1 General Design Considerations

2 Analysis

3 Systems Not Specifically Detailed for Seismic Resistance

4 Moment Frames

5 Braced Frames

6 Composite Moment Frames

7 Composite Braced Frames and Shear Walls

8 Diaphragms, Collectors and Chords

9 Provisions and Standards

10 Engineered Damping Systems

Index

AMERICAN INSTITUTE OF STEEL CONSTRUCTION

SEISMIC
DESIGN

MANUAL

**AMERICAN INSTITUTE
OF
STEEL CONSTRUCTION**

SECOND EDITION

AISC © 2012

by

American Institute of Steel Construction

ISBN 1-56424-061-4

All rights reserved. This book or any part thereof must not be reproduced in any form without the written permission of the publisher. The AISC logo is a registered trademark of AISC.

The information presented in this publication has been prepared in accordance with recognized engineering principles and is for general information only. While it is believed to be accurate, this information should not be used or relied upon for any specific application without competent professional examination and verification of its accuracy, suitability and applicability by a licensed professional engineer, designer, or architect. The publication of the material contained herein is not intended as a representation or warranty on the part of the American Institute of Steel Construction or of any other person named herein, that this information is suitable for any general or particular use or of freedom from infringement of any patent or patents. Anyone making use of this information assumes all liability arising from such use.

Caution must be exercised when relying upon other specifications and codes developed by other bodies and incorporated by reference herein since such material may be modified or amended from time to time subsequent to the printing of this edition. The Institute bears no responsibility for such material other than to refer to it and incorporate it by reference at the time of the initial publication of this edition.

Printed in the United States of America

First Printing: September 2012

DEDICATION

Clarkson ("Pinky") W. Pinkham

This edition of the AISC *Seismic Design Manual* is dedicated to the memory of Clarkson W. Pinkham, a long-time member of the AISC Committee on Specifications and Task Committee 9—Seismic Design. Mr. Pinkham, or Pinky as he was affectionately known to his professional colleagues, was president and member of the Los Angeles consulting structural engineering firm, S.B. Barnes and Associates, for 62 years. He served on the AISC Committee on Specifications from the mid-1970s until the year 2000, and Task Committee 9—Seismic Design from the mid-1990s until 2010. As a member of Task Committee 9 and technical secretary for the 1997 AISC *Seismic Provisions for Structural Steel Buildings*, he was a major contributor and instrumental in the early development of that standard. Pinky was one of the first proponents of including composite systems in the AISC *Seismic Provisions* and, for the first time, this *Seismic Design Manual* includes two chapters on such systems. Pinky received the AISC Lifetime Achievement Award in 1999. Through a career that spanned more than six decades, he spent a lifetime sharing his expertise with others in the field. He was passionate about learning up until his death in 2012 at the age of 92. Pinky was generous in sharing his abundance of structural engineering experience and knowledge through committee involvement and with those who requested it on the subjects of structural steel, concrete and masonry design, cold-formed structures, and timber. By providing solutions and recommendations in this way, Pinky improved the integrity of numerous structures; in particular, their capacity to resist seismic-generated forces. He was elected president of the Structural Engineers Association of Southern California (SEAOSC) in 1971, and later served as president of the Structural Engineers Association of California (SEAOC) in 1975. He was twice given the S.B. Barnes Award for Research, and in 1994 was inducted into the SEAOC College of Fellows, the highest honor awarded by SEAOC. In 2009, the Structural Engineering Institute of the American Society of Civil Engineers awarded Pinky the Walter P. Moore, Jr., Award in recognition of his dedication to and technical expertise in the development of structural codes and standards. AISC will always remember his dedication to the development of standards related to the design and construction of structural steel and it is especially fitting that the 2nd Edition of the AISC *Seismic Design Manual* be dedicated to the memory of Clarkson W. Pinkham.

FOREWORD

The American Institute of Steel Construction, founded in 1921, is the nonprofit technical specifying and trade organization for the fabricated structural steel industry in the United States. Executive and engineering headquarters of AISC are maintained in Chicago. The Institute is supported by four classes of membership: Active Members engaged in the fabrication, production and sale of structural steel; Associate Members, who include Erectors, Detailers, Service Consultants, Software Developers, and Steel Product Manufacturers; Professional Members, who are individuals or firms engaged in the practice of architecture or engineering, including architectural and engineering educators; and Affiliate Members, who include General Contractors, Building Inspectors and Code Officials. The continuing financial support and active participation of Members in the engineering, research and development activities of the Institute make possible the publishing of this *Seismic Design Manual*.

The Institute's objective is to make structural steel the material of choice, by being the leader in structural-steel-related technical and market-building activities, including: specification and code development, research, education, technical assistance, quality certification, standardization, and market development.

To accomplish this objective, the Institute publishes manuals, design guides and specifications. Best known and most widely used is the *Steel Construction Manual*, which holds a highly respected position in engineering literature. The Manual is based on the *Specification for Structural Steel Buildings* and the *Code of Standard Practice for Steel Buildings and Bridges*. Both standards are included in the Manual for easy reference.

The Institute also publishes technical information and timely articles in its *Engineering Journal*, Design Guide series, *Modern Steel Construction* magazine, and other design aids, research reports, and journal articles. Nearly all of the information AISC publishes is available for download from the AISC web site at **www.aisc.org**.

PREFACE

This is the second edition of the AISC *Seismic Design Manual*, intended to assist designers in properly applying AISC standards and provisions in the design of steel frames to resist high-seismic loadings. This Manual is intended for use in conjunction with the AISC *Steel Construction Manual*, 14th Edition.

The following consensus standards are printed in Part 9 of this Manual:

- 2010 *Seismic Provisions for Structural Steel Buildings* (ANSI/AISC 341-10)
- 2010 *Prequalified Connections for Special and Intermediate Steel Moment Frames for Seismic Applications* including Supplement No. 1 (ANSI/AISC 358-10 and ANSI/AISC 358s1-11)

The design examples contained in this Manual demonstrate an approach to design, and are not intended to suggest that the approach presented is the only approach. The committee responsible for the development of these design examples recognizes that designers have alternate approaches that work best for them and their projects. Design approaches that differ from those presented in these examples are considered viable as long as the AISC *Specification* and AISC *Seismic Provisions*, sound engineering, and project specific requirements are satisfied.

The following major changes and improvements have been made in this revision:

- More thorough and comprehensive design examples, updated for the 2010 AISC *Seismic Provisions*
- Side-by-side LRFD and ASD design methodologies for design examples
- Addition of Part 2: Analysis
- Addition of column base plate design examples
- Extended discussion in Part 8 on collector elements
- Addition of Part 10: Engineered Damping Systems
- Addition of buckling-restrained braced frame systems and examples
- Addition of new chapters on composite moment frames and composite braced frames

By the AISC Committee on Manuals and Textbooks,

Mark V. Holland, Chairman
Gary C. Violette, Vice-Chairman
Abbas Aminmansour
Charles J. Carter
Harry A. Cole
Brad Davis
Bo Dowswell
Lanny J. Flynn
Patrick J. Fortney
W. Scott Goodrich
Christopher M. Hewitt
W. Steven Hofmeister
William P. Jacobs
Bill R. Lindley, II

Ronald L. Meng
Larry Muir
Thomas M. Murray
Rafael Sabelli
Clifford W. Schwinger
William N. Scott
William T. Segui
Victor Shneur
Marc L. Sorenson
William A. Thornton
Michael A. West
Ronald G. Yeager
Cynthia J. Duncan, Secretary

and the AISC Subcommittee on Seismic Design,

Rafael Sabelli, Chairman	Brett R. Manning
Thomas A. Sabol, Vice-Chairman	Heath E. Mitchell
Allen Adams	Kevin S. Moore
Scott M. Adan	Larry Muir
William A. Andrews	Clinton O. Rex
Richard M. Drake	John A. Rolfes
Michael D. Engelhardt	William N. Scott
Patrick J. Fortney	Victor Shneur
Timothy P. Fraser	Harold O. Sprague
John L. Harris, III	Amit H. Varma
James O. Malley	Leigh Arber, Secretary

The committee gratefully acknowledges the contributions made to this Manual by the following individuals: Eric Bolin, Areti Carter, Maria E. Chumbita, Janet Cummins, Thomas Dehlin, Richard Drake, Erica Fischer, Louis Geschwindner, Amir Gilani, Keith Grubb, Jerome Hajjar, Amit Kanvinde, Richard Kaehler, Ryan Kersting, Zhichao Lai, Dawn Lehman, Brent Leu, Kit Miyamoto, Keith Palmer, Davis Parsons II, Paul Richards, Kimberly Robinson, Charles Roeder, Brandt Saxey, Thomas Schlafly, Bahram Shahrooz, Chia-Ming Uang, and Jie Zuo.

SCOPE

The specification requirements and other design recommendations and considerations summarized in this Manual apply in general to the design and construction of seismic force resisting systems in steel buildings and other structures. The AISC *Seismic Design Manual* is intended to be applied in conjunction with the AISC *Steel Construction Manual*, which provides guidance on the use of the AISC *Specification for Structural Steel Buildings*.

In addition to the requirements of the AISC *Specification*, the design of seismic force resisting systems must meet the requirements in the AISC *Seismic Provisions for Structural Steel Buildings*, except in the following cases for which use of the AISC *Seismic Provisions* is not required:

- Buildings and other structures in Seismic Design Category (SDC) A
- Buildings and other structures in SDC B or C with $R = 3$ systems (steel systems not specifically detailed for seismic resistance per ASCE/SEI 7 Table 12.2-1)
- Nonbuilding structures similar to buildings with $R = 1½$ braced-frame systems or $R = 1$ moment-frame systems; see ASCE/SEI 7 Table 15.4-1
- Nonbuilding structures not similar to buildings (see ASCE/SEI 7 Table 15.4-2), which are designed to meet the requirements in other standards entirely

Conversely, use of the AISC *Seismic Provisions* is required in the following cases:

- Buildings and other structures in SDC B or C when one of the exemptions for steel seismic force resisting systems above does not apply
- Buildings and other structures in SDC B or C that use composite seismic force resisting systems (those containing composite steel-and-concrete members and those composed of steel members in combination with reinforced concrete members)
- Buildings in SDC D, E or F
- Nonbuilding structures in SDC D, E or F when the exemption above does not apply

The *Seismic Design Manual* consists of ten parts addressing various topics related to the design and construction of seismic force resisting systems of structural steel and structural steel acting compositely with reinforced concrete. Part 1 stipulates the specific editions of the specifications, codes and standards referenced in this Manual, and provides a discussion of general design considerations related to seismic design. Part 2 provides some guidance on structural analysis procedures employed. For the design of systems not detailed for seismic resistance, see Part 3. Parts 4 through 7 apply to the various types of seismic force resisting systems, including design examples. Part 8 discusses other systems, such as diaphragm chords and collectors, which are important in seismic design. Part 10 addresses engineering damping systems. For applicable AISC seismic standards, see Part 9.

PART 1

GENERAL DESIGN CONSIDERATIONS

1.1   SCOPE 1–4
1.2   APPLICABLE SPECIFICATIONS, CODES AND OTHER REFERENCES 1–4
    Specifications, Codes and Standards for Structural Steel Buildings 1–4
    Other AISC Reference Documents 1–5
1.3   SEISMIC DESIGN OVERVIEW AND DESIGN CONSIDERATIONS 1–5
    Performance Goals 1–5
    Applicable Building Code 1–6
    Risk Category and Seismic Design Category 1–7
    Earthquake Ground Motion and Response Spectrum 1–7
    Maximum Considered Earthquake and Design Basis Earthquake 1–10
    Systems Defined in ASCE/SEI 7 1–10
    Seismic Performance Factors 1–12
        Response Modification Coefficient, R 1–12
        $R = 3$ Applications 1–13
        Deflection Amplification Factor, C_d 1–14
        Overstrength Factor, Ω_o 1–14
        Redundancy Factor, ρ 1–15
    Maximum Force Delivered by the System 1–16
    Building Joints 1–16
        Expansion Joints 1–16
        Seismic Joints 1–17
        Building Separations 1–17
    Building Drift 1–18
        Deflection Compatibility 1–18
    Lowest Anticipated Service Temperature 1–18
    Quality Control and Quality Assurance 1–19
    Design Drawing Requirements 1–21
        Structural Design Drawing Requirements 1–21
        SFRS Member and Connection Material Specifications 1–21
        Demand Critical Welds 1–21

Locations and Dimensions of Protected Zones 1–22
Additional Structural Design Drawing Detail Requirements
in the Provisions 1–22
AWS D1.8 Structural Welding Code—Seismic Supplement 1-23
Composite Systems 1–23
1.4 DESIGN TABLE DISCUSSION 1–25
Seismic Weld Access Hole Configuration 1–25
Member Ductility Requirements 1–25
Local Buckling Requirements 1–25
Table 1-A. Limiting Width-to-Thickness Ratios for W-Shape Flanges
and Webs in Compression 1–26
Table 1-B. Limiting Width-to-Thickness Ratios for Angle Legs
in Compression 1–27
Table 1-C. Limiting Width-to-Thickness Ratios for Rectangular
HSS Walls in Compression 1–28
Table 1-D. Limiting Width-to-Thickness Ratios for Round HSS
and Pipe Walls in Compression 1–29
Strength of Steel Headed Stud Anchors 1–30
ASCE/SEI 7 Design Coefficients and Factors for SFRS 1–30
PART 1 REFERENCES 1–31
DESIGN TABLES 1–33
Table 1-1. Workable Seismic Weld Access Hole Configurations 1–33
Table 1-2. Summary of Member Ductility Requirements 1–34
Table 1-3. Sections That Satisfy Seismic Width-to-Thickness
Requirements, W-Shapes 1–36
Table 1-4. Sections That Satisfy Seismic Width-to-Thickness
Requirements, Angles 1–52
Table 1-5a. Sections That Satisfy Seismic Width-to-Thickness
Requirements, Rectangular HSS 1–53
Table 1-5b. Sections That Satisfy Seismic Width-to-Thickness
Requirements, Square HSS 1–54
Table 1-6. Sections That Satisfy Seismic Width-to-Thickness
Requirements, Round HSS 1–55
Table 1-7. Sections That Satisfy Seismic Width-to-Thickness
Requirements, Pipe 1–57
Table 1-8. Shear Stud Anchor Nominal Horizontal Shear Strength
and 25% Reduced Nominal Horizontal Shear Strength for
Steel Headed Stud Anchors 1–58

Table 1-9a. Design Coefficients and Factors for Steel and Steel and
Concrete Composite Seismic Force Resisting Systems 1–59

Table 1-9b. Design Coefficients and Factors for Nonbuilding Structures
Similar to Buildings 1–62

1.1 SCOPE

The design considerations summarized in this Part apply in general to the design and construction of steel buildings for seismic applications. The specific editions of specifications, codes and other references listed below are referenced throughout this Manual.

1.2 APPLICABLE SPECIFICATIONS, CODES AND OTHER REFERENCES

Specifications, Codes and Standards for Structural Steel Buildings

Subject to the requirements in the applicable building code and the contract documents, the design, fabrication and erection of structural steel buildings is governed as indicated in the AISC *Specification* Sections A1 and B2, and AISC *Seismic Provisions* Sections A2 and B2 as follows:

1. ASCE/SEI 7: *Minimum Design Loads for Buildings and Other Structures*, ASCE/SEI 7-10 (ASCE, 2010). Available from the American Society of Civil Engineers, ASCE/SEI 7 provides the general requirements for loads, load factors and load combinations.
2. AISC *Specification*: *Specification for Structural Steel Buildings*, ANSI/AISC 360-10 (AISC, 2010a). This standard provides the general requirements for design and construction of structural steel buildings, and is included in Part 16 of the AISC *Steel Construction Manual* and is also available at **www.aisc.org**.
3. AISC *Seismic Provisions*: *Seismic Provisions for Structural Steel Buildings*, ANSI/AISC 341-10 (AISC, 2010b). This standard provides the design and construction requirements for seismic force resisting systems in structural steel buildings, and is included in Part 9 of this Manual and is also available at **www.aisc.org**.
4. ANSI/AISC 358: AISC *Prequalified Connections for Special and Intermediate Steel Moment Frames for Seismic Applications*, ANSI/AISC 358-10 (AISC, 2010c). This standard specifies design, detailing, fabrication and quality criteria for connections that are prequalified in accordance with the AISC *Seismic Provisions* for use with special and intermediate moment frames. It is included in Part 9 of this Manual and is also available at **www.aisc.org**.
5. AISC *Code of Standard Practice*: AISC *Code of Standard Practice for Steel Buildings and Bridges* (AISC, 2010d). This document provides the standard of custom and usage for the fabrication and erection of structural steel, and is included in Part 16 of the AISC *Steel Construction Manual* and is also available at **www.aisc.org**.

Other referenced standards include:

1. RCSC *Specification*: *Specification for Structural Joints Using High-Strength Bolts* (RCSC, 2009), reprinted in Part 16 of the AISC *Steel Construction Manual* with the permission of the Research Council on Structural Connections and available at **www.boltcouncil.org**, provides the additional requirements specific to bolted joints with high-strength bolts.

2. AWS D1.1: *Structural Welding Code—Steel*, AWS D1.1/D1.1M:2010 (AWS, 2010). Available from the American Welding Society, AWS D1.1 provides additional requirements specific to welded joints. Requirements for the proper specification of welds can be found in AWS A2.4: *Standard Symbols for Welding, Brazing, and Nondestructive Examination* (AWS, 2007).
3. AWS D1.8: *Structural Welding Code—Seismic Supplement*, AWS D1.8/D1.8M:2009 (AWS, 2009). Available from the American Welding Society, AWS D1.8 acts as a supplement to AWS D1.1 and provides additional requirements specific to welded joints in seismic applications.
4. ACI 318: *Building Code Requirements for Structural Concrete*, ACI 318-08 (ACI, 2008). Available from the American Concrete Institute, ACI 318 provides additional requirements for reinforced concrete, including composite design and the design of steel-to-concrete anchorage.

Other AISC Reference Documents

The AISC *Steel Construction Manual* (AISC, 2011), referred to as the AISC *Manual* is available from AISC at **www.aisc.org**. This publication provides design recommendations and specification requirements for various topics related to steel building design and construction.

1.3 SEISMIC DESIGN OVERVIEW AND DESIGN CONSIDERATIONS

Performance Goals

Seismic design is the practice of proportioning and detailing a structure so that it can withstand shaking from an earthquake event with acceptable performance. The AISC *Seismic Provisions for Structural Steel Buildings* are intended to provide a means of designing structures constructed to respond to maximum considered earthquake ground shaking, as defined in ASCE/SEI 7, with a low probability of collapse, while potentially sustaining significant inelastic behavior and structural damage. Fundamental to seismic design is the practice of proportioning and detailing the structure so that it can withstand large deformation demands, accommodated through inelastic behavior in structural elements that have been specifically designed to withstand this behavior acceptably. This requires careful proportioning of the structural system so that inelastic behavior occurs in pre-selected elements that have appropriate section properties to sustain large inelastic deformation demands without loss of strength, and assuring that connections of structural elements are adequate to develop the strength of the connected members.

Performance appropriate to the function of the structure is a fundamental consideration for seismic design. Potential considerations are post-earthquake reparability and serviceability for earthquakes of different severity. Most structures are designed only with an expectation of protecting life safety, rather than assuring either the feasibility of repair or post-earthquake utility. Buildings assigned to Risk Categories III and IV, as defined in ASCE/SEI 7, are expected to withstand severe earthquakes with limited levels of damage, and in some cases, allow post-earthquake occupancy. The criteria of the AISC *Seismic Provisions*, when used together with the requirements of ASCE/SEI 7, are intended to provide performance appropriate to the structure's risk category[1]. For some buildings, performance that exceeds these expectations may be appropriate. In those cases, designers must develop supplementary criteria to those contained in the AISC *Seismic Provisions* and ASCE/SEI 7.

Building performance is not a function of the structural system alone. Many building structures have exhibited ill effects from damage to nonstructural components, including breaks in fire protection systems and impaired egress, which have precluded building functions and thus impaired performance. Proper consideration of the behavior of nonstructural components is essential to enhanced building performance. Industrial and nonbuilding structures often contain elements that require some measure of protection from large deformations.

Generally, seismic force resisting systems (SFRS) are classified into three levels of inelastic response capability, designated as ordinary, intermediate or special, depending on the level of ductility that the system is expected to provide. A system designated as ordinary is designed and detailed to provide limited ability to exhibit inelastic response without failure or collapse. The design requirements for such systems, including limits on proportioning and detailing, are not as stringent as those systems classified as intermediate or special. Ordinary systems provide seismic resistance primarily through their strength. Structures such as these must be designed for higher force demands with commensurately less stringent ductility and member stability requirements. Some steel structures achieve acceptable seismic performance by providing ductility in specific structural elements that are designed to undergo nonlinear deformation without strength loss and dissipate seismic energy. Examples of ductile steel structures include special moment frames, eccentrically braced frames, and buckling-restrained braced frames. The ability of these structures to deform inelastically, without strength loss or instability, permits them to be designed for lower forces than structures with ordinary detailing.

Enhanced performance, relative to that provided by conformance to the AISC *Seismic Provisions* and ASCE/SEI 7, can be a required consideration for certain nuclear structures and critical military structures, but is beyond the scope of this Manual. Critical structures generally are designed to remain elastic, even for large infrequent seismic events.

Applicable Building Code

National model building codes are published so that state and local authorities may adopt the code's prescriptive provisions to standardize design and construction practices in their jurisdiction. The currently used model code in the U.S. for the structural design of buildings and nonbuilding structures is the International Building Code (IBC), published by the International Code Council (ICC) (ICC, 2012). Oftentimes the adopted provisions are amended based on jurisdictional requirements to develop local building codes (e.g., California Building Code and the Building Code of New York City). Local codes are then enforced by law and any deviation must be approved by the local building authority. As the local code provisions may change between jurisdictions, the AISC *Specification* and AISC *Seismic Provisions* refer to this code as the applicable building code.

---

[1] Codes have historically used occupancy category. This classification was changed to risk category in ASCE/SEI 7-10 and IBC 2012. Where classification by occupancy category is still employed, the more stringent of the two is used.

The primary performance objective of these model codes is that of "life safety" for building occupants for all the various demands to which the building will be subjected. To satisfy this objective for structures required to resist strong ground motions from earthquakes, these codes reference ASCE/SEI 7 for seismic analysis and design provisions. Seismic design criteria in this standard prescribe minimum requirements for both the strength and stiffness of SFRS and the structural elements they include. The seismic design criteria in ASCE/SEI 7 for the most part are based on the *NEHRP Recommended Provisions for Seismic Regulations for New Buildings and Other Structures* (FEMA, 2009).

The seismic design of nonbuilding structures is addressed separately in ASCE/SEI 7 in Chapter 15. Nonbuilding structures are defined as all self-supporting structures that carry gravity loads and that may be required to resist the effects of seismic loads, with certain exclusions. ASCE/SEI 7 develops an appropriate interface with building structures for those types of nonbuilding structures that have dynamic behaviors similar to buildings. There are other nonbuilding structures that have little similarity to buildings in terms of dynamic response, which are not specifically covered by AISC documents.

Risk Category and Seismic Design Category

In ASCE/SEI 7, the expected performance of a structure is determined by assigning it to a risk category. There are four risk categories (I, II, III and IV), based on the risk posed to society as a consequence of structural failure or loss of function. In seismic design, the risk category is used in conjunction with parameters that define the intensity of design ground shaking in determining the importance factor and the seismic design category (SDC) for which a structure must be designed. There are six SDC, designated by the letters A through F. Structures assigned to Seismic Design Category A are not anticipated to experience ground shaking of sufficient intensity to cause unacceptable performance, even if they are not specifically designed for seismic resistance. Structures in Seismic Design Categories B or C can experience motion capable of producing unacceptable damage when the structures have not been designed for seismic resistance. Structures in Seismic Design Category D are expected to experience intense ground shaking, capable of producing unacceptable performance in structures that have unfavorable structural systems and which have not been detailed to provide basic levels of inelastic deformation response without failure. Structures assigned to Seismic Design Categories E and F are located within a few miles of major active faults capable of producing large magnitude earthquakes and ground motions with peak ground accelerations exceeding $0.6g$. Even well-designed structures with extensive inelastic response capability can be severely damaged under such conditions, requiring careful selection and proportioning of structures.

Earthquake Ground Motion and Response Spectrum

An earthquake causes ground motions that may propagate from the hypocenter in any direction. These motions produce horizontal and vertical ground accelerations at the earth's surface, which in turn cause structural accelerations. While it is possible to use earthquake ground motions recorded in past earthquakes to simulate the behavior of structures, the required analysis procedures are complex, and the analysis results are sensitive to the characteristics of the individual ground motions selected, which may not actually be similar to those a structure will experience in the future. To simplify the uncertainty and complexity associated with using recorded motions to predict a structure's response, response earthquake spectra are used. A response spectrum for a given earthquake ground motion indicates the maximum (absolute value), expressed either as acceleration, velocity or displacement, that an elastic single-degree-of-freedom (SDOF) oscillator will experience as a function of the structure's period and equivalent damping factor. Figure 1-1a shows an example of an acceleration response spectrum. On average, low-rise buildings (Figure 1-1b) tend to have short periods,

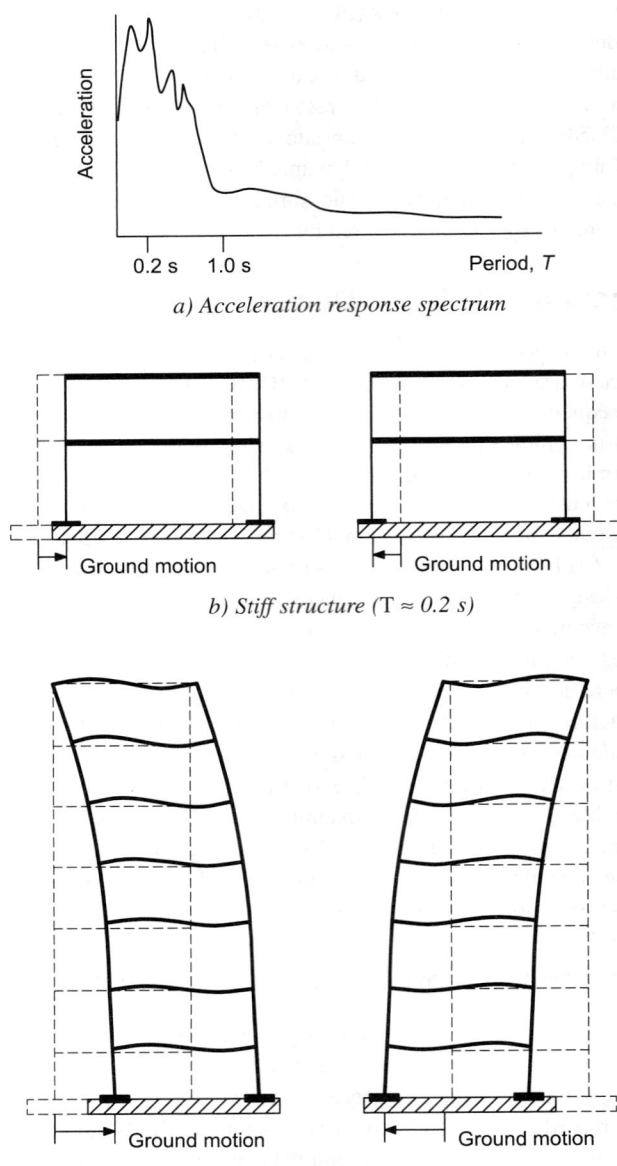

Fig. 1-1. Earthquake acceleration and structure response.

while tall structures tend to be flexible with longer periods (Figure 1-1c). For a given ground motion, short period structures tend to experience higher acceleration, and therefore, higher inertial force (mass times acceleration), than do longer period structures. However, long period structures generally experience greater displacement.

Multi-story buildings are multi-degree-of-freedom systems with multiple modes of vibration. Each mode has a characteristic deflected shape and period. Since earthquake ground motion contains energy caused by vibration across an entire spectrum of frequencies, each acceleration frequency that corresponds to a mode imparts energy into the structure. Figure 1-2 shows an example of a two-dimensional five-story building frame and the modal information for the first four modes. Although the mode shapes are shown separately, the actual building motion will consist of combined response in each of the several modes. Using the modal shape of the structure for each mode and the effective percentage of the structure's mass mobilized when vibrating in that mode, it is possible to use the same SDOF response spectrum discussed above to determine the maximum response for each mode. These maxima are then combined to estimate the total maximum response based on the participation of each mode. These maxima for the various modes will generally occur at different points in time. Modal combination rules approximately account for this effect. Detailed information about structural response using modal analysis can be found in Chopra (2007).

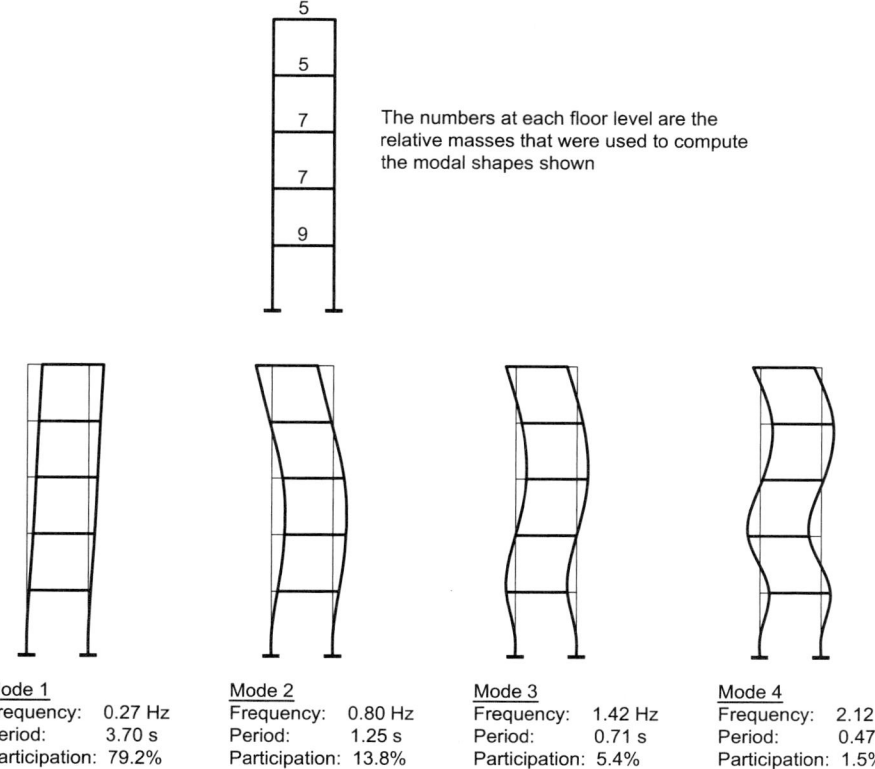

Fig. 1-2. Vibration modes for a multi-degree-of-freedom building caused by application of a typical earthquake acceleration design spectrum.

Maximum Considered Earthquake and Design Basis Earthquake

Ground motion hazards in ASCE/SEI 7 are defined as maximum considered earthquake ground motions. They are based on the proximity of the site to active faults, the activity of these faults, magnitude of the event these faults can produce, and the regional and local geology at a site. The design intent of ASCE/SEI 7 is to assure that ordinary occupancy structures (structures assigned to Risk Categories I and II) have not greater than a 10% chance of collapse should they experience maximum considered earthquake shaking. Except for regions located within a few miles of major active faults, such as some sites in coastal California, the maximum considered earthquake is selected with an annual frequency that will provide a uniform collapse risk of 1% probability in 50 years (denoted MCE_R). In regions close to major active faults probabilistic estimates of ground shaking at these annual frequencies are very intense and impractical for use in design. Therefore, on such sites, the MCE_R is capped by a conservative deterministic estimate of the ground motion resulting from a maximum magnitude earthquake on the nearby fault, resulting in a higher collapse risk. This is a change from prior editions of ASCE/SEI 7. See ASCE (2010) commentary and FEMA (2009) for further information. The MCE_R is represented by a generalized elastic acceleration response spectrum (see previous). This response spectrum is subsequently reduced by two-thirds to represent the elastic response for the design basis earthquake for which a structure is designed. Detailed information about this reduction can be found in FEMA (2009).

Systems Defined in ASCE/SEI 7

A steel SFRS is generally classified into three levels of expected inelastic response capability, designated as ordinary, intermediate or special, depending on the level of ductility that the system is expected to provide. Systems designated as ordinary are designed and detailed to provide limited ductility, but the requirements are not as stringent as those systems classified as intermediate or special. In some cases, an SFRS can be classified as a "structure not specifically detailed for seismic resistance" in accordance with the applicable building code. Each classification is characterized by the following seismic performance factors:

- Response modification coefficient, R
- Overstrength factor, Ω_o
- Deflection amplification factor, C_d

When used in combination, these factors quantitatively outline the expected performance of an SFRS. Other factors that influence the performance are the importance factor, I_e, and redundancy factor, ρ. These factors are discussed in the following.

Designing to meet the seismic requirements of the AISC *Seismic Provisions* is mandatory for structures where they have been specifically referenced in Table 12.2-1 of ASCE/SEI 7. For steel structures, typically this occurs in SDC D and higher where R is greater than 3. However, there are instances where an R less than 3 is assigned to a system and the *Provisions* are still required. These limited cases occur in ASCE/SEI 7 Table 12.2-1 for cantilevered column systems and Table 15.4-1 for nonbuilding structures similar to buildings. For composite steel-concrete structures, there are cases where the *Provisions* are required in SDC B and C, as specified in Table 12.2-1 of ASCE/SEI 7. This typically occurs for composite systems designated as ordinary where the counterpart reinforced concrete systems have designated values of R and design requirements for SDC B and C.

Applications where R is greater than 3 are intended for buildings that are designed to meet the requirements of both the AISC *Seismic Provisions* and the AISC *Specification*. The use of R greater than 3 in the calculation of the seismic base shear requires the use of a seismically designed and detailed system that is able to provide the level of ductility commensurate with the value of R selected in the design. This level of ductility is achieved through a combination of proper material and section selection, the use of low width-to-thickness members for the energy dissipating elements of the SFRS, detailing member connections to resist the local demands at the capacity of the system, and providing for system lateral stability at the large deformations expected in a major earthquake. Consider the following three examples:

1. Special concentrically braced fame (SCBF) systems—SCBF systems are generally configured so that energy dissipation will occur by tension yielding and/or compression buckling in the braces. The connections of the braces to the columns and beams and between the columns and beams themselves must then be proportioned to remain essentially elastic as they undergo these deformations. See Figure 1-3.
2. Eccentrically braced frame (EBF) systems—EBF systems are generally configured so that energy dissipation will occur by shear and/or flexural yielding in the link. The beam outside the link, connections, braces and columns must then be proportioned to remain essentially elastic as they undergo these deformations. See Figure 1-4.
3. Special moment frame (SMF) systems—SMF systems are generally configured so that energy dissipation will occur by flexural yielding in the girders near, but away from, the connection of the girders to the columns. The connections of the girders to the columns and the columns themselves must then be proportioned to remain essentially elastic as they undergo these deformations. See Figure 1-5.

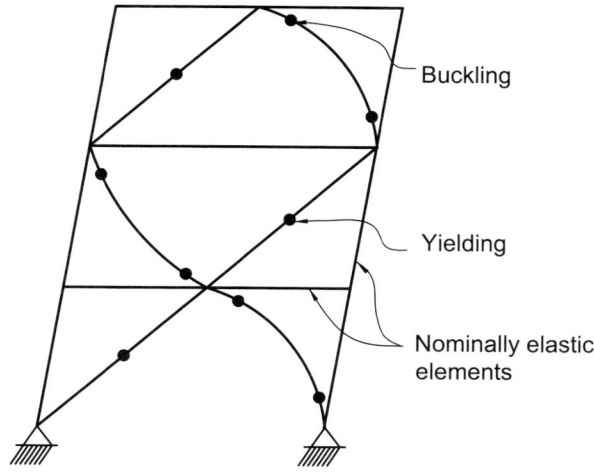

Fig. 1-3. Ductile braced frames.

Seismic Performance Factors

Response Modification Coefficient, R

The SDC is used, along with the SFRS type, to establish a minimum level of inelastic, ductile response that is required of a structure. The corresponding expected system behavior is codified in the form of an *R*-factor, which is a response modification factor applied to the lateral force to adjust a structure's required lateral strength considering its inelastic response capability.

The response modification coefficient, *R*, sets the minimum fraction ($1/R$) of the strength required to resist design earthquake shaking elastically for which it is permissible to design

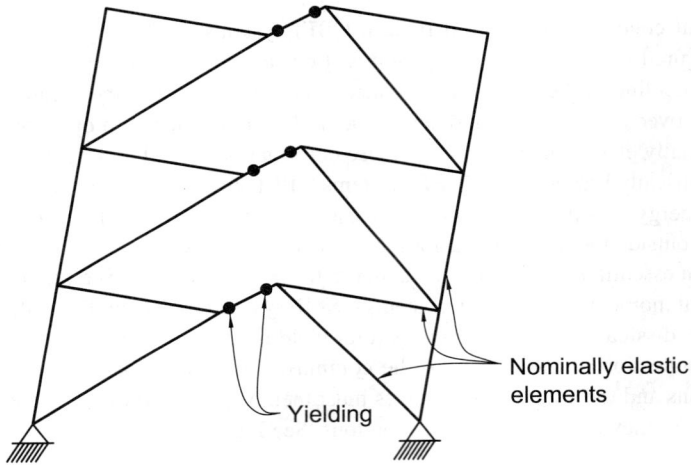

Fig. 1-4. Ductile eccentrically braced frames.

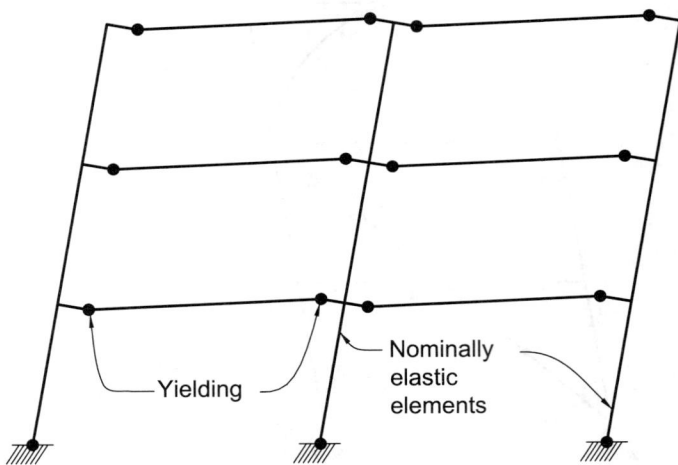

Fig. 1-5. Ductile moment frames.

1.3 SEISMIC DESIGN OVERVIEW AND DESIGN CONSIDERATIONS

a structure. Structures designed with a large value of R must have extensive capability to withstand large inelastic deformation demands during design shaking. Structures designed with an R approximating 1 are anticipated to experience design shaking while remaining essentially elastic. Figure 1-6 shows the relationship between R and the design-level forces, along with the corresponding lateral deformation of the structural system (FEMA, 2009).

Factors that determine the magnitude of the response modification factor are the vulnerability of the gravity load resisting system to a failure of elements in the SFRS, the level and reliability of the inelasticity the system can attain, and potential backup frame resistance such as that which is provided by dual-frame systems. As illustrated in Figure 1-6, in order for a system to utilize a higher value of R, other elements of the system must have adequate strength and deformation capacity to remain stable at the maximum lateral deflection levels. If the system redundancy and system overstrength cannot be achieved, a lower value of R should be incorporated in the design and detailing of the structure. Values of R for all structural systems are defined in Table 12.2-1 of ASCE/SEI 7. Tables 1-9a and 1-9b in this Manual summarize the R-Factors and other factors specified in ASCE/SEI 7 for steel and composite systems. More detailed discussion on the system design parameters can be found in FEMA (2009).

R = 3 Applications

For structures assigned to SDC B and C in ASCE/SEI 7 the designer is given a choice to either solely use the AISC *Specification* to design and detail the structure (typically assigned an R of 3) or to assign a higher value of R to a system detailed for seismic resistance and follow the requirements of the AISC *Seismic Provisions*. The resulting systems have ductility associated with conventional steel framing not specifically detailed for high seismic resistance. It is important to note, however, that even steel structures not specifically designed or

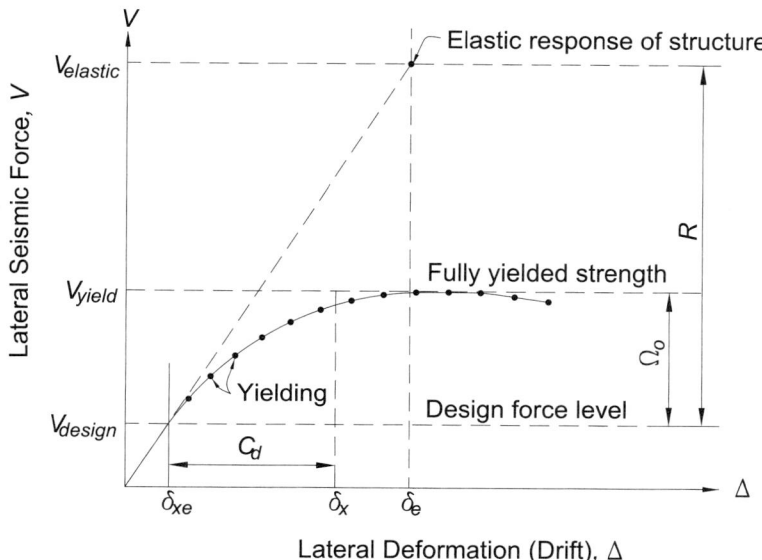

Fig. 1-6. Relationship between R, *design level forces, and lateral deformation.*

detailed for seismic resistance possess some inherent amount of seismic resistance, which may be adequate to resist a limited amount of seismic demand.

It is recognized that when the designer has the option to design a building to meet the AISC *Specification* with $R = 3$, such a design will generally be more cost effective than the same structure designed in accordance with the AISC *Seismic Provisions* using a higher value of R. The extra fabrication, erection and inspection costs needed to achieve the high ductility commensurate with the higher R more than offset the additional steel tonnage required by the $R = 3$ system.

The $R = 3$ option is not generally available for composite steel-concrete systems. For composite systems, the designer must follow the requirements outlined in Table 12.2-1 of ASCE/SEI 7.

Deflection Amplification Factor, C_d

The elastic story drifts calculated under reduced lateral forces are multiplied by the deflection amplification factor, C_d, to better estimate the total story drifts likely to result from the design earthquake ground motion. These amplified story drifts are used to verify compliance with the allowable story drift and to determine seismic demands on elements of the structure that are not part of the SFRS and on nonstructural components within the structure.

Overstrength Factor, Ω_o

Most seismic force resisting systems rely on dissipation of earthquake energy through varying levels of inelastic behavior. Steel seismic system definitions in the AISC *Seismic Provisions* designate the elements intended to dissipate the majority of this energy through ductile response, and those that are intended to remain essentially elastic. The application of an overstrength factor, Ω_o, is applied to some seismic load combinations in ASCE/SEI 7 and in certain cases in the AISC *Seismic Provisions* to provide a design force that will result in essentially elastic response for specific components. These load combinations are invoked for members or connections whose inelastic behavior may cause poor system performance. They generally meet the following criteria: They are critical elements on the load path, and they are not likely to perform well in the inelastic range.

Members and connections requiring the special seismic load combinations incorporating the overstrength factor, Ω_o, in ASCE/SEI 7, include the following (the applicable section of ASCE/SEI 7 is provided in parentheses):

1. Elements supporting discontinuous walls or frames (Section 12.3.3.3)
2. Collectors for structures in SDC C through F (Section 12.10.2.1)
3. Batter piles (Section 12.13.6.4)
4. Pile anchorage (Section 12.13.6.5)
5. Pile splices (Section 12.13.6.6)

In the AISC *Seismic Provisions*, this concept is addressed using the term, amplified seismic load. In some cases, the amplified seismic load defers to the use of the ASCE/SEI 7 load combinations that include Ω_o, while for other situations, the amplified seismic load is a term defined in the AISC *Seismic Provisions* to meet a capacity design requirement. Sections of the AISC *Seismic Provisions* where the amplified seismic load is invoked for the design of certain elements or connections include:

1.3 SEISMIC DESIGN OVERVIEW AND DESIGN CONSIDERATIONS

Section D1.4a—Required compressive and tensile strength of columns
Section D2.5b—Required strength of column splices
Section D2.6a—Required axial strength of column bases
Section D2.6b—Required shear strength of column bases
Section D2.6c—Required flexural strength of column bases
Section E1.6b—Required shear strength of beam-to-column connections for ordinary moment frames
Sections E2.6d and G2.6d—Required shear strength of beam-to-column connections for intermediate moment frames and composite intermediate moment frames
Sections E3.4a and G3.4a—Moment ratio check for special moment frames and composite special moment frames (also referred to as the strong-column-weak-beam calculation)
Sections E3.4c and G3.4c—Required column strength at unbraced beam-to-column connections for special moment frames and composite special moment frames
Section E3.6d and G3.6d—Required shear strength of beam-to-column connections for special moment frames and composite special moment frames
Section E4.3b—Required strength of nonspecial segment members and connections for special truss moment frames
Section E5.4a—Required strength of columns in ordinary cantilever column systems
Section E6.4a—Required strength of columns in special cantilever column systems
Section F1.2—Determination of eccentric moments in members for ordinary concentrically braced frames, if an eccentricity is present
Section F1.4a—Determination of required strength of beams in V-braced and inverted-V-braced ordinary concentrically braced frames
Section F1.6—Diagonal brace connections in ordinary concentrically braced frames
Section F2.3—Required strength of columns, beams and connections in special concentrically braced frames
Section F2.4a—Provides an exception to the lateral force distribution requirement in special concentrically braced frames
Sections F3.3 and F3.6c—Required strength of diagonal braces and their connections, beams outside links, and columns for eccentrically braced frames
Sections F4.3 and F4.6c—Required strength of beams, columns and connections in buckling-restrained braced frames
Sections F5.3 and F5.6b—Required strength of horizontal and vertical boundary elements and connections in special plate shear walls

See the applicable sections of the AISC *Seismic Provisions* for specific requirements.

Redundancy Factor, ρ

Redundancy is an important property for structures designed with the expectation that damage will occur. Redundant structures have alternative load paths so that if some elements are severely damaged and lose load carrying capacity, other elements will be able to continue to provide a safe load path. Adequate redundancy is ensured when a large number of plastic hinges must form throughout the structure in a progressive manner before formation of a mechanism and when no one element is required to provide the full seismic resistance of the structure. To encourage provision of a minimum level of redundancy in the structure, ASCI/SEI 7 Section 12.3.4 stipulates a redundancy factor, ρ, based on the structure's configuration and the number of independent seismic force resisting elements present. When structures do not satisfy minimum criteria, this factor amplifies the required strength of the lateral system. The elastic analysis of the SFRS is performed using V, the total design lateral force, based on the tabulated value of R, and ρ is applied to the resulting Q_E member force effects, where Q_E is the effect of horizontal seismic forces.

Maximum Force Delivered by the System

The maximum force delivered by the system is a concept used in several applications in the practice of seismic design. The maximum force delivered by the system is often one of the limits for required strength of a seismic resisting element. For example, a thorough discussion of how this force may be determined for SCBF brace connections is contained in the AISC *Seismic Provisions* Commentary Section F2.6c.

Building Joints

Expansion Joints

Expansion joints in a structure are provided to avoid impairing the function of the facility or causing damage to the structural or architectural components. The number and location of building expansion joints is a design issue not fully treated in technical literature.

- The AISC *Specification* considers expansion joints a serviceability issue, and Section L7 states that "The effects of thermal expansion and contraction of a building shall be considered. Damage to building cladding can cause water penetration and may lead to corrosion."
- ASCE/SEI 7 also considers expansion joints a serviceability issue indicating in Section 1.3.2 that "Structural systems, and members thereof, shall be designed to have adequate stiffness to limit deflections, lateral drift, vibration, or any other deformations that adversely affect the intended use and performance of buildings and other structures."

Typical locations of expansion joints include:

- Where steel framing changes direction
- Separating wings of L, U and T shaped buildings
- At additions to existing buildings
- At locations where interior heating conditions change, such as where heated offices abut an unheated warehouse
- To break very long structures into shorter structures

The width of an expansion joint is determined from the basic thermal expansion expression for the material used for the structural frame:

$$\Delta_L = \alpha L \Delta_T \tag{1-1}$$

where
Δ_L = change in length
α = 6.5×10^{-6}, coefficient of linear expansion for steel structures

1.3 SEISMIC DESIGN OVERVIEW AND DESIGN CONSIDERATIONS

L = length subject to the temperature change
Δ_T = design temperature change

See AISC *Manual* Table 17-11 for additional information on coefficients of expansion.

Seismic Joints

Seismic joints are similar in form to expansion joints but are the result of very different structural considerations. They must accommodate movement in both orthogonal directions simultaneously and their spacing is not typically affected by building length or size. Seismic joints are used to separate an irregular structure into multiple regular structures in an effort to provide better seismic performance of the overall building.

The design of seismic joints is complex and includes efforts by all members of the design team to assure that the joint is properly sized, adequately sealed from weather, and safe to walk on, as well as to provide for adequate movement of other systems crossing the joint and means to maintain the fire ratings of the floor, roof and wall systems. Seismic joints are costly and architecturally undesirable, so they should be incorporated with discretion.

When seismic joints are determined to be necessary or desirable for a particular building, the locations of the joints are often obvious and inherent. Many of the locations appropriate for expansion joints are also appropriate for seismic joints. Requirements for determining the seismic separation between buildings are prescribed in ASCE/SEI 7.

The width of seismic joints in modern buildings can vary from just a few inches to several feet, depending on building height and stiffness. Joints in more recent buildings tend to be much wider than their predecessors. This is due to several major factors, the most important of which is changes in the codes. Other contributing factors are the lower lateral stiffness of many modern buildings and the greater recognition by engineers of the magnitude of real lateral deformations induced by an earthquake.

Seismic joints often result in somewhat complicated structural framing conditions. In the simplest of joints, separate columns are placed at either side of the joint to provide the necessary structural support. This is common in parking structures. When double columns are not acceptable, the structure must either be cantilevered from more widely spaced columns or seated connections must be used. In the case of seated connections, there is the temptation to limit the travel of the sliding element, because longer sliding surfaces using Teflon sliders or similar devices are costly and the seat element may interfere with other elements of the building. It is strongly recommended that seated connections be designed to allow for movements that exceed those calculated for the design basis earthquake to allow for the effects of greater earthquakes and because the consequences of the structure falling off of the seat may be disastrous. Where this is not possible, restraint cables such as those often used on bridges should be considered.

Building Separations

Separations between adjacent buildings that are constructed at different times, have different ownership, or are otherwise not compatible with each other may be necessary and unavoidable if both buildings are located at or near the common property line. ASCE/SEI 7 prescribes setbacks for property lines. An exception can be made where justified by rational analysis based on inelastic response to design ground motions.

Building Drift

Story drift is the maximum lateral displacement within a story (i.e., the displacement of one floor relative to the floor below caused by the effects of seismic loads). Buildings subjected to earthquakes need drift control to limit damage to fragile nonstructural elements, and to minimize differential movement demands on the structure. It is expected that the design of moment resisting frames, and the design of tall, narrow shear-wall or braced-frame buildings will be governed at least in part by drift considerations.

The allowable story drift limits are defined in ASCE/SEI 7 Table 12.12-1 and are a function of the seismic lateral force resisting system and the building risk category. The prescribed story drift limits are applicable to each story. They must not be exceeded in any story even though the drift in other stories may be well below the limit.

Deflection Compatibility

ASCE/SEI 7 prescribes requirements for deformation compatibility for Seismic Design Categories D through F to ensure that the SFRS provides adequate deformation control to protect elements of the structure that are not part of the seismic force resisting system. This is intended to ensure that components designed as gravity supporting only can also resist P-Δ moments, based on total story drifts.

Lowest Anticipated Service Temperature

Most structural steels can fracture either in a ductile or in a brittle manner. The mode of fracture is governed by the temperature at fracture, the rate at which the loads are applied, and the magnitude of the constraints that would prevent plastic deformation. Fracture toughness is a measure of the energy required to cause an element to fracture; the more energy that is required, the tougher the material, i.e., it takes more energy to fracture a ductile material than a brittle material. Additionally, lower temperatures have an adverse impact on material ductility. Fracture toughness for materials can be established by using fracture-mechanics test methods.

Traditionally, the fracture toughness for structural steels has been primarily characterized by testing Charpy V-notch (CVN) specimens at different temperatures [ASTM E23 (ASTM, 2007)]. The CVN test produces failures at very high strain rates. If testing is carried out over a range of temperatures, the results of energy absorbed versus temperature can be plotted to give an S-curve as shown in Figure 1-7. Usually, three specimens are tested at a given temperature and the results averaged.

Carbon and low alloy steels exhibit a change in fracture behavior as the temperature falls with the failure mode changing from ductile to brittle. At high temperatures, the fracture is characterized by pure ductile tearing. At low temperatures, the fracture surface is characterized by cleavage fractures. The decrease in fracture toughness at low temperatures decreases the fracture capacity of the member, resulting in poorer cyclic behavior. (Austenitic stainless steels do not show this change in fracture behavior, with the fracture remaining ductile even to very low temperatures. This is one reason why these types of alloys are used in cryogenic applications.)

The AISC *Seismic Provisions* Commentary Section A3.4 acknowledges that in structures with exposed structural steel, demand critical welds may be subject to service temperatures less than 50 °F on a regular basis. In these cases, the AISC *Seismic Provisions* Commentary suggests that the minimum qualification temperature provided in AWS D1.8 Annex A be adjusted such that the test temperature for the CVN toughness qualification tests be no more than 20 °F above the lowest anticipated service temperature (LAST).

It is recognized that the LAST is defined differently in different industries. For example, the current AASHTO CVN toughness requirements are specified to avoid brittle fracture in steel bridges above the LAST, which is defined in terms of three temperature zones. In arctic offshore applications the LAST can be either the minimum design temperature or a selected value below the design temperature, depending upon the consequences of failure.

The AISC *Seismic Provisions* are intended to ensure ductile performance for a low probability earthquake event. The LAST is normally defined to ensure ductile performance for a low probability temperature extreme. The direct combination of two low probability events would be statistically very unlikely. As a result, the definition of LAST need not be excessively restrictive for seismic applications. For purposes of the AISC *Seismic Provisions*, the LAST may be considered to be the lowest one-day mean temperature compiled from National Oceanic and Atmospheric Administration data. For more information, go to **www.noaa.gov** and **www.climate.gov**.

Quality Control and Quality Assurance

The *International Building Code* (ICC, 2012) refers to the 2010 AISC *Specification* and the 2010 AISC *Seismic Provisions* for all quality requirements for structural steel. The scope statement in Section J1 of the AISC *Seismic Provisions* gives the following explanation for quality control and quality assurance:

> Quality control (QC) as specified in this chapter shall be provided by the fabricator, erector, or other responsible contractor as applicable. Quality assurance (QA) as specified in this chapter shall be provided by others when required by the authority having jurisdiction, applicable building code, purchaser, owner, or engineer of record (EOR).

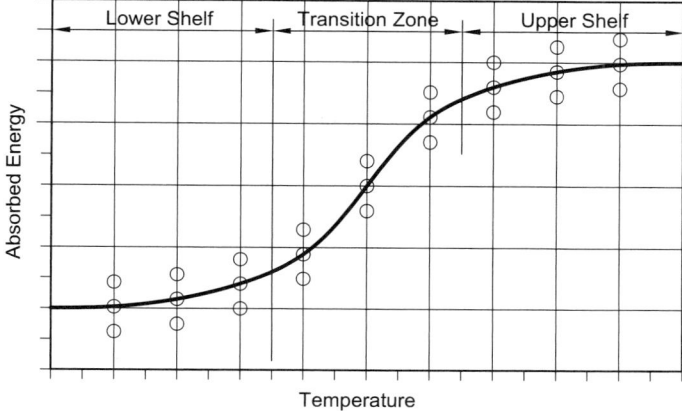

Fig. 1-7. Typical Charpy V-notch test results.

When ductile seismic response should be assured and the AISC *Seismic Provisions* govern the design, fabrication and erection, steel framing needs to meet special quality requirements as appropriate for the various components of the structure. These requirements, applicable only to members of the SFRS, are provided in:

- ANSI/AISC 341-10, *Seismic Provisions for Structural Steel Buildings* (AISC, 2010b)
- AWS D1.8/D1.8M:2009, *Structural Welding Code—Seismic Supplement* (AWS, 2009)
- ANSI/AISC 358-10, *Prequalified Connections for Special and Intermediate Steel Moment Frames for Seismic Applications* (AISC, 2010c)
- 2012 *International Building Code*, Chapter 17 (ICC, 2012)

Additional quality requirements are specified in:

- ANSI/AISC 360-10, *Specification for Structural Steel Buildings* (AISC, 2010a)
- AISC 303-10, *Code of Standard Practice for Steel Buildings and Bridges* (AISC, 2010d)
- AWS D1.1/D1.1M:2010, *Structural Welding Code—Steel* (AWS, 2010)
- 2009 RCSC *Specification for Structural Joints Using High-Strength Bolts* (RCSC, 2009)

The requirements of AISC *Seismic Provisions* Chapter J specify QC and QA special requirements for all responsible parties related to the following:

- Fabricator and erector documents
- Quality assurance agency documents
- Inspection and nondestructive testing personnel
- Inspection tasks
- Welding inspection and nondestructive testing
- Inspection of high-strength bolting
- Other steel structure inspections
- Inspection of composite structures
- Inspection of piling

To meet the requirements of the *International Building Code*, as part of the contract documents, the registered design professional in responsible charge must prepare a "statement of special inspections," which is termed the quality assurance plan (QAP) in the AISC *Seismic Provisions*. The QAP should be prepared by the engineer of record and made a part of the contract documents. The plan should contain, at a minimum, a written description of qualifications, procedures, quality inspections, resources and records to be used to provide assurance that the structure complies with the engineer's quality requirements, specifications and contract documents. Chapter J of the AISC *Seismic Provisions* provides the minimum acceptable requirements for a QAP for the SFRS, including requirements for the contract documents, quality assurance agency documents, inspection points, and frequencies, along with special requirements for weld and bolt inspections.

AISC *Seismic Provisions* Chapter J has specific requirements for nondestructive testing of welds, in addition to those in AISC *Specification* Section N4.5, which must be shown on the contract documents. Quality assurance requirements for bolting include verifying that faying surfaces meet the specification requirements and that the bolts are properly tensioned per the RCSC *Specification*.

Design Drawing Requirements

Structural Design Drawing Requirements

For systems not requiring seismic detailing, structural design drawings are to meet the requirements in the AISC *Code of Standard Practice* as stipulated in AISC *Specification* Section A4. Shop and erection drawings should follow design documents to convey specified information for fabrication and erection. For systems designed to meet the AISC *Seismic Provisions*, additional requirements are provided in AISC *Seismic Provisions* Section A4 with supplementary discussion in the *Provisions* Commentary Section A4. It is important to define all structural elements in the building that resist seismic loads, including struts, collectors, chords, diaphragms and trusses. Also, the SFRS members should be indicated in both plan and elevation drawings. If the SFRS includes other materials, these elements should be defined as such where the steel connects to them.

SFRS Member and Connection Material Specifications

SFRS material requirements are discussed in AISC *Seismic Provisions* Section A3.1 and in the material sections of the various prequalified connections in ANSI/AISC 358. Wide flange shapes will generally be ASTM A992 material. ASTM A992 has specified maximum yield stress and maximum yield-to-tensile ratio to ensure ductility along with a limit on the carbon equivalent to ensure weldability. Material requirements for the connection elements must be consistent with the prequalified details in ANSI/AISC 358. Bolt material grade, size, location and tensioning must be shown on the design drawings. Bolts typically are designed as bearing-type connections with standard holes and all bolts are required to be pretensioned and have Class A faying surfaces. AISC *Seismic Provisions* Section D2.3 on welded joints, references AISC *Specification* Chapter J. AISC *Specification* Section J2 stipulates that all requirements from AWS D1.1, including weld procedure specifications, are applicable except for the specific AWS D1.1 provisions cited. The AISC *Seismic Provisions* Section A3.4 requires that all welds in the SFRS be made with filler metals meeting the requirements specified in clause 6.3 of AWS D1.8. Clause 6.3 requires that all welds provide a minimum Charpy V-notch toughness of 20 ft-lb at 0 °F, either by test or manufacturer's certification. There are additional requirements for demand critical welds as noted below.

Demand Critical Welds

Welds are designated demand critical in the AISC *Seismic Provisions* based on consideration of the inelastic strain demand and the consequence of failure. The location of these demand critical welds is given in the AISC *Seismic Provisions* and in ANSI/AISC 358 in the section applicable to the designated SFRS. As specified in AISC *Seismic Provisions* Section A3.4b, demand critical welds shall be made with filler metals meeting the requirements of AWS D1.8 clause 6.3. Clause 6.3 requires a minimum Charpy V-notch toughness of 40 ft-lb at 70 °F for a LAST of 50 °F or greater. See AWS D1.8 for detailed testing requirements and for a LAST less than 50 °F.

There are a number of other quality control and quality assurance items associated with demand critical welds that are covered in the AISC *Seismic Provisions* and AWS D1.8. Items such as use of backing bars and run-off tabs, including requirements for trimming and finishing of run-off tabs, are specifically addressed.

Locations and Dimensions of Protected Zones

Protected zones are designated by the AISC *Seismic Provisions* for different systems and generally are areas encompassing the plastic hinging region. The FEMA/SAC testing has demonstrated the sensitivity of these areas to fracture caused by discontinuities resulting from welding, penetrations, changes in section, or construction-caused notches (Ricles et al., 2003). Fabrication and erection work, and the subsequent work by other trades, have the potential to cause discontinuities in the SFRS. AISC *Seismic Provisions* Sections D1.3 and I2.1 provide detailed requirements for the protected zone.

The locations and dimensions of these protected zones for moment connections are specified in the AISC *Seismic Provisions* and in ANSI/AISC 358 for each SFRS. For example, according to AISC *Seismic Provisions* Section F2.5c, the protected zone for special concentrically braced frames includes "the center one-quarter of the brace length and a zone adjacent to each connection equal to the brace depth in the plane of buckling" as well as "elements that connect braces to beams and columns." For eccentrically braced frames, AISC *Seismic Provisions* Section F3.5c defines the protected zone as the link. In any case, the requirements in AISC *Seismic Provisions* Sections D1.3 and I2.1 must be satisfied.

When located in the protected zone these discontinuities are required to be repaired by the responsible contractor to the satisfaction of the engineer of record. The AISC *Seismic Provisions* require that the protected zones be shown on the design drawings. The contractor needs to use this information to control construction activities in this area.

Additional Structural Design Drawing Detail Requirements in the Provisions

Following are some of the additional requirements from the AISC *Seismic Provisions* that may affect structural design drawing details:

1. SFRS column splice requirements are given in AISC *Seismic Provisions* Section D2.5a. The splices need to be located away from beam-to-column connections, with the provisions stipulating 4 ft or more away from the connection; however, in general, splices should be in the middle third of the column (see Exceptions in Section D2.5a). Because of the splice strength requirements in Section D2.5, it is important that the splice be fully detailed on the design drawings. Where bolted splices are used there must be plates or channels on both sides of the web.
2. Column splice requirements for columns that are not part of the SFRS are given in the AISC *Seismic Provisions* Section D2.5c. The minimum shear forces required to be developed in these splices will require a special column splice and this detail should also be shown on the design drawings.
3. SFRS column bases must meet the requirements of AISC *Seismic Provisions* Section D2.6 and anchor rod embedment and reinforcing steel should be designed according to ACI 318 Appendix D. Anchor rod sizes and locations, along with washer requirements, hole sizes and base plate welds must meet these design requirements and must be shown. Special embedment used for base fixity must also be shown on the structural design drawings. The Commentary to Section D2.6 gives a good discussion along with examples of how to develop these forces. For column bases that are not part of

1.3 SEISMIC DESIGN OVERVIEW AND DESIGN CONSIDERATIONS

SFRS, some consideration should be given to developing a limited amount of base shear either by embedment or by bearing on the anchor rods. AISC *Seismic Provisions* Section D2.6b stipulates the required shear strength for column bases, including those not designated as part of the SFRS.

4. Width-to-thickness ratios of SFRS members must be less than those that are resistant to local buckling in order to achieve the required inelastic deformations required. While the width-to-thickness ratios given in the AISC *Specification* Table B4.1 for compact sections are adequate to prevent buckling before the onset of strain hardening, tests have shown that they are not adequate for the required inelastic performance in several SFRS. AISC *Seismic Provisions* Table D1.1 gives the limiting width-to-thickness ratios for moderately ductile and highly ductile members. Classification of members as moderately or highly ductile may govern member size for the various systems.

5. Requirements for stability bracing of beams are provided for each system. The bracing required is stipulated in AISC *Seismic Provisions* Section D1.2 and depends on whether the beam is moderately or highly ductile. Special bracing is required adjacent to plastic hinge locations. If the bracing requirement cannot be met by the floor slab and the elements of the moment connection, then the required bracing member(s) and connection(s) should be shown. For example, special moment frame beams require bracing that satisfy the provisions for highly ductile members as given in AISC *Seismic Provisions* D1.2b. While the floor slab typically will brace the top flange, additional braces should be shown where required with the necessary connections.

AWS D1.8 Structural Welding Code—Seismic Supplement

AWS D1.8, subclause 1.2.1 lists the information that the engineer of record is required to provide on the contract documents specifically related to welding of the SFRS. Additionally, gouges and notches are not permitted and while grinding to a flush condition is not required, the contour should provide a smooth transition. AWS D1.8 provides recommended details for these areas.

AWS D1.8 contains a number of other special requirements that should be specifically referenced in the contract documents. In addition to the filler metal requirements mentioned previously, demand critical welds have the following requirements:

- Manufacturer's certificates of conformance for filler metals
- Special restrictions on care and exposure of electrodes
- Supplemental welder qualification for restricted access welding for bottom flange welding through access holes
- Special weld sequence for bottom flange welding through access holes
- Supplementary requirements for qualification of ultrasonic testing technicians

Composite Systems

For buildings with composite members and/or composite SFRS, an important change in the 2010 AISC *Seismic Provisions* is the integration of what were formerly presented separately in Parts I (steel) and II (composite) into a combined set of provisions. This edition of the *Seismic Design Manual* follows that approach by adding examples for composite systems.

The 2010 AISC *Seismic Provisions* for the seismic design of composite structural steel and reinforced concrete buildings are based upon the 1994 *NEHRP Provisions* (FEMA, 1994) and subsequent modifications made in the 1997, 2000, 2003 and 2009 *NEHRP Provisions* (FEMA, 2009) and in ASCE/SEI 7. Because composite systems are comprised of integrated steel and concrete components, both the AISC *Specification* and ACI 318 form an important basis for provisions related to composite construction.

There is, at present, limited experience in the U.S. with composite building systems subjected to extreme seismic loads. Extensive design and performance experience with this type of construction in Japan clearly indicates that composite systems, due to their inherent rigidity and toughness, can equal or exceed the performance of buildings comprised of reinforced concrete systems or structural steel systems (Deierlein and Noguchi, 2004; Yamanouchi et al., 1998). Composite systems have been extensively used in tall buildings throughout the world.

Careful attention to all aspects of the design is necessary in the design of composite systems, particularly with respect to the general building layout and detailing of members and connections. Composite connection details are illustrated throughout this Manual to convey the basic character of the force transfer in composite systems. However, these details should not necessarily be treated as design standards. The cited references provide more specific information on the design of composite connections. For a general discussion of these issues and some specific design examples, refer to Viest et al. (1997).

The design and construction of composite elements and systems continues to evolve in practice. Except where explicitly stated, the AISC *Seismic Provisions* are not intended to limit the application of new systems for which testing and analysis demonstrates that the structure has adequate strength, ductility and toughness. It is generally anticipated that the overall behavior of the composite systems herein will be similar to that for counterpart structural steel systems or reinforced concrete systems and that inelastic deformations will occur in conventional ways, such as flexural yielding of beams in fully restrained moment frames or axial yielding and/or buckling of braces in braced frames.

When systems have both ductile and nonductile elements, the relative stiffness of each should be properly modeled; the ductile elements can deform inelastically while the nonductile elements remain nominally elastic. When using elastic analysis, member stiffness should be reduced to account for the degree of cracking at the onset of significant yielding in the structure. Additionally, it is necessary to account for material overstrength that may alter relative strength and stiffness.

Parts 6 and 7 of this Manual provide discussion and example problems for the design of members and connections for composite moment frame and braced frame systems, respectively, as well as guidelines for traversing through the AISC *Seismic Provisions* and AISC *Specification* relative to each specific building system. Where possible, the example problems presented were developed to be companions to the example problems presented in other parts of this Manual. For instance, the example problem for the composite special moment frame system illustrates the application of the composite requirements when a concrete-filled tube column replaces the steel column of the special moment frame structure illustrated in the example problems in Part 4 of this Manual.

1.4 DESIGN TABLE DISCUSSION

Seismic Weld Access Hole Configurations

Table 1-1. Workable Seismic Weld Access Hole Configurations

Fourteen configurations are given based upon the minimum seismic weld access hole profile. This table is intended to be used in conjunction with Table 1-3 for quick selection of weld access hole geometry for wide-flange beams when the special seismic weld access hole is used. A workable seismic access hole configuration from Table 1-1 is given in Table 1-3 for each shape listed. Where a dash is shown, no configuration shown in Table 1-1 meets all criteria.

AISC *Specification* Section J1.6 provides general requirements for weld access holes. It should be noted that the geometries shown in Table 1-1 represent only one set of configurations that satisfy the dimensions and tolerances in AWS D1.8 Figure 6.2. Other configurations that comply with AWS D1.8 Figure 6.2 may also be used. The special seismic weld access hole is required for beams in ordinary moment frames per AISC *Seismic Provisions* Section E1.6b(c), and for beams in welded unreinforced flange-welded web (WUF-W) moment connections per ANSI/AISC 358.

Member Ductility Requirements

Table 1-2. Summary of Member Ductility Requirements

Ductility requirements are summarized for SFRS members per Chapters E, F, G and H of the AISC *Seismic Provisions*.

Local Buckling Requirements

Table 1-3. Sections That Satisfy Seismic Width-to-Thickness Requirements, W-Shapes

W-shapes with $F_y = 50$ ksi (ASTM A992) that satisfy the moderately or highly ductile width-to-thickness requirements per the AISC *Seismic Provisions* Table D1.1 are indicated with a "•" in the column corresponding to the member requirements for specific SFRS. This includes W-shapes that incorporate reduced beam section moment connections. See Table 1-2 for a summary of the member ductility requirements for the SFRS in the AISC *Seismic Provisions*. A wide-flange section satisfies these requirements if its flange and web width-to-thickness ratios are less than or equal to the corresponding limits listed in Table 1-A, which is summarized from the requirements in Table D1.1 of the AISC *Seismic Provisions*. For cases where the limiting web width-to-thickness ratio is a function of the member's required axial strength, P_u or P_a, the member will satisfy the width-to-thickness requirements if P_u or P_a is less than or equal to the value tabulated for $P_{u\ max}$ or $P_{a\ max}$, respectively. The nominal axial yield strength of a member, P_y, is calculated as $F_y A_g$. Note that in these cases it is assumed that $C_a = P_u/\phi_c P_y > 0.125$ or $\Omega_c P_a/P_y > 0.125$. Exceptions for intermediate moment frame and special moment frame beams with $C_a < 0.125$ are indicated in the footnotes of Table 1-A. Where a dash is shown, there is no limitation on the values of P_u or P_a.

Also provided is the maximum spacing of beam bracing for moderately ductile and highly ductile beams, $L_{b\,max}$, where for moderately ductile beams, $L_{b\,max} = 0.17 r_y\, E/F_y$, and highly ductile beams, $L_{b\,max} = 0.086 r_y\, E/F_y$. Note that W-shapes that do not satisfy either moderately or highly ductile width-to-thickness ratios are not included in Table 1-3.

Table 1-A
Limiting Width-to-Thickness Ratios for W-Shape Flanges and Webs in Compression

	Member	Limiting Width-to-Thickness Ratio Flange, b/t	Limiting Width-to-Thickness Ratio Web, h/t_w
Moderately Ductile	Diagonal Brace	$0.38\sqrt{E/F_y}$	$1.49\sqrt{E/F_y}$
	Beam,[a] Column, EBF Link[b]	$0.38\sqrt{E/F_y}$	For $C_a \leq 0.125$ $3.76\sqrt{E/F_y}\,(1 - 2.75 C_a)$ For $C_a > 0.125$ $1.12\sqrt{E/F_y}\,(2.33 - C_a) \geq 1.49\sqrt{E/F_y}$ where $C_a = \dfrac{P_u}{\phi_c P_y}$ (LRFD) $C_a = \dfrac{\Omega_c P_a}{P_y}$ (ASD)
Highly Ductile	Diagonal Brace	$0.30\sqrt{E/F_y}$	$1.49\sqrt{E/F_y}$
	Beam,[c] Column, Chords in STMF Special Segment, EBF Link, SPSW VBE & HBE	$0.30\sqrt{E/F_y}$	For $C_a \leq 0.125$ $2.45\sqrt{E/F_y}\,(1 - 0.93 C_a)$ For $C_a > 0.125$ $0.77\sqrt{E/F_y}\,(2.93 - C_a) \geq 1.49\sqrt{E/F_y}$ where $C_a = \dfrac{P_u}{\phi_c P_y}$ (LRFD) $C_a = \dfrac{\Omega_c P_a}{P_y}$ (ASD)

[a] For W-shape beams in SMF systems where C_a is less than or equal to 0.125, the limiting ratio h/t_w shall not exceed $3.76\sqrt{E/F_y}$.
[b] Applies to EBF links meeting the exception in Section F3.5b(1).
[c] For W-shape beams in SMF systems where C_a is less than or equal to 0.125, the limiting width-to-thickness ratio h/t_w shall not exceed $2.45\sqrt{E/F_y}$.

Table 1-4. Sections That Satisfy Seismic Width-to-Thickness Requirements, Angles

Angles with $F_y = 36$ ksi (A36), including both single and double angle configurations, that satisfy AISC *Seismic Provisions* local buckling requirements for use as diagonal braces in SCBF, OCBF, EBF, and the special segment of STMF chords are indicated with a "•" in the corresponding column. An angle satisfies these requirements if the greatest leg width-to-thickness ratio is less than or equal to the corresponding limits listed in Table 1-B, which is summarized from the requirements in Table D1.1 of the AISC *Seismic Provisions*. Note that angles that do not satisfy either moderately or highly ductile width-to-thickness ratios are not included in Table 1-4.

Table 1-B
Limiting Width-to-Thickness Ratios for Angle Legs in Compression

	Member	Width-to-Thickness Ratio	Limiting Width-to-Thickness Ratio
Moderately Ductile	Diagonal Brace	b/t	$0.38\sqrt{E/F_y}$
Highly Ductile	Diagonal Brace, Chords in STMF Special Segment	b/t	$0.30\sqrt{E/F_y}$

Table 1-5a. Sections That Satisfy Seismic Width-to-Thickness Requirements, Rectangular HSS

Table 1-5b. Sections That Satisfy Seismic Width-to-Thickness Requirements, Square HSS

Rectangular and square HSS with $F_y = 46$ ksi (ASTM A500 Grade B) that satisfy the AISC *Seismic Provisions* local buckling requirements for use as diagonal braces or columns in SCBF, and braces in OCBF and EBF are indicated with a "•" in the corresponding column. A rectangular or square HSS satisfies these requirements if its flange and web width-to-thickness ratios are less than or equal to the corresponding limits listed in Table 1-C, which is summarized from the requirements of Table D1.1 of the AISC *Seismic Provisions*. Note that HSS sections that do not satisfy either moderately or highly ductile width-to-thickness ratios are not included in Tables 1-5a or 1-5b.

Table 1-C
Limiting Width-to-Thickness Ratios for Rectangular and Square HSS Walls in Compression

	Member	Width-to-Thickness Ratio	Limiting Width-to-Thickness Ratio
Moderately Ductile	Diagonal Brace, Beam, Column	b/t	$0.64\sqrt{E/F_y}$ [a]
Highly Ductile	Diagonal Brace, Beam, Column	b/t	$0.55\sqrt{E/F_y}$

[a] The limiting width-to-thickness ratio of walls of rectangular and square HSS members used as beams or columns shall not exceed $1.12\sqrt{E/F_y}$.

Table 1-6. Sections That Satisfy Seismic Width-to-Thickness Requirements, Round HSS

Round HSS sections with $F_y = 42$ ksi (ASTM A500 Grade B) that satisfy the AISC *Seismic Provisions* local buckling requirements for use as braces or columns in SCBF and braces in OCBF and EBF are indicated with a "•" in the corresponding column. A round HSS satisfies these requirements if its width-to-thickness ratio is less than or equal to the corresponding limit listed in Table 1-D. Note that round HSS sections that do not satisfy either moderately or highly ductile width-to-thickness ratios are not included in Table 1-6.

Table 1-D
Limiting Width-to-Thickness Ratios for Round HSS and Pipe Walls in Compression

	Member	Width-to-Thickness Ratio	Limiting Width-to-Thickness Ratio
Moderately Ductile	Diagonal Brace, Beam, Column	D/t	$0.044\,(E/F_y)$ [a]
Highly Ductile	Diagonal Brace, Beam, Column	D/t	$0.038\,(E/F_y)$

[a] The limiting diameter-to-thickness ratio of walls of round HSS members used as beams or columns shall not exceed $0.07E/F_y$.

Table 1-7. Sections That Satisfy Seismic Width-to-Thickness Requirements, Pipe

Pipes with F_y = 35 ksi (ASTM A53 Grade B) that satisfy AISC *Seismic Provisions* local buckling requirements for use as braces or columns in SCBF and braces in OCBF and EBF are indicated with a "•" in the corresponding column. A pipe satisfies these requirements if its width-to-thickness ratio, D/t, is less than or equal to the corresponding limit listed in Table 1-D. Note that pipe that do not satisfy either moderately or highly ductile width-to-thickness ratios are not included in Table 1-7.

Strength of Steel Headed Stud Anchors

Table 1-8. Nominal Horizontal Shear Strength and 25% Reduced Nominal Horizontal Shear Strength for One Steel Headed Stud Anchor

The nominal shear strength of steel headed stud anchors is given in Table 1-8, in accordance with AISC *Specification* Chapter I. This table provides the nominal shear strength for one steel headed stud anchor embedded in a solid concrete slab or in a composite slab with decking, as given in AISC *Specification* Section I8.2a. The nominal shear strength with the 25% reduction as specified in AISC *Seismic Provisions* Section D2.8 for intermediate or special SFRS of Sections G2, G3, G4, H2, H3, H5 and H6 is also given in Table 1-8. According to the User Note in AISC *Seismic Provisions* Section D2.8, the 25% reduction is not necessary for gravity or collector components in structures with intermediate or special seismic force resisting systems designed for the amplified seismic load. Nominal horizontal shear strength values are presented based upon the position of the steel anchor, profile of the deck, and orientation of the deck relative to the steel anchor. See AISC *Specification* Commentary Figure C-I8.1.

ASCE/SEI 7 Design Coefficients and Factors for SFRS

Table 1-9a. Design Coefficients and Factors for Steel and Steel and Concrete Composite Seismic Force Resisting Systems

This table is based on ASCE/SEI 7 Table 12.2-1 and provides design coefficients and factors for steel and composite seismic force resisting systems (ASCE, 2010).

Table 1-9b. Design Coefficients and Factors for Nonbuilding Structures Similar to Buildings

This table is based on ASCE/SEI 7 Table 15.4-1 and provides design coefficients and factors for steel and composite seismic force resisting systems in nonbuilding structures similar to buildings (ASCE, 2010).

PART 1 REFERENCES

ACI (2008), *Building Code Requirements for Structural Concrete*, ACI 318-08, American Concrete Institute, Farmington Hills, MI.

AISC (2010a), *Specification for Structural Steel Buildings*, ANSI/AISC 360-10, American Institute of Steel Construction, Chicago, IL.

AISC (2010b), *Seismic Provisions for Structural Steel Buildings*, ANSI/AISC 341-10, American Institute of Steel Construction, Chicago, IL.

AISC (2010c), *Prequalified Connections for Special and Intermediate Steel Moment Frames for Seismic Applications*, ANSI/AISC 358-10, American Institute of Steel Construction, Chicago, IL.

AISC (2010d), *Code of Standard Practice for Steel Buildings and Bridges*, American Institute of Steel Construction, Chicago, IL.

AISC (2011), *Steel Construction Manual*, 14th Ed., American Institute of Steel Construction, Chicago, IL.

ASCE (2010), *Minimum Design Loads for Buildings and Other Structures*, ASCE/SEI 7-10, American Society of Civil Engineers, Reston, VA.

ASTM (2007), *Standard Test Methods for Notched Bar Impact Testing of Metallic Materials*, ASTM E23-07ae1, ASTM International, West Conshohocken, PA.

AWS (2007), *Standard Symbols for Welding, Brazing, and Nondestructive Examination*, AWS A2.4, American Welding Society, Miami, FL.

AWS (2009), *Structural Welding Code—Seismic Supplement*, AWS D1.8/D1.8M:2009, American Welding Society, Miami, FL.

AWS (2010), *Structural Welding Code—Steel*, AWS D1.1/D1.1M:2010, American Welding Society, Miami, FL.

Chopra, A.K. (2007), *Dynamics of Structures: Theory and Applications to Earthquake Engineering*, 3rd Ed., Prentice Hall, Upper Saddle River, NJ.

Deierlein, G.G. and Noguchi, H. (2004), "Overview of US-Japan Research on the Seismic Design of Composite Reinforced Concrete and Steel Moment Frame Structures," *Journal of Structural Engineering*, ASCE, Vol. 130, No. 2, pp. 361–367, Reston, VA.

FEMA (1994), *NEHRP Recommended Provisions for Seismic Regulations for New Buildings and Other Structures*, Washington, DC.

FEMA (2009), *NEHRP Recommended Provisions for Seismic Regulations for New Buildings and Other Structures*, FEMA P-750, Washington, DC.

ICC (2012), *International Building Code*, International Code Council, Falls Church, VA.

RCSC (2009), *Specification for Structural Joints Using High-Strength Bolts*, Research Council on Structural Connections, American Institute of Steel Construction, Chicago, IL.

Ricles, J.M., Mao, C., Lu, L.W. and Fisher, J.W. (2003), "Ductile Details For Welded Unreinforced Moment Connections Subject To Inelastic Cyclic Loading," *Journal of Engineering Structures*, Elsevier, Vol. 25, pp. 667–680.

Viest, I.M., Colaco, J.P., Furlong, R.W., Griffis, L.G., Leon, R.T. and Wyllie, L.A., Jr. (1997), *Composite Construction: Design for Buildings*, McGraw-Hill/ASCE, Reston, VA.

Yamanouchi, H., Nishiyama, I. and Kobayashi, J. (1998), "Development and Usage of Composite and Hybrid Building Structure in Japan," *ACI SP-174*, American Concrete Institute, pp. 151–174.

Table 1-1
Workable Seismic Weld Access Hole Configurations

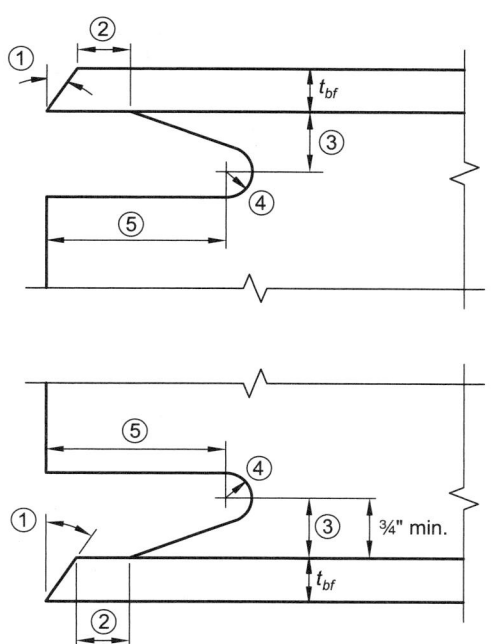

Access Hole Type	Dimension for weld access hole geometry in accordance with AWS D1.8/D1.8M subclause 6.10.1				
	①	②	③	④	⑤
	degrees	in.	in.	in.	in.
A	30	1/2	3/4	1/2	1 1/4
B	↓	1/2	3/4	↓	1 1/2
C		3/4	1		2 1/2
D		1	1 1/4		3 1/2
E		1 1/4	1 1/2		4 1/2
F		1 1/2	1 3/4		5 1/2
G		1 3/4	2		6 1/2
H		2	2 1/4		7 1/2
I		2 1/4	2 1/2		8 1/2
J		2 1/2	2 3/4		9 1/2
K		3	3		11
L		3 1/4	3 1/2		12 1/2
M		3 3/4	4		14
N	↓	4	4 1/4	↓	15

Table 1-2
Summary of Member Ductility Requirements

System	Highly Ductile λ_{hd}	Moderately Ductile λ_{md}	No Ductility Requirements per *Seismic Provisions*	*Seismic Provisions* Section Reference
Ordinary Moment Frame (OMF)			•	E1.5a
Intermediate Moment Frame (IMF)				
• Beams		•		E2.5a
• Columns		•		E2.5a
Special Moment Frames (SMF)				
• Beams	•			E3.5a
• Columns	•			E3.5a
Special Truss Moment Frames (STMF)				
• Chords in Special Segment	•			E4.5c
• Special Segment Diagonal Webs	•			E4.5c
Ordinary Cantilever Column Systems (OCCS)			•	E5.5a
Special Cantilever Column Systems (SCCS)				
• Columns	•			E6.5a
Ordinary Concentrically Braced Frames (OCBF)				
• Diagonal Braces		•		F1.5a
Special Concentrically Braced Frames (SCBF)				
• Diagonal Braces	•			F2.5a
• Beams		•		F2.5a
• Columns	•			F2.5a
Eccentrically Braced Frames (EBF)				
• Diagonal Braces		•		F3.5a
• Columns	•			F3.5a
• Link Beams	•[a]			F3.5b(1)
• Beams outside of the Link		•		F3.5a
Buckling-Restrained Braced Frames (BRBF)				
• Beams		•		F4.5a
• Columns		•		F4.5a
Special Plate Shear Walls (SPSW)				
• Horizontal Boundary Element		•		F5.5a
• Vertical Boundary Element		•		F5.5a
• Intermediate Boundary Elements		•		F5.5a
Composite Ordinary Moment Frames (C-OMF)			•	G1.5

[a] See exceptions in Section F3.5b(1).
[b] See exception in Section G3.5a.

Table 1-2 (continued)
Summary of Member Ductility Requirements

System	Highly Ductile λ_{hd}	Moderately Ductile λ_{md}	No Ductility Requirements per *Seismic Provisions*	*Seismic Provisions* Section Reference
Composite Intermediate Moment Frames (C-IMF)				
• Steel and Composite Beams		•		G2.5a
• Steel and Composite Columns		•		G2.5a
Composite Special Moment Frames (C-SMF)				
• Steel and Composite Beams	•			G3.5a
• Steel and Composite Columns	•			G3.5a
• Reinforced Concrete-Encased Beams	•[b]			G3.5a
Composite Partially Restrained Moment Frames (C-PRMF)				
• Steel Columns		•		G4.5a
• Composite Beams		•		G4.5b
Composite Ordinary Braced Frames (C-OBF)			•	H1.5a
Composite Special Concentrically Braced Frames (C-SCBF)				
• Composite Columns		•		H2.5a
• Steel Braces or Composite Braces		•		H2.5a
• Steel or Composite Beams			•	H2.5a
Composite Eccentrically Braced Frames (C-EBF)				
• Diagonal Braces			•	H3.5 & F3.5a
• Columns		•		H3.5 & F3.5a
• Link Beams	•[a]			H3.5 & F3.5b(1)
• Beams outside of the Link			•	H3.5 & F3.5a
Composite Ordinary Shear Walls (C-OSW)				
• Steel Coupling Beams			•	H4.5b(1)
• Encased Composite Coupling Beams			•	H4.5b(1)&(2)
Composite Special Shear Walls (C-SSW)				
• Unencased Structural Steel Columns		•		H5.5b
• Concrete Encased Structural Steel Columns		•		H5.5b
• Steel Coupling Beams	•[a]			H5.5c, F3.5b(1)
• Encased Composite Coupling Beams	•[a]			H5.5c,d, F3.5b(1)
Composite Plate Shear Walls (C-PSW)				
• Steel and Composite Horizontal Boundary Elements		•		H6.5a
• Steel and Composite Vertical Boundary Elements		•		H6.5a

[a] See exceptions in Section F3.5b(1).
[b] See exception in Section G3.5a.

Table 1-3
Sections That Satisfy Seismic Width-to-Thickness Requirements
W-Shapes

$F_y = 50$ ksi

Shape	IMF	SMF	STMF	SCCS	OCBF	SCBF			$L_{b\,max}$, ft	
	Beams and Columns	Beams and Columns	Chord Segment	Columns	Diagonal Braces	Diagonal Braces	Columns	Beams	λ_{hd}	λ_{md}
W44×335	•	•	•	•			•	•	14.5	28.7
×290	•	•	•	•			•	•	14.5	28.7
×262	•	•	•	•			•	•	14.4	28.5
×230	•	•	•	•			•	•	14.3	28.2
W40×593	•	•	•	•	•	•	•	•	15.8	31.2
×503	•	•	•	•	•	•	•	•	15.5	30.6
×431	•	•	•	•	•	•	•	•	15.2	30.0
×397	•	•	•	•	•	•	•	•	15.1	29.9
×372	•	•	•	•	•	•	•	•	15.0	29.6
×362	•	•	•	•	•	•	•	•	15.0	29.6
×324	•	•	•	•	•	•	•	•	14.9	29.4
×297	•	•	•	•			•	•	14.7	29.1
×277	•	•	•	•			•	•	14.9	29.4
×249	•	•	•	•			•	•	14.8	29.2
×215	•	•	•	•			•	•	14.7	29.1
×199	•							•	14.3	28.3
W40×392	•	•	•	•	•	•	•	•	11.0	21.7
×331	•	•	•	•	•	•	•	•	10.7	21.1
×327	•	•	•	•	•	•	•	•	10.7	21.2
×294	•	•	•	•	•	•	•	•	10.6	21.0
×278	•	•	•	•	•	•	•	•	10.5	20.7
×264	•	•	•	•	•	•	•	•	10.5	20.7
×235	•	•	•	•			•	•	10.6	20.9
×211	•	•	•	•			•	•	10.4	20.6
×183	•	•	•	•			•	•	10.4	20.5
×167	•	•	•	•			•	•	9.98	19.7
×149	•	•	•	•			•	•	9.52	18.8

Table 1-3 (continued)
Sections That Satisfy Seismic Width-to-Thickness Requirements
W-Shapes

F_y = 50 ksi

Shape	EBF			BRBF	SPSW	$P_{a\,max}$, kips ASD		$P_{u\,max}$, kips LRFD		Web Access Holes
	Diagonal Braces	Columns	Links	Beams and Columns	HBE and VBE	λ_{hd}	λ_{md}	λ_{hd}	λ_{md}	
W44×335		•	•	•	•	2600	2720	3900	4080	F
×290		•	•	•	•	1290	1690	1930	2540	E
×262		•	•	•	•	590	1140	887	1710	E
×230		•	•	•	•	156	606	234	910	D
W40×593	•	•	•	•	•	—	—	—	—	J
×503	•	•	•	•	•	—	—	—	—	I
×431	•	•	•	•	•	—	—	—	—	H
×397	•	•	•	•	•	—	—	—	—	G
×372	•	•	•	•	•	—	—	—	—	G
×362	•	•	•	•	•	—	—	—	—	G
×324	•	•	•	•	•	—	—	—	—	F
×297		•	•	•	•	2470	2520	3710	3790	E
×277		•	•	•	•	1730	1960	2600	2940	E
×249		•	•	•	•	1040	1410	1560	2110	E
×215		•	•	•	•	222	722	333	1090	D
×199			1		•	205	669	309	1010	D
W40×392	•	•	•	•	•	—	—	—	—	H
×331	•	•	•	•	•	—	—	—	—	G
×327	•	•	•	•	•	—	—	—	—	G
×294	•	•	•	•	•	—	—	—	—	F
×278	•	•	•	•	•	—	—	—	—	F
×264	•	•	•	•	•	—	—	—	—	F
×235		•	•	•	•	1470	1660	2200	2500	E
×211		•	•	•	•	876	1190	1320	1790	E
×183		•	•	•	•	186	606	280	911	D
×167		•	•	•	•	172	561	259	843	D
×149		•	•	•	•	112	416	169	625	C

Note 1: Links in EBF that meet the exception in the AISC *Seismic Provisions* Section F3.5b(1) need only meet the limits for moderately ductile members.

Table 1-3 (continued)
Sections That Satisfy Seismic Width-to-Thickness Requirements
W-Shapes

$F_y = 50$ ksi

Shape	IMF Beams and Columns	SMF Beams and Columns	STMF Chord Segment	SCCS Columns	OCBF Diagonal Braces	SCBF Diagonal Braces	SCBF Columns	SCBF Beams	$L_{b\,max}$, ft λ_{hd}	$L_{b\,max}$, ft λ_{md}
W36×652	•	•	•	•	•	•	•	•	17.0	33.7
×529	•	•	•	•	•	•	•	•	16.6	32.9
×487	•	•	•	•	•	•	•	•	16.5	32.5
×441	•	•	•	•	•	•	•	•	16.3	32.2
×395	•	•	•	•	•	•	•	•	16.1	31.9
×361	•	•	•	•	•	•	•	•	16.0	31.6
×330	•	•	•	•	•	•	•	•	15.9	31.5
×302	•	•	•	•	•	•	•	•	15.9	31.4
×282	•	•	•	•	•		•	•	15.8	31.2
×262	•	•	•	•	•		•	•	15.6	30.9
×247	•	•	•	•	•		•	•	15.5	30.7
×231	•	•	•	•			•	•	15.4	30.5
W36×256	•	•	•	•	•	•	•	•	11.0	21.8
×232	•	•	•	•			•	•	10.9	21.5
×210	•	•	•	•			•	•	10.7	21.2
×194	•	•	•	•			•	•	10.6	21.0
×182	•	•	•	•			•	•	10.6	21.0
×170	•	•	•	•			•	•	10.5	20.8
×160	•	•	•	•			•	•	10.4	20.5
×150	•	•	•	•			•	•	10.3	20.3
×135	•							•	9.89	19.6
W33×387	•	•	•	•	•	•	•	•	15.7	31.0
×354	•	•	•	•	•	•	•	•	15.5	30.7
×318	•	•	•	•	•	•	•	•	15.4	30.5
×291	•	•	•	•	•	•	•	•	15.3	30.2
×263	•	•	•	•	•	•	•	•	15.2	30.1
×241	•	•	•	•			•	•	15.0	29.7
×221	•	•	•	•			•	•	14.9	29.5
×201	•	•	•	•			•	•	14.8	29.3
W33×169	•	•	•	•			•	•	10.4	20.5
×152	•	•	•	•			•	•	10.3	20.3
×141	•	•	•	•			•	•	10.1	20.0
×130	•	•	•	•			•	•	9.93	19.6
×118	•							•	9.64	19.1

Table 1-3 (continued)
Sections That Satisfy Seismic Width-to-Thickness Requirements
W-Shapes

F_y = 50 ksi

Shape	EBF			BRBF	SPSW	$P_{a\,max}$, kips ASD		$P_{u\,max}$, kips LRFD		Web Access Holes
	Diagonal Braces	Columns	Links	Beams and Columns	HBE and VBE	λ_{hd}	λ_{md}	λ_{hd}	λ_{md}	
W36×652	•	•	•	•	•	—	—	—	—	K
×529	•	•	•	•	•	—	—	—	—	I
×487	•	•	•	•	•	—	—	—	—	I
×441	•	•	•	•	•	—	—	—	—	H
×395	•	•	•	•	•	—	—	—	—	G
×361	•	•	•	•	•	—	—	—	—	G
×330	•	•	•	•	•	—	—	—	—	F
×302	•	•	•	•	•	—	—	—	—	F
×282		•	•	•	•	2430	2450	3650	3690	E
×262		•	•	•	•	2010	2110	3020	3170	E
×247		•	•	•	•	1670	1830	2500	2750	E
×231		•	•	•	•	1340	1560	2010	2350	D
W36×256	•	•	•	•	•	—	—	—	—	F
×232		•	•	•	•	1870	1930	2810	2900	E
×210		•	•	•	•	1520	1630	2290	2450	E
×194		•	•	•	•	1100	1290	1650	1940	D
×182		•	•	•	•	825	1070	1240	1610	D
×170		•	•	•	•	536	841	805	1260	D
×160		•	•	•	•	336	675	506	1020	D
×150		•	•	•	•	174	538	262	809	C
×135			1			107	387	160	582	C
W33×387	•	•	•	•	•	—	—	—	—	G
×354	•	•	•	•	•	—	—	—	—	G
×318	•	•	•	•	•	—	—	—	—	F
×291	•	•	•	•	•	—	—	—	—	F
×263	•	•	•	•	•	—	—	—	—	E
×241		•	•	•	•	2120	—	3180	—	E
×221		•	•	•	•	1670	1760	2510	2650	D
×201		•	•	•	•	1210	1390	1810	2090	D
W33×169		•	•	•	•	770	997	1160	1500	D
×152		•	•	•	•	517	780	777	1170	D
×141		•	•	•	•	317	610	477	917	C
×130		•	•	•	•	163	474	245	712	C
×118			1			85.3	322	128	483	C

Note 1: Links in EBF that meet the exception in the AISC *Seismic Provisions* Section F3.5b(1) need only meet the limits for moderately ductile members.

AMERICAN INSTITUTE OF STEEL CONSTRUCTION

Table 1-3 (continued)
Sections That Satisfy Seismic Width-to-Thickness Requirements
W-Shapes

$F_y = 50$ ksi

Shape	IMF	SMF	STMF	SCCS	OCBF		SCBF		$L_{b\,max}$, ft	
	Beams and Columns	Beams and Columns	Chord Segment	Columns	Diagonal Braces	Diagonal Braces	Columns	Beams	λ_{hd}	λ_{md}
W30×391	•	•	•	•	•	•	•	•	15.3	30.2
×357	•	•	•	•	•	•	•	•	15.1	29.9
×326	•	•	•	•	•	•	•	•	15.0	29.6
×292	•	•	•	•	•	•	•	•	14.9	29.4
×261	•	•	•	•	•	•	•	•	14.7	29.0
×235	•	•	•	•	•	•	•	•	14.6	28.8
×211	•	•	•	•			•	•	14.5	28.7
×191	•	•	•	•			•	•	14.4	28.4
×173	•	•	•	•			•	•	14.2	28.1
W30×148	•	•	•	•			•	•	9.48	18.7
×132	•	•	•	•			•	•	9.35	18.5
×124	•	•	•	•			•	•	9.27	18.3
×116	•	•	•	•			•	•	9.10	18.0
×108	•	•	•	•			•	•	8.94	17.7
×99	•							•	8.73	17.3
×90	•							•	8.69	17.2
W27×539	•	•	•	•	•	•	•	•	15.2	30.0
×368	•	•	•	•	•	•	•	•	14.5	28.6
×336	•	•	•	•	•	•	•	•	14.3	28.3
×307	•	•	•	•	•	•	•	•	14.2	28.0
×281	•	•	•	•	•	•	•	•	14.1	27.9
×258	•	•	•	•	•	•	•	•	14.0	27.6
×235	•	•	•	•	•	•	•	•	13.8	27.4
×217	•	•	•	•	•	•	•	•	13.8	27.3
×194	•	•	•	•	•	•	•	•	13.7	27.0
×178	•	•	•	•	•	•	•	•	13.5	26.7
×161	•	•	•	•			•	•	13.4	26.5
×146	•	•	•	•			•	•	13.3	26.3
W27×129	•	•	•	•			•	•	9.19	18.2
×114	•	•	•	•			•	•	9.06	17.9
×102	•	•	•	•			•	•	8.94	17.7
×94	•	•	•	•			•	•	8.81	17.4
×84	•							•	8.60	17.0

Table 1-3 (continued)
Sections That Satisfy Seismic Width-to-Thickness Requirements
W-Shapes

F_y = 50 ksi

Shape	EBF			BRBF	SPSW	$P_{a\,max}$, kips ASD		$P_{u\,max}$, kips LRFD		Web Access Holes
	Diagonal Braces	Columns	Links	Beams and Columns	HBE and VBE	λ_{hd}	λ_{md}	λ_{hd}	λ_{md}	
W30×391	•	•	•	•	•	—	—	—	—	H
×357	•	•	•	•	•	—	—	—	—	G
×326	•	•	•	•	•	—	—	—	—	G
×292	•	•	•	•	•	—	—	—	—	F
×261	•	•	•	•	•	—	—	—	—	E
×235	•	•	•	•	•	—	—	—	—	E
×211	•	•	•	•	•	—	—	—	—	D
×191		•	•	•	•	1510	1570	2260	2350	D
×173			•	•	•	1110	1250	1670	1870	D
W30×148		•	•	•	•	896	1030	1350	1550	D
×132		•	•	•	•	654	816	982	1230	C or D
×124		•	•	•	•	479	674	720	1010	C
×116		•	•	•	•	361	571	542	859	C
×108		•	•	•	•	242	466	364	701	C
×99			1			114	352	171	530	C
×90			1			21.6	156	32.4	235	B
W27×539	•	•	•	•	•	—	—	—	—	K
×368	•	•	•	•	•	—	—	—	—	H
×336	•	•	•	•	•	—	—	—	—	G
×307	•	•	•	•	•	—	—	—	—	G
×281	•	•	•	•	•	—	—	—	—	F
×258	•	•	•	•	•	—	—	—	—	F
×235	•	•	•	•	•	—	—	—	—	E
×217	•	•	•	•	•	—	—	—	—	E
×194	•	•	•	•	•	—	—	—	—	E
×178	•	•	•	•	•	—	—	—	—	D
×161		•	•	•	•	1400	1410	2110	2120	D
×146		•	•	•	•	1040	1120	1570	1690	C
W27×129		•	•	•	•	893	971	1340	1460	D
×114		•	•	•	•	642	759	965	1140	C
×102		•	•	•	•	350	524	527	788	C
×94		•	•	•	•	215	409	324	615	C
×84			1		•	85.0	278	128	418	B

Note 1: Links in EBF that meet the exception in the AISC *Seismic Provisions* Section F3.5b(1) need only meet the limits for moderately ductile members.

Table 1-3 (continued)
Sections That Satisfy Seismic Width-to-Thickness Requirements
W-Shapes

$F_y = 50$ ksi

Shape	IMF Beams and Columns	SMF Beams and Columns	STMF Chord Segment	SCCS Columns	OCBF Diagonal Braces	SCBF Diagonal Braces	SCBF Columns	SCBF Beams	$L_{b\,max}$, ft λ_{hd}	λ_{md}
W24×370	•	•	•	•	•	•	•	•	13.6	26.9
×335	•	•	•	•	•	•	•	•	13.4	26.5
×306	•	•	•	•	•	•	•	•	13.3	26.3
×279	•	•	•	•	•	•	•	•	13.2	26.0
×250	•	•	•	•	•	•	•	•	13.1	25.8
×229	•	•	•	•	•	•	•	•	12.9	25.6
×207	•	•	•	•	•	•	•	•	12.8	25.3
×192	•	•	•	•	•	•	•	•	12.8	25.2
×176	•	•	•	•	•	•	•	•	12.6	25.0
×162	•	•	•	•	•	•	•	•	12.7	25.1
×146	•	•	•	•	•	•	•	•	12.5	24.7
×131	•	•	•	•	•	•	•	•	12.3	24.4
×117	•							•	12.2	24.2
×104	•							•	12.1	23.9
W24×103	•	•	•	•			•	•	8.27	16.4
×94	•	•	•	•			•	•	8.23	16.3
×84	•	•	•	•			•	•	8.11	16.0
×76	•	•	•	•			•	•	7.98	15.8
×68	•							•	7.77	15.4
W24×62	•	•	•	•			•	•	5.74	11.3
×55	•						•	•	5.57	11.0
W21×201	•	•	•	•	•	•	•	•	12.6	24.8
×182	•	•	•	•	•	•	•	•	12.5	24.7
×166	•	•	•	•	•	•	•	•	12.4	24.6
×147	•	•	•	•	•	•	•	•	12.3	24.2
×132	•	•	•	•	•	•	•	•	12.2	24.1
×122	•	•	•	•	•	•	•	•	12.1	24.0
×111	•	•	•	•	•	•	•	•	12.1	23.8
×101	•							•	12.0	23.7
W21×93	•	•	•	•	•	•	•	•	7.65	15.1
×83	•	•	•	•			•	•	7.61	15.0
×73	•	•	•	•			•	•	7.52	14.9
×68	•	•	•	•			•	•	7.48	14.8
×62	•	•	•	•			•	•	7.36	14.5
×55	•						•	•	7.19	14.2

Table 1-3 (continued)
Sections That Satisfy Seismic Width-to-Thickness Requirements
W-Shapes

$F_y = 50$ ksi

Shape	EBF			BRBF	SPSW	$P_{a\,max}$, kips ASD		$P_{u\,max}$, kips LRFD		Web Access Holes
	Diagonal Braces	Columns	Links	Beams and Columns	HBE and VBE	λ_{hd}	λ_{md}	λ_{hd}	λ_{md}	
W24×370	•	•	•	•	•	—	—	—	—	I
×335	•	•	•	•	•	—	—	—	—	H
×306	•	•	•	•	•	—	—	—	—	G
×279	•	•	•	•	•	—	—	—	—	G
×250	•	•	•	•	•	—	—	—	—	F
×229	•	•	•	•	•	—	—	—	—	F
×207	•	•	•	•	•	—	—	—	—	E
×192	•	•	•	•	•	—	—	—	—	E
×176	•	•	•	•	•	—	—	—	—	E
×162	•	•	•	•	•	—	—	—	—	D
×146	•	•	•	•	•	—	—	—	—	D
×131	•	•	•	•	•	—	—	—	—	C
×117			1	•	•	841	903	1260	1360	C
×104			1	•	•	557	673	837	1010	C
W24×103		•	•	•	•	740	795	1110	1200	C
×94		•	•	•	•	556	644	836	968	C
×84		•	•	•	•	336	465	506	698	C
×76		•	•	•	•	193	344	290	517	C
×68			1			75.7	242	114	364	B
W24×62		•	•	•	•	124	258	187	387	B
×55		•	•	•	•	38.9	148	58.5	223	A or B
W21×201	•	•	•	•	•	—	—	—	—	E
×182	•	•	•	•	•	—	—	—	—	E
×166	•	•	•	•	•	—	—	—	—	E
×147	•	•	•	•	•	—	—	—	—	D
×132	•	•	•	•	•	—	—	—	—	D
×122	•	•	•	•	•	—	—	—	—	C
×111	•	•	•	•	•	—	—	—	—	C
×101			1			810	838	1220	1260	C
W21×93	•	•	•	•	•	—	—	—	—	C
×83		•	•	•	•	707	716	1060	1080	C
×73		•	•	•	•	456	517	685	776	C
×68		•	•	•	•	347	427	521	642	C
×62		•	•	•	•	220	324	330	487	B
×55			1			113	231	170	347	A or B

Note 1: Links in EBF that meet the exception in the AISC *Seismic Provisions* Section F3.5b(1) need only meet the limits for moderately ductile members.

Table 1-3 (continued)
Sections That Satisfy Seismic Width-to-Thickness Requirements
W-Shapes

$F_y = 50$ ksi

Shape	IMF Beams and Columns	SMF Beams and Columns	STMF Chord Segment	SCCS Columns	OCBF Diagonal Braces	SCBF Diagonal Braces	SCBF Columns	SCBF Beams	$L_{b\,max}$, ft λ_{hd}	λ_{md}
W21×57	•	•	•	•			•	•	5.61	11.1
×50	•	•	•	•			•	•	5.40	10.7
×44	•	•	•	•			•	•	5.24	10.4
W18×311	•	•	•	•	•	•	•	•	12.3	24.2
×283	•	•	•	•	•	•	•	•	12.1	23.9
×258	•	•	•	•	•	•	•	•	12.0	23.7
×234	•	•	•	•	•	•	•	•	11.8	23.4
×211	•	•	•	•	•	•	•	•	11.7	23.2
×192	•	•	•	•	•	•	•	•	11.6	22.9
×175	•	•	•	•	•	•	•	•	11.5	22.7
×158	•	•	•	•	•	•	•	•	11.4	22.5
×143	•	•	•	•	•	•	•	•	11.3	22.3
×130	•	•	•	•	•	•	•	•	11.2	22.2
×119	•	•	•	•	•	•	•	•	11.2	22.1
×106	•	•	•	•	•	•	•	•	11.1	21.9
×97	•	•	•	•	•	•	•	•	11.0	21.8
×86	•	•	•	•	•	•	•	•	10.9	21.6
×76	•							•	10.8	21.4
W18×71	•	•	•	•	•	•	•	•	7.07	14.0
×65	•	•	•	•	•	•	•	•	7.02	13.9
×60	•	•	•	•			•	•	6.98	13.8
×55	•	•	•	•			•	•	6.94	13.7
×50	•	•	•	•			•	•	6.86	13.6
W18×46	•	•	•	•			•	•	5.36	10.6
×40	•	•	•	•			•	•	5.28	10.4
×35	•	•	•	•			•	•	5.07	10.0
W16×100	•	•	•	•	•	•	•	•	10.4	20.6
×89	•	•	•	•	•	•	•	•	10.4	20.5
×77	•	•	•	•	•	•	•	•	10.3	20.3
×67	•							•	10.2	20.2
W16×57	•	•	•	•	•	•	•	•	6.65	13.1
×50	•	•	•	•			•	•	6.61	13.1
×45	•	•	•	•			•	•	6.53	12.9
×40	•	•	•	•			•	•	6.53	12.9
×36	•							•	6.32	12.5

Table 1-3 (continued)
Sections That Satisfy Seismic Width-to-Thickness Requirements
W-Shapes

F_y = 50 ksi

| Shape | EBF | | | BRBF | SPSW | $P_{a\,max}$, kips ASD | | $P_{u\,max}$, kips LRFD | | Web Access Holes |
	Diagonal Braces	Columns	Links	Beams and Columns	HBE and VBE	λ_{hd}	λ_{md}	λ_{hd}	λ_{md}	
W21×57		•	•	•	•	217	307	326	461	B
×50		•	•	•	•	117	219	176	330	A or B
×44		•	•	•	•	38.3	133	57.6	201	A or B
W18×311	•	•	•	•	•	—	—	—	—	I
×283	•	•	•	•	•	—	—	—	—	H
×258	•	•	•	•	•	—	—	—	—	G
×234	•	•	•	•	•	—	—	—	—	G
×211	•	•	•	•	•	—	—	—	—	F
×192	•	•	•	•	•	—	—	—	—	F
×175	•	•	•	•	•	—	—	—	—	E
×158	•	•	•	•	•	—	—	—	—	E
×143	•	•	•	•	•	—	—	—	—	D
×130	•	•	•	•	•	—	—	—	—	D
×119	•	•	•	•	•	—	—	—	—	D
×106	•	•	•	•	•	—	—	—	—	C
×97	•	•	•	•	•	—	—	—	—	C
×86	•	•	•	•	•	—	—	—	—	C
×76			1			595	620	895	932	C
W18×71	•	•	•	•	•	—	—	—	—	C
×65	•	•	•	•	•	—	—	—	—	C
×60		•	•	•	•	444	472	668	709	C
×55		•	•	•	•	346	391	520	588	B
×50		•	•	•	•	217	288	326	433	A or B
W18×46		•	•	•	•	212	273	319	411	B
×40			•	•	•	65.4	156	98.3	235	A or B
×35			•	•	•	30.9	107	46.5	161	A or B
W16×100	•	•	•	•	•	—	—	—	—	C
×89	•	•	•	•	•	—	—	—	—	C
×77	•	•	•	•	•	—	—	—	—	C
×67			1			583	586	877	881	B or C
W16×57	•	•	•	•	•	—	—	—	—	C
×50		•	•	•	•	402	415	604	624	B
×45		•	•	•	•	284	321	427	483	A or B
×40		•	•	•	•	149	214	224	322	A or B
×36			1			107	174	160	261	A or B

Note 1: Links in EBF that meet the exception in the AISC *Seismic Provisions* Section F3.5b(1) need only meet the limits for moderately ductile members.

Table 1-3 (continued)
Sections That Satisfy Seismic Width-to-Thickness Requirements
W-Shapes

F_y = 50 ksi

Shape	IMF	SMF	STMF	SCCS	OCBF	SCBF			$L_{b\,max}$, ft	
	Beams and Columns	Beams and Columns	Chord Segment	Columns	Diagonal Braces	Diagonal Braces	Columns	Beams	λ_{hd}	λ_{md}
W16×31	•	•	•	•			•	•	4.86	9.61
×26	•							•	4.66	9.20
W14×730	•	•	•	•	•	•	•	•	19.5	38.5
×665	•	•	•	•	•	•	•	•	19.2	38.0
×605	•	•	•	•	•	•	•	•	18.9	37.4
×550	•	•	•	•	•	•	•	•	18.7	36.9
×500	•	•	•	•	•	•	•	•	18.4	36.4
×455	•	•	•	•	•	•	•	•	18.2	36.0
×426	•	•	•	•	•	•	•	•	18.0	35.7
×398	•	•	•	•	•	•	•	•	17.9	35.4
×370	•	•	•	•	•	•	•	•	17.7	35.1
×342	•	•	•	•	•	•	•	•	17.6	34.8
×311	•	•	•	•	•	•	•	•	17.5	34.5
×283	•	•	•	•	•	•	•	•	17.3	34.3
×257	•	•	•	•	•	•	•	•	17.2	33.9
×233	•	•	•	•	•	•	•	•	17.0	33.7
×211	•	•	•	•	•	•	•	•	16.9	33.4
×193	•	•	•	•	•	•	•	•	16.8	33.3
×176	•	•	•	•	•	•	•	•	16.7	33.0
×159	•	•	•	•	•	•	•	•	16.6	32.9
×145	•	•	•	•	•	•	•	•	16.5	32.7
W14×132	•	•	•	•	•	•	•	•	15.6	30.9
×120	•				•			•	15.5	30.7
×109	•				•			•	15.5	30.6
W14×82	•	•	•	•	•	•	•	•	10.3	20.4
×74	•	•	•	•	•	•	•	•	10.3	20.4
×68	•	•	•	•	•	•	•	•	10.2	20.2
×61	•				•			•	10.2	20.1
W14×53	•	•	•	•	•	•	•	•	7.98	15.8
×48	•	•	•	•	•	•		•	7.94	15.7
×43	•							•	7.86	15.5
W14×38	•	•	•	•			•	•	6.44	12.7
×34	•							•	6.36	12.6
×30	•							•	6.19	12.2

Table 1-3 (continued)
Sections That Satisfy Seismic Width-to-Thickness Requirements
W-Shapes

F_y = 50 ksi

Shape	EBF			BRBF	SPSW	$P_{a\,max}$, kips ASD		$P_{u\,max}$, kips LRFD		Web Access Holes
	Diagonal Braces	Columns	Links	Beams and Columns	HBE and VBE	λ_{hd}	λ_{md}	λ_{hd}	λ_{md}	
W16×31		•	•	•	•	40.3	114	60.6	171	A or B
×26			1			9.23	51.6	13.9	77.5	A or B
W14×730	•	•	•	•	•	—	—	—	—	N
×665	•	•	•	•	•	—	—	—	—	M
×605	•	•	•	•	•	—	—	—	—	L
×550	•	•	•	•	•	—	—	—	—	K
×500	•	•	•	•	•	—	—	—	—	K
×455	•	•	•	•	•	—	—	—	—	J
×426	•	•	•	•	•	—	—	—	—	J
×398	•	•	•	•	•	—	—	—	—	I
×370	•	•	•	•	•	—	—	—	—	H
×342	•	•	•	•	•	—	—	—	—	H
×311	•	•	•	•	•	—	—	—	—	G
×283	•	•	•	•	•	—	—	—	—	G
×257	•	•	•	•	•	—	—	—	—	F
×233	•	•	•	•	•	—	—	—	—	F
×211	•	•	•	•	•	—	—	—	—	E
×193	•	•	•	•	•	—	—	—	—	E
×176	•	•	•	•	•	—	—	—	—	D
×159	•	•	•	•	•	—	—	—	—	D
×145	•	•	•	•	•	—	—	—	—	D
W14×132	•	•	•	•	•	—	—	—	—	D
×120	•	•	1			—	—	—	—	C
×109	•	•	1			—	—	—	—	C
W14×82	•	•	•	•	•	—	—	—	—	C
×74	•	•	•	•	•	—	—	—	—	C
×68	•	•	•	•	•	—	—	—	—	C
×61	•	•	1			—	—	—	—	B
W14×53	•	•	•	•	•	—	—	—	—	B
×48	•	•	•	•	•	—	—	—	—	B
×43			1			344	356	518	535	A or B
W14×38		•	•	•	•	266	289	400	434	A or B
×34			1			181	219	273	329	A or B
×30			1			128	171	192	258	A or B

Note 1: Links in EBF that meet the exception in the AISC *Seismic Provisions* Section F3.5b(1) need only meet the limits for moderately ductile members.

Table 1-3 (continued)
Sections That Satisfy Seismic Width-to-Thickness Requirements
W-Shapes

$F_y = 50$ ksi

Shape	IMF Beams and Columns	SMF Beams and Columns	STMF Chord Segment	STMF Columns	SCCS	OCBF Diagonal Braces	SCBF Diagonal Braces	SCBF Columns	SCBF Beams	$L_{b\,max}$, ft λ_{hd}	λ_{md}
W14×26	•	•	•	•				•	•	4.49	8.87
×22	•								•	4.32	8.55
W12×336	•	•	•	•	•	•	•	•	•	14.4	28.5
×305	•	•	•	•	•	•	•	•	•	14.2	28.1
×279	•	•	•	•	•	•	•	•	•	14.0	27.8
×252	•	•	•	•	•	•	•	•	•	13.9	27.4
×230	•	•	•	•	•	•	•	•	•	13.8	27.2
×210	•	•	•	•	•	•	•	•	•	13.6	27.0
×190	•	•	•	•	•	•	•	•	•	13.5	26.7
×170	•	•	•	•	•	•	•	•	•	13.4	26.5
×152	•	•	•	•	•	•	•	•	•	13.3	26.2
×136	•	•	•	•	•	•	•	•	•	13.1	26.0
×120	•	•	•	•	•	•	•	•	•	13.0	25.7
×106	•	•	•	•	•	•	•	•	•	12.9	25.6
×96	•	•	•	•		•		•	•	12.8	25.4
×87	•					•		•	•	12.8	25.2
×79	•					•		•	•	12.7	25.1
×72	•					•		•	•	12.6	25.0
W12×58	•					•			•	10.4	20.6
×53	•					•			•	10.3	20.4
W12×50	•	•	•	•	•	•	•		•	8.15	16.1
×45	•	•	•	•	•	•	•		•	8.11	16.0
×40	•				•				•	8.06	15.9
W12×35	•	•	•	•				•	•	6.40	12.7
×30	•							•	•	6.32	12.5
×26	•							•	•	6.28	12.4
W12×22	•	•	•	•				•	•	3.52	6.97
×19	•	•	•	•				•	•	3.42	6.75
×16	•								•	3.21	6.35
×14	•								•	3.13	6.19

Table 1-3 (continued)
Sections That Satisfy Seismic Width-to-Thickness Requirements
W-Shapes

F_y = 50 ksi

Shape	EBF			BRBF	SPSW	$P_{a\,max}$, kips ASD		$P_{u\,max}$, kips LRFD		Web Access Holes
	Diagonal Braces	Columns	Links	Beams and Columns	HBE and VBE	λ_{hd}	λ_{md}	λ_{hd}	λ_{md}	
W14×26		•	•	•	•	77.4	126	116	189	A or B
×22			1			20.2	68.8	30.4	103	A or B
W12×336	•	•	•	•	•	—	—	—	—	I
×305	•	•	•	•	•	—	—	—	—	I
×279	•	•	•	•	•	—	—	—	—	H
×252	•	•	•	•	•	—	—	—	—	G
×230	•	•	•	•	•	—	—	—	—	G
×210	•	•	•	•	•	—	—	—	—	F
×190	•	•	•	•	•	—	—	—	—	F
×170	•	•	•	•	•	—	—	—	—	E
×152	•	•	•	•	•	—	—	—	—	E
×136	•	•	•	•	•	—	—	—	—	D
×120	•	•	•	•	•	—	—	—	—	D
×106	•	•	•	•	•	—	—	—	—	C
×96	•	•	•	•	•	—	—	—	—	C
×87	•		1			—	—	—	—	C
×79	•		1			—	—	—	—	C
×72	•		1			—	—	—	—	C
W12×58	•		1			—	—	—	—	B
×53	•		1			—	—	—	—	A or B
W12×50	•	•	•	•	•	—	—	—	—	B
×45	•	•	•	•	•	—	—	—	—	A or B
×40	•		1			—	—	—	—	A or B
W12×35		•	•	•	•	302	305	453	458	A or B
×30			1	•	•	178	205	267	309	A or B
×26			1			88.1	133	132	200	A or B
W12×22		•	•	•	•	131	151	197	228	A or B
×19		•	•	•	•	73.1	103	110	155	A or B
×16			1			37.5	70.3	56.4	106	A
×14			1			10.7	39.5	16.0	59.3	—

Note 1: Links in EBF that meet the exception in the AISC *Seismic Provisions* Section F3.5b(1) need only meet the limits for moderately ductile members.

Table 1-3 (continued)
Sections That Satisfy Seismic Width-to-Thickness Requirements
W-Shapes

$F_y = 50$ ksi

Shape	IMF	SMF	STMF	SCCS	OCBF	SCBF			$L_{b\,max}$, ft	
	Beams and Columns	Beams and Columns	Chord Segment	Columns	Diagonal Braces	Diagonal Braces	Columns	Beams	λ_{hd}	λ_{md}
W10×112	•	•	•	•	•	•	•	•	11.1	22.0
×100	•	•	•	•	•	•	•	•	11.0	21.8
×88	•	•	•	•	•	•	•	•	10.9	21.6
×77	•	•	•	•	•	•	•	•	10.8	21.4
×68	•	•	•	•	•	•	•	•	10.8	21.3
×60	•				•			•	10.7	21.1
×54	•				•			•	10.6	21.0
×49	•				•			•	10.6	20.9
W10×45	•	•	•	•	•	•	•	•	8.35	16.5
×39	•				•			•	8.23	16.3
×33	•				•			•	8.06	15.9
W10×30	•	•	•	•	•	•	•	•	5.69	11.3
×26	•	•	•	•	•	•	•	•	5.65	11.2
×22	•							•	5.53	10.9
W10×19	•	•	•	•			•	•	3.63	7.18
×17	•	•	•	•				•	3.51	6.94
×15	•							•	3.37	6.66
×12									3.26	6.45
W8×67	•	•	•	•	•	•	•	•	8.81	17.4
×58	•	•	•	•	•	•	•	•	8.73	17.3
×48	•	•	•	•	•	•	•	•	8.65	17.1
×40	•	•	•	•	•	•	•	•	8.48	16.8
×35	•				•			•	8.44	16.7
×31								•	8.40	16.6
W8×28	•	•	•	•	•	•	•	•	6.73	13.3
×24	•				•			•	6.69	13.2
W8×21	•	•	•	•	•	•	•	•	5.24	10.4
×18	•				•			•	5.11	10.1

Table 1-3 (continued)
Sections That Satisfy Seismic Width-to-Thickness Requirements
W-Shapes

$F_y = 50$ ksi

Shape	EBF			BRBF	SPSW	$P_{a\,max}$, kips ASD		$P_{u\,max}$, kips LRFD		Web Access Holes
	Diagonal Braces	Columns	Links	Beams and Columns	HBE and VBE	λ_{hd}	λ_{md}	λ_{hd}	λ_{md}	
W10×112	•	•	•	•	•	—	—	—	—	D
×100	•	•	•	•	•	—	—	—	—	D
×88	•	•	•	•	•	—	—	—	—	C
×77	•	•	•	•	•	—	—	—	—	C
×68	•	•	•	•	•	—	—	—	—	C
×60	•		1			—	—	—	—	C
×54	•		1			—	—	—	—	B
×49	•		1			—	—	—	—	A or B
W10×45	•	•	•	•	•	—	—	—	—	B
×39	•		1			—	—	—	—	A or B
×33	•		1			—	—	—	—	A or B
W10×30	•	•	•	•	•	—	—	—	—	A or B
×26	•	•	•	•	•	—	—	—	—	A or B
×22			1			183	187	275	281	A or B
W10×19	•	•	•	•	•	—	—	—	—	A or B
×17		•	•	•	•	140	144	211	216	A
×15			1			113	119	169	179	A
×12						44.2	63.8	66.4	96.0	—
W8×67	•	•	•	•	•	—	—	—	—	C
×58	•	•	•	•	•	—	—	—	—	C
×48	•	•	•	•	•	—	—	—	—	C
×40	•	•	•	•	•	—	—	—	—	A or B
×35	•		1			—	—	—	—	A or B
×31						—	—	—	—	A or B
W8×28	•	•	•	•	•	—	—	—	—	A or B
×24	•		1			—	—	—	—	A or B
W8×21	•	•	•	•	•	—	—	—	—	A or B
×18	•		1			—	—	—	—	A

Note 1: Links in EBF that meet the exception in the AISC *Seismic Provisions* Section F3.5b(1) need only meet the limits for moderately ductile members.

Table 1-4
Sections That Satisfy Seismic Width-to-Thickness Requirements
Angles

$F_y = 36$ ksi

Shape	STMF Chords	OCBF and EBF Diagonal Braces	SCBF Diagonal Braces	Shape	STMF Chords	OCBF and EBF Diagonal Braces	SCBF Diagonal Braces
L8×8×1^1/$_8$	•	•	•	L4×3^1/$_2$×^{1}/$_2$		•	•
×1	•	•	•	×^{3}/$_8$		•	
×^{7}/$_8$		•		L4×3×^{5}/$_8$	•	•	•
×^{3}/$_4$		•		×^{1}/$_2$	•	•	•
L8×6×1	•	•	•	×^{3}/$_8$		•	
×^{7}/$_8$		•		L3^1/$_2$×3^1/$_2$×^{1}/$_2$	•	•	•
×^{3}/$_4$		•		×^{7}/$_{16}$	•	•	•
L8×4×1	•	•	•	×^{3}/$_8$		•	
×^{7}/$_8$		•		L3^1/$_2$×3×^{1}/$_2$	•	•	•
×^{3}/$_4$		•		×^{7}/$_{16}$	•	•	•
L7×4×^{3}/$_4$		•		×^{3}/$_8$		•	
L6×6×1	•	•	•	L3^1/$_2$×2^1/$_2$×^{1}/$_2$	•	•	•
×^{7}/$_8$	•	•	•	×^{3}/$_8$		•	
×^{3}/$_4$	•	•	•	L3×3×^{1}/$_2$	•	•	•
×^{5}/$_8$		•		×^{7}/$_{16}$	•	•	•
×^{9}/$_{16}$		•		×^{3}/$_8$	•	•	•
L6×4×^{7}/$_8$	•	•	•	×^{5}/$_{16}$		•	
×^{3}/$_4$	•	•	•	L3×2^1/$_2$×^{1}/$_2$	•	•	•
×^{5}/$_8$		•		×^{7}/$_{16}$	•	•	•
×^{9}/$_{16}$		•		×^{3}/$_8$	•	•	•
L5×5×^{7}/$_8$	•	•	•	×^{5}/$_{16}$		•	
×^{3}/$_4$		•		L3×2×^{1}/$_2$	•	•	•
×^{5}/$_8$	•	•	•	×^{3}/$_8$	•	•	•
×^{1}/$_2$		•		×^{5}/$_{16}$		•	
L5×3^1/$_2$×^{3}/$_4$	•	•	•				
×^{5}/$_8$	•	•	•				
×^{1}/$_2$		•					
L5×3×^{1}/$_2$		•					
L4×4×^{3}/$_4$	•	•	•				
×^{5}/$_8$	•	•	•				
×^{1}/$_2$	•	•	•				
×^{7}/$_{16}$		•					
×^{3}/$_8$		•					

Table 1-5a
Sections That Satisfy Seismic Width-to-Thickness Requirements
Rectangular HSS

F_y = 46 ksi

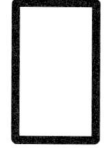

Shape	OCBF and EBF — Diagonal Braces	SCBF — Diagonal Braces	SCCS and SCBF[a] — Columns	Shape	OCBF and EBF — Diagonal Braces	SCBF — Diagonal Braces	SCCS and SCBF[a] — Columns
HSS10×8×5/8	•			HSS4×3×3/8	•	•	•
HSS10×6×5/8	•			×5/16	•	•	•
HSS10×4×5/8	•			×1/4	•		
HSS9×7×5/8	•	•	•	HSS4×2½×3/8	•	•	•
HSS9×5×5/8	•	•	•	HSS4×2½×5/16	•	•	•
HSS8×6×5/8	•	•	•	×1/4	•		
HSS8×6×1/2	•			HSS4×2×3/8	•	•	•
HSS8×4×5/8	•	•	•	×5/16	•	•	•
HSS8×4×1/2	•			×1/4	•		
HSS8×3×1/2	•			HSS3½×2½×3/8	•	•	•
HSS7×5×1/2	•	•	•	×5/16	•	•	•
HSS7×4×1/2	•	•	•	×1/4	•	•	•
HSS7×3×1/2	•	•	•	HSS3½×2×1/4	•	•	•
HSS6×5×1/2	•	•	•	HSS3½×1½×1/4	•	•	•
×3/8	•			HSS3×2½×5/16	•	•	•
HSS6×4×1/2	•	•	•	×1/4	•	•	•
×3/8	•			HSS3×2×5/16	•	•	•
HSS6×3×1/2	•	•	•	×1/4	•	•	•
×3/8	•			×3/16	•		
HSS6×2×3/8	•			HSS3×1½×1/4	•	•	•
HSS5×4×1/2	•	•	•	×3/16	•		
×3/8	•	•	•	HSS3×1×3/16	•		
×5/16	•			HSS2½×2×1/4	•	•	•
HSS5×3×1/2	•	•	•	×3/16	•	•	•
×3/8	•	•	•	HSS2½×1½×1/4	•	•	•
×5/16	•			×3/16	•	•	•
HSS5×2×3/8	•	•	•	HSS2½×1×3/16	•	•	•
×5/16	•			HSS2¼×2×3/16	•	•	•
				HSS2×1½×3/16	•	•	•
				×1/8	•		

[a] Sections also satisfy STMF truss chords.

Table 1-5b
Sections That Satisfy Seismic Width-to-Thickness Requirements
Square HSS

$F_y = 46$ ksi

Shape	OCBF and EBF Diagonal Braces	SCBF Diagonal Braces	SCCS and SCBF[a] Columns	Shape	OCBF and EBF Diagonal Braces	SCBF Diagonal Braces	SCCS and SCBF[a] Columns
HSS10×10×5/8	•			HSS4×4×1/2	•	•	•
HSS9×9×5/8	•	•	•	×3/8	•	•	•
HSS8×8×5/8	•	•	•	×5/16	•	•	•
×1/2	•			×1/4	•		
HSS7×7×5/8	•	•	•	HSS3 1/2×3 1/2×3/8	•	•	•
×1/2	•	•	•	×5/16	•	•	•
HSS6×6×5/8	•	•	•	×1/4	•	•	•
×1/2	•	•	•	HSS3×3×3/8	•	•	•
×3/8	•			×5/16	•	•	•
HSS5 1/2×5 1/2×3/8	•	•	•	×1/4	•	•	•
×5/16	•			×3/16	•		
HSS5×5×1/2	•	•	•	HSS2 1/2×2 1/2×5/16	•	•	•
×3/8	•	•	•	×1/4	•	•	•
×5/16	•			×3/16	•	•	•
HSS4 1/2×4 1/2×1/2	•	•	•	HSS2 1/4×2 1/4×1/4	•	•	•
×3/8	•	•	•	×3/16	•	•	•
×5/16	•	•	•	HSS2×2×1/4	•	•	•
				×3/16	•	•	•

[a] Sections also satisfy STMF truss chord requirements.

Table 1-6
Sections That Satisfy Seismic Width-to-Thickness Requirements
Round HSS

$F_y = 42$ ksi

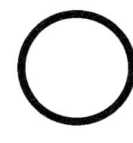

Shape	OCBF and EBF Diagonal Braces	SCBF Diagonal Braces	SCCS and SCBF[a] Columns	Shape	OCBF and EBF Diagonal Braces	SCBF Diagonal Braces	SCCS and SCBF[a] Columns
HSS16×0.625	•			HSS6×0.500	•	•	•
HSS14×0.625	•	•	•	×0.375	•	•	•
×0.500	•			×0.312	•	•	•
HSS12.750×0.500	•			×0.280	•	•	•
HSS10.750×0.500	•	•	•	×0.250	•	•	•
HSS10×0.625	•	•	•	HSS5.563×0.500	•	•	•
×0.500	•	•	•	×0.375	•	•	•
×0.375	•			×0.258	•	•	•
HSS9.625×0.500	•	•	•	HSS5.500×0.500	•	•	•
×0.375	•			×0.375	•	•	•
HSS8.625×0.625	•	•	•	×0.258	•	•	•
×0.500	•	•	•	HSS5×0.500	•	•	•
×0.375	•	•	•	×0.375	•	•	•
×0.322	•			×0.312	•	•	•
HSS7.625×0.375	•	•	•	×0.258	•	•	•
×0.328	•	•	•	×0.250	•	•	•
HSS7.500×0.500	•	•	•	×0.188	•		
×0.375	•	•	•	HSS4.500×0.375	•	•	•
×0.312	•	•	•	×0.337	•	•	•
HSS7×0.500	•	•	•	×0.237	•	•	•
×0.375	•	•	•	×0.188	•	•	•
×0.312	•	•	•	HSS4×0.313	•	•	•
×0.250	•			×0.250	•	•	•
HSS6.875×0.500	•	•	•	×0.237	•	•	•
×0.375	•	•	•	×0.226	•	•	•
×0.312	•	•	•	×0.220	•	•	•
×0.250	•			×0.188	•	•	•
HSS6.625×0.500	•	•	•	HSS3.500×0.313	•	•	•
×0.432	•	•	•	×0.300	•	•	•
×0.375	•	•	•	×0.250	•	•	•
×0.312	•	•	•	×0.216	•	•	•
×0.280	•	•	•	×0.203	•	•	•
×0.250	•			×0.188	•	•	•
				×0.125	•		

[a] Sections also satisfy STMF truss chord requirements.

Table 1-6 (continued)
Sections That Satisfy Seismic Width-to-Thickness Requirements
Round HSS

$F_y = 42$ ksi

Shape	OCBF and EBF Diagonal Braces	SCBF Diagonal Braces	SCCS and SCBF[a] Columns	Shape	OCBF and EBF Diagonal Braces	SCBF Diagonal Braces	SCCS and SCBF[a] Columns
HSS3×0.250	•	•	•	HSS2.375×0.250	•	•	•
×0.216	•	•	•	×0.218	•	•	•
×0.203	•	•	•	×0.188	•	•	•
×0.188	•	•	•	×0.154	•	•	•
×0.152	•	•	•	×0.125	•	•	•
×0.134	•	•	•	HSS1.900×0.188	•	•	•
×0.125	•	•	•	×0.145	•	•	•
HSS2.875×0.250	•	•	•	×0.120	•	•	•
×0.203	•	•	•	HSS1.660×0.140	•	•	•
×0.188	•	•	•				
×0.125	•	•	•				
HSS2.500×0.250	•	•	•				
×0.188	•	•	•				
×0.125	•	•	•				

[a] Sections also satisfy STMF truss chord requirements.

Table 1-7
Sections That Satisfy Seismic Width-to-Thickness Requirements
Pipe

F_y = 35 ksi

Shape	OCBF and EBF Diagonal Braces	SCBF Diagonal Braces	SCCS and SCBF[a] Columns	Shape	OCBF and EBF Diagonal Braces	SCBF Diagonal Braces	SCCS and SCBF[a] Columns
Standard Weight (Std.)				**Extra Strong (x-Strong)**			
Pipe 10 Std.	•			Pipe 12 x-Strong	•	•	•
Pipe 8 Std.	•	•	•	Pipe 10 x-Strong	•	•	•
Pipe 6 Std.	•	•	•	Pipe 8 x-Strong	•	•	•
Pipe 5 Std.	•	•	•	Pipe 6 x-Strong	•	•	•
Pipe 4 Std.	•	•	•	Pipe 5 x-Strong	•	•	•
Pipe 3½ Std.	•	•	•	Pipe 4 x-Strong	•	•	•
Pipe 3 Std.	•	•	•	Pipe 3½ x-Strong	•	•	•
Pipe 2½ Std.	•	•	•	Pipe 3 x-Strong	•	•	•
Pipe 2 Std.	•	•	•	Pipe 2½ x-Strong	•	•	•
Pipe 1½ Std.	•	•	•	Pipe 2 x-Strong	•	•	•
Pipe 1¼ Std.	•	•	•	Pipe 1½ x-Strong	•	•	•
Pipe 1 Std.	•	•	•	Pipe 1¼ x-Strong	•	•	•
Pipe ¾ Std.	•	•	•	Pipe 1 x-Strong	•	•	•
Pipe ½ Std.	•	•	•	Pipe ¾ x-Strong	•	•	•
				Pipe ½ x-Strong	•	•	•
				Double-Extra-Strong (xx-Strong)			
				Pipe 8 xx-Strong	•	•	•
				Pipe 6 xx-Strong	•	•	•
				Pipe 5 xx-Strong	•	•	•
				Pipe 4 xx-Strong	•	•	•
				Pipe 3 xx-Strong	•	•	•
				Pipe 2½ xx-Strong	•	•	•
				Pipe 2 xx-Strong	•	•	•

[a] Sections also satisfy STMF truss chord requirements.

Table 1-8
Shear Stud Anchor
$F_u = 65$ ksi — Nominal Horizontal Shear Strength and 25% Reduced Nominal Horizontal Shear Strength for Steel Headed Stud Anchors, kips

Q_n

Deck Condition			Stud Diameter in.	Normal Weight Concrete $w_c = 145$ pcf				Lightweight Concrete $w_c = 110$ pcf			
				$f'_c = 3$ ksi		$f'_c = 4$ ksi		$f'_c = 3$ ksi		$f'_c = 4$ ksi	
				Nominal	25% Reduced	Nominal	25% Reduced	Nominal	25% Reduced	Nominal	25% Reduced
No Deck			3/8	5.26	3.95	5.38	4.04	4.28	3.21	5.31	3.98
			1/2	9.35	7.01	9.57	7.18	7.60	5.70	9.43	7.07
			5/8	14.6	11.0	15.0	11.3	11.9	8.93	14.7	11.0
			3/4	21.0	15.8	21.5	16.1	17.1	12.8	21.2	15.9
Deck Parallel	$\frac{w_r}{h_r} \geq 1.5$		3/8	5.26	3.95	5.38	4.04	4.28	3.21	5.31	3.98
			1/2	9.35	7.01	9.57	7.18	7.60	5.70	9.43	7.07
			5/8	14.6	11.0	15.0	11.3	11.9	8.93	14.7	11.0
			3/4	21.0	15.8	21.5	16.1	17.1	12.8	21.2	15.9
	$\frac{w_r}{h_r} < 1.5$		3/8	4.58	3.44	4.58	3.44	4.28	3.21	4.58	3.44
			1/2	8.14	6.11	8.14	6.11	7.60	5.70	8.14	6.11
			5/8	12.7	9.53	12.7	9.53	11.9	8.93	12.7	9.53
			3/4	18.3	13.7	18.3	13.7	17.1	12.8	18.3	13.7
Deck Perpendicular	Weak studs, per rib	1	3/8	4.31	3.23	4.31	3.23	4.28	3.21	4.31	3.23
			1/2	7.66	5.75	7.66	5.75	7.60	5.70	7.66	5.75
			5/8	12.0	9.00	12.0	9.00	11.9	8.93	12.0	9.00
			3/4	17.2	12.9	17.2	12.9	17.1	12.8	17.2	12.9
		2	3/8	3.66	2.75	3.66	2.75	3.66	2.75	3.66	2.75
			1/2	6.51	4.88	6.51	4.88	6.51	4.88	6.51	4.88
			5/8	10.2	7.65	10.2	7.65	10.2	7.65	10.2	7.65
			3/4	14.6	11.0	14.6	11.0	14.6	11.0	14.6	11.0
		3	3/8	3.02	2.27	3.02	2.27	3.02	2.27	3.02	2.27
			1/2	5.36	4.02	5.36	4.02	5.36	4.02	5.36	4.02
			5/8	8.38	6.29	8.38	6.29	8.38	6.29	8.38	6.29
			3/4	12.1	9.08	12.1	9.08	12.1	9.08	12.1	9.08
	Strong studs, per rib	1	3/8	5.26	3.95	5.38	4.04	4.28	3.21	5.31	3.98
			1/2	9.35	7.01	9.57	7.18	7.60	5.70	9.43	7.07
			5/8	14.6	11.0	15.0	11.3	11.9	8.93	14.7	11.0
			3/4	21.0	15.8	21.5	16.1	17.1	12.8	21.2	15.9
		2	3/8	4.58	3.44	4.58	3.44	4.28	3.21	4.58	3.44
			1/2	8.14	6.11	8.14	6.11	7.60	5.70	8.14	6.11
			5/8	12.7	9.53	12.7	9.53	11.9	8.93	12.7	9.53
			3/4	18.3	13.7	18.3	13.7	17.1	12.8	18.3	13.7
		3	3/8	3.77	2.83	3.77	2.83	3.77	2.83	3.77	2.83
			1/2	6.70	5.03	6.70	5.03	6.70	5.03	6.70	5.03
			5/8	10.5	7.88	10.5	7.88	10.5	7.88	10.5	7.88
			3/4	15.1	11.3	15.1	11.3	15.1	11.3	15.1	11.3

Note: Tabulated values are applicable only to concrete made with ASTM C33 aggregates for normal weight concrete and ASTM C330 aggregates for lightweight concrete.
After-weld shear stud lengths assumed to be ≥ deck height + 1.5 in.
All symbols shown are defined in AISC *Specification* Chapter I.

Table 1-9a
Design Coefficients and Factors for Steel and Steel and Concrete Composite Seismic Force Resisting Systems[j]

Seismic Force Resisting System	Resp. Mod. Coeff., R[a]	Over-strength Factor, Ω_o	Deflection Amp. Factor, C_d[b]	Structural System Limitations Including Structural Height, h_n, Limits in ft[c]				
				Seismic Design Category				
				B	C	D[d]	E[d]	F[e]
STEEL SYSTEMS								
Steel eccentrically braced frames (EBF)	8	2	4	NL	NL	160	160	100
Steel special concentrically braced frames (SCBF)	6	2	5	NL	NL	160	160	100
Steel ordinary concentrically braced frames (OCBF)	3¼	2	3¼	NL	NL	35[g]	35[g]	NP[g]
Steel buckling-restrained braced frames (BRBF)	8	2½	5	NL	NL	160	160	100
Steel special plate shear walls (SPSW)	7	2	6	NL	NL	160	160	100
Steel special moment frames (SMF)	8	3	5½	NL	NL	NL	NL	NL
Steel special truss moment frames (STMF)	7	3	5½	NL	NL	160	100	NP
Steel intermediate moment frames (IMF)	4½	3	4	NL	NL	35[h]	NP[h]	NP[h]
Steel ordinary moment frames (OMF)	3½	3	3	NL	NL	NP[i]	NP[i]	NP[i]
Steel special cantilever column systems (SCCS)	2½	1¼	2½	35	35	35	35	35
Steel ordinary cantilever column systems (OCCS)	1¼	1¼	1¼	35	35	NP[i]	NP[i]	NP[i]
Steel systems not specifically detailed for seismic resistance	3	3	3	NL	NL	NP	NP	NP
COMPOSITE SYSTEMS								
Steel and concrete composite eccentrically braced frames (C-EBF)	8	2½	4	NL	NL	160	160	100
Steel and concrete composite special concentrically braced frames (C-SCBF)	5	2	4½	NL	NL	160	160	100
Steel and concrete composite ordinary braced frames (C-OBF)	3	2	3	NL	NL	NP	NP	NP

[a] Response modification coefficient, R, used throughout ASCE/SEI 7 (ASCE, 2010).
[b] Deflection amplification factor, C_d, for use in ASCE/SEI 7 Sections 12.8.6, 12.8.7 and 12.9.2
[c] NL = not limited and NP = not permitted.
[d] See ASCE/SEI 7 Section 12.2.5.4 for a description of seismic force resisting systems limited to buildings with a structural height, h_n, of 240 ft or less
[e] See ASCE/SEI 7 Section 12.2.5.4 for a description of seismic force resisting systems limited to buildings with a structural height, h_n, of 160 ft or less
[f] Ordinary moment frame is permitted to be used in lieu of intermediate moment frame for Seismic Design Categories B or C.
[g] Steel ordinary concentrically braced frames are permitted in single-story buildings up to a structural height, h_n, of 60 ft (18.3 m) where the dead load of the roof does not exceed 20 psf.
[h] See ASCE/SEI 7 Section 12.2.5.7 for limitations in structures assigned to Seismic Design Categories D, E or F.
[i] See ASCE/SEI 7 Section 12.2.5.6 for limitations in structures assigned to Seismic Design Categories D, E or F.
[j] This table is based on ASCE/SEI Table 12.2-1 and is reprinted with permission from ASCE.

Table 1-9a (continued)
Design Coefficients and Factors for Steel and Steel and Concrete Composite Seismic Force Resisting Systems[j]

Seismic Force Resisting System	Resp. Mod. Coeff., R[a]	Over-strength Factor, Ω_0	Deflection Amp. Factor, C_d[b]	Structural System Limitations Including Structural Height, h_n, Limits in ft[c]				
				Seismic Design Category				
				B	C	D[d]	E[d]	F[e]
COMPOSITE SYSTEMS								
Steel and concrete composite plate shear walls (C-PSW)	6½	2½	5½	NL	NL	160	160	100
Steel and concrete composite special shear walls (C-SSW)	6	2½	5	NL	NL	160	160	100
Steel and concrete composite ordinary shear walls (C-OSW)	5	2½	4½	NL	NL	NP	NP	NP
Steel and concrete composite special moment frames (C-SMF)	8	3	5½	NL	NL	NL	NL	NL
Steel and concrete composite intermediate moment frames (C-IMF)	5	3	4½	NL	NL	NP	NP	NP
Steel and concrete composite partially restrained moment frames (C-PRMF)	6	3	5½	160	160	100	NP	NP
Steel and concrete composite ordinary moment frames (C-OMF)	3	3	2½	NL	NP	NP	NP	NP
DUAL SYSTEMS								
Dual Systems with SMF capable of resisting at least 25% of prescribed seismic forces								
Steel eccentrically braced frames	8	2½	4	NL	NL	NL	NL	NL
Steel special concentrically braced frames	7	2½	5½	NL	NL	NL	NL	NL
Steel buckling-restrained braced frames	8	2½	5	NL	NL	NL	NL	NL
Steel special plate shear walls	8	2½	6½	NL	NL	NL	NL	NL
Dual Systems with IMF capable of resisting at least 25% of prescribed seismic forces								

[a] Response modification coefficient, R, used throughout ASCE/SEI 7 (ASCE, 2010).
[b] Deflection amplification factor, C_d, for use in ASCE/SEI 7 Sections 12.8.6, 12.8.7 and 12.9.2
[c] NL = not limited and NP = not permitted.
[d] See ASCE/SEI 7 Section 12.2.5.4 for a description of seismic force resisting systems limited to buildings with a structural height, h_n, of 240 ft or less
[e] See ASCE/SEI 7 Section 12.2.5.4 for a description of seismic force resisting systems limited to buildings with a structural height, h_n, of 160 ft or less
[f] Ordinary moment frame is permitted to be used in lieu of intermediate moment frame for Seismic Design Categories B or C.
[g] Steel ordinary concentrically braced frames are permitted in single-story buildings up to a structural height, h_n, of 60 ft (18.3 m) where the dead load of the roof does not exceed 20 psf.
[h] See ASCE/SEI 7 Section 12.2.5.7 for limitations in structures assigned to Seismic Design Categories D, E or F.
[i] See ASCE/SEI 7 Section 12.2.5.6 for limitations in structures assigned to Seismic Design Categories D, E or F.
[j] This table is based on ASCE/SEI 7 Table 12.2-1 and is reprinted with permission from ASCE.

Table 1-9a (continued)
Design Coefficients and Factors for Steel and Steel and Concrete Composite Seismic Force Resisting Systems[j]

Seismic Force Resisting System	Resp. Mod. Coeff., R[a]	Over-strength Factor, Ω_o	Deflection Amp. Factor, C_d[b]	Structural System Limitations Including Structural Height, h_n, Limits in ft[c]				
				Seismic Design Category				
				B	C	D[d]	E[d]	F[e]
DUAL SYSTEMS								
Steel special concentrically braced frames[f]	6	2¹/₂	5	NL	NL	35	NP	NP
DUAL COMPOSITE SYSTEMS								
Dual Composite Systems with SMF capable of resisting at least 25% of prescribed seismic forces								
Steel and concrete composite eccentrically braced frames	8	2¹/₂	4	NL	NL	NL	NL	NL
Steel and concrete composite special concentrically braced frames	6	2¹/₂	5	NL	NL	NL	NL	NL
Steel and concrete composite plate shear walls	7¹/₂	2¹/₂	6	NL	NL	NL	NL	NL
Steel and concrete composite special shear walls	7	2¹/₂	6	NL	NL	NL	NL	NL
Steel and concrete composite ordinary shear walls	6	2¹/₂	5	NL	NL	NP	NP	NP
Dual Composite Systems with IMF capable of resisting at least 25% of prescribed seismic forces								
Steel and concrete composite special concentrically braced frames	5¹/₂	2¹/₂	4¹/₂	NL	NL	160	100	NP
Steel and concrete composite ordinary braced frames	3¹/₂	2¹/₂	3	NL	NL	NP	NP	NP
Steel and concrete composite ordinary shear walls	5	3	4¹/₂	NL	NL	NP	NP	NP

[a] Response modification coefficient, R, used throughout ASCE/SEI 7 (ASCE, 2010).
[b] Deflection amplfication factor, C_d, for use in ASCE/SEI 7 Sections 12.8.6, 12.8.7 and 12.9.2
[c] NL = not limited and NP = not permitted.
[d] See ASCE/SEI 7 Section 12.2.5.4 for a description of seismic force resisting systems limited to buildings with a structural height, h_n, of 240 ft or less
[e] See ASCE/SEI 7 Section 12.2.5.4 for a description of seismic force resisting systems limited to buildings with a structural height, h_n, of 160 ft or less
[f] Ordinary moment frame is permitted to be used in lieu of intermediate moment frame for Seismic Design Categories B or C.
[g] Steel ordinary concentrically braced frames are permitted in single-story buildings up to a structural height, h_n, of 60 ft (18.3 m) where the dead load of the roof does not exceed 20 psf.
[h] See ASCE/SEI 7 Section 12.2.5.7 for limitations in structures assigned to Seismic Design Categories D, E or F.
[i] See ASCE/SEI 7 Section 12.2.5.6 for limitations in structures assigned to Seismic Design Categories D, E or F.
[j] This table is based on ASCE/SEI 7 Table 12.2-1 and is reprinted with permission from ASCE.

Table 1-9b
Design Coefficients and Factors for Nonbuilding Structures Similar to Buildings[e]

Nonbuilding Structure Type	Response Mod. Coeff., R	Over-strength Factor, Ω_o	Deflection Amp. Factor, C_d	Structural System Limitations Including Structural Height Limits, h_n, in ft[a]				
				Seismic Design Category				
				B	C	D	E	F
Steel Storage racks	4	2	3^1/$_2$	NL	NL	NL	NL	NL
Building frame systems:								
Steel special concentrically braced frames (SCBF)	6	2	5	NL	NL	160	160	100
Steel ordinary concentrically braced frames (OCBF)	3^1/$_4$	2	3^1/$_4$	NL	NL	35[b]	35[b]	NP[b]
With permitted height increase	2^1/$_2$	2	2^1/$_2$	NL	NL	160	160	100
With unlimited height	1^1/$_2$	1	1^1/$_2$	NL	NL	NL	NL	NL
Moment-resisting frame systems:								
Steel special moment frames (SMF)	8	3	5^1/$_2$	NL	NL	NL	NL	NL
Steel intermediate moment frames (IMF)	4^1/$_2$	3	4	NL	NL	35[c,d]	NP[c,d]	NP[c,d]
With permitted height increase	2^1/$_2$	2	2^1/$_2$	NL	NL	160	160	100
With unlimited height	1^1/$_2$	1	1^1/$_2$	NL	NL	NL	NL	NL
Steel ordinary moment frames (OMF)	3^1/$_2$	3	3	NL	NL	NP[c,d]	NP[c,d]	NP[c,d]
With permitted height increase	2^1/$_2$	2	2^1/$_2$	NL	NL	100	100	NP[c,d]
With unlimited height	1	1	1	NL	NL	NL	NL	NL

[a] NL = not limited and NP = not permitted.
[b] Steel ordinary braced frames are permitted in pipe racks up to 65 ft.
[c] Steel ordinary moment frames and intermediate moment frames are permitted in pipe racks up to a height of 65 ft where the moment joints of field connections are constructed of bolted end plates.
[d] Steel ordinary moment frames and intermediate moment frames are permitted in pipe racks up to a height of 35 ft.
[e] This table is based on ASCE/SEI 7 Table 15.4-1 and is reprinted with permission from ASCE.

PART 2
ANALYSIS

2.1 SCOPE 2–2
2.2 ROLE OF STRUCTURAL ANALYSIS IN DESIGN 2–2
    Ductile Design Mechanism 2–2
    Capacity Design 2–3
2.3 ANALYSIS PROCEDURES 2–3
    Elastic, Inelastic and Plastic Analysis 2–4
    Stability Design Methods in the AISC *Specification* 2–5
        Direct Analysis Method 2–5
        Effective Length Method 2–5
        First-Order Analysis Method 2–6
    Analysis Methods in ASCE/SEI 7 and the Direct Analysis Method 2–6
        Equivalent Lateral Force Analysis and the Direct Analysis Method 2–6
        Modal Response Spectrum Analysis and the Direct Analysis Method ... 2–7
2.4 STRUCTURAL MODELING 2–7
    Strength of Structural Elements 2–7
    Stiffness of Structural Elements 2–8
        Steel Elements 2–8
        Composite Elements 2–8
        Connections and Panel Zones 2–9
        Column Bases and Foundations 2–11
        Diaphragms for Three-Dimensional Analysis 2–12
    Gravity Loads 2–13
        Gravity Loads in Diagonal Braces and Special Plate Shear Walls 2–13
PART 2 REFERENCES 2–15

2.1 SCOPE

This Part provides an overview of the analysis provisions in ASCE/SEI 7, the AISC *Specification*, and the AISC *Seismic Provisions*, and how they are applied to seismic design.

2.2 ROLE OF STRUCTURAL ANALYSIS IN DESIGN

The basic role of analysis in seismic design is to provide the engineer with an understanding of the structure's behavior under design earthquakes. In its most simple form, analysis will consist of simple static linear methods and will provide information on the required design strength and system deformation under specified loading. For some structures, analysis may include static or dynamic nonlinear methods that provide information on the nonlinear deformation of individual elements, patterns of mechanism formation, and the peak demands that can be delivered to individual structural elements and their connections. The method of analysis selected must as a minimum conform to the requirements of the applicable building code. Since the results of seismic analysis inherently depend on the assumed properties of the structural elements, seismic analysis must often be performed in an iterative manner, initiating with assumed member sizes and configurations, and refined as member selection is confirmed.

Chapter C of the AISC *Seismic Provisions* requires that analysis of a structure for strength design of seismic force resisting components conforms to the applicable building code and the AISC *Specification*, as well as additional system-level requirements prescribed in the respective system sections.

Ductile Design Mechanism[1]

Structures required to resist the effects of earthquake ground motions should be designed to allow controlled inelastic, ductile deformations of the system. Accepted design practice is to limit these inelastic actions to certain components of the seismic force resisting system (SFRS) in order to develop a reliable ductile design mechanism that dissipates energy. Components of the ductile design mechanism are then designed and detailed to maintain the structural integrity of the system at large inelastic deformations. How this energy dissipation occurs depends on the structural system type used as the SFRS. Each SFRS in the AISC *Seismic Provisions* includes a "Basis of Design" section that defines the locations where inelastic actions are intended to occur. Accordingly, the provisions in ASCE/SEI 7, the AISC *Specification*, the AISC *Seismic Provisions*, and ANSI/AISC 358 are intended to work together to ensure that the resulting frames can undergo controlled deformations in a ductile manner and that those deformations are distributed throughout the frame. Clearly identifying the intended ductile design mechanism will provide insight on which aspects of the structural model may need detailed consideration. Many of the ductile design mechanisms shown in Part 1 were identified from structural behavior at large deformations from nonlinear static analyses using lateral forces that approximate the fundamental elastic mode shape. Real structures in earthquakes exhibit variability in the formation of ductile

---

[1] The term, ductile design mechanism, is intended to capture all possible system-specific mechanisms that are discussed in the AISC *Seismic Provisions*.

design mechanisms. Thus, the design and detailing requirements of the AISC *Seismic Provisions* and ANSI/AISC 358 are intended to desensitize the structure to earthquake characteristics so that multiple mechanisms do not lead to undesirable modes of failure.

Capacity Design

Capacity design is a design philosophy wherein inelastic actions under strong ground motion are presumed to be concentrated in predetermined critical zones of the SFRS. The AISC *Seismic Provisions* employs this methodology by stipulating that the required strength of certain elements of the SFRS be defined by forces corresponding to the expected capacity (based on available strength) of certain designated yielding members. The adjacent nonyielding members and connections are then protected because they are designed to remain nominally elastic regardless of the magnitude of ground shaking; in essence, these protected components are designed to be insensitive to the characteristics of the earthquake, ensuring that the desired ductile design mechanism(s) can develop. See AISC *Seismic Provisions* Commentary Section A3.1.

ASCE/SEI 7 addresses the concept of capacity design by using a system overstrength factor, Ω_o (see Part 1). ASCE/SEI 7 Section 12.4 modifies some of the basic load combinations to address load conditions where the overstrength factor is required, but does not explicitly provide guidance on application to steel frames. The AISC *Seismic Provisions* explicitly prescribe where to apply the overstrength factor or, alternatively, an estimated maximum seismic load determined from a capacity design analysis outlined in the respective chapter for each SFRS.

In many instances, ASCE/SEI 7 and the AISC *Seismic Provisions* explicitly prescribe when amplified seismic loads are to be used. Amplified seismic loads are defined in ASCE/SEI 7 as

$$E_m = E_{mh} \pm E_v \qquad \text{(ASCE/SEI 7 Eq. 12.4-5 and 12.4-6)}$$

where
$E_{mh} = \Omega_o Q_E$ = horizontal seismic load effect including overstrength factor
Ω_o = overstrength factor as defined in Tables 12.2-1, 15.4-1 and 15.4-2 of ASCE/SEI 7
Q_E = effect of horizontal seismic (earthquake-induced) forces
E_v = vertical seismic load effect

The load effect, E_{mh}, is based on code-specified loads and the code-specified overstrength factor. However, the AISC *Seismic Provisions* sometimes redefines E_{mh} as the forces resulting from the expected strengths of the designated yielding members of the SFRS.

2.3 ANALYSIS PROCEDURES

To determine the required strength of structural steel systems, members and connections, AISC *Specification* Section B3.1 permits design forces to be determined by elastic, inelastic or plastic analysis. Note that AISC *Specification* Appendix 1, Inelastic Design, is not intended for seismic design. For a discussion of the application of the AISC *Specification*, AISC *Seismic Provisions*, and ASCE/SEI 7 in seismic analysis, see Nair et al. (2011).

While non-SFRS members and connections may be analytically assumed not to resist horizontal ground motion (i.e., $\pm\rho Q_E$ from ASCE/SEI 7), they must be reliable in resisting the vertical inertial forces induced by vertical ground motion (i.e., $\pm 0.2S_{DS}D$ from ASCE/SEI 7). Non-SFRS members must also be designed to ensure deformation compatibility at large lateral displacements to maintain structural integrity of the structure. Equally, the destabilizing effect that non-SFRS framing can have on a structure (e.g., leaning column effects) must be addressed in the analysis and design of the stabilizing SFRS. The SFRS also consists of diaphragms, chords and collectors.

Elastic, Inelastic and Plastic Analysis

Elastic seismic analysis procedures in ASCE/SEI 7 reduce the seismic response by a factor of $1/R$, where R is the response modification coefficient. The intent of this reduction is to target the elastic response at the onset of the first significant yield (e.g., plastic hinge in a beam or compression buckling of a brace). Consequently, inelastic or plastic analysis as outlined in Appendix 1 of the AISC *Specification* is not permitted for determining the component design forces from seismic effects—see the AISC *Specification* Commentary to Appendix 1 for further discussion. Therefore, analytical consistency with the AISC *Specification* and the AISC *Seismic Provisions* is primarily maintained using an elastic analysis procedure. Although a nonlinear response history analysis is permitted, it is not commonly used to determine member design forces, but as an assessment tool to judge acceptance of a design. In specific cases, a nonlinear static analysis may be used to capture the nonlinear elastic response of a component or connection, such as when rotational springs are used to represent partially restrained connections.

AISC *Specification* Chapter C requires that a rigorous second-order analysis be used to determine the required strengths of components using the appropriate load combinations. The analysis must include consideration of certain effects that can influence the stability of the structure and its elements, including second order effects (both P-Δ and P-δ). Additional discussion can be found in Wilson and Habibullah (1987), White and Hajjar (1991), and Geschwindner (2002). There are different methods by which to address second-order effects, including iterative or noniterative solutions with either stationary or incremental loading. For example, some computer programs use a vertical load combination in conjunction with the approximate geometric stiffness matrix to reduce the structural stiffness to account for geometric nonlinearities. The resulting structural stiffness from this initial analysis is used for all subsequent load analyses (e.g., dead, live, lateral). This method is advantageous as it allows superposition of individual load effects because the stiffness is held constant. This approach typically captures only the P-Δ effect, and P-δ is either neglected or approximated by segmenting members into two or more sections. Some programs can iterate by ramping the gravity loads in conjunction with the geometric stiffness matrix to more accurately capture the change in system stiffness during each load step. Lateral loads can then be iterated by ramping them so that the analysis captures the additional changes to the system stiffness during each step. In this method, superposition of individual load effects is not appropriate and the vertical loads would therefore need to be included in the analysis.

With reference to seismic analysis, the structural stiffness is constant (reduced based on an initial analysis) when using the ASCE/SEI 7 modal response spectrum analysis (MRSA) to determine the total lateral seismic forces and linear response history analysis. The results of these analyses are then combined with other load effects based on the same reduced stiffness. This procedure is not applicable to a nonlinear response time history analysis as the structural stiffness would need to be updated at each time step based on all load effects included in the analysis.

Gravity loads should be included in the seismic analysis in order to accurately address second-order effects, including the destabilizing effect generated by non-SFRS framing, and the effect of these loads on the periods of a structure. A three-dimensional mathematical model can be developed that captures all loading conditions or, in the case of a two-dimensional analysis, an ancillary P-Delta column, as a minimum, can be modeled as a substitute for the gravity (non-SFRS) framing system. The P-Delta column is commonly modeled to provide no lateral stiffness to the SFRS, but could be calibrated to provide the same stiffness as that provided by the gravity system.

As an alternative to a rigorous second-order analysis, second order effects can be approximated by amplifying the axial forces and moments in members and connections from a first-order analysis through an approximate second-order analysis outlined in AISC *Specification* Appendix 8. The provisions for performing this amplified first-order analysis were developed on the basis of elastic theory and are not appropriate for inelastic analysis.

Stability Design Methods in the AISC *Specification*

The AISC *Specification* outlines three stability design methods and corresponding elastic analysis requirements (see Table 2-2 in the AISC *Manual*) as follows:

- Direct analysis method (AISC *Specification* Sections C2 and C3)
- Effective length method (AISC *Specification* Appendix 7, Section 7.2)
- First-order analysis method (AISC *Specification* Appendix 7, Section 7.3)

The use of each of these methods in seismic design is explained in the following discussions. Additional information on each of the methods can be found in the Commentary to the applicable sections in the AISC *Specification*.

Direct Analysis Method

Provisions for the direct analysis method (DM) are outlined in AISC *Specification* Sections C2 and C3. This analysis procedure is permitted for all steel structures and is required when the ratio of maximum second-order drift to maximum first-order drift, which can be taken as B_2 in Appendix 8 using nominal stiffness properties, exceeds 1.5. The DM requires P-Delta effects to be considered either directly through a second-order elastic analysis or through an amplified first-order analysis.

Effective Length Method

Provisions for the effective length method (ELM) are outlined in AISC *Specification* Appendix 7, Section 7.2. When permitted by Section 7.2.1, there are no deviations from the elastic analysis provisions in ASCE/SEI 7. The ELM addresses P-Delta effects either directly through a second-order elastic analysis or through an amplified first-order analysis.

In the ELM procedure, interaction between frame behavior and that of its members is approximated by the effective length factor, K. This factor is used to represent the influence of the system on the strength of an individual member. Where the flexural stiffness of a column is considered to contribute to the lateral stability and resistance to lateral loads, K for that member is determined from a sidesway buckling analysis. Alternatively, the effective length factor may be computed using the alignment charts as discussed in detail in the Commentary to AISC *Specification* Appendix 7. It is permitted to use $K = 1.0$ for design for compression effects if $B_2 \leq 1.1$.

First-Order Analysis Method

Provisions for the first-order analysis method (FOM) are outlined in AISC *Specification* Appendix 7, Section 7.3. With this approach, second-order effects are captured through the application of an additional lateral load equal to at least 0.42% of the story gravity load applied in each load case. No further second-order analysis is necessary. The required strengths are taken as the forces and moments obtained from the analysis and the effective length factor is $K = 1.0$.

Analysis Methods in ASCE/SEI 7 and the Direct Analysis Method

ASCE/SEI 7 Section 12.6 outlines three seismic analytical procedures as follows:

- Equivalent lateral force analysis (ELF) (ASCE/SEI 7 Section 12.8)
- Modal response spectrum analysis (MRSA) (ASCE/SEI 7 Section 12.8)
- Seismic response history procedures, linear and nonlinear (ASCE/SEI 7 Chapter 16)

Detailed information can be found in the commentary to Section 12.6 of ASCE/SEI 7 and in the *NEHRP Recommended Seismic Provisions for New Buildings and Other Structures* (FEMA, 2009a). The following discussion summarizes the ELF and MRSA analysis methods and how they relate to the direct analysis method of the AISC *Specification*.

Equivalent Lateral Force Analysis and the Direct Analysis Method

The provisions for the DM are consistent with the elastic analysis provisions given in ASCE/SEI 7 Section 12.8 for the ELF, provided that the following conditions are maintained throughout the analysis:

- The mathematical model for analysis considers all forms of deformation of the structural components, including stiffness reductions and geometric imperfections in accordance with AISC *Specification* Chapter C. The stability coefficient, θ, will generally limit B_2 to less than 1.7, permitting geometric imperfections to be neglected in the analysis for seismic load combinations. Consequently, notional loads should be applied in the mathematical model for gravity-only load combinations (if the same model is used) in lieu of modeling the out-of-plumbness by shifting work points.
- The fundamental period of the structure, T, is limited to T_a or $C_u T_a$ if T is computed by analytical methods. If the computed value for T is less than $C_u T_a$ then T is used as

the fundamental period. This is because T_a has been statistically derived from actual building responses therefore capturing all influential factors. See ASCE/SEI 7 Section 12.9.4.
- Forces and deformations resulting from analysis with seismic forces reduced by a factor of $1/R$, where R is the response modification coefficient, include second-order effects either through a second-order analysis, an amplified first-order analysis, or a hybrid combination of the two methods, independent of the stability coefficient, θ, in ASCE/SEI 7.

The AISC *Specification* and the AISC *Seismic Provisions* deal directly with strength design of members and connections. Verification of seismic drift limits and potential post-earthquake instability are addressed in the applicable building code. As such, some of the provisions for the DM are not directly applicable for a drift analysis. However, they can be conservatively applied for drift analysis.

Other methodologies for applying the DM have been proposed by Nair et al. (2011).

Modal Response Spectrum Analysis and the Direct Analysis Method

The provisions for the DM are consistent with elastic analysis provisions in ASCE/SEI 7 for MRSA, provided that the following conditions are maintained throughout the analysis:

- All the requirements listed previously for the ELF are maintained.
- Forces and drifts are scaled as required by ASCE/SEI 7 Section 12.9.4. Note that T used in this scaling is limited as discussed previously for the ELF.

The same procedure is followed in regards to a drift analysis. Though the scaling of drifts is not required unless assigned to a certain seismic design category, allowable drift limits and stability provisions of ASCE/SEI 7 are applicable.

2.4 STRUCTURAL MODELING

A mathematical model used for structural analysis is simply an interpretation of what configuration of components, mechanical characteristics, and mass distribution is significant to the distribution of forces and deformations in the system. Models can be simple (such as a two-dimensional finite element model based on centerline dimensions) or highly sophisticated (such as a three-dimensional continuum model that can explicitly capture material nonlinearity and buckling). Both strength and stiffness are required to characterize the mechanical properties of a component.

Strength of Structural Elements

The strength of structural elements is typically not a modeling consideration for elastic analysis. Information on modeling component strengths for nonlinear dynamic analysis can be found in NCJV (2010), Deierlein et al. (2010), PEER (2010), PEER/ATC (2010), FEMA (2009b), and ASCE (2006).

Stiffness of Structural Elements

AISC *Seismic Provisions* Chapter C states that stiffness properties of components for an elastic analysis should be based on the elastic sections and that the effects of cracked sections shall be considered for composite components. AISC *Specification* Chapter C and the commentary to AISC *Seismic Provisions* Chapter C give recommendations for effective stiffness values to be used in analysis.

Steel Elements

The stiffness properties of steel beams, columns and braces used in the mathematical model will depend upon the stability design method selected and, potentially, the magnitude of straining the member undergoes. Reduced stiffness for all members contributing to the lateral stability of the structure is required when using the DM to determine design forces. It is important to note that the stiffness reduction terms in the DM include a component representing material nonlinearity (e.g., accounting for residual stresses) and a component representing member out-of-straightness and other uncertainties. Consequently, stiffness reduction is separated into a load-dependent factor and load-independent factor, complicating its direct application to dynamic analysis.

Research has demonstrated that residual stresses have a lesser effect on shear stiffness than flexural stiffness. For simplicity, the shear modulus, G, can be reduced in proportion to the reduction in the modulus of elasticity, E, with no further reduction to account for axial load effects.

It is common to model steel beams that are part of the SFRS without composite action because the reliability of the composite stiffness at large inelastic deformations is questionable due to the potential for failure of steel headed stud anchors. If composite action is taken into account, the following applicable effects should be considered.

Composite Elements

The stiffness properties of steel members acting compositely with concrete should include the following applicable effects: concrete cracking of the section, steel reinforcement ratio, section configuration, material properties of the concrete, and variations of these factors along the member length. The flexural stiffness, EI_{eff}, and axial stiffness, EA_{eff}, based on a transformed cracked section analysis (that also accounts for variations along the member length) should be used in lieu of EI and EA in all analysis methods. Recommendations are provided in AISC *Seismic Provisions* Chapter C Commentary based on ACI 318 provisions.

For steel beams with a composite slab, composite action can be included where the slab and shear connection to the beam have been designed and detailed to provide acceptable behavior (see Commentary to Chapter G in the AISC *Seismic Provisions*). For concrete-encased steel beams and beams acting compositely with a concrete slab, a plastic stress distribution corresponding to the ultimate nominal strengths of each component can be used to compute a lower-bound elastic moment of inertia, I_{LB}. For a steel beam with a composite slab in a moment frame with double curvature bending, the effective flexural stiffness, EI_{eff}, can be taken as the average of the stiffness in the positive and negative bending regions, as follows:

STRUCTURAL MODELING

$$EI_{eff} = 0.5 \left(\overbrace{E_s I_s}^{negative} + \overbrace{E_s I_{LB}}^{positive} \right) \qquad (2\text{-}1)$$

where
- E_s = modulus of elasticity of steel, ksi
- I_{LB} = $I_s + A_s(Y_{ENA} - d_3)^2 + (\Sigma Q_n/F_y)(2d_3 + d_1 - Y_{ENA})^2$, in.^{4} (*Spec.* Eq. C-I3-1)
- A_s = area of steel cross section, in.^{2}
- ΣQ_n = sum of the nominal shear strength of steel anchors between the point of maximum positive moment and the point of zero moment to either side, kips
- I_s = moment of inertia of steel cross section, in.^{4}
- Y_{ENA} = distance from bottom of the steel section to the elastic neutral axis, in.
  = $[A_s d_3 + (\Sigma Q_n/F_y)(2d_3 + d_1)]/[A_s + (\Sigma Q_n/F_y)]$, in.   (*Spec.* Eq. C-I3-2)
- d_1 = distance from the compression force in the concrete to the top of the steel section, in.
- d_3 = distance from the resultant steel tension force for full section tension yield ($P_y = F_y A_s$) to the top of the steel, in.

I_{LB}, based on a plastic stress distribution, is recommended in lieu of 75% of I_{equiv} (see AISC *Specification* Chapter I commentary), where

$$I_{equiv} = I_s + \sqrt{(\Sigma Q_n / C_f)}(I_{tr} - I_s) \qquad (\text{*Spec.* Eq. C-I3-4})$$

where
- C_f = compression force in concrete slab for fully composite beam; smaller of $A_s F_y$ and $0.85 f_c' A_c$, kips
- I_{tr} = moment of inertia for the fully composite uncracked transformed section, in.^{4}

AISC *Seismic Provisions* Commentary Chapter G discusses limitations on using partially composite beams in certain composite systems.

The flexural stiffness of composite columns and braces (encased or filled) can be taken as EI_{eff} prescribed in AISC *Specification* Chapter I. The axial stiffness can be taken as

$$EA_{eff} = E_s A_s + c_3 E_c A_c \qquad (2\text{-}2)$$

where
- A_c = area of concrete slab within the effective width, in.^{2}
- E_c = modulus of elasticity of concrete, ksi
- c_3 = 0.4 for filled sections and 0.2 for encased sections

Equation 2-2 is taken from the LRFD *Specification for Structural Steel Buildings* (AISC, 2000).

Connections and Panel Zones

Connections and panel zones can contribute significantly to the overall lateral flexibility of a system and the resulting deformations are required to be addressed in the analysis for determining the distribution of design forces and story drifts. In modeling moment or braced frames, the impact of connection size and stiffness should be considered.

Research (FEMA, 2000a) has demonstrated that panel zone deformations in steel moment frames can have significant impact on earthquake-induced lateral drift. However, modeling framing using center line-to-center line dimensions for the framing elements can approximate the effects of panel zone flexibility reasonably well for elastic analysis (see Figure 2-1). Zero-stiffness end offsets may be modeled to analytically provide forces at the panel zone faces but not influence the periods of vibration. Alternatively, panel zone models that include web doubler plates and continuity plates can be explicitly modeled or implicitly included by modeling partially rigid end offsets. Fully rigid offsets alone should not be assumed to be the only source of panel zone stiffness (Tsai and Popov, 1990). Several panel zones models are illustrated in FEMA 355C (FEMA, 2000a).

Explicit connection modeling by rotational springs is permitted when based on analytical and experimental test data. Such an approach may be warranted when accounting for the effects of partially restrained connections or other mechanical characteristics of a connection such as bolt slip. Alternatively, beams can be modeled with an equivalent flexural stiffness, EI_{eff}.

Beams with reduced beam sections (RBS) can be addressed by physically modeling a prismatic or parabolic tapered section at the RBS location. If a prismatic section is used, one possibility is to take the moment of inertia at the outer edge of the center two-thirds of the RBS (ANSI/AISC 358 Chapter 5). The flange width, $b_{f,RBS}$, is:

$$b_{f,RBS} = 2(R-c) + b_f - 2\sqrt{R^2 - \left(\frac{b}{3}\right)^2} \qquad (2\text{-}3)$$

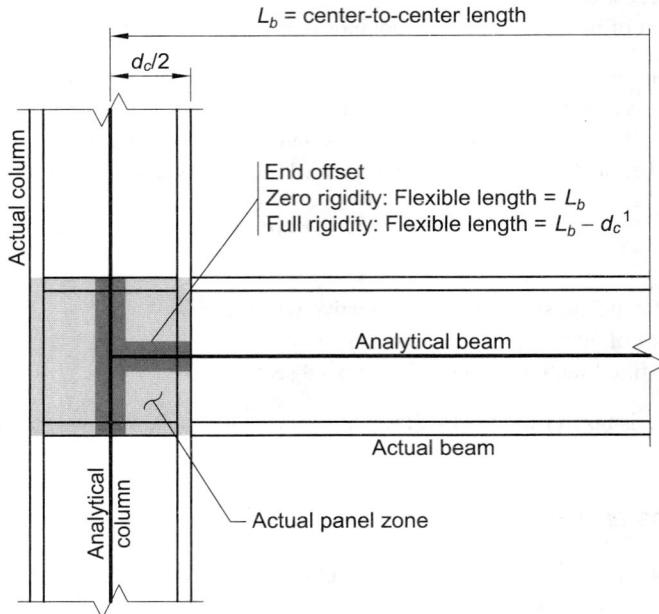

[1] Assumes same column depth at both ends of beam

Fig. 2–1. Modeling end offsets at panel zones.

where

$$R = \frac{4c^2 + b^2}{8c}$$

= radius of cut from ANSI/AISC 358 Figure 5.1

This approach may be counteracted by neglecting composite action with the concrete slab between protected zones. It is also common not to explicitly model the RBS and to use either an EI_{eff} for the beam or simply to amplify the elastic story drifts to account for the reduced stiffness, as shown in Example 4.3.1 of this Manual. Additional information on steel moment frames can be found in ANSI/AISC 358, FEMA 350 (FEMA, 2000b), and NEHRP Seismic Design Technical Brief No. 2 (Hamburger et al., 2009). For composite frames, the effects of cracking on the beam-to-column joint stiffness should be included.

A common question regarding connection deformations in braced frames is whether the ends of a brace should be considered as a moment resisting or pinned connection. The answer will depend on the gusset connection detailing. Fundamentally, a brace-end connection at a beam-to-column joint or at a beam interior segment can be assumed pinned out-of-plane and fixed in-plane, because the out-of-plane stiffness of the gusset plate is significantly smaller than the in-plane stiffness.

Similar to beam-to-column joints in moment frames, partially restrained end zones or ancillary stub members can be modeled at the ends of braces to represent the increased in-plane flexural stiffness provided by the gusset connections. The flexural stiffness at these sections typically ranges from 2 to 4 times that of the brace. The beam-to-column connection where a brace member intersects can be modeled as a fully restrained connection; otherwise the connection can be modeled as a simple connection depending on project specific requirements. Additional information concerning steel braced frames can be found in NCJV (2010) and Carter (2009).

AISC Design Guide 20, *Steel Plate Shear Walls* (Sabelli and Bruneau, 2006) provides information regarding modeling practices for special plate shear walls. For composite construction, the effects of cracking on the beam-to-column joint stiffness should be included.

Column Bases and Foundations

ASCE/SEI 7 Section 12.7 states that for the purpose of determining seismic loads, the structure can be considered fixed at the base. That is, the base where seismic motions are introduced into the structure is globally restrained horizontally, vertically and rotationally about the horizontal axes. Alternatively, flexibility of the supporting soil (including deformations of the foundation components) or soil-structure interaction may be included. The theoretical derivation of soil-structure interaction effects was developed on the basis of a rigid foundation. Therefore, support flexibility and soil-structure interaction cannot be applied concurrently.

Flexibility of the supporting soil is commonly modeled using soil springs assuming the foundation component is rigid. Alternatively, foundation components may be explicitly modeled to address their flexibility. For nonlinear response history analysis, soil springs should directly model the nonlinear behavior of the supporting soil.

Column base modeling is a function of frame mechanics, detailing and rigidity of the foundation components, and is not related to the global restraint of the seismic base. Partially restrained base models may be used to more accurately capture rotational characteristics of base plate connections based on experimental results. Alternatively, pinned bases may be modeled to account for connection, foundation and soil flexibility, although the column base may be detailed to be fixed to the foundation component.

Diaphragms for Three-Dimensional Analysis

Diaphragms, chords, collectors and associated elements distribute seismic forces to the SFRS. The diaphragm model used in analysis should realistically model the diaphragm's in-plane stiffness and the distribution of lateral forces. ASCE/SEI 7 Section 12.3.1 classifies a diaphragm as rigid, semi-rigid or flexible depending on its in-plane stiffness. A diaphragm made up of a composite slab can be modeled as rigid when the diaphragm's span-to-depth ratio is 3 or less in structures with no horizontal irregularities. This assumption simplifies calculations because the diaphragm moves as a rigid body about the center of rigidity of a given story and the total mass can be assumed to be lumped at the center of mass. Alternatively, a semi-rigid diaphragm explicitly models the diaphragm's in-plane stiffness. In either model, lateral forces are distributed to the various SFRS in proportion to their relative elastic lateral stiffness and distance from the center of rigidity. For flexible diaphragms, an SFRS is assumed to resist forces proportional to the mass that is tributary to the SFRS.

Diaphragm slabs can be modeled using either membrane or shell elements. In-plane stiffness reduction factors should be applied to account for cracking of the concrete and other factors that decrease the stiffness of the diaphragm. Membranes differ from shells in that membranes do not provide out-of-plane or rotational stiffness which can increase the computational demand and the flexural stiffness at joints. However, membrane edges have to be supported by framing.

The axial forces developed in horizontal members on a given floor are dependent on the in-plane stiffness of the diaphragm model assigned to that floor. Caution should be exercised in assigning diaphragm models where horizontal members are designed to transmit or redistribute seismic forces to and between SFRS. In many cases, these members are required to be designed for amplified seismic loads including the overstrength factor, Ω_o, and thereby, are intended to remain essentially elastic.

A rigid diaphragm model prevents relative in-plane movement between nodes on a given floor. Thus, axial forces will not develop in horizontal members connected to the diaphragm, inhibiting the design of members that transmit forces to or between SFRS or chords of a vertical truss spanning between levels (e.g., outriggers). The effect of this node lock will increase forces carried by diagonal members between diaphragms. Alternatively, a semi-rigid diaphragm can be modeled. A disadvantage of this model is that the magnitude of the axial force in a horizontal member will depend on the in-plane stiffness at the node and how the diaphragm is modeled along the length of that member.

Another alternative is to release nodes from the diaphragm constraint. This may also include restructuring the extents of the rigid diaphragm so that a core area is a rigid diaphragm and the surrounding areas are semi-rigid based on structural properties assigned to the diaphragm system.

It is possible to model the diaphragm by decoupling a three-dimensional structure into multiple two-dimensional analyses where lateral forces are applied as point forces at nodes or as uniform or triangular distributed loads along horizontal members. Capturing the required magnitude of the axial force in a three-dimensional analysis can be more challenging as zero to very low stiffness diaphragm models can lead to increases in $P\text{-}\Delta$ forces transferred to the SFRS and/or modeling errors. It is recommended that the analyst perform a parametric study with various diaphragm assignments and assemblies to determine the most efficient model to adequately capture a reasonable estimate of the diaphragm behavior and required axial force.

Gravity Loads

All gravity loads should be modeled in the analysis in order to accurately address second-order effects and to capture the distribution of gravity load effects on vertical force-resisting members. A mathematical model is commonly analyzed as a fully constructed, cohesive structure for each load effect or load combination. This practice is not, however, consistent with how a structure is built, where some load effects are distributed based on construction sequence. This is particularly true for the distribution of self-weight in braced frames and structures with outriggers or hat trusses where installation of diagonal members may be completed after the surrounding framing and floor system is constructed and at different story elevations. In the latter case, dead load effects created during construction in exterior vertical force-resisting members can be underestimated because these members can in effect hang from the stiffer outrigger/truss system in the analysis, increasing the forces in the interior vertical support system. Similarly, gravity effects can be distributed to diagonal braces in proportion to their contribution to joint stiffness.

For cases when the distribution of dead load effects is a concern, a staged construction analysis can be performed. In its fundamental form, the mathematical model of a complete system is broken down into assemblies, commonly story levels, and the analysis will incrementally add each group and superimpose the results onto the previous analysis. Care should be exercised when gravity effects produce sway and when geometric imperfections are included in the analysis (by either method discussed previously) as a staged analysis cannot handle sway of different assemblies and superposition must be applicable. Alternatively, some analysis programs allow the analyst to automatically not include diagonal members during the gravity load analysis. Though this is more related to analysis than modeling, an alternative modeling technique would be to provide axial force releases in the diagonal members for the gravity load analysis. In these cases, the method used to address geometric nonlinearities within the analysis program is of critical importance, and this will dictate which technique can be used. Another consideration is differential settlement of vertical force-resisting systems under dead load effects.

Gravity Loads in Diagonal Braces and Special Plate Shear Walls

The AISC *Seismic Provisions* stipulate that the gravity forces be neglected in braces in buckling-restrained braced frames and web plates in special plate shear walls. These provisions are intended to restrict the use of SFRS components that are required to dissipate significant amounts of energy by inelastic actions to simultaneously provide structural integrity of the structure under gravity loads. Many of the capacity design analysis provisions have been developed based on this concept.

This approach can be a concern for complex structures that contain purposely sloped or stepped non-SFRS columns or where diagonal braces are required to stabilize a structure that undergoes sidesway from gravity loads (e.g., sloping structural system) or are required to directly participate in carrying gravity loads (e.g., diagrid system). A three-dimensional nonlinear dynamic analysis may be necessary to verify the seismic performance of complex structures. If lateral support for other load effects is provided by an ancillary non-SFRS back-up system, it should not be excluded from the mathematical model used for seismic analysis.

PART 2 REFERENCES

AISC (2000), *Load and Resistance Factor Design Specification for Structural Steel Buildings*, American Institute of Steel Construction, Chicago, IL.

ASCE (2006), *Seismic Rehabilitation of Existing Buildings*, ASCE 41-06, American Society of Civil Engineers, Reston, VA.

Carter, C.J. (2009), "Origins of $R = 3$," *Proceedings of the 2009 Structures Congress*, ASCE, Austin, TX, April 30 – May 2, 2009, pp. 1–10.

Deierlein, G.G., Reinhorn, A.M. and Willford, M.R. (2010), "Nonlinear Structural Analysis for Seismic Design," NIST GCR 10-917-5, NEHRP Seismic Design Technical Brief No. 4, NEHRP Consultants Joint Venture, partnership of the Applied Technology Council and the Consortium of Universities for Research in Earthquake Engineering, National Institute of Standards and Technology, Gaithersburg, MD.

FEMA (2000a), *State of the Art Report on Systems Performance of Steel Moment Frames Subject to Earthquake Ground Shaking*, FEMA 355c, prepared by the SAC Joint Venture for the Federal Emergency Management Agency, Washington, DC.

FEMA (2000b), *Recommended Seismic Design Criteria for New Steel Moment-Frame Buildings*, FEMA 350, Federal Emergency Management Agency, Washington, DC.

FEMA (2009a), *NEHRP Recommended Seismic Provisions for New Buildings and Other Structures*, FEMA P-750, Federal Emergency Management Agency, Washington, DC.

FEMA (2009b), *Quantification of Building Seismic Performance Factors*, FEMA P-695, Federal Emergency Management Agency, Washington, DC.

Geschwindner, L.F. (2002), "A Practical Approach to Frame Analysis, Stability and Leaning Columns," *Engineering Journal*, AISC, Vol. 39, No. 4, 4th Quarter, pp. 167–181.

Hamburger, R.O., Krawinkler, H., Malley, J.O. and Adan, S.M. (2009), "Seismic Design of Steel Special Moment Frames: a Guide for Practicing Engineers," NIST GCR 09-917-3, NEHRP Seismic Design Technical Brief No. 2, NEHRP Consultants Joint Venture, partnership of the Applied Technology Council and the Consortium of Universities for Research in Earthquake Engineering, National Institute of Standards and Technology, Gaithersburg, MD.

Nair, S., Malley, J.O. and Hooper, J.D. (2011), "Design of Steel Buildings for Earthquake and Stability by Application of ASCE 7 and AISC 360," *Engineering Journal*, AISC, Vol. 48, No. 3, 3rd Quarter, pp. 199–204.

NCJV (2010), *Evaluation of the FEMA P-695 Methodology for Quantification of Building Seismic Performance Factors*, NIST GCR 10-917-8, NEHRP Consultants Joint Venture, partnership of the Applied Technology Council and the Consortium of Universities for Research in Earthquake Engineering, National Institute of Standards and Technology, Gaithersburg, MD.

PEER (2010), *Seismic Design Guidelines for Tall Buildings*, Pacific Earthquake Engineering Research Center, University of California at Berkeley, Berkeley, CA.

PEER/ATC (2010), *Modeling and Acceptance Criteria for Seismic Design and Analysis of Tall Buildings*, PEER/ATC 72-1 Report, Applied Technology Council, Redwood City, CA, October.

Sabelli, R. and Bruneau, M. (2006), *Steel Plate Shear Walls*, Design Guide 20, AISC, Chicago, IL.

Tsai, K.C. and Popov, E.P. (1990), "Seismic Panel Zone Design Effect on Elastic Story Drift in Steel Frames," *Journal of Structural Engineering*, ASCE, Vol. 116, No. 12, pp. 3235–3301.

White, D. and Hajjar, J. (1991), "Application of Second-Order Elastic Analysis in LRFD: Research to Practice," *Engineering Journal*, AISC, Vol. 28, No. 4, 4th Quarter, pp. 133–148.

Wilson, E.L. and Habibullah, A. (1987), "Static and Dynamic Analysis of Multi-Story Buildings Including P-Delta Effects," *Earthquake Spectra*, Earthquake Engineering Research Institute, Vol. 3, Issue 3.

PART 3

SYSTEMS NOT SPECIFICALLY DETAILED FOR SEISMIC RESISTANCE

3.1 SCOPE 3–2
3.2 GENERAL DISCUSSION 3–2
3.3 DESIGN EXAMPLE PLAN AND ELEVATIONS 3–2
3.4 MOMENT FRAMES 3–4
    Example 3.4.1. Moment Frame Story Drift Check 3–5
    Example 3.4.2. Moment Frame Column Design 3–6
    Example 3.4.3. Moment Frame Beam Design 3–10
    Example 3.4.4. Moment Frame Beam-to-Column Connection Design 3–13
3.5 BRACED FRAMES 3–21
    Example 3.5.1. Braced Frame Brace Design 3–22
    Example 3.5.2. Braced Frame Column Design 3–24
    Example 3.5.3. Braced Frame Brace-to-Beam/Column Connection Design 3–25
PART 3 REFERENCES 3–53

3.1 SCOPE

This Part shows member and connection designs for braced and moment frame systems that are not specifically detailed for seismic resistance. Seismic design of the seismic force resisting system in accordance with the AISC *Seismic Provisions* is referred to as "seismic detailing" by the applicable building code. The systems in this Part are designed according to the requirements of the AISC *Specification*. The Scope statement at the front of this Manual discusses the differentiation between seismic force resisting systems that require special detailing for seismic resistance and those that do not.

3.2 GENERAL DISCUSSION

Systems requiring structural steel design in accordance with the AISC *Specification* only are addressed in this Part. It is a common misconception that when seismic detailing of the seismic force resisting system is not required, there are no other seismic design requirements. Regardless of the seismic detailing requirements, structures assigned to Seismic Design Categories B through F are subject to many other seismic design considerations prescribed in the applicable building code. For example, ASCE/SEI 7 contains numerous requirements, such as:

- Table 12.3-1, Horizontal Structural Irregularities
- Table 12.3-2, Vertical Structural Irregularities
- Section 12.4, Seismic Load Effects and Combinations
- Section 12.5, Direction of Loading
- Section 12.8.4.3, Amplification of Accidental Torsional Moment
- Section 12.10.2, Collector Elements
- Section 12.13, Foundation Design

3.3 DESIGN EXAMPLE PLAN AND ELEVATIONS

The following sections consist of design examples for a typical building not requiring seismic detailing. See Figure 3-1 for a typical floor plan for this building with composite flooring. Design Examples 3.4.1 through 3.4.4 demonstrate the design of a typical moment frame for the building. See Figure 3-2 for an elevation of the moment frame. Design Examples 3.5.1 through 3.5.3 demonstrate the design of a typical braced frame for the building. See Figure 3-3 for an elevation of the braced bay.

The code specified loading is as follows:

D_{floor} = 85 psf
D_{roof} = 68 psf
L_{floor} = 80 psf
S = 20 psf
Curtain wall = 175 lb/ft

Wind loads are determined according to Chapter 28, Part 2 of ASCE/SEI 7. The assumed parameters are: Basic Wind Speed is 115 miles per hour (3 second gust), Wind Exposure Category is B, topographic factor K_{zt} is 1.0, and the building is in Risk Category II. Required

3.3 DESIGN EXAMPLE PLAN AND ELEVATIONS

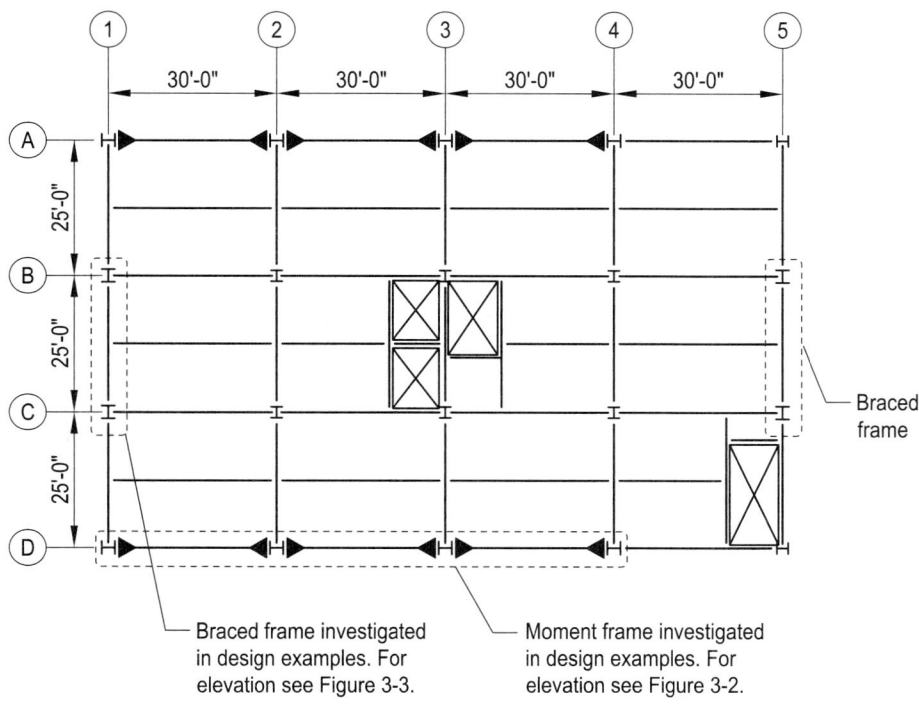

Fig. 3-1. Floor plan for Part 3 design examples.

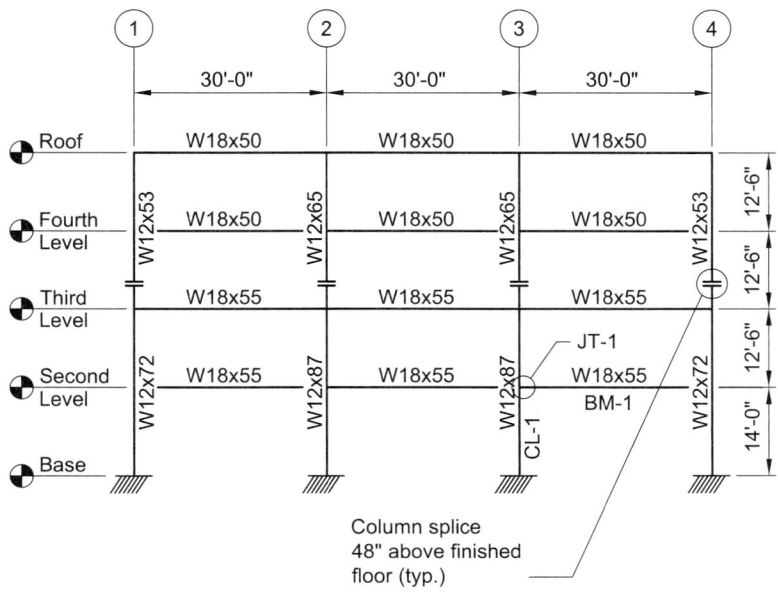

Fig. 3-2. Moment frame elevation for Examples 3.4.1, 3.4.2, 3.4.3 and 3.4.4.
For floor plan, see Figure 3-1.

AMERICAN INSTITUTE OF STEEL CONSTRUCTION

strengths from load combinations that include wind loads were shown not to govern over load combinations that include seismic loads for both the braced frame and the moment frame. Therefore, wind loads are not included in the design examples in Part 3.

The necessary parameters for determining seismic loading are given with each design example.

3.4 MOMENT FRAMES

Moment frames resist lateral forces and displacements through flexure and shear in the beams and columns. The necessary restraint must be provided by the moment connections between the beam and the columns.

Moment frames tend to have larger and heavier beam and column sizes than braced frames. The increase in member sizes and related costs is often accepted to gain the increased flexibility provided in the architectural and mechanical layout in the structure. The absence of diagonal bracing members can provide greater freedom in the configuration of walls and in the routing of mechanical ductwork and piping. Moment frames are often positioned at the perimeter of the structure, allowing maximum flexibility of the interior spaces. Drift control is required by the applicable building code to help limit damage to both the structural and nonstructural systems.

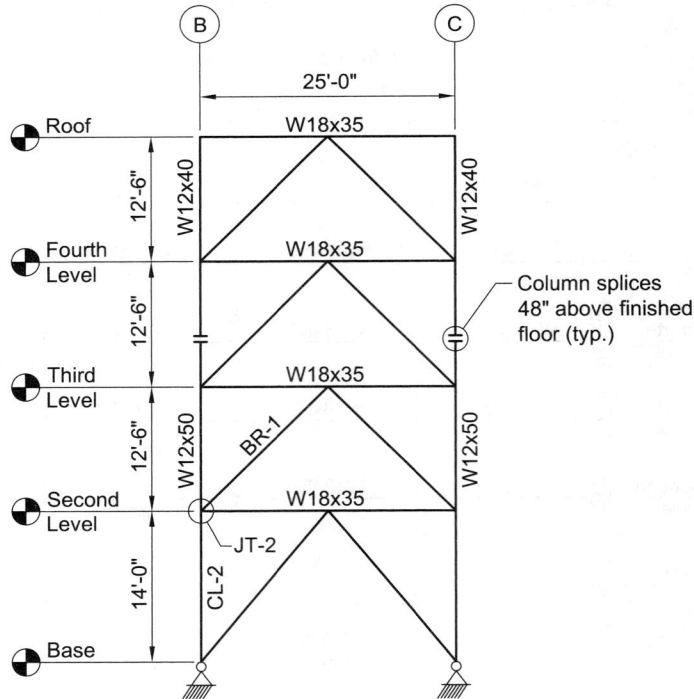

Fig. 3-3. Braced frame elevation for Examples 3.5.1, 3.5.2 and 3.5.3. For floor plan, see Figure 3-1.

3.4 MOMENT FRAMES

Because the moment frame in the following examples does not require seismic detailing, it is designed in accordance with the provisions of the AISC *Specification*.

Example 3.4.1. Moment Frame Story Drift Check

Given:
Determine if the moment frame satisfies the ASCE/SEI 7 seismic story drift requirements.

Refer to the moment frame elevation shown in Figure 3-2. The applicable building code specifies the use of ASCE/SEI 7 for seismic story drift requirements. In accordance with ASCE/SEI 7:

Risk Category: II
Seismic Design Category: C
Deflection Amplification Factor, C_d: 3
Seismic Importance Factor, I_e: 1.0
Allowable Story Drift, Δ_a: $0.020 h_{sx}$

Solution:
From a second-order elastic analysis of the structure, the elastic displacement computed under strength-level design earthquake forces at each level are:

$\delta_{re} = 1.87$ in.
$\delta_{4e} = 1.54$ in.
$\delta_{3e} = 1.03$ in.
$\delta_{2e} = 0.477$ in.
$\delta_{be} = 0$ in.

The deflection at level x is:

$$\delta_x = \frac{C_d \delta_{xe}}{I_e} \quad \text{(ASCE/SEI 7 Eq. 12.8-15)}$$

The allowable story drift at level x, from ASCE/SEI 7 Table 12.12-1, is:

$$\Delta_a = 0.020 h_{sx}$$

where

h_{sx} = story height below level x, ft

Between the roof level and level 4:

$$\begin{aligned} \delta_r &= \frac{C_d (\delta_{re} - \delta_{4e})}{I_e} \\ &= \frac{3(1.87 \text{ in.} - 1.54 \text{ in.})}{1.0} \\ &= 0.990 \text{ in.} \end{aligned}$$

$$\Delta_a = 0.020(12.5 \text{ ft})(12.0 \text{ in./ft})$$
$$= 3.00 \text{ in.} > 0.990 \text{ in.} \quad \textbf{o.k.}$$

Between level 4 and level 3:

$$\delta_4 = \frac{C_d(\delta_{4e} - \delta_{3e})}{I_e}$$
$$= \frac{3(1.54 \text{ in.} - 1.03 \text{ in.})}{1.0}$$
$$= 1.53 \text{ in.}$$
$$\Delta_a = 0.020(12.5 \text{ ft})(12.0 \text{ in./ft})$$
$$= 3.00 \text{ in.} > 1.53 \text{ in.} \quad \textbf{o.k.}$$

Between level 3 and level 2:

$$\delta_3 = \frac{C_d(\delta_{3e} - \delta_{2e})}{I_e}$$
$$= \frac{3(1.03 \text{ in.} - 0.477 \text{ in.})}{1.0}$$
$$= 1.66 \text{ in.}$$
$$\Delta_a = 0.020(12.5 \text{ ft})(12.0 \text{ in./ft})$$
$$= 3.00 \text{ in.} > 1.66 \text{ in.} \quad \textbf{o.k.}$$

Between level 2 and the base level:

$$\delta_2 = \frac{C_d(\delta_{2e} - \delta_{be})}{I_e}$$
$$= \frac{3(0.477 \text{ in.} - 0 \text{ in.})}{1.0}$$
$$= 1.43 \text{ in.}$$
$$\Delta_a = 0.020(14.0 \text{ ft})(12.0 \text{ in./ft})$$
$$= 3.36 \text{ in.} > 1.43 \text{ in.} \quad \textbf{o.k.}$$

Comment:

In this case, the member sizes resulted from strength requirements. The seismic story drift requirements do not always govern the design of moment frames.

Example 3.4.2. Moment Frame Column Design

Given:

Refer to Column CL-1 in Figure 3-2. Verify that a W12×87 ASTM A992 W-shape is sufficient to resist the following required strengths between the base and second levels. The applicable building code specifies the use of ASCE/SEI 7 for calculation of loads.

The load combinations that include seismic effects are:

3.4 MOMENT FRAMES

LRFD	ASD
LRFD Load Combination 5 from ASCE/SEI 7 Section 12.4.2.3 $(1.2+0.2S_{DS})D+\rho Q_E+0.5L+0.2S$ (including the 0.5 load factor on L permitted in ASCE/SEI 7 Section 12.4.2.3)	ASD Load Combination 5 from ASCE/SEI 7 Section 12.4.2.3 $(1.0+0.14S_{DS})D+H+F+0.7\rho Q_E$

From ASCE/SEI 7, this structure is assigned to Seismic Design Category C ($\rho = 1.0$) and $S_{DS} = 0.352$.

The required strengths of Column CL-1 determined by a second-order analysis including the effects of P-δ and P-Δ with reduced stiffness as required by the direct analysis method are:

LRFD	ASD
P_u = 233 kips V_u = 35.0 kips $M_{u\ top}$ = 201 kip-ft $M_{u\ bot}$ = −320 kip-ft	P_a = 165 kips V_a = 23.4 kips $M_{a\ top}$ = 131 kip-ft $M_{a\ bot}$ = −210 kip-ft

There are no transverse loadings between the floors in the plane of bending, and the beams framing into the column weak axis are pin-connected and produce negligible moments.

Solution:

From AISC *Manual* Table 2-4, the material properties are as follows:

ASTM A992
F_y = 50 ksi
F_u = 65 ksi

From AISC *Manual* Table 1-1, the geometric properties are as follows:

W12×87
r_x = 5.38 in. r_y = 3.07 in.

Available Compressive Strength of Column CL-1

Because the member is being designed using the direct analysis method, K is taken as 1.0.

$$\frac{KL_x}{r_x} = \frac{1.0(14.0 \text{ ft})(12.0 \text{ in./ft})}{5.38 \text{ in.}}$$
$$= 31.2$$
$$\frac{KL_y}{r_y} = \frac{1.0(14.0 \text{ ft})(12.0 \text{ in./ft})}{3.07 \text{ in.}}$$
$$= 54.7 \quad \textbf{governs}$$

From AISC *Manual* Table 4-1, the available compressive strength is:

LRFD	ASD
$\phi_c P_n = 925$ kips	$\dfrac{P_n}{\Omega_c} = 616$ kips

Available Flexural Strength of Column CL-1

Check the unbraced length for flexure
From AISC *Manual* Table 3-2:

$L_p = 10.8$ ft
$L_r = 43.1$ ft
$L_p < L_b = 14.0$ ft $< L_r$

Therefore, the member is subject to lateral-torsional buckling.

Calculate C_b using AISC *Specification* Equation F1-1.

LRFD	ASD
$M_{u\ top} = 201$ kip-ft $\\$ $M_{u\ bot} = -320$ kip-ft $\\$ $M(x) = M_{top} - \left(\dfrac{M_{top} - M_{bot}}{L}\right)x$ $\\$ $= 201$ kip-ft $- \left(\dfrac{201\text{ kip-ft} + 320\text{ kip-ft}}{14.0\text{ ft}}\right)x$ $\\$ $= 201$ kip-ft $- (37.2$ kips$)x$	$M_{a\ top} = 131$ kip-ft $\\$ $M_{a\ bot} = -210$ kip-ft $\\$ $M(x) = M_{top} - \left(\dfrac{M_{top} - M_{bot}}{L}\right)x$ $\\$ $= 131$ kip-ft $- \left(\dfrac{131\text{ kip-ft} + 210\text{ kip-ft}}{14.0\text{ ft}}\right)x$ $\\$ $= 131$ kip-ft $- (24.4$ kips$)x$
Quarter point moments are: $\\$ $\lvert M(x = 3.50\text{ ft})\rvert = M_A$ $\\$ $\quad = \lvert 201$ kip-ft $-(37.2$ kips$)(3.50$ ft$)\rvert$ $\\$ $\quad = 70.8$ kip-ft $\\$ $\lvert M(x = 7.00\text{ ft})\rvert = M_B$ $\\$ $\quad = \lvert 201$ kip-ft $-(37.2$ kips$)(7.00$ ft$)\rvert$ $\\$ $\quad = 59.4$ kip-ft $\\$ $\lvert M(x = 10.5\text{ ft})\rvert = M_C$ $\\$ $\quad = \lvert 201$ kip-ft $-(37.2$ kips$)(10.5$ ft$)\rvert$ $\\$ $\quad = 190$ kip-ft $\\$ $M_{max} = 320$ kip-ft	Quarter point moments are: $\\$ $\lvert M(x = 3.50\text{ ft})\rvert = M_A$ $\\$ $\quad = \lvert 131$ kip-ft $-(24.4$ kips$)(3.50$ ft$)\rvert$ $\\$ $\quad = 45.6$ kip-ft $\\$ $\lvert M(x = 7.00\text{ ft})\rvert = M_B$ $\\$ $\quad = \lvert 131$ kip-ft $-(24.4$ kips$)(7.00$ ft$)\rvert$ $\\$ $\quad = 39.8$ kip-ft $\\$ $\lvert M(x = 10.5\text{ ft})\rvert = M_C$ $\\$ $\quad = \lvert 131$ kip-ft $-(24.4$ kips$)(10.5$ ft$)\rvert$ $\\$ $\quad = 125$ kip-ft $\\$ $M_{max} = 210$ kip-ft

3.4 MOMENT FRAMES

LRFD	ASD
$C_b = \dfrac{12.5 M_{max}}{2.5 M_{max} + 3 M_A + 4 M_B + 3 M_C}$	$C_b = \dfrac{12.5 M_{max}}{2.5 M_{max} + 3 M_A + 4 M_B + 3 M_C}$
$= [12.5(320 \text{ kip-ft})]/[2.5(320 \text{ kip-ft})$	$= [12.5(210 \text{ kip-ft})]/[2.5(210 \text{ kip-ft})$
$+ 3(70.8 \text{ kip-ft})$	$+ 3(45.6 \text{ kip-ft})$
$+ 4(59.4 \text{ kip-ft}) + 3(190 \text{ kip-ft})]$	$+ 4(39.8 \text{ kip-ft}) + 3(125 \text{ kip-ft})]$
$= 2.20$	$= 2.19$

From AISC *Manual* Table 3-10, with $L_b = 14.0$ ft, the available flexural strength of a W12×87 is:

LRFD	ASD
$\phi_b M_n = 2.20(477 \text{ kip-ft})$ $= 1{,}050 \text{ kip-ft}$	$\dfrac{M_n}{\Omega_b} = 2.19(318 \text{ kip-ft})$ $= 696 \text{ kip-ft}$
Check yielding (plastic moment) limit state, using AISC *Manual* Table 3-2,	Check yielding (plastic moment) limit state, using AISC *Manual* Table 3-2,
$\phi_b M_p = 495 \text{ kip-ft} < 1{,}050 \text{ kip-ft}$	$\dfrac{M_p}{\Omega_b} = 329 \text{ kip-ft} < 696 \text{ kip-ft}$

Therefore, the yielding limit state governs.

Interaction of Flexure and Compression in Column CL-1

Using AISC *Specification* Section H1, check the interaction of compression and flexure in Column CL-1, as follows:

LRFD	ASD
$P_c = \phi_c P_n$, as determined previously $= 925 \text{ kips}$ $\dfrac{P_r}{P_c} = \dfrac{233 \text{ kips}}{925 \text{ kips}}$ $= 0.252$ Because $P_r/P_c \geq 0.2$, use AISC *Specification* Equation H1-1a,	$P_c = \dfrac{P_n}{\Omega_c}$, as determined previously $= 616 \text{ kips}$ $\dfrac{P_r}{P_c} = \dfrac{165 \text{ kips}}{616 \text{ kips}}$ $= 0.268$ Because $P_r/P_c \geq 0.2$, use AISC *Specification* Equation H1-1a,

LRFD	ASD
$\dfrac{P_r}{P_c} + \dfrac{8}{9}\left(\dfrac{M_{rx}}{M_{cx}} + \dfrac{M_{ry}}{M_{cy}}\right) \leq 1.0$	$\dfrac{P_r}{P_c} + \dfrac{8}{9}\left(\dfrac{M_{rx}}{M_{cx}} + \dfrac{M_{ry}}{M_{cy}}\right) \leq 1.0$
$0.252 + \dfrac{8}{9}\left(\dfrac{320 \text{ kip-ft}}{495 \text{ kip-ft}} + 0\right) = 0.827$	$0.268 + \dfrac{8}{9}\left(\dfrac{210 \text{ kip-ft}}{329 \text{ kip-ft}} + 0\right) = 0.835$
$0.827 < 1.0$ **o.k.**	$0.835 < 1.0$ **o.k.**

Available Shear Strength of Column CL-1

From AISC *Manual* Table 3-2, the available shear strength of a W12×87 is:

LRFD	ASD
$\phi_v V_n = 193$ kips > 35.0 kips **o.k.**	$V_n / \Omega_v = 129$ kips > 23.4 kips **o.k.**

The W12×87 is adequate to resist the required strengths given for Column CL-1.

Note: Load combinations that do not include seismic effects must also be investigated.

Example 3.4.3. Moment Frame Beam Design

Given:

Refer to Beam BM-1 in Figure 3-2. Verify that a W18×55 ASTM A992 W-shape is sufficient to resist the following required strengths. The applicable building code specifies the use of ASCE/SEI 7 for calculation of loads. See the Design Example Plan and Elevation section for code specified loading.

The load combinations that include seismic effects are:

LRFD	ASD
LRFD Load Combination 5 from ASCE/SEI 7 Section 12.4.2.3 $(1.2 + 0.2 S_{DS})D + \rho Q_E + 0.5L + 0.2S$ (including the 0.5 load factor on L permitted in ASCE/SEI 7 Section 12.4.2.3)	ASD Load Combination 5 from ASCE/SEI 7 Section 12.4.2.3 $(1.0 + 0.14 S_{DS})D + H + F + 0.7 \rho Q_E$

From ASCE/SEI 7, this structure is assigned to Seismic Design Category C ($\rho = 1.0$) and $S_{DS} = 0.352$.

The required strengths determined by a second-order analysis including the effects of $P\text{-}\delta$ and $P\text{-}\Delta$ with reduced stiffness as required by the direct analysis method are:

3.4 MOMENT FRAMES 3–11

LRFD	ASD
$P_u = 0$ kips	$P_a = 0$ kips
$V_u = 33.9$ kips	$V_a = 23.1$ kips
$M_{u\ left} = -316$ kip-ft	$M_{a\ left} = -212$ kip-ft
$M_{CL} = 58.6$ kip-ft	$M_{CL} = 40.6$ kip-ft
$M_{u\ right} = 167$ kip-ft	$M_{a\ right} = 106$ kip-ft

Solution:

From AISC *Manual* Table 2-4, the material properties are as follows:

ASTM A992
$F_y = 50$ ksi
$F_u = 65$ ksi

From AISC *Manual* Table 1-1, the geometric properties are as follows:

W18×55
$d = 18.1$ in. $t_w = 0.390$ in. $r_y = 1.67$ in. $S_x = 98.3$ in.^{3} $Z_x = 112$ in.^{3}
$r_{ts} = 2.00$ in. $J = 1.66$ in.^{4} $h_o = 17.5$ in.

Assume that the beam flanges are braced at the columns.

Available Flexural Strength of Beam BM-1

From AISC *Manual* Table 3-2:

$L_p = 5.90$ ft
$L_r = 17.6$ ft
$L_r < L_b = 30.0$ ft

The limit states of yielding and lateral-torsional buckling are applicable, as given in AISC *Specification* Section F2.

Calculate C_b using AISC *Specification* Commentary Equation C-F1-5, which applies to gravity loaded beams with the top flange laterally restrained; the top flange is restrained by the composite slab.

LRFD	ASD
$M_o = M_{u\ left} = -316$ kip-ft	$M_o = M_{a\ left} = -212$ kip-ft
$M_1 = M_{u\ right} = 167$ kip-ft	$M_1 = M_{a\ right} = 106$ kip-ft
$M_{CL} = 58.6$ kip-ft	$M_{CL} = 40.6$ kip-ft

AMERICAN INSTITUTE OF STEEL CONSTRUCTION

LRFD	ASD
$(M_o + M_1)^* = M_o$ $= -316$ kip-ft because M_1 is positive $C_b = 3.0 - \frac{2}{3}\left(\frac{M_1}{M_o}\right) - \frac{8}{3}\left[\frac{M_{CL}}{(M_o + M_1)^*}\right]$ $= 3.0 - \frac{2}{3}\left(\frac{167 \text{ kip-ft}}{-316 \text{ kip-ft}}\right) - \frac{8}{3}\left(\frac{58.6 \text{ kip-ft}}{-316 \text{ kip-ft}}\right)$ $= 3.85$	$(M_o + M_1)^* = M_o$ $= -212$ kip-ft because M_1 is positive $C_b = 3.0 - \frac{2}{3}\left(\frac{M_1}{M_o}\right) - \frac{8}{3}\left[\frac{M_{CL}}{(M_o + M_1)^*}\right]$ $= 3.0 - \frac{2}{3}\left(\frac{106 \text{ kip-ft}}{-212 \text{ kip-ft}}\right) - \frac{8}{3}\left(\frac{40.6 \text{ kip-ft}}{-212 \text{ kip-ft}}\right)$ $= 3.84$

Per the User Note in AISC *Specification* Section F2, the **W18×55** is compact for $F_y = 50$ ksi.

Because AISC *Manual* Table 3-10 does not provide a strength for a **W18×55** with an unbraced length of 30 ft, calculate the strength from the AISC *Specification*. From AISC *Specification* Section F2, with compact flanges and web and $L_b > L_r$, the applicable limit states are yielding and lateral-torsional buckling.

$$M_n = F_{cr}S_x \leq M_p \qquad \qquad (Spec. \text{ Eq. F2-3})$$

$$F_{cr} = \frac{C_b \pi^2 E}{\left(\frac{L_b}{r_{ts}}\right)^2} \sqrt{1 + 0.078 \frac{Jc}{S_x h_o}\left(\frac{L_b}{r_{ts}}\right)^2} \qquad \qquad (Spec. \text{ Eq. F2-4})$$

$$\frac{Jc}{S_x h_o} = 0.000965$$

LRFD	ASD
$F_{cr} = \frac{3.85\pi^2(29,000 \text{ ksi})}{\left[\frac{30.0 \text{ ft}(12 \text{ in./ft})}{2.00 \text{ in.}}\right]^2}$ $\times \sqrt{1 + 0.078(0.000965)\left[\frac{30.0 \text{ ft}(12 \text{ in./ft})}{2.00 \text{ in.}}\right]^2}$ $= 63.1$ ksi	$F_{cr} = \frac{3.84\pi^2(29,000 \text{ ksi})}{\left[\frac{30.0 \text{ ft}(12 \text{ in./ft})}{2.00 \text{ in.}}\right]^2}$ $\times \sqrt{1 + 0.078(0.000965)\left[\frac{30.0 \text{ ft}(12 \text{ in./ft})}{2.00 \text{ in.}}\right]^2}$ $= 62.9$ ksi

3.4 MOMENT FRAMES

LRFD	ASD
$M_n = 63.1 \text{ ksi}(98.3 \text{ in.}^3)$ $\quad = 6{,}200 \text{ kip-in.}$ $\quad = 517 \text{ kip-ft} \leq M_p$ $M_p = F_y Z_x$ $\quad = 50 \text{ ksi}(112 \text{ in.}^3)(1 \text{ ft}/12 \text{ in.})$ $\quad = 467 \text{ kip-ft} \quad \textbf{controls}$ $\phi_b M_n = 0.90(467 \text{ kip-ft})$ $\quad = 420 \text{ kip-ft} > 316 \text{ kip-ft} \quad \textbf{o.k.}$	$M_n = 62.9 \text{ ksi}(98.3 \text{ in.}^3)$ $\quad = 6{,}180 \text{ kip-in.}$ $\quad = 515 \text{ kip-ft} \leq M_p$ $M_p = F_y Z_x$ $\quad = 50 \text{ ksi}(112 \text{ in.}^3)(1 \text{ ft}/12 \text{ in.})$ $\quad = 467 \text{ kip-ft} \quad \textbf{controls}$ $\dfrac{M_n}{\Omega_b} = \dfrac{467 \text{ kip-ft}}{1.67}$ $\quad = 280 \text{ kip-ft} > 212 \text{ kip-ft} \quad \textbf{o.k.}$

Available Shear Strength of Beam BM-1

From AISC *Manual* Table 3-2, the available shear strength of the W18×55 is:

LRFD	ASD
$\phi_v V_n = 212 \text{ kips} > 33.9 \text{ kips} \quad \textbf{o.k.}$	$V_n / \Omega_v = 141 \text{ kips} > 23.1 \text{ kips} \quad \textbf{o.k.}$

The W18×55 is adequate to resist the loads given for Beam BM-1.

Note: Load combinations that do not include seismic effects must also be investigated.

Example 3.4.4. Moment Frame Beam-to-Column Connection Design

Given:

Refer to Joint JT-1 in Figure 3-2. Design a bolted flange-plated fully restrained moment connection between Beam BM-1 and Column CL-1. The beam and column are ASTM A992 W-shapes and ASTM A36 is used for the connecting material. Use ASTM A325-N bolts and 70-ksi electrodes.

From Example 3.4.3, the required strengths are:

LRFD	ASD
$V_u = 33.9 \text{ kips}$ $M_u = 316 \text{ kip-ft}$	$V_a = 23.1 \text{ kips}$ $M_a = 212 \text{ kip-ft}$

Solution:

From AISC *Manual* Table 2-4, the material properties are as follows:

ASTM A36
$F_y = 36$ ksi
$F_u = 58$ ksi

ASTM A992
$F_y = 50$ ksi
$F_u = 65$ ksi

From AISC *Manual* Table 1-1, the geometric properties are as follows:

W18×55
$d = 18.1$ in. $t_w = 0.390$ in. $t_f = 0.630$ in. $S_x = 98.3$ in.^{3} $b_f = 7.53$ in.

Available Flexural Strength of Beam BM-1

AISC *Specification* Section F13 requires that tensile rupture of the tension flange be investigated if

$$F_u A_{fn} < Y_t F_y A_{fg}$$

Since $F_y/F_u = 50$ ksi/65 ksi $= 0.77 < 0.8$:

$$Y_t = 1.0$$

For two rows of ⅞-in.-diameter ASTM A325-N bolts in standard holes in the beam tension flange, using AISC *Specification* Section B4.3b:

$$\begin{aligned}
A_{fg} &= b_f t_f \\
&= 7.53 \text{ in.}(0.630 \text{ in.}) \\
&= 4.74 \text{ in.}^2 \\
A_{fn} &= A_{fg} - 2(d_h + \tfrac{1}{16} \text{ in.})t_f \\
&= 4.74 \text{ in.}^2 - 2(1.00 \text{ in.})(0.630 \text{ in.}) \\
&= 3.48 \text{ in.}^2 \\
Y_t F_y A_{fg} &= 1.0(50 \text{ ksi})(4.74 \text{ in.}) \\
&= 237 \text{ kips} \\
F_u A_{fn} &= (65 \text{ ksi})(3.48 \text{ in.}^2) \\
&= 226 \text{ kips}
\end{aligned}$$

Since $F_u A_{fn} < Y_t F_y A_{fg}$, the limit state of tensile rupture of the flange applies.

$$M_n = \frac{F_u A_{fn}}{A_{fg}} S_x \qquad \text{(Spec. Eq. F13-1)}$$

$$\begin{aligned}
&= \frac{226 \text{ kips}}{4.74 \text{ in.}^2}(98.3 \text{ in.}^3) \\
&= 4,690 \text{ kip-in.} \\
&= 391 \text{ kip-ft}
\end{aligned}$$

3.4 MOMENT FRAMES

The available flexural strength of the W18×55 is:

LRFD	ASD
$\phi M_n = 0.90(391 \text{ kip-ft})$ $= 352 \text{ kip-ft} > 316 \text{ kip-ft}$ **o.k.**	$\dfrac{M_n}{\Omega} = \dfrac{391 \text{ kip-ft}}{1.67}$ $= 234 \text{ kip-ft} > 212 \text{ kip-ft}$ **o.k.**

Single-Plate Web Connection

The single plate connection in an FR moment connection need not be designed for eccentricity on the bolts; however, AISC *Manual* Table 10-10a is applied here for simplicity. Conservatively, using AISC *Manual* Table 10-10a, select a ⁵⁄₁₆-in.-thick ASTM A36 plate with (3) ⅞-in.-diameter ASTM A325-N bolts (Group A) in standard holes connected to the beam web, and a ¼-in. fillet weld to the column flange. The available strength of the single-plate connection is:

LRFD	ASD
$\phi R_n = 48.9 \text{ kips} > 33.9 \text{ kips}$ **o.k.**	$\dfrac{R_n}{\Omega} = 32.6 \text{ kips} > 23.1 \text{ kips}$ **o.k.**

Because the bolt bearing limit state is included in Table 10-10a, the beam web is acceptable by inspection, as the beam web thickness of 0.390 in. is greater than the plate thickness of ⁵⁄₁₆ in.

Use a ⁵⁄₁₆-in.-thick, single-plate connection with (3) ⅞-in.-diameter ASTM A325-N bolts in standard holes to the beam web and ¼-in. fillet weld to the column flange.

Flange Plate Connection

Determine the required number of bolts in the flange plate

The flange force is:

LRFD	ASD
$P_{uf} = \dfrac{M_u}{d}$ $= \dfrac{316 \text{ kip-ft}(12.0 \text{ in./ft})}{18.1 \text{ in.}}$ $= 210 \text{ kips}$	$P_{af} = \dfrac{M_a}{d}$ $= \dfrac{212 \text{ kip-ft}(12.0 \text{ in./ft})}{18.1 \text{ in.}}$ $= 141 \text{ kips}$

From AISC *Manual* Table 7-1 for bolt shear, the required number of ⅞-in.-diameter ASTM A325-N bolts is:

LRFD	ASD
$n_{min} = \dfrac{P_{uf}}{\phi r_n}$ $= \dfrac{210 \text{ kips}}{24.3 \text{ kips/bolt}}$ $= 8.64 \text{ bolts}$	$n_{min} = \dfrac{P_{af}}{r_n/\Omega}$ $= \dfrac{141 \text{ kips}}{16.2 \text{ kips/bolt}}$ $= 8.70 \text{ bolts}$

Try ten bolts on a 4-in. gage. Using AISC *Manual* Tables 7-4 and 7-5 for bearing strength with $L_e = 2$ in. and $s = 3$ in., the available bearing strength of the beam flange is:

LRFD	ASD
$\phi R_n = n(\phi r_n)t_f$ $= 8(102 \text{ kip/in.})(0.630 \text{ in.})$ $+ 2(89.6 \text{ kip/in.})(0.630 \text{ in.})$ $= 627 \text{ kips} > 210 \text{ kips}$ **o.k.**	$\dfrac{R_n}{\Omega} = n\left(\dfrac{r_n}{\Omega}\right)t_f$ $= 8(68.3 \text{ kip/in.})(0.630 \text{ in.})$ $+ 2(59.7 \text{ kip/in.})(0.630 \text{ in.})$ $= 419 \text{ kips} > 141 \text{ kips}$ **o.k.**

Size the flange plate for the tension force

The minimum thickness of a 7-in.-wide plate for tension yielding is:

LRFD	ASD
$t_{min} = \dfrac{P_{uf}}{\phi F_y b_p}$ $= \dfrac{210 \text{ kips}}{0.90(36 \text{ ksi})(7.00 \text{ in.})}$ $= 0.926 \text{ in.}$	$t_{min} = \dfrac{P_{af}}{F_y b_p/\Omega}$ $= \dfrac{141 \text{ kips}}{(36 \text{ ksi})(7.00 \text{ in.})/1.67}$ $= 0.934 \text{ in.}$

Try a 1 in. × 7 in. plate. The available tensile rupture strength of the plate is determined according to AISC *Specification* Section D2 as follows:

3.4 MOMENT FRAMES

LRFD	ASD
$\phi R_n = \phi_t F_u A_e$ $= \phi_t F_u A_n U$ $= 0.75(58 \text{ ksi})(1.00 \text{ in.})$ $\times [7.00 \text{ in.} - 2(^{15}\!/_{16} \text{ in.} + ^1\!/_{16} \text{ in.})](1.0)$ $= 218 \text{ kips} > 210 \text{ kips}$ **o.k.**	$\dfrac{R_n}{\Omega} = \dfrac{F_u A_e}{\Omega_t}$ $= \dfrac{F_u A_n U}{\Omega_t}$ $= (1/2.00)(58 \text{ ksi})(1.00 \text{ in.})$ $\times [7.00 \text{ in.} - 2(^{15}\!/_{16} \text{ in.} + ^1\!/_{16} \text{ in.})]$ $\times (1.0)$ $= 145 \text{ kips} > 141 \text{ kips}$ **o.k.**

Using AISC *Manual* Tables 7-4 and 7-5 with $L_e = 2$ in. and $s = 3$ in., the bearing strength of the flange plate is:

LRFD	ASD
$\phi R_n = n(\phi r_n) t_p$ $= 8(91.4 \text{ kip/in.})(1.00 \text{ in.})$ $+ 2(79.9 \text{ kip/in.})(1.00 \text{ in.})$ $= 891 \text{ kips} > 210 \text{ kips}$ **o.k.**	$\dfrac{R_n}{\Omega} = n\left(\dfrac{r_n}{\Omega}\right) t_p$ $= 8(60.9 \text{ kip/in.})(1.00 \text{ in.})$ $+ 2(53.3 \text{ kip/in.})(1.00 \text{ in.})$ $= 594 \text{ kips} > 141 \text{ kips}$ **o.k.**

Check the flange plate and beam flange for block shear rupture

The two cases for which block shear must be considered in the flange plate are shown in Figure 3-4.

Case 1 involves the tearout of the two blocks outside of the two rows of bolt holes in the flange plate. For this case, the tension area has a width of 2(1½ in.). Case 2 involves the tearout of the block between the two rows of holes in the flange plate. For this case, the tension area has a width of 4 in. Because the shear areas are the same in both cases, Case 1 governs for the flange plate. The beam flange must also be checked for Case 1, but need not be checked for Case 2 due to the presence of the web.

The nominal strength for the limit state of block shear rupture is given by AISC *Specification* Equation J4-5:

$$R_n = 0.60 F_u A_{nv} + U_{bs} F_u A_{nt} \le 0.60 F_y A_{gv} + U_{bs} F_u A_{nt}$$

Check the flange plate for Case 1

From AISC *Specification* Equation J4-5:

LRFD	ASD
$\phi R_n = \phi U_{bs} F_u A_{nt}$ $+ \min(\phi 0.60 F_y A_{gv},\ \phi 0.60 F_u A_{nv})$	$\dfrac{R_n}{\Omega} = \dfrac{U_{bs} F_u A_{nt}}{\Omega}$ $+ \min\left(\dfrac{0.60 F_y A_{gv}}{\Omega},\ \dfrac{0.60 F_u A_{nv}}{\Omega}\right)$
$U_{bs} = 1.0$ for uniform tension stress	$U_{bs} = 1.0$ for uniform tension stress
Tension rupture component from AISC *Manual* Table 9-3a:	Tension rupture component from AISC *Manual* Table 9-3a:
$\phi U_{bs} F_u A_{nt} = 2(1.0)(43.5 \text{ kip/in.})(1 \text{ in.})$ $= 87.0$ kips	$\dfrac{U_{bs} F_u A_{nt}}{\Omega} = 2(1.0)(29.0 \text{ kip/in.})(1 \text{ in.})$ $= 58.0$ kips
Shear yielding component from AISC *Manual* Table 9-3b:	Shear yielding component from AISC *Manual* Table 9-3b:
$\phi 0.6 F_y A_{gv} = 2(227 \text{ kip/in.})(1 \text{ in.})$ $= 454$ kips	$\dfrac{0.6 F_y A_{gv}}{\Omega} = 2(151 \text{ kip/in.})(1 \text{ in.})$ $= 302$ kips

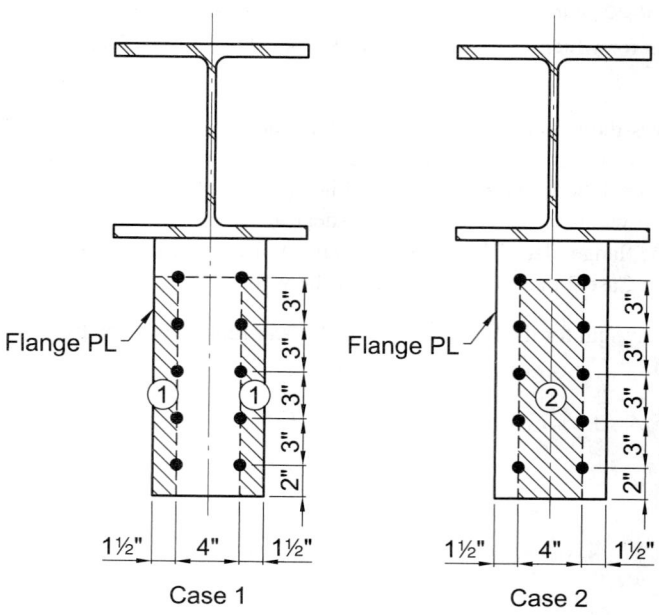

Fig. 3-4. Block shear failure paths for the flange plate in Example 3.4.4.

3.4 MOMENT FRAMES

LRFD	ASD
Shear rupture component from AISC Manual Table 9-3c: $$\phi 0.6F_u A_{nv} = 2(248 \text{ kip/in.})(1 \text{ in.})$$ $$= 496 \text{ kips}$$ $$\phi R_n = 87.0 \text{ kips} + 454 \text{ kips}$$ $$= 541 \text{ kips} > 210 \text{ kips} \quad \textbf{o.k.}$$	Shear rupture component from AISC Manual Table 9-3c: $$\frac{0.6F_u A_{nv}}{\Omega} = 2(165 \text{ kip/in.})(1 \text{ in.})$$ $$= 330 \text{ kips}$$ $$\frac{R_n}{\Omega} = 58.0 \text{ kips} + 302 \text{ kips}$$ $$= 360 \text{ kips} > 141 \text{ kips} \quad \textbf{o.k.}$$

Check the beam flange for Case 1

From AISC *Manual* Tables 9-3a, 9-3b and 9-3c for L_{eh} = 1¾ in. and L_{ev} = 2 in.:

U_{bs} = 1.0 for uniform tension stress

LRFD	ASD
$$\phi U_{bs} F_u A_{nt} = 2(1.0)(60.9 \text{ kip/in.})(0.630 \text{ in.})$$ $$= 76.7 \text{ kips}$$ $$\phi 0.6F_y A_{gv} = 2(315 \text{ kip/in.})(0.630 \text{ in.})$$ $$= 397 \text{ kips}$$ $$\phi 0.6F_u A_{nv} = 2(278 \text{ kip/in.})(0.630 \text{ in.})$$ $$= 350 \text{ kips}$$ $$\phi R_n = 76.7 \text{ kips} + 350 \text{ kips}$$ $$= 427 \text{ kips} > 210 \text{ kips} \quad \textbf{o.k.}$$	$$\frac{U_{bs} F_u A_{nt}}{\Omega} = 2(1.0)(40.6 \text{ kip/in.})(0.630 \text{ in.})$$ $$= 51.2 \text{ kips}$$ $$\frac{0.6F_y A_{gv}}{\Omega} = 2(210 \text{ kip/in.})(0.630 \text{ in.})$$ $$= 265 \text{ kips}$$ $$\frac{0.6F_u A_{nv}}{\Omega} = 2(185 \text{ kip/in.})(0.630 \text{ in.})$$ $$= 233 \text{ kips}$$ $$\frac{R_n}{\Omega} = 51.2 \text{ kips} + 233 \text{ kips}$$ $$= 284 \text{ kips} > 141 \text{ kips} \quad \textbf{o.k.}$$

Use (5) rows of ⅞-in.-diameter ASTM A325-N bolts in standard holes at a 4-in. gage to connect each flange plate to the beam flange. Use 2-in. edge distance and 3-in. spacing for the bolts.

Check the flange plate for the compression force

$$r = \frac{t}{\sqrt{12}}$$
$$= \frac{1.00 \text{ in.}}{\sqrt{12}}$$
$$= 0.289 \text{ in.}$$

From AISC *Specification* Commentary Table C-A-7.1, use $K = 0.65$, and $l = 2\frac{1}{2}$ in. (2-in. edge distance plus ½-in. beam setback):

$$\frac{Kl}{r} = \frac{0.65(2.50 \text{ in.})}{0.289 \text{ in.}}$$
$$= 5.62$$

According to AISC *Specification* Section J4.4, because $Kl/r \leq 25$, $F_{cr} = F_y$ and the compressive strength of the flange plate is:

$P_n = F_y A_g$ (*Spec.* Eq. J4-6)
   = (36 ksi)(7.00 in.)(1.00 in.)
   = 252 kips

LRFD	ASD
$\phi P_n = 0.90(252 \text{ kips})$ $= 227 \text{ kips} > 210 \text{ kips}$ **o.k.**	$\dfrac{P_n}{\Omega} = \dfrac{252 \text{ kips}}{1.67}$ $= 151 \text{ kips} > 141 \text{ kips}$ **o.k.**

Use 1 in. × 7 in. ASTM A36 flange plates.

Design the weld between the flange plates and column flange

The directional strength increase is used in determining the required weld size. The length of the weld, l_w, is taken to be the width of the 7-in. plate less twice the weld size.

Determine the weld size

Solving for D_{min} from AISC *Manual* Equation 8-2 and applying the directional strength increase of AISC *Specification* Equation J2-5:

LRFD	ASD
$D_{min} = \dfrac{P_{uf}}{2(1.5)(1.392 \text{ kip/in.})l_w}$ $= \dfrac{210 \text{ kips}}{2(1.5)(1.392 \text{ kip/in.})(5.88 \text{ in.})}$ $= 8.55$ sixteenths	$D_{min} = \dfrac{P_{af}}{2(1.5)(0.928 \text{ kip/in.})l_w}$ $= \dfrac{141 \text{ kips}}{2(1.5)(0.928 \text{ kip/in.})(5.88 \text{ in.})}$ $= 8.61$ sixteenths

Use ⁹⁄₁₆-in. fillet welds on both sides to connect the flange plates to the column flange.

Comment:

The column must be checked for panel zone and stiffening requirements. For further information, see AISC Design Guide No. 13, *Stiffening of Wide-Flange Columns at Moment Connections: Wind and Seismic Applications* (Carter, 1999).

3.5 BRACED FRAMES

The final connection design and geometry is shown in Figure 3-5.

3.5 BRACED FRAMES

Braced frames gain their strength and their resistance to lateral forces and displacements primarily from the axial strength and stiffness of the bracing members. Braced frames are arranged such that the centerlines of the framing members (braces, columns and beams) coincide or nearly coincide, thus eliminating the majority of flexure that might occur due to lateral forces.

Braced-frame systems tend to be more economical than moment-resisting frames when material, fabrication and erection costs are considered. These efficiencies are often offset by reduced flexibility in floor plan layout, space planning, and electrical and mechanical routing encountered as a result of the space requirements of the brace members.

Braced frames typically are located in walls that stack vertically between floor levels. In the typical office building, these walls generally occur in the "core" area around stair and elevator shafts, central restrooms, and mechanical and electrical rooms. This generally allows for greater architectural flexibility in placement and configuration of exterior windows and

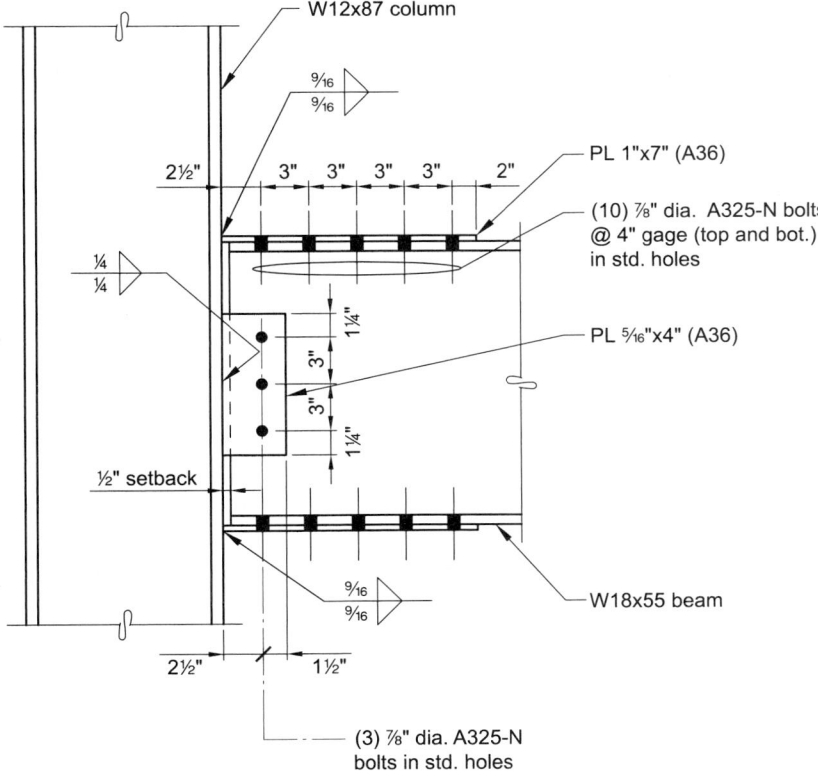

Fig. 3-5. Connection as designed in Example 3.4.4.

cladding. Depending on the plan location and the size of the core area of the building, the torsional resistance offered by the braced frames may become a controlling design parameter. Differential drift between stories at the building perimeter must be considered with this type of layout, as rotational displacements of the floor diaphragms may impose deformation demands on the cladding system and other nonstructural elements of the building.

Because the braced frame in the following examples does not require seismic detailing, it is designed in accordance with the provisions of the AISC *Specification*.

Example 3.5.1. Braced Frame Brace Design

Given:

Select an ASTM A36 double-angle section to act as Brace BR-1 in Figure 3-3 and resist the following axial forces. The applicable building code specifies the use of ASCE/SEI 7 for calculation of required strength. See the Design Example Plan and Elevation section for code specified loading.

The governing load combinations include seismic effects as follows:

LRFD	ASD
Maximum brace compression from LRFD Load Combination 5 from ASCE/SEI 7 Section 12.4.2.3 $$(1.2 + 0.2S_{DS})D + \rho Q_E + 0.5L + 0.2S$$ (including the 0.5 factor on L permitted by ASCE/SEI 7 Section 12.4.2.3)	Maximum brace compression from ASD Load Combination 5 from ASCE/SEI 7 Section 12.4.2.3 $$(1.0 + 0.14S_{DS})D + H + F + 0.7\rho Q_E$$
Maximum brace tension from LRFD Load Combination 6 from ASCE/SEI 7 Section 12.4.2.3 $$(0.9 - 0.2S_{DS})D + \rho Q_E + 1.6H$$	Maximum brace tension from ASD Load Combination 8 from ASCE/SEI 7 Section 12.4.2.3 $$(0.6 - 0.14S_{DS})D + 0.7\rho Q_E + H$$

From ASCE/SEI 7, this structure is assigned to Seismic Design Category C ($\rho = 1.0$) and $S_{DS} = 0.352$.

The required strengths of Brace BR-1 determined by a second-order analysis including the effects of P-δ and P-Δ with reduced stiffness as required by the direct analysis method are:

LRFD	ASD
Maximum Compression $P_u = 127$ kips Maximum Tension $P_u = 89.6$ kips	Maximum Compression $P_a = 83.4$ kips Maximum Tension $P_a = 60.2$ kips

3.5 BRACED FRAMES

Assume that the ends of the brace are pinned and braced against translation.

Solution:

From AISC *Manual* Table 2-4, the material properties are as follows:

ASTM A36
$F_y = 36$ ksi
$F_u = 58$ ksi

The effective length of the brace is:

$$KL = (1.0)\sqrt{(12.5 \text{ ft})^2 + (12.5 \text{ ft})^2}$$
$$= 17.7 \text{ ft}$$

This unbraced length has been conservatively determined by calculating the distance between the work points based on the intersection of the centerlines of the brace, column and beams, and using $K = 1.0$. Shorter unbraced lengths may be used if justified by the engineer of record.

Brace Selection

Select a trial brace size based on the effective length and the compressive strength of the brace. Based on the discussion in AISC *Specification* Commentary Section J1.7, it is assumed that the effect of the load eccentricity with respect to the center of gravity of the brace is negligible and can be ignored. Use AISC *Manual* Table 4-9 to select trial brace sections. Possible double angle braces include 2L5×5×⅝, 2L6×6×⅜, or 2L6×4×⅝ LLBB. Use a 2L6×4×⅝ LLBB for the trial design due to architectural needs. From AISC *Manual* Table 4-9, the available strength of the 2L6×4×⅝ LLBB brace (⅜-in. separation) in compression with $KL = 17.7$ ft is controlled by the *y-y* axis. By interpolation:

LRFD	ASD
$\phi_c P_n = 143$ kips > 127 kips **o.k.**	$\dfrac{P_n}{\Omega_c} = 94.6$ kips > 83.4 kips **o.k.**

The 2L6×4×⅝ LLBB is adequate.

Element Slenderness

Table 4-9 considers the AISC *Specification* Section E6.2 requirement that the effective slenderness ratio, Ka/r_i, of each of the component shapes between fasteners may not exceed three-fourths times the governing slenderness ratio of the built-up member. Per AISC *Manual* Table 4-9, at least two welded or pretensioned bolted intermediate connectors must be provided.

Available Tensile Strength of Brace

From AISC *Manual* Table 5-8, the available strength of the 2L6×4×⅝ brace for tensile yielding on the gross section is:

LRFD	ASD
$\phi_t P_n = 379$ kips > 89.6 kips **o.k.**	$\dfrac{P_n}{\Omega_t} = 252$ kips > 60.2 kips **o.k.**

The 2L6×4×⅝ is adequate for tensile yielding on the gross area.

See Example 3.5.3 for calculations confirming that the tensile rupture strength on the effective net section of the brace is adequate with a single row of (5) ¾-in. bolts spaced at 3 in. connecting the double-angle brace to a gusset plate.

Use 2L6×4×⅝ LLBB with a ⅜-in. separation, assuming a ⅜-in. gusset plate, and two intermediate connectors for Brace BR-1.

Note that the intermediate connectors can be fastened by welding or with pretensioned bolts. If bolted intermediate connectors are used, a net section tensile rupture check at the connectors is also required.

Example 3.5.2. Braced Frame Column Design

Given:

Refer to Column CL-2 in Figure 3-3. Select an ASTM A992 W-shape with a nominal depth of 12 in. to resist the following required strengths. The applicable building code specifies the use of ASCE/SEI 7 for the calculation of the required strength. See the Design Example Plan and Elevation section for code specified loading.

The load combinations that include seismic effects are:

LRFD	ASD
Maximum column compression from LRFD Load Combination 5 from ASCE/SEI 7 Section 12.4.2.3	Maximum column compression from ASD Load Combination 6 from ASCE/SEI 7 Section 12.4.2.3
$(1.2 + 0.2S_{DS})D + \rho Q_E + 0.5L + 0.2S$	$(1.0 + 0.10S_{DS})D + H + F + 0.525\rho Q_E + 0.75L + 0.75S$
(including the 0.5 load factor on *L* permitted by ASCE/SEI 7 Section 12.4.2.3)	
Maximum column tension from LRFD Load Combination 6 from ASCE/SEI 7 Section 12.4.2.3	Maximum column tension from ASD Load Combination 8 from ASCE/SEI 7 Section 12.4.2.3
$(0.9 - 0.2S_{DS})D + \rho Q_E + 1.6H$	$(0.6 - 0.14S_{DS})D + 0.7\rho Q_E + H$

3.5 BRACED FRAMES

From ASCE/SEI 7, this structure is assigned to Seismic Design Category C ($\rho = 1.0$) and $S_{DS} = 0.352$.

The required strengths of Column CL-2 determined by a second-order analysis including the effects of P-δ and P-Δ with reduced stiffness as required by the direct analysis method are:

LRFD	ASD
Maximum Compression $P_u = 351$ kips Maximum Tension $P_u = 42.1$ kips	Maximum Compression $P_a = 253$ kips Maximum Tension $P_a = 28.7$ kips

Consider that the ends of the column are pinned and braced against translation for both the x-x and y-y axes.

Solution:

From AISC *Manual* Table 2-4, the material properties are as follows:

ASTM A992
$F_y = 50$ ksi
$F_u = 65$ ksi

Using AISC *Manual* Table 4-1 with $KL = 14$ ft, select a W12×50.

LRFD	ASD
$\phi_c P_n = 384$ kips > 351 kips **o.k.**	$\dfrac{P_n}{\Omega_c} = 255$ kips > 253 kips **o.k.**

The W12×50 is adequate.

There is net tension (uplift) on the column. Using AISC *Manual* Table 5-1, the available strength of the W12×50 in axial tension is adequate.

Use a W12×50 for braced-frame Column CL-2.

Example 3.5.3. Braced Frame Brace-to-Beam/Column Connection Design

Given:

Refer to Joint JT-2 in Figure 3-3. Design the connection between the brace, beam and column. Use a gusset plate concentric to the brace and welded to the beam with 70-ksi electrodes. Connect the gusset and the beam to the column using a bolted single-plate connection. Use ASTM A36 for all plate material, use the brace and column as designed in

Examples 3.5.1 and 3.5.2, respectively, and use an ASTM A992 W18×35 for the beam, as required for strength and connection geometry. The applicable building code specifies the use of ASCE/SEI 7 for calculation of the required strengths. See the Design Example Plan and Elevation section for code specified loading.

The required strengths are:

LRFD	ASD
Beam Shear	Beam Shear
$V_u = 4.00$ kips	$V_a = 2.63$ kips
Brace Compression	Brace Compression
$P_u = 127$ kips	$P_a = 83.4$ kips
Brace Tension	Brace Tension
$P_u = 89.6$ kips	$P_a = 60.2$ kips

From Examples 3.5.1 and 3.5.2, the brace is an ASTM A36 2L6×4×⅝ LLBB section with ⅜-in. separation for a ⅜-in.-thick gusset plate, and the column is an ASTM A992 W12×50.

Solution:

From AISC *Manual* Table 2-4, the material properties are as follows:

ASTM A36
$F_y = 36$ ksi
$F_u = 58$ ksi

ASTM A992
$F_y = 50$ ksi
$F_u = 65$ ksi

From AISC *Manual* Tables 1-1, 1-7 and 1-15, the geometric properties are as follows:

Beam
W18×35
$d = 17.7$ in. $t_w = 0.300$ in. $t_f = 0.425$ in. $k_{des} = 0.827$ in.

Column
W12×50
$d = 12.2$ in. $t_w = 0.370$ in. $t_f = 0.640$ in. $k_{des} = 1.14$ in.

Brace
2L6×4×⅝ LLBB
$A_g = 11.7$ in.^{2}
$\bar{x} = 1.03$ in. for single angle $\bar{y} = 2.03$ in.

3.5 BRACED FRAMES

Brace-to-Gusset Connection Design

Choose to use oversized holes in the gusset plate and standard holes in the brace. In this example, decisions related to hole sizes should be carefully considered. Oversized holes could be used in all plies if required for extra tolerance. Providing different sized holes in various plies can make squaring and plumbing the structure difficult and is therefore generally avoided. Providing consistent hole sizes in all plies allows drift pins to be used in squaring and plumbing the structure. Providing oversized hole sizes allows for more fit-up tolerance but requires connections to be designed as slip critical. Providing standard holes limits the available fit-up tolerance but generally will result in fewer bolts designed based on the bearing values.

Using AISC *Manual* Table 7-3 for ¾-in.-diameter ASTM A325 (Group A) slip-critical bolts in double shear, Class B faying surfaces, oversized holes in the gusset, and standard holes in the brace, the available shear strength and the required number of bolts is:

LRFD	ASD
$\phi R_n = 1.67(16.1 \text{ kips})$ $= 26.9 \text{ kips/bolt}$	$\dfrac{R_n}{\Omega} = 1.67(10.8 \text{ kips})$ $= 18.0 \text{ kips/bolt}$
$n_{req'd} = \dfrac{P_u}{\phi R_n}$ $= \dfrac{127 \text{ kips}}{26.9 \text{ kips/bolt}}$ $= 4.72 \text{ bolts}$	$n_{req'd} = \dfrac{P_a}{R_n/\Omega}$ $= \dfrac{83.4 \text{ kips}}{18.0 \text{ kips/bolt}}$ $= 4.63 \text{ bolts}$

Try (5) ¾-in.-diameter bolts at 3-in. spacing.

Check brace net section for tensile rupture strength

The net area of the brace is:

$A_n = A_g - 2(d_h + \frac{1}{16} \text{ in.})t$
$= 11.7 \text{ in.}^2 - 2\,(⅞ \text{ in.})(⅝ \text{ in.})$
$= 10.6 \text{ in.}^2$

From AISC *Specification* Table D3.1:

$U = 1 - \dfrac{\bar{x}}{l}$
$= 1 - \dfrac{1.03 \text{ in.}}{4(3.00 \text{ in.})}$
$= 0.914$

$A_e = A_n U$ (Spec. Eq. D3-1)
$= (10.6 \text{ in.}^2)(0.914)$
$= 9.69 \text{ in.}^2$

$P_n = F_u A_e$ (*Spec.* Eq. D2-2)

$= (58 \text{ ksi})(9.69 \text{ in.}^2)$

$= 562 \text{ kips}$

The available tensile strength of the brace due to the limit state of tensile rupture is determined from AISC *Specification* Section D2, as follows:

LRFD	ASD
$\phi_t P_n = 0.75 P_n$ $= 0.75(562 \text{ kips})$ $= 422 \text{ kips} > 89.6 \text{ kips}$ **o.k.**	$\dfrac{P_n}{\Omega_t} = \dfrac{P_n}{2.00}$ $= \dfrac{562 \text{ kips}}{2.00}$ $= 281 \text{ kips} > 60.2 \text{ kips}$ **o.k.**

Check bolt bearing on the brace and shear strength of the bolts

According to the User Note in AISC *Specification* Section J3.6, the strength of the bolt group is taken as the sum of the effective strengths of the individual fasteners. In the following calculations, the available bearing strength and tearout strength limit states from AISC *Specification* Equation J3-6a are separated for clarity. Assume that bolt hole deformation is a design consideration.

LRFD	ASD
Design shear strength per bolt from AISC *Manual* Table 7-1 is: $\phi R_n = 35.8 \text{ kips/bolt}$ Design bearing strength on angles $\phi R_n = \phi 2.4 dt F_u$ $= 0.75(2.4)(\tfrac{3}{4} \text{ in.})(2)(\tfrac{5}{8} \text{ in.})(58 \text{ ksi})$ $= 97.9 \text{ kips}$ Design bearing strength on gusset $\phi R_n = \phi 2.4 dt F_u$ $= 0.75(2.4)(\tfrac{3}{4} \text{ in.})(\tfrac{3}{8} \text{ in.})(58 \text{ ksi})$ $= 29.4 \text{ kips}$	Allowable shear strength per bolt from AISC *Manual* Table 7-1 is: $\dfrac{R_n}{\Omega} = 23.9 \text{ kips/bolt}$ Allowable bearing strength on angles $\dfrac{R_n}{\Omega} = \dfrac{2.4 dt F_u}{\Omega}$ $= \dfrac{(2.4)(\tfrac{3}{4} \text{ in.})(2)(\tfrac{5}{8} \text{ in.})(58 \text{ ksi})}{2.00}$ $= 65.3 \text{ kips}$ Allowable bearing strength on gusset $\dfrac{R_n}{\Omega} = \dfrac{2.4 dt F_u}{\Omega}$ $= \dfrac{(2.4)(\tfrac{3}{4} \text{ in.})(\tfrac{3}{8} \text{ in.})(58 \text{ ksi})}{2.00}$ $= 19.6 \text{ kips}$

3.5 BRACED FRAMES   3–29

LRFD	ASD
Tearout strength on angles, edge (assuming 1.50 in. edge distance): $$\phi R_n = \phi 1.2 t L_c F_u$$ $$= 0.75(1.2)(2)(\text{⅝ in.})$$ $$\times [1.50 \text{ in.} - \tfrac{1}{2}(\tfrac{13}{16} \text{ in.})](58 \text{ ksi})$$ $$= 71.4 \text{ kips}$$	Tearout strength on angles, edge (assuming 1.50 in. edge distance): $$\frac{R_n}{\Omega} = \frac{1.2 t L_c F_u}{\Omega}$$ $$= (1/2.00)(1.2)(2)(\text{⅝ in.})$$ $$\times [1.50 \text{ in.} - \tfrac{1}{2}(\tfrac{13}{16} \text{ in.})](58 \text{ ksi})$$ $$= 47.6 \text{ kips}$$
Tearout strength on angles, spacing between bolts: $$\phi R_n = \phi 1.2 t L_c F_u$$ $$= 0.75(1.2)(2)(\text{⅝ in.})$$ $$\times (3.00 \text{ in.} - \tfrac{13}{16} \text{ in.})(58 \text{ ksi})$$ $$= 143 \text{ kips}$$	Tearout strength on angles, spacing between bolts: $$\frac{R_n}{\Omega} = \frac{1.2 t L_c F_u}{\Omega}$$ $$= (1/2.00)(1.2)(2)(\text{⅝ in.})$$ $$\times (3.00 \text{ in.} - \tfrac{13}{16} \text{ in.})(58 \text{ ksi})$$ $$= 95.2 \text{ kips}$$
Tear-out strength on gusset, edge—assume 2-in. edge distance: $$\phi R_n = \phi 1.2 t L_c F_u$$ $$= 0.75(1.2)(\text{⅜ in.})$$ $$\times (2.00 \text{ in.} - \tfrac{15}{32} \text{ in.})(58 \text{ ksi})$$ $$= 30.0 \text{ kips}$$	Tear-out strength on gusset, edge—assume 2-in. edge distance: $$\frac{R_n}{\Omega} = \frac{1.2 t L_c F_u}{\Omega}$$ $$= (1/2.00)(1.2)(\text{⅜ in.})$$ $$\times (2.00 \text{ in.} - \tfrac{15}{32} \text{ in.})(58 \text{ ksi})$$ $$= 20.0 \text{ kips}$$
Tear-out strength on gusset, spacing between bolts: $$\phi R_n = \phi 1.2 t L_c F_u$$ $$= 0.75(1.2)(\text{⅜ in.})$$ $$\times (3.00 \text{ in.} - \tfrac{15}{16} \text{ in.})(58 \text{ ksi})$$ $$= 40.4 \text{ kips}$$	Tear-out strength on gusset, spacing between bolts: $$\frac{R_n}{\Omega} = \frac{1.2 t L_c F_u}{\Omega}$$ $$= (1/2.00)(1.2)(\text{⅜ in.})$$ $$\times (3.00 \text{ in.} - \tfrac{15}{16} \text{ in.})(58 \text{ ksi})$$ $$= 26.9 \text{ kips}$$
Since all bearing limit state strengths exceed the slip-critical strength of 26.9 kips/bolt, bearing does not govern.	Since all bearing limit state strengths exceed the slip-critical strength of 18.0 kips/bolt, bearing does not govern.

Check block shear strength of brace

$n = 5$
$L_{ev} = 1\frac{1}{2}$ in.
$L_{eh} = 2\frac{1}{2}$ in.

From AISC *Specification* Equation J4-5:

LRFD	ASD
$\phi R_n = \phi U_{bs} F_u A_{nt}$ $+ \min(\phi 0.60 F_y A_{gv},\ \phi 0.60 F_u A_{nv})$	$\dfrac{R_n}{\Omega} = \dfrac{U_{bs} F_u A_{nt}}{\Omega}$ $+ \min\left(\dfrac{0.60 F_y A_{gv}}{\Omega},\ \dfrac{0.60 F_u A_{nv}}{\Omega}\right)$
$U_{bs} = 1.0$	$U_{bs} = 1.0$
Tension rupture component from AISC *Manual* Table 9-3a:	Tension rupture component from AISC *Manual* Table 9-3a:
$\phi U_{bs} F_u A_{nt} = 2(1.0)(89.7 \text{ kip/in.})(\frac{5}{8} \text{ in.})$ $= 112 \text{ kips}$	$\dfrac{U_{bs} F_u A_{nt}}{\Omega} = 2(1.0)(59.8 \text{ kip/in.})(\frac{5}{8} \text{ in.})$ $= 74.8 \text{ kips}$
Shear yielding component from AISC *Manual* Table 9-3b:	Shear yielding component from AISC *Manual* Table 9-3b:
$\phi 0.60 F_y A_{gv} = 2(219 \text{ kip/in.})(\frac{5}{8} \text{ in.})$ $= 274 \text{ kips}$	$\dfrac{0.60 F_y A_{gv}}{\Omega} = 2(146 \text{ kip/in.})(\frac{5}{8} \text{ in.})$ $= 183 \text{ kips}$
Shear rupture component from AISC *Manual* Table 9-3c:	Shear rupture component from AISC *Manual* Table 9-3c:
$\phi 0.60 F_u A_{nv} = 2(250 \text{ kip/in.})(\frac{5}{8} \text{ in.})$ $= 313 \text{ kips}$	$\dfrac{0.60 F_u A_{nv}}{\Omega} = 2(166 \text{ kip/in.})(\frac{5}{8} \text{ in.})$ $= 208 \text{ kips}$
$\phi R_n = 112 \text{ kips} + 274 \text{ kips}$ $= 386 \text{ kips} > 89.6 \text{ kips}$ **o.k.**	$\dfrac{R_n}{\Omega} = 74.8 \text{ kips} + 183 \text{ kips}$ $= 258 \text{ kips} > 60.2 \text{ kips}$ **o.k.**

Use (5) ¾-in. ASTM A325-SC bolts to connect the brace angle to the gusset plate. Use Class B faying surfaces, standard holes in the brace, and oversized holes in the gusset.

Try a ⅜-in. trial gusset plate thickness.

3.5 BRACED FRAMES

Check the gusset compression buckling strength

Using the Whitmore section as discussed in the AISC *Manual* Part 9, the available width is greater than the Whitmore width determined as follows:

$l_w = 2l \tan 30°$
$ = 2(4)(3.00 \text{ in.})\tan 30°$
$ = 13.9 \text{ in.}$

The radius of gyration of the gusset plate buckling in the weak direction is:

$r = \dfrac{t}{\sqrt{12}}$
$ = \dfrac{\tfrac{3}{8} \text{ in.}}{\sqrt{12}}$
$ = 0.108 \text{ in.}$

The average length of the gusset plate beyond the connection on the Whitmore width is approximately 6.5 in. For a fixed-fixed buckling condition, $K = 0.65$ [see Dowswell (2006)], and

$\dfrac{KL}{r} = \dfrac{0.65(6.50 \text{ in.})}{0.108 \text{ in.}}$
$\phantom{\dfrac{KL}{r}} = 39.1$

From AISC *Manual* Table 4-22 for $F_y = 36$ ksi, the available critical stress is:

LRFD	ASD
$\phi_c F_{cr} = 29.9$ ksi	$\dfrac{F_{cr}}{\Omega_c} = 19.9$ ksi
The design compressive strength is:	The allowable compressive strength is:
$\phi R_n = \phi F_{cr} A_g$ $= (29.9 \text{ ksi})(13.9 \text{ in.})(\tfrac{3}{8} \text{ in.})$ $= 156 \text{ kips} > 127 \text{ kips}$ **o.k.**	$\dfrac{R_n}{\Omega} = \dfrac{F_{cr} A_g}{\Omega}$ $= (19.9 \text{ ksi})(13.9 \text{ in.})(\tfrac{3}{8} \text{ in.})$ $= 104 \text{ kips} > 83.4 \text{ kips}$ **o.k.**

The ⅜-in. gusset plate is o.k. Additional checks are required as follows.

Connection Interface Forces

The forces resulting from the applied brace force at the gusset-to-beam, gusset-to-column, and beam-to-column interfaces are determined using the Uniform Force Method (UFM). The planes of uniform forces will be set as the vertical bolt line and the gusset/beam interface. The assumption of a plane of uniform force at the vertical bolt line allows the bolts at the column connection to be designed for shear only (no eccentricity). However, this convenient assumption for connection design requires that a corresponding moment be resolved in the design of the members. In this case, the moment will be assigned to the beam. It should be noted that this assumption is different than that made for the typical cases of the UFM shown in the AISC *Manual* and is not a requirement for this type of connection. Appropriate work points and uniform force planes can often be selected conveniently to balance engineering, fabrication and erection economy. As is demonstrated in the following, the application of the UFM in terms of equations used will remain unchanged despite the change in interface location to the column bolt line.

Using the connection geometry given in Figure 3-6 and using the UFM described in AISC *Manual* Part 13, determine the connection interface forces as follows.

The beam eccentricity to the plane of uniform force is:

$e_b = 0.5 d_b$
$ = 0.5(17.7 \text{ in.})$
$ = 8.85 \text{ in.}$

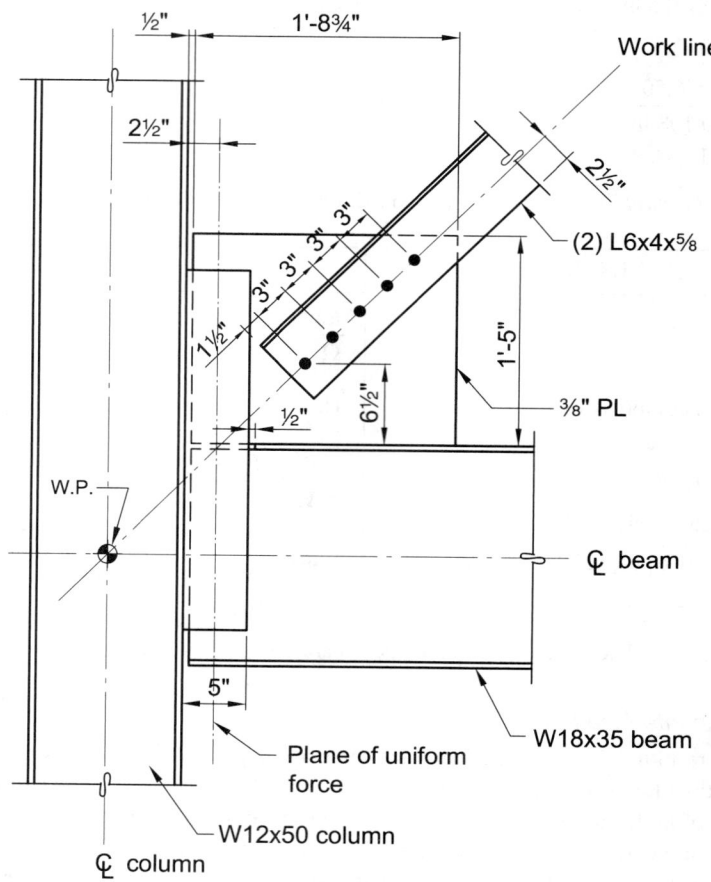

Fig. 3-6. Initial connection geometry for Example 3.5.3.

3.5 BRACED FRAMES

where d_b is the depth of the beam.

The column eccentricity to the plane of uniform force is:

$e_c = 0.5d_c + 2.5$ in.
$ = 0.5(12.2$ in.$) + 2.5$ in.
$ = 8.60$ in.

where d_c is the depth of the column.

The horizontal eccentricity from the plane of uniform force to the centroid of the beam-to-gusset connection is:

$\bar{\alpha} = 0.5(20.75$ in.$) - 2.50$ in. $+ 0.500$ in.
$\phantom{\bar{\alpha}} = 8.38$ in.

Assuming four bolts are used in the gusset-to-single plate connections spaced at 3 in. starting 3½ in. from the top of the beam, the vertical eccentricity from the plane of uniform force to the centroid of the gusset-to-column connection is:

$\bar{\beta} = 3.50$ in. $+ \dfrac{3(3.00 \text{ in.})}{2}$
$\phantom{\bar{\beta}} = 8.00$ in.
$\theta = 45°$

Since the gusset-to-beam connection is more rigid than the gusset-to-column connection, the beam can be assumed to resist the moment generated by eccentricity between the actual gusset centroids and the ideal centroids calculated using the UFM. Therefore:

$\beta = \bar{\beta} = 8.00$ in.

$\alpha = K + \bar{\beta} \tan \theta$ \hfill (*Manual* Eq. 13-15)

where

$K = e_b \tan \theta - e_c$ \hfill (*Manual* Eq. 13-16)

Therefore:

$\alpha = (e_b + \beta) \tan \theta - e_c$
$ = (8.85$ in. $+ 8.00$ in.$) \tan(45°) - 8.60$ in.
$ = 8.25$ in.

The distance from work point to centroid of gusset is:

$r = \sqrt{(\alpha + e_c)^2 + (\beta + e_b)^2}$ \hfill (*Manual* Eq. 13-6)
$ = \sqrt{(8.25 \text{ in.} + 8.60 \text{ in.})^2 + (8.00 \text{ in.} + 8.85 \text{ in.})^2}$
$ = 23.8$ in.

AMERICAN INSTITUTE OF STEEL CONSTRUCTION

The free body diagram forces are determined as follows.

From AISC *Manual* Equation 13-2:

LRFD	ASD
$V_{uc} = \dfrac{\beta}{r} P_u$ $= \dfrac{8.00 \text{ in.}}{23.8 \text{ in.}}(127 \text{ kips})$ $= 42.7 \text{ kips}$	$V_{ac} = \dfrac{\beta}{r} P_a$ $= \dfrac{8.00 \text{ in.}}{23.8 \text{ in.}}(83.4 \text{ kips})$ $= 28.0 \text{ kips}$

From AISC *Manual* Equation 13-3:

LRFD	ASD
$H_{uc} = \dfrac{e_c}{r} P_u$ $= \dfrac{8.60 \text{ in.}}{23.8 \text{ in.}}(127 \text{ kips})$ $= 45.9 \text{ kips}$	$H_{ac} = \dfrac{e_c}{r} P_a$ $= \dfrac{8.60 \text{ in.}}{23.8 \text{ in.}}(83.4 \text{ kips})$ $= 30.1 \text{ kips}$

From AISC *Manual* Equation 13-4:

LRFD	ASD
$V_{ub} = \dfrac{e_b}{r} P_u$ $= \dfrac{8.85 \text{ in.}}{23.8 \text{ in.}}(127 \text{ kips})$ $= 47.2 \text{ kips}$	$V_{ab} = \dfrac{e_b}{r} P_a$ $= \dfrac{8.85 \text{ in.}}{23.8 \text{ in.}}(83.4 \text{ kips})$ $= 31.0 \text{ kips}$

From AISC *Manual* Equation 13-5:

LRFD	ASD
$H_{ub} = \dfrac{\alpha}{r} P_u$ $= \dfrac{8.25 \text{ in.}}{23.8 \text{ in.}}(127 \text{ kips})$ $= 44.0 \text{ kips}$	$H_{ab} = \dfrac{\alpha}{r} P_a$ $= \dfrac{8.25 \text{ in.}}{23.8 \text{ in.}}(83.4 \text{ kips})$ $= 28.9 \text{ kips}$

3.5 BRACED FRAMES

The ⅛-in. difference between the ideal centroid, α, and the actual centroid, $\bar{\alpha}$, determined previously, could be neglected but is included here to illustrate the UFM procedure. From AISC *Manual* Equation 13-17:

LRFD	ASD
$M_{ub} = V_{ub}\|\alpha - \bar{\alpha}\|$ $= 47.2 \text{ kips}\|8.25 \text{ in.} - 8.38 \text{ in.}\|$ $= 6.14 \text{ kip-in.}$	$M_{ab} = V_{ab}\|\alpha - \bar{\alpha}\|$ $= 31.0 \text{ kips}\|8.25 \text{ in.} - 8.38 \text{ in.}\|$ $= 4.03 \text{ kip-in.}$

The moments at the column-gusset plate interface and the column-beam interface due to the plane of uniform force set at the vertical bolt line are as follows:

LRFD	ASD
$M_{ucg} = V_{uc}e$ $= 42.7 \text{ kips}(2.50 \text{ in.})$ $= 107 \text{ kip-in.}$ $M_{ucb} = V_{ub}e$ $= 47.2 \text{ kips }(2.50 \text{ in.})$ $= 118 \text{ kip-in.}$	$M_{acg} = V_{ac}e$ $= 28.0 \text{ kips}(2.50 \text{ in.})$ $= 70.0 \text{ kip-in.}$ $M_{acb} = V_{ab}e$ $= 31.0 \text{ kips }(2.50 \text{ in.})$ $= 77.5 \text{ kip-in.}$

The LRFD and ASD geometry and required strengths are shown in Figures 3-7a and 3-7b, respectively.

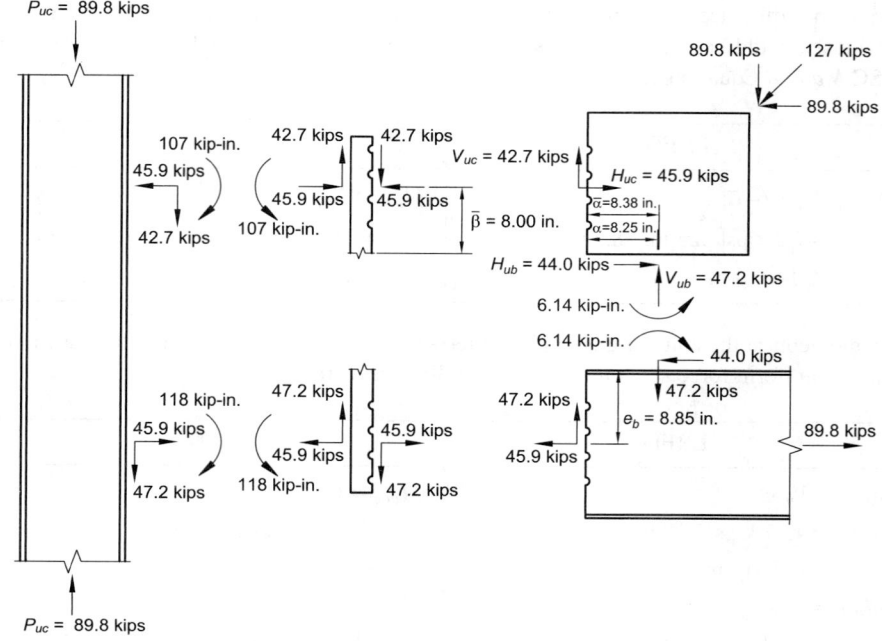

Fig. 3-7a. LRFD free body forces and moments.

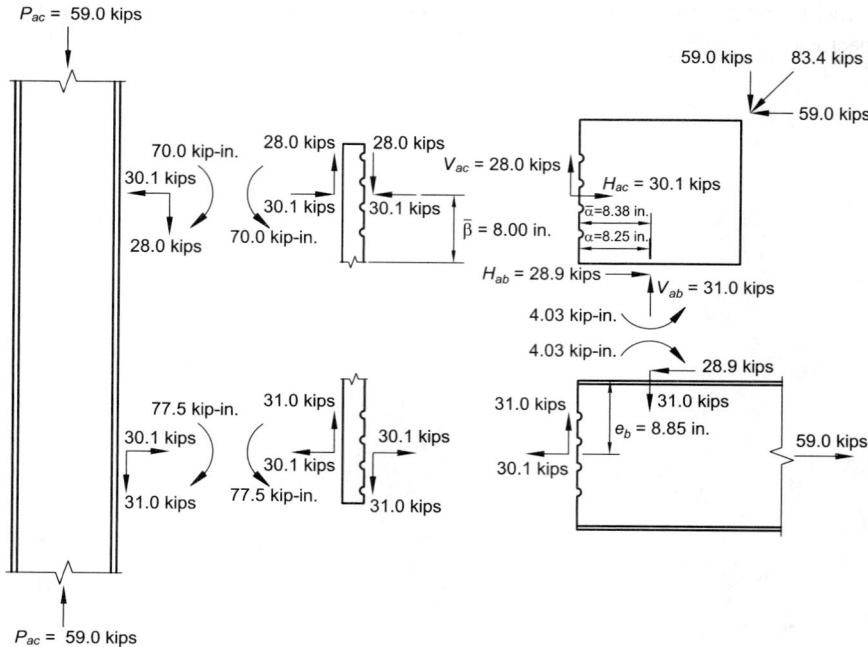

Fig. 3-7b. ASD free body forces and moments.

3.5 BRACED FRAMES

Gusset-to-Beam Interface

Design the gusset-to-beam weld

Treating the welds as a line:

$l_w = 20.75$ in.

$$Z_w = \frac{(20.75 \text{ in.})^2}{4}$$

$= 108$ in.^{3}/in.

The forces along the gusset-to-beam interface are:

LRFD	ASD
$f_{uv} = \dfrac{H_{ub}}{l_w}$ $= \dfrac{44.0 \text{ kips}}{20.75 \text{ in.}}$ $= 2.12$ kip/in.	$f_{av} = \dfrac{H_{ab}}{l_w}$ $= \dfrac{28.9 \text{ kips}}{20.75 \text{ in.}}$ $= 1.39$ kip/in.
$f_{ua} = \dfrac{V_{ub}}{l_w}$ $= \dfrac{47.2 \text{ kips}}{20.75 \text{ in.}}$ $= 2.27$ kip/in.	$f_{aa} = \dfrac{V_{ab}}{l_w}$ $= \dfrac{31.0 \text{ kips}}{20.75 \text{ in.}}$ $= 1.49$ kip/in.
$f_{ub} = \dfrac{M_{ub}}{Z_w}$ $= \dfrac{6.14 \text{ kip-in.}}{108 \text{ in.}^3/\text{in.}}$ $= 0.0569$ kip/in.	$f_{ab} = \dfrac{M_{ab}}{Z_w}$ $= \dfrac{4.03 \text{ kip-in.}}{108 \text{ in.}^3/\text{in.}}$ $= 0.0373$ kip/in.
The resultant force is: $f_{u,\,peak} = \sqrt{f_{uv}^2 + (f_{ua} + f_{ub})^2}$ $= \sqrt{\begin{array}{l}(2.12 \text{ kip/in.})^2 \\ + (2.27 \text{ kip/in.} + 0.0569 \text{ kip/in.})^2\end{array}}$ $= 3.15$ kip/in.	The resultant force is: $f_{a,\,peak} = \sqrt{f_{av}^2 + (f_{aa} + f_{ab})^2}$ $= \sqrt{\begin{array}{l}(1.39 \text{ kip/in.})^2 \\ + (1.49 \text{ kip/in.} + 0.0373 \text{ kip/in.})^2\end{array}}$ $= 2.07$ kip/in.

LRFD	ASD
$f_{u,avg} = 0.5\left(f_{u,peak} + \sqrt{f_{uv}^2 + (f_{ua} - f_{ub})^2}\right)$	$f_{a,avg} = 0.5\left(f_{a,peak} + \sqrt{f_{av}^2 + (f_{aa} - f_{ab})^2}\right)$
$= 0.5\left(\begin{array}{l}3.15 \text{ kip/in.} \\ + \sqrt{(2.12 \text{ kip/in.})^2 \\ + (2.27 \text{ kip/in.} - 0.0569 \text{ kip/in.})^2}\end{array}\right)$	$= 0.5\left(\begin{array}{l}2.07 \text{ kip/in.} \\ + \sqrt{(1.39 \text{ kip/in.})^2 \\ + (1.49 \text{ kip/in.} - 0.0373 \text{ kip/in.})^2}\end{array}\right)$
$= 3.11 \text{ kip/in.}$	$= 2.04 \text{ kip/in.}$
$\dfrac{f_{u,peak}}{f_{u,avg}} = \dfrac{3.15 \text{ kip/in.}}{3.11 \text{ kip/in.}}$	$\dfrac{f_{a,peak}}{f_{a,avg}} = \dfrac{2.07 \text{ kip/in.}}{2.04 \text{ kip/in.}}$
$= 1.01$	$= 1.01$
Since $f_{peak}/f_{avg} < 1.25$, the weld ductility factor of 1.25 will be applied. For a discussion of the weld ductility factor, see AISC *Manual* Part 13.	Since $f_{peak}/f_{avg} < 1.25$, the weld ductility factor of 1.25 will be applied. For a discussion of the weld ductility factor, see AISC *Manual* Part 13.
Load angle:	Load angle:
$\theta = \tan^{-1}\left(\dfrac{f_{ua} + f_{ub}}{f_v}\right)$	$\theta = \tan^{-1}\left(\dfrac{f_{aa} + f_{ab}}{f_{av}}\right)$
$= \tan^{-1}\left(\dfrac{2.27 \text{ kip/in.} + 0.0569 \text{ kip/in.}}{2.12 \text{ kip/in.}}\right)$	$= \tan^{-1}\left(\dfrac{1.49 \text{ kip/in.} + 0.0373 \text{ kip/in.}}{1.39 \text{ kip/in.}}\right)$
$= 47.7°$	$= 47.7°$
Required weld leg, D, including the weld ductility factor and directional weld strength increase:	Required weld leg, D, including the weld ductility factor and directional weld strength increase:
$D \geq 1.25\left[\dfrac{f_{u,avg}}{2\phi R_n\left[1 + 0.5\sin^{1.5}(\theta)\right]}\right]$	$D \geq 1.25\left[\dfrac{f_{a,avg}}{2(R_n/\Omega)\left[1 + 0.5\sin^{1.5}(\theta)\right]}\right]$
$= 1.25 \times \left[\dfrac{3.11 \text{ kip/in.}}{2(1.392 \text{ kip/in.})\left[1 + 0.5\sin^{1.5}(47.7°)\right]}\right]$	$= 1.25 \times \left[\dfrac{2.04 \text{ kip/in.}}{2(0.928 \text{ kip/in.})\left[1 + 0.5\sin^{1.5}(47.7°)\right]}\right]$
$= 1.06$ sixteenths	$= 1.04$ sixteenths
For a derivation of the weld shear strength, $\phi R_n = 1.392$ kip/in., see AISC *Manual* Part 8.	For a derivation of the weld shear strength, $\dfrac{R_n}{\Omega} = 0.928$ kip/in., see AISC *Manual* Part 8.

3.5 BRACED FRAMES

The weld size is controlled by the minimum size of fillet weld given in AISC *Specification* Table J2.4.

Use a 20.75-in. -long, double-sided ³⁄₁₆-in. fillet weld to connect the gusset plate to the beam.

A conservative method to determine the minimum gusset plate thickness is to set the shear rupture strength of the base metal of the gusset plate equal to the required shear rupture strength of the weld. From AISC *Manual* Equation 9-3:

LRFD	ASD
$t_{min} = \dfrac{6.19D}{F_u}$ $= \dfrac{6.19(1.07 \text{ sixteenths})}{58 \text{ ksi}}$ $= 0.114 \text{ in.} \leq \text{⅜ in.}$ **o.k.**	$t_{min} = \dfrac{6.19D}{F_u}$ $= \dfrac{6.19(1.06 \text{ sixteenths})}{58 \text{ ksi}}$ $= 0.113 \text{ in.} \leq \text{⅜ in.}$ **o.k.**

Therefore, the gusset plate thickness of ⅜ in. is acceptable.

Check the beam web at the beam-to-gusset interface

The normal and flexural forces at the gusset-to-beam interface can be converted into an effective normal force in order to facilitate the web local yielding and web local crippling checks. The effective normal force for use with the full length of the gusset can be conservatively calculated as:

LRFD	ASD
$N_{eff} = V_{ub} + \dfrac{4M_{ub}}{L}$ $= 47.2 \text{ kips} + \dfrac{4(6.14 \text{ kip-in.})}{20.75 \text{ in.}}$ $= 48.4 \text{ kips}$	$N_{eff} = V_{ab} + \dfrac{4M_{ab}}{L}$ $= 31.0 \text{ kips} + \dfrac{4(4.03 \text{ kip-in.})}{20.75 \text{ in.}}$ $= 31.8 \text{ kips}$

Check beam web local yielding

The beam force is applied at $\alpha = 8.25$ in. from the beam end. Because $\alpha < d = 17.7$ in.,

$R_n = F_{yw} t_w (2.5k + l_b)$ \hfill (*Spec.* Eq. J10-3)

$ = (50 \text{ ksi})(0.300 \text{ in.})[2.5(0.827 \text{ in.}) + 20.75 \text{ in.}]$

$ = 342 \text{ kips}$

LRFD	ASD
$\phi R_n = 1.00(342 \text{ kips})$ $= 342 \text{ kips} > 48.4 \text{ kips}$ **o.k.**	$\dfrac{R_n}{\Omega} = \dfrac{342 \text{ kips}}{1.50}$ $= 228 \text{ kips} > 31.8 \text{ kips}$ **o.k.**

Check beam web local crippling

Since the framed beam-to-column connection will provide significant restraint to the web relative to crippling, AISC *Specification* Equation J10-4 is used despite the fact that the force is applied less than $d/2$ from the end of the beam.

Using AISC *Manual* Table 9-4 and Equations 9-49a and 9-49b:

LRFD	ASD
$\phi R_3 = 38.7 \text{ kips}$ $\phi R_4 = 3.89 \text{ kip/in.}$ $\phi R_n = 2[\phi R_3 + l_b(\phi R_4)]$ $= 2[38.7 \text{ kips} + 20.75 \text{ in.}(3.89 \text{ kip/in.})]$ $= 239 \text{ kips} > 48.4 \text{ kips}$ **o.k.**	$R_3/\Omega = 25.8 \text{ kips}$ $R_4/\Omega = 2.59 \text{ kip/in.}$ $\dfrac{R_n}{\Omega} = 2[R_3/\Omega + l_b(R_4/\Omega)]$ $= 2[25.8 \text{ kips} + 20.75 \text{ in.}(2.59 \text{ kip/in.})]$ $= 159 \text{ kips} > 31.8 \text{ kips}$ **o.k.**

Gusset-to-Column Interface

Check the gusset at the gusset-to-column interface

Try a length above the top of beam of 14 in. to allow for shaping of the gusset if desired.

LRFD	ASD
Forces at interface $V_{uc} = 42.7 \text{ kips}$ $H_{uc} = 45.9 \text{ kips}$ Shear yielding on gross section, from AISC *Specification* Equation J4-3: $\phi R_n = \phi 0.6 F_y A_{gv}$ $= 1.00(0.6)(36 \text{ ksi})(14.0 \text{ in.})(\tfrac{3}{8} \text{ in.})$ $= 113 \text{ kips} > 42.7 \text{ kips}$ **o.k.**	Forces at interface $V_{ac} = 28.0 \text{ kips}$ $H_{ac} = 30.1 \text{ kips}$ Shear yielding on gross section from AISC *Specification* Equation J4-3: $\dfrac{R_n}{\Omega} = \dfrac{0.6 F_y A_{gv}}{\Omega}$ $= \dfrac{(0.6)(36 \text{ ksi})(14.0 \text{ in.})(\tfrac{3}{8} \text{ in.})}{1.50}$ $= 75.6 \text{ kips} > 28.0 \text{ kips}$ **o.k.**

3.5 BRACED FRAMES

3–41

LRFD	ASD
Tension yielding on gross section, from AISC *Specification* Equation J4-1: $$\phi R_n = \phi F_y A_g$$ $$= 0.90(36 \text{ ksi})(14.0 \text{ in.})(\tfrac{3}{8} \text{ in.})$$ $$= 170 \text{ kips} > 45.9 \text{ kips} \quad \textbf{o.k.}$$	Tension yielding on gross section, from AISC *Specification* Equation J4-1: $$\frac{R_n}{\Omega} = \frac{F_y A_g}{\Omega}$$ $$= \frac{(36 \text{ ksi})(14.0 \text{ in.})(\tfrac{3}{8} \text{ in.})}{1.67}$$ $$= 113 \text{ kips} > 30.1 \text{ kips} \quad \textbf{o.k.}$$
Check block shear relative to shear load, $n = 4$, $L_{ev} = 1\tfrac{1}{2}$ in., $L_{eh} = 2$ in.	Check block shear relative to shear load, $n = 4$, $L_{ev} = 1\tfrac{1}{2}$ in., $L_{eh} = 2$ in.
From AISC *Specification* Equation J4-5: $$\phi R_n = \phi U_{bs} F_u A_{nt}$$ $$+ \min(\phi 0.60 F_y A_{gv}, \phi 0.60 F_u A_{nv})$$	From AISC *Specification* Equation J4-5: $$\frac{R_n}{\Omega} = \frac{U_{bs} F_u A_{nt}}{\Omega}$$ $$+ \min\left(\frac{0.60 F_y A_{gv}}{\Omega}, \frac{0.60 F_u A_{nv}}{\Omega}\right)$$
$U_{bs} = 1.0$	$U_{bs} = 1.0$
Tension rupture component from AISC *Manual* Table 9-3a: $$\phi U_{bs} F_u A_{nt} = (1.0)(68.0 \text{ kip/in.})(\tfrac{3}{8} \text{ in.})$$ $$= 25.5 \text{ kips}$$	Tension rupture component from AISC *Manual* Table 9-3a: $$\frac{U_{bs} F_u A_{nt}}{\Omega} = (1.0)(45.3 \text{ kip/in.})(\tfrac{3}{8} \text{ in.})$$ $$= 17.0 \text{ kips}$$
Shear yielding component from AISC *Manual* Table 9-3b: $$\phi 0.60 F_y A_{gv} = (170 \text{ kip/in.})(\tfrac{3}{8} \text{ in.})$$ $$= 63.8 \text{ kips}$$	Shear yielding component from AISC *Manual* Table 9-3b: $$\frac{0.60 F_y A_{gv}}{\Omega} = (113 \text{ kip/in.})(\tfrac{3}{8} \text{ in.})$$ $$= 42.4 \text{ kips}$$
Shear rupture component from AISC *Manual* Table 9-3c: $$\phi 0.60 F_u A_{nv} = (194 \text{ kip/in.})(\tfrac{3}{8} \text{ in.})$$ $$= 72.8 \text{ kips}$$	Shear rupture component from AISC *Manual* Table 9-3c: $$\frac{0.60 F_u A_{nv}}{\Omega} = (129 \text{ kip/in.})(\tfrac{3}{8} \text{ in.})$$ $$= 48.4 \text{ kips}$$
$$\phi R_n = 25.5 \text{ kip/in.} + 63.8 \text{ kip/in.}$$ $$= 89.3 \text{ kips} > 42.7 \text{ kips} \quad \textbf{o.k.}$$	$$\frac{R_n}{\Omega} = 17.0 \text{ kip/in.} + 42.4 \text{ kip/in.}$$ $$= 59.4 \text{ kips} > 28.0 \text{ kips} \quad \textbf{o.k.}$$

LRFD	ASD
Block shear relative to normal load	Block shear relative to normal load
Calculate tension rupture component:	Calculate tension rupture component:
$\phi U_{bs} F_u A_{nt} = 0.75(1.0)(58 \text{ kips})(\tfrac{3}{8} \text{ in.})$ $\times [10.5 \text{ in.} - 3.5(0.875 \text{ in.})]$ $= 121 \text{ kips}$	$\dfrac{U_{bs} F_u A_{nt}}{\Omega} = (1/2.00)(1.0)(58 \text{ kips})(\tfrac{3}{8} \text{ in.})$ $\times [10.5 \text{ in.} - 3.5(0.875 \text{ in.})]$ $= 80.9 \text{ kips}$
Calculate shear yielding component;	Calculate shear yielding component:
$\phi 0.60 F_y A_{gv} = 0.75(0.60)(36 \text{ ksi})(\tfrac{3}{8} \text{ in.})$ $\times (2.00 \text{ in.})$ $= 12.2 \text{ kips}$	$\dfrac{0.60 F_y A_{gv}}{\Omega} = (1/2.00)(0.60)(36 \text{ ksi})$ $\times (\tfrac{3}{8} \text{ in.})(2.00 \text{ in.})$ $= 8.10 \text{ kips}$
Calculate shear rupture component:	Calculate shear rupture component:
$\phi 0.60 F_u A_{nv} = 0.75(0.60)(58 \text{ ksi})(\tfrac{3}{8} \text{ in.})$ $\times [2.00 \text{ in.} - 0.5(0.875 \text{ in.})]$ $= 15.3 \text{ kips}$	$\dfrac{0.60 F_u A_{nv}}{\Omega} = (1/2.00)(0.60)(58 \text{ ksi})(\tfrac{3}{8} \text{ in.})$ $\times [2.00 \text{ in.} - 0.5(0.875 \text{ in.})]$ $= 10.2 \text{ kips}$
$\phi R_n = 121 \text{ kips} + 12.2 \text{ kips}$ $= 133 \text{ kips} > 45.9 \text{ kips}$ **o.k.**	$\dfrac{R_n}{\Omega} = 80.9 \text{ kips} + 8.10 \text{ kips}$ $= 89.0 \text{ kips} > 30.1 \text{ kips}$ **o.k.**
Combined shear and normal block shear:	Combined shear and normal block shear:
$\left(\dfrac{45.9 \text{ kips}}{133 \text{ kips}}\right)^2 + \left(\dfrac{42.7 \text{ kips}}{89.3 \text{ kips}}\right)^2$ $= 0.348 < 1.0$ **o.k.**	$\left(\dfrac{30.1 \text{ kips}}{89.0 \text{ kips}}\right)^2 + \left(\dfrac{28.0 \text{ kips}}{59.4 \text{ kips}}\right)^2$ $= 0.337 < 1.0$ **o.k.**

Gusset-to-single-plate connection design

The resultant forces that will be resisted by the bolts in the gusset plate are:

LRFD	ASD
$R_u = \sqrt{(V_{uc})^2 + (H_{uc})^2}$ $= \sqrt{(42.7 \text{ kips})^2 + (45.9 \text{ kips})^2}$ $= 62.7 \text{ kips}$	$R_a = \sqrt{(V_{ac})^2 + (H_{ac})^2}$ $= \sqrt{(28.0 \text{ kips})^2 + (30.1 \text{ kips})^2}$ $= 41.1 \text{ kips}$

3.5 BRACED FRAMES

LRFD	ASD
From AISC *Manual* Table 7-1, (4) ¾-in.-diameter ASTM A325-N bolts are required. $\phi R_n = (4)(17.9 \text{ kips})$ $= 71.6 \text{ kips} > 62.7 \text{ kips}$ **o.k.**	From AISC *Manual* Table 7-1, (4) ¾-in.-diameter ASTM A325-N bolts are required. $\dfrac{R_n}{\Omega} = (4)(11.9 \text{ kips})$ $= 47.6 \text{ kips} > 41.1 \text{ kips}$ **o.k.**

Use (4) ¾-in.-diameter ASTM A325-N bolts to connect the gusset plate to the column.

Using AISC *Manual* Tables 7-4 and 7-5 to check bolt bearing on the gusset plate with $s = 3$ in. and $L_e = 2$ in., the available bearing strength based on one bolt is:

LRFD	ASD
$\phi R_n = \phi r_n t$ $= (78.3 \text{ kip/in.})(\text{⅜ in.})$ $= 29.4 \text{ kips} > 17.9 \text{ kips}$ **o.k.**	$\dfrac{R_n}{\Omega} = \left(\dfrac{r_n}{\Omega}\right) t$ $= (52.2 \text{ kip/in.})(\text{⅜ in.})$ $= 19.6 \text{ kips} > 11.9 \text{ kips}$ **o.k.**

Therefore, bolt shear governs over bolt bearing.

Single plate design

Check single plate—assume ⅜-in.-thick plate

LRFD	ASD
Shear yielding on gross section, from AISC *Specification* Equation J4-3: $\phi R_n = \phi 0.60 F_y A_{gv}$ $= 1.00(0.60)(36 \text{ ksi})(12.0 \text{ in.})(\text{⅜ in.})$ $= 97.2 \text{ kips} > 42.7 \text{ kips}$ **o.k.**	Shear yielding on gross section, from AISC *Specification* Equation J4-3: $\dfrac{R_n}{\Omega} = \dfrac{0.60 F_y A_{gv}}{\Omega}$ $= \dfrac{(0.60)(36 \text{ ksi})(12.0 \text{ in.})(\text{⅜ in.})}{1.50}$ $= 64.8 \text{ kips} > 28.0 \text{ kips}$ **o.k.**

LRFD	ASD
Shear rupture on net section, from AISC *Specification* Equation J4-4:	Shear rupture on net section, from AISC *Specification* Equation J4-4:

LRFD:

$\phi R_n = \phi 0.60 F_u A_{nv}$
$= 0.75(0.60)(58 \text{ ksi})$
$\times [12.0 \text{ in.} - 4(0.875 \text{ in.})](\tfrac{3}{8} \text{ in.})$
$= 83.2 \text{ kips} > 42.7 \text{ kips}$ **o.k.**

Tensile yielding on gross section, from AISC *Specification* Equation J4-1:

$\phi R_n = \phi F_y A_g$
$= 0.90(36 \text{ ksi})(12.0 \text{ in.})(\tfrac{3}{8} \text{ in.})$
$= 146 \text{ kips} > 45.9 \text{ kips}$ **o.k.**

Tensile rupture on net section, from AISC *Specification* Equation J4-2:

$\phi R_n = \phi F_u A_e$
$= 0.75(58 \text{ ksi})$
$\times [12.0 \text{ in.} - 4(0.875 \text{ in.})](\tfrac{3}{8} \text{ in.})$
$= 139 \text{ kips} > 45.9 \text{ kips}$ **o.k.**

Block shear on single plate relative to shear load, with $n = 4$, $L_{ev} = 1\tfrac{1}{2}$ in., $L_{eh} = 2\tfrac{1}{2}$ in.

From AISC *Specification* Equation J4-5:

$\phi R_n = \phi U_{bs} F_u A_{nt}$
$\quad + \min(\phi 0.60 F_y A_{gv}, \phi 0.60 F_u A_{nv})$

$U_{bs} = 1.0$

Tension rupture component from AISC *Manual* Table 9-3a:

$\phi U_{bs} F_u A_{nt} = (1.0)(89.7 \text{ kip/in.})(\tfrac{3}{8} \text{ in.})$
$= 33.6 \text{ kips}$

ASD:

$\dfrac{R_n}{\Omega} = \dfrac{0.60 F_u A_{nv}}{\Omega}$
$= (1/2.00)(0.60)(58 \text{ ksi})$
$\times [12.0 \text{ in.} - 4(0.875 \text{ in.})](\tfrac{3}{8} \text{ in.})$
$= 55.5 \text{ kips} > 28.0 \text{ kips}$ **o.k.**

Tensile yielding on gross section, from AISC *Specification* Equation J4-1:

$\dfrac{R_n}{\Omega} = \dfrac{F_y A_g}{\Omega}$
$= \dfrac{(36 \text{ ksi})(12.0 \text{ in.})(\tfrac{3}{8} \text{ in.})}{1.67}$
$= 97.0 \text{ kips} > 30.1 \text{ kips}$ **o.k.**

Tensile rupture on net section, from AISC *Specification* Equation J4-2:

$\dfrac{R_n}{\Omega} = \dfrac{F_u A_e}{\Omega}$
$= (1/2.00)(58 \text{ ksi})$
$\times [12.0 \text{ in.} - 4(0.875 \text{ in.})](\tfrac{3}{8} \text{ in.})$
$= 92.4 \text{ kips} > 30.1 \text{ kips}$ **o.k.**

Block shear on single plate relative to shear load, with $n = 4$, $L_{ev} = 1\tfrac{1}{2}$ in., $L_{eh} = 2\tfrac{1}{2}$ in.

From AISC *Specification* Equation J4-5:

$\dfrac{R_n}{\Omega} = \dfrac{U_{bs} F_u A_{nt}}{\Omega}$
$\quad + \min\left(\dfrac{0.60 F_y A_{gv}}{\Omega}, \dfrac{0.60 F_u A_{nv}}{\Omega}\right)$

$U_{bs} = 1.0$

Tension rupture component from AISC *Manual* Table 9-3a:

$\dfrac{U_{bs} F_u A_{nt}}{\Omega} = (1.0)(59.8 \text{ kip/in.})(\tfrac{3}{8} \text{ in.})$
$= 22.4 \text{ kips}$

3.5 BRACED FRAMES   3-45

LRFD	ASD
Shear yielding component from AISC *Manual* Table 9-3b: $\phi 0.60 F_y A_{gv} = 63.8$ kips, from gusset-to-column interface check Shear rupture component from AISC *Manual* Table 9-3c: $\phi 0.60 F_u A_{nv} = 72.8$ kips, from gusset-to-column interface check $\phi R_n = 33.6$ kips + 63.8 kips $\quad\;\; = 97.4$ kips > 42.7 kips **o.k.** Block shear on single plate relative to normal load Calculate tension rupture component: $\phi U_{bs} F_u A_{nt} = 121$ kips, from gusset-to-column interface check Calculate shear yielding component: $\phi 0.60 F_y A_{gv} = 0.75(0.60)(36 \text{ ksi})(\text{\textfrac{3}{8} in.})$ $\qquad\qquad\quad \times (2.50 \text{ in.})$ $\qquad\qquad\; = 15.2$ kips Calculate shear rupture component: $\phi 0.60 F_u A_{nv} = 0.75(0.60)(58 \text{ ksi})(\text{\textfrac{3}{8} in.})$ $\qquad\qquad\quad \times [2.50 \text{ in.} - 0.5(0.875 \text{ in.})]$ $\qquad\qquad\; = 20.2$ kips $\phi R_n = 121$ kips + 15.2 kips $\quad\;\; = 136$ kips > 45.9 kips **o.k.**	Shear yielding component from AISC *Manual* Table 9-3b: $\dfrac{0.60 F_y A_{gv}}{\Omega} = 42.4$ kips, from gusset-to-column interface check Shear rupture component from AISC *Manual* Table 9-3c: $\dfrac{0.60 F_u A_{nv}}{\Omega} = 48.4$ kips, from gusset-to-column interface check $\dfrac{R_n}{\Omega} = 22.4$ kips + 42.4 kips $\quad\;\; = 64.8$ kips > 28.0 kips **o.k.** Block shear on single plate relative to normal load Calculate tension rupture component: $\dfrac{U_{bs} F_u A_{nt}}{\Omega} = 80.9$ kips, from gusset-to-column interface check Calculate shear yielding component: $\dfrac{0.60 F_y A_{gv}}{\Omega} = \dfrac{0.60(36 \text{ ksi})(\text{\textfrac{3}{8} in.})(2.50 \text{ in.})}{2.00}$ $\qquad\qquad = 10.1$ kips Calculate shear rupture component: $\dfrac{0.6 F_u A_{nv}}{\Omega} = (1/2.00)(0.60)(58 \text{ ksi})(\text{\textfrac{3}{8} in.})$ $\qquad\qquad\quad \times [2.50 \text{ in.} - 0.5(0.875 \text{ in.})]$ $\qquad\qquad = 13.5$ kips $\dfrac{R_n}{\Omega} = 80.9$ kips + 10.1 kips $\quad\;\; = 91.0$ kips > 30.1 kips **o.k.**

AMERICAN INSTITUTE OF STEEL CONSTRUCTION

LRFD	ASD
Combined shear and normal block shear:	Combined shear and normal block shear:
$\left(\dfrac{42.7 \text{ kips}}{97.4 \text{ kips}}\right)^2 + \left(\dfrac{45.9 \text{ kips}}{136 \text{ kips}}\right)^2$ $= 0.306 < 1.0$ **o.k.**	$\left(\dfrac{28.0 \text{ kips}}{64.8 \text{ kips}}\right)^2 + \left(\dfrac{30.1 \text{ kips}}{91.0 \text{ kips}}\right)^2$ $= 0.296 < 1.0$ **o.k.**

Use a ⅜-in.-thick single plate.

Beam-to-Column Single Plate Connection

The forces on the connection are:

LRFD	ASD

$V_u = R_u + V_{ub}$
   $= 4.00$ kips $+ 47.2$ kips
   $= 51.2$ kips
	$V_a = R_a + V_{ab}$
	   $= 2.63$ kips $+ 31.0$ kips
	   $= 33.6$ kips

$H_u = H_{uc}$
   $= 45.9$ kips
	$H_a = H_{ac}$
	   $= 30.1$ kips

The resultant force that will be resisted by the bolts is:
$R_u = \sqrt{(51.2 \text{ kips})^2 + (45.9 \text{ kips})^2}$
   $= 68.8$ kips
	The resultant force that will be resisted by the bolts is:
	$R_a = \sqrt{(33.6 \text{ kips})^2 + (30.1 \text{ kips})^2}$
	   $= 45.1$ kips

From AISC *Manual* Table 7-1, (4) ¾-in.-diameter ASTM A325-N bolts are required.	From AISC *Manual* Table 7-1, (4) ¾-in.-diameter ASTM A325-N bolts are required.

$\phi R_n = (4)(17.9 \text{ kips})$
   $= 71.6$ kips > 68.8 kips **o.k.**
	$\dfrac{R_n}{\Omega} = (4)(11.9 \text{ kips})$
	   $= 47.6$ kips > 45.1 kips **o.k.**

Use (4) ASTM A325-N bolts to connect the beam to the column.

Using AISC *Manual* Tables 7-4 and 7-5 for bolt bearing on the single plate, with $s = 3$ in. and $L_e = 2$ in. (note that $L_e = 2$ in. is used conservatively to employ Table 7-5). The available bearing strength of the plate based on one bolt is:

3.5 BRACED FRAMES

LRFD	ASD
$\phi R_n = \phi r_n t_p$ $= (78.3 \text{ kip/in.})(\sfrac{3}{8} \text{ in.})$ $= 29.4 \text{ kips} > 17.9 \text{ kips}$ **o.k.**	$\dfrac{R_n}{\Omega} = \left(\dfrac{r_n}{\Omega}\right) t_p$ $= (52.2 \text{ kip/in.})(\sfrac{3}{8} \text{ in.})$ $= 19.6 \text{ kips} > 11.9 \text{ kips}$ **o.k.**

Therefore, bolt shear governs over bearing.

Beam single-plate-to-column connection weld

Treating the welds as a line:

$l_w = 12.0 \text{ in.}$

$Z_w = \dfrac{(12.0 \text{ in.})^2}{4}$

$= 36.0 \text{ in.}^3/\text{in.}$

The forces along the beam-to-column interface are:

LRFD	ASD
$f_v = \dfrac{51.2 \text{ kips}}{12.0 \text{ in.}}$ $= 4.27 \text{ kip/in.}$	$f_v = \dfrac{33.6 \text{ kips}}{12.0 \text{ in.}}$ $= 2.80 \text{ kip/in.}$
$f_a = \dfrac{45.9 \text{ kips}}{12.0 \text{ in.}}$ $= 3.83 \text{ kip/in.}$	$f_a = \dfrac{30.1 \text{ kips}}{12.0 \text{ in.}}$ $= 2.51 \text{ kip/in.}$
$f_b = \dfrac{118 \text{ kip-in.}}{36.0 \text{ in.}^3/\text{in.}}$ $= 3.28 \text{ kip/in.}$	$f_b = \dfrac{77.5 \text{ kip-in.}}{36.0 \text{ in.}^3/\text{in.}}$ $= 2.15 \text{ kip/in.}$
$f_{peak} = \sqrt{f_v^2 + (f_a + f_b)^2}$ $= \sqrt{4.27^2 + (3.83 + 3.28)^2}$ $= 8.29 \text{ kip/in.}$	$f_{peak} = \sqrt{f_v^2 + (f_a + f_b)^2}$ $= \sqrt{2.80^2 + (2.51 + 2.15)^2}$ $= 5.44 \text{ kip/in.}$

LRFD	ASD
Load Angle $$\theta = \tan^{-1}\left(\frac{3.83 \text{ kip/in.} + 3.28 \text{ kip/in.}}{4.27 \text{ kip/in.}}\right)$$ $$= 59.0°$$	Load Angle $$\theta = \tan^{-1}\left(\frac{2.51 \text{ kip/in.} + 2.15 \text{ kip/in.}}{2.80 \text{ kip/in.}}\right)$$ $$= 59.0°$$
$$D \geq \frac{8.29 \text{ kip/in.}}{2(1.392 \text{ kip/in.})\left[1 + 0.5\sin^{1.5}(59.0°)\right]}$$ $$= 2.13 \text{ sixteenths}$$	$$D \geq \frac{5.44 \text{ kip/in.}}{2(0.928 \text{ kip/in.})\left[1 + 0.5\sin^{1.5}(59.0°)\right]}$$ $$= 2.10 \text{ sixteenths}$$

A ³⁄₁₆-in. fillet weld on both sides of the single plate is adequate.

Determine the single plate thickness

Try a ⅜-in.-thick plate.

LRFD	ASD
Shear yielding on gross section, from AISC *Specification* Equation J4-3: $$\phi R_n = \phi 0.60 F_y A_{gv}$$ $$= 1.00(0.60)(36 \text{ ksi})(12.0 \text{ in.})(\tfrac{3}{8} \text{ in.})$$ $$= 97.2 \text{ kips} > 51.2 \text{ kips} \quad \textbf{o.k.}$$	Shear yielding on gross section, from AISC *Specification* Equation J4-3: $$\frac{R_n}{\Omega} = \frac{0.60 F_y A_{gv}}{\Omega}$$ $$= \frac{(0.60)(36 \text{ ksi})(12.0 \text{ in.})(\tfrac{3}{8} \text{ in.})}{1.50}$$ $$= 64.8 \text{ kips} > 33.6 \text{ kips} \quad \textbf{o.k.}$$
Shear rupture on net section, from AISC *Specification* Equation J4-4: $$\phi R_n = \phi 0.60 F_u A_{nv}$$ $$= 0.75(0.60)(58 \text{ ksi})$$ $$\times [12.0 \text{ in.} - 4(0.875 \text{ in.})](\tfrac{3}{8} \text{ in.})$$ $$= 83.2 \text{ kips} > 47.2 \text{ kips} \quad \textbf{o.k.}$$	Shear rupture on net section, from AISC *Specification* Equation J4-4: $$\frac{R_n}{\Omega} = \frac{0.60 F_u A_{nv}}{\Omega}$$ $$= (1/2.00)(0.60)(58 \text{ ksi})$$ $$\times [12.0 \text{ in.} - 4(0.875 \text{ in.})](\tfrac{3}{8} \text{ in.})$$ $$= 55.5 \text{ kips} > 31.0 \text{ kips} \quad \textbf{o.k.}$$
Tensile yielding on gross section, from AISC *Specification* Equation J4-1 (use stresses calculated for weld): $$\frac{f_{ua} + f_{ub}}{\phi F_y t_p} = \frac{3.83 \text{ kip/in.} + 3.28 \text{ kip/in.}}{0.90(36 \text{ ksi})(\tfrac{3}{8} \text{ in.})}$$ $$= 0.585 < 1.0 \quad \textbf{o.k.}$$	Tensile yielding on gross section, from AISC *Specification* Equation J4-1 (use stresses calculated for weld): $$\frac{f_{aa} + f_{ab}}{F_y t_p / \Omega} = \frac{2.51 \text{ kip/in.} + 2.15 \text{ kip/in.}}{(36 \text{ ksi})(\tfrac{3}{8} \text{ in.})/1.67}$$ $$= 0.576 < 1.0 \quad \textbf{o.k.}$$

3.5 BRACED FRAMES

LRFD	ASD
Tensile rupture on net section from AISC *Specification* Equation J4-2: $\phi R_n = \phi F_u A_e$ $= 0.75(58 \text{ ksi})$ $\times [12.0 \text{ in.} - 4(0.875 \text{ in.})](\tfrac{3}{8} \text{ in.})$ $= 139 \text{ kips} > 45.9 \text{ kips}$ **o.k.** Combined shear and normal block shear strengths from gusset-to-column check— using values from gusset-to-column single plate: $\left(\dfrac{47.2 \text{ kips}}{97.4 \text{ kips}}\right)^2 + \left(\dfrac{45.9 \text{ kips}}{136 \text{ kips}}\right)^2$ $= 0.349 < 1.0$ **o.k.**	Tensile rupture on net section from AISC *Specification* Equation J4-2: $\dfrac{R_n}{\Omega} = \dfrac{F_u A_e}{\Omega}$ $= (1/2.00)(58 \text{ ksi})$ $\times [12.0 \text{ in.} - 4(0.875 \text{ in.})](\tfrac{3}{8} \text{ in.})$ $= 92.4 \text{ kips} > 30.1 \text{ kips}$ **o.k.** Combined shear and normal block shear strengths from gusset-to-column check— using values from gusset-to-column single plate: $\left(\dfrac{31.0 \text{ kips}}{64.8 \text{ kips}}\right)^2 + \left(\dfrac{30.1 \text{ kips}}{91.0 \text{ kips}}\right)^2$ $= 0.338 < 1.0$ **o.k.**

Check block shear at beam web

With beam flange intact, only axial force will cause block shear.

LRFD	ASD
Block shear relative to normal load: $\phi R_n = \phi U_{bs} F_u A_{nt}$ $\quad + \min(\phi 0.60 F_y A_{gv}, \phi 0.60 F_u A_{nv})$ $\phi U_{bs} F_u A_{nt} = 0.75(1.0)(65 \text{ ksi})$ $\quad \times [9.00 \text{ in.} - 3(0.875 \text{ in.})]$ $\quad \times (0.300 \text{ in.})$ $\quad = 93.2 \text{ kips}$ $\phi 0.60 F_y A_{gv} = 2(0.75)(0.60)(50 \text{ ksi})$ $\quad \times (0.300 \text{ in.})(2.00 \text{ in.})$ $\quad = 27.0 \text{ kips}$	Block shear relative to normal load: $\dfrac{R_n}{\Omega} = \dfrac{U_{bs} F_u A_{nt}}{\Omega}$ $\quad + \min\left(\dfrac{0.60 F_y A_{gv}}{\Omega}, \dfrac{0.60 F_u A_{nv}}{\Omega}\right)$ $\dfrac{U_{bs} F_u A_{nt}}{\Omega} = (1/2.00)(1.0)(65 \text{ ksi})$ $\quad \times [9.00 \text{ in.} - 3(0.875 \text{ in.})]$ $\quad \times (0.300 \text{ in.})$ $\quad = 62.2 \text{ kips}$ $\dfrac{0.60 F_y A_{gv}}{\Omega} = 2(1/2.00)(0.60)(50 \text{ ksi})$ $\quad \times (0.300 \text{ in.})(2.00 \text{ in.})$ $\quad = 18.0 \text{ kips}$

LRFD	ASD
$\phi 0.60 F_u A_{nv} = 2(0.75)(0.60)(65 \text{ ksi})$ $\times [2.00 \text{ in.} - 0.5(0.875 \text{ in.})]$ $\times (0.300 \text{ in.})$ $= 27.4 \text{ kips}$	$\dfrac{0.60 F_u A_{nv}}{\Omega} = 2(1/2.00)(0.60)(65 \text{ ksi})$ $\times [2.00 \text{ in.} - 0.5(0.875 \text{ in.})]$ $\times (0.300 \text{ in.})$ $= 18.3 \text{ kips}$
$\phi R_n = 93.2 \text{ kips} + 27.0 \text{ kips}$ $= 120 \text{ kips} > 45.9 \text{ kips}$ **o.k.**	$\dfrac{R_n}{\Omega} = 62.2 \text{ kips} + 18.0 \text{ kips}$ $= 80.2 \text{ kips} > 30.1 \text{ kips}$ **o.k.**

Since the gusset-to-column and the beam-to-column single plates are treated as identical plates, several checks related to these can be combined.

Single plate to column weld design

The beam-to-column and gusset-to-column single plates will be treated as separate connections. Conservatively, each single plate will be assumed to be 12 in. long to maintain symmetry relative to the actual loads.

Consider only the portion of the single plate attached to the gusset, design the single plate to column weld. Treating the welds as a line:

$l_w = 12.0$ in.

$Z_w = \dfrac{(12.0 \text{ in.})^2}{4}$

$= 36.0$ in.^{3}/in.

The forces along the gusset-to-single plate interface are:

LRFD	ASD
$f_{uv} = \dfrac{V_{uc}}{l_w}$ $= \dfrac{42.7 \text{ kips}}{12.0 \text{ in.}}$ $= 3.56$ kip/in.	$f_{av} = \dfrac{V_{ac}}{l_w}$ $= \dfrac{28.0 \text{ kips}}{12.0 \text{ in.}}$ $= 2.33$ kip/in.
$f_{ua} = \dfrac{H_{uc}}{l_w}$ $= \dfrac{45.9 \text{ kips}}{12.0 \text{ in.}}$ $= 3.83$ kip/in.	$f_{aa} = \dfrac{H_{ac}}{l_w}$ $= \dfrac{30.1 \text{ kips}}{12.0 \text{ in.}}$ $= 2.51$ kip/in.

3.5 BRACED FRAMES

LRFD	ASD
$f_{ub} = \dfrac{M_{ucg}}{Z_w}$	$f_{ab} = \dfrac{M_{acg}}{Z_w}$
$= \dfrac{107 \text{ kip-in.}}{36.0 \text{ in.}^3/\text{in.}}$	$= \dfrac{70.0 \text{ kip-in.}}{36.0 \text{ in.}^3/\text{in.}}$
$= 2.97$ kip/in.	$= 1.94$ kip/in.
$f_{u,\,peak} = \sqrt{f_{uv}^2 + (f_{ua} + f_{ub})^2}$	$f_{a,\,peak} = \sqrt{f_{av}^2 + (f_{aa} + f_{ab})^2}$
$= \sqrt{3.56^2 + (3.83 + 2.97)^2}$	$= \sqrt{2.33^2 + (2.51 + 1.94)^2}$
$= 7.68$ kip/in.	$= 5.02$ kip/in.
$\theta = \tan^{-1}\left(\dfrac{f_{ua} + f_{ub}}{f_{uv}}\right)$	$\theta = \tan^{-1}\left(\dfrac{f_{aa} + f_{ab}}{f_{av}}\right)$
$= \tan^{-1}\left(\dfrac{3.83 \text{ kip/in.} + 2.97 \text{ kip/in.}}{3.56 \text{ kip/in.}}\right)$	$= \tan^{-1}\left(\dfrac{2.51 \text{ kip/in.} + 1.94 \text{ kip/in.}}{2.33 \text{ kip/in.}}\right)$
$= 62.4°$	$= 62.4°$
$D \geq \dfrac{7.68 \text{ kip/in.}}{2(1.392 \text{ kip/in.})\left[1 + 0.5\sin^{1.5}(62.4°)\right]}$	$D \geq \dfrac{5.02 \text{ kip/in.}}{2(0.928 \text{ kip/in.})\left[1 + 0.5\sin^{1.5}(62.4°)\right]}$
$= 1.95$ sixteenths	$= 1.91$ sixteenths

A $\tfrac{3}{16}$-in. fillet weld on both sides of the single plate is adequate.

Note: Since the bolts in the single plate will add ductility to the connection and also make this interface less rigid than the gusset-to-beam interface, the weld ductility factor applied to the gusset-to-beam interface need not be applied here.

Regarding the design of the weld to the single plate, from AISC *Specification* Table J2.4, the minimum size fillet weld allowed for the parts being connected is $\tfrac{3}{16}$ in. The AISC *Manual* Part 10 recommends developing the strength of the plate to ensure plastic yielding of the plate, instead of fracture in the fillet weld. A minimum fillet weld of $\tfrac{5}{8}$ times the plate thickness for both sides of the plate is needed to develop the plate strength. Since this requirement is intended to ensure that the simple beam end rotation can be accommodated in a ductile manner, it need not be applied to $R = 3$ bracing connections. Use a $\tfrac{3}{16}$-in. fillet weld.

The final connection design and geometry is shown in Figure 3-8.

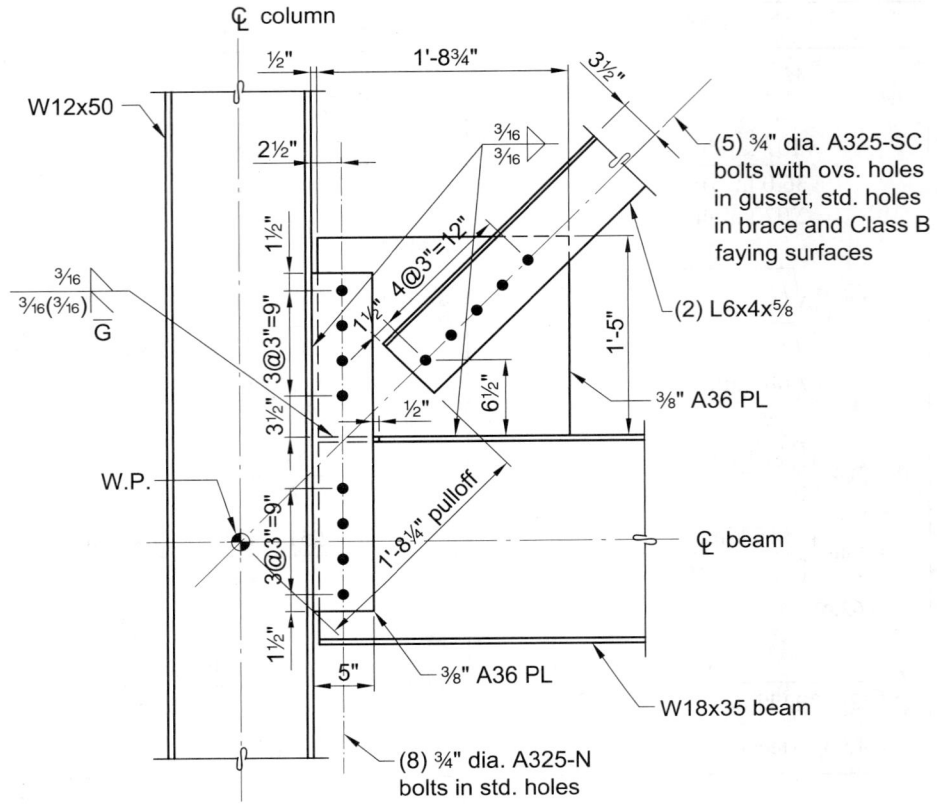

Fig. 3-8. Connection as designed in Example 3.5.3.

PART 3 REFERENCES

Carter, C.J. (1999), *Stiffening of Wide-Flange Column at Moment Connections: Wind and Seismic Applications*, Design Guide 13, AISC, Chicago, IL.

Dowswell, B. (2006), "Effective Length Factors for Gusset Plate Buckling," *Engineering Journal*, AISC, Vol. 43, No. 2, 2nd Quarter, pp. 91–101.

PART 4
MOMENT FRAMES

4.1 SCOPE 4–2
4.2 ORDINARY MOMENT FRAMES (OMF) 4–2
    OMF Design Example Plan and Elevation 4–3
    Example 4.2.1. OMF Story Drift and Stability Check 4–5
    Example 4.2.2. OMF Column Strength Check 4–7
    Example 4.2.3. OMF Beam Strength Check 4–12
    Example 4.2.4. OMF Beam-Column Connection Design 4–16
4.3 SPECIAL MOMENT FRAMES (SMF) AND INTERMEDIATE
    MOMENT FRAMES (IMF) 4–33
    SMF Design Example Plan and Elevation 4–36
    Example 4.3.1. SMF Story Drift and Stability Check 4–36
    Example 4.3.2. SMF Column Strength Check 4–42
    Example 4.3.3. SMF Beam Strength Check 4–46
    Example 4.3.4. SMF Beam-Column Connection Design 4–57
4.4 COLUMN SPLICE AND COLUMN BASE DESIGN EXAMPLES 4–76
    Example 4.4.1. Gravity Column Splice Design in a Moment
    Frame Building 4–76
    Example 4.4.2. SMF Column Splice Design 4–90
    Example 4.4.3. SMF Column Base Design 4–94
    Example 4.4.4. SMF Embedded Column Base Design 4–113
4.5 DESIGN TABLE DISCUSSION 4–120
    DESIGN TABLES 4–122
    Table 4-1. Comparison of Requirements for SMF, IMF and OMF 4–122
    Table 4-2. SMF Design Values 4–124
PART 4 REFERENCES 4–138

AMERICAN INSTITUTE OF STEEL CONSTRUCTION

4.1 SCOPE

The following types of moment frames are addressed in this Part: ordinary moment frame (OMF) systems, intermediate moment frame (IMF) systems, and special moment frame (SMF) systems. The AISC *Seismic Provisions* requirements and other design considerations summarized in this Part apply to the design of the members and connections in moment frames that require seismic detailing according to the AISC *Seismic Provisions*.

Moment frame systems resist lateral forces through the flexural and shear strengths of the beams and columns. Lateral displacement is resisted primarily through the flexural stiffness of the framing members and the restraint of relative rotation between the beams and columns at the connections, or "frame action." Moment frame systems tend to have larger and heavier beam and column sizes than in braced frame systems, because the beams and columns are often sized for drift control rather than strength. The increase in member sizes and related costs, however, may be acceptable because of the increased flexibility in the architectural and mechanical layout in the structure. The absence of diagonal bracing members can provide greater freedom in the configuration of walls and in the routing of mechanical ductwork and piping. On the other hand, the flexible nature of the frames does warrant some additional consideration of the interaction between the steel frame and more rigid architectural cladding systems. AISC Design Guide 3, *Serviceability Design Considerations for Steel Buildings* (West and Fisher, 2003), discusses recommended drift limits for various cladding systems.

4.2 ORDINARY MOMENT FRAMES (OMF)

The only system-specific requirements for an OMF pertain to the beam-to-column moment connections. The general intent of the OMF design provisions provided in AISC *Seismic Provisions* Section E1 is that connection failure should not be the first significant inelastic event in the response of the frame to earthquake loading, recognizing that a connection failure is typically one of the least ductile failure modes of a steel frame. Thus the basic design requirement is to provide a frame with strong moment connections. In accordance with AISC *Seismic Provisions* Section E1.6, two connection types are permitted when designing OMF systems—fully restrained (FR) and partially restrained (PR), as defined in AISC *Specification* Section B3.6b.

All FR connections in OMF systems must satisfy at least one of the following three options given in AISC *Seismic Provisions* Section E1.6b.

(a) FR moment connections are designed for a required flexural strength equal to the expected flexural strength of the beam multiplied by 1.1, as follows:

$$M_u = 1.1 R_y M_p \quad \text{(LRFD)} \tag{4-1a}$$

or

$$M_a = (1.1/1.5) R_y M_p \quad \text{(ASD)} \tag{4-1b}$$

where
R_y = ratio of the expected yield stress to the specified minimum yield stress, F_y
M_p = nominal plastic flexural strength of the beam

The required shear strength of the connection is determined using a shear force due to earthquake loads associated with the development of these expected flexural moments simultaneously at each end of the beam.

(b) FR moment connections are designed for a required flexural strength and shear strength equal to the maximum moment and corresponding shear that can be transferred to the connection by the system, including the effects of material overstrength and strain hardening. As discussed in AISC *Seismic Provisions* Commentary Section E1.6b, specific examples of potentially limiting aspects of the system include:

- Flexural yielding of the column when the flexural strength of the column is less than that of the beam
- The panel zone shear strength of the column, in recognition of the fact that testing has shown that panel zone shear yielding provides a fairly ductile response in this joint
- The foundation uplift
- The earthquake force determined using an R value of 1

(c) FR moment connections between wide flange beams and the flange of wide flange columns are designed according to the connection design requirements of the IMF (AISC *Seismic Provisions* Section E2.6) or SMF (AISC *Seismic Provisions* Section E3.6), or a connection is used that resembles the tested WUF-W or WUF-B connections that are included in ANSI/AISC 358. See AISC *Seismic Provisions* Section E1.6b(c) for detailed requirements.

As described in AISC *Seismic Provisions* Section E1.6c, PR moment connections are required to develop available strengths similar to those of FR moment connections. In addition, PR moment connections must have a nominal flexural strength no less than $0.50M_p$ of the connected beam (or $0.50M_p$ of the column for one-story structures). The strength and flexibility of the connection must be considered in the design, including the effect on overall frame stability.

OMF systems are not required to have any special detailing of the panel zones, and have no special requirements for the relationship between beam and column strength. This is indicative of the overall OMF system, where the detailing requirements are reduced and the seismic forces are larger than moment frame systems intended to provide higher ductility. This basic design philosophy for OMF systems allows for their use as an economical moment frame system when OMF systems are permitted by the applicable building code.

According to ASCE/SEI 7 Section 12.2.5.6, OMF frames are permitted to be used in Seismic Design Categories D, E and F for one-story structures under certain height and loading limitations.

OMF Design Example Plan and Elevation

The following section consists of four design examples for an OMF system. See Figure 4-1 for the roof plan and Figure 4-2 for the elevation of the building moment frames.

The code-specified gravity loading is as follows:

$D = 15$ psf
$S = 20$ psf

From ASCE/SEI 7, the following parameters apply: Risk Category II, Seismic Design Category D, $R = 3\frac{1}{2}$, $\Omega_o = 3$, $C_d = 3$, $I_e = 1.00$, $S_{DS} = 0.528$, and $\rho = 1.0$. According to ASCE/SEI 7 Section 12.3.4.2, $\rho = 1.0$ if each story resists more than 35% of the base shear in the direction of interest and loss of moment resistance at the beam-to-column connections at both ends of a beam will not result in more than a 33% reduction in story strength, nor does the resulting system have an extreme torsional irregularity. ρ is taken as 1.0 for this reason.

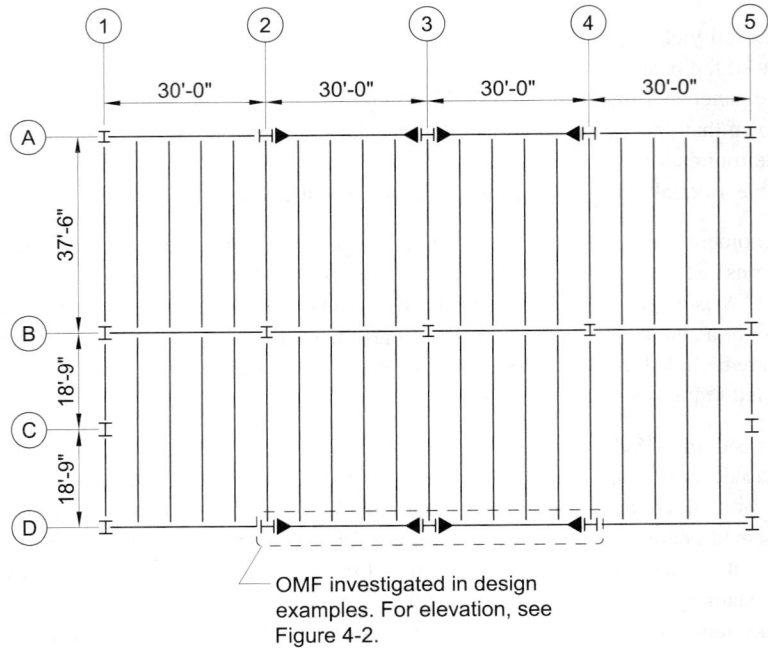

Fig. 4-1. OMF roof plan.

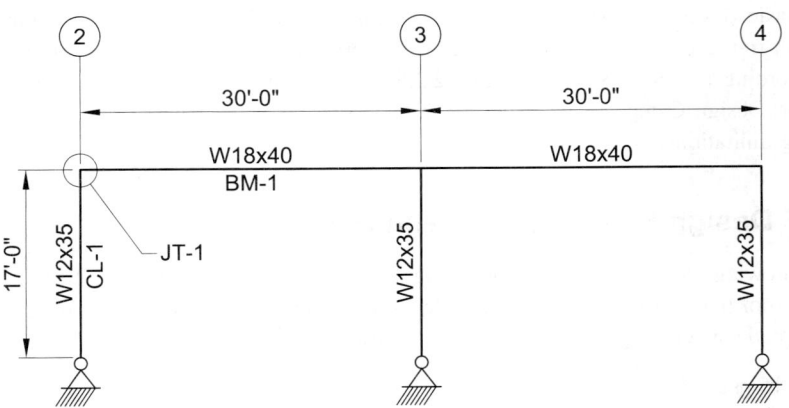

Fig. 4-2. OMF elevation.

Example 4.2.1. OMF Story Drift and Stability Check

Given:
Refer to the roof plan shown in Figure 4-1 and the OMF elevation shown in Figure 4-2. Determine if the frame satisfies the drift and stability requirements. The applicable building code specifies the use of ASCE/SEI 7 for calculation of loads. The loading and applicable ASCE parameters are as given previously.

The seismic design story shear, V_x, is 11.5 kips.

From an elastic analysis of the structure that includes second-order effects and accounts for panel-zone deformations, the elastic drift at the top of the story is:

$\delta_{te} = 0.980$ in.

At the base of the structure:

$\delta_{be} = 0$ in.

Solution:

Drift Check
Section 12.8.6 of ASCE/SEI 7 defines the design story drift, Δ, as the largest difference of the deflections of vertically aligned points at the top and bottom of the story under consideration along any of the edges of the structure. This calculated deflection includes the effects of elastic and inelastic drift, which in this example includes second-order effects. From ASCE/SEI 7 Equation 12.8-15:

$$\Delta = \frac{C_d(\delta_{te} - \delta_{be})}{I_e}$$
$$= \frac{3(0.980 \text{ in.} - 0 \text{ in.})}{1.00}$$
$$= 2.94 \text{ in.}$$

From ASCE/SEI 7 Table 12.12-1, the allowable story drift at level x, Δ_a, is $0.020h_{sx}$, where h_{sx} is the story height below level x. It is assumed in this example that Δ_a can be increased to $0.025h_{sx}$ because interior walls, partitions, ceilings, and exterior wall systems are designed to accommodate these increased story drifts. ASCE/SEI 7 Section 12.12.1.1 requires, for seismic force resisting systems comprised solely of moment frames in structures assigned to Seismic Design Category D, E or F, that the design story drift not exceed Δ_a/ρ for any story. Determine the allowable story drift as follows:

$$\frac{\Delta_a}{\rho} = \frac{0.025h_{sx}}{1.0}$$
$$= \frac{0.025(17.0 \text{ ft})(12 \text{ in./ft})}{1.0}$$
$$= 5.10 \text{ in.} > 2.94 \text{ in.} \quad \textbf{o.k.}$$

Frame Stability Check

ASCE/SEI 7 Section 12.8.7 investigates potential for instability by use of a stability coefficient, θ, calculated as:

$$\theta = \frac{P_x \Delta I_e}{V_x h_{sx} C_d} \quad \text{(ASCE/SEI 7 Eq. 12.8-16)}$$

where
- P_x = total vertical design load at and above level x
- Δ = design story drift occurring simultaneously with V_x
- I_e = seismic importance factor
- V_x = seismic design story shear acting between levels x and $x-1$
- h_{sx} = story height below level x
- C_d = deflection amplification factor

ASCE/SEI 7 does not explicitly specify load factors to be used on the gravity loads for determining P_x, except Section 12.8.7 does specify that no individual load factor need exceed 1.0. For this example, the load combination used to compute the total vertical load on a given story, P_x, acting simultaneously with the horizontal earthquake force, V_x, is $1.0D + 0.2S$, taken from ASCE/SEI 7 Section 2.3 with the dead load factor limited to 1.0 as explained. Note that consistent with this, the same combination was used in the second-order analysis as used for this example for the purpose of computing the fundamental period, base shear, and design story drift.

The total vertical design load is:

$$P_x = 120 \text{ ft}(75.0 \text{ ft})[1.0(15 \text{ psf}) + 0.2(20 \text{ psf})] / 1{,}000 \text{ lb/kip}$$
$$= 171 \text{ kips}$$

The stability coefficient, θ, from ASCE/SEI 7 Equation 12.8-16 is:

$$\theta = \frac{171 \text{ kips}(2.94 \text{ in.})(1.00)}{11.5 \text{ kips}(17.0 \text{ ft})(12 \text{ in./ft})(3)}$$
$$= 0.0714$$

Because a second-order analysis was used to compute the story drift, θ is adjusted as follows according to ASCE/SEI 7 Section 12.8.7 before checking θ_{max}.

$$\frac{\theta}{1+\theta} = \frac{0.0714}{1+0.0714}$$
$$= 0.0666$$

Per ASCE/SEI 7, if θ from a first-order analysis or $\theta/(1+\theta)$ from a second-order analysis is less than or equal to 0.10, second-order effects need not be considered for computing story drift. Note that whether or not second-order effects on member forces must be considered per ASCE/SEI 7 has to be verified, as it was in this example; however, Chapter C of the AISC *Specification* requires second-order effects be considered in all cases.

4.2 ORDINARY MOMENT FRAMES

Check the maximum permitted θ

The stability coefficient may not exceed θ_{max}. The ratio of shear demand to shear capacity for the story between levels x and $x-1$ is β. Conservatively, using a value of 1.0 for β:

$$\theta_{max} = \frac{0.5}{\beta C_d} \leq 0.25 \quad \text{(ASCE/SEI 7 Eq. 12.8-17)}$$

$$= \frac{0.5}{1.0(3)}$$

$$= 0.167 \leq 0.25$$

The adjusted stability coefficient satisfies the maximum:

$0.0666 < 0.167$ **o.k.**

The moment frame meets the allowable story drift and stability requirements for seismic loading.

Example 4.2.2. OMF Column Strength Check

Given:

Refer to Column CL-1 in Figure 4-2. Determine the adequacy of the ASTM A992 W12×35 column for the following loading. The required strength of columns should be determined in accordance with AISC *Seismic Provisions* Section D1.4a. The applicable building code specifies the use of ASCE/SEI 7 for calculation of loads.

The governing load combinations that include seismic effects are:

LRFD	ASD
LRFD Load Combination 5 from ASCE/SEI 7 Section 12.4.2.3 $(1.2+0.2S_{DS})D+\rho Q_E +0.5L+0.2S$ (including the permitted 0.5 factor L in Section 12.4.2.3)	ASD Load Combination 6 from ASCE/SEI 7 Section 12.4.2.3 $(1.0+0.10S_{DS})D+0.525\rho Q_E$ $+H+F+0.75L$ $+0.75S$

From a second-order analysis including the effects of P-Δ and P-δ effects as well as the reduced stiffness required by the direct analysis method, the column required strengths are:

LRFD	ASD
$P_u = 15.2$ kips $V_u = 3.78$ kips $M_{u\ top} = 64.3$ kip-ft $M_{u\ bot} = 0$ kip-ft	$P_a = 17.5$ kips $V_a = 2.57$ kips $M_{a\ top} = 43.7$ kip-ft $M_{a\ bot} = 0$ kip-ft

The higher ASD required axial strength compared to LRFD could be explained by the higher load factor on snow load, S, of 0.75 for ASD versus 0.2 for LRFD.

According to ASCE/SEI 7, the load combinations including amplified seismic loads (including overstrength factor, Ω_o) are:

LRFD	ASD
LRFD Load Combinations 5 and 7 from ASCE/SEI 7 Section 12.4.3.2	ASD Load Combinations 5, 6 and 8 from ASCE/SEI 7 Section 12.4.3.2
$(1.2+0.2S_{DS})D+\Omega_o Q_E +0.5L+0.2S$	$(1.0+0.14S_{DS})D+H+F+0.7\Omega_o Q_E$
$(0.9-0.2S_{DS})D+\Omega_o Q_E +1.6H$	$(1.0+0.105S_{DS})D+H+F+0.525\Omega_o Q_E$
(including the 0.5 factor on L permitted by Section 12.4.3.2)	$+0.75L+0.75S$
	$(0.6-0.14S_{DS})D+0.7\Omega_o Q_E +H$

From the frame analysis, the maximum required axial strength in this column from the governing load combination that includes the amplified seismic load is:

LRFD	ASD
$P_u = 21.0$ kips	$P_a = 20.6$ kips

There are no transverse loadings between the column supports in the plane of bending and the columns are considered to be pinned at the base.

Solution:

From AISC *Manual* Table 2-4, the material properties are as follows:

ASTM A992
$F_y = 50$ ksi
$F_u = 65$ ksi

From AISC *Manual* Table 1-1, the geometric properties are as follows:

W12×35
$r_x = 5.25$ in. $r_y = 1.54$ in.

Section E1.5a of the AISC *Seismic Provisions* states that there are no limitations on width-to-thickness ratios of members of an OMF, beyond those in the AISC *Specification*.

Available Flexural Strength

Per the User Note in AISC *Specification* Section F2, the column has compact flanges and web. The available flexural strength is the lower value obtained according to the limit states of lateral-torsional buckling and yielding.

With no interior brace points, the unbraced column length is $L_b = 17.0$ ft.

4.2 ORDINARY MOMENT FRAMES

From AISC *Manual* Table 3-2:

$L_p = 5.44$ ft
$L_r = 16.6$ ft

$L_b > L_p$; therefore, the limit state of lateral-torsional buckling applies.

Calculate C_b using AISC *Specification* Equation F1-1.

LRFD	ASD
$M_{u\ top} = 64.3$ kip-ft $M_{u\ bot} = 0$ kip-ft	$M_{a\ top} = 43.7$ kip-ft $M_{a\ bot} = 0$ kip-ft
Quarter point moments are: $$\begin{aligned} M(x = 4.25 \text{ ft}) &= M_A \\ &= 0.25(64.3 \text{ kip-ft}) \\ &= 16.1 \text{ kip-ft} \\ M(x = 8.50 \text{ ft}) &= M_B \\ &= 0.50(64.3 \text{ kip-ft}) \\ &= 32.2 \text{ kip-ft} \\ M(x = 12.75 \text{ ft}) &= M_C \\ &= 0.75(64.3 \text{ kip-ft}) \\ &= 48.2 \text{ kip-ft} \\ M_{max} &= 64.3 \text{ kip-ft} \end{aligned}$$ $$\begin{aligned} C_b &= \frac{12.5 M_{max}}{\left(\begin{array}{c} 2.5 M_{max} + 3 M_A \\ + 4 M_B + 3 M_C \end{array}\right)} \\ &= \frac{12.5(64.3 \text{ kip-ft})}{\left[\begin{array}{c} 2.5(64.3 \text{ kip-ft}) + 3(16.1 \text{ kip-ft}) \\ + 4(32.2 \text{ kip-ft}) + 3(48.2 \text{ kip-ft}) \end{array}\right]} \\ &= 1.67 \end{aligned}$$	Quarter point moments are: $$\begin{aligned} M(x = 4.25 \text{ ft}) &= M_A \\ &= 0.25(43.7 \text{ kip-ft}) \\ &= 10.9 \text{ kip-ft} \\ M(x = 8.50 \text{ ft}) &= M_B \\ &= 0.50(43.7 \text{ kip-ft}) \\ &= 21.9 \text{ kip-ft} \\ M(x = 12.75 \text{ ft}) &= M_C \\ &= 0.75(43.7 \text{ kip-ft}) \\ &= 32.8 \text{ kip-ft} \\ M_{max} &= 43.7 \text{ kip-ft} \end{aligned}$$ $$\begin{aligned} C_b &= \frac{12.5 M_{max}}{\left(\begin{array}{c} 2.5 M_{max} + 3 M_A \\ + 4 M_B + 3 M_C \end{array}\right)} \\ &= \frac{12.5(43.7 \text{ kip-ft})}{\left[\begin{array}{c} 2.5(43.7 \text{ kip-ft}) + 3(10.9 \text{ kip-ft}) \\ + 4(21.9 \text{ kip-ft}) + 3(32.8 \text{ kip-ft}) \end{array}\right]} \\ &= 1.67 \end{aligned}$$

Check lateral-torsional buckling using AISC *Manual* Table 6-1 with $L_b = 17.0$ ft and the modification to b_x for when $C_b > 1.0$ (AISC *Manual* Equation 6-5).

LRFD	ASD
$b_x = 0.00766 \text{ (kip-ft)}^{-1}$	$b_x = 0.0115 \text{ (kip-ft)}^{-1}$
$\phi_b M_n = C_b \left(\dfrac{8}{9}\right)\left(\dfrac{1}{b_x}\right)$	$\dfrac{M_n}{\Omega_b} = C_b \left(\dfrac{8}{9}\right)\left(\dfrac{1}{b_x}\right)$
$= 1.67 \left(\dfrac{8}{9}\right)\left(\dfrac{1}{0.00766 (\text{kip-ft})^{-1}}\right)$	$= 1.67 \left(\dfrac{8}{9}\right)\left(\dfrac{1}{0.0115 (\text{kip-ft})^{-1}}\right)$
$= 194 \text{ kip-ft}$	$= 129 \text{ kip-ft}$

Check yielding using AISC *Manual* Table 3-2.

LRFD	ASD
$\phi_b M_n = \phi_b M_p$	$\dfrac{M_n}{\Omega_b} = \dfrac{M_p}{\Omega_b}$
$= 192 \text{ kip-ft} < 194 \text{ kip-ft}$	$= 128 \text{ kip-ft} < 129 \text{ kip-ft}$
Use $\phi_b M_n = 192 \text{ kip-ft}$	Use $\dfrac{M_n}{\Omega_b} = 128 \text{ kip-ft}$

Available Axial Compressive Strength

The unbraced length of the column for buckling about both the strong and weak axis is 17.0 ft. The column has slender elements according to AISC *Manual* Table 1-1.

The direct analysis method described in Section C of the AISC *Specification* states that the effective length factor, K, of all members shall be taken as unity unless a smaller value can be justified by rational analysis.

Therefore:

$K_x = 1.0$

$K_y = 1.0$

The slenderness ratios about the strong and weak axis are:

$$\dfrac{K_x L_x}{r_x} = \dfrac{1.0(17.0 \text{ ft})(12.0 \text{ in./ft})}{5.25 \text{ in.}}$$

$$= 38.9$$

$$\dfrac{K_y L_y}{r_y} = \dfrac{1.0(17.0 \text{ ft})(12.0 \text{ in./ft})}{1.54 \text{ in.}}$$

$$= 132 \quad \textbf{governs}$$

4.2 ORDINARY MOMENT FRAMES

Because the W12×35 is not included in AISC *Manual* Table 4-1, use AISC *Manual* Table 6-1 to determine the available compressive strength, with $KL_y = 17.0$ ft:

LRFD	ASD
$p = 0.00754 \text{ kips}^{-1}$ $$\phi_c P_n = \frac{1}{p}$$ $$= \frac{1}{0.00754 \text{ kips}^{-1}}$$ $= 133 \text{ kips}$	$p = 0.0113 \text{ kips}^{-1}$ $$\frac{P_n}{\Omega_c} = \frac{1}{p}$$ $$= \frac{1}{0.0113 \text{ kips}^{-1}}$$ $= 88.5 \text{ kips}$

Combined Loading

Using AISC *Specification* Section H1, determine whether the applicable interaction equation is satisfied, as follows:

LRFD	ASD
$$\frac{P_r}{P_c} = \frac{15.2 \text{ kips}}{133 \text{ kips}}$$ $= 0.114$	$$\frac{P_r}{P_c} = \frac{17.5 \text{ kips}}{88.5 \text{ kips}}$$ $= 0.198$
Because $P_r/P_c < 0.2$, use AISC *Specification* Equation H1-1b:	Because $P_r/P_c < 0.2$, use AISC *Specification* Equation H1-1b:
$$\frac{P_r}{2P_c} + \left(\frac{M_{rx}}{M_{cx}} + \frac{M_{ry}}{M_{cy}}\right) \leq 1.0$$ $$\frac{0.114}{2} + \left(\frac{64.3 \text{ kip-ft}}{192 \text{ kip-ft}} + 0\right) = 0.392$$ $0.392 < 1.0$ **o.k.**	$$\frac{P_r}{2P_c} + \left(\frac{M_{rx}}{M_{cx}} + \frac{M_{ry}}{M_{cy}}\right) \leq 1.0$$ $$\frac{0.198}{2} + \left(\frac{43.7 \text{ kip-ft}}{128 \text{ kip-ft}} + 0\right) = 0.440$$ $0.440 < 1.0$ **o.k.**

Alternatively, Section H1.3 of the AISC *Specification* may be used for the interaction check for this column since the column is only subject to bending about a single axis. The interaction equations in Section H1.3 would result in a higher column strength than demonstrated by this procedure.

Required Axial Strength of Column Including Amplified Seismic Loads

Determine the required axial compressive strength using load combinations including amplified seismic loads per Section D1.4a(2) of the AISC *Seismic Provisions*.

Per Section D.1.4a(2) of the AISC *Seismic Provisions*, it is permitted to neglect moments in the column for determination of required strength because the column moments do not result from loads applied between points of lateral support.

LRFD	ASD
$\phi_c P_n = 133$ kips $> P_u = 21.0$ kips	$\dfrac{P_n}{\Omega_c} = 88.5$ kips $> P_a = 20.6$ kips

Available Shear Strength

Using AISC *Manual* Table 3-2, the available shear strength for a W12×35 is:

LRFD	ASD
$\phi_v V_{nx} = 113$ kips $> V_u = 3.78$ kips **o.k.**	$\dfrac{V_{nx}}{\Omega_v} = 75.0$ kips $> V_a = 2.57$ kips **o.k.**

The W12×35 is adequate to resist the required strengths given for Column CL-1.

Example 4.2.3. OMF Beam Strength Check

Given:

Refer to Beam BM-1 in Figure 4-2. Determine the adequacy of the ASTM A992 W18×40 for the following loading. The applicable building code specifies the use of ASCE/SEI 7 for calculation of loads. The governing load combinations which include seismic effects are:

LRFD	ASD
LRFD Load Combination 5 from ASCE/SEI 7 Section 12.4.2.3 $(1.2 + 0.2 S_{DS})D + \rho Q_E + 0.5L + 0.2S$ (including the 0.5 factor on L permitted by Section 12.4.2.3)	ASD Load Combination 6 from ASCE/SEI 7 Section 12.4.2.3 $(1.0 + 0.10 S_{DS})D + H + F + 0.525 \rho Q_E + 0.75L + 0.75S$

From a second-order analysis considering P-Δ and P-δ effects as well as the reduced stiffness required by the direct analysis method, the beam required strengths are:

LRFD	ASD
$P_u = 2.54$ kips $M_u = 82.9$ kip-ft $V_u = 10.9$ kips	$P_a = 0.784$ kips $M_a = 78.0$ kip-ft $V_a = 11.8$ kips

The top and bottom beam flanges are braced every 6 ft by infill beams.

4.2 ORDINARY MOMENT FRAMES

Solution:

From AISC *Manual* Table 2-4, the material properties are as follows:

ASTM A992
F_y = 50 ksi
F_u = 65 ksi

From AISC *Manual* Table 1-1, the geometric properties are as follows:

Beam
W18×40
r_x = 7.21 in. r_y = 1.27 in.

AISC *Seismic Provisions* Section E1.5a states that there are no limitations on width-to-thickness ratios of members of an OMF, beyond those in the AISC *Specification*.

AISC *Seismic Provisions* Section E1.5a also states that there are no requirements for stability bracing of beams or joints in OMF, beyond those in the AISC *Specification*.

Available Flexural Strength

Per the User Note in AISC *Specification* Section F2, the beam has compact flanges and web. The available flexural strength is the lower value obtained according to the limit states of lateral-torsional buckling and yielding.

Note: The infill beams or joists are not described in this example. It is presumed that the combination of these members (with suitable connections) and a roof deck diaphragm will provide an adequate lateral brace for the top flange of this beam. With appropriate detailing, the bottom flange of the beam could also be braced by the infill beams or joists. This is assumed to be the case in this example.

The unbraced beam length is:

L_b (top flange in compression) = 6.00 ft (spacing of infill beams)
L_b (bottom flange in compression) = 6.00 ft

From Manual Table 3-2 for a W18×40:

L_p = 4.49 ft L_r = 13.1 ft

$L_b > L_p$; therefore, the limit state of lateral-torsional buckling applies.

Calculate C_b using AISC *Specification* Equation F1-1.

LRFD	ASD
M_{max} = 82.9 kip-ft M_A = 67.0 kip-ft M_B = 52.2 kip-ft M_C = 38.3 kip-ft	M_{max} = 78.0 kip-ft M_A = 61.0 kip-ft M_B = 45.3 kip-ft M_C = 30.9 kip-ft

LRFD	ASD
$C_b = \dfrac{12.5 M_{max}}{\begin{pmatrix} 2.5 M_{max} + 3 M_A \\ + 4 M_B + 3 M_C \end{pmatrix}}$ $= \dfrac{12.5(82.9 \text{ kip-ft})}{\begin{bmatrix} 2.5(82.9 \text{ kip-ft}) + 3(67.0 \text{ kip-ft}) \\ + 4(52.2 \text{ kip-ft}) + 3(38.3 \text{ kip-ft}) \end{bmatrix}}$ $= 1.42$	$C_b = \dfrac{12.5 M_{max}}{\begin{pmatrix} 2.5 M_{max} + 3 M_A \\ + 4 M_B + 3 M_C \end{pmatrix}}$ $= \dfrac{12.5(78.0 \text{ kip-ft})}{\begin{bmatrix} 2.5(78.0 \text{ kip-ft}) + 3(61.0 \text{ kip-ft}) \\ + 4(45.3 \text{ kip-ft}) + 3(30.9 \text{ kip-ft}) \end{bmatrix}}$ $= 1.50$

Compute the lateral-torsional buckling strength using AISC *Manual* Table 3-10 with $L_b = 6.00$ ft:

LRFD	ASD
$\phi_b M_n = C_b (274 \text{ kip-ft})$ $= 1.42(274 \text{ kip-ft})$ $= 389 \text{ kip-ft}$	$\dfrac{M_n}{\Omega_b} = C_b (183 \text{ kip-ft})$ $= 1.50(183 \text{ kip-ft})$ $= 275 \text{ kip-ft}$

Check yielding using AISC *Manual* Table 3-2:

LRFD	ASD
$\phi_b M_n = \phi_b M_{px}$ $= 294 \text{ kip-ft} < 389 \text{ kip-ft}$ Use $\phi_b M_n = 294$ kip-ft	$\dfrac{M_n}{\Omega_b} = \dfrac{M_{px}}{\Omega_b}$ $= 196 \text{ kip-ft} < 275 \text{ kip-ft}$ Use $\dfrac{M_n}{\Omega_b} = 196$ kip-ft

Available Axial Compressive Strength

The infill beams provide bracing in the beam's weak axis and the unbraced length, L_y, is 6.00 ft. The beam is not braced in the strong axis.

$$\dfrac{K_x L_x}{r_x} = \dfrac{1.0(30.0 \text{ ft})(12.0 \text{ in./ft})}{7.21 \text{ in.}}$$
$$= 49.9$$

$$\dfrac{K_y L_y}{r_y} = \dfrac{1.0(6.00 \text{ ft})(12.0 \text{ in./ft})}{1.27 \text{ in.}}$$
$$= 56.7 \quad \textbf{governs}$$

4.2 ORDINARY MOMENT FRAMES

Using the AISC *Manual* Table 6-1 with an unbraced length of 6.00 ft in the weak axis, the available axial strength is:

LRFD	ASD
$p = 0.00252 \text{ kips}^{-1}$ $\phi_c P_n = \dfrac{1}{p}$ $= \dfrac{1}{0.00252 \text{ kips}^{-1}}$ $= 397 \text{ kips}$	$p = 0.00379 \text{ kips}^{-1}$ $\dfrac{P_n}{\Omega_c} = \dfrac{1}{p}$ $= \dfrac{1}{0.00379 \text{ kips}^{-1}}$ $= 264 \text{ kips}$

Combined Loading

Using AISC *Specification* H1, determine whether the applicable interaction equation is satisfied, as follows:

LRFD	ASD
$\dfrac{P_r}{P_c} = \dfrac{2.54 \text{ kips}}{397 \text{ kips}}$ $= 0.00640$ Because $P_r/P_c < 0.2$, use AISC *Specification* Equation H1-1b: $\dfrac{P_r}{2P_c} + \left(\dfrac{M_{rx}}{M_{cx}} + \dfrac{M_{ry}}{M_{cy}}\right) \leq 1.0$ $\dfrac{0.00640}{2} + \left(\dfrac{82.9 \text{ kip-ft}}{294 \text{ kip-ft}} + 0\right) = 0.285$ $0.285 < 1.0$ **o.k.**	$\dfrac{P_r}{P_c} = \dfrac{0.784 \text{ kips}}{264 \text{ kips}}$ $= 0.00297$ Because $P_r/P_c < 0.2$, use AISC *Specification* Equation H1-1b: $\dfrac{P_r}{2P_c} + \left(\dfrac{M_{rx}}{M_{cx}} + \dfrac{M_{ry}}{M_{cy}}\right) \leq 1.0$ $\dfrac{0.00297}{2} + \left(\dfrac{78.0 \text{ kip-ft}}{196 \text{ kip-ft}} + 0\right) = 0.399$ $0.399 < 1.0$ **o.k.**

Available Shear Strength of Beam

From AISC *Manual* Table 3-2, the available shear strength for a W18×40 is:

LRFD	ASD
$\phi_v V_n = 169 \text{ kips} > V_u = 10.9 \text{ kips}$ **o.k.**	$\dfrac{V_n}{\Omega_v} = 113 \text{ kips} > V_a = 11.8 \text{ kips}$ **o.k.**

The W18×40 is adequate to resist the required strengths given for Beam BM-1.

Note that load combinations that do not include seismic effects must also be investigated.

Example 4.2.4. OMF Beam-Column Connection Design

Given:
Refer to Joint JT-1 in Figure 4-2. Design a fully restrained (FR) moment connection for the configuration shown in Figure 4-3. The beam and column are ASTM A992 W-shapes and the plate material is ASTM A36. Use 70-ksi electrodes and ASTM A325 bolts.

To avoid the field welding requirements associated with the prescriptive connection described in AISC *Seismic Provisions* Section E1.6b(c), an eight-bolt stiffened end-plate connection is used.

The required shear strengths for the column based on a second-order analysis are given in Example 4.2.2. The other shear forces acting at the beam end simultaneously with E_{mh} are:

$V_D = 4.86$ kips
$V_S = 6.49$ kips
$V_{E_v} = 0.2S_{DS}D$
$\quad\quad = 0.2(0.528)\,(4.86 \text{ kips})$
$\quad\quad = 0.513$ kips

Solution:
From AISC *Manual* Table 2-4, the material properties are as follows:

ASTM A992
$F_y = 50$ ksi
$F_u = 65$ ksi

From AISC *Manual* Table 1-1, the geometric properties are as follows:

Column
W12×35
$A = 10.3$ in.^{2} $\quad d = 12.5$ in. $\quad t_w = 0.300$ in. $\quad b_f = 6.56$ in. $\quad t_f = 0.520$ in.
$k_{des} = 0.820$ in. $\quad k_1 = \frac{3}{4}$ in. $\quad Z_x = 51.2$ in.^{3} $\quad h/t_w = 36.2$

$h = d - 2k_{des}$
$\quad = 12.5$ in. $- 2(0.820$ in.$)$
$\quad = 10.9$ in.

Beam
W18×40
$d = 17.9$ in. $\quad t_w = 0.315$ in. $\quad b_f = 6.02$ in. $\quad t_f = 0.525$ in. $\quad Z_x = 78.4$ in.^{3}

Determine the appropriate force and flexural strength levels for the design of this connection detail according to AISC *Seismic Provisions* Section E1.6b(b). This section stipulates that the connection design should be based on the maximum moment that can be transferred to the connection by the system, including the effects of material overstrength and strain

4.2 ORDINARY MOMENT FRAMES

hardening. In this example, the flexural strength that can be transferred is based on the smaller of the expected flexural strength, including 1.1, of the beam or column, or the flexural strength resulting from panel zone shear. The Commentary to this section of the AISC *Seismic Provisions* notes that column yielding and panel zone shear strength are two factors that could limit the forces developed by the system.

For the W18×40 beam, with $R_y = 1.1$ from AISC *Seismic Provisions* Table A3.1 for ASTM A992 material, the expected flexural strength is:

$$M_{p,\,exp} = 1.1 R_y M_p$$
$$= 1.1(1.1)(50 \text{ ksi})(78.4 \text{ in.}^3)$$
$$= 4,740 \text{ kip-in.}$$

The column flexural strength, accounting for overstrength and strain hardening, is equal to $1.1 R_y M_p$. For the W12×35 column, with $R_y = 1.1$, the expected flexural strength is:

$$M_{p,\,exp} = 1.1 R_y M_p$$
$$= 1.1(1.1)(50 \text{ ksi})(51.2 \text{ in.}^3)$$
$$= 3,100 \text{ kip-in.}$$

The column panel zone shear strength is evaluated using AISC *Specification* Section J10.6. Panel zone deformations were included in the analysis of the structure. Using required strengths from Example 4.2.2, check the limit given in Section J10.6 to determine the applicable equation, as follows:

LRFD	ASD
$P_r = 15.2$ kips $$\frac{P_r}{P_c} = \frac{P_r}{P_y}$$ $$= \frac{15.2 \text{ kips}}{50 \text{ ksi}(10.3 \text{ in.}^2)}$$ $$= 0.0295 < 0.75$$	$P_r = 17.5$ kips $$\frac{P_r}{P_c} = \frac{P_r}{0.60 P_y}$$ $$= \frac{17.5 \text{ kips}}{0.60(50 \text{ ksi})(10.3 \text{ in.}^2)}$$ $$= 0.0566 < 0.75$$

Therefore, use AISC *Specification* Equation J10-11 to calculate the panel zone yielding strength, as follows:

$$R_n = 0.60 F_y d_c t_w \left(1 + \frac{3 b_{cf} t_{cf}^2}{d_b d_c t_w}\right) \qquad (\textit{Spec. Eq. J10-11})$$

Including a strain hardening factor of 1.1 and R_y as recommended in AISC *Seismic Provisions* Commentary Section E1.6b(b), the force transferred to the connection due to panel zone yielding is:

$$R_{ne} = 0.60(1.1)R_y F_y d_c t_w \left(1 + \frac{3b_{cf} t_{cf}^2}{d_b d_c t_w}\right)$$

$$= 0.60(1.1)(1.1)(50 \text{ ksi})(12.5 \text{ in.})(0.300 \text{ in.})$$

$$\times \left[1 + \frac{3(6.56 \text{ in.})(0.520 \text{ in.})^2}{17.9 \text{ in.}(12.5 \text{ in.})(0.300 \text{ in.})}\right]$$

$$= 147 \text{ kips}$$

LRFD	ASD
$V_{ue} = R_{ne}$ $= 147 \text{ kips}$	$V_{ae} = R_{ne}/1.5$ $= 147 \text{ kips}/1.5$ $= 98.0 \text{ kips}$

The required flexural strength is:

LRFD	ASD
$M_{ue} = V_{ue}(d_b - t_f)$ $= 147 \text{ kips}(17.9 \text{ in.} - 0.525 \text{ in.})$ $= 2,550 \text{ kip-in.}$	$M_{ae} = V_{ae}(d_b - t_f)$ $= 98.0 \text{ kips}(17.9 \text{ in.} - 0.525 \text{ in.})$ $= 1,700 \text{ kip-in.}$

There is also shear in the column due to story shear.

LRFD	ASD
$V_{uc} = \dfrac{M_{ue}}{H}$ $= \dfrac{2,550 \text{ kip-in.}}{(17.0 \text{ ft})(12 \text{ in./ft})}$ $= 12.5 \text{ kips}$	$V_{ac} = \dfrac{M_{ae}}{H}$ $= \dfrac{1,700 \text{ kip-in.}}{(17.0 \text{ ft})(12 \text{ in./ft})}$ $= 8.33 \text{ kips}$

This shear should be added to the panel zone strength to recalculate the required flexural strength, as follows:

LRFD	ASD
$M_{ue} = (V_{ue} + V_{uc})(d_b - t_f)$ $= (147 \text{ kips} + 12.5 \text{ kips})$ $\times (17.9 \text{ in.} - 0.525 \text{ in.})$ $= 2,770 \text{ kip-in.}$	$M_{ae} = (V_{ae} + V_{ac})(d_b - t_f)$ $= (98.0 \text{ kips} + 8.33 \text{ kips})$ $\times (17.9 \text{ in.} - 0.525 \text{ in.})$ $= 1,850 \text{ kip-in.}$

The process could be iterated until the shear and moment values converge, but the difference is negligible between the initial calculation and the convergence. For simplicity, use this value as the required flexural strength.

Therefore, the column panel zone shear strength controls the maximum force that can be delivered by the system to the connection, in accordance with AISC *Seismic Provisions* Section E1.6b(b) and Commentary Section E1.6b(b).

Calculate the corresponding shear for the beam-to-column connection design using AISC *Seismic Provisions* Section E1.6b(b). The required shear strength of the connection is based on the load combinations in the applicable building code that include the amplified seismic load. In determining the amplified seismic load, the effect of horizontal forces including overstrength, E_{mh}, is determined from:

$$E_{mh} = 2[1.1R_y M_p]/L_{cf} \qquad \text{(\textit{Provisions} Eq. E1-1)}$$

where

L_{cf} = clear length of the beam
= 30.0 ft (12 in./ft) − 12.5 in.
= 348 in.

Because AISC *Seismic Provisions* Section E1.6b(b) is used, the term $1.1R_y M_p$ is substituted with M_{ue} (LRFD) or M_{ae} (ASD) based on the panel zone strength as calculated.

The shear in the column is:

LRFD	ASD
V due to $E_{mh} = \dfrac{2M_{ue}}{L_{cf}}$ $= \dfrac{2(2{,}770 \text{ kip-in.})}{348 \text{ in.}}$ $= 15.9 \text{ kips}$	V due to $E_{mh} = \dfrac{2M_{ae}}{L_{cf}}$ $= \dfrac{2(1{,}850 \text{ kip-in.})}{348 \text{ in.}}$ $= 10.6 \text{ kips}$

The controlling load combinations from ASCE/SEI 7 are:

LRFD	ASD
Load Combination 5 from Section 12.4.3.2, including the 0.5 factor on L permitted for certain occupancies and $\Omega_o Q_E = E_{mh}$ $$\begin{aligned}V_u &= (1.2+0.2S_{DS})D + E_{mh} + 0.5L + 0.2S \\ &= [1.2+0.2(0.528)](4.86 \text{ kips}) \\ &\quad + 15.9 \text{ kips} + 0 \text{ kips} + 0.2(6.49 \text{ kips}) \\ &= 23.5 \text{ kips}\end{aligned}$$	Load Combination 6 from Section 12.4.3.2 with $\Omega_o Q_E = E_{mh}$ $$\begin{aligned}V_a &= (1.0+0.105S_{DS})D + H + F \\ &\quad + 0.525E_{mh} + 0.75L + 0.75S \\ &= [1.0+0.105(0.528)](4.86 \text{ kips}) \\ &\quad + 0 \text{ kips} \\ &\quad + 0 \text{ kips} + 0.525(10.6 \text{ kips}) \\ &\quad + 0 \text{ kips} + 0.75(6.49 \text{ kips}) \\ &= 15.6 \text{ kips}\end{aligned}$$

End Plate Design

The design methodology used for the moment end-plate connections is taken from AISC Design Guide 4, *Extended End-Plate Moment Connections—Seismic and Wind Applications* (Murray and Sumner, 2003). ANSI/AISC 358 outlines requirements and design methodology for prequalified moment end-plate connections for special and intermediate moment frames. However, for an ordinary moment frame, the basic design equations and methodology described in AISC Design Guide 4 can be used. Note that Design Guide 4 includes only the LRFD method and the equations are modified here for ASD.

Based upon preliminary calculations, it was determined that an eight-bolt stiffened end-plate connection would be required to make the column flange work in bending.

Figure 4-3 illustrates the configuration and key dimensions associated with this type of connection.

Determine the required bolt diameter, $d_{b,reqd}$, from AISC Design Guide 4 Equation 3.6 using the bolt spacing provided in Figure 4-4 and ASTM A325-N bolts, as follows:

LRFD	ASD
$$d_{b,\,reqd} = \sqrt{\dfrac{2M_{ue}}{\pi \phi F_{nt}(\Sigma d_n)}}$$ $$= \sqrt{\dfrac{2(2{,}770 \text{ kip-in.})}{\begin{bmatrix}\pi(0.75)(90 \text{ ksi}) \\ \times \begin{pmatrix}22.6 \text{ in.}+19.6 \text{ in.} \\ +15.1 \text{ in.}+12.1 \text{ in.}\end{pmatrix}\end{bmatrix}}}$$ $$= 0.614 \text{ in.}$$	$$d_{b,\,reqd} = \sqrt{\dfrac{2\Omega M_{ae}}{\pi F_{nt}(\Sigma d_n)}}$$ $$= \sqrt{\dfrac{2(2.00)(1{,}850 \text{ kip-in.})}{\begin{bmatrix}\pi(90 \text{ ksi}) \\ \times \begin{pmatrix}22.6 \text{ in.}+19.6 \text{ in.} \\ +15.1 \text{ in.}+12.1 \text{ in.}\end{pmatrix}\end{bmatrix}}}$$ $$= 0.614 \text{ in.}$$

4.2 ORDINARY MOMENT FRAMES

The value of F_{nt}, the nominal tensile strength of the bolt, is from AISC *Specification* Table J3.2 and Σd_n is the sum of h_1 through h_4.

Use ¾-in.-diameter ASTM A325-N bolts in standard holes.

Calculate M_{np} based on the ¾-in.-diameter A325-N bolt strength with $A_b = 0.442$ in.² from AISC *Manual* Table 7-1, as follows:

$P_t = F_{nt} A_b$

$= 90$ ksi $(0.442$ in.²$)$

$= 39.8$ kips

From AISC Design Guide 4 Equation 3.8, the flexural design strength of the connection is:

LRFD	ASD
$\phi M_{np} = \phi\left[2P_t(\Sigma d_n)\right]$ $= 0.75 \begin{bmatrix} 2(39.8 \text{ kips}) \\ \times \begin{pmatrix} 22.6 \text{ in.} + 19.6 \text{ in.} \\ +15.1 \text{ in.} + 12.1 \text{ in.} \end{pmatrix} \end{bmatrix}$ $= 4{,}140$ kip-in. 4,140 kip-in. > 2,770 kip-in. **o.k.**	$\dfrac{M_{np}}{\Omega} = \dfrac{2P_t(\Sigma d_n)}{\Omega}$ $= \dfrac{\begin{bmatrix} 2(39.8 \text{ kips}) \\ \times \begin{pmatrix} 22.6 \text{ in.} + 19.6 \text{ in.} \\ +15.1 \text{ in.} + 12.1 \text{ in.} \end{pmatrix} \end{bmatrix}}{2.00}$ $= 2{,}760$ kip-in. 2,760 kip-in. > 1,850 kip-in. **o.k.**

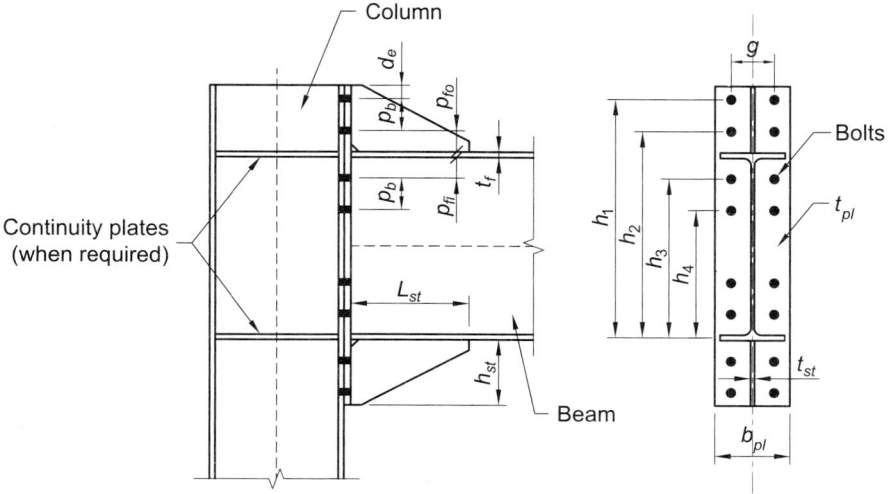

Fig. 4-3. Configuration for eight-bolt stiffened end-plate connection.

Determine the required end plate thickness

The required end plate thickness is determined from AISC Design Guide 4 Equation 3.10. The necessary parameters are determined as follows based on Figure 4-4. From Table 3.3 of AISC Design Guide 4:

$$s = \frac{1}{2}\sqrt{b_p g}$$

$$= \frac{1}{2}\sqrt{7.00 \text{ in.}(4.00 \text{ in.})}$$

$$= 2.65 \text{ in.}$$

$p_b = 3.00$ in.

$p_{fo} = 2.00$ in.

$p_{fi} = 2.00$ in.

$d_e = 1.25$ in.

Because $d_e < s$, Case 1 of AISC Design Guide 4 applies. From Table 3.3 of AISC Design Guide 4:

$$Y_p = \frac{b_p}{2}\left[h_1\left(\frac{1}{2d_e}\right) + h_2\left(\frac{1}{p_{fo}}\right) + h_3\left(\frac{1}{p_{fi}}\right) + h_4\left(\frac{1}{s}\right)\right]$$

$$+ \frac{2}{g}\left[h_1\left(d_e + \frac{p_b}{4}\right) + h_2\left(p_{fo} + \frac{3p_b}{4}\right) + h_3\left(p_{fi} + \frac{p_b}{4}\right) + h_4\left(s + \frac{3p_b}{4}\right) + p_b^2\right] + g$$

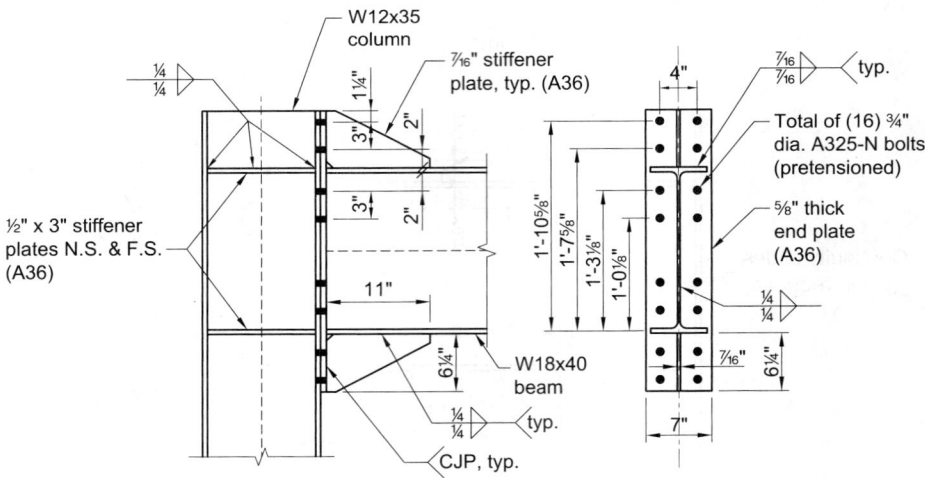

Fig. 4-4. Detailed OMF connection.

$$= \frac{7.00 \text{ in.}}{2}\left[22.6 \text{ in.}\left(\frac{1}{2(1.25 \text{ in.})}\right) + 19.6 \text{ in.}\left(\frac{1}{2.00 \text{ in.}}\right) + 15.1 \text{ in.}\left(\frac{1}{2.00 \text{ in.}}\right) + 12.1 \text{ in.}\left(\frac{1}{2.65 \text{ in.}}\right)\right]$$

$$+ \frac{2}{4.00 \text{ in.}}\left[\begin{array}{l}22.6 \text{ in.}\left(1.25 \text{ in.} + \dfrac{(3.00 \text{ in.})}{4}\right) + 19.6 \text{ in.}\left(2.00 \text{ in.} + \dfrac{3(3.00 \text{ in.})}{4}\right) \\ + 15.1 \text{ in.}\left(2.00 \text{ in.} + \dfrac{(3.00 \text{ in.})}{4}\right) + 12.1 \text{ in.}\left(2.65 \text{ in.} + \dfrac{3(3.00 \text{ in.})}{4}\right) \\ + (3.00 \text{ in.})^2 \end{array}\right] + 4.00 \text{ in.}$$

$= 232$ in.

From AISC Design Guide 4 Equation 3.10, the required end plate thickness is:

LRFD	ASD
$t_{pl,\,reqd} = \sqrt{\dfrac{1.11\phi M_{ue}}{\phi_b F_{yp} Y_p}}$	$t_{pl,\,reqd} = \sqrt{\dfrac{1.11\Omega_b M_{ae}}{\Omega F_y Y_p}}$
$= \sqrt{\dfrac{1.11(0.75)(2{,}770 \text{ kip-in.})}{0.90(36 \text{ ksi})(232 \text{ in.})}}$	$= \sqrt{\dfrac{1.11(1.67)(1{,}850 \text{ kip-in.})}{2.00(36 \text{ ksi})(232 \text{ in.})}}$
$= 0.554$ in.	$= 0.453$ in.

Use a ⅝-in.-thick ASTM A36 end plate.

Size the end-plate stiffener

Match the stiffener strength to the beam web strength using AISC Design Guide 4 Equation 3.15.

$$t_{st,\,reqd} = t_{wb}\left(\frac{F_{yb}}{F_{ys}}\right)$$

$$= 0.315 \text{ in.}\left(\frac{50 \text{ ksi}}{36 \text{ ksi}}\right)$$

$$= 0.438 \text{ in.}$$

Use a ⁷⁄₁₆-in.-thick ASTM A36 plate for the stiffener.

The height of the stiffener is:

$$h_{st} = p_{fo} + p_b + d_e$$
$$= 2.00 \text{ in.} + 3.00 \text{ in.} + 1.25 \text{ in.}$$
$$= 6.25 \text{ in.}$$

The length of the stiffener is determined from AISC Design Guide 4 Equation 2.11.

$$L_{st} = \frac{h_{st}}{\tan 30°}$$
$$= \frac{6.25 \text{ in.}}{\tan 30°}$$
$$= 10.8 \text{ in.}$$

Use $L_{st} = 11.0$ in.

Check for local buckling of the stiffener. The stiffener slenderness ratio is:

$$\lambda = \frac{h_{st}}{t_{st}}$$
$$= \frac{6.25 \text{ in.}}{7/16 \text{ in.}}$$
$$= 14.3$$

The limiting slenderness ratio from AISC *Specification* Table B4.1a Case 1 is:

$$\lambda_r = 0.56\sqrt{\frac{E}{F_y}}$$
$$= 0.56\sqrt{\frac{29,000 \text{ ksi}}{36 \text{ ksi}}}$$
$$= 15.9$$

14.3 < 15.9; therefore, the stiffener is not subject to local buckling.

Determine type and size of stiffener welds

AISC Design Guide 4 states that the weld of this stiffener plate to the end plate should be a complete-joint-penetration groove weld if $t_{st} > 3/8$ in.; therefore, weld the 7/16-in.-thick stiffener plate to the end plate with a complete-joint-penetration groove weld.

AISC Design Guide 4 states that the weld of this stiffener plate to the beam flange should be capable of developing the shear yield strength of the stiffener plate.

For the 7/16-in.-thick ASTM A36 plate:

$$V_n = 0.60 F_y A_{gv} \quad \text{(Spec. Eq. J4-3)}$$
$$= 0.60(36 \text{ ksi})(7/16 \text{ in.})$$
$$= 9.45 \text{ kip/in.}$$

For a two-sided fillet weld, calculate the required leg size, w_{req}, by setting the available shear yield strength of the plate equal to the available shear strength of the weld and solving for w_{req}.

4.2 ORDINARY MOMENT FRAMES

LRFD	ASD
$w_{req} = \dfrac{\phi_v V_n}{2\phi(0.60)(F_{EXX})/\sqrt{2}}$ $= \dfrac{1.00(9.45 \text{ kip/in.})}{2(0.75)(0.60)(70 \text{ ksi})/\sqrt{2}}$ $= 0.212 \text{ in.}$	$w_{req} = \dfrac{\Omega V_n/\Omega_v}{2(0.60)(F_{EXX})/\sqrt{2}}$ $= \dfrac{2.00(9.45 \text{ kip/in.})/1.50}{2(0.60)(70 \text{ ksi})/\sqrt{2}}$ $= 0.212 \text{ in.}$

Use ¼-in. fillet welds.

Check end-plate bolts for beam shear transfer

Per AISC Design Guide 4, a conservative check is to assume that only the bolts opposite the compression flange of the beam transfer the shear loads. In this case, this would be a total of (8) ¾-in.-diameter ASTM A325-N bolts. From AISC *Manual* Table 7-1, the available shear strength of the bolts is:

LRFD	ASD
$\phi V_n = n(\phi r_n)$ $= 8(17.9 \text{ kips})$ $= 143 \text{ kips}$ $143 \text{ kips} > 23.5 \text{ kips}$ **o.k.**	$\dfrac{V_n}{\Omega} = n\left(\dfrac{r_n}{\Omega}\right)$ $= 8(11.9 \text{ kips})$ $= 95.5 \text{ kips}$ $95.5 \text{ kips} > 15.6 \text{ kips}$ **o.k.**

Check compression bolts bearing/tearout per AISC Specification Section J3.10

For all eight bolts, the bearing strength when deformation at the bolt hole at service load is a consideration is:

$R_n = 2.4 dt F_u$
$= 2.4(¾ \text{ in.})(⅝ \text{ in.})(58 \text{ ksi})$
$= 65.3 \text{ kips/bolt}$

For the six inner bolts, the tearout strength when deformation at the bolt hole at service load is a consideration, with $l_c = 3 \text{ in.} - {}^{13}\!/_{16} \text{ in.} = 2.19 \text{ in.}$, is:

$R_n = 1.2 l_c t F_u$
$= 1.2(2.19 \text{ in.})(⅝ \text{ in.})(58 \text{ ksi})$
$= 95.3 \text{ kips/bolt}$

For the two outside bolts, the tearout strength when deformation at the bolt hole at service load is a consideration, with $l_c = 1¼ \text{ in.} - ({}^{13}\!/_{16} \text{ in.})/2 = 0.844 \text{ in.}$, is:

$R_n = 1.2l_c t F_u$
$= 1.2(0.844 \text{ in.})(\text{⅝ in.})(58 \text{ ksi})$
$= 36.7$ kips/bolt

LRFD	ASD
$\phi R_n = 0.75[6(65.3 \text{ kips}) + 2(36.7 \text{ kips})]$ $= 349$ kips 349 kips > 23.5 kips **o.k.**	$\dfrac{R_n}{\Omega} = \dfrac{6(65.3 \text{ kips}) + 2(36.7 \text{ kips})}{2.00}$ $= 233$ kips 233 kips > 15.6 kips **o.k.**

Design of Beam Flange-to-End-Plate Weld

The beam flange-to-end-plate weld is designed based on the recommendations of AISC Design Guide 4. Design the weld for the flange force, but no less than ϕR_n or R_n/Ω given as follows:

LRFD	ASD
$\phi R_n = \phi 0.6 F_{yp} b_f t_f < F_{fu}$ $= 0.90(0.6)(50 \text{ ksi})(6.02 \text{ in.})$ $\times (0.525 \text{ in.})$ $= 85.3$ kips The flange force is: $F_{fu} = \dfrac{M_{ue}}{d - t_f}$ $= \dfrac{2{,}770 \text{ kip-in.}}{17.9 \text{ in.} - 0.525 \text{ in.}}$ $= 159$ kips Design beam flange-to-end-plate welds for a required strength, $F_{fu} = 159$ kips	$\dfrac{R_n}{\Omega} = \dfrac{0.6 F_{yp} b_f t_f}{\Omega} < F_{fa}$ $= \dfrac{(0.6)(50 \text{ ksi})(6.02 \text{ in.})(0.525 \text{ in.})}{1.67}$ $= 56.8$ kips The flange force is: $F_{fa} = \dfrac{M_{ae}}{d - t_f}$ $= \dfrac{1{,}850 \text{ kip-in.}}{17.9 \text{ in.} - 0.525 \text{ in.}}$ $= 106$ kips Design beam flange-to-end-plate welds for a required strength, $F_{fa} = 106$ kips

Effective length of weld available, l_e, on both sides of flanges:

$l_e = b_f + (b_f - t_w)$
$= 6.02 \text{ in.} + (6.02 \text{ in.} - 0.315 \text{ in.})$
$= 11.7$ in.

A factor of 1.5 is applied to the weld strength because the weld is at a 90° angle to the load, according to AISC *Specification* Section J2.4, as follows:

4.2 ORDINARY MOMENT FRAMES

LRFD	ASD
$w_{req} = \dfrac{F_{fu}}{\phi 0.60\left(F_{EXX}/\sqrt{2}\right)1.5l_e}$ $= \dfrac{159 \text{ kips}}{0.75(0.60)\left(70 \text{ ksi}/\sqrt{2}\right)(1.5)(11.7 \text{ in.})}$ $= 0.407 \text{ in.}$	$w_{req} = \dfrac{\Omega F_{fa}}{0.60\left(F_{EXX}/\sqrt{2}\right)1.5l_e}$ $= \dfrac{2.00(106 \text{ kips})}{(0.60)\left(70 \text{ ksi}/\sqrt{2}\right)(1.5)(11.7 \text{ in.})}$ $= 0.407 \text{ in.}$

Use 7/16-in. fillet welds (two-sided) for the beam flange-to-end-plate weld.

Design of Beam Web-to-End-Plate Weld

AISC Design Guide 4 requires that the beam web-to-end-plate weld develop the available tensile yield strength of the web in the vicinity of the tension bolts.

The available tensile yield strength of the beam web and required weld size are:

LRFD	ASD
$\phi_t T_n = \phi_t F_{yw} t_w$ $= 0.90(50 \text{ ksi})(0.315 \text{ in.})$ $= 14.2 \text{ kip/in.}$	$\dfrac{T_n}{\Omega_t} = \dfrac{F_{yw} t_w}{\Omega_t}$ $= \dfrac{(50 \text{ ksi})(0.315 \text{ in.})}{1.67}$ $= 9.43 \text{ kip/in.}$
$w_{req} = \dfrac{\phi_t T_n}{2\phi\left(0.60 F_{EXX}/\sqrt{2}\right)1.5}$ $= \dfrac{14.2 \text{ kip/in.}}{2(0.75)\left[0.60(70 \text{ ksi})/\sqrt{2}\right]1.5}$ $= 0.213 \text{ in.}$	$w_{req} = \dfrac{\Omega(T_n/\Omega_t)}{2\left(0.60 F_{EXX}/\sqrt{2}\right)1.5}$ $= \dfrac{2.00(9.43 \text{ kip/in.})}{2\left[0.60(70 \text{ ksi})/\sqrt{2}\right]1.5}$ $= 0.212 \text{ in.}$

Use 1/4-in. fillet welds (two-sided) for the beam web-to-end-plate weld.

Column Flange Flexural Strength

With no column flange stiffeners, AISC Design Guide 4 Table 3.5 gives the following:

$$s = \dfrac{1}{2}\sqrt{b_{fc} g}$$
$$= \dfrac{1}{2}\sqrt{6.56 \text{ in.}(4.00 \text{ in.})}$$
$$= 2.56 \text{ in.}$$

$p_b = 3.00$ in.
$p_{so} = 2.00$ in.
$p_{si} = 2.00$ in.
$c = p_{so} + p_{si} + t_f$
$= 2.00$ in. $+ 2.00$ in. $+ 0.525$ in.
$= 4.53$ in.

From AISC Design Guide 4 Table 3.5:

$$Y_c = \frac{b_{fc}}{2}\left[h_1\left(\frac{1}{s}\right) + h_4\left(\frac{1}{s}\right)\right]$$
$$+ \frac{2}{g}\left[h_1\left(p_b + \frac{c}{2} + s\right) + h_2\left(\frac{p_b}{2} + \frac{c}{4}\right) + h_3\left(\frac{p_b}{2} + \frac{c}{2}\right) + h_4(s)\right] + \frac{g}{2}$$

$$= \frac{6.56 \text{ in.}}{2}\left[22.6 \text{ in.}\left(\frac{1}{2.56 \text{ in.}}\right) + 12.1 \text{ in.}\left(\frac{1}{2.56 \text{ in.}}\right)\right]$$

$$+ \frac{2}{4.00 \text{ in.}}\begin{bmatrix} 22.6 \text{ in.}\left(3.00 \text{ in.} + \frac{4.53 \text{ in.}}{2} + 2.56 \text{ in.}\right) + 19.6 \text{ in.}\left(\frac{3.00 \text{ in.}}{2} + \frac{4.53 \text{ in.}}{4}\right) \\ + 15.1 \text{ in.}\left(\frac{3.00 \text{ in.}}{2} + \frac{4.53 \text{ in.}}{2}\right) + 12.1 \text{ in.}(2.56 \text{ in.}) \end{bmatrix}$$

$$+ \frac{4.00 \text{ in.}}{2}$$

$= 205$ in.

From AISC Design Guide 4 Equation 3.21, the column flange flexural strength is:

LRFD	ASD
$\phi M_{cf} = \phi_b F_{yc} Y_c t_{fc}^2$	$\dfrac{M_{cf}}{\Omega} = \dfrac{F_{yc} Y_c t_{fc}^2}{\Omega_b}$
$= 0.90(50 \text{ ksi})(205 \text{ in.})(0.520 \text{ in.})^2$	
$= 2{,}490$ kip-in.	$= \dfrac{(50 \text{ ksi})(205 \text{ in.})(0.520 \text{ in.})^2}{1.67}$
	$= 1{,}660$ kip-in.
2,490 kip-in. < 2,770 kip-in. **n.g.**	1,660 kip-in. < 1,850 kip-in. **n.g.**

Therefore, column stiffeners must be added.

Assume the stiffener thickness is $t_s = \frac{1}{2}$ in. Then:

$p_{so} = p_{si}$
$= \dfrac{c - t_s}{2}$
$= \dfrac{4.53 \text{ in.} - \frac{1}{2} \text{ in.}}{2}$
$= 2.02$ in.

4.2 ORDINARY MOMENT FRAMES

With stiffeners added, AISC Design Guide 4 Table 3.5 provides the following equation:

$$Y_c = \frac{b_{fc}}{2}\left[h_1\left(\frac{1}{s}\right)+h_2\left(\frac{1}{p_{so}}\right)+h_3\left(\frac{1}{p_{si}}\right)+h_4\left(\frac{1}{s}\right)\right]$$

$$+\frac{2}{g}\left[h_1\left(s+\frac{p_b}{4}\right)+h_2\left(p_{so}+\frac{3p_b}{4}\right)+h_3\left(p_{si}+\frac{p_b}{4}\right)+h_4\left(s+\frac{3p_b}{4}\right)+p_b^2\right]+g$$

$$= \frac{6.56 \text{ in.}}{2}\begin{bmatrix} 22.6 \text{ in.}\left(\dfrac{1}{2.56 \text{ in.}}\right)+19.6 \text{ in.}\left(\dfrac{1}{2.02 \text{ in.}}\right)+15.1 \text{ in.}\left(\dfrac{1}{2.02 \text{ in.}}\right) \\ +12.1 \text{ in.}\left(\dfrac{1}{2.56 \text{ in.}}\right) \end{bmatrix}$$

$$+\frac{2}{4.00 \text{ in.}}\begin{bmatrix} 22.6 \text{ in.}\left(2.56 \text{ in.}+\dfrac{3.00 \text{ in.}}{4}\right)+19.6 \text{ in.}\left(2.02 \text{ in.}+\dfrac{3(3.00 \text{ in.})}{4}\right) \\ +15.1 \text{ in.}\left(2.02 \text{ in.}+\dfrac{3.00 \text{ in.}}{4}\right)+12.1 \text{ in.}\left(2.56 \text{ in.}+\dfrac{3(3.00 \text{ in.})}{4}\right) \\ +(3.00 \text{ in.})^2 \end{bmatrix}$$

$$+ 4.00 \text{ in.}$$
$$= 239 \text{ in.}$$

From AISC Design Guide 4 Equation 3.21, the available strength of the stiffened column flange is:

LRFD	ASD
$\phi M_{cf} = \phi_b F_{yc} Y_c t_{fc}^2$ $= 0.90(50 \text{ ksi})(239 \text{ in.})(0.520 \text{ in.})^2$ $= 2{,}910 \text{ kip-in.}$	$\dfrac{M_{cf}}{\Omega} = \dfrac{F_{yc} Y_c t_{fc}^2}{\Omega_b}$ $= \dfrac{(50 \text{ ksi})(239 \text{ in.})(0.520 \text{ in.})^2}{1.67}$ $= 1{,}930 \text{ kip-in.}$
2,910 kip-in. > 2,770 kip-in. **o.k.**	1,930 kip-in. > 1,850 kip-in. **o.k.**

Therefore, the connection will be adequate if stiffeners are added as designed in the following.

Column Stiffener Plates and Welds

The stiffener design is based on the minimum strength determined from flange local bending, column web local yielding, and column web local crippling. The minimum available strength based on these limit states will then be subtracted from the required flange force, F_{fu} or F_{fa}, to determine the stiffener required strength.

Calculate the available flexural strength of the flange using the available flexural strength of the unstiffened column determined previously.

LRFD	ASD
$\phi R_n = \dfrac{\phi M_{cf}}{d - t_{fb}}$ $= \dfrac{2{,}490 \text{ kip-in.}}{17.9 \text{ in.} - 0.525 \text{ in.}}$ $= 143 \text{ kips}$	$\dfrac{R_n}{\Omega} = \dfrac{M_{cf}}{\Omega(d - t_{fb})}$ $= \dfrac{1{,}660 \text{ kip-in.}}{17.9 \text{ in.} - 0.525 \text{ in.}}$ $= 95.5 \text{ kips}$

Calculate the available column web local yielding strength opposite the beam flange from AISC Design Guide 4 Equation 3.24. The parameter, C_t, is 1.0 because the distance from the top of the beam to the top of the column is greater than or equal to $d/2$ of the column.

$$R_n = \left[C_t\left(6k_{c,des} + 2t_p\right) + N\right]F_{ywc}t_{wc}$$
$$= \{1.0[6(0.820 \text{ in.}) + 2(\tfrac{5}{8} \text{ in.})] + [0.525 \text{ in.} + 2(\tfrac{3}{8} \text{ in.})]\}(50 \text{ ksi})(0.300 \text{ in.})$$
$$= 112 \text{ kips}$$

The available column web local yielding strength is:

LRFD	ASD
$\phi R_n = 1.00(112 \text{ kips})$ $= 112 \text{ kips}$	$\dfrac{R_n}{\Omega} = \dfrac{112 \text{ kips}}{1.50}$ $= 74.7 \text{ kips}$

Calculate the column web local crippling available strength opposite the beam flange force. The flange force applied from the top of the beam is located more than the half the column depth from the end of the column; therefore use AISC *Specification* Equation J10-4.

$$R_n = 0.80t_w^2\left[1 + 3\left(\dfrac{l_b}{d}\right)\left(\dfrac{t_w}{t_f}\right)^{1.5}\right]\sqrt{\dfrac{EF_{yw}t_f}{t_w}} \quad \text{(Spec. Eq. J10-4)}$$

$$= 0.80(0.300 \text{ in.})^2\left[1 + 3\left(\dfrac{0.525 \text{ in.}}{12.5 \text{ in.}}\right)\left(\dfrac{0.300 \text{ in.}}{0.520 \text{ in.}}\right)^{1.5}\right]\sqrt{\dfrac{29{,}000 \text{ ksi}(50 \text{ ksi})(0.520 \text{ in.})}{0.300 \text{ in.}}}$$

$$= 120 \text{ kips}$$

The available column web local crippling strength is:

LRFD	ASD
$\phi R_n = 0.75(120 \text{ kips})$ $= 90.0 \text{ kips}$	$\dfrac{R_n}{\Omega} = \dfrac{120 \text{ kips}}{2.00}$ $= 60.0 \text{ kips}$

Determine the stiffener required strength.

4.2 ORDINARY MOMENT FRAMES 4–31

LRFD	ASD
$F_{cu} = F_{fu} - \min(\phi R_n)$ $= 159 \text{ kips}$ $\quad - \min(143, 112, 90.0) \text{ kips}$ $= 69.0 \text{ kips}$	$F_{cu} = F_{fa} - \min\left(\dfrac{R_n}{\Omega}\right)$ $= 106 \text{ kips}$ $\quad - \min(95.5, 74.7, 60.0) \text{ kips}$ $= 46.0 \text{ kips}$

Use ½ in. × 3 in. ASTM A36 stiffener plates with ¾-in. clips along the flange on both sides of the column web and at the beam top and bottom flange.

The required axial strength per stiffener is:

LRFD	ASD
$P_u = \dfrac{F_{cu}}{2}$ $= \dfrac{69.0 \text{ kips}}{2}$ $= 34.5 \text{ kips}$	$P_a = \dfrac{F_{ca}}{2}$ $= \dfrac{46.0 \text{ kips}}{2}$ $= 23.0 \text{ kips}$

From AISC *Specification* Equation J4-6, the available axial strength per stiffener with a ¾-in. clip is:

LRFD	ASD
$\phi P_n = \phi F_y t_p b_p$ $= 0.90(36 \text{ ksi})(½ \text{ in.})(3.00 \text{ in.} - ¾ \text{ in.})$ $= 36.5 \text{ kips}$ 36.5 kips > 34.5 kips **o.k.**	$\dfrac{P_n}{\Omega} = \dfrac{F_y t_p b_p}{\Omega}$ $= \dfrac{(36 \text{ ksi})(½ \text{ in.})(3.00 \text{ in.} - ¾ \text{ in.})}{1.67}$ $= 24.3 \text{ kips}$ 24.3 kips > 23.0 kips **o.k.**

From AISC *Specification* Equation J4-3, the available shear strength along the column web is:

LRFD	ASD
$\phi V_n = \phi 0.60 F_{yp} t_p l_p$ $= 0.90(0.60)(36 \text{ ksi})(½ \text{ in.})(10.0 \text{ in.})$ $= 97.2 \text{ kips}$ 97.2 kips > 34.5 kips **o.k.**	$\dfrac{V_n}{\Omega} = \dfrac{0.60 F_{yp} t_p l_p}{\Omega}$ $= \dfrac{(0.60)(36 \text{ ksi})(½ \text{ in.})(10.0 \text{ in.})}{1.67}$ $= 64.7 \text{ kips}$ 64.7 kips > 23.0 kips **o.k.**

The value of $l_p = 10.0$ in. is based on the length of contact of the stiffener plates, including a reduction for the corner clips required to avoid the k-area of the column.

Weld of Stiffener to Column Flange
According to AISC *Specification* Section J2.4, with $b_p = 3.00$ in. $- ¾$ in.$= 2.25$ in.:

LRFD	ASD
$w_{reqd} = \dfrac{P_u}{2\phi\left(\dfrac{0.6(F_{EXX})}{\sqrt{2}}\right)b_p(1.5)}$	$w_{reqd} = \dfrac{\Omega P_a}{2\left(\dfrac{0.6(F_{EXX})}{\sqrt{2}}\right)b_p(1.5)}$
$= \dfrac{34.5 \text{ kips}}{2(0.75)\left[\dfrac{0.6(70 \text{ ksi})}{\sqrt{2}}\right](2.25 \text{ in.})(1.5)}$	$= \dfrac{2.00(23.0 \text{ kips})}{2\left[\dfrac{0.6(70 \text{ ksi})}{\sqrt{2}}\right](2.25 \text{ in.})(1.5)}$
$= 0.229$ in.	$= 0.229$ in.

Use ¼-in. fillet welds (two sided).

Weld of Stiffener to Column Web
According to AISC *Specification* Section J2.4:

LRFD	ASD
$w_{reqd} = \dfrac{P_u}{2\phi\left(\dfrac{0.60(F_{EXX})}{\sqrt{2}}\right)(1.0)l_p}$	$w_{reqd} = \dfrac{\Omega P_a}{2\left(\dfrac{0.60(F_{EXX})}{\sqrt{2}}\right)(1.0)l_p}$
$= \dfrac{34.5 \text{ kips}}{2(0.75)\left[\dfrac{0.60(70 \text{ ksi})}{\sqrt{2}}\right](1.0)(10.0 \text{ in.})}$	$= \dfrac{2.00(23.0 \text{ kips})}{2\left[\dfrac{0.60(70 \text{ ksi})}{\sqrt{2}}\right](1.0)(10.0 \text{ in.})}$
$= 0.0774$ in.	$= 0.0774$ in.

Use ¼-in. fillet welds (two sided). Based on AISC *Specification* Table J2.4, a 3⁄16-in. fillet weld is acceptable; however, ¼-in. fillet welds are used to be consistent with the stiffener-to-column flange welds.

The fully detailed end-plate connection is shown in Figure 4-4.

4.3 SPECIAL MOMENT FRAMES (SMF) AND INTERMEDIATE MOMENT FRAMES (IMF)

Special moment frame (SMF) and intermediate moment frame (IMF) systems, which are addressed in AISC *Seismic Provisions* Sections E3 and E2, respectively, resist lateral forces and displacements through the flexural and shear strengths of the beams and columns. Lateral displacement is resisted primarily through the flexural stiffness of the framing members and the restraint of relative rotation between the beams and columns at the connections, or "frame action." SMF and IMF systems must be capable of providing a story drift angle of at least 0.04 rad per AISC *Seismic Provisions* Section E3.6b and 0.02 rad per AISC *Seismic Provisions* Section E2.6b, respectively. An overview of SMF behavior and design issues is provided by Hamburger et al. (2009).

SMF and IMF systems tend to have larger and heavier beam and column sizes than braced-frame systems, as the beams and columns are often sized for drift control rather than for strength. The increase in member sizes and related costs, however, may be acceptable based on the increased flexibility in the architectural and mechanical layout in the structure. The absence of diagonal bracing members can provide greater freedom in configuring walls and routing mechanical ductwork and piping. As with other moment-frame systems, SMF and IMF systems are often located at the perimeter of the structure, allowing maximum flexibility in interior spaces without complicating the routing of building services such as mechanical ducts beneath the frame girders. The flexible nature of the frames, however, warrants additional consideration of the interaction between the steel frame and architectural cladding systems.

Current requirements for SMF and IMF systems are the result of research and analysis completed by various groups, including the Federal Emergency Management Agency (FEMA), AISC, the National Institute of Standards and Technology (NIST), the National Science Foundation (NSF), and the SAC Joint Venture. These requirements include prequalification of the connections used, per Section K1 of the AISC *Seismic Provisions*, or qualification through testing in accordance with Section K2 of the AISC *Seismic Provisions*. Design and detailing requirements for moment connections prequalified in accordance with AISC *Seismic Provisions* Section K1 may be found in AISC *Prequalified Connections for Special and Intermediate Steel Moment Frames for Seismic Applications*, herein referred to as ANSI/AISC 358. ANSI/AISC 358 is included in Part 9.2 of this Manual.

A primary focus point of the testing requirements lies in the measurement of inelastic deformations of beam-to-column moment connections. Plastic rotation of the specimen was used initially as the basis for qualification; however, this quantity is dependent on the selection of plastic hinge locations and member span. To avoid confusion, it was decided to use the centerline dimensions of the frame to define the total drift angle, which includes both elastic and inelastic deformations of the connections.

Most beam-to-column moment connections for SMF and IMF systems develop inelasticity in the beams and in the column panel zone, as shown in Figure 4-5. Panel zone deformation, while more difficult to predict, can contribute a significant amount of ductility to the frame. There are various factors that must be considered when accounting for panel zone deformation including continuity plates, doubler plates, and toughness of the *k*-area. In regard to these two areas of inelastic deformation—beam and panel zone—the AISC *Seismic Provisions* Section K2 requires that at least 75% of the observed inelastic deformation under testing procedures be as intended in the design of a prototype connection. This means that if the connection is anticipated to achieve inelasticity through plastic rotation in the beam, at least 75% of the actual deformation must occur in the beam-hinge locations when tested.

Currently, there are two primary methods used to move plastic hinging of the beam away from the column. These two methods focus on either reducing the cross-sectional properties of the beam at a defined location away from the column, or special detailing of the beam-to-column connection in order to provide adequate strength and toughness in the connection to force inelasticity into the beam just adjacent to the column flange. Reduced beam section (RBS) connections are typically fabricated by trimming the flanges of the beams at a short distance away from the face of the column in order to reduce the beam section properties at a defined location for formation of the plastic hinge (Figure 4-6). Research has included a straight reduced segment, an angularly tapered segment, and a circular reduced segment. A higher level of ductility was noted in the latter, and the RBS is typically fabricated using a circular reduced segment.

ANSI/AISC 358 includes six prequalified SMF and IMF connections, including the reduced beam section illustrated in the examples. Each of these prequalified connections has a design procedure similar to that employed in Example 4.3.4. Designers should evaluate the requirements of their project, the abilities of local fabricators and erectors, and the relative cost-effectiveness of different beam-to-column connections to determine the most appropriate connection for a given project.

Special connection detailing for added toughness and strength takes many forms using both welded and bolted connections. In many of the connections, both proprietary and non-proprietary, such factors as welding procedures, weld-access-hole detailing, web-plate attachment and flange-plate usage have been considered. For additional information on the specification of these connections, see ANSI/AISC 358 in Part 9.2 of this Manual.

Panel zone behavior is difficult to predict and is complicated by the presence of continuity plates and doubler plates, as well as k-area toughness. Three basic approaches are most

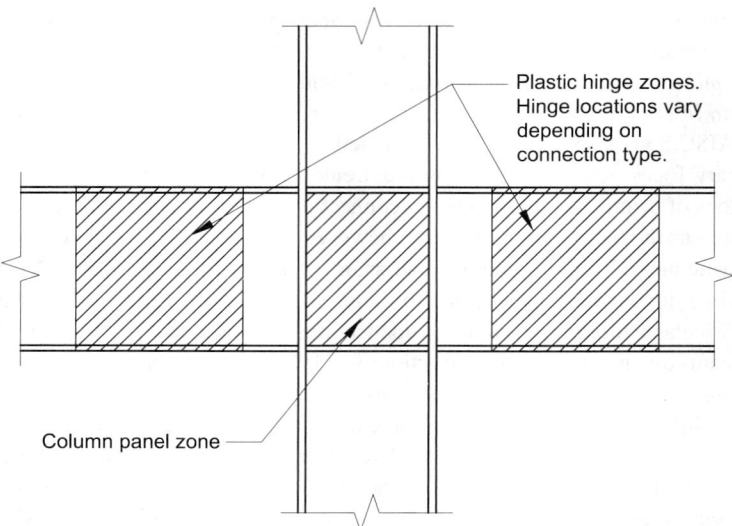

Fig. 4-5. Areas where inelastic deformation may be expected.

4.3 SPECIAL MOMENT FRAMES AND INTERMEDIATE MOMENT FRAMES

commonly used: "strong panel," "balanced panel" and "weak panel." These three terms relate the strength and inelastic behavior of the panel in relation to the strength and inelastic behavior of the framing members in the connection. In a "strong panel," the panel-zone strength is greater than the surrounding framing components to the point where the vast majority of the inelastic deformation of the frame occurs in the beam. In a "weak panel," the strength of the panel-zone is low enough relative to the framing members such that the majority of the inelastic deformation of the connection and frame occurs in the panel zone. A "balanced panel" falls between the strong and weak panel, where inelastic deformation in the framing members and panel zone are similar. The requirements in the AISC *Seismic Provisions* generally provide for strong or balanced panel zone designs in SMF. The full range of panel zone designs are permitted for IMF and OMF.

Another consideration in the design of SMF systems is the concept of "strong column-weak beam." The AISC *Seismic Provisions* provide for the proper proportioning of the frame elements in Equation E3-1.

$$\frac{\Sigma M^*_{pc}}{\Sigma M^*_{pb}} > 1.0 \qquad (\textit{Provisions Eq. E3-1})$$

where

ΣM^*_{pc} = sum of the projections of the nominal flexural strengths of the columns (including haunches where used) above and below the joint to the beam centerline with a reduction for the axial force in the column

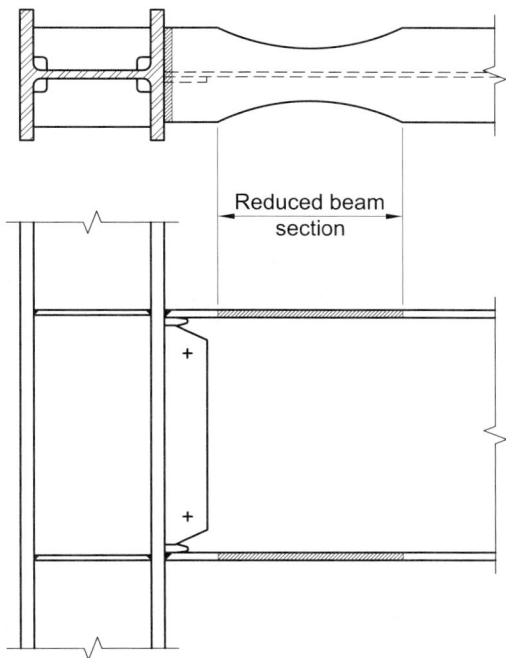

Fig. 4-6. Reduced beam section (RBS) connection.

AMERICAN INSTITUTE OF STEEL CONSTRUCTION

ΣM_{pb}^* = sum of the projections of the expected flexural strengths of the beam at the plastic hinge locations to the column centerline

This provision is not intended to eliminate all yielding in the columns. Rather, as described in AISC *Seismic Provisions* Commentary Section E3.4a, it is intended to result in framing systems that have distributed inelasticity in large seismic events, and discourages story mechanisms.

The primary difference between SMF systems and IMF systems is the interstory drift angle capacities. While this requirement differs for SMF and IMF systems, there are many requirements that are similar between the two frame types. This comparison is summarized in Table 4-1 of this Manual located at the end of this Part.

SMF Design Example Plan and Elevation

The following examples illustrate the design of special moment frames (SMF) based on AISC *Seismic Provisions* Section E3. Design of intermediate moment frames (IMF) reflects requirements outlined in AISC *Seismic Provisions* Section E2 that are, in most instances, similar to those in Section E3 or that do not vary from frame design requirements in the AISC *Specification*. For this reason, Part 4 does not present examples that focus exclusively on IMF, although these examples should prove useful when designing IMF frames as well. Table 4-1 in this Manual compares the significant design requirements for OMF, IMF and SMF systems, and clarifies which portions of the SMF examples apply to IMF design.

The plan and elevation are shown in Figure 4-7 and Figure 4-8, respectively. The code-specified gravity loading is as follows:

D_{floor} = 85 psf
D_{roof} = 68 psf
$L_{o,floor}$ = 80 psf
L_{floor} = 50 psf (reduced)
S = 20 psf
Curtain wall = 175 lb/ft along building perimeter at every level

For the SMF examples, it has been determined from ASCE/SEI 7 that the following factors are applicable: Risk Category I, Seismic Design Category D, $R = 8$, $\Omega_o = 3$, $C_d = 5\frac{1}{2}$, $I_e = 1.00$, $S_{DS} = 1.0$, and $\rho = 1.0$ (per ASCE/SEI 7 Section 12.3.4.2, $\rho = 1.0$ if the story resists more than 35% of the base shear in the direction of interest, loss of one bay of SMF will not result in more than a 33% reduction in story strength, nor does the resulting system have an extreme torsional irregularity).

Example 4.3.1. SMF Story Drift and Stability Check

Given:

Refer to the floor plan shown in Figure 4-7 and the SMF elevation shown in Figure 4-8. Determine if the frame satisfies the ASCE/SEI 7 drift and stability requirements based on the given loading.

The applicable building code specifies the use of ASCE/SEI 7 for calculation of loads.

4.3 SPECIAL MOMENT FRAMES AND INTERMEDIATE MOMENT FRAMES

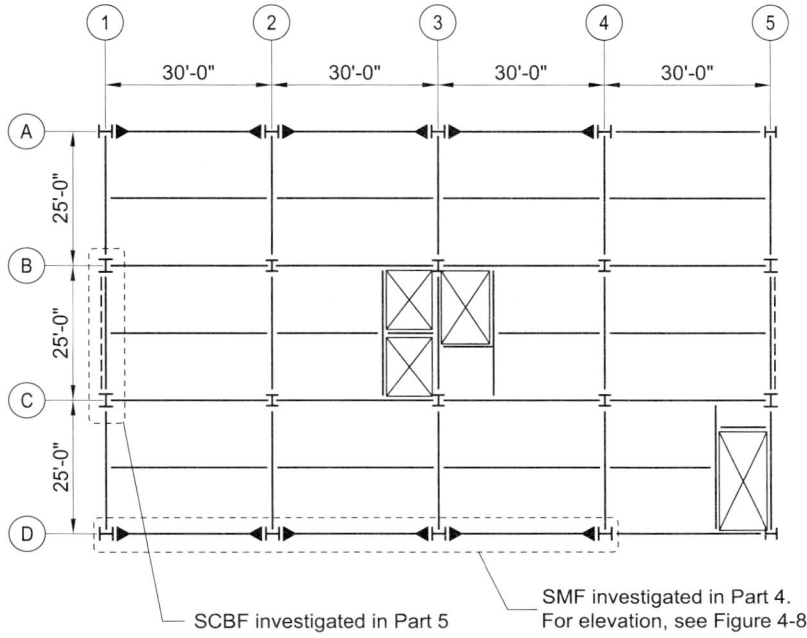

Fig. 4-7. SMF floor plan.

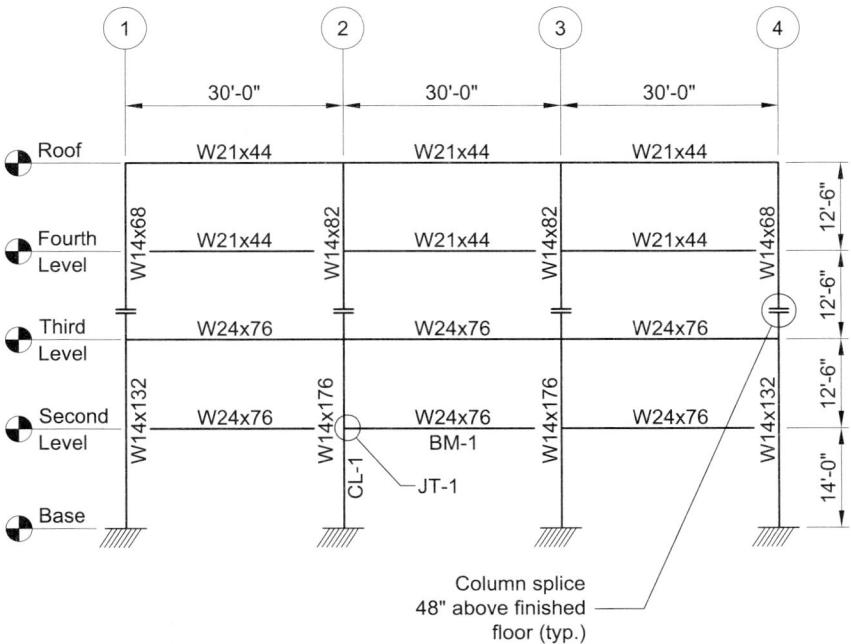

Fig. 4-8. SMF elevation.

The seismic design story shear at the third levels, V_x, is 140 kips as defined in ASCE/SEI 7 Section 12.8.4. From an elastic analysis of the structure that includes second-order effects and accounts for panel-zone deformations, the maximum interstory drift occurs between the third and fourth levels: $\delta_{xe} = \delta_{4e} - \delta_{3e} = 0.482$ in.

In this example, the stability check will be performed for the *third* level. This checks the stability of the columns supporting the *third* level. The story drift between the *second* and *third* levels is $\delta_{3e} - \delta_{2e} = 0.365$ in..

Solution:

From AISC *Manual* Table 1-1, the geometric properties are as follows:

W24×76
$b_f = 8.99$ in.

Reduced beam section (RBS) connections are used at the frame beam-to-column connections and the flange cut will reduce the stiffness of the beam. Example 4.3.3 illustrates the design of the RBS geometry and the flange cut on one side of the web is $c = 2$ in. Section 5.8, Step 1, of ANSI/AISC 358 states that the calculated elastic drift, based on gross beam section properties, may be multiplied by 1.1 for flange reductions up to 50% of the beam flange width in lieu of specific calculations of effective stiffness. Amplification of drift values for cuts less than the maximum may be linearly interpolated.

For $b_f = 8.99$ in., the maximum cut is:

0.5(8.99 in.) = 4.50 in.

Thus, the total 4-in. cut is:

$$\left(\frac{4.00 \text{ in.}}{4.50 \text{ in.}}\right) 100 = 88.9\% \text{ of the maximum cut}$$

The calculated elastic drift needs to be amplified by 8.89% (say 9%).

Drift Check

From an elastic analysis of the structure that includes second-order effects, the maximum interstory drift occurs between the third and fourth levels. The effective elastic drift is:

$$\begin{aligned}
\delta_{xe} &= \delta_{4e} - \delta_{3e} \\
&= 0.482 \text{ in.} \\
\delta_{xe\,RBS} &= 1.09 \delta_{xe} \\
&= 1.09(0.482 \text{ in.}) \\
&= 0.525 \text{ in.}
\end{aligned}$$

Per the AISC *Seismic Provisions* Section B1, the design story drift and the story drift limits are those stipulated by the applicable building code. ASCE/SEI 7 Section 12.8.6 defines the design story drift, Δ, computed from δ_x, as the difference in the deflections at the center of

4.3 SPECIAL MOMENT FRAMES AND INTERMEDIATE MOMENT FRAMES

mass at the top and bottom of the story under consideration, which in this case is the third level.

$$\Delta = \frac{C_d \delta_{xe}}{I_e} \quad \text{(ASCE/SEI 7 Eq. 12.8-15)}$$

$$= \frac{5\frac{1}{2}(0.525 \text{ in.})}{1.00}$$

$$= 2.89 \text{ in.}$$

From ASCE/SEI 7 Table 12.12-1, the allowable story drift at level x, Δ_a, is $0.020h_{sx}$, where h_{sx} is the story height below level x. Although not assumed in this example, Δ_a can be increased to $0.025h_{sx}$ if interior walls, partitions, ceilings and exterior wall systems are designed to accommodate these increased story drifts. ASCE/SEI 7 Section 12.12.1.1 requires for seismic force resisting systems comprised solely of moment frames in structures assigned to Seismic Design Category D, E or F, that the design story drift not exceed Δ_a/ρ for any story. Determine the allowable story drift as follows:

$$\frac{\Delta_a}{\rho} = \frac{0.020 h_{sx}}{\rho}$$

$$= \frac{0.020(12.5 \text{ ft})(12 \text{ in./ft})}{1.0}$$

$$= 3.00 \text{ in.}$$

$\Delta = 2.89$ in. $< \Delta_a$ **o.k.**

The frame satisfies the drift requirements.

Frame Stability Check

ASCE/SEI Section 12.8.7 provides a method for the evaluation of the P-Δ effects on moment frames based on a stability coefficient, θ, which should be checked for each floor. For the purposes of illustration, this example checks the stability coefficient only for the third level. The stability coefficient, θ, is determined as follows:

$$\theta = \frac{P_x \Delta I_e}{V_x h_{sx} C_d} \quad \text{(ASCE/SEI 7 Eq. 12.8-16)}$$

$A_{floor} = A_{roof} \approx 75 \text{ ft}(120 \text{ ft}) = 9,000 \text{ ft}^2$

$D_{floor} = 9,000 \text{ ft}^2 (85 \text{ psf}) / 1,000 \text{ lb/kip}$
$\phantom{D_{floor}} = 765 \text{ kips}$

$D_{roof} = 9,000 \text{ ft}^2 (68 \text{ psf}) / (1,000 \text{ lb/kip})$
$\phantom{D_{roof}} = 612 \text{ kips}$

$D_{wall} = 175 \text{ lb/ft}[2(75 \text{ ft}+120 \text{ ft})] / (1,000 \text{ lb/kip})$
$\phantom{D_{wall}} = 68.3 \text{ kips per level}$

$L_{floor} = 9,000 \text{ ft}^2 (50 \text{ psf}) / (1,000 \text{ lb/kip})$
$\phantom{L_{floor}} = 450 \text{ kips}$

$$L_{roof} = 9,000 \text{ ft}^2 (20 \text{ psf}) / (1,000 \text{ lb/kip})$$
$$= 180 \text{ kips}$$

ASCE/SEI 7 does not explicitly specify load factors to be used on the gravity loads for determining P_x, except that Section 12.8.7 does specify that no individual load factor need exceed 1.0. This means that if the combinations of ASCE/SEI 7 Section 2.3 are used, a factor of 1.0 can be used for dead load rather than the usual 1.2 factor used in the LRFD load combination, for example. This also means that the vertical component $0.2S_{DS}D$ need not be considered here. Therefore, for this example, the load combination used to compute the total vertical load on a given story, P_x, acting simultaneously with the seismic design story shear, V_x, is $1.0D + 0.5L$ based on ASCE/SEI 7 Section 2.3 including the 0.5 factor on L permitted by Section 2.3, where L is the reduced live load. Note that consistent with this, the same combination was used in the second order analysis for this example for the purpose of computing the fundamental period, base shear, and design story drift.

The total dead load in the columns supporting the third level, assuming that the columns support two floors of curtain wall in addition to other dead loads, is:

$$1.0P_D = 1.0[612 \text{ kips} + 2(765 \text{ kips}) + 2(68.3 \text{ kips})]$$
$$= 2,280 \text{ kips}$$

The total live load in the columns supporting the third level is:

$$0.5P_L = 0.5[(2)(450 \text{ kips}) + 180 \text{ kips}]$$
$$= 540 \text{ kips}$$

Therefore, the total vertical design load carried by these columns is:

$$P_x = 2,280 \text{ kips} + 540 \text{ kips}$$
$$= 2,820 \text{ kips}$$

The seismic design story between the second and third level, including the 9% amplification on the drift, is:

$$\Delta = \frac{C_d \delta_{xe}}{I_e} \quad \text{(from ASCE/SEI 7 Eq. 12.8-15)}$$
$$= \frac{5\tfrac{1}{2}(1.09)(0.365 \text{ in.})}{1.00}$$
$$= 2.19 \text{ in.}$$

From an elastic analysis of the structure, the seismic design story shear at the third level under the story drift loading using the equivalent lateral force procedure is $V_x = 140$ kips and the floor-to-floor height below the third level is $h_{sx} = 12.5$ ft.

4.3 SPECIAL MOMENT FRAMES AND INTERMEDIATE MOMENT FRAMES

Therefore, the stability coefficient is:

$$\theta = \frac{2{,}820 \text{ kips}(2.19 \text{ in.})(1.0)}{140 \text{ kips}(12.5 \text{ ft})(12 \text{ in./ft})(5\frac{1}{2})}$$
$$= 0.0535$$

Because a second-order analysis was used to compute the story drift, θ is adjusted as follows to verify compliance with θ_{max}, per ASCE/SEI 7 Section 12.8.7.

$$\frac{\theta}{1+\theta} = \frac{0.0535}{1+0.0535}$$
$$= 0.0508$$

According to ASCE/SEI 7, if θ is less than or equal to 0.10, second-order effects need not be considered for computing story drift. Note that whether or not second-order effects on member forces must be considered per ASCE/SEI 7 has to be verified, as it was in this example; however, Chapter C of the AISC *Specification* requires second-order effects be considered in all cases in the analysis used for member design.

Check the maximum permitted θ

The stability coefficient may not exceed θ_{max}. In determining θ_{max}, β is the ratio of shear demand to shear capacity for the level being analyzed, and may be conservatively taken as 1.0.

$$\theta_{max} = \frac{0.5}{\beta C_d} \qquad \text{(ASCE/SEI 7 Eq. 12.8-17)}$$
$$= \frac{0.5}{1.0(5\frac{1}{2})}$$
$$= 0.0909 \leq 0.25$$

The adjusted stability coefficient satisfies the maximum:

$0.0508 < 0.0909$ **o.k.**

The moment frame meets the allowable story drift and stability requirements for seismic loading.

Comments:

There are a total of six bays of SMF in this example. Considering the relative expense of SMF connections and because the drift and stability limits are met, it may be more cost-effective to reduce the number of bays to four, and increase member sizes to satisfy the strength and stiffness requirements.

Example 4.3.2. SMF Column Strength Check

Given:

Refer to Column CL-1 on the first level in Figure 4-8. Determine the adequacy of the ASTM A992 W14×176 to resist the loads given.

There is no transverse loading between the column supports in the plane of bending.

The applicable building code specifies the use of ASCE/SEI 7 for calculation of loads. The required strengths are determined by a second-order analysis including the effects of P-δ and P-Δ with reduced stiffness as required by the direct analysis method. The governing load combinations for shear that include seismic effects are:

LRFD	ASD
LRFD Load Combination 5 from ASCE/SEI 7 Section 12.4.2.3 $V_u = (1.2 + 0.2S_{DS})D + \rho Q_E$ $+ 0.5L + 0.2S$ $= 32.0$ kips (including the 0.5 factor on L permitted in ASCE/SEI 7 Section 12.4.2.3)	ASD Load Combination 5 from ASCE/SEI 7 Section 12.4.2.3 $V_a = (1.0 + 0.14S_{DS})D + H + F + 0.7\rho Q_E$ $= 22.4$ kips

AISC *Seismic Provisions* Section D1.4a requires, with limited exceptions, that the amplified seismic load (i.e., the seismic load multiplied by the overstrength factor, Ω_o) be used to calculate required column axial strength. Moment need not be combined simultaneously with the amplified seismic axial load in this case because there is no transverse loading between the column supports. The redundancy factor, ρ, and the overstrength factor need not be applied simultaneously.

The governing load combinations for axial strength that include the amplified seismic load from ASCE/SEI 7 are:

LRFD	ASD
LRFD Load Combination 5 from ASCE/SEI 7 Section 12.4.3.2 $P_u = (1.2 + 0.2S_{DS})D + \Omega_o Q_E$ $+ 0.5L + 0.2S$ $= 249$ kips	ASD Load Combination 6 from ASCE/SEI 7 Section 12.4.3.2 $P_a = (1.0 + 0.105S_{DS})D + H + F$ $+ 0.525\Omega_o Q_E + 0.75L + 0.75S$ $= 218$ kips

4.3 SPECIAL MOMENT FRAMES AND INTERMEDIATE MOMENT FRAMES

The governing load combinations for axial and flexural strength that include seismic effects from ASCE/SEI 7 are:

LRFD	ASD
LRFD Load Combination 5 from ASCE/SEI 7 Section 12.4.2.3 $$\begin{aligned}P_u &= (1.2+0.2S_{DS})D+\rho Q_E \\ &\quad +0.5L+0.2S \\ &= 243 \text{ kips} \\ M_u &= (1.2+0.2S_{DS})D+\rho Q_E \\ &\quad +0.5L+0.2S \\ M_{u\,top} &= 125 \text{ kip-ft} \\ M_{u\,bot} &= -298 \text{ kip-ft}\end{aligned}$$	ASD Load Combination 6 from ASCE/SEI 7 Section 12.4.2.3 $$\begin{aligned}P_a &= (1.0+0.10S_{DS})D+H+F \\ &\quad +0.525\rho Q_E+0.75L+0.75S \\ &= 214 \text{ kips} \\ M_a &= (1.0+0.10S_{DS})D+0.525\rho Q_E \\ &\quad +0.75L+0.75S \\ M_{a\,top} &= 67.0 \text{ kip-ft} \\ M_{a\,bot} &= -158 \text{ kip-ft}\end{aligned}$$

Solution:

From AISC *Manual* Table 2-4, the material properties are as follows:

ASTM A992
$F_y = 50$ ksi
$F_u = 65$ ksi

From AISC *Manual* Table 1-1, the geometric properties are as follows:

Column
W14×176
$d = 15.2$ in. $t_w = 0.830$ in. $I_x = 2{,}140$ in.^{4} $I_y = 838$ in.^{4}
$A = 51.8$ in.^{2} $r_x = 6.43$ in. $r_y = 4.02$ in. $S_x = 281$ in.^{3}
$Z_x = 320$ in.^{3} $b_f = 15.7$ in. $t_f = 1.31$ in. $k_{des} = 1.91$ in.
$h/t_w = 13.7$ $b_f/2t_f = 5.97$

Beam
W24×76
$I_x = 2{,}100$ in.^{4}

Column Element Slenderness

AISC *Seismic Provisions* Section E3.5a requires that the stiffened and unstiffened elements of SMF columns satisfy the requirements of Section D1.1 for highly ductile members.

From the AISC *Seismic Provisions* Table D1.1, for flanges of highly ductile members:

$$\lambda_{hd} = 0.30\sqrt{\frac{E}{F_y}}$$
$$= 0.30\sqrt{\frac{29{,}000 \text{ ksi}}{50 \text{ ksi}}}$$
$$= 7.22$$

Because $\lambda = b_f/2t_f = 5.97 < \lambda_{hd}$, the flanges satisfy the requirements for highly ductile elements.

The limiting width-to-thickness ratio for webs of highly ductile members is determined as follows from Table D1.1 using the governing load case for axial load, including the amplified seismic load, as stipulated in AISC *Seismic Provisions* Section D1.4a:

LRFD	ASD
$C_a = \dfrac{P_u}{\phi_c P_y}$	$C_a = \dfrac{\Omega_c P_a}{P_y}$
$= \dfrac{P_u}{0.90 F_y A_g}$	$= \dfrac{1.67 P_a}{F_y A_g}$
$= \dfrac{249 \text{ kips}}{0.90(50 \text{ ksi})(51.8 \text{ in.}^2)}$	$= \dfrac{1.67(218 \text{ kips})}{(50 \text{ ksi})(51.8 \text{ in.}^2)}$
$= 0.107$	$= 0.141$
Because $C_a \leq 0.125$,	Because $C_a > 0.125$,
$\lambda_{hd} = 2.45\sqrt{\dfrac{E}{F_y}}(1-0.93C_a)$	$\lambda_{hd} = 0.77\sqrt{\dfrac{E}{F_y}}(2.93-C_a) \geq 1.49\sqrt{\dfrac{E}{F_y}}$
$= 2.45\sqrt{\dfrac{29{,}000 \text{ ksi}}{50 \text{ ksi}}}[1-0.93(0.107)]$	$= 0.77\sqrt{\dfrac{29{,}000 \text{ ksi}}{50 \text{ ksi}}}(2.93-0.141)$
$= 53.1$	$\geq 1.49\sqrt{\dfrac{29{,}000 \text{ ksi}}{50 \text{ ksi}}}$
	$= 51.7 \geq 35.9$
	Use $\lambda_{hd} = 51.7$.

Because $\lambda = h/t_w = 13.7 < \lambda_{hd}$, the web satisfies the requirements for highly ductile elements.

Alternatively, Table 1-3 in this Manual can be used to confirm that members satisfy the requirements for highly ductile members.

4.3 SPECIAL MOMENT FRAMES AND INTERMEDIATE MOMENT FRAMES

Effective Length Factor

The direct analysis method in AISC *Specification* Section C3 states that the effective length factor K of all members shall be taken as unity unless a smaller value can be justified by rational analysis. Therefore,

$K_x = 1.0$
$K_y = 1.0$

Available Compressive Strength

Determine what the controlling slenderness ratio of the column is:

$$\frac{K_x L_x}{r_x} = \frac{1.0(14.0 \text{ ft})(12.0 \text{ in./ft})}{6.43 \text{ in.}}$$
$$= 26.1$$

$$\frac{K_y L_y}{r_y} = \frac{1.0(14.0 \text{ ft})(12 \text{ in./ft})}{4.02 \text{ in.}}$$
$$= 41.8 \quad \textbf{governs}$$

Using AISC *Manual* Table 4-1, with $K_y L_y = 14.0$ ft, the available compressive strength of the **W14×176** column is:

LRFD	ASD
$\phi_c P_n = 2{,}050$ kips > 249 kips **o.k.**	$\dfrac{P_n}{\Omega_c} = 1{,}360$ kips > 218 kips **o.k.**

Available Flexural Strength

From AISC *Manual* Table 3-2, determine for the **W14×176** whether the limit state of lateral-torsional buckling applies for flexural strength, i.e., $L_b > L_p$.

$L_p = 14.2$ ft
$L_r = 73.2$ ft
$L_b = 14.0$ ft $< L_p$

From AISC *Specification* Section F2, with compact flanges and web and $L_b \leq L_p$, the applicable limit state is yielding. Using AISC *Manual* Table 3-2, the available flexural strength of the **W14×176** column is:

LRFD	ASD
$M_{cx} = \phi_b M_{px}$ $= 1{,}200$ kip-ft	$M_{cx} = \dfrac{M_{px}}{\Omega_b}$ $= 798$ kip-ft

Combined Loading

Check the interaction of compression and flexure using AISC *Specification* Section H1.1, and the governing load case for combined loading.

LRFD	ASD
$\dfrac{P_r}{P_c} = \dfrac{243 \text{ kips}}{2{,}050 \text{ kips}}$ $= 0.119 < 0.2$	$\dfrac{P_r}{P_c} = \dfrac{214 \text{ kips}}{1{,}360 \text{ kips}}$ $= 0.157 < 0.2$
Therefore, use AISC *Specification* Equation H1-1b	Therefore, use AISC *Specification* Equation H1-1b
$\dfrac{P_r}{2P_c} + \left(\dfrac{M_{rx}}{M_{cx}} + \dfrac{M_{ry}}{M_{cy}}\right) \leq 1.0$	$\dfrac{P_r}{2P_c} + \left(\dfrac{M_{rx}}{M_{cx}} + \dfrac{M_{ry}}{M_{cy}}\right) \leq 1.0$
$\dfrac{0.119}{2} + \left(\dfrac{298 \text{ kip-ft}}{1{,}200 \text{ kip-ft}} + 0\right) = 0.308$	$\dfrac{0.157}{2} + \left(\dfrac{158 \text{ kip-ft}}{798 \text{ kip-ft}} + 0\right) = 0.276$
$0.308 \leq 1.0$ **o.k.**	$0.276 \leq 1.0$ **o.k.**

Available Shear Strength

Using AISC *Manual* Table 3-2 for the W14×176 column:

LRFD	ASD
$\phi V_n = 378$ kips > 32.0 kips **o.k.**	$\dfrac{V_n}{\Omega} = 252$ kips > 22.4 kips **o.k.**

The W14×176 is adequate to resist the loads given for Column CL-1.

Comments:

The beam and column sizes selected were based on a least-weight solution for drift control; thus, the column size is quite conservative for strength.

Example 4.3.3. SMF Beam Strength Check

Given:

Refer to Beam BM-1 in Figure 4-8. Determine the adequacy of the W24×76 ASTM A992 W-shape to resist the following loading. The beam end connections utilize the reduced beam section (RBS) prequalified in accordance with ANSI/AISC 358 and shown in Figure 4-9. Also, design the lateral bracing for the beam using ASTM A36 angles. Assume that the beam flanges are braced at the columns.

4.3 SPECIAL MOMENT FRAMES AND INTERMEDIATE MOMENT FRAMES

The applicable building code specifies the use of ASCE/SEI 7 for calculation of loads. The required strengths at the face of the column and the centerline of the RBS are determined by a second-order analysis including the effects of P-δ and P-Δ with reduced stiffness as required by the direct analysis method.

The governing load combinations for the required flexural and shear strength at the face of the column are:

LRFD	ASD
LRFD Load Combination 5 from ASCE/SEI 7 Section 12.4.2.3 $$M_u = (1.2 + 0.2S_{DS})D + \rho Q_E$$ $$+ 0.5L + 0.2S$$ $$= -273 \text{ kip-ft}$$ $$V_u = (1.2 + 0.2S_{DS})D + \rho Q_E$$ $$+ 0.5L + 0.2S$$ $$= 33.8 \text{ kips}$$ (including the 0.5 factor on L permitted in ASCE/SEI 7 Section 12.4.2.3)	ASD Load Combination 5 from ASCE/SEI 7 Section 12.4.2.3 $$M_a = (1.0 + 0.14S_{DS})D + H + F$$ $$+ 0.7\rho Q_E$$ $$= -136 \text{ kip-ft}$$ $$V_a = (1.0 + 0.14S_{DS})D + H + F$$ $$+ 0.7\rho Q_E$$ $$= 22.8 \text{ kips}$$

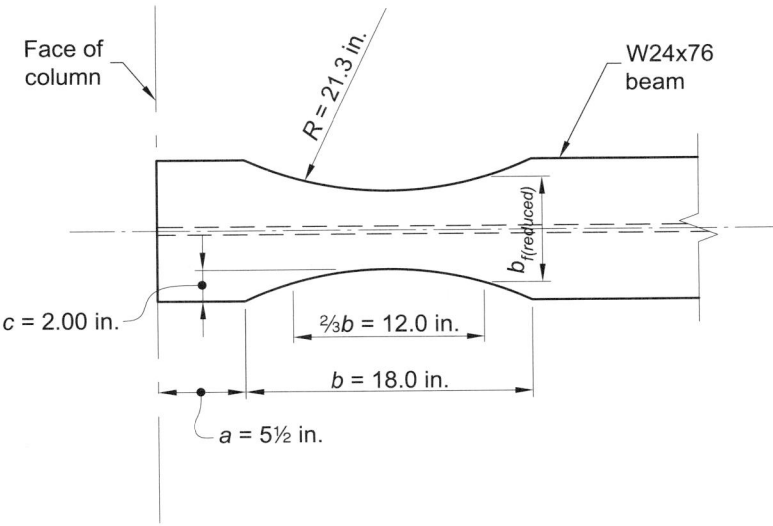

Fig. 4-9. Initial RBS detail for Examples 4.3.3 and 4.3.4.

The governing load combinations for the required flexural and shear strength at the centerline of the RBS are:

LRFD	ASD
$M_u = (1.2 + 0.2S_{DS})D + \rho Q_E$ $+ 0.5L + 0.2S$ $= -246$ kip-ft	$M_a = (1.0 + 0.14S_{DS})D + H + F + 0.7\rho Q_E$ $= -168$ kip-ft

The required shear strength at the RBS is not given because the shear at the face of the column is greater than at the RBS and the available shear strength is the same at each location since the web is not modified by the RBS cut.

Solution:

From AISC *Manual* Table 2-4, the beam material properties are as follows:

ASTM A992
$F_y = 50$ ksi
$F_u = 65$ ksi

From AISC *Manual* Table 1-1, the beam geometric properties are as follows:

W24×76
$d = 23.9$ in. $b_f = 8.99$ in. $t_f = 0.680$ in. $t_w = 0.440$ in.
$k_{des} = 1.18$ in. $h/t_w = 49.0$ $S_x = 176$ in.^{3} $Z_x = 200$ in.^{3}
$r_y = 1.92$ in. $h_o = 23.2$ in.

RBS Dimensions

According to the requirements of ANSI/AISC 358 Section 5.8, Step 1, the designer must choose a section that satisfies the dimensional constraints listed below. For this example, trial values of a, b and c are chosen as shown in Figure 4-9. Example 4.3.4 demonstrates that these dimensions are acceptable. Other dimensions that satisfy the requirements of ANSI/AISC 358 could have been selected. Dimensions that satisfy the dimensional constraints listed below may still require adjustment to satisfy all of the requirements of ANSI/AISC 358 Section 5.8.

$0.5b_{bf} \leq a \leq 0.75b_{bf}$ (ANSI/AISC 358 Eq. 5.8-1)
$0.65d \leq b \leq 0.85d$ (ANSI/AISC 358 Eq. 5.8-2)
$0.1b_{bf} \leq c \leq 0.25b_{bf}$ (ANSI/AISC 358 Eq. 5.8-3)

Check Beam Element Slenderness

AISC *Seismic Provisions* Section E3.5a requires that the stiffened and unstiffened elements of SMF beams satisfy the requirements of AISC *Seismic Provisions* Section D1.1 for highly ductile members.

4.3 SPECIAL MOMENT FRAMES AND INTERMEDIATE MOMENT FRAMES

ANSI/AISC 358 Section 5.3.1 permits calculation of the width-to-thickness ratio for the flanges based on a value of b_f not less than the flange width at the ends of the center two-thirds of the reduced section provided that gravity loads do not shift the location of the plastic hinge a significant distance from the center of the RBS. Assuming this is the case here, the RBS radius of cut from ANSI/AISC 358 Figure 5.1 and the dimensions given in Figure 4-9 is:

$$R = \frac{4c^2 + b^2}{8c}$$

$$= \frac{4(2.00 \text{ in.})^2 + (18.0 \text{ in.})^2}{8(2.00 \text{ in.})}$$

$$= 21.3 \text{ in.}$$

At the edge of the center two-thirds of the RBS, the reduced flange width is, from geometry:

$$b_{f,RBS} = 2(R-c) + b_f - 2\sqrt{R^2 - \left(\frac{b}{3}\right)^2} \quad (2\text{-}3)$$

$$= 2(21.3 \text{ in.} - 2.00 \text{ in.}) + (8.99 \text{ in.}) - 2\sqrt{(21.3 \text{ in.})^2 - \left(\frac{18.0 \text{ in.}}{3}\right)^2}$$

$$= 6.72 \text{ in.}$$

$$\lambda_f = \frac{b_{f,RBS}}{2t_f}$$

$$= \frac{6.72 \text{ in.}}{2(0.680 \text{ in.})}$$

$$= 4.94$$

From AISC *Seismic Provisions* Table D1.1, the limiting flange width-to-thickness ratio for highly ductile members is:

$$\lambda_{hd} = 0.30\sqrt{\frac{E}{F_y}}$$

$$= 0.30\sqrt{\frac{29,000 \text{ ksi}}{50 \text{ ksi}}}$$

$$= 7.22$$

Because $\lambda_f < \lambda_{hd}$, the flanges satisfy the requirements for highly ductile members.

From AISC *Seismic Provisions* Table D1.1, for webs of rolled I-shaped sections used as beams or columns, recognizing that $C_a = P_u/(\phi P_n)$ is assumed to be zero because no axial force is present for the beam, the limiting width-to-thickness ratio is:

$$\lambda_{hd} = 2.45\sqrt{\frac{E}{F_y}}$$

$$= 2.45\sqrt{\frac{29{,}000 \text{ ksi}}{50 \text{ ksi}}}$$

$$= 59.0$$

Because $\lambda_w = h/t_w = 49.0 < \lambda_{hd}$, the web satisfies the requirements for highly ductile members.

Alternatively, using Table 4-2 of this Manual, it can be seen that a W24×76 will satisfy the width-to-thickness requirements for an SMF beam.

Spacing of Lateral Bracing

AISC *Seismic Provisions* Section D1.2b requires that both flanges be laterally braced at intervals not to exceed:

$$0.086 r_y \left(\frac{E}{F_y}\right) = 0.086(1.92 \text{ in.})\left(\frac{29{,}000 \text{ ksi}}{50 \text{ ksi}}\right)\left(\frac{1}{12 \text{ in./ft}}\right)$$

$$= 7.98 \text{ ft}$$

Alternatively, using Table 4-2 for a W24×76, it can be seen that L_{bmax} is equal to 7.98 ft.

The composite concrete and metal deck diaphragm provides continuous lateral support to the top flange of the beam; however, the only lateral supports for the bottom flange occur at the end connections. Therefore, a bottom flange brace must be provided at least every 7.98 ft. The distance between column centerlines is 30.0 ft. If three braces are provided along the length, the unbraced length of the beam, L_b, would be:

$$L_b = \frac{30.0 \text{ ft}}{4}$$

$$= 7.50 \text{ ft} < 7.98 \text{ ft}$$

Therefore, provide lateral bracing of the bottom flange at 7.50 ft intervals.

Available Flexural Strength

Check the available flexural strength of the beam (including the reduced section) as stipulated in ANSI/AISC 358 Section 5.8, Step 1.

First, check the unbraced length using AISC *Manual* Table 3-2:

$L_p = 6.78 \text{ ft}$ $\qquad L_r = 19.5 \text{ ft}$

Therefore, $L_p < L_b < L_r$.

This suggests that bracing must be provided more closely than 7.50 ft on center to develop M_p in the frame beam but, as discussed in the following, recognizing that $C_b > 1.0$ helps establish that M_p can be developed with bracing intervals further apart than 6.78 ft.

When designing an RBS connection, it is assumed that the flexural strength of the member at the reduced section will control the moment strength of the beam. According to AISC *Specification* Section F2, where $L_b \leq L_p$, beam strength is controlled by M_p. When the RBS section is proportioned and located according to the provisions of ANSI/AISC 358, the flexural strength of the RBS will control beam strength and this assumption does not need to be verified. In these cases, the flexural strength of the unreduced section is limited by $M_p = F_y Z_x$ and the flexural strength of the reduced beam section will be $M_{pRBS} = F_y Z_{RBS}$, where Z_{RBS} is the plastic section modulus at the center of the reduced beam section, as defined in ANSI/AISC 358 Equation 5.8-4, and Z_x is the plastic section modulus of the unreduced beam section. However, in cases where $L_b > L_p$, which is the case in this example, the assumption will have to be verified. Note that as a practical matter, the typical value of C_b is greater than 1.0 for moment frame beams and when the limits imposed by the AISC *Seismic Provisions* on unbraced length are considered, lateral-torsional buckling typically will not reduce the flexural strength of the unreduced section below M_p.

For the unreduced section, from AISC *Specification* Section F2, with compact flanges and web and $L_p < L_b \leq L_r$, the applicable flexural strength limit states are yielding and lateral-torsional buckling. For the limit state of yielding and lateral-torsional buckling, the following equation applies.

$$M_n = C_b \left[M_p - \left(M_p - 0.7 F_y S_x \right) \left(\frac{L_b - L_p}{L_r - L_p} \right) \right] \leq M_p \qquad (Spec.\ Eq.\ F2\text{-}2)$$

$$C_b = \frac{12.5 M_{max}}{2.5 M_{max} + 3 M_A + 4 M_B + 3 M_C} \qquad (Spec.\ Eq.\ F1\text{-}1)$$

If bracing is provided at 7.50 ft on center, there are four unbraced segments along the beam, although the two segments on each side of the beam midspan are symmetric assuming that the seismic load case on the beam is considered. The moment diagram from the elastic analysis has an approximately constant slope such that the values of M_{max}, M_A, M_B and M_C can be obtained by proportioning the moment diagram shown in Figure 4-10. This approximation assumes that the impact of gravity load is such that it does not significantly influence the shape of the moment diagram resulting from lateral load.

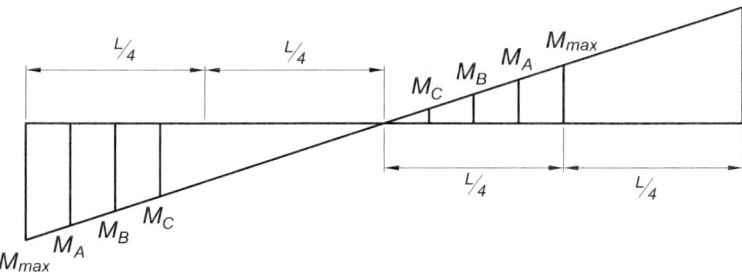

Fig. 4-10. Moment diagram for Beam BM-1.

For the exterior segments of the beam, where M is the moment at the end of the beam:

$M_{max} = M$
$M_A = |0.875M|$ $M_B = |0.75\ M|$ $M_C = |0.625M|$

$$C_b = \frac{12.5M}{2.5M + 3(0.875M) + 4(0.75M) + 3(0.625M)}$$
$$= 1.25$$

For the interior segments of the beam:

$M_{max} = 0.5M$
$M_A = |0.375M|$ $M_B = |0.25\ M|$ $M_C = |0.125M|$

$$C_b = \frac{12.5(0.5M)}{2.5(0.5M) + 3(0.375M) + 4(0.25M) + 3(0.125M)}$$
$$= 1.67$$

The available flexural strength of the beam end segment is determined in the following. The end segment is the governing case because the ratio of C_b values for the exterior and interior segments is less than the ratio of the maximum moments for the segments. From AISC *Specification* Section F2.2, for the limit state of lateral-torsional buckling, with $L_p < L_b \leq L_r$:

$$M_n = C_b\left[M_p - (M_p - 0.7F_y S_x)\left(\frac{L_b - L_p}{L_r - L_p}\right)\right] \leq M_p \qquad (Spec.\ Eq.\ F2\text{-}2)$$

where

$M_p = F_y Z_x$ $(Spec.\ Eq.\ F2\text{-}1)$
$\quad = 50\ \text{ksi}(200\ \text{in.}^3)(1\ \text{ft}/12\ \text{in.})$
$\quad = 833\ \text{kip-ft}$

$0.7F_y S_x = 0.7(50\ \text{ksi})(176\ \text{in.}^3)(1\ \text{ft}/12\ \text{in.})$
$\quad = 513\ \text{kip-ft}$

For the end segment:

$$M_n = 1.25\left[833\ \text{kip-ft} - (833\ \text{kip-ft} - 513\ \text{kip-ft})\left(\frac{7.50\ \text{ft} - 6.78\ \text{ft}}{19.5\ \text{ft} - 6.78\ \text{ft}}\right)\right]$$
$$= 1{,}020\ \text{kip-ft}$$

Therefore, $M_n = M_p = 833$ kip-ft because M_n cannot be greater than M_p (as indicated in AISC *Specification* Equation F2-1) regardless of the value of C_b and bracing may be provided at 7.5 ft on center to achieve M_p.

4.3 SPECIAL MOMENT FRAMES AND INTERMEDIATE MOMENT FRAMES

Plastic Section Modulus at the Center of the RBS

At the centerline of the reduced beam section, using ANSI/AISC 358 Section 5.8, the plastic section modulus is:

(ANSI/AISC 358 Eq. 5.8-4)
$$Z_{RBS} = Z_x - 2ct_{bf}(d - t_{bf})$$
$$= 200 \text{ in.}^3 - 2(2.00 \text{ in.})(0.680 \text{ in.})(23.9 \text{ in.} - 0.680 \text{ in.})$$
$$= 137 \text{ in.}^3$$

Available and Required Flexural Strength at Centerline of RBS and Face of Column

As determined previously, the nominal flexural strength is the plastic moment of the beam, M_p. At the centerline of the RBS, the nominal and available flexural strengths are:

$$M_{n@RBS} = F_y Z_{RBS}$$
$$= 50 \text{ ksi}(137 \text{ in.}^3)$$
$$= 6{,}850 \text{ kip-in.}$$
$$= 571 \text{ kip-ft}$$

LRFD	ASD
$\phi_b M_{n@RBS} = 0.90(571 \text{ kip-ft})$ $= 514 \text{ kip-ft}$	$\dfrac{M_{n@RBS}}{\Omega_b} = \dfrac{571 \text{ kip-ft}}{1.67}$ $= 342 \text{ kip-ft}$
$M_{u@RBS} = 246 \text{ kip-ft} < 514 \text{ kip-ft}$ **o.k.**	$M_{a@RBS} = 168 \text{ kip-ft} < 342 \text{ kip-ft}$ **o.k.**

At the face of the column, the nominal and available flexural strengths are:

LRFD	ASD
$\phi M_n = \phi_b M_p$ $= 0.90(833 \text{ kip-ft})$ $= 750 \text{ kip-ft}$	$\dfrac{M_n}{\Omega} = \dfrac{M_p}{\Omega_b}$ $= \dfrac{833 \text{ kip-ft}}{1.67}$ $= 499 \text{ kip-ft}$
$M_u = 273 \text{ kip-ft} < 750 \text{ kip-ft}$ **o.k.**	$M_a = 136 \text{ kip-ft} < 499 \text{ kip-ft}$ **o.k.**

Available Shear Strength
Using AISC *Manual* Table 3-2 for the W24×76 beam:

LRFD	ASD
$\phi_v V_n = 315$ kips	$\dfrac{V_n}{\Omega_v} = 210$ kips
$V_u = 33.8$ kips < 315 kips **o.k.**	$V_a = 22.8$ kips < 210 kips **o.k.**

The W24×76 is adequate to resist the loads given for Beam BM-1.

Comment:
The preceding flexural check could have been conservatively made using the required strength at the face of the column compared to the available strength at the centerline of the RBS. This approach might be useful if there is uncertainty regarding the geometry of the RBS, particularly the values of a and b since these are needed to determine the location of the RBS centerline.

Lateral Bracing
According to the AISC *Seismic Provisions* Section D1.2b, which references AISC *Specification* Appendix 6, the required strength of nodal lateral bracing away from an expected plastic hinge location is determined from AISC *Specification* Appendix 6 as follows:

$$P_{rb} = \frac{0.02 M_r C_d}{h_o} \quad\quad (Spec.\ Eq.\ A\text{-}6\text{-}7)$$

$R_y = 1.1$ from AISC *Seismic Provisions* Table A3.1

$C_d = 1.0$

where, according to AISC *Seismic Provisions* D1.2a(a)(2):

LRFD	ASD
$M_r = R_y F_y Z$ $= 1.1(50\text{ ksi})(200\text{ in.}^3)$ $= 11{,}000$ kip-in.	$M_r = \dfrac{R_y F_y Z}{1.5}$ $= \dfrac{1.1(50\text{ ksi})(200\text{ in.}^3)}{1.5}$ $= 7{,}330$ kip-in.

4.3 SPECIAL MOMENT FRAMES AND INTERMEDIATE MOMENT FRAMES

Alternatively, Table 4-2 of this Manual can be used to determine M_r. The required brace force using AISC *Specification* Equation A-6-7 is:

LRFD	ASD
$P_{urb} = \dfrac{0.02(11,000 \text{ kip-in.})(1.0)}{23.2 \text{ in.}}$ $= 9.48 \text{ kips}$	$P_{arb} = \dfrac{0.02(7,330 \text{ kip-in.})(1.0)}{23.2 \text{ in.}}$ $= 6.32 \text{ kips}$

The length of the brace is assumed to extend from the centerline of the bottom flange of the W24×76 SMF beam to the centerline of the top flange of the adjacent gravity beam. The size of the adjacent gravity beam is unknown, but assume for this calculation that the flange thickness is the same as the W24×76. The center-to-center spacing of the beams is 12 ft 6 in., as indicated in Figure 4-7. Therefore, the length of the brace is approximately:

$$L = \sqrt{[12.5 \text{ ft}(12 \text{ in./ft})]^2 + (23.9 \text{ in.} - 0.680 \text{ in.})^2} \, (1/12 \text{ in./ft})$$
$$= 12.6 \text{ ft}$$

From AISC *Manual* Table 4-12 for eccentrically loaded single angles with the eccentricity equal to or less than 0.75 times the angle thickness, try a L5×5×5/16 with $K = 1.0$. For ASTM A36, the available axial strength of the single angle is found through interpolation using $KL = 12.6$ ft.

LRFD	ASD
$\phi_c P_n = 22.9$ kips	$\dfrac{P_n}{\Omega_c} = 15.0$ kips
$P_{urb} = 9.48$ kips < 22.9 kips **o.k.**	$P_{arb} = 6.32$ kips < 15.0 kips **o.k.**

AISC *Seismic Provisions* Section D1.2b also specifies a minimum stiffness for lateral bracing according to Appendix 6 of the AISC *Specification*. The kicker brace selected in this example is considered a nodal brace. Assuming a rigid brace support, from AISC *Specification* Equation A-6-8, the required brace stiffness is:

LRFD	ASD
$\beta_{br} = \dfrac{1}{\phi}\left(\dfrac{10 M_r C_d}{L_b h_o}\right)$	$\beta_{br} = \Omega\left(\dfrac{10 M_r C_d}{L_b h_o}\right)$

LRFD	ASD
where $\phi = 0.75$ $M_r = 11{,}000$ kip-in. $C_d = 1.0$ $L_b = 7.50$ ft(12 in./ft) $ = 90.0$ in. $h_o = 23.2$ in. $\beta_{br} = \dfrac{1}{0.75}\left[\dfrac{10(11{,}000 \text{ kip-in.})(1.0)}{(90.0 \text{ in.})(23.2 \text{ in.})}\right]$ $\phantom{\beta_{br}} = 70.2$ kip/in.	where $\Omega = 2.00$ $M_r = 7{,}330$ kip-in. $C_d = 1.0$ $L_b = 7.50$ ft(12 in./ft) $ = 90.0$ in. $h_o = 23.2$ in. $\beta_{br} = 2.00\left[\dfrac{10(7{,}330 \text{ kip-in.})(1.0)}{(90.0 \text{ in.})(23.2 \text{ in.})}\right]$ $\phantom{\beta_{br}} = 70.2$ kip/in.

The stiffness of the L5×5×5/16 brace, with $A = 3.07$ in.^{2}, in the horizontal plane is:

$$k = \frac{AE}{L}\cos^2(\theta)$$

$$\theta = \tan^{-1}\left(\frac{23.2 \text{ in.}}{12.5 \text{ ft}(12 \text{ in./ft})}\right)$$

$$ = 8.79°$$

$$k = \frac{3.07 \text{ in.}^2 (29{,}000 \text{ ksi})}{152 \text{ in.}}\cos^2(8.79°)$$

$$ = 572 \text{ kip/in.} > \beta_{br} = 70.2 \text{ kip/in.}$$

$k > \beta_{br}$ **o.k.**

L5×5×5/16 ASTM A36 kickers will be provided to brace the beam bottom flange at a spacing of 7.50 ft. The brace at midspan can be designed in a similar manner with $C_d = 2.0$, because it is the brace closest to the inflection point.

Note that because this connection features a prequalified RBS moment connection supporting a concrete structural slab, according to ANSI/AISC 358 Section 5.3.1(7) the slab plus the typical lateral stability bracing provides sufficient stability so that additional bracing adjacent to the plastic hinges is not required, provided that shear connectors are provided at a minimum spacing of 12 in. (but omitted in the RBS protected zone).

Comment:

In addition to checking that the beam available flexural strength is greater than the required flexural strength from code-specified load combinations at the center of the RBS, the maximum probable moment, M_{pr}, at the column face needs to be checked against the expected moment strength of the unreduced beam section. This will be done in Example 4.3.4.

Example 4.3.4. SMF Beam-Column Connection Design

The SMF beam-column connection design presented in this example has been chosen to demonstrate the application of the design provisions for prequalified RBS connections in accordance with ANSI/AISC 358. This example demonstrates that the RBS geometry developed below is satisfactory. Some of the results from this example are used in Example 4.3.3. The geometry of an RBS connection is not unique and alternative configurations of the RBS geometry are possible.

Given:

Refer to Joint JT-1 in Figure 4-8. Design the connection between Beam BM-1 and Column CL-1 using the reduced beam section (RBS) shown in Figure 4-9. All beams and columns are ASTM A992 W-shapes. Plate material is ASTM A572 Grade 50. The gravity loads on the beam are:

$w_D = 0.84$ kip/ft $\qquad w_L = 0.60$ kip/ft

Procedure:

The procedure outlined below follows the order of the design procedure outlined in ANSI/AISC 358 Section 5.8. The term "Step n" indicates the actual step number in ANSI/AISC 358 Section 5.8. The steps from ANSI/AISC 358 are augmented with some additional checks in this example. Some of the steps listed in Table 4-A are executed in detail in Example 4.3.3, the SMF beam strength check. The procedure is defined for LRFD only.

In addition, panel zone and bracing requirements are checked.

Solution:

From AISC *Manual* Table 2-4, the W-shape material properties are as follows:

ASTM A992
$F_y = 50$ ksi
$F_u = 65$ ksi

From AISC *Manual* Table 2-5, the plate material properties are as follows:

ASTM A572 Grade 50
$F_y = 50$ ksi
$F_u = 65$ ksi

From AISC *Manual* Table 1-1, the geometric properties are as follows:

Column
W14×176
$A = 51.8$ in.^{2} $\qquad d = 15.2$ in. $\qquad t_w = 0.830$ in. $\qquad b_f = 15.7$ in.
$t_f = 1.31$ in. $\qquad t_{fdet} = 1\tfrac{5}{16}$ in. $\qquad k_{det} = 2\tfrac{5}{8}$ in. $\qquad k_1 = 1\tfrac{5}{8}$ in.
$Z_x = 320$ in.^{3}

Table 4-A
RBS Design Procedure Per ANSI/AISC 358

Check system limitations per Section 5.2.
Check prequalification limits per Section 5.3.
Step 1. Choose trial values for the RBS dimensions a, b and c. See also Example 4.3.3.
Step 2. Compute plastic section modulus at the center of RBS, Z_e. See Example 4.3.3.
Step 3. Compute the probable maximum moment at the center of RBS.
Step 4. Compute the shear force at the center of the RBS at each end of beam.
Step 5. Compute the probable maximum moment at the face of the column.
Step 6. Compute the plastic moment of the beam based on expected yield stress.
Step 7. Check that moment at the face of the column, M_f, does not exceed available strength, $\phi_d M_{pe}$.
Step 8. Determine the required shear strength, V_u, of beam and beam web-to-column connection from Equation 5.8-9.
Step 9. Design the beam web-to-column connection per Section 5.6.
Step 10. Check continuity plate requirements per Chapter 2.
Step 11. Check column-beam relationship limitations according to Section 5.4.

Beam
W24×76
$A = 22.4$ in.^{2} $d = 23.9$ in. $t_w = 0.440$ in. $b_f = 8.99$ in.
$t_f = 0.680$ in. $Z_x = 200$ in.^{3} $r_y = 1.92$ in.

System limitations per ANSI/AISC 358 Section 5.2

The frame is a special moment frame and the RBS connection is prequalified for SMF and IMF systems.

Prequalification limits per ANSI/AISC 358 Section 5.3

Check beam requirements

The W24×76 beam satisfies the requirements of ANSI/AISC 358 Section 5.3.1 as a rolled wide flange member, with depth less than a W36, weight less than 300 lb/ft, and flange thickness less than 1.75 in. The clear span-to-depth ratio of the beam is at least 7 as required for an SMF system:

$$\text{Clear span/depth} = \frac{(360 \text{ in.} - 15.2 \text{ in.})}{23.9 \text{ in.}}$$
$$= 14.4 \geq 7 \quad \textbf{o.k.}$$

The beam also satisfies the maximum width-to-thickness ratios for the flange, measured at the edge of the center two-thirds of the RBS, and the web specified by ANSI/AISC 358 Section 5.3.1(6), as shown in Example 4.3.3.

Beam lateral bracing must be provided in conformance with the AISC *Seismic Provisions*. This beam supports a concrete structural slab that is connected between the protected zones with welded shear connectors spaced at a maximum of 12 in. Consequently, according to the Exception in Section 5.3.1(7) of ANSI/AISC 358, supplemental lateral bracing is not required at the reduced section. Minimum spacing between the face of the column and the first beam lateral support and minimum spacing between lateral supports is shown in Example 4.3.3.

The protected zone consists of the portion of the beam between the face of the column and the end of the reduced beam section farthest from the face of the column. Figure 5.1 of ANSI/AISC 358 shows the location of the protected zone. This information should be clearly identified on the structural design drawings, on shop drawings, and on erection drawings.

Check column requirements
The W14×176 column satisfies the requirements of Section 5.3.2 as a rolled wide flange member, with the frame beam connected to the column flange and with a column depth less than a W36.

The column also satisfies the maximum width-to-thickness ratios for the flanges and the web specified by Section 5.3.2(6), as shown in Example 4.3.2.

Column lateral bracing must conform to the requirements of the AISC *Seismic Provisions*. Section E3.4c allows the use of a strong-column/weak-beam ratio (AISC *Seismic Provisions* Equation E3-1) greater than 2.0 to show that a column remains elastic outside of the panel zone at restrained beam-to-column connections. If it can be demonstrated that the column remains elastic outside of the panel zone, Section E3.4c(1) requires the column flanges to be braced at the level of the beam top flanges only. With a column-beam moment ratio of 1.72 in this example (see calculations following), the column cannot be assumed to remain elastic and bracing is required at both the top and bottom flanges of the beam. Column flange bracing at these locations may be provided by continuity plates and a full-depth shear plate between the continuity plates at the connection of the girder framing into the weak axis of the column.

ANSI/AISC 358 provides only an LRFD design procedure for the RBS connection; therefore, the RBS connection must be designed using LRFD, even in the case where ASD was used for the remainder of the design. The following calculations illustrate the LRFD procedure.

Trial Values for the RBS Dimensions a, b and c
(Step 1 in ANSI/AISC 358 Section 5.8)
The dimensions of the RBS cut will be determined so that the RBS has sufficient strength to resist the flexural loads prescribed by the building code as well as so that the probable

maximum moment in the beam at the face of the column does not exceed the expected plastic moment capacity of the beam. The former check is performed in Example 4.3.3, while the latter check is performed in the following.

For the trial values of the RBS dimensions, use the values in Figure 4-9 and check per ANSI/AISC 358 Equations 5.8-1 to 5.8-3.

$a = 5.50$ in.
$0.5b_f \leq a \leq 0.75b_f$ (ANSI/AISC 358 Eq. 5.8-1)
$0.5b_f = 0.5(8.99$ in.$)$
$\quad = 4.50$ in.
$0.75b_f = 0.75(8.99$ in.$)$
$\quad = 6.74$ in.
4.50 in. $\leq$ 5.50 in. $\leq$ 6.74 in.   **o.k.**

$b = 18.0$ in.
$0.65d \leq b \leq 0.85d$ (ANSI/AISC 358 Eq. 5.8-2)
$0.65d = 0.65(23.9$ in.$)$
$\quad = 15.5$ in.
$0.85d = 0.85(23.9$ in.$)$
$\quad = 20.3$ in.
15.5 in. $\leq$ 18.0 in. $\leq$ 20.3 in.   **o.k.**

$c = 2.00$ in.
$0.1b_{bf} \leq c \leq 0.25b_f$ (ANSI/AISC 358, Eq. 5.8-3)
$0.1b_{bf} = 0.1(8.99$ in.$)$
$\quad = 0.899$ in.
$0.25b_{bf} = 0.25(8.99$ in.$)$
$\quad = 2.25$ in.
0.899 in. $\leq$ 2.00 in. $\leq$ 2.25 in.   **o.k.**

Plastic Section Modulus at the Center of the Reduced Beam Section
(Step 2 in ANSI/AISC 358 Section 5.8)

The value of the plastic section modulus at the center of the RBS, $Z_{RBS} = 137$ in.^{3}, is computed in Example 4.3.3.

Probable maximum moment at the center of the reduced beam section
(Step 3 in ANSI/AISC 358 Section 5.8)

From Example 4.3.3, $Z_{RBS} = 137$ in.^{3}, therefore:

$$C_{pr} = \frac{F_y + F_u}{2F_y} \leq 1.2$$ (ANSI/AISC 358 Eq. 2.4.3-2)

$$= \frac{50 \text{ ksi} + 65 \text{ ksi}}{2(50 \text{ ksi})}$$

$$= 1.15 \leq 1.2$$

$R_y = 1.1$ from AISC *Seismic Provisions* Table A3.1

$M_{pr} = C_{pr} R_y F_y Z_{RBS}$ (ANSI/AISC 358 Eq. 5.8-5)

$= 1.15(1.1)(50 \text{ ksi})(137 \text{ in.}^3)$

$= 8{,}670$ kip-in.

The value of M_{pr} is intended to represent the maximum moment that can occur at the center of the RBS cut when the reduced section has yielded and strain hardened.

Shear force at the center of the reduced beam sections at each end of the beam (Step 4 in ANSI/AISC 358 Section 5.8)

The shear force at the center of the RBS at each end of the beam is computed from a free body diagram of the portion of the beam between the RBS centers. For this free body diagram, assume the moment at the center of each RBS is equal to M_{pr} as computed in Step 3.

The gravity load on the beam is computed from the load combination provided in ANSI/AISC 358 Section 5.8, Step 4, as follows:

$w_u = 1.2D + 0.5L + 0.2S$

$= 1.2(0.840 \text{ kip/ft}) + 0.5(0.600 \text{ kip/ft})$

$\quad + 0.2(0 \text{ kip/ft})$

$= 1.31$ kip/ft

The distance from the column face to the center of the RBS cut is determined from ANSI/AISC 358 Figure 5.2 as follows:

$S_h = a + (b/2)$

$= 5.50 \text{ in.} + (18.0 \text{ in.}/2)$

$= 14.5$ in.

The distance between centers of RBS cuts is:

$L_h = L - 2(d_{col}/2) - 2S_h$

$= 360 \text{ in.} - 2(15.2 \text{ in.}/2) - 2(14.5 \text{ in.})$

$= 316$ in.

Figure 4-11 shows the key beam dimensions. Figure 4-12 shows a free body diagram of the portion of the beam between RBS cuts.

As shown in Figure 4-12, V_{RBS} and V'_{RBS} are the symbols used for the shear at the center of the RBS cuts. V_{RBS} is the larger of the two shear forces and V'_{RBS} is the smaller of the two. By summing moments about the right end of this free body diagram, the shear forces can be computed as follows:

$$V_{RBS} = \frac{2M_{pr}}{L_h} + \frac{w_u L_h}{2}$$

$$= \frac{2(8{,}670 \text{ kip-in.})}{316 \text{ in.}} + \frac{1.31 \text{ kip/ft}(1 \text{ ft}/12 \text{ in.})(316 \text{ in.})}{2}$$

$$= 72.1 \text{ kips}$$

Summing moments about the left end:

$$V'_{RBS} = \frac{2M_{pr}}{L_h} - \frac{w_u L_h}{2}$$

$$= \frac{2(8{,}670 \text{ kip-in.})}{316 \text{ in.}} - \frac{1.31 \text{ kip/ft}(1 \text{ ft}/12 \text{ in.})(316 \text{ in.})}{2}$$

$$= 37.6 \text{ kips}$$

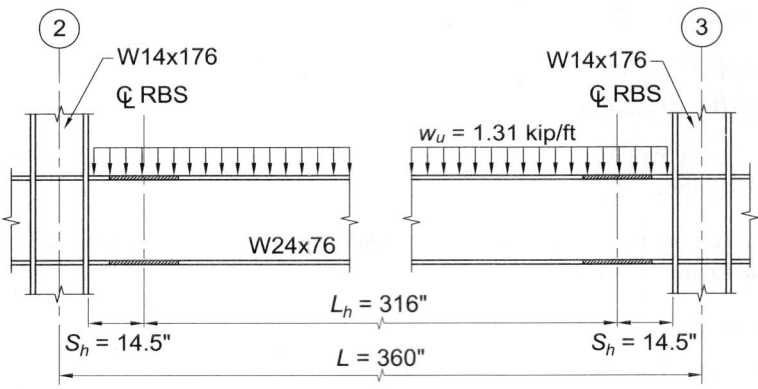

Fig. 4-11. Beam dimensions.

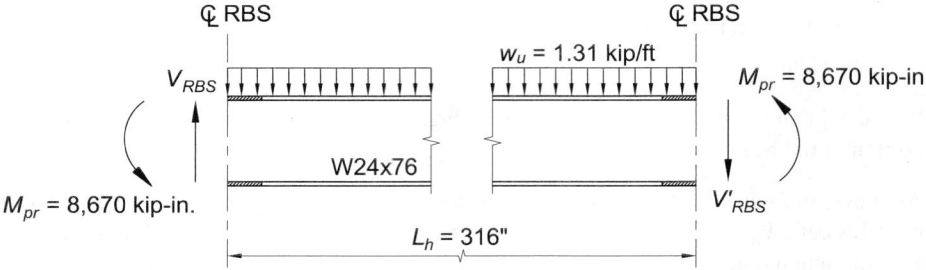

Fig. 4-12. Free body diagram of portion of beam between RBS cuts.

4.3 SPECIAL MOMENT FRAMES AND INTERMEDIATE MOMENT FRAMES

If the gravity load on the beam is something other than a uniform load, the correct shear forces at the centers of the RBS cuts are still obtained from equilibrium of the portion of the beam between the centers of the RBS cuts (i.e., by summing moments about each end of the free body diagram).

If the gravity load on the beam is very large, there is a possibility that the location of the plastic hinge may shift a significant distance outside of the RBS. If this is the case, the design procedure in ANSI/AISC 358 would require some modification, since the design procedure assumes the plastic hinge forms within the RBS. The possibility of the plastic hinge shifting outside of the RBS can be checked by drawing the moment diagram for the portion of the beam between RBS cuts. If the point of maximum moment is outside of the RBS and exceeds M_p of the full beam cross section, the plastic hinge location will not form in the RBS, and the ANSI/AISC 358 design procedure must be modified. This is unlikely to occur for typical spans and gravity loads, but may be a possibility for cases of very long beam spans and/or very large gravity loads. Figure 4-13 shows the moment diagram for the portion of the beam between RBS cuts for this example. This moment diagram confirms that the maximum moments occur at the RBS cuts, and therefore the plastic hinges will form in the RBS cuts, as assumed in the ANSI/AISC 358 design procedure.

Probable maximum moment at the face of the column
(Step 5 in ANSI/AISC 358 Section 5.8)

The probable maximum moment at the face of the column, M_f, is computed by taking a free body diagram of the portion of the beam between the center of the RBS cut and the face of the column. Summing moments for the free body diagram results in Equation 5.8-6 in ANSI/AISC 358. The probable maximum moment at the face of each column is:

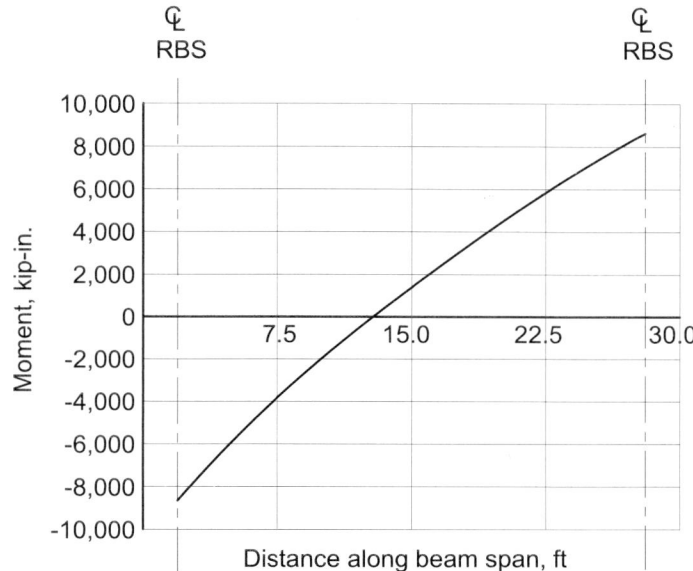

Fig. 4-13. Probable moment diagram for portion of beam between centers of RBS cuts.

$$M_f = M_{pr} + V_{RBS}S_h \quad \text{(ANSI/AISC 358 Eq. 5.8-6)}$$
$$= 8{,}670 \text{ kip-in.} + 72.1 \text{ kips}(14.5 \text{ in.})$$
$$= 9{,}720 \text{ kip-in.}$$

$$M'_f = M_{pr} + V'_{RBS}S_h$$
$$= 8{,}670 \text{ kip-in.} + 37.6 \text{ kips}(14.5 \text{ in.})$$
$$= 9{,}220 \text{ kip-in.}$$

The free body diagram corresponding to Equation 5.8-6 is shown in Figure 4-14 for the left side of the beam.

As noted in ANSI/AISC 358, this free body diagram and Equation 5.8-6 neglect the gravity load on the beam between the center of the RBS and the face of the column. Neglecting this gravity load introduces little error. For this example, if the gravity load of 1.31 kip/ft was included in the free body diagram in Figure 4-14, the value of M_f would increase by 11.5 kip-in.

Plastic moment of the beam based on the expected yield stress
(Step 6 in ANSI/AISC 358 Section 5.8)

$$M_{pe} = R_y F_y Z_x \quad \text{(ANSI/AISC 358 Eq. 5.8-7)}$$
$$= 1.1(50 \text{ ksi})\left(200 \text{ in.}^3\right)$$
$$= 11{,}000 \text{ kip-in.}$$

Alternatively, using AISC *Seismic Manual* Table 4-2 for the W24×76 beam, $R_y M_p = 917$ kip-ft = 11,000 kip-in.

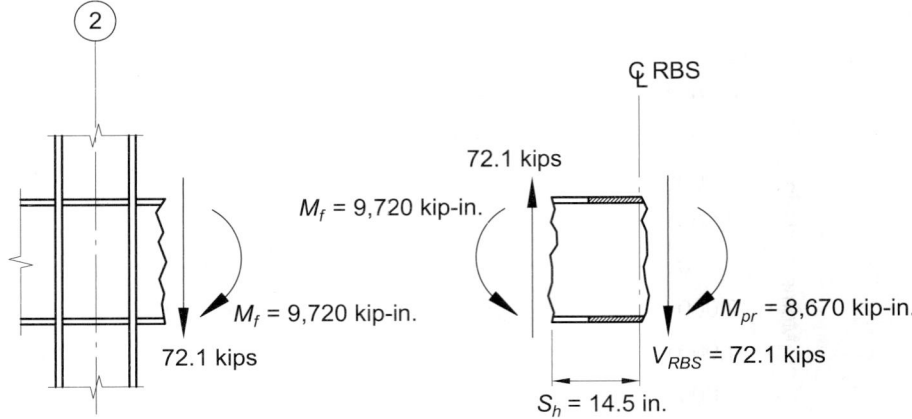

Fig. 4-14. Free body diagram of portion of beam between center of RBS and face of column.

4.3 SPECIAL MOMENT FRAMES AND INTERMEDIATE MOMENT FRAMES

Check that moment at the face of the column, M_f, does not exceed $\phi_d M_{pe}$ (Step 7 in ANSI/AISC 358 Section 5.8)

From ANSI/AISC 358 Section 2.4.1

$\phi_d = 1.00$

$\phi_d M_{pe} = 1.0(11{,}000 \text{ kip-in.})$
$= 11{,}000 \text{ kip-in.}$

$M_f = 9{,}720 \text{ kip-in.}$

$M_f \leq \phi_d M_{pe}$ \hfill (ANSI/AISC 358 Eq. 5.8-8)

9,720 kip-in. < 11,000 kip-in. **o.k.**

Because Equation 5.8-8 is satisfied, the preliminary values of $a = 5.50$ in., $b = 18.0$ in., and $c = 2.00$ in. are acceptable.

Because there is a significant difference between M_f and $\phi_d M_{pe}$, it may be possible to reduce the depth of the RBS cut. Reducing the RBS cut (the c dimension) from 2.0 in. to 1.5 in. will still satisfy Equation 5.8-8, and will result in a smaller increase in story drift ratio due to the presence of the RBS cut. On the other hand, increasing the RBS cut would reduce the shear demand on the panel zone, as discussed in Step 9 of this example. For the purpose of this example, continue with the RBS dimensions of $a = 5.50$ in., $b = 18.0$ in., and $c = 2.00$ in.

Required shear strength, V_u, of the beam and beam web-to-column connection (Step 8 in ANSI/AISC 358 Section 5.8)

The required shear strength of the beam and the beam-to-column connection, V_u, can be calculated by taking the previously computed value of V_{RBS} and adding the shear due to the gravity load on the portion of the beam between the center of the RBS and the face of the column:

$V_u = V_{RBS} + w_u S_h$
$= 72.1 \text{ kips} + 1.31 \text{ kip/ft}(1 \text{ ft}/12 \text{ in.})(14.5 \text{ in.})$
$= 73.7 \text{ kips}$

Note that there is little error in taking $V_u = V_{RBS}$.

The design shear strength of the W24×76 beam, ϕV_n, is 316 kips from AISC *Manual* Table 3-6.

$V_u = 73.7 \text{ kips} \leq \phi V_n$ **o.k.**

Design the beam web-to-column connection according to ANSI/AISC 358 Section 5.6 (Step 9 in ANSI/AISC 358 Section 5.8)

The required shear force at the column face is $V_u = 73.7$ kips as determined previously.

Select a single-plate connection with a plate at least ⅜ in. thick to support erection loads, per ANSI/AISC 358 Section 5.6(2)(a). The same section requires that the beam web be welded to the column flange using a complete-joint-penetration (CJP) groove weld.

With the single plate as backing, use a CJP groove weld to connect the beam web to the column flange.

From the AISC *Specification* Section G2.1, the required minimum remaining web depth between weld access holes for the 73.7 kips shear force is:

$$d_{min} = \frac{73.7 \text{ kips}}{\phi 0.6 F_y t_w C_v}$$

$$= \frac{73.7 \text{ kips}}{1.00(0.6)(50 \text{ ksi})(0.440 \text{ in.})(1.0)}$$

$$= 5.58 \text{ in.}$$

By inspection, sufficient web depth remains. **o.k.**

Continuity plate requirements according to ANSI/AISC 358 Chapter 2 (Step 10 in ANSI/AISC 358 Section 5.8)

ANSI/AISC 358 requires continuity plates for the prequalified RBS connection unless the exceptions of AISC *Seismic Provisions* Section E3.6f are met and both Equations E3-8 and E3-9 are satisfied.

$t_{cf} = 1.31$ in.
$R_{yb} = R_{yc} = 1.1$ from AISC *Seismic Provisions* Table A3.1

$$t_{cf} \geq 0.4\sqrt{1.8 b_{bf} t_{bf} \frac{R_{yb} F_{yb}}{R_{yc} F_{yc}}} \qquad (Provisions \text{ Eq. E3-8})$$

$$\geq 0.4\sqrt{1.8(8.99 \text{ in.})(0.680 \text{ in.})\frac{(1.1)(50 \text{ ksi})}{(1.1)(50 \text{ ksi})}}$$

1.31 in. < 1.33 in.   **n.g.**

$$t_{cf} \geq \frac{b_{bf}}{6} \qquad (Provisions \text{ Eq. E3-9})$$

$$t_{cf} \geq \frac{8.99 \text{ in.}}{6}$$

1.31 < 1.50 in.   **n.g.**

Neither Equation E3-8 nor Equation E3-9 is satisfied, so the minimum thickness requirements of Section E3.6f are not met. Therefore, continuity plates are required.

For this two-sided connection, the thickness of the continuity plates is required to be at least equal to the thicker beam flange on either side of the column according to AISC *Seismic Provisions* E3.6f(2). Therefore the minimum continuity plate thickness is 0.680 in.

Use ¾-in.-thick ASTM A572 Grade 50 continuity plates on both sides of the web.

Alternatively, the W14×176 column could be upsized to a W14×211 to avoid the need for a continuity plate. The decision to upsize the column should also consider the need to provide a doubler plate for the panel zone, as discussed in the following. For the purposes of this example, the column size will not be changed and ¾-in.-thick continuity plates are required.

Welds between the continuity plate and the column flanges are required to be CJP groove welds according to AISC *Seismic Provisions* Section E3.6f(3).

The AISC *Seismic Provisions* do not specify the width of the continuity plate. AISC *Specification* Section J10.8 says that the minimum width of each continuity plate plus $t_{cw}/2$ must be greater than $b_{bf}/3$. As shown below, however, this width does not appear to be sufficient to stiffen the column flanges due to the significant clip in the plate resulting from the column fillet.

From AISC *Specification* Section J10.8, the minimum continuity plate width is:

$$\frac{8.99 \text{ in.}}{3} - \frac{0.830 \text{ in.}}{2} = 2.58 \text{ in.}$$

While a 2⅝-in.-wide continuity plate is the minimum width permitted, this is too narrow because the resulting contact width is only 0.870 in. as shown in Figure 4-15, once the impact of the fillet is considered. AISC *Seismic Provisions* Section I2.4, which references AWS D1.8 clause 4.1, limits the corner clip to not more than ½ in. beyond the published k_1 dimension, where k_1 for a W14×176 is 1⅝ in. Thus, the length of contact between each continuity plate and the column flange is:

$$2.58 \text{ in.} - \left[(1⅝ \text{ in.} + ½ \text{ in.}) - \frac{0.830 \text{ in.}}{2} \right] = 0.870 \text{ in.}$$

The typical practice, therefore, is to set the continuity plate so that it is at least as wide as the edge of the frame beam flange:

$$\frac{8.99 \text{ in.}}{2} - \frac{0.830 \text{ in.}}{2} = 4.08 \text{ in.}$$

or it is as wide as the edge of the column flange:

$$\frac{15.7 \text{ in.}}{2} - \frac{0.830 \text{ in.}}{2} = 7.44 \text{ in.}$$

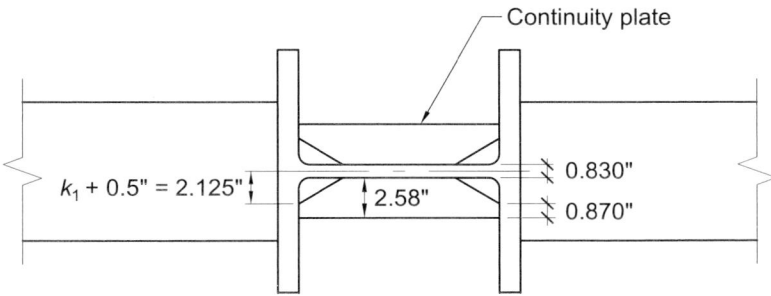

Fig. 4-15. Contact area between minimum-width continuity plate and column flange.

For illustration in this example, use 6.00 in. as the plate width, so that the contact width is:

$$6.00 \text{ in.} - \left[(1\tfrac{5}{8} \text{ in.} + \tfrac{1}{2} \text{ in.}) - \frac{0.830 \text{ in.}}{2}\right] = 4.29 \text{ in.}$$

According to AISC *Seismic Provisions* Section E3.6f(3), the strength of the sum of the welded joints of the continuity plates to the column web weld is the smallest of:

(a) Sum of design strengths in tension of continuity plate contact area with the column flanges
(b) Design strength in shear of continuity plate contact area with the web
(c) Design strength in shear of the column panel zone
(d) Sum of expected yield strengths of the beam flanges transmitting force to the continuity plates

Note that (a) and (b) can be analyzed for each continuity plate to column web on either side of the web, whereas (c) and (d) apply to the welds of both continuity plates.

Assuming a ¾-in.-thick plate, for continuity plate requirement (a), and using AISC *Specification* Section J4.1(a), the design tensile strength is:

$\phi_t T_n = \phi_t F_y (\text{contact area})$
$\quad = 0.90(50 \text{ ksi})(2)(4.29 \text{ in.})(¾ \text{ in.})$
$\quad = 290 \text{ kips}$

For continuity plate requirement (b):

AISC *Seismic Provisions* Section I2.4 states that continuity plates are to be detailed in accordance with AWS D1.8 clause 4.1. The corner clip of the continuity plate along the column web is equal to $k_{det} + 1.5$ in. $= 4.13$ in. The contact width with the web is:

Contact width $= 15.2$ in. $- 2(4.13$ in.$)$
$\qquad\qquad\qquad = 6.94$ in.

The design shear strength of the continuity plate based on the contact area with the web, from AISC *Specification* Section J4.2(a), is:

$\phi_v V_n = \phi_v 0.60 F_y (\text{contact area})$
$\quad = 1.00(0.60)(50 \text{ ksi})(6.94 \text{ in.})(¾ \text{ in.})$
$\quad = 156 \text{ kips}$

For continuity plate requirement (c) and assuming that $P_r \le 0.75 P_c$, the design strength of the panel zone, ϕR_n, is (note that AISC *Seismic Provisions* Section E3.6e(1) revises the value of ϕ to 1.00):

4.3 SPECIAL MOMENT FRAMES AND INTERMEDIATE MOMENT FRAMES

$$\phi R_n = \phi 0.60 F_y d_c t_w \left(1 + \frac{3 b_{cf} t_{cf}^2}{d_b d_c t_w}\right) \quad \textit{(Spec. Eq. J10-11)}$$

$$= 1.00(0.60)(50 \text{ ksi})(15.2 \text{ in.})(0.830 \text{ in.}) \times \left[1 + \frac{3(15.7 \text{ in.})(1.31 \text{ in.})^2}{23.9 \text{ in.}(15.2 \text{ in.})(0.830 \text{ in.})}\right]$$

$$= 480 \text{ kips}$$

Since this requirement applies to the entire panel zone, it will be divided by 2 when compared with requirements (a) and (b).

For continuity plate requirement (d):

$$T_n = 2 R_y F_y b_f t_f$$
$$= 2(1.1)(50 \text{ ksi})(8.99 \text{ in.})(0.680 \text{ in.})$$
$$= 672 \text{ kips}$$

Since this requirement applies to the entire panel zone, it will be divided by 2 when compared with requirements (a) and (b).

The continuity plate to column web weld is based on the smallest of 290 kips, 156 kips, 480 kips/2 = 240 kips, or 672 kips/2 = 336 kips. Thus, the design should be based on 156 kips.

To transfer 156 kips, a ½-in. double-sided fillet weld is required over the contact length. From AISC *Manual* Equation 8-2a:

$$D = \frac{156 \text{ kips}}{2(1.392 \text{ kip/in. per sixteenth})(6.94 \text{ in.})}$$
$$= 8.07 \text{ sixteenths}$$

Check column-beam relationships per ANSI/AISC 358 Section 5.4 (Step 11 in ANSI/AISC 358 Section 5.8)

AISC *Seismic Provisions* Section E3.4a requires that SMF connections satisfy the following strong-column-weak-beam criterion, assuming that the exceptions stated in Section E3.4a are not met.

$$\frac{\Sigma M_{pc}^*}{\Sigma M_{pb}^*} > 1.0 \quad \textit{(Provisions Eq. E3-1)}$$

The value of M_{pc}^* in this example is based on projecting M_{pc} to the beam centerline assuming that the column shear, V_c, is in equilibrium with the column moment, M_{pc}. This is consistent with the definition of M_{pc}^* in AISC *Seismic Provisions* Section E3.4a. Alternatively, the column shear could be computed to be in equilibrium with the beam moment, M_{pr}. The latter approach will result in a smaller value of M_{pc}^* and, when applied to Equation E3-1, will produce a slightly more conservative result.

The axial load on the column must also be considered when determining the flexural strength of the column at the beam centerline. (For simplicity, the same axial load will be used above and below the joint although this is not quite accurate.) Using $P_{uc} = 249$ kips as given in Example 4.3.2, and the height of the column to its assumed points of inflection above [$h_t = (12.5\text{ ft}/2)(12\text{ in./ft}) = 75.0$ in.] and below [$h_b = (14\text{ ft}/2)(12\text{ in./ft}) = 84.0$ in.] the beam centerline, ΣM_{pc}^* is determined as follows:

$$\Sigma M_{pc}^* = Z_{xt}\left(F_y - \frac{P_{uc}}{A_g}\right)\left(\frac{h_t}{h_t - d_b/2}\right) \quad \textit{(Provisions Eq. E3-2a)}$$

$$+ Z_{xb}\left(F_y - \frac{P_{uc}}{A_g}\right)\left(\frac{h_b}{h_b - d_b/2}\right)$$

$$= 320\text{ in.}^3\left(50\text{ ksi} - \frac{249\text{ kips}}{51.8\text{ in.}^2}\right)$$

$$\times\left\{\left[\frac{75.0\text{ in.}}{75.0\text{ in.} - (23.9\text{ in.}/2)}\right] + \left[\frac{84.0\text{ in.}}{84.0\text{ in.} - (23.9\text{ in.}/2)}\right]\right\}$$

$$= 34{,}100\text{ kip-in.}$$

The expected flexural demand of the beam at the column centerline is defined in ANSI/AISC 358 Section 5.4 as:

$$\Sigma M_{pb}^* = \Sigma(M_{pr} + M_{uv})$$

where

$$\Sigma M_{uv} = \Sigma\left[V_{RBS}\left(a + \frac{b}{2} + \frac{d_c}{2}\right)\right]$$

ΣM_{pr} = summation of the probable maximum moment at the center of each RBS determined previously

The term ΣM_{uv} is the sum of the moments produced at the column centerline by the shear at the plastic hinges. Recalling the values of V_{RBS} and V'_{RBS} computed in Step 4 of this example and the values of the RBS cut confirmed in Step 1, ΣM_{uv} is:

$$\Sigma M_{uv} = (V_{RBS} + V'_{RBS})\left(a + \frac{b}{2} + \frac{d_c}{2}\right)$$

$$= (72.1\text{ kips} + 37.6\text{ kips})\left(5.50\text{ in.} + \frac{18.0\text{ in.}}{2} + \frac{15.2\text{ in.}}{2}\right)$$

$$= 2{,}420\text{ kip-in.}$$

Therefore, the expected flexural demand of the beam at the column centerline is:

$$\Sigma M_{pb}^* = 2M_{pr} + \Sigma M_{uv}$$

$$= 2(8{,}670\text{ kip-in.}) + 2{,}420\text{ kip-in.}$$

$$= 19{,}800\text{ kip-in.}$$

4.3 SPECIAL MOMENT FRAMES AND INTERMEDIATE MOMENT FRAMES

$$\frac{\Sigma M_{pc}^*}{\Sigma M_{pb}^*} = \frac{34,100 \text{ kip-in.}}{19,800 \text{ kip-in.}}$$

$$= 1.72$$

$$1.72 > 1.0 \quad \textbf{o.k.}$$

Therefore, the strong-column-weak-beam check is satisfied.

Panel Zone Check

AISC *Seismic Provisions* Section E3.6e specifies that the required panel zone shear strength be calculated by summing the moments at the column faces as determined by projecting the expected moments at the plastic hinge points to the column faces; in this example, M_f and M_f'.

From statics, it can be seen that column panel-zone shear based on the summation of the expected moments at the column faces should be reduced by the column shear, V_c. The column shear, V_c, is not from the code-specified loads but is, instead, the column shear developed from the plastic hinging of the RBS. Assuming points of inflection at the mid-height of the columns above and below the panel zone and as previously determined in this example, $M_f = 9,720$ kip-in. on the left side of the beam and $M_f' = 9,220$ kip-in. on the right side, the value of V_c, ignoring the small effect of gravity loads between the expected plastic hinge location and the face of the column, is:

$$V_c = \frac{M_f + M_f'}{\frac{h_b}{2} + \frac{h_t}{2}}$$

$$= \frac{9,720 \text{ kip-in.} + 9,220 \text{ kip-in.}}{\left[\frac{(14.0 \text{ ft} + 12.5 \text{ ft})(12 \text{ in.}/1 \text{ ft})}{2}\right]}$$

$$= 119 \text{ kips}$$

where

h_b = story height above the joint, in.

h_t = story height below the joint, in.

The required strength of the panel zone is:

$$R_u = \frac{\Sigma M_f}{d_b - t_f} - V_c$$

$$= \frac{9,720 \text{ kip-in.} + 9,220 \text{ kip-in.}}{23.9 \text{ in.} - 0.680 \text{ in.}} - 119 \text{ kips}$$

$$= 697 \text{ kips}$$

AISC *Seismic Provisions* Section E3.6e(1) requires that the design shear strength of the panel zone be determined in accordance with the limit state of shear yielding in AISC *Specification* Section J10.6 with $\phi_v = 1.00$. Specifically, AISC *Specification* Section J10.6(b) is applicable, because frame stability, including plastic panel-zone deformation, is considered in the analysis. Determine the applicable equation as follows:

$P_r = 243$ kips from Example 4.3.2

$P_r < 0.75 P_c$
$P_r < 0.75 F_y A_g$
$P_r < 0.75(50 \text{ ksi})(51.8 \text{ in.}^2)$
243 kips < 1,940 kips   **o.k.**

Therefore, the shear strength of the panel zone is given by AISC *Specification* Equation J10-11:

$$R_n = 0.60 F_y d_c t_w \left[1 + \frac{3 b_{cf} t_{cf}^2}{d_b d_c t_w}\right] \quad \text{(Spec. Eq. J10-11)}$$

$\phi R_n = 1.00(0.60)(50 \text{ ksi})(15.2 \text{ in.})(0.830 \text{ in.})$

$$\times \left[1 + \frac{3(15.7 \text{ in.})(1.31 \text{ in.})^2}{(23.9 \text{ in.})(15.2 \text{ in.})(0.830 \text{ in.})}\right]$$

$= 480$ kips

Alternatively, using Table 4-2 of this Manual for the **W14×176** column:

$0.75 P_y = 1,940$ kips
$\phi R_{v1} = 378$ kips
$\phi R_{v2} = 2,420$ kip-in.

$\phi R_n = \phi R_{v1} + \dfrac{\phi R_{v2}}{d_b}$

$= 378 \text{ kips} + \dfrac{2,420 \text{ kip-in.}}{23.9 \text{ in.}}$

$= 479$ kips

Because $R_u = 697$ kips $> \phi R_n$, a column-web doubler plate is required. Note that if V_c had not been subtracted, the required panel zone strength would have been approximately 816 kips, which is a 17% increase in demand.

Comments:

Tests and analyses have shown that the actual shear strength of the panel zone might be considerably larger than the shear force that causes global shear yielding, because of strain hardening and the additional resistance provided by the column flanges bounding the panel zone. Therefore, AISC *Specification* Section J10.6(b) permits utilization of this additional shear resistance in design when the flexibility of the panel zone is considered in analysis. Designers should be aware, however, that significant inelastic deformations might be associated with this increase in resistance. For connections in which beam flanges are directly welded to column flanges, large inelastic shear distortion of the panel zone might contribute

4.3 SPECIAL MOMENT FRAMES AND INTERMEDIATE MOMENT FRAMES

to the initiation and propagation of fracture at welded beam-to-column connections. In such cases, sharing of inelastic deformations between beams and panel zones is not encouraged. See Hamburger et al. (2009) for additional information.

It has already been pointed out in this example that reducing the RBS cut (i.e., dimension c) might be possible to bring M_f closer to $\phi_d M_{pe}$ and reduce the impact of the RBS on frame stiffness. Also, increasing the RBS cut dimensions would reduce shear demand on the panel zone and, in some cases, eliminate the need to install doubler plates.

Size Web Doubler Plate

The minimum thickness of each component of the panel zone, without the aid of intermediate plug welds between the column web and the doubler is:

$$t \geq \frac{(d_z + w_z)}{90} \qquad \text{(\textit{Provisions} Eq. E3-7)}$$

From Table 4-2 of this Manual, for the W24×76 beam:

$$\frac{d_z}{90} = 0.250 \text{ in}$$

From Table 4-2 of this Manual, for the W14×176 column:

$$\frac{w_z}{90} = 0.140 \text{ in.}$$

$$t \geq 0.250 \text{ in.} + 0.140 \text{ in.} = 0.390 \text{ in.}$$

The column web satisfies this requirement:

$$t_w = 0.830 \text{ in.} > 0.390 \text{ in.} \qquad \textbf{o.k.}$$

If the doubler plate satisfies this minimum thickness, it is permitted to be applied directly to the column web or spaced away from the web, without the use of plug welds.

The available shear strength of the panel zone is checked using AISC *Specification* Equation J10-11 with the thickness, t_w, taken as the combined thickness of the column web and doubler plate.

$$R_n = 0.60 F_y d_c t_w \left[1 + \frac{3 b_{cf} t_{cf}^2}{d_b d_c t_w} \right] \qquad \text{(\textit{Spec.} Eq. J10-11)}$$

Where t_w used in two places is replaced by $t_w + t_p$.

Rearranging to solve for t_p:

$$t_w + t_p \geq \left[R_u - \frac{0.60 F_y \left(3 b_{cf} t_{cf}^2 \right)}{d_b} \right] \left[\frac{1}{0.6 F_y d_c} \right]$$

$$t_p \geq \left\{ 697 \text{ kips} - \frac{0.60(50 \text{ ksi})\left[3(15.7 \text{ in.})(1.31 \text{ in.})^2\right]}{(23.9 \text{ in.})} \right\}$$

$$\times \left[\frac{1}{0.60(50 \text{ ksi})(15.2 \text{ in.})} \right] - 0.830 \text{ in.}$$

$t_p \geq 0.476$

Use a ½-in.-thick doubler plate.

Because the doubler plate meets the minimum thickness required by AISC *Seismic Provisions* Equation E3-7 (0.390 in.), plug welds between the doubler and the column web are not required. The length of the clip of the continuity plate at the doubler plate is not required to meet AWS D1.8 clause 4.1. Use a 1 in. × 1 in. clip.

Extend the doubler plate 6 in. above and below the beams. Attach the doubler plate to the column flanges using complete-joint-penetration groove welds, as stipulated in AISC *Seismic Provisions* Section E3.6e(3)(2). A minimum-sized fillet weld may be used across the top and bottom of the doubler plate to avoid free edges, but is not required.

Alternatively, two doubler plates spaced away from the column web can be used as shown in AISC *Seismic Provisions* Commentary Figure C-E3.3(c).

Figure 4-16 shows the final configuration of the panel zone.

Installing doubler plates can be costly, so selecting a larger column may also be considered. Upsizing the column may potentially eliminate the need for continuity plates. The amount that the column can be upsized without losing the savings associated with eliminating the doubler plate varies significantly depending on the project and geographic region, but a general rule of thumb suggests that upsizing the column between 50 to 100 lb/ft might still be more cost-effective than installing doubler plates and continuity plates. The column would have to be upsized to a W14×257 to eliminate both continuity plates and doubler plates—a weight increase of 81 lb/ft from the W14×176 used in this example. This weight increase is in the middle of the 50 to 100 lb/ft range where upsizing might be cost effective and a discussion with a fabricator is recommended. Nevertheless, for the purposes of this example, the W14×176 column size will be retained to illustrate the design of the doubler plate.

Column Bracing Requirements

AISC *Seismic Provisions* Section E3.4c allows the use of a strong-column/weak-beam ratio (AISC *Seismic Provisions* Equation E3-1) greater than 2.0 to show that a column remains elastic outside of the panel zone at restrained beam-to-column connections. If it can be demonstrated that the column remains elastic outside of the panel zone, AISC *Seismic Provisions* Section E3.4c(1) requires the column flanges to be braced at the level of the beam top flanges only. With a ratio of 1.72 in this example, the column cannot be assumed to remain elastic and bracing is required at both the top and bottom flanges of the beam.

4.3 SPECIAL MOMENT FRAMES AND INTERMEDIATE MOMENT FRAMES

Column flange restraint at these locations can be provided by continuity plates and a full-depth shear plate between the continuity plates at the connection of the girder framing into the weak axis of the column.

Specify Beam Flange-to-Column Flange Connection

Per AISC *Seismic Provisions* Section E3.6c, the connection configuration must comply with the requirements of the prequalified connection, or provisions of qualifying cyclic test results in accordance with Section K2. ANSI/AISC 358 Section 5.5(1) requires a complete-joint-penetration groove weld,

Use a complete-joint-penetration groove weld to connect the beam flanges to the column flange. The weld access hole geometry is required to comply with AISC *Specification* Section J1.6. The welds are also considered demand critical.

The final connection design and geometry is shown in Figure 4-16.

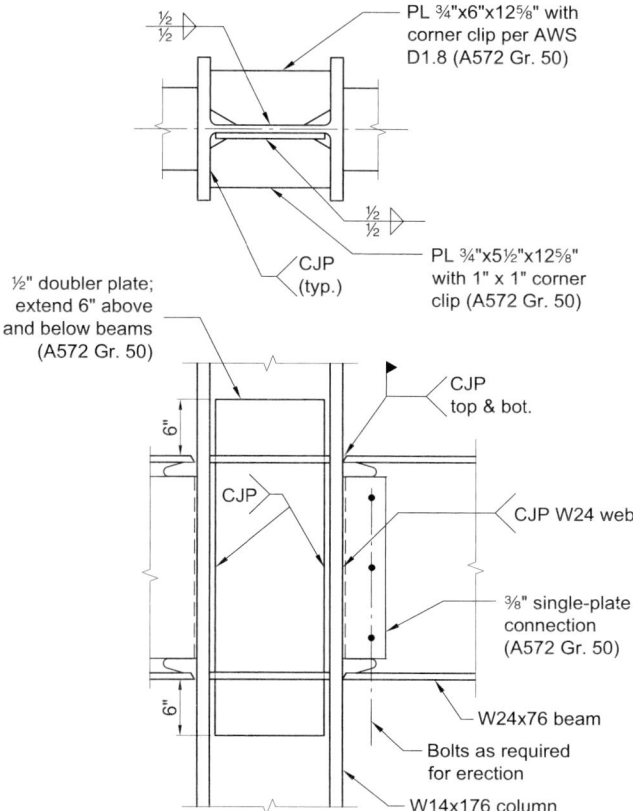

Note: For weld backing requirements, see ANSI/AISC 358 Chapter 3.

Fig. 4-16. Design Example 4.3.4 connection geometry.

4.4 COLUMN SPLICE AND COLUMN BASE DESIGN EXAMPLES

The following design examples address the design of gravity column splices, SMF column splices, SMF column bases, and SMF embedded column bases.

Example 4.4.1. Gravity Column Splice Design in a Moment Frame Building

Given:
Refer to the floor plan shown in Figure 4-7 and the SMF elevation shown in Figure 4-8. Design a splice using bolted flange plates between the third and fourth levels for the gravity column located at the intersection of grids 2 and B. Use ASTM A572 Grade 50 for all splice material. The column sizes above and below the splice are ASTM A992 W12×40 and W12×58, respectively.

Solution:
From AISC *Manual* Table 2-4, the beam and column material properties are as follows:

ASTM A992
$F_y = 50$ ksi
$F_u = 65$ ksi

From AISC *Manual* Table 2-5, the splice material properties are as follows:

ASTM A572 Grade 50
$F_y = 50$ ksi
$F_u = 65$ ksi

From AISC *Manual* Table 1-1, the column geometric properties are as follows:

W12×58—Lower shaft
$d = 12.2$ in. $d_{det} = 12\tfrac{1}{4}$ in. $t_f = 0.640$ in. $b_f = 10.0$ in.
$Z_x = 86.4$ in.^{3} $Z_y = 32.5$ in.^{3}

W12×40—Upper shaft
$d = 11.9$ in. $d_{det} = 12$ in. $t_f = 0.515$ in. $b_f = 8.01$ in.
$Z_x = 57.0$ in.^{3} $Z_y = 16.8$ in.^{3}

AISC *Seismic Provisions* Sections D2.1, D2.5a and D2.5c have requirements for gravity column splices. Note that these gravity column splice provisions are equally applicable to gravity column splices in braced-frame buildings.

Check splice location
AISC *Seismic Provisions* Section D2.5a requires that the splice be located a minimum of 4 ft from the beam-to-column connections. The three exceptions to this requirement do not apply for this building.

4.4 COLUMN SPLICE AND COLUMN BASE DESIGN EXAMPLES

Assume that the gravity column splices are at the same vertical elevation as the SMF column splices shown in Figure 4-8. This location satisfies AISC *Seismic Provisions* Section D2.5a.

Required Shear Strength of Splice in Weak Axis of Column

AISC *Seismic Provisions* Section D2.5c requires that, with respect to both orthogonal axes, the column splice be able to develop a required shear strength equal to:

LRFD	ASD
$V_u = \dfrac{M_{pc}}{H}$	$V_a = \dfrac{M_{pc}}{1.5H}$

In the weak axis of the column, the required shear strength of the splice is:

LRFD	ASD
$V_{uy} = \dfrac{F_y Z_y}{H}$ $= \dfrac{50 \text{ ksi}(16.8 \text{ in.}^3)}{12.5 \text{ ft}(12 \text{ in./ft})}$ $= 5.60 \text{ kips}$	$V_{ay} = \dfrac{F_y Z_y}{1.5H}$ $= \dfrac{50 \text{ ksi}(16.8 \text{ in.}^3)}{1.5(12.5 \text{ ft})(12 \text{ in./ft})}$ $= 3.73 \text{ kips}$

The shear force to be resisted by each flange splice plate is half of M_{pc}/H. Therefore, for one splice plate:

LRFD	ASD
$V_{uy} = 5.60 \text{ kips}/2$ $= 2.80 \text{ kips}$	$V_{ay} = 3.73 \text{ kips}/2$ $= 1.87 \text{ kips}$

Note that the smaller column, the W12×40, controls the required shear strength, as is stipulated in AISC *Seismic Provisions* Section D2.5c.

Conservatively ignoring frictional resistance between the upper and lower shafts due to column dead load, this force will be resisted by the splice material.

Required Compressive Strength of Splice

With the upper shaft centered on the lower shaft, the dimensions of the upper shaft are such that it will achieve full contact bearing on the lower shaft. Therefore, the splice will not be required to transfer any compressive loads if the upper shaft is finished to bear on the lower shaft. Because a note stating, "finish to bear," is provided on the detail, Case I-A applies from AISC *Manual* Part 14, Table 14-3.

Splice Geometry

Try the column splice detail from AISC *Manual* Part 14, Table 14-3, Case I-A.

W12×40

$d_u = d_{det}$
  $= 12$ in.

W12×58

$d_l = d_{det}$
  $= 12\frac{1}{4}$ in.

$d_u + \frac{1}{4}$ in. $\leq d_l \leq d_u + \frac{5}{8}$ in.
$d_u + \frac{1}{4}$ in. $= 12\frac{1}{4}$ in.
$d_u + \frac{5}{8}$ in. $= 12\frac{5}{8}$ in.
$12\frac{1}{4}$ in. $\leq 12\frac{1}{4}$ in. $\leq 12\frac{5}{8}$ in. **o.k.**

From Case 1-A of AISC *Manual* Table 14-3, use Type 2 flange plates.

PL$\frac{3}{8}$ in. × 8 in. × 1 ft 0$\frac{1}{2}$ in.
$g_u = g_l = 5\frac{1}{2}$ in.

Splice Bolts

Because the centroid of each bolt group is eccentric to the column ends, there will be a moment on each bolt group. Using the geometry shown in Table 14-3, Case 1-A of the AISC *Manual* and considering the eccentricity from the center of the bolt group to the column interface, this moment is:

LRFD	ASD

$M_u = V_{uy}e$
  $= 2.80$ kips$\left[\frac{1}{2}(3 \text{ in.}) + 1\frac{3}{4} \text{ in.}\right]$
  $= 9.10$ kip-in.
	$M_a = V_{ay}e$
	  $= 1.87$ kips$\left[\frac{1}{2}(3 \text{ in.}) + 1\frac{3}{4} \text{ in.}\right]$
	  $= 6.08$ kip-in.

The geometry of each bolt group is such that the bolts are all equidistant from the centroid of their bolt group. Therefore, the moment will be shared equally between the bolts. The x-, y- and radial distances from the center of gravity of the bolt group to the center of each bolt following the procedure and definitions in AISC *Manual* Part 7, are:

$c_x = \frac{1}{2}(5\frac{1}{2}$ in.$)$
  $= 2.75$ in.
$c_y = \frac{1}{2}(3$ in.$)$
  $= 1.50$ in.
$c = \sqrt{(2.75 \text{ in.})^2 + (1.50 \text{ in.})^2}$
  $= 3.13$ in.

The polar moment of inertia of the bolt group is:

$$I_y \approx \Sigma c_x^2 = 4(2.75 \text{ in.}^2)^2 (1/\text{in.}^2)$$
$$= 30.3 \text{ in.}^4/\text{in.}^2$$
$$I_x \approx \Sigma c_y^2 = 4(1.50 \text{ in.}^2)^2 (1/\text{in.}^2)$$
$$= 9.00 \text{ in.}^4/\text{in.}^2$$
$$I_p \approx I_x + I_y$$
$$\approx 30.3 \text{ in.}^4/\text{in.}^2 + 9.00 \text{ in.}^4/\text{in.}^2 = 39.3 \text{ in.}^4/\text{in.}^2$$

From AISC *Manual* Equation 7-2a, the direct shear force on each bolt due to the concentric force, V_{uy} and V_{ay}, applied at 90° with respect to the vertical is:

LRFD	ASD
From AISC *Manual* Equation 7-3a	From AISC *Manual* Equation 7-3b
$r_{pxu} = r_{pu}\sin\theta$ $= \dfrac{V_{uy}\sin 90°}{n}$ $= \dfrac{2.80 \text{ kips}(1.00)}{4}$ $= 0.700 \text{ kips}$	$r_{pxa} = r_{pa}\sin\theta$ $= \dfrac{V_{ay}\sin 90°}{n}$ $= \dfrac{1.87 \text{ kips}(1.00)}{4}$ $= 0.468 \text{ kips}$
From AISC *Manual* Equation 7-4a	From AISC *Manual* Equation 7-4b
$r_{pyu} = r_{pu}\cos\theta$ $= \dfrac{V_{uy}\cos 90°}{n}$ $= \dfrac{2.80 \text{ kips}(0)}{4}$ $= 0 \text{ kips}$	$r_{pya} = r_{pa}\cos\theta$ $= \dfrac{V_{ay}\cos 90°}{n}$ $= \dfrac{1.87 \text{ kips}(0)}{4}$ $= 0 \text{ kips}$

The additional shear force on each of the four bolts in the bolt group due to the moment caused by eccentricity is:

LRFD	ASD
From AISC *Manual* Equation 7-6a	From AISC *Manual* Equation 7-6b
$r_{mxu} = \left(\dfrac{M_u c_y}{I_p}\right)$ $= \dfrac{9.10 \text{ kip-in.}(1.50 \text{ in.})}{\left(39.3 \dfrac{\text{in.}^4}{\text{in.}^2}\right)}$ $= 0.347 \text{ kips}$	$r_{mxa} = \left(\dfrac{M_a c_y}{I_p}\right)$ $= \dfrac{6.08 \text{ kip-in.}(1.50 \text{ in.})}{\left(39.3 \dfrac{\text{in.}^4}{\text{in.}^2}\right)}$ $= 0.232 \text{ kips}$
From AISC *Manual* Equation 7-7a	From AISC *Manual* Equation 7-7b
$r_{myu} = \left(\dfrac{M_u c_x}{I_p}\right)$ $= \dfrac{9.10 \text{ kip-in.}(2.75 \text{ in.})}{\left(39.3 \dfrac{\text{in.}^4}{\text{in.}^2}\right)}$ $= 0.637 \text{ kips}$	$r_{mya} = \left(\dfrac{M_a c_x}{I_p}\right)$ $= \dfrac{6.08 \text{ kip-in.}(2.75 \text{ in.})}{\left(39.3 \dfrac{\text{in.}^4}{\text{in.}^2}\right)}$ $= 0.425 \text{ kips}$

The required strength per bolt is then:

LRFD	ASD
From AISC *Manual* Equation 7-8a	From AISC *Manual* Equation 7-8b
$r_u = \sqrt{(r_{pxu} + r_{mxu})^2 + (r_{pyu} + r_{myu})^2}$ $= \sqrt{(0.700 \text{ kips} + 0.347 \text{ kips})^2 + (0 \text{ kips} + 0.637 \text{ kips})^2}$ $= 1.23 \text{ kips}$	$r_u = \sqrt{(r_{pxa} + r_{mxa})^2 + (r_{pya} + r_{mya})^2}$ $= \sqrt{(0.468 \text{ kips} + 0.232 \text{ kips})^2 + (0 \text{ kips} + 0.425 \text{ kips})^2}$ $= 0.819 \text{ kips}$

From AISC *Manual* Table 7-1 for a ¾-in.-diameter ASTM A325-N bolt (Group A):

LRFD	ASD
$\phi r_n = 17.9 \text{ kips}$ $\phi r_n > r_u$ **o.k.**	$\dfrac{r_n}{\Omega} = 11.9 \text{ kips}$ $\dfrac{r_n}{\Omega} > r_a$ **o.k.**

Use ¾-in.-diameter ASTM A325-N bolts in standard holes.

4.4 COLUMN SPLICE AND COLUMN BASE DESIGN EXAMPLES

Bearing Strength of Splice Plate

Using AISC *Manual* Table 7-5 with $L_e = 1\frac{1}{4}$ in., hole type = STD, $F_u = 65$ ksi:

LRFD	ASD
$\phi r_n = 49.4$ kip/in.($\frac{3}{8}$ in.) $= 18.5$ kips $\phi r_n > r_u$ **o.k.**	$\dfrac{r_n}{\Omega} = 32.9$ kip/in.($\frac{3}{8}$ in.) $= 12.3$ kips $\dfrac{r_n}{\Omega} > r_a$ **o.k.**

Bearing Strength of the Column Flanges

Since the column flanges are thicker and wider than the splice plates and their tensile strength is equal to the splice material, the bearing strength of the column flanges is adequate.

Block Shear Rupture of the Splice Plates

A block shear failure path is assumed as shown in Figure 4-17. The available strength for the limit state of block shear rupture is given in AISC *Specification* Section J4.3 as follows:

$R_n = 0.60 F_u A_{nv} + U_{bs} F_u A_{nt} \leq 0.60 F_y A_{gv} + U_{bs} F_u A_{nt}$ (*Spec.* Eq. J4-5)

$U_{bs} = 1.0$

$A_{nt} = (3 \text{ in.} + 1\frac{1}{2} \text{ in.})(\frac{3}{8} \text{ in.}) - 1.5(^{13}/_{16} \text{ in.} + ^{1}/_{16} \text{ in.})(\frac{3}{8} \text{ in.})$
$\quad = 1.20 \text{ in.}^2$

$A_{gv} = (5\frac{1}{2} \text{ in.} + 1\frac{1}{4} \text{ in.})(\frac{3}{8} \text{ in.})$
$\quad = 2.53 \text{ in.}^2$

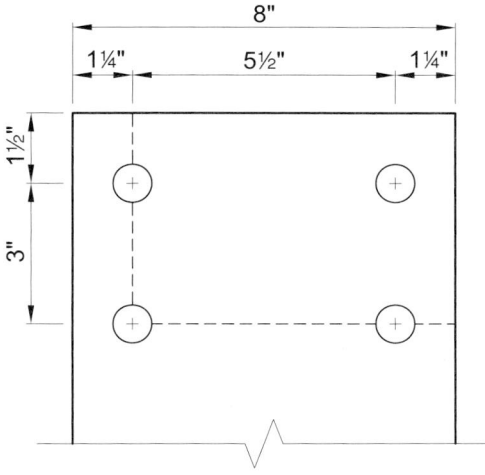

Fig. 4-17. Block shear failure path for splice plate.

$$A_{nv} = 2.53 \text{ in.}^2 - (1.5)(^{13}\!/_{16} \text{ in.} + ^1\!/_{16} \text{ in.})(^3\!/_8 \text{ in.})$$
$$= 2.04 \text{ in.}^2$$

$$\begin{aligned}F_u A_{nt} &= 65 \text{ ksi}\left(1.20 \text{ in.}^2\right) \\ &= 78.0 \text{ kips}\end{aligned}$$

$$\begin{aligned}0.60 F_u A_{nv} &= 0.60(65 \text{ ksi})\left(2.04 \text{ in.}^2\right) \\ &= 79.6 \text{ kips}\end{aligned}$$

$$\begin{aligned}0.60 F_y A_{gv} &= 0.60(50 \text{ ksi})\left(2.53 \text{ in.}^2\right) \\ &= 75.9 \text{ kips}\end{aligned}$$

$$\begin{aligned}R_n &= 0.60 F_u A_{nv} + U_{bs} F_u A_{nt} \leq 0.60 F_y A_{gv} + U_{bs} F_u A_{nt} \quad \text{(Spec. Eq. J4-5)}\\ &= 79.6 \text{ kips} + 1.0(78.0 \text{ kips}) \leq 75.9 \text{ kips} + 1.0(78.0 \text{ kips}) \\ &= 158 \text{ kips} > 154 \text{ kips}\end{aligned}$$

Use $R_n = 154$ kips.

The available strength for the limit state of block shear rupture is:

LRFD	ASD
$\phi R_n = 0.75(154 \text{ kips})$ $= 116 \text{ kips}$ 116 kips $\geq$ 2.80 kips **o.k.**	$\dfrac{R_n}{\Omega} = \dfrac{154 \text{ kips}}{2.00}$ $= 77.0 \text{ kips}$ 77.0 kips $\geq$ 1.87 kips **o.k.**

Shear Yielding of the Splice Plates

From AISC *Specification* Section J4.2, the available shear strength due to the limit state of shear yielding of one splice plate is:

LRFD	ASD
$\phi V_n = \phi 0.60 F_y A_g$ $= 1.00(0.60)(50 \text{ ksi})(^3\!/_8 \text{ in.})(8.00 \text{ in.})$ $= 90.0 \text{ kips} > 2.80 \text{ kips}$	$\dfrac{V_n}{\Omega} = \dfrac{0.60 F_y A_g}{\Omega}$ $= \dfrac{0.60(50 \text{ ksi})(^3\!/_8 \text{ in.})(8.00 \text{ in.})}{1.50}$ $= 60.0 \text{ kips} > 1.87 \text{ kips}$ **o.k.**

Shear Yielding of the Column Flanges

Because the column flanges are thicker and wider than the splice plates and their yield strength is equal to the splice material, the shear yielding strength of the column flanges is adequate.

Shear Rupture of the Splice Plates

The net area of one splice plate is:

$$A_n = [8.00 \text{ in.} - 2(^{13}\!/_{16} \text{ in.} + ^{1}\!/_{16} \text{ in.})](^{3}\!/_{8} \text{ in.})$$
$$= 2.34 \text{ in.}^2$$

From AISC *Specification* Equation J4-4, the available strength due to the limit state of shear rupture for each splice plate is:

LRFD	ASD
$\phi V_n = \phi 0.60 F_u A_{nv}$ $= 0.75(0.60)(65 \text{ ksi})(2.34 \text{ in.}^2)$ $= 68.4 \text{ kips} > 2.80 \text{ kips}$ **o.k.**	$\dfrac{V_n}{\Omega} = \dfrac{0.60 F_u A_{nv}}{\Omega}$ $= \dfrac{0.60(65 \text{ ksi})(2.34 \text{ in.}^2)}{2.00}$ $= 45.6 \text{ kips} > 1.87 \text{ kips}$ **o.k.**

Required Shear Strength of the Splice in the Strong Axis of the Column

AISC *Seismic Provisions* Section D2.5c requires that the column splice be able to develop a required shear strength in the strong axis of the column equal to:

LRFD	ASD
$V_{ux} = \dfrac{M_{pcx}}{H}$ $= \dfrac{F_y Z_x}{H}$ $= \dfrac{50 \text{ ksi}(57.0 \text{ in.}^3)}{12.5 \text{ ft}(12 \text{ in./ft})}$ $= 19.0 \text{ kips}$	$V_{ax} = \dfrac{M_{pcx}}{1.5H}$ $= \dfrac{F_y Z_x}{1.5H}$ $= \dfrac{50 \text{ ksi}(57.0 \text{ in.}^3)}{1.5(12.5 \text{ ft})(12 \text{ in./ft})}$ $= 12.7 \text{ kips}$

Bolted splice plates could be provided on the column web, but it may be possible to resist the strong-axis shear through weak axis bending of the flange plates.

Since there are two flange splice plates, the applied force on each plate is one half of the shear calculated for the strong axis of the column.

LRFD	ASD
$V_{ux} = \dfrac{19.0 \text{ kips}}{2}$ $= 9.50 \text{ kips}$	$V_{ax} = \dfrac{12.7 \text{ kips}}{2}$ $= 6.35 \text{ kips}$

Weak-Axis Flexural Yielding of the Splice Plate

Assuming the column is rigid enough to force all deformation into the splice plate, the relative movement between the columns will cause weak-axis plate bending. The bending behavior in the plate is that of a beam fixed at one end, free to deflect vertically but not rotate at the other (Case 23 of Table 3-23 in the AISC *Manual*).

The limit states checked are flexural yielding of the splice plate, shear yielding of the splice plate, shear rupture of the splice plate, and prying action on the innermost bolts.

The length of bending is the distance between the bearing plane of the columns and the innermost bolt line, which is 1.75 in. according to Figure 4-18.

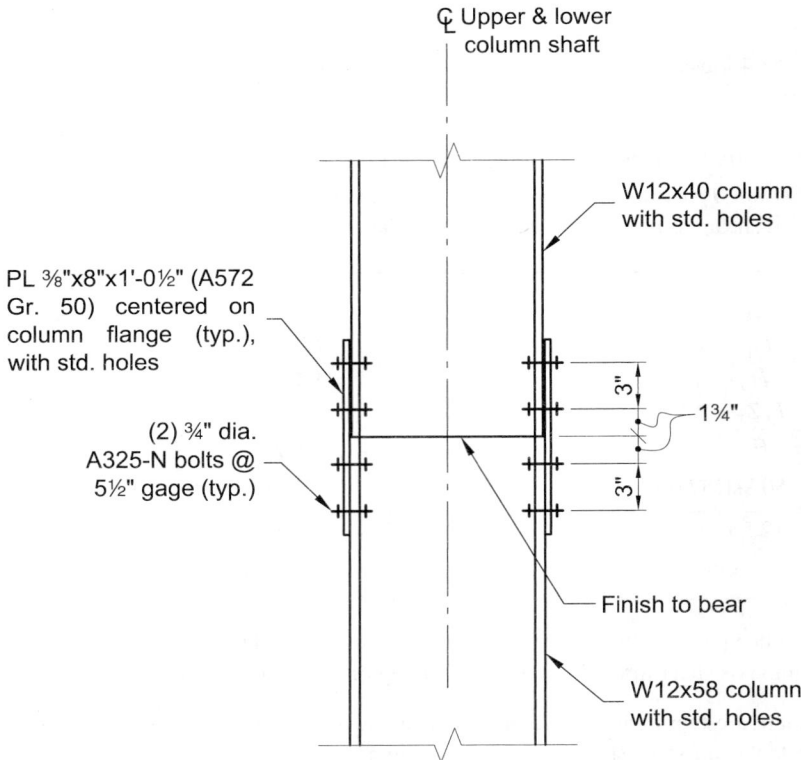

Fig. 4-18. Connection as designed in Example 4.4.1.

The required flexural strength of the plate, from AISC *Manual* Table 3-23 Case 23, is:

LRFD	ASD
$M_u = \dfrac{V_{ux}L}{2}$ $= \dfrac{9.50 \text{ kips}(1.75 \text{ in.})}{2}$ $= 8.31 \text{ kip-in.}$	$M_a = \dfrac{V_{ax}L}{2}$ $= \dfrac{6.35 \text{ kips}(1.75 \text{ in.})}{2}$ $= 5.56 \text{ kip-in.}$

As determined previously, the splice plates are PL³⁄₈ in. × 8 in. × 1 ft 0¹⁄₂ in. Using AISC *Specification* Section F11, determine the available flexural yielding strength of the plate. Note that the dimension t used in AISC *Specification* Section F11 is parallel to the axis of bending, and therefore $t = 8.00$ in. for weak-axis bending of the splice plate in this example.

Check the limit on $L_b d/t^2$:

$$\frac{L_b d}{t^2} = \frac{1.75 \text{ in.}(\tfrac{3}{8} \text{ in.})}{(8.00 \text{ in.})^2}$$
$$= 0.0103$$

$$\frac{0.08E}{F_y} = \frac{0.08(29{,}000 \text{ ksi})}{50 \text{ ksi}}$$
$$= 46.4$$

Because $\dfrac{L_b d}{t^2} < \dfrac{0.08E}{F_y}$, AISC *Specification* Equation F11-1 applies. The nominal flexural yielding strength of the plate from Equation F11-1 is:

$$M_n = F_y Z \leq 1.6 M_y$$
$$= 50 \text{ ksi} \frac{(8.00 \text{ in.})(\tfrac{3}{8} \text{ in.})^2}{4}$$
$$= 14.1 \text{ kip-in.}$$

$$1.6 M_y = 1.6 F_y S_x$$
$$= 1.6(50 \text{ ksi}) \left[\frac{(8.00 \text{ in.})(\tfrac{3}{8} \text{ in.})^2}{6} \right]$$
$$= 15.0 \text{ kip-in.}$$

14.1 kip-in. ≤ 15.0 kip-in., therefore, $M_n = 14.1$ kip-in.

The available flexural yielding strength is:

LRFD	ASD
$\phi_b M_n = \phi_b F_y Z \leq 1.6 M_y$ $= 0.90(14.1 \text{ kip-in.})$ $= 12.7 \text{ kip-in.} > 8.31 \text{ kip-in.}$ **o.k.**	$\dfrac{M_n}{\Omega_b} = \dfrac{F_y Z}{\Omega_b}$ $= \dfrac{14.1 \text{ kip-in.}}{1.67}$ $= 8.44 \text{ kip-in.} > 5.56 \text{ kip-in.}$ **o.k.**

Shear Yielding of the Splice Plate
Using AISC *Specification* Equation J4-3:

LRFD	ASD
$\phi R_n = \phi 0.60 F_y A_{gv}$ $= 1.00(0.60)(50 \text{ ksi})(\frac{3}{8} \text{ in.})(8.00 \text{ in.})$ $= 90.0 \text{ kips} > 9.50 \text{ kips}$ **o.k.**	$\dfrac{R_n}{\Omega} = \dfrac{0.60 F_y A_{gv}}{\Omega}$ $= \dfrac{(0.60)(50 \text{ ksi})(\frac{3}{8} \text{ in.})(8.00 \text{ in.})}{1.50}$ $= 60.0 \text{ kips} > 6.35 \text{ kips}$ **o.k.**

Shear Rupture of the Splice Plate

$$A_{nv} = 8.00 \text{ in.}(\tfrac{3}{8} \text{ in.}) - 2(^{13}/_{16} \text{ in.} + \tfrac{1}{16} \text{ in.})(\tfrac{3}{8} \text{ in.})$$
$$= 2.34 \text{ in.}^2$$

Using AISC *Specification* Equation J4-4:

LRFD	ASD
$\phi R_n = \phi 0.60 F_u A_{nv}$ $= 0.75(0.60)(65 \text{ ksi})(2.34 \text{ in.}^2)$ $= 68.4 \text{ kips} > 9.50 \text{ kips}$ **o.k.**	$\dfrac{R_n}{\Omega} = \dfrac{0.60 F_u A_{nv}}{\Omega}$ $= \dfrac{0.60(65 \text{ ksi})(2.34 \text{ in.}^2)}{2.00}$ $= 45.6 \text{ kips} > 6.35 \text{ kips}$ **o.k.**

Prying Action on the Splice Plates

Because the innermost bolts will dominate the resistance to the tension force, only the two bolts closest to the interface are considered. The required strength per bolt, T, is taken as half of the shear force at each flange plate, therefore:

4.4 COLUMN SPLICE AND COLUMN BASE DESIGN EXAMPLES

LRFD	ASD
$T = \dfrac{9.50 \text{ kips}}{2}$ $= 4.75$ kips The available tensile strength per bolt before prying action effects are considered, B, is 29.8 kips from AISC *Manual* Table 7-2.	$T = \dfrac{6.35 \text{ kips}}{2}$ $= 3.18$ kips The available tensile strength per bolt before prying action effects are considered, B, is 19.9 kips from AISC *Manual* Table 7-2.

The parameters required for checking prying action are defined in AISC *Manual* Part 9 and given in Figure 4-19 for this example.

$b = 1.75$ in.

$d_b = \sfrac{3}{4}$ in.

$d' = \sfrac{13}{16}$ in.

$b' = b - d_b/2$ (*Manual* Eq. 9-21)

    $= 1.75$ in. $- \sfrac{3}{4}$ in./2

    $= 1.38$ in.

$a = 4.50$ in.

$a' = a + d_b/2 \leq 1.25b + d_b/2$ (*Manual* Eq. 9-27)

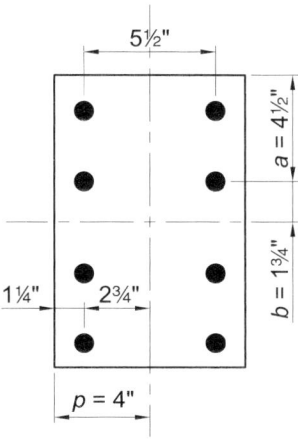

Fig. 4-19. Prying action terminology.

where

$a + d_b/2 = 4.50$ in. $+ \frac{3}{4}$ in./2
$= 4.88$ in.

and

$1.25b + d_b/2 = 1.25(1.75$ in.$) + \frac{3}{4}$ in./2
$= 2.56$ in.
4.88 in. > 2.56 in.

Use $a' = 2.56$ in.

To calculate the tributary length, p, the AISC *Manual* refers to Dowswell (2011) as one method to calculate the length. According to this reference, the tributary length, p_e, can be taken as $p_e = 4\sqrt{bc}$ (Dowswell, 2011, Equation 33) where b is as defined above and where $c = a + b$, and a is limited to $1.25b$. For this calculation:

$a = 4.50$ in. $\leq 1.25b = 2.19$ in. (Use $a = 2.19$ in.)

$c = a + b$
$= 2.19$ in. $+ 1.75$ in.
$= 3.94$ in.

$p_e = 4\sqrt{bc}$
$= 4\sqrt{1.75 \text{ in.}(3.94 \text{ in.})}$
$= 10.5$ in.

This tributary width is limited by the geometry of the plate. The tributary width cannot be greater than the actual edge distance to the end of the plate on one side and half of the bolt gage in the other direction. Therefore, use:

$p = 1.25$ in. $+ 2.75$ in.
$= 4.00$ in.

The remaining variables from AISC *Manual* Part 9 are as follows:

$\delta = 1 - d'/p$ (*Manual* Eq. 9-24)
$= 1 - {}^{13}\!/_{16}$ in./4.00 in.
$= 0.797$

$\rho = b'/a'$ (*Manual* Eq. 9-26)
$= 1.38$ in./2.56 in.
$= 0.539$

4.4 COLUMN SPLICE AND COLUMN BASE DESIGN EXAMPLES

From AISC *Manual* Equation 9-25, β is:

LRFD	ASD
$\beta = \dfrac{1}{\rho}\left(\dfrac{B}{T} - 1\right)$	$\beta = \dfrac{1}{\rho}\left(\dfrac{B}{T} - 1\right)$
$= \dfrac{1}{0.539}\left(\dfrac{29.8 \text{ kips}}{4.75 \text{ kips}} - 1\right)$	$= \dfrac{1}{0.539}\left(\dfrac{19.9 \text{ kips}}{3.18 \text{ kips}} - 1\right)$
$= 9.78$	$= 9.75$

The required plate thickness to develop the available strength of the bolt, B, with no prying action, is calculated from AISC *Manual* Equation 9-20 as:

LRFD	ASD
$t_c = \sqrt{\dfrac{4Bb'}{\phi p F_u}}$	$t_c = \sqrt{\dfrac{\Omega 4Bb'}{p F_u}}$
$= \sqrt{\dfrac{4(29.8 \text{ kips})(1.38 \text{ in.})}{0.90(4.00 \text{ in.})(65 \text{ ksi})}}$	$= \sqrt{\dfrac{1.67(4)(19.9 \text{ kips})(1.38 \text{ in.})}{(4.00 \text{ in.})(65 \text{ ksi})}}$
$= 0.838$ in.	$= 0.840$ in.

Because the splice plate is thinner than t_c, prying on the bolts will occur at the bolt ultimate strength.

Because the fitting geometry is known, the available tensile strength of the bolt including the effects of prying action can be determined as:

$$T_{avail} = BQ \qquad \text{(Manual Eq. 9-31)}$$

where Q is based on α' determined from AISC *Manual* Equation 9-35.

LRFD	ASD
$\alpha' = \dfrac{1}{\delta(1+\rho)}\left[\left(\dfrac{t_c}{t}\right)^2 - 1\right]$	$\alpha' = \dfrac{1}{\delta(1+\rho)}\left[\left(\dfrac{t_c}{t}\right)^2 - 1\right]$
$= \dfrac{1}{0.797(1+0.539)}\left[\left(\dfrac{0.838 \text{ in.}}{\text{3/8 in.}}\right)^2 - 1\right]$	$= \dfrac{1}{0.797(1+0.539)}\left[\left(\dfrac{0.840 \text{ in.}}{\text{3/8 in.}}\right)^2 - 1\right]$
$= 3.26$	$= 3.28$

LRFD	ASD
Because $\alpha' > 1$, use AISC *Manual* Equation 9-34:	Because $\alpha' > 1$, use AISC *Manual* Equation 9-34:
$Q = \left(\dfrac{t}{t_c}\right)^2 (1+\delta)$	$Q = \left(\dfrac{t}{t_c}\right)^2 (1+\delta)$
$= \left(\dfrac{\text{⅜ in.}}{0.838 \text{ in.}}\right)^2 (1+0.797)$	$= \left(\dfrac{\text{⅜ in.}}{0.840 \text{ in.}}\right)^2 (1+0.797)$
$= 0.360$	$= 0.358$
The available tensile strength of each bolt is:	The available tensile strength of each bolt is:
$T_{avail} = BQ$ $= 29.8 \text{ kips}(0.360)$ $= 10.7 \text{ kips} > 4.75 \text{ kips}$ **o.k.**	$T_{avail} = BQ$ $= 19.9 \text{ kips}(0.358)$ $= 7.12 \text{ kips} > 3.18 \text{ kips}$ **o.k.**

The final connection design and geometry for the flange connection is shown in Figure 4-18.

Example 4.4.2. SMF Column Splice Design

Given:

Design a splice for the SMF column located on grid 4 in Figure 4-8. The column material is ASTM A992.

The applicable building code specifies the use of ASCE/SEI 7 for calculation of loads. The required column strengths between the third and fourth levels were determined by a second-order analysis including the effects of $P\text{-}\delta$ and $P\text{-}\Delta$ with reduced stiffness as required by the direct analysis method. The governing load combinations in ASCE/SEI 7, including the overstrength factor (referred to as the amplified seismic load in the AISC *Seismic Provisions*), follow.

The required compressive strength of the column is:

4.4 COLUMN SPLICE AND COLUMN BASE DESIGN EXAMPLES 4-91

LRFD	ASD
From ASCE/SEI 7 Section 12.4.3.2 LRFD Load Combination 5 $P_u = (1.2 + 0.2S_{DS})D + \Omega_o Q_E$ $\quad + 0.5L + 0.2S$ $\quad = 140$ kips (including the 0.5 factor on L permitted in Section 12.4.3.2)	From ASCE/SEI 7 Section 12.4.3.2 ASD Load Combination 6 $P_a = (1.0 + 0.105S_{DS})D + 0.525\Omega_o Q_E$ $\quad + 0.75L + 0.75S$ $\quad = 109$ kips

The required tensile strength of the column is:

LRFD	ASD
From ASCE/SEI 7 Section 12.4.3.2 LRFD Load Combination 7 $T_u = (0.9 - 0.2S_{DS})D + \Omega_o Q_E$ $\quad = 15.3$ kips	From ASCE/SEI 7 Section 12.4.3.2 ASD Load Combination 8 $T_a = (0.6 - 0.14S_{DS})D + 0.7\Omega_o Q_E$ $\quad = 8.64$ kips

The required shear strength of the column is:

LRFD	ASD
From ASCE/SEI 7 Section 12.4.3.2 LRFD Load Combination 5 $V_u = (1.2 + 0.2S_{DS})D + \Omega_o Q_E$ $\quad + 0.5L + 0.2S$ $\quad = 47.2$ kips	From ASCE/SEI 7 Section 12.4.3.2 ASD Load Combination 6 $V_a = (1.0 + 0.105S_{DS})D + 0.525\Omega_o Q_E$ $\quad + 0.75L + 0.75S$ $\quad = 26.9$ kips

From ASCE/SEI 7, use Seismic Design Category D, $\Omega_o = 3.0$, $\rho = 1.0$, and $S_{DS} = 1.0$.

Assume that there is no transverse loading between the column supports in the plane of bending and that the connections into the column weak-axis produce negligible moments on the column.

Solution:

From AISC *Manual* Table 2-4, the column material properties are as follows:

ASTM A992
$F_y = 50$ ksi
$F_u = 65$ ksi

From AISC Manual Table 1-1, the column geometric properties are as follows:

W14×68—Upper Shaft
$A = 20.0$ in.^{2} $d = 14.0$ in. $b_f = 10.0$ in. $t_f = 0.720$ in.
$t_w = 0.415$ in. $Z_x = 115$ in.^{3}

W14×132—Lower Shaft
$Z_x = 234$ in.^{3}

There is no net tensile load effect on the column; therefore, the requirements of AISC *Seismic Provisions* Section D2.5b(1), (2) and (3) do not apply.

Splice Connection

According to AISC *Seismic Provisions* Section E3.6g, welded splices in SMF columns shall be made with complete-joint-penetration (CJP) groove welds. The use of CJP groove welds ensures that the required axial strength and the required flexural strength of the splice will be achieved.

Use CJP groove welds to splice the column webs and flanges directly as shown in Figure 4-20.

Required Shear Strength of the Web Splice

Per AISC *Seismic Provisions* Sections D2.5b, D2.5c and E3.6g, the required shear strength of the web splice is equal to the greater of the required strength determined using the load combinations in the applicable building code, including the amplified seismic load, and the following:

LRFD	ASD
$V_u = \dfrac{\Sigma M_{pc}}{H}$	$V_a = \dfrac{\Sigma M_{pc}}{1.5H}$

where ΣM_{pc} is the sum of the nominal plastic flexural strengths of the columns above and below the splice. Because this requirement is for web splices, ΣM_{pc} in the strong axis of the column will be considered.

LRFD	ASD
$V_u = \dfrac{\Sigma M_{pc}}{H}$	$V_a = \dfrac{\Sigma M_{pc}}{1.5H}$
$= \dfrac{F_y(Z_{xtop} + Z_{xbot})}{H}$	$= \dfrac{F_y(Z_{xtop} + Z_{xbot})}{1.5H}$
$= \dfrac{(50\text{ ksi})(115\text{ in.}^3 + 234\text{ in.}^3)}{12.5\text{ ft}(12\text{ in./ft})}$	$= \dfrac{(50\text{ ksi})(115\text{ in.}^3 + 234\text{ in.}^3)}{1.5(12.5\text{ ft})(12\text{ in./ft})}$
$= 116$ kips	$= 77.6$ kips

Using the load combinations in ASCE/SEI 7 including the amplified seismic load, the required shear strength is given as:

LRFD	ASD
$V_u = 47.2$ kips	$V_a = 26.9$ kips
Therefore $\dfrac{\Sigma M_{pc}}{H}$ governs in determining the required shear strength of the splice.	Therefore $\dfrac{\Sigma M_{pc}}{1.5H}$ governs in determining the required shear strength of the splice.

Using AISC *Specification* Equation G2-1, the required web depth to develop this force through shear yielding of the web is:

LRFD	ASD
$d_w = \dfrac{V_u}{\phi_v 0.6 F_y t_w C_v}$ $= \dfrac{116 \text{ kips}}{1.00(0.6)(50 \text{ ksi})(0.415 \text{ in.})(1.0)}$ $= 9.32$ in.	$d_w = \dfrac{\Omega_v V_a}{0.6 F_y t_w C_v}$ $= \dfrac{1.50(77.6 \text{ kips})}{(0.6)(50 \text{ ksi})(0.415 \text{ in.})(1.0)}$ $= 9.35$ in.

Therefore, the maximum length of each weld access hole, l_h, permitted in the direction of the web is:

LRFD	ASD
$l_h = \frac{1}{2}[d - 2t_f - d_w]$ $= \frac{1}{2}[14.0 \text{ in.} - 2(0.720 \text{ in.}) - 9.32 \text{ in.}]$ $= 1.62$ in.	$l_h = \frac{1}{2}[d - 2t_f - d_w]$ $= \frac{1}{2}[14.0 \text{ in.} - 2(0.720 \text{ in.}) - 9.35 \text{ in.}]$ $= 1.61$ in.

Therefore, specify that the access holes for the flange splice welds may not extend more than 1½ in. measured perpendicular to the inside flange surface as shown in Figure 4-20.

Location of Splice

AISC *Seismic Provisions* Section D2.5a requires that splices be located 4 ft away from the beam-to-column flange connection. The clear distance between the beam-to-column connections is approximately 10.8 ft. Because the webs and flanges are joined by CJP welds, AISC *Seismic Provisions* Section D2.5a(2) permits the splice to be located a minimum of the column depth (14.0 in.) from the beam-to-column flange connection.

The column splice location shown in Figure 4-8 is acceptable.

Additional Weld Requirements

Per AISC *Seismic Provisions* Section A3.4b, the filler metal used to make the splice welds must satisfy AWS D1.8/D1.8M clause 6.3. Additionally, AISC *Seismic Provisions* Section D2.5d requires that weld tabs be removed.

AISC *Specification* Section J1.6 provides additional requirements for weld access hole geometry. The final connection design is shown in Figure 4-20.

Example 4.4.3. SMF Column Base Design

Given:

Refer to Column CL-1 in Figure 4-8. Design a fixed column base plate for the ASTM A992 W-shape. The base and other miscellaneous plate material is ASTM A572 Grade 50. The anchor rod material is ASTM F1554 Grade 105. The 2¼-in.-diameter anchor rods have an embedment length, h_{ef}, of at least 25 in. The column is centered on a reinforced concrete foundation. The foundation concrete compressive strength, f'_c, is 4 ksi with ASTM A615 Grade 60 reinforcement. The anchor rod concrete edge distances, c_{a1} and c_{a2}, are both greater than 37.5 in.

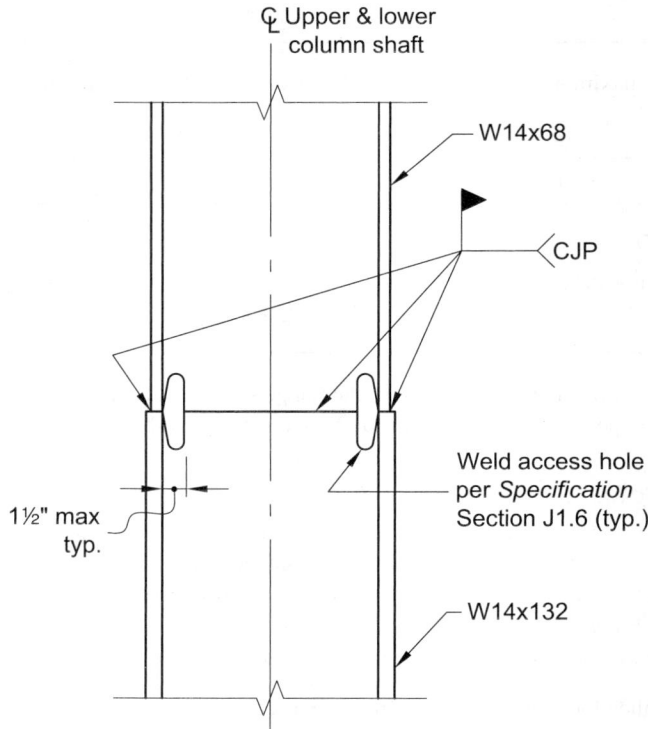

Fig. 4-20. Connection as designed in Example 4.4.2

4.4 COLUMN SPLICE AND COLUMN BASE DESIGN EXAMPLES

The applicable building code specifies the use of ASCE/SEI 7 for calculation of loads. The required column strengths at the base level were determined by a second-order analysis including the effects of P-δ and P-Δ with reduced stiffness as required by the direct analysis method. The governing load combinations in ASCE/SEI 7, including the overstrength factor (referred to as the amplified seismic load in the AISC *Seismic Provisions*), follow. In this example, two of the controlling limit states are tensile yielding in the anchor rods and bending in the base plate. For these limit states, the axial force needs to be minimized as this will increase the overturning (bending) in the base plate and increase the tensile force in the anchor rods; therefore, the required axial compressive strength is determined from:

LRFD	ASD
LRFD Load Combination 7 from ASCE/SEI 7 Section 12.4.3.2 $P_u = (0.9 - 0.2S_{DS})D + \Omega_o Q_E$ $= 98.8$ kips	ASD Load Combination 8 from ASCE/SEI 7 Section 12.4.3.2 $P_a = (0.6 - 0.14S_{DS})D + 0.7\Omega_o Q_E$ $= 64.5$ kips

The required flexural strength is determined from:

LRFD	ASD
LRFD Load Combination 7 from ASCE/SEI 7 Section 12.4.3.2 $M_u = (0.9 - 0.2S_{DS})M_D + \Omega_o M_{Q_E}$ $= 946$ kip-ft	ASD Load Combination 8 from ASCE/SEI 7 Section 12.4.3.2 $M_a = (0.6 - 0.14S_{DS})M_D + 0.7\Omega_o M_{Q_E}$ $= 662$ kip-ft

The required shear strength is determined from:

LRFD	ASD
LRFD Load Combination 5 from ASCE/SEI 7 Section 12.4.3.2 $V_u = (1.2 + 0.2S_{DS})D + \Omega_o Q_E$ $= 96.0$ kips	ASD Load Combination 5 from ASCE/SEI 7 Section 12.4.3.2 $V_a = (1.0 + 0.14S_{DS})D + 0.7\Omega_o Q_E$ $= 67.2$ kips

Assume that the connection into the column weak-axis produces negligible moments on the column.

From ASCE/SEI 7, use Seismic Design Category D, $\Omega_o = 3.0$, $\rho = 1.0$, and $S_{DS} = 1.0$.

Use LRFD provisions for the concrete design.

Solution:

From AISC *Manual* Table 2-4, the column material properties are as follows:

ASTM A992
$F_y = 50$ ksi
$F_u = 65$ ksi

From AISC *Manual* Table 2-5, the base plate material properties are as follows:

ASTM A572 Grade 50
$F_y = 50$ ksi
$F_u = 65$ ksi

From AISC *Manual* Table 2-6, the anchor rod material properties are as follows:

ASTM F1554 Grade 105
$F_u, f_{uta} = 125$ ksi

From ASTM A615, the concrete reinforcement properties are as follows:

ASTM A615 Grade 60
$F_y = 60$ ksi

From AISC *Manual* Table 1-1, the column and beam geometric properties are as follows:

W14×176
$A = 51.8$ in.^{2} $d = 15.2$ in. $b_f = 15.7$ in. $t_f = 1.31$ in.
$t_w = 0.830$ in. $k_{des} = 1.91$ in. $Z_x = 320$ in.^{3}

W24×76
$d = 23.9$ in.

From AISC *Manual* Table 7-17, the 2¼-in.-diameter anchor rod has an area of $A = 3.98$ in.^{2}

Required Strengths at Column Base

AISC *Seismic Provisions* Section D2.6a.(a) defines the required axial strength as the required strength calculated using the load combinations of the applicable building code, including the amplified seismic load.

By reference to AISC *Seismic Provisions* Section D2.5c, which references Section D2.5b, AISC *Seismic Provisions* Section D2.6b indirectly stipulates that the required shear strength of the column base be the greater of the required shear strength determined from load combinations including the amplified seismic load (Section D2.5b(b)) or the required column strength as stipulated in the system chapters (Section D2.5b(a)). Here, the provisions of Section E3.6g apply, as follows:

LRFD	ASD
$V_u = \dfrac{\Sigma M_{pc}}{H}$	$V_a = \dfrac{\Sigma M_{pc}}{1.5H}$

where ΣM_{pc} is the sum of the nominal plastic flexural strengths of the columns above and below the splice, or in this case, the base.

Therefore:

LRFD	ASD
$V_u = \dfrac{\Sigma M_{pc}}{H}$ $= \dfrac{2(50 \text{ ksi})(320 \text{ in.}^3)}{(12 \text{ in./ft})(14.0 \text{ ft})}$ $= 190 \text{ kips} > 96.0 \text{ kips}$ Use $V_u = 190$ kips.	$V_a = \dfrac{\Sigma M_{pc}}{1.5H}$ $= \dfrac{2(50 \text{ ksi})(320 \text{ in.}^3)}{1.5(12 \text{ in./ft})(14.0 \text{ ft})}$ $= 127 \text{ kips} > 67.2 \text{ kips}$ Use $V_a = 127$ kips.

AISC *Seismic Provisions* Section D2.6c(b) requires that the flexural strength equal or exceed the lesser of the load combination of the applicable building code, including the amplified seismic load, or the following:

LRFD	ASD
$M_u = 1.1 R_y F_y Z_x$ $= \dfrac{1.1(1.1)(50 \text{ ksi})(320 \text{ in.}^3)}{(12 \text{ in./ft})}$ $= 1{,}610 \text{ kip-ft} > 946 \text{ kip-ft}$ Use $M_u = 946$ kip-ft.	$M_a = \dfrac{1.1 R_y F_y Z_x}{1.5}$ $= \dfrac{1.1(1.1)(50 \text{ ksi})(320 \text{ in.}^3)}{1.5(12 \text{ in./ft})}$ $= 1{,}080 \text{ kip-ft} > 662 \text{ kip-ft}$ Use $M_a = 662$ kip-ft.

Initial Size of Base Plate

The base plate dimensions shall be large enough for the installation of at least four anchor rods, as required by the Occupational Safety and Health Administration (OSHA, 2008).

Try a plate with: $N = 32$ in., $B = 32$ in., and anchor rod edge distance $= 4$ in. Try two rows of four equally spaced rods, as shown in Figure 4-23.

Using the recommendations from AISC Design Guide 1, *Base Plate and Anchor Rod Design* (Fisher and Kloiber, 2010), determine the required base plate thickness and anchor rod tension force.

Base Plate Eccentricity and Critical Eccentricity

For the calculation of the base plate eccentricity, e, from AISC Design Guide 1 Equation 3.3.6:

LRFD	ASD
$e = \dfrac{M_u}{P_u}$ $= \dfrac{(946 \text{ kip-ft})(12 \text{ in./ft})}{98.8 \text{ kips}}$ $= 115 \text{ in.}$	$e = \dfrac{M_a}{P_a}$ $= \dfrac{(662 \text{ kip-ft})(12 \text{ in./ft})}{64.5 \text{ kips}}$ $= 123 \text{ in.}$

For the calculation of the critical eccentricity, e_{crit}:

$$e_{crit} = \frac{N}{2} - \frac{P_r}{2q_{max}} \quad \text{(AISC Design Guide 1 Eq. 3.3.7)}$$

For the calculation of the maximum plate bearing stress, q_{max}:

$$q_{max} = f_{p(max)}B \quad \text{(AISC Design Guide 1 Eq. 3.3.4)}$$

For the calculation, assume the concrete bearing frustum area ratio equals 2.0 from ACI 318 Section 10.14.1:

$$\sqrt{\frac{A_2}{A_1}} = 2.0$$

The available bearing strength is determined from AISC *Specification* Equation J8-2.

LRFD	ASD
$f_{p(max)} = \phi(0.85f'_c)\sqrt{\dfrac{A_2}{A_1}}$ $= 0.65(0.85)(4 \text{ ksi})(2.0)$ $= 4.42 \text{ ksi}$	$f_{p(max)} = \dfrac{0.85f'_c}{\Omega}\sqrt{\dfrac{A_2}{A_1}}$ $= \dfrac{0.85(4 \text{ ksi})(2.0)}{2.31}$ $= 2.94 \text{ ksi}$
$q_{max} = f_{p(max)}B$ $= 4.42 \text{ ksi}(32.0 \text{ in.})$ $= 141 \text{ kip/in.}$	$q_{max} = f_{p(max)}B$ $= 2.94 \text{ ksi}(32.0 \text{ in.})$ $= 94.1 \text{ kip/in.}$

4.4 COLUMN SPLICE AND COLUMN BASE DESIGN EXAMPLES

LRFD	ASD
$e_{crit} = \dfrac{N}{2} - \dfrac{P_u}{2q_{max}}$ $= \dfrac{32.0 \text{ in.}}{2} - \dfrac{98.8 \text{ kips}}{2(141 \text{ kip/in.})}$ $= 15.6 \text{ in.}$	$e_{crit} = \dfrac{N}{2} - \dfrac{P_a}{2q_{max}}$ $= \dfrac{32.0 \text{ in.}}{2} - \dfrac{64.5 \text{ kips}}{2(94.1 \text{ kip/in.})}$ $= 15.7 \text{ in.}$

With $e > e_{crit}$, the eccentricity meets the AISC Design Guide 1 criteria for a base plate with a large moment (Figure 4-21).

Per AISC Design Guide 1 Section 3.4, the following inequality must be satisfied:

$$\left(f + \frac{N}{2}\right)^2 \geq \frac{2P_r(e+f)}{q_{max}} \qquad \text{(AISC Design Guide 1 Eq. 3.4.4)}$$

For the calculation of f:

$$f = \frac{N}{2} - \text{edge distance}$$
$$= \frac{32.0 \text{ in.}}{2} - 4.00 \text{ in.}$$
$$= 12.0 \text{ in.}$$

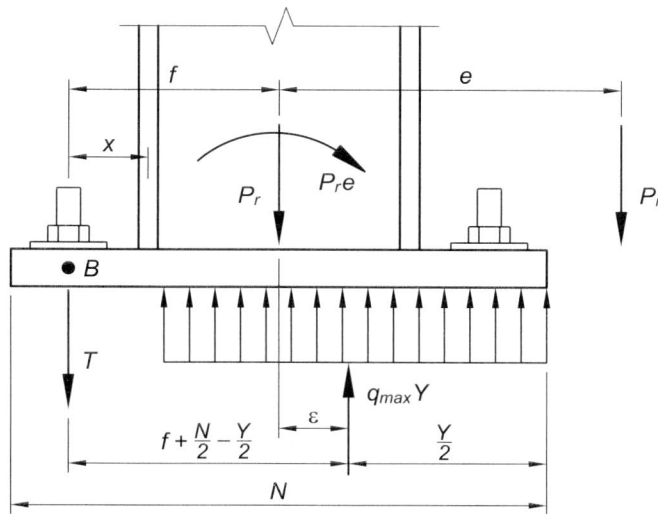

Fig. 4-21. Base plate with large moment (Fisher and Kloiber, 2010).

Therefore:

$$\left(f+\frac{N}{2}\right)^2 = \left(12.0 \text{ in.} + \frac{32.0 \text{ in.}}{2}\right)^2$$
$$= 784 \text{ in.}^2$$

LRFD	ASD
$\dfrac{2P_u(e+f)}{q_{max}} = \dfrac{2(98.8 \text{ kips})(115 \text{ in.}+12.0 \text{ in.})}{141 \text{ kip/in.}}$	$\dfrac{2P_a(e+f)}{q_{max}} = \dfrac{2(64.5 \text{ kips})(123 \text{ in.}+12.0 \text{ in.})}{94.1 \text{ kip/in.}}$
$= 178 \text{ in.}^2$	$= 185 \text{ in.}^2$

With $\left(f+\dfrac{N}{2}\right)^2 > \dfrac{2P_r(e+f)}{q_{max}}$, the inequality is satisfied and a real solution is possible.

Base Plate Bearing Length
From AISC Design Guide 1 Equation 3.4.3, the base plate bearing length is:

LRFD	ASD
$Y = \left(f+\dfrac{N}{2}\right) \pm \sqrt{\left(f+\dfrac{N}{2}\right)^2 - \dfrac{2P_u(e+f)}{q_{max}}}$	$Y = \left(f+\dfrac{N}{2}\right) \pm \sqrt{\left(f+\dfrac{N}{2}\right)^2 - \dfrac{2P_a(e+f)}{q_{max}}}$
$= \sqrt{784 \text{ in.}^2} - \sqrt{784 \text{ in.}^2 - 178 \text{ in.}^2}$	$= \sqrt{784 \text{ in.}^2} - \sqrt{784 \text{ in.}^2 - 185 \text{ in.}^2}$
$= 3.38 \text{ in.}$	$= 3.53 \text{ in.}$

Required Rod Tensile Strength
From AISC Design Guide 1 Equation 3.4.2, the required rod tensile strength for the anchor group on one side of the base plate is:

LRFD	ASD
$N_{ua} = q_{max}Y - P_u$	$N_{aa} = q_{max}Y - P_u$
$= 141 \text{ kip/in.}(3.38 \text{ in.}) - 98.8 \text{ kips}$	$= 94.1 \text{ kip/in.}(3.53 \text{ in.}) - 64.5 \text{ kips}$
$= 378 \text{ kips}$	$= 268 \text{ kips}$

Base Plate Thickness
Check the base plate for flexural yielding at both the bearing and tension interfaces. At the bearing interface, the bearing pressures between the concrete and the plate will cause bending for the cantilever lengths m and n as shown in Figure 4-22. At the tension interface, the anchor rods cause bending for the cantilever length, x, as shown in Figure 4-21.

4.4 COLUMN SPLICE AND COLUMN BASE DESIGN EXAMPLES

For the calculation of the assumed bending lines at the bearing interface, from AISC Design Guide 1 Section 3.1.2:

$$m = \frac{N - 0.95d}{2}$$
$$= \frac{32.0 \text{ in.} - 0.95(15.2 \text{ in.})}{2}$$
$$= 8.78 \text{ in.}$$

$$n = \frac{B - 0.8b_f}{2}$$
$$= \frac{32.0 \text{ in.} - 0.8(15.7 \text{ in.})}{2}$$
$$= 9.72 \text{ in.}$$

For the calculation of the base plate cantilever bending line distance at the tension interface:

$$x = f - \frac{d}{2} + \frac{t_f}{2} \qquad \text{(AISC Design Guide 1 Eq. 3.4.6)}$$
$$= 12.0 \text{ in.} - \frac{15.2 \text{ in.}}{2} + \frac{1.31 \text{ in.}}{2}$$
$$= 5.06 \text{ in.}$$

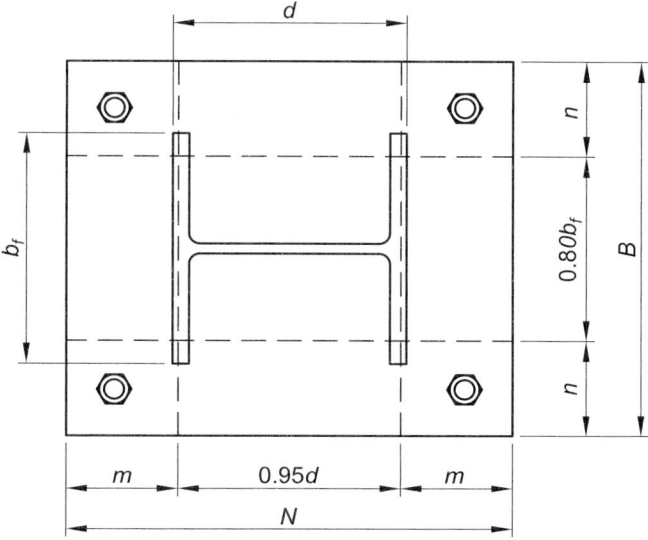

Fig. 4-22. Assumed bending lines (Fisher and Kloiber, 2010).

For flexural yielding at the bearing interface and $Y < \max(m, n)$, from AISC Design Guide 1 Equation 3.3.15:

LRFD	ASD
$t_{p(req)} = 2.11\sqrt{\dfrac{f_{p(max)}Y\left(\max(m,n) - \dfrac{Y}{2}\right)}{F_y}}$	$t_{p(req)} = 2.58\sqrt{\dfrac{f_{p(max)}Y\left(\max(m,n) - \dfrac{Y}{2}\right)}{F_y}}$
$= 2.11\sqrt{\dfrac{4.42 \text{ ksi}(3.38 \text{ in.}) \times \left(9.72 \text{ in.} - \dfrac{3.38 \text{ in.}}{2}\right)}{50 \text{ ksi}}}$	$= 2.58\sqrt{\dfrac{2.94 \text{ ksi}(3.53 \text{ in.}) \times \left(9.72 \text{ in.} - \dfrac{3.53 \text{ in.}}{2}\right)}{50 \text{ ksi}}}$
$= 3.27$ in.	$= 3.32$ in.

For flexural yielding at the tension interface, from AISC Design Guide 1 Equation 3.4.7:

LRFD	ASD
$t_{p(req)} = 2.11\sqrt{\dfrac{N_{ua}x}{BF_y}}$	$t_{p(req)} = 2.58\sqrt{\dfrac{N_{aa}x}{BF_y}}$
$= 2.11\sqrt{\dfrac{378 \text{ kips}(5.06 \text{ in.})}{32.0 \text{ in.}(50 \text{ ksi})}}$	$= 2.58\sqrt{\dfrac{268 \text{ kips}(5.06 \text{ in.})}{32.0 \text{ in.}(50 \text{ ksi})}}$
$= 2.31$ in.	$= 2.38$ in.

Use a PL3½ in. × 32 in. × 2 ft 8 in. ASTM A572 Grade 50 for the base plate.

Plate Washer Bearing Strength

According to AISC *Manual* Table 14-2, use a ⅞ in. × 5¼ in. × 5¼ in. plate washer, welded to the top of the base plate, to transfer the shear to the anchor rods. Also, interpolating from Table 14-2, use a 3½-in.-diameter hole for the 2¼-in.-diameter anchor rods.

Determine the available bearing strength assuming deformation at the bolt hole is not a design consideration.

The clear distance to the edge of the bearing plate, l_c, is taken as:

$$l_c = \dfrac{5\frac{1}{4} \text{ in.} - 2\frac{1}{4} \text{ in.}}{2}$$
$$= 1.50 \text{ in.}$$

$$R_n = 1.5l_c t F_u n_v \leq 3.0 dt F_u n_v \quad \text{(Spec. Eq. J3-6b)}$$
$$= 1.5(1.50 \text{ in.})(\tfrac{7}{8} \text{ in.})(65 \text{ ksi})(8)$$
$$\leq 3.0(2\tfrac{1}{4} \text{ in.})(\tfrac{7}{8} \text{ in.})(65 \text{ ksi})(8)$$
$$= 1{,}020 \text{ kips} < 3{,}070 \text{ kips}$$

LRFD	ASD
$\phi R_n = 0.75(1{,}020 \text{ kips})$ $= 765 \text{ kips}$ $\phi R_n > V_u = 190 \text{ kips}$ **o.k.**	$\dfrac{R_n}{\Omega} = \dfrac{1{,}020 \text{ kips}}{2.00}$ $= 510 \text{ kips}$ $\dfrac{R_n}{\Omega} > V_a = 127 \text{ kips}$ **o.k.**

Anchor Rod Combined Tension and Shear

Using the recommendations from AISC Design Guide 1 and AISC *Specification* Section J3.7, the available tensile stress of the anchor rod subject to combined tensile and shear loads is checked, including the effects of bending.

Based on testing performed by Gomez et al. (2010), this approach was determined to provide a reasonable and conservative strength estimate for earthquake design. Therefore, given the comprehensive testing and design approach, the general anchor strength requirement of ACI 318 Section D4.3 for resistance to combined tensile and shear loads can be satisfied.

The anchor rod nominal tensile stress, from AISC *Specification* Table J3.2:

$$F_{nt} = 0.75 F_u$$
$$= 0.75(125 \text{ ksi})$$
$$= 93.8 \text{ ksi}$$

The anchor rod nominal shear stress with threads not excluded from the shear plane from AISC *Specification* Table J3.2:

$$F_{nv} = 0.450 F_u$$
$$= 0.450(125 \text{ ksi})$$
$$= 56.3 \text{ ksi}$$

The anchor rod required shear stress, f_{rv}:

LRFD	ASD
$f_{rv} = \dfrac{V_u}{n_v A_g}$ $= \dfrac{190 \text{ kips}}{8(3.98 \text{ in.}^2)}$ $= 5.97 \text{ ksi}$	$f_{rv} = \dfrac{V_a}{n_v A_g}$ $= \dfrac{127 \text{ kips}}{8(3.98 \text{ in.}^2)}$ $= 3.99 \text{ ksi}$

Therefore, the nominal tensile stress from AISC *Specification* Equation J3-3 is:

LRFD	ASD
$F'_{nt} = 1.3F_{nt} - \dfrac{F_{nt}}{\phi F_{nv}} f_{rv} < F_{nt}$	$F'_{nt} = 1.3F_{nt} - \dfrac{\Omega F_{nt}}{F_{nv}} f_{rv} < F_{nt}$
$= 1.3(93.8 \text{ ksi})$	$= 1.3(93.8 \text{ ksi})$
$\quad - \dfrac{93.8 \text{ ksi}}{0.75(56.3 \text{ ksi})}(5.97 \text{ ksi})$	$\quad - \dfrac{2.00(93.8 \text{ ksi})}{56.3 \text{ ksi}}(3.99 \text{ ksi})$
$= 109 \text{ ksi} > 93.8 \text{ ksi}$	$= 109 \text{ ksi} > 93.8 \text{ ksi}$
Therefore use $F'_{nt} = 93.8$ ksi	Therefore use $F'_{nt} = 93.8$ ksi
$\phi F'_{nt} = 0.75(93.8 \text{ ksi})$	$\dfrac{F'_{nt}}{\Omega} = \dfrac{93.8 \text{ ksi}}{2.00}$
$= 70.4$ ksi	$= 46.9$ ksi

The anchor rod combined tensile and bending stresses, f_t, is:

$$f_t = f_{ta} + f_{tb}$$

From AISC Design Guide 1 (Fisher and Kloiber, 2010), the anchor rod bending moment lever arm, l, is taken as:

$$l = \frac{t_p}{2} + \frac{t_{washer}}{2}$$
$$= \frac{3\tfrac{1}{2} \text{ in.}}{2} + \frac{\tfrac{7}{8} \text{ in.}}{2}$$
$$= 2.19 \text{ in.}$$

The anchor rod plastic section modulus, Z, is:

$$Z = \frac{d_b^3}{6}$$
$$= \frac{(2\tfrac{1}{4} \text{ in.})^3}{6}$$
$$= 1.90 \text{ in.}^3$$

Determine the anchor rod tensile stress, assuming that only the rods on one side of the base plate are in tension at any time.

LRFD	ASD
$f_{ta} = \dfrac{N_{ua}}{n_t A_g}$ $= \dfrac{378 \text{ kips}}{4(3.98 \text{ in.}^2)}$ $= 23.7 \text{ ksi}$	$f_{ta} = \dfrac{N_{aa}}{n_t A_g}$ $= \dfrac{268 \text{ kips}}{4(3.98 \text{ in.}^2)}$ $= 16.8 \text{ ksi}$
Anchor rod bending stress $M_{tb} = \dfrac{V_u l}{n_v}$ $= \dfrac{190 \text{ kips}(2.19 \text{ in.})}{8}$ $= 52.0 \text{ kip-in.}$	Anchor rod bending stress $M_{tb} = \dfrac{V_a l}{n_v}$ $= \dfrac{127 \text{ kips}(2.19 \text{ in.})}{8}$ $= 34.8 \text{ kip-in.}$
$f_{tb} = \dfrac{M_{tb}}{Z}$ $= \dfrac{52.0 \text{ kip-in.}}{1.90 \text{ in.}^3}$ $= 27.4 \text{ ksi}$	$f_{tb} = \dfrac{M_{tb}}{Z}$ $= \dfrac{34.8 \text{ kip-in.}}{1.90 \text{ in.}^3}$ $= 18.3 \text{ ksi}$
Combined stress $f_t = f_{ta} + f_{tb}$ $= 23.7 \text{ ksi} + 27.4 \text{ ksi}$ $= 51.1 \text{ ksi} < 70.4 \text{ ksi}$ **o.k.**	Combined stress $f_t = f_{ta} + f_{tb}$ $= 16.8 \text{ ksi} + 18.3 \text{ ksi}$ $= 35.1 \text{ ksi} < 46.9 \text{ ksi}$ **o.k.**

Concrete Anchorage Strengths

The available strengths of the column base concrete elements are checked in accordance with ACI 318 Appendix D. Section D.3.3.3 requires the anchor design strength associated with concrete failure modes be reduced by a factor of 0.75 for structures assigned to Seismic Design Category C, D, E or F. The same section requires that the concrete be assumed cracked unless it can be demonstrated otherwise. Section D.3.3.6 permits the use of a 0.4 factor when not designing to fail either the anchor rod or the connection to the anchor rod per Sections D.3.3.4 and D.3.3.5, respectively. Although longer embedment depths are permitted, with respect to the basic strength equation, ACI 318 Section D.5.2.2 and this example limit the minimum effective embedment depth, h_{ef}, of the anchor rods to 25 in.

Design Requirements for Tensile Loading

Although checked previously in accordance with AISC provisions, the following illustrates the anchor tensile loading checks in accordance with ACI 318 Appendix D provisions. Per Section D3.3.4, to ensure anchor rod ductile behavior, the design steel tensile strength, ϕN_{sa}, must be less than the concrete breakout, $0.75\phi N_{cbg}$, pullout, $0.75\phi N_{pn}$, and side-face blowout, $0.75\phi N_{sb}$, strengths. By inspection, the side-face blowout limit state is not applicable.

The steel tensile strength of the anchor rod group of four (on one side of the base plate):

$$\phi N_{sa} = \phi n \, A_{se,N} f_{uta}$$ (from ACI 318 Eq. D-3)

where
ϕ = 0.75 from ACI 318 Section D4.4(a)(i)

$$A_{se,N} = \frac{\pi}{4}\left(d_a - \frac{0.9743}{n_t}\right)^2$$ from ACI 318 Section RD.5.1.2

n_t = 4.5 threads/in. from AISC *Manual* Table 7-17

Therefore:

$$A_{se,N} = \frac{\pi}{4}\left(2\tfrac{1}{4}\text{ in.} - \frac{0.9743}{4.50 \text{ in.}}\right)^2$$
$$= 3.25 \text{ in.}^2$$

$$\phi N_{sa} = 0.75(4)(3.25 \text{ in.}^2)(125 \text{ ksi})$$
$$= 1{,}220 \text{ kips} > N_{ua} = 378 \text{ kips} \quad \textbf{o.k.}$$

For the design tensile concrete breakout strength of the anchor group:

$$N_{cbg} = \left(\frac{A_{Nc}}{A_{Nco}}\right)\psi_{ec,N}\psi_{ed,N}\psi_{c,N}\psi_{cp,N}N_b$$ (ACI 318 Eq. D-5)

where the following values are assumed.

$\psi_{ec,N}$ = 1.0 from ACI 318 Section D.5.2.4

$\psi_{ed,N}$ = 1.0 from ACI 318 D.5.2.5

$\psi_{c,N}$ = 1.0 from ACI 318 D.5.2.6

$\psi_{cp,N}$ = 1.0 from ACI 318 Section D.5.2.7

$$A_{Nc} = [(n-1)s + 2(1.5)h_{ef}]2(1.5)h_{ef}$$ from ACI 318 Figure RD.5.2.1

s = $[B - 2(\text{Edge Distance})]/(n-1)$
$= [32.0 \text{ in.} - 2(4.00 \text{ in.})]/(4-1)$
$= 8.00 \text{ in.}$

Therefore:

$$A_{Nc} = [(4-1)8.00 \text{ in.} + 2(1.5)(25.0 \text{ in.})](2)(1.5)(25.0 \text{ in.})$$
$$= 7{,}430 \text{ in.}^2$$

4.4 COLUMN SPLICE AND COLUMN BASE DESIGN EXAMPLES

For the calculation of A_{Nco}:

$$A_{Nco} = 9h_{ef}^2 \qquad \text{(ACI 318 Eq. D-6)}$$
$$= 9(25.0 \text{ in.})^2$$
$$= 5{,}630 \text{ in.}^2$$

For the calculation of N_b:

$$N_b = 16\lambda\sqrt{f_c'}\, h_{ef}^{5/3} \qquad \text{(ACI 318 Eq. D-8)}$$
$$= \frac{16(1)\sqrt{4{,}000 \text{ psi}}\,(25.0 \text{ in.})^{5/3}}{1{,}000 \text{ lb/kip}}$$
$$= 216 \text{ kips}$$

Therefore:

$$N_{cbg} = \left(\frac{7{,}430 \text{ in.}^2}{5{,}630 \text{ in.}^2}\right)(1.0)(1.0)(1.0)(1.0)(216 \text{ kips})$$
$$= 285 \text{ kips}$$

$$0.4(0.75)\phi N_{cbg} = 0.4(0.75)(0.75)(285 \text{ kips})$$
$$= 64.1 \text{ kips} < 378 \text{ kips} \qquad \textbf{n.g.}$$

Per ACI 318 Section D.4.2.1 provide supplemental reinforcement to restrain the concrete breakout. From ACI 318 Section D.5.2.9:

$$A_s = \frac{T_u}{0.75\phi f_y}$$

$$\phi = 0.75$$

$$A_s = \frac{378 \text{ kips}}{0.75(0.75)(60 \text{ ksi})}$$
$$= 11.2 \text{ in.}^2$$

Provide at least 11.2 in.^{2} of vertical reinforcing stirrups spaced within $0.5h_{ef}$ of each anchor rod group per ACI 318 Section RD.5.2.9.

For the design pullout strength of the anchor group, including the additional 0.75 factor stipulated in ACI 318 Section D.5.2.9 and 0.4 factor stipulated in D.3.3.6:

$$0.4(0.75)\phi N_{pn} = 0.4(0.75)\phi n \psi_{c,P} N_p \qquad \text{(from ACI 318 Eq. D-14)}$$

where
  ϕ = 0.7 from ACI 318 Section D.4.4(c)ii for Condition B
  $\psi_{c,P}$ = 1.0 from ACI 318 Section D.5.3.6

For the calculation of N_p,

$$N_p = 8A_{brg} f'_c \qquad \text{(ACI 318 Eq. D-15)}$$

For calculation of the anchor head bearing area, A_{brg}, try a 1 in. × 4½ in. × 4½ in. plate washer with a double heavy hex nut head on the embedded end of the anchor rod.

$$A_{brg} = A_{plate} - A_{se}$$

$$A_{brg} = (4½ \text{ in.})^2 - 3.25 \text{ in.}^2$$

$$= 17.0 \text{ in.}^2$$

$$N_p = 8(17.0 \text{ in.}^2)(4 \text{ ksi})$$

$$= 544 \text{ kips}$$

Therefore:

$$0.4(0.75)\phi N_{pn} = 0.4(0.75)(0.7)(4)(1.0)(544 \text{ kips})$$
$$= 457 \text{ kips} > 378 \text{ kips} \qquad \textbf{o.k.}$$

Anchor Rod Head Plate Washer Flexural Strength

The plastic section modulus per unit width, Z, of the plate washer is:

$$Z = \frac{bd^2}{4}$$

$$= \frac{1.00 \text{ in.}(1.00 \text{ in.})^2}{4}$$

$$= 0.250 \text{ in.}^3$$

The nominal flexural strength of the plate washer is:

$$M_n = F_y Z \qquad \text{(Spec. Eq. F11-1)}$$

$$= 50 \text{ ksi}(0.250 \text{ in.}^3)$$

$$= 12.5 \text{ kip-in.}$$

Therefore, from AISC *Specification* Section F11.1 and ACI 318 Section D.3.3.6, the available flexural strength of the plate washer is:

LRFD	ASD
$0.4\phi M_n = 0.4(0.90)(12.5 \text{ kip-in.})$ $= 4.50 \text{ kip-in.}$	$\dfrac{0.4 M_n}{\Omega} = \dfrac{0.4(12.5 \text{ kip-in.})}{1.67}$ $= 2.99 \text{ kip-in.}$

For the calculation of the plate washer cantilever bending moment, the plate washer cantilever distance, l, is:

$$l = \frac{(B_{washer} - B_{nut\ head})}{2}$$
$$= \frac{(4\frac{1}{2}\ \text{in.} - 3\frac{1}{2}\ \text{in.})}{2}$$
$$= 0.500\ \text{in.}$$

where $B_{nut\ head}$ is the heavy hex nut F dimension given in AISC *Manual* Table 7-19.

Therefore:

LRFD	ASD
For the plate washer load, w_u,	For the plate washer load, w_a,
$w_u = \dfrac{N_{ua}}{A_{brg}}$ $= \dfrac{378\ \text{kips}}{17.0\ \text{in.}^2}$ $= 22.2\ \text{ksi}$	$w_a = \dfrac{N_{aa}}{A_{brg}}$ $= \dfrac{268\ \text{kips}}{17.0\ \text{in.}^2}$ $= 15.8\ \text{ksi}$
For a 1-in. strip of plate:	For a 1-in. strip of plate:
$M_u = \dfrac{w_u l^2}{2}$ $= \dfrac{(22.2\ \text{kip/in.})(0.500\ \text{in.})^2}{2}$ $= 2.78\ \text{kip-in.} < 4.50\ \text{kip-in.}$ **o.k.**	$M_a = \dfrac{w_a l^2}{2}$ $= \dfrac{(15.8\ \text{kip/in.})(0.500\ \text{in.})^2}{2}$ $= 1.98\ \text{kip-in.} < 2.99\ \text{kip-in.}$ **o.k.**

Design Requirements for Shear Loading

Although checked previously in accordance with AISC provisions, the following illustrates the shear loading checks in accordance with ACI 318 Appendix D provisions. Frictional shear resistance developed between the base plate and the concrete is neglected in consideration of earthquake loading. By inspection, the concrete breakout strength of the anchor group in shear is not applicable.

The design steel shear strength of the entire anchor group, including the grout pad factor of 0.80 (ACI 318 Section D.6.1.3) is:

$\phi V_{sa} = \phi\ 0.80 n 0.6 A_{se,V} f_{uta}$ (ACI 318 Eq. D-20)

where
 $\phi = 0.65$ from ACI 318 Section D.4.4(a)ii

Therefore:

$$\phi V_{sa} = 0.65(0.8)(8)(0.6)(3.25 \text{ in.}^2)(125 \text{ ksi})$$
$$= 1{,}010 \text{ kips} > 190 \text{ kips} \quad \textbf{o.k.}$$

For the interaction of tensile and shear forces, from ACI 318 Section D.7:

$$\frac{V_u}{\phi V_{sa}} = \frac{190 \text{ kips}}{1{,}010 \text{ kips}}$$
$$= 0.188$$

$$\frac{N_{ua}}{\phi N_{sa}} = \frac{378 \text{ kips}}{1{,}220 \text{ kips}}$$
$$= 0.310$$

Because $V_u \leq 0.2\phi V_{sa}$, the full strength in tension is permitted according to ACI 318 Section D.7.1.

For the design pryout strength of the anchor group, ACI 318 Section D.3.3.5 requires that the strength be greater than the shear associated with a ductile failure of the attachment if the requirements of ACI 318 Section D.3.3.4 are not met. Because the shear strength is based on hinging in the column, the ductile failure requirement is met if the design strength exceeds the column shear strength.

As indicated previously, the anchor rods on both sides of the base plate are provided with supplemental reinforcement. In the region between each anchor group, the supplemental reinforcement may overlap, contributing to either group. In consideration of the concrete breakout strength for prying, a conservative estimate considers only 75% of the supplemental total for both groups. Also, the N_{cbg} term is now determined based on the area of supplemental reinforcement instead of ACI 318 Appendix D, Equation D-5. Therefore, the revised design pryout strength is:

$$0.75\phi V_{cpg} = 0.75\phi \, k_{cp} \, N_{cbg} \quad \text{(from ACI 318 Eq. D-31)}$$

where
 ϕ = 0.70 from ACI 318 Section D.4.4(c)ii Condition B
 k_{cp} = 2.0 from ACI 318 Section D.6.3.1
 N_{cbg} = 0.75(2)(11.2 in.²)(60 ksi)
  = 1,010 kips

Therefore:

$$0.75\phi V_{cpg} = 0.75(0.70)(2.0)(1{,}010 \text{ kips})$$
$$= 1{,}060 \text{ kips} > 190 \text{ kips} \quad \textbf{o.k.}$$

Recheck the interaction of tensile and shear forces, using ACI 318 Section D.7.1, with $\phi V_{sa} = 0.75\,\phi V_{cpg}$, as follows:

$$\frac{V_u}{\phi V_{sa}} = \frac{190 \text{ kips}}{1{,}060 \text{ kips}}$$
$$= 0.179 < 0.20$$

Because $V_u \leq 0.2\phi V_{sa}$, the full strength in tension is permitted according to ACI 318 Section D.7.1. Therefore, ϕV_n is controlled by a ductile steel element.

Design of Column Web-to-Base Plate Weld

The effective length of weld available, l_e, on both sides of web, holding welds back from the "k" region, is:

$$l_e = d - 2k_{des}$$
$$= 15.2 \text{ in.} - 2(1.91 \text{ in.})$$
$$= 11.4 \text{ in.}$$

From AISC *Manual* Equation 8-2, the weld size in sixteenths of an inch is:

LRFD	ASD
$D_{req} = \dfrac{V_u}{1.392(2l_e)}$	$D_{req} = \dfrac{V_a}{0.928(2l_e)}$
$= \dfrac{190 \text{ kips}}{1.392 \text{ kip/in.}(2)(11.4 \text{ in.})}$	$= \dfrac{127 \text{ kips}}{0.928 \text{ kip/in.}(2)(11.4 \text{ in.})}$
$= 5.99$ sixteenths	$= 6.00$ sixteenths

Conservatively use 7/16-in. fillet welds (two-sided) for the column web-to-base plate weld.

Design of Washer Plate to Base Plate Weld

The effective length of weld available, l_e, on each of the eight plates (two sides), is:

$$l_e = 2(5\tfrac{1}{4} \text{ in.})$$
$$= 10.5 \text{ in.}$$

From AISC *Manual* Equation 8-2, the weld size in sixteenths of an inch is:

LRFD	ASD
$D_{req} = \dfrac{V_u}{1.392(8l_e)}$	$D_{req} = \dfrac{V_u}{0.928(8l_e)}$
$= \dfrac{190 \text{ kips}}{1.392 \text{ kip/in.}(8)(10.5 \text{ in.})}$	$= \dfrac{127 \text{ kips}}{0.928 \text{ kip/in.}(8)(10.5 \text{ in.})}$
$= 1.62$ sixteenths	$= 1.63$ sixteenths

The minimum weld size based on the thinner part joined from AISC *Specification* Table J2.4 controls. Based on the 0.830-in. web, use 5/16-in. fillet welds (two sides) for the washer plate-to-base plate weld.

The final connection design and geometry for the moment frame column base is shown in Figure 4-23.

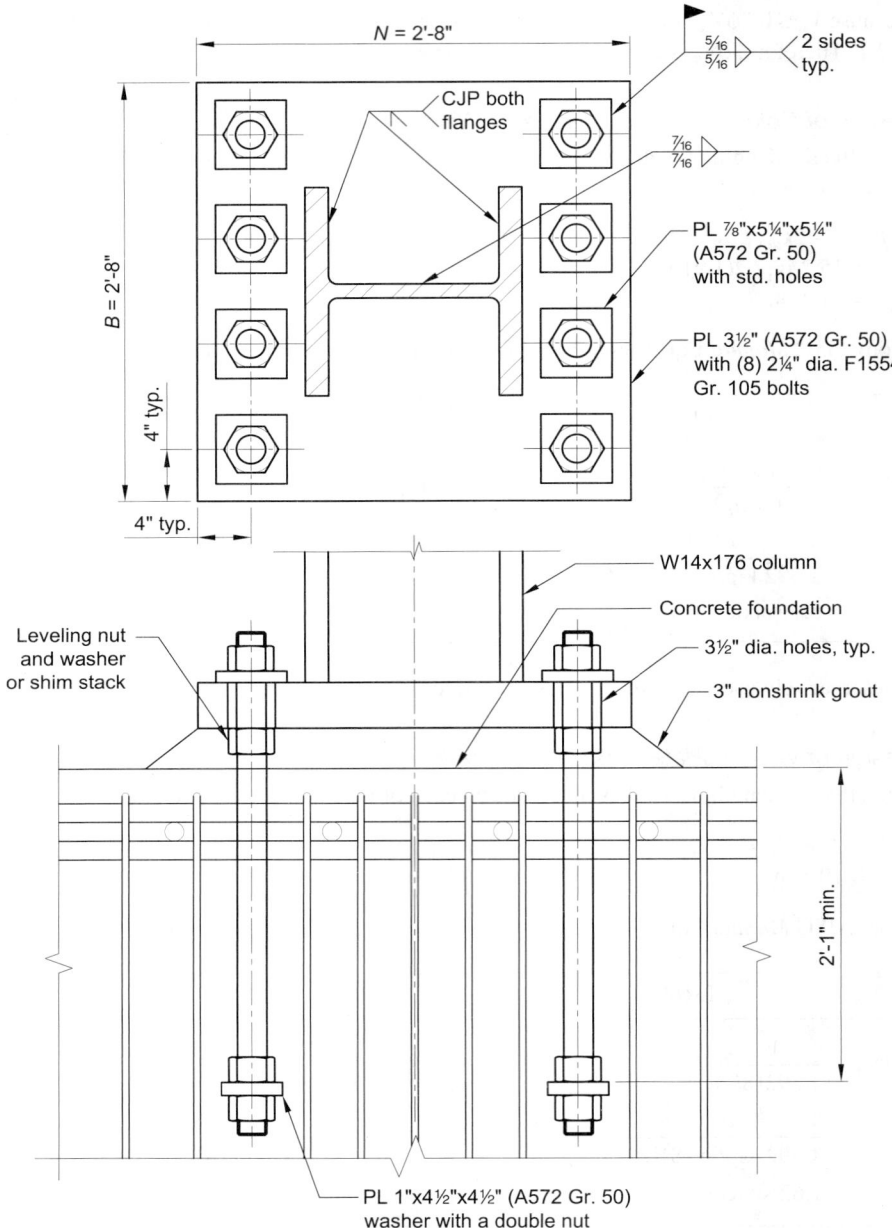

Fig. 4-23. Connection cross section as designed in Example 4.4.3.

Example 4.4.4. SMF Embedded Column Base Design

Given:
Refer to Column CL-1 in Figure 4-8. Design an embedded column base plate for the ASTM A992 W-shape. The column is centered on a 72-in.-wide reinforced concrete foundation. The foundation concrete compressive strength, f'_c, is 4 ksi with ASTM A615 Grade 60 reinforcement. Use ASTM A572 Grade 50 plate material.

The applicable building code specifies the use of ASCE/SEI 7 for calculation of loads. The required column strengths at the base level were determined by a second-order analysis including the effects of $P\text{-}\delta$ and $P\text{-}\Delta$ with reduced stiffness as required by the direct analysis method. The governing load combinations in ASCE/SEI 7, including the overstrength factor (referred to as the amplified seismic load in the AISC *Seismic Provisions*), follow.

In this example, the controlling limit state is yielding of the face plates. For this limit state, the axial force needs to be maximized as this will increase the bearing force and subsequent bending (yielding) in the plates. Therefore, the required axial strength is determined from:

LRFD	ASD
LRFD Load Combination 5 from ASCE/SEI 7 Section 12.4.3.2 $$P_u = (1.2 + 0.2S_{DS})D + \Omega_o Q_E$$ $$+ 0.5L + 0.2S$$ $$= 250 \text{ kips}$$ (including the 0.5 factor on L permitted in Section 12.4.3.2)	ASD Load Combination 6 from ASCE/SEI 7 Section 12.4.3.2 $$P_a = (1.0 + 0.105S_{DS})D + 0.525\Omega_o Q_E$$ $$+ 0.75L + 0.75S$$ $$= 215 \text{ kips}$$

The required flexural strength is determined from:

LRFD	ASD
LRFD Load Combination 7 from ASCE/SEI 7 Section 12.4.3.2 $$M_u = (0.9 - 0.2S_{DS})D + \Omega_o Q_E$$ $$= 946 \text{ kip-ft}$$	ASD Load Combination 8 from ASCE/SEI 7 Section 12.4.3.2 $$M_a = (0.6 - 0.14S_{DS})D + 0.7\Omega_o Q_E$$ $$= 662 \text{ kip-ft}$$

The required shear strength is determined from:

LRFD	ASD
LRFD Load Combination 5 from ASCE/SEI 7 Section 12.4.3.2 $$V_u = (1.2 + 0.2S_{DS})D + \Omega_o Q_E$$ $$= 96.0 \text{ kips}$$	ASD Load Combination 5 from ASCE/SEI 7 Section 12.4.3.2 $$V_a = (1.0 + 0.14S_{DS})D + 0.7\Omega_o Q_E$$ $$= 67.2 \text{ kips}$$

Consider that the connection into the column weak-axis produces negligible moments on the column. With respect to the foundation, consider that the ACI 318 reinforcement requirements are adequate for all applicable concrete limit states including punching shear.

From ASCE/SEI 7, use Seismic Design Category D, $\Omega_o = 3.0$, $\rho = 1.0$ and $S_{DS} = 1.0$.

Use LRFD provisions for the concrete design. The final connection design and geometry for the embedded column base is shown in Figure 4-24.

Solution:

From AISC *Manual* Table 2-4, the column material properties are as follows:

ASTM A992
$F_y = 50$ ksi
$F_u = 65$ ksi

From AISC *Manual* Table 2-5, the plate material properties are as follows:

ASTM A572 Grade 50
$F_y = 50$ ksi
$F_u = 65$ ksi

From ASTM A615, the concrete reinforcement properties are as follows:

ASTM A615 Grade 60
$F_y = 60$ ksi

From AISC *Manual* Table 1-1, the geometric properties are as follows:

Column
W14×176
$A = 51.8$ in.^{2} $d = 15.2$ in. $b_f = 15.7$ in. $t_w = 0.830$ in.
$t_f = 1.31$ in. $Z_x = 320$ in.^{3}

Beam
W24×76
$d = 23.9$ in.

Required Strengths at the Column Base

AISC *Seismic Provisions* Section D2.6a requires that the axial strength equals or exceeds the required strength calculated using the load combinations of the applicable building code, including the amplified seismic load.

By reference to AISC *Seismic Provisions* Section D2.5c, which references Section D2.5b, AISC *Seismic Provisions* Section D2.6b indirectly stipulates that the required shear strength of the column base be the greater of the required shear strength determined from load combinations including the amplified seismic load (Section D2.5b(b)) or the required column strength as stipulated in the system chapters (Section D2.5b(a)). Here, the provisions of Section E3.6g apply, as follows:

LRFD	ASD
$V_u = \dfrac{\Sigma M_{pc}}{H}$	$V_a = \dfrac{\Sigma M_{pc}}{1.5H}$

where ΣM_{pc} is the sum of the nominal plastic flexural strengths of the columns above and below the splice, or in this case, the base.

For the calculation of M_{pc}:

$M_{pc} = F_y Z_x$

$= (50 \text{ ksi})(320 \text{ in.}^3)(1 \text{ ft}/12 \text{ in.})$

$= 1{,}330$ kip-ft

Therefore:

LRFD	ASD
$V_u = \dfrac{\Sigma M_{pc}}{H}$ $= \dfrac{2(1{,}330 \text{ kip-ft})}{14.0 \text{ ft}}$ $= 190 \text{ kips} > 96.0 \text{ kips}$ Use $V_u = 190$ kips.	$V_a = \dfrac{\Sigma M_{pc}}{1.5H}$ $= \dfrac{2(1{,}330 \text{ kip-ft})}{1.5(14.0 \text{ ft})}$ $= 127 \text{ kips} > 67.2 \text{ kips}$ Use $V_a = 127$ kips.

AISC *Seismic Provisions* Section D2.6c(b) requires that the flexural strength equals or exceeds the lesser of the load combination of the applicable building code, including the amplified seismic load, or $1.1 R_y F_y Z$.

LRFD	ASD
$M_u = 1.1 R_y F_y Z_x$ $= \dfrac{1.1(1.1)(50 \text{ ksi})(320 \text{ in.}^3)}{(12 \text{ in./ft})}$ $= 1{,}610 \text{ kip-ft} > 946 \text{ kip-ft}$	$M_a = \dfrac{1.1 R_y F_y Z_x}{1.5}$ $= \dfrac{1.1(1.1)(50 \text{ ksi})(320 \text{ in.}^3)}{1.5(12 \text{ in./ft})}$ $= 1{,}080 \text{ kip-ft} > 662 \text{ kip-ft}$
Use $M_u = 946$ kip-ft.	Use $M_a = 662$ kip-ft.

Required Column Embedment Depth

Consider the base condition similar to a structural steel coupling beam embedded in a composite special shear wall, per AISC *Seismic Provisions* Section H5.5c. Therefore, Section H4.5b(1)(3) will be used to determine the embedment length. For the calculation of the embedment length, L_e:

$$V_n = 1.54\sqrt{f'_c}\left(\frac{b_w}{b_f}\right)^{0.66} \beta_1 b_f L_e \left[\frac{0.58 - 0.22\beta_1}{0.88 + \dfrac{g}{2L_e}}\right] \quad \text{(\textit{Provisions} Eq. H4-2)}$$

where

$\beta_1 = 0.85$ from ACI 318 Section 10.2.7.3
$g = H$
$= 14.0 \text{ ft}(12 \text{ in./ft})$
$= 168 \text{ in.}$

Try an embedment length, L_e, of 22 in.

Therefore:

$$V_n = 1.54\sqrt{4.0 \text{ ksi}}\left(\frac{72.0 \text{ in.}}{15.7 \text{ in.}}\right)^{0.66}(0.85)(15.7 \text{ in.})(22.0 \text{ in.})\left[\frac{0.58 - 0.22(0.85)}{0.88 + \dfrac{168 \text{ in.}}{2(22.0 \text{ in.})}}\right]$$

$= 207 \text{ kips} > 190 \text{ kips}$ **o.k.**

As indicated in AISC *Seismic Provisions* Section H4.5b(1)(3), the embedment is considered to begin inside the first layer of confining reinforcement in the foundation.

Longitudinal Foundation Reinforcement

AISC *Seismic Provisions* Section H4.5b(1)(4) requires that longitudinal foundation reinforcement with nominal axial strength equal to the expected shear strength of the column be placed over the embedment length.

$$A_s = \frac{V_u}{F_y}$$
$$= \frac{190 \text{ kips}}{60 \text{ ksi}}$$
$$= 3.17 \text{ in.}^2$$

AISC *Seismic Provisions* Section H4.5b(1)(4) requires two-thirds of this reinforcement in the top layer. It is permitted to use reinforcement placed for other purposes as part of the required longitudinal reinforcement.

AISC *Seismic Provisions* Section H5.5c requires that this reinforcement be confined by transverse reinforcement that meets the requirements for boundary members of ACI 318 Section 21.9.6. For this example, as stated above, the foundation reinforcing requirements are considered adequate per ACI 318.

Minimum Face Bearing Plate Thickness

AISC *Seismic Provisions* Section H5.5c requires face bearing plates on both sides of the column at the face of the foundation and near the end of the embedded region. At a minimum, the stiffener thickness should meet the detailing requirements of AISC *Seismic Provisions* Section F3.5b(4) where,

$t_{min} = 0.75 t_w > \frac{3}{8}$ in.
$= 0.75(0.830 \text{ in.})$
$= 0.623 \text{ in.} > \frac{3}{8}$ in.

Yielding in the Face Bearing Plates

The column axial force is distributed from the column to the face bearing plates and then to the foundation in direct bearing. As outlined in AISC *Manual* Part 14, the critical face plate cantilever dimension, l, is determined as the larger of m, n or $\lambda n'$ (as depicted in Figure 4-22), where:

$$m = \frac{N - 0.95d}{2} \qquad \text{(\textit{Manual} Eq. 14-2)}$$

$$n = \frac{B - 0.8b_f}{2} \qquad \text{(\textit{Manual} Eq. 14-3)}$$

$$\lambda n' = \frac{\lambda \sqrt{db_f}}{4} \qquad \text{(from \textit{Manual} Eq. 14-4)}$$

$N = d$

$B = b_f$

$\lambda = 1.0$ (conservative per AISC *Manual* Part 14)

Therefore:

$$m = \frac{15.2 \text{ in.} - 0.95(15.2 \text{ in.})}{2}$$
$$= 0.380 \text{ in.}$$

$$n = \frac{15.7 \text{ in.} - 0.8(15.7 \text{ in.})}{2}$$
$$= 1.57 \text{ in.}$$

$$\lambda n' = \frac{1.0\sqrt{15.2 \text{ in.}(15.7 \text{ in.})}}{4}$$
$$= 3.86 \text{ in.}$$

For the yielding limit state, the required minimum thickness is determined from AISC *Manual* Equations 14-7a and 14-7b:

LRFD	ASD
$t_{min} = l\sqrt{\dfrac{2P_u}{0.9F_y BN}}$	$t_{min} = l\sqrt{\dfrac{3.33P_a}{F_y BN}}$
$= 3.86 \text{ in.}$	$= 3.86 \text{ in.}$
$\times \sqrt{\dfrac{2(250 \text{ kips})}{0.9(50 \text{ ksi})(15.7 \text{ in.})(15.2 \text{ in.})}}$	$\times \sqrt{\dfrac{3.33(215 \text{ kips})}{50 \text{ ksi}(15.7 \text{ in.})(15.2 \text{ in.})}}$
$= 0.833 \text{ in.}$	$= 0.946 \text{ in.}$

Due to the different load combinations used for LRFD versus ASD, there is a slight discrepancy between the LRFD and ASD results for the required shear strength. Typically, one method should be chosen and used consistently throughout an entire design. For the purposes of this example, the LRFD result will be used.

Because flexural yielding at the bearing interface controls the face plate design, the fillet weld connection provisions of AISC *Seismic Provisions* Section F3.5b(4) are not applicable and the thickness should be fully developed. Therefore, the face plates are welded to the columns with complete-joint-penetration groove welds.

Use ⅞-in.-thick ASTM A572 Grade 50 face bearing plates.

Required Transfer Reinforcement

AISC *Seismic Provisions* Section H5.5c requires two regions of transfer reinforcement attached to both the embedded flanges. The area of transfer reinforcement is:

$A_{tb} \geq 0.03 f'_c L_e b_f / F_{ysr}$ (*Provisions* Eq. H5-1)

$F_{ysr} = 60$ ksi (deformed bar anchor)

4.4 COLUMN SPLICE AND COLUMN BASE DESIGN EXAMPLES

Therefore:

$$A_{tb} = 0.03(4 \text{ ksi})(22.0 \text{ in.})(15.7 \text{ in.})/60 \text{ ksi}$$
$$= 0.691 \text{ in.}^2$$

The provision requires that all transfer bars be fully developed where they engage the embedded flange. For this example, consider a bar length of 36 in. fully developed per ACI 318.

Use (2) ¾ in. × 36 in. bars in each region.

$$A_{tb} = \frac{(2)\pi(¾ \text{ in.})^2}{4}$$
$$= 0.884 \text{ in.}^2 > 0.691 \text{ in.}^2 \quad \textbf{o.k.}$$

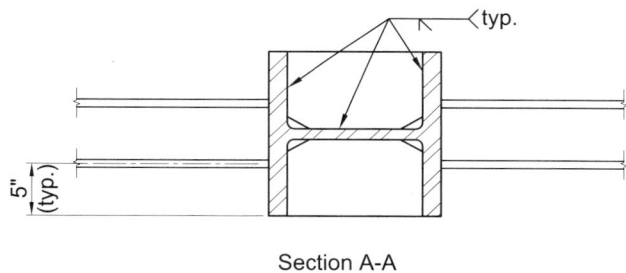

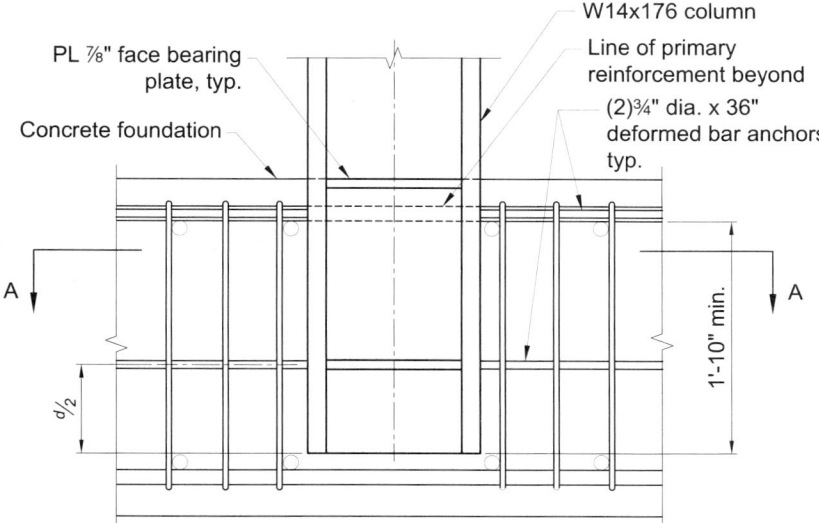

Note: The deformed bar anchor-to-column flange connection should match the strength of the bar.

Fig. 4-24. Connection cross section as designed in Example 4.4.4.

The weld of the deformed bar to the column flange should be a flux-filled material using an electric arc welding process, that develops the strength of the rebar according to AWS D1.1 clause 7.

AISC *Seismic Provisions* Section H5.5c also requires that the not-to-exceed transfer reinforcement area is:

$$\Sigma A_{tb} < 0.08 L_e b_w - A_s \qquad \text{(Provisions Eq. H5-2)}$$
$$< 0.08 \ (22.0 \text{ in.})(72.0 \text{ in.}) - A_s$$
$$< 127 \text{ in.}^2 - A_s$$

In AISC *Seismic Provisions* Equation H5-2, A_s is the longitudinal area of reinforcement provided over the embedment length. As noted in the Given statement, the foundation reinforcing requirements are considered adequate per ACI 318. Therefore, this check is provided for illustrative purposes only.

The final connection design and geometry for the embedded column base is shown in Figure 4-24.

4.5 DESIGN TABLE DISCUSSION

Table 4-1. Comparison of Requirements for SMF, IMF and OMF

Several categories of connection and design criteria are listed in Table 4-1. The *Seismic Provisions* requirements for each category are given for OMF, IMF and SMF.

Table 4-2. SMF Design Tables

Various values useful in the design of SMF are tabulated. Values are given for W-shapes that meet the width-to-thickness requirements for SMF beams and columns with $F_y = 50$ ksi (ASTM A992).

For cases where the limiting web width-to-thickness ratio is a function of the member's required axial strength, P_u or P_a, according to AISC *Seismic Provisions* Table D1.1, the member will satisfy the width-to-thickness requirements for highly ductile members if P_u or P_a is less than or equal to the value tabulated for $P_{u\,max}$ or $P_{a\,max}$, respectively. The nominal axial yield strength of a member, P_y, is calculated as $F_y A_g$. Note that it is assumed that $C_a = P_u/\phi_c P_y > 0.125$ or $C_a = \Omega_c P_a/P_y > 0.125$. Where a dash is shown, there is no limitation on the values of P_u or P_a.

The value $1.1 R_y M_p$ is given to aid in several calculations, including the determination of the required shear strength of SMF connections and the SMF column-beam moment ratio.

Several values are tabulated to enable quick determination of column panel zone shear strength. To determine if AISC *Specification* Equations J10-11 or J10-12 are applicable, $0.75 P_c$ is given for comparison with the required axial strength, P_r. If P_r is less than or equal

4.5 DESIGN TABLE DISCUSSION

to $0.75P_c$, then the values of ϕR_{v1} and ϕR_{v2} or R_{v1}/Ω and R_{v2}/Ω can be used to calculate the available panel zone shear strength. Considering strength of a column without doubler plates:

$$R_n = 0.60 F_y d_c t_{cw} \left(1 + \frac{3 b_{cf} t_{cf}^2}{d_b d_c t_w} \right) \quad \text{(Spec. Eq. J10-11)}$$

where
    F_y = specified minimum yield stress of the column web, ksi
    b_{cf} = width of column flange, in.
    d_b = depth of beam, in.
    d_c = depth of column, in.
    t_{cf} = thickness of column flange, in.
    t_w = thickness of column web, in.

Expanding AISC *Specification* Equation J10-11 yields:

$$R_n = 0.60 F_y d_c t_w + 0.60 F_y d_c t_w \left(\frac{3 b_{cf} t_{cf}^2}{d_b d_c t_w} \right)$$

R_{v1} and R_{v2} are defined as:

$$R_{v1} = 0.60 F_y d_c t_w$$

$$R_{v2} = 0.60 F_y d_c t_w \left(\frac{3 b_{cf} t_{cf}^2}{d_c t_w} \right)$$

Substituting into the expanded version of AISC *Specification* Equation J10-11, the available panel zone shear strength is:

LRFD	ASD
$\phi R_v = \phi R_{v1} + \dfrac{\phi R_{v2}}{d_b}$	$\dfrac{R_v}{\Omega} = \dfrac{R_{v1}}{\Omega} + \dfrac{R_{v2}}{\Omega d_b}$

To aid in the determination of the minimum panel-zone element thicknesses, $w_z/90$ or $d_z/90$ are also tabulated. Therefore, the sum of the corresponding $w_z/90$ or $d_z/90$ values for the SMF beam and column will determine the minimum panel-zone element thicknesses per AISC *Seismic Provisions* Equation E3-7:

$$t \geq (d_z + w_z)/90$$

Values are also tabulated to aid in the determination of lateral bracing requirements. The value given for $L_{b\ max}$ is the maximum distance between lateral braces specified in AISC *Seismic Provisions* Section D1.2b. The required brace strength at beam-to-column connections stipulated in AISC *Seismic Provisions* Section E3.4c(1), equal to $0.02 F_y b_f t_f$, is also given. All lateral bracing is also required to have a minimum stiffness based on a moment equal to $R_y M_p = R_y F_y Z$. The value of this moment is tabulated.

Table 4-1
Comparison of Requirements for SMF, IMF and OMF

	Special Moment Frame (SMF)	Intermediate Moment Frame (IMF)	Ordinary Moment Frame (OMF)
Story Drift Angle	0.04 rad	0.02 rad	No specified minimum
Connection Flexural Strength	Performance confirmed by testing per AISC *Seismic Provisions* Chapter K; connection achieves minimum 80% of nominal plastic moment of the beam at story drift angle of 0.04 rad	Performance confirmed by testing per AISC *Seismic Provisions* Chapter K; connection achieves minimum 80% of nominal plastic moment of the beam at story drift angle of 0.02 rad	FR: Develop $1.1R_yM_p$ of beam, maximum moment developed by system or satisfy requirements in AISC *Seismic Provisions* Section E1.6b, E2.6 and E3.6
Connection Shear Strength	V for load combination including overstrength plus shear from application of $E_{mh} = 2[1.1R_yM_p]/L_h$	V for load combination including overstrength plus shear from application of $E_{mh} = 2[1.1R_yM_p]/L_h$	V for load combination including overstrength plus shear from application of $E_{mh} = 2[1.1R_yM_p]/L_{cf}$
	– or –	– or –	– or –
	Lesser V permitted if justified by analysis. See also the exception provided in AISC *Seismic Provisions* Section E3.6d	Lesser V permitted if justified by analysis. See also the exception provided in AISC *Seismic Provisions* Section E2.6d	Lesser V permitted if justified by analysis
Panel Zone Shear Strength	For $P_r \leq 0.75P_c$, compute strength per AISC *Specification* Eq. J10-11 using $\phi_v = 1.00$ (LRFD) or $\Omega_v = 1.50$ (ASD)	No additional requirements beyond AISC *Specification*	No additional requirements beyond AISC *Specification*
	For $P_r > 0.75P_c$, compute strength per AISC *Specification* Eq. J10-12 using $\phi_v = 1.00$ (LRFD) or $\Omega_v = 1.50$ (ASD)		
Panel Zone Thickness	$t \geq (d_z + w_z)/90$	No additional requirements beyond AISC *Specification*	No additional requirements beyond AISC *Specification*

Table 4-1 (continued)
Comparison of Requirements for SMF, IMF and OMF

	Special Moment Frame (SMF)	Intermediate Moment Frame (IMF)	Ordinary Moment Frame (OMF)
Continuity Plates	To match tested condition or ANSI/AISC 358 Section 2.4.4	To match tested condition or ANSI/AISC 358 Section 2.4.4	Provide continuity plates as required by AISC *Seismic Provisions* Section E1.6b
Beam-Column Proportion	$\dfrac{\Sigma M^*_{pc}}{\Sigma M^*_{pb}} > 1.0$	No additional requirements beyond AISC *Specification*	No additional requirements beyond AISC *Specification*
Width-to-Thickness Limitations	Beams and columns to satisfy the AISC *Seismic Provisions* Section D1.1 for highly ductile members	Beams and columns to satisfy the AISC *Seismic Provisions* Section D1.1 for moderately ductile members	No additional requirements beyond AISC *Specification*
Stability Bracing of Beams	Beam bracing required to satisfy AISC *Seismic Provisions* Section D1.2b for highly ductile members	Beam bracing required to satisfy AISC *Seismic Provisions* Section D1.2a for moderately ductile members	No additional requirements beyond AISC *Specification*
Column Splice	Splices shall satisfy AISC *Seismic Provisions* Section D2.5 and E3.6g; bolts or CJP groove welds	Splices shall satisfy AISC *Seismic Provisions* Sections D2.5 and E2.6g; bolts or CJP groove welds	No additional requirements beyond AISC *Specification*
Protected Zone	As established by ANSI/AISC 358 for each prequalified connection; generally, one-half beam depth beyond centerline of plastic hinge	As established by ANSI/AISC 358 for each prequalified connection; generally, one-half beam depth beyond centerline of plastic hinge	None

Table 4-2
SMF Design Values
W-Shapes

$R_y = 1.1$ $F_y = 50$ ksi

Shape	$P_{a\,max}$ (ASD)	$P_{u\,max}$ (LRFD)	$1.1R_yM_p$	Panel Zone ASD ($\Omega = 1.50$)			Panel Zone LRFD ($\phi = 1.00$)		
				R_{v1}/Ω	R_{v2}/Ω	$0.75P_c$	ϕR_{v1}	ϕR_{v2}	$0.75P_c$
	kips	kips	kip-ft	kips	kip-in.	kips	kips	kip-in.	kips
W44×335	2600	3900	8170	906	2990	2220	1360	4480	3690
×290	1290	1930	7110	754	2370	1920	1130	3550	3200
×262	590	887	6400	680	1910	1740	1020	2870	2900
×230	156	234	5550	609	1410	1530	914	2120	2540
W40×593	—	—	13900	1540	10500	3920	2310	15700	6530
×503	—	—	11700	1300	7500	3330	1950	11200	5550
×431	—	—	9880	1110	5410	2860	1660	8120	4760
×397	—	—	9080	1000	4680	2630	1500	7010	4390
×372	—	—	8470	942	4060	2480	1410	6090	4130
×362	—	—	8270	909	3880	2390	1360	5820	3980
×324	—	—	7360	804	3130	2140	1210	4690	3570
×297	2470	3710	6710	740	2580	1960	1110	3870	3270
×277	1730	2600	6300	659	2370	1830	989	3550	3060
×249	1040	1560	5650	591	1910	1650	887	2870	2760
×215	222	333	4860	507	1410	1430	761	2120	2380
W40×392	—	—	8620	1180	4720	2610	1770	7090	4350
×331	—	—	7210	996	3320	2200	1490	4980	3660
×327	—	—	7110	963	3290	2160	1440	4940	3600
×294	—	—	6400	856	2680	1940	1280	4020	3230
×278	—	—	6000	828	2360	1850	1240	3540	3090
×264	—	—	5700	768	2140	1740	1150	3210	2900
×235	1470	2200	5090	659	1780	1550	989	2670	2590
×211	876	1320	4570	591	1430	1400	887	2140	2330
×183	186	280	3900	507	1020	1200	761	1530	2000
×167	172	259	3490	502	751	1110	753	1130	1850
×149	112	169	3010	481	488	986	722	732	1640

Table 4-2
SMF Design Values
W-Shapes

$F_y = 50$ ksi $R_y = 1.1$

Shape	Panel Zone $\dfrac{w_z}{90}$ or $\dfrac{d_z}{90}$	$L_{b\,max}$	Lateral Bracing ASD $\dfrac{0.02F_y b_f t_f}{1.5}$	ASD $\dfrac{0.02M_r C_d}{h_o}$	LRFD $0.02F_y b_f t_f$	LRFD $\dfrac{0.02M_r C_d}{h_o}$	$R_y M_p$
	in.	ft	kips	kips	kips	kips	kip-ft
W44×335	0.450	14.5	18.8	28.2	28.1	42.2	7430
×290	0.449	14.5	16.6	24.6	25.0	36.9	6460
×262	0.450	14.4	15.0	22.2	22.4	33.3	5820
×230	0.450	14.3	12.9	19.3	19.3	29.0	5040
W40×593	0.406	15.8	36.0	50.9	53.9	76.3	12700
×503	0.406	15.5	30.2	43.3	45.3	64.9	10600
×431	0.406	15.2	25.5	36.9	38.2	55.4	8980
×397	0.407	15.1	23.6	34.0	35.4	51.0	8250
×372	0.406	15.0	22.0	31.9	33.0	47.9	7700
×362	0.406	15.0	21.4	31.2	32.2	46.7	7520
×324	0.406	14.9	19.2	27.9	28.8	41.8	6690
×297	0.406	14.7	17.4	25.5	26.1	38.3	6100
×277	0.406	14.9	16.6	24.1	25.0	36.1	5730
×249	0.406	14.8	15.0	21.6	22.4	32.4	5130
×215	0.406	14.7	12.9	18.7	19.3	28.1	4420
W40×392	0.406	11.0	20.8	32.1	31.2	48.1	7840
×331	0.406	10.7	17.3	27.1	26.0	40.6	6550
×327	0.406	10.7	17.2	26.7	25.8	40.1	6460
×294	0.406	10.6	15.4	24.2	23.2	36.3	5820
×278	0.406	10.5	14.5	22.7	21.7	34.1	5450
×264	0.406	10.5	13.7	21.6	20.6	32.5	5180
×235	0.406	10.6	12.5	19.4	18.8	29.2	4630
×211	0.406	10.4	11.2	17.5	16.8	26.2	4150
×183	0.407	10.4	9.44	15.0	14.2	22.5	3550
×167	0.406	9.98	8.10	13.5	12.2	20.3	3180
×149	0.406	9.52	6.53	11.7	9.79	17.6	2740

AMERICAN INSTITUTE OF STEEL CONSTRUCTION

Table 4-2 (continued)
SMF Design Values
W-Shapes

$R_y = 1.1$ $F_y = 50$ ksi

Shape	$P_{a\,max}$ (ASD)	$P_{u\,max}$ (LRFD)	$1.1R_yM_p$	Panel Zone ASD ($\Omega = 1.50$)			Panel Zone LRFD ($\phi = 1.00$)		
				R_{v1}/Ω	R_{v2}/Ω	$0.75P_c$	ϕR_{v1}	ϕR_{v2}	$0.75P_c$
	kips	kips	kip-ft	kips	kip-in.	kips	kips	kip-in.	kips
W36×652	—	—	14700	1620	13200	4320	2430	19900	7200
×529	—	—	11700	1280	8740	3510	1920	13100	5850
×487	—	—	10700	1180	7370	3220	1770	11100	5360
×441	—	—	9630	1060	6070	2930	1590	9110	4880
×395	—	—	8620	937	4880	2610	1410	7320	4350
×361	—	—	7810	851	4050	2390	1280	6070	3980
×330	—	—	7110	769	3410	2180	1150	5110	3630
×302	—	—	6450	705	2830	2000	1060	4240	3340
×282	2430	3650	6000	657	2460	1870	985	3680	3110
×262	2010	3020	5550	620	2070	1740	930	3100	2900
×247	1670	2500	5190	587	1800	1630	881	2710	2720
×231	1340	2010	4860	555	1570	1530	832	2360	2560
W36×256	—	—	5240	718	2190	1690	1080	3290	2820
×232	1870	2810	4720	646	1790	1530	968	2680	2550
×210	1520	2290	4200	609	1350	1390	914	2030	2320
×194	1100	1650	3870	558	1150	1280	838	1730	2140
×182	825	1240	3620	526	1010	1210	790	1520	2010
×170	536	805	3370	492	871	1130	738	1310	1880
×160	336	506	3150	468	749	1060	702	1120	1760
×150	174	262	2930	449	636	997	673	954	1660
W33×387	—	—	7870	907	5050	2570	1360	7580	4280
×354	—	—	7160	826	4220	2340	1240	6330	3900
×318	—	—	6400	732	3430	2110	1100	5140	3510
×291	—	—	5850	668	2860	1930	1000	4280	3210
×263	—	—	5240	600	2340	1740	900	3510	2900
×241	2120	3180	4740	568	1870	1600	852	2800	2670
×221	1670	2510	4320	525	1550	1470	788	2330	2450
×201	1210	1810	3900	482	1250	1330	723	1870	2220
W33×169	770	1160	3170	453	1030	1110	679	1540	1860
×152	517	777	2820	425	782	1010	638	1170	1680
×141	317	477	2590	403	636	934	604	954	1560
×130	163	245	2350	384	504	862	576	757	1440

Table 4-2 (continued)
SMF Design Values
W-Shapes

$F_y = 50$ ksi $R_y = 1.1$

Shape	Panel Zone $\frac{w_z}{90}$ or $\frac{d_z}{90}$	$L_{b\,max}$	ASD $0.02F_y b_f t_f$ / 1.5	ASD $\frac{0.02M_r C_d}{h_o}$	LRFD $0.02F_y b_f t_f$	LRFD $\frac{0.02M_r C_d}{h_o}$	$R_y M_p$
	in.	ft	kips	kips	kips	kips	kip-ft
W36×652	0.378	17.0	41.5	56.8	62.3	85.1	13300
×529	0.378	16.6	33.4	46.3	50.1	69.5	10700
×487	0.377	16.5	30.6	42.7	45.8	64.0	9760
×441	0.378	16.3	27.7	38.4	41.5	57.6	8750
×395	0.378	16.1	24.6	34.6	37.0	52.0	7840
×361	0.378	16.0	22.4	31.6	33.6	47.4	7100
×330	0.378	15.9	20.5	28.8	30.7	43.2	6460
×302	0.377	15.9	18.7	26.4	28.1	39.6	5870
×282	0.377	15.8	17.4	24.6	26.1	36.9	5450
×262	0.378	15.6	15.9	22.7	23.9	34.1	5040
×247	0.378	15.5	14.9	21.3	22.3	32.0	4720
×231	0.378	15.4	13.9	20.1	20.8	30.1	4410
W36×256	0.377	11.0	14.1	21.4	21.1	32.0	4770
×232	0.377	10.9	12.7	19.3	19.0	29.0	4290
×210	0.378	10.7	11.1	17.3	16.6	26.0	3820
×194	0.378	10.6	10.2	16.0	15.2	24.0	3520
×182	0.377	10.6	9.52	15.0	14.3	22.5	3290
×170	0.378	10.5	8.80	14.0	13.2	20.9	3060
×160	0.377	10.4	8.16	13.1	12.2	19.6	2860
×150	0.378	10.3	7.52	12.2	11.3	18.3	2660
W33×387	0.349	15.7	24.6	33.9	36.9	50.9	7150
×354	0.349	15.5	22.4	31.1	33.6	46.6	6510
×318	0.349	15.4	20.2	28.0	30.2	42.0	5820
×291	0.348	15.3	18.3	25.7	27.5	38.5	5320
×263	0.348	15.2	16.5	23.2	24.8	34.8	4770
×241	0.349	15.0	14.8	21.0	22.3	31.5	4310
×221	0.348	14.9	13.5	19.3	20.2	28.9	3930
×201	0.349	14.8	12.0	17.4	18.1	26.1	3540
W33×169	0.348	10.4	9.35	14.1	14.0	21.2	2880
×152	0.349	10.3	8.20	12.7	12.3	19.0	2560
×141	0.349	10.1	7.36	11.7	11.0	17.5	2360
×130	0.349	9.93	6.56	10.6	9.83	16.0	2140

Table 4-2 (continued)
SMF Design Values
W-Shapes

$R_y = 1.1$ $F_y = 50$ ksi

Shape	$P_{a\,max}$ (ASD)	$P_{u\,max}$ (LRFD)	$1.1R_yM_p$	Panel Zone ASD ($\Omega = 1.50$)			Panel Zone LRFD ($\phi = 1.00$)		
				R_{v1}/Ω	R_{v2}/Ω	$0.75P_c$	ϕR_{v1}	ϕR_{v2}	$0.75P_c$
	kips	kips	kip-ft	kips	kip-in.	kips	kips	kip-in.	kips
W30×391	—	—	7310	903	5570	2590	1350	8360	4310
×357	—	—	6660	813	4670	2360	1220	7000	3940
×326	—	—	6000	739	3880	2160	1110	5820	3600
×292	—	—	5340	653	3140	1940	979	4710	3230
×261	—	—	4750	588	2480	1730	882	3720	2890
×235	—	—	4270	520	2040	1560	779	3060	2600
×211	—	—	3790	479	1580	1400	718	2370	2340
×191	1510	2260	3400	436	1270	1260	654	1910	2100
×173	1110	1670	3060	398	1030	1150	597	1550	1910
W30×148	896	1350	2520	399	877	981	599	1320	1640
×132	654	982	2200	373	630	873	559	945	1460
×124	479	720	2060	353	545	821	530	817	1370
×116	361	542	1910	339	455	770	509	683	1280
×108	242	364	1740	325	364	713	487	546	1190
W27×539	—	—	9530	1280	11500	3580	1920	17300	5960
×368	—	—	6250	839	5420	2450	1260	8140	4090
×336	—	—	5700	756	4550	2230	1130	6830	3720
×307	—	—	5190	687	3770	2030	1030	5660	3380
×281	—	—	4720	621	3220	1870	932	4830	3120
×258	—	—	4300	568	2690	1710	853	4030	2850
×235	—	—	3890	522	2210	1560	784	3310	2600
×217	—	—	3580	471	1900	1440	707	2860	2400
×194	—	—	3180	422	1510	1280	632	2260	2140
×178	—	—	2870	403	1200	1180	605	1800	1970
×161	1400	2110	2600	364	980	1070	546	1470	1790
×146	1040	1570	2340	332	799	972	497	1200	1620
W27×129	893	1340	1990	337	726	851	505	1090	1420
×114	642	965	1730	311	524	756	467	786	1260
×102	350	527	1540	279	413	675	419	620	1130
×94	215	324	1400	264	333	621	395	500	1040

Table 4-2 (continued)
SMF Design Values
W-Shapes

F_y = 50 ksi R_y = 1.1

Shape	Panel Zone $\dfrac{w_z}{90}$ or $\dfrac{d_z}{90}$	$L_{b\,max}$	Lateral Bracing				$R_y M_p$
			ASD		LRFD		
			$\dfrac{0.02 F_y b_f t_f}{1.5}$	$\dfrac{0.02 M_r C_d}{h_o}$	$0.02 F_y b_f t_f$	$\dfrac{0.02 M_r C_d}{h_o}$	
	in.	ft	kips	kips	kips	kips	kip-ft
W30×391	0.315	15.3	25.4	34.5	38.1	51.8	6650
×357	0.315	15.1	23.1	31.6	34.7	47.5	6050
×326	0.314	15.0	21.0	28.7	31.6	43.1	5450
×292	0.314	14.9	18.9	25.7	28.3	38.6	4860
×261	0.314	14.7	16.7	23.1	25.1	34.6	4320
×235	0.314	14.6	15.1	20.8	22.7	31.3	3880
×211	0.314	14.5	13.3	18.6	19.9	27.9	3440
×191	0.315	14.4	11.9	16.8	17.9	25.2	3090
×173	0.314	14.2	10.7	15.2	16.1	22.8	2780
W30×148	0.315	9.48	8.26	12.4	12.4	18.6	2290
×132	0.314	9.35	7.00	10.9	10.5	16.4	2000
×124	0.315	9.27	6.51	10.2	9.77	15.3	1870
×116	0.314	9.10	5.95	9.49	8.93	14.2	1730
×108	0.314	8.94	5.32	8.75	7.98	13.1	1590
W27×539	0.282	15.2	36.1	47.8	54.2	71.7	8660
×368	0.283	14.5	24.3	32.6	36.5	48.9	5680
×336	0.283	14.3	22.2	29.9	33.3	44.9	5180
×307	0.282	14.2	20.1	27.5	30.1	41.2	4720
×281	0.283	14.1	18.5	25.1	27.8	37.6	4290
×258	0.283	14.0	16.9	23.0	25.3	34.5	3910
×235	0.283	13.8	15.2	20.9	22.9	31.3	3540
×217	0.282	13.8	14.1	19.4	21.2	29.1	3260
×194	0.282	13.7	12.5	17.3	18.8	25.9	2890
×178	0.282	13.5	11.2	15.7	16.8	23.6	2610
×161	0.283	13.4	10.1	14.3	15.1	21.4	2360
×146	0.283	13.3	9.10	12.9	13.7	19.3	2130
W27×129	0.282	9.19	7.33	10.9	11.0	16.4	1810
×114	0.283	9.06	6.26	9.53	9.39	14.3	1570
×102	0.283	8.94	5.53	8.50	8.30	12.8	1400
×94	0.282	8.81	4.97	7.78	7.45	11.7	1270

Table 4-2 (continued)
SMF Design Values
W-Shapes

$R_y = 1.1$ $F_y = 50$ ksi

Shape	$P_{a\,max}$ (ASD)	$P_{u\,max}$ (LRFD)	$1.1R_yM_p$	Panel Zone					
				ASD ($\Omega = 1.50$)			LRFD ($\phi = 1.00$)		
				R_{v1}/Ω	R_{v2}/Ω	$0.75P_c$	ϕR_{v1}	ϕR_{v2}	$0.75P_c$
	kips	kips	kip-ft	kips	kip-in.	kips	kips	kip-in.	kips
W24×370	—	—	5700	851	6080	2450	1280	9120	4090
×335	—	—	5140	759	4980	2210	1140	7470	3690
×306	—	—	4650	683	4180	2020	1020	6270	3360
×279	—	—	4210	619	3490	1840	929	5230	3070
×250	—	—	3750	547	2830	1650	821	4240	2760
×229	—	—	3400	499	2350	1510	749	3530	2520
×207	—	—	3060	447	1920	1370	671	2880	2280
×192	—	—	2820	413	1660	1270	620	2490	2120
×176	—	—	2580	378	1390	1160	567	2080	1940
×162	—	—	2360	353	1160	1080	529	1740	1790
×146	—	—	2110	321	920	968	482	1380	1610
×131	—	—	1870	296	713	869	445	1070	1450
W24×103	740	1110	1410	270	519	682	404	778	1140
×94	556	836	1280	250	417	623	375	625	1040
×84	336	506	1130	227	321	556	340	481	926
×76	193	290	1010	210	249	504	315	374	840
W24×62	124	187	771	204	147	410	306	221	683
×55	38.9	58.5	676	186	107	365	280	161	608
W21×201	—	—	2670	419	2010	1330	628	3010	2220
×182	—	—	2400	377	1640	1210	565	2460	2010
×166	—	—	2180	338	1380	1100	506	2060	1830
×147	—	—	1880	318	992	972	477	1490	1620
×132	—	—	1680	283	805	873	425	1210	1460
×122	—	—	1550	260	686	808	391	1030	1350
×111	—	—	1410	237	565	734	355	848	1220
W21×93	—	—	1110	251	437	614	376	655	1020
×83	707	1060	988	220	350	549	331	525	915
×73	456	685	867	193	273	484	289	409	806
×68	347	521	807	181	233	450	272	349	750
×62	220	330	726	168	187	412	252	280	686
W21×57	217	326	650	171	166	376	256	249	626
×50	117	176	555	158	112	331	237	168	551
×44	38.3	57.6	481	145	79.0	293	217	118	488

Table 4-2 (continued)
SMF Design Values
W-Shapes

$F_y = 50$ ksi $R_y = 1.1$

Shape	Panel Zone $\dfrac{w_z}{90}$ or $\dfrac{d_z}{90}$	$L_{b\,max}$	ASD $0.02F_y b_f t_f$ 1.5	ASD $\dfrac{0.02M_r C_d}{h_o}$	LRFD $0.02F_y b_f t_f$	LRFD $\dfrac{0.02M_r C_d}{h_o}$	$R_y M_p$
	in.	ft	kips	kips	kips	kips	kip-ft
W24×370	0.251	13.6	24.8	32.8	37.3	49.1	5180
×335	0.250	13.4	22.3	29.9	33.5	44.9	4680
×306	0.250	13.3	20.4	27.3	30.6	40.9	4230
×279	0.250	13.2	18.5	24.9	27.8	37.3	3830
×250	0.250	13.1	16.6	22.4	24.9	33.5	3410
×229	0.250	12.9	15.1	20.4	22.7	30.6	3090
×207	0.251	12.8	13.6	18.4	20.4	27.7	2780
×192	0.251	12.8	12.7	17.1	19.0	25.6	2560
×176	0.250	12.6	11.5	15.7	17.3	23.5	2340
×162	0.251	12.7	10.6	14.4	15.9	21.6	2150
×146	0.250	12.5	9.37	13.0	14.1	19.5	1920
×131	0.251	12.3	8.26	11.5	12.4	17.3	1700
W24×103	0.250	8.27	5.88	8.74	8.82	13.1	1280
×94	0.251	8.23	5.29	7.96	7.94	11.9	1160
×84	0.251	8.11	4.63	7.05	6.95	10.6	1030
×76	0.250	7.98	4.08	6.32	6.11	9.48	917
W24×62	0.250	5.74	2.77	4.86	4.15	7.29	701
×55	0.251	5.57	2.36	4.25	3.54	6.38	614
W21×201	0.219	12.6	13.7	18.2	20.5	27.2	2430
×182	0.219	12.5	12.3	16.5	18.5	24.7	2180
×166	0.220	12.4	11.2	15.0	16.9	22.5	1980
×147	0.220	12.3	9.58	13.0	14.4	19.5	1710
×132	0.219	12.2	8.60	11.7	12.9	17.6	1530
×122	0.220	12.1	7.94	10.9	11.9	16.3	1410
×111	0.219	12.1	7.18	9.93	10.8	14.9	1280
W21×93	0.219	7.65	5.22	7.83	7.83	11.7	1010
×83	0.219	7.61	4.65	6.98	6.98	10.5	898
×73	0.219	7.52	4.09	6.15	6.14	9.23	788
×68	0.219	7.48	3.78	5.75	5.66	8.63	733
×62	0.220	7.36	3.38	5.18	5.07	7.76	660
W21×57	0.220	5.61	2.84	4.66	4.26	6.99	591
×50	0.219	5.40	2.33	3.99	3.49	5.99	504
×44	0.220	5.24	1.95	3.45	2.93	5.17	437

Table 4-2 (continued)
SMF Design Values
W-Shapes

$R_y = 1.1$ $F_y = 50$ ksi

Shape	$P_{a\,max}$ (ASD)	$P_{u\,max}$ (LRFD)	$1.1 R_y M_p$	Panel Zone					
				ASD ($\Omega = 1.50$)			LRFD ($\phi = 1.00$)		
				R_{v1}/Ω	R_{v2}/Ω	$0.75 P_c$	ϕR_{v1}	ϕR_{v2}	$0.75 P_c$
	kips	kips	kip-ft	kips	kip-in.	kips	kips	kip-in.	kips
W18×311	—	—	3800	678	5410	2060	1020	8110	3440
×283	—	—	3410	613	4460	1870	920	6690	3120
×258	—	—	3080	550	3750	1710	826	5620	2850
×234	—	—	2770	490	3130	1540	734	4690	2570
×211	—	—	2470	439	2540	1400	658	3810	2340
×192	—	—	2230	392	2110	1260	588	3170	2110
×175	—	—	2010	356	1730	1160	534	2590	1930
×158	—	—	1790	319	1410	1040	479	2110	1740
×143	—	—	1620	285	1170	945	427	1760	1580
×130	—	—	1460	259	968	862	388	1450	1440
×119	—	—	1320	249	762	790	373	1140	1320
×106	—	—	1160	221	594	700	331	891	1170
×97	—	—	1060	199	504	641	299	756	1070
×86	—	—	938	177	395	569	265	592	949
W18×71	—	—	736	183	301	470	275	451	784
×65	—	—	671	166	256	430	248	384	716
×60	444	668	620	151	219	396	227	329	660
×55	346	520	565	141	179	365	212	269	608
×50	217	326	509	128	146	331	192	219	551
W18×46	212	319	457	130	133	304	195	200	506
×40	65.4	98.3	395	113	99.6	266	169	149	443
×35	30.9	46.5	335	106	65.0	232	159	97.5	386
W16×100	—	—	998	199	605	662	298	908	1100
×89	—	—	882	176	478	590	265	717	983
×77	—	—	756	150	357	509	225	535	848
W16×57	—	—	529	141	218	378	212	328	630
×50	402	604	464	124	168	331	186	253	551
×45	284	427	415	111	135	299	167	202	499
×40	149	224	368	97.6	107	266	146	161	443
W16×31	40.3	60.6	272	87.5	64.2	205	131	96.4	342

Table 4-2 (continued)
SMF Design Values
W-Shapes

F_y = 50 ksi R_y = 1.1

Shape	Panel Zone $\dfrac{w_z}{90}$ or $\dfrac{d_z}{90}$	$L_{b\,max}$	Lateral Bracing				$R_y M_p$
			ASD		LRFD		
			$\dfrac{0.02 F_y b_f t_f}{1.5}$	$\dfrac{0.02 M_r C_d}{h_o}$	$0.02 F_y b_f t_f$	$\dfrac{0.02 M_r C_d}{h_o}$	
	in.	ft	kips	kips	kips	kips	kip-ft
W18×311	0.187	12.3	21.9	28.2	32.9	42.3	3460
×283	0.188	12.1	19.8	25.6	29.8	38.3	3100
×258	0.188	12.0	18.1	23.3	27.1	35.0	2800
×234	0.188	11.8	16.5	21.2	24.7	31.8	2520
×211	0.188	11.7	14.8	19.1	22.2	28.7	2250
×192	0.188	11.6	13.4	17.3	20.1	26.0	2030
×175	0.187	11.5	12.1	15.9	18.1	23.8	1820
×158	0.187	11.4	10.8	14.3	16.3	21.4	1630
×143	0.187	11.3	9.86	13.0	14.8	19.5	1480
×130	0.188	11.2	8.96	11.7	13.4	17.6	1330
×119	0.188	11.2	7.99	10.7	12.0	16.1	1200
×106	0.187	11.1	7.02	9.48	10.5	14.2	1050
×97	0.187	11.0	6.44	8.74	9.66	13.1	967
×86	0.187	10.9	5.70	7.75	8.55	11.6	853
W18×71	0.188	7.07	4.13	6.05	6.19	9.07	669
×65	0.188	7.02	3.80	5.51	5.69	8.27	610
×60	0.187	6.98	3.50	5.15	5.25	7.73	564
×55	0.187	6.94	3.16	4.69	4.74	7.04	513
×50	0.187	6.86	2.85	4.26	4.28	6.39	463
W18×46	0.188	5.36	2.44	3.80	3.67	5.70	416
×40	0.187	5.28	2.11	3.30	3.16	4.96	359
×35	0.187	5.07	1.70	2.82	2.55	4.23	305
W16×100	0.167	10.4	6.83	9.08	10.2	13.6	908
×89	0.167	10.4	6.07	8.07	9.10	12.1	802
×77	0.166	10.3	5.22	7.01	7.83	10.5	688
W16×57	0.166	6.65	3.39	4.90	5.09	7.36	481
×50	0.167	6.61	2.97	4.30	4.45	6.45	422
×45	0.166	6.53	2.65	3.89	3.98	5.84	377
×40	0.167	6.53	2.36	3.45	3.54	5.18	335
W16×31	0.167	4.86	1.62	2.55	2.43	3.83	248

Table 4-2 (continued)
SMF Design Values
W-Shapes

$R_y = 1.1$ $F_y = 50$ ksi

Shape	$P_{a\,max}$ (ASD)	$P_{u\,max}$ (LRFD)	$1.1R_yM_p$	Panel Zone ASD ($\Omega = 1.50$)			Panel Zone LRFD ($\phi = 1.00$)		
				R_{v1}/Ω	R_{v2}/Ω	$0.75P_c$	ϕR_{v1}	ϕR_{v2}	$0.75P_c$
	kips	kips	kip-ft	kips	kip-in.	kips	kips	kip-in.	kips
W14×730	—	—	8370	1380	25900	4840	2060	38800	8060
×665	—	—	7460	1220	21700	4410	1830	32500	7350
×605	—	—	6660	1090	18100	4010	1630	27100	6680
×550	—	—	5950	962	15100	3650	1440	22600	6080
×500	—	—	5290	858	12500	3310	1290	18700	5510
×455	—	—	4720	768	10400	3020	1150	15600	5030
×426	—	—	4380	703	9260	2810	1050	13900	4690
×398	—	—	4040	648	8090	2630	972	12100	4390
×370	—	—	3710	594	7000	2450	891	10500	4090
×342	—	—	3390	539	6000	2270	809	9000	3790
×311	—	—	3040	482	4960	2060	723	7450	3430
×283	—	—	2730	431	4140	1870	646	6210	3120
×257	—	—	2460	387	3430	1700	581	5140	2840
×233	—	—	2200	342	2820	1540	514	4230	2570
×211	—	—	1970	308	2310	1400	462	3460	2330
×193	—	—	1790	276	1950	1280	414	2930	2130
×176	—	—	1610	252	1620	1170	378	2420	1940
×159	—	—	1450	224	1330	1050	335	1990	1750
×145	—	—	1310	201	1100	961	302	1660	1600
W14×132	—	—	1180	190	936	873	284	1400	1460
W14×82	—	—	701	146	443	540	219	665	900
×74	—	—	635	128	373	491	192	560	818
×68	—	—	580	116	311	450	174	467	750
W14×53	—	—	439	103	211	351	154	316	585
×48	—	—	395	93.8	171	317	141	256	529
W14×38	266	400	310	87.4	108	252	131	162	420
W14×26	77.4	116	203	70.9	53.2	173	106	79.9	288

Table 4-2 (continued)
SMF Design Values
W-Shapes

F_y = 50 ksi R_y = 1.1

Shape	Panel Zone $\dfrac{w_z}{90}$ or $\dfrac{d_z}{90}$	Lateral Bracing					$R_y M_p$
		$L_{b\,max}$	ASD		LRFD		
			$\dfrac{0.02 F_y b_f t_f}{1.5}$	$\dfrac{0.02 M_r C_d}{h_o}$	$0.02 F_y b_f t_f$	$\dfrac{0.02 M_r C_d}{h_o}$	
	in.	ft	kips	kips	kips	kips	kip-ft
W14×730	0.140	19.5	58.6	69.6	87.9	104	7610
×665	0.140	19.2	53.3	63.5	80.0	95.2	6780
×605	0.140	18.9	48.3	58.0	72.4	86.9	6050
×550	0.140	18.7	43.8	52.8	65.7	79.1	5410
×500	0.140	18.4	39.7	47.8	59.5	71.7	4810
×455	0.140	18.2	36.0	43.4	53.9	65.2	4290
×426	0.140	18.0	33.8	40.6	50.8	60.9	3980
×398	0.140	17.9	31.5	37.9	47.3	56.8	3670
×370	0.140	17.7	29.3	35.5	43.9	53.3	3370
×342	0.140	17.6	27.0	32.9	40.5	49.3	3080
×311	0.140	17.5	24.4	29.9	36.6	44.8	2760
×283	0.140	17.3	22.2	27.2	33.3	40.8	2480
×257	0.140	17.2	20.2	24.6	30.2	36.9	2230
×233	0.140	17.0	18.2	22.4	27.3	33.5	2000
×211	0.140	16.9	16.4	20.3	24.6	30.4	1790
×193	0.140	16.8	15.1	18.5	22.6	27.7	1630
×176	0.140	16.7	13.7	16.9	20.6	25.3	1470
×159	0.140	16.6	12.4	15.3	18.6	22.9	1320
×145	0.140	16.5	11.3	13.9	16.9	20.9	1190
W14×132	0.140	15.6	10.1	12.5	15.1	18.8	1070
W14×82	0.140	10.3	5.76	7.61	8.64	11.4	637
×74	0.140	10.3	5.29	6.90	7.93	10.3	578
×68	0.140	10.2	4.80	6.34	7.20	9.51	527
W14×53	0.140	7.98	3.55	4.84	5.32	7.26	399
×48	0.140	7.94	3.19	4.36	4.78	6.53	359
W14×38	0.145	6.44	2.32	3.32	3.49	4.97	282
W14×26	0.145	4.49	1.41	2.18	2.11	3.28	184

Table 4-2 (continued)
SMF Design Values
W-Shapes

$R_y = 1.1$ $F_y = 50$ ksi

Shape	$P_{a\,max}$ (ASD)	$P_{u\,max}$ (LRFD)	$1.1R_yM_p$	Panel Zone ASD ($\Omega = 1.50$)			Panel Zone LRFD ($\phi = 1.00$)		
				R_{v1}/Ω	R_{v2}/Ω	$0.75P_c$	ϕR_{v1}	ϕR_{v2}	$0.75P_c$
	kips	kips	kip-ft	kips	kip-in.	kips	kips	kip-in.	kips
W12×336	—	—	3040	598	7040	2230	897	10600	3710
×305	—	—	2710	531	5820	2010	797	8720	3360
×279	—	—	2430	487	4800	1840	730	7190	3070
×252	—	—	2160	431	3950	1670	647	5920	2780
×230	—	—	1950	390	3320	1520	584	4970	2540
×210	—	—	1750	347	2770	1390	520	4160	2320
×190	—	—	1570	305	2310	1260	458	3460	2100
×170	—	—	1390	269	1840	1130	403	2760	1880
×152	—	—	1230	238	1470	1010	358	2210	1680
×136	—	—	1080	212	1160	898	318	1740	1500
×120	—	—	938	186	909	792	279	1360	1320
×106	—	—	827	157	717	702	236	1080	1170
×96	—	—	741	140	593	635	210	889	1060
W12×50	—	—	362	90.3	199	329	135	298	548
×45	—	—	324	81.1	160	295	122	240	491
W12×35	302	453	258	75.0	106	232	113	160	386
W12×22	131	197	148	64.0	43.7	146	95.9	65.5	243
×19	73.1	110	125	57.3	29.5	125	86.0	44.2	209
W10×112	—	—	741	172	975	740	258	1460	1230
×100	—	—	655	151	775	659	226	1160	1100
×88	—	—	570	131	606	585	196	909	975
×77	—	—	492	112	463	511	169	695	851
×68	—	—	430	97.8	359	448	147	539	746
W10×45	—	—	277	70.7	185	299	106	277	499
W10×30	—	—	185	63.0	90.7	199	94.5	136	332
×26	—	—	158	53.6	67.0	171	80.3	101	285
W10×19	—	—	109	51.0	37.6	126	76.5	56.4	211
×17	140	211	94.3	48.5	26.2	112	72.7	39.3	187

Table 4-2 (continued)
SMF Design Values
W-Shapes

F_y = 50 ksi R_y = 1.1

Shape	Panel Zone $\dfrac{w_z}{90}$ or $\dfrac{d_z}{90}$	$L_{b\,max}$	ASD $\dfrac{0.02F_y b_f t_f}{1.5}$	ASD $\dfrac{0.02M_r C_d}{h_o}$	LRFD $0.02F_y b_f t_f$	LRFD $\dfrac{0.02M_r C_d}{h_o}$	$R_y M_p$
	in.	ft	kips	kips	kips	kips	kip-ft
W12×336	0.121	14.4	26.4	32.0	39.7	48.1	2760
×305	0.121	14.2	23.8	29.0	35.8	43.4	2460
×279	0.122	14.0	21.6	26.3	32.4	39.5	2200
×252	0.121	13.9	19.5	23.8	29.3	35.7	1960
×230	0.122	13.8	17.8	21.8	26.7	32.7	1770
×210	0.121	13.6	16.2	19.9	24.3	29.9	1600
×190	0.121	13.5	14.7	18.0	22.1	26.9	1430
×170	0.121	13.4	13.1	16.3	19.7	24.4	1260
×152	0.121	13.3	11.7	14.5	17.5	21.7	1110
×136	0.121	13.1	10.3	12.9	15.5	19.3	981
×120	0.121	13	9.10	11.4	13.7	17.1	853
×106	0.121	12.9	8.05	10.1	12.1	15.2	752
×96	0.121	12.8	7.32	9.14	11.0	13.7	674
W12×50	0.121	8.15	3.45	4.55	5.17	6.82	330
×45	0.122	8.11	3.09	4.09	4.63	6.14	294
W12×35	0.127	6.40	2.27	3.13	3.41	4.69	235
W12×22	0.127	3.52	1.14	1.81	1.71	2.71	134
×19	0.128	3.42	0.936	1.52	1.40	2.28	113
W10×112	0.0989	11.1	8.67	10.6	13.0	15.9	674
×100	0.0984	11.0	7.69	9.53	11.5	14.3	596
×88	0.098	10.9	6.80	8.45	10.2	12.7	518
×77	0.0984	10.8	5.92	7.36	8.87	11.0	447
×68	0.0984	10.8	5.18	6.50	7.78	9.74	391
W10×45	0.0984	8.35	3.31	4.25	4.97	6.37	252
W10×30	0.105	5.69	1.98	2.69	2.96	4.03	168
×26	0.105	5.65	1.69	2.33	2.54	3.49	143
W10×19	0.105	3.63	1.06	1.61	1.59	2.42	99
×17	0.105	3.51	0.882	1.40	1.32	2.11	85.7

PART 4 REFERENCES

Dowswell, B. (2011), "A Yield Line Component Method for Bolted Flange Connections," *Engineering Journal*, American Institute of Steel Construction, Vol. 48, No. 2, 2nd Quarter, pp. 93–116.

Fisher, J.M. and Kloiber, L.A. (2010), *Base Plate and Anchor Rod Design*, Design Guide 1, 2nd Ed., AISC, Chicago, IL.

Gomez, I., Smith, C., Deierlein, G. and Kanvinde, A. (2010), "Shear Transfer in Exposed Column Base Plates," **http://nees.org/resources/837**.

Hamburger, R., Krawinkler, H., Malley, J. and Adan, S. (2009), *Seismic Design of Steel Special Moment Frames: A Guide for Practicing Engineers*, NEHRP Seismic Design Technical Brief No. 2, National Institute of Standards and Technology.

Murray, T.M. and Sumner, E.A. (2003), *Extended End-Plate Moment Connections—Seismic and Wind Applications*, Design Guide 4, 2nd Ed., AISC, Chicago, IL.

OSHA (2008), Occupational Safety and Health Regulations, Title 29, Code of Federal Regulations, U.S. Government Printing Office, Washington, DC.

West, M.A. and Fisher, J.M. (2003), *Serviceability Design Considerations for Steel Buildings*, Design Guide 3, 2nd Ed., AISC, Chicago, IL.

PART 5
BRACED FRAMES

5.1 SCOPE 5–3
5.2 ORDINARY CONCENTRICALLY BRACED FRAMES (OCBF) 5–3
    OCBF Design Example Plan and Elevation 5–4
    Example 5.2.1 OCBF Diagonal Brace Design 5–6
    Example 5.2.2 OCBF Column Design 5–15
    Example 5.2.3 OCBF Beam Design 5–18
    Example 5.2.4 OCBF Brace-to-Beam/Column Connection Design 5–25
    Example 5.2.5 OCBF Tension-Only Diagonal Brace Design 5–75
5.3 SPECIAL CONCENTRICALLY BRACED FRAMES (SCBF) 5–82
    SCBF Design Example Plan and Elevation 5–86
    Example 5.3.1 SCBF Brace Design 5–87
    Example 5.3.2 SCBF Analysis 5–93
    Example 5.3.3 SCBF Column Design 5–98
    Example 5.3.4 SCBF Beam Design 5–104
    Example 5.3.5 SCBF Beam Design 5–119
    Example 5.3.6 SCBF Column Splice Design 5–129
    Example 5.3.7 SCBF Maximum Force Limited by Foundation Uplift 5–136
    Example 5.3.8 SCBF Brace-to-Beam Connection Design 5–140
    Example 5.3.9 SCBF Brace-to-Beam Connection Design 5–178
    Example 5.3.10 SCBF Brace-to-Beam/Column Connection Design 5–202
    Example 5.3.11 SCBF Brace-to-Beam/Column Connection Design
    with Elliptical Clearance and Fixed Beam-to-Column Connection 5–269
    Example 5.3.12 SCBF Brace-to-Beam/Column Connection Design—
    In Plane Brace Buckling 5–299
5.4 ECCENTRICALLY BRACED FRAMES (EBF) 5–334
    EBF Design Example Plan and Elevation 5–338
    Example 5.4.1 EBF Story Drift Check 5–339
    Example 5.4.2 EBF Link Design 5–340
    Example 5.4.3 EBF Beam Outside of the Link Design 5–353
    Example 5.4.4 EBF Brace Design 5–362

    Example 5.4.5 EBF Column Design 5–367
    Example 5.4.6 EBF Brace-to-Link Connection Design 5–372
    Example 5.4.7 EBF Brace-to-Beam/Column Connection Design 5–379
5.5 BUCKLING-RESTRAINED BRACED FRAMES (BRBF) 5–413
    BRBF Design Example Plan and Elevation 5–418
    Example 5.5.1 BRBF Brace Design 5–419
    Example 5.5.2 BRBF Column Design 5–425
    Example 5.5.3 BRBF Beam Design 5–430
5.6 NONBUILDING STRUCTURES: A SPECIAL CASE 5–443

5.1 SCOPE

The AISC *Seismic Provisions* requirements and other design considerations summarized in this Part apply to the design of the members and connections in braced frames that require seismic detailing according to the AISC *Seismic Provisions*.

5.2 ORDINARY CONCENTRICALLY BRACED FRAMES (OCBF)

Ordinary concentrically braced frame (OCBF) systems, like other concentrically braced frame systems, resist lateral forces and displacements primarily through the axial strength and stiffness of the brace members. The design of OCBF systems is addressed in AISC *Seismic Provisions* Section F1. Concentrically braced frames are arranged such that the centerlines of the framing members (braces, columns and beams) coincide or nearly coincide, thus minimizing flexural behavior. While special concentrically braced frame (SCBF) systems have numerous detailing requirements to ensure greater ductility, OCBF systems anticipate little inelastic deformation and are designed using a higher seismic force level to account for their limited system ductility. OCBF systems, with their relatively simple design and construction procedures, can be an attractive choice for smaller buildings and nonbuilding structures. OCBF systems may be less desirable in larger buildings and buildings with a higher seismic performance objective.

Concentrically braced frame systems tend to be more economical than moment resisting frames and eccentrically braced frames in terms of material, fabrication and erection costs. They do, however, often have reduced flexibility in floor-plan layout, space planning, and electrical and mechanical routing as a result of the presence of braces. In certain circumstances, however, braced frames are exposed and featured in the architecture of the building. Several configurations of braced frames may be considered, including those shown in AISC *Seismic Provisions* Commentary Figures C-F2.1 and C-F2.2.

Braced frames typically are located in walls that stack vertically between floor levels. In the typical office building, these walls generally occur in the core area around stair and elevator shafts, central restrooms, and mechanical and electrical rooms. This generally allows for greater architectural flexibility in placement and configuration of exterior windows and cladding. Depending on the plan location and the size of the core area of the building, the torsional resistance offered by the braced frames may become a controlling design parameter. Differential drift between stories at the exterior perimeter must be considered with this type of layout, as rotational displacements of the floor diaphragms may impose forces on the cladding system and other nonstructural elements of the building.

In designing and detailing OCBF systems, there are few special considerations. The design of OCBF members is mostly based upon typical steel design procedures, as outlined in the AISC *Specification*. The requirements for OCBF systems in the AISC *Seismic Provisions* include the following:

- Braces are moderately ductile members as given in Section F1.5a
- The required strength of bracing connections is given in Section F1.6a
- The brace slenderness limit of $KL/r \leq 4\sqrt{E/F_y}$ for V or inverted-V configurations is given in Section F1.5b
- The requirements for beams in V or inverted-V frames are given in Section F1.4a

The connection strength requirement of AISC *Seismic Provisions* Section F1.6a is intended to ensure that the brace member acts as the ductile link (brace yielding) in the frame prior to the connections failing, thus providing more reliability to the system. The limit on the slenderness in V-type and inverted V-type braced frames is intended to limit the unbalanced force that develops on the braced frame beam when the compression brace buckles and its strength degrades while the tension brace yields. The buckling of the compression brace results in a significant reduction in the frame shear resistance. This slenderness limit does not apply to braces in two-story X-braced frames, because that configuration prevents or reduces the magnitude of unbalanced forces on the beam.

K-braced frames, as defined by the AISC *Seismic Provisions* Glossary, where a brace frames to a column at a location where there is no out-of-plane support, are not permitted in OCBF systems. The definition of K-braced frames precludes the use of braces framing to columns between diaphragm levels or locations of out-of-plane lateral support for the columns. This definition also precludes multi-tiered concentric braced frames where there are two or more levels of bracing between diaphragm levels or locations of out-of-plane lateral support for the columns.

OCBF Design Example Plan and Elevation

The following examples illustrate the design of an OCBF system based on the AISC *Seismic Provisions* Section F1. The plan and elevation are shown in Figure 5-1 and Figure 5-2.

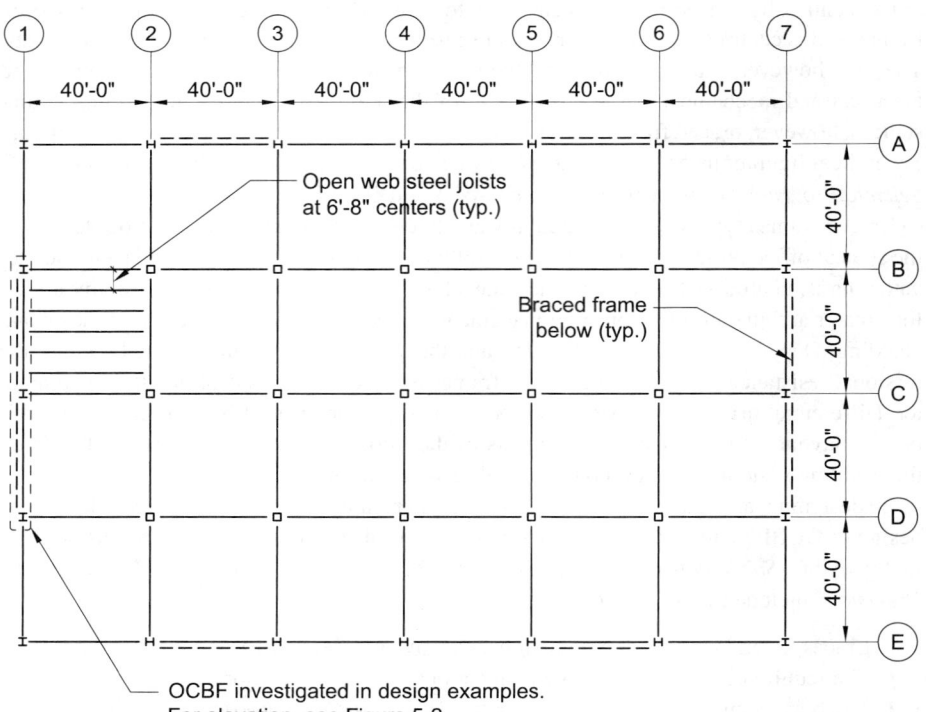

Fig. 5-1. OCBF roof plan.

5.2 ORDINARY CONCENTRICALLY BRACED FRAMES

The gravity loading is as follows:

$D = 18$ psf
$L = 0$ psf
$S = 30$ psf

The vertical load of the exterior wall is supported at grade. The seismic weight of the wall that is tributary to the roof level is 140 lb/ft on all four sides of the building perimeter. The lateral earthquake force, E, acting at the roof level along grid 1 is 65.8 kips as calculated per ASCE/SEI 7 Section 12.8.

The applicable building code specifies the use of ASCE/SEI 7 for calculation of loads. From ASCE/SEI 7, the following parameters apply: Seismic Design Category D, $R = 3\frac{1}{4}$, $\Omega_o = 2$, $I_e = 1.0$, $S_{DS} = 0.528$, and $\rho = 1.0$. ASCE/SEI 7 does not permit an $R = 3$ system in Seismic Design Category D; therefore, an OCBF system is used for this building and designed according to the AISC *Seismic Provisions*. The structural framing is regular and has two bays of seismic force resisting perimeter framing on each side in each orthogonal direction. Therefore, ASCE/SEI 7 Section 12.3.4.2b permits the redundancy factor, ρ, to be taken as 1.0.

The vertical seismic load effect, E_v, based on LRFD load combinations in ASCE/SEI 7 Section 12.4.2.3, is:

$$0.2S_{DS}D = 0.2(0.528)D \qquad \text{(ASCE/SEI 7 Eq. 12.4-4)}$$
$$= 0.106D$$

For ASD load combinations in ASCE/SEI 7 Section 12.4.2.3, E_v is:

$$0.14S_{DS}D = 0.14(0.528)D$$
$$= 0.0739D$$

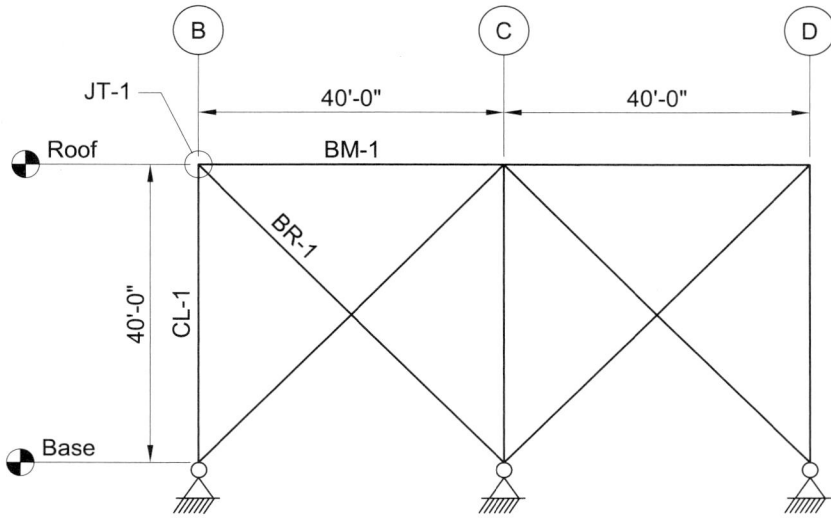

Fig. 5-2. OCBF elevation.

and

$$0.10S_{DS}D = 0.10(0.528)D$$
$$= 0.0528D$$

Note that according to ASCE/SEI 7 Table 12.2-1, buildings with OCBF frames in Seismic Design Categories D and E are only permitted up to a structural height of 35 ft. An exception applies for Seismic Design Categories D, E and F that allows the maximum structural height to be increased to 60 ft for single-story buildings where the dead load of the roof does not exceed 20 psf, which is the case here.

Assume that the ends of the diagonal braces are pinned and braced against translation for both the *x-x* and *y-y* axes. The loads given for each example are from a first-order analysis. Assume that the effective length method of AISC *Specification* Appendix 7 is used for stability design. AISC *Specification* Appendix 8 will be applied to approximate a second-order analysis.

Example 5.2.1. OCBF Diagonal Brace Design

Given:
Refer to the roof plan shown in Figure 5-1 and the Brace BR-1 shown in Figure 5-2. Select an ASTM A992 W-shape for the diagonal braces to resist the loads given.

The axial loads and moments on the brace due to a first-order analysis are:

$P_D = 5.54$ kips $\qquad P_S = 6.70$ kips $\qquad P_{Q_E} = \pm 22.3$ kips $\qquad M_D = 2.34$ kip-ft

The dead load bending moment indicated above is due to the self-weight of the brace assuming a member that weighs 33 lb/ft. Sometimes this self-weight loading is ignored in the design of vertical diagonal braces where judgment would indicate that the loading is minimal and only uses a small percentage of the member strength. However, in this example, considering the relatively long length of the diagonal brace and that the self-weight moment is resisted by the weak axis flexural strength of the brace, the dead load moment is included in this design check. There are no bending moments due to live loads or snow loads.

P_{story} is the total vertical load on the story calculated using the following governing load combination. From the analysis:

LRFD	ASD
LRFD Load Combination 5 from ASCE/SEI 7 Section 12.4.2.3 (including the 0.5 factor on *L* permitted in Section 12.4.2.3)	ASD Load Combination 5 from ASCE/SEI 7 Section 12.4.2.3
$(1.2 + 0.2S_{DS})D + \rho Q_E + 0.5L + 0.2S$	$(1.0 + 0.14S_{DS})D + H + F + 0.7\rho Q_E$
$P_{story} = 1{,}130$ kips	$P_{story} = 740$ kips

5.2 ORDINARY CONCENTRICALLY BRACED FRAMES

The story shear from the analysis is 136 kips. The first order interstory drift due to this shear force without the C_d factor applied from the analysis model is $\Delta_H = 0.0941$ in.

Solution:

From AISC *Manual* Table 2-4, the material properties are:

ASTM A992
$F_y = 50$ ksi
$F_u = 65$ ksi

Required Strength

Determine the required strength

The load combinations that include seismic effects are:

LRFD	ASD
LRFD Load Combinations 5 and 6 from ASCE/SEI 7 Section 12.4.2.3 (including the 0.5 factor on L permitted in ASCE/SEI 7 Section 12.4.2.3) $(1.2 + 0.2S_{DS})D + \rho Q_E + 0.5L + 0.2S$ $(0.9 - 0.2S_{DS})D + \rho Q_E + 1.6H$	ASD Load Combinations 5 and 8 from ASCE/SEI 7 Section 12.4.2.3 $(1.0 + 0.14S_{DS})D + H + F + 0.7\rho Q_E$ $(0.6 - 0.14S_{DS})D + 0.7\rho Q_E + H$

Considering the load combinations given in ASCE/SEI 7, the maximum compressive axial force in the diagonal brace is:

LRFD	ASD
LRFD Load Combination 5 from ASCE/SEI 7 Section 12.4.2.3 $P_u = [1.2 + 0.2(0.528)](5.54 \text{ kips})$ $+ 1.0(22.3 \text{ kips}) + 0.5(0 \text{ kips})$ $+ 0.2(6.70 \text{ kips})$ $= 30.9$ kips	ASD Load Combination 5 from ASCE/SEI 7 Section 12.4.2.3 $P_a = [1.0 + 0.14(0.528)](5.54 \text{ kips})$ $+ 0 \text{ kips} + 0 \text{ kips}$ $+ 0.7(1.0)(22.3 \text{ kips})$ $= 21.6$ kips

The maximum bending moment in the brace concurrent with the above load combination is:

LRFD	ASD
$M_u = [1.2 + 0.2(0.528)](2.34 \text{ kip-ft})$ $\quad + 1.0(0 \text{ kip-ft}) + 0.5(0 \text{ kip-ft})$ $\quad + 0.2(0 \text{ kip-ft})$ $\quad = 3.06 \text{ kip-ft}$	$M_a = [1.0 + 0.14(0.528)](2.34 \text{ kip-ft})$ $\quad + 0 \text{ kip-ft} + 0 \text{ kip-ft}$ $\quad + 0.7(1.0)(0 \text{ kip-ft})$ $\quad = 2.51 \text{ kip-ft}$

The ASCE/SEI 7 load combination that results in the maximum axial tensile force in the diagonal brace is:

LRFD	ASD
LRFD Load Combination 6 from ASCE/SEI 7 Section 12.4.2.3 $P_u = [0.9 - 0.2(0.528)](5.54 \text{ kips})$ $\quad + 1.0(-22.3 \text{ kips}) + 1.6(0 \text{ kips})$ $\quad = -17.9 \text{ kips}$	ASD Load Combination 8 from ASCE/SEI 7 Section 12.4.2.3 $P_a = [0.6 - 0.14(0.528)](5.54 \text{ kips})$ $\quad + 0.7(1.0)(-22.3 \text{ kips}) + 0 \text{ kips}$ $\quad = -12.7 \text{ kips}$

The maximum bending moment in the brace concurrent with the above load combination is:

LRFD	ASD
$M_u = [0.9 - 0.2(0.528)](2.34 \text{ kip-ft})$ $\quad + 1.0(0 \text{ kip-ft}) + 1.6(0 \text{ kip-ft})$ $\quad = 1.86 \text{ kip-ft}$	$M_a = [0.6 - 0.14(0.528)](2.34 \text{ kip-ft})$ $\quad + 0 \text{ kip-ft} + 0 \text{ kip-ft}$ $\quad + 0.7(1.0)(0 \text{ kip-ft})$ $\quad = 1.23 \text{ kip-ft}$

Try a **W10×33** with its flanges oriented parallel to the plane of the braced frame.

From AISC *Manual* Table 2-4, the material properties are as follows:

ASTM A992
$F_y = 50$ ksi
$F_u = 65$ ksi

From AISC *Manual* Table 1-1, the geometric properties for the **W10×33** are as follows:

$A = 9.71$ in.^{2} $b_f = 7.96$ in. $d = 9.73$ in. $t_w = 0.290$ in.
$t_f = 0.435$ in. $k_{des} = 0.935$ in. $b_f/2t_f = 9.15$ $h/t_w = 27.1$
$r_x = 4.19$ in. $I_y = 36.6$ in.^{4} $r_y = 1.94$ in.

5.2 ORDINARY CONCENTRICALLY BRACED FRAMES

Brace Slenderness

Check brace element width-to-thickness ratios

According to AISC *Seismic Provisions* Sections F1.5a, braces are required to satisfy the requirements for moderately ductile members. Elements in the brace members must not exceed λ_{md} width-to-thickness requirements given in Section D1.1.

From Table D1.1 of the AISC *Seismic Provisions*:

$$\lambda_{md} = 0.38\sqrt{\frac{E}{F_y}}$$
$$= 0.38\sqrt{\frac{29{,}000 \text{ ksi}}{50 \text{ ksi}}}$$
$$= 9.15$$

Because $b_f/2t_f \leq \lambda_{md}$, the flanges meet the requirements for moderately ductile members.

From Table D1.1 of the AISC *Seismic Provisions* for webs of rolled I-shaped sections used as diagonal braces:

$$\lambda_{md} = 1.49\sqrt{\frac{E}{F_y}}$$
$$= 1.49\sqrt{\frac{29{,}000 \text{ ksi}}{50 \text{ ksi}}}$$
$$= 35.9$$

Because $h/t_w \leq \lambda_{md}$, the web meets the requirements for moderately ductile members.

Alternatively, Table 1-3 can be used to verify that the member satisfies the local width-to-thickness requirements for OCBF diagonal braces.

Additionally, the W10×33 does not contain slender compression elements according to AISC *Specification* Table B4.1a.

Available Compressive Strength

Determine K

As stated in the OCBF Design Example Plan and Elevation section, the effective length method in AISC *Specification* Appendix 7 is used for stability design. According to AISC *Specification* Appendix 7, Section 7.2.3(a), for braced frame systems, the effective length factor, K, for members subject to compression is taken as 1.0, unless a rational analysis indicates that a lower value is appropriate.

The length of the brace diagonal in each bay, based on the geometry in Figure 5-2, is:

$$L = \sqrt{(40.0 \text{ ft})^2 + (40.0 \text{ ft})^2}$$
$$= 56.6 \text{ ft}$$

This length has been determined by calculating the distance between the work points based on the intersection of the centerlines of the brace, column and beams. Shorter unbraced lengths of the brace may be used if justified by the engineer of record. By inspection, the laterally unbraced length of the diagonal brace in the in-plane (about the y-y axis) direction is half of the overall length. For buckling out-of-plane (about the x-x axis), if both of the diagonals are continuous for their full length and are connected at the intersection point, then the effective length factor, K, is 0.5 (El-Tayem and Goel, 1986; Picard and Beaulieu, 1987). This requires a connection between the diagonal members at their intersection that is rigid in flexure out-of-plane. The available axial compressive strength of diagonals in X-bracing where one of the diagonal braces is not continuous through the intersection can be determined by an energy method (Nair, 1997).

Assume that the connection of the half brace sections at the X-brace intersection is rigid out-of-plane. The braces are oriented such that buckling about the y-y axis of the brace occurs in the plane of the frame.

$L_x = 56.6$ ft

$L_y = 0.5L$

$\quad = 0.5(56.6 \text{ ft})$

$\quad = 28.3$ ft

$K_x = 0.5$

$K_y = 1.0$

$\dfrac{K_x L_x}{r_x} = \dfrac{0.5(56.6 \text{ ft})(12 \text{ in./ft})}{4.19 \text{ in.}}$

$\quad = 81.1$

$\dfrac{K_y L_y}{r_y} = \dfrac{1.0(28.3 \text{ ft})(12 \text{ in./ft})}{1.94 \text{ in.}}$

$\quad = 175$ (governs)

The slenderness, KL/r, is less than 200 and therefore meets the recommendation of the User Note in Section E2 of the AISC *Specification*.

Using AISC *Specification* Equation E3-1 and AISC *Manual* Table 4-22 with $KL/r = 175$, the available compressive strength is determined as follows:

5.2 ORDINARY CONCENTRICALLY BRACED FRAMES

LRFD	ASD
$\phi_c F_{cr} = 7.38$ ksi $\phi_c P_n = \phi_c F_{cr} A_g$ $\quad = 7.38 \text{ ksi}(9.71 \text{ in.}^2)$ $\quad = 71.7$ kips	$\dfrac{F_{cr}}{\Omega_c} = 4.91$ ksi $\dfrac{P_n}{\Omega_c} = \left(\dfrac{F_{cr}}{\Omega_c}\right) A_g$ $\quad = 4.91 \text{ ksi}(9.71 \text{ in.}^2)$ $\quad = 47.7$ kips

Second-order effects and interaction between axial force and flexure are checked in the following.

Available Flexural Strength

Because there is no bending moment in the strong axis, $M_{cx} = 0$.

From AISC *Manual* Table 3-4, the available strength in the weak axis is:

LRFD	ASD
$\phi_b M_{ny} = 52.5$ kip-ft	$\dfrac{M_{ny}}{\Omega_b} = 34.9$ kip-ft

Second-Order Effects

Second-order effects are addressed using the procedure in AISC *Specification* Appendix 8, as follows:

$$M_r = B_1 M_{nt} + B_2 M_{lt} \qquad \text{(Spec. Eq. A-8-1)}$$
$$P_r = P_{nt} + B_2 P_{lt} \qquad \text{(Spec. Eq. A-8-2)}$$

Calculate B_1

$C_m = 1.0$ as moment is due to self-weight applied between supports
$\alpha = 1.00$ (LRFD); $\alpha = 1.60$ (ASD)

The elastic critical buckling strength, P_{e1}, is calculated in the plane of bending. For this calculation, the plane of bending will be in the plane of the frame, about the brace's y-y axis.

$K_1 L_y = 1.0(28.3 \text{ ft})$
$\quad\quad = 28.3$ ft

$$P_{e1} = \frac{\pi^2 EI_y^*}{(K_1 L)^2} \quad \text{(Spec. Eq. A-8-5)}$$

$$= \frac{\pi^2 (29{,}000 \text{ ksi})(36.6 \text{ in.}^4)}{[28.3 \text{ ft}(12 \text{ in./ft})]^2}$$

$$= 90.8 \text{ kips}$$

From AISC *Specification* Equation A-8-3:

LRFD	ASD
$B_1 = \dfrac{C_m}{1 - \alpha P_r / P_{e1}} \geq 1$	$B_1 = \dfrac{C_m}{1 - \alpha P_r / P_{e1}} \geq 1$
$= \dfrac{1.0}{1 - [1.00(30.9 \text{ kips})/90.8 \text{ kips}]}$	$= \dfrac{1.0}{1 - [1.60(21.6 \text{ kips})/90.8 \text{ kips}]}$
$= 1.52 \geq 1$ **o.k.**	$= 1.61 \geq 1$ **o.k.**

Calculate B_2

P_{story} is given as 1,130 kips (LRFD) and 740 kips (ASD) and H is given as 136 kips.

$$P_{e\,story} = R_M \frac{HL}{\Delta_H} \quad \text{(Spec. Eq. A-8-7)}$$

$$= 1.00 \frac{136 \text{ kips}(40.0 \text{ ft})}{(0.0941 \text{ in.})(1 \text{ ft}/12 \text{ in.})}$$

$$= 694{,}000 \text{ kips}$$

Using AISC *Specification* Equation A-8-6:

LRFD	ASD
$B_2 = \dfrac{1}{1 - \dfrac{\alpha P_{story}}{P_{e\,story}}} \geq 1$	$B_2 = \dfrac{1}{1 - \dfrac{\alpha P_{story}}{P_{e\,story}}} \geq 1$
$= \dfrac{1}{1 - \dfrac{1.00(1{,}130 \text{ kips})}{694{,}000 \text{ kips}}}$	$= \dfrac{1}{1 - \dfrac{1.60(740 \text{ kips})}{694{,}000 \text{ kips}}}$
$= 1.00$	$= 1.00$

Because $B_2 \leq 1.5$, the effective length method is a valid way to check stability according to AISC *Specification* Appendix 7.

5.2 ORDINARY CONCENTRICALLY BRACED FRAMES

The required flexural strength of the brace including second-order effects, using AISC *Specification* Equation A-8-1, is:

LRFD	ASD
$M_{nt} = M_u$ $\quad = 3.06$ kip-ft $M_{lt} = 0$ kip-ft $M_r = B_1 M_{nt} + B_2 M_{lt}$ $\quad = 1.52(3.06 \text{ kip-ft}) + 1.00(0 \text{ kip-ft})$ $\quad = 4.65$ kip-ft	$M_{nt} = M_a$ $\quad = 2.51$ kip-ft $M_{lt} = 0$ kip-ft $M_r = B_1 M_{nt} + B_2 M_{lt}$ $\quad = 1.61(2.51 \text{ kip-ft}) + 1.00(0 \text{ kip-ft})$ $\quad = 4.04$ kip-ft

Because $B_2 = 1.00$, the required axial compressive strength of the brace including second-order effects, based on AISC *Specification* Equation A-8-2, is:

LRFD	ASD
LRFD Load Combination 5 from ASCE/SEI 7 Section 12.4.2.3 $(1.2 + 0.2 S_{DS}) D + B_1 (\rho Q_E) + 0.5L + 0.2S$ $P_u = [1.2 + 0.2(0.528)](5.54 \text{ kips})$ $\quad + 1.00(1.0)(22.3 \text{ kips}) + 0.5(0 \text{ kips})$ $\quad + 0.2(6.70 \text{ kips})$ $\quad = 30.9$ kips	ASD Load Combination 5 from ASCE/SEI 7 Section 12.4.2.3 $(1.0 + 0.14 S_{DS}) D + H + F + B_1 (0.7 \rho Q_E)$ $P_a = [1.0 + 0.14(0.528)](5.54 \text{ kips})$ $\quad + 0 \text{ kips} + 0 \text{ kips}$ $\quad + 1.00(0.7)(1.0)(22.3 \text{ kips})$ $\quad = 21.6$ kips

Combined Loading (Compression and Flexure)

Check combined loading of the W10×33 brace

Determine the applicable equation, using AISC *Specification* Section H1:

LRFD	ASD
$\dfrac{P_r}{P_c} = \dfrac{30.9 \text{ kips}}{71.7 \text{ kips}}$ $\quad = 0.431$	$\dfrac{P_r}{P_c} = \dfrac{21.6 \text{ kips}}{47.7 \text{ kips}}$ $\quad = 0.453$

Because $P_r/P_c \geq 0.2$, the brace design is controlled by the equation:

$$\frac{P_r}{P_c} + \frac{8}{9}\left(\frac{M_{rx}}{M_{cx}} + \frac{M_{ry}}{M_{cy}}\right) \leq 1.0 \qquad \text{(Spec. Eq. H1-1a)}$$

LRFD	ASD
$\dfrac{30.9 \text{ kips}}{71.7 \text{ kips}} + \dfrac{8}{9}\left(0 + \dfrac{4.65 \text{ kip-ft}}{52.5 \text{ kip-ft}}\right) = 0.510$ $0.510 < 1.0$ **o.k.**	$\dfrac{21.6 \text{ kips}}{47.7 \text{ kips}} + \dfrac{8}{9}\left(0 + \dfrac{4.04 \text{ kip-ft}}{34.9 \text{ kip-ft}}\right) = 0.556$ $0.556 < 1.0$ **o.k.**

Note that the weak axis bending moment from the self-weight of the diagonal brace utilizes about 8% of the member available strength.

Available Tensile Strength

From AISC *Manual* Table 5-1, the available strength of the W10×33 brace in axial tension for yielding on the gross section is:

LRFD	ASD
$\phi_t P_n = 437$ kips > 17.9 kips **o.k.**	$\dfrac{P_n}{\Omega_t} = 291$ kips > 12.7 kips **o.k.**

Combined Loading (Tension and Flexure)

Check combined loading of the W10×33

As previously determined:

LRFD	ASD
$M_{ry} = M_u$ $\quad = 1.86$ kip-ft $P_r = T_u$ $\quad = 17.9$ kips $P_c = \phi_t P_n$ $\quad = 437$ kips	$M_{ry} = M_a$ $\quad = 1.23$ kip-ft $P_r = T_a$ $\quad = 12.7$ kips $P_c = \dfrac{P_n}{\Omega_t}$ $\quad = 291$ kips

Consider second-order effects per Appendix 8 of the AISC *Specification*. As previously calculated, $B_2 = 1.0$. According to Appendix 8, Section 8.2 of the AISC *Specification*, B_1 should be taken as 1.0 for members not subject to compression. Given that both B_1 and B_2 are equal to 1, there is no amplification required for second-order effects for the loads on the member when the diagonal brace is in tension.

LRFD	ASD
$\dfrac{P_r}{P_c} = \dfrac{17.9 \text{ kips}}{437 \text{ kips}}$ $= 0.0410$	$\dfrac{P_r}{P_c} = \dfrac{12.7 \text{ kips}}{291 \text{ kips}}$ $= 0.0436$

Because $P_r/P_c < 0.2$, the brace design is controlled by the equation:

$$\frac{P_r}{2P_c} + \left(\frac{M_{rx}}{M_{cx}} + \frac{M_{ry}}{M_{cy}}\right) \leq 1.0 \qquad \text{(Spec. Eq. H1-1b)}$$

LRFD	ASD
$\dfrac{17.9 \text{ kips}}{2(437 \text{ kips})} + \left(0 + \dfrac{1.86 \text{ kip-ft}}{52.5 \text{ kip-ft}}\right) = 0.0559$ < 1.0 **o.k.**	$\dfrac{12.7 \text{ kips}}{2(291 \text{ kips})} + \left(0 + \dfrac{1.23 \text{ kip-ft}}{34.9 \text{ kip-ft}}\right) = 0.0571$ < 1.0 **o.k.**

The W10×33 is adequate for the OCBF diagonal brace BR-1. The brace is oriented with the flanges parallel to the plane of the braced frame.

Example 5.2.2. OCBF Column Design

Given:

Refer to Column CL-1 in Figure 5-2. Select a 40-ft-long ASTM A992 W-shape to resist the loads given for the column.

The loads on Column CL-1 due to a first-order analysis are:

$P_D = 16.4$ kips $P_S = 19.9$ kips $P_{Q_E} = \pm 15.8$ kips

Assume that the ends of the columns are pinned and braced against translation for both the x-x and y-y axes. The loading in the columns is from a first-order analysis. Appendix 8 of the AISC *Specification* can be applied to approximate a second-order analysis.

Solution:

From AISC *Manual* Table 2-4, the material properties are:

ASTM A992
$F_y = 50$ ksi
$F_u = 65$ ksi

Required Strength

AISC *Seismic Provisions* Section D1.4a requires that the axial compressive and tensile strength be determined using the amplified seismic load; that is, the seismic load multiplied by the overstrength factor, Ω_o.

The governing load combinations, including the overstrength factor, for the required axial compressive strength are:

LRFD	ASD
LRFD Load Combination 5 from ASCE/SEI 7 Section 12.4.3.2 (including the 0.5 factor on L permitted in Section 12.4.3.2)	ASD Load Combination 5 from ASCE/SEI 7 Section 12.4.3.2
$P_u = (1.2 + 0.2S_{DS})P_D + \Omega_o P_{Q_E}$ $+ 0.5P_L + 0.2P_S$ $= [1.2 + 0.2(0.528)](16.4 \text{ kips})$ $+ 2(15.8 \text{ kips}) + 0.5(0 \text{ kips})$ $+ 0.2(19.9 \text{ kips})$ $= 57.0 \text{ kips}$	$P_a = (1.0 + 0.14S_{DS})P_D + P_H$ $+ P_F + 0.7\Omega_o P_{Q_E}$ $= [1.0 + 0.14(0.528)](16.4 \text{ kips})$ $+ 0 \text{ kips} + 0 \text{ kips}$ $+ 0.7(2)(15.8 \text{ kips})$ $= 39.7 \text{ kips}$

The governing load combinations, including the overstrength factor, for the required axial tensile strength is:

LRFD	ASD
LRFD Load Combination 7 from ASCE/SEI 7 Section 12.4.3.2	ASD Load Combination 8 from ASCE/SEI 7 Section 12.4.3.2
$P_u = (0.9 - 0.2S_{DS})P_D + \Omega_o P_{Q_E} + 1.6P_H$ $= [0.9 - 0.2(0.528)](16.4 \text{ kips})$ $+ 2(-15.8 \text{ kips}) + 1.6(0 \text{ kips})$ $= -18.6 \text{ kips}$	$P_a = (0.6 - 0.14S_{DS})P_D$ $+ 0.7\Omega_o P_{Q_E} + P_H$ $P_a = [0.6 - 0.14(0.528)](16.4 \text{ kips})$ $+ 0.7(2)(-15.8 \text{ kips}) + 0 \text{ kips}$ $= -13.5 \text{ kips}$

Second-Order Effects

Use the procedure of AISC *Specification* Appendix 8 to determine the second-order effects on the required strengths, where the required flexural strength and required axial strength are given as:

$M_r = B_1 M_{nt} + B_2 M_{lt}$ (Spec. Eq. A-8-1)

$P_r = P_{nt} + B_2 P_{lt}$ (Spec. Eq. A-8-2)

There is no bending moment in the column due to either vertical loading or lateral translation. Consequently there is no requirement to determine multipliers for the required flexural strength due to second-order effects. The lateral drift is minimal. As calculated in Example 5.2.1, $B_2 = 1.0$. Therefore there is no amplification of the axial load in the column due to P-Δ. In summary, no adjustments to the member forces calculated by a first-order analysis are required due to second-order effects.

5.2 ORDINARY CONCENTRICALLY BRACED FRAMES

Try a W10×49.

From AISC *Manual* Table 1-1, the geometric properties are as follows:

$A = 14.4$ in.^{2}	$d = 10.0$ in.	$t_w = 0.340$ in.	$b_f = 10.0$ in.
$t_f = 0.560$ in.	$r_x = 4.35$ in.	$r_y = 2.54$ in.	

Column Slenderness

There are no specific requirements for member ductility for columns in OCBF systems in Section F1 of the AISC *Seismic Provisions*. Therefore, check width-to-thickness ratios for element slenderness according to Table B4.1a of the AISC *Specification*. As indicated in AISC *Manual* Table 1-1, the W10×49 section is not slender for compression.

Available Compressive Strength

Determine K

According to AISC *Specification* Appendix 7, Section 7.2.3(a), for braced frame systems, the effective length factor for members subject to compression shall be taken as 1.0.

Therefore:

$K_x = 1.0$ $K_y = 1.0$
$L_x = 40.0$ ft $L_y = 40.0$ ft

$$\frac{K_x L_x}{r_x} = \frac{1.0(40.0 \text{ ft})(12 \text{ in./ft})}{4.35 \text{ in.}}$$
$$= 110$$

$$\frac{K_y L_y}{r_y} = \frac{1.0(40.0 \text{ ft})(12 \text{ in./ft})}{2.54 \text{ in.}}$$
$$= 189 \text{ (governs)}$$

From AISC *Manual* Table 4-22 with $KL/r = 189$ and using AISC *Specification* Equation E3-1, the available compressive strength is:

LRFD	ASD
$\phi_c F_{cr} = 6.32$ ksi $\phi_c P_n = \phi_c F_{cr} A_g$ $= 6.32 \text{ ksi}(14.4 \text{ in.}^2)$ $= 91.0 \text{ kips} > 57.0 \text{ kips}$ **o.k.**	$\dfrac{F_{cr}}{\Omega_c} = 4.21$ ksi $\dfrac{P_n}{\Omega_c} = \left(\dfrac{F_{cr}}{\Omega_c}\right) A_g$ $= 4.21 \text{ ksi}(14.4 \text{ in.}^2)$ $= 60.6 \text{ kips} > 39.7 \text{ kips}$ **o.k.**

Available Tensile Strength

From AISC *Manual* Table 5-1, the available strength of the W10×49 column in axial tension for yielding on the gross section is:

LRFD	ASD
$\phi_t P_n = 648$ kips > 18.6 kips **o.k.**	$\dfrac{P_n}{\Omega_t} = 431$ kips > 13.5 kips **o.k.**

The W10×49 for OCBF Column CL-1 is adequate.

Example 5.2.3. OCBF Beam Design

Given:

Refer to Beam BM-1 in Figure 5-2. Select a 40-ft-long ASTM A992 W-shape to resist the loads shown below.

The loads on the beam due to a first-order analysis are:

$P_D = 3.92$ kips (tens.) $P_L = 0$ kips $P_S = 4.74$ kips (tens.) $P_{Q_E} = \pm 16.5$ kips
$M_D = 72.0$ kip-ft $M_S = 120$ kip-ft
$V_D = 7.20$ kips $V_S = 12.0$ kips

Assume that the ends of the beam are pinned and braced against translation for both the *x-x* and *y-y* axes.

Solution:

From AISC *Manual* Table 2-4, the material properties are:

ASTM A992
$F_y = 50$ ksi
$F_u = 65$ ksi

Required Strength

The beam is a collector element transferring diaphragm shear to the OCBF braces. According to Section 12.10.2.1 of ASCE/SEI 7, the forces in the collector are calculated using the seismic load effects including the overstrength factor. The axial force in the beam from dead and snow load is in tension.

The governing load combinations in ASCE/SEI 7 used for determining the required flexural strength of the beam are used to determine the required axial strengths.

5.2 ORDINARY CONCENTRICALLY BRACED FRAMES

The required axial compressive strength of the beam, with axial tension shown as negative, is:

LRFD	ASD
LRFD Load Combination 5 from ASCE/SEI 7 Section 12.4.3.2 (including the 0.5 factor on L permitted in Section 12.4.3.2)	ASD Load Combination 5 from ASCE/SEI 7 Section 12.4.3.2
$P_u = (1.2+0.2S_{DS})P_D + \Omega_o P_{Q_E}$ $+ 0.5P_L + 0.2P_S$ $= [1.2+0.2(0.528)](-3.92 \text{ kips})$ $+ 2(16.5 \text{ kips}) + 0.5(0 \text{ kips})$ $+ 0.2(-4.74 \text{ kips})$ $= 26.9 \text{ kips}$	$P_a = (1.0+0.14S_{DS})P_D$ $+ P_H + P_F + 0.7\Omega_o P_{Q_E}$ $= [1.0+0.14(0.528)](-3.92 \text{ kips})$ $+ 0 \text{ kips} + 0 \text{ kips}$ $+ 0.7(2)(16.5 \text{ kips})$ $= 18.9 \text{ kips}$

The required axial tensile strength of the beam is:

LRFD	ASD
$P_u = [1.2+0.2(0.528)](-3.92 \text{ kips})$ $+ 2(-16.5 \text{ kips}) + 0.5(0 \text{ kips})$ $+ 0.2(-4.74 \text{ kips})$ $= -39.1 \text{ kips}$	$P_a = [1.0+0.14(0.528)](-3.92 \text{ kips})$ $+ 0 \text{ kips} + 0 \text{ kips}$ $+ 0.7(2)(-16.5 \text{ kips})$ $= -27.3 \text{ kips}$

The required shear strength of the beam is:

LRFD	ASD
$V_u = [1.2+0.2(0.528)](7.20 \text{ kips})$ $+ 2(0 \text{ kips}) + 0.5(0 \text{ kips})$ $+ 0.2(12.0 \text{ kips})$ $= 11.8 \text{ kips}$	$V_a = [1.0+0.14(0.528)](7.20 \text{ kips})$ $+ 0 \text{ kips} + 0 \text{ kips}$ $+ 0.7(2)(0 \text{ kips})$ $= 7.73 \text{ kips}$

The required flexural strength of the beam is:

LRFD	ASD
$M_u = [1.2 + 0.2(0.528)](72.0 \text{ kip-ft})$ $\quad + 2(0 \text{ kip-ft}) + 0.5(0 \text{ kip-ft})$ $\quad + 0.2(120 \text{ kip-ft})$ $\quad = 118 \text{ kip-ft}$	$M_a = [1.0 + 0.14(0.528)](72.0 \text{ kip-ft})$ $\quad + 0 \text{ kip-ft} + 0 \text{ kip-ft}$ $\quad + 0.7(2)(0 \text{ kip-ft})$ $\quad = 77.3 \text{ kip-ft}$

Try a W18×50.

From AISC *Manual* Table 1-1, the geometric properties are as follows:

$A = 14.7 \text{ in.}^2$ $\quad d = 18.0 \text{ in.}$ $\quad b_f = 7.50 \text{ in.}$ $\quad t_f = 0.570 \text{ in.}$
$t_w = 0.355 \text{ in.}$ $\quad k_{des} = 0.972 \text{ in.}$ $\quad h/t_w = 45.2$ $\quad r_x = 7.38 \text{ in.}$
$r_y = 1.65 \text{ in.}$ $\quad I_x = 800 \text{ in.}^4$ $\quad S_x = 88.9 \text{ in.}^3$ $\quad Z_x = 101 \text{ in.}^3$

Beam Slenderness

There are no specific requirements for member ductility for beams in OCBF systems in Section F1 of the AISC *Seismic Provisions*. Therefore, check width-to-thickness ratios for element slenderness according to Table B4.1a and Table B4.1b of the AISC *Specification*.

As indicated in AISC *Manual* Table 1-1, the W18×50 is slender for compression and compact for flexure.

Available Compressive Strength

Determine K

According to AISC *Specification* Appendix 7, Section 7.2.3(a), for braced frame systems, the effective length factor for members subject to compression shall be taken as 1.0. Consider the open web steel joists at the top flange of the beam to provide the strength and stiffness required by AISC *Specification* Appendix 6 to stabilize the top flange of the beam in the y-y axis at 6 ft 8 in. centers. Consider that the bottom flange of the beam is stabilized in the y-y axis at midspan by a bottom chord extension from the open web steel joist. Consider the effective length of the beam in compression about the y-y axis to be based on the unsupported length of the bottom flange.

Therefore:

$K_x = 1.0$ $\quad K_y = 1.0$
$L_x = 40.0 \text{ ft}$ $\quad L_y = 20.0 \text{ ft}$

$$\frac{K_x L_x}{r_x} = \frac{1.0(40.0 \text{ ft})(12 \text{ in./ft})}{7.38 \text{ in.}}$$
$$= 65.0$$

$$\frac{K_y L_y}{r_y} = \frac{1.0(20.0 \text{ ft})(12 \text{ in./ft})}{1.65 \text{ in.}}$$

$$= 145 \text{ (governs)}$$

The combination of the top flange bracing and the bottom flange bracing from the open web steel joist at midspan creates a torsional brace. This example uses a simplified calculation of the available compressive strength according to AISC *Specification* Section E7 that considers the limit state of flexural buckling using the minor axis unbraced length of the member that is based on the bottom flange unbraced length. A greater compressive strength may be available due to the additional minor axis constraint at the top flange. See Section 8.3 of this Manual for a method to determine the available torsional buckling strength considering constraint at the top flange.

Because the web is considered a slender element for axial compression $\left(h/t_w > 1.49\sqrt{E/F_y}\right.$ $= 1.49\sqrt{29{,}000 \text{ ksi}/50 \text{ ksi}} = 35.9\big)$ a reduction for slenderness is required for calculating the available compressive strength per Section E7.2 of the AISC *Specification*.

This reduction is included in AISC *Manual* Table 6-1; therefore, use AISC *Manual* Table 6-1 to determine the available compressive strength of the W18×50. From Table 6-1, for $K_y L_y = 20$ ft:

LRFD	ASD
$p \times 10^3 = 6.37 (\text{kips})^{-1}$	$p \times 10^3 = 9.58 (\text{kips})^{-1}$
$\phi_c P_n = \dfrac{1}{p}$	$P_n/\Omega_c = \dfrac{1}{p}$
$= \dfrac{1}{6.37 \times 10^{-3} (\text{kips})^{-1}}$	$= \dfrac{1}{9.58 \times 10^{-3} (\text{kips})^{-1}}$
$= 157$ kips	$= 104$ kips

Available Flexural Strength

Because the beam is bending about its major axis, and has both compact flanges and a compact web in flexure, the available flexural strength is determined in accordance with AISC *Specification* Section F2.

The open web steel joists provide lateral support of the compression flange at 6 ft 8 in. centers.

$L_b = 6.67$ ft

According to AISC *Manual* Table 3-2:

$L_p = 5.83$ ft
$L_r = 16.9$ ft

Therefore $L_p < L_b \le L_r$ and the limit state of lateral-torsional buckling applies. Conservatively, use $C_b = 1.0$.

From AISC *Manual* Table 3-10, the available flexural strength of the beam is:

LRFD	ASD
$\phi_b M_n = 368$ kip-ft	$\dfrac{M_n}{\Omega_b} = 245$ kip-ft

Second-Order Effects
Following the procedure of AISC *Specification* Appendix 8:

$M_r = B_1 M_{nt} + B_2 M_{lt}$ (*Spec.* Eq. A-8-1)

$P_r = P_{nt} + B_2 P_{lt}$ (*Spec.* Eq. A-8-2)

Calculate B_1

$C_m = 1.0$ as the beam is subject to transverse loading between supports
$\alpha = 1.00$ (LRFD); $\alpha = 1.60$ (ASD)

$K_1 L_x = 1.0(40.0 \text{ ft})$
$\quad\quad = 40.0 \text{ ft}$

$$P_{e1} = \frac{\pi^2 EI_x}{(K_1 L)^2} \quad\quad (\textit{Spec. Eq. A-8-5})$$

$$= \frac{\pi^2 (29{,}000 \text{ ksi})(800 \text{ in.}^4)}{[40.0 \text{ ft}(12 \text{ in./ft})]^2}$$

$= 994$ kips

LRFD	ASD
$B_1 = \dfrac{C_m}{1 - \alpha P_r / P_{e1}}$	$B_1 = \dfrac{C_m}{1 - \alpha P_r / P_{e1}}$
$= \dfrac{1.0}{1 - \dfrac{1.00(26.9 \text{ kips})}{994 \text{ kips}}}$	$= \dfrac{1.0}{1 - \dfrac{1.60(18.9 \text{ kips})}{994 \text{ kips}}}$
$= 1.03$	$= 1.03$

5.2 ORDINARY CONCENTRICALLY BRACED FRAMES

Calculate B_2

$B_2 = 1.00$ as calculated in Example 5.2.1
$P_{nt} = 0$ kips
$P_{lt} = P_u$ or P_a as determined previously
$M_{nt} = M_u$ or M_a as determined previously
$M_{lt} = 0$ kip-ft because there is no moment due to seismic loading

From AISC *Specification* Equation A-8-2 and the applicable ASCE/SEI 7 load combination, the required axial compressive strength is:

LRFD	ASD
LRFD Load Combination 5 from ASCE/SEI 7 Section 12.4.3.2 (including the 0.5 factor on L permitted in Section 12.4.3.2)	ASD Load Combination 5 from ASCE/SEI 7 Section 12.4.3.2
$P_u = (1.2 + 0.2S_{DS})P_D + B_2\Omega_o P_{Q_E}$ $\quad + 0.5P_L + 0.2P_S$ $\quad = [1.2 + 0.2(0.528)](-3.92 \text{ kips})$ $\quad + 1.00(2)(16.5 \text{ kips}) + 0.5(0 \text{ kips})$ $\quad + 0.2(-4.74 \text{ kips})$ $\quad = 26.9 \text{ kips}$	$P_a = (1.0 + 0.14S_{DS})P_D$ $\quad + P_H + P_F + B_2(0.7\Omega_o P_{Q_E})$ $\quad = [1.0 + 0.14(0.528)](-3.92 \text{ kips})$ $\quad + 0 \text{ kips} + 0 \text{ kips}$ $\quad + 1.00(0.7)(2)(16.5 \text{ kips})$ $\quad = 18.9 \text{ kips}$

From AISC *Specification* Equation A-8-1, the required flexural strength is:

LRFD	ASD
$M_{rx} = B_1 M_{nt} + B_2 M_{lt}$ $\quad = 1.03(118 \text{ kip-ft}) + 1.00(0 \text{ kip-ft})$ $\quad = 122 \text{ kip-ft}$	$M_{rx} = B_1 M_{nt} + B_2 M_{lt}$ $\quad = 1.03(77.3 \text{ kip-ft}) + 1.00(0 \text{ kip-ft})$ $\quad = 79.6 \text{ kip-ft}$

Combined Loading (Flexure and Compression)
Determine the applicable equation in AISC *Specification* Section H1.1:

LRFD	ASD
$\dfrac{P_r}{P_c} = \dfrac{26.9 \text{ kips}}{157 \text{ kips}}$ $\quad = 0.171$	$\dfrac{P_r}{P_c} = \dfrac{18.9 \text{ kips}}{104 \text{ kips}}$ $\quad = 0.182$

Because $P_r/P_c < 0.2$, the beam design is controlled by the equation:

$$\frac{P_r}{2P_c} + \left(\frac{M_{rx}}{M_{cx}} + \frac{M_{ry}}{M_{cy}}\right) \leq 1.0 \qquad \text{(Spec. Eq. H1-1b)}$$

LRFD	ASD
$\dfrac{26.9 \text{ kips}}{2(157 \text{ kips})} + \left(\dfrac{122 \text{ kip-ft}}{368 \text{ kip-ft}} + 0\right) = 0.417$ 0.417 < 1.0 **o.k.**	$\dfrac{18.9 \text{ kips}}{2(104 \text{ kips})} + \left(\dfrac{79.6 \text{ kip-ft}}{245 \text{ kip-ft}} + 0\right) = 0.416$ 0.416 < 1.0 **o.k.**

Available Shear Strength

From AISC *Manual* Table 3-6, the available shear strength of the W18×50 beam is:

LRFD	ASD
$\phi_v V_n = 192$ kips > 11.8 kips **o.k.**	$\dfrac{V_n}{\Omega_v} = 128$ kips > 7.73 kips **o.k.**

Available Tensile Strength

From AISC *Manual* Table 5-1, the available strength of the W18×50 beam in axial tension for yielding on the gross section is:

LRFD	ASD
$\phi_t P_n = 662$ kips > 39.1 kips	$\dfrac{P_n}{\Omega_t} = 440$ kips > 27.3 kips **o.k.**

Consider second-order effects (tension loading)

Consider second order effects according to Appendix 8 of the AISC *Specification*. As previously calculated, $B_2 = 1.0$. According to AISC *Specification* Appendix 8, Section 8.2, B_1 is taken as 1.0 for members not subject to compression. Given that both B_1 and B_2 are equal to 1.0, there is no amplification required for second-order effects for the loads on the member when the diagonal brace is in tension.

Combined Loading (Flexure and Tension)

Because the axial tensile force is greater than the axial compressive force, interaction will be checked. As previously determined:

5.2 ORDINARY CONCENTRICALLY BRACED FRAMES

LRFD	ASD
$M_{rx} = M_u$ $= 118$ kip-ft $P_r = T_u$ $= 39.1$ kips $P_c = \phi_t P_n$ $= 662$ kips	$M_{rx} = M_a$ $= 77.3$ kip-ft $P_r = T_a$ $= 27.3$ kips $P_c = P_n / \Omega_t$ $= 440$ kips

Determine the applicable equation in AISC *Specification* Section H1.1:

LRFD	ASD
$\dfrac{P_r}{P_c} = \dfrac{39.1 \text{ kips}}{662 \text{ kips}}$ $= 0.0591$	$\dfrac{P_r}{P_c} = \dfrac{27.3 \text{ kips}}{440 \text{ kips}}$ $= 0.0620$

Since $P_r/P_c < 0.2$, the beam design is controlled by the equation:

$$\frac{P_r}{2P_c} + \left(\frac{M_{rx}}{M_{cx}} + \frac{M_{ry}}{M_{cy}}\right) \leq 1.0 \qquad (Spec.\ Eq.\ H1\text{-}1b)$$

LRFD	ASD
$\dfrac{39.1 \text{ kips}}{2(662 \text{ kips})} + \left(\dfrac{118 \text{ kip-ft}}{368 \text{ kip-ft}} + 0\right) = 0.350$ $0.350 < 1.0$ **o.k.**	$\dfrac{27.3 \text{ kips}}{2(440 \text{ kips})} + \left(\dfrac{77.3 \text{ kip-ft}}{245 \text{ kip-ft}} + 0\right) = 0.347$ $0.347 < 1.0$ **o.k.**

Note that the available flexural strength was conservatively based on $C_b = 1.0$. Determining C_b and applying it would have resulted in a higher available flexural strength.

The W18×50 is adequate for use as the OCBF Beam BM-1.

Example 5.2.4. OCBF Brace-to-Beam/Column Connection Design

Given:

Refer to Joint JT-1 in Figure 5-2. Design the connection between the brace, beam and column. Use a bolted connection for the brace-to-gusset connection. Use a single-plate connection to connect the beam and gusset to the column and a welded connection between the beam and gusset plate. Use ASTM A36 for all plate and angle material. Assume the member sizes are as determined in the previous OCBF examples. Use ¾-in.-diameter ASTM A325-N bolts and 70-ksi weld electrodes.

From Example 5.2.1, the loads on the connection from the brace based on a first order analysis are:

$P_D = 5.54$ kips $P_S = 6.70$ kips $P_{Q_E} = \pm 22.3$ kips

From Example 5.2.3, the loads on the connection from the beam (collector element), based on a first-order analysis are:

$P_D = 3.92$ kips (tens.) $P_L = 0$ kips $P_S = 4.74$ kips (tens.) $P_{Q_E} = \pm 16.5$ kips
$M_D = 72.0$ kip-ft $M_L = 0$ kip-ft $M_S = 120$ kip-ft $M_{Q_E} = 0$ kip-ft
$V_D = 7.20$ kips $V_L = 0$ kips $V_S = 12.0$ kips $V_{Q_E} = 0$ kips

Solution:

From AISC *Manual* Table 2-5, the material properties are as follows:

ASTM A36
$F_y = 36$ ksi
$F_u = 58$ ksi

From AISC *Manual* Table 1-1, the geometric properties are as follows:

Beam
W18×50
$A = 14.7$ in.^{2} $d = 18.0$ in. $b_f = 7.50$ in. $t_f = 0.570$ in.
$t_w = 0.355$ in. $T = 15\tfrac{1}{2}$ in. $k_{des} = 0.972$ in. $I_x = 800$ in.^{4}
$S_x = 88.9$ in.^{3} $r_x = 7.38$ in. $Z_x = 101$ in.^{3} $r_y = 1.65$ in.

Column
W10×49
$d = 10.0$ in. $t_f = 0.560$ in. $t_w = 0.340$ in. $k_{des} = 1.06$ in.

Brace
W10×33
$A = 9.71$ in.^{2} $d = 9.73$ in. $t_f = 0.435$ in. $t_w = 0.290$ in.
$b_f = 7.96$ in.

Required Strength
The governing load combinations for the collector force are:

5.2 ORDINARY CONCENTRICALLY BRACED FRAMES

LRFD	ASD
LRFD Load Combination 5 from ASCE/SEI 7 Section 12.4.3.2 (including the 0.5 factor on L permitted in Section 12.4.3.2) $(1.2 + 0.2S_{DS})P_D + \Omega_o P_{Q_E} + 0.5P_L + 0.2P_S$	ASD Load Combination 5 from ASCE/SEI 7 Section 12.4.3.2 $(1.0 + 0.14S_{DS})P_D + P_H + P_F + 0.7\Omega_o P_{Q_E}$

The required axial compressive strength of the collector at the beam-to-column connection is, from the loads given in Example 5.2.3:

LRFD	ASD
$P_u = [1.2 + 0.2(0.528)](0 \text{ kips})$ $+ 2(16.5 \text{ kips}) + 0.5(0 \text{ kips})$ $+ 0.2(0 \text{ kips})$ $= 33.0 \text{ kips}$	$P_a = [1.0 + 0.14(0.528)](0 \text{ kips})$ $+ 0 \text{ kips} + 0 \text{ kips} + 0.7(2)(16.5 \text{ kips})$ $= 23.1 \text{ kips}$

Note: The above load results from the transfer of the collector force from the beam in the adjacent bay. The axial components from snow and gravity axial loads used in Example 5.2.3 are transferred from the brace gusset directly into the braced frame beam.

According to AISC *Seismic Provisions* Section F1.6a, the required strength of diagonal brace connections is the load effect based upon the amplified seismic load. Based on the loads given for the brace from Example 5.2.1, the maximum axial tensile force in the diagonal brace based upon the amplified seismic load, is:

LRFD	ASD
LRFD Load Combination 7 from ASCE/SEI 7 Section 12.4.3.2 $(0.9 - 0.2S_{DS})P_D + \Omega_o P_{Q_E} + 1.6P_H$ $P_u = [0.9 - 0.2(0.528)](5.54 \text{ kips})$ $+ 2(-22.3 \text{ kips}) + 1.6(0 \text{ kips})$ $= -40.2 \text{ kips}$	ASD Load Combination 8 from ASCE/SEI 7 Section 12.4.3.2 $(0.6 - 0.14S_{DS})P_D + P_H + P_F + 0.7\Omega_o P_{Q_E}$ $P_a = [0.6 - 0.14(0.528)](5.54 \text{ kips})$ $+ 0 \text{ kips} + 0 \text{ kips}$ $+ 0.7(2)(-22.3 \text{ kips})$ $= -28.3 \text{ kips}$

According to the exception in AISC *Seismic Provisions* Section F1.6a, the required axial tension strength need not exceed the expected yield strength multiplied by 1.00 (LRFD) or divided by 1.50 (ASD):

LRFD	ASD
$T_{u,exp} = 1.00(R_y F_y A_g)$ $= 1.00(1.1)(50 \text{ ksi})(9.71 \text{ in.}^2)$ $= 534 \text{ kips}$	$T_{a,exp} = R_y F_y A_g / 1.50$ $= 1.1(50 \text{ ksi})(9.71 \text{ in.}^2)/1.50$ $= 356 \text{ kips}$

Therefore, the required strength of the brace connection in tension is P_u = 40.2 kips and P_a = 28.3 kips.

The required shear strength of the beam concurrent with axial tension in the brace is:

LRFD	ASD
LRFD Load Combination 7 from ASCE/SEI 7 Section 12.4.3.2 $(0.9 - 0.2S_{DS})V_D + \Omega_o V_{Q_E} + 1.6V_H$ $V_u = [0.9 - 0.2(0.528)](7.20 \text{ kips})$ $+ 2(0 \text{ kips}) + 1.6(0 \text{ kips})$ $= 5.72 \text{ kips}$	ASD Load Combination 8 from ASCE/SEI 7 Section 12.4.3.2 $(0.6 - 0.14S_{DS})V_D + V_H + V_F + 0.7\Omega_o V_{Q_E}$ $V_a = [0.6 - 0.14(0.528)](7.20 \text{ kips})$ $+ 0 \text{ kips} + 0 \text{ kips} + 0.7(2)(0 \text{ kips})$ $= 3.79 \text{ kips}$

The above shear force is concurrent with the maximum tension force in the diagonal brace.

Considering the load combinations given in ASCE/SEI 7, the maximum compressive axial force in the diagonal brace based upon the amplified seismic load is:

LRFD	ASD
LRFD Load Combination 5 from ASCE/SEI 7 Section 12.4.3.2 (including the 0.5 factor on L from Section 12.4.3.2) $(1.2 + 0.2S_{DS})P_D + \Omega_o P_{Q_E} + 0.5P_L$ $+ 0.2P_S$ $P_u = [1.2 + 0.2(0.528)](5.54 \text{ kips})$ $+ 2(22.3 \text{ kips}) + 0.5(0 \text{ kips})$ $+ 0.2(6.70 \text{ kips})$ $= 53.2 \text{ kips}$	ASD Load Combination from ASCE/SEI 5 Section 12.4.3.2 $(1.0 + 0.14S_{DS})P_D + P_H + P_F$ $+ 0.7\Omega_o P_{Q_E}$ $P_a = [1.0 + 0.14(0.528)](5.54 \text{ kips})$ $+ 0 \text{ kips} + 0 \text{ kips}$ $+ 0.7(2)(22.3 \text{ kips})$ $= 37.2 \text{ kips}$

5.2 ORDINARY CONCENTRICALLY BRACED FRAMES

According to the Exception in AISC *Seismic Provisions* Section F1.6a, the required axial strength of the brace connection in compression need not exceed the lesser of the expected yield strength and $1.14F_{cre}A_g$, where F_{cre} is based on the expected yield stress, R_yF_y.

As determined in Example 5.2.1, the available compressive strength of the brace is:

LRFD	ASD
$\phi P_n = 71.7$ kips	$\dfrac{P_n}{\Omega} = 47.7$ kips

The available compressive strength is greater than the maximum compressive axial force calculated using the amplified seismic load. Therefore, the exception limiting the required axial compressive strength to the expected yield strength and $1.14F_{cre}A_g$ will not govern. The required strength of the brace connection in compression is $P_u = 53.2$ kips and $P_a = 37.2$ kips.

The required shear strength of the beam that is concurrent with maximum axial compression in the brace is, as calculated in Example 5.2.3:

LRFD	ASD
$V_u = 11.8$ kips	$V_a = 7.73$ kips

Brace-to-Gusset Connection

Using AISC *Manual* Table 7-1 for ¾-in.-diameter A325-N bolts (Group A) in double shear:

LRFD	ASD
$\phi r_n = 35.8$ kips	$\dfrac{r_n}{\Omega} = 23.9$ kips

For the limit state of bolt shear, the minimum number of bolts required in the brace-to-gusset connection is:

LRFD	ASD
$n = \dfrac{P_u}{\phi r_n}$ $= \dfrac{53.2 \text{ kips}}{35.8 \text{ kips}}$ $= 1.49$ bolts	$n = \dfrac{P_a}{r_n/\Omega}$ $= \dfrac{37.2 \text{ kips}}{23.9 \text{ kips}}$ $= 1.56$ bolts

To facilitate erection, use oversized holes in one ply of the connection as permitted in AISC *Seismic Provisions* Section D2.2(3).

When oversized holes are used in the diagonal brace connection, the required strength for the limit state of bolt slip need not exceed the load effect calculated using the load combinations not including the amplified seismic load, according to AISC *Seismic Provisions* Section F1.6a(3). These correspond to the required strengths calculated for the member design in Example 5.2.1.

Therefore, the required strength for the limit state of bolt slip need not exceed:

LRFD	ASD
$P_{u\;slip} = 30.9$ kips	$P_{a\;slip} = 21.6$ kips

From AISC *Manual* Table 7-3 for ¾-in.-diameter A325-SC bolts (Group A) in double shear, Class A faying surfaces, oversized holes in the diagonal brace web and standard holes in the gusset and angles:

LRFD	ASD
$\phi r_n = 16.1$ kips	$\dfrac{r_n}{\Omega} = 10.8$ kips

For the limit state of bolt slip, the minimum number of bolts required in the brace-to-gusset connection is:

LRFD	ASD
$n = \dfrac{P_u}{\phi r_n}$ $= \dfrac{30.9 \text{ kips}}{16.1 \text{ kips}}$ $= 1.92$ bolts	$n = \dfrac{P_a}{r_n/\Omega}$ $= \dfrac{21.6 \text{ kips}}{10.8 \text{ kips}}$ $= 2.00$ bolts

Use four claw angles to connect the brace to the gusset as shown in Figure 5-5. Try (4) L3½×3½×5/16 claw angles each connected to the gusset with (2) ¾-in.-diameter ASTM A325 bolts in double shear and to the brace web with (2) ¾-in.-diameter ASTM A325 bolts in double shear. Therefore, the total number of bolts at the brace-to-angle connection and at the angle-to-gusset connection, $N_b = 4$, is greater than the minimum number of bolts, n, calculated above.

From AISC *Manual* Tables 1-7 and 1-7a:

Claw Angles
L3½×3½×5/16

$A = 2.10$ in.² $\qquad t_f = 0.435$ in. $\qquad \bar{x} = 0.979$ in. $\qquad g = 2$ in.

5.2 ORDINARY CONCENTRICALLY BRACED FRAMES

For short claw angle connections, eccentricity may be an issue. For angles with the ratio $L/g \geq 4$, the eccentricity effect of connections to opposite angle legs can safely be ignored (Thornton, 1996). L is the distance between the centers of bolt groups on opposite legs of the angle, and g is the bolt gage in the angle leg. See Figure 5-3.

Consider a 2.00-in. edge distance on the brace and the gusset, ½-in. space between the end of the brace and the end of the gusset, and 4-in. spacing between bolts.

$$L = 2\left(\frac{4.00 \text{ in.}}{2} + 2.00 \text{ in.} + \frac{0.500 \text{ in.}}{2}\right)$$
$$= 8.50 \text{ in.}$$

$g = 2.00$ in.

$$\frac{L}{g} = \frac{8.50 \text{ in.}}{2.00 \text{ in.}}$$
$$= 4.25 > 4 \qquad \textbf{o.k.}$$

Check tensile yielding of the angles

$A_g =$ gross area of four angles
$= 4A$
$= 4(2.10 \text{ in.}^2)$
$= 8.40 \text{ in.}^2$

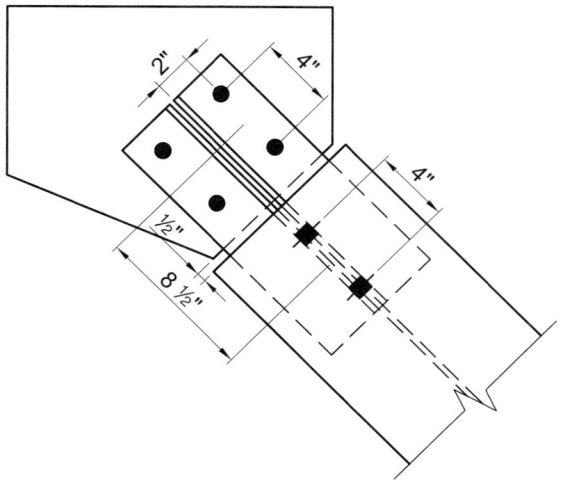

Fig. 5-3. Single claw angle dimensions for check of eccentric effect.

For tensile yielding of connecting elements, the nominal strength is:

$R_n = F_y A_g$ (*Spec.* Eq. J4-1)

$= 36 \text{ ksi}(8.40 \text{ in.}^2)$

$= 302 \text{ kips}$

The available tensile strength (yielding) of the four angles is:

LRFD	ASD
$\phi R_n = 0.90(302 \text{ kips})$ $= 272 \text{ kips} > 40.2 \text{ kips}$ **o.k.**	$\dfrac{R_n}{\Omega} = (302 \text{ kips})/1.67$ $= 181 \text{ kips} > 28.3 \text{ kips}$ **o.k.**

Check tensile rupture of the angles

From AISC *Specification* Table D3.1, the shear lag factor is:

$U = 1 - \dfrac{\bar{x}}{l}$

$= 1 - \dfrac{0.979 \text{ in.}}{4.00 \text{ in.}}$

$= 0.755$

Use standard holes in the angles. For calculation of net area, AISC *Specification* Section B4.3b defines the width of the bolt hole as $\frac{1}{16}$ in. greater than the nominal dimension of the hole, where the nominal hole dimension is given in Table J3.3.

$A_n = A_g - 4t d_h$

$= 8.40 \text{ in.}^2 - 4(\frac{5}{16} \text{ in.})(\frac{13}{16} \text{ in.} + \frac{1}{16} \text{ in.})$

$= 7.31 \text{ in.}^2$

$A_e = A_n U$ (*Spec.* Eq. D3-1)

$= 7.31 \text{ in.}^2 (0.755)$

$= 5.52 \text{ in.}^2$

For tensile rupture of connecting elements, the nominal strength is:

$R_n = F_u A_e$ (*Spec.* Eq. J4-2)

$= 58 \text{ ksi}(5.52 \text{ in.}^2)$

$= 320 \text{ kips}$

5.2 ORDINARY CONCENTRICALLY BRACED FRAMES

LRFD	ASD
$\phi R_n = 0.75(320 \text{ kips})$ $= 240 \text{ kips} > 40.2 \text{ kips}$ **o.k.**	$\dfrac{R_n}{\Omega} = \dfrac{320 \text{ kips}}{2.00}$ $= 160 \text{ kips} > 28.3 \text{ kips}$ **o.k.**

Check block shear rupture of the angles

AISC *Manual* Tables 9-3a, 9-3b and 9-3c for block shear may be used here for accurately calculating the tension rupture component. For the shear components, the values in the tables are based on a bolt spacing of 3.00 in., whereas this connection uses 4.00-in. bolt spacing. For this reason, the tables are not used here for calculating shear components (but could have been used as a conservative check).

The horizontal edge distance along the tension plane, L_{eh}, is calculated as the angle leg less the gage:

$L_{eh} = 3\frac{1}{2}$ in. $- 2.00$ in.
$\qquad = 1.50$ in.

Use an edge distance, L_{ev}, of 1.50 in. at the ends of the angles.

The nominal strength for the limit state of block shear rupture is:

$R_n = U_{bs} F_u A_{nt} + \min(0.60 F_y A_{gv}, 0.60 F_u A_{nv})$ (from *Spec*. Eq. J4-5)

Tension rupture component for one angle:

From AISC *Manual* Table 9-3a with $F_u = 58$ ksi, $L_{eh} = 1.50$ in., and $U_{bs} = 1.0$:

LRFD	ASD
$\dfrac{\phi F_u A_{nt}}{t} = 46.2 \text{ kip/in.}$ $\phi F_u A_{nt} U_{bs} = 46.2 \text{ kip/in.}(\frac{5}{16} \text{ in.})(1.0)$ $\qquad = 14.4 \text{ kips}$	$\dfrac{F_u A_{nt}}{t\Omega} = 30.8 \text{ kip/in.}$ $\dfrac{F_u A_{nt} U_{bs}}{\Omega} = 30.8 \text{ kip/in.}(\frac{5}{16} \text{ in.})(1.0)$ $\qquad = 9.63 \text{ kips}$

Shear yielding component for one angle:

$0.60 F_y A_{gv} = 0.60(36 \text{ ksi})(4.00 \text{ in.} + 1.50 \text{ in.})(\frac{5}{16} \text{ in.})$
$\qquad\quad = 37.1$ kips

LRFD	ASD
$\phi 0.60 F_y A_{gv} = 0.75(37.1 \text{ kips})$ $= 27.8 \text{ kips}$	$\dfrac{0.60 F_y A_{gv}}{\Omega} = \dfrac{37.1 \text{ kips}}{2.00}$ $= 18.6 \text{ kips}$

Shear rupture component for one angle:

$$0.60 F_u A_{nv} = 0.60(58 \text{ ksi})[4.00 \text{ in.} + 1.50 \text{ in.} - 1.5(^{13}\!/_{16} \text{ in.} + {}^{1}\!/_{16} \text{ in.})]({}^{5}\!/_{16} \text{ in.})$$
$$= 45.5 \text{ kips}$$

LRFD	ASD
$\phi 0.60 F_u A_{nv} = 0.75(45.5 \text{ kips})$ $= 34.1 \text{ kips}$	$\dfrac{0.60 F_u A_{nv}}{\Omega} = \dfrac{45.5 \text{ kips}}{2.00}$ $= 22.8 \text{ kips}$

Shear yielding governs over shear rupture. The available strength for the limit state of block shear rupture for the four angles is:

LRFD	ASD
$\phi R_n = 4(14.4 \text{ kips} + 27.8 \text{ kips})$ $= 169 \text{ kips} > 40.2 \text{ kips}$ **o.k.**	$\dfrac{R_n}{\Omega} = 4(9.63 \text{ kips} + 18.6 \text{ kips})$ $= 113 \text{ kips} > 28.3 \text{ kips}$ **o.k.**

Check tension rupture of the brace

The claw angles are connected only to the web of the W10×33 brace and not to the flanges. Therefore shear lag may reduce the effective area. The bolt holes in the web of the brace are oversized for erection tolerance.

Because the tension load is transferred only at the web of the wide flange brace, Case 2 of AISC *Specification* Table D3.1 is applicable. However to simplify calculation of the net section, consider the tensile rupture capacity of the web element only. This is similar to Case 3 of Table D3.1, which applies to members with transverse welds to some but not all of the cross-sectional elements.

From AISC *Specification* Table J3.3, the diameter of an oversized hole for a ¾-in.-diameter bolt is $^{15}\!/_{16}$ in. From AISC *Specification* Section B4.3b, when computing the net area the width of the bolt hole is taken as $^{1}\!/_{16}$ in. greater than the nominal dimension of the hole.

$$d_h = {}^{15}\!/_{16} \text{ in.} + {}^{1}\!/_{16} \text{ in.}$$
$$= 1.00 \text{ in.}$$

5.2 ORDINARY CONCENTRICALLY BRACED FRAMES

Effective net area:

$U = 1.0$

$A_n = (d - 2d_h)t_w$
$= [9.73 \text{ in.} - 2(1.00 \text{ in.})](0.290 \text{ in.})$
$= 2.24 \text{ in.}^2$

$A_e = A_n U$ (*Spec.* Eq. D3-1)
$= 2.24 \text{ in.}^2 (1.0)$
$= 2.24 \text{ in.}^2$

For tensile rupture of the brace web, the nominal strength is:

$R_n = F_u A_e$ (*Spec.* Eq. J4-2)
$= 65 \text{ ksi}(2.24 \text{ in.}^2)$
$= 146 \text{ kips}$

The available tensile rupture strength of the brace web is:

LRFD	ASD
$\phi R_n = 0.75(146 \text{ kips})$ $= 110 \text{ kips} > 40.2 \text{ kips}$ **o.k.**	$\dfrac{R_n}{\Omega} = \dfrac{146 \text{ kips}}{2.00}$ $= 73.0 \text{ kips} > 28.3 \text{ kips}$ **o.k.**

For this lightly loaded member, this conservative and simplified calculation indicates that the available tensile rupture strength is adequate.

Alternatively, the effective net area could be calculated for the entire section as follows. Calculate U, the shear lag factor, in accordance with Table D3.1, Case 2, of the AISC *Specification*. AISC *Specification* Commentary Figure C-D3.1 recommends treating half of the flange and a portion of the web as an angle. This is shown in Figure 5-4.

First, calculate $\bar{x}$ of the angle, where $\bar{x}$ is measured from the centerline of the web (this calculation ignores the fillets):

$A = \dfrac{b_f}{2}(t_f) + \left(\dfrac{d}{2} - t_f\right)\left(\dfrac{t_w}{2}\right)$

$= \dfrac{7.96 \text{ in.}}{2}(0.435 \text{ in.}) + \left(\dfrac{9.73 \text{ in.}}{2} - 0.435 \text{ in.}\right)\left(\dfrac{0.290 \text{ in.}}{2}\right)$

$= 2.37 \text{ in.}^2$

$$\bar{x} = \frac{\Sigma \bar{x}_i A_i}{A}$$

$$= \frac{1}{2.37 \text{ in.}^2} \left[\left(\frac{0.290 \text{ in.}}{4} \right) \left(\frac{0.290 \text{ in.}}{2} \right) \left(\frac{9.73 \text{ in.}}{2} - 0.435 \text{ in.} \right) + \left(\frac{3.98 \text{ in.}}{2} \right) (3.98 \text{ in.})(0.435 \text{ in.}) \right]$$

$$= 1.47 \text{ in.}$$

AISC *Specification* Commentary Section D3 states that $\bar{x}$ is the perpendicular distance from the connection plane to the centroid of the member section. Therefore, the $\bar{x}$ used in the tension rupture calculation is the calculated $\bar{x}$ of 1.47 in. minus half the web thickness. From AISC *Specification* Table D3.1:

$$U = 1 - \frac{\bar{x}}{l}$$

$$= 1 - \frac{1.47 \text{ in.} - \frac{1}{2}(0.290 \text{ in.})}{4.00 \text{ in.}}$$

$$= 0.669$$

For a **W10×33** brace, with $A = 9.71$ in.^{2} and using oversized holes in the brace web ($d_h = 1.00$ in.), the effective net area is:

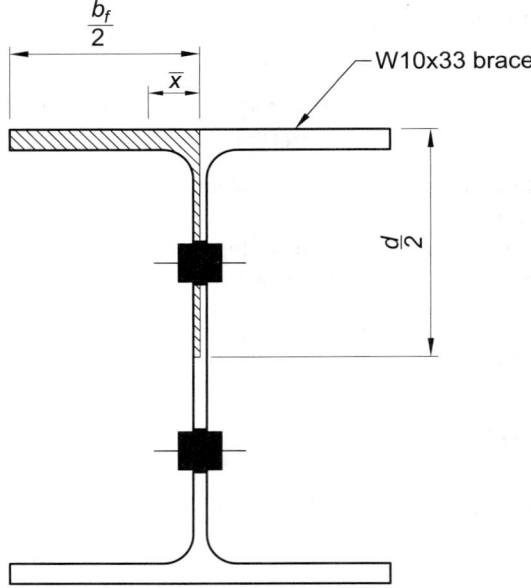

Fig. 5-4. Tension rupture on brace.

5.2 ORDINARY CONCENTRICALLY BRACED FRAMES

$$A_e = A_n U \quad \text{(Spec. Eq. D3-1)}$$
$$= (A - 2d_h t_w)U$$
$$= \left[9.71 \text{ in.}^2 - 2(1.00 \text{ in.})(0.290 \text{ in.})\right](0.669)$$
$$= 6.11 \text{ in.}^2$$

For tensile rupture of the beam web, the nominal strength is:

$$R_n = F_u A_e \quad \text{(Spec. Eq. J4-2)}$$
$$= 65 \text{ ksi}\left(6.11 \text{ in.}^2\right)$$
$$= 397 \text{ kips}$$

The available tensile rupture strength of the brace web is:

LRFD	ASD
$\phi R_n = 0.75(397 \text{ kips})$ $= 298 \text{ kips} > 40.2 \text{ kips}$ **o.k.**	$\dfrac{R_n}{\Omega} = \dfrac{397 \text{ kips}}{2.00}$ $= 199 \text{ kips} > 28.3 \text{ kips}$ **o.k.**

As shown, the available strength of the W-shape brace for the limit state of tensile rupture as calculated per the simplified calculation (with only the brace web considered effective) is adequate for the applied loads. However, if additional capacity were required, the available strength as calculated per AISC *Specification* Table D3.1, Case 2, is much greater.

Check block shear rupture of the brace web

The portion of the brace web between the bolt lines is checked for block shear. Assume a gusset plate thickness, t_g, of ⅜ in.

$U_{bs} = 1.0$ for uniform tensile stress

$$A_{nt} = (2g + t_g - d_h)t_w$$
$$= [2(2.00 \text{ in.}) + \text{⅜ in.} - 1.00 \text{ in.}](0.290 \text{ in.})$$
$$= 0.979 \text{ in.}^2$$

$L_{ev} = 2.00$ in.

$s \phantom{_{ev}} = 4.00$ in.

$$A_{gv} = 2(L_{ev} + s)t_w$$
$$= 2(2.00 \text{ in.} + 4.00 \text{ in.})(0.290 \text{ in.})$$
$$= 3.48 \text{ in.}^2$$

$$A_{nv} = 2(L_{ev} + s - 1.5d_h)t_w$$
$$= 2[2.00 \text{ in.} + 4.00 \text{ in.} - 1.5(1.00 \text{ in.})](0.290 \text{ in.})$$
$$= 2.61 \text{ in.}^2$$

The nominal strength for the limit state of block shear rupture is given by:

$$R_n = 0.60F_u A_{nv} + U_{bs} F_u A_{nt} \leq 0.60F_y A_{gv} + U_{bs} F_u A_{nt} \quad \text{(Spec. Eq. J4-5)}$$

AISC *Specification* Equation J4-5 can be expressed as follows to determine the available strength of the brace web for the limit state of block shear rupture:

LRFD	ASD
$\phi R_n = \phi U_{bs} F_u A_{nt}$ $+ \min(\phi 0.60 F_y A_{gv},\ \phi 0.60 F_u A_{nv})$	$\dfrac{R_n}{\Omega} = \dfrac{U_{bs} F_u A_{nt}}{2.00}$ $+ \min\left(\dfrac{0.60 F_y A_{gv}}{2.00},\ \dfrac{0.60 F_u A_{nv}}{2.00}\right)$
$\phi U_{bs} F_u A_{nt} = 0.75(1.0)(65\text{ ksi})$ $\times (0.979 \text{ in.}^2)$ $= 47.7 \text{ kips}$ $\phi 0.60 F_y A_{gv} = 0.75(0.60)(50 \text{ ksi})$ $\times (3.48 \text{ in.}^2)$ $= 78.3 \text{ kips}$ $\phi 0.60 F_u A_{nv} = 0.75(0.60)(65 \text{ ksi})$ $\times (2.61 \text{ in.}^2)$ $= 76.3 \text{ kips}$ $\phi R_n = 47.7 \text{ kips} + 76.3 \text{ kips}$ $= 124 \text{ kips} > 40.2 \text{ kips} \quad \textbf{o.k.}$	$\dfrac{U_{bs} F_u A_{nt}}{2.00} = \dfrac{(1.0)(65 \text{ ksi})(0.979 \text{ in.}^2)}{2.00}$ $= 31.8 \text{ kips}$ $\dfrac{0.60 F_y A_{gv}}{2.00} = \dfrac{(0.60)(50 \text{ ksi})(3.48 \text{ in.}^2)}{2.00}$ $= 52.2 \text{ kips}$ $\dfrac{0.60 F_u A_{nv}}{2.00} = \dfrac{(0.60)(65 \text{ ksi})(2.61 \text{ in.}^2)}{2.00}$ $= 50.9 \text{ kips}$ $\dfrac{R_n}{\Omega} = 31.8 \text{ kips} + 50.9 \text{ kips}$ $= 82.7 \text{ kips} > 28.3 \text{ kips} \quad \textbf{o.k.}$

Check block shear rupture of the gusset plate

With an assumed gusset thickness, $t_g = \frac{3}{8}$ in., and standard holes in the gusset:

$d_h = {}^{13}\!/_{16}$ in. $+ {}^{1}\!/_{16}$ in.
$\quad\ = 0.875$ in.

$U_{bs} = 1.0$

$A_{nt} = (2g + t_w - d_h) t_g$
$\quad\ \ = [2(2.00 \text{ in.}) + 0.290 \text{ in.} - 0.875 \text{ in.}](\frac{3}{8} \text{ in.})$
$\quad\ \ = 1.28 \text{ in.}^2$

$L_{ev} = 2.00$ in.

$s \quad = 4.00$ in.

5.2 ORDINARY CONCENTRICALLY BRACED FRAMES

$$A_{nv} = 2(L_{ev} + s - 1.5d_h)t_g$$
$$= 2[2.00 \text{ in.} + 4.00 \text{ in.} - 1.5(0.875 \text{ in.})](\text{⅜ in.})$$
$$= 3.52 \text{ in.}^2$$

$$A_{gv} = 2(L_{ev} + s)t_g$$
$$= 2(2.00 \text{ in.} + 4.00 \text{ in.})(\text{⅜ in.})$$
$$= 4.50 \text{ in.}^2$$

The nominal strength for the limit state of block shear rupture is given by:

$$R_n = 0.60F_u A_{nv} + U_{bs}F_u A_{nt} \leq 0.60F_y A_{gv} + U_{bs}F_u A_{nt} \qquad (Spec. \text{ Eq. J4-5})$$

AISC *Specification* Equation J4-5 can be expressed as follows to determine the available strength of the gusset plate for the limit state of block shear rupture:

LRFD	ASD
$\phi R_n = \phi U_{bs}F_u A_{nt}$ $+ \min(\phi 0.60F_y A_{gv}, \phi 0.60F_u A_{nv})$	$\dfrac{R_n}{\Omega} = \dfrac{U_{bs}F_u A_{nt}}{2.00}$ $+ \min\left(\dfrac{0.60F_y A_{gv}}{2.00}, \dfrac{0.60F_u A_{nv}}{2.00}\right)$
$\phi U_{bs}F_u A_{nt} = 0.75(1.0)(58 \text{ ksi})$ $\times (1.28 \text{ in.}^2)$ $= 55.7 \text{ kips}$ $\phi 0.60F_y A_{gv} = 0.75(0.60)(36 \text{ ksi})$ $\times (4.50 \text{ in.}^2)$ $= 72.9 \text{ kips}$ $\phi 0.60F_u A_{nv} = 0.75(0.60)(58 \text{ ksi})$ $\times (3.52 \text{ in.}^2)$ $= 91.9 \text{ kips}$ $\phi R_n = 55.7 \text{ kips} + 72.9 \text{ kips}$ $= 129 \text{ kips} > 40.2 \text{ kips}$ **o.k.**	$\dfrac{U_{bs}F_u A_{nt}}{2.00} = \dfrac{(1.0)(58 \text{ ksi})(1.28 \text{ in.}^2)}{2.00}$ $= 37.1 \text{ kips}$ $\dfrac{0.60F_y A_{gv}}{2.00} = \dfrac{(0.60)(36 \text{ ksi})(4.50 \text{ in.}^2)}{2.00}$ $= 48.6 \text{ kips}$ $\dfrac{0.60F_u A_{nv}}{2.00} = \dfrac{(0.60)(58 \text{ ksi})(3.52 \text{ in.}^2)}{2.00}$ $= 61.2 \text{ kips}$ $\dfrac{R_n}{\Omega} = 37.1 \text{ kips} + 48.6 \text{ kips}$ $= 85.7 \text{ kips} > 28.3 \text{ kips}$ **o.k.**

Check the gusset plate for buckling on the Whitmore section

The "Whitmore section" is discussed in AISC *Manual* Part 9 (Figure 9-1) and in Thornton and Lini (2011), and is shown for this example in Figure 5-5.

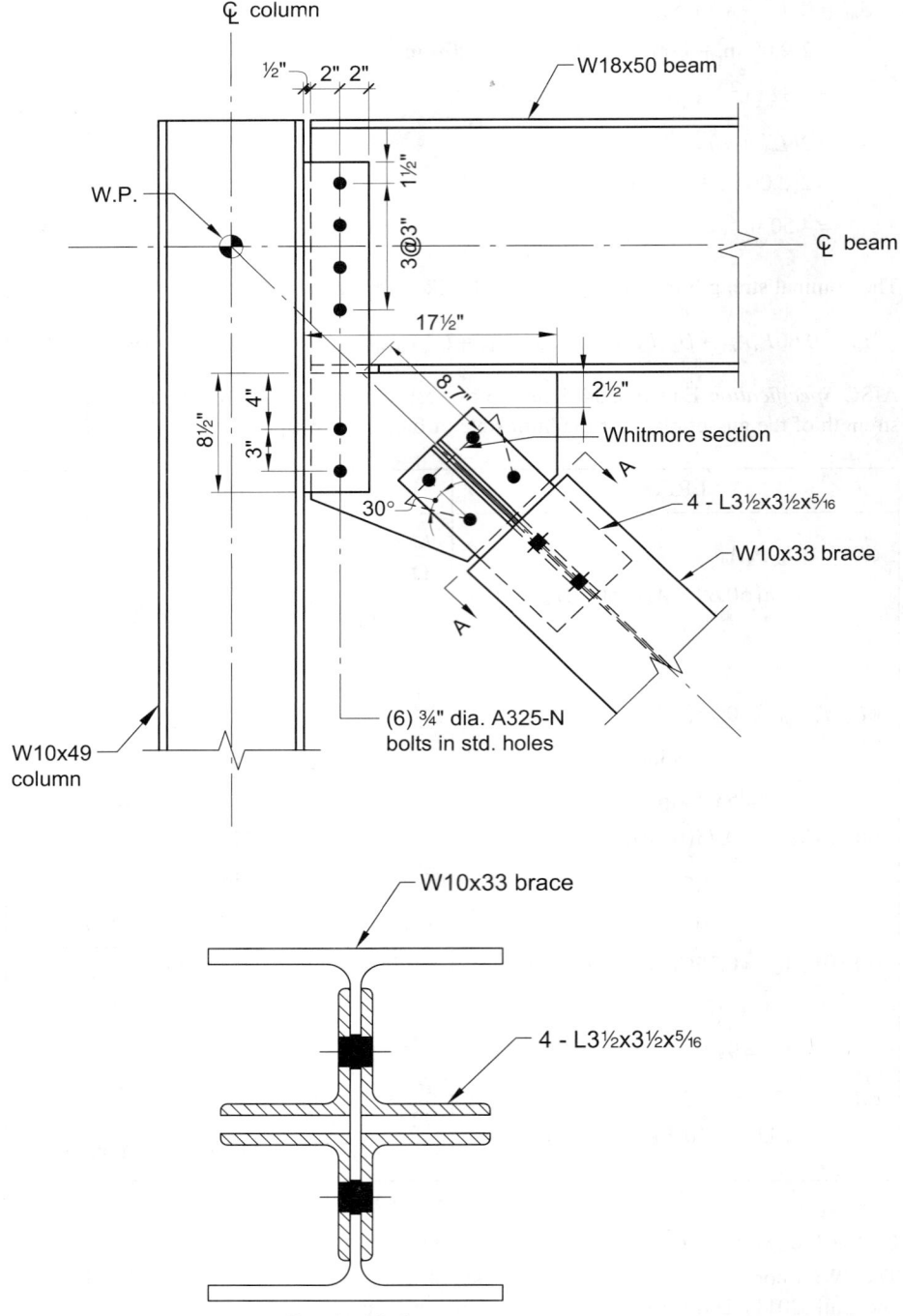

Fig. 5-5. Assumed initial geometry for Examples 5.2.1 through 5.2.4.

5.2 ORDINARY CONCENTRICALLY BRACED FRAMES

On the gusset plate, the space between the bolt lines of the angles is:

$$2g + t_w = 2(2.00 \text{ in.}) + 0.290 \text{ in.}$$
$$= 4.29 \text{ in.}$$

The Whitmore width is:

$$l_w = 2l \tan 30° + s$$
$$= 2(4.00 \text{ in.}) \tan 30° + 4.29 \text{ in.}$$
$$= 8.91 \text{ in.}$$

$$r = \frac{t_g}{\sqrt{12}}$$
$$= \frac{\text{⅜ in.}}{\sqrt{12}}$$
$$= 0.108 \text{ in.}$$

Use the effective length factor, K, of 0.50 as established by full scale tests on bracing connections (Gross, 1990) and as referenced in the AISC *Design Examples* V14.0. Note that this K value requires the gusset to be supported on both edges. Alternatively, the effective length factor for gusset buckling could be determined according to Dowswell (2006).

From Figure 5-5, the unbraced length of the gusset plate along the axis of the brace is $L = 8.70$ in. (Example 5.3.10 provides an equation for calculating the length of buckling; here it is determined graphically.)

$$\frac{KL}{r} = \frac{0.50(8.70 \text{ in.})}{0.108 \text{ in.}}$$
$$= 40.3$$

From AISC *Manual* Table 4-22 with $F_y = 36$ ksi and $\frac{KL}{r} = 40.3$:

LRFD	ASD
$\phi_c F_{cr} = 29.8$ ksi	$\dfrac{F_{cr}}{\Omega_c} = 19.8$ ksi

Therefore, from AISC *Specification* Equation E3-1, the available compressive strength based on flexural buckling is:

LRFD	ASD
$\phi P_n = \phi_c F_{cr} A_g$ $= 29.8 \text{ ksi}(8.91 \text{ in.})(\text{⅜ in.})$ $= 99.6 \text{ kips} > 53.2 \text{ kips}$ **o.k.**	$\dfrac{P_n}{\Omega} = \left(\dfrac{F_{cr}}{\Omega_c}\right) A_g$ $= 19.8 \text{ ksi}(8.91 \text{ in.})(\text{⅜ in.})$ $= 66.2 \text{ kips} > 37.2 \text{ kips}$ **o.k.**

Note: An alternative is to use a reduced unbraced buckling length for the gusset calculated from the average values from the end and center of the Whitmore section. See AISC *Design Examples* V14.0, Example II.C-2 (AISC, 2011).

Because the absolute value of the required strength of the connection in tension is less than the required strength of the connection in compression, tension yielding on the Whitmore section will not control.

Check bolt bearing on the angles

Standard holes are used in the angles. From AISC *Specification* Table J3.3, for a ¾-in.-diameter bolt, $d_h = {}^{13}/_{16}$ in.

The bearing strength requirements per bolt are given by AISC *Specification* Section J3.10.

For the interior bolt with a bolt spacing of 4.00 in., the bearing strength per bolt is:

$$r_n = 1.2 l_c t F_u \leq 2.4 d t F_u \qquad (Spec.\ Eq.\ J3\text{-}6a)$$

$$1.2 l_c t F_u = 1.2(4.00 \text{ in.} - {}^{13}/_{16} \text{ in.})({}^{5}/_{16} \text{ in.})(58 \text{ ksi})$$
$$= 69.3 \text{ kips}$$

$$2.4 d t F_u = 2.4({}^{3}/_{4} \text{ in.})({}^{5}/_{16} \text{ in.})(58 \text{ ksi})$$
$$= 32.6 \text{ kips}$$

Therefore, nominal bearing strength for the interior bolt is $r_n = 32.6$ kips. The available bearing strength of the interior bolt is:

LRFD	ASD
$\phi r_n = 0.75(32.6 \text{ kips})$ $= 24.5 \text{ kips}$	$\dfrac{r_n}{\Omega} = \dfrac{32.6 \text{ kips}}{2.00}$ $= 16.3 \text{ kips}$

Note that AISC *Manual* Table 7-4 could also have been used; however, it is based on smaller bolt spacing than 4.00 in.

For the end bolt, with $L_e = 1.50$ in., the nominal bearing strength per bolt for each angle is:

$$r_n = 1.2 l_c t F_u \leq 2.4 d t F_u \qquad (Spec.\ Eq.\ J3\text{-}6a)$$

$$= 1.2\left[1.50 \text{ in.} - \tfrac{1}{2}({}^{13}/_{16} \text{ in.})\right]({}^{5}/_{16} \text{ in.})(58 \text{ ksi})$$

$$\leq 2.4({}^{3}/_{4} \text{ in.})({}^{5}/_{16} \text{ in.})(58 \text{ ksi})$$

$$= 23.8 \text{ kips} \leq 32.6 \text{ kips}$$

The nominal bearing strength for the end bolt is $r_n = 23.8$ kips. The available bearing strength of the end bolt is:

LRFD	ASD
$\phi r_n = 0.75(23.8 \text{ kips})$ $= 17.9 \text{ kips}$	$\dfrac{r_n}{\Omega} = \dfrac{23.8 \text{ kips}}{2.00}$ $= 11.9 \text{ kips}$

Considering four angles, each with one end bolt and one interior bolt:

LRFD	ASD
$\phi R_n = 4 \begin{bmatrix} 1 \text{ bolt}(24.5 \text{ kips}) \\ + 1 \text{ bolt}(17.9 \text{ kips}) \end{bmatrix}$ $= 170 \text{ kips} > 53.2 \text{ kips}$ **o.k.**	$\dfrac{R_n}{\Omega} = 4 \begin{bmatrix} 1 \text{ bolt}(16.3 \text{ kips}) \\ + 1 \text{ bolt}(11.9 \text{ kips}) \end{bmatrix}$ $= 113 \text{ kips} > 37.2 \text{ kips}$ **o.k.**

Check bolt bearing on brace web

Oversized holes are used in the brace. From AISC *Specification* Table J3.3, for a ¾-in.-diameter bolt, $d_h = {}^{15}\!/_{16}$ in.

For the interior bolt with a bolt spacing of 4.00 in., the bearing strength per bolt is:

$$r_n = 1.2 L_c t F_u < 2.4 dt F_u \qquad (Spec.\text{ Eq. J3-6a})$$

$$1.2 L_c t F_u = 1.2(4.00 \text{ in.} - {}^{15}\!/_{16} \text{ in.})(0.290 \text{ in.})(65 \text{ ksi})$$
$$= 69.3 \text{ kips}$$

$$2.4 dt F_u = 2.4(¾ \text{ in.})(0.290 \text{ in.})(65 \text{ ksi})$$
$$= 33.9 \text{ kips}$$

Therefore, the nominal bearing strength of the interior bolts is 33.9 kips. The available bearing strength of the interior bolts is:

LRFD	ASD
$\phi r_n = 0.75(33.9 \text{ kips})$ $= 25.4 \text{ kips}$	$\dfrac{r_n}{\Omega} = \dfrac{33.9 \text{ kips}}{2.00}$ $= 17.0 \text{ kips}$

Note that AISC *Manual* Table 7-4 could have been used, but the table is based on smaller bolt spacing than the 4.00 in. used in this example.

Use AISC *Manual* Table 7-5 for the end bolts. For $L_e = 2.00$ in., the bearing strength per end bolt is:

LRFD	ASD
$\phi r_n = 87.8$ kip/in.	$\dfrac{r_n}{\Omega} = 58.5$ kip/in.

The available strength of the end bolt is:

LRFD	ASD
$\phi r_n = 87.8$ kip/in.$(0.290$ in.$)$ $= 25.5$ kips	$\dfrac{r_n}{\Omega} = 58.5$ kip/in.$(0.290$ in.$)$ $= 17.0$ kips

Considering two interior bolts and two end bolts on the brace:

LRFD	ASD
$\phi R_n = \begin{bmatrix} 2 \text{ bolts}(25.4 \text{ kips}) \\ + 2 \text{ bolts}(25.5 \text{ kips}) \end{bmatrix}$ $= 102$ kips > 53.2 kips **o.k.**	$\dfrac{R_n}{\Omega} = \begin{bmatrix} 2 \text{ bolts}(17.0 \text{ kips}) \\ + 2 \text{ bolts}(17.0 \text{ kips}) \end{bmatrix}$ $= 68.0$ kips > 37.2 kips **o.k.**

Check bolt bearing on the gusset

Standard holes are used in the gusset. From AISC *Specification* Table J3.3, for a ¾-in.-diameter bolt, $d_h = {}^{13}\!/_{16}$ in.

For the interior bolt with a bolt spacing of 4.00 in., the bearing strength per bolt is:

$r_n = 1.2 L_c t F_u < 2.4 d t F_u$ (*Spec.* Eq. J3-6a)

$1.2 L_c t F_u = 1.2(4.00 \text{ in.} - {}^{13}\!/_{16} \text{ in.})(\text{⅜ in.})(58 \text{ ksi})$
$\qquad\qquad = 83.2$ kips

$2.4 d t F_u = 2.4(\text{¾ in.})(\text{⅜ in.})(58 \text{ ksi})$
$\qquad\qquad = 39.2$ kips

Therefore, the nominal bearing strength of the interior bolt is 39.2 kips. The available bearing strength of the interior bolt is:

LRFD	ASD
$\phi r_n = 0.75(39.2 \text{ kips})$ $= 29.4$ kips	$\dfrac{r_n}{\Omega} = \dfrac{39.2 \text{ kips}}{2.00}$ $= 19.6$ kips

5.2 ORDINARY CONCENTRICALLY BRACED FRAMES

Note that AISC *Manual* Table 7-4 could also have been used. However, it is based on smaller bolt spacing than 4.00 in.

Use AISC *Manual* Table 7-5 for end bolts. For $L_e = 2.00$ in., the bearing strength per end bolt is:

LRFD	ASD
$\phi r_n = 78.3$ kip/in.	$\dfrac{r_n}{\Omega} = 52.2$ kip/in.

The available strength of the end bolt is:

LRFD	ASD
$\phi r_n = 78.3$ kip/in.($\tfrac{3}{8}$ in.) $= 29.4$ kips	$\dfrac{r_n}{\Omega} = 52.2$ kip/in.($\tfrac{3}{8}$ in.) $= 19.6$ kips

Considering two end bolts and two interior bolts:

LRFD	ASD
$\phi R_n = \begin{bmatrix} 2\text{ bolts}(29.4\text{ kips}) \\ + 2\text{ bolts}(29.4\text{ kips}) \end{bmatrix}$ $= 118$ kips > 53.2 kips **o.k.**	$\dfrac{R_n}{\Omega} = \begin{bmatrix} 2\text{ bolts}(19.6\text{ kips}) \\ + 2\text{ bolts}(19.6\text{ kips}) \end{bmatrix}$ $= 78.4$ kips > 37.2 kips **o.k.**

Use (4) ASTM A325-SC bolts in double shear to connect (4) L3½×3½×5/16 to the brace web. Use standard holes in the angles and gusset, and oversized holes in the brace web. Use (4) ASTM A325-N bolts in double shear to connect the (4) L3½×3½×5/16 to the gusset.

Connection Interface Forces

The forces at the gusset-to-beam and gusset-to-column interfaces are determined using the Uniform Force Method. The planes of uniform forces will be set at the column bolt line and the gusset/beam interface. The assumption of a plane of uniform force at the column bolt line allows the bolts at the column connection to be designed for shear and axial load only (no eccentricity) and therefore simplifies the design.

It should be noted that this assumption is different than that made for the typical cases of the Uniform Force Method discussed in the AISC *Manual* where the uniform force at the column is at the face of the column flange. Appropriate work points and uniform force planes can often be selected conveniently to balance engineering, fabrication and erection economy.

As previously determined, the maximum brace force according to ASCE/SEI 7 load combinations is 53.2 kips (LRFD) or 37.2 kips (ASD) acting in compression. The maximum brace force in tension is 40.2 kips (LRFD) or 28.3 kips (ASD). Consider the larger compression force to act in both directions in order to simplify calculations.

Assume an initial connection geometry as shown in Figure 5-5. Using the analysis found in AISC *Manual* Part 13:

$$e_b = \frac{d_b}{2}$$
$$= \frac{18.0 \text{ in.}}{2}$$
$$= 9.00 \text{ in.}$$

$$e_c = \frac{d_c}{2} + 2.50 \text{ in.}$$
$$= \frac{10.0 \text{ in.}}{2} + 2.50 \text{ in.}$$
$$= 7.50 \text{ in.}$$

Set β as the distance from the bottom of the beam to the center of the two bolts connecting the single plate to the gusset.

$$\beta = 5.50 \text{ in.}$$

Use a shared single-plate connection to connect the beam and gusset to the column. Therefore, the bottom flange of the beam must be either coped or blocked flush to clear the single-plate shear connection. Consider no weld between the gusset and the beam for 5 in. to allow for a 4½- in.-wide single plate with a ½-in. clearance between the plate and the start of the blocked beam flange. Assume a 17.0-in.-long gusset with a ½-in. clearance to the column flange. Consider the gusset-to-beam weld length as 12.5 in. Because the bolt line is used as the plane of uniform force, the distance to the center of the gusset-to-beam weld, $\bar{\alpha}$, must be set from the bolt line.

$$\bar{\alpha} = \frac{12.5 \text{ in.}}{2} + 4.50 \text{ in.} + 0.500 \text{ in.} - 2.50 \text{ in.}$$
$$= 8.75 \text{ in.}$$

Note: Alternatively, where the beam flange is blocked flush to lap the shear tab, the gusset could be welded to the beam with a one-sided fillet weld on the far side of the gusset, and a flush partial-joint-penetration groove weld on the near side. This would allow the full length of the gusset along the beam to be included in the design at this interface.

Setting $\beta = \bar{\beta}$, the value of α required for the uniform forces is:

$$\alpha = e_b \tan\theta - e_c + \beta \tan\theta \qquad \text{(from \textit{Manual} Eq. 13-1)}$$
$$= 9.00 \text{ in.} (\tan 45°) - 7.50 \text{ in.} + 5.50 \text{ in.} (\tan 45°)$$
$$= 7.00 \text{ in.}$$

Because the α required for uniform forces does not equal $\bar{\alpha}$ based on this initial geometry, uniform forces at the interfaces are not possible with the current configuration. The connection geometry can be adjusted by an iterative process to achieve the uniform distribution. Alternatively, the connection can be analyzed with an additional moment per

5.2 ORDINARY CONCENTRICALLY BRACED FRAMES

the method described as "Analysis of Existing Diagonal Bracing Connections" in Part 13 of the AISC *Manual*.

Because the gusset-to-beam connection is more rigid than the gusset-to-column connection, the beam can be assumed to resist the moment generated by the eccentricity between the actual gusset connection centroids and the ideal centroids calculated using the Uniform Force Method.

Using $\alpha = 7.00$ in. and $\beta = 5.50$ in.:

$$r = \sqrt{(\alpha + e_c)^2 + (\beta + e_b)^2} \qquad \text{(Manual Eq. 13-6)}$$

$$= \sqrt{(7.00 \text{ in.} + 7.50 \text{ in.})^2 + (5.50 \text{ in.} + 9.00 \text{ in.})^2}$$

$$= 20.5 \text{ in.}$$

The required shear force at the gusset-to-column connection is determined as:

$$V_c = \frac{\beta}{r} P \qquad \text{(Manual Eq. 13-2)}$$

LRFD	ASD
$V_{uc} = \dfrac{\beta}{r} P_u$	$V_{ac} = \dfrac{\beta}{r} P_a$
$= \dfrac{5.50 \text{ in.}}{20.5 \text{ in.}}(53.2 \text{ kips})$	$= \dfrac{5.50 \text{ in.}}{20.5 \text{ in.}}(37.2 \text{ kips})$
$= 14.3$ kips	$= 9.98$ kips

The required axial force at the gusset-to-column connection is determined as:

$$H_c = \frac{e_c}{r} P \qquad \text{(Manual Eq. 13-3)}$$

LRFD	ASD
$H_{uc} = \dfrac{e_c}{r} P_u$	$H_{ac} = \dfrac{e_c}{r} P_a$
$= \dfrac{7.50 \text{ in.}}{20.5 \text{ in.}}(53.2 \text{ kips})$	$= \dfrac{7.50 \text{ in.}}{20.5 \text{ in.}}(37.2 \text{ kips})$
$= 19.5$ kips	$= 13.6$ kips

The required shear force at the gusset-to-beam connection is determined as:

$$H_b = \frac{\alpha}{r} P \qquad \text{(Manual Eq. 13-5)}$$

LRFD	ASD
$H_{ub} = \dfrac{\alpha}{r} P_u$ $= \dfrac{7.00 \text{ in.}}{20.5 \text{ in.}}(53.2 \text{ kips})$ $= 18.2 \text{ kips}$	$H_{ab} = \dfrac{\alpha}{r} P_a$ $= \dfrac{7.00 \text{ in.}}{20.5 \text{ in.}}(37.2 \text{ kips})$ $= 12.7 \text{ kips}$

The required axial force at the gusset-to-beam connection is determined as:

$$V_b = \frac{e_b}{r} P \qquad \text{(Manual Eq. 13-4)}$$

LRFD	ASD
$V_{ub} = \dfrac{e_b}{r} P_u$ $= \dfrac{9.00 \text{ in.}}{20.5 \text{ in.}}(53.2 \text{ kips})$ $= 23.4 \text{ kips}$	$V_{ab} = \dfrac{e_b}{r} P_a$ $= \dfrac{9.00 \text{ in.}}{20.5 \text{ in.}}(37.2 \text{ kips})$ $= 16.3 \text{ kips}$

The moment at the gusset-to-beam interface is:

$$M_b = V_b |\alpha - \bar{\alpha}| \qquad \text{(Manual Eq. 13-17)}$$

LRFD	ASD
$M_{ub} = V_{ub} |\alpha - \bar{\alpha}|$ $= 23.4 \text{ kips}|7.00 \text{ in.} - 8.75 \text{ in.}|$ $= 41.0 \text{ kip-in.}$	$M_{ab} = V_{ab} |\alpha - \bar{\alpha}|$ $= 16.3 \text{ kips}|7.00 \text{ in.} - 8.75 \text{ in.}|$ $= 28.5 \text{ kip-in.}$

These forces are illustrated symbolically in Figure 5-6.

Gusset-to-Beam Connection

Design gusset-to-beam weld

The gusset-to-beam weld will be determined by applying the Elastic Method discussed in AISC *Manual* Part 8.

To accommodate the bottom flange block, which extends ½ in. past the single plate, the maximum length of weld along the gusset-to-beam interface is:

$l_{wb} = 17.0 \text{ in.} + 0.500 \text{ in.} - 4.50 \text{ in.} - 0.500 \text{ in.}$
$= 12.5 \text{ in.}$

5.2 ORDINARY CONCENTRICALLY BRACED FRAMES

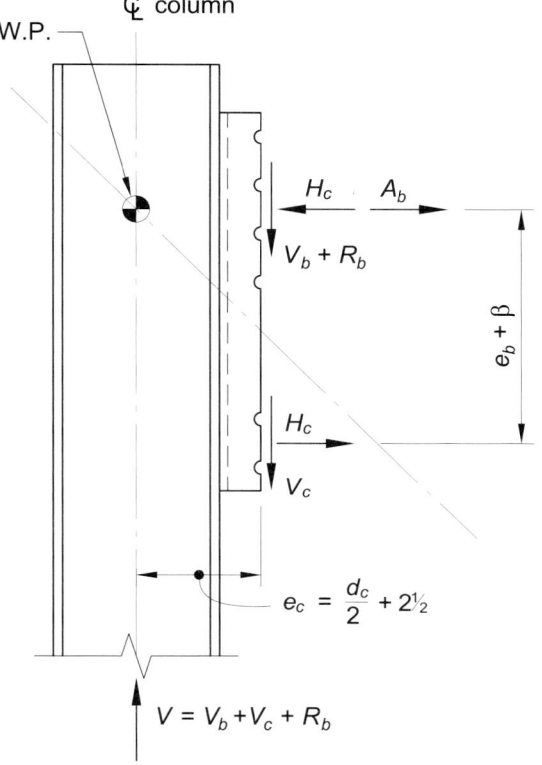

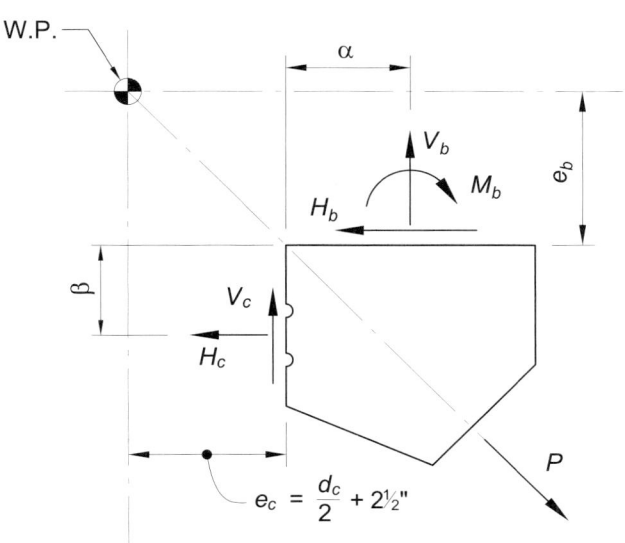

Fig. 5-6. Free-body diagrams for Example 5.2.4.

Treating the weld as a line:

$$S_w = \frac{(12.5 \text{ in.})^2}{6}$$

$$= 26.0 \text{ in.}^3/\text{in.}$$

The shear force, axial force, and force due to flexure per linear inch of weld are:

LRFD	ASD
$f_{uv} = \dfrac{H_{ub}}{l_{wb}}$ $= \dfrac{18.2 \text{ kips}}{12.5 \text{ in.}}$ $= 1.46 \text{ kip/in.}$	$f_{av} = \dfrac{H_{ab}}{l_{wb}}$ $= \dfrac{12.7 \text{ kips}}{12.5 \text{ in.}}$ $= 1.02 \text{ kip/in.}$
$f_{ua} = \dfrac{V_{ub}}{l_{wb}}$ $= \dfrac{23.4 \text{ kips}}{12.5 \text{ in.}}$ $= 1.87 \text{ kip/in.}$	$f_{aa} = \dfrac{V_{ab}}{l_{wb}}$ $= \dfrac{16.3 \text{ kips}}{12.5 \text{ in.}}$ $= 1.30 \text{ kip/in.}$
$f_{ub} = \dfrac{M_{ub}}{S_w}$ $= \dfrac{41.0 \text{ kip-in.}}{26.0 \text{ in.}^3/\text{in.}}$ $= 1.58 \text{ kip/in.}$	$f_{ab} = \dfrac{M_{ab}}{S_w}$ $= \dfrac{28.5 \text{ kip-in.}}{26.0 \text{ in.}^3/\text{in.}}$ $= 1.10 \text{ kip/in.}$

The force on the weld due to bending is determined using elastic section properties as per the Elastic Method indicated in Part 8 of the AISC *Manual*. Generally this method is considered conservative. The Instantaneous Center of Rotation Method, also found in Part 8, often results in smaller required weld sizes for welds subject to eccentricity. In the examples in this Manual employing special concentrically braced frame connections, a plastic stress distribution is used to determine the forces at the beam-to-gusset interface. This example will use the more conservative elastic method.

Use a vector sum (square root of the sum of the squares) to combine the shear, axial and bending stresses on the gusset-to-beam interface. Because the bending stress acts in opposite directions over each half of the length, this creates both a maximum (peak) and a minimum stress. The average stress is determined based on the maximum (peak) stress and the minimum stress. All stress units below are in kip/in.

5.2 ORDINARY CONCENTRICALLY BRACED FRAMES

LRFD	ASD
$f_{u,peak} = \sqrt{(f_{ua} + f_{ub})^2 + f_{uv}^2}$ $= \sqrt{(1.87 + 1.58)^2 + (1.46)^2}$ $= 3.75 \text{ kip/in.}$	$f_{a,peak} = \sqrt{(f_{aa} + f_{ab})^2 + f_{av}^2}$ $= \sqrt{(1.30 + 1.10)^2 + (1.02)^2}$ $= 2.61 \text{ kip/in.}$
$f_{u,avg} = \frac{1}{2}\left[\sqrt{(f_{ua} - f_{ub})^2 + f_{uv}^2} + \sqrt{(f_{ua} + f_{ub})^2 + f_{uv}^2}\right]$ $= \frac{1}{2}\left[\sqrt{(1.87 - 1.58)^2 + (1.46)^2} + \sqrt{(1.87 + 1.58)^2 + (1.46)^2}\right]$ $= 2.62 \text{ kip/in.}$	$f_{a,avg} = \frac{1}{2}\left[\sqrt{(f_{aa} - f_{ab})^2 + f_{av}^2} + \sqrt{(f_{aa} + f_{ab})^2 + f_{av}^2}\right]$ $= \frac{1}{2}\left[\sqrt{(1.30 - 1.10)^2 + (1.02)^2} + \sqrt{(1.30 + 1.10)^2 + (1.02)^2}\right]$ $= 1.82 \text{ kip/in.}$
$\dfrac{f_{u,peak}}{f_{u,avg}} = \dfrac{3.75 \text{ kip/in.}}{2.62 \text{ kip/in.}}$ $= 1.43$	$\dfrac{f_{a,peak}}{f_{a,avg}} = \dfrac{2.61 \text{ kip/in.}}{1.82 \text{ kip/in.}}$ $= 1.43$
$\theta = \tan^{-1}\left(\dfrac{f_{ua} + f_{ub}}{f_{uv}}\right)$ $= \tan^{-1}\left(\dfrac{1.87 \text{ kip/in.} + 1.58 \text{ kip/in.}}{1.46 \text{ kip/in.}}\right)$ $= 67.1°$	$\theta = \tan^{-1}\left(\dfrac{f_{aa} + f_{ab}}{f_{av}}\right)$ $= \tan^{-1}\left(\dfrac{1.30 \text{ kip/in.} + 1.10 \text{ kip/in.}}{1.02 \text{ kip/in.}}\right)$ $= 67.0°$

According to the AISC *Manual* Part 13, because the gusset is directly welded to the beam, the weld is designed for the larger of the peak stress and 1.25 times the average stress. Because $f_{peak}/f_{avg} > 1.25$, the weld ductility factor need not be applied. For a discussion of the weld ductility factor of 1.25, see AISC *Manual* Part 13.

Therefore, $f_r = f_{peak} = 3.75$ kip/in. (LRFD) and 2.61 kip/in. (ASD).

The strength of fillet welds defined in AISC *Specification* Section J2 can be simplified, as explained in Part 8 of the AISC *Manual*, to AISC *Manual* Equations 8-2a and 8-2b:

LRFD	ASD
$\phi R_n = (1.392 \text{ kip/in.}) Dl$	$\dfrac{R_n}{\Omega} = (0.928 \text{ kip/in.}) Dl$

The required weld size at the gusset-to-beam interface is:

LRFD	ASD
$D \geq \dfrac{f_{u,\,peak}}{2(1.392 \text{ kip/in.})\left(1+0.50\sin^{1.5}\theta\right)}$	$D \geq \dfrac{f_{a,\,peak}}{2(0.928 \text{ kip/in.})\left(1+0.50\sin^{1.5}\theta\right)}$
$\geq \dfrac{3.75 \text{ kip/in.}}{2(1.392 \text{ kip/in.})\left[1+0.50\sin^{1.5}(67.1°)\right]}$	$\geq \dfrac{2.61 \text{ kip/in.}}{2(0.928 \text{ kip/in.})\left[1+0.50\sin^{1.5}(67.0°)\right]}$
$= 0.934$ sixteenths	$= 0.975$ sixteenths

From AISC *Specification* Table J2.4, the minimum size fillet weld allowed for the parts being connected is 3/16 in.

Use two-sided 3/16-in. fillet welds to connect the gusset plate to the beam.

Check gusset plate rupture at beam weld

A conservative method to determine the minimum gusset plate thickness required to transfer the shear and tension forces is to set the shear rupture strength of the weld (based on the resultant force) equal to the shear rupture strength of the gusset plate. Using AISC *Manual* Equation 9-3:

LRFD	ASD
$t_{min} = \dfrac{6.19D}{F_u}$	$t_{min} = \dfrac{6.19D}{F_u}$
$= \dfrac{6.19 \text{ kip/in.}(0.934)}{58 \text{ ksi}}$	$= \dfrac{6.19 \text{ kip/in.}(0.975)}{58 \text{ ksi}}$
$= 0.0997$ in.	$= 0.104$ in.
3/8 in. > 0.0977 in. **o.k.**	3/8 in. > 0.104 in. **o.k.**

Use a 3/8-in.-thick gusset plate to connect the brace to the beam and column.

Alternatively, the required thickness of the gusset plate could be determined by checking the strength of gusset plate directly.

Check gusset plate yielding at beam weld

It can be shown that since the gusset plate satisfies the minimum thickness criteria for rupture based on weld size, it also satisfies the tension yielding criteria.

Check beam web local yielding

The maximum stress per unit length on the gusset-to-beam interface along the weld due to moment M_b is $M_b/(l^2/4)$ assuming a plastic stress distribution. Conservatively neglecting the portion of this stress distribution that acts in the reverse direction, and considering the total force to be applied at the center of the bearing length, the resultant compressive force is:

LRFD	ASD
$R_u = V_{ub} + \dfrac{M_{ub}}{\left(\dfrac{l^2}{4}\right)}(l)$	$R_a = V_{ab} + \dfrac{M_{ab}}{\left(\dfrac{l^2}{4}\right)}(l)$
$= V_{ub} + 4\left(\dfrac{M_{ub}}{l}\right)$	$= V_{ab} + 4\left(\dfrac{M_{ab}}{l}\right)$
$= 23.4 \text{ kips} + 4\left(\dfrac{41.0 \text{ kip-in.}}{12.5 \text{ in.}}\right)$	$= 16.3 \text{ kips} + 4\left(\dfrac{28.5 \text{ kip-in.}}{12.5 \text{ in.}}\right)$
$= 36.5$ kips	$= 25.4$ kips

The beam is checked for the limit state of web local yielding due to the force from the plate welded to the beam flange.

The force is applied a distance α from the beam end. Because $\alpha < d_b = 18.0$ in, AISC *Specification* Equation J10-3 is applicable.

For a force applied at a distance less than the depth of the member:

$R_n = F_{yw}t_w(2.5k + l_b)$ (*Spec.* Eq. J10-3)

$= (50 \text{ ksi})(0.355 \text{ in.})[2.5(0.972 \text{ in.}) + 12.5 \text{ in.}]$

$= 265$ kips

LRFD	ASD
$\phi R_n = 1.00(265 \text{ kips})$ $= 265 \text{ kips} > 36.5 \text{ kips}$ **o.k.**	$\dfrac{R_n}{\Omega} = \dfrac{265 \text{ kips}}{1.50}$ $= 177 \text{ kips} > 25.4 \text{ kips}$ **o.k.**

Alternatively, the available strength for web yielding can be determined from AISC *Manual* Table 9-4.

Check beam web local crippling

A portion of the force is applied within $d/2$ of the member end; therefore, use AISC *Specification* Section J10.3(b). Check the length of bearing relative to the beam depth:

$\dfrac{l_b}{d} = \dfrac{12.5 \text{ in.}}{18.0 \text{ in.}}$

$= 0.694 > 0.2$

Therefore, use AISC *Specification* Equation J10-5b to determine the available strength, through use of AISC *Manual* Table 9-4.

From AISC *Manual* Table 9-4 for the **W18×50**:

LRFD	ASD
$\phi R_5 = 52.0$ kips	$\dfrac{R_5}{\Omega} = 34.7$ kips
$\phi R_6 = 6.30$ kip/in.	$\dfrac{R_6}{\Omega} = 4.20$ kip/in.

From AISC *Manual* Equation 9-48:

LRFD	ASD
$\phi R_n = \phi R_5 + l_b(\phi R_6)$	$\dfrac{R_n}{\Omega} = \dfrac{R_5}{\Omega} + l_b \dfrac{R_6}{\Omega}$
$= 52.0 \text{ kips} + 12.5 \text{ in.}(6.30 \text{ kip/in.})$	$= 34.7 \text{ kips} + 12.5 \text{ in.}(4.20 \text{ kip/in.})$
$= 131 \text{ kips} > 36.5 \text{ kips}$ **o.k.**	$= 87.2 \text{ kips} > 25.4 \text{ kips}$ **o.k.**

Beam and Gusset-to-Column Connection

Use a single-plate connection that combines the connections of the beam and gusset to the column. Design the bolted connections of the gusset to the single plate and of the beam to the single plate individually. Design the weld of the single plate to the column considering the combined plate length. The forces used to design the single-plate will be those derived per the Uniform Force Method. Additional forces beyond those calculated by this method may occur in the connection of the beam/gusset connection to the column due to the rotation of the beam relative to the column. While forces in the connections due to rotation from seismic drift are opposite the forces determined by the Uniform Force Method, the beam and gusset connection to the column will be designed following the single plate design philosophy in Part 10 of the AISC *Manual* to provide additional rotational ductility to address both rotation from seismic drift and simple-beam end rotation. The eccentricity on the single plate due to the braced frame shear is addressed by the Uniform Force Method which applies a couple based on the H_c axial forces applied at the center of the beam and the center of the gusset-to-column connection.

Design gusset-to-column bolted connection

The resultant force on the bolts in the gusset plate is:

5.2 ORDINARY CONCENTRICALLY BRACED FRAMES

LRFD	ASD
$R_u = \sqrt{V_{uc}^2 + H_{uc}^2}$ $= \sqrt{(14.3 \text{ kips})^2 + (19.5 \text{ kips})^2}$ $= 24.2$ kips	$R_a = \sqrt{V_{ac}^2 + H_{ac}^2}$ $= \sqrt{(9.98 \text{ kips})^2 + (13.6 \text{ kips})^2}$ $= 16.9$ kips

Try two bolts connecting the gusset to a single plate. The required shear strength per bolt is:

LRFD	ASD
$V_u = \dfrac{R_u}{2}$ $= \dfrac{24.2 \text{ kips}}{2}$ $= 12.1$ kips/bolt	$V_a = \dfrac{R_a}{2}$ $= \dfrac{16.9 \text{ kips}}{2}$ $= 8.45$ kips/bolt

From AISC *Manual* Table 7-1, the shear strength of a ¾-in.-diameter ASTM A325-N bolt in single shear is:

LRFD	ASD
$\phi r_n = 17.9$ kips > 12.1 kips **o.k.**	$\dfrac{r_n}{\Omega} = 11.9$ kips > 8.45 kips **o.k.**

From AISC *Manual* Table 7-4 with 3 in. bolt spacing, the bearing strength per inch of single-plate thickness is:

LRFD	ASD
$\phi r_n = 78.3$ kip/in.	$\dfrac{r_n}{\Omega} = 52.2$ kip/in.

Assume a ⁵⁄₁₆-in.-thick single plate.

For the interior bolt, the available bearing strength of the single plate is:

LRFD	ASD
$\phi r_n = 78.3$ kip/in.(⁵⁄₁₆ in.) $= 24.5$ kips/bolt	$\dfrac{r_n}{\Omega} = 52.2$ kip/in.(⁵⁄₁₆ in.) $= 16.3$ kips/bolt

The edge distances in the single plate are 1½ in. vertically and 2 in. horizontally. Conservatively, use the lesser of these edge distances. A more refined check would calculate the edge distance in the direction of the force. For the end bolt, with $L_v = 1½$ in., the nominal bearing strength per bolt is:

$$r_n = \frac{1.2 l_c F_u}{t} \leq \frac{2.4 d F_u}{t} \qquad \text{(from } Spec. \text{ Eq. J3-6a)}$$

$$\frac{1.2 l_c F_u}{t} = \frac{1.2 [1.50 \text{ in.} - ½(^{13}/_{16} \text{ in.})](58 \text{ ksi})}{t}$$

$$= 76.1 \text{ kip/in.}$$

$$\frac{2.4 d F_u}{t} = \frac{2.4(¾ \text{ in.})(58 \text{ ksi})}{t}$$

$$= 104 \text{ kip/in.}$$

Therefore, the nominal bearing strength of the end bolt is $r_n = 76.1$ kip/in. The available bearing strength of the end bolt is:

LRFD	ASD
$\phi r_n = 0.75(76.1 \text{ kip/in.})$ $= 57.1$ kip/in.	$\dfrac{r_n}{\Omega} = \dfrac{76.1 \text{ kip/in.}}{2.00}$ $= 38.1$ kip/in.

For the end bolt, the available bearing strength of the single plate is:

LRFD	ASD
$\phi r_n = 57.1$ kip/in.($^5/_{16}$ in.) $= 17.8$ kips/bolt > 12.1 kips/bolt **o.k.**	$\dfrac{r_n}{\Omega} = 38.1$ kip/in.($^5/_{16}$ in.) $= 11.9$ kips/bolt > 8.45 kips/bolt **o.k.**

The available bolt shear strength and the bearing strength for the end and interior bolts exceeds the required shear strength per bolt.

The gusset is ⅜-in.-thick and will have greater bearing strength than the $^5/_{16}$-in. single plate; therefore, the gusset plate is not checked for bearing strength.

Block shear rupture in the gusset-to-column single-plate connection

Check block shear relative to normal force on the single plate.

According to AISC *Specification* Section B4.3b, in computing net area for tension and shear, the width of a bolt hole is taken as $^1/_{16}$ in. larger than the nominal dimension of the hole. The nominal diameter of the hole from Table J3.3 is $^{13}/_{16}$ in.

The nominal strength for the limit state of block shear rupture relative to the normal force on the single plate is:

$$R_n = 0.60F_u A_{nv} + U_{bs} F_u A_{nt} \leq 0.60 F_y A_{gv} + U_{bs} F_u A_{nt} \qquad (Spec.\ Eq.\ J4\text{-}5)$$

where

d_h = 13/16 in. + 1/16 in.
   = 0.875 in.

$U_{bs} = 1.0$

$A_{nv} = 2(L_{eh} - 0.5 d_h) t_p$
   = 2[2.00 in. − 0.5(0.875 in.)](5/16 in.)
   = 0.977 in.^{2}

$A_{gv} = 2 L_{eh} t_p$
   = 2(2.00 in.)(5/16 in.)
   = 1.25 in.^{2}

$A_{nt} = (s - d_h) t_p$
   = (3.00 in. − 0.875 in.)(5/16 in.)
   = 0.664 in.^{2}

$0.60 F_u A_{nv} + U_{bs} F_u A_{nt}$ = 0.60(58 ksi)(0.977 in.^{2}) + 1.0(58 ksi)(0.664 in.^{2})
   = 72.5 kips

$0.60 F_y A_{gv} + U_{bs} F_u A_{nt}$ = 0.60(36 ksi)(1.25 in.^{2}) + 1.0(58 ksi)(0.664 in.^{2})
   = 65.5 kips

Therefore, the nominal strength for the limit state of block shear rupture is R_n = 65.5 kips. The available strength for the limit state of block shear rupture on the single plate is:

LRFD	ASD
$\phi R_n = 0.75(65.5\ \text{kips})$ = 49.1 kips > 19.5 kips **o.k.**	$\dfrac{R_n}{\Omega} = \dfrac{65.5\ \text{kips}}{2.00}$ = 32.8 kips > 13.6 kips **o.k.**

Check block shear relative to shear force on the single plate.

In this case, the AISC *Manual* tables will be used to determine the available strength for the limit state of block shear rupture, because the parameters fall within the limits of the tables provided. For the single plate at the gusset-to-column connection:

$n = 2$
$L_{ev} = 1\frac{1}{2}$ in.
$L_{eh} = 2.00$ in.
$U_{bs} = 1.0$

From AISC *Specification* Equation J4-5, the available strength for the limit state of block shear rupture can be written as:

LRFD	ASD
$\phi R_n = \phi U_{bs} F_u A_{nt}$ $+ \min(\phi 0.60 F_y A_{gv}, \phi 0.60 F_u A_{nv})$	$\dfrac{R_n}{\Omega} = \dfrac{U_{bs} F_u A_{nt}}{\Omega}$ $+ \min\left(\dfrac{0.60 F_y A_{gv}}{\Omega}, \dfrac{0.60 F_u A_{nv}}{\Omega}\right)$
Tension rupture component from AISC *Manual* Table 9-3a: $\phi U_{bs} F_u A_{nt} = 1.0(68.0 \text{ kip/in.})(5/16 \text{ in.})$ $= 21.3$ kips	Tension rupture component from AISC *Manual* Table 9-3a: $\dfrac{U_{bs} F_u A_{nt}}{\Omega} = 1.0(45.3 \text{ kip/in.})(5/16 \text{ in.})$ $= 14.2$ kips
Shear yielding component from AISC *Manual* Table 9-3b: $\phi 0.60 F_y A_{gv} = 72.9 \text{ kip/in}(5/16 \text{ in.})$ $= 22.8$ kips	Shear yielding component from AISC *Manual* Table 9-3b: $\dfrac{0.60 F_y A_{gv}}{\Omega} = (48.6 \text{ kip/in.})(5/16 \text{ in.})$ $= 15.2$ kips
Shear rupture component from AISC *Manual* Table 9-3c: $\phi 0.60 F_u A_{nv} = 83.2 \text{ kip/in.}(5/16 \text{ in.})$ $= 26.0$ kips	Shear rupture component from AISC *Manual* Table 9-3c: $\dfrac{0.60 F_u A_{nv}}{\Omega} = 55.5 \text{ kip/in.}(5/16 \text{ in.})$ $= 17.3$ kips
The available strength for the limit state of block shear rupture relative to the shear force on the single plate is: $\phi R_n = 21.3$ kips $\quad + \min(22.8 \text{ kips}, 26.0 \text{ kips})$ $\quad = 44.1$ kips > 14.3 kips **o.k.**	The available strength for the limit state of block shear rupture relative to the shear force on the single plate is: $\dfrac{R_n}{\Omega} = 14.2$ kips $\quad + \min(15.2 \text{ kips}, 17.3 \text{ kips})$ $\quad = 29.4$ kips > 9.98 kips **o.k.**

Combined shear and normal block shear design check using an elliptical equation

For the single plate at the gusset-to-column connection, the interaction of shear and normal block shear is considered as follows:

LRFD	ASD
$\left(\dfrac{V_r}{V_c}\right)^2 + \left(\dfrac{P_r}{P_c}\right)^2 \leq 1.0$	$\left(\dfrac{V_r}{V_c}\right)^2 + \left(\dfrac{P_r}{P_c}\right)^2 \leq 1.0$
$\left(\dfrac{14.3 \text{ kips}}{44.1 \text{ kips}}\right)^2 + \left(\dfrac{19.5 \text{ kips}}{49.1 \text{ kips}}\right)^2$	$\left(\dfrac{9.98 \text{ kips}}{29.4 \text{ kips}}\right)^2 + \left(\dfrac{13.6 \text{ kips}}{32.8 \text{ kips}}\right)^2$
$= 0.263 \leq 1.0$ **o.k.**	$= 0.287 \leq 1.0$ **o.k.**

Block shear rupture in the ⅜-in.-thick gusset plate is also adequate as the gusset is thicker than the single plate.

Tensile rupture in the gusset-to-column single plate

Conservatively consider only a 6.00-in. length of single plate under axial tension from the gusset. The nominal tensile rupture strength is:

$$R_n = F_u A_e \qquad \text{(Spec. Eq. J4-2)}$$

where

$d_h = 0.875$ in.

$U = 1.0$

$A_n = (l - 2d_h)t_p$
$= [6.00 \text{ in.} - 2(0.875 \text{ in.})](\text{⁵⁄₁₆ in.})$
$= 1.33 \text{ in.}^2$

$A_e = A_n U \qquad \text{(Spec. Eq. D3-1)}$
$= 1.33 \text{ in.}^2 (1.0)$
$= 1.33 \text{ in.}^2$

Therefore:

$R_n = 58 \text{ ksi}(1.33 \text{ in.}^2)$
$= 77.1$ kips

The available tensile rupture strength is:

LRFD	ASD
$\phi R_n = 0.75(77.1 \text{ kips})$ $= 57.8 \text{ kips} > 19.5 \text{ kips}$	$\dfrac{R_n}{\Omega} = \dfrac{77.1 \text{ kips}}{2.00}$ $= 38.6 \text{ kips} > 13.6 \text{ kips}$

Tensile rupture in the ⅜-in.-thick gusset is also okay because of its greater thickness.

Tensile yielding in the gusset-to-column single plate

Again, conservatively consider only a 6.00-in. length of single plate under axial tension from the gusset. The nominal tensile yielding strength is:

$$R_n = F_y A_g \qquad \text{(Spec. Eq. J4-1)}$$

where

$A_g = l t_p$
$= 6.00 \text{ in.}(\text{⁵⁄₁₆ in.})$
$= 1.88 \text{ in.}^2$

Therefore:

$R_n = 36 \text{ ksi}(1.88 \text{ in.}^2)$
$= 67.7 \text{ kips}$

The available tensile yielding strength is:

LRFD	ASD
$\phi R_n = 0.90(67.7 \text{ kips})$ $= 60.9 \text{ kips} > 19.5 \text{ kips}$ **o.k.**	$\dfrac{R_n}{\Omega} = \dfrac{67.7 \text{ kips}}{1.67}$ $= 40.5 \text{ kips} > 13.6 \text{ kips}$ **o.k.**

Tensile yielding in the ⅜-in.-thick gusset is also okay because of its greater thickness.

Shear rupture in the gusset-to-column single plate

Check the available shear rupture strength at the net section through the bolt line. Conservatively, consider only a 6.00-in. length of single plate.

$A_{nv} = (l - 2d_h)t_p$
$= [6.00 \text{ in.} - 2(0.875 \text{ in.})](\text{⁵⁄₁₆ in.})$
$= 1.33 \text{ in.}^2$

$R_n = 0.60 F_u A_{nv}$ (*Spec.* Eq. J4-4)

$ = 0.60(58 \text{ ksi})(1.33 \text{ in.}^2)$

$ = 46.3$ kips

LRFD	ASD
$\phi R_n = 0.75(46.3 \text{ kips})$ $= 34.7 \text{ kips} > 14.3 \text{ kips}$ **o.k.**	$\dfrac{R_n}{\Omega} = \dfrac{46.3 \text{ kips}}{2.00}$ $= 23.2 \text{ kips} > 9.98 \text{ kips}$ **o.k.**

Shear rupture in the ⅜-in.-thick gusset is also okay because of its greater thickness.

Shear yielding in the gusset-to-column single plate

Check the available shear yielding strength at the net section through the bolt line.

$A_{gv} = l t_p$

$\phantom{A_{gv}} = 6.00 \text{ in.}(\text{⁵⁄₁₆ in.})$

$\phantom{A_{gv}} = 1.88 \text{ in.}^2$

$R_n = 0.60 F_y A_{gv}$ (*Spec.* Eq. J4-3)

$ = 0.60(36 \text{ ksi})(1.88 \text{ in.}^2)$

$ = 40.6$ kips

LRFD	ASD
$\phi R_n = 1.00(40.6 \text{ kips})$ $= 40.6 \text{ kips} > 14.3 \text{ kips}$ **o.k.**	$\dfrac{R_n}{\Omega} = \dfrac{40.6 \text{ kips}}{1.50}$ $= 27.1 \text{ kips} > 9.98 \text{ kips}$ **o.k.**

Shear yielding in the ⅜-in.-thick gusset is also okay because of its greater thickness.

Use a ⁵⁄₁₆-in.-thick single plate with (2) ¾-in.-diameter ASTM A325-N bolts in standard holes to connect the ⅜-in.-thick gusset to the column.

Design the beam-to-column single plate connection

The beam-to-column joint transfers both vertical shear and horizontal force. The horizontal forces acting at the beam-to-column interface are the uniform force component, $H - H_b = H_c$, and the collector force, A_b. For this particular connection at this location in the structure, when the diagonal brace is in tension, the resultant horizontal force between the beam and the column is a compression force with a magnitude of H_c. However, when the diagonal brace is in compression, the collector force between the beam and the column will be in tension. Therefore, the collector and H_c forces act in opposite directions. Conservatively, use the greater of H_c and the collector force, A_b, for the design of the single plate.

LRFD	ASD
$P_u = \max \begin{Bmatrix} H - H_{ub} = H_{uc} \\ A_{ub} \end{Bmatrix}$	$P_a = \max \begin{Bmatrix} H - H_{ab} = H_{ac} \\ A_{ab} \end{Bmatrix}$
$= \max \begin{Bmatrix} 19.5 \text{ kips} \\ 33.0 \text{ kips} \end{Bmatrix}$	$= \max \begin{Bmatrix} 13.6 \text{ kips} \\ 23.1 \text{ kips} \end{Bmatrix}$
$= 33.0$ kips	$= 23.1$ kips

Note that the determination of the relative directions of the collector force and H_c forces at the column face may not always be as apparent as in this single-story structure. A conservative approach is to add the absolute values of the two components.

The vertical force on the beam web-to-column connection is, as shown in Figure 5-6:

LRFD	ASD
$R_{ub} = 11.8$ kips	$R_{ab} = 7.73$ kips
$V_{ub} = 23.4$ kips	$V_{ab} = 16.3$ kips
$V_u = R_{ub} + V_{ub}$	$V_a = R_{ab} + V_{ab}$
$= 11.8 \text{ kips} + 23.4 \text{ kips}$	$= 7.73 \text{ kips} + 16.3 \text{ kips}$
$= 35.2$ kips	$= 24.0$ kips

Note that the vertical shear force calculated above is conservative as the analysis has been simplified by considering the maximum brace force as equal in magnitude in either tension or compression. A more exact analysis would include the actual tension and compression forces combined with the respective beam reactions with consideration of the direction of loading of each force component. For this structure, the larger diagonal brace force which acts in compression, and its resultant V_b component which acts upwards, would be counteracted by the beam reaction acting downwards. So to remedy the shortfall of this simplification, the vertical force, V_u (LRFD) and V_a (ASD), could be calculated for both the maximum force due to compression in the brace with its concurrent reaction and the maximum reaction resulting from tension force in the brace with the vertical beam reaction.

For the case where the brace is in compression:

LRFD	ASD
$R_u = 11.8$ kips	$R_a = 7.73$ kips
$V_{ub} = -23.4$ kips	$V_{ab} = -16.3$ kips
$V_u = R_u + V_{ub}$	$V_a = R_a + V_{ab}$
$= 11.8 \text{ kips} + (-23.4 \text{ kips})$	$= 7.73 \text{ kips} + (-16.3 \text{ kips})$
$= -11.6$ kips	$= -8.57$ kips

5.2 ORDINARY CONCENTRICALLY BRACED FRAMES

For the case where the brace is in tension:

The maximum shear at the beam-to-column interface will occur when the diagonal brace is in tension based on ASCE/SEI 7 Section 12.4.3.2 Load Combination 5 (LRFD and ASD). The beam reaction, V_u or V_a, is the concurrent force.

LRFD	ASD
LRFD Load Combination 5 from ASCE/SEI 7 Section 12.4.3.2 (including the 0.5 factor on L permitted by Section 12.4.3.2)	ASD Load Combination 5 from ASCE/SEI 7 Section 12.4.3.2
$(1.2 + 0.2S_{DS})T_D + \Omega_o T_{Q_E} + 0.5T_L$ $+ 0.2T_S$	$(1.0 + 0.14S_{DS})T_D + T_H + T_F$ $+ 0.7\Omega_o T_{Q_E}$
$T_u = [1.2 + 0.2(0.528)](5.54 \text{ kips})$ $+ 2(-22.3 \text{ kips}) + 0.5(0 \text{ kips})$ $+ 0.2(6.70 \text{ kips})$ $= -36.0 \text{ kips}$	$T_a = [1.0 + 0.14(0.528)](5.54 \text{ kips})$ $+ 0 \text{ kips} + 0 \text{ kips}$ $+ 0.7(2)(-22.3 \text{ kips})$ $= -25.3 \text{ kips}$

Calculate V_b concurrent with tension in the brace by prorating the tensile force in the brace calculated above to the maximum compressive force in the brace calculated at the beginning of this example.

LRFD	ASD
$V_u = R_u + \dfrac{T_u}{P_{uc}} V_{ub}$ $= 11.8 \text{ kips} + \left(\dfrac{-36.0 \text{ kips}}{53.2 \text{ kips}}\right)(-23.4 \text{ kips})$ $= 27.6 \text{ kips}$	$V_a = R_a + \dfrac{T_a}{P_a} V_{ab}$ $= 7.73 \text{ kips} + \left(\dfrac{-25.3 \text{ kips}}{37.2 \text{ kips}}\right)(-16.3 \text{ kips})$ $= 18.8 \text{ kips}$

Therefore, the maximum vertical force in the beam-to-column connection is $V_u = 27.6$ kips (LRFD) or $V_a = 18.8$ kips (ASD).

Combine the maximum vertical force with the horizontal force at the beam-to-column interface as follows:

LRFD	ASD
$R_u = \sqrt{V_u^2 + P_u^2}$ $= \sqrt{(27.6 \text{ kips})^2 + (33.0 \text{ kips})^2}$ $= 43.0 \text{ kips}$	$R_a = \sqrt{V_a^2 + P_a^2}$ $= \sqrt{(18.8 \text{ kips})^2 + (23.1 \text{ kips})^2}$ $= 29.8 \text{ kips}$

Try (4) ¾-in.-diameter A325-N bolts in the single plate connecting the beam and the column.

Available shear strength of the bolt group
From the check of the gusset-to-column single plate design, the available strength of the ¾-in.-diameter ASTM A325-N bolt in the ⁵⁄₁₆-in.-thick plate is 17.9 kips (LRFD) and 11.9 kips (ASD) for bolt shear and 17.8 kips (LRFD) and 11.9 kips (ASD) for bolt bearing. The required number of bolts is:

LRFD	ASD
$n_{min} = \dfrac{R_u}{\phi r_n}$ $= \dfrac{43.0 \text{ kips}}{17.8 \text{ kips}}$ $= 2.42$	$n_{min} = \dfrac{R_a}{(r_n/\Omega)}$ $= \dfrac{29.8 \text{ kips}}{11.9 \text{ kips}}$ $= 2.50$

Use four bolts so that the connection is at least half the depth of the beam.

The beam web thickness is 0.355 in., which is slightly thicker than the single plate. Additionally, the beam specified minimum tensile strength, F_u, of 65 ksi is greater than the F_u of the single plate. Therefore, the bolt available bearing strength on the beam web is greater than that of the single plate, and the bolt bearing strength of the beam web is adequate.

Block shear rupture in the beam-to-column single-plate connection
Check block shear relative to the normal force in the single plate.

According to AISC *Specification* Section B4.3b, in computing net area for tension and shear, the width of a bolt hole is taken as ¹⁄₁₆ in. larger than the nominal dimension of the hole. The nominal diameter of the hole from Table J3.3 is ¹³⁄₁₆ in.

The available strength for the limit state of block shear rupture is:

$$R_n = 0.60 F_u A_{nv} + U_{bs} F_u A_{nt} \leq 0.60 F_y A_{gv} + U_{bs} F_u A_{nt} \qquad (Spec.\ Eq.\ J4\text{-}5)$$

where

$U_{bs} = 1.0$

$A_{nv} = 2(L_{eh} - 0.5 d_h) t_p$
$\quad = 2[2.00 \text{ in.} - 0.5(0.875 \text{ in.})](⁵⁄₁₆ \text{ in.})$
$\quad = 0.977 \text{ in.}^2$

$A_{gv} = 2 L_{eh} t_p$
$\quad = 2(2.00 \text{ in.})(⁵⁄₁₆ \text{ in.})$
$\quad = 1.25 \text{ in.}^2$

5.2 ORDINARY CONCENTRICALLY BRACED FRAMES

$$A_{nt} = [3s - (\text{no. bolt holes})d_h]t_p$$
$$= [3(3.00 \text{ in.}) - (2 + \frac{1}{2} + \frac{1}{2})(0.875 \text{ in.})](\frac{5}{16} \text{ in.})$$
$$= 1.99 \text{ in.}^2$$

$$0.60F_u A_{nv} + U_{bs}F_u A_{nt} = 0.60(58 \text{ ksi})(0.977 \text{ in.}^2) + 1.0(58 \text{ ksi})(1.99 \text{ in.}^2)$$
$$= 149 \text{ kips}$$

$$0.60F_y A_{gv} + U_{bs}F_u A_{nt} = 0.60(36 \text{ ksi})(1.25 \text{ in.}^2) + 1.0(58 \text{ ksi})(1.99 \text{ in.}^2)$$
$$= 142 \text{ kips}$$

Therefore, the nominal strength for the limit state of block shear rupture is $R_n = 142$ kips. The available strength for the limit state of block shear rupture on the single plate is:

LRFD	ASD
$\phi R_n = 0.75(142 \text{ kips})$ $= 107 \text{ kips} > 33.0 \text{ kips}$ **o.k.**	$\dfrac{R_n}{\Omega} = \dfrac{142 \text{ kips}}{2.00}$ $= 71.0 \text{ kips} > 23.1 \text{ kips}$ **o.k.**

Use the AISC *Manual* tables to determine the available strength of the single plate for the limit state of block shear rupture relative to the shear force on the single plate. For the single plate at the gusset-to-column connection:

$n = 4$
$L_{ev} = 1\frac{1}{2}$ in.
$L_{eh} = 2.00$ in.
$U_{bs} = 1.0$

From AISC *Specification* Equation J4-5, the available strength for the limit state of block shear rupture can be written as:

LRFD	ASD
$\phi R_n = \phi U_{bs} F_u A_{nt}$ $\quad + \min(\phi 0.60 F_y A_{gv}, \phi 0.60 F_u A_{nv})$	$\dfrac{R_n}{\Omega} = \dfrac{U_{bs} F_u A_{nt}}{\Omega}$ $\quad + \min\left(\dfrac{0.60 F_y A_{gv}}{\Omega}, \dfrac{0.60 F_u A_{nv}}{\Omega}\right)$
Tension rupture component from AISC *Manual* Table 9-3a:	Tension rupture component from AISC *Manual* Table 9-3a:
$\phi U_{bs} F_u A_{nt} = 1.0(68.0 \text{ kip/in.})(\frac{5}{16} \text{ in.})$ $= 21.3 \text{ kips}$	$\dfrac{U_{bs} F_u A_{nt}}{\Omega} = 1.0(45.3 \text{ kip/in.})(\frac{5}{16} \text{ in.})$ $= 14.2 \text{ kips}$

LRFD	ASD
Shear yielding component from AISC *Manual* Table 9-3b: $$\phi 0.60 F_y A_{gv} = (170 \text{ kip/in.})(\tfrac{5}{16} \text{ in.})$$ $$= 53.1 \text{ kips}$$ Shear rupture component from AISC *Manual* Table 9-3c: $$\phi 0.60 F_u A_{nv} = (194 \text{ kip/in.})(\tfrac{5}{16} \text{ in.})$$ $$= 60.6 \text{ kips}$$ The total available block shear rupture strength of the single plate at the beam-to-column connection is: $$\phi R_n = 21.3 \text{ kips}$$ $$+ \min(53.1 \text{ kips}, 60.6 \text{ kips})$$ $$= 74.4 \text{ kips} > 27.6 \text{ kips} \quad \textbf{o.k.}$$	Shear yielding component from AISC *Manual* Table 9-3b: $$\frac{0.60 F_y A_{gv}}{\Omega} = (113 \text{ kip/in.})(\tfrac{5}{16} \text{ in.})$$ $$= 35.3 \text{ kips}$$ Shear rupture component from AISC *Manual* Table 9-3c: $$\frac{0.60 F_u A_{nv}}{\Omega} = (129 \text{ kip/in.})(\tfrac{5}{16} \text{ in.})$$ $$= 40.3 \text{ kips}$$ The total available block shear rupture strength of the single plate at the beam-to-column connection is: $$\frac{R_n}{\Omega} = 14.2 \text{ kips}$$ $$+ \min(35.3 \text{ kips}, 40.3 \text{ kips})$$ $$= 49.5 \text{ kips} > 18.8 \text{ kips} \quad \textbf{o.k.}$$

Block shear rupture in the beam web is also okay, based on the greater thickness and the higher F_y and F_u values.

Combined shear and normal block shear design check using an elliptical equation

For the single-plate at the beam-to-column connection, the interaction of shear and normal block shear rupture is considered as follows:

LRFD	ASD
$$\left(\frac{V_r}{V_c}\right)^2 + \left(\frac{P_r}{P_c}\right)^2 \le 1.0$$ $$\left(\frac{27.6 \text{ kips}}{74.4 \text{ kips}}\right)^2 + \left(\frac{33.0 \text{ kips}}{107 \text{ kips}}\right)^2 = 0.233$$ $$0.233 \le 1.0 \quad \textbf{o.k.}$$	$$\left(\frac{V_r}{V_c}\right)^2 + \left(\frac{P_r}{P_c}\right)^2 \le 1.0$$ $$\left(\frac{18.8 \text{ kips}}{49.5 \text{ kips}}\right)^2 + \left(\frac{23.1 \text{ kips}}{71.0 \text{ kips}}\right)^2 = 0.250$$ $$0.250 \le 1.0 \quad \textbf{o.k.}$$

Tensile yielding in the beam-to-column single plate

Consider 12.0 in. of the plate to be effective.

5.2 ORDINARY CONCENTRICALLY BRACED FRAMES

$$A_g = lt_p$$
$$= (12.0 \text{ in.})(\tfrac{5}{16} \text{ in.})$$
$$= 3.75 \text{ in.}^2$$

The nominal strength due to tensile yielding is:

$$R_n = F_y A_g \hspace{2cm} (Spec. \text{ Eq. D2-1})$$
$$= (36 \text{ ksi})(3.75 \text{ in.}^2)$$
$$= 135 \text{ kips}$$

The available strength due to tensile yielding in the beam-to-column single plate is:

LRFD	ASD
$\phi R_n = 0.90(135 \text{ kips})$ $= 122 \text{ kips} > 33.0 \text{ kips}$ **o.k.**	$\dfrac{R_n}{\Omega} = \dfrac{135 \text{ kips}}{1.67}$ $= 80.8 \text{ kips} > 23.1 \text{ kips}$ **o.k.**

The beam web has a greater thickness (0.355 in.) and a higher specified minimum yield stress of $F_y = 50$ ksi; therefore, the available tensile strength due to yielding in the beam web is also adequate.

Tensile rupture in the beam-to-column single plate

Consider 12.0 in. of the plate to be effective.

$$d_h = 0.875 \text{ in.}$$

$$A_n = (l - 4d_h)t_p$$
$$= [12.0 \text{ in.} - 4(0.875 \text{ in.})](\tfrac{5}{16} \text{ in.})$$
$$= 2.66 \text{ in.}^2$$

$$U = 1.0$$

$$A_e = A_n U \hspace{2cm} (Spec. \text{ Eq. D3-1})$$
$$= (2.66 \text{ in.}^2)(1.0)$$
$$= 2.66 \text{ in.}^2$$

The nominal strength due to tensile rupture is:

$$R_n = F_u A_e \hspace{2cm} (Spec. \text{ Eq. J4-2})$$
$$= (58 \text{ ksi})(2.66 \text{ in.}^2)$$
$$= 154 \text{ kips}$$

The available strength due to tensile rupture in the beam-to-column single plate is:

LRFD	ASD
$\phi R_n = 0.75(154 \text{ kips})$ $= 116 \text{ kips} > 33.0 \text{ kips}$ **o.k.**	$\dfrac{R_n}{\Omega} = \dfrac{154 \text{ kips}}{2.00}$ $= 77.0 \text{ kips} > 23.1 \text{ kips}$ **o.k.**

The beam web has a greater thickness (0.355 in.) and a higher specified minimum tensile strength than the single plate, therefore, the available strength due to tensile rupture in the beam web is also adequate.

Shear rupture in the beam-to-column single plate

Check the available shear rupture strength at the net section through the bolt line. Conservatively consider only a 12.0 in. length of single plate.

$$A_{nv} = (l - 4d_h)t_p$$
$$= [12.0 \text{ in.} - 4(0.875 \text{ in.})](\tfrac{5}{16} \text{ in.})$$
$$= 2.66 \text{ in.}^2$$

The nominal strength due to shear rupture is:

$$R_n = 0.60 F_u A_{nv} \qquad \qquad (Spec. \text{ Eq. J4-4})$$
$$= 0.60(58 \text{ ksi})(2.66 \text{ in.}^2)$$
$$= 92.6 \text{ kips}$$

The available strength due to shear rupture is:

LRFD	ASD
$\phi R_n = 0.75(92.6 \text{ kips})$ $= 69.5 \text{ kips} > 27.6 \text{ kips}$ **o.k.**	$\dfrac{R_n}{\Omega} = \dfrac{92.6 \text{ kips}}{2.00}$ $= 46.3 \text{ kips} > 18.8 \text{ kips}$ **o.k.**

The beam web is thicker (0.355 in.) and has a higher specified minimum tensile strength (65 ksi) than the single plate; therefore, the available strength of the beam web due to shear rupture is also adequate.

Shear yielding in the beam-to-column single plate

Check the available shear yielding strength at the gross section through the bolt line. Conservatively consider only a 12.0 in. length of single plate.

5.2 ORDINARY CONCENTRICALLY BRACED FRAMES

$$A_{gv} = lt_p$$
$$= 12.0 \text{ in.}(\tfrac{5}{16} \text{ in.})$$
$$= 3.75 \text{ in.}^2$$

The nominal strength due to shear yielding is:

$$R_n = 0.60 F_y A_{gv} \qquad \text{(Spec. Eq. J4-3)}$$
$$= 0.60(36 \text{ ksi})(3.75 \text{ in.}^2)$$
$$= 81.0 \text{ kips}$$

The available strength due to shear yielding is:

LRFD	ASD
$\phi R_n = 1.00(81.0 \text{ kips})$ $= 81.0 \text{ kips} > 27.6 \text{ kips}$ **o.k.**	$\dfrac{R_n}{\Omega} = \dfrac{81.0 \text{ kips}}{1.50}$ $= 54.0 \text{ kips} > 18.8 \text{ kips}$ **o.k.**

The beam web is thicker (0.355 in.) with a higher specified minimum tensile strength (65 ksi) than the single plate; therefore, the available strength of the beam web due to shear yielding is also adequate.

Use a minimum 5/16-in.-thick single plate with (4) ¾-in.-diameter ASTM A325-N bolts in standard holes to connect the beam to the column.

Design the weld of the combined single plate to the column face

The weld of the single plate could be determined assuming two individual single plates. However this neglects the increased bending capacity of a 22-in.-long plate relative to the summation of bending capacities of a 12.0-in.-long single plate and a 6.00-in.-long single plate. Therefore, design the weld based on a 23.5-in.-long single plate.

When the collector force acts in tension on the column face, the H_c force on the gusset-to-column interface is also in tension. The collector force in the beam, A_b, acts 5.75 in. above the neutral axis of the single plate, and the H_c force at the gusset-to-column interface acts 8.75 in. below the neutral axis of the single plate, as determined in the following.

Eccentricity of A_b on the single plate:

$$e_{A_b} = (23.5 \text{ in.}/2) - 1.5 \text{ in.} - 3.0 \text{ in} - 1.5 \text{ in.}$$
$$= 5.75 \text{ in.}$$

Eccentricity of H_c on the single plate:

$$e_{H_c} = (23.5 \text{ in.}/2) - 1.5 \text{ in.} - 1.5 \text{ in.}$$
$$= 8.75 \text{ in.}$$

Eccentricity of vertical shear on the column face: $e_c = 2.50$ in.

The total normal force at the column face is:

LRFD	ASD
$H_u = A_{ub} + H_{uc}$ $= 33.0 \text{ kips} + 19.5 \text{ kips}$ $= 52.5 \text{ kips}$	$H_a = A_{ab} + H_{ac}$ $= 23.1 \text{ kips} + 13.6 \text{ kips}$ $= 36.7 \text{ kips}$

The total shear force at the column face is:

LRFD	ASD
$V_u = R_{ub} + V_{ub} + V_{uc}$ $= 11.8 \text{ kips} + 23.4 \text{ kips} + 14.3 \text{ kips}$ $= 49.5 \text{ kips}$	$V_a = R_{ab} + V_{ab} + V_{ac}$ $= 7.73 \text{ kips} + 16.3 \text{ kips} + 9.98 \text{ kips}$ $= 34.0 \text{ kips}$

For moment on a weld group, sum moments about the mid-height centerline of the single plate at the face of the column:

LRFD	ASD
$M_u = V_u e_c + A_{ub} e_{A_b} - H_{uc} e_{H_c}$ $= 49.5 \text{ kips}(2.50 \text{ in.})$ $+ 33.0 \text{ kips}(5.75 \text{ in.})$ $- 19.5 \text{ kips}(8.75 \text{ in.})$ $= 143 \text{ kip-in.}$	$M_a = V_a e_c + A_{ab} e_{A_b} - H_{ac} e_{H_c}$ $= 34.0 \text{ kips}(2.50 \text{ in.})$ $+ 23.1 \text{ kips}(5.75 \text{ in.})$ $- 13.6 \text{ kips}(8.75 \text{ in.})$ $= 98.8 \text{ kip-in.}$

The stresses at the single plate-to-column interface are determined as follows:

$l = 23.5 \text{ in.}$

$$Z_w = \frac{l^2}{4}$$
$$= \frac{(23.5 \text{ in.})^2}{4}$$
$$= 138 \text{ in.}^2$$

5.2 ORDINARY CONCENTRICALLY BRACED FRAMES

LRFD	ASD
$f_{uv} = \dfrac{V_u}{l}$ $= \dfrac{49.5 \text{ kips}}{23.5 \text{ in.}}$ $= 2.11 \text{ kip/in.}$	$f_{av} = \dfrac{V_{aa}}{l}$ $= \dfrac{34.0 \text{ kips}}{23.5 \text{ in.}}$ $= 1.45 \text{ kip/in.}$
$f_{ua} = \dfrac{H_u}{l}$ $= \dfrac{52.5 \text{ kips}}{23.5 \text{ in.}}$ $= 2.23 \text{ kip/in.}$	$f_a = \dfrac{H_{aa}}{l}$ $= \dfrac{36.7 \text{ kips}}{23.5 \text{ in.}}$ $= 1.56 \text{ kip/in.}$
$f_{ub} = \dfrac{M_u}{Z_w}$ $= \dfrac{143 \text{ kip-in.}}{138 \text{ in.}^2}$ $= 1.04 \text{ kip/in.}$	$f_{ab} = \dfrac{M_{aa}}{Z_w}$ $= \dfrac{98.8 \text{ kip-in.}}{138 \text{ in.}^2}$ $= 0.716 \text{ kip/in.}$
$f_{ur} = \sqrt{f_{uv}^2 + (f_{ua} + f_{ub})^2}$ $= \sqrt{\begin{array}{c}(2.11 \text{ kip/in.})^2 \\ +(2.23 \text{ kip/in.} + 1.04 \text{ kip/in.})^2\end{array}}$ $= 3.89 \text{ kip/in.}$	$f_{ar} = \sqrt{f_{av}^2 + (f_{aa} + f_{ab})^2}$ $= \sqrt{\begin{array}{c}(1.45 \text{ kip/in.})^2 \\ +(1.56 \text{ kip/in.} + 0.716 \text{ kip/in.})^2\end{array}}$ $= 2.70 \text{ kip/in.}$
Using the conservative solution (adding the flexural stress), the angle of the resultant load with respect to the weld is: $\theta = \tan^{-1}\left(\dfrac{f_{ua} + f_{ub}}{f_{uv}}\right)$ $= \tan^{-1}\left(\dfrac{2.23 \text{ kip/in.} + 1.04 \text{ kip/in.}}{2.11 \text{ kip/in.}}\right)$ $= 57.2°$	Using the conservative solution (adding the flexural stress), the angle of the resultant load with respect to the weld is: $\theta = \tan^{-1}\left(\dfrac{f_{aa} + f_{ab}}{f_{av}}\right)$ $= \tan^{-1}\left(\dfrac{1.56 \text{ kip/in.} + 0.716 \text{ kip/in.}}{1.45 \text{ kip/in.}}\right)$ $= 57.5°$

The weld size is determined from AISC *Manual* Equation 8-2a (LRFD) and 8-2b (ASD):

LRFD	ASD
$D = \dfrac{f_{ur}}{2(1.392 \text{ kip/in.})(1.0+0.50\sin^{1.5}\theta)}$ $= \dfrac{3.89 \text{ kip/in.}}{2(1.392 \text{ kip/in.})}$ $\times \dfrac{1}{\left[1.0+0.50\sin^{1.5}(57.2°)\right]}$ $= 1.01 \text{ sixteenths}$	$D = \dfrac{f_{ar}}{2(0.928 \text{ kip/in.})(1.0+0.50\sin^{1.5}\theta)}$ $= \dfrac{2.70 \text{ kip/in.}}{2(0.928 \text{ kip/in.})}$ $\times \dfrac{1}{\left[1.0+0.50\sin^{1.5}(57.5°)\right]}$ $= 1.05 \text{ sixteenths}$

Considering the column flange thickness and the single-plate thickness, the minimum fillet weld size from AISC *Specification* Table J2.4 is ³⁄₁₆ in. However, according to the AISC *Manual* Part 10 discussion of single-plate connections, the weld between a single plate and the support should be sized as:

$⅝ t_p = ⅝(⁵⁄₁₆ \text{ in.})$
$= 0.195 \text{ in.}$

The use of the above minimum weld size combined with the single plate requirement for connection plate thicknesses to be less than $d_b - ¹⁄₁₆$ in. according to AISC *Manual* Table 10-9 facilitates ductile behavior in the connection.

Use two-sided ¼-in. fillet welds at the single plate to column connection.

Check single-plate shear rupture at weld to column

One method to determine the minimum single-plate thickness required to transfer the shear and tension forces is to set the weld strength (based on the resultant force) equal to the shear rupture strength of the single plate. From AISC *Manual* Part 9, the minimum required single-plate thickness is:

$$t_{min} = \dfrac{6.19D}{F_u} \hspace{3cm} (\textit{Manual Eq. 9-3})$$

LRFD	ASD
$t_{min} = \dfrac{6.19 \text{ kip/in.}(1.01)}{58 \text{ ksi}}$ $= 0.108 \text{ in.} < ⁵⁄₁₆ \text{ in.}$ **o.k.**	$t_{min} = \dfrac{6.19 \text{ kip/in.}(1.05)}{58 \text{ ksi}}$ $= 0.112 \text{ in.} < ⁵⁄₁₆ \text{ in.}$ **o.k.**

Check compression on the single plate

When the brace force is in compression, the beam-to-column axial force is in compression. The unit force on the single plate in compression results from axial and bending forces combined.

5.2 ORDINARY CONCENTRICALLY BRACED FRAMES

Check the plate for the limit state of buckling using the double-coped beam procedure given in AISC *Manual* Part 9.

$$F_{cr} = QF_y \qquad \text{(\textit{Manual} Eq. 9-14)}$$

Calculate Q for the single plate:

$$\lambda = \frac{h_o\sqrt{F_y}}{10t_p\sqrt{475 + 280\left(\frac{h_o}{c}\right)^2}} \qquad \text{(\textit{Manual} Eq. 9-18)}$$

$$= \frac{(23.5 \text{ in.})\sqrt{36 \text{ ksi}}}{10(\sfrac{5}{16} \text{ in.})\sqrt{475 + 280\left(\frac{23.5 \text{ in.}}{2.50 \text{ in.}}\right)^2}}$$

$$= 0.284$$

Because $\lambda \leq 0.7$,

$$Q = 1 \qquad \text{(\textit{Manual} Eq. 9-15)}$$

$F_{cr} = F_y$; therefore, plate buckling does not control.

Use a $\sfrac{5}{16}$-in.-thick single plate 23.5 in. long.

Check column web local yielding

The peak unit bending force, f_b, is less than the axial unit bending force, f_a. Therefore, the bending forces do not affect the overall concentrated force on the gusset nor do they affect the length of force applied on the interface. A portion of the concentrated force is applied within a distance less than the depth of the column.

For a force applied at a distance less than the depth of the member:

$$R_n = F_{yw}t_w(2.5k + l_b) \qquad \text{(\textit{Spec.} Eq. J10-3)}$$

$$= (50 \text{ ksi})(0.340 \text{ in.})[2.5(1.06 \text{ in.}) + 23.5 \text{ in.}]$$

$$= 445 \text{ kips}$$

LRFD	ASD
$\phi R_n = 1.00(445 \text{ kips})$ $= 445 \text{ kips} > 33.0 \text{ kips}$ **o.k.**	$\dfrac{R_n}{\Omega} = \dfrac{445 \text{ kips}}{1.50}$ $= 297 \text{ kips} > 23.1 \text{ kips}$ **o.k.**

Alternatively, the available strength for web yielding can be determined per Part 9 of the AISC *Manual*, and Table 9-4.

Check column web local crippling

A portion of the concentrated force is applied at a distance less than $d/2$ from the end of the column; therefore, use AISC *Specification* Section J10.3(b). Check the length of bearing relative to the column depth:

$$\frac{l_b}{d} = \frac{23.5 \text{ in.}}{10.0 \text{ in.}}$$
$$= 2.35 > 0.2$$

Therefore, use AISC *Specification* Equation J10-5b to determine the available strength, through use of AISC *Manual* Table 9-4.

From AISC *Manual* Table 9-4 for the **W10×49**:

LRFD	ASD
$\phi R_5 = 48.5$ kips	$\dfrac{R_5}{\Omega} = 32.3$ kips
$\phi R_6 = 10.1$ kip/in.	$\dfrac{R_6}{\Omega} = 6.76$ kip/in.

From AISC *Manual* Equations 9-48a and 9-48b:

LRFD	ASD
$\phi R_n = \phi R_5 + l_b (\phi R_6)$ $= 48.5 \text{ kips} + 23.5 \text{ in.}(10.1 \text{ kip/in.})$ $= 286 \text{ kips} > 33.0 \text{ kips}$ **o.k.**	$\dfrac{R_n}{\Omega} = \dfrac{R_5}{\Omega} + l_b \dfrac{R_6}{\Omega}$ $= 32.3 \text{ kips} + 23.5 \text{ in.}(6.76 \text{ kip/in.})$ $= 191 \text{ kips} > 23.1 \text{ kips}$ **o.k.**

The final connection design and geometry is shown in Figure 5-7.

5.2 ORDINARY CONCENTRICALLY BRACED FRAMES

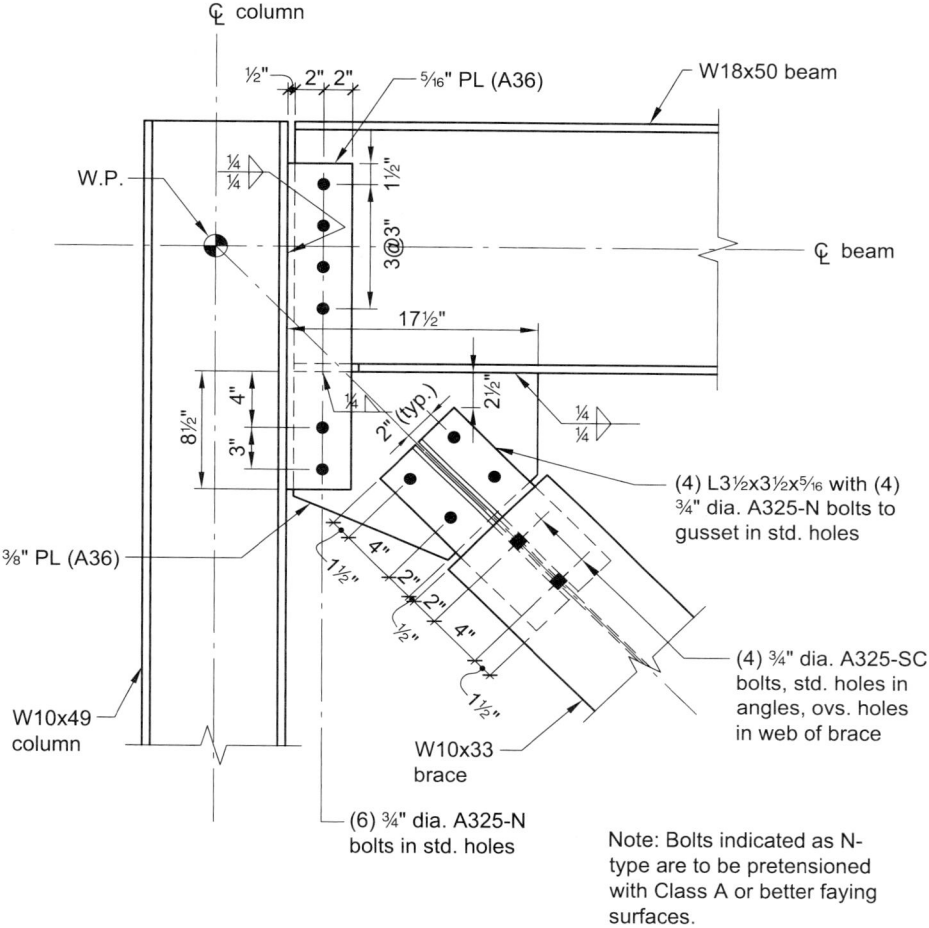

Fig. 5-7. Connection designed in Example 5.2.4.

Example 5.2.5. OCBF Tension-Only Diagonal Brace Design

Given:

Unlike special systems, tension-only bracing is permitted in OCBF systems; therefore, this example demonstrates a tension-only brace design for the same configuration as Example 5.2.4. Refer to Brace BR-1 shown in Figure 5-2. Select an ASTM A36 single-angle section for the diagonal brace to resist the loads shown below as a tension-only bracing configuration.

The applicable building code specifies the use of ASCE/SEI 7 for calculation of loads. From a first-order analysis, the loads on the brace are:

$P_D = 0$ kips $P_S = 0$ kips $P_{Q_E} = \pm 51.1$ kips
$P_H = 0$ kips $P_L = 0$ kips $M_D = 1.13$ kip-ft

The dead load bending moment indicated above is due to the self-weight of the brace assuming a member that weighs 16 lb/ft. Sometimes this self-weight loading is ignored in the design of vertical diagonal braces where judgment would indicate that the loading is minimal and only uses a small percentage of the available member strength. However, in this example, considering the relatively long length of the diagonal brace, the dead load moment is included in this design check. There are no bending moments due to live loads or snow loads.

The story shear, H, from the first-order analysis is 136 kips and the first-order interstory drift due to that load without the C_d factor applied from the analysis model is:

$\Delta_H = 0.761$ in.

Solution:

From AISC *Manual* Table 2-5, the material properties are:

ASTM A36
$F_y = 36$ ksi
$F_u = 58$ ksi

Determine the required strength of the diagonal brace

Considering the load combinations given in ASCE/SEI 7, the governing load combination and resultant maximum axial tension and bending moment in the diagonal brace are:

LRFD	ASD
LRFD Load Combination 5 from ASCE/SEI 7 Section 12.4.2.3 (including the 0.5 factor on L permitted in Section 12.4.2.3)	ASD Load Combination 5 from ASCE/SEI 7 Section 12.4.2.3
$(1.2 + 0.2S_{DS})D + \rho Q_E + 0.5L + 0.2S$	$(1.0 + 0.14S_{DS})D + H + F + 0.7\rho Q_E$
$P_u = [1.2 + 0.2(0.528)](0 \text{ kips})$ $\quad + 1.0(-51.1 \text{ kips}) + 0.5(0 \text{ kips})$ $\quad + 0.2(0 \text{ kips})$ $\quad = -51.1$ kips	$P_a = [1.0 + 0.14(0.528)](0 \text{ kips})$ $\quad + 0 \text{ kips} + 0 \text{ kips}$ $\quad + 0.7(1.0)(-51.1 \text{ kips})$ $\quad = -35.8$ kips
$M_u = [1.2 + 0.2(0.528)](1.13 \text{ kip-ft})$ $\quad + 1.0(0 \text{ kip-ft}) + 0.5(0 \text{ kip-ft})$ $\quad + 0.2(0 \text{ kip-ft})$ $\quad = 1.48$ kip-ft	$M_a = [1.0 + 0.14(0.528)](1.13 \text{ kip-ft})$ $\quad + 0 \text{ kip-ft} + 0 \text{ kip-ft}$ $\quad + 0.7(1.0)(0 \text{ kip-ft})$ $\quad = 1.21$ kip-ft

Try an **L5×5×½** for the brace member.

DESIGN OF PARTIALLY-RESTRAINED MOMENT CONNECTIONS

From AISC *Manual* Table 1-7, the geometric properties are as follows:

L5×5×¹/₂

$A = 4.79$ in.² $b = 5.00$ in. $d = 5.00$ in.

$r_x = r_y = 1.53$ in. $t = 0.500$ in. $r_z = 0.980$ in.

$S_x = S_y = 3.15$ in.³

Check brace element width-to-thickness ratios

The braces must satisfy the requirements for moderately ductile members, as stipulated in Sections F1.5a and D1.1 of the AISC *Seismic Provisions*. Elements of the brace members must not exceed λ_{md} width-to-thickness ratios.

From Table 1-4 of this Manual, the L5×5×¹/₂ satisfies width-to-thickness ratios for OCBF diagonal braces (moderately ductile members).

Determine the effective slenderness ratio

The available compressive strength of a tension-only brace is ignored in the design of the bracing. Therefore in order to ensure the brace will buckle in compression under relatively minor loading, use a tension-only brace with a slenderness ratio greater than the recommended maximum effective slenderness ratio, KL/r, of 200 as indicated in the User Note in Section E2 of the AISC *Specification*. According to the User Note in AISC *Specification* Section D1, KL/r of members designed on the basis of tension should preferably not exceed 300. Therefore the effective slenderness ratio, KL/r, is selected to be greater than 200, but less than 300.

Determine K

According to AISC *Specification* Appendix 7, Section 7.2.3(a), for braced frame systems the effective length factor for members subject to compression shall be taken as 1.0, unless a rational analysis indicates that a lower value is appropriate.

The overall length of the brace diagonal in each bay is:

$$L = \sqrt{(40.0 \text{ ft})^2 + (40.0 \text{ ft})^2}$$
$$= 56.6 \text{ ft}$$

This length has been determined by calculating the distance between the work points based on the intersection of the centerlines of the diagonal braces, columns and beam. Shorter lengths may be used if justified by the engineer of record.

Single angles in X-bracing are normally continuous for the full diagonal length of the bay with the orientation of each brace reversed as shown in Figure 5-8, permitting the braces to be connected to each other by bolting at mid-length. The effective length in this arrangement is 0.85 times the half diagonal length considering the radius of gyration in the z-axis, r_z (El-Tayem and Goel, 1986).

$L_z = 0.5L$
    $= 0.5(56.6 \text{ ft})$
    $= 28.3 \text{ ft}$

$K_z = 0.85$

$\dfrac{K_z L_z}{r_z} = \dfrac{0.85(28.3 \text{ ft})(12 \text{ in./ft})}{0.980 \text{ in.}}$

    $= 295$

The slenderness, $\dfrac{KL}{r}$, is greater than 200, but less than 300, and therefore meets the desired range based on the User Notes in Sections D1 and E2 of the AISC *Specification*.

Note that the suggested slenderness limit of 300 does not apply to rod bracing, nor does the 0.85 effective length factor.

Determine the available tensile strength

For tensile yielding on the gross section, the nominal tensile strength is:

$P_n = F_y A_g$ (*Spec.* Eq. D2-1)

    $= 36 \text{ ksi}(4.79 \text{ in.}^2)$

    $= 172 \text{ kips}$

The available tensile strength is:

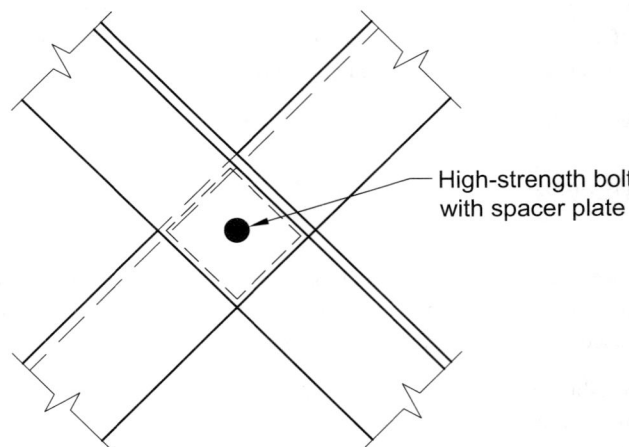

Fig. 5-8. Connection of single-angle diagonal braces at mid-point.

5.2 ORDINARY CONCENTRICALLY BRACED FRAMES

LRFD	ASD
$\phi_t P_n = 0.90(172 \text{ kips})$ $= 155 \text{ kips} > 51.1 \text{ kips}$ **o.k.**	$\dfrac{P_n}{\Omega_t} = \dfrac{172 \text{ kips}}{1.67}$ $= 103 \text{ kips} > 35.8 \text{ kips}$ **o.k.**

The limit state of tension rupture on the effective area should also be checked; however, by inspection, it would not control.

Determine the available flexural strength

During the governing seismic load conditions, the bracing is subject to significant axial tension with some minor flexure due to self-weight. The large axial tension loading provides a stabilizing effect to the brace and negates the effect of lateral-torsional buckling due to flexure. Therefore, even though the member is not laterally restrained along the length, when consideration is given to the significant axial tension load in the member, flexural strength can be based on the limit state of yielding only. This assumes that the single-angle has continuous lateral restraint along the length; therefore, the lateral-torsional buckling limit state does not apply. Additionally, because the section is compact, the limit state of leg local buckling does not apply.

The nominal flexural strength due to yielding is:

$M_n = 1.5 M_y$ (*Spec*. Eq. F10-1)

$M_n = 1.5 S_x F_y$

$\quad = 1.5(3.15 \text{ in.}^3)(36 \text{ ksi})(1 \text{ ft}/12 \text{ in.})$

$\quad = 14.2 \text{ kip-ft}$

The available flexural strength is:

LRFD	ASD
$\phi_b M_n = 0.90(14.2 \text{ kip-ft})$ $= 12.8 \text{ kip-ft} > 1.48 \text{ kip-ft}$ **o.k.**	$\dfrac{M_n}{\Omega_b} = \dfrac{14.2 \text{ kip-ft}}{1.67}$ $= 8.50 \text{ kip-ft} > 1.21 \text{ kip-ft}$ **o.k.**

Consider second-order effects

Follow the calculation procedure of AISC *Specification* Appendix 8.

$M_r = B_1 M_{nt} + B_2 M_{lt}$ (*Spec*. Eq. A-8-1)

$P_r = P_{nt} + B_2 P_{lt}$ (*Spec*. Eq. A-8-2)

Calculate B_1

$B_1 = 1.00$ according to Section 8.2 of AISC *Specification* Appendix 8, as the member is not subject to compression.

Calculate B_2

$\alpha = 1.00$ (LRFD); $\alpha = 1.60$ (ASD)

P_{story} is the total vertical load on the story calculated using the applicable load case. As calculated in Example 5.2.1:

LRFD	ASD
$P_{story} = 1{,}130$ kips	$P_{story} = 740$ kips

$R_M = 1.0$ (braced frame)

$$P_{e\ story} = R_M \frac{HL}{\Delta_H} \qquad \text{(Spec. Eq. A-8-7)}$$

$$= 1.0 \frac{136 \text{ kips}(40.0 \text{ ft})}{(0.761 \text{ in.})(1 \text{ ft}/12 \text{ in.})}$$

$$= 85{,}800 \text{ kips}$$

Using AISC *Specification* Equation A-8-6:

LRFD	ASD
$B_2 = \dfrac{1}{1 - \dfrac{\alpha P_{story}}{P_{e\ story}}} \geq 1$	$B_2 = \dfrac{1}{1 - \dfrac{\alpha P_{story}}{P_{e\ story}}} \geq 1$
$= \dfrac{1}{1 - \dfrac{1.00(1{,}130 \text{ kips})}{85{,}800 \text{ kips}}}$	$= \dfrac{1}{1 - \dfrac{1.60(740 \text{ kips})}{85{,}800 \text{ kips}}}$
$= 1.01$	$= 1.01$

First-order bending moments with the structure restrained against lateral translation (gravity loads in this case), and due to lateral translation of the story are, respectively:

LRFD	ASD
$M_{nt} = M_u$	$M_{nt} = M_a$
$= 1.48$ kip-ft	$= 1.21$ kip-ft
$M_{lt} = 0$ kip-ft	$M_{lt} = 0$ kip-ft

The required flexural strength of the brace including second-order effects is, from AISC *Specification* Equation A-8-1:

5.2 ORDINARY CONCENTRICALLY BRACED FRAMES

LRFD	ASD
$M_r = B_1 M_{nt} + B_2 M_{lt}$ $= 1.00(1.48 \text{ kip-ft}) + 1.01(0 \text{ kip-ft})$ $= 1.48 \text{ kip-ft}$	$M_r = B_1 M_{nt} + B_2 M_{lt}$ $= 1.00(1.21 \text{ kip-ft}) + 1.01(0 \text{ kip-ft})$ $= 1.21 \text{ kip-ft}$

First-order axial forces with the structure restrained against lateral translation (gravity loads in this case), and due to lateral translation of the story from seismic loading are, respectively:

LRFD	ASD
$P_{nt} = 0$ kips $P_{lt} = 51.1$ kips	$P_{nt} = 0$ kips $P_{lt} = 35.8$ kips

The required strength of the brace including second-order effects is, from AISC *Specification* Equation A-8-2:

LRFD	ASD
$P_r = P_{nt} + B_2 P_{lt}$ $= 0 \text{ kips} + 1.01(51.1 \text{ kips})$ $= 51.6 \text{ kips}$	$P_r = P_{nt} + B_2 P_{lt}$ $= 0 \text{ kips} + 1.01(35.8 \text{ kips})$ $= 36.2 \text{ kips}$

Check combined loading of the L5×5×½ brace:

LRFD	ASD
$\dfrac{P_r}{P_c} = \dfrac{51.6 \text{ kips}}{155 \text{ kips}}$ $= 0.333$	$\dfrac{P_r}{P_c} = \dfrac{36.2 \text{ kips}}{103 \text{ kips}}$ $= 0.351$

Because $P_r/P_c \geq 0.2$, the brace design is controlled by the equation:

$$\frac{P_r}{P_c} + \frac{8}{9}\left(\frac{M_{rx}}{M_{cx}} + \frac{M_{ry}}{M_{cy}}\right) \leq 1.0 \qquad \text{(Spec. Eq. H1-1a)}$$

LRFD	ASD
$\dfrac{51.6 \text{ kips}}{155 \text{ kips}} + \dfrac{8}{9}\left(0 + \dfrac{1.48 \text{ kip-ft}}{12.8 \text{ kip-ft}}\right) = 0.436$ $0.436 < 1.0$ **o.k.**	$\dfrac{36.2 \text{ kips}}{103 \text{ kips}} + \dfrac{8}{9}\left(0 + \dfrac{1.21 \text{ kip-ft}}{8.50 \text{ kip-ft}}\right) = 0.478$ $0.478 < 1.0$ **o.k.**

Note that the y-y axis bending moment from the self-weight of the diagonal brace utilizes about 11% of the member capacity.

Use an L5×5×¹/₂ in the tension-only configuration for OCBF diagonal Brace BR-1.

Braces must be continuous through and bolted to each other at the intersecting joint as shown in Figure 5-8.

5.3 SPECIAL CONCENTRICALLY BRACED FRAMES (SCBF)

Special concentrically braced frame (SCBF) systems, like other concentrically braced frames, resist lateral forces and displacements primarily through the axial strength and stiffness of the brace members. In concentrically braced frames, the centerlines of the framing members (braces, columns, and beams) coincide or nearly coincide, eliminating or minimizing flexure in the system. The design of SCBF systems is addressed in AISC *Seismic Provisions* Section F2. While the general layout of an SCBF is very similar to that of an ordinary concentrically braced frame (OCBF), there are additional detailing requirements to focus ductile behavior of the frames into the braces. These detailing requirements provide for greater energy dissipation and ductility, allowing SCBF systems to be designed using a lower force level in comparison to that of OCBF systems.

Concentrically braced frame systems tend to be more economical than moment resisting frames and eccentrically braced frames in terms of material, fabrication and erection costs. They do, however, often have reduced flexibility in floor-plan layout, space planning, and electrical and mechanical routing as a result of the presence of braces. In certain circumstances, however, braced frames are exposed and featured in the architecture of the building.

Braced frames typically are located in walls that stack vertically between floor levels. In the typical office building, these walls generally occur in the core area around stair and elevator shafts, central restrooms, and mechanical and electrical rooms. This generally allows for greater architectural flexibility in placement and configuration of exterior windows and cladding.

In considering the configuration of a braced frame system, both in plan and elevation, it is important to note the requirements for redundancy in the system. The AISC *Seismic Provisions* require that a braced frame system balance the compression and tension braces. AISC *Seismic Provisions* Section F2.4a requires that along any line of bracing, the braces are oriented to resist at least 30% but not more than 70% of the total horizontal force in tension unless the exception in Section F2.4a is met.

The AISC *Seismic Provisions* limit member slenderness, compressive strength, and width-to-thickness ratios, in addition to requiring special detailing for gusset plates. The cumulative effect of these requirements is intended to result in braces that maintain a high level of ductility and hysteretic damping when subjected to severe seismic forces.

Brace slenderness is limited to ensure adequate compressive strength and resistance to the cyclic degradation of the brace. The post-buckling performance of the brace is dependent on the compactness of the members used. Members with a higher width-to-thickness ratio are more susceptible to local buckling, which may lead to tearing of the brace material in the buckled areas prior to the dissipation of a significant amount of energy. This behavior results in a system with significantly lower energy dissipation capability.

The last of the predominant issues relating to the bracing members is the spacing of intermediate connectors of double angle, double channel or similar built-up braces. AISC *Seismic Provisions* Section F2.5b notes that connectors should be placed such that the a/r_i

value for the individual components of the brace is less than 40% of the governing slenderness of the built-up member. Additionally, it is required that the connectors have a shear strength that develops the tensile strength of individual components of the brace. These two provisions are intended to ensure that the brace buckles as a unit, thus allowing more reliable behavior. The connector requirements are reduced when it can be shown that the brace assembly can buckle as a single element without inducing shear forces in the connectors between the individual members. In any case, no fewer than two connectors are allowed with uniform spacing, and bolted connectors are not permitted in the middle one-fourth of the clear brace length. The limitation on the location of bolted attachments is included to guard against premature fracture due to the formation of a plastic hinge in the buckled brace.

In order to increase ductility and energy dissipation of the system the connections must be detailed to accommodate the effects of brace buckling. Currently, there are two approaches used in the design of these connections; these are stated in AISC *Seismic Provisions* Sections F2.6c(3)(a) and F2.6c(3)(b). The first approach creates enough strength and rigidity in the connections to force the brace to form plastic hinges at the ends and middle of the brace under compressive forces. The second approach utilizes out-of-plane buckling of the gusset plate such that plastic hinges occur in the gusset plate at the brace ends with a hinge still occurring at the midpoint of the brace. This usually is accommodated in one of two ways. As one option, the connection can be detailed such that the end of the brace is located a distance of at least two times the thickness of the gusset from the intersection of the gusset and the beam or column. This configuration is shown in AISC *Seismic Provisions* Commentary Figure C-F2.9. The value of two times the thickness of the gusset has been developed through research and analysis. Alternatively, an elliptical yield line approach can be used (Lehman et al., 2008). AISC *Seismic Provisions* Section F2.6b addresses beam-to-column connection issues related to the accommodation of large seismic drifts associated with the yielding and buckling of the braces. This provision is discussed in greater detail in the following.

The design requirements for most basic frame configurations are covered by the conditions listed earlier in this section. V-type and inverted V-type frames, however, are required to meet additional criteria, as noted in AISC *Seismic Provisions* Section F2.4b.

These requirements are intended to reduce the effect of a loss in strength of the compression brace relative to the tension brace in the post-buckling range, as shown in Figure 5-9. As the compression brace buckles under load, its capability to resist the vertical load is diminished relative to the strength of the tension brace. This results in an unbalanced vertical load between the two members, which exerts additional vertical force on the beam. Braced frame configurations utilizing zipper columns and two-story X configurations, as shown in Figures 5-9(b) and 5-9(c), distribute this unbalanced vertical load to other levels that are not experiencing high seismic demands, providing for better overall frame performance.

Another check covered in the AISC *Seismic Provisions* relates to columns that are part of the SCBF system. Columns are required to meet the highly ductile width-to-thickness criteria according to AISC *Seismic Provisions* Section F2.5a, and have special considerations for their splices. According to AISC *Seismic Provisions* Section F2.6d, column splices must develop a required shear strength equal to $\Sigma M_{pc}/H_c$ for LRFD and $\Sigma M_{pc}/(1.5H_c)$ for ASD. This requirement is intended to account for the possibility of the columns sharing some of the lateral force demand through frame action as the brace elements deform inelastically, deflecting the frames beyond what elastic calculations might predict. Additionally, it is noted that the column splices must be located at least 4 ft from the beam-to-column flange connections in AISC *Seismic Provisions* Section D2.5a.

Design of Gusseted Beam-to-Column Connections to Accommodate Large Drifts

AISC *Seismic Provisions* Section F2.6b requires that gusseted beam-to-column connections be designed to accommodate demands corresponding to large drifts. In the context of this provision, the connection consists of the gusset plate, the affected parts of the beam and column, and any other connection material, such as angles and plates, interconnecting these elements.

Two methods of accommodating demands corresponding to large drifts are provided. First, as described in AISC *Seismic Provisions* Section F2.6b(a), the connection may be detailed to provide sufficient rotation capacity such that the beam and column are not constrained to rotate together as the frame deforms. The provision defines this required relative rotation as 0.025 rad. Connections similar to the simple connections presented in Part 10 of

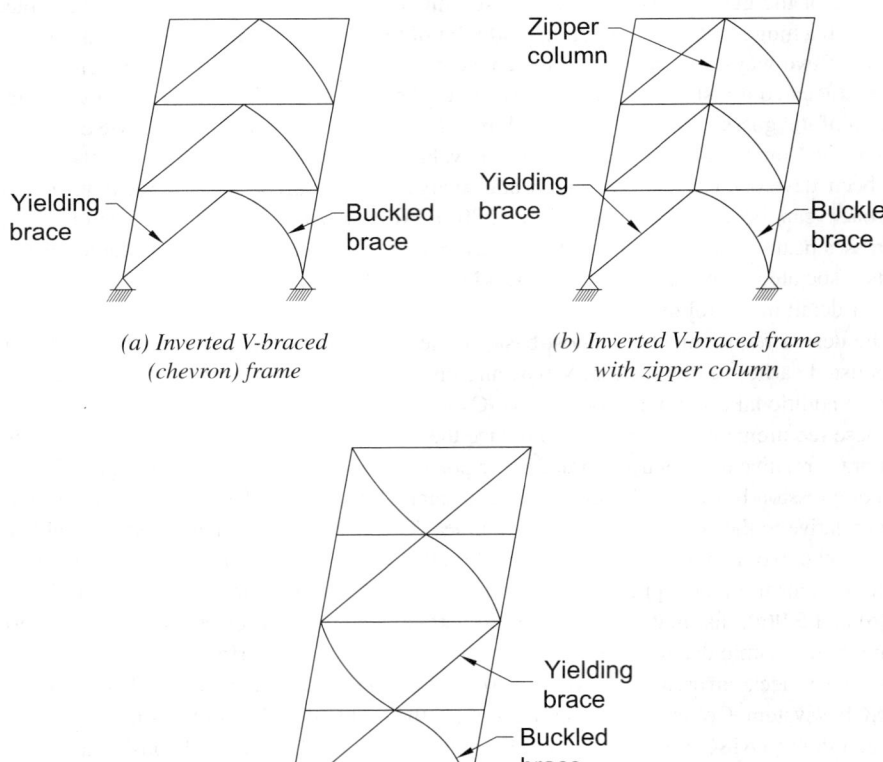

Fig. 5-9. Assumed inelastic deformation of various braced frame configurations.

5.3 SPECIAL CONCENTRICALLY BRACED FRAMES

the AISC *Manual* and meeting the rotational ductility checks described in Part 9 of the AISC *Manual* can be assumed to provide a minimum of 0.03 rad and satisfy the intent of the AISC *Specification* Section B3.6a for simple connections. The Part 9 rotational ductility checks are intended for use with connections between 6 in. and 36 in. deep and with geometries similar to those shown in the AISC *Manual*. The use of deeper connections, smaller set-off distances between the supported and supporting members, or smaller edge distances can affect the ability of connections to accommodate large rotations in a ductile manner.

It is important to note that these bounds apply to the connection as a whole. For example, if the connection at the column face consists of a double-angle connection from column flange-to-gusset and a double-angle connection from column flange-to-beam web, the two double-angle connections should not be considered as separate; they should be considered as rotating about a single point and the entire depth of the assembly should not exceed 36 in. in order for the rotation requirements to be deemed satisfied in the absence of further demonstration. Physical tests can also be used to demonstrate adequate rotation capacity.

The second method of accommodating demands corresponding to large drifts is described in AISC *Seismic Provisions* Section F2.6b(b). Rather than attempting to determine the actual demand placed on gusseted connections by seismic drifts, this method establishes an upper bound demand based on flexural yielding of either the beam or the column. It is assumed that these members have sufficient rotational ductility to maintain their function as braced-frame members when subjected to inelastic rotation. The connection is designed to resist a moment corresponding to the lesser of 1.1 times the expected beam flexural strength and 1.1 times the sum of the expected column flexural strength above and below the connection. This moment is considered in conjunction with the brace forces corresponding to the brace expected strength. Connection assemblies may be designed to resist this moment in one of two ways. The entire assembly may be analyzed with the required moment and axial force applied and all connection elements designed for the corresponding forces. Connecting the beam itself to the column by a fully restrained moment connection capable of resisting the expected flexural strength of the beam is another option. With this option the gusset plate and related connection elements may be designed for forces derived considering the brace connection required strength.

Thus, there are three methods of complying with AISC *Seismic Provisions* Section F2.6b presented in this Manual. Each of these methods is presented in a different connection example—Examples 5.3.10, 5.3.11 and 5.3.12. These examples also illustrate three different methods of accommodating the rotation associated with brace buckling as required by Section F2.6c(3). There is no correlation between the method of accommodating frame drift and the method of accommodating brace rotation due to buckling, i.e., any method of complying with Section F2.6b may be used in conjunction with any method of complying with Section F2.6c(3). Examples 5.3.10, 5.3.11 and 5.3.12 are configured as follows:

Example	Method of complying with AISC *Seismic Provisions* Section F2.6b	Method of complying with AISC *Seismic Provisions* Section F2.6c(3)
5.3.10	Detailed to provide rotation per Section F2.6b(a)	Linear hinge zone
5.3.11	Detailed as FR connection per Section F2.6b(b)(i)	Elliptical hinge zone
5.3.12	Designed to resist moments per Section F2.6b(b)	Hinge plate for in-plane brace buckling

Examples 5.3.1 through 5.3.6 address analysis and SCBF member design issues. Example 5.3.7 demonstrates how to determine the maximum force on the system when limited by foundation uplift. Examples 5.3.8 and 5.3.9 address brace-to-beam connection design.

SCBF Design Example Plan and Elevation

The following examples illustrate the design of SCBF systems based on AISC *Seismic Provisions* Section F2. The plan and elevation are shown in Figure 5-10 and Figure 5-11. The lateral forces shown in Figure 5-11 are the seismic forces from the equivalent lateral force procedure of ASCE/SEI 7 Section 12.8 and apply to the entire frame.

The code-specified gravity loading is as follows:

D_{floor} = 85 psf
D_{roof} = 68 psf
L_{floor} = 80 psf (50 psf reduced)
S = 20 psf
Curtain wall = 175 lb/ft along building perimeter at every level

From ASCE/SEI 7, the Seismic Design Category is D, Ω_o = 2.0, R = 6, ρ = 1.3, and S_{DS} = 1.0. Assume that the effective length method of AISC *Specification* Appendix 7 is used for stability design.

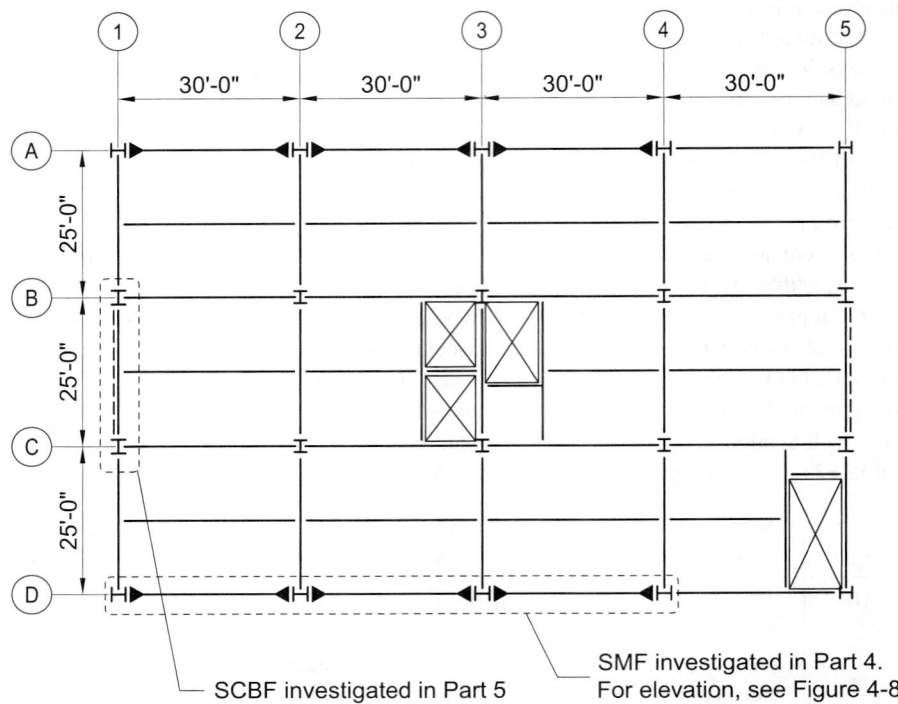

Fig. 5-10. SCBF plan for SCBF member examples.

Example 5.3.1. SCBF Brace Design

Given:

Refer to Brace BR-1 in Figure 5-11. Select an ASTM A500 Grade B round HSS to resist the following axial loads.

$P_D = 18.0$ kips $P_L = 9.50$ kips $P_{Q_E} = \pm 197$ kips

The applicable building code specifies the use of ASCE/SEI 7 for calculation of loads. The axial force due to the snow load is negligible.

Relevant seismic design parameters were given in the SCBF Design Example Plan and Elevation section.

From an elastic analysis, the first-order interstory drift between the base and the second level is $\Delta_H = 0.200$ in.

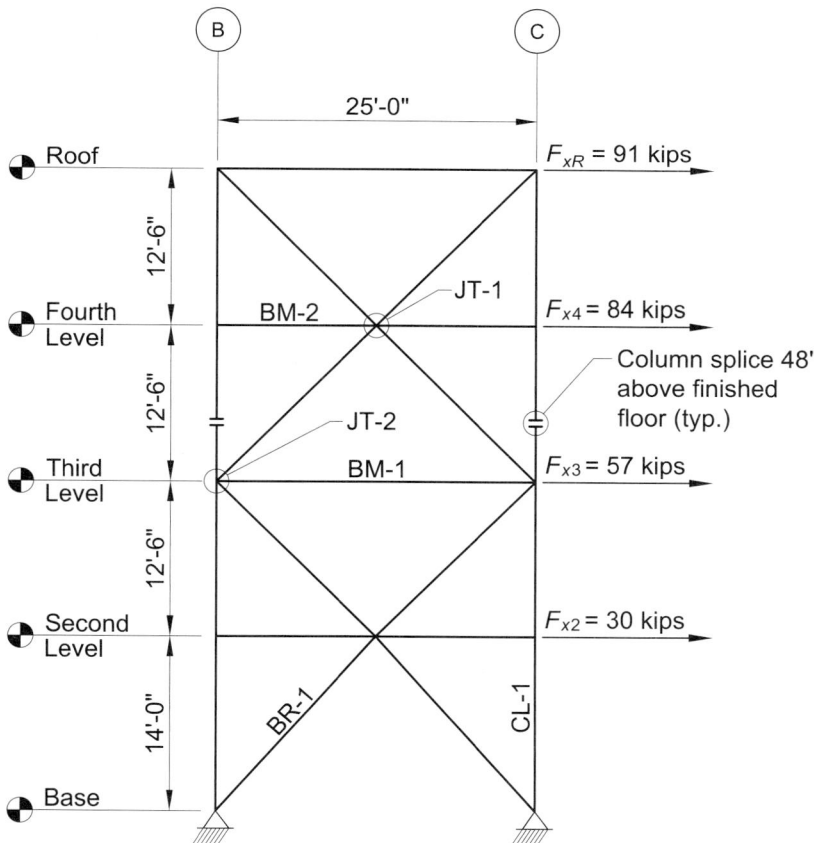

Fig. 5-11. SCBF elevation for SCBF member examples.

Assume that the ends of the brace are pinned and braced against translation for both the x-x and y-y axes.

Solution:

From AISC *Manual* Table 2-4, the material properties are as follows:

ASTM A500 Grade B
$F_y = 42$ ksi
$F_u = 58$ ksi

Required Strength

Determine the required strength

The governing load combinations that include seismic effects are:

LRFD	ASD
LRFD Load Combinations 5 and 6 from ASCE/SEI 7 Section 12.4.2.3 (including the 0.5 factor on L permitted by Section 12.4.2.3)	ASD Load Combinations 5 and 8 from ASCE/SEI 7 Section 12.4.2.3
$(1.2 + 0.2S_{DS})D + \rho Q_E + 0.5L + 0.2S$	$(1.0 + 0.14S_{DS})D + H + F + 0.7\rho Q_E$
$(0.9 - 0.2S_{DS})D + \rho Q_E + 1.6H$	$(0.6 - 0.14S_{DS})D + 0.7\rho Q_E + H$

The required axial compressive strength of the brace is:

LRFD	ASD
LRFD Load Combination 5 from ASCE/SEI 7 Section 12.4.2.3	ASD Load Combination 5 from ASCE/SEI 7 Section 12.4.2.3
$P_u = [1.2 + 0.2(1.0)](18.0 \text{ kips})$ $+ 1.3(197 \text{ kips})$ $+ 0.5(9.50 \text{ kips}) + 0.2(0 \text{ kips})$ $= 286 \text{ kips}$	$P_a = [1.0 + 0.14(1.0)](18.0 \text{ kips})$ $+ 0 \text{ kips} + 0 \text{ kips}$ $+ 0.7(1.3)(197 \text{ kips})$ $= 200 \text{ kips}$

5.3 SPECIAL CONCENTRICALLY BRACED FRAMES

The required axial tensile strength of the brace is:

LRFD	ASD
LRFD Load Combination 6 from ASCE/SEI 7 Section 12.4.2.3 $$P_u = [0.9 - 0.2(1.0)](18.0 \text{ kips})$$ $$+ 1.3(-197 \text{ kips}) + 1.6(0 \text{ kips})$$ $$= -244 \text{ kips}$$	ASD Load Combination 8 from ASCE/SEI 7 Section 12.4.2.3 $$P_a = [0.6 - 0.14(1.0)](18.0 \text{ kips})$$ $$+ 0.7(1.3)(-197 \text{ kips}) + 0 \text{ kips}$$ $$= -171 \text{ kips}$$

The unbraced length of the brace from work point-to-work point is:

$$L = \sqrt{(14.0 \text{ ft})^2 + (12.5 \text{ ft})^2}$$
$$= 18.8 \text{ ft}$$

This length has been determined by calculating the distance between the work points based on the intersection of the centerlines of the brace, column and beams. Shorter unbraced lengths of the brace may be used if justified by the engineer of record.

AISC *Seismic Provisions* Section F2.4a requires that between 30% and 70% of the total horizontal force is resisted by braces in tension. From analysis, the total horizontal force in the line of the braced frame is 91 kips + 84 kips + 57 kips + 30 kips = 262 kips. The horizontal component of the axial force due to earthquake force in Brace BR-1, when it is in tension is:

$$\left(\frac{12.5 \text{ ft}}{18.8 \text{ ft}}\right)(197 \text{ kips}) = 131 \text{ kips}$$

which is 50% of the total horizontal force in the line of the braced frame. Therefore, it meets the lateral force distribution requirements in AISC *Seismic Provisions* Section F2.4a.

Try a round **HSS8.625×0.500** for the brace.

From AISC *Manual* Table 1-13, the geometric properties are as follows:

$D = 8.625$ in. $t_{nom} = 0.500$ in. $t_{des} = 0.465$ in. $A = 11.9$ in.^{2}
$I = 100$ in.^{4} $r = 2.89$ in.

Width-to-Thickness Limitations

According to AISC *Seismic Provisions* Section F2.5a, braces must satisfy the requirements for highly ductile members. Elements in the brace members must not exceed λ_{hd} width-to-thickness ratios in AISC *Seismic Provisions* Table D1.1.

From Table D1.1:

$$\frac{D}{t_{des}} = \frac{8.625 \text{ in.}}{0.465 \text{ in.}}$$
$$= 18.5$$

$$\lambda_{hd} = 0.038 \frac{E}{F_y}$$
$$= 0.038 \left(\frac{29,000 \text{ ksi}}{42 \text{ ksi}} \right)$$
$$= 26.2$$

Since $\frac{D}{t_{des}} \leq \lambda_{hd}$, the HSS8.625×0.500 satisfies the width-to-thickness limitation for highly ductile members.

Alternatively, using Table 1-6, it can be seen that the HSS8.625×0.500 will satisfy the width-to-thickness requirements for an SCBF brace.

Brace Slenderness

Use $K = 1.0$ for both the x-x and y-y axes. According to AISC *Seismic Provisions* Section F2.5b(1), braces must have a slenderness ratio $\frac{KL}{r} \leq 200$.

$$\frac{KL}{r} = \frac{1.0(18.8 \text{ ft})(12 \text{ in./ft})}{2.89 \text{ in.}}$$
$$= 78.1 < 200 \quad \textbf{o.k.}$$

Second-Order Effects

Follow the procedure of AISC *Specification* Appendix 8. Because there are no moments, only the following equation need be checked.

$$P_r = P_{nt} + B_2 P_{lt} \quad \quad \quad (Spec. \text{ Eq. A-8-2})$$

Calculate B_2

To determine P_{story}, use an area of 9,000 ft² on each floor and the gravity loads given in the SCBF Design Example Plan and Elevation section. Use load combinations that include seismic effects; in this case, Load Combination 5 from ASCE/SEI 7 Section 12.4.2.3 for LRFD and ASD governs.

5.3 SPECIAL CONCENTRICALLY BRACED FRAMES

LRFD	ASD
$P_{story} = 9{,}000 \text{ ft}^2 \begin{Bmatrix} [1.2 + 0.2(1.0)] \\ \times [68 \text{ psf} + 3(85 \text{ psf})] \\ + 0 \text{ psf} + 0.5(3)(50 \text{ psf}) \\ + 0.2(20 \text{ psf}) \end{Bmatrix}$ $\times (1 \text{ kip}/1{,}000 \text{ lb})$ $+ \begin{Bmatrix} [1.2 + 0.2(1.0)] \\ \times [175 \text{ lb/ft}(4)(390 \text{ ft})] \\ \times (1 \text{ kip}/1{,}000 \text{ lb}) \end{Bmatrix}$ $= 5{,}160 \text{ kips}$	$P_{story} = 9{,}000 \text{ ft}^2 \begin{Bmatrix} [1.0 + 0.14(1.0)] \\ \times [68 \text{ psf} + 3(85 \text{ psf})] \\ + 0 \text{ psf} + 0 \text{ psf} + 0 \text{ psf} \end{Bmatrix}$ $\times (1 \text{ kip}/1{,}000 \text{ lb})$ $+ \begin{Bmatrix} [1.0 + 0.14(1.0)] \\ \times [175 \text{ lb/ft}(4)(390 \text{ ft})] \\ \times (1 \text{ kip}/1{,}000 \text{ lb}) \end{Bmatrix}$ $= 3{,}630 \text{ kips}$

The total story shear, H, with two bays of bracing in the direction under consideration, where each braced frame is designed to resist the seismic loads shown in Figure 5-11, is determined as follows. From an elastic analysis, the first-order interstory drift is $\Delta_H = 0.200$ in.

$H = 2(91 \text{ kips} + 84 \text{ kips} + 57 \text{ kips} + 30 \text{ kips})$
$\quad = 524 \text{ kips}$

$L = 14.0 \text{ ft}$

$R_M = 1.0$ for a braced frame

$P_{e\ story} = R_M \dfrac{HL}{\Delta_H}$ *(Spec. Eq. A-8-7)*

$\quad = 1.0 \dfrac{524 \text{ kips}(14.0 \text{ ft})}{(0.200 \text{ in.})(1 \text{ ft}/12 \text{ in.})}$

$\quad = 440{,}000 \text{ kips}$

Using AISC *Specification* Equation A-8-6:

LRFD	ASD
$\alpha = 1.00$ $B_2 = \dfrac{1}{1 - \dfrac{\alpha P_{story}}{P_{e\ story}}} \geq 1$ $= \dfrac{1}{1 - \dfrac{1.00(5{,}160 \text{ kips})}{440{,}000 \text{ kips}}}$ $= 1.01 \geq 1$	$\alpha = 1.60$ $B_2 = \dfrac{1}{1 - \dfrac{\alpha P_{story}}{P_{e\ story}}} \geq 1$ $= \dfrac{1}{1 - \dfrac{1.60(3{,}630 \text{ kips})}{440{,}000 \text{ kips}}}$ $= 1.01 \geq 1$

Because $B_2 \le 1.5$, the effective length method is a valid way to check stability according to AISC *Specification* Appendix 7.

The required axial compressive strength of the brace including second order effects is, from AISC *Specification* Equation A-8-2:

$$P_r = P_{nt} + B_2 P_{lt} \qquad (Spec.\ Eq.\ A\text{-}8\text{-}2)$$

LRFD	ASD
$P_u = (1.2 + 0.2 S_{DS}) P_D + B_2 \rho Q_E$ $+ 0.5L + 0.2S$ $= [1.2 + 0.2(1.0)](18.0 \text{ kips})$ $+ 1.01(1.3)(197 \text{ kips})$ $+ 0.5(9.50 \text{ kips}) + 0.2(0 \text{ kips})$ $= 289 \text{ kips}$	$P_a = [1.0 + 0.14(1.0)] P_D$ $+ P_H + P_F + 0.7 \rho B_2 P_{Q_E}$ $= [1.0 + 0.14(1.0)](18.0 \text{ kips})$ $+ 0 \text{ kips} + 0 \text{ kips}$ $+ 0.7(1.3)(1.01)(197 \text{ kips})$ $= 202 \text{ kips}$

Available Compressive Strength

As stated previously, use $L = 18.8$ ft for the unbraced length of the brace.

From AISC *Manual* Table 4-5 for the **HSS8.625×0.500** brace with $KL = 18.8$ ft (using interpolation), the available compressive strength is:

LRFD	ASD
$\phi_c P_n = 309 \text{ kips} > 289 \text{ kips}$ **o.k.**	$\dfrac{P_n}{\Omega_c} = 206 \text{ kips} > 202 \text{ kips}$ **o.k.**

Available Tensile Strength

From AISC *Manual* Table 5-6 for the **HSS8.625×0.500** brace, the available tensile yielding strength is:

LRFD	ASD
$\phi_t P_n = 450 \text{ kips} > 244 \text{ kips}$ **o.k.**	$\dfrac{P_n}{\Omega_t} = 299 \text{ kips} > 171 \text{ kips}$ **o.k.**

Tensile rupture on the net section must also be checked at the connection; see Examples 5.3.8 and 5.3.10 for illustration of this check.

Use an **HSS8.625×0.500** for SCBF Brace BR-1.

Comments:

The engineer of record may be able to justify a shorter unbraced length for the brace. In this example, if an unbraced length of 14 ft could be justified, an **HSS7.500×0.500** could have been used for the brace. Because the end connections may be designed to resist the expected yield strength of the brace in tension, a 13% decrease in brace area would reduce the required connection strength.

Example 5.3.2. SCBF Analysis

Given:
Refer to the braced frame elevation and sizes shown in Figure 5-12. All braces are ASTM A500 Grade B round HSS. Perform an analysis to determine the expected strengths in tension and compression of the braces according to AISC *Seismic Provisions* Section F2.3. Some engineers may choose not to change the brace size at every level, but they are different at every level in these design examples to fully illustrate the AISC *Seismic Provisions* requirements.

Solution:
From AISC *Manual* Table 2-4, the material properties are as follows:

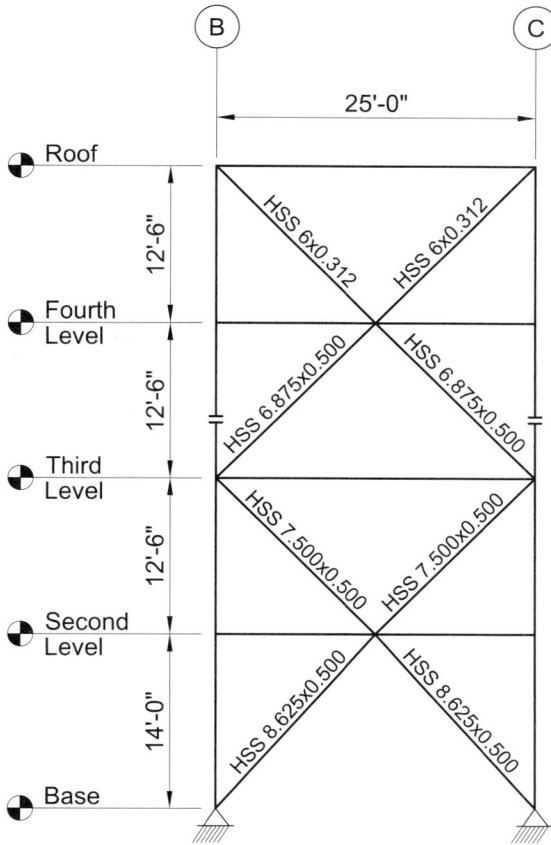

Fig. 5-12. SCBF elevation for Example 5.3.2.

ASTM A500 Grade B
$F_y = 42$ ksi
$F_u = 58$ ksi

From AISC *Manual* Table 1-13, the geometric properties of the braces are:

HSS6×0.312
$A = 5.22$ in.^{2} $r = 2.02$ in.

HSS6.875×0.500
$A = 9.36$ in.^{2} $r = 2.27$ in.

HSS7.500×0.500
$A = 10.3$ in.^{2} $r = 2.49$ in.

HSS8.625×0.500
$A = 11.9$ in.^{2} $r = 2.89$ in.

According to AISC *Seismic Provisions* Section F2.3, the required strengths of columns, beams and connections are based on the load combinations in the applicable building code, where the amplified seismic load, E_{mh}, is based on the larger force determined from the following two analyses:

(i) An analysis in which all braces are assumed to resist forces corresponding to their expected strength in compression or in tension
(ii) An analysis in which all braces in tension are assumed to resist forces corresponding to their expected strength and all braces in compression are assumed to resist their expected post-buckling strength

In order to study the effects of analyses (i) and (ii) on the rest of the frame, the expected strengths in tension and compression and the post-buckling strength in compression must be determined for all of the braces.

For determining the expected strength in compression, AISC *Seismic Provisions* Section F2.3 requires that the brace length used shall not exceed the distance from brace end-to-brace end. The work point-to-work point length of the typical brace above the base level is:

$$L = \sqrt{(12.5 \text{ ft})^2 + (12.5 \text{ ft})^2}$$
$$= 17.7 \text{ ft}$$

The work point-to-work point length of the brace at the base level is:

$$L = \sqrt{(14.0 \text{ ft})^2 + (12.5 \text{ ft})^2}$$
$$= 18.8 \text{ ft}$$

The brace length will be less than this distance because of the column and beam depth and because the gusset will accommodate brace buckling [AISC *Seismic Provisions* Section

5.3 SPECIAL CONCENTRICALLY BRACED FRAMES

F2.6c(3)(b)] by allowing a $2t$ clearance between the end of the brace and the line of restraint. It is likely that the actual length from brace end-to-brace end between the connections will be significantly less than the work point-to-work point distance calculated previously. Example 5.3.8 verifies that the actual length of the brace is approximately 12 to 13 ft; therefore, use a length of 12 ft for determining the expected strength in compression for all braces.

The following Tables 5-1 and 5-2 show the expected strengths in tension and the expected and post-buckling strengths in compression of all braces. A sample calculation is given for the HSS6×0.312, and a similar procedure is used to determine the strengths of the other braces. From AISC *Seismic Provisions* Table A3.1:

$R_y = 1.4$

From AISC *Seismic Provisions* Section F2.3, the expected strength of the brace in tension is:

$$P_{tension} = R_y F_y A_g$$
$$= 1.4(42 \text{ ksi})(5.22 \text{ in.}^2)$$
$$= 307 \text{ kips}$$

In compression, $R_y F_y$ is used in lieu of F_y for the determination of F_{cre} according to AISC *Seismic Provisions* Section F2.3, where F_{cre} is determined from AISC *Specification* Chapter E, using the equations for F_{cr}.

$$\frac{KL}{r} = \frac{1.0(12.0 \text{ ft})(12 \text{ in./ft})}{2.02 \text{ in.}}$$
$$= 71.3$$

$$4.71\sqrt{\frac{E}{R_y F_y}} = 4.71\sqrt{\frac{29,000 \text{ ksi}}{1.4(42 \text{ ksi})}}$$
$$= 105$$

Because $71.3 < 105$, AISC *Specification* Equation E3-2 applies, and F_{cre} is determined as follows:

$$F_e = \frac{\pi^2 E}{\left(\dfrac{KL}{r}\right)^2} \qquad \text{(Spec. Eq. E3-4)}$$
$$= \frac{\pi^2 (29,000 \text{ ksi})}{(71.3)^2}$$
$$= 56.3 \text{ ksi}$$

$$F_{cre} = \left[0.658^{\frac{R_y F_y}{F_e}}\right] R_y F_y \qquad \text{(from } Spec. \text{ Eq. E3-2)}$$

$$= \left[0.658^{\frac{1.4(42 \text{ ksi})}{(56.3 \text{ ksi})}}\right](1.4)(42 \text{ ksi})$$

$$= 38.0 \text{ ksi}$$

From AISC *Seismic Provisions* Section F2.3, the expected strength of the brace in compression is:

$$P_{compression} = 1.14 F_{cre} A_g$$

$$= 1.14(38.0 \text{ ksi})(5.22 \text{ in.}^2)$$

$$= 226 \text{ kips}$$

Table 5-1
Expected Brace Strength in Tension

Brace Member	A in.^{2}	$R_y F_y A_g$ kips
HSS6×0.312	5.22	307
HSS6.875×0.500	9.36	550
HSS7.500×0.500	10.3	606
HSS8.625×0.500	11.9	700

Table 5-2
Expected Brace Strength and Post-Buckling Brace Strength in Compression

Brace Member	$A = A_g$ in.^{2}	r in.	Length ft	KL/r	F_{cre} ksi	Expected Strength in Compression $1.14 F_{cre} A_g$ kips	Expected Post-Buckling Strength in Compression $0.3(1.14 F_{cre} A_g)$ kips
HSS6×0.312	5.22	2.02	12.0	71.3	38.0	226	67.8
HSS6.875×0.500	9.36	2.27	12.0	63.4	41.6	444	133
HSS7.500×0.500	10.3	2.49	12.0	57.8	44.1	518	155
HSS8.625×0.500	11.9	2.89	12.0	49.8	47.5	644	193

5.3 SPECIAL CONCENTRICALLY BRACED FRAMES

In Examples 5.3.3 through 5.3.6, the forces generated in this analysis will be considered in the design of the beam, column and column splice connection. The diagram in Figure 5-13 shows the forces imposed on the frame from buckling and yielding of the braces. For the analysis provisions of AISC *Seismic Provisions* F2.3(ii), the expected strengths of the braces in compression shown in Figure 5-13a are multiplied by 0.3 (expected post-buckling brace strength) and shown in Figure 5-13b.

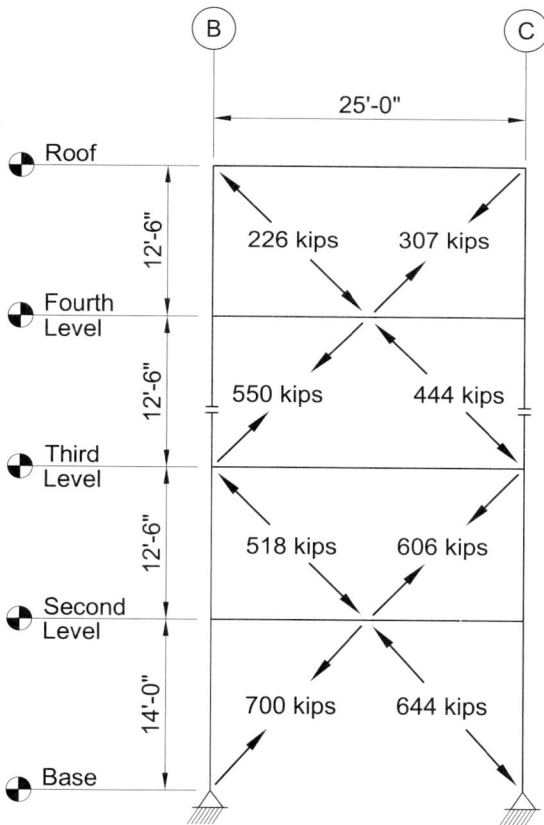

Fig. 5-13a. Forces imposed on frame from brace buckling/yielding according to AISC Seismic Provisions Section F2.3(i).

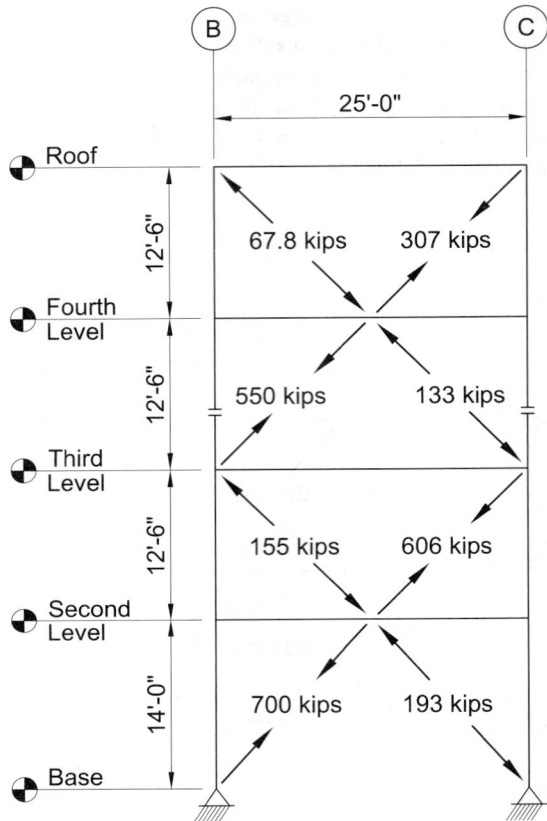

Fig. 5-13b. Forces imposed on frame from brace buckling/yielding according to AISC Seismic Provisions Section F2.3(ii).

Example 5.3.3. SCBF Column Design

Given:

Refer to Column CL-1 in Figure 5-11. Select an ASTM A992 W-shape with the available strength required by the AISC *Seismic Provisions*.

Relevant seismic parameters were given in the SCBF Design Example Plan and Elevation section. The column forces from gravity and snow loads are the following:

$P_D = 147$ kips $\qquad P_L = 60.0$ kips $\qquad P_S = 7.00$ kips

The seismic force in Column CL-1 from the seismic forces stipulated by the applicable building code using an equivalent lateral force analysis, not including the Ω_o amplification, was determined from analysis to be $P_{Q_E} = 248$ kips.

The forces resulting from the expected strengths of the braces defined in AISC *Seismic Provisions* Section F2.3 and calculated in Example 5.3.2 must be considered. The Exception

5.3 SPECIAL CONCENTRICALLY BRACED FRAMES

2(a) in Section F2.3 will also be considered. There are two additional Exceptions: 2(b) forces corresponding to the resistance of the foundation to overturning uplift, and 2(c) forces determined from nonlinear analysis. Exception 2(b) will be considered in Example 5.3.6.

Assume that the ends of the column are pinned and braced against translation for both the x-x and y-y axes.

Solution:

From AISC *Manual* Table 2-4, the material properties are as follows:

ASTM A992
$F_y = 50$ ksi
$F_u = 65$ ksi

Required Strength

Determine the required strength of the column from AISC Seismic Provisions Section F2.3 (Mechanism Analysis)

According to AISC *Seismic Provisions* Section F2.3, the required strengths of columns are based on the load combinations in the applicable building code, where the amplified seismic load, E_{mh}, is based on an analysis in which all braces are assumed to resist forces corresponding to their expected strengths in compression or in tension. The analysis in which the compression braces are at their post-buckling strength does not govern here.

Figure 5-14 shows the forces from the expected strengths of the braces as determined in Example 5.3.2. These forces can be considered as applied loads acting on the columns and as applied loads on the beam, which are shown here as beam shears acting on the column. Because seismic forces must be considered in both directions, both columns in the frame must be designed both for the maximum tension, shown for the column on gridline B, and for the maximum compression, shown for the column on gridline C.

The axial compression force in the column from this analysis is, with forces that produce compression in the column shown as positive:

$$P_{E_{mh}} = (307 \text{ kips} + 444 \text{ kips} + 606 \text{ kips})\sin 45° + (8.84 \text{ kips} - 11.3 \text{ kips})$$
$$= 957 \text{ kips (compression)}$$

The axial tension force in the column from this analysis is, with forces that produce tension in the column shown as negative:

$$T_{E_{mh}} = (-226 \text{ kips} - 550 \text{ kips} - 518 \text{ kips})\sin 45° + (8.84 \text{ kips} - 11.3 \text{ kips})$$
$$= -917 \text{ kips (tension)}$$

Note that since the expected strength from the brace at the lowest level is not included, the forces in tension and compression are not exactly equal.

Using the load combinations in ASCE/SEI 7 including the overstrength factor as required by AISC *Seismic Provisions* Section F2.3 where the amplified seismic load is substituted with the analysis described in Section F2.3, the required axial compressive strength of the column is:

LRFD	ASD
LRFD Load Combination 5 from ASCE/SEI 7 Section 12.4.3.2 (including the 0.5 factor on L permitted in Section 12.4.3.2) $$P_u = (1.2 + 0.2S_{DS})P_D + P_{E_{mh}} + 0.5P_L$$ $$+ 0.2P_S$$ $$= [1.2 + 0.2(1.0)](147 \text{ kips}) + 957 \text{ kips}$$ $$+ 0.5(60.0 \text{ kips}) + 0.2(7.00 \text{ kips})$$ $$= 1,190 \text{ kips}$$	ASD Load Combination 5 from ASCE/SEI 7 Section 12.4.3.2 $$P_a = (1.0 + 0.14S_{DS})P_D + P_H + P_F$$ $$+ 0.7P_{E_{mh}}$$ $$= [1.0 + 0.14(1.0)](147 \text{ kips}) + 0 \text{ kips}$$ $$+ 0 \text{ kips} + 0.7(957 \text{ kips})$$ $$= 837 \text{ kips}$$

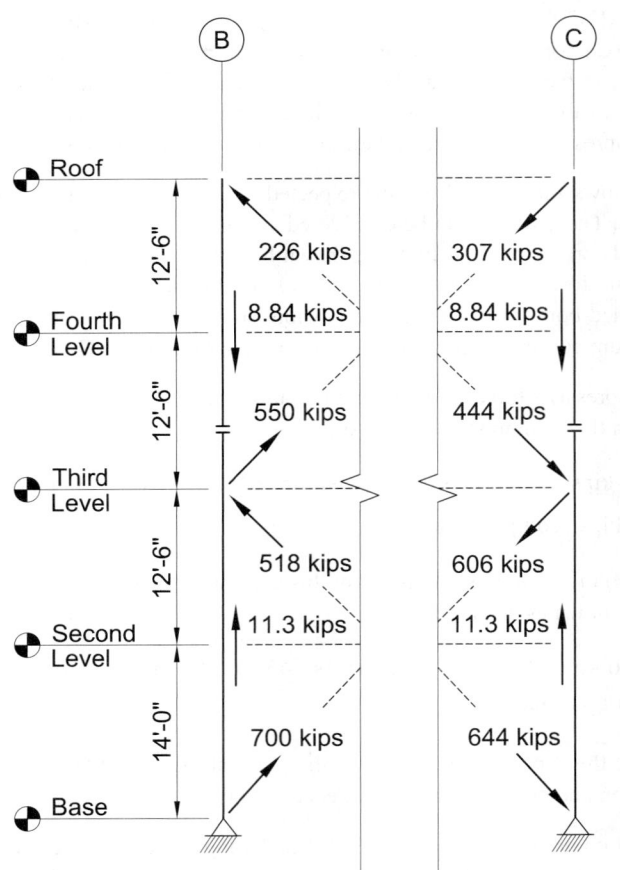

Fig. 5-14. SCBF applied column forces for Example 5.3.3.

The required axial tensile strength of the column is:

LRFD	ASD
LRFD Load Combination 7 from ASCE/SEI 7 Section 12.4.3.2	ASD Load Combination 8 from ASCE/SEI 7 Section 12.4.3.2
$P_u = (0.9 - 0.2S_{DS})P_D + P_{E_{mh}}$ $= [0.9 - 0.2(1.0)](147 \text{ kips})$ $+ (-917 \text{ kips})$ $= -814 \text{ kips}$	$P_a = (0.6 - 0.14S_{DS})P_D + 0.7P_{E_{mh}} + P_H$ $= [0.6 - 0.14(1.0)](147 \text{ kips})$ $+ 0.7(-917 \text{ kips}) + 0 \text{ kips}$ $= -574 \text{ kips}$

Determine the required strength of the column from AISC Seismic Provisions Section F2.3 Exception (2)(a)

AISC *Seismic Provisions* Section F2.3 Exception (2)(a) states that the required strength of columns need not exceed the forces determined using load combinations stipulated by the applicable building code including the amplified seismic load, applied to a building frame model in which all compression braces have been removed. For exterior columns (columns at the ends of a braced frame or at the ends of several bays of bracing), forces determined using this exception may be lower than those required by AISC *Seismic Provisions* Section D1.4a(2), in which case it might not be prudent to use this exception. In this example with a 2-story X configuration, the required strength of the column determined from a model in which compression braces have been removed results in required strengths in tension that are significantly less than forces determined from the analysis provisions of AISC *Seismic Provisions* Section F2.3.

For this example and for other typical frames, a model that includes all braces can be used with the load combinations including the code-based amplified seismic load to determine the appropriate required strength of exterior columns. However, it should be noted that for interior columns in multi-bay braces, a building frame model in which all compression braces have been removed should be used.

Figure 5-11 shows the forces from an equivalent lateral force analysis, before the overstrength factor is applied. The seismic force in Column CL-1 from the seismic forces stipulated by the applicable building code using an equivalent lateral force analysis, not including the Ω_o amplification, was determined from analysis to be $P_{Q_E} = 248$ kips.

$P_{Q_E} = 248$ kips (compression)

$P_{Q_E} = 248$ kips (tension)

Using these forces in the full model which includes the compression braces as an approximation of Exception (2)(a) of AISC *Seismic Provisions* Section F2.3, the load combinations in ASCE/SEI 7 give a required axial compressive strength for the column as follows:

LRFD	ASD
LRFD Load Combination 5 from ASCE/SEI 7 Section 12.4.3.2 $P_u = (1.2 + 0.2S_{DS})P_D + \Omega_o P_{QE} + 0.5P_L$ $+ 0.2P_S$ $= [1.2 + 0.2(1.0)](147 \text{ kips})$ $+ 2.0(248 \text{ kips}) + 0.5(60.0 \text{ kips})$ $+ 0.2(7.00 \text{ kips})$ $= 733 \text{ kips}$	ASD Load Combination 5 from ASCE/SEI 7 Section 12.4.3.2 $P_a = (1.0 + 0.14S_{DS})P_D + P_H + P_F$ $+ 0.7\Omega_o P_{QE}$ $= [1.0 + 0.14(1.0)](147 \text{ kips})$ $+ 0 \text{ kips} + 0 \text{ kips}$ $+ 0.7(2.0)(248 \text{ kips})$ $= 515 \text{ kips}$

The required axial tensile strength of the column is:

LRFD	ASD
LRFD Load Combination 7 from ASCE/SEI 7 Section 12.4.3.2 $P_u = (0.9 - 0.2S_{DS})P_D + \Omega_o P_{QE} + 1.6P_H$ $= [0.9 - 0.2(1.0)](147 \text{ kips})$ $+ 2.0(-248 \text{ kips}) + 1.6(0 \text{ kips})$ $= -393 \text{ kips}$	ASD Load Combination 8 from ASCE/SEI 7 Section 12.4.3.2 $P_a = (0.6 - 0.14S_{DS})P_D + 0.7\Omega_o P_{QE} + P_H$ $= [0.6 - 0.14(1.0)](147 \text{ kips})$ $+ 0.7(2.0)(-248 \text{ kips}) + 0 \text{ kips}$ $= -280 \text{ kips}$

Because these required strengths are less than those determined from the mechanism analysis of AISC *Seismic Provisions* Section F2.3, they will be used for the design of the column.

Second-Order Effects

Because the column is designed for code-based forces rather than the mechanism analysis requirements of AISC *Seismic Provisions* Section F2.3, second-order effects should be considered. From Example 5.3.1 for the brace at this level, $B_2 = 1.01$. Because the column does not have moments, there is no need to calculate B_1 factors.

Therefore, the required axial compressive strength of the column including second-order effects is, from AISC *Specification* Equation A-8-2:

$P_r = P_{nt} + B_2 P_{lt}$ (*Spec.* Eq. A-8-2)

5.3 SPECIAL CONCENTRICALLY BRACED FRAMES

LRFD	ASD
$P_u = (1.2 + 0.2S_{DS})P_D + B_2\Omega_o P_{QE}$ $\quad + 0.5P_L + 0.2P_S$ $= [1.2 + 0.2(1.0)](147 \text{ kips})$ $\quad + 1.01(2.0)(248 \text{ kips})$ $\quad + 0.5(60.0 \text{ kips})$ $\quad + 0.2(7.00 \text{ kips})$ $= 738 \text{ kips}$	$P_a = (1.0 + 0.14S_{DS})P_D + P_H + P_F$ $\quad + B_2(0.7)\Omega_o P_{QE}$ $= [1.0 + 0.14(1.0)](147 \text{ kips})$ $\quad + 0 \text{ kips} + 0 \text{ kips}$ $\quad + 1.01(0.7)(2.0)(248 \text{ kips})$ $= 518 \text{ kips}$

For comparison, Table 5-3 provides a summary of the required axial strengths of the column based on the two different analyses considered.

Try a W12×96.

Available Compressive Strength

Use $K = 1.0$ for both the *x-x* and *y-y* axes. From AISC *Manual* Table 4-1, the available strength in axial compression for a W12×96 with $KL = 14$ ft:

LRFD	ASD
$\phi_c P_n = 1{,}020 \text{ kips} > 738 \text{ kips}$ **o.k.**	$\dfrac{P_n}{\Omega_c} = 680 \text{ kips} > 518 \text{ kips}$ **o.k.**

Available Tensile Strength

From AISC *Manual* Table 5-1, the available strength of the W12×96 column in axial tension for yielding on the gross section is:

LRFD	ASD
$\phi_t P_n = 1{,}270 \text{ kips} > 393 \text{ kips}$ **o.k.**	$\dfrac{P_n}{\Omega_t} = 844 \text{ kips} > 280 \text{ kips}$ **o.k.**

Width-to-Thickness Limitations

According to AISC *Seismic Provisions* Section F2.5a, the stiffened and unstiffened elements of columns must satisfy the requirements for highly ductile members in Section D1.1.

From Table 1-3 of this Manual, it can be seen that a W12×96 will satisfy the width-to-thickness limits for an SCBF column (note that any value of $P_{u\,max}$ and $P_{a\,max}$ is permissible, as shown in Table 1-3).

Use a W12×96 for SCBF Column CL-1.

Table 5-3
Required Axial Strength of Column CL-1 According to the Requirements of AISC Seismic Provisions Section F2.3 and D1.4a

Analysis with braces at expected strengths in tension and compression [AISC *Seismic Provisions* Section F2.3(i)]				Analysis with code-specified amplified seismic loads [AISC *Seismic Provisions* Section D1.4a(2)]			
Compression		Tension		Compression		Tension	
LRFD	ASD	LRFD	ASD	LRFD	ASD	LRFD	ASD
1,190	837	−814	−574	738	518	−393	−280

Example 5.3.4. SCBF Beam Design

Given:

Refer to Beam BM-2 in Figure 5-11. Select an ASTM A992 W-shape with a maximum depth of 36 in. Design the beam as a noncomposite beam for strength, although the composite deck can be considered to brace the beam as discussed later in this example. The applicable building code specifies the use of ASCE/SEI 7 for calculation of loads.

Assume the brace sizes are as shown in Figure 5-12. Relevant seismic parameters were given in the SCBF Design Example Plan and Elevation section. The gravity shears and moments on the beam, assuming a simple span from column line B to C, are:

$V_D = 11.2$ kips $V_L = 8.50$ kips $M_D = 120$ kip-ft $M_L = 100$ kip-ft

Solution:

From AISC *Manual* Table 2-4, the material properties are as follows:

ASTM A992
$F_y = 50$ ksi
$F_u = 65$ ksi

As required by AISC *Seismic Provisions* Section F2.3, the required strength of the beams shall be based on the load combinations in the applicable building code, including the amplified seismic load. The amplified seismic load is determined from the larger of:

(i) An analysis in which all braces are assumed to resist forces corresponding to their expected strength in compression or in tension
(ii) An analysis in which all braces in tension are assumed to resist forces corresponding to their expected strength and all braces in compression are assumed to resist their expected post-buckling strength

5.3 SPECIAL CONCENTRICALLY BRACED FRAMES

These forces are shown in Tables 5-1 and 5-2, and the forces acting on beam BM-2 are shown in Figure 5-15.

Required Strength

Determine the required axial strength of the beam based on AISC Seismic Provisions Section F2.3(i)

From AISC *Seismic Provisions* Section F2.3(i), the required axial strength of the beam is based on the braces at their expected strengths in tension and compression. The "unbalanced" vertical force is determined from the vertical component of all four brace forces.

$$P_y = (307 \text{ kips} - 226 \text{ kips} + 444 \text{ kips} - 550 \text{ kips})\sin 45°$$
$$= -17.7 \text{ kips}$$

This unbalanced vertical force can be considered as a load acting at the midpoint of the beam, and produces the following shear and moment:

$$V_{E_{mh}} = \frac{P_y}{2}$$
$$= \frac{17.7 \text{ kips}}{2}$$
$$= 8.85 \text{ kips}$$

$$M_{E_{mh}} = \frac{P_y L}{4}$$
$$= \frac{17.7 \text{ kips}(25.0 \text{ ft})}{4}$$
$$= 111 \text{ kip-ft}$$

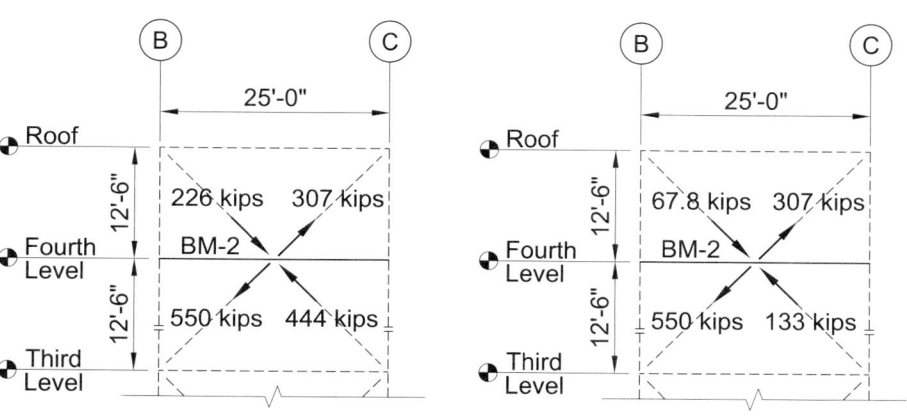

(a) Forces from AISC Seismic Provisions Section F2.3(i)

(b) Forces from AISC Seismic Provisions Section F2.3(ii)

Fig. 5-15. Forces acting on Beam BM-2 from a mechanism analysis of AISC Seismic Provisions Section F2.3 as carried out in Example 5.3.2.

Note that the unbalanced vertical force from the braces is considered to act at a single point for the purpose of evaluating member limit states in the beam. In the connection design presented in Example 5.3.8, beam local limit states are evaluated using internal forces determined in the brace connection design.

To determine the required axial force of the beam, the horizontal component of the difference between the sum of the expected strengths of the braces below the beam and the sum of the expected strengths of the braces above the beam can be thought of as a "story force" which the beam must deliver to the braces. Since the braced frame is in the middle bay of a three-bay building, half of this story force can be considered to enter the braces from each side, and is carried by Beam BM-2 to the braces connected to the beam midspan. This force could act in either direction and is shown as positive.

$$P_x = \cos 45° \left[\frac{\Sigma(\text{Braces below beam}) - \Sigma(\text{Braces above beam})}{2} \right]$$

$$= \cos 45° \left[\frac{(550 \text{ kips} + 444 \text{ kips}) - (226 \text{ kips} + 307 \text{ kips})}{2} \right]$$

$$= 163 \text{ kips}$$

Using the load combinations in ASCE/SEI 7, the required axial strength of Beam BM-2 according to the analysis requirements of AISC *Seismic Provisions* Section F2.3(i) is:

LRFD	ASD
LRFD Load Combination 5 from ASCE/SEI 7 Section 12.4.3.2 (including the 0.5 factor on L permitted in Section 12.4.3.2)	ASD Load Combination 5 from ASCE/SEI 7 Section 12.4.3.2
$P_u = (1.2 + 0.2S_{DS})P_D + P_{E_{mh}}$ $\quad + 0.5P_L + 0.2P_S$ $= [1.2 + 0.2(1.0)](0 \text{ kips}) + 163 \text{ kips}$ $\quad + 0.5(0 \text{ kips}) + 0.2(0 \text{ kips})$ $= 163 \text{ kips}$	$P_a = (1.0 + 0.14S_{DS})P_D + P_H$ $\quad + P_F + 0.7P_{E_{mh}}$ $= [1.0 + 0.14(1.0)](0 \text{ kips}) + 0 \text{ kips}$ $\quad + 0 \text{ kips} + 0.7(163 \text{ kips})$ $= 114 \text{ kips}$

The required shear strength of Beam BM-2 according to the analysis requirements of AISC *Seismic Provisions* Section F2.3(i) is:

5.3 SPECIAL CONCENTRICALLY BRACED FRAMES

LRFD	ASD
LRFD Load Combination 5 from ASCE/SEI 7 Section 12.4.3.2 $$V_u = (1.2 + 0.2S_{DS})V_D + V_{E_{mh}}$$ $$+ 0.5V_L + 0.2V_S$$ $$= [1.2 + 0.2(1.0)](11.2 \text{ kips})$$ $$+ 8.85 \text{ kips} + 0.5(8.50 \text{ kips})$$ $$+ 0.2(0 \text{ kips})$$ $$= 28.8 \text{ kips}$$	ASD Load Combination 5 from ASCE/SEI 7 Section 12.4.3.2 $$V_a = (1.0 + 0.14S_{DS})V_D + V_H$$ $$+ V_F + 0.7V_{E_{mh}}$$ $$= [1.0 + 0.14(1.0)](11.2 \text{ kips})$$ $$+ 0 \text{ kips} + 0 \text{ kips} + 0.7(8.85 \text{ kips})$$ $$= 19.0 \text{ kips}$$

The required flexural strength of Beam BM-2 according to the analysis requirements of AISC *Seismic Provisions* Section F2.3(i) is:

LRFD	ASD
LRFD Load Combination 5 from ASCE/SEI 7 Section 12.4.3.2 $$M_u = (1.2 + 0.2S_{DS})M_D + M_{E_{mh}}$$ $$+ 0.5M_L + 0.2M_S$$ $$= [1.2 + 0.2(1.0)](120 \text{ kip-ft})$$ $$+ 111 \text{ kip-ft} + 0.5(100 \text{ kip-ft})$$ $$+ 0.2(0 \text{ kip-ft})$$ $$= 329 \text{ kip-ft}$$	ASD Load Combination 5 from ASCE/SEI 7 Section 12.4.3.2 $$M_a = (1.0 + 0.14S_{DS})M_D + M_H$$ $$+ M_F + 0.7M_{E_{mh}}$$ $$= [1.0 + 0.14(1.0)](120 \text{ kip-ft})$$ $$+ 0 \text{ kip-ft} + 0 \text{ kip-ft}$$ $$+ 0.7(111 \text{ kip-ft})$$ $$= 215 \text{ kip-ft}$$

Determine the required axial strength of the beam based on AISC Seismic Provisions Section F2.3(ii)

From AISC *Seismic Provisions* Section F2.3(ii), the required axial strength of the beam is based on the braces at their expected strength in tension and post-buckling strengths in compression. For this analysis, the expected strengths of the braces in compression must be multiplied by 0.3 to approximate their post-buckling strength as shown in Table 5-2.

The "unbalanced" vertical force is determined from the vertical component of all four brace forces.

$$P_y = (307 \text{ kips} - 67.8 \text{ kips} + 133 \text{ kips} - 550 \text{ kips})\sin 45°$$
$$= -126 \text{ kips}$$

This unbalanced vertical force can be considered as a load acting on the beam, and produces the following shear and moment:

$$V_{E_{mh}} = \frac{P_y}{2}$$
$$= \frac{126 \text{ kips}}{2}$$
$$= 63.0 \text{ kips}$$

$$M_{E_{mh}} = \frac{P_y L}{4}$$
$$= \frac{126 \text{ kips}(25.0 \text{ ft})}{4}$$
$$= 788 \text{ kip-ft}$$

To determine the required axial force of the beam, the horizontal component of the difference between the sum of the expected strengths of the braces below the beam and the sum of the expected strengths of the braces above the beam can be thought of as a "story force" which the beam must deliver to the braces. Since the braced frame is in the middle bay of a three-bay building, half of this story force can be considered to enter the braces from each side.

$$P_x = (\cos 45°) \left[\frac{\Sigma(\text{Braces below beam}) - \Sigma(\text{Braces above beam})}{2} \right]$$
$$= (\cos 45°) \left[\frac{(550 \text{ kips} + 133 \text{ kips}) - (67.8 \text{ kips} + 307 \text{ kips})}{2} \right]$$
$$= 109 \text{ kips}$$

Using the load combinations in ASCE/SEI 7, the required axial strength of BM-2 according to the analysis requirements of AISC *Seismic Provisions* Section F2.3(ii) is:

LRFD	ASD
LRFD Load Combination 5 from ASCE/SEI 7 Section 12.4.3.2 $P_u = (1.2 + 0.2S_{DS})P_D + P_{E_{mh}}$ $+ 0.5P_L + 0.2P_S$ $= [1.2 + 0.2(1.0)](0 \text{ kips}) + 109 \text{ kips}$ $+ 0.5(0 \text{ kips}) + 0.2(0 \text{ kips})$ $= 109 \text{ kips}$	ASD Load Combination 5 from ASCE/SEI 7 Section 12.4.3.2 $P_a = (1.0 + 0.14S_{DS})P_D + P_H$ $+ P_F + 0.7P_{E_{mh}}$ $= [1.0 + 0.14(1.0)](0 \text{ kips}) + 0 \text{ kips}$ $+ 0 \text{ kips} + 0.7(109 \text{ kips})$ $= 76.3 \text{ kips}$

5.3 SPECIAL CONCENTRICALLY BRACED FRAMES

The required shear strength of BM-2 according to the analysis requirements of AISC *Seismic Provisions* Section F2.3(ii) is:

LRFD	ASD
LRFD Load Combination 5 from ASCE/SEI 7 Section 12.4.3.2 $$V_u = (1.2+0.2S_{DS})V_D + V_{E_{mh}} \\ + 0.5V_L + 0.2V_S \\ = [1.2+0.2(1.0)](11.2 \text{ kips}) \\ + 63.0 \text{ kips} + 0.5(8.50 \text{ kips}) \\ + 0.2(0 \text{ kips}) \\ = 82.9 \text{ kips}$$	ASD Load Combination 5 from ASCE/SEI 7 Section 12.4.3.2 $$V_a = (1.0+0.14S_{DS})V_D + V_H \\ + V_F + 0.7V_{E_{mh}} \\ = [1.0+0.14(1.0)](11.2 \text{ kips}) \\ + 0 \text{ kips} + 0 \text{ kips} + 0.7(63.0 \text{ kips}) \\ = 56.9 \text{ kips}$$

The required flexural strength of Beam BM-2 according to the analysis requirements of AISC *Seismic Provisions* Section F2.3(ii) is:

LRFD	ASD
LRFD Load Combination 5 from ASCE/SEI 7 Section 12.4.3.2 $$M_u = (1.2+0.2S_{DS})M_D + M_{E_{mh}} \\ + 0.5M_L + 0.2M_S \\ = [1.2+0.2(1.0)](120 \text{ kip-ft}) \\ + 788 \text{ kip-ft} + 0.5(100 \text{ kip-ft}) \\ + 0.2(0 \text{ kip-ft}) \\ = 1{,}010 \text{ kip-ft}$$	ASD Load Combination 5 from ASCE/SEI 7 Section 12.4.3.2 $$M_a = (1.0+0.14S_{DS})M_D + M_H \\ + M_F + 0.7M_{E_{mh}} \\ = [1.0+0.14(1.0)](120 \text{ kip-ft}) \\ + 0 \text{ kip-ft} + 0 \text{ kip-ft} \\ + 0.7(788 \text{ kip-ft}) \\ = 688 \text{ kip-ft}$$

Note that the analysis of AISC *Seismic Provisions* Section F2.3(ii), with the braces at post-buckling strength in compression, gives significantly higher required shear and moment for the beam, and a moderately lower required axial force. The shear and moment resulting from the analysis of Section F2.3(ii) do not act simultaneously with the axial force resulting from Section F2.3(i).

In summary, the required strength of Beam BM-2 determined by the analysis provisions of AISC *Seismic Provisions* Section F2.3(i) is:

LRFD	ASD
$P_u = 163$ kips $V_u = 28.8$ kips $M_u = 329$ kip-ft	$P_a = 114$ kips $V_a = 19.0$ kips $M_a = 215$ kip-ft

The required strength of Beam BM-2 determined by the analysis provisions of AISC *Seismic Provisions* Section F2.3(ii) is:

LRFD	ASD
$P_u = 109$ kips	$P_a = 76.3$ kips
$V_u = 82.9$ kips	$V_a = 56.9$ kips
$M_u = 1{,}010$ kip-ft	$M_a = 688$ kip-ft

Beam Size Selection

The beam is subject to axial and flexural forces. The discussion in Part 8 and Table 8-1 of this Manual regarding the design of collector beams is applicable to the design of beams within a braced frame.

Try a W27×114.

From AISC *Manual* Table 1-1, the geometric properties are as follows:

W27×114
$A = 33.6$ in.^{2} $d = 27.3$ in. $t_w = 0.570$ in. $b_f = 10.1$ in.
$t_f = 0.930$ in. $k_{des} = 1.53$ in. $h/t_w = 42.5$ $I_x = 4{,}080$ in.^{4}
$S_x = 299$ in.^{3} $r_x = 11.0$ in. $Z_x = 343$ in.^{3} $I_y = 159$ in.^{4}
$r_y = 2.18$ in. $h_o = 26.4$ in. $J = 7.33$ in.^{4} $C_w = 27{,}600$ in.^{6}

In order to determine which limit states apply, the beam bracing requirements must be investigated.

Bracing Requirements

According to AISC *Seismic Provisions* Section F2.4b(2), beams in SCBF using V- and inverted-V configurations must satisfy the bracing requirements for moderately ductile members. This beam is considered part of such a configuration because it is intersected by braces at its midspan. AISC *Seismic Provisions* Section D1.2a requires that beam bracing in moderately ductile members have a maximum spacing of:

$$L_b = 0.17 r_y E/F_y \qquad \text{(\textit{Provisions} Eq. D1-2)}$$
$$= 0.17(2.18 \text{ in.})(1 \text{ ft}/12 \text{ in.})(29{,}000 \text{ ksi})/(50 \text{ ksi})$$
$$= 17.9 \text{ ft}$$

The beam span is 25 ft; therefore it is economical to provide bracing at midspan of the beam ($L_b = 12.5$ ft).

AISC *Seismic Provisions* Section D1.2a(a)(1) requires that both flanges of the beam be laterally braced or the cross section be torsionally braced. Assume the beam shown in Figure 5-10, spanning between column lines 1 and 2, at midspan of the SCBF frame will be used to provide lateral bracing.

5.3 SPECIAL CONCENTRICALLY BRACED FRAMES

Determine lateral bracing requirements

Beam bracing requirements are given in AISC *Specification* Appendix 6. The required strength of lateral nodal bracing is:

$$P_{rb} = 0.02 M_r C_d / h_o \qquad \text{(Spec. Eq. A-6-7)}$$

where

$$C_d = 1.0$$

From AISC *Seismic Provisions* Equation D1-1, the required flexural strength is:

LRFD	ASD
$M_r = R_y F_y Z$	$M_r = R_y F_y Z / 1.5$
$= 1.1(50 \text{ ksi})(343 \text{ in.}^3)$	$= 1.1(50 \text{ ksi})(343 \text{ in.}^3)/1.5$
$= 18{,}900$ kip-in.	$= 12{,}600$ kip-in.

From AISC *Specification* Equation A-6-7, the required strength of lateral nodal bracing is:

LRFD	ASD
$P_{rb} = 0.02(R_y F_y Z) C_d / h_o$	$P_{rb} = 0.02(R_y F_y Z) C_d / h_o$
$= 0.02(18{,}900 \text{ kip-in.})(1.0)/26.4 \text{ in.}$	$= 0.02(12{,}600 \text{ kip-in.})(1.0)/26.4 \text{ in.}$
$= 14.3$ kips	$= 9.55$ kips

The required stiffness of lateral nodal bracing is, according to AISC *Specification* Equation A-6-8:

LRFD	ASD
$\beta_{br} = \dfrac{1}{\phi}\left(\dfrac{10 M_r C_d}{L_b h_o}\right)$	$\beta_{br} = \Omega\left(\dfrac{10 M_r C_d}{L_b h_o}\right)$
$= \dfrac{1}{0.75}\left[\dfrac{10(18{,}900 \text{ kip-in.})(1.0)}{(12.5 \text{ ft})(12 \text{ in./ft})(26.4 \text{ in.})}\right]$	$= 2.00\left[\dfrac{10(12{,}600 \text{ kip-in.})(1.0)}{(12.5 \text{ ft})(12 \text{ in./ft})(26.4 \text{ in.})}\right]$
$= 63.6$ kip/in.	$= 63.6$ kip/in.

The axial stiffness of the member providing bracing to the beam is:

$$k = \frac{AE}{L}$$

The required area of the brace is:

$$k \geq \beta_{br} = 63.6 \text{ kip/in.}$$

$$A \geq \beta_{br}\left(\frac{L}{E}\right)$$

$$\geq 63.6 \text{ kip/in.} \left[\frac{30.0 \text{ ft}(12 \text{ in./ft})}{29{,}000 \text{ ksi}}\right]$$

$$= 0.790 \text{ in.}^2$$

Provide beam lateral bracing of both flanges at midspan of the beam (12.5 ft) with a minimum area of 0.790 in.^{2} and with an available axial compressive strength of 14.3 kips (LRFD) and 9.55 kips (ASD).

Note: The gravity beam shown (but not sized) in Figure 5-10 should be able to provide this lateral bracing, depending on the depth of the beam and the connection type.

Available Flexural Strength

Beam lateral bracing will be provided at 12.5 ft. However, the composite slab can be considered to continuously brace the beam, and therefore the limit state of lateral-torsional buckling does not apply and the available flexural strength is based on the plastic moment of the beam. From AISC *Manual* Table 3-6, the available flexural strength of the beam is:

LRFD	ASD
$\phi_b M_p = 1{,}290$ kip-ft	$\dfrac{M_p}{\Omega_b} = 856$ kip-ft

Available Compressive Strength

In compression, the beam is considered continuously braced by the slab so minor-axis flexural buckling about the y-y axis does not apply. For major-axis flexural buckling about the x-x axis, the beam is assumed unbraced ($KL = 25$ ft). As explained in Part 8 for collectors, torsional buckling is considered because the torsional unbraced length is not the same as the minor-axis flexural buckling unbraced length. Because the top flange is constrained by the composite slab, the applicable torsional limit state is constrained-axis flexural-torsional buckling, as discussed in Part 8 of this Manual.

For torsional buckling, the beam is considered unbraced between torsional brace points. In this example, the lateral braces of both flanges at midspan are assumed to provide a torsional braced point. Therefore the unbraced length for torsional buckling is taken as 12.5 ft. To summarize:

$L_x = 25.0$ ft (flexural buckling about x-x axis)
$L_y = 0$ ft (flexural buckling about y-y axis does not apply)
$L_z = 12.5$ ft (constrained-axis flexural-torsional buckling)

From AISC *Manual* Table 1-1 and AISC *Specification* Table B4.1, the web is slender for compression. Therefore the reduction factor for slender stiffened elements, Q_a, must be determined.

5.3 SPECIAL CONCENTRICALLY BRACED FRAMES

Determine the critical buckling strength for flexural buckling about the x-x axis, assuming Q = 1

$K_x = 1.0$

$L_x = 25.0$ ft

$$\frac{K_x L_x}{r_x} = \frac{1.0(25.0 \text{ ft})(12 \text{ in./ft})}{11.0 \text{ in.}}$$

$= 27.3$

The elastic buckling stress is:

$$F_e = \frac{\pi^2 E}{\left(\frac{KL}{r}\right)^2} \qquad \text{(Spec. Eq. E3-4)}$$

$$= \frac{\pi^2 (29,000 \text{ ksi})}{(27.3)^2}$$

$= 384$ ksi

The value of F_{cr} before local buckling effects are considered is determined as follows:

$$\frac{F_y}{F_e} = \frac{50 \text{ ksi}}{384 \text{ ksi}}$$

$= 0.130$

Because $0.130 < 2.25$, use AISC *Specification* Equation E3-2 to determine the critical buckling stress.

$$F_{cr} = \left(0.658^{\frac{F_y}{F_e}}\right) F_y \qquad \text{(Spec. Eq. E3-2)}$$

$$= \left(0.658^{\frac{50 \text{ ksi}}{384 \text{ ksi}}}\right) 50 \text{ ksi}$$

$= 47.3$ ksi

Determine the critical buckling strength for constrained-axis flexural-torsional buckling, assuming Q = 1

For the limit state of constrained-axis flexural-torsional buckling, the unbraced length is 12.5 ft and the top flange of the beam is considered continuously braced by the slab as described in Part 8 of this Manual.

$$F_e = \left\{ \frac{\pi^2 E \left[C_w + I_y (d/2)^2 \right]}{(K_z L)^2} + GJ \right\} \left[\frac{1}{I_x + I_y + (d/2)^2 A_g} \right] \quad (8\text{-}3)$$

$$= \left\{ \frac{\pi^2 (29{,}000 \text{ ksi}) \left[27{,}600 \text{ in.}^6 + 159 \text{ in.}^4 \left(\frac{27.3 \text{ in.}}{2} \right)^2 \right]}{[1.0(12.5 \text{ ft})(12 \text{ in./ft})]^2} + 11{,}200 \text{ ksi} \left(7.33 \text{ in.}^4 \right) \right\}$$

$$\times \left[\frac{1}{4{,}080 \text{ in.}^4 + 159 \text{ in.}^4 + \left(\frac{27.3 \text{ in.}}{2} \right)^2 \left(33.6 \text{ in.}^2 \right)} \right]$$

$$= 77.2 \text{ ksi}$$

The value of F_{cr} before local buckling effects are considered is determined as follows:

$$\frac{F_y}{F_e} = \frac{50 \text{ ksi}}{77.2 \text{ ksi}}$$

$$= 0.648$$

Because $0.648 < 2.25$, use Equation E3-2 to determine the critical buckling stress.

$$F_{cr} = \left[0.658^{\frac{F_y}{F_e}} \right] F_y \quad \text{(Spec. Eq. E3-2)}$$

$$= \left[0.658^{\frac{50 \text{ ksi}}{77.2 \text{ ksi}}} \right] 50 \text{ ksi}$$

$$= 38.1 \text{ ksi}$$

Because F_{cr} is lower for constrained-axis flexural-torsional buckling, this limit state governs over major axis flexural buckling.

Determine the reduction factor, Q, for slender elements

To determine the reduction factor, Q, use AISC *Specification* Section E7.2, with $f = F_{cr}$, and the minimum F_{cr} from the two preceding limit states. The reduced effective width of the slender web is determined as follows:

$$b = h$$
$$= d - 2k_{des}$$
$$= 27.3 \text{ in.} - 2(1.53 \text{ in.})$$
$$= 24.2 \text{ in.}$$

$$f = F_{cr}$$
$$= 38.1 \text{ ksi}$$

5.3 SPECIAL CONCENTRICALLY BRACED FRAMES

$$b_e = 1.92t\sqrt{\frac{E}{f}}\left[1 - \frac{0.34}{(b/t)}\sqrt{\frac{E}{f}}\right] \leq b \quad \text{(Spec. Eq. E7-17)}$$

$$= 1.92(0.570 \text{ in.})\sqrt{\frac{29,000 \text{ ksi}}{38.1 \text{ ksi}}}\left[1 - \frac{0.34}{42.5}\sqrt{\frac{29,000 \text{ ksi}}{38.1 \text{ ksi}}}\right] \leq 24.2 \text{ in.}$$

$$= 23.5 \text{ in.} \leq 24.2 \text{ in.}$$

$$Q_a = \frac{A_e}{A_g} \quad \text{(Spec. Eq. E7-16)}$$

$$= \frac{A_g - t_w(h - b_e)}{A_g}$$

$$= \frac{33.6 \text{ in.}^2 - 0.570 \text{ in.}(24.2 \text{ in.} - 23.5 \text{ in.})}{33.6 \text{ in.}^2}$$

$$= 0.988$$

$$Q_s = 1.0$$

$$Q = Q_s Q_a$$
$$= 1.0(0.988)$$
$$= 0.988$$

For the governing limit state of constrained-axis flexural-torsional buckling, accounting for slender elements, the available strength is determined as follows from AISC *Specification* Section E7:

$$\frac{QF_y}{F_e} = \frac{0.988(50 \text{ ksi})}{(77.2 \text{ ksi})}$$
$$= 0.640$$

Because $0.640 < 2.25$, use AISC *Specification* Equation E7-2.

$$F_{cr} = Q\left[0.658^{\frac{QF_y}{F_e}}\right]F_y \quad \text{(Spec. Eq. E7-2)}$$

$$= 0.988\left[0.658^{\frac{0.988(50 \text{ ksi})}{77.2 \text{ ksi}}}\right](50 \text{ ksi})$$

$$= 37.8 \text{ ksi}$$

$$P_n = F_{cr}A_g \quad \text{(Spec. Eq. E7-1)}$$
$$= 37.8 \text{ ksi}(33.6 \text{ in.}^2)$$
$$= 1,270 \text{ kips}$$

LRFD	ASD
$\phi_c P_n = 0.90(1,270 \text{ kips})$ $= 1,140 \text{ kips}$	$\dfrac{P_n}{\Omega_c} = \dfrac{1,270 \text{ kips}}{1.67}$ $= 760 \text{ kips}$

Second-Order Effects

Because the seismic component of the beam required strength comes from the mechanism analysis of AISC *Seismic Provisions* Section F2.3 and is based on the expected strengths of the braces, P-Δ effects need not be considered and B_2 from AISC *Specification* Appendix 8 need not be applied. P-Δ effects do not increase the forces corresponding to the expected brace strengths in compression and tension; instead, they may be thought of as contributing to the system reaching that state. P-δ effects do apply, however. The effective length method is used.

$$B_1 = \frac{C_m}{1 - \alpha P_r / P_{e1}} \geq 1 \qquad \text{(Spec. Eq. A-8-3)}$$

$$P_{e1} = \frac{\pi^2 EI^*}{(K_1 L)^2} \qquad \text{(Spec. Eq. A-8-5)}$$

$$= \frac{\pi^2 (29,000 \text{ ksi})(4,080 \text{ in.}^4)}{[1.0(25.0 \text{ ft})(12 \text{ in./ft})]^2}$$

$$= 13,000 \text{ kips}$$

$C_m = 1.0$, because there is transverse loading between supports

LRFD	ASD
$B_1 = \dfrac{1.0}{1 - [1.00(163 \text{ kips})/13,000 \text{ kips}]}$ $= 1.01$	$B_1 = \dfrac{1.0}{1 - [1.60(114 \text{ kips})/13,000 \text{ kips}]}$ $= 1.01$

The B_1 factor (P-δ effect) need only be applied to the first-order moment with the structure restrained against translation. The required flexural strength of Beam BM-2 according to the analysis requirements of AISC *Seismic Provisions* Section F2.3(i) and including second-order effects is determined from ASCE/SEI 7 Section 12.4.3.2 Load Combination 5 for LRFD and ASD:

LRFD	ASD
$M_u = B_1(1.2+0.2S_{DS})M_D + M_{E_{mh}}$ $\quad + B_1 0.5M_L + 0.2M_S$ $\quad = 1.01[1.2+0.2(1.0)](120 \text{ kip-ft})$ $\quad + 111 \text{ kip-ft} + 1.01(0.5)(100 \text{ kip-ft})$ $\quad + 0.2(0 \text{ kip-ft})$ $\quad = 331 \text{ kip-ft}$	$M_a = B_1(1.0+0.14S_{DS})M_D + M_H$ $\quad + M_F + 0.7M_{E_{mh}}$ $\quad = 1.01[1.0+0.14(1.0)](120 \text{ kip-ft})$ $\quad + 0 \text{ kip-ft} + 0 \text{ kip-ft}$ $\quad + 0.7(111 \text{ kip-ft})$ $\quad = 216 \text{ kip-ft}$

The required flexural strength of Beam BM-2 according to the analysis requirements of AISC *Seismic Provisions* Section F2.3(ii) and including second-order effects is:

LRFD	ASD
$M_u = B_1(1.2+0.2S_{DS})M_D + M_{E_{mh}}$ $\quad + B_1 0.5M_L + 0.2M_S$ $\quad = 1.01[1.2+0.2(1.0)](120 \text{ kip-ft})$ $\quad + 788 \text{ kip-ft} + 1.01(0.5)(100 \text{ kip-ft})$ $\quad + 0.2(0 \text{ kip-ft})$ $\quad = 1,010 \text{ kip-ft}$	$M_a = B_1(1.0+0.14S_{DS})M_D + M_H$ $\quad + M_F + 0.7M_{E_{mh}}$ $\quad = 1.01[1.0+0.14(1.0)](120 \text{ kip-ft})$ $\quad + 0 \text{ kip-ft} + 0 \text{ kip-ft}$ $\quad + 0.7(788 \text{ kip-ft})$ $\quad = 690 \text{ kip-ft}$

In summary, including second-order effects, the required strength of Beam BM-2 determined by the analysis provisions of AISC *Seismic Provisions* Section F2.3(i) is:

LRFD	ASD
$P_u = 163 \text{ kips}$ $V_u = 28.8 \text{ kips}$ $M_u = 331 \text{ kip-ft}$	$P_a = 114 \text{ kips}$ $V_a = 19.0 \text{ kips}$ $M_a = 216 \text{ kip-ft}$

Including second-order effects, the required strength of Beam BM-2 determined by the analysis provisions of AISC *Seismic Provisions* Section F2.3(ii) is:

LRFD	ASD
$P_u = 109 \text{ kips}$ $V_u = 82.9 \text{ kips}$ $M_u = 1,010 \text{ kip-ft}$	$P_a = 76.3 \text{ kips}$ $V_a = 56.9 \text{ kips}$ $M_a = 690 \text{ kip-ft}$

Combined Loading

For the analysis provisions of AISC *Seismic Provisions* Section F2.3(i):

LRFD	ASD
$\dfrac{P_r}{P_c} = \dfrac{163 \text{ kips}}{1{,}140 \text{ kips}}$ $= 0.143$	$\dfrac{P_r}{P_c} = \dfrac{114 \text{ kips}}{760 \text{ kips}}$ $= 0.150$

Because $P_r/P_c < 0.2$, the beam-column design is controlled by the equation:

$$\frac{P_r}{2P_c} + \left(\frac{M_{rx}}{M_{cx}} + \frac{M_{ry}}{M_{cy}}\right) \leq 1.0 \qquad (Spec.\ Eq.\ H1\text{-}1b)$$

LRFD	ASD
$\dfrac{0.143}{2} + \dfrac{331 \text{ kip-ft}}{1{,}290 \text{ kip-ft}} + 0 = 0.328$ $0.328 < 1.0 \quad \textbf{o.k.}$	$\dfrac{0.150}{2} + \dfrac{216 \text{ kip-ft}}{856 \text{ kip-ft}} + 0 = 0.327$ $0.327 < 1.0 \quad \textbf{o.k.}$

For the analysis provisions of AISC *Seismic Provisions* Section F2.3(ii):

LRFD	ASD
$\dfrac{P_r}{P_c} = \dfrac{109 \text{ kips}}{1{,}140 \text{ kips}}$ $= 0.0956$	$\dfrac{P_r}{P_c} = \dfrac{76.3 \text{ kips}}{760 \text{ kips}}$ $= 0.100$

Because $P_r/P_c < 0.2$, the beam-column design is controlled by the equation:

$$\frac{P_r}{2P_c} + \left(\frac{M_{rx}}{M_{cx}} + \frac{M_{ry}}{M_{cy}}\right) \leq 1.0 \qquad (Spec.\ Eq.\ H1\text{-}1b)$$

LRFD	ASD
$\dfrac{0.0956}{2} + \dfrac{1{,}010 \text{ kip-ft}}{1{,}290 \text{ kip-ft}} + 0 = 0.831$ $0.831 < 1.0 \quad \textbf{o.k.}$	$\dfrac{0.100}{2} + \dfrac{690 \text{ kip-ft}}{856 \text{ kip-ft}} + 0 = 0.856$ $0.856 < 1.0 \quad \textbf{o.k.}$

5.3 SPECIAL CONCENTRICALLY BRACED FRAMES

Check shear strength of the W27×114
From AISC *Manual* Table 3-2:

LRFD	ASD
$\phi_v V_n = 467$ kips > 82.9 kips **o.k.**	$\dfrac{V_n}{\Omega_v} = 311$ kips > 56.9 kips **o.k.**

Check width-to-thickness limits of the W27×114
According to AISC *Seismic Provisions* Section F2.5a, beams in SCBF must satisfy the requirements for moderately ductile members. From Table 1-3 of this Manual, the W27×114 satisfies the limiting width-to-thickness ratios and P_u and P_a are less than the maximum permitted.

Example 5.3.5. SCBF Beam Design

Given:

Refer to Beam BM-1 in Figure 5-11. Select an ASTM A992 W-shape with a maximum depth of 36 in. Design the beam as a noncomposite beam for strength, although the composite deck can be considered to brace the beam. The applicable building code specifies the use of ASCE/SEI 7 for calculation of loads.

Assume the brace sizes are as shown in Figure 5-12. Relevant seismic design parameters were given in the SCBF Design Example Plan and Elevation section. The gravity shears and moments on the beam are:

$V_D = 11.2$ kips $V_L = 8.50$ kips
$M_D = 120$ kip-ft $M_L = 100$ kip-ft

Note that in Example 5.3.10, the bracing connections at the third level use a splice in the beam away from the gusset plate. Based on the connection configuration, a shorter length could have been used for the beam design here. In this example, the full 25-ft bay width is used as the length of the beam.

Solution:

From AISC *Manual* Table 2-4, the material properties are as follows:

ASTM A992
$F_y = 50$ ksi
$F_u = 65$ ksi

As required by AISC *Seismic Provisions* Section F2.3, the required strength of the beams are based on the load combinations in the applicable building code, including the amplified seismic loads. The amplified seismic loads are determined from the larger of:

(i) An analysis in which all braces are assumed to resist forces corresponding to their expected strength in compression or in tension

(ii) An analysis in which all braces in tension are assumed to resist forces corresponding to their expected strength and all braces in compression are assumed to resist their expected post-buckling strength

These forces are shown in Tables 5-1 and 5-2, and the forces acting on Beam BM-1 are shown in Figure 5-16.

Unlike Beam BM-2 designed in Example 5.3.4, these forces do not cause shears and moments on the beam; the only shears and moments are from gravity loads.

Required Strength

Determine the required axial strength of the beam based on AISC Seismic Provisions Section F2.3(i)

From AISC *Seismic Provisions* Section F2.3(i), the required axial strength of the beam is based on the braces at their expected strengths in tension and compression. To determine the required axial force on the beam, the horizontal component of the difference between the sum of the expected strengths of the braces below the beam and the sum of the expected strengths of the braces above the beam can be thought of as a "story force." The story force for the analysis in AISC *Seismic Provisions* Section F2.3(i) with tension and compression braces at their expected strengths is:

$$P_x = (\cos 45°)[\Sigma(\text{Braces below beam}) - \Sigma(\text{Braces above beam})]$$
$$= (\cos 45°)[(518 \text{ kips} + 606 \text{ kips}) - (550 \text{ kips} + 444 \text{ kips})]$$
$$= 91.9 \text{ kips}$$

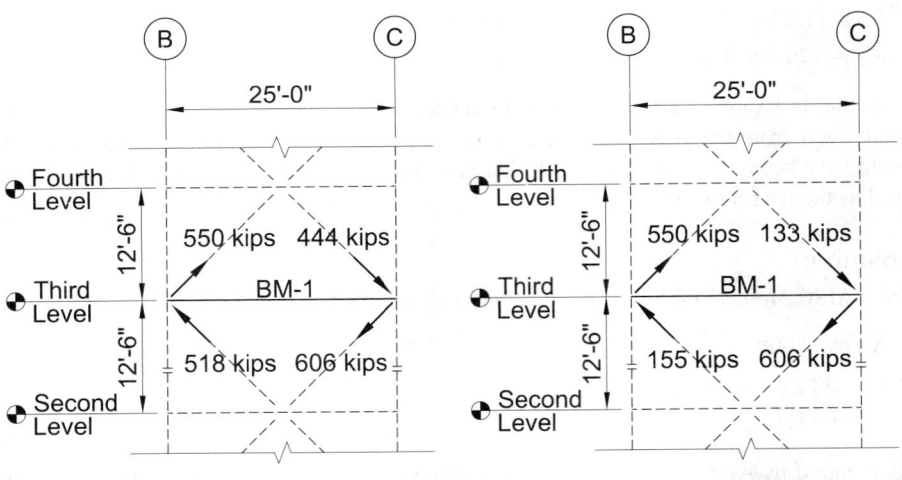

(a) Forces from AISC Seismic Provisions Section F2.3(i)

(b) Forces from AISC Seismic Provisions Section F2.3(ii)

Fig. 5-16. Forces on Beam BM-1 from a mechanism analysis of AISC Seismic Provisions Section F2.3 as carried out in Example 5.3.2.

5.3 SPECIAL CONCENTRICALLY BRACED FRAMES

Because the braced frame is in the middle bay of a three-bay building, half of this story force, or 46.0 kips, can be considered to enter the braced bay from each side. From equilibrium of the joints at each end of the beam, this results in an axial force in the beam of $E_{mh} = 68.6$ kips, as shown in Figure 5-17.

Determine the required axial strength of the beam based on AISC Seismic Provisions Section F2.3(ii)

For this analysis, the expected strength of the braces in compression must be multiplied by 0.3 to approximate their post-buckling strength, as shown in Table 5-2.

Figure 5-16(b) shows the forces corresponding to the tension braces at their expected strengths and the compression braces at their post-buckling strength. Similar to Beam BM-2 in Example 5.3.4, an equivalent "story force" can be determined as:

$$P_x = (\cos 45°)[\Sigma(\text{Braces below beam}) - \Sigma(\text{Braces above beam})]$$
$$= (\cos 45°)[(155 \text{ kips} + 606 \text{ kips}) - (550 \text{ kips} + 133 \text{ kips})]$$
$$= 55.2 \text{ kips}$$

Since the braced frame is in the middle bay of a three-bay building, half of this story force, or 27.6 kips, can be considered to enter the braced bay from each side. From equilibrium of the joints at each end of the beam, this results in a axial force in the beam of $E_{mh} = 307$ kips, as shown in Figure 5-18.

The analysis of AISC *Seismic Provisions* Section F2.3(ii) governs, in which tension braces are at their expected strengths and compression braces are at their post-buckling strengths.

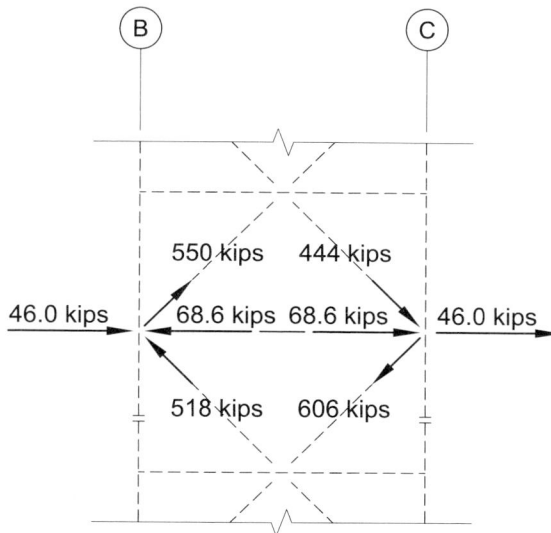

Fig. 5-17. Axial force in Beam BM-1 from the mechanism analysis of AISC Seismic Provisions Section F2.3(i).

The required shear and flexural strength of the beam comes from gravity loads only, and therefore are the same for both analysis cases.

Using the load combinations in ASCE/SEI 7, the required shear strength of Beam BM-1 according to the analysis requirements of AISC *Seismic Provisions* Section F2.3(ii) is:

LRFD	ASD
LRFD Load Combination 5 from ASCE/SEI 7 Section 12.4.3.2 (including the 0.5 load factor on L permitted by Section 12.4.3.2) $$V_u = (1.2 + 0.2S_{DS})V_D + V_{E_{mh}}$$ $$+ 0.5V_L + 0.2V_S$$ $$= [1.2 + 0.2(1.0)](11.2 \text{ kips}) + 0 \text{ kips}$$ $$+ 0.5(8.50 \text{ kips}) + 0.2(0 \text{ kips})$$ $$= 19.9 \text{ kips}$$	ASD Load Combination 5 from ASCE/SEI 7 Section 12.4.3.2 $$V_a = (1.0 + 0.14S_{DS})V_D + V_H$$ $$+ V_F + 0.7V_{E_{mh}}$$ $$= [1.0 + 0.14(1.0)](11.2 \text{ kips}) + 0 \text{ kips}$$ $$+ 0 \text{ kips} + 0.7(0 \text{ kips})$$ $$= 12.8 \text{ kips}$$

The required flexural strength of Beam BM-1 according to the analysis requirements of AISC *Seismic Provisions* Section F2.3(ii) is:

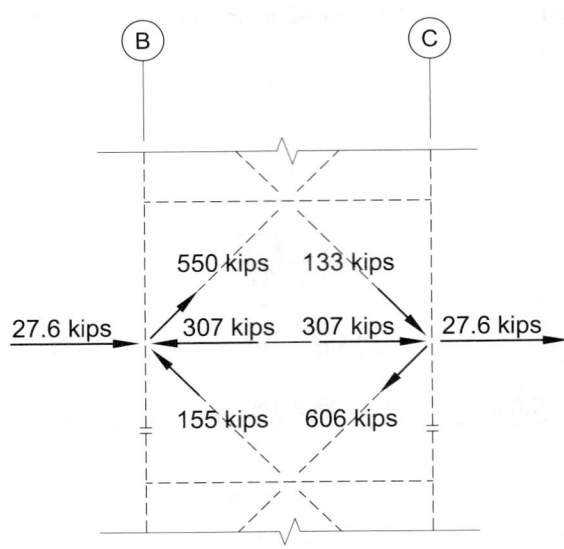

Fig. 5-18. Axial force in Beam BM-1 from the mechanism analysis of AISC Seismic Provisions F2.3(ii).

5.3 SPECIAL CONCENTRICALLY BRACED FRAMES

LRFD	ASD
LRFD Load Combination 5 from ASCE/SEI 7 Section 12.4.3.2 (including the 0.5 load factor on L permitted by Section 12.4.3.2) $M_u = (1.2 + 0.2S_{DS})M_D + M_{E_{mh}}$ $\quad + 0.5M_L + 0.2M_S$ $\quad = [1.2 + 0.2(1.0)](120 \text{ kip-ft})$ $\quad + 0 \text{ kip-ft} + 0.5(100 \text{ kip-ft})$ $\quad + 0.2(0 \text{ kip-ft})$ $\quad = 218 \text{ kip-ft}$	ASD Load Combination 5 from ASCE/SEI 7 Section 12.4.3.2 $M_a = (1.0 + 0.14S_{DS})M_D + M_H$ $\quad + M_F + 0.7M_{E_{mh}}$ $\quad = [1.0 + 0.14(1.0)](120 \text{ kip-ft})$ $\quad + 0 \text{ kip-ft} + 0 \text{ kip-ft} + 0.7(0 \text{ kip-ft})$ $\quad = 137 \text{ kip-ft}$

The required axial strength of Beam BM-1 according to the analysis requirements of AISC *Seismic Provisions* Section F2.3(ii) is:

LRFD	ASD
LRFD Load Combination 5 from ASCE/SEI 7 Section 12.4.3.2 $P_u = (1.2 + 0.2S_{DS})P_D + P_{E_{mh}}$ $\quad + 0.5P_L + 0.2P_S$ $\quad = [1.2 + 0.2(1.0)](0 \text{ kips}) + 307 \text{ kips}$ $\quad + 0.5(0 \text{ kips}) + 0.2(0 \text{ kips})$ $\quad = 307 \text{ kips}$	ASD Load Combination 5 from ASCE/SEI 7 Section 12.4.3.2 $P_a = (1.0 + 0.14S_{DS})P_D + P_H$ $\quad + P_F + 0.7P_{E_{mh}}$ $\quad = [1.0 + 0.14(1.0)](0 \text{ kips}) + 0 \text{ kips}$ $\quad + 0 \text{ kips} + 0.7(307 \text{ kips})$ $\quad = 215 \text{ kips}$

The beam is subject to axial and flexural forces. See Part 8 and Table 8-1 of this Manual for a discussion of collector beams, which also generally applies to beams within a braced frame.

In flexure, the beam is considered continuously braced by the slab and lateral-torsional buckling does not apply.

In compression, the beam is considered continuously braced by the slab in the y-y direction so minor-axis flexural buckling does not apply. For major-axis flexural buckling, the beam is assumed unbraced. As explained in Part 8 for collectors, torsional buckling is considered because the torsional unbraced length is not equal to the minor-axis flexural buckling unbraced length. For torsional buckling, the beam is considered braced by the gravity beam and its connection at midspan. Since the top flange is constrained by the composite slab, the applicable torsional limit state is constrained-axis flexural-torsional buckling, as discussed in Part 8 of this Manual.

Try a W24×68.

Beam Size Selection
From AISC *Manual* Table 1-1, the geometric properties are as follows:

W24×68

$A = 20.1$ in.^{2}	$d = 23.7$ in.	$t_w = 0.415$ in.	$b_f = 8.97$ in.
$t_f = 0.585$ in.	$k_{des} = 1.09$ in.	$h/t_w = 52.0$	$I_x = 1,830$ in.^{4}
$S_x = 154$ in.^{3}	$r_x = 9.55$ in.	$Z_x = 177$ in.^{3}	$I_y = 70.4$ in.^{4}
$r_y = 1.87$ in.	$h_o = 23.1$ in.	$J = 1.87$ in.^{4}	$C_w = 9,430$ in.^{6}

Lateral Bracing Requirements
Because this beam is not part of a V- or inverted-V-braced frame (there is no brace connection at the midspan of the beam), there are no lateral bracing requirements in the AISC *Seismic Provisions*, other than what may be required for strength. However, there is a gravity beam framing into the beam at midspan. The gravity beam at midspan and its connection will be considered to provide a torsional brace point for the limit state of constrained-axis flexural-torsional buckling.

Available Flexural Strength
The composite slab can be considered to continuously brace the beam, and therefore the limit state of lateral-torsional buckling does not apply and the available flexural strength is based on the plastic moment. From AISC *Manual* Table 3-6, the available flexural strength of the beam is:

LRFD	ASD
$\phi_b M_p = 664$ kip-ft	$\dfrac{M_p}{\Omega_b} = 442$ kip-ft

Available Compressive Strength
The unbraced lengths for flexural buckling were discussed previously. To summarize:

$L_x = 25.0$ ft

$L_y = 0$ ft (lateral movement is braced by the slab)

$L_z = 12.5$ ft (torsion with top flange restrained by the slab)

From AISC *Manual* Table 1-1 and AISC *Specification* Table B4.1, the web is slender for compression. Therefore the reduction factor for slender elements, Q, must be determined.

Determine the critical buckling strength for flexural buckling about the x-x axis, assuming Q = 1

$K_x = 1.0$

$L_x = 25.0$ ft

$\dfrac{K_x L_x}{r_x} = \dfrac{1.0(25.0 \text{ ft})(12 \text{ in./ft})}{9.55 \text{ in.}}$

$= 31.4$

$$F_e = \frac{\pi^2 E}{\left(\dfrac{KL}{r}\right)^2} \qquad \text{(Spec. Eq. E3-4)}$$

$$= \frac{\pi^2 (29{,}000 \text{ ksi})}{(31.4)^2}$$

$$= 290 \text{ ksi}$$

The value of F_{cr} before local buckling effects are considered is determined as follows:

$$\frac{F_y}{F_e} = \frac{50 \text{ ksi}}{290 \text{ ksi}}$$

$$= 0.172$$

Because $0.172 < 2.25$, AISC *Specification* Equation E3-2 applies.

$$F_{cr} = \left(0.658^{\frac{F_y}{F_e}}\right) F_y \qquad \text{(Spec. Eq. E3-2)}$$

$$= \left(0.658^{\frac{50 \text{ ksi}}{290 \text{ ksi}}}\right)(50 \text{ ksi})$$

$$= 46.5 \text{ ksi}$$

Determine the critical buckling strength for constrained-axis flexural-torsional buckling, assuming Q = 1

For the limit state of constrained-axis flexural-torsional buckling, the unbraced length is 12.5 ft and the top flange of the beam is considered continuously braced by the slab as described in Part 8 of this Manual.

$$F_e = \left\{ \frac{\pi^2 E\left[C_w + I_y (d/2)^2\right]}{(K_z L)^2} + GJ \right\} \left[\frac{1}{I_x + I_y + (d/2)^2 A_g} \right] \qquad (8\text{-}3)$$

$$= \left\{ \frac{\pi^2 (29{,}000 \text{ ksi})\left[9{,}430 \text{ in.}^6 + 70.4 \text{ in.}^4 \left(\dfrac{23.7 \text{ in.}}{2}\right)^2\right]}{[1.0(12.5 \text{ ft})(12 \text{ in./ft})]^2} + 11{,}200 \text{ ksi}\left(1.87 \text{ in.}^4\right) \right\}$$

$$\times \left[\frac{1}{1{,}830 \text{ in.}^4 + 70.4 \text{ in.}^4 + \left(\dfrac{23.7 \text{ in.}}{2}\right)^2 \left(20.1 \text{ in.}^2\right)} \right]$$

$$= 56.5 \text{ ksi}$$

The value of F_{cr} before local buckling effects are considered, with $\dfrac{F_y}{F_e} \leq 2.25$, is:

$$F_{cr} = \left[0.658^{\frac{F_y}{F_e}}\right] F_y \qquad \text{(Spec. Eq. E3-2)}$$

$$= \left[0.658^{\frac{50 \text{ ksi}}{56.5 \text{ ksi}}}\right] 50 \text{ ksi}$$

$$= 34.5 \text{ ksi}$$

Because F_{cr} is lower for constrained-axis flexural-torsional buckling, this limit state governs over major axis flexural buckling.

Determine the reduction factor, Q, for slender elements

To determine the reduction factor Q, use AISC *Specification* Section E7.2, with $f = F_{cr}$, using the minimum F_{cr} from the two preceding limit states. Determine the effective width, b_e, as follows:

$b = h$
$\quad = d - 2k_{des}$
$\quad = 23.7 \text{ in.} - 2(1.09 \text{ in.})$
$\quad = 21.5 \text{ in.}$

$f = F_{cr}$

$$b_e = 1.92t\sqrt{\dfrac{E}{f}}\left[1 - \dfrac{0.34}{(b/t)}\sqrt{\dfrac{E}{f}}\right] \leq b \qquad \text{(Spec. Eq. E7-17)}$$

$$= 1.92(0.415 \text{ in.})\sqrt{\dfrac{29{,}000 \text{ ksi}}{34.5 \text{ ksi}}}\left[1 - \dfrac{0.34}{(52.0)}\sqrt{\dfrac{29{,}000 \text{ ksi}}{34.5 \text{ ksi}}}\right] \leq 21.5 \text{ in.}$$

$$= 18.7 \text{ in.} \leq 21.5 \text{ in.}$$

$$Q_a = \dfrac{A_e}{A_g} \qquad \text{(Spec. Eq. E7-16)}$$

$$= \dfrac{A_g - t_w(h - b_e)}{A_g}$$

$$= \dfrac{20.1 \text{ in.}^2 - 0.415 \text{ in.}(21.5 \text{ in.} - 18.7 \text{ in.})}{20.1 \text{ in.}^2}$$

$$= 0.942$$

$Q_s = 1.0$

$Q = Q_s Q_a$
$\quad = 1.0(0.942)$
$\quad = 0.942$

5.3 SPECIAL CONCENTRICALLY BRACED FRAMES

Determine the available compressive strength for the governing limit state of constrained-axis flexural-torsional buckling, accounting for slender elements

$$\frac{QF_y}{F_e} = \frac{0.942(50 \text{ ksi})}{(56.5 \text{ ksi})}$$

$$= 0.834$$

Because $0.834 < 2.25$, use AISC *Specification* Equation E7-2.

$$F_{cr} = Q \left(0.658^{\frac{QF_y}{F_e}} \right) F_y \quad \text{(Spec. Eq. E7-2)}$$

$$= 0.942 \left[0.658^{\frac{0.942(50 \text{ ksi})}{56.5 \text{ ksi}}} \right] (50 \text{ ksi})$$

$$= 33.2 \text{ ksi}$$

$$P_n = F_{cr} A_g \quad \text{(Spec. Eq. E7-1)}$$

$$= 33.2 \text{ ksi} (20.1 \text{ in.}^2)$$

$$= 667 \text{ kips}$$

The available compressive strength is:

LRFD	ASD
$\phi_c P_n = 0.90(667 \text{ kips})$ $= 600 \text{ kips}$	$\dfrac{P_n}{\Omega_c} = \dfrac{667 \text{ kips}}{1.67}$ $= 399 \text{ kips}$

Second-Order Effects

Because the seismic component of the required strength of the beam comes from the mechanism analysis of AISC *Seismic Provisions* Section F2.3 and is based on the expected strengths of the braces, P-Δ effects need not be considered and B_2 from AISC *Specification* Appendix 8 need not be applied. P-Δ effects do not increase the forces corresponding to the expected brace strengths in compression and tension; instead, they may be thought of as contributing to the system reaching that state. P-δ effects do apply, however, and B_1 is determined as follows. The effective length method is used.

$$B_1 = \frac{C_m}{1 - \alpha P_r / P_{e1}} \geq 1 \quad \text{(Spec. Eq. A-8-3)}$$

$$P_{e1} = \frac{\pi^2 EI^*}{(K_1 L)^2} \qquad \text{(Spec. Eq. A-8-5)}$$

$$= \frac{\pi^2 (29{,}000 \text{ ksi})(1{,}830 \text{ in.}^4)}{[1.0(25.0 \text{ ft})(12 \text{ in./ft})]^2}$$

$$= 5{,}820 \text{ kips}$$

$C_m = 1.0$ because there is transverse loading

LRFD	ASD
$B_1 = \dfrac{1.0}{1-[1.00(307 \text{ kips})/5{,}820 \text{ kips}]}$ $= 1.06$	$B_1 = \dfrac{1.0}{1-[1.60(215 \text{ kips})/5{,}820 \text{ kips}]}$ $= 1.06$

The B_1 factor (P-δ effect) need only be applied to the first-order moment with the structure restrained against translation. The required flexural strength of Beam BM-1 according to the analysis requirements of AISC *Seismic Provisions* Section F2.3(ii) and including second-order effects is determined from ASCE/SEI 7 Section 12.4.3.2 Load Combination 5 for LRFD and ASD:

LRFD	ASD
$M_u = B_1(1.2+0.2S_{DS})M_D + M_{E_{mh}}$ $+ B_1 0.5M_L + 0.2M_S$ $= 1.06[1.2+0.2(1.0)](120 \text{ kip-ft})$ $+ 0 \text{ kip-ft} + 1.06(0.5)(100 \text{ kip-ft})$ $+ 0.2(0 \text{ kip-ft})$ $= 231 \text{ kip-ft}$	$M_a = B_1(1.0+0.14S_{DS})M_D + M_H$ $+ M_F + 0.7M_{E_{mh}}$ $= 1.06[1.0+0.14(1.0)](120 \text{ kip-ft})$ $+ 0 \text{ kip-ft} + 0 \text{ kip-ft} + 0.7(0 \text{ kip-ft})$ $= 145 \text{ kip-ft}$

Combined Loading

LRFD	ASD
$\dfrac{P_r}{P_c} = \dfrac{307 \text{ kips}}{600 \text{ kips}}$ $= 0.512$	$\dfrac{P_r}{P_c} = \dfrac{215 \text{ kips}}{399 \text{ kips}}$ $= 0.539$

Because $P_r/P_c \geq 0.2$, the beam-column design is controlled by the equation:

$$\frac{P_r}{P_c} + \frac{8}{9}\left(\frac{M_{rx}}{M_{cx}} + \frac{M_{ry}}{M_{cy}}\right) \leq 1.0 \qquad \text{(Spec. Eq. H1-1a)}$$

5.3 SPECIAL CONCENTRICALLY BRACED FRAMES

LRFD	ASD
$0.512 + \dfrac{8}{9}\left(\dfrac{231 \text{ kip-ft}}{664 \text{ kip-ft}} + 0\right) = 0.821$	$0.539 + \dfrac{8}{9}\left(\dfrac{145 \text{ kip-ft}}{442 \text{ kip-ft}} + 0\right) = 0.831$
$0.821 < 1.0$ **o.k.**	$0.831 < 1.0$ **o.k.**

Available Shear Strength

From AISC *Manual* Table 3-2, the available shear strength is:

LRFD	ASD
$\phi_v V_n = 295$ kips > 19.9 kips **o.k.**	$\dfrac{V_n}{\Omega_v} = 197$ kips > 12.8 kips **o.k.**

Width-to-Thickness Limitations

According to AISC *Seismic Provisions* Section F2.5a, beams in SCBF shall satisfy the requirements for moderately ductile members. From Table 1-3 of this Manual, the W24×68 satisfies the limiting width-to-thickness ratios and P_u and P_a are less than the maximum permitted.

Example 5.3.6. SCBF Column Splice Design

Given:

Design a fully welded splice between the third and fourth levels for the SCBF column located on grid C in Figure 5-11. The column material is ASTM A992, the upper shaft is a W12×45 and the lower shaft is a W12×96. The applicable building code specifies the use of ASCE/SEI 7 for calculation of loads.

The relevant seismic parameters were given in the SCBF Design Example Plan and Elevation section.

The required axial strengths of the columns due to dead (including curtain wall), live and snow loads at the splice location are:

$P_D = 66.3$ kips $P_L = 18.8$ kips $P_S = 7.00$ kips

The seismic component of the required axial strength of the column due to code-specified seismic loads from the applicable building code is:

$P_{Q_E} = 45.5$ kips

Assume that the ends of the column are pinned and braced against translation for both the *x-x* and *y-y* axes and the column moment produced by the gravity framing connections is negligible.

Solution:

From AISC *Manual* Table 2-4, the material properties are as follows:

ASTM A992
$F_y = 50$ ksi
$F_u = 65$ ksi

From AISC *Manual* Table 1-1, the geometric properties are as follows:

W12×45
$A = 13.1$ in.^{2} $d = 12.1$ in. $b_f = 8.05$ in. $t_f = 0.575$ in.
$t_w = 0.335$ in. $Z_x = 64.2$ in.^{3}

W12×96
$Z_x = 147$ in.^{3}

Required Strength

AISC *Seismic Provisions* Section F2.6d requires that SCBF column splices comply with Section D2.5, which states that the required strength of column splices is the greater of (a) the required strength of the columns, including that determined from Chapters E, F, G and H, and Section D1.4a, or (b) the required strength determined using the load combinations stipulated in the applicable building code, including the amplified seismic load, but need not exceed the maximum loads that can be transferred to the splice by the system. Also, for columns with net tension, three other specific conditions must be satisfied, as stipulated in Section D2.5b.

The required axial strength of columns in SCBF frames is based on the expected strength of the braces, as defined in AISC *Seismic Provisions* Section F2.3. Example 5.3.2 provides a description of this analysis. For the column at the lowest story, Example 5.3.3 illustrates the determination of the column force. For the splice location, only the braces at the top two stories need to be considered.

From Example 5.3.2, with brace forces shown in Figure 5-13 and Tables 5-1 and 5-2, the expected tensile strength of the HSS6×0.312 brace between level 4 and the roof is:

$P_{tension} = 307$ kips

From Example 5.3.2, in Table 5-2, the expected compressive strength of the HSS6×0.312 brace between level 4 and the roof is given as:

$P_{compression} = 226$ kips

The vertical components of these brace expected strengths are transferred to the column. At the fourth level, the brace forces at the beam midpoint connection are carried across in beam shear. The forces acting on the columns due to the expected strengths of the braces are as shown in Figure 5-19.

The axial force in the column at the splice location due to seismic load effects (including the amplified seismic load) is:

5.3 SPECIAL CONCENTRICALLY BRACED FRAMES

$P_{E_{mh}} = 307 \text{ kips}(\sin 45°) + 8.84 \text{ kips}$
$= 226 \text{ kips (compression)}$

$P_{E_{mh}} = 226 \text{ kips}(\sin 45°) - 8.84 \text{ kips}$
$= 151 \text{ kips (tension)}$

At this level, Exception 2(a) for the column in AISC *Seismic Provisions* Section F2.3 can be shown not to result in reduced forces; therefore the exception is not used.

For comparison, the seismic component of the required axial strength of the column due to code-specified seismic loads from the applicable building code is given as:

$P_{Q_E} = 45.5 \text{ kips}$

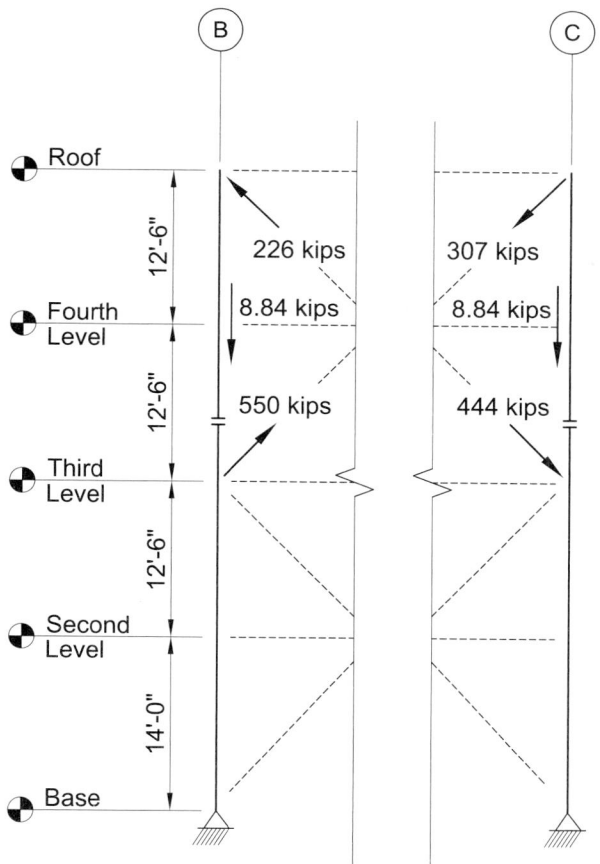

Fig. 5-19. SCBF column forces for splice design from Example 5.3.3.

Using amplified seismic forces, this becomes:

$$P_{E_{mh}} = \Omega_o P_{QE}$$
$$= 2.0(45.5 \text{ kips})$$
$$= 91.0 \text{ kips}$$

The seismic component of the required strength of the column using the analysis requirements of AISC *Seismic Provisions* Section F2.3 (226 kips compression and 151 kips tension) is greater than that determined from the code-specified loads (91.0 kips tension or compression). Therefore, use the analysis requirements of AISC *Seismic Provisions* Section F2.3 for design of the splice.

Using the load combinations in ASCE/SEI 7, the required axial compressive strength of the column is:

LRFD	ASD
LRFD Load Combination 5 from ASCE/SEI 7 Section 12.4.3.2 (including the 0.5 factor on L permitted in Section 12.4.3.2)	ASD Load Combination 5 from ASCE/SEI 7 Section 12.4.3.2
$P_u = (1.2 + 0.2 S_{DS}) P_D + P_{E_{mh}}$ $\quad + 0.5 P_L + 0.2 P_S$ $= [1.2 + 0.2(1.0)](66.3 \text{ kips})$ $\quad + 226 \text{ kips} + 0.5(18.8 \text{ kips})$ $\quad + 0.2(7.00 \text{ kips})$ $= 330 \text{ kips}$	$P_a = (1.0 + 0.14 S_{DS}) P_D + P_H$ $\quad + P_F + 0.7 P_{E_{mh}}$ $= [1.0 + 0.14(1.0)](66.3 \text{ kips})$ $\quad + 0 \text{ kips} + 0 \text{ kips} + 0.7(226 \text{ kips})$ $= 234 \text{ kips}$

The required axial tensile strength of the column is:

LRFD	ASD
LRFD Load Combination 7 from ASCE/SEI 7 Section 12.4.3.2	ASD Load Combination 8 from ASCE/SEI 7 Section 12.4.3.2
$P_u = (0.9 - 0.2 S_{DS}) P_D + P_{E_{mh}} + 1.6 P_H$ $= [0.9 - 0.2(1.0)](66.3 \text{ kips})$ $\quad + (-151 \text{ kips}) + 1.6(0 \text{ kips})$ $= -105 \text{ kips}$	$P_a = (0.6 - 0.14 S_{DS}) P_D + 0.7 P_{E_{mh}} + P_H$ $= [0.6 - 0.14(1.0)](66.3 \text{ kips})$ $\quad + 0.7(-151 \text{ kips}) + 0 \text{ kips}$ $= -75.2 \text{ kips}$

As stated above, this splice is to be a welded splice. AISC *Seismic Provisions* Section F2.6d requires that groove welds must be complete-joint-penetration (CJP) groove welds.

Use CJP groove welds to splice the column flanges and web.

5.3 SPECIAL CONCENTRICALLY BRACED FRAMES

Required Flexural Strength

AISC *Seismic Provisions* Section F2.6d requires the column splice to develop 50% of the lesser available flexural strength of the connected members. For simplicity, use the plastic flexural strength, $\phi_b M_p$ (LRFD) or M_p/Ω_b (ASD).

For the smaller column, W12×45, determine the available flexural strength from AISC *Manual* Table 3-6:

LRFD	ASD
$\phi_b M_p = 241$ kip-ft	$\dfrac{M_p}{\Omega_b} = 160$ kip-ft

The required flexural strength of the splice is:

LRFD	ASD
$M_u = 0.50(\phi_b M_p)$ $= 0.50(241 \text{ kip-ft})$ $= 121$ kip-ft	$M_a = 0.50\left(\dfrac{M_p}{\Omega_b}\right)$ $= 0.50(160 \text{ kip-ft})$ $= 80.0$ kip-ft

Assuming that the entire moment is taken through the flange splices, the required strength of each flange splice is:

LRFD	ASD
$R_u = \dfrac{M_u}{d - t_f}$ $= \dfrac{121 \text{ kip-ft}(12 \text{ in./ft})}{12.1 \text{ in.} - 0.575 \text{ in.}}$ $= 126$ kips	$R_a = \dfrac{M_a}{d - t_f}$ $= \dfrac{80.0 \text{ kip-ft}(12 \text{ in./ft})}{12.1 \text{ in.} - 0.575 \text{ in.}}$ $= 83.3$ kips

The available strength of each CJP groove welded flange splice is controlled by the base metal strength according to AISC *Specification* Table J2.5. Thus, based on tension yielding of the flange from AISC *Specification* Section D2(a), the available strength of the CJP groove weld is:

LRFD	ASD
$\phi R_n = 0.90 F_y b_f t_f$ $= 0.90(50 \text{ ksi})(8.05 \text{ in.})(0.575 \text{ in.})$ $= 208$ kips > 126 kips **o.k.**	$\dfrac{R_n}{\Omega} = F_y b_f t_f / 1.67$ $= (50 \text{ ksi})(8.05 \text{ in.})(0.575 \text{ in.})/1.67$ $= 139$ kips > 83.3 kips **o.k.**

Required Shear Strength

AISC *Seismic Provisions* Section F2.6d defines the required shear strength of the splice as at least $\Sigma M_{pc}/H_c$ (LRFD) or $\Sigma M_{pc}/(1.5H_c)$ (ASD), where ΣM_{pc} is the sum of the nominal plastic flexural strengths of the columns above and below the splice, and H_c is the clear height of the column between beam connections. A CJP groove weld will be used.

Assume that the 12.5-ft story height is from top of steel to top of steel. The beam at the story above the splice is a W27. Therefore, the approximate value for H_c is:

$$H_c = 12.5 \text{ ft} - (27 \text{ in.})(1 \text{ ft}/12 \text{ in.})$$
$$= 10.3 \text{ ft}$$

$$\Sigma M_{pc} = F_y \left(Z_{x\,bot} + Z_{x\,top} \right)$$
$$= 50 \text{ ksi} \left(147 \text{ in.}^3 + 64.2 \text{ in.}^3 \right)(1 \text{ ft}/12 \text{ in.})$$
$$= 880 \text{ kip-ft}$$

The required shear strength of the splice is:

LRFD	ASD
$\dfrac{\Sigma M_{pc}}{H_c} = \dfrac{880 \text{ kip-ft}}{10.3 \text{ ft}}$ $= 85.4 \text{ kips}$	$\dfrac{\Sigma M_{pc}}{1.5H_c} = \dfrac{880 \text{ kip-ft}}{1.5(10.3 \text{ ft})}$ $= 57.0 \text{ kips}$

For the limit state of shear yielding according to AISC *Specification* Section G2, the available shear strength of the W12×45 column is:

LRFD	ASD
$\phi R_n = \phi 0.6 F_y A_w C_v$ $= 1.00(0.6)(50 \text{ ksi})$ $\times (12.1 \text{ in.})(0.335 \text{ in.})(1.0)$ $= 122 \text{ kips} > 85.4 \text{ kips}$ **o.k.**	$\dfrac{R_n}{\Omega} = \dfrac{0.6 F_y A_w C_v}{1.50}$ $= (1/1.50)0.60(50 \text{ ksi})$ $\times (12.1 \text{ in.})(0.335 \text{ in.})(1.0)$ $= 81.1 \text{ kips} > 57.0 \text{ kips}$ **o.k.**

For the shear in the weak axis of the column, the column flanges of the smaller member will easily be able to meet the required shear strength, since the M_p values for the columns are smaller in this direction and the flange area is significantly larger than the web area in this case.

Additional Requirements for Columns Subject to a Net Tensile Load Effect

AISC *Seismic Provisions* Section D2.5b has additional requirements for welded column splices in which any portion of the column is subjected to a net tensile load effect determined

5.3 SPECIAL CONCENTRICALLY BRACED FRAMES

using the load combinations stipulated in the applicable building code, including the amplified seismic load. These additional requirements are:

(1) The available strength of partial-joint-penetration (PJP) groove welded joints, if used, shall be at least equal to 200% of the required strength.
(2) The available strength for each flange splice shall be at least equal to $0.5R_yF_yb_ft_f$ (LRFD) or $(0.5/1.5)R_yF_yb_ft_f$ (ASD).
(3) Where butt joints in column splices are made with CJP groove welds, when the tension stress at any location in the smaller flange exceeds $0.30F_y$ (LRFD) or $0.20F_y$ (ASD), tapered transitions are required between flanges of unequal thickness or width.

As determined previously, the column is subjected to a net tensile load effect.

Since there is net tension, the additional requirements must be met.

(1) AISC *Seismic Provisions* Section D2.5b(1) does not apply because partial-joint-penetration (PJP) welds are not used.
(2) AISC *Seismic Provisions* Section D2.5b(2) requires that the available strength of each flange splice be at least $0.5R_yF_yb_ft_f$ (LRFD) or $(0.5/1.5)R_yF_yb_ft_f$ (ASD). With a CJP groove weld, the available strength of the smaller flange can be developed, so this requirement will be met.
(3) AISC *Seismic Provisions* Section D2.5b(3) requires tapered transitions when the tension stress in the smaller flange exceeds $0.30F_y$ (LRFD) and $0.20F_y$ (ASD) for butt joints with CJP groove welds. The tension stress over the cross section is:

LRFD	ASD
$\dfrac{T_u}{A_g} = \dfrac{-105 \text{ kips}}{13.1 \text{ in.}^2}$ $= 8.02$ ksi	$\dfrac{T_a}{A_g} = \dfrac{-75.2 \text{ kips}}{13.1 \text{ in.}^2}$ $= 5.74$ ksi
$0.3F_y = 0.3(50 \text{ ksi})$ $= 15.0$ ksi	$0.2F_y = 0.2(50 \text{ ksi})$ $= 10.0$ ksi
8.02 ksi < 15.0 ksi	5.74 ksi < 10.0 ksi

Therefore, the requirements in AISC *Seismic Provisions* Section D2.5b(3) need not be met.

Check Splice Location

The splice location satisfies the requirement in AISC *Seismic Provisions* Section D2.5a that the splice be located 4 ft or more away from the beam-to-column flange connection.

The final connection design is shown in Figure 5-20.

Example 5.3.7. SCBF Maximum Force Limited by Foundation Uplift

Given:
Some of the sections in the AISC *Seismic Provisions* allow the required strength of certain members or components to be limited by the forces corresponding to a maximum force that can be delivered by the system. One example is AISC *Seismic Provisions* Section F2.3, Exception (2)(b), which states that the required strength of the column need not exceed the forces corresponding to the resistance of the foundation to overturning uplift. The maximum force that can be delivered is the force required to overturn the foundation. The use of Section F2.3 Exception (2)(b) will be illustrated in this example.

Refer to the SCBF elevation shown in Figure 5-21. Determine the maximum force that can be delivered to Column CL-1 based on the foundation uplift resistance of the system. The seismic loads at each floor are given in Figure 5-21. Assume a concrete density equal to 150 lb/ft^3 and a soil density equal to 100 lb/ft.^{3} As given in Example 5.3.3 for the SCBF column design, the column forces at the base from gravity and snow loads are: $P_D = 147$ kips, $P_L = 60.0$ kips, $P_S = 7.00$ kips. The relevant seismic parameters were given in the SCBF Design Example Plan and Elevation section.

Solution:

Dead Load Resistance to Overturning

The volumes of the mat, soil and slab are:

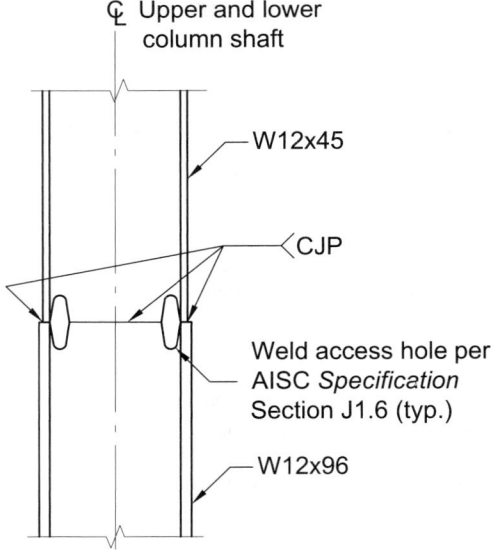

Note: Erection aids not shown for clarity.

Fig. 5-20. SCBF column splice designed in Example 5.3.6.

5.3 SPECIAL CONCENTRICALLY BRACED FRAMES

$$V_{mat} = 4.00 \text{ ft}(6.50 \text{ ft})(35.0 \text{ ft})$$
$$= 910 \text{ ft}^3$$

$$V_{soil\ over\ mat} = \left(\frac{8}{12} \text{ ft}\right)(6.50 \text{ ft})(35.0 \text{ ft})$$
$$= 152 \text{ ft}^3$$

$$V_{slab\ over\ mat} = \left(\frac{4}{12} \text{ ft}\right)(6.50 \text{ ft})(35.0 \text{ ft})$$
$$= 75.8 \text{ ft}^3$$

Using the densities given, the weights of the mat, soil and slab are:

$$W_{mat} = 910 \text{ ft}^3 \left(150 \text{ lb/ft}^3\right) / (1,000 \text{ lb/kip})$$
$$= 137 \text{ kips}$$

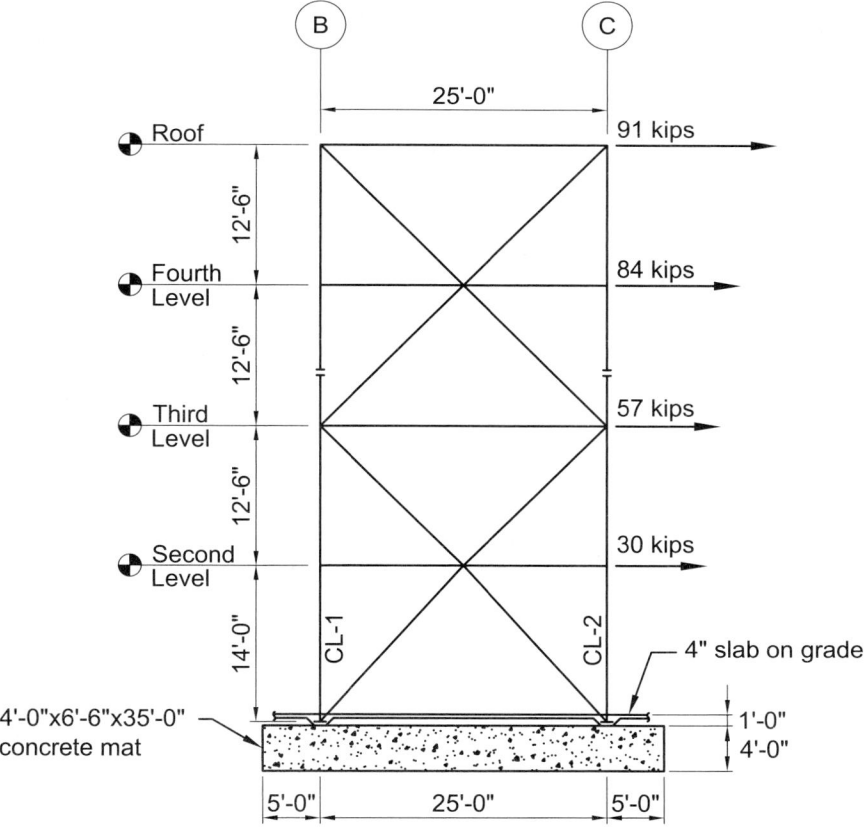

Fig. 5-21. SCBF elevation and foundation.

$$W_{soil\ over\ mat} = 152\ \text{ft}^3\left(100\ \text{lb/ft}^3\right)/(1{,}000\ \text{lb/kip})$$
$$= 15.2\ \text{kips}$$

$$W_{slab\ over\ mat} = 75.8\ \text{ft}^3\left(150\ \text{lb/ft}^3\right)/(1{,}000\ \text{lb/kip})$$
$$= 11.4\ \text{kips}$$

Summing moments at the extreme lower right corner of the mat allows estimation of the maximum moment required to overturn the foundation.

Element	Dead Load kips	Live Load kips	Snow Load kips	Moment Arm ft	Resisting Moment (D) kip-ft	Resisting Moment (L) kip-ft	Resisting Moment (S) kip-ft
Conc. Mat	137	–	–	17.5	2,400	–	–
Soil over Mat	15.2	–	–	17.5	266	–	–
Slab over Mat	11.4	–	–	17.5	200	–	–
Column CL-2	147	60.0	7.00	5.00	735	300	35.0
Column CL-1	147	60.0	7.00	30.0	4,410	1,800	210
Σ	–	–	–	–	8,010	2,100	245

The controlling load combinations (for compression in the column) that include seismic load are ASCE/SEI 7 Section 12.4.3.2 Load Combination 5 for LRFD and Load Combination 6 for ASD. Setting the overturning moment, M_E, equal to the resisting moment in these load combinations, the overturning moment required to cause uplift of the foundation is:

LRFD	ASD
$(1.2+0.2S_{DS})M_D + M_E + 0.5M_L + 0.2M_S = 0$	$(1.0+0.105S_{DS})M_D + 0.525M_E + 0.75M_L + 0.75M_S = 0$
$M_E = \left\vert \begin{array}{c} -[1.2+0.2(1.0)]M_D - 0.5M_L \\ -0.2M_S \end{array} \right\vert$	$M_E = \left\vert \begin{array}{c} \{-[1.0+0.105(1.0)]M_D \\ -0.75M_L - 0.75M_S\}/0.525 \end{array} \right\vert$
$= \left\vert \begin{array}{c} -[1.2+0.2(1.0)](8{,}010\ \text{kip-ft}) \\ -0.5(2{,}100\ \text{kip-ft}) \\ -0.2(245\ \text{kip-ft}) \end{array} \right\vert$	$= \left\vert \begin{array}{c} \{-1.105(8{,}010\ \text{kip-ft}) \\ -0.75(2{,}100\ \text{kip-ft}) \\ -0.75(245\ \text{kip-ft})\}/0.525 \end{array} \right\vert$
$= 12{,}300\ \text{kip-ft}$	$= 20{,}200\ \text{kip-ft}$

The overturning moment required to cause uplift of the foundation, M_E, should be compared to the overturning moment caused by the seismic loads including overstrength, $M_{E_{mh}}$. If M_E is less than $M_{E_{mh}}$, then the seismic component of the required column force in Example 5.3.3 could be reduced by the ratio of those moments.

5.3 SPECIAL CONCENTRICALLY BRACED FRAMES

The overturning moment caused by the seismic loads given in Figure 5-21, and including overstrength, is:

$$M_{E_{mh}} = \Omega_o \sum_{i=1}^{n} F_i h_i$$

$$= 2.0 \begin{bmatrix} 30.0 \text{ kips}(14.0 \text{ ft}) + 57.0 \text{ kips}(26.5 \text{ ft}) + 84.0 \text{ kips}(39.0 \text{ ft}) \\ + 91.0 \text{ kips}(51.5 \text{ ft}) \end{bmatrix}$$

$$= 19,800 \text{ kip-ft}$$

For convenience, use the concept of an effective overstrength factor, Ω'_o, determined as follows:

LRFD	ASD
$\Omega'_o = \dfrac{M_E}{M_{E_{mh}}} \Omega_o$	$\Omega'_o = \dfrac{M_E}{M_{E_{mh}}} \Omega_o$
$= \dfrac{12,300 \text{ kip-ft}}{19,800 \text{ kip-ft}}(2.0)$	$= \dfrac{20,200 \text{ kip-ft}}{19,800 \text{ kip-ft}}(2.0)$
$= 1.24$	$= 2.04$

From Example 5.3.3, the seismic component of the required column strength based on the seismic loads is $P_{Q_E} = 248$ kips in tension or compression. Rather than amplifying this force by $\Omega_o = 2.0$ as shown in Example 5.3.3, it could instead be amplified by 1.24 (LRFD) or 2.04 (ASD) as allowed by AISC *Seismic Provisions* Section F2.3 Exception (2)(b). For determining the required axial compressive strength, the controlling load combinations that include seismic load are ASCE/SEI 7 Section 12.4.3.2 Load Combination 5 for LRFD and Load Combination 5 for ASD.

LRFD	ASD
The required axial compressive strength of the column is:	The required axial compressive strength of the column is:
$P_u = (1.2 + 0.2S_{DS})P_D + \Omega'_o P_{Q_E}$ $\quad + 0.5P_L + 0.2P_s$ $= [1.2 + 0.2(1.0)](147 \text{ kips})$ $\quad + 1.24(248 \text{ kips}) + 0.5(60.0 \text{ kips})$ $\quad + 0.2(7.00 \text{ kips})$ $= 545 \text{ kips}$	$P_a = (1.0 + 0.105S_{DS})P_D + 0.525\Omega'_o P_{Q_E}$ $\quad + 0.75P_L + 0.75P_s$ $= [1.0 + 0.105(1.0)](147 \text{ kips})$ $\quad + 0.525(2.04)(248 \text{ kips})$ $\quad + 0.75(60.0 \text{ kips}) + 0.75(7.00 \text{ kips})$ $= 478 \text{ kips}$

For determining the required axial tensile strength of the column, ASCE/SEI 7 Section 12.4.3.2 Load Combination 7 for LRFD and Load Combination 8 for ASD apply. A similar approach is used to calculate the maximum tension force in the column due to foundation uplift. Re-calculating M_E for the governing load combination for tension in the column:

LRFD	ASD
$(0.9-0.2S_{DS})M_D + M_E = 0$	$(0.6-0.14S_{DS})M_D + 0.7M_E = 0$
$M_E = \lvert -[0.9-0.2(1.0)]M_D \rvert$	$M_E = \lvert -[0.6-0.14(1.0)]M_D/0.7 \rvert$
$\quad = \lvert -[0.9-0.2(1.0)](8{,}010 \text{ kip-ft}) \rvert$	$\quad = \lvert -0.46(8{,}010 \text{ kip-ft})/0.7 \rvert$
$\quad = 5{,}610 \text{ kip-ft}$	$\quad = 5{,}260 \text{ kip-ft}$

Use an effective overstrength factor, similar to the compression case, calculated as:

LRFD	ASD
$\Omega_o'' = \dfrac{M_E}{M_{E_{mh}}}\Omega_o$	$\Omega_o'' = \dfrac{M_E}{M_{E_{mh}}}\Omega_o$
$\quad = \dfrac{5{,}610 \text{ kip-ft}}{19{,}800 \text{ kip-ft}}(2.0)$	$\quad = \dfrac{5{,}260 \text{ kip-ft}}{19{,}800 \text{ kip-ft}}(2.0)$
$\quad = 0.567$	$\quad = 0.531$

LRFD	ASD
The required axial tensile strength of the column is:	The required axial tensile strength of the column is:
$P_u = (0.9-0.2S_{DS})P_D + \Omega_o' P_{QE}$	$P_a = (0.6-0.14S_{DS})P_D + 0.7\Omega_o' P_{QE}$
$\quad = [0.9-0.2(1.0)](147 \text{ kips})$	$\quad = [0.6-0.14(1.0)](147 \text{ kips})$
$\quad\quad + 0.567(-248 \text{ kips})$	$\quad\quad + 0.7(0.531)(-248 \text{ kips})$
$\quad = -37.7 \text{ kips}$	$\quad = -24.6 \text{ kips}$

As stated in the Exception to AISC *Seismic Provisions* Section F2.3, the required strength of the columns does not need to exceed the forces corresponding to the resistance of the foundation to overturning uplift. These forces are smaller than the required strengths of the column as determined in Example 5.3.3, and could have been used as the required strengths for the design of the column.

Example 5.3.8. SCBF Brace-to-Beam Connection Design

Given:

Refer to Joint JT-1 in Figure 5-11. Design the connection between the braces and the beam. Use an ASTM A36 welded gusset plate concentric to the braces and 70-ksi electrodes to connect the braces to the beam. Use ASTM A572 Grade 50 material for brace reinforcement. All braces are ASTM A500 Grade B round HSS and the beam is an ASTM A992 W27×114. The applicable building code specifies the use of ASCE/SEI 7 for calculation of loads.

5.3 SPECIAL CONCENTRICALLY BRACED FRAMES

Relevant seismic design parameters were given in the SCBF Design Example Plan and Elevation section.

The complete connection design is shown in Figure 5-22.

Solution:

From AISC *Manual* Tables 2-4 and 2-5, the material properties are as follows:

ASTM A36
$F_y = 36$ ksi
$F_u = 58$ ksi

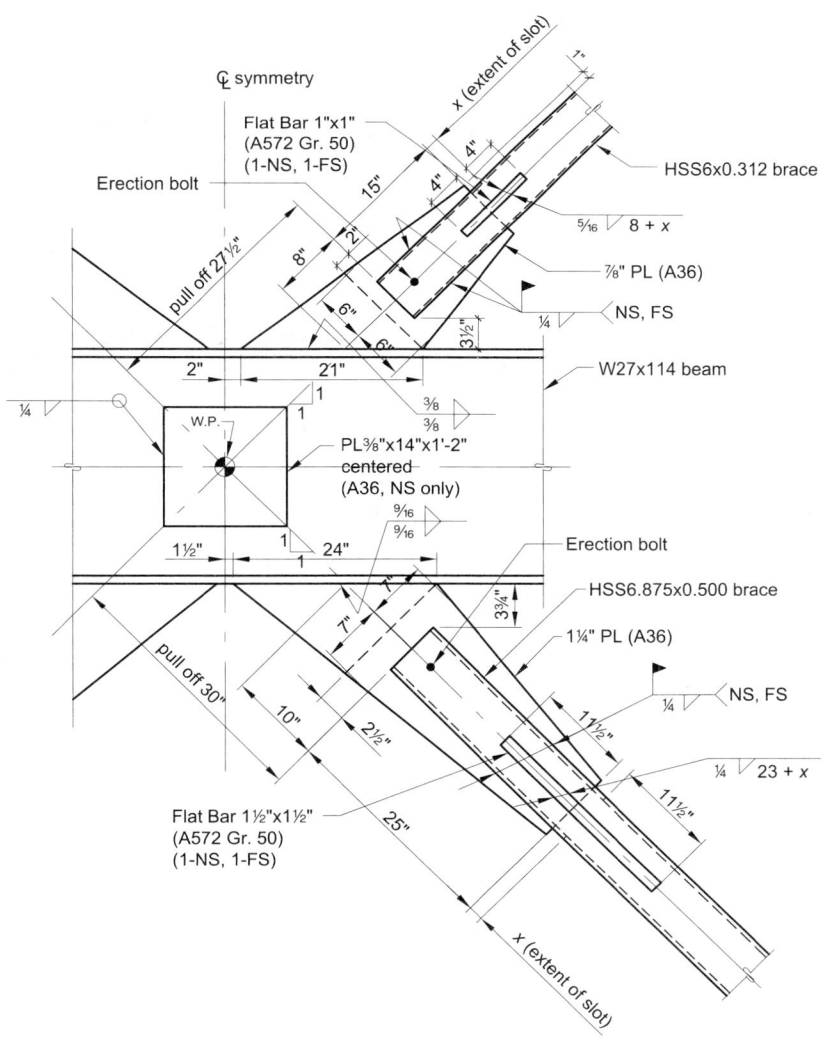

Fig. 5-22. Final connection design for Example 5.3.8.

ASTM A500 Grade B
$F_y = 42$ ksi
$F_u = 58$ ksi

ASTM A992
$F_y = 50$ ksi
$F_u = 65$ ksi

From AISC *Manual* Tables 1-1 and 1-13, the geometric properties are as follows:

Brace (above the beam)
HSS6×0.312
$t_{nom} = 0.312$ in. $t_{des} = 0.291$ in. $A = 5.22$ in.^{2} $r = 2.02$ in.

Brace (below the beam)
HSS6.875×0.500
$t_{nom} = 0.500$ in. $t_{des} = 0.465$ in. $A = 9.36$ in.^{2} $r = 2.27$ in.

Beam
W27×114
$d = 27.3$ in. $t_w = 0.570$ in. $t_f = 0.930$ in. $k_{des} = 1.53$ in.

AISC *Seismic Provisions* Sections F2.3(i) and F2.3(ii) define the two mechanism analyses that must be considered in determining the required strength of beams, columns and connections. AISC *Seismic Provisions* Section F2.6c specifies the required strength of bracing connections. For the required compressive strength, Section F2.6c(2) has an additional 1.1 factor (relative to the requirements of Section F2.3) applied to the required strength of the connection.

For these SCBF connection examples, the requirements of AISC *Seismic Provisions* Sections B2 and F2.3 will be used for both LRFD and ASD, except for the limit state of compression buckling on the Whitmore section, which will use the 1.1 factor specified in AISC *Seismic Provisions* Section F2.6c(2).

The required strength of the bracing connection due to seismic loading is based on ASCE/SEI 7 Section 12.4.3.2 Load Combination 5 (LRFD and ASD) with $\Omega_o Q_E = E_{mh}$, as discussed in AISC *Seismic Provisions* Section F2.3.

Determine the expected tensile strength of the braces

The brace connections must be designed to develop the larger forces determined from the two analyses specified in AISC *Seismic Provisions* Section F2.3. The required tensile strength of the connection is based on the expected strength.

For the HSS6×0.312 brace above the beam:

From AISC *Seismic Provisions* Table A3.1:

$R_y = 1.4$

5.3 SPECIAL CONCENTRICALLY BRACED FRAMES

Therefore:

$$P_{tension} = R_y F_y A_g$$
$$= 1.4(42 \text{ ksi})(5.22 \text{ in.}^2)$$
$$= 307 \text{ kips}$$

For the braces above the beam, the required strength of the bracing connection when the brace is in tension is:

LRFD	ASD
$P_u = 1.0 E_{mh}$ $= 1.0 (307 \text{ kips})$ $= 307 \text{ kips}$	$P_a = 0.7 E_{mh}$ $= 0.7 (307 \text{ kips})$ $= 215 \text{ kips}$

For the HSS6.875×0.500 brace below the beam:

$$P_{tension} = R_y F_y A_g$$
$$= 1.4(42 \text{ ksi})(9.36 \text{ in.}^2)$$
$$= 550 \text{ kips}$$

For the braces below the beam, the required strength of the bracing connection when the brace is in tension is:

LRFD	ASD
$P_u = 1.0 E_{mh}$ $= 1.0 (550 \text{ kips})$ $= 550 \text{ kips}$	$P_a = 0.7 E_{mh}$ $= 0.7 (550 \text{ kips})$ $= 385 \text{ kips}$

Determine the expected strength in compression of the braces

For determining the expected strength of the brace in compression, $R_y F_y$ is used in lieu of F_y for the determination of F_{cre} according to AISC *Seismic Provisions* Section F2.3. The brace length used for the determination of F_{cre} must not exceed the distance from brace end to brace end. Estimate that the length of the connections will reduce the brace length to approximately 12 ft. This will be verified once the connection is complete. Therefore, a length of 12 ft will be used to determine the brace expected strength in compression.

For the HSS6×0.312 brace above the beam:

Use AISC *Specification* Chapter E with $F_y = R_y F_y$ to determine F_{cre}, as follows:

$$\frac{KL}{r} = \frac{1.0(12.0 \text{ ft})(12 \text{ in./ft})}{2.02 \text{ in.}}$$
$$= 71.3$$

$$4.71\sqrt{\frac{E}{R_y F_y}} = 4.71\sqrt{\frac{29{,}000 \text{ ksi}}{1.4(42 \text{ ksi})}}$$

$$= 105$$

When: $\dfrac{KL}{r} \leq 4.71\sqrt{\dfrac{E}{R_y F_y}}$:

$$F_e = \frac{\pi^2 E}{\left(\dfrac{KL}{r}\right)^2} \qquad \text{(Spec. Eq. E3-4)}$$

$$= \frac{\pi^2 (29{,}000 \text{ ksi})}{(71.3)^2}$$

$$= 56.3 \text{ ksi}$$

$$F_{cre} = \left(0.658^{\frac{R_y F_y}{F_e}}\right) R_y F_y \qquad \text{(from Spec. Eq. E3-2)}$$

$$= \left[0.658^{\frac{1.4(42 \text{ ksi})}{56.3 \text{ ksi}}}\right](1.4)(42 \text{ ksi})$$

$$= 38.0 \text{ ksi}$$

The expected compressive strength of the braces above the beam is:

$$P_{compression} = 1.14 F_{cre} A_g$$

$$= 1.14(38.0 \text{ ksi})(5.22 \text{ in.}^2)$$

$$= 226 \text{ kips}$$

And the expected post-buckling strength is:

$$0.3 P_{compression} = 0.3(226 \text{ kips})$$

$$= 67.8 \text{ kips}$$

For the braces above the beam, the required strength of the bracing connection when the brace is in compression is based on E_{mh} equal to the lesser of $R_y F_y A_g$ and $1.14 F_{cre} A_g$ according to AISC *Seismic Provisions* Section F2.3; therefore, use $E_{mh} = 226$ kips.

The required strength is:

LRFD	ASD
$P_u = 1.0 E_{mh}$	$P_a = 0.7 E_{mh}$
$= 1.0 \, (226 \text{ kips})$	$= 0.7(226 \text{ kips})$
$= 226 \text{ kips}$	$= 158 \text{ kips}$

5.3 SPECIAL CONCENTRICALLY BRACED FRAMES

For the braces above the beam, the required strength of the bracing connection when the brace is in compression at its post-buckling strength is:

LRFD	ASD
$P_u = 1.0 E_{mh}$ $= 1.0\,(67.8 \text{ kips})$ $= 67.8 \text{ kips}$	$P_a = 0.7 E_{mh}$ $= 0.7(67.8 \text{ kips})$ $= 47.5 \text{ kips}$

For the HSS6.875×0.500 brace below the beam:

$$\frac{KL}{r} = \frac{1.0(12.0 \text{ ft})(12 \text{ in./ft})}{2.27 \text{ in.}}$$
$$= 63.4$$

As calculated previously, $4.71\sqrt{\dfrac{E}{R_y F_y}} = 105$.

When $\dfrac{KL}{r} \leq 4.71\sqrt{\dfrac{E}{R_y F_y}}$:

$$F_e = \frac{\pi^2 E}{\left(\dfrac{KL}{r}\right)^2} \qquad \text{(Spec. Eq. E3-4)}$$

$$= \frac{\pi^2 (29{,}000 \text{ ksi})}{(63.4)^2}$$

$$= 71.2 \text{ ksi}$$

$$F_{cre} = \left(0.658^{\frac{R_y F_y}{F_e}}\right) R_y F_y \qquad \text{(from Spec. Eq. E3-2)}$$

$$= \left[0.658^{\frac{1.4(42 \text{ ksi})}{71.2 \text{ ksi}}}\right](1.4)(42 \text{ ksi})$$

$$= 41.6 \text{ ksi}$$

The expected compressive strength of the braces below the beam is:

$$P_{compression} = 1.14 F_{cre} A_g$$
$$= 1.14(41.6 \text{ ksi})(9.36 \text{ in.}^2)$$
$$= 444 \text{ kips}$$

And the expected post-buckling strength is:

$$0.3 P_{compression} = 0.3(444 \text{ kips})$$
$$= 133 \text{ kips}$$

For the braces below the beam, the required strength of the bracing connection when the brace is in compression is based on E_{mh} equal to the lesser of $R_y F_y A_g$ and $1.14 F_{cre} A_g$ according to AISC *Seismic Provisions* Section F2.3; therefore, use $E_{mh} = 444$ kips.

LRFD	ASD
$P_u = 1.0 E_{mh}$ $= 1.0\,(444 \text{ kips})$ $= 444 \text{ kips}$	$P_a = 0.7 E_{mh}$ $= 0.7(444 \text{ kips})$ $= 311 \text{ kips}$

For the braces below the beam, the required strength of the bracing connection when the brace is in compression at its post-buckling strength is:

LRFD	ASD
$P_u = 1.0 E_{mh}$ $= 1.0\,(133 \text{ kips})$ $= 133 \text{ kips}$	$P_a = 0.7 E_{mh}$ $= 0.7(133 \text{ kips})$ $= 93.1 \text{ kips}$

The two sets of forces are shown in Figures 5-23 and 5-24.

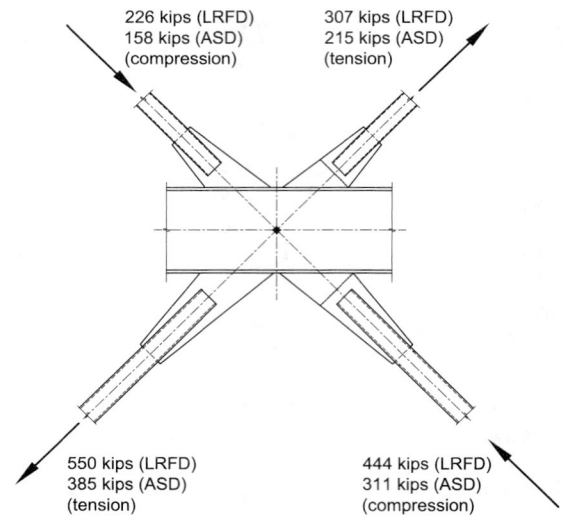

Fig. 5-23. Required strength of bracing connections according to AISC Seismic Provisions Section F2.3(i).

Top Brace-to-Beam Connection

The required tensile strength of the connection is based upon $R_y F_y A_g$ of the braces as stipulated in AISC *Seismic Provisions* Section F2.6c(1). All limit states applicable to tension or compression in the brace must be checked.

Determine the minimum length, l, required for the brace-gusset lap

The limit state of shear rupture in the brace wall is used to determine the minimum brace-gusset lap length. Note that the expected brace rupture strength, $R_t F_u$, may be used in the determination of the available strength according to AISC *Seismic Provisions* Section A3.2.

Using AISC *Specification* Section J4.2, including R_t from AISC *Seismic Provisions* Table A3.1:

$R_t = 1.3$

$R_n = 0.60 R_t F_u A_{nv}$ (from *Spec.* Eq. J4-4)

In this equation, A_{nv} is taken as the cross-sectional area of the four walls of the brace, $A_{nv} = 4lt_{des}$. Therefore:

$R_n = 0.60 R_t F_u (4lt_{des})$

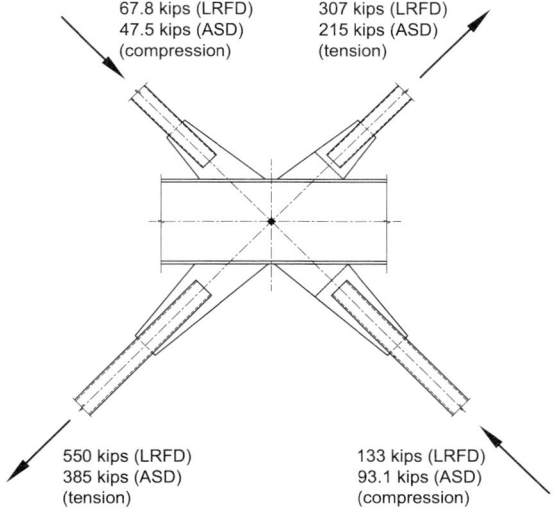

Fig. 5-24. Required strength of bracing connections according to AISC Seismic Provisions Section F2.3(ii).

Setting the available shear rupture strength equal to the required tensile strength and solving for the minimum lap length, l:

LRFD	ASD
$l \geq \dfrac{P_u}{\phi(0.60)R_t F_u(4t_{des})}$ $\geq \dfrac{307 \text{ kips}}{0.75(0.60)(1.3)(58 \text{ ksi})(4)(0.291 \text{ in.})}$ $\geq 7.77 \text{ in.}$	$l \geq \dfrac{\Omega P_a}{0.60 R_t F_u(4t_{des})}$ $\geq \dfrac{2.00(215 \text{ kips})}{0.60(1.3)(58 \text{ ksi})(4)(0.291 \text{ in.})}$ $\geq 8.17 \text{ in.}$

Note that this length is the minimum required for the limit state of shear rupture in the brace wall. A longer length may be used when designing the fillet welds between the brace and the gusset plate, if desired, to allow a smaller fillet weld size as is implemented in the following.

Size the weld between the brace and the gusset plate

The strength of fillet welds defined in AISC *Specification* Section J2 can be simplified, as explained in Part 8 of the AISC *Manual*, to AISC *Manual* Equations 8-2a and 8-2b:

LRFD	ASD
$\phi R_n = 1.392 Dl$	$\dfrac{R_n}{\Omega} = 0.928 Dl$

Try ¼-in. fillet welds for the four lines of weld, which can be made in a single pass:

LRFD	ASD
$4(1.392)Dl \geq P_u$ $l \geq \dfrac{307 \text{ kips}}{4(1.392 \text{ kip/in.})(4 \text{ sixteenths})}$ $\geq 13.8 \text{ in.}$	$4(0.928)Dl \geq P_a$ $l \geq \dfrac{215 \text{ kips}}{4(0.928 \text{ kip/in.})(4 \text{ sixteenths})}$ $\geq 14.5 \text{ in.}$

The designs in LRFD and ASD give slightly different required lengths of weld. For convenience, the more conservative result will be used in subsequent calculations and in the final design. In practice, designers should consistently use one method or the other.

Use (4) 15-in. long, ¼-in. fillet welds to connect the brace above the beam to the gusset plate.

Check block shear rupture of the gusset plate

The available strength for the limit state of block shear rupture is:

$$R_n = 0.60 F_u A_{nv} + U_{bs} F_u A_{nt} \leq 0.60 F_y A_{gv} + U_{bs} F_u A_{nt} \quad \quad (Spec.\ Eq.\ J4\text{-}5)$$

5.3 SPECIAL CONCENTRICALLY BRACED FRAMES

Since the gross shear area, A_{gv}, and the net shear area, A_{nv}, are equal in this case, the shear yielding component, $0.60F_yA_{gv}$, is smaller than the shear rupture component, $0.60F_uA_{nv}$, and the right side of the equation controls.

$U_{bs} = 1.0$

$$0.60F_yA_{gv} = 0.60F_y(2)lt_p$$
$$= 0.60(36 \text{ ksi})(2)(15.0 \text{ in.})(t_p)$$
$$= 648t_p$$

$$U_{bs}F_uA_{nt} = 1.0F_uD_{brace}t_p$$
$$= 1.0(58 \text{ ksi})(6.00 \text{ in.})(t_p)$$
$$= 348t_p$$

LRFD	ASD
$\phi R_n = 0.75(0.60F_yA_{gv} + U_{bs}F_uA_{nt}) \geq P_u$	$\dfrac{R_n}{\Omega} = \dfrac{0.60F_yA_{gv} + U_{bs}F_uA_{nt}}{\Omega} \geq P_a$
$t_p \geq \dfrac{307 \text{ kips}}{0.75(648 \text{ kip/in.} + 348 \text{ kip/in.})}$	$t_p \geq \dfrac{2.00(215 \text{ kips})}{(648 \text{ kip/in.} + 348 \text{ kip/in.})}$
≥ 0.411 in.	≥ 0.432 in.

Check required gusset width and thickness based on the limit state of tensile yielding

Tensile yielding is checked on a section of the gusset plate commonly referred to as the Whitmore section. This section is explained in AISC *Manual* Part 9 (Figure 9-1) and in Thornton and Lini (2011). Because the width and thickness of the gusset plate have not yet been chosen, the minimum area will be determined for this limit state. The nominal tensile yielding strength is:

$$R_n = F_yA_g \qquad \qquad (\textit{Spec.} \text{ Eq. J4-1})$$

LRFD	ASD
$\phi R_n \geq P_u$	$\dfrac{R_n}{\Omega} \geq P_a$
$t_pw_p \geq \dfrac{P_u}{\phi F_y}$	$t_pw_p \geq \dfrac{\Omega P_a}{F_y}$
$\geq \dfrac{307 \text{ kips}}{0.90(36 \text{ ksi})}$	$\geq \dfrac{1.67(215 \text{ kips})}{36 \text{ ksi}}$
≥ 9.48 in.^{2}	≥ 9.97 in.^{2}

A gusset plate will be chosen that has a width on the Whitmore section, w_p, of 12 in. and a thickness, t_p, of ⅞ in. This meets the minimum required gusset plate thickness for the limit state of block shear rupture calculated previously.

Check that the bracing connection can accommodate brace buckling according to AISC Seismic Provisions Section F2.6c(3)

The requirements of AISC *Seismic Provisions* Section F2.6c(3) are met through the use of option (b): rotation capacity. As explained in the User Note of that section and in the Commentary Figure C-F2.9, accommodation of inelastic rotation is accomplished with the brace terminating before the line of restraint. Figure 5-22 shows the $2t$ clearance beyond the end of the brace.

The choice of a relatively small Whitmore section results in a tapered gusset, which is beneficial because it allows the brace to be located closer to the beam while still accommodating brace rotation by providing a $2t$ clearance per AISC *Seismic Provisions* Section F2.6c(3) and Commentary.

Check the maximum Whitmore section

As explained in AISC *Manual* Part 9, the maximum width of the Whitmore section that can be considered effective is defined by a 30° spread to each side, but not exceeding the actual width of the gusset plate. To make sure that the 12.0 in. width chosen previously can be considered effective in tension, check the maximum Whitmore section.

$$w_p \leq D_{brace} + 2l \tan 30°$$
$$\leq 6.00 \text{ in.} + 2(15.0 \text{ in.}) \tan 30° = 23.3 \text{ in.}$$

12.0 in. ≤ 23.3 in.   **o.k.**

Therefore, the 12 in. width and ⅞ in. thickness for the gusset plate chosen previously is acceptable.

The actual angle of the gusset edge, measured relative to the centerline of the brace, is:

$$\phi = \tan^{-1}\left[\frac{\frac{1}{2}(w_p - D_{brace})}{l}\right]$$
$$= \tan^{-1}\left[\frac{\frac{1}{2}(12.0 \text{ in.} - 6.00 \text{ in.})}{15.0 \text{ in.}}\right]$$
$$= 11.3°$$

Note: this angle that is smaller than the Whitmore section angle of 30° provides a more compact gusset.

Check brace effective net area

From AISC *Seismic Provisions* Section F2.5b(3), the brace effective net area, A_e, shall not be less than the brace gross area, A_g.

$$A_n = A_g - 2[t_p + 2(gap)]t_{des}$$

5.3 SPECIAL CONCENTRICALLY BRACED FRAMES

Using a gap of $\frac{1}{16}$ in. on each side of the brace slot to allow clearance for erection:

$$A_n = 5.22 \text{ in.}^2 - 2[\tfrac{7}{8} \text{ in.} + 2(\tfrac{1}{16} \text{ in.})](0.291 \text{ in.})$$
$$= 4.64 \text{ in.}^2$$

From AISC *Specification* Table D3.1, Case 5, because $l > 1.3D$, $U = 1.0$, and the effective net area is:

$$A_e = 1.0(4.64 \text{ in.}^2)$$
$$= 4.64 \text{ in.}^2$$

Because $A_e < A_g$, brace reinforcement is required. The approximate area of reinforcement required, A_{rn}, is the area removed, but the position of the reinforcement will reduce U to less than 1.0 because of its position. The required area of reinforcement can be obtained from: $(A_n + A_{rn})U \geq A_g$.

Assuming a value of $U = 0.80$:

$$A_{rn} = \frac{A_g}{0.80} - A_n$$
$$= \frac{5.22 \text{ in.}^2}{0.80} - 4.64 \text{ in.}^2$$
$$= 1.89 \text{ in.}^2$$

Try two 1 in. × 1 in. flat bars, with a total area of 2.00 in.² AISC *Seismic Provisions* Section F2.5b(3)(i) requires that the specified minimum yield strength of the reinforcement be at least that of the brace; therefore, use ASTM A572 Grade 50 material for the flat bar. The cross-sectional geometry is shown in Figure 5-25.

$$r_1 = \frac{D_{brace}}{2} - \frac{t_{des}}{2}$$
$$= \frac{6.00 \text{ in.}}{2} - \frac{0.291 \text{ in.}}{2}$$
$$= 2.85 \text{ in.}$$

$$r_2 = \frac{D_{brace}}{2} + \frac{1.00 \text{ in.}}{2}$$
$$= \frac{6.00 \text{ in.}}{2} + \frac{1.00 \text{ in.}}{2}$$
$$= 3.50 \text{ in.}$$

The distance to the centroid of a partial circle is given by:

$$\bar{x} = \frac{r_1 \sin \theta}{\theta}$$

where the total arc of the partial circle is 2θ, and θ is measured in radians. Although the brace is slightly less than a full half-circle because of the slot as shown in Figure 5-25, use an angle, θ, of $\pi/2$ for simplicity. This is slightly unconservative for calculating the value of the shear lag factor, U. A more precise calculation could be performed using the exact angle.

$$\bar{x}_{brace} = 2.85 \text{ in.} \left(\frac{\sin(\pi/2) \text{ rad}}{(\pi/2) \text{ rad}} \right)$$

$$= 1.81 \text{ in.}$$

$$\bar{x}_{re} = r_2$$

$$= 3.50 \text{ in.}$$

Determine $\bar{x}$ for the composite cross section.

Part	$\bar{x}$ in.	A in.^{2}	$\bar{x}A$ in.^{3}
Half of brace	1.81	2.32	4.20
One flat bar	3.50	1.00	3.50
Σ	–	3.32	7.70

$$\bar{x} = \frac{\Sigma \bar{x} A}{\Sigma A}$$

$$= \frac{7.70 \text{ in.}^3}{3.32 \text{ in.}^2}$$

$$= 2.32 \text{ in.}$$

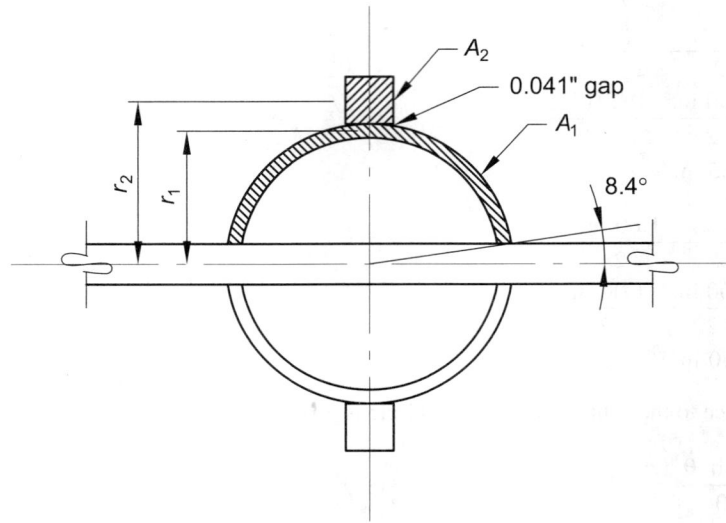

Fig. 5-25. Cross section of brace above beam at net section.

From AISC *Specification* Table D3.1, Case 2, which applies to round HSS with reinforcement added:

$$U = 1 - \frac{\bar{x}}{l}$$
$$= 1 - \frac{2.32 \text{ in.}}{15.0 \text{ in.}}$$
$$= 0.845$$

$$A_n = A_{n(brace)} + A_{rn}$$
$$= 4.64 \text{ in.}^2 + 2(1.00 \text{ in.})(1.00 \text{ in.})$$
$$= 6.64 \text{ in.}^2$$

$$A_e = UA_n$$
$$= 0.845(6.64 \text{ in.}^2)$$
$$= 5.61 \text{ in.}^2 > 5.22 \text{ in.}^2 \quad \textbf{o.k.}$$

Design welds connecting flat bars to brace

According to AISC *Seismic Provisions* Section F2.5b(3)(ii), the flat bar must be connected to the pipe brace to develop the expected strength of the flat bar on each side of the reduced section (the expected yield strength, $R_y F_y$, is used here). The reduced section is the length of the HSS from the extent of the slot (dimension x of Figure 5-22) to the start of the HSS-to-gusset weld. The required strength of the weld is based on the expected flat bar yield strength, using R_y from AISC *Seismic Provisions* Table A3.1 for ASTM A572 Grade 50 bars. For ASD, use 1.0/1.5 of the expected strength of the flat bar reinforcement.

LRFD	ASD
$R_y F_y A_{fb} = 1.1(50 \text{ ksi})(1.00 \text{ in.}^2)$ $= 55.0$ kips	$R_y F_y A_{fb} / 1.5 = 1.1(50 \text{ ksi})(1.00 \text{ in.}^2)/1.5$ $= 36.7$ kips

There is a small gap of approximately 0.041 in. between the face of the pipe brace and the edge of the flat bar, as indicated in Figure 5-25. Since this is less than $\frac{1}{16}$ in., it can be neglected according to AWS D1.1 clause 5.22.1. A single-pass $\frac{5}{16}$-in. fillet weld can be used.

With two welds, the length of $\frac{5}{16}$-in. fillet welds connecting the flat bar to the brace is determined from AISC *Manual* Equations 8-2a and 8-2b as follows:

LRFD	ASD
$l_w = \dfrac{55.0 \text{ kips}}{2(1.392 \text{ kip/in.})(5 \text{ sixteenths})}$ $= 3.95$ in.	$l_w = \dfrac{36.7 \text{ kips}}{2(0.928 \text{ kip/in.})(5 \text{ sixteenths})}$ $= 3.95$ in.

Use a 1 in. × 1 in. flat bar with 5/16-in. fillet welds; the detail extends past both sides of the reduced section of the brace.

The flat bar fillet weld develops the expected strength of the bar on each side of the end of the brace slot. The brace slot may be longer than the slot length by a maximum erection clearance of x inches (see Figure 5-22), as determined by the fabricator. The length of the flat bar will be 4.00 in. + 4.00 in. + x in. = 8.00 in. + x in.

Check the gusset plate for buckling on the Whitmore section

From Figure 5-22, the buckling length, which is taken along the brace centerline (Dowswell, 2006), is l_b = 8.00 in. (Example 5.3.10 provides an equation for calculating the length of buckling; here it is determined graphically.) AISC *Seismic Provisions* Section F2.6c(2) specifies a required compressive strength for buckling limit states that is at least equal to 1.1 times the expected brace strength in compression for LRFD. The stress over the Whitmore section, using the Whitmore width of 12.0 in., is:

LRFD	ASD
$f_{ua} = \dfrac{1.1(226 \text{ kips})}{12.0 \text{ in.}(\text{⅞ in.})}$ $= 23.7 \text{ ksi}$	$f_{aa} = \dfrac{1.1(158 \text{ kips})}{12.0 \text{ in.}(\text{⅞ in.})}$ $= 16.6 \text{ ksi}$

The radius of gyration of the gusset plate is:

$$r = \frac{t_p}{\sqrt{12}}$$
$$= \frac{\text{⅞ in.}}{\sqrt{12}}$$
$$= 0.253 \text{ in.}$$

Recommended values for the effective length factor, K, are given in Dowswell (2006). However, that paper does not address the case of a single gusset plate with the $2t$ clearance to accommodate brace buckling [called an "extended" gusset plate in Dowswell (2006)]. Therefore, in this case, use K = 1.2 from AISC *Specification* Commentary Table C-A-7.1 assuming that the gusset plate is fixed at one end and free to translate but not rotate at the other. With $l_b = L$:

$$\frac{KL}{r} = \frac{1.2(8.00 \text{ in.})}{0.253 \text{ in.}}$$
$$= 37.9$$

Interpolating from AISC *Manual* Table 4-22:

LRFD	ASD
$\phi_c F_{cr}$ = 30.0 ksi > 23.7 ksi　　**o.k.**	$\dfrac{F_{cr}}{\Omega_c}$ = 20.0 ksi > 16.6 ksi　　**o.k.**

5.3 SPECIAL CONCENTRICALLY BRACED FRAMES

Determine the forces at the gusset-to-beam interface

Based on the required tensile strength of the bracing connection (which is larger than the required compressive strength), the shear force at the interface of the gusset with the beam flange is:

LRFD	ASD
$V_u = 307 \text{ kips}(\cos 45°)$ $= 217 \text{ kips}$	$V_a = 215 \text{ kips}(\cos 45°)$ $= 152 \text{ kips}$

The normal (tension) force is:

LRFD	ASD
$N_u = 307 \text{ kips}(\sin 45°)$ $= 217 \text{ kips}$	$N_a = 215 \text{ kips}(\sin 45°)$ $= 152 \text{ kips}$

The contact length between the gusset plate and the beam top flange, as shown in Figure 5-22, is 21.0 in. and the brace line of action misses the centroid of the contact length by 1.5 in. due to the increasing width of the gusset plate. Therefore, the moment on the contact surface is:

LRFD	ASD
$M_u = N_u(1.5 \text{ in.})$ $= 217 \text{ kips}(1.5 \text{ in.})$ $= 326 \text{ kip-in.}$	$M_a = N_a(1.5 \text{ in.})$ $= 152 \text{ kips}(1.5 \text{ in.})$ $= 228 \text{ kip-in.}$

Check the gusset and design the weld at the gusset-to-beam flange interface

The forces are:

LRFD	ASD
Shear V_u = 217 kips Normal N_u = 217 kips Moment M_u = 326 kip-in.	Shear V_a = 152 kips Normal N_a = 152 kips Moment M_a = 228 kip-in.

The moment, M_u or M_a, and the normal force, N_u or N_a, can be combined to give an equivalent normal force, $N_{u\,equiv}$ or $N_{a\,equiv}$. This equivalent tension assumes a plastic stress distribution for the moment, M_u or M_a, which is similar to the stress distribution in the weld assumed in the inelastic method of the AISC *Manual* Part 8 eccentrically loaded weld group tables. On one half of the contact length, the stress due to the normal force, N_u or N_a, and the stress due to the moment are additive. On the other half, the stresses are in opposite directions. For convenience of calculations, one of the forces in the force couple (due to the moment, M_u or M_a) is imagined reversed so the entire contact surface is in uniform tension or compression. The equivalent normal force is:

LRFD	ASD
$N_{u\,equiv} = N_u + \dfrac{4M_u}{l_b}$ $= 217 \text{ kips} + \dfrac{4(326 \text{ kip-in.})}{21.0 \text{ in.}}$ $= 279 \text{ kips}$	$N_{a\,equiv} = N_a + \dfrac{4M_a}{l_b}$ $= 152 \text{ kips} + \dfrac{4(228 \text{ kip-in.})}{21.0 \text{ in.}}$ $= 195 \text{ kips}$

The gusset stresses are:

LRFD	ASD
In tension $f_{ua} = \dfrac{N_{u\,equiv}}{t_p l_b}$ $= \dfrac{279 \text{ kips}}{(\text{⅞ in.})(21.0 \text{ in.})}$ $= 15.2 \text{ ksi}$ The design tensile yielding stress from AISC *Specification* Section J4.1 is: $\phi F_y = 0.90(36 \text{ ksi})$ $= 32.4 \text{ ksi} > 15.2 \text{ ksi}$ **o.k.**	In tension $f_{aa} = \dfrac{N_{a\,equiv}}{t_p l_b}$ $= \dfrac{195 \text{ kips}}{(\text{⅞ in.})(21.0 \text{ in.})}$ $= 10.6 \text{ ksi}$ The allowable tensile yielding stress from AISC *Specification* Section J4.1 is: $\dfrac{F_y}{\Omega} = \dfrac{36 \text{ ksi}}{1.67}$ $= 21.6 \text{ ksi} > 10.6 \text{ ksi}$ **o.k.**

LRFD	ASD
In shear $f_{uv} = \dfrac{V_u}{t_p l_b}$ $= \dfrac{217 \text{ kips}}{(\text{⅞ in.})(21.0 \text{ in.})}$ $= 11.8 \text{ ksi}$	In shear $f_{av} = \dfrac{V_a}{t_p l_b}$ $= \dfrac{152 \text{ kips}}{(\text{⅞ in.})(21.0 \text{ in.})}$ $= 8.27 \text{ ksi}$

5.3 SPECIAL CONCENTRICALLY BRACED FRAMES

LRFD	ASD
The design shear yielding stress from AISC *Specification* Section J4.2 is: $\phi 0.60 F_y = 1.00(0.60)(36 \text{ ksi})$ $= 21.6 \text{ ksi} > 11.8 \text{ ksi}$ **o.k.**	The allowable shear yielding stress from AISC *Specification* Section J4.2 is: $\dfrac{0.60 F_y}{\Omega} = \dfrac{0.60(36 \text{ ksi})}{1.50}$ $= 14.4 \text{ ksi} > 8.27 \text{ ksi}$ **o.k.**

Size gusset-to-beam weld

The angle of the resultant force can be calculated and used in the directional strength increase of fillet welds according to AISC *Specification* Equation J2-5 as follows:

LRFD	ASD
$\theta = \tan^{-1}\left(\dfrac{N_{u\,equiv}}{V_u}\right)$ $= \tan^{-1}\left(\dfrac{279 \text{ kips}}{217 \text{ kips}}\right)$ $= 52.1°$	$\theta = \tan^{-1}\left(\dfrac{N_{a\,equiv}}{V_a}\right)$ $= \tan^{-1}\left(\dfrac{195 \text{ kips}}{152 \text{ kips}}\right)$ $= 52.1°$

AISC *Specification* Section J2.4 allows an increase in the available strength of fillet welds when the angle of loading is not along the weld longitudinal axis, which is used in the following calculation.

The weld ductility factor, equal to 1.25, which is explained in AISC *Manual* Part 13, is applied here. Using AISC *Manual* Equations 8-2a and 8-2b, the number of sixteenths of fillet weld required is:

LRFD	ASD
$D_{req'd} \geq$ $\dfrac{1.25\sqrt{N_{u\,equiv}^2 + V_u^2}}{2(1.392 \text{ kip/in.})(1+0.5\sin^{1.5}\theta)(l_b)}$ $= \dfrac{1.25\sqrt{(279 \text{ kips})^2 + (217 \text{ kips})^2}}{2(1.392 \text{ kip/in.})(1.35)(21.0 \text{ in.})}$ $= 5.60 \text{ sixteenths}$	$D_{req'd} \geq$ $\dfrac{1.25\sqrt{N_{a\,equiv}^2 + V_a^2}}{2(0.928 \text{ kip/in.})(1+0.5\sin^{1.5}\theta)(l_b)}$ $= \dfrac{1.25\sqrt{(195 \text{ kips})^2 + (152 \text{ kips})^2}}{2(0.928 \text{ kip/in.})(1.35)(21.0 \text{ in.})}$ $= 5.87 \text{ sixteenths}$

An alternative fully plastic approach to the gusset-to-beam stresses is shown in the following calculations and presented in the *IBC Structural/Seismic Design Manual* (SEAOC, 2006), where the normal and bending stresses are assumed to act over separate portions of the contact length, l_b, and are set equal to each other in order to result in a uniform stress distribution as shown in Figure 5-26.

From Figure 5-26, the moment about the center of the contact length, l_b, is:

$$M = F'\left(\frac{a-e}{2} + e\right) \quad (2)$$
$$= F'(a+e)$$

so,

$$F' = \frac{M}{a+e}$$

and

$$f_b = \frac{F'}{(a-e)t_p}$$
$$= \frac{M}{(a^2 - e^2)t_p}$$

Likewise, from Figure 5-26:

$$f_a = \frac{N}{2et_p}$$

Setting $f_a = f_b$ and solving for e:

$$e = \sqrt{\left(\frac{M}{N}\right)^2 + a^2} - \left(\frac{M}{N}\right)$$

For this example:

$$a = \frac{21.0 \text{ in.}}{2}$$
$$= 10.5 \text{ in.}$$

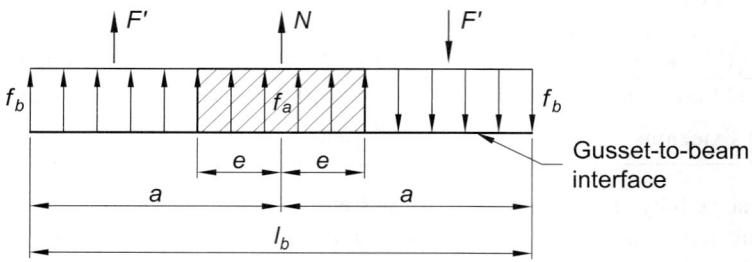

Fig. 5-26. Fully plastic stress distribution on gusset-to-beam interface.

5.3 SPECIAL CONCENTRICALLY BRACED FRAMES

LRFD	ASD
$M_u = 326$ kip-in. $N_u = 217$ kips $e = \sqrt{\left(\dfrac{326 \text{ kip-in.}}{217 \text{ kips}}\right)^2 + (10.5 \text{ in.})^2}$ $\quad - \left(\dfrac{326 \text{ kip-in.}}{217 \text{ kips}}\right)$ $= 9.10$ in.	$M_a = 228$ kip-in. $N_a = 152$ kips $e = \sqrt{\left(\dfrac{228 \text{ kip-in.}}{152 \text{ kips}}\right)^2 + (10.5 \text{ in.})^2}$ $\quad - \left(\dfrac{228 \text{ kip-in.}}{152 \text{ kips}}\right)$ $= 9.11$ in.

Therefore:

LRFD	ASD
$f_{ua} = \dfrac{217 \text{ kips}}{2(9.10 \text{ in.})(\text{7/8 in.})}$ $= 13.6$ ksi $f_{ub} = \dfrac{326 \text{ kip-in.}}{\left[(10.5 \text{ in.})^2 - (9.10 \text{ in.})^2\right](\text{7/8 in.})}$ $= 13.6$ ksi	$f_{aa} = \dfrac{152 \text{ kips}}{2(9.11 \text{ in.})(\text{7/8 in.})}$ $= 9.53$ ksi $f_{ab} = \dfrac{228 \text{ kip-in.}}{\left[(10.5 \text{ in.})^2 - (9.11 \text{ in.})^2\right](\text{7/8 in.})}$ $= 9.56$ ksi

As required, $f_a = f_b = 13.6$ ksi (LRFD) and approximately 9.53 (ASD).

Compare this to 15.2 ksi (LRFD) and 10.6 ksi (ASD) using the simpler method. The simpler method is about 12% (LRFD) or 11% (ASD) conservative.

Using the plastic method to check the required weld size, define an equivalent tensile (normal) force as:

LRFD	ASD
$N'_{u\,equiv} = (13.6 \text{ ksi})(21.0 \text{ in.})(\text{7/8 in.})$ $\quad\quad\quad = 250$ kips $\theta = \tan^{-1}\left(\dfrac{N_{u\,equiv}}{V_u}\right)$ $\quad = \tan^{-1}\left(\dfrac{250 \text{ kips}}{217 \text{ kips}}\right)$ $\quad = 49.0°$	$N'_{a\,equiv} = (9.53 \text{ ksi})(21.0 \text{ in.})(\text{7/8 in.})$ $\quad\quad\quad = 175$ kips $\theta = \tan^{-1}\left(\dfrac{N_{a\,equiv}}{V_a}\right)$ $\quad = \tan^{-1}\left(\dfrac{175 \text{ kips}}{152 \text{ kips}}\right)$ $\quad = 49.0°$

LRFD	ASD
$D_{req'd} = \dfrac{1.25\sqrt{(250 \text{ kips})^2 + (217 \text{ kips})^2}}{2(21.0 \text{ in.})(1.392 \text{ kip/in.})}$ $\times \dfrac{1}{\left(1.0 + 0.5\sin^{1.5} 49.0°\right)}$ $= 5.33$ sixteenths	$D_{req'd} = \dfrac{1.25\sqrt{(175 \text{ kips})^2 + (152 \text{ kips})^2}}{2(21.0 \text{ in.})(0.928 \text{ kip/in.})}$ $\times \dfrac{1}{\left(1.0 + 0.5\sin^{1.5} 49.0°\right)}$ $= 5.60$ sixteenths

This result is within 5% of the simpler method and both will require a ⅜-in. fillet weld.

The plastic method is slightly less conservative than the more common simple method. It can always be used for these calculations but will not be pursued further in this Manual.

Use a ⅜-in. fillet weld on each side of the gusset at the gusset-to-beam connection for the brace above the beam.

Check beam web local yielding

For a force applied at a distance from the end that is greater than the depth of the member:

$R_n = F_{yw} t_w (5k + l_b)$ *(Spec. Eq. J10-2)*

$= 50 \text{ ksi}(0.570 \text{ in.})[5(1.53 \text{ in.}) + 21.0 \text{ in.}]$

$= 817$ kips

LRFD	ASD
$\phi R_n = 1.00 \, (817 \text{ kips})$ $= 817$ kips $> N_{u\,equiv} = 279 \text{ kips}$ **o.k.**	$\dfrac{R_n}{\Omega} = \dfrac{817 \text{ kips}}{1.50}$ $= 545$ kips $> N_{a\,equiv} = 195 \text{ kips}$ **o.k.**

Web local yielding applies to both tension and compression loads. Web local crippling applies only to the compression loads, but the larger tension load is used here for convenience. If desired, the compressive strength of the brace could be used for web local crippling.

Check beam web local crippling

For a force applied greater than a distance of $d/2$ from the beam end:

$$R_n = 0.80t_w^2 \left[1 + 3\left(\frac{l_b}{d}\right)\left(\frac{t_w}{t_f}\right)^{1.5}\right] \sqrt{\frac{EF_{yw}t_f}{t_w}} \qquad \text{(Spec. Eq. J10-4)}$$

$$= 0.80(0.570 \text{ in.})^2 \left[1 + 3\left(\frac{21.0 \text{ in.}}{27.3 \text{ in.}}\right)\left(\frac{0.570 \text{ in.}}{0.930 \text{ in.}}\right)^{1.5}\right]$$

$$\times \sqrt{\frac{29,000 \text{ ksi}(50 \text{ ksi})(0.930 \text{ in.})}{0.570 \text{ in.}}}$$

$$= 842 \text{ kips}$$

LRFD	ASD
$\phi R_n = 0.75(842 \text{ kips})$ $= 632 \text{ kips}$ $> N_{u\,equiv} = 279 \text{ kips}$ **o.k.**	$\dfrac{R_n}{\Omega} = \dfrac{842 \text{ kips}}{2.00}$ $= 421 \text{ kips}$ $> N_{a\,equiv} = 195 \text{ kips}$ **o.k.**

This completes the design of the top brace to the beam. Figure 5-22 shows the configuration.

Bottom Brace-to-Beam Connection

The required tensile strength of the connection is based on $R_y F_y A_g$ of the braces as stipulated in AISC *Seismic Provisions* Section F2.6c(1). All limit states applicable to tension or compression in the brace must be checked.

For the HSS6.875×0.500 below the beam, the required strength of the bracing connections was determined at the beginning of this example.

Determine the minimum length, l, required for the brace-gusset lap

The limit state of shear rupture in the brace wall is used to determine the minimum brace-gusset lap length. Note that the expected brace rupture strength, $R_t F_u$, may be used according to AISC *Seismic Provisions* Section A3.2.

Using AISC *Specification* Section J4.2, including R_t from AISC *Seismic Provisions* Table A3.1:

$R_t = 1.3$

$R_n = 0.60 R_t F_u A_{nv}$ (from *Spec.* Eq. J4-4)

In this equation, A_{nv} is taken as the cross-sectional area of the four walls of the brace, $A_{nv} = 4lt_{des}$. Therefore:

$R_n = 0.60 R_t F_u (4lt_{des})$

Setting the available shear rupture strength equal to the required tensile strength and solving for the minimum lap length, *l*:

LRFD	ASD
$l \geq \dfrac{P_u}{\phi(0.60)R_t F_u(4t_{des})}$	$l \geq \dfrac{\Omega P_a}{0.60 R_t F_u(4t_{des})}$
$\geq \dfrac{550 \text{ kips}}{0.75(0.60)(1.3)(58 \text{ ksi})(4)(0.465 \text{ in.})}$	$\geq \dfrac{2.00(385 \text{ kips})}{0.60(1.3)(58 \text{ ksi})(4)(0.465 \text{ in.})}$
≥ 8.71 in.	≥ 9.15 in.

Note that this length is the minimum required for the limit state of shear rupture in the brace wall. A longer length may be used when designing the fillet welds between the brace and the gusset plate, if desired, to allow a smaller fillet weld size as is implemented in the following.

Size the weld between the brace and the gusset plate

The strength of fillet welds defined in AISC *Specification* Section J2 can be simplified, as explained in Part 8 of the AISC *Manual*, to Equations 8-2a and 8-2b:

LRFD	ASD
$\phi R_n = 1.392 Dl$	$\dfrac{R_n}{\Omega} = 0.928 Dl$

Using ¼-in. fillet welds for the four lines of weld so that they can be made in a single pass:

LRFD	ASD
$4(1.392)Dl \geq T_u$	$4(0.928)Dl \geq T_a$
$l \geq \dfrac{550 \text{ kips}}{4(1.392 \text{ kip/in.})(4 \text{ sixteenths})}$	$l \geq \dfrac{385 \text{ kips}}{4(0.928 \text{ kip/in.})(4 \text{ sixteenths})}$
≥ 24.7 in.	≥ 25.9 in.

The designs in LRFD and ASD give slightly different required lengths of weld. Use the LRFD result in this example. In practice, designers should consistently use one method or the other.

Use (4) 25-in. long, ¼-in. fillet welds to connect the brace below the beam to the gusset plate.

Check block shear rupture of the gusset plate

The available strength for the limit state of block shear rupture is:

$$R_n = 0.60 F_u A_{nv} + U_{bs} F_u A_{nt} \leq 0.60 F_y A_{gv} + U_{bs} F_u A_{nt} \qquad \text{(Spec. Eq. J4-5)}$$

Since the gross shear area, A_{gv}, and the net shear area, A_{nv}, are equal in this case, the shear yielding component, $0.60F_y A_{gv}$, is smaller than the shear rupture component, $0.60F_u A_{nv}$, and the right side of the equation controls.

$U_{bs} = 1.0$

$$0.60F_y A_{gv} = 0.60F_y(2)lt_p$$
$$= 0.60(36 \text{ ksi})(2)(25.0 \text{ in.})t_p$$
$$= 1,080 t_p$$

$$U_{bs} F_u A_{nt} = 1.0 F_u D_{brace} t_p$$
$$= 1.0(58 \text{ ksi})(6.875 \text{ in.})t_p$$
$$= 399 t_p$$

LRFD	ASD
$\phi R_n = 0.75(0.60 F_y A_{gv} + U_{bs} F_u A_{nt}) \geq P_u$	$\dfrac{R_n}{\Omega} = \dfrac{0.6 F_y A_{gv} + U_{bs} F_u A_{nt}}{\Omega} \geq P_a$
$t_p \geq \dfrac{550 \text{ kips}}{0.75(1,080 \text{ kip/in.} + 399 \text{ kip/in.})}$	$t_p \geq \dfrac{2.00(385 \text{ kips})}{(1,080 \text{ kip/in.} + 399 \text{ kip/in.})}$
≥ 0.496 in.	≥ 0.521 in.

Check required gusset width and thickness based on the limit state of tensile yielding

Tensile yielding is checked on a section of the gusset plate commonly referred to as the Whitmore section. This section is explained in AISC *Manual* Part 9 (Figure 9-1) and in Thornton and Lini (2011). Because the width and thickness of the gusset plate have not yet been chosen, the minimum area will be determined for this limit state. The nominal tensile yielding strength is:

$R_n = F_y A_g$ (*Spec.* Eq. J4-1)

LRFD	ASD
$\phi R_n \geq P_u$	$\dfrac{R_n}{\Omega} \geq P_a$
$A_g \geq \dfrac{P_u}{\phi F_y}$	$A_g \geq \dfrac{\Omega P_a}{F_y}$
$\geq \dfrac{550 \text{ kips}}{0.90(36 \text{ ksi})}$	$\geq \dfrac{1.67(385 \text{ kips})}{36 \text{ ksi}}$
≥ 17.0 in.^{2}	≥ 17.9 in.^{2}

Choose a reduced Whitmore width, w_p, of approximately twice the brace width. This does not exceed the maximum Whitmore width described in AISC *Manual* Part 9. Therefore, $w_p = 14.0$ in. and the gusset plate thickness is:

LRFD	ASD
$t_p \geq \dfrac{A_g}{w_p}$ $= \dfrac{17.0 \text{ in.}^2}{14.0 \text{ in.}}$ $= 1.21$ in.	$t_p \geq \dfrac{A_g}{w_p}$ $= \dfrac{17.9 \text{ in.}^2}{14.0 \text{ in.}}$ $= 1.28$ in.

The minimum required gusset thickness for this limit state is higher than the minimum required for the limit state of block shear as calculated previously.

Using the LRFD solution, a 1¼-in.-thick plate is selected for the gusset plate below the beam.

Check brace effective net area

From AISC *Seismic Provisions* Section F2.5b(3), the brace effective net area, A_e, should not be less than the brace gross area, A_g. Thus:

$$A_n = A_g - 2[t_p + 2(gap)]t_{des}$$

Using a gap of 1/16 in. on each side of the slot to allow clearance for erection:

$$A_n = 9.36 \text{ in.}^2 - 2[1¼ \text{ in.} + 2(\text{1/16 in.})](0.465 \text{ in.})$$
$$= 8.08 \text{ in.}^2$$

From AISC *Specification* Table D3.1, because $l > 1.3D$, $U = 1.0$, and the effective net area is:

$$A_e = UA_n$$
$$= 1.0(8.08 \text{ in.}^2)$$
$$= 8.08 \text{ in.}^2$$

Because $A_e < A_g$, brace reinforcement is required. The approximate area of reinforcement required, A_{rn}, is the area removed, but reinforcement will reduce U to less than 1.0 because of its position. The required area of reinforcement can be obtained from $(A_n + A_{rn})U \geq A_g$.

Assuming a value of $U = 0.80$:

$$A_{rn} = \dfrac{A_g}{0.80} - A_n$$
$$= \dfrac{9.36 \text{ in.}^2}{0.80} - 8.08 \text{ in.}^2$$
$$= 3.62 \text{ in.}^2$$

Try two 1½ in. × 1½ in. flat bars, with a total area of 4.50 in.² AISC *Seismic Provisions* Section F2.5b(3)(i) requires that the specified minimum yield strength of the reinforcement be at least that of the brace; therefore, use ASTM A572 Grade 50 material for the flat bar. The geometry is shown in Figure 5-27.

$$r_1 = \frac{D_{brace}}{2} - \frac{t_{des}}{2}$$
$$= \frac{6.875 \text{ in.}}{2} - \frac{0.465 \text{ in.}}{2}$$
$$= 3.21 \text{ in.}$$

$$r_2 = \frac{D_{brace}}{2} + \frac{1½ \text{ in.}}{2}$$
$$= \frac{6.875 \text{ in.}}{2} + \frac{1½ \text{ in.}}{2}$$
$$= 4.19 \text{ in.}$$

The distance to the centroid of a partial circle is given by:

$$\bar{x} = \frac{r_1 \sin \theta}{\theta}$$

where the total arc of the partial circle is 2θ, and θ is measured in radians. Although the brace is slightly less than a full half-circle because of the slot as shown in Figure 5-27, use an angle, θ, of $\pi/2$ for simplicity. This is slightly unconservative. A more precise calculation could be performed using the exact angle.

$$\bar{x}_{brace} = 3.21 \text{ in.} \left(\frac{\sin(\pi/2) \text{ rad}}{(\pi/2) \text{ rad}} \right)$$
$$= 2.04 \text{ in.}$$

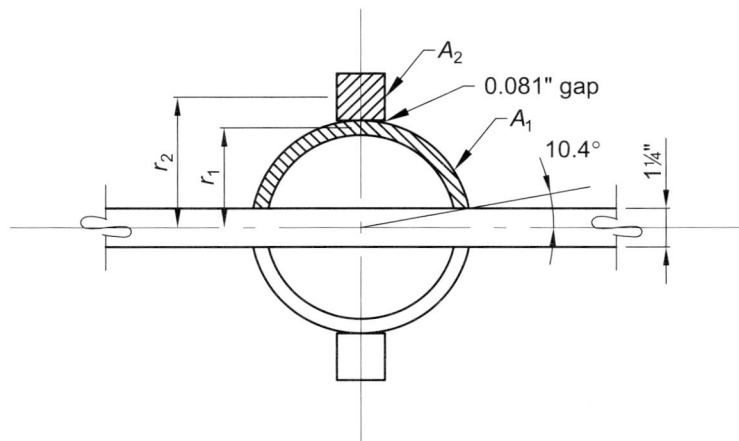

Fig. 5-27. Cross section of the brace below the beam at the net section.

$\bar{x}_{re} = r_2$
$= 4.19$ in.

Determine $\bar{x}$ for the composite cross section.

Part	$\bar{x}$ in.	A in.^{2}	$\bar{x}A$ in.^{3}
Half of brace	2.04	4.04	8.24
One flat bar	4.19	2.25	9.43
Σ	–	6.29	17.7

$\bar{x} = \dfrac{\Sigma \bar{x} A}{\Sigma A}$

$= \dfrac{17.7 \text{ in.}^3}{6.29 \text{ in.}^2}$

$= 2.81$ in.

From AISC *Specification* Table D3.1, Case 2, which applies to round HSS with reinforcement added:

$U = 1 - \dfrac{\bar{x}}{l}$

$= 1 - \dfrac{2.81 \text{ in.}}{26.0 \text{ in.}}$

$= 0.892$

$A_n = A_{n(brace)} + A_{rn}$

$= 8.08 \text{ in.}^2 + 2(2.25 \text{ in.}^2)$

$= 12.6 \text{ in.}^2$

$A_e = U A_n$

$= 0.892(12.6 \text{ in.}^2)$

$= 11.2 \text{ in.}^2 > A_g = 9.36 \text{ in.}^2$ **o.k.**

Design welds connecting flat bars to brace

According to AISC *Seismic Provisions* Section F2.5b(3)(ii), the flat bar must be connected to the brace to develop the expected strength of the flat bar on each side of the reduced section (the expected yield strength, $R_y F_y$, is used here). The reduced section is the length of the HSS from the extent of the slot (dimension x of Figure 5-22) to the start of the HSS-to-gusset weld. The required strength of the weld is based on the expected flat bar yield strength, using R_y from AISC *Seismic Provisions* Table A3.1 for ASTM A572 Grade 50 bars. For ASD, use 1.0/1.5 of the expected strength of the flat bar reinforcement.

5.3 SPECIAL CONCENTRICALLY BRACED FRAMES

LRFD	ASD
$R_y F_y A_{fb} = 1.1(50 \text{ ksi})(2.25 \text{ in.}^2)$ $= 124 \text{ kips}$	$R_y F_y A_{fb}/1.5 = 1.1(50 \text{ ksi})(2.25 \text{ in.}^2)/1.5$ $= 82.5 \text{ kips}$

There is a small gap of approximately 0.081 in. between the face of the brace and the edge of the flat bar as shown in Figure 5-27. Since this is more than $\frac{1}{16}$ in., the fillet weld of the flat bar to the brace would be increased by the fabricator by $\frac{1}{16}$ in. (see AWS D1.1 clause 5.22.1). Thus, to keep the weld as a single pass weld, $\frac{1}{4}$-in. fillet welds can be used and the flat bar length on each side of the reduced section of the brace, x, is determined using AISC *Manual* Equations 8-2a or 8-2b as follows:

LRFD	ASD
$l_w = \dfrac{124 \text{ kips}}{2(1.392 \text{ kip/in.})(4 \text{ sixteenths})}$ $= 11.1 \text{ in.}$	$l_w = \dfrac{82.5 \text{ kips}}{2(0.928 \text{ kip/in.})(4 \text{ sixteenths})}$ $= 11.1 \text{ in.}$

Use a 1½ in. × 1½ in. flat bar with ¼-in. fillet welds; the detail extends 11½ in. past both sides of the reduced section of the brace. Figure 5-22 shows a ¼-in. fillet weld for the flat bar to HSS weld. Note that fabricators typically assume the weld given on detail drawings to be the weld required for strength. Therefore, the gap between the edge of the flat bar and the curved brace, which is greater than $\frac{1}{16}$ in., will cause the shop to increase the fillet weld size to $\frac{5}{16}$ in.

The flat bar fillet weld develops the expected yield strength of the bar on each side of the reduced section of the brace. The brace slot may be longer than the slot length by an erection clearance of x inches (see Figure 5-22) as determined by the fabricator. The length of the flat bar will be a minimum of 11½ in. + 11½ in. + x in. = 23.0 in. + x in.

Check the gusset plate for buckling on the Whitmore section

From Figure 5-22, the buckling length, which is taken along the brace centerline (Dowswell, 2006), is $l_b = 10.0$ in. (Example 5.3.10 provides an equation for calculating the length of buckling; here it is determined graphically.) AISC *Seismic Provisions* Section F2.6c(2) specifies a required compressive strength for buckling limit states that is at least equal to 1.1 times the expected brace strength in compression. The stress over the Whitmore section, using the Whitmore width of 14.0 in., is:

LRFD	ASD
$f_{ua} = \dfrac{1.1(444 \text{ kips})}{14.0 \text{ in.}(1\frac{1}{4} \text{ in.})}$ $= 27.9 \text{ ksi}$	$f_{aa} = \dfrac{1.1(311 \text{ kips})}{14.0 \text{ in.}(1\frac{1}{4} \text{ in.})}$ $= 19.5 \text{ ksi}$

The radius of gyration of the gusset plate is:

$$r = \frac{t_p}{\sqrt{12}}$$

$$= \frac{1\tfrac{1}{4}}{\sqrt{12}}$$

$$= 0.361 \text{ in.}$$

Recommended values for the effective length factor, K, are given in Dowswell (2006). However, that paper does not address the case of a single gusset plate with the $2t$ clearance to accommodate brace buckling [called an "extended" gusset in Dowswell (2006)]. Therefore, in this case, use $K = 1.2$ from AISC *Specification* Commentary Table C-A-7.1 assuming that the gusset plate is fixed at one end, and free to translate but not rotate at the other. With $l_b = L$:

$$\frac{KL}{r} = \frac{1.2(10.0 \text{ in.})}{0.361 \text{ in.}}$$

$$= 33.2$$

Interpolating from AISC *Manual* Table 4-22:

LRFD	ASD
$\phi_c F_{cr} = 30.6$ ksi > 27.9 ksi **o.k.**	$\dfrac{F_{cr}}{\Omega_c} = 20.4$ ksi > 19.5 ksi **o.k.**

Determine the forces at the gusset-to-beam interface

Based on the required tensile strength of the bracing connection (which is larger than the required compressive strength) for the brace below the beam, the shear force at the interface of the gusset with the beam bottom flange is:

LRFD	ASD
$V_u = 550$ kips$(\cos 45°)$ $= 389$ kips	$V_a = 385$ kips$(\cos 45°)$ $= 272$ kips

The normal (tension) force is:

LRFD	ASD
$N_u = 550$ kips$(\sin 45°)$ $= 389$ kips	$N_a = 385$ kips$(\sin 45°)$ $= 272$ kips

The contact length between the gusset plate and the beam bottom flange, as shown in Figure 5-22, is 24.0 in. and the brace line of action misses the centroid of the contact length

5.3 SPECIAL CONCENTRICALLY BRACED FRAMES

by 1 in. due to the increasing width of the gusset plate. Therefore, the moment on the contact surface is:

LRFD	ASD
$M_u = N_u (1.0 \text{ in.})$ $= 389 \text{ kips}(1.0 \text{ in.})$ $= 389 \text{ kip-in.}$	$M_a = N_a (1.0 \text{ in.})$ $= 272 \text{ kips}(1.0 \text{ in.})$ $= 272 \text{ kip-in.}$

Combining the moment, M_u or M_a, and the normal force, N_u or N_a, into an equivalent normal force as explained for the design of the brace above the beam:

LRFD	ASD
$N_{u\,equiv} = N_u + \dfrac{4M_u}{l_b}$ $= 389 \text{ kips} + \dfrac{4(389 \text{ kip-in.})}{24.0 \text{ in.}}$ $= 454 \text{ kips}$	$N_{a\,equiv} = N_a + \dfrac{4M_a}{l_b}$ $= 272 \text{ kips} + \dfrac{4(272 \text{ kip-in.})}{24.0 \text{ in.}}$ $= 317 \text{ kips}$

The gusset stresses are:

LRFD	ASD
In tension $f_{ua} = \dfrac{N_{u\,equiv}}{t_p l_b}$ $= \dfrac{454 \text{ kips}}{(1\tfrac{1}{4} \text{ in.})(24.0 \text{ in.})}$ $= 15.1 \text{ ksi}$ The design tensile yielding stress from AISC *Specification* Section J4.1 is: $\phi F_y = 0.90(36 \text{ ksi})$ $= 32.4 \text{ ksi} > 15.1 \text{ ksi}$ **o.k.**	In tension $f_{aa} = \dfrac{N_{a\,equiv}}{t_p l_b}$ $= \dfrac{317 \text{ kips}}{(1\tfrac{1}{4} \text{ in.})(24.0 \text{ in.})}$ $= 10.6 \text{ ksi}$ The allowable tensile yielding stress from AISC *Specification* Section J4.1 is: $\dfrac{F_y}{\Omega} = \dfrac{36 \text{ ksi}}{1.67}$ $= 21.6 \text{ ksi} > 10.6 \text{ ksi}$ **o.k.**

LRFD	ASD
In shear	In shear
$f_{uv} = \dfrac{V_u}{t_p l_b}$	$f_{av} = \dfrac{V_a}{t_p l_b}$
$= \dfrac{389 \text{ kips}}{(1\frac{1}{4} \text{ in.})(24.0 \text{ in.})}$	$= \dfrac{272 \text{ kips}}{(1\frac{1}{4} \text{ in.})(24.0 \text{ in.})}$
$= 13.0$ ksi	$= 9.07$ ksi
The design shear yielding stress from AISC *Specification* Section J4.2 is:	The allowable shear yielding stress from AISC *Specification* Section J4.2 is:
$\phi 0.60 F_y = 1.00(0.60)(36 \text{ ksi})$ $= 21.6$ ksi > 13.0 ksi **o.k.**	$\dfrac{0.60 F_y}{\Omega} = \dfrac{0.60(36 \text{ ksi})}{1.50}$ $= 14.4$ ksi > 9.07 ksi **o.k.**

Size gusset-to-beam weld

The angle of the resultant force can be calculated and used in the directional strength increase of fillet welds according to AISC *Specification* Equation J2-5.

LRFD	ASD
$\theta = \tan^{-1}\left(\dfrac{N_{u\,equiv}}{V_u}\right)$	$\theta = \tan^{-1}\left(\dfrac{N_{a\,equiv}}{V_a}\right)$
$= \tan^{-1}\left(\dfrac{454 \text{ kips}}{389 \text{ kips}}\right)$	$= \tan^{-1}\left(\dfrac{317 \text{ kips}}{272 \text{ kips}}\right)$
$= 49.4°$	$= 49.4°$

AISC *Specification* Section J2.4 allows an increase in the available strength of fillet welds when the angle of loading is not along the weld longitudinal axis, which is used in the following calculation.

The weld ductility factor, equal to 1.25, which is explained in AISC *Manual* Part 13, is applied here. Using AISC *Manual* Equations 8-2a and 8-2b, the number of sixteenths of fillet weld required is:

5.3 SPECIAL CONCENTRICALLY BRACED FRAMES

LRFD	ASD
$D_{req'd} \geq$ $$\frac{1.25\sqrt{N_{u\,equiv}^2 + V_u^2}}{2(1.392 \text{ kip/in.})(1+0.5\sin^{1.5}\theta)(l_b)}$$ $$= \frac{1.25\sqrt{(454 \text{ kips})^2 + (389 \text{ kips})^2}}{2(1.392 \text{ kip/in.})(1.33)(24.0 \text{ in.})}$$ $$= 8.41 \text{ sixteenths}$$	$D_{req'd} \geq$ $$\frac{1.25\sqrt{N_{a\,equiv}^2 + V_a^2}}{2(0.928 \text{ kip/in.})(1+0.5\sin^{1.5}\theta)(l_b)}$$ $$= \frac{1.25\sqrt{(317 \text{ kips})^2 + (272 \text{ kips})^2}}{2(0.928 \text{ kip/in.})(1.33)(24.0 \text{ in.})}$$ $$= 8.81 \text{ sixteenths}$$

Use a ⁹⁄₁₆-in. fillet weld on each side of the gusset at the gusset-to-beam connection for the brace below the beam.

Check beam web local yielding

For a force applied at a distance from the end that is greater than the depth of the member:

$$R_n = F_{yw}t_w(5k + l_b) \qquad \text{(Spec. Eq. J10-2)}$$
$$= 50 \text{ ksi}(0.570 \text{ in.})[5(1.53 \text{ in.}) + 24.0 \text{ in.}]$$
$$= 902 \text{ kips}$$

LRFD	ASD
$\phi R_n = 1.00(902 \text{ kips})$ $= 902 \text{ kips}$ $> N_{u\,equiv} = 454 \text{ kips}$ **o.k.**	$\dfrac{R_n}{\Omega} = \dfrac{902 \text{ kips}}{1.50}$ $= 601 \text{ kips}$ $> N_{a\,equiv} = 317 \text{ kips}$ **o.k.**

Check beam web local crippling

Web local yielding applies to both tension and compression loads. Web local crippling applies only to the compression loads, but the larger tension load is used here for convenience. If desired, the compressive strength of the brace could be used for web local crippling.

For a force applied greater than a distance of $d/2$ from the beam end:

$$R_n = 0.80t_w^2\left[1 + 3\left(\frac{l_b}{d}\right)\left(\frac{t_w}{t_f}\right)^{1.5}\right]\sqrt{\frac{EF_{yw}t_f}{t_w}} \qquad \text{(Spec. Eq. J10-4)}$$

$$= 0.80(0.570 \text{ in.})^2\left[1 + 3\left(\frac{24.0 \text{ in.}}{27.3 \text{ in.}}\right)\left(\frac{0.570 \text{ in.}}{0.930 \text{ in.}}\right)^{1.5}\right]$$

$$\times \sqrt{\frac{29,000 \text{ ksi}(50 \text{ ksi})(0.930 \text{ in.})}{0.570 \text{ in.}}}$$

$$= 906 \text{ kips}$$

LRFD	ASD
$\phi R_n = 0.75(906 \text{ kips})$ $= 680 \text{ kips}$ $> N_{u\,equiv} = 454 \text{ kips}$ **o.k.**	$\dfrac{R_n}{\Omega} = \dfrac{906 \text{ kips}}{2.00}$ $= 453 \text{ kips}$ $> N_{a\,equiv} = 317 \text{ kips}$ **o.k.**

Figure 5-22 shows the final configuration.

Beam Web Available Shear Strength

Figure 5-28 shows the shear distribution in the beam web due to the gusset vertical components. The design of the top gusset shown earlier in this example was based on the expected strength in tension, but this force is not simultaneous with tension in the bottom brace (as shown in Figure 5-23). Therefore, for checking the beam web, the brace below the beam will be considered to be in tension and the brace above the beam will be considered to be in compression. For the brace in tension below the beam, the forces at the gusset-beam interface have already been calculated. For the brace in compression above the beam, the forces at the gusset-beam interface are:

LRFD	ASD
Normal $N_u = 226 \text{ kips}(\sin 45°)$ $\qquad = 160 \text{ kips}$	Normal $N_a = 158 \text{ kips}(\sin 45°)$ $\qquad = 112 \text{ kips}$
Shear $V_u = 226 \text{ kips}(\cos 45°)$ $\qquad = 160 \text{ kips}$	Shear $V_a = 158 \text{ kips}(\cos 45°)$ $\qquad = 112 \text{ kips}$
Moment $M_u = N_u(1.50 \text{ in.})$ $\qquad = 160 \text{ kips}(1.50 \text{ in.})$ $\qquad = 240 \text{ kip-in.}$	Moment $M_a = N_a(1.50 \text{ in.})$ $\qquad = 112 \text{ kips}(1.50 \text{ in.})$ $\qquad = 168 \text{ kip-in.}$

The braces to the left of the beam centerline, where the brace below the beam is in tension as shown in Figure 5-23, result in the highest shear force. The total vertical shear at the beam centerline is:

LRFD	ASD
$V_u = (226 \text{ kips} + 550 \text{ kips})(\sin 45°)$ $\quad = 549 \text{ kips}$	$V_a = (158 \text{ kips} + 385 \text{ kips})(\sin 45°)$ $\quad = 384 \text{ kips}$

5.3 SPECIAL CONCENTRICALLY BRACED FRAMES

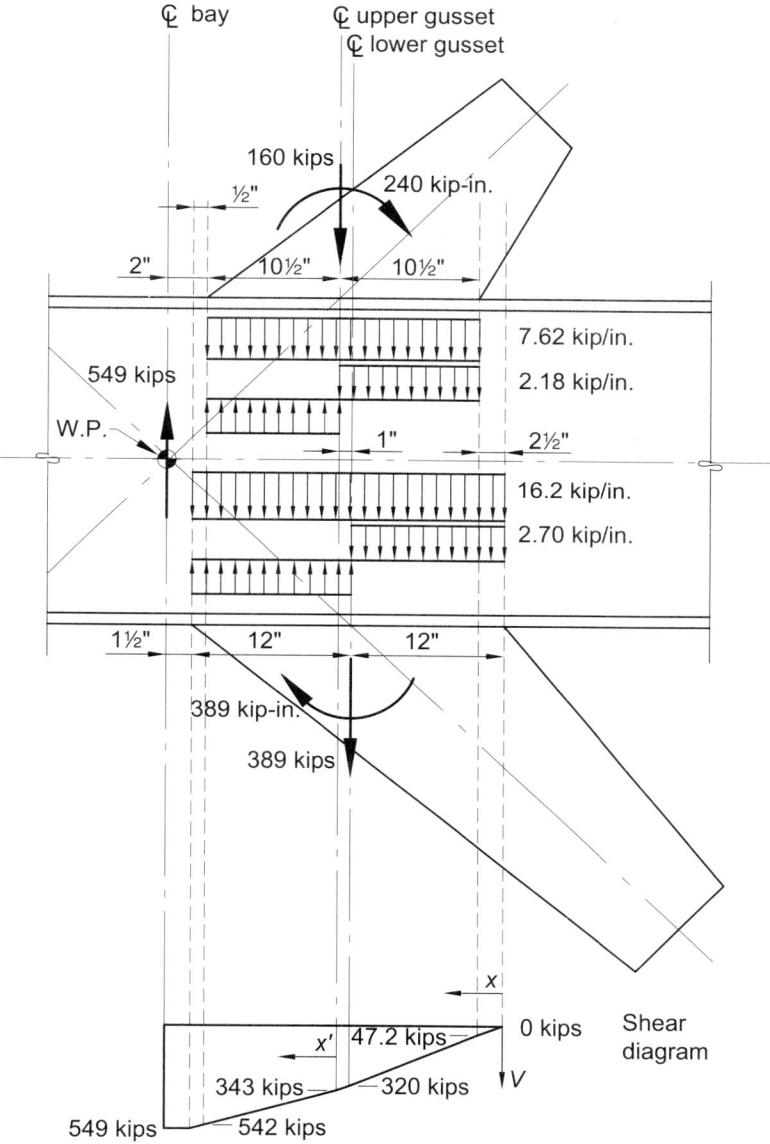

Fig. 5-28a. Beam shear due to gusset forces (LRFD).

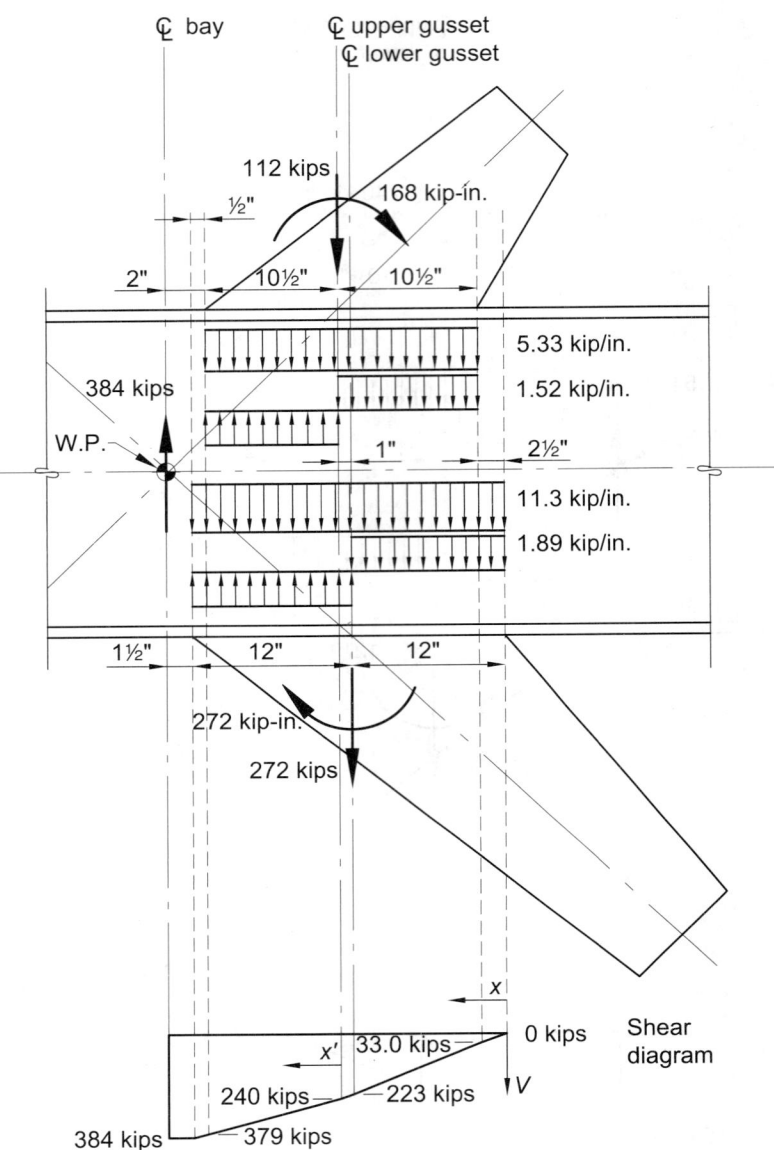

Fig. 5-28b. Beam shear due to gusset forces (ASD).

From AISC *Manual* Table 3-6, the available shear strength of the W27×114 is:

LRFD	ASD
$\phi_v V_n = 467$ kips < 549 kips **n.g.**	$\dfrac{V_n}{\Omega_v} = 311$ kips < 384 kips **n.g.**

Because the beam available shear strength is not adequate, a doubler plate is required on the web. From Figure 3-28a and 3-28b, measuring x' as shown, the equation for the shear, V, in the region of interest is:

LRFD	ASD
$V_u(x') = 343 \text{ kips} + \dfrac{(542 \text{ kips} - 343 \text{ kips})}{10.5 \text{ in.}} x'$ $= 343 \text{ kips} + (19.0 \text{ kip/in.}) x'$ (for $0 \le x' \le 10.5$ in.)	$V_a(x') = 240 \text{ kips} + \dfrac{(379 \text{ kips} - 240 \text{ kips})}{10.5 \text{ in.}} x'$ $= 240 \text{ kips} + (13.2 \text{ kip/in.}) x'$ (for $0 \le x' \le 10.5$ in.)

Setting this shear equal to the available shear strength of the beam and solving for x' to find out where the beam available strength is adequate results in:

LRFD	ASD
$V_u(x') = 467$ kips $x' = \dfrac{467 \text{ kips} - 343 \text{ kips}}{19.0 \text{ kip/in.}}$ $= 6.53$ in.	$V_a(x') = 311$ kips $x' = \dfrac{311 \text{ kips} - 240 \text{ kips}}{13.2 \text{ kip/in.}}$ $= 5.38$ in.

The distance from the beam centerline to this point is:

LRFD	ASD
12.5 in. − 6.53 in. = 5.97 in.	12.5 in. − 5.38 in. = 7.12 in.

Use a 14 in. × 14 in. doubler plate centered on the web as shown in Figure 5-22.

Based on shear yielding and AISC *Specification* Section J4.2, the plate thickness required is determined as follows:

LRFD	ASD
$\phi R_n = 1.00(0.60 F_y A_{gv})$ $= (1.00)(0.60)(36 \text{ ksi})(14.0 \text{ in.})(t_d)$ $= 302 t_d \text{ kips}$	$\dfrac{R_n}{\Omega} = \dfrac{0.60 F_y A_{gv}}{1.50}$ $= \dfrac{0.60(36 \text{ ksi})(14.0 \text{ in.})(t_d)}{1.50}$ $= 202 t_d \text{ kips}$
The required doubler plate shear strength is:	The required doubler plate shear strength is:
549 kips − 467 kips = 82.0 kips	384 kips − 311 kips = 73.0 kips
Setting the available strength equal to the required strength and solving for t_d:	Setting the available strength equal to the required strength and solving for t_d:
$t_d \geq \dfrac{82.0 \text{ kips}}{302 \text{ kip/in.}}$ $= 0.272 \text{ in.}$	$t_d \geq \dfrac{73.0 \text{ kips}}{202 \text{ kip/in.}}$ $= 0.361 \text{ in.}$

Using the instantaneous center of rotation method of AISC *Manual* Part 8, use Table 8-8 with $\theta = 0°$ to determine the strength of welds on the doubler plate. For a channel-shaped weld on half of the doubler plate, the geometric variables are:

$l = 14.0$ in.

$k = 7.00$ in./14.0 in.
$= 0.500$

$kl = 7.00$ in.

$xl = 0.125(14.0 \text{ in.}) = 1.75$ in.

$e_x = al$
$= 7.00 \text{ in.} - 1.75 \text{ in.}$
$= 5.25 \text{ in.}$

$a = 0.375$

By interpolation, AISC *Manual* Table 8-8 with $\theta = 0°$ gives:

$C = 3.35$

From AISC *Manual* Table 8-3, $C_1 = 1.00$. The fillet weld size required is:

5.3 SPECIAL CONCENTRICALLY BRACED FRAMES

LRFD	ASD
$D_{min} = \dfrac{V_{doubler}}{\phi CC_1 l}$	$D_{min} = \dfrac{\Omega V_{doubler}}{CC_1 l}$
$= \dfrac{82.0 \text{ kips}}{0.75(3.35)(1.00)(14.0 \text{ in.})}$	$= \dfrac{2.00(73.0 \text{ kips})}{(3.35)(1.00)(14.0 \text{ in.})}$
$= 2.33$	$= 3.11$

Use a PL ³⁄₈ in. × 14 in. × 1 ft 2 in., with ¼-in. fillet welds.

Comment:

The length of the brace used to determine the expected strength of both braces was 12 ft. This length should be verified once the connection design is complete. For both braces, the length of the brace with a 12.5-ft story height and a 25-ft bay as shown in Figure 5-11 is:

$$L = \sqrt{(12.5 \text{ ft})^2 + (25.0 \text{ ft}/2)^2}$$
$$= 17.7 \text{ ft}$$

From Figure 5-22, with a pull-off dimension of 27½ in. at both ends of the top brace, the actual unbraced length of the brace is:

$$L = 17.7 \text{ ft} - 2(27\tfrac{1}{2} \text{ in.})(1 \text{ ft}/12 \text{ in.})$$
$$= 13.1 \text{ ft}$$

For the bottom brace, with a pull-off dimension of 30 in. at both ends of the brace, the actual unbraced length of the brace is:

$$L = 17.7 \text{ ft} - 2(30.0 \text{ in.})(1 \text{ ft}/12 \text{ in.})$$
$$= 12.7 \text{ ft}$$

Therefore, the length of 12 ft used for the determination of the expected compressive strength of the brace is adequate because it does not exceed the actual length from brace end-to-brace end as required by AISC *Seismic Provisions* Section F2.3.

The reduced Whitmore section used in this example greatly reduced the size of the gusset plates, at the expense of the gusset thickness. This is usually a good trade-off, but if a lighter beam were used it is possible that the beam web local yielding or local crippling checks would have failed. A beam web doubler plate would then be necessary for one or both of these limit states. This doubler requirement can usually be avoided by increasing the gusset-to-beam interface length by increasing the Whitmore section (the maximum width that can be considered effective, as explained in AISC *Manual* Part 9, is limited to the width on a 30° spread).

The reader should keep in mind that there is not just one way to design these connections. Any method that satisfies equilibrium and the applicable limit states is an acceptable method.

Example 5.3.9. SCBF Brace-to-Beam Connection Design

Given:
An alternative design for Example 5.3.8 at Joint JT-1 of Figure 5-11 is presented here. Example 5.3.8 used separate gusset plates for each brace, while this example uses continuous "chevron type" gusset plates for the top and bottom braces as shown in Figure 5-29. This is a common arrangement for this situation, and the example provides an admissible internal force distribution that differs from the distribution used for the four separate gusset plates. All braces are ASTM A500 Grade B round HSS and the beam is an ASTM A992 W27×114. For the connection, ASTM A36 plate material and 70-ksi electrodes are used.

Solution:
From AISC *Manual* Table 1-1 and 1-13, the geometric properties are:

HSS6×0.312 (brace above the beam)
$A = 5.22$ in.^{2} $\quad t_{nom} = 0.312$ in. $\quad t_{des} = 0.291$ in. $\quad r = 2.02$ in.

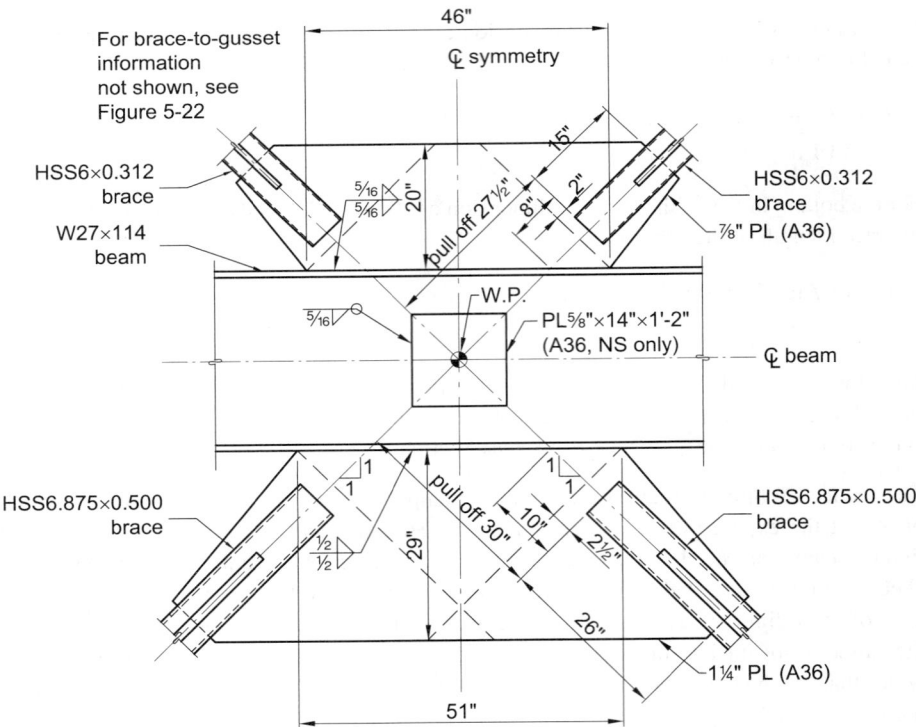

Fig. 5-29. Connection with chevron gusset plates designed in Exaample 5.3.9.

5.3 SPECIAL CONCENTRICALLY BRACED FRAMES

HSS6.875×0.500 (brace below the beam)
$A = 9.36$ in.² $\quad t_{nom} = 0.500$ in. $\quad t_{des} = 0.465$ in. $\quad r = 2.27$ in.

Beam
W27×114
$d = 27.3$ in. $\quad t_w = 0.570$ in. $\quad t_f = 0.930$ in. $\quad k_{des} = 1.53$ in.

From AISC *Manual* Table 2-4 and 2-5, the material properties are:

ASTM A36
$F_y = 36$ ksi
$F_u = 58$ ksi

ASTM A500 Grade B
$F_y = 42$ ksi
$F_u = 58$ ksi

ASTM A992
$F_y = 50$ ksi
$F_u = 65$ ksi

The complete connection design is shown in Figure 5-29.

AISC *Seismic Provisions* Sections F2.3(i) and F2.3(ii) define the two mechanism analyses that must be considered in determining the required strength of beams, columns and connections. For this example, only the mechanism analysis of Section F2.3(i) will be considered. AISC *Seismic Provisions* Sections F2.6c specifies the required strength of bracing connections. For the required compressive strength, Section F2.6c(2) has an additional 1.1 factor (relative to the requirements of Section F2.3) applied to the required strength of the connection.

For these SCBF connection examples, the requirements of AISC *Seismic Provisions* Sections B2 and F2.3 will be used for both LRFD and ASD.

Determine the expected strengths of the braces

The calculations for brace expected strengths were shown in Example 5.3.8 and are not repeated here. The required strengths are given in the following and shown in Figure 5-30.

For the HSS6×0.312 brace above the beam:

For the braces above the beam, the required strength of the bracing connection when the brace is in tension is:

LRFD	ASD
$P_u = 1.0 E_{mh}$ $= 307$ kips	$P_a = 0.7 E_{mh}$ $= 0.7(307$ kips$)$ $= 215$ kips

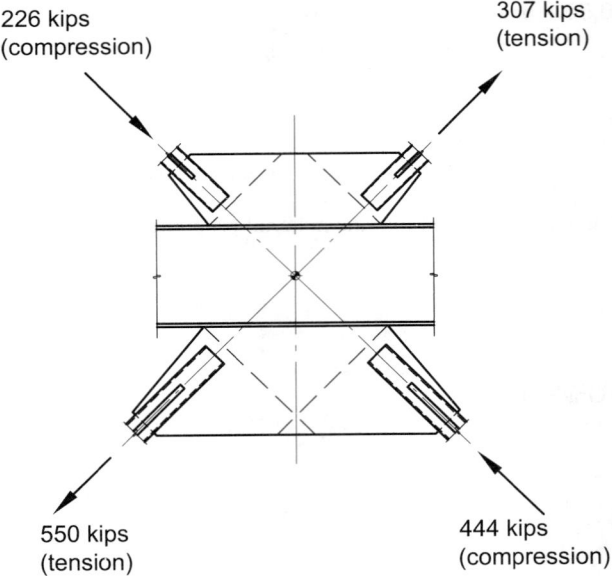

Fig. 5-30a. Required strength of bracing connections according to AISC Seismic Provisions Section F2.3(i) for LRFD design.

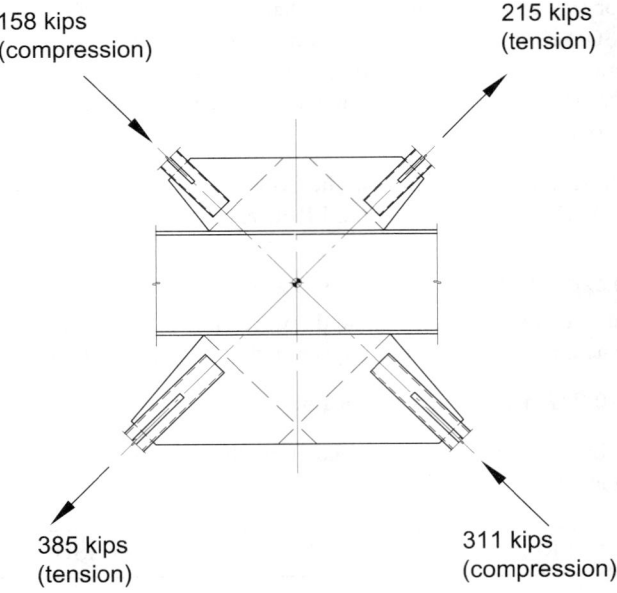

Fig. 5-30b. Required strength of bracing connections according to AISC Seismic Provisions Section F2.3(i) for ASD design.

5.3 SPECIAL CONCENTRICALLY BRACED FRAMES

For the braces above the beam, the required strength of the bracing connection when the brace is in compression is:

LRFD	ASD
$P_u = 1.0E_{mh}$ $= 226$ kips	$P_a = 0.7E_{mh}$ $= 0.7(226$ kips$)$ $= 158$ kips

For the HSS6.875×0.500 brace below the beam:

For the braces below the beam, the required strength of the bracing connection when the brace is in tension is:

LRFD	ASD
$P_u = 1.0E_{mh}$ $= 550$ kips	$P_a = 0.7E_{mh}$ $= 0.7(550$ kips$)$ $= 385$ kips

For the braces below the beam, the required strength of the bracing connection when the brace is in compression is:

LRFD	ASD
$P_u = 1.0E_{mh}$ $= 444$ kips	$P_a = 0.7E_{mh}$ $= 0.7(444$ kips$)$ $= 311$ kips

Brace-to-gusset design

The brace-to-gusset design is exactly the same as for the design shown in Figure 5-22 and is not repeated here. The gusset checks involving the Whitmore sections in Example 5.3.8 are conservative for the geometry in this example, as shown in Figure 5-29, because the gusset plates in this condition have a wider Whitmore width with the same plate thickness. Therefore, the gusset thicknesses previously chosen, ⅞-in.-thick plate for the top gusset and 1¼-in.-thick plate for the bottom gusset, are satisfactory. For the case of chevron gusset plates in this example, an effective length factor K of 0.65 (Dowswell, 2012) can be used in lieu of the value of 1.2 used in Example 5.3.8. If these gusset buckling limit states control the gusset thickness, the reduced K may allow thinner gussets to be used. For simplicity, the thickness of the gusset plates will be kept the same as in Example 5.3.8, although a thinner gusset may be acceptable here and may be more economical.

The usual controlling interface for this arrangement is Section a-a of Figure 5-31a. Figure 5-31a gives the general geometry and the sign convention. Figure 5-31b shows the interface forces on section a-a. Subsequently, the gusset is cut at its centerline at section b-b, as shown in Figure 5-31c, which also shows the forces on this interface. Horizontal and vertical force components and moments without a prime symbol act on section a-a. Components and moments with a prime symbol act on section b-b.

Because there are four braces with four different loads and two gussets with different geometry, Figure 5-31 introduces a sign convention as a means of keeping track of all the quantities. All of the force quantities are shown in the positive directions in these figures. If a quantity calculates as negative, it acts in a direction opposite to that shown. Alternatively, the designer can work with basic problem-specific free body diagrams. For the moments M_1 and M_2, which act on section a-a, M_1 is considered positive when it acts in a clockwise direction, whereas M_2 is considered positive when it acts in a counter-clockwise direction. Consequently, the total moment on section a-a is the difference between those two moments rather than the sum, where

$$M_1 = H_1 e + V_1 \Delta$$
$$M_2 = H_2 e - V_2 \Delta$$
$$\Delta = \tfrac{1}{2}(L_2 - L_1) \text{(Note: } \Delta \text{ is negative if } L_2 < L_1)$$

The moments M_1 and M_2 are due to the brace forces P_1 and P_2, respectively, and are taken about the midpoint of section a-a, with brace forces resolved at the common work point. The moments M_1' and M_2', also due to the brace forces P_1 and P_2, are taken at the midpoint of section b-b, as illustrated in Figure 5-31c, where

$$M_1' = \tfrac{1}{8} L V_1 - \tfrac{1}{4} h H_1 - \tfrac{1}{2} M_1$$
$$M_2' = \tfrac{1}{8} L V_2 - \tfrac{1}{4} h H_2 - \tfrac{1}{2} M_2$$

The derivation of M_1' and M_2' can be illustrated by considering half of the gusset plate, as shown in Figure 5-31c. Internal forces from both brace forces are considered evenly distributed across section a-a.

The sign convention used in this example is as follows:

- P_1 and P_2 are positive for tension and negative for compression
- If P_1 is positive, V_1 and H_1 are positive also
- If P_1 is negative, V_1 and H_1 are negative also
- If P_2 is positive, V_2 and H_2 are positive also
- If P_2 is negative, V_2 and H_2 are negative also

5.3 SPECIAL CONCENTRICALLY BRACED FRAMES

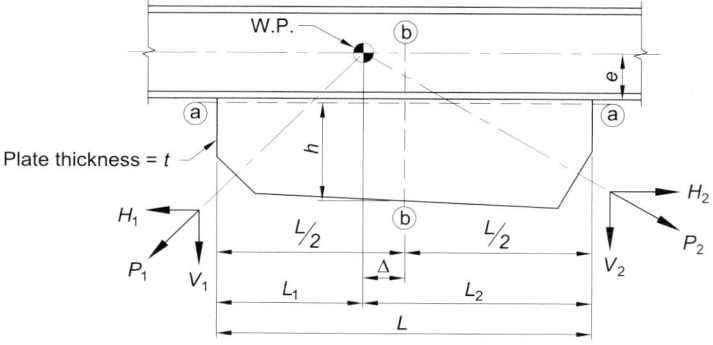

Fig. 5-31a. Chevron brace gusset forces.

Forces on section a-a
  Normal: $N = V_1 + V_2$
  Shear: $V = H_1 - H_2$
  Moment: $M = M_1 - M_2$

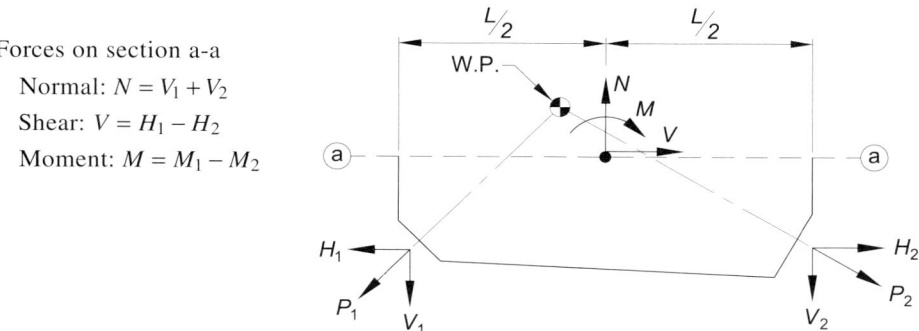

Fig. 5-31b. Forces on section a-a (positive direction shown).

Forces on section b-b
  Normal: $N' = \frac{1}{2}(H_1 + H_2)$
  Shear: $V' = \frac{1}{2}(V_1 - V_2) - \frac{2}{L}M$
  Moment: $M' = M'_1 + M'_2$

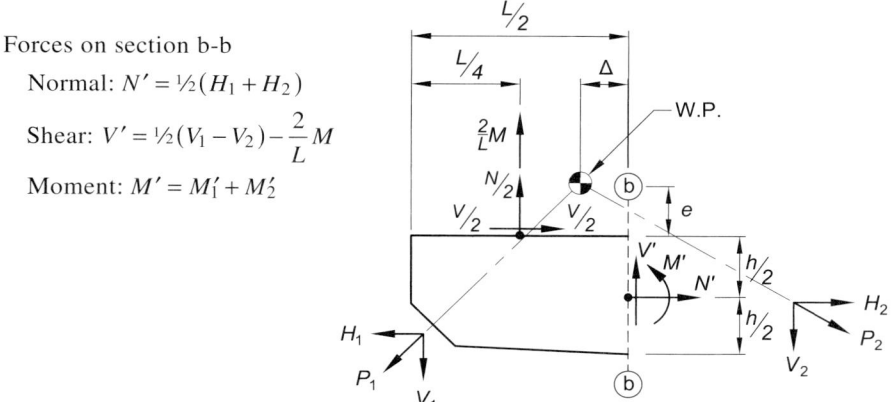

Fig. 5-31c. Forces on section b-b (positive directions shown).

From Figure 5-31c, the forces on section b-b can be derived (note that the brace force P_2 and its components do not act directly on the half-gusset shown as the P_1 force does, but is shown for completeness). For horizontal equilibrium:

$$N' + \frac{V}{2} - H_1 = 0$$

Therefore:

$$N' = -\frac{V}{2} + H_1$$

$$= -\frac{1}{2}(H_1 - H_2) + H_1$$

$$= \frac{1}{2}(H_1 + H_2)$$

For vertical equilibrium:

$$V' + \frac{2M}{L} + \frac{N}{2} - V_1 = 0$$

Therefore:

$$V' = -\frac{2M}{L} - \frac{N}{2} + V_1$$

$$= -\frac{2M}{L} - \frac{1}{2}(V_1 + V_2) + V_1$$

$$= -\frac{2M}{L} - \frac{1}{2}V_1 - \frac{1}{2}V_2 + V_1$$

$$= \frac{1}{2}(V_1 - V_2) - \frac{2M}{L}$$

For moment equilibrium, with moments taken about the midpoint of section b-b and brace forces acting at the work point:

$$M' + H_1\left(e + \frac{h}{2}\right) + V_1\Delta - \frac{V}{2}\left(\frac{h}{2}\right) - \frac{N}{2}\left(\frac{L}{4}\right) - \frac{2}{L}M\left(\frac{L}{4}\right) = 0$$

Therefore:

5.3 SPECIAL CONCENTRICALLY BRACED FRAMES

$$M' = -H_1\left(e + \frac{h}{2}\right) - V_1\Delta + \frac{1}{4}Vh + \frac{1}{8}NL + \frac{1}{2}M$$

$$= -H_1e - H_1\left(\frac{h}{2}\right) - V_1\Delta + \frac{1}{4}h(H_1 - H_2) + \frac{1}{8}L(V_1 + V_2) + \frac{1}{2}(M_1 - M_2)$$

$$= \frac{1}{8}L(V_1 + V_2) - \frac{1}{2}hH_1 + \frac{1}{4}hH_1 - \frac{1}{4}hH_2 + \frac{1}{2}(M_1 - M_2) - H_1e - V_1\Delta$$

$$= \frac{1}{8}L(V_1 + V_2) - \frac{1}{4}hH_1 - \frac{1}{4}hH_2 + \frac{1}{2}M_1 - \frac{1}{2}M_2 - (H_1e + V_1\Delta)$$

$$= \frac{1}{8}L(V_1 + V_2) - \frac{1}{4}h(H_1 + H_2) + \frac{1}{2}M_1 - \frac{1}{2}M_2 - M_1$$

$$= M_1' + M_2'$$

$$M_1' = \frac{1}{8}LV_1 - \frac{1}{4}hH_1 - \frac{1}{2}M_1$$

$$M_2' = \frac{1}{8}LV_2 - \frac{1}{4}hH_2 - \frac{1}{2}M_2$$

Design of gusset above the beam

The required strengths shown in Figure 5-30 represent external loads acting in a left-to-right motion. With external loads reversed, the brace expected strengths in compression and tension will be reversed but the effect on the gusset plate will be equivalent.

For the geometry shown in Figure 5-29 for the gusset plate above the beam and Figure 5-30 for general terminology:

$\Delta = 0$

$L_1 = L_2$
$\quad = 23.0$ in.

$L = L_1 + L_2$
$\quad = 46.0$ in.

$e = d/2$
$\quad = 27.3$ in. $/2$
$\quad = 13.7$ in.

$h = 20.0$ in.

For the forces given in Figures 5-30a and 5-30b:

LRFD	ASD
$P_1 = -226$ kips	$P_1 = -158$ kips
$H_1 = (-226 \text{ kips})(\cos 45°)$ $= -160$ kips	$H_1 = (-158 \text{ kips})(\cos 45°)$ $= -112$ kips
$V_1 = (-226 \text{ kips})(\sin 45°)$ $= -160$ kips	$V_1 = (-158 \text{ kips})(\sin 45°)$ $= -112$ kips
$P_2 = +307$ kips	$P_2 = +215$ kips
$H_2 = (307 \text{ kips})(\cos 45°)$ $= 217$ kips	$H_2 = (215 \text{ kips})(\cos 45°)$ $= 152$ kips
$V_2 = (307 \text{ kips})(\sin 45°)$ $= 217$ kips	$V_2 = (215 \text{ kips})(\sin 45°)$ $= 152$ kips

The moments are:

LRFD	ASD
$M_1 = H_1 e + V_1 \Delta$ $= [-160 \text{ kips}(13.7 \text{ in.})] +$ $[-160 \text{ kips}(0 \text{ in.})]$ $= -2,190$ kip-in.	$M_1 = H_1 e + V_1 \Delta$ $= [-112 \text{ kips}(13.7 \text{ in.})] +$ $[-112 \text{ kips}(0 \text{ in.})]$ $= -1,530$ kip-in.
$M_2 = H_2 e - V_2 \Delta$ $= [217 \text{ kips}(13.7 \text{ in.})]$ $-[217 \text{ kips}(0 \text{ in.})]$ $= 2,970$ kip-in.	$M_2 = H_2 e - V_2 \Delta$ $= [152 \text{ kips}(13.7 \text{ in.})]$ $-[152 \text{ kips}(0 \text{ in.})]$ $= 2,080$ kip-in.
$M_1' = \frac{1}{8}LV_1 - \frac{1}{4}hH_1 - \frac{1}{2}M_1$ $= \frac{1}{8}(46.0 \text{ in.})(-160 \text{ kips})$ $- \frac{1}{4}(20.0 \text{ in.})(-160 \text{ kips})$ $- \frac{1}{2}(-2,190 \text{ kip-in.})$ $= +975$ kip-in.	$M_1' = \frac{1}{8}LV_1 - \frac{1}{4}hH_1 - \frac{1}{2}M_1$ $= \frac{1}{8}(46.0 \text{ in.})(-112 \text{ kips})$ $- \frac{1}{4}(20.0 \text{ in.})(-112 \text{ kips})$ $- \frac{1}{2}(-1,530 \text{ kip-in.})$ $= +681$ kip-in.

5.3 SPECIAL CONCENTRICALLY BRACED FRAMES

LRFD	ASD
$M_2' = \dfrac{1}{8}LV_2 - \dfrac{1}{4}hH_2 - \dfrac{1}{2}M_2$	$M_2' = \dfrac{1}{8}LV_2 - \dfrac{1}{4}hH_2 - \dfrac{1}{2}M_2$
$= \dfrac{1}{8}(46.0 \text{ in.})(217 \text{ kips})$	$= \dfrac{1}{8}(46.0 \text{ in.})(152 \text{ kips})$
$- \dfrac{1}{4}(20.0 \text{ in.})(217 \text{ kips})$	$- \dfrac{1}{4}(20.0 \text{ in.})(152 \text{ kips})$
$- \dfrac{1}{2}(2{,}970 \text{ kip-in.})$	$- \dfrac{1}{2}(2{,}080 \text{ kip-in.})$
$= -1{,}320 \text{ kip-in.}$	$= -926 \text{ kip-in.}$

For the gusset above the beam, the forces on section a-a are:

LRFD	ASD
Normal $N_u = V_1 + V_2$ $= -160 \text{ kips} + 217 \text{ kips}$ $= 57.0 \text{ kips}$	Normal $N_a = V_1 + V_2$ $= -112 \text{ kips} + 152 \text{ kips}$ $= 40.0 \text{ kips}$
Shear $V_u = H_1 - H_2$ $= -160 \text{ kips} - 217 \text{ kips}$ $= -377 \text{ kips}$	Shear $V_a = H_1 - H_2$ $= -112 \text{ kips} - 152 \text{ kips}$ $= -264 \text{ kips}$
Moment $M_u = M_1 - M_2$ $= -2{,}190 \text{ kip-in.}$ $-2{,}970 \text{ kip-in.}$ $= -5{,}160 \text{ kip-in.}$	Moment $M_a = M_1 - M_2$ $= -1{,}530 \text{ kip-in.}$ $-2{,}080 \text{ kip-in.}$ $= -3{,}610 \text{ kip-in.}$

Check available strength of the gusset plate above the beam on section a-a

The available shear strength of the gusset plate on section a-a is:

$$V_n = 0.60 F_y A_{gv} \qquad \qquad (Spec.\ Eq.\ J4\text{-}3)$$

LRFD	ASD
$\phi V_n = 1.00(0.60)(36 \text{ ksi})$ $\quad \times (\text{\textonesuperior/\textsubscript{8}} \text{ in.})(46.0 \text{ in.})$ $\quad = 869 \text{ kips} > 377 \text{ kips} \quad$ **o.k.**	$\dfrac{V_n}{\Omega} = 0.60(36 \text{ ksi})(\text{\textonesuperior/\textsubscript{8}} \text{ in.})$ $\quad \times (46.0 \text{ in.})/1.50$ $\quad = 580 \text{ kips} > 264 \text{ kips} \quad$ **o.k.**

The normal force involves both N and M. It is convenient to introduce an equivalent normal force as:

$$N_{equiv} = |N| + \left|\frac{4M}{L}\right|$$

This is determined as follows. Over half the gusset, the normal force is $\frac{N}{2} + \frac{2M}{L}$, and over the other half it is $\frac{N}{2} - \frac{2M}{L}$. For simplicity in calculations, one of the moment forces, $\frac{2M}{L}$, is reversed so that a uniform equivalent normal force exists over the entire gusset section a-a. This is also convenient for use in the beam web local yielding and web local crippling equations of the AISC *Specification* that assume a uniform compression over the contact area. The equivalent normal force is:

LRFD	ASD
$N_{u\,equiv} = \|57.0 \text{ kips}\| + \left\|\dfrac{4(5{,}160 \text{ kip-in.})}{46.0 \text{ in.}}\right\|$ $= 506 \text{ kips}$	$N_{a\,equiv} = \|40.0 \text{ kips}\| + \left\|\dfrac{4(3{,}610 \text{ kip-in.})}{46.0 \text{ in.}}\right\|$ $= 354 \text{ kips}$

The available strength of the gusset plate to resist this force is determined for the limit state of tensile yielding:

$N_n = F_y A_g$ (Spec. Eq. J4-1)

LRFD	ASD
$\phi N_n = \phi F_y t_p L$ $= 0.90(36 \text{ ksi})(\frac{7}{8} \text{ in.})(46.0 \text{ in.})$ $= 1{,}300 \text{ kips} > 506 \text{ kips}$ **o.k.**	$\dfrac{N_n}{\Omega} = \dfrac{F_y t_p L}{\Omega}$ $= \dfrac{36 \text{ ksi}(\frac{7}{8} \text{ in.})(46.0 \text{ in.})}{1.67}$ $= 868 \text{ kips} > 354 \text{ kips}$ **o.k.**

The gusset shear and normal forces and strengths calculated previously do not consider interaction. Interaction seldom controls at this location because the gusset thickness is usually controlled by the limit states associated with the brace-to-gusset connection. If interaction is to be considered here, the following interaction equation from plasticity theory is recommended (Astaneh, 1998):

LRFD	ASD
$\left(\dfrac{M_u}{\phi M_n}\right) + \left(\dfrac{N_u}{\phi N_n}\right)^2 + \left(\dfrac{V_u}{\phi V_n}\right)^4 \leq 1$	$\left(\dfrac{\Omega M_a}{M_n}\right) + \left(\dfrac{\Omega N_a}{N_n}\right)^2 + \left(\dfrac{\Omega V_a}{V_n}\right)^4 \leq 1$

For the present problem, the required strengths have been calculated in the preceding text, as have the available strengths in shear and tension. The available flexural strength of the gusset plate is calculated using the plastic section modulus of the gusset plate at section a-a:

5.3 SPECIAL CONCENTRICALLY BRACED FRAMES

LRFD	ASD
$\phi M_n = \phi F_y Z$ $= 0.90(36 \text{ ksi})(\text{⅞ in.})\dfrac{(46.0 \text{ in.})^2}{4}$ $= 15{,}000 \text{ kip-in.}$	$\dfrac{M_n}{\Omega} = \dfrac{F_y Z}{\Omega}$ $= \dfrac{36 \text{ ksi}(\text{⅞ in.})\dfrac{(46.0 \text{ in.})^2}{4}}{1.67}$ $= 9{,}980 \text{ kip-in.}$

Therefore, the interaction equation from plasticity theory is:

LRFD	ASD
$\left(\dfrac{5{,}160 \text{ kip-in.}}{15{,}000 \text{ kip-in.}}\right) + \left(\dfrac{57.0 \text{ kips}}{1{,}300 \text{ kips}}\right)^2$ $+ \left(\dfrac{377 \text{ kips}}{869 \text{ kips}}\right)^4$ $= 0.381 \leq 1.0$ **o.k.**	$\left(\dfrac{3{,}610 \text{ kip-in.}}{9{,}980 \text{ kip-in.}}\right) + \left(\dfrac{40.0 \text{ kips}}{868 \text{ kips}}\right)^2$ $+ \left(\dfrac{264 \text{ kips}}{580 \text{ kips}}\right)^4$ $= 0.407 \leq 1.0$ **o.k.**

Design the weld at the gusset-to-beam flange interface

The use of a plastic distribution for the moment is convenient for calculation as mentioned previously, but requires sufficient ductility. The gusset and the beam can be assumed to be sufficiently ductile, but the fillet welds or PJP groove welds generally used to connect the gusset to the beam are well-known to have less ductility when loaded at angles significantly different from the longitudinal axis, which often is the case with the moment forces. Therefore, it is prudent to use the weld ductility factor originally derived from Richard (1986) as a value of 1.4, and modified by Hewitt and Thornton (2004) to a 90% confidence limit and the value of 1.25. This value, which is explained in AISC *Manual* Part 13, is used in these calculations. The original 1.4 factor is from Richard's work on corner gussets. Nevertheless, it is reasonable to use some "ductility factor" here because the weld is assumed to be uniformly loaded over each half width, even though the actual distribution can vary. The use of a CJP groove weld avoids this issue, but likely at greater cost.

The resultant force on the weld is:

LRFD	ASD
$R_u = \sqrt{N_{u\,equiv}^2 + V_u^2}$ $= \sqrt{(506 \text{ kips})^2 + (377 \text{ kips})^2}$ $= 631 \text{ kips}$	$R_a = \sqrt{N_{a\,equiv}^2 + V_a^2}$ $= \sqrt{(354 \text{ kips})^2 + (264 \text{ kips})^2}$ $= 442 \text{ kips}$

The angle of the resultant force can be calculated and used in the directional strength increase for fillet welds as follows:

LRFD	ASD
$\theta = \tan^{-1}\left(\dfrac{N_{u\,equiv}}{V_u}\right)$	$\theta = \tan^{-1}\left(\dfrac{N_{a\,equiv}}{V_a}\right)$
$= \tan^{-1}\left(\dfrac{506 \text{ kips}}{377 \text{ kips}}\right)$	$= \tan^{-1}\left(\dfrac{354 \text{ kips}}{264 \text{ kips}}\right)$
$= 53.3°$	$= 53.3°$

AISC *Specification* Section J2.4 allows an increase in the available strength of fillet welds when the angle of loading is not along the weld longitudinal axis. The directional strength increase is determined from the following portion of AISC *Specification* Equation J2-5:

$1.0 + 0.50\sin^{1.5}53.3° = 1.36$

Using AISC *Manual* Equations 8-2a and 8-2b, the number of sixteenths of fillet weld required is:

LRFD	ASD
$D_{req'd} \geq$	$D_{req'd} \geq$
$\dfrac{1.25R_u}{2(1.392 \text{ kip/in.})(1.36)l}$	$\dfrac{1.25R_a}{2(0.928 \text{ kip/in.})(1.36)l}$
$= \dfrac{1.25(631 \text{ kips})}{2(1.392 \text{ kip/in.})(1.36)(46.0 \text{ in.})}$	$= \dfrac{1.25(442 \text{ kips})}{2(0.928 \text{ kip/in.})(1.36)(46.0 \text{ in.})}$
$= 4.53$ sixteenths	$= 4.76$ sixteenths

Based on the thickness of the thinner connected part, the minimum fillet weld size required by AISC *Specification* Table J2.4 is 5/16 in.

Use double-sided 5/16-in. fillet welds to connect the top gusset plate to the beam.

Check beam web local yielding

For a force applied at a distance from the beam end that is greater than the depth of the member:

$R_n = F_{yw}t_w(5k + l_b)$ (*Spec.* Eq. J10-2)

$= (50 \text{ ksi})(0.570 \text{ in.})[5(1.53 \text{ in.}) + 46.0 \text{ in.}]$

$= 1,530$ kips

LRFD	ASD
$\phi R_n = 1.00(1{,}530 \text{ kips})$ $= 1{,}530 \text{ kips} > 506 \text{ kips}$ **o.k.**	$\dfrac{R_n}{\Omega} = \dfrac{1{,}530 \text{ kips}}{1.50}$ $= 1{,}020 \text{ kips} > 354 \text{ kips}$ **o.k.**

Check beam web local crippling

For a force applied greater than a distance of $d/2$ from the beam end:

$$R_n = 0.80 t_w^2 \left[1 + 3\left(\dfrac{l_b}{d}\right)\left(\dfrac{t_w}{t_f}\right)^{1.5}\right]\sqrt{\dfrac{EF_{yw}t_f}{t_w}} \qquad (\textit{Spec. Eq. J10-4})$$

$$= (0.80)(0.570 \text{ in.})^2 \left[1 + 3\left(\dfrac{46.0 \text{ in.}}{27.3 \text{ in.}}\right)\left(\dfrac{0.570 \text{ in.}}{0.930 \text{ in.}}\right)^{1.5}\right]$$

$$\times \sqrt{\dfrac{29{,}000 \text{ ksi}(50 \text{ ksi})(0.930 \text{ in.})}{0.570 \text{ in.}}}$$

$$= 1{,}370 \text{ kips}$$

LRFD	ASD
$\phi R_n = 0.75(1{,}370 \text{ kips})$ $= 1{,}030 \text{ kips} > 506 \text{ kips}$ **o.k.**	$\dfrac{R_n}{\Omega} = \dfrac{1{,}370 \text{ kips}}{2.00}$ $= 685 \text{ kips} > 354 \text{ kips}$ **o.k.**

This completes the design of the top gusset for the forces on section a-a.

The forces on gusset section b-b are:

LRFD	ASD
Normal $N_u' = \dfrac{1}{2}(H_1 + H_2)$ $\quad = \dfrac{1}{2}(-160 \text{ kips} + 217 \text{ kips})$ $\quad = 28.5 \text{ kips}$ Shear $V_u' = \dfrac{1}{2}(V_1 - V_2) - \dfrac{2}{L}(M_u)$ $\quad = \dfrac{1}{2}(-160 \text{ kips} - 217 \text{ kips})$ $\quad \quad - \dfrac{2(-5{,}160 \text{ kip-in.})}{46.0 \text{ in.}}$ $\quad = 35.8 \text{ kips}$	Normal $N_a' = \dfrac{1}{2}(H_1 + H_2)$ $\quad = \dfrac{1}{2}(-112 \text{ kips} + 152 \text{ kips})$ $\quad = 20.0 \text{ kips}$ Shear $V_a' = \dfrac{1}{2}(V_1 - V_2) - \dfrac{2}{L}(M_a)$ $\quad = \dfrac{1}{2}(-112 \text{ kips} - 152 \text{ kips})$ $\quad \quad - \dfrac{2(-3{,}610 \text{ kip-in.})}{46.0 \text{ in.}}$ $\quad = 25.0 \text{ kips}$

LRFD	ASD
Moment $M'_u = M'_1 + M'_2$ $= 975$ kip-in. $+(-1,320$ kip-in.$)$ $= -345$ kip-in.	Moment $M'_a = M'_1 + M'_2$ $= 681$ kip-in. $+(-926$ kip-in.$)$ $= -245$ kip-in.

Check gusset available strength on section b-b
The available shear strength of the gusset plate on section b-b is:

$V_n = 0.60 F_y A_{gv}$ (*Spec.* Eq. J4-3)

LRFD	ASD
$\phi V_n = 1.00(0.60)(36 \text{ ksi})$ $\times (\frac{7}{8} \text{ in.})(20.0 \text{ in.})$ $= 378$ kips > 35.8 kip **o.k.**	$\dfrac{V_n}{\Omega} = 0.60(36 \text{ ksi})(\frac{7}{8} \text{ in.})$ $\times (20.0 \text{ in.})/1.50$ $= 252$ kips > 25.0 kips **o.k.**

The normal force involves both N and M. It is convenient to introduce an equivalent normal force, as before, using the governing condition where N and the component of M are additive. This can be written as:

$N'_{equiv} = |N'| + \left|\dfrac{4M'}{h}\right|$

LRFD	ASD
$N'_{u\,equiv} = \|28.5 \text{ kips}\| + \left\|\dfrac{4(-345 \text{ kip-in.})}{20.0 \text{ in.}}\right\|$ $= 97.5$ kips	$N'_{a\,equiv} = \|20.0 \text{ kips}\| + \left\|\dfrac{4(-245 \text{ kip-in.})}{20.0 \text{ in.}}\right\|$ $= 69.0$ kips

The available strength of the gusset plate to resist this force is determined for the limit state of tensile yielding:

$N_n = F_y A_g$ (*Spec.* Eq. J4-1)

LRFD	ASD
$\phi N_n = \phi F_y t_p L$ $= 0.90(36 \text{ ksi})(\frac{7}{8} \text{ in.})(20.0 \text{ in.})$ $= 567$ kips > 97.5 kips **o.k.**	$\dfrac{N_n}{\Omega} = \dfrac{F_y t_p L}{\Omega}$ $= \dfrac{36 \text{ ksi}(\frac{7}{8} \text{ in.})(20.0 \text{ in.})}{1.67}$ $= 377$ kips > 69.0 kips **o.k.**

5.3 SPECIAL CONCENTRICALLY BRACED FRAMES

Design of gusset below the beam

For the geometry shown in Figure 5-29 for the gusset plate above the beam and using Figure 5-31 for the general terminology:

$\Delta = 0$

$L_1 = L_2$
   $= 25.5$ in.

$L = L_1 + L_2$
   $= 51.0$ in.

$e = d/2$
   $= 27.3$ in./2
   $= 13.7$ in.

$h = 29.0$ in.

For the forces given in Figures 5-30a and 5-30b:

LRFD	ASD
$P_1 = 550$ kips	$P_1 = 385$ kips

$H_1 = (550 \text{ kips})(\cos 45°)$
    $= 389$ kips
	$H_1 = (385 \text{ kips})(\cos 45°)$
	    $= 272$ kips

$V_1 = (550 \text{ kips})(\sin 45°)$
    $= 389$ kips
	$V_1 = (385 \text{ kips})(\sin 45°)$
	    $= 272$ kips

$P_2 = -444$ kips	$P_2 = -311$ kips

$H_2 = (-444 \text{ kips})(\cos 45°)$
    $= -314$ kips
	$H_2 = (-311 \text{ kips})(\cos 45°)$
	    $= -220$ kips

$V_2 = (-444 \text{ kips})(\sin 45°)$
    $= -314$ kips
	$V_2 = (-311 \text{ kips})(\sin 45°)$
	    $= -220$ kips

The moments are:

LRFD	ASD

$M_1 = H_1 e + V_1 \Delta$
   $= [389 \text{ kips}(13.7 \text{ in.})]$
   $+ [389 \text{ kips}(0 \text{ in.})]$
   $= 5{,}330$ kip-in.
	$M_1 = H_1 e + V_1 \Delta$
	   $= [272 \text{ kips}(13.7 \text{ in.})]$
	   $+ [272 \text{ kips}(0 \text{ in.})]$
	   $= 3{,}730$ kip-in.

LRFD	ASD
$M_2 = H_2 e - V_2 \Delta$ $= [-314 \text{ kips}(13.7 \text{ in.})]$ $\quad -[-314 \text{ kips}(0 \text{ in.})]$ $= -4{,}300 \text{ kip-in.}$	$M_2 = H_2 e - V_2 \Delta$ $= [-220 \text{ kips}(13.7 \text{ in.})]$ $\quad -[-220 \text{ kips}(0 \text{ in.})]$ $= -3{,}010 \text{ kip-in.}$
$M_1' = \frac{1}{8} L V_1 - \frac{1}{4} h H_1 - \frac{1}{2} M_1$ $= \frac{1}{8}(51.0 \text{ in.})(389 \text{ kips})$ $\quad - \frac{1}{4}(29.0 \text{ in.})(389 \text{ kips})$ $\quad - \frac{1}{2}(5{,}330 \text{ kip-in.})$ $= -3{,}010 \text{ kip-in.}$	$M_1' = \frac{1}{8} L V_1 - \frac{1}{4} h H_1 - \frac{1}{2} M_1$ $= \frac{1}{8}(51.0 \text{ in.})(272 \text{ kips})$ $\quad - \frac{1}{4}(29.0 \text{ in.})(272 \text{ kips})$ $\quad - \frac{1}{2}(3{,}730 \text{ kip-in.})$ $= -2{,}100 \text{ kip-in.}$
$M_2' = \frac{1}{8} L V_2 - \frac{1}{4} h H_2 - \frac{1}{2} M_2$ $= \frac{1}{8}(51.0 \text{ in.})(-314 \text{ kips})$ $\quad - \frac{1}{4}(29.0 \text{ in.})(-314 \text{ kips})$ $\quad - \frac{1}{2}(-4{,}300 \text{ kip-in.})$ $= 2{,}420 \text{ kip-in.}$	$M_2' = \frac{1}{8} L V_2 - \frac{1}{4} h H_2 - \frac{1}{2} M_2$ $= \frac{1}{8}(51.0 \text{ in.})(-220 \text{ kips})$ $\quad - \frac{1}{4}(29.0 \text{ in.})(-220 \text{ kips})$ $\quad - \frac{1}{2}(-3{,}010 \text{ kip-in.})$ $= 1{,}700 \text{ kip-in.}$

For the bottom gusset, the forces on section a-a are:

LRFD	ASD
Normal $N_u = V_1 + V_2$ $= 389 \text{ kips} - 314 \text{ kips}$ $= 75.0 \text{ kips}$	Normal $N_a = V_1 + V_2$ $= 272 \text{ kips} + (-220 \text{ kips})$ $= 52.0 \text{ kips}$
Shear $V_u = H_1 - H_2$ $= 389 \text{ kips} - (-314 \text{ kips})$ $= 703 \text{ kips}$	Shear $V_a = H_1 - H_2$ $= 272 \text{ kips} - (-220 \text{ kips})$ $= 492 \text{ kips}$
Moment $M_u = M_1 - M_2$ $= 5{,}330 \text{ kip-in.}$ $\quad -(-4{,}300 \text{ kip-in.})$ $= 9{,}630 \text{ kip-in.}$	Moment $M_a = M_1 - M_2$ $= 3{,}730 \text{ kip-in.}$ $\quad -(-3{,}010 \text{ kip-in.})$ $= 6{,}740 \text{ kip-in.}$

5.3 SPECIAL CONCENTRICALLY BRACED FRAMES

Check available strength of bottom gusset below the beam on section a-a

The available shear strength of the gusset plate on section a-a is:

$$V_n = 0.60 F_y A_{gv}$$ (Spec. Eq. J4-3)

LRFD	ASD
$\phi V_n = 1.00(0.60)(36 \text{ ksi})$ $\times (1\frac{1}{4} \text{ in.})(51.0 \text{ in.})$ $= 1{,}380 \text{ kips} > 703 \text{ kips}$ **o.k.**	$\dfrac{V_n}{\Omega} = 0.60(36 \text{ ksi})(1\frac{1}{4} \text{ in.})$ $\times (51.0 \text{ in.})/1.50$ $= 918 \text{ kips} > 492 \text{ kips}$ **o.k.**

The normal force involves both N_u or N_a and M_u or M_a. It is convenient to introduce an equivalent normal force, as before, as:

LRFD	ASD
$N_{u\,equiv} = \left\|75.0 \text{ kips}\right\| + \left\|\dfrac{4(9{,}630 \text{ kip-in.})}{51.0 \text{ in.}}\right\|$ $= 830 \text{ kips}$	$N_{a\,equiv} = \left\|52.0 \text{ kips}\right\| + \left\|\dfrac{4(6{,}740 \text{ kip-in.})}{51.0 \text{ in.}}\right\|$ $= 581 \text{ kips}$

The available strength of the gusset plate to resist this force is determined for the limit state of tensile yielding:

$$N_n = F_y A_g$$ (Spec. Eq. J4-1)

LRFD	ASD
$\phi N_n = \phi F_y t_p L$ $= 0.90(36 \text{ ksi})(1\frac{1}{4} \text{ in.})(51.0 \text{ in.})$ $= 2{,}070 \text{ kips} > 830 \text{ kips}$ **o.k.**	$\dfrac{N_n}{\Omega} = \dfrac{F_y t_p L}{\Omega}$ $= \dfrac{36 \text{ ksi}(1\frac{1}{4} \text{ in.})(51.0 \text{ in.})}{1.67}$ $= 1{,}370 \text{ kips} > 581 \text{ kips}$ **o.k.**

Interaction as calculated for the top gusset above the beam is not repeated here.

Design the weld at the gusset-to-beam flange interface for the gusset below the beam

As discussed for the gusset above the beam, the 1.25 ductility factor is used here.

The resultant force on the weld is:

LRFD	ASD
$R_u = \sqrt{N_{u\,equiv}^2 + V_u^2}$ $= \sqrt{(830 \text{ kips})^2 + (703 \text{ kips})^2}$ $= 1{,}090 \text{ kips}$	$R_a = \sqrt{N_{a\,equiv}^2 + V_a^2}$ $= \sqrt{(581 \text{ kips})^2 + (492 \text{ kips})^2}$ $= 761 \text{ kips}$

The angle of the resultant force can be calculated and used in the directional strength increase for fillet welds as follows:

LRFD	ASD
$\theta = \tan^{-1}\left(\dfrac{N_{u\,equiv}}{V_u}\right)$ $= \tan^{-1}\left(\dfrac{830 \text{ kips}}{703 \text{ kips}}\right)$ $= 49.7°$	$\theta = \tan^{-1}\left(\dfrac{N_{a\,equiv}}{V_a}\right)$ $= \tan^{-1}\left(\dfrac{581 \text{ kips}}{492 \text{ kips}}\right)$ $= 49.7°$

AISC *Specification* Section J2.4 allows an increase in the available strength of fillet welds when the angle of loading is not along the weld longitudinal axis. The directional strength increase is determined from the following portion of AISC *Specification* Equation J2-5:

$1.0 + 0.50\sin^{1.5}49.7° = 1.33$

Using AISC *Manual* Equations 8-2a and 8-2b, the number of sixteenths of fillet weld required is:

LRFD	ASD
$D_{req'd} \geq$ $\dfrac{1.25 R_u}{2(1.392 \text{ kip/in.})(1.33)l}$ $= \dfrac{1.25(1{,}090 \text{ kips})}{2(1.392 \text{ kip/in.})(1.33)(51.0 \text{ in.})}$ $= 7.22 \text{ sixteenths}$	$D_{req'd} \geq$ $\dfrac{1.25 R_a}{2(0.928 \text{ kip/in.})(1.33)l}$ $= \dfrac{1.25(761 \text{ kips})}{2(0.928 \text{ kip/in.})(1.33)(51.0 \text{ in.})}$ $= 7.56 \text{ sixteenths}$

Use double-sided ½-in. fillet welds to connect the bottom gusset plate to the beam.

Check beam web local yielding

For a force applied at a distance from the beam end that is greater than the depth of the member:

5.3 SPECIAL CONCENTRICALLY BRACED FRAMES

$$R_n = F_{yw}t_w(5k + l_b) \quad \text{(Spec. Eq. J10-2)}$$
$$= (50 \text{ ksi})(0.570 \text{ in.})[5(1.53 \text{ in.}) + 51.0 \text{ in.}]$$
$$= 1{,}670 \text{ kips}$$

LRFD	ASD
$\phi R_n = 1.00(1{,}670 \text{ kips})$ $= 1{,}670 \text{ kips} > 830 \text{ kips}$ **o.k.**	$\dfrac{R_n}{\Omega} = \dfrac{1{,}670 \text{ kips}}{1.50}$ $= 1{,}110 \text{ kips} > 581 \text{ kips}$ **o.k.**

Check beam web local crippling

For a force applied greater than a distance of $d/2$ from the beam end:

$$R_n = 0.80 t_w^2 \left[1 + 3\left(\frac{l_b}{d}\right)\left(\frac{t_w}{t_f}\right)^{1.5}\right]\sqrt{\frac{EF_{yw}t_f}{t_w}} \quad \text{(Spec. Eq. J10-4)}$$

$$= (0.80)(0.570 \text{ in.})^2 \left[1 + 3\left(\frac{51.0 \text{ in.}}{27.3 \text{ in.}}\right)\left(\frac{0.570 \text{ in.}}{0.930 \text{ in.}}\right)^{1.5}\right]$$
$$\times \sqrt{\frac{29{,}000 \text{ ksi}(50 \text{ ksi})(0.930 \text{ in.})}{0.570 \text{ in.}}}$$

$$= 1{,}470 \text{ kips}$$

LRFD	ASD
$\phi R_n = 0.75(1{,}470 \text{ kips})$ $= 1{,}100 \text{ kips} > 830 \text{ kips}$ **o.k.**	$\dfrac{R_n}{\Omega} = \dfrac{1{,}470 \text{ kips}}{2.00}$ $= 735 \text{ kips} > 581 \text{ kips}$ **o.k.**

This completes the design of the bottom gusset for the forces on section a-a.

The forces on gusset section b-b are:

LRFD	ASD
Normal $N_u' = \dfrac{1}{2}(H_1 + H_2)$ $\dfrac{1}{2}[389 \text{ kips} + (-314 \text{ kips})]$ $= 37.5 \text{ kips}$	Normal $N_a' = \dfrac{1}{2}(H_1 + H_2)$ $= \dfrac{1}{2}[272 \text{ kips} + (-220 \text{ kips})]$ $= 26.0 \text{ kips}$

LRFD	ASD
Shear $V'_u = \frac{1}{2}(V_1 - V_2) - \frac{2}{L}(M_u)$	Shear $V'_a = \frac{1}{2}(V_1 - V_2) - \frac{2}{L}(M_a)$
$= \frac{1}{2}[389 \text{ kips} - (-314 \text{ kips})]$	$= \frac{1}{2}[272 \text{ kips} - (-220 \text{ kips})]$
$\quad - \frac{2(9{,}630 \text{ kip-in.})}{51.0 \text{ in.}}$	$\quad - \frac{2(6{,}740 \text{ kip-in.})}{51.0 \text{ in.}}$
$= -26.1 \text{ kips}$	$= -18.3 \text{ kips}$
Moment $M'_u = M'_1 + M'_2$	Moment $M'_a = M'_1 + M'_2$
$= -3{,}010 \text{ kip-in.}$	$= -2{,}100 \text{ kip-in.}$
$\quad +2{,}420 \text{ kip-in.}$	$\quad +1{,}700 \text{ kip-in.}$
$= -590 \text{ kip-in.}$	$= -400 \text{ kip-in.}$

Check gusset available strength on section b-b

The available shear strength of the gusset plate on section b-b is:

$V_n = 0.60 F_y A_{gv}$ \hfill (*Spec.* Eq. J4-3)

LRFD	ASD
$\phi V_n = 1.00(0.60)(36 \text{ ksi})$	$\frac{V_n}{\Omega} = 0.60(36 \text{ ksi})(1\frac{1}{4} \text{ in.})$
$\quad \times (1\frac{1}{4} \text{ in.})(29.0 \text{ in.})$	$\quad \times (29.0 \text{ in.})/1.50$
$= 783 \text{ kips} > 26.1 \text{ kips}$ **o.k.**	$= 522 \text{ kips} > 18.3 \text{ kips}$ **o.k.**

The normal force involves both N and M. It is convenient to introduce an equivalent normal force, as before, as:

$N'_{equiv} = |N'| + \left|\frac{4M'}{h}\right|$

LRFD	ASD
$N'_{u\,equiv} = |37.5 \text{ kips}| + \left|\frac{4(-590 \text{ kip-in.})}{29.0 \text{ in.}}\right|$	$N'_{a\,equiv} = |26.0 \text{ kips}| + \left|\frac{4(-400 \text{ kip-in.})}{29.0 \text{ in.}}\right|$
$= 119 \text{ kips}$	$= 81.2 \text{ kips}$

The available strength of the gusset plate to resist this force is determined for the limit state of tensile yielding:

$N_n = F_y A_g$ \hfill (*Spec.* Eq. J4-1)

5.3 SPECIAL CONCENTRICALLY BRACED FRAMES

LRFD	ASD
$\phi N_n = \phi F_y t_p L$ $= 0.90(36 \text{ ksi})(1\tfrac{1}{4} \text{ in.})(29.0 \text{ in.})$ $= 1{,}170 \text{ kips} > 119 \text{ kips}$ **o.k.**	$\dfrac{N_n}{\Omega} = \dfrac{F_y t_p L}{\Omega}$ $= \dfrac{36 \text{ ksi}(1\tfrac{1}{4} \text{ in.})(29.0 \text{ in.})}{1.67}$ $= 781 \text{ kips} > 81.2 \text{ kips}$ **o.k.**

Check beam web shear at the centerline of the connection

From Figure 5-32:

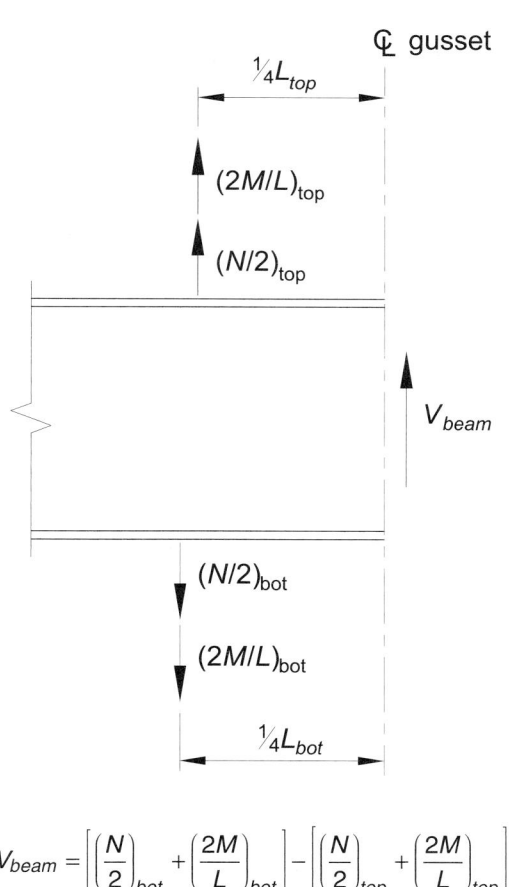

$$V_{beam} = \left[\left(\dfrac{N}{2}\right)_{bot} + \left(\dfrac{2M}{L}\right)_{bot} \right] - \left[\left(\dfrac{N}{2}\right)_{top} + \left(\dfrac{2M}{L}\right)_{top} \right]$$

Fig. 5-32. Free body diagram to calculate beam shear, V_{beam}, (forces are shown in assumed positive direction).

LRFD	ASD
$V_{beam} = \left[\dfrac{75.0 \text{ kips}}{2} + \dfrac{2(9{,}630 \text{ kip-in.})}{51.0 \text{ in.}}\right]$ $- \left[\dfrac{57.0 \text{ kips}}{2} + \dfrac{2(-5{,}160 \text{ kip-in.})}{46.0 \text{ in.}}\right]$ $= 611 \text{ kips}$	$V_{beam} = \left[\dfrac{52.0 \text{ kips}}{2} + \dfrac{2(6{,}740 \text{ kip-in.})}{51.0 \text{ in.}}\right]$ $- \left[\dfrac{40.0 \text{ kips}}{2} + \dfrac{2(-3{,}610 \text{ kip-in.})}{46.0 \text{ in.}}\right]$ $= 427 \text{ kips}$

From AISC *Manual* Table 3-6, the available shear strength of the W27×114 is:

LRFD	ASD
$\phi_v V_n = 467 \text{ kips} < 611 \text{ kips}$ **n.g.**	$\dfrac{V_n}{\Omega_v} = 311 \text{ kips} < 427 \text{ kips}$ **n.g.**

The available shear strength is less than the required shear strength. Thus, a web doubler plate is required, as in Example 5.3.8.

The normal force per inch for the bottom gusset is:

LRFD	ASD
$p_{bot} = \left[\left(\dfrac{N}{2}\right)_{bot} + \left(\dfrac{2M}{L}\right)_{bot}\right]/(L/2)$ $= \left[\left(\dfrac{75.0 \text{ kips}}{2}\right) + \left(\dfrac{2(9{,}630 \text{ kip-in.})}{51.0 \text{ in.}}\right)\right]$ $/(51.0/2)$ $= 16.3 \text{ kip/in.}$	$p_{bot} = \left[\left(\dfrac{N}{2}\right)_{bot} + \left(\dfrac{2M}{L}\right)_{bot}\right]/(L/2)$ $= \left[\left(\dfrac{52.0 \text{ kips}}{2}\right) + \left(\dfrac{2(6{,}740 \text{ kip-in.})}{51.0 \text{ in.}}\right)\right]$ $/(51.0/2)$ $= 11.4 \text{ kip/in.}$

The normal force per inch for the top gusset is:

LRFD	ASD
$p_{top} = \left[\left(\dfrac{N}{2}\right)_{top} + \left(\dfrac{2M}{L}\right)_{top}\right]/(L/2)$ $= \left[\left(\dfrac{57.0 \text{ kips}}{2}\right) + \left(\dfrac{2(-5{,}160 \text{ kip-in.})}{46.0 \text{ in.}}\right)\right]$ $/(46.0/2)$ $= -8.52 \text{ kip/in.}$	$p_{top} = \left[\left(\dfrac{N}{2}\right)_{top} + \left(\dfrac{2M}{L}\right)_{top}\right]/(L/2)$ $= \left[\left(\dfrac{40.0 \text{ kips}}{2}\right) + \left(\dfrac{2(-3{,}610 \text{ kip-in.})}{46.0 \text{ in.}}\right)\right]$ $/(46.0/2)$ $= -5.95 \text{ kip/in.}$

5.3 SPECIAL CONCENTRICALLY BRACED FRAMES

Starting at the beam centerline, the shear per inch is:

LRFD	ASD
$v(x) = 611$ kips $\quad -(8.52 \text{ kip/in.} + 16.3 \text{ kip/in.})x$ $\quad = 611 \text{ kips} - (24.8 \text{ kip/in.})x$	$v(x) = 427$ kips $\quad -(5.95 \text{ kip/in.} + 11.4 \text{ kip/in.})x$ $\quad = 427 \text{ kips} - (17.4 \text{ kip/in.})x$

where x is measured from the beam centerline.

Setting $v(x)$ equal to the available strength of the beam and solving for x:

LRFD	ASD
$611 \text{ kips} - (24.8 \text{ kip/in.})x = 467$ kips $x = 5.81$ in.	$427 \text{ kips} - (17.4 \text{ kip/in.})x = 311$ kips $x = 6.67$ in.

Try a 14 in.×14 in. doubler plate. Using the available shear strength of the beam and AISC *Specification* Equation J4-3 and solving for $t_{doubler}$:

LRFD	ASD
$\phi V_{n\,(beam)} + \phi V_{n\,(doubler)} \geq V_u$ $467 \text{ kips} + 1.00(t_{doubler})(14.0 \text{ in})$ $\quad \times (0.60)(36 \text{ ksi})$ ≥ 611 kips $t_{doubler} \geq 0.476$ in.	$V_{n\,(beam)}/\Omega + V_{n\,(doubler)}/\Omega \geq V_a$ $311 \text{ kips} + \{0.60(t_{doubler})(14.0 \text{ in})$ $\quad \times (36 \text{ ksi})/1.50\}$ ≥ 427 kips $t_{doubler} \geq 0.575$ in.

Typically, one method should be chosen and used consistently throughout an entire design. In design by LRFD, a ½-in. doubler plate would be adequate. For this example, the ASD result will be used.

Use a doubler plate ⅝ in.×14 in.×1'-2", near side only.

The required shear strength of the doubler plate is:

LRFD	ASD
$V_{beam} - \phi V_{n\,(beam)} = 611 \text{ kips} - 467$ kips $= 144$ kips	$V_{beam} - V_{n\,(beam)}/\Omega = 427 \text{ kips} - 311$ kips $= 116$ kips

Using the instantaneous center of rotation method of AISC *Manual* Part 8, use Table 8-8 with $\theta = 0°$ to determine the strength of welds on the doubler plate. For a channel-shaped weld on half of the doubler plate, the geometric variables are:

$l = 14.0$ in.

$k = 7.00$ in. / 14.0 in.
$ = 0.500$

$kl = 7.00$ in.

$xl = 0.125(14.0 \text{ in.}) = 1.75$ in.

$e_x = al$
$ = 7.00 \text{ in.} - 1.75 \text{ in.}$
$ = 5.25$ in.

$a = 0.375$

By interpolation, AISC *Manual* Table 8-8 with $\theta = 0°$ gives:

$C = 3.35$

From AISC *Manual* Table 8-3, $C_1 = 1.00$. The fillet weld size required is:

LRFD	ASD
$D_{min} = \dfrac{V_{beam}}{\phi CC_1 l}$	$D_{min} = \dfrac{\Omega V_{beam}}{CC_1 l}$
$\phantom{D_{min}} = \dfrac{144 \text{ kips}}{0.75(3.35)(1.00)(14.0 \text{ in.})}$	$\phantom{D_{min}} = \dfrac{2.00(116 \text{ kips})}{(3.35)(1.00)(14.0 \text{ in.})}$
$\phantom{D_{min}} = 4.09$	$\phantom{D_{min}} = 4.95$

Use a ⁵⁄₁₆-in. all around fillet weld, as shown in Figure 5-29. All details are shown in Figure 5-29.

Example 5.3.10. SCBF Brace-to-Beam/Column Connection Design

Given:

Refer to Joint JT-2 at level 3 in Figure 5-11. Design the connection between braces, beam and column using splices in the beam away from the gusset plates. The brace is designed to buckle out-of-plane. Use ASTM A36 welded gusset plates concentric to the braces and 70-ksi electrodes to connect the braces to the gusset plates and the gusset plates to the beam and column. As designed in Examples 5.3.1, 5.3.3 and 5.3.5, the braces are ASTM A500 Grade B round HSS sections, the column is an ASTM A992 W12×96, and the beam is an ASTM A992 W24×68. The brace reinforcing bars are ASTM A572 Grade 50 material. As noted in

5.3 SPECIAL CONCENTRICALLY BRACED FRAMES

Example 5.3.5, this connection uses ASTM A572 Grade 50 splices in the beam away from the connection. ASTM A992 W24×146 beam stubs are used at the beam ends to meet the high shear demand from the braces over the connection. Use ASTM A325-X bolts and 70-ksi weld electrodes. The applicable building code specifies the use of ASCE/SEI 7 for calculation of loads. The gravity shears and moments on the beam are:

$V_D = 11.2$ kips $V_L = 8.50$ kips $M_D = 120$ kip-ft $M_L = 100$ kip-ft

The relevant seismic parameters are given in the SCBF Design Example Plan and Elevation section.

Solution:

This connection design uses splices in the beam to provide a simple beam-to-column connection satisfying AISC *Seismic Provisions* Section F2.6b(a).

From AISC *Manual* Tables 2-4 and 2-5, the material properties are as follows:

ASTM A36
$F_y = 36$ ksi
$F_u = 58$ ksi

ASTM A572 Grade 50
$F_y = 50$ ksi
$F_u = 65$ ksi

ASTM A500 Grade B
$F_y = 42$ ksi
$F_u = 58$ ksi

ASTM A992
$F_y = 50$ ksi
$F_u = 65$ ksi

From AISC *Manual* Tables 1-1 and 1-13, the geometric properties are as follows:

Brace (above the beam)
HSS6.875×0.500
$t_{nom} = 0.500$ in. $t_{des} = 0.465$ in. $A = 9.36$ in.^{2} $r = 2.27$ in.

Brace (below the beam)
HSS7.500×0.500
$t_{nom} = 0.500$ in. $t_{des} = 0.465$ in. $A = 10.3$ in.^{2} $r = 2.49$ in.

Beam
W24×68
$d = 23.7$ in. $t_w = 0.415$ in. $t_f = 0.585$ in. $k_{des} = 1.09$ in.

Beam stub
W24×146

| $A = 43.0$ in.² | $d = 24.7$ in. | $t_w = 0.650$ in. | $t_f = 1.09$ in. |
| $k_{des} = 1.59$ in. | $T = 20\frac{3}{4}$ in. | $r_y = 3.01$ in. | |

Column
W12×96

| $d = 12.7$ in. | $t_w = 0.550$ in. | $t_f = 0.900$ in. | $k_{des} = 1.50$ in. |

The complete connection design is shown in Figure 5-33. The connection geometry and member forces are as shown in Figures 5-34 and 5-35. These were originally determined in Example 5.3.5. The calculations will be shown again here.

See the discussion under "Solution" in Example 5.3.8 for a discussion of the analysis forces required by the AISC *Seismic Provisions* and of the LRFD and ASD approaches.

In Example 5.3.8, there were two braces above the beam and two braces below, so the direction of loading did not affect the connection design. In this corner connection, since the braces above and below the beam are not the same size, the direction of loading affects the amount of force that must be considered in the connection design. Two design cases will be considered.

AISC *Seismic Provisions* Sections F2.3(i) and F2.3(ii) define the two mechanism analyses that must be considered in determining the required strength of beams, columns and connections. AISC *Seismic Provisions* Section F2.6c specifies the required strength of bracing connections. For the required compressive strength based on buckling limit states, Section F2.6c(2) has an additional 1.1 factor (relative to the requirements of Section F2.3) applied to the required strength of the connection.

For this SCBF connection example, the requirements of AISC *Seismic Provisions* Section F2.3 will be used for both LRFD and ASD, except for the limit state of compression buckling on the Whitmore section, which will use the 1.1 factor specified in AISC *Seismic Provisions* Section F2.6c(2).

Design Case I

Design Case I shows brace strengths which correspond to lateral forces applied in the positive *x*-direction. The brace above the beam is at its expected strength in tension, and the brace below the beam is at its expected strength (or its post-buckling strength) in compression. These forces above and below must be considered simultaneously.

Determine the expected tensile strength of the HSS6.875×0.500 brace above the beam for Design Case I

From AISC *Seismic Provisions* Table A3.1:

$R_y = 1.4$

$$P_{tension} = R_y F_y A_g$$
$$= 1.4(42 \text{ ksi})(9.36 \text{ in.}^2)$$
$$= 550 \text{ kips}$$

5.3 SPECIAL CONCENTRICALLY BRACED FRAMES

The required strength of the bracing connection due to seismic loading is based on ASCE/SEI 7 Section 12.4.3.2 Load Combination 5 (LRFD and ASD) with $\Omega_o Q_E = E_{mh}$, as discussed in AISC *Seismic Provisions* Section F2.3.

The required strength of the bracing connection when the brace is in tension is:

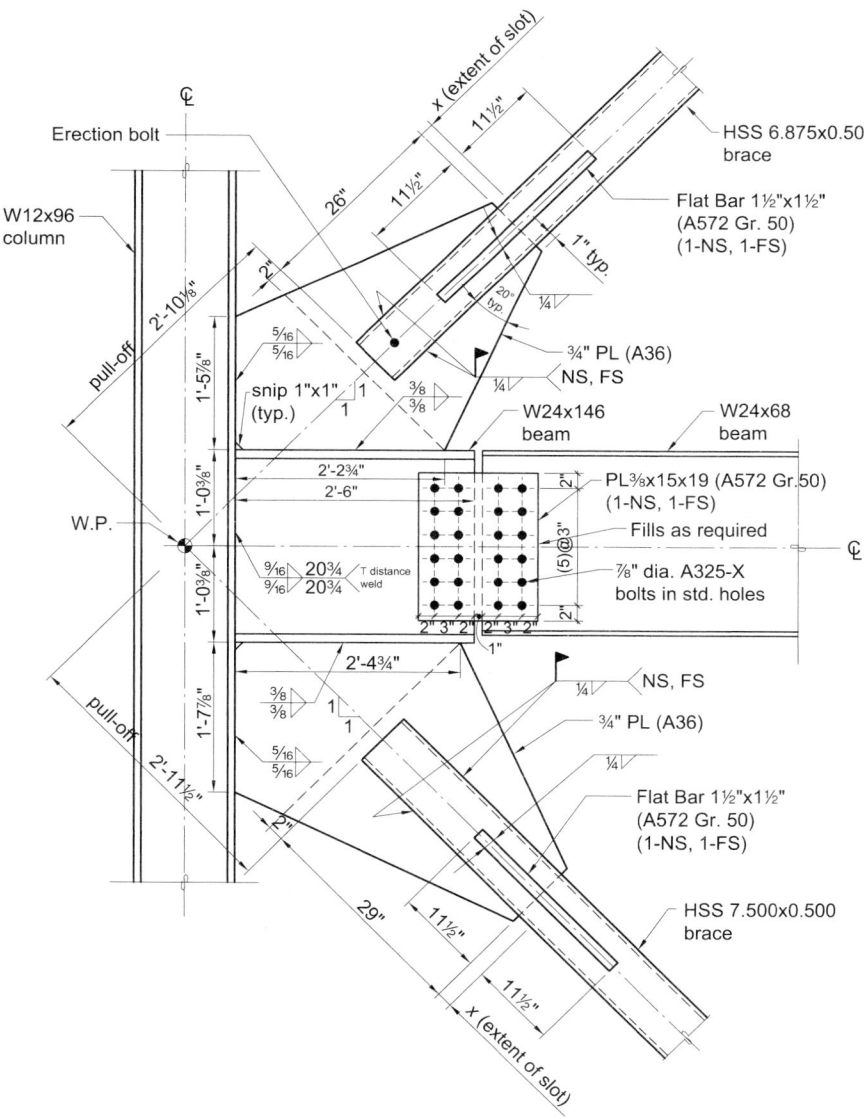

Note: A325-X bolts are to be pretensioned with slip-critical faying surfaces.

Fig. 5-33. Geometry and completed design for Example 5.3.10.

LRFD	ASD
$P_u = 1.0 E_{mh}$ $= 1.0(550 \text{ kips})$ $= 550 \text{ kips}$	$P_a = 0.7 E_{mh}$ $= 0.7(550 \text{ kips})$ $= 385 \text{ kips}$

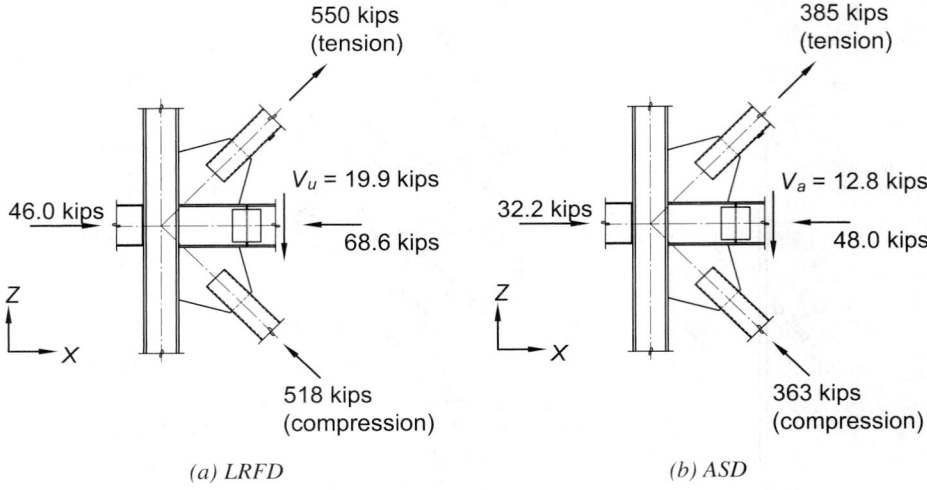

Fig. 5-34. Required strength of bracing connections according to AISC Seismic Provisions Section F2.3(i) for Design Case I.

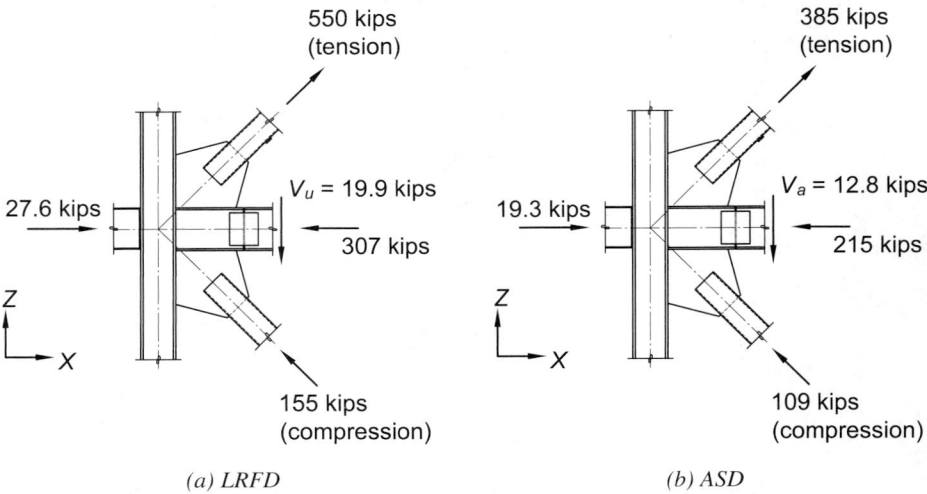

Fig. 5-35. Required strength of bracing connections according to AISC Seismic Provisions Section F2.3(ii) for Design Case I.

Determine the expected compressive strength of the HSS7.500×0.500 brace below the beam for Design Case I

For determining the expected strength of the brace in compression, $R_y F_y$ is used in lieu of F_y for the determination of F_{cre} according to AISC *Seismic Provisions* Section F2.3. The brace length used for the determination of F_{cre} must not exceed the distance from brace end to brace end; therefore a length of 12.0 ft will be used as previously stated in Example 5.3.2.

$$\frac{KL}{r} = \frac{1.0(12.0 \text{ ft})(12 \text{ in./ft})}{2.49 \text{ in.}}$$

$$= 57.8$$

$$4.71\sqrt{\frac{E}{R_y F_y}} = 4.71\sqrt{\frac{29{,}000 \text{ ksi}}{1.4(42 \text{ ksi})}}$$

$$= 105$$

With $\dfrac{KL}{r} \le 4.71\sqrt{\dfrac{E}{R_y F_y}}$:

$$F_e = \frac{\pi^2 E}{\left(\dfrac{KL}{r}\right)^2} \qquad \text{(Spec. Eq. E3-4)}$$

$$= \frac{\pi^2 (29{,}000 \text{ ksi})}{(57.8)^2}$$

$$= 85.7 \text{ ksi}$$

$$F_{cre} = \left[0.658^{\frac{R_y F_y}{F_e}}\right] R_y F_y \qquad \text{(from Spec. Eq. E3-2)}$$

$$= \left[0.658^{\frac{1.4(42 \text{ ksi})}{85.7 \text{ ksi}}}\right](1.4)(42 \text{ ksi})$$

$$= 44.1 \text{ ksi}$$

The expected compressive strength of the brace below the beam, based on AISC *Seismic Provisions* Section F2.3, is taken as the lesser of $R_y F_y A_g$ and $1.14 F_{cre} A_g$.

$$P_{compression} = \min\{1.14 F_{cre} A_g,\ R_y F_y A_g\}$$

$$= \min\left\{\left[1.14(44.1 \text{ ksi})(10.3 \text{ in.}^2)\right],\ \left[1.4(42 \text{ ksi})(10.3 \text{ in.}^2)\right]\right\}$$

$$= \min\{518 \text{ kips},\ 606 \text{ kips}\}$$

$$= 518 \text{ kips}$$

The required strength due to seismic loading of the bracing connection when the brace is in compression is:

LRFD	ASD
$P_u = 1.0E_{mh}$ $= 1.0(518 \text{ kips})$ $= 518 \text{ kips}$	$P_a = 0.7E_{mh}$ $= 0.7(518 \text{ kips})$ $= 363 \text{ kips}$

And the expected post-buckling brace strength from AISC *Seismic Provisions* Section F2.3 is:

$$0.3P_{compression} = 0.3(518 \text{ kips})$$
$$= 155 \text{ kips}$$

The required strength due to seismic loading of the bracing connection when the brace is in compression, based on post-buckling strength, is:

LRFD	ASD
$P_u = 1.0E_{mh}$ $= 1.0(155 \text{ kips})$ $= 155 \text{ kips}$	$P_a = 0.7E_{mh}$ $= 0.7(155 \text{ kips})$ $= 109 \text{ kips}$

The two sets of forces to be considered in Design Case I are shown in Figures 5-34 and 5-35. Determination of the collector force from outside the frame and the axial forces in the beam are shown for the beam design in Example 5.3.5 (see Figures 5-17 and 5-18).

Design Case II

Design Case II shows brace strengths corresponding to lateral forces applied in the negative *x*-direction. The brace above the beam is at its expected strength (or post-buckling strength) in compression, and the brace below the beam is at its expected strength in tension. These forces must be considered simultaneously.

Determine the expected compressive strength of the HSS6.875×0.500 brace above the beam for Design Case II

$$\frac{KL}{r} = \frac{1.0(12.0 \text{ ft})(12 \text{ in./ft})}{2.27 \text{ in.}}$$
$$= 63.4$$

$$4.71\sqrt{\frac{E}{R_y F_y}} = 4.71\sqrt{\frac{29{,}000 \text{ ksi}}{1.4(42 \text{ ksi})}}$$
$$= 105$$

With $\dfrac{KL}{r} \leq 4.71\sqrt{\dfrac{E}{R_y F_y}}$:

5.3 SPECIAL CONCENTRICALLY BRACED FRAMES

$$F_e = \frac{\pi^2 E}{\left(\frac{KL}{r}\right)^2} \quad \text{(Spec. Eq. E3-4)}$$

$$= \frac{\pi^2 (29{,}000 \text{ ksi})}{(63.4)^2}$$

$$= 71.2 \text{ ksi}$$

$$F_{cre} = \left[0.658^{\frac{R_y F_y}{F_e}}\right] R_y F_y \quad \text{(from Spec. Eq. E3-2)}$$

$$= \left[0.658^{\frac{1.4(42 \text{ ksi})}{71.2 \text{ ksi}}}\right] (1.4)(42 \text{ ksi})$$

$$= 41.6 \text{ ksi}$$

The expected compressive strength of the brace above the beam, based on AISC *Seismic Provisions* Section F2.3, is taken as the lesser of $R_y F_y A_g$ or $1.14 F_{cre} A_g$:

$$\begin{aligned} P_{compression} &= \min\{1.14 F_{cre} A_g,\, R_y F_y A_g\} \\ &= \min\left\{\left[1.14(41.6 \text{ ksi})(9.36 \text{ in.}^2)\right],\, \left[1.4(42 \text{ ksi})(9.36 \text{ in.}^2)\right]\right\} \\ &= \min\{444 \text{ kips},\, 550 \text{ kips}\} \\ &= 444 \text{ kips} \end{aligned}$$

The required strength due to seismic loading of the bracing connection when the brace is in compression is:

LRFD	ASD
$P_u = 1.0 E_{mh}$ $= 1.0(444 \text{ kips})$ $= 444 \text{ kips}$	$P_a = 0.7 E_{mh}$ $= 0.7(444 \text{ kips})$ $= 311 \text{ kips}$

And the expected post-buckling strength is:

$$\begin{aligned} 0.3 P_{compression} &= 0.3(444 \text{ kips}) \\ &= 133 \text{ kips} \end{aligned}$$

The required strength due to seismic loading of the bracing connection when the brace is in compression, based on post-buckling dtrength, is:

LRFD	ASD
$P_u = 1.0 E_{mh}$ $= 1.0(133 \text{ kips})$ $= 133 \text{ kips}$	$P_a = 0.7 E_{mh}$ $= 0.7(133 \text{ kips})$ $= 93.1 \text{ kips}$

Determine the expected tensile strength of the HSS7.500×0.500 brace below the beam for Design Case II

From AISC *Seismic Provisions* Table A3.1:

$R_y = 1.4$

$P_{tension} = R_y F_y A_g$

$= 1.4(42 \text{ ksi})(10.3 \text{ in.}^2)$

$= 606 \text{ kips}$

The required strength of the bracing connection due to seismic loading when the brace is in tension is:

LRFD	ASD
$P_u = 1.0 E_{mh}$ $= 1.0(606 \text{ kips})$ $= 606 \text{ kips}$	$P_a = 0.7 E_{mh}$ $= 0.7(606 \text{ kips})$ $= 424 \text{ kips}$

The two sets of forces to be considered in Design Case II are shown in Figures 5-36 and 5-37 (also see Figures 5-17 and 5-18 of Example 5.3.5).

There is no shear in the beam due to seismic loads. The required shear strength of the beam due to gravity loads is:

LRFD	ASD
LRFD Load Combination 5 from ASCE/SEI 7 Section 12.4.3.2 $V_u = (1.2 + 0.2 S_{DS}) V_D + \Omega_o V_{QE}$ $\quad + 0.5 V_L + 0.2 V_S$ $= 1.4(11.2 \text{ kips}) + 2.0(0 \text{ kips})$ $\quad + 0.5(8.50 \text{ kips}) + 0.2(0 \text{ kips})$ $= 19.9 \text{ kips}$	ASD Load Combination 5 from ASCE/SEI 7 Section 12.4.3.2 $V_a = (1.0 + 0.14 S_{DS}) V_D + V_H$ $\quad + V_F + 0.7 \Omega_o V_{QE}$ $= 1.14(11.2 \text{ kips}) + 0 \text{ kips} + 0 \text{ kips}$ $\quad + 0.7(2.0)(0 \text{ kips})$ $= 12.8 \text{ kips}$

Main Member Design Considerations

Considering Design Cases I and II, the total maximum vertical shear is the sum of the vertical components of the expected strength of the braces above and below the beam.

For Design Case I, this shear is:

LRFD	ASD
$(550 \text{ kips} + 518 \text{ kips}) \sin 45° = 755 \text{ kips}$	$(385 \text{ kips} + 363 \text{ kips}) \sin 45° = 529 \text{ kips}$

5.3 SPECIAL CONCENTRICALLY BRACED FRAMES

For Design Case II, this shear is:

LRFD	ASD
$(444 \text{ kips} + 606 \text{ kips})\sin 45° = 742 \text{ kips}$	$(311 \text{ kips} + 424 \text{ kips})\sin 45° = 520 \text{ kips}$

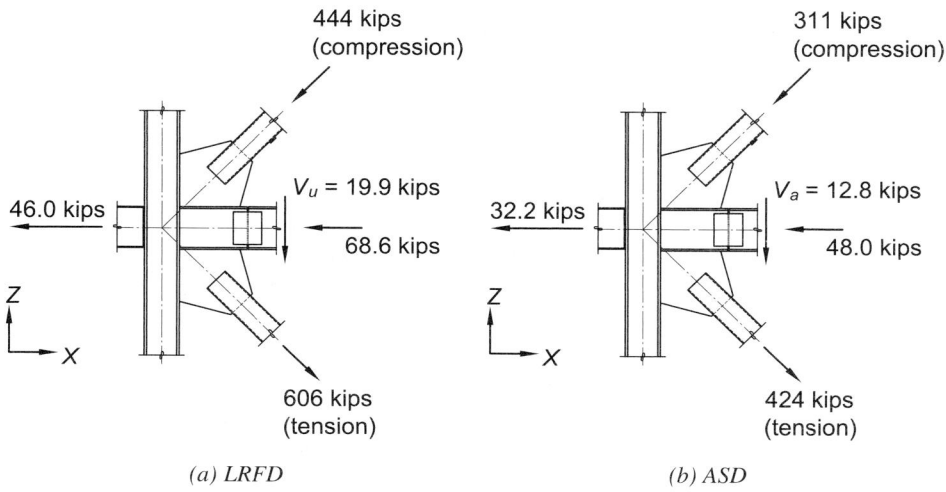

Fig. 5-36. Required strength of bracing connections according to AISC Seismic Provisions Section F2.3(i) for Design Case II.

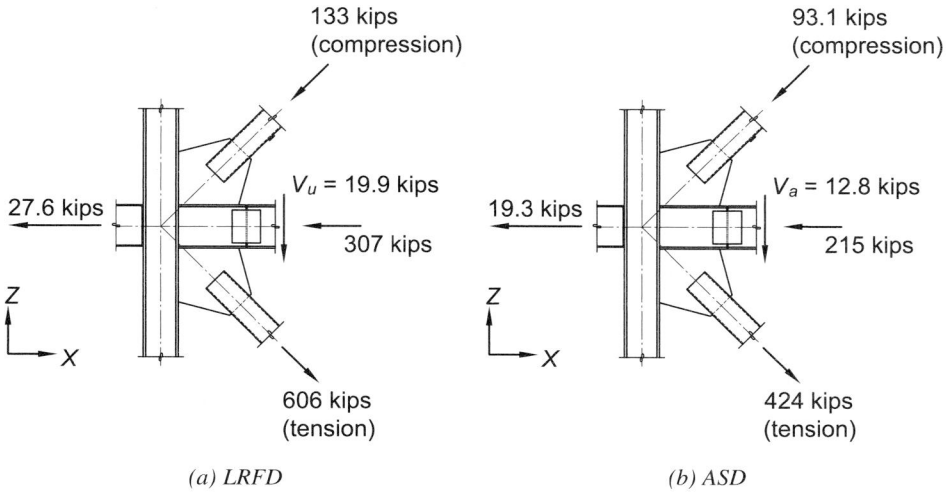

Fig. 5-37. Required strength of bracing connections according to AISC Seismic Provisions Section F2.3(ii) for Design Case II.

Design Case I controls. In the usual computer or manual analysis of this design problem where all members intersect at a common gravity axis work point, the beam does not participate in the carrying of this shear, and is designed for gravity loads and the axial load due to the mechanism analysis required by AISC *Seismic Provisions* Section F2.3. In reality, however, the beam participates with the gusset plates as the principal carrier of the shear due to the brace force vertical components. The total vertical shear in this case is 755 kips (LRFD) and 529 kips (ASD). As a "rule of thumb," the beam should be able to carry one half or more of this shear, plus the specified gravity shear, to avoid the need for doubler plates. The chosen W24×68 beam, with an available shear strength of 295 kips (LRFD) and 197 kips (ASD) from AISC *Manual* Table 3-6, will require doubler plates.

To avoid the use of doubler plates, use a W24×146. This is an increase in weight of (146 lb/ft − 68 lb/ft)(25 ft) = 1,950 lb.

Alternatively, the beam stubs shown in Figure 5-33 can use the heavier W24×146 section and the original W24×68 can be used between the splices. As yet another possibility, a continuous plate can be used in lieu of the W24×146 stub, and the W24×68 can be connected to this plate. This option is shown in Figure 5-43 as an illustration, without calculations. The option using the W24×146 and the W24×68 infill piece will be used here.

Brace-to-Gusset Connection Design

The design approach used here will follow that of Example 5.3.8, with single pass, ¼-in. field welds between the brace and the gusset. The weld length required is determined using AISC *Manual* Equations 8-2a and 8-2b. For the top gusset, the maximum force is 550 kips (LRFD) and 385 kips (ASD), thus:

LRFD	ASD
$\phi R_n = 1.392 Dl$	$\dfrac{R_n}{\Omega} = 0.928 Dl$
4 welds$(1.392 \text{ kip/in.}) Dl > P_u$	4 welds$(0.928 \text{ kip/in.}) Dl > P_a$
$l = \dfrac{550 \text{ kips}}{4 \text{ welds}(1.392 \text{ kip/in.})(4 \text{ sixteenths})}$	$l = \dfrac{385 \text{ kips}}{4 \text{ welds}(0.928 \text{ kip/in.})(4 \text{ sixteenths})}$
= 24.7 in.	= 25.9 in.

Use (4) 26-in.-long ¼-in. fillet welds to connect the brace above the beam to the gusset plate.

For the bottom gusset, the maximum force is 606 kips (LRFD) and 424 kips (ASD) and the required weld length is:

5.3 SPECIAL CONCENTRICALLY BRACED FRAMES

LRFD	ASD
4 welds$(1.392)Dl > P_u$ $$l = \frac{606 \text{ kips}}{4 \text{ welds}(1.392 \text{ kip/in.})(4 \text{ sixteenths})}$$ $= 27.2$ in.	4 welds$(0.928)Dl > P_u$ $$l = \frac{424 \text{ kips}}{4 \text{ welds}(0.928 \text{ kip/in.})(4 \text{ sixteenths})}$$ $= 28.6$ in.

Use (4) 29-in.-long ¼-in. fillet welds to connect the brace below the beam to the gusset plate.

Determine the minimum length, l, required for the brace-to-gusset lap

The limit state of shear rupture in the brace wall is used to determine the minimum brace-to-gusset lap length. Note that the expected brace rupture strength, $R_t F_u$, may be used in the determination of the available strength according to the User Note in AISC *Seismic Provisions* Section A3.2.

Using AISC *Specification* Section J4.2, including R_t from AISC *Seismic Provisions* Table A3.1:

$R_t = 1.3$

$R_n = 0.60 R_t F_u A_{nv}$ (from *Spec.* Eq. J4-4)

In this equation, A_{nv} is taken as the cross-sectional area of the four walls of the brace, $A_{nv} = 4lt_{des}$. Therefore:

$R_n = 0.60 R_t F_u (4lt_{des})$

Solving for the minimum lap length, *l*, for the brace above the beam:

LRFD	ASD
$$l \geq \frac{P_u}{\phi(0.60) R_t F_u (4t_{des})}$$ $$\geq \frac{550 \text{ kips}}{0.75(0.60)(1.3)(58 \text{ ksi})(4)(0.465 \text{ in.})}$$ ≥ 8.71 in.	$$l \geq \frac{\Omega P_a}{0.60 R_t F_u (4t_{des})}$$ $$\geq \frac{2.00(385 \text{ kips})}{0.60(1.3)(58 \text{ ksi})(4)(0.465 \text{ in.})}$$ ≥ 9.15 in.

The 26 in. required for the ¼-in. fillet weld controls.

Solving for the minimum lap length, *l*, for the brace below the beam:

LRFD	ASD
$l \geq \dfrac{606 \text{ kips}(8.71 \text{ in.})}{550 \text{ kips}}$ ≥ 9.60 in.	$l \geq \dfrac{424 \text{ kips}(9.15 \text{ in.})}{385 \text{ kips}}$ ≥ 10.1 in.

The 29 in. length required for the ¼-in. fillet welds controls.

Check that the bracing connection can accommodate brace buckling according to AISC Seismic Provisions Section F2.6c(3)

The requirements of AISC *Seismic Provisions* Section F2.6c(3) are met through the use of option (b)—rotation capacity. As explained in the User Note of that section and in the Commentary Figure C-F2.9, accommodation of inelastic rotation is accomplished with the brace terminating before the line of restraint. Figures 5-37 and 5-38 show the $2t$ clearance beyond the end of the brace.

The choice of a relatively small Whitmore section results in a tapered gusset, which is beneficial because it allows the brace to be located closer to the beam while still accommodating brace rotation by providing a $2t$ clearance according to AISC *Seismic Provisions* Section F2.6c(3) and Commentary.

Determine gusset plate thickness for the limit state of tensile yielding on the Whitmore section

To keep the gussets compact, choose an angle ϕ, as shown in Figure 5-38, of 20°. Example 5.3.8 used $\phi = 10°$, but in this example a smaller angle will result in shorter gusset interfaces and larger welds and may result in concentrated forces that cause yielding or crippling in the beam and column.

With $\phi = 20°$, the gusset thickness can be estimated.

For the top brace, the width of the gusset at the Whitmore section is:

$$w_p = D_{brace} + 2l \tan \phi$$
$$= 6.875 \text{ in.} + 2(26.0 \text{ in.})(\tan 20°)$$
$$= 25.8 \text{ in.}$$

Find the minimum gusset plate thickness based on the limit state of tensile yielding.

$$R_n = F_y A_g \hspace{4cm} (Spec. \text{ Eq. J4-1})$$
$$= F_y t_p w_p$$

5.3 SPECIAL CONCENTRICALLY BRACED FRAMES

LRFD	ASD
$t_p \geq \dfrac{P_u}{\phi F_y w_p}$	$t_p \geq \dfrac{\Omega P_a}{F_y w_p}$
$= \dfrac{550 \text{ kips}}{0.90(36 \text{ ksi})(25.8 \text{ in.})}$	$= \dfrac{1.67(385 \text{ kips})}{36 \text{ ksi}(25.8 \text{ in.})}$
$= 0.658$ in.	$= 0.692$ in.

Use a ¾-in.-thick gusset plate for the brace above the beam.

For the brace below the beam, the width of the gusset on the Whitmore section is:

$w_p = D_{brace} + 2l \tan \phi$

$= 7.50 \text{ in.} + 2(29.0 \text{ in.})(\tan 20°)$

$= 28.6 \text{ in.}$

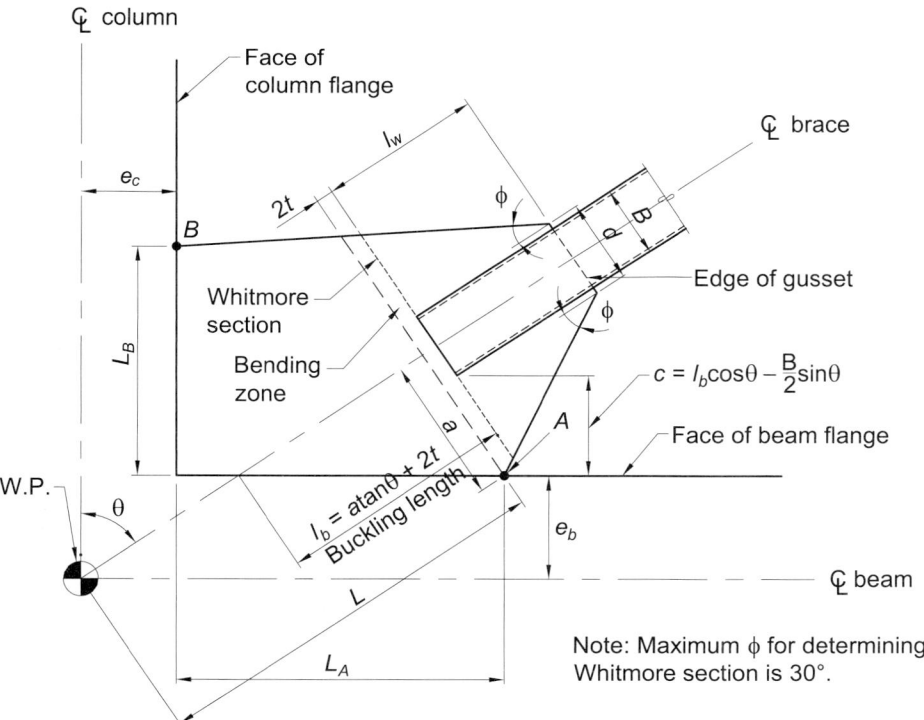

Fig. 5-38. Geometry of gusset to accommodate bending zone.

Find the minimum gusset plate thickness based on the limit state of tensile yielding.

$$R_n = F_y A_g$$
$$= F_y t_p w_p$$
(Spec. Eq. J4-1)

LRFD	ASD
$t_p \geq \dfrac{P_u}{\phi F_y w_p}$	$t_p \geq \dfrac{\Omega P_a}{F_y w_p}$
$= \dfrac{606 \text{ kips}}{0.90(36 \text{ ksi})(28.6 \text{ in.})}$	$= \dfrac{1.67(424 \text{ kips})}{36 \text{ ksi}(28.6 \text{ in.})}$
$= 0.654$ in.	$= 0.688$ in.

Use a ¾-in.-thick gusset plate for the brace below the beam.

Check block shear rupture of the gusset plate

The available strength for the limit state of block shear rupture is:

$$R_n = 0.60 F_u A_{nv} + U_{bs} F_u A_{nt} \leq 0.60 F_y A_{gv} + U_{bs} F_u A_{nt}$$
(Spec. Eq. J4-5)

For the gusset plate above the beam:

$$U_{bs} = 1.0$$

$$A_{nt} = 6.875 \text{ in.}(\text{¾ in.})$$
$$= 5.16 \text{ in.}^2$$

$$A_{gv} = A_{nv}$$
$$= 2(26.0 \text{ in.})(\text{¾ in.})$$
$$= 39.0 \text{ in.}^2$$

Since the gross shear area, A_{gv}, and the net shear area, A_{nv}, are equal in this case, the shear yielding component, $0.60 F_y A_{gv}$, is smaller than the shear rupture component, $0.60 F_u A_{nv}$.

$$0.60 F_y A_{gv} = 0.60(36 \text{ ksi})(39.0 \text{ in.}^2)$$
$$= 842 \text{ kips}$$

$$U_{bs} F_u A_{nt} = 1.0(58 \text{ ksi})(5.16 \text{ in.}^2)$$
$$= 299 \text{ kips}$$

5.3 SPECIAL CONCENTRICALLY BRACED FRAMES

LRFD	ASD
$\phi R_n = \phi(0.60 F_y A_{gv} + U_{bs} F_u A_{nt})$ $= 0.75(842 \text{ kips} + 299 \text{ kips})$ $= 856 \text{ kips} > 550 \text{ kips}$ **o.k.**	$\dfrac{R_n}{\Omega} = \dfrac{0.60 F_y A_{gv} + U_{bs} F_u A_{nt}}{\Omega}$ $= \dfrac{842 \text{ kips} + 299 \text{ kips}}{2.00}$ $= 571 \text{ kips} > 385 \text{ kips}$ **o.k.**

For the gusset plate below the beam:

$U_{bs} = 1.0$

$A_{nt} = 7.50 \text{ in.}(\sfrac{3}{4} \text{ in.})$
$\phantom{A_{nt}} = 5.63 \text{ in.}^2$

$A_{gv} = A_{nt}$
$\phantom{A_{gv}} = 2(29.0 \text{ in.})(\sfrac{3}{4} \text{ in.})$
$\phantom{A_{gv}} = 43.5 \text{ in.}^2$

$0.60 F_y A_{gv} = 0.60(36 \text{ ksi})(43.5 \text{ in.}^2)$
$\phantom{0.60 F_y A_{gv}} = 940 \text{ kips}$

$U_{bs} F_u A_{nt} = 1.0(58 \text{ ksi})(5.63 \text{ in.}^2)$
$\phantom{U_{bs} F_u A_{nt}} = 327 \text{ kips}$

LRFD	ASD
$\phi R_n = \phi(0.60 F_y A_{gv} + U_{bs} F_u A_{nt})$ $= 0.75(940 \text{ kips} + 327 \text{ kips})$ $= 950 \text{ kips} > 606 \text{ kips}$ **o.k.**	$\dfrac{R_n}{\Omega} = \dfrac{0.60 F_y A_{gv} + U_{bs} F_u A_{nt}}{\Omega}$ $= \dfrac{940 \text{ kips} + 327 \text{ kips}}{2.00}$ $= 634 \text{ kips} > 424 \text{ kips}$ **o.k.**

Determine gusset geometry

From Figure 5-38, the gusset geometry can be determined as follows:

$L = \sqrt{(e_b \tan\theta + a\sin\theta\tan\theta)^2 + (e_b + a\sin\theta)^2}$

$a = \dfrac{d}{2} + (l_w + 2t)\tan\phi$

$L_A = \dfrac{a}{\cos\theta} + e_b \tan\theta - e_c$

$L_B = (L + l_w + 2t)\cos\theta + \dfrac{d}{2}\sin\theta - e_b - \left[(L + l_w + 2t)\sin\theta - \dfrac{d}{2}\cos\theta - e_c\right]\tan(90° - \theta - \phi)$

For the gusset above the beam:

The width of the gusset edge, d, is 2.00 in. wider than the brace diameter to allow clearance between the HSS and the gusset corner, i.e., 1.00 in. on each side of the HSS.

$d = 6.875$ in. $+ 2.00$ in.
$\quad = 8.88$ in.

$e_b = \dfrac{24.7 \text{ in.}}{2}$
$\quad = 12.4$ in.

$e_c = \dfrac{12.7 \text{ in.}}{2}$
$\quad = 6.35$ in.

$\theta = 45°$

$\phi = 20°$

AISC *Seismic Provisions* Section F2.6c(3) requires that the brace connection accommodate the flexural forces or rotation imposed by brace buckling. This can be achieved either by option (a) designing the connection to have an available flexural strength of the expected brace flexural strength, $R_y M_p$, multiplied by 1.1 (LRFD) or by 1.1/1.5 (ASD) or option (b) providing rotation capacity to accommodate the required rotation. This brace configuration satisfies option (b) as it provides rotation capacity by providing the minimum $2t$ offset distance recommended in AISC *Seismic Provisions* Commentary Section F2.6c(3). Using a ¾-in.-thick gusset plate, $2t = 2(¾$ in.$) = 1.50$ in., but use 2.00 in. to allow for a possible gusset thickness increase as the calculations proceed. With $l_w = 26.0$ in.:

$a = \dfrac{d}{2} + (l_w + 2t)\tan\phi$

$\quad = \dfrac{8.88 \text{ in.}}{2} + (26.0 \text{ in.} + 2.00 \text{ in.})(\tan 20°)$

$\quad = 14.6$ in.

$L = \sqrt{(e_b \tan\theta + a\sin\theta \tan\theta)^2 + (e_b + a\sin\theta)^2}$

$\quad = \sqrt{[12.4 \text{ in.}(\tan 45°) + 14.6 \text{ in.}(\sin 45°)(\tan 45°)]^2 + [12.4 \text{ in.} + 14.6 \text{ in.}(\sin 45°)]^2}$

$\quad = 32.1$ in. Use $L = $ 2 ft 8⅛ in.

$L_A = \dfrac{a}{\cos\theta} + e_b \tan\theta - e_c$

$\quad = \dfrac{14.6 \text{ in.}}{\cos 45°} + 12.4 \text{ in.}(\tan 45°) - 6.35$ in.

$\quad = 26.7$ in. Use $L_A = 2$ ft 2¾ in.

$$L_B = (L + l_w + 2t)\cos\theta + \frac{d}{2}\sin\theta - e_b - \left[(L + l_w + 2t)\sin\theta - \frac{d}{2}\cos\theta - e_c\right]\tan(90° - \theta - \phi)$$

$$= (32.1 \text{ in.} + 26.0 \text{ in.} + 2.00 \text{ in.})(\cos 45°) + \frac{8.88 \text{ in.}}{2}(\sin 45°) - 12.4 \text{ in.}$$

$$-\left[(32.1 \text{ in.} + 26.0 \text{ in.} + 2.00 \text{ in.})(\sin 45°) - \frac{8.88 \text{ in.}}{2}(\cos 45°) - 6.35 \text{ in.}\right]$$

$$\times \tan(90° - 45° - 20°)$$

$$= 17.8 \text{ in.} \quad \text{Use } L_B = 1 \text{ ft } 5\% \text{ in.}$$

$$l_b = a\tan\theta + 2t$$
$$= 14.6 \text{ in.}(\tan 45°) + 2.00 \text{ in.}$$
$$= 16.6 \text{ in.}$$

For the bottom gusset:

$$d = 7.50 \text{ in.} + 2.00 \text{ in.}$$
$$= 9.50 \text{ in.}$$

$l_w = 29.0$ in.

$2t = 2.00$ in.

$$a = \frac{d}{2} + (l_w + 2t)\tan\phi$$
$$= \frac{9.50 \text{ in.}}{2} + (29.0 \text{ in.} + 2.00 \text{ in.})(\tan 20°)$$
$$= 16.0 \text{ in.}$$

$$L = \sqrt{(e_b\tan\theta + a\sin\theta\tan\theta)^2 + (e_b + a\sin\theta)^2}$$
$$= \sqrt{\left[12.4 \text{ in.}(\tan 45°) + 16.0 \text{ in.}(\sin 45°)(\tan 45°)\right]^2 + \left[12.4 \text{ in.} + 16.0 \text{ in.}(\sin 45°)\right]^2}$$
$$= 33.5 \text{ in.} \quad \text{Use } L = 2 \text{ ft } 9\frac{1}{2} \text{ in.}$$

$$L_A = \frac{a}{\cos\theta} + e_b\tan\theta - e_c$$
$$= \frac{16.0 \text{ in.}}{\cos 45°} + 12.4 \text{ in.}(\tan 45°) - 6.35 \text{ in.}$$
$$= 28.7 \text{ in.} \quad \text{Use } L_A = 2 \text{ ft } 4\frac{3}{4} \text{ in.}$$

$$L_B = (L + l_w + 2t)\cos\theta + \frac{d}{2}\sin\theta - e_b - \left[(L + l_w + 2t)\sin\theta - \frac{d}{2}\cos\theta - e_c\right]\tan(90° - \theta - \phi)$$

$$= (33.5 \text{ in.} + 29.0 \text{ in.} + 2.00 \text{ in.})(\cos 45°) + \frac{9.50 \text{ in.}}{2}(\sin 45°) - 12.4 \text{ in.}$$

$$-\left[(33.5 \text{ in.} + 29.0 \text{ in.} + 2.00 \text{ in.})(\sin 45°) - \frac{9.50 \text{ in.}}{2}(\cos 45°) - 6.35 \text{ in.}\right]$$

$$\times \tan(90° - 45° - 20°)$$

$$= 19.8 \text{ in.} \quad \text{Use } L_B = 1 \text{ ft } 7\% \text{ in.}$$

$$l_b = a\tan\theta + 2t$$
$$= 16.0 \text{ in.}(\tan 45°) + 2.00 \text{ in.}$$
$$= 18.0 \text{ in.}$$

This completes the gusset geometry, and the basic gusset geometry of Figure 5-33 can be generated.

Top Brace-to-Gusset Connection

The design of the top brace-to-gusset connection in Example 5.3.8 is very similar. The gusset plate there is 1¼ in. thick while it is ¾ in. thick here due to the wider gusset plate used in this example. For the limit state of tensile rupture of the brace, the check in Example 5.3.8 is adequate and need not be repeated here. Because the gusset plate is thinner in this example, less area is removed from the brace to accommodate the gusset plate.

Check the top gusset plate for buckling on the Whitmore section

Because the gusset geometry is different from the gusset in Example 5.3.8, gusset plate buckling must be investigated. Determine the available compressive strength using an effective length factor, $K = 0.6$, for the extended corner gusset, from Dowswell (2006). As noted in the beginning of Example 5.3.8, the additional 1.1 factor specified in AISC *Seismic Provisions* Section F2.6c(2) is applied to the expected brace strength for determining the required compressive strength based on buckling limit states.

$$\frac{Kl_b}{r} = \frac{0.6(16.6 \text{ in.})}{¾ \text{ in.}/\sqrt{12}}$$
$$= 46.0$$

From AISC *Manual* Table 4-22, with $KL/r = 46.0$, the available critical stress is:

LRFD	ASD
$\phi_c F_{cr} = 29.0$ ksi	$\dfrac{F_{cr}}{\Omega_c} = 19.3$ ksi

From AISC *Specification* Equation E3-1, using the width at the Whitmore section, the available compressive strength of the top gusset plate is:

LRFD	ASD
$\phi_c P_n = \phi_c F_{cr} A_g$ $= \phi_c F_{cr} t_p w_p$ $= 29.0 \text{ ksi}(¾ \text{ in.})(25.8 \text{ in.})$ $= 561$ kips	$\dfrac{P_n}{\Omega_c} = \dfrac{F_{cr} A_g}{\Omega_c}$ $= \dfrac{F_{cr} t_p w_p}{\Omega_c}$ $= 19.3 \text{ ksi}(¾ \text{ in.})(25.8 \text{ in.})$ $= 373$ kips

The required compressive strength of the gusset plate is:

LRFD	ASD
P_u = 1.1(444 kips) = 488 kips 561 kips > 488 kips **o.k.**	P_a = 1.1(311 kips) = 342 kips 373 kips > 342 kips **o.k.**

Bottom Brace-to-Gusset Connection

Check the brace effective net area

From AISC *Seismic Provisions* Section F2.5b(3), the brace effective net area, $A_{n(brace)}$, should not be less than the brace gross area, A_g. Thus the net area is:

$$A_n = A_g - 2[t_p + 2(gap)]t_{des}$$

Using a gap of $\frac{1}{16}$ in. on each side of the brace slot to allow clearance for erection:

$$A_n = 10.3 \text{ in.}^2 - 2[\tfrac{3}{4} \text{ in.} + 2(\tfrac{1}{16} \text{ in.})](0.465 \text{ in.})$$
$$= 9.49 \text{ in.}^2$$

From AISC *Specification* Table D3.1, Case 5, if l = 29.0 in. > 1.3(7.50 in.) = 9.75 in., U = 1.0, thus:

$$A_e = UA_n$$
$$= 1.0(9.49 \text{ in.}^2)$$
$$= 9.49 \text{ in.}^2$$

Since $A_e < A_g$, reinforcement is required. The approximate area of reinforcement required, A_{rn}, is the area removed, but the position of the reinforcement will reduce U to less than 1.0. The required area of reinforcement can be obtained from:

$$(A_n + A_{rn})U \geq A_g$$

Try U = 0.80, then:

$$A_{rn} = \frac{10.3 \text{ in.}^2}{0.80} - 9.49 \text{ in.}^2$$
$$= 3.39 \text{ in.}^2$$

Try two flat bars of ASTM A572 Grade 50 steel 1½ in. × 1½ in., with a total area of 2(1½ in.)² = 4.50 in.² With F_y = 50 ksi, ASTM A572 Grade 50 material satisfies the requirement in AISC *Seismic Provisions* Section F2.5b(3), that the yield strength of the reinforcement be at least the specified minimum yield strength of the member.

The arrangement is shown in Figure 5-39.

From Figure 5-39:

$$r_1 = \frac{D_{brace}}{2} - \frac{t_{des}}{2}$$
$$= \frac{7.50 \text{ in.}}{2} - \frac{0.465 \text{ in.}}{2}$$
$$= 3.52 \text{ in.}$$

$$r_2 = \frac{D_{brace}}{2} + \frac{1\frac{1}{2} \text{ in.}}{2}$$
$$= \frac{7.50 \text{ in.}}{2} + \frac{1\frac{1}{2} \text{ in.}}{2}$$
$$= 4.50 \text{ in.}$$

The distance to the centroid of a partial circle is given by:

$$\bar{x} = \frac{r_1 \sin\theta}{\theta}$$

where the total arc of the partial circle is 2θ, and θ is measured in radians. Although the brace is slightly less than a full half-circle because of the slot as shown in Figure 5-39, use an angle, θ, of $\pi/2$ for simplicity.

$$\bar{x}_{brace} = 3.52 \text{ in.} \left[\frac{\sin(\pi/2) \text{ rad}}{(\pi/2) \text{ rad}} \right]$$
$$= 2.24 \text{ in.}$$

$$\bar{x}_{re} = r_2$$
$$= 4.50 \text{ in.}$$

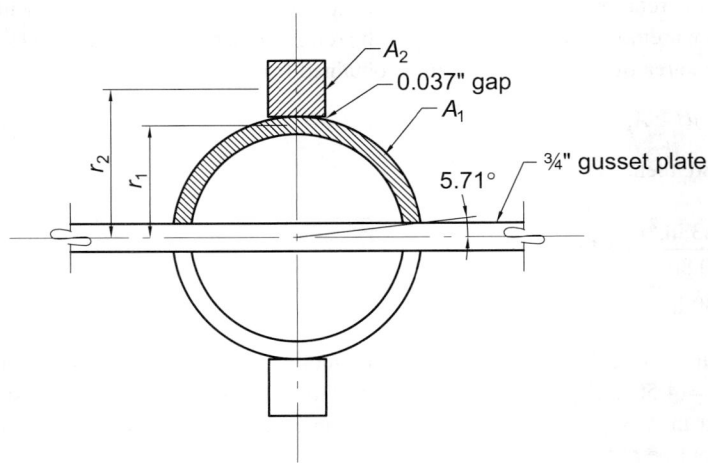

Fig. 5-39. Cross section of brace below the beam at net section.

5.3 SPECIAL CONCENTRICALLY BRACED FRAMES

Determine $\bar{x}$ for the composite cross section.

Part	$\bar{x}$ in.	A in.^{2}	$\bar{x}A$ in.^{3}
Half of brace	2.24	4.75	10.6
One Flat Bar	4.50	2.25	10.1
Σ	–	7.00	20.7

$$\bar{x} = \frac{\Sigma \bar{x} A}{\Sigma A}$$

$$= \frac{20.7 \text{ in.}^3}{7.00 \text{ in.}^2}$$

$$= 2.96 \text{ in.}$$

From AISC *Specification* Table D3.1, Case 2, which applies to HSS with reinforcement added:

$$U = 1 - \frac{\bar{x}}{l}$$

$$= 1 - \frac{2.96 \text{ in.}}{29.0 \text{ in.}}$$

$$= 0.898$$

$$A_n = A_{n(brace)} + A_{rn}$$

$$= 9.49 \text{ in.}^2 + 4.50 \text{ in.}^2$$

$$= 14.0 \text{ in.}^2$$

$$A_e = U A_n$$

$$= 0.898 \left(14.0 \text{ in.}^2\right)$$

$$= 12.6 \text{ in.}^2 > 10.3 \text{ in.}^2 \quad \textbf{o.k.}$$

Design welds connecting flat bars to brace

According to AISC *Seismic Provisions* Section F2.5b(3)(ii), the flat bar is connected to the HSS brace to develop the expected strength of the flat bar on each side of the reduced section (the expected yield strength, $R_y F_y$, is used here). The reduced section is the length of the HSS from the extent of the slot (dimension x of Figure 5-33) to the start of the HSS-to-gusset weld. The required strength of the weld is based on the expected flat bar yield strength using R_y from AISC *Seismic Provisions* Table A3.1 for ASTM A572 Grade 50 bars. For ASD, use 1.0/1.5 of the expected strength of the flat bar reinforcement.

LRFD	ASD
$R_y F_y A_{fb} = 1.1(50 \text{ ksi})\left(2.25 \text{ in.}^2\right)$ $= 124 \text{ kips}$	$R_y F_y A_{fb}/1.5 = 1.1(50 \text{ ksi})\left(2.25 \text{ in.}^2\right)/1.5$ $= 82.5 \text{ kips}$

Using two single pass ¼-in. fillet welds, from AISC *Manual* Equations 8-2a and 8-2b, the weld length required is:

LRFD	ASD
$1.392Dl > P_u$ $$l_w = \frac{124 \text{ kips}}{2 \text{ welds}(1.392 \text{ kip/in.})(4 \text{ sixteenths})}$$ $= 11.1$ in.	$0.928Dl > P_u$ $$l_w = \frac{82.5 \text{ kips}}{2 \text{ welds}(0.928 \text{ kip/in.})(4 \text{ sixteenths})}$$ $= 11.1$ in.

Use (2) 11½-in.-long ¼-in. fillet welds on each side of the reduced section of the brace.

Since the gap between the edge of the 1½ in. × 1½ in. flat bar and the brace is 0.037 in., as shown in Figure 5-39, and is less than ¹⁄₁₆ in. (see AWS D1.1 clause 5.22.1), the ¼-in. fillet welds are adequate. Note that the flat bar reinforcement needs to extend 11½ in. on each side of the end of the actual slot, which includes the dimension x that may be required for erection.

Check the bottom gusset plate for buckling on the Whitmore section

Determine the available compressive strength using an effective length factor, $K = 0.6$, for the extended corner gusset, from Dowswell (2006). As noted in the beginning of Example 5.3.8, the additional 1.1 factor specified in AISC *Seismic Provisions* Section F2.6c(2) is applied to compression buckling limit states.

$$\frac{Kl_b}{r} = \frac{0.6(18.0 \text{ in.})}{\text{¾ in.}/\sqrt{12}}$$
$$= 49.9$$

From AISC *Manual* Table 4-22, with $KL/r = 49.9$, the available critical stress is:

LRFD	ASD
$\phi_c F_{cr} = 28.4$ ksi	$\dfrac{F_{cr}}{\Omega_c} = 18.9$ ksi

From AISC *Specification* Equation E3-1, using the width at the Whitmore section, the available compressive strength of the bottom gusset plate is:

LRFD	ASD
$\phi_c P_n = \phi_c F_{cr} A_g$ $= \phi_c F_{cr} t_p w_p$ $= 28.4$ ksi$(¾$ in.$)(28.6$ in.$)$ $= 609$ kips	$\dfrac{P_n}{\Omega_c} = \dfrac{F_{cr} A_g}{\Omega_c}$ $= \dfrac{F_{cr} t_p w_p}{\Omega_c}$ $= 18.9$ ksi$(¾$ in.$)(28.6$ in.$)$ $= 405$ kips

5.3 SPECIAL CONCENTRICALLY BRACED FRAMES

The required compressive strength of the gusset plate is:

LRFD	ASD
$P_u = 1.1(518 \text{ kips})$ $= 570$ kips $609 \text{ kips} > 570 \text{ kips}$ **o.k.**	$P_a = 1.1(363 \text{ kips})$ $= 399$ kips $405 \text{ kips} > 399 \text{ kips}$ **o.k.**

Connection Interface Forces

The Uniform Force Method (UFM) requires that a constraint on the locations of the interface centroids be satisfied in order to eliminate moments on the gusset-to-beam and gusset-to-column interfaces, M_b and M_c, respectively. When this constraint is not satisfied, moments will be introduced on the connection interfaces. This is discussed in Part 13 of the AISC *Manual*, and the terminology used there is repeated here. Let $\bar{\alpha}$ and $\bar{\beta}$ represent the distance from the column flange to the actual centroids of the gusset-to-beam and gusset-to-column connections, respectively. When the calculated $\alpha > \bar{\alpha}$ or the calculated $\beta > \bar{\beta}$, the additional shear induced in the beam or column due to the moment may add to the shear, V_b, in the beam and H_c in the column. Thus, for the beam:

When $\alpha > \bar{\alpha}$:

$$\text{Total beam shear} = \max\left\{\left|\frac{V_b}{2}\right| + \left|\frac{M_b}{\bar{\alpha}-s}\right|, |V_b|\right\} + R_b$$

When $\alpha < \bar{\alpha}$:

$$\text{Total beam shear} = \max\left\{\left|\frac{V_b}{2}\right| - \left|\frac{M_b}{\bar{\alpha}-s}\right|, |V_b|\right\} + R_b$$

where
 R_b = beam end reaction
 s = snip size in the gusset where the top flange of the beam connects to the column flange

For the column:

When $\beta > \bar{\beta}$:

$$\text{Total column shear} = \max\left\{\left|\frac{H_c}{2}\right| + \left|\frac{M_c}{\bar{\beta}-s}\right|, |H_c|\right\}$$

When $\beta < \bar{\beta}$:

$$\text{Total column shear} = \max\left\{\left|\frac{H_c}{2}\right| - \left|\frac{M_c}{\bar{\beta}-s}\right|, |H_c|\right\}$$

In nonseismic and low-seismic design, this is not an issue because the brace forces are more closely matched to the beam and column sizes and calculated loads are used. In some structures detailed for seismic resistance, the connections are not designed for calculated loads but rather must be designed for the expected tensile strength of the brace, $R_y F_y A_g$. This is normally larger than the actual design load from the applicable building code. For instance, the HSS6.875×0.500 brace would normally be designed for point-to-point buckling with a length of $\sqrt{(12.5 \text{ ft})^2 + (12.5 \text{ ft})^2} = 17.7$ ft. The available compressive strength of this brace is 207 kips (LRFD) and 137 kips (ASD) from AISC *Manual* Table 4-5, and the actual brace load will be less than this. But, we are designing the connections of this member for 550 kips (LRFD) and 385 kips (ASD), which is at least 550 kips/207 kips = 2.7 times the maximum possible required strength. This puts a great demand on the gusset, beam and column which must be accommodated. So, it is important to distribute this high demand in the most optimal manner.

Top Gusset—Design Case I

From the geometry shown in Figure 5-33 and the Uniform Force Method variables in AISC *Manual* Part 13:

$$\bar{\alpha} = \frac{26\tfrac{3}{4} \text{ in.} - 1.00 \text{ in.}}{2} + 1.00 \text{ in.}$$

$$= 13.9 \text{ in.}$$

$$\bar{\beta} = \frac{17\tfrac{7}{8} \text{ in.} - 1.00 \text{ in.}}{2} + 1.00 \text{ in.}$$

$$= 9.45 \text{ in.}$$

Choosing $\beta = \bar{\beta} = 9.45$ in., the constraint between α and β given by AISC *Manual* Equation 13-1, $\alpha - \beta \tan\theta = e_b \tan\theta = e_c$, gives:

$$\alpha = \beta \tan\theta + e_b \tan\theta - e_c$$

$$= 9.45 \text{ in.}(\tan 45°) + 12.4 \text{ in.}(\tan 45°) - 6.35 \text{ in.}$$

$$= 15.5 \text{ in.}$$

Since $\alpha > \bar{\alpha}$, the moment $M_b = V_b(\alpha - \bar{\alpha})$ may add to the beam shear. Choose $\alpha = \bar{\alpha} = 13.9$ in., then:

$$\alpha - \beta \tan\theta = e_b \tan\theta - e_c \quad \text{(\textit{Manual} Eq. 13-1)}$$

$$13.9 \text{ in.} - \beta(\tan 45°) = 12.4 \text{ in.}(\tan 45°) - 6.35 \text{ in.}$$

$$\beta = 7.85 \text{ in.} < \bar{\beta} = 9.45 \text{ in.}$$

The column shear will not be increased by the moment $M_c = H_c(\bar{\beta} - \beta)$ because $\beta \leq \bar{\beta}$. Therefore, use $\alpha = \bar{\alpha} = 13.9$ in. and $\beta = 7.85$ in. Then:

$$r = \sqrt{(\alpha + e_c)^2 + (\beta + e_b)^2} \quad \text{(\textit{Manual} Eq. 13-6)}$$

$$= \sqrt{(13.9 \text{ in.} + 6.35 \text{ in.})^2 + (7.85 \text{ in.} + 12.4 \text{ in.})^2}$$

$$= 28.6 \text{ in.}$$

5.3 SPECIAL CONCENTRICALLY BRACED FRAMES

The controlling brace forces for the top gusset interface forces are:

LRFD	ASD
From AISC *Manual* Equation 13-3: $$H_{uc} = \frac{e_c}{r} P_u$$ $$= \frac{6.35 \text{ in.}}{28.6 \text{ in.}}(550 \text{ kips})$$ $$= 122 \text{ kips}$$	From AISC *Manual* Equation 13-3: $$H_{ac} = \frac{e_c}{r} P_a$$ $$= \frac{6.35 \text{ in.}}{28.6 \text{ in.}}(385 \text{ kips})$$ $$= 85.5 \text{ kips}$$
From AISC *Manual* Equation 13-5: $$H_{ub} = \frac{\alpha}{r} P_u$$ $$= \frac{13.9 \text{ in.}}{28.6 \text{ in.}}(550 \text{ kips})$$ $$= 267 \text{ kips}$$	From AISC *Manual* Equation 13-5: $$H_{ab} = \frac{\alpha}{r} P_a$$ $$= \frac{13.9 \text{ in.}}{28.6 \text{ in.}}(385 \text{ kips})$$ $$= 187 \text{ kips}$$
From AISC *Manual* Equation 13-2: $$V_{uc} = \frac{\beta}{r} P_u$$ $$= \frac{7.85 \text{ in.}}{28.6 \text{ in.}}(550 \text{ kips})$$ $$= 151 \text{ kips}$$	From AISC *Manual* Equation 13-2: $$V_{ac} = \frac{\beta}{r} P_a$$ $$= \frac{7.85 \text{ in.}}{28.6 \text{ in.}}(385 \text{ kips})$$ $$= 106 \text{ kips}$$
From AISC *Manual* Equation 13-4: $$V_{ub} = \frac{e_b}{r} P_u$$ $$= \frac{12.4 \text{ in.}}{28.6 \text{ in.}}(550 \text{ kips})$$ $$= 238 \text{ kips}$$	From AISC *Manual* Equation 13-4: $$V_{ab} = \frac{e_b}{r} P_a$$ $$= \frac{12.4 \text{ in.}}{28.6 \text{ in.}}(385 \text{ kips})$$ $$= 167 \text{ kips}$$
From AISC *Manual* Equation 13-19: $$M_{uc} = H_{uc}(\overline{\beta} - \beta)$$ $$= 122 \text{ kips}(9.45 \text{ in.} - 7.85 \text{ in.})$$ $$= 195 \text{ kip-in.}$$	From AISC *Manual* Equation 13-19: $$M_{ac} = H_{ac}(\overline{\beta} - \beta)$$ $$= 85.5 \text{ kips}(9.45 \text{ in.} - 7.85 \text{ in.})$$ $$= 137 \text{ kip-in.}$$

Note that the sum of the horizontal gusset forces must equal the brace horizontal component. The sum of the vertical gusset forces must equal the brace vertical component.

Bottom Gusset—Design Case I

From the geometry shown in Figure 5-33:

$$\bar{\alpha} = \frac{28.8 \text{ in.} - 1.00 \text{ in.}}{2} + 1.00 \text{ in.}$$
$$= 14.9 \text{ in.}$$

$$\bar{\beta} = \frac{19.9 \text{ in.} - 1.00 \text{ in.}}{2} + 1.00 \text{ in.}$$
$$= 10.5 \text{ in.}$$

Choose $\alpha = \bar{\alpha} = 14.9$ in., then:

$$\alpha - \beta \tan\theta = e_b \tan\theta - e_c \quad \text{(Manual Eq. 13-1)}$$
$$14.9 \text{ in.} - \beta(\tan 45°) = 12.4 \text{ in.}(\tan 45°) - 6.35 \text{ in.}$$
$$\beta = 8.85 \text{ in.} < \bar{\beta} = 10.5 \text{ in.} \quad \textbf{o.k.}$$

Use $\alpha = \bar{\alpha} = 14.9$ in. and $\beta = 8.85$ in.

$$r = \sqrt{(\alpha + e_c)^2 + (\beta + e_b)^2} \quad \text{(Manual Eq. 13-6)}$$
$$= \sqrt{(14.9 \text{ in.} + 6.35 \text{ in.})^2 + (8.85 \text{ in.} + 12.4 \text{ in.})^2}$$
$$= 30.1 \text{ in.}$$

LRFD	ASD
From AISC *Manual* Equation 13-3: $$H_{uc} = \frac{e_c}{r} P_u$$ $$= \frac{6.35 \text{ in.}}{30.1 \text{ in.}}(518 \text{ kips})$$ $$= 109 \text{ kips}$$	From AISC *Manual* Equation 13-3: $$H_{ac} = \frac{e_c}{r} P_a$$ $$= \frac{6.35 \text{ in.}}{30.1 \text{ in.}}(363 \text{ kips})$$ $$= 76.6 \text{ kips}$$
From AISC *Manual* Equation 13-5: $$H_{ub} = \frac{\alpha}{r} P_u$$ $$= \frac{14.9 \text{ in.}}{30.1 \text{ in.}}(518 \text{ kips})$$ $$= 256 \text{ kips}$$	From AISC *Manual* Equation 13-5: $$H_{ab} = \frac{\alpha}{r} P_a$$ $$= \frac{14.9 \text{ in.}}{30.1 \text{ in.}}(363 \text{ kips})$$ $$= 180 \text{ kips}$$

5.3 SPECIAL CONCENTRICALLY BRACED FRAMES

LRFD	ASD
From AISC *Manual* Equation 13-2: $$V_{uc} = \frac{\beta}{r}P_u$$ $$= \frac{8.85 \text{ in.}}{30.1 \text{ in.}}(518 \text{ kips})$$ $$= 152 \text{ kips}$$	From AISC *Manual* Equation 13-2: $$V_{ac} = \frac{\beta}{r}P_a$$ $$= \frac{8.85 \text{ in.}}{30.1 \text{ in.}}(363 \text{ kips})$$ $$= 107 \text{ kips}$$
From AISC *Manual* Equation 13-4: $$V_{ub} = \frac{e_b}{r}P_u$$ $$= \frac{12.4 \text{ in.}}{30.1 \text{ in.}}(518 \text{ kips})$$ $$= 213 \text{ kips}$$	From AISC *Manual* Equation 13-4: $$V_{ab} = \frac{e_b}{r}P_a$$ $$= \frac{12.4 \text{ in.}}{30.1 \text{ in.}}(363 \text{ kips})$$ $$= 150 \text{ kips}$$
From AISC *Manual* Equation 13-19: $$M_{uc} = H_{uc}(\overline{\beta} - \beta)$$ $$= 109 \text{ kips}(10.5 \text{ in.} - 8.85 \text{ in.})$$ $$= 180 \text{ kip-in.}$$	From AISC *Manual* Equation 13-19: $$M_{ac} = H_{ac}(\overline{\beta} - \beta)$$ $$= 76.6 \text{ kips}(10.5 \text{ in.} - 8.85 \text{ in.})$$ $$= 126 \text{ kip-in.}$$

Figures 5-40a and 5-40b show the force distribution for Design Case I. The total column shear when $\beta < \overline{\beta}$ is discussed in the previous Connection Interface Forces section. In this example, the column shear, H_c, is greater than the combined shear, $\left|\frac{H_c}{2}\right| - \left|\frac{M_c}{\overline{\beta}-s}\right|$. Therefore, Figures 5-40a and 5-40b show only the H_c forces.

Top gusset—Design Case II

The geometry is the same as Design Case I, only the loads have changed in magnitude and direction as shown in Figures 5-36 and 5-37.

LRFD	ASD
From AISC *Manual* Equation 13-3: $$H_{uc} = \frac{e_c}{r}P_u$$ $$= \frac{6.35 \text{ in.}}{28.6 \text{ in.}}(444 \text{ kips})$$ $$= 98.6 \text{ kips}$$	From AISC *Manual* Equation 13-3: $$H_{ac} = \frac{e_c}{r}P_a$$ $$= \frac{6.35 \text{ in.}}{28.6 \text{ in.}}(311 \text{ kips})$$ $$= 69.1 \text{ kips}$$

LRFD	ASD
From AISC *Manual* Equation 13-5:	From AISC *Manual* Equation 13-5:
$H_{ub} = \dfrac{\alpha}{r} P_u$	$H_{ab} = \dfrac{\alpha}{r} P_a$
$= \dfrac{13.9 \text{ in.}}{28.6 \text{ in.}} (444 \text{ kips})$	$= \dfrac{13.9 \text{ in.}}{28.6 \text{ in.}} (311 \text{ kips})$
$= 216 \text{ kips}$	$= 151 \text{ kips}$
From AISC *Manual* Equation 13-2:	From AISC *Manual* Equation 13-2:
$V_{uc} = \dfrac{\beta}{r} P_u$	$V_{ac} = \dfrac{\beta}{r} P_a$
$= \dfrac{7.85 \text{ in.}}{28.6 \text{ in.}} (444 \text{ kips})$	$= \dfrac{7.85 \text{ in.}}{28.6 \text{ in.}} (311 \text{ kips})$
$= 122 \text{ kips}$	$= 85.4 \text{ kips}$

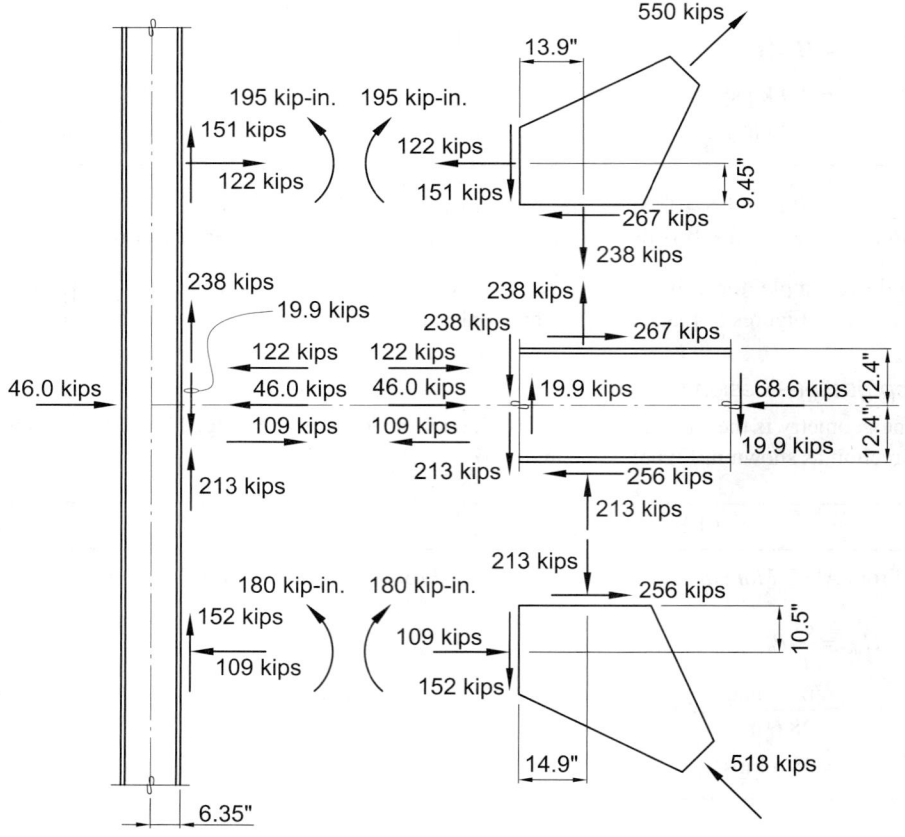

Fig. 5-40a. Design Case I gusset interface forces (LRFD).

5.3 SPECIAL CONCENTRICALLY BRACED FRAMES

LRFD	ASD
From AISC *Manual* Equation 13-4: $$V_{ub} = \frac{e_b}{r} P_u$$ $$= \frac{12.4 \text{ in.}}{28.6 \text{ in.}}(444 \text{ kips})$$ $$= 193 \text{ kips}$$ From AISC *Manual* Equation 13-19: $$M_{uc} = H_{uc}(\overline{\beta} - \beta)$$ $$= 98.6 \text{ kips}(9.45 \text{ in.} - 7.85 \text{ in.})$$ $$= 158 \text{ kip-in.}$$	From AISC *Manual* Equation 13-4: $$V_{ab} = \frac{e_b}{r} P_a$$ $$= \frac{12.4 \text{ in.}}{28.6 \text{ in.}}(311 \text{ kips})$$ $$= 135 \text{ kips}$$ From AISC *Manual* Equation 13-19: $$M_{ac} = H_{ac}(\overline{\beta} - \beta)$$ $$= 69.1 \text{ kips}(9.45 \text{ in.} - 7.85 \text{ in.})$$ $$= 111 \text{ kip-in.}$$

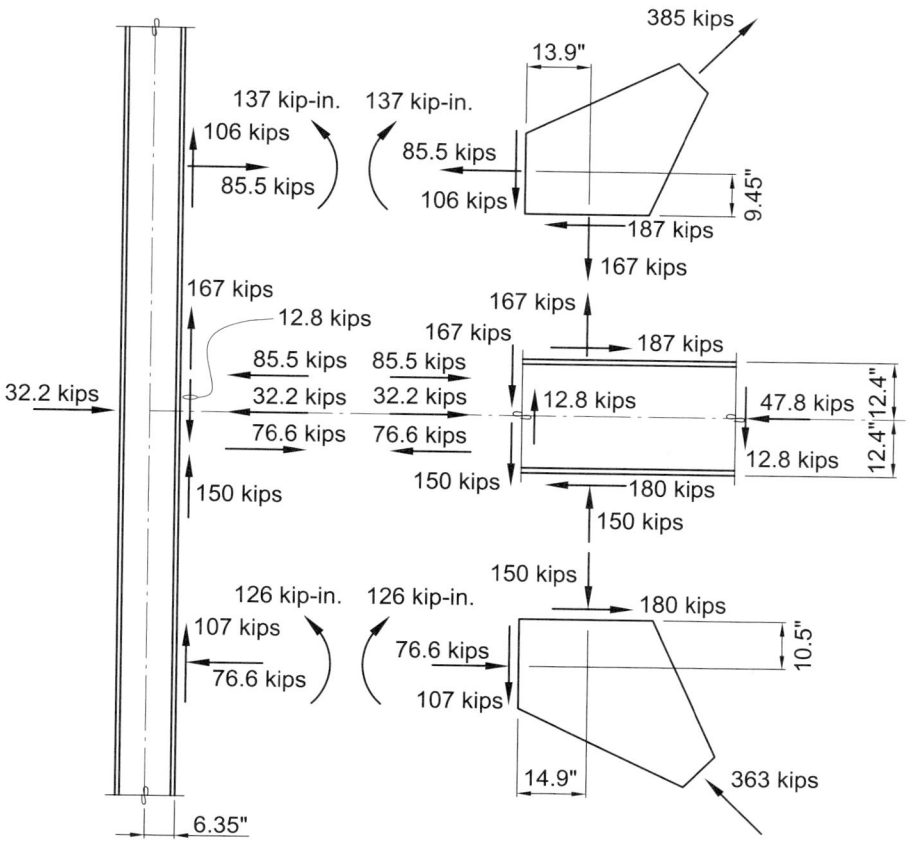

Fig 5-40b. Design Case I gusset interface forces (ASD).

Gusset Below the Beam—Design Case II

LRFD	ASD
From AISC *Manual* Equation 13-3:	From AISC *Manual* Equation 13-3:
$H_{uc} = \dfrac{e_c}{r} P_u$	$H_{ac} = \dfrac{e_c}{r} P_a$
$= \dfrac{6.35 \text{ in.}}{30.1 \text{ in.}} (606 \text{ kips})$	$= \dfrac{6.35 \text{ in.}}{30.1 \text{ in.}} (424 \text{ kips})$
$= 128 \text{ kips}$	$= 89.4 \text{ kips}$
From AISC *Manual* Equation 13-5:	From AISC *Manual* Equation 13-5:
$H_{ub} = \dfrac{\alpha}{r} P_u$	$H_{ab} = \dfrac{\alpha}{r} P_a$
$= \dfrac{14.9 \text{ in.}}{30.1 \text{ in.}} (606 \text{ kips})$	$= \dfrac{14.9 \text{ in.}}{30.1 \text{ in.}} (424 \text{ kips})$
$= 300 \text{ kips}$	$= 210 \text{ kips}$
From AISC *Manual* Equation 13-2:	From AISC *Manual* Equation 13-2:
$V_{uc} = \dfrac{\beta}{r} P_u$	$V_{ac} = \dfrac{\beta}{r} P_a$
$= \dfrac{8.85 \text{ in.}}{30.1 \text{ in.}} (606 \text{ kips})$	$= \dfrac{8.85 \text{ in.}}{30.1 \text{ in.}} (424 \text{ kips})$
$= 178 \text{ kips}$	$= 125 \text{ kips}$
From AISC *Manual* Equation 13-4:	From AISC *Manual* Equation 13-4:
$V_{ub} = \dfrac{e_b}{r} P_u$	$V_{ab} = \dfrac{e_b}{r} P_a$
$= \dfrac{12.4 \text{ in.}}{30.1 \text{ in.}} (606 \text{ kips})$	$= \dfrac{12.4 \text{ in.}}{30.1 \text{ in.}} (424 \text{ kips})$
$= 250 \text{ kips}$	$= 175 \text{ kips}$
From AISC *Manual* Equation 13-19:	From AISC *Manual* Equation 13-19:
$M_{uc} = H_{uc}(\bar{\beta} - \beta)$	$M_{ac} = H_{ac}(\bar{\beta} - \beta)$
$= 128 \text{ kips}(10.5 \text{ in.} - 8.85 \text{ in.})$	$= 89.4 \text{ kips}(10.5 \text{ in.} - 8.85 \text{ in.})$
$= 211 \text{ kip-in.}$	$= 148 \text{ kip-in.}$

Figures 5-41a and 5-41b show the force distribution for Design Case II. The total column shear when $\beta < \bar{\beta}$ is discussed in the previous Connection Interface Forces section.

In this example, the column shear, H_c, is greater than the combined shear, $\left| \dfrac{H_c}{2} \right| - \left| \dfrac{M_c}{\bar{\beta} - s} \right|$. Therefore, Figures 5-41a and 5-41b show only the H_c forces.

5.3 SPECIAL CONCENTRICALLY BRACED FRAMES

Each of the Design Cases I and II has a subsidiary case in which the compression brace post-buckling strength is considered. This affects the design of the main members but not, in this case, the gusset connection.

Ductility Requirements

AISC *Seismic Provisions* Section F2.6b and Commentary require that connections that involve a beam, a column and a brace satisfy option (a) or (b) in that section. This example will use option (a)—a simple beam-to-column connection.

To satisfy option (a), a splice can be provided in the beam just outside of the connection region as is done in this example. If the beam splice were a perfect pin, then $(1.1R_yM_p)_{splice} = 0$. As long as the splice can accommodate 0.025 rad of rotation without binding (i.e., no fouling of parts), AISC *Seismic Provisions* Section F2.6b(a) will be satisfied. The simple connections presented in Parts 9 and 10 of the AISC *Manual* are deemed to comply with Section F2.6b(a).

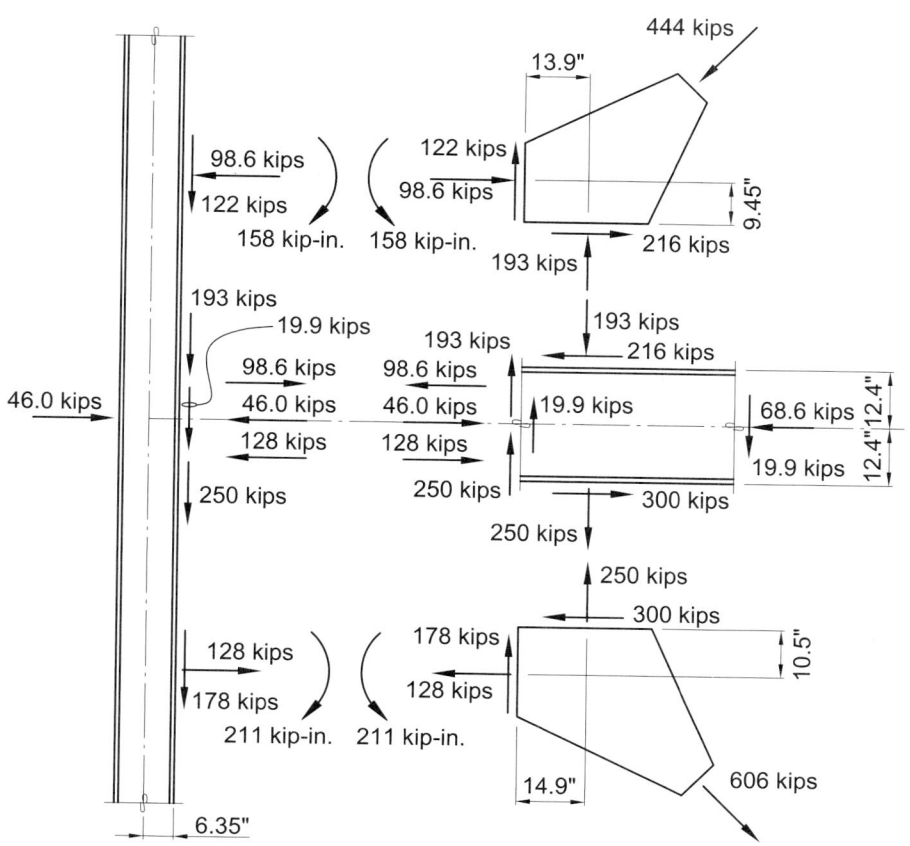

Fig. 5-41a. Design Case II gusset interface forces (LRFD).

Design of Splice

Since the beam splice is in a main member, the design of the member needs to be considered. In normal design practice, the beam will be considered to be continuous from column-to-column. In this example, a splice is inserted 2 ft 6½ in. from the face of the column. This splice must carry the loads that would exist at this point in the continuous beam. Therefore, the splice must carry the beam shear, beam axial force, and a moment equal to the beam shear times 2 ft 6½ in. This moment is the moment that would exist in the beam as designed, without the splice.

The extended shear tab presented in Part 10 of the AISC *Manual* will be used for the splice design. The splice must be designed before it can be checked for ductility.

From Example 5.3.5 the required strength of the beam, and therefore the splice connection, is as follows. These forces are also shown in Figures 5-35 and 5-37.

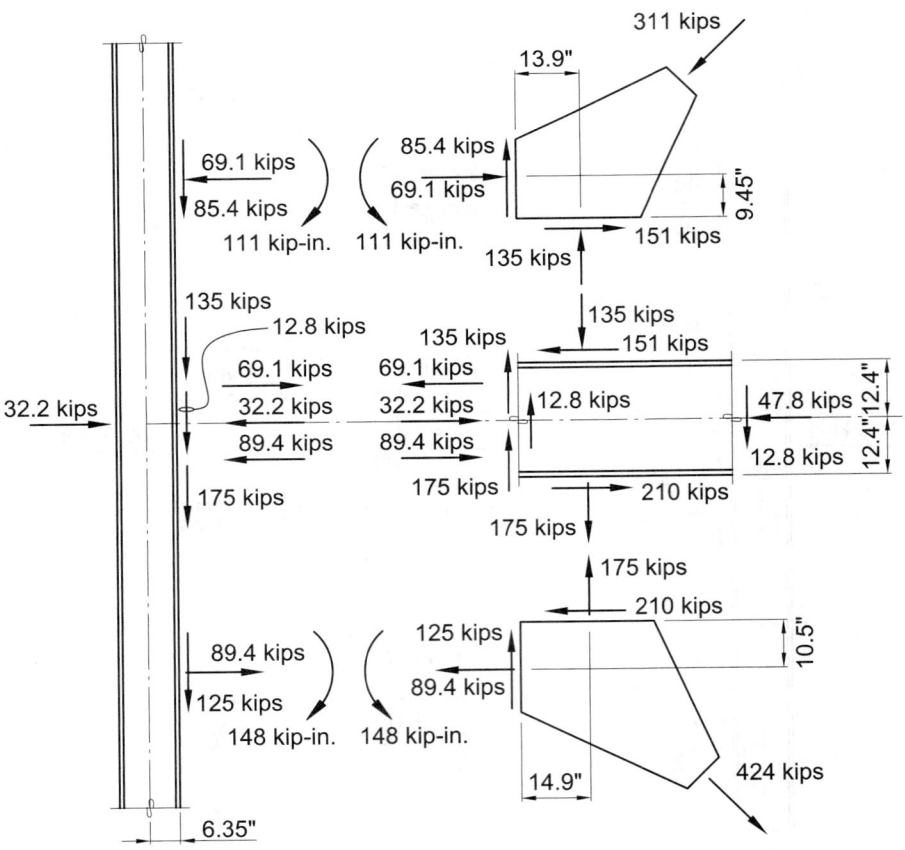

Fig. 5-41b. Design Case II gusset interface forces (ASD).

5.3 SPECIAL CONCENTRICALLY BRACED FRAMES

LRFD	ASD
Shear: $V_u = 19.9$ kips (gravity)	Shear: $V_a = 12.8$ kips (gravity)
Axial: $P_u = 307$ kips (compression)	Axial: $P_a = 215$ kips (compression)

Figure 5-42 shows the beam splice (one plate on the near face and one on the far face). In order to prevent binding at a 0.025 rad story drift, the clearance between the beam and the beam stub at the splice must be at least 12.4 in.(0.025 rad) = 0.310 in., where 12.4 in. is half of the W24×146 beam stub depth. Use a 1.00 in. clearance as shown in Figure 5-33.

The splice is on the beam centerline. Use ⅞-in.-diameter ASTM A325-X bolts.

For gravity load alone—the connection to the W24×68 is designed as follows:

The gravity shear force needs to be delivered from the centroid of the W24×68 bolt group to the face of the column. Therefore:

$e_x = 30.5$ in. $+ (½$ in. $+ 2$ in. $+ 1½$ in.$)$
$\quad = 34.5$ in.

Interpolating from AISC *Manual* Table 7-7 for angle = 0° with $s = 3$ in., $e_x = 34.5$ in., and $n = 6$:

$C = 1.56$

Using AISC *Manual* Table 7-1 for ⅞-in.-diameter ASTM A325-X (Group A) bolts in double shear, the available shear strength is:

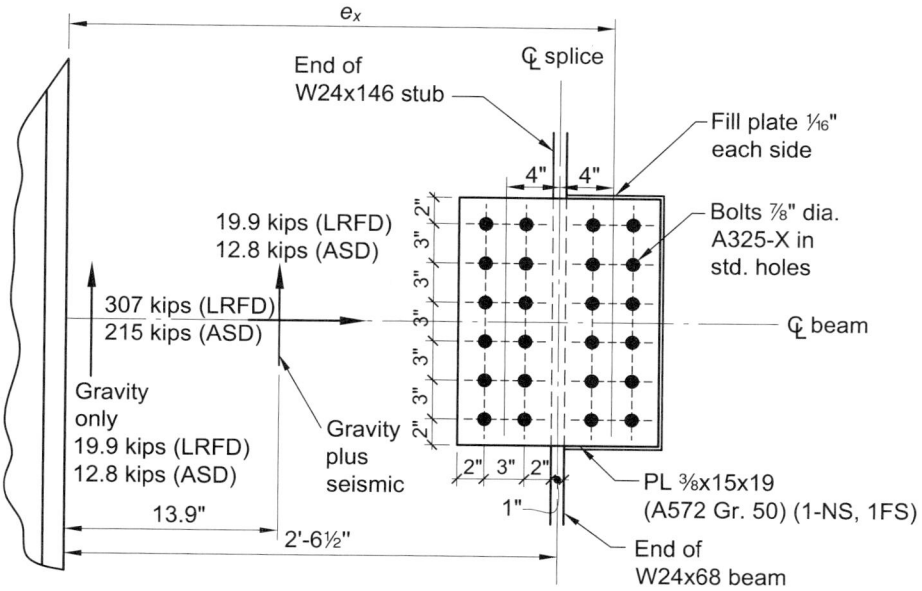

Fig. 5-42. Beam splice.

LRFD	ASD
$\phi r_n = 61.3$ kips	$\dfrac{r_n}{\Omega} = 40.9$ kips
$\phi R_n = C\phi r_n$	$\dfrac{R_n}{\Omega} = C\left(\dfrac{r_n}{\Omega}\right)$
$= 1.56(61.3 \text{ kips})$	$= 1.56(40.9 \text{ kips})$
$= 95.6$ kips > 19.9 kips **o.k.**	$= 63.8$ kips > 12.8 kips **o.k.**

For gravity load alone—the connection to the W24×146 is designed as follows:

The gravity shear force needs to be delivered from the centroid of the W24×146 bolt group to the face of the column. Therefore:

$e_x = 30.5$ in. $- 4.00$ in.
   $= 26.5$ in.

Interpolating from AISC *Manual* Table 7-7 for angle $= 0°$ with $s = 3$ in., $e_x = 26.5$ in., and $n = 6$:

$C = 2.02$

The available shear strength of the W24×146 bolts is:

LRFD	ASD
$\phi R_n = C\phi r_n$	$\dfrac{R_n}{\Omega} = C\left(\dfrac{r_n}{\Omega}\right)$
$= 2.02(61.3 \text{ kips})$	$= 2.02(40.9 \text{ kips})$
$= 124$ kips > 19.9 kips **o.k.**	$= 82.6$ kips > 12.8 kips **o.k.**

For gravity plus seismic forces:

The majority of the horizontal seismic force is resolved into the gussets and does not reach the column face. The average gusset-to-beam connection length, from the geometry of Figure 5-33 and subtracting 1 in. for the snip, is:

(25¾ in. + 27¾ in.)/2 = 26.8 in.

Assume for calculation purposes a point (26.8 in. + 1.00 in.)/2 = 13.9 in. from the column face, as shown in Figure 5-42, can be used as a reference point to check the splice under gravity plus seismic loading.

The resultant of the beam shear and axial forces and the load angle from the horizontal axis of the beam, γ, are found as follows:

5.3 SPECIAL CONCENTRICALLY BRACED FRAMES

LRFD	ASD
$R_u = \sqrt{P_u^2 + V_u^2}$ $= \sqrt{(307 \text{ kips})^2 + (19.9 \text{ kips})^2}$ $= 308 \text{ kips}$ $\gamma = \tan^{-1}\left(\dfrac{V_u}{P_u}\right)$ $= \tan^{-1}\left(\dfrac{19.9 \text{ kips}}{307 \text{ kips}}\right)$ $= 3.71°$	$R_a = \sqrt{P_a^2 + V_a^2}$ $= \sqrt{(215 \text{ kips})^2 + (12.8 \text{ kips})^2}$ $= 215 \text{ kips}$ $\gamma = \tan^{-1}\left(\dfrac{V_a}{P_a}\right)$ $= \tan^{-1}\left(\dfrac{12.8 \text{ kips}}{215 \text{ kips}}\right)$ $= 3.41°$

The distance from the gravity plus seismic resultant force to the centroid of the W24×146 bolts is:

$e_x = 30.5 \text{ in.} - 13.9 \text{ in.} - 4.00 \text{ in.}$
$= 12.6 \text{ in.}$

Use AISC *Manual* Table 7-7 with the angle from the vertical equal to 90° − 3.71° = 86.3° (LRFD) and 90° − 3.41° = 86.6° (ASD). Interpolating from AISC *Manual* Table 7-7 for an angle equal to 75° with $s = 3$ in., $e_x = 12.6$ in., and $n = 6$:

$C = 8.06$

LRFD	ASD
$\phi R_n = C\phi r_n$ $= 8.06(61.3 \text{ kips})$ $= 494 \text{ kips} > 308 \text{ kips}$ **o.k.**	$\dfrac{R_n}{\Omega} = C\left(\dfrac{r_n}{\Omega}\right)$ $= 8.06(40.9 \text{ kips})$ $= 330 \text{ kips} > 215 \text{ kips}$ **o.k.**

For the W24×68 bolts:

$e_x = 30.5 \text{ in.} - 13.9 \text{ in.} + 4.00 \text{ in.}$
$= 20.6 \text{ in.}$

Interpolating from AISC *Manual* Table 7-7 for an angle equal to 75° with $s = 3$ in., $e_x = 20.6$ in., and $n = 6$:

$C = 6.54$

LRFD	ASD
$\phi R_n = C\phi r_n$ $= 6.54(61.3 \text{ kips})$ $= 401 \text{ kips} > 308 \text{ kips}$ **o.k.**	$\dfrac{R_n}{\Omega} = C\left(\dfrac{r_n}{\Omega}\right)$ $= 6.54(40.9 \text{ kips})$ $= 267 \text{ kips} > 215 \text{ kips}$ **o.k.**

Check bolt bearing on the W24×146 and W24×68

Because the force is eccentric and the AISC *Manual* eccentrically loaded bolt group tables are used, the lowest tearout strength of the bolt group should be used; in this case, the edge bolt controls.

The bearing and tearout strength limit states are separated for clarity in the following. For SCBF systems, where large deformations are anticipated, it is appropriate to assume that deformation at the bolt hole is not a design consideration for the seismic loading, and the tearout strength is checked at the end bolt with the 2.00 in. edge distance using AISC *Specification* Equation J3-6b. AISC *Seismic Provisions* Section D2.2(1) limits the nominal bearing strength at bolt holes to $2.4dtF_u$; therefore the available bearing and tearout strengths for the W24×146 web are:

LRFD	ASD
Bearing: Gravity $\phi r_n = \phi 2.4 dt F_u$ $= 0.75(2.4)(\text{⅞ in.})(0.650 \text{ in.})$ $\quad \times (65 \text{ ksi})$ $= 66.5 \text{ kips/bolt}$ $\phi R_n = C\phi r_n$ $= 2.02(66.5 \text{ kips/bolt})$ $= 134 \text{ kips} > 19.9 \text{ kips}$ **o.k.** Bearing: Gravity plus seismic $\phi r_n = 66.5 \text{ kips/bolt}$ $\phi R_n = C\phi r_n$ $= 8.06(66.5 \text{ kips/bolt})$ $= 536 \text{ kips} > 308 \text{ kips}$ **o.k.**	Bearing: Gravity $\dfrac{r_n}{\Omega} = \dfrac{2.4 dt F_u}{\Omega}$ $= \dfrac{2.4(\text{⅞ in.})(0.650 \text{ in.})(65 \text{ ksi})}{2.00}$ $= 44.4 \text{ kips/bolt}$ $\dfrac{R_n}{\Omega} = C\left(\dfrac{r_n}{\Omega}\right)$ $= 2.02(44.4 \text{ kips/bolt})$ $= 89.7 \text{ kips} > 12.8 \text{ kips}$ **o.k.** Bearing: Gravity plus seismic $\dfrac{r_n}{\Omega} = 44.4 \text{ kips/bolt}$ $\dfrac{R_n}{\Omega} = C\left(\dfrac{r_n}{\Omega}\right)$ $= 8.06(44.4 \text{ kips/bolt})$ $= 358 \text{ kips} > 215 \text{ kips}$ **o.k.**

5.3 SPECIAL CONCENTRICALLY BRACED FRAMES

LRFD	ASD
Tearout: Gravity	Tearout: Gravity
$\phi r_n = \phi 1.2 l_c t F_u$ $= 0.75(1.2)\left[2.00 \text{ in.} - \tfrac{1}{2}(^{15}\!/_{16} \text{ in.})\right]$ $\times (0.650 \text{ in.})(65 \text{ ksi})$ $= 58.2$ kips/bolt	$\dfrac{r_n}{\Omega} = \dfrac{1.2 l_c t F_u}{\Omega}$ $= 1.2\left[2.00 \text{ in.} - \tfrac{1}{2}(^{15}\!/_{16} \text{ in.})\right]$ $\times (0.650 \text{ in.})(65 \text{ ksi})/2.00$ $= 38.8$ kips/bolt
$\phi R_n = C \phi r_n$ $= 2.02(58.2 \text{ kips/bolt})$ $= 118$ kips > 19.9 kips **o.k.**	$\dfrac{R_n}{\Omega} = C\left(\dfrac{r_n}{\Omega}\right)$ $= 2.02(38.8 \text{ kips/bolt})$ $= 78.4$ kips > 12.8 kips **o.k.**
Tearout: Gravity plus seismic	Tearout: Gravity plus seismic
$\phi r_n = 58.2$ kips/bolt $(1.5/1.2)$ $= 72.8$ kips/bolt	$\dfrac{r_n}{\Omega} = 38.8$ kips/bolt $(1.5/1.2)$ $= 48.5$ kips/bolt
$\phi R_n = C \phi r_n$ $= 8.06(72.8 \text{ kips/bolt})$ $= 587$ kips > 308 kips **o.k.**	$\dfrac{R_n}{\Omega} = C\left(\dfrac{r_n}{\Omega}\right)$ $= 8.06(48.5 \text{ kips/bolt})$ $= 391$ kips > 215 kips **o.k.**

For the W24×68:

LRFD	ASD
Bearing: Gravity	Bearing: Gravity
$\phi r_n = \phi 2.4 d t F_u$ $= 0.75(2.4)(\tfrac{7}{8} \text{ in.})(0.415 \text{ in.})$ $\times (65 \text{ ksi})$ $= 42.5$ kips/bolt	$\dfrac{r_n}{\Omega} = \dfrac{2.4 d t F_u}{\Omega}$ $= \dfrac{2.4(\tfrac{7}{8} \text{ in.})(0.415 \text{ in.})(65 \text{ ksi})}{2.00}$ $= 28.3$ kips/bolt
$\phi R_n = C \phi r_n$ $= 1.56(42.5 \text{ kips/bolt})$ $= 66.3$ kips > 19.9 kips **o.k.**	$\dfrac{R_n}{\Omega} = C\left(\dfrac{r_n}{\Omega}\right)$ $= 1.56(28.3 \text{ kips/bolt})$ $= 44.1$ kips > 12.8 kips **o.k.**

LRFD	ASD
Bearing: Gravity plus seismic	Bearing: Gravity plus seismic
$\phi r_n = 42.5$ kips/bolt	$\dfrac{r_n}{\Omega} = 28.3$ kips/bolt
$\phi R_n = C\phi r_n$ $= 6.54(42.5 \text{ kips/bolt})$ $= 278$ kips < 308 kips **n.g.**	$\dfrac{R_n}{\Omega} = C\left(\dfrac{r_n}{\Omega}\right)$ $= 6.54(28.3 \text{ kips/bolt})$ $= 185$ kips < 215 kips **n.g.**

A web doubler can be used to increase the W24×68 web thickness, or a less approximate analysis of the bolt group can be used. Entering Table 7-7 of the AISC *Manual* at $\theta = 75°$ when the true angle is 86.3° (LRFD) and 86.6° (ASD) is very conservative. A computer program based on the instantaneous center of rotation method of Part 7 of the AISC *Manual* yields a C value equal to 9.76. This value of C, rather than the value of 6.54 from AISC *Manual* Table 7-7 at 75°, will be used in subsequent calculations. Thus, the available bearing and tearout strengths are:

LRFD	ASD
Bearing: Gravity plus seismic	Bearing: Gravity plus seismic
$\phi R_n = C\phi r_n$ $= 9.76(42.5 \text{ kips/bolt})$ $= 415$ kips > 308 kips **o.k.**	$\dfrac{R_n}{\Omega} = C\left(\dfrac{r_n}{\Omega}\right)$ $= 9.76(28.3 \text{ kips/bolt})$ $= 276$ kips > 215 kips **o.k.**
Tearout: Gravity	Tearout: Gravity
$\phi r_n = \phi 1.2 l_c t F_u$ $= 0.75(1.2)$ $\times [2.00 \text{ in.} - \frac{1}{2}(\frac{15}{16} \text{ in.})]$ $\times (0.415 \text{ in.})(65 \text{ ksi})$ $= 37.2$ kips/bolt	$\dfrac{r_n}{\Omega} = \dfrac{1.2 l_c t F_u}{\Omega}$ $= 1.2[2.00 \text{ in.} - \frac{1}{2}(\frac{15}{16} \text{ in.})]$ $\times (0.415 \text{ in.})(65 \text{ ksi})/2.00$ $= 24.8$ kips/bolt
$\phi R_n = C\phi r_n$ $= 1.56(37.2 \text{ kips/bolt})$ $= 58.0$ kips > 19.9 kips **o.k.**	$\dfrac{R_n}{\Omega} = C\left(\dfrac{r_n}{\Omega}\right)$ $= 1.56(24.8 \text{ kips/bolt})$ $= 38.7$ kips > 12.8 kips **o.k.**

5.3 SPECIAL CONCENTRICALLY BRACED FRAMES

LRFD	ASD
Tearout: Gravity plus seismic	Tearout: Gravity plus seismic
$\phi r_n = 37.2$ kips/bolt$(1.5/1.2)$ $\quad = 46.5$ kips/bolt $\phi R_n = C\phi r_n$ $\quad = 9.76(46.5$ kips/bolt$)$ $\quad = 454$ kips > 308 kips **o.k.**	$\dfrac{r_n}{\Omega} = 24.8$ kips/bolt$(1.5/1.2)$ $\quad = 31.0$ kips/bolt $\dfrac{R_n}{\Omega} = 9.76(31.0$ kips/bolt$)$ $\quad = 303$ kips > 215 kips **o.k.**

Size splice plates

Choose plates of ASTM A572 Grade 50 steel and a total thickness that exceeds the web thickness of the lighter beam. Try (2) ⅜-in.-thick plates. The length, l, is the horizontal distance between the last bolt on the W24×68 beam and the first bolt on the W24×146 beam stub, which is 5.00 in.

Check axial compression of splice plates

As the axial force in the beam due to seismic loads is always in compression, net tension is not a limit state. With $K = 1.2$ from AISC *Specification* Commentary Table C-A-7.1:

$$\frac{Kl}{r} = \frac{1.2(5.00 \text{ in.})}{\text{⅜ in.}/\sqrt{12}}$$
$$= 55.4$$

From AISC *Manual* Table 4-22 for $F_y = 50$ ksi, the available critical stress is:

LRFD	ASD
$\phi_c F_{cr} = 36.0$ ksi	$\dfrac{F_{cr}}{\Omega_c} = 23.9$ ksi
The design compressive strength of the two plates is:	The allowable compressive strength of the two plates is:
$\phi_c R_n = \phi_c F_{cr} A_g$ $\quad = 36.0$ ksi$($⅜ in.$)(19.0$ in.$)(2)$ $\quad = 513$ kips > 307 kips **o.k.**	$\dfrac{R_n}{\Omega_c} = \dfrac{F_{cr} A_g}{\Omega_c}$ $\quad = 23.9$ ksi$($⅜ in.$)(19.0$ in.$)(2)$ $\quad = 341$ kips > 215 kips **o.k.**

Check splice gross section for shear and flexural yielding for gravity-only forces

The required shear strength due to gravity load only is:

LRFD	ASD
Shear: $V_u = 19.9$ kips	Shear: $V_a = 12.8$ kips

Moment at critical section:

The critical section is at the first line of bolts in the W24×68 side of the splice; 33.0 in. from the column face. The required moment is:

LRFD	ASD
$M_{u\ splice} = 19.9$ kips$(33.0$ in.$)$ $= 657$ kip-in.	$M_{a\ splice} = 12.8$ kips$(33.0$ in.$)$ $= 422$ kip-in.

From AISC *Specification* Equation J4-3, the available shear strength of both splice plates is:

LRFD	ASD
$\phi R_n = \phi 0.60 F_y A_{gv}$ $= 1.00(0.60)(50 \text{ ksi})(\sfrac{3}{8} \text{ in.})$ $\times (19.0 \text{ in.})(2)$ $= 428 \text{ kips} > 19.9 \text{ kips}$ **o.k.**	$\dfrac{R_n}{\Omega} = \dfrac{0.60 F_y A_{gv}}{\Omega}$ $= \dfrac{0.60(50 \text{ ksi})(\sfrac{3}{8} \text{ in.})(19.0 \text{ in.})(2)}{1.50}$ $= 285 \text{ kips} > 12.8 \text{ kips}$ **o.k.**

From AISC *Specification* Section J4.5 and Section F11, the available flexural strength is:

LRFD	ASD
$\phi_b M_n = \phi_b F_y Z$ $= 0.90(50 \text{ ksi})(2)\dfrac{(\sfrac{3}{8} \text{ in.})(19.0 \text{ in.})^2}{4}$ $= 3{,}050$ kip-in. > 657 kip-in. **o.k.**	$\dfrac{M_n}{\Omega_b} = \dfrac{F_y Z}{\Omega_b}$ $= 50 \text{ ksi}(2)\dfrac{(\sfrac{3}{8} \text{ in.})(19.0 \text{ in.})^2}{4}\left(\dfrac{1}{1.67}\right)$ $= 2{,}030$ kip-in. > 422 kip-in. **o.k.**

Check splice net section for shear and flexural rupture for gravity-only forces

From AISC *Specification* Equation J4-4, the available shear strength is:

5.3 SPECIAL CONCENTRICALLY BRACED FRAMES

LRFD	ASD
$\phi R_n = \phi 0.60 F_u A_{nv}$ $= 0.75(0.60)(65 \text{ ksi})$ $\times [19.0 \text{ in.} - 6(^{15}/_{16} \text{ in.} + ^{1}/_{16} \text{ in.})]$ $\times (^{3}/_{8} \text{ in.})(2)$ $= 285 \text{ kips} > 19.9 \text{ kips}$ **o.k.**	$\dfrac{R_n}{\Omega} = \dfrac{0.60 F_u A_{nv}}{\Omega}$ $= 0.60(65 \text{ ksi})$ $\times [19.0 \text{ in.} - 6(^{15}/_{16} \text{ in.} + ^{1}/_{16} \text{ in.})]$ $\times (^{3}/_{8} \text{ in.})(2)/2.00$ $= 190 \text{ kips} > 12.8 \text{ kips}$ **o.k.**

From AISC *Manual* Equation 9-4, the available flexural strength is determined as follows:

$$Z_{net} = \frac{2(^{3}/_{8} \text{ in.})(19.0 \text{ in.})^2}{4}$$
$$- 2(^{3}/_{8} \text{ in.})(1.0 \text{ in.})$$
$$\times (1.5 \text{ in.} + 4.5 \text{ in.} + 7.5 \text{ in.})(2)$$
$$= 47.4 \text{ in.}^3$$

LRFD	ASD
$\phi_b M_n = \phi_b F_u Z_{net}$ $= 0.75(65 \text{ ksi})(47.4 \text{ in.}^3)$ $= 2{,}310 \text{ kip-in.} > 657 \text{ kip-in.}$ **o.k.**	$\dfrac{M_n}{\Omega_b} = \dfrac{F_u Z_{net}}{2.00}$ $= \dfrac{65 \text{ ksi}(47.4 \text{ in.}^3)}{2.00}$ $= 1{,}540 \text{ kip-in.} > 422 \text{ kip-in.}$ **o.k.**

Check splice for shear and flexural yielding for gravity and seismic forces

There is no shear in the splice due to seismic loads. From previous calculations, for gravity loading, the available shear strength is as follows:

LRFD	ASD
$\phi R_n = 428 \text{ kips} > 19.9 \text{ kips}$ **o.k.**	$\dfrac{R_v}{\Omega} = 285 \text{ kips} > 12.8 \text{ kips}$ **o.k.**

Moment at critical section:

The critical section is at the first line of bolts in the W24×68 side of the splice; 19.1 in. from the gravity plus seismic resultant force. The required moment at the critical section is:

LRFD	ASD
$M_{u\ splice} = 19.9$ kips(19.1 in.) $= 380$ kip-in.	$M_{a\ splice} = 12.8$ kips(19.1 in.) $= 244$ kip-in.

From previous calculations, the available flexural strength is:

LRFD	ASD
$\phi M_n = 3,050$ kip-in. > 380 kip-in. **o.k.**	$\dfrac{M_n}{\Omega} = 2,030$ kip-in. > 244 kip-in. **o.k.**

Check splice net section for shear and flexural rupture for gravity and seismic forces

LRFD	ASD
$\phi R_n = 285$ kips > 19.9 kips **o.k.**	$\dfrac{R_n}{\Omega} = 190$ kips > 12.8 kips **o.k.**
$\phi M_n = 2,310$ kip-in. > 380 kip-in. **o.k.**	$\dfrac{M_n}{\Omega} = 1,540$ kip-in. > 244 kip-in. **o.k.**

The splice is satisfactory for the required strengths.

Check the ductility of the splice

The procedure used for the extended single-plate connection in AISC *Manual* Part 10 can be used to check the ductility of the splice. From AISC *Manual* Part 10, the maximum splice plate thickness permitted is:

$$t_{max} = \frac{6M_{max}}{F_y d^2} \qquad \text{(\textit{Manual} Eq. 10-3)}$$

where

$$M_{max} = \frac{F_v}{0.90}(A_b C') \qquad \text{(\textit{Manual} Eq. 10-4)}$$

For the splice plate and bolts, the nominal shear stress of ASTM A325-X bolts from AISC *Specification* Table J3.2 is:

$F_{nv} = F_v = 68$ ksi

The area of a ⅞-in.-diameter bolt, from AISC *Manual* Table 7-1, is:

$A_b = 0.601$ in.^{2}

From AISC *Manual* Table 7-7 for angle = 0° with $s = 3$ in. and $n = 6$:

$C' = 54.2$ in.

The nominal flexural strength of the bolt group is:

$$M_{max} = \frac{F_v}{0.90}(A_b C') \quad \text{(Manual Eq. 10-4)}$$

$$= \frac{68 \text{ ksi}}{0.90}\left[0.601 \text{ in.}^2 (54.2 \text{ in.})\right]$$

$$= 2{,}460 \text{ kip-in.}$$

$$t_{max} = \frac{6M_{max}}{F_y d^2}$$

$$= \frac{6(2{,}460 \text{ kip-in.})}{50 \text{ ksi}(19.0 \text{ in.})^2}$$

$$= 0.818 \text{ in.}$$

Since $2t = 2(\text{\textthreeeighths}$ in.$) = 0.750$ in. $< t_{max} = 0.818$ in., the splice satisfies the ductility requirement.

Beam-to-Column Interface—Design Case I

The forces at the beam-to-column interface, shown in Figures 5-40a and 5-40b, are:

LRFD	ASD
Normal: $N_u = \lvert 109 \text{ kips} - 46.0 \text{ kips} - 122 \text{ kips}\rvert$ $= 59.0$ kips (compression) Shear: $V_u = 238 \text{ kips} + 213 \text{ kips} - 19.9 \text{ kips}$ $= 431$ kips	Normal: $N_a = \lvert 76.6 \text{ kips} - 32.2 \text{ kips} - 85.5 \text{ kips}\rvert$ $= 41.1$ kips (compression) Shear: $V_a = 167 \text{ kips} + 150 \text{ kips} - 12.8 \text{ kips}$ $= 304$ kips

Check beam stub gross section for shear and tension yielding

From AISC *Manual* Table 3-6, the available shear strength of the W24×146 beam stub is:

LRFD	ASD
$\phi V_n = 482$ kips > 431 kips **o.k.**	$\dfrac{V_n}{\Omega} = 321$ kips > 304 kips **o.k.**

Check the available compressive strength of the beam stub. Treating the beam stub as a connecting element, determine whether the available compressive strength can be determined using AISC *Specification* Section J4.4:

$$\frac{KL}{r} = \frac{1.0(30 \text{ in.})}{3.01 \text{ in.}}$$

$= 9.97 < 25$; therefore, AISC *Specification* Section J4.4 is applicable

$P_n = F_y A_g$ \hfill (*Spec.* Eq. J4-6)

$= 50 \text{ ksi}(43.0 \text{ in.}^2)$

$= 2,150 \text{ kips}$

The available compressive strength is:

LRFD	ASD
$\phi P_n = \phi F_y A_g$ $= 0.90(2,150 \text{ kips})$ $= 1,940 \text{ kips} > 59.0 \text{ kips}$ **o.k.**	$\dfrac{P_n}{\Omega} = \dfrac{2,150 \text{ kips}}{1.67}$ $= 1,290 \text{ kips} > 41.1 \text{ kips}$ **o.k.**

Design of beam stub web-to-column weld

The resultant force to be resisted by the weld is:

LRFD	ASD
$R_u = \sqrt{V_u^2 + N_u^2}$ $= \sqrt{(431 \text{ kips})^2 + (59.0 \text{ kips})^2}$ $= 435 \text{ kips}$	$R_a = \sqrt{V_a^2 + N_a^2}$ $= \sqrt{(304 \text{ kips})^2 + (41.1 \text{ kips})^2}$ $= 307 \text{ kips}$

The angle of the resultant force can be calculated and used in the directional strength increase of fillet welds according to AISC *Specification* Equation J2-5. The angle of the resultant with respect to the vertical along the column is:

LRFD	ASD
$\theta = \tan^{-1}\left(\dfrac{N_u}{V_u}\right)$ $= \tan^{-1}\left(\dfrac{59.0 \text{ kips}}{431 \text{ kips}}\right)$ $= 7.79°$	$\theta = \tan^{-1}\left(\dfrac{N_a}{V_a}\right)$ $= \tan^{-1}\left(\dfrac{41.1 \text{ kips}}{304 \text{ kips}}\right)$ $= 7.70°$

5.3 SPECIAL CONCENTRICALLY BRACED FRAMES

The directional strength increase is calculated as follows:

LRFD	ASD
$1.0 + 0.5\sin^{1.5} 7.79° = 1.02$	$1.0 + 0.5\sin^{1.5} 7.70° = 1.02$

The required weld size is calculated as follows from AISC *Manual* Equations 8-2a and 8-2b:

LRFD	ASD
$D_{req'd} = \dfrac{435 \text{ kips}}{2(1.392 \text{ kip/in.})(20\frac{3}{4} \text{ in.})(1.02)}$ $= 7.38$ sixteenths	$D_{req'd} = \dfrac{307 \text{ kips}}{2(0.928 \text{ kip/in.})(20\frac{3}{4} \text{ in.})(1.02)}$ $= 7.82$ sixteenths

½-in. double-sided fillet welds are required for Design Case I on the beam *T*-distance of 20¾ in.

The normal force of 59.0 kips (LRFD) or 41.1 kips (ASD) on the column indicates that web local crippling and local yielding checks should be made on the column as follows.

Check column web local yielding

From AISC *Specification* Equation J10-2, because the load is applied greater than *d* from the end of the column, the available web local yielding strength of the column is:

LRFD	ASD
$\phi R_n = \phi F_{yw} t_w (5k_{des} + l_b)$ $= 1.00(50 \text{ ksi})(0.550 \text{ in.})$ $\times [5(1.50 \text{ in.}) + 20\frac{3}{4} \text{ in.}]$ $= 777 \text{ kips} > 59.0 \text{ kips}$ **o.k.**	$\dfrac{R_n}{\Omega} = \dfrac{F_{yw} t_w (5k_{des} + l_b)}{\Omega}$ $= 50 \text{ ksi}(0.550 \text{ in.})$ $\times [5(1.50 \text{ in.}) + 20\frac{3}{4} \text{ in.}]$ $\times (1/1.50)$ $= 518 \text{ kips} > 41.1 \text{ kips}$ **o.k.**

Check column web local crippling

From AISC *Specification* Equation J10-4, because the load is applied greater than *d*/2 from the end of the column, the available web local crippling strength of the column is:

LRFD	ASD
$\phi R_n = \phi 0.80 t_w^2 \left[1 + 3\left(\dfrac{l_b}{d}\right)\left(\dfrac{t_w}{t_f}\right)^{1.5}\right]$ $\times \sqrt{\dfrac{EF_{yw}t_f}{t_w}}$ $= 0.75(0.80)(0.550 \text{ in.})^2$ $\times \left[1 + 3\left(\dfrac{20\frac{3}{4} \text{ in.}}{12.7 \text{ in.}}\right)\left(\dfrac{0.550 \text{ in.}}{0.900 \text{ in.}}\right)^{1.5}\right]$ $\times \sqrt{\dfrac{29{,}000 \text{ ksi}(50 \text{ ksi})(0.900 \text{ in.})}{0.550 \text{ in.}}}$ $= 934 \text{ kips} > 59.0 \text{ kips}$ **o.k.**	$\dfrac{R_n}{\Omega} = 0.80 t_w^2 \left[1 + 3\left(\dfrac{l_b}{d}\right)\left(\dfrac{t_w}{t_f}\right)^{1.5}\right]$ $\times \sqrt{\dfrac{EF_{yw}t_f}{t_w}}\left(\dfrac{1}{\Omega}\right)$ $= 0.80(0.550 \text{ in.})^2$ $\times \left[1 + 3\left(\dfrac{20\frac{3}{4} \text{ in.}}{12.7 \text{ in.}}\right)\left(\dfrac{0.550 \text{ in.}}{0.900 \text{ in.}}\right)^{1.5}\right]$ $\times \sqrt{\dfrac{29{,}000 \text{ ksi}(50 \text{ ksi})(0.900 \text{ in.})}{0.550 \text{ in.}}}$ $\times \left(\dfrac{1}{2.00}\right)$ $= 623 \text{ kips} > 41.1 \text{ kips}$ **o.k.**

The limit state of column web compression buckling is not checked here because only the beam stub web is attached to the column flange. Therefore, pinching of the column web would not occur, as it would if the beam stub flanges were also connected.

Beam-to-Column Interface—Design Case II

The forces at the beam-to-column interface, shown in Figures 5-41a and 5-41b, are:

LRFD	ASD
Normal: $N_u = \lvert 98.6 \text{ kips} - 128 \text{ kips} - 46.0 \text{ kips} \rvert$ $= 75.4 \text{ kips (compression)}$ Shear: $V_u = 193 \text{ kips} + 250 \text{ kips} + 19.9 \text{ kips}$ $= 463 \text{ kips}$	Normal: $N_a = \lvert 69.1 \text{ kips} - 32.2 \text{ kips} - 89.4 \text{ kips} \rvert$ $= 52.5 \text{ kips (compression)}$ Shear: $V_a = 135 \text{ kips} + 175 \text{ kips} + 12.8 \text{ kips}$ $= 323 \text{ kips}$

Check beam stub for shear and tension yielding

The available shear yielding strength determined previously for Design Case I is:

LRFD	ASD
$\phi V_n = 482 \text{ kips} > 463 \text{ kips}$ **o.k.**	$\dfrac{V_n}{\Omega} = 321 \text{ kips} < 323 \text{ kips}$ **n.g.**

5.3 SPECIAL CONCENTRICALLY BRACED FRAMES

Due to the different load combinations used for LRFD versus ASD, there is a slight discrepancy between the LRFD and ASD results for the required shear strength. Typically, one method should be chosen and used consistently throughout an entire design. For the purposes of this example, the LRFD result will be used here.

The available compressive strength determined previously for Design Case I is:

LRFD	ASD
$\phi P_n = 1{,}940 \text{ kips} > 75.4 \text{ kips}$ **o.k.**	$\dfrac{P_n}{\Omega} = 1{,}290 \text{ kips} > 52.5 \text{ kips}$ **o.k.**

Design of beam stub web-to-column weld

The resultant force at the beam-to-column interface is:

LRFD	ASD
$R_u = \sqrt{V_u^2 + N_u^2}$ $= \sqrt{(463 \text{ kips})^2 + (75.4 \text{ kips})^2}$ $= 469 \text{ kips}$	$R_a = \sqrt{V_a^2 + N_a^2}$ $= \sqrt{(323 \text{ kips})^2 + (52.5 \text{ kips})^2}$ $= 327 \text{ kips}$

The beam stub web-to-column weld size is determined from AISC *Manual* Equations 8-2a and 8-2b, including the directional strength increase of AISC *Specification* Equation J2-5, as follows:

LRFD	ASD
Load angle $\theta = \tan^{-1}\left(\dfrac{N_u}{V_u}\right)$ $= \tan^{-1}\left(\dfrac{75.4 \text{ kips}}{463 \text{ kips}}\right)$ $= 9.25°$ Directional strength increase $1.0 + 0.5\sin^{1.5} 9.25° = 1.03$	Load angle $\theta = \tan^{-1}\left(\dfrac{N_a}{V_a}\right)$ $= \tan^{-1}\left(\dfrac{52.5 \text{ kips}}{323 \text{ kips}}\right)$ $= 9.23°$ Directional strength increase $1.0 + 0.5\sin^{1.5} 9.23° = 1.03$

LRFD	ASD
Number of sixteenths inch of weld required $$D_{req'd} = \frac{469 \text{ kips}}{2(1.392 \text{ kip/in.})(20\frac{3}{4} \text{ in.})(1.03)}$$ $$= 7.88 \text{ sixteenths}$$	Number of sixteenths inch of weld required $$D_{req'd} = \frac{327 \text{ kips}}{2(0.928 \text{ kip/in.})(20\frac{3}{4} \text{ in.})(1.03)}$$ $$= 8.24 \text{ sixteenths}$$

Note that this Case controls (Design Case II) requiring double-sided $\frac{9}{16}$-in. fillet welds.

The column must also be checked for web local crippling and web local yielding. These limit states will not control here for Design Case II. The calculations were shown for Design Case I.

Top Gusset-to-Beam Interface—Design Case I

The forces at the top gusset-to-beam interface, shown in Figures 5-40a and 5-40b, are:

LRFD	ASD
Normal: $N_u = 238$ kips	Normal: $N_a = 167$ kips
Shear: $V_u = 267$ kips	Shear: $V_a = 187$ kips

Check top gusset for shear yielding and tension yielding along the beam flange

The available shear yielding strength of the gusset plate is determined from AISC *Specification* Equation J4-3, and the available tensile yielding strength is determined from AISC *Specification* Equation J4-1 as follows:

LRFD	ASD
$\phi V_n = \phi 0.60 F_y A_{gv}$ $= 1.00(0.60)(36 \text{ ksi})(\frac{3}{4} \text{ in.})$ $\times (26\frac{3}{4} \text{ in.} - 1.00 \text{ in.})$ $= 417 \text{ kips} > 267 \text{ kips}$ **o.k.** $\phi P_n = \phi F_y A_g$ $= 0.90(36 \text{ ksi})(\frac{3}{4} \text{ in.})$ $\times (26\frac{3}{4} \text{ in.} - 1.00 \text{ in.})$ $= 626 \text{ kips} > 238 \text{ kips}$ **o.k.**	$\frac{V_n}{\Omega} = \frac{0.60 F_y A_{gv}}{\Omega}$ $= 0.60(36 \text{ ksi})$ $\times (\frac{3}{4} \text{ in.})(26\frac{3}{4} \text{ in.} - 1.00 \text{ in.})$ $\times (1/1.50)$ $= 278 \text{ kips} > 187 \text{ kips}$ **o.k.** $\frac{P_n}{\Omega} = \frac{F_y A_g}{\Omega}$ $= \frac{36 \text{ ksi}(\frac{3}{4} \text{ in.})(26\frac{3}{4} \text{ in.} - 1.00 \text{ in.})}{1.67}$ $= 416 \text{ kips} > 167 \text{ kips}$ **o.k.**

5.3 SPECIAL CONCENTRICALLY BRACED FRAMES

Although it seldom controls, interaction can be checked here using the interaction formula of Example 5.3.8.

Design of beam flange-to-top gusset weld

The beam flange-to-top gusset plate weld is determined as follows using the directional strength increase of AISC *Specification* Equation J2-5, the 1.25 ductility factor discussed in AISC *Manual* Part 13, and AISC *Manual* Equations 8-2a and 8-2b:

LRFD	ASD
Resultant force $$R_u = \sqrt{V_u^2 + N_u^2}$$ $$= \sqrt{(267 \text{ kips})^2 + (238 \text{ kips})^2}$$ $$= 358 \text{ kips}$$	Resultant force $$R_a = \sqrt{V_a^2 + N_a^2}$$ $$= \sqrt{(187 \text{ kips})^2 + (167 \text{ kips})^2}$$ $$= 251 \text{ kips}$$
Load angle $$\theta = \tan^{-1}\left(\frac{N_u}{V_u}\right)$$ $$= \tan^{-1}\left(\frac{238 \text{ kips}}{267 \text{ kips}}\right)$$ $$= 41.7°$$	Load angle $$\theta = \tan^{-1}\left(\frac{N_a}{V_a}\right)$$ $$= \tan^{-1}\left(\frac{167 \text{ kips}}{187 \text{ kips}}\right)$$ $$= 41.8°$$
Directional strength increase $$1.0 + 0.50\sin^{1.5} 41.7° = 1.27$$ $$D_{req'd} = \frac{1.25(358 \text{ kips})}{2(1.392 \text{ kip/in.})}$$ $$\times \frac{1}{(26\tfrac{3}{4} \text{ in.} - 1.00 \text{ in.})(1.27)}$$ $$= 4.92 \text{ sixteenths}$$	Directional strength increase $$1.0 + 0.50\sin^{1.5} 41.8° = 1.27$$ $$D_{req'd} = \frac{1.25(251 \text{ kips})}{2(0.928 \text{ kip/in.})}$$ $$\times \frac{1}{(26\tfrac{3}{4} \text{ in.} - 1.00 \text{ in.})(1.27)}$$ $$= 5.17 \text{ sixteenths}$$

Use double-sided ⅜-in. fillet welds.

The 1.25 factor in the numerator is a ductility factor from the work of Richard (1986) as modified by Hewitt and Thornton (2004). Richard found that the ratio of the maximum stress to the average stress on corner gussets was 1.4, which was reduced to a 90% confidence limit by Hewitt and Thornton resulting in the 1.25 factor. The UFM assumes a uniform distribution of stress on the gusset edge, but the actual distribution is not likely to be uniform; the 1.25 factor accounts for this.

Beam stub strength

Check web local yielding

From AISC *Specification* Equation J10-3, because the load is applied less than or equal to the beam stub depth, d, from the end of the beam stub, the available web local yielding strength of the beam stub is:

LRFD	ASD
$\phi R_n = \phi F_{yw} t_w (2.5 k_{des} + l_b)$ $= 1.00 (50 \text{ ksi})(0.650 \text{ in.})$ $\times \begin{bmatrix} 2.5(1.59 \text{ in.}) \\ +(26\frac{3}{4} \text{ in.} - 1.00 \text{ in.}) \end{bmatrix}$ $= 966 \text{ kips} > 238 \text{ kips}$ **o.k.**	$\dfrac{R_n}{\Omega} = \dfrac{F_{yw} t_w (2.5 k_{des} + l_b)}{\Omega}$ $= 50 \text{ ksi}(0.650 \text{ in.})$ $\times [2.5(1.59 \text{ in.}) + (26\frac{3}{4} \text{ in.} - 1.00 \text{ in.})]$ $\times (1/1.50)$ $= 644 \text{ kips} > 167 \text{ kips}$ **o.k.**

Check web local crippling

Because the compressive force is applied at the centroid of the beam stub-to-gusset interface, which is a distance from the beam stub end that is greater than $d/2$, the nominal web local crippling strength is:

$$R_n = 0.80 t_w^2 \left[1 + 3\left(\dfrac{l_b}{d}\right)\left(\dfrac{t_w}{t_f}\right)^{1.5}\right]\sqrt{\dfrac{EF_{yw}t_f}{t_w}} \qquad \text{(Spec. Eq. J10-4)}$$

The available web local crippling strength is:

LRFD	ASD
$\phi R_n = 0.75(0.80)(0.650 \text{ in.})^2$ $\times \begin{bmatrix} 1 + 3\left(\dfrac{26\frac{3}{4} \text{ in.} - 1.00 \text{ in.}}{24.7 \text{ in.}}\right) \\ \times \left(\dfrac{0.650 \text{ in.}}{1.09 \text{ in.}}\right)^{1.5} \end{bmatrix}$ $\times \sqrt{\dfrac{29,000 \text{ ksi}(50 \text{ ksi})(1.09 \text{ in.})}{0.650 \text{ in.}}}$ $= 965 \text{ kips} > 238 \text{ kips}$ **o.k.**	$\dfrac{R_n}{\Omega} = 0.80(0.650 \text{ in.})^2$ $\times \begin{bmatrix} 1 + 3\left(\dfrac{26\frac{3}{4} \text{ in.} - 1.00 \text{ in.}}{24.7 \text{ in.}}\right) \\ \times \left(\dfrac{0.650 \text{ in.}}{1.09 \text{ in.}}\right)^{1.5} \end{bmatrix}$ $\times \sqrt{\dfrac{29,000 \text{ ksi}(50 \text{ ksi})(1.09 \text{ in.})}{0.650 \text{ in.}}}$ $\times \left(\dfrac{1}{2.00}\right)$ $= 643 \text{ kips} > 167 \text{ kips}$ **o.k.**

5.3 SPECIAL CONCENTRICALLY BRACED FRAMES

Top Gusset-to-Beam Interface—Design Case II
The forces at the top gusset-to-beam interface, shown in Figures 5-41a and 5-41b, are:

LRFD	ASD
Normal: $N_u = 193$ kips	Normal: $N_a = 135$ kips
Shear: $V_u = 216$ kips	Shear: $V_a = 151$ kips

Check top gusset gross section for shear and tension yielding
From Design Case I:

LRFD	ASD
$\phi V_n = 417$ kips > 216 kips **o.k.**	$\dfrac{V_n}{\Omega} = 0.60(36\text{ ksi})$ $\times (\sqrt[3]{4}\text{ in.})(26\sqrt[3]{4}\text{ in.} - 1.00\text{ in.})$ $\times (1/1.50)$ $= 278$ kips > 151 kips **o.k.**
$\phi P_n = \phi F_y A_g$ $= 0.90(36\text{ ksi})(\sqrt[3]{4}\text{ in.})$ $\times (26\sqrt[3]{4}\text{ in.} - 1.00\text{ in.})$ $= 626$ kips > 193 kips **o.k.**	$\dfrac{P_n}{\Omega} = \dfrac{F_y A_g}{\Omega}$ $= \dfrac{36\text{ ksi}(\sqrt[3]{4}\text{ in.})(26\sqrt[3]{4}\text{ in.} - 1.00\text{ in.})}{1.67}$ $= 416$ kips > 135 kips **o.k.**

Beam flange-to-top gusset weld
The beam flange-to-top gusset plate weld is determined as follows using the directional strength increase of AISC *Specification* Equation J2-5, the 1.25 ductility factor discussed in AISC *Manual* Part 13, and AISC *Manual* Equations 8-2a and 8-2b:

LRFD	ASD
Resultant force $R_u = \sqrt{V_u^2 + N_u^2}$ $= \sqrt{(216\text{ kips})^2 + (193\text{ kips})^2}$ $= 290$ kips	Resultant force $R_a = \sqrt{V_a^2 + N_a^2}$ $= \sqrt{(151\text{ kips})^2 + (135\text{ kips})^2}$ $= 203$ kips

LRFD	ASD
Load angle $$\theta = \tan^{-1}\left(\frac{N_u}{V_u}\right)$$ $$= \tan^{-1}\left(\frac{193 \text{ kips}}{216 \text{ kips}}\right)$$ $$= 41.8°$$	Load angle $$\theta = \tan^{-1}\left(\frac{N_a}{V_a}\right)$$ $$= \tan^{-1}\left(\frac{135 \text{ kips}}{151 \text{ kips}}\right)$$ $$= 41.8°$$
Directional strength increase $$1.0 + 0.50\sin^{1.5}41.8° = 1.27$$	Directional strength increase $$1.0 + 0.50\sin^{1.5}41.8° = 1.27$$
$$D_{req'd} = \frac{1.25(290 \text{ kips})}{2(1.392 \text{ kip/in.})}$$ $$\times \frac{1}{(26\tfrac{3}{4} \text{ in.} - 1.00 \text{ in.})(1.27)}$$ $$= 3.98 \text{ sixteenths}$$	$$D_{req'd} = \frac{1.25(203 \text{ kips})}{2(0.928 \text{ kip/in.})}$$ $$\times \frac{1}{(26\tfrac{3}{4} \text{ in.} - 1.00 \text{ in.})(1.27)}$$ $$= 4.18 \text{ sixteenths}$$

This requires double-sided 5⁄16-in. fillet welds. Note that Design Case I controls, however, requiring a 3⁄8-in. fillet weld.

Beam stub strength

Check beam stub web local yielding
From Design Case I calculation:

LRFD	ASD
$\phi R_n = 966$ kips > 193 kips **o.k.**	$\dfrac{R_n}{\Omega} = 644$ kips > 135 kips **o.k.**

Check beam stub web local crippling
From Design Case I calculation:

LRFD	ASD
$\phi R_n = 965$ kips > 193 kips **o.k.**	$\dfrac{R_n}{\Omega} = 643$ kips > 135 kips **o.k.**

5.3 SPECIAL CONCENTRICALLY BRACED FRAMES

Top Gusset-to-Column Interface—Design Case I
The forces at the top gusset-to-column interface, shown in Figures 5-40a and 5-40b, are:

LRFD	ASD
Normal: $N_u = 122$ kips	Normal: $N_a = 85.5$ kips
Shear: $V_u = 151$ kips	Shear: $V_a = 106$ kips
Moment: $M_{uc} = 195$ kip-in.	Moment: $M_{ac} = 137$ kip-in.

Combine the axial force and the moment by converting the moment into an equivalent axial force derived from the moment equation for a simply supported member with a concentrated load at midspan (s is the snip dimension of the gusset and $\overline{\beta}$ is the distance to the centroid of the column-to-gusset connection, determined previously):

LRFD	ASD
$N_{u\,equiv} = N_u + \dfrac{4M_{uc}}{2(\overline{\beta}-s)}$	$N_{a\,equiv} = N_a + \dfrac{4M_{ac}}{2(\overline{\beta}-s)}$
$= 122 \text{ kips} + \dfrac{2(195 \text{ kip-in.})}{9.45 \text{ in.} - 1.00 \text{ in.}}$	$= 85.5 \text{ kips} + \dfrac{2(137 \text{ kip-in.})}{9.45 \text{ in.} - 1.00 \text{ in.}}$
$= 168$ kips	$= 118$ kips

This is not a real load but results in the same demand on the gusset and weld as working with N and M_c separately and allows the direct use of AISC *Specification* Section J10.

Design of column flange-to-top gusset weld
The column flange-to-top gusset plate weld is determined as follows using the directional strength increase of AISC *Specification* Equation J2-5, the 1.25 ductility factor discussed in AISC *Manual* Part 13, and AISC *Manual* Equations 8-2a and 8-2b:

LRFD	ASD
Resultant force	Resultant force
$R_u = \sqrt{V_u^2 + N_{u\,equiv}^2}$	$R_a = \sqrt{V_a^2 + N_{a\,equiv}^2}$
$= \sqrt{(151 \text{ kips})^2 + (168 \text{ kips})^2}$	$= \sqrt{(106 \text{ kips})^2 + (118 \text{ kips})^2}$
$= 226$ kips	$= 159$ kips

LRFD	ASD
Load angle	Load angle
$\theta = \tan^{-1}\left(\dfrac{N_{u\,equiv}}{V_u}\right)$	$\theta = \tan^{-1}\left(\dfrac{N_{a\,equiv}}{V_a}\right)$
$= \tan^{-1}\left(\dfrac{168 \text{ kips}}{151 \text{ kips}}\right)$	$= \tan^{-1}\left(\dfrac{118 \text{ kips}}{106 \text{ kips}}\right)$
$= 48.1°$	$= 48.1°$
Directional strength increase	Directional strength increase
$1.0 + 0.50\sin^{1.5} 48.1° = 1.32$	$1.0 + 0.50\sin^{1.5} 48.1° = 1.32$
$D_{req'd} = \dfrac{1.25(226 \text{ kips})}{2(1.392 \text{ kip/in.})}$	$D_{req'd} = \dfrac{1.25(159 \text{ kips})}{2(0.928 \text{ kip/in.})}$
$\times \dfrac{1}{(17\text{⅞ in.} - 1.00 \text{ in.})(1.32)}$	$\times \dfrac{1}{(17\text{⅞ in.} - 1.00 \text{ in.})(1.32)}$
$= 4.56$ sixteenths	$= 4.81$ sixteenths

Use a double-sided ⁵⁄₁₆-in. fillet weld.

Check top gusset for shear yielding and tension yielding along the column flange

The available shear yielding strength of the gusset plate at the column flange interface is determined from AISC *Specification* Equation J4-3, and the available tensile yielding strength at the column flange interface is determined from AISC *Specification* Equation J4-1 as follows:

LRFD	ASD
$\phi V_n = \phi 0.60 F_y A_{gv}$	$\dfrac{V_n}{\Omega} = \dfrac{0.60 F_y A_{gv}}{\Omega}$
$= 1.00(0.60)(36 \text{ ksi})(\text{¾ in.})$	$= \dfrac{0.60(36 \text{ ksi})(\text{¾ in.})(17\text{⅞ in.} - 1.00 \text{ in.})}{1.50}$
$\times (17\text{⅞ in.} - 1.00 \text{ in.})$	
$= 273$ kips > 151 kips **o.k.**	$= 182$ kips > 106 kips **o.k.**
$\phi P_n = \phi F_y A_g$	$\dfrac{P_n}{\Omega} = \dfrac{F_y A_g}{\Omega}$
$= 0.90(36 \text{ ksi})(\text{¾ in.})$	$= \dfrac{36 \text{ ksi}(\text{¾ in.})(17\text{⅞ in.} - 1.00 \text{ in.})}{1.67}$
$\times (17\text{⅞ in.} - 1.00 \text{ in.})$	
$= 410$ kips > 168 kips **o.k.**	$= 273$ kips > 118 kips **o.k.**

5.3 SPECIAL CONCENTRICALLY BRACED FRAMES

Check column web local yielding

From AISC *Specification* Equation J10-2, because the load is applied greater than the column depth, d, from the end of the column, the available web local yielding strength of the column is:

LRFD	ASD
$\phi R_n = \phi F_{yw} t_w (5k_{des} + l_b)$ $= 1.00(50 \text{ ksi})(0.550 \text{ in.})$ $\times [5(1.50 \text{ in.}) + (17\frac{7}{8} \text{ in.} - 1.00 \text{ in.})]$ $= 670 \text{ kips} > 168 \text{ kips}$ **o.k.**	$\dfrac{R_n}{\Omega} = \dfrac{F_{yw} t_w (5k_{des} + l_b)}{\Omega}$ $= 50 \text{ ksi}(0.550 \text{ in.})$ $\times [5(1.50 \text{ in.}) + (17\frac{7}{8} \text{ in.} - 1.00 \text{ in.})]$ $\times (1/1.50)$ $= 447 \text{ kips} > 118 \text{ kips}$ **o.k.**

Check column web local crippling

Because the load is applied greater than $d/2$ from the end of the column, the available web local crippling strength of the column is determined from AISC *Specification* Equation J10-4 as follows:

LRFD	ASD
$\phi R_n = \phi 0.80 t_w^2 \left[1 + 3\left(\dfrac{l_b}{d}\right)\left(\dfrac{t_w}{t_f}\right)^{1.5}\right]$ $\times \sqrt{\dfrac{E F_{yw} t_f}{t_w}}$ $= 0.75(0.80)(0.550 \text{ in.})^2$ $\times \left[1 + 3\left(\dfrac{17\frac{7}{8} \text{ in.} - 1.00 \text{ in.}}{12.7 \text{ in.}}\right) \times \left(\dfrac{0.550 \text{ in.}}{0.900 \text{ in.}}\right)^{1.5}\right]$ $\times \sqrt{\dfrac{29{,}000 \text{ ksi}(50 \text{ ksi})(0.900 \text{ in.})}{0.550 \text{ in.}}}$ $= 812 \text{ kips} > 168 \text{ kips}$ **o.k.**	$\dfrac{R_n}{\Omega} = 0.80 t_w^2 \left[1 + 3\left(\dfrac{l_b}{d}\right)\left(\dfrac{t_w}{t_f}\right)^{1.5}\right]$ $\times \sqrt{\dfrac{E F_{yw} t_f}{t_w}} \left(\dfrac{1}{\Omega}\right)$ $= 0.80(0.550 \text{ in.})^2$ $\times \left[1 + 3\left(\dfrac{17\frac{7}{8} \text{ in.} - 1.00 \text{ in.}}{12.7 \text{ in.}}\right) \times \left(\dfrac{0.550 \text{ in.}}{0.900 \text{ in.}}\right)^{1.5}\right]$ $\times \sqrt{\dfrac{29{,}000 \text{ ksi}(50 \text{ ksi})(0.900 \text{ in.})}{0.550 \text{ in.}}}$ $\times \left(\dfrac{1}{2.00}\right)$ $= 541 \text{ kips} > 118 \text{ kips}$ **o.k.**

Check column web shear strength

From AISC *Manual* Table 3-6, for a W12×96, the available shear strength is:

LRFD	ASD
$\phi V_n = 210$ kips > 122 kips **o.k.**	$\dfrac{V_n}{\Omega} = 140$ kips > 85.5 kips **o.k.**

Top Gusset-to-Column Interface — Design Case II

The forces at the top gusset-to-column interface, shown in Figures 5-41a and 5-41b, are:

LRFD	ASD
Normal: $N_u = 98.6$ kips	Normal: $N_a = 69.1$ kips
Shear: $V_u = 122$ kips	Shear: $V_a = 85.4$ kips
Moment: $M_{uc} = 158$ kip-in.	Moment: $M_{ac} = 111$ kip-in.

Comparing these loads with those of Design Case I, it can be seen that Design Case I controls.

This completes the top gusset design.

Bottom Gusset-to-Beam Interface—Design Case I

The forces at the bottom gusset-to-beam interface, shown in Figures 5-40a and 5-40b, are:

LRFD	ASD
Normal: $N_u = 213$ kips	Normal: $N_a = 150$ kips
Shear: $V_u = 256$ kips	Shear: $V_a = 180$ kips

Check bottom gusset for shear and tension yielding along the beam flange

The available shear yielding strength of the gusset plate is determined from AISC *Specification* Equation J4-3, and the available tensile yielding strength is determined from AISC *Specification* Equation J4-1 as follows:

LRFD	ASD
$\phi V_n = \phi 0.60 F_y A_{gv}$ $\quad = 1.00(0.60)(36\text{ ksi})(\tfrac{3}{4}\text{ in.})$ $\quad \times (28\tfrac{3}{4}\text{ in.} - 1.00\text{ in.})$ $\quad = 450$ kips > 256 kips **o.k.**	$\dfrac{V_n}{\Omega} = \dfrac{0.60 F_y A_{gv}}{\Omega}$ $\quad = 0.60(36\text{ ksi})$ $\quad \times (\tfrac{3}{4}\text{ in.})(28\tfrac{3}{4}\text{ in.} - 1.00\text{ in.})$ $\quad \times (1/1.50)$ $\quad = 300$ kips > 180 kips **o.k.**

LRFD	ASD
$\phi P_n = \phi F_y A_g$ $= 0.90(36 \text{ ksi})(\tfrac{3}{4} \text{ in.})$ $\times (28\tfrac{3}{4} \text{ in.} - 1.00 \text{ in.})$ $= 674 \text{ kips} > 213 \text{ kips}$ **o.k.**	$\dfrac{P_n}{\Omega} = \dfrac{F_y A_g}{\Omega}$ $= \dfrac{36 \text{ ksi}(\tfrac{3}{4} \text{ in.})(28\tfrac{3}{4} \text{ in.} - 1.00 \text{ in.})}{1.67}$ $= 449 \text{ kips} > 150 \text{ kips}$ **o.k.**

Design of beam flange-to-bottom gusset weld

The beam flange-to-bottom gusset plate weld is determined as follows using the directional strength increase of AISC *Specification* Equation J2-5, the 1.25 ductility factor discussed in AISC *Manual* Part 13, and AISC *Manual* Equations 8-2a and 8-2b:

LRFD	ASD
Resultant force $R_u = \sqrt{V_u^2 + N_u^2}$ $= \sqrt{(256 \text{ kips})^2 + (213 \text{ kips})^2}$ $= 333 \text{ kips}$	Resultant load $R_a = \sqrt{V_a^2 + N_a^2}$ $= \sqrt{(180 \text{ kips})^2 + (150 \text{ kips})^2}$ $= 234 \text{ kips}$
Load angle $\theta = \tan^{-1}\left(\dfrac{N_u}{V_u}\right)$ $= \tan^{-1}\left(\dfrac{213 \text{ kips}}{256 \text{ kips}}\right)$ $= 39.8°$	Load angle $\theta = \tan^{-1}\left(\dfrac{N_a}{V_a}\right)$ $= \tan^{-1}\left(\dfrac{150 \text{ kips}}{180 \text{ kips}}\right)$ $= 39.8°$
Directional strength increase $1.0 + 0.50 \sin^{1.5} 39.8° = 1.26$ $D_{req'd} = \dfrac{1.25(333 \text{ kips})}{2(1.392 \text{ kip/in.})}$ $\times \dfrac{1}{(28\tfrac{3}{4} \text{ in.} - 1.00 \text{ in.})(1.26)}$ $= 4.28 \text{ sixteenths}$	Directional strength increase $1.0 + 0.50 \sin^{1.5} 39.8° = 1.26$ $D_{req'd} = \dfrac{1.25(234 \text{ kips})}{2(0.928 \text{ kip/in.})}$ $\times \dfrac{1}{(28\tfrac{3}{4} \text{ in.} - 1.00 \text{ in.})(1.26)}$ $= 4.51 \text{ sixteenths}$

Use a double-sided 5/16-in. fillet weld. Design Case II must also be investigated.

Beam stub strength

Check beam stub web local yielding

Because the normal force acts at the centroid of the bottom gusset-to-beam interface, which is less than the depth of the beam stub, d, the available web local yielding strength of the beam stub is determined from AISC *Specification* Equation J10-3 as follows:

LRFD	ASD
$\phi R_n = \phi F_{yw} t_w (2.5 k_{des} + l_b)$ $= 1.00(50 \text{ ksi})(0.650 \text{ in.})$ $\times [2.5(1.59 \text{ in.}) + (28\tfrac{3}{4} \text{ in.} - 1.00 \text{ in.})]$ $= 1{,}030 \text{ kips} > 213 \text{ kips}$ **o.k.**	$\dfrac{R_n}{\Omega} = \dfrac{F_{yw} t_w (2.5 k_{des} + l_b)}{\Omega}$ $= 50 \text{ ksi}(0.650 \text{ in.})$ $\times [2.5(1.59 \text{ in.}) + (28\tfrac{3}{4} \text{ in.} - 1.00 \text{ in.})]$ $\times (1/1.50)$ $= 687 \text{ kips} > 150 \text{ kips}$ **o.k.**

Check beam stub web local crippling

The normal force acts at the centroid of the bottom gusset-to-beam interface, which is greater than $d/2$ from the end of the beam. The available web local crippling strength of the beam stub is determined from AISC *Specification* Equation J10-4:

LRFD	ASD
$\phi R_n = \phi 0.80 t_w^2 \left[1 + 3 \left(\dfrac{l_b}{d} \right) \left(\dfrac{t_w}{t_f} \right)^{1.5} \right]$ $\times \sqrt{\dfrac{E F_{yw} t_f}{t_w}}$ $= 0.75(0.80)(0.650 \text{ in.})^2$ $\times \left[1 + 3 \left(\dfrac{28\tfrac{3}{4} \text{ in.} - 1.00 \text{ in.}}{24.7 \text{ in.}} \right) \right.$ $\left. \times \left(\dfrac{0.650 \text{ in.}}{1.09 \text{ in.}} \right)^{1.5} \right]$ $\times \sqrt{\dfrac{29{,}000 \text{ ksi}(50 \text{ ksi})(1.09 \text{ in.})}{0.650 \text{ in.}}}$ $= 1{,}010 \text{ kips} > 213 \text{ kips}$ **o.k.**	$\dfrac{R_n}{\Omega} = 0.80 t_w^2 \left[1 + 3 \left(\dfrac{l_b}{d} \right) \left(\dfrac{t_w}{t_f} \right)^{1.5} \right]$ $\times \sqrt{\dfrac{E F_{yw} t_f}{t_w}} \left(\dfrac{1}{\Omega} \right)$ $= 0.80(0.650 \text{ in.})^2$ $\times \left[1 + 3 \left(\dfrac{28\tfrac{3}{4} \text{ in.} - 1.00 \text{ in.}}{24.7 \text{ in.}} \right) \right.$ $\left. \times \left(\dfrac{0.650 \text{ in.}}{1.09 \text{ in.}} \right)^{1.5} \right]$ $\times \sqrt{\dfrac{29{,}000 \text{ ksi}(50 \text{ ksi})(1.09 \text{ in.})}{0.650 \text{ in.}}}$ $\times \left(\dfrac{1}{2.00} \right)$ $= 673 \text{ kips} > 150 \text{ kips}$ **o.k.**

5.3 SPECIAL CONCENTRICALLY BRACED FRAMES

Bottom Gusset-to-Beam Interface—Design Case II

The forces at the bottom gusset-to-beam interface, shown in Figures 5-41a and 5-41b, are:

LRFD	ASD
Normal: $N_u = 250$ kips	Normal: $N_a = 175$ kips
Shear: $V_u = 300$ kips	Shear: $V_a = 210$ kips

Check bottom gusset for shear and tension yielding along the beam flange

From previous calculations for Design Case I:

LRFD	ASD
$\phi P_n = 674$ kips > 250 kips **o.k.**	$\dfrac{P_n}{\Omega} = 449$ kips > 175 kips **o.k.**
$\phi V_n = 450$ kips > 300 kips **o.k.**	$\dfrac{V_n}{\Omega} = 300$ kips > 210 kips **o.k.**

Beam flange-to-bottom gusset weld

The beam flange-to-bottom gusset plate weld is determined as follows using the directional strength increase of AISC *Specification* Equation J2-5, the 1.25 ductility factor discussed in AISC *Manual* Part 13, and AISC *Manual* Equations 8-2a and 8-2b:

LRFD	ASD
Resultant force $R_u = \sqrt{V_u^2 + N_u^2}$ $= \sqrt{(300 \text{ kips})^2 + (250 \text{ kips})^2}$ $= 391$ kips	Resultant force $R_a = \sqrt{V_a^2 + N_a^2}$ $= \sqrt{(210 \text{ kips})^2 + (175 \text{ kips})^2}$ $= 273$ kips
Load angle $\theta = \tan^{-1}\left(\dfrac{N_u}{V_u}\right)$ $= \tan^{-1}\left(\dfrac{250 \text{ kips}}{300 \text{ kips}}\right)$ $= 39.8°$	Load angle $\theta = \tan^{-1}\left(\dfrac{N_a}{V_a}\right)$ $= \tan^{-1}\left(\dfrac{175 \text{ kips}}{210 \text{ kips}}\right)$ $= 39.8°$

LRFD	ASD
Directional strength increase $1.0 + 0.50\sin^{1.5} 39.8° = 1.26$	Directional strength increase $1.0 + 0.50\sin^{1.5} 39.8° = 1.26$
$D_{req'd} = \dfrac{1.25(391 \text{ kips})}{2(1.392 \text{ kip/in.})}$ $\times \dfrac{1}{(28\frac{3}{4} \text{ in.} - 1.00 \text{ in.})(1.26)}$ $= 5.02$ sixteenths	$D_{req'd} = \dfrac{1.25(273 \text{ kips})}{2(0.928 \text{ kip/in.})}$ $\times \dfrac{1}{(28\frac{3}{4} \text{ in.} - 1.00 \text{ in.})(1.26)}$ $= 5.26$ sixteenths

Use a double-sided ⅜-in. fillet weld. Design Case II controls.

Check beam stub web local yielding
The available web local yielding strength of the beam is (from previous calculations):

LRFD	ASD
$\phi R_n = 1{,}030$ kips > 250 kips **o.k.**	$\dfrac{R_n}{\Omega} = 687$ kips > 175 kips **o.k.**

Check beam stub web local crippling
The available web local crippling strength of the beam is (from previous calculations):

LRFD	ASD
$\phi R_n = 1{,}010$ kips > 250 kips **o.k.**	$\dfrac{R_n}{\Omega} = 673$ kips > 175 kips **o.k.**

Bottom Gusset-to-Column Interface—Design Case I
The forces at the bottom gusset-to-column interface, shown in Figures 5-40a and 5-40b, are:

LRFD	ASD
Normal: $N_u = 109$ kips	Normal: $N_a = 76.6$ kips
Shear: $V_u = 152$ kips	Shear: $V_a = 107$ kips
Moment: $M_{uc} = 180$ kip-in.	Moment: $M_{ac} = 126$ kip-in.

Combine the axial force and the moment by converting the moment into an equivalent axial force derived from the moment equation for a simply supported member with a concentrated load at midspan (s is the snip dimension of the gusset and $\overline{\beta}$ is the distance to the centroid of the column-to-gusset connection, determined previously):

5.3 SPECIAL CONCENTRICALLY BRACED FRAMES

LRFD	ASD
$N_{u\,equiv} = T_u + \dfrac{4M_{uc}}{2(\overline{\beta}-s)}$	$N_{a\,equiv} = T_a + \dfrac{4M_{ac}}{2(\overline{\beta}-s)}$
$= 109 \text{ kips} + \dfrac{2(180 \text{ kip-in.})}{10.5 \text{ in.} - 1.00 \text{ in.}}$	$= 76.6 \text{ kips} + \dfrac{2(126 \text{ kip-in.})}{10.5 \text{ in.} - 1.00 \text{ in.}}$
$= 147 \text{ kips}$	$= 103 \text{ kips}$

Column flange-to-bottom gusset plate weld

The column flange-to-bottom gusset plate weld is determined as follows using the directional strength increase of AISC *Specification* Equation J2-5, the 1.25 ductility factor discussed in AISC *Manual* Part 13, and AISC *Manual* Equations 8-2a and 8-2b:

LRFD	ASD
Resultant force	Resultant force
$R_u = \sqrt{V_u^2 + N_{u\,equiv}^2}$	$R_a = \sqrt{V_a^2 + N_{a\,equiv}^2}$
$= \sqrt{(152 \text{ kips})^2 + (147 \text{ kips})^2}$	$= \sqrt{(107 \text{ kips})^2 + (103 \text{ kips})^2}$
$= 211 \text{ kips}$	$= 149 \text{ kips}$
Load angle	Load angle
$\theta = \tan^{-1}\left(\dfrac{N_{u\,equiv}}{V_u}\right)$	$\theta = \tan^{-1}\left(\dfrac{N_{a\,equiv}}{V_a}\right)$
$= \tan^{-1}\left(\dfrac{147 \text{ kips}}{152 \text{ kips}}\right)$	$= \tan^{-1}\left(\dfrac{103 \text{ kips}}{107 \text{ kips}}\right)$
$= 44.0°$	$= 43.9°$
Directional strength increase	Directional strength increase
$1.0 + 0.50\sin^{1.5} 44.0° = 1.29$	$1.0 + 0.50\sin^{1.5} 43.9° = 1.29$
$D_{req'd} = \dfrac{1.25(211 \text{ kips})}{2(1.392 \text{ kip/in.})}$	$D_{req'd} = \dfrac{1.25(149 \text{ kips})}{2(0.928 \text{ kip/in.})}$
$\times \dfrac{1}{(19\tfrac{7}{8} \text{ in.} - 1.00 \text{ in.})(1.29)}$	$\times \dfrac{1}{(19\tfrac{7}{8} \text{ in.} - 1.00 \text{ in.})(1.29)}$
$= 3.89 \text{ sixteenths}$	$= 4.12 \text{ sixteenths}$

Use a double-sided $\tfrac{5}{16}$-in. fillet weld. Design Case II must also be investigated.

Check bottom gusset plate for shear and tensile yielding along the column flange

The available shear yielding strength of the gusset plate at the column flange interface is determined from AISC *Specification* Equation J4-3, and the available tensile yielding strength at the column flange interface is determined from AISC *Specification* Equation J4-1 as follows:

LRFD	ASD
Shear: $$\phi V_n = \phi 0.60 F_y A_{gv}$$ $$= 1.00(0.60)(36 \text{ ksi})(\tfrac{3}{4} \text{ in.})$$ $$\times (19\tfrac{7}{8} \text{ in.} - 1.00 \text{ in.})$$ $$= 306 \text{ kips} > 152 \text{ kips} \quad \textbf{o.k.}$$	Shear: $$\frac{V_n}{\Omega} = \frac{0.60 F_y A_{gv}}{\Omega}$$ $$= 0.60(36 \text{ ksi})$$ $$\times (\tfrac{3}{4} \text{ in.})(19\tfrac{7}{8} \text{ in.} - 1.00 \text{ in.})$$ $$\times (1/1.50)$$ $$= 204 \text{ kips} > 107 \text{ kips} \quad \textbf{o.k.}$$
Normal: $$\phi P_n = \phi F_y A_g$$ $$= 0.90(36 \text{ ksi})(\tfrac{3}{4} \text{ in.})$$ $$\times (19\tfrac{7}{8} \text{ in.} - 1.00 \text{ in.})$$ $$= 459 \text{ kips} > 147 \text{ kips} \quad \textbf{o.k.}$$	Normal: $$\frac{P_n}{\Omega} = \frac{F_y A_g}{\Omega}$$ $$= \frac{36 \text{ ksi}(\tfrac{3}{4} \text{ in.})(19\tfrac{7}{8} \text{ in.} - 1.00 \text{ in.})}{1.67}$$ $$= 305 \text{ kips} > 103 \text{ kips} \quad \textbf{o.k.}$$

Check column web local yielding

Because the normal force is applied at a distance from the column end that is greater than or equal to the column depth, d, the available web local yielding strength of the column from AISC *Specification* Equation J10-2 is:

LRFD	ASD
$$\phi R_n = \phi F_{yw} t_w (5k_{des} + l_b)$$ $$= 1.00(50 \text{ ksi})(0.550 \text{ in.})$$ $$\times [5(1.50 \text{ in.}) + (19\tfrac{7}{8} \text{ in.} - 1.00 \text{ in.})]$$ $$= 725 \text{ kips} > 147 \text{ kips} \quad \textbf{o.k.}$$	$$\frac{R_n}{\Omega} = \frac{F_{yw} t_w (5k_{des} + l_b)}{\Omega}$$ $$= 50 \text{ ksi}(0.550 \text{ in.})$$ $$\times [5(1.50 \text{ in.}) + (19\tfrac{7}{8} \text{ in.} - 1.00 \text{ in.})]$$ $$\times (1/1.50)$$ $$= 484 \text{ kips} > 103 \text{ kips} \quad \textbf{o.k.}$$

5.3 SPECIAL CONCENTRICALLY BRACED FRAMES

Check column web local crippling

Because the normal force is applied at a distance from the column end that is greater than or equal to $d/2$, the available web local crippling strength of the column from AISC *Specification* Equation J10-4 is:

LRFD	ASD
$\phi R_n = \phi 0.80 t_w^2 \left[1 + 3\left(\dfrac{l_b}{d}\right)\left(\dfrac{t_w}{t_f}\right)^{1.5}\right]$ $\times \sqrt{\dfrac{EF_{yw}t_f}{t_w}}$ $= 0.75(0.80)(0.550 \text{ in.})^2$ $\times \left[1 + 3\left(\dfrac{19\% \text{ in.} - 1.00 \text{ in.}}{12.7 \text{ in.}}\right) \times \left(\dfrac{0.550 \text{ in.}}{0.900 \text{ in.}}\right)^{1.5}\right]$ $\times \sqrt{\dfrac{29{,}000 \text{ ksi}(50 \text{ ksi})(0.900 \text{ in.})}{0.550 \text{ in.}}}$ $= 875 \text{ kips} > 147 \text{ kips}$ **o.k.**	$\dfrac{R_n}{\Omega} = 0.80 t_w^2 \left[1 + 3\left(\dfrac{l_b}{d}\right)\left(\dfrac{t_w}{t_f}\right)^{1.5}\right]$ $\times \sqrt{\dfrac{EF_{yw}t_f}{t_w}} \left(\dfrac{1}{\Omega}\right)$ $= 0.80(0.550 \text{ in.})^2$ $\times \left[1 + 3\left(\dfrac{19\% \text{ in.} - 1.00 \text{ in.}}{12.7 \text{ in.}}\right) \times \left(\dfrac{0.550 \text{ in.}}{0.900 \text{ in.}}\right)^{1.5}\right]$ $\times \sqrt{\dfrac{29{,}000 \text{ ksi}(50 \text{ ksi})(0.900 \text{ in.})}{0.550 \text{ in.}}}$ $\times \left(\dfrac{1}{2.00}\right)$ $= 583 \text{ kips} > 103 \text{ kips}$ **o.k.**

Check column web shear strength

From AISC *Manual* Table 3-6, the available shear strength of the W12×96 column is:

LRFD	ASD
$\phi V_n = 210 \text{ kips} > 109 \text{ kips}$ **o.k.**	$\dfrac{V_n}{\Omega} = 140 \text{ kips} > 76.6 \text{ kips}$ **o.k.**

Bottom Gusset to Column Interface—Design Case II

The forces at the bottom gusset-to-column interface, shown in Figures 5-41a and 5-41b, are:

LRFD	ASD
Normal: $N_u = 128$ kips	Normal: $N_a = 89.4$ kips
Shear: $V_u = 178$ kips	Shear: $V_a = 125$ kips
Moment: $M_{uc} = 211$ kip-in.	Moment: $M_{ac} = 148$ kip-in.

Similar to previous calculations, the axial force and moment are combined by converting the moment into an equivalent axial force:

LRFD	ASD
$N_{u\,equiv} = N_u + \dfrac{4M_{uc}}{2(\overline{\beta}-s)}$	$N_{a\,equiv} = N_a + \dfrac{4M_{ac}}{2(\overline{\beta}-s)}$
$= 128 \text{ kips} + \dfrac{2(211 \text{ kip-in.})}{10.5 \text{ in.} - 1.00 \text{ in.}}$	$= 89.4 \text{ kips} + \dfrac{2(148 \text{ kip-in.})}{10.5 \text{ in.} - 1.00 \text{ in.}}$
$= 172 \text{ kips}$	$= 121 \text{ kips}$

Design of column flange-to-bottom gusset weld

The column flange-to-bottom gusset plate weld is determined as follows using the directional strength increase of AISC *Specification* Equation J2-5, the 1.25 ductility factor discussed in AISC *Manual* Part 13, and AISC *Manual* Equations 8-2a and 8-2b:

LRFD	ASD
Resultant force	Resultant force
$R_u = \sqrt{V_u^2 + N_{u\,equiv}^2}$	$R_a = \sqrt{V_a^2 + N_{a\,equiv}^2}$
$= \sqrt{(178 \text{ kips})^2 + (172 \text{ kips})^2}$	$= \sqrt{(125 \text{ kips})^2 + (121 \text{ kips})^2}$
$= 248 \text{ kips}$	$= 174 \text{ kips}$
Load angle	Load angle
$\theta = \tan^{-1}\left(\dfrac{N_{u\,equiv}}{V_u}\right)$	$\theta = \tan^{-1}\left(\dfrac{N_{a\,equiv}}{V_a}\right)$
$= \tan^{-1}\left(\dfrac{172 \text{ kips}}{178 \text{ kips}}\right)$	$= \tan^{-1}\left(\dfrac{121 \text{ kips}}{125 \text{ kips}}\right)$
$= 44.0°$	$= 44.1°$
Directional strength increase	Directional strength increase
$1.0 + 0.50\sin^{1.5} 44.0° = 1.29$	$1.0 + 0.50\sin^{1.5} 44.1° = 1.29$
$D_{req'd} = \dfrac{1.25(248 \text{ kips})}{2(1.392 \text{ kip/in.})}$	$D_{req'd} = \dfrac{1.25(174 \text{ kips})}{2(0.928 \text{ kip/in.})}$
$\times \dfrac{1}{(19\frac{7}{8} \text{ in.} - 1.00 \text{ in.})(1.29)}$	$\times \dfrac{1}{(19\frac{7}{8} \text{ in.} - 1.00 \text{ in.})(1.29)}$
$= 4.57 \text{ sixteenths}$	$= 4.81 \text{ sixteenths}$

Use a double-sided 5/16-in. fillet weld.

Check bottom gusset shear and tensile yielding along the column flange

From previous calculations for Design Case I, the available shear yielding and available tensile yielding strengths of the bottom gusset are:

LRFD	ASD
Shear:	Shear:
$\phi V_n = 306$ kips > 178 kips **o.k.**	$\dfrac{V_n}{\Omega} = 204$ kips > 125 kips **o.k.**
Normal:	Normal:
$\phi N_n = 459$ kips > 172 kips **o.k.**	$\dfrac{N_n}{\Omega} = 305$ kips > 121 kips **o.k.**

Check column web local yielding

From previous calculations for Design Case I, the available column web local yielding strength is:

LRFD	ASD
$\phi R_n = 725$ kips > 172 kips **o.k.**	$\dfrac{R_n}{\Omega} = 484$ kips > 121 kips **o.k.**

Check column web local crippling

From previous calculations for Design Case I, the available column web local crippling strength is:

LRFD	ASD
$\phi R_n = 875$ kips > 172 kips **o.k.**	$\dfrac{R_n}{\Omega} = 583$ kips > 121 kips **o.k.**

Check column web shear strength

From previous calculations for Design Case I, the available shear strength of the W12×96 column is:

LRFD	ASD
$\phi V_n = 210$ kips > 128 kips **o.k.**	$\dfrac{V_n}{\Omega} = 140$ kips > 89.4 kips **o.k.**

The complete design is shown in Figure 5-33.

Alternate Detail Using a Continuous Gusset Plate

An alternate detail using a continuous gusset plate instead of a beam stub is shown in Figure 5-43. This alternate uses a ¾-in.-thick gusset plate with plate reinforcement in lieu of the W24×146 beam stub and eliminates many welds. Note that the horizontal dimension 2α is used to set the gusset horizontal dimension.

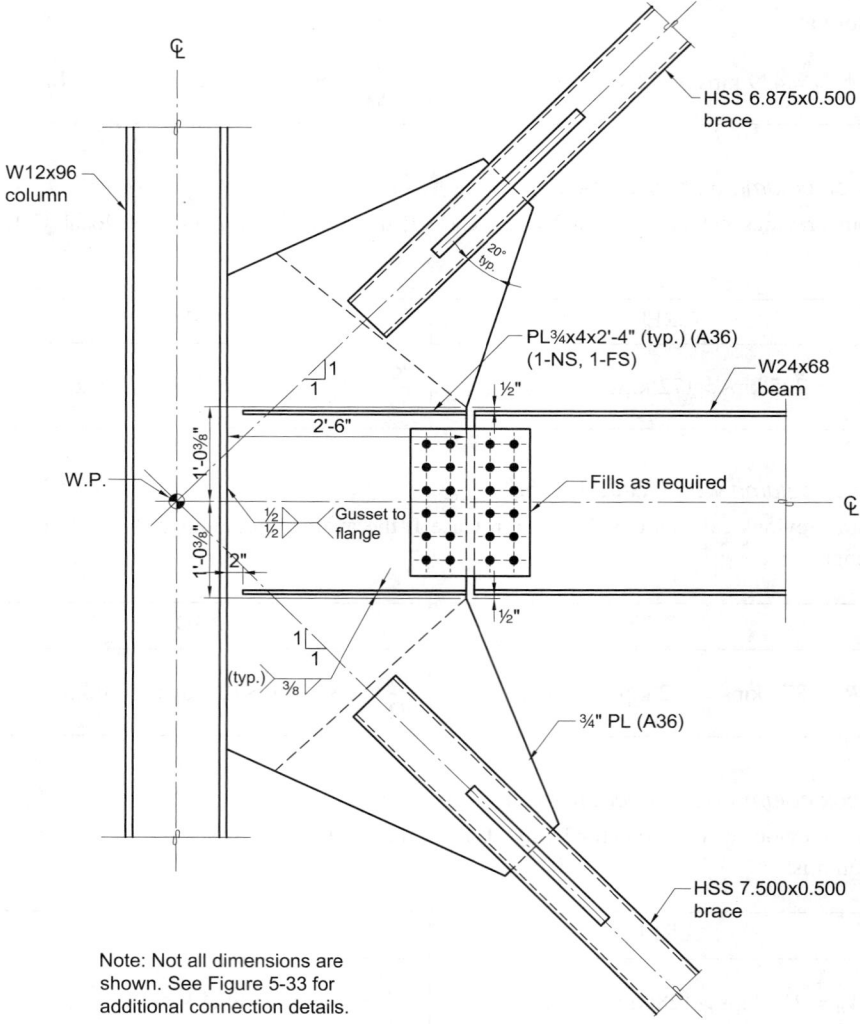

Fig. 5-43. Alternate design using continuous gusset plate.

Example 5.3.11. SCBF Brace-to-Beam/Column Connection Design with Elliptical Clearance and Fixed Beam-to-Column Connection

Given:

Refer to Joint JT-1 at the third level in Figure 5-44 (the plan is given in Figure 5-10). Design the connection between brace, beam and column. Use an ASTM A36 welded gusset plate concentric to the braces and 70-ksi electrodes to connect the brace to the gusset plate and the gusset plate to the beam and column. Use ASTM A572 Grade 50 continuity plates. The brace is an ASTM A500 Grade B round HSS, the beam is an ASTM A992 W24×68, and the column is a W12×96. The applicable building code specifies the use of ASCE/SEI 7 for calculation of loads. The gravity shear forces at the end of the beam are:

$V_D = 4.50$ kips
$V_L = 3.00$ kips

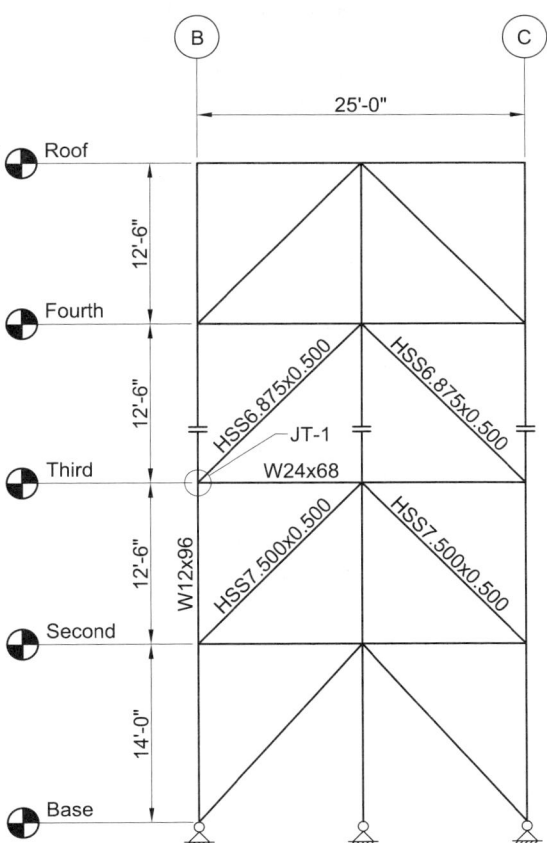

Fig. 5-44. Frame elevation.

This example illustrates an alternative method for gusset plate design to that provided in the AISC *Seismic Provisions* Commentary Figure C-F2.9, proposed by Lehman et al. (2008), to accommodate brace buckling as required by AISC *Seismic Provisions* Section F2.6c(3). In lieu of the $2t_p$ linear brace offset, an $8t_p$ elliptical offset may be used. In particular, for rectangular gusset plates, the $8t_p$ elliptical offset offers a more compact gusset plate with equivalent deformation capacity to accommodate the out-of-plane brace movement. However, for gusset plates that have significant taper, as is the case in Example 5.3.10, the $8t_p$ elliptical offset and the $2t_p$ linear offset offer similar results. This example, as shown in Figure 5-45, illustrates a connection design using the elliptical offset method applied to a rectangular gusset plate.

In addition to illustrating the application of the elliptical clearance methodology, this example uses a fixed beam-to-column connection to satisfy the requirements of item (b) of AISC *Seismic Provisions* Section F2.6b. In the design, the beam web and flanges are welded to the column flange with CJP groove welds. The flange weld requires a substantial corner clip in the gusset plate for access. This clip is detailed as 1.5 in. In this example, the clip is considered for rupture limit states, but it is ignored for yielding limit states.

Some features of this example, including the elliptical clearance, the fixed beam-to-column connection, and the sizing of welds at the gusset plate interfaces are provided as an alternative to Example 5.3.10. The brace-to-gusset calculations are not shown in this example because they are similar to Example 5.3.10.

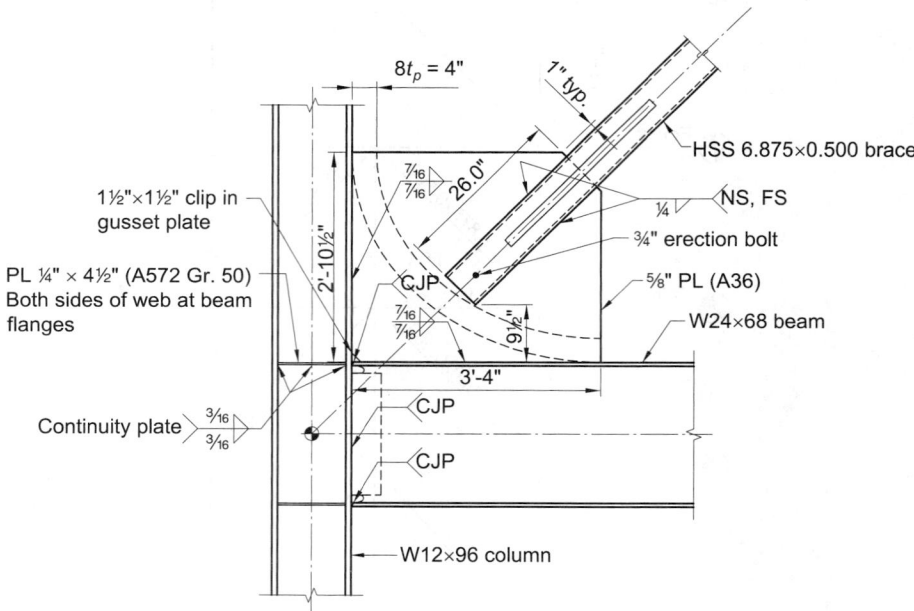

Notes:
Welds of web and doubler/shear plate to column flange are demand critical.

Fig. 5-45. Rectangular gusset plate with $8t_p$ elliptical brace offset addressed in Example 5.3.11.

5.3 SPECIAL CONCENTRICALLY BRACED FRAMES

Solution:

From AISC *Manual* Tables 2-4 and 2-5 and AISC *Seismic Provisions* Table A3.1, the material properties are as follows:

ASTM A36
$F_y = 36$ ksi
$F_u = 58$ ksi

ASTM A572 Grade 50
$F_y = 50$ ksi
$F_u = 65$ ksi

ASTM A500 Grade B
$F_y = 42$ ksi
$F_u = 58$ ksi
$R_y = 1.4$
$R_t = 1.3$

ASTM A992
$F_y = 50$ ksi
$F_u = 65$ ksi

From AISC *Manual* Table 1-1, the geometric properties are as follows:

Brace (above the beam)
HSS 6.875×0.500
$t_{nom} = 0.500$ in. $t_{des} = 0.465$ in. $A = 9.36$ in.² $r = 2.27$ in.

Brace (below the beam)
HSS 7.50×0.500
$t_{nom} = 0.500$ in. $t_{des} = 0.465$ in. $A = 10.3$ in.² $r = 2.49$ in.

Beam
W24×68
$d = 23.7$ in. $t_w = 0.415$ in. $b_f = 8.97$ in. $t_f = 0.585$ in.
$k_{des} = 1.09$ in.

Column
W12×96
$d = 12.7$ in. $t_w = 0.550$ in. $b_f = 12.2$ in. $t_f = 0.900$ in.
$k_{des} = 1.50$ in.

Required Strength
For the HSS6.875×0.500 brace above the beam

According to AISC *Seismic Provisions* Section F2.3(i), the amplified seismic load is determined from the expected strengths of the brace in compression and in tension. The expected tensile strength of the brace is determined as follows.

From AISC *Seismic Provisions* Section F2.3 and Table A3.1:

$R_y = 1.4$

$$P_{tension} = R_y F_y A_g$$
$$= 1.4(42 \text{ ksi})(9.36 \text{ in.}^2)$$
$$= 550 \text{ kips}$$

The required tensile strength due to seismic loading is:

LRFD	ASD
$P_u = 1.0 E_{mh}$	$P_a = 0.7 E_{mh}$
$= 550$ kips	$= 0.7(550 \text{ kips})$
	$= 385$ kips

The expected strength of the brace in compression (using $L = 12$ ft as the actual length of the brace) is determined as follows:

$$\frac{KL}{r} = \frac{1.0(12.0 \text{ ft})(12 \text{ in./ft})}{2.27 \text{ in.}}$$
$$= 63.4$$

$$4.71\sqrt{\frac{E}{R_y F_y}} = 4.71\sqrt{\frac{29,000 \text{ ksi}}{1.4(42 \text{ ksi})}}$$
$$= 105$$

With $\frac{KL}{r} \leq 4.71\sqrt{\frac{E}{R_y F_y}}$:

$$F_e = \frac{\pi^2 E}{\left(\frac{KL}{r}\right)^2} \qquad \text{(\textit{Spec.} Eq. E3-4)}$$

$$= \frac{\pi^2 (29,000 \text{ ksi})}{(63.4)^2}$$
$$= 71.2 \text{ ksi}$$

$$F_{cre} = \left[0.658^{\frac{R_y F_y}{F_e}}\right] R_y F_y \qquad \text{(from \textit{Spec.} Eq. E3-2)}$$

$$= \left[0.658^{\frac{1.4(42 \text{ ksi})}{71.2 \text{ ksi}}}\right](1.4)(42 \text{ ksi})$$
$$= 41.6 \text{ ksi}$$

5.3 SPECIAL CONCENTRICALLY BRACED FRAMES

The expected compressive strength of the brace is:

$$P_{compression} = 1.14 F_{cre} A_g$$
$$= 1.14(41.6 \text{ ksi})(9.36 \text{ in.}^2)$$
$$= 444 \text{ kips}$$

The required compressive strength due to seismic loading is:

LRFD	ASD
$P_u = 1.0 E_{mh}$ $= 444 \text{ kips}$	$P_a = 0.7 E_{mh}$ $= 0.7(444 \text{ kips})$ $= 311 \text{ kips}$

The expected strength of the brace in compression (using $L = 12$ ft) at its post-buckling strength is:

$$0.3 P_{compression} = 0.3(444 \text{ kips})$$
$$= 133 \text{ kips}$$

The required compressive strength based on post-buckling strength is:

LRFD	ASD
$P_u = 1.0 E_{mh}$ $= 133 \text{ kips}$	$P_a = 0.7 E_{mh}$ $= 0.7(133 \text{ kips})$ $= 93.1 \text{ kips}$

For the HSS7.500×0.500 brace below the beam

The connection of the brace below the beam is not designed as part of this example for Joint JT-1, but the brace member size is important when considering the analysis provisions of AISC *Seismic Provisions* Section F2.3.

The expected strength of the brace in tension is determined as follows.

From AISC *Seismic Provisions* Table A3.1:

$R_y = 1.4$

$$P_{tension} = R_y F_y A_g$$
$$= 1.4(42 \text{ ksi})(10.3 \text{ in.}^2)$$
$$= 606 \text{ kips}$$

The required tensile strength due to seismic loading is:

LRFD	ASD
$P_u = 1.0 E_{mh}$ $= 606$ kips	$P_a = 0.7 E_{mh}$ $= 0.7(606 \text{ kips})$ $= 424$ kips

The expected strength of the brace in compression (using $L = 12$ ft) is determined as follows:

$$\frac{KL}{r} = \frac{1.0(12.0 \text{ ft})(12 \text{ in./ft})}{2.49 \text{ in.}}$$
$$= 57.8$$

$$4.71\sqrt{\frac{E}{R_y F_y}} = 4.71\sqrt{\frac{29,000 \text{ ksi}}{1.4(42 \text{ ksi})}}$$
$$= 105$$

With $\dfrac{KL}{r} \leq 4.71\sqrt{\dfrac{E}{R_y F_y}}$:

$$F_e = \frac{\pi^2 E}{\left(\dfrac{KL}{r}\right)^2} \qquad (Spec.\ Eq.\ E3\text{-}4)$$

$$= \frac{\pi^2 (29,000 \text{ ksi})}{(57.8)^2}$$
$$= 85.7 \text{ ksi}$$

$$F_{cre} = \left[0.658^{\frac{R_y F_y}{F_e}}\right] R_y F_y \qquad (\text{from } Spec.\ Eq.\ E3\text{-}2)$$

$$= \left[0.658^{\frac{1.4(42 \text{ ksi})}{85.7 \text{ ksi}}}\right] (1.4)(42 \text{ ksi})$$
$$= 44.1 \text{ ksi}$$

The expected compressive strength of the brace below the beam is:

$$P_{compression} = 1.14 F_{cre} A_g$$
$$= 1.14(44.1 \text{ ksi})(10.3 \text{ in.}^2)$$
$$= 518 \text{ kips}$$

The required compressive strength is:

5.3 SPECIAL CONCENTRICALLY BRACED FRAMES

LRFD	ASD
$P_u = 1.0 E_{mh}$ $= 518$ kips	$P_a = 0.7 E_{mh}$ $= 0.7(518$ kips$)$ $= 363$ kips

The expected strength of the brace in compression (using $L = 12$ ft) at its post-buckling strength is:

$$0.3 P_{compression} = 0.3(518 \text{ kips})$$
$$= 155 \text{ kips}$$

The required compressive strength based on post-buckling strength is:

LRFD	ASD
$P_u = 1.0 E_{mh}$ $= 155$ kips	$P_a = 0.7 E_{mh}$ $= 0.7(155$ kips$)$ $= 109$ kips

The brace-to-gusset connection and brace reinforcement will not be addressed in this example. As in Example 5.3.10, the brace-to-gusset weld will be ¼-in. fillet welds that are 26 in. long.

For reference, the final design using these methodologies is shown in Figure 5-45. The symbols used are shown in Figure 5-46.

Gusset Plate Design

The geometry of the gusset plate and location of the end of the brace are established using the approach described in Lehman et al. (2008). The calculations for the brace connection are shown in the following. The horizontal gusset dimension, a, has been chosen as 40 in. and the vertical dimension is calculated. These values result in an economical gusset plate thickness and weld sizes. The value of a is based on iterations using the method outlined in Lehman et al. and allows for a brace-to-gusset weld length of 26 in.

From the geometry in Figure 5-46 and based on the choice of $a = 40$ in., the gusset length along the column flange is:

$$b = (a + e_c)\tan\gamma - e_b$$
$$= [40 \text{ in.} + \tfrac{1}{2}(12.7 \text{ in.})]\tan 45° - \tfrac{1}{2}(23.7 \text{ in.})$$
$$= 34.5 \text{ in.}$$

Where b is the vertical gusset dimension, $\gamma = 45°$ is the angle between the brace and the horizontal as shown in Figure 5-46 and determined from the elevation geometry in Figure 5-44, and e_c and e_b are the eccentricities of the gusset edges from the column and beam centerlines, respectively (that is, half the member depth).

Half of the lengths of the major and minor axis of the ellipse are then calculated using a gusset plate thickness of ⅝ in. based on yielding on the Whitmore section.

Check required gusset plate thickness based on the limit state of tensile yielding

Tension yielding is checked on a section of the gusset plate commonly referred to as the Whitmore section. This section is explained in AISC *Manual* Part 9 (Figure 9-1) and in Thornton and Lini (2011). The width of the Whitmore section is determined based on a 30° spread.

$$w_p = 2l_w \tan 30° + D_{brace}$$
$$= 2(26 \text{ in.})(\tan 30°) + 6.875 \text{ in.}$$
$$= 36.9 \text{ in.}$$

From AISC *Specification* Equation J4-1, the available tensile yielding strength is:

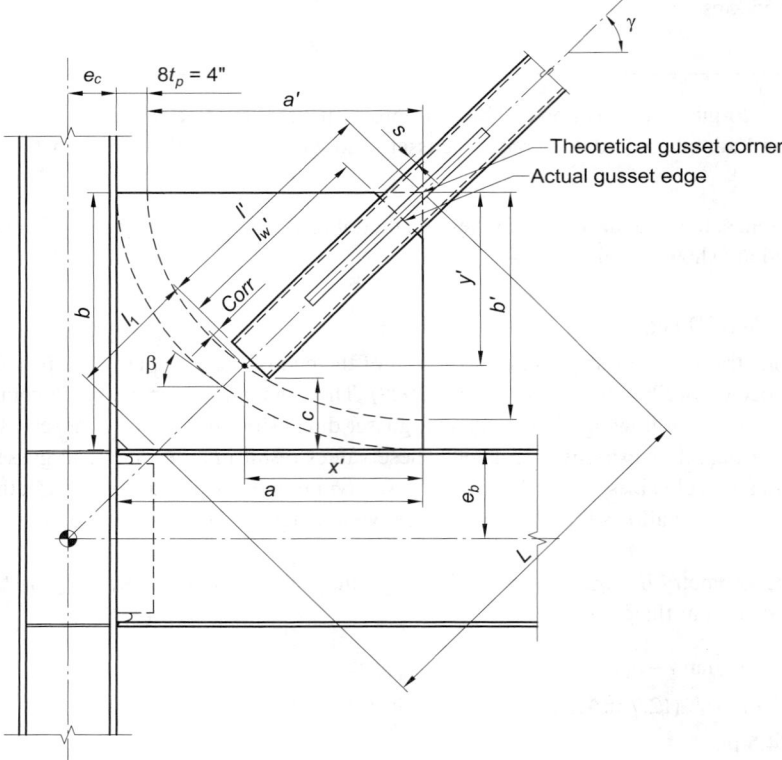

Fig. 5-46. Illustration of symbols used for lengths and angles.

5.3 SPECIAL CONCENTRICALLY BRACED FRAMES

LRFD	ASD
$\phi R_n = \phi F_y A_g$	$\dfrac{R_n}{\Omega} = \dfrac{F_y A_g}{\Omega}$
Setting this equal to the required tensile strength of the brace connection, and with $A_g = t_p w_p$, the gusset plate thickness is:	Setting this equal to the required tensile strength of the brace connection, and with $A_g = t_p w_p$, the gusset plate thickness is:
$t_p = \dfrac{P_u}{\phi F_y w_p}$ $= \dfrac{550 \text{ kips}}{0.90(36 \text{ ksi})(36.9 \text{ in.})}$ $= 0.460 \text{ in.}$	$t_p = \dfrac{\Omega P_a}{F_y w_p}$ $= \dfrac{1.67(385 \text{ kips})}{(36 \text{ ksi})(36.9 \text{ in.})}$ $= 0.484 \text{ in.}$

Try a ⅝-in.-thick gusset plate.

This calculation does not include any reduction considering that the Whitmore width extends into the web of the column or beam. If the Whitmore width enters into a beam or column web that is substantially thinner than the gusset, there is a potential for web local yielding.

In the configuration selected, the Whitmore width does not intrude into the beam or column web. This can be demonstrated by a geometric evaluation.

Determine geometry of the gusset plate

The determination of the location of the end of the brace, as determined in the following, is based on the methodology described in Lehman et al. (2008); the equations in the following are updated from the reference. The location may also be determined from Kotulka (2007). Note that the determination of the final dimensions of the gusset plate based on either method is iterative.

$b' = b - 8t_p$
 $= 34.5 \text{ in.} - 8(⅝ \text{ in.})$
 $= 29.5 \text{ in.}$

$a' = a - 8t_p$
 $= 40.0 \text{ in.} - 8(⅝ \text{ in.})$
 $= 35.0 \text{ in.}$

The aspect ratio of the ellipse is:

$$\rho = \frac{a'}{b'}$$
$$= \frac{35.0 \text{ in.}}{29.5 \text{ in.}}$$
$$= 1.19$$

The dimension y' defines the y-coordinate of the intersection of the brace axis with the ellipse:

$$y' = a'\sqrt{\frac{1}{\cot^2\gamma + \rho^2}}$$
$$= 35.0 \text{ in.}\sqrt{\frac{1}{\cot^2 45° + (1.19)^2}}$$
$$= 22.5 \text{ in.}$$

The x-coordinate of the ellipse is then found from:

$$x' = a'\sqrt{1 - \left(\frac{y'}{b'}\right)^2}$$
$$= 35.0 \text{ in.}\sqrt{1 - \left(\frac{22.5 \text{ in.}}{29.5 \text{ in.}}\right)^2}$$
$$= 22.6 \text{ in.}$$

To ensure that the entire brace cross section remains clear of the elliptical zone, the brace is shifted from the x' and y' coordinates using the correction factor, $Corr$, calculated in the following:

$$\beta = \tan^{-1}\left(\frac{x'}{y'\rho^2}\right)$$
$$= \tan^{-1}\left[\frac{22.6 \text{ in.}}{(22.5 \text{ in.})(1.19)^2}\right]$$
$$= 35.3°$$

$$Corr = \frac{D_{brace}}{2}\tan(90° - \beta - \gamma)$$
$$= \frac{6.875 \text{ in.}}{2}\tan(90° - 35.3° - 45.0°)$$
$$= 0.588$$

In the preceding equation, $D_{brace}/2$ is expressed as c in Lehman et al. (2008), and is defined as the distance from the brace centroidal axis to the extreme fiber of the brace ($D/2$ in this case).

The maximum distance from the theoretical gusset corner to the end of the brace is l':

$$l' = \sqrt{(x')^2 + (y')^2} - Corr$$
$$= \sqrt{(22.6 \text{ in.})^2 + (22.5 \text{ in.})^2} - 0.588 \text{ in.}$$
$$= 31.3 \text{ in.}$$

The brace length overlapping the gusset plate must then be checked to ensure that there is adequate length for the required weld:

$$l'_w = l' - \left(\frac{D_{brace}}{2} + s\right)\cot\gamma$$
$$= 31.3 \text{ in.} - \left(\frac{6.875 \text{ in.}}{2} + 1.00 \text{ in.}\right)\cot 45°$$
$$= 26.9 \text{ in.}$$

where s is the "shoulder" of the gusset at the brace as shown in Figures 5-45 and 5-46.

This is greater than the 26 in. required for the ¼-in. fillet welds (determined in Example 5.3.10). Therefore, the geometry of the gusset plate is now set. If l'_w were less than 26 in., then the gusset plate height and width would have to be increased.

The thickness of the gusset plate was tentatively assumed to be ⅝ in. and needs to be verified for the limit states of block shear rupture and compression buckling.

Check compression buckling on the Whitmore section

The limit state of compression buckling is checked using AISC *Specification* Section J4.4. First determine KL/r as follows.

The length of the brace centerline from the theoretical gusset corner to the intersection with the beam flange is calculated as:

$$L = \frac{b}{\sin\gamma}$$
$$= \frac{34.5 \text{ in.}}{\sin 45°}$$
$$= 48.8 \text{ in.}$$

The centerline length of buckling, l_1, is:

$$l_1 = L - l'$$
$$= 48.8 \text{ in.} - 31.3 \text{ in.}$$
$$= 17.5 \text{ in.}$$

The radius of gyration of the gusset plate is:

$$r = \frac{t_p}{\sqrt{12}}$$

$$= \frac{\text{⅝ in.}}{\sqrt{12}}$$

$$= 0.180 \text{ in.}$$

The elliptical clearance provided in this example results in an extended corner gusset plate, therefore, from Dowswell (2006), use $K = 0.6$.

$$\frac{KL}{r} = \frac{0.6(17.5 \text{ in.})}{0.180 \text{ in.}}$$

$$= 58.3$$

From AISC *Manual* Table 4-22 with $F_y = 36$ ksi and $\frac{KL}{r} = 58.3$:

LRFD	ASD
$\phi_c F_{cr} = 27.1$ ksi	$\frac{F_{cr}}{\Omega_c} = 18.0$ ksi

AISC *Seismic Provisions* Sections F2.3(i) and F2.3(ii) define the two mechanism analyses that must be considered in determining the required strength of connections. AISC *Seismic Provisions* Section F2.6c specifies the required strength of bracing connections, in particular. For the required compressive strength, Section F2.6c(2) has an additional 1.1 factor (relative to the requirements of Section F2.3) applied to the required strength of the connection. The requirements of AISC *Seismic Provisions* Section F2.3 will be used for both LRFD and ASD, except for the limit state of compression buckling on the Whitmore section, which will use the 1.1 factor specified in AISC *Seismic Provisions* Section F2.6c(2).

Therefore, the required compressive strength of the bracing connection is based on the expected compressive strength of the brace due to seismic loading as determined previously, with the 1.1 factor applied. From AISC *Specification* Equation E3-1, the available compressive strength at the Whitmore section, based on flexural buckling, is:

LRFD	ASD
$\phi P_n = \phi_c F_{cr} A_g$ $= 27.1$ ksi$(36.9$ in.$)($⅝ in.$)$ $= 625$ kips 625 kips > 1.1 $(444$ kips$) = 488$ kips **o.k.**	$\frac{P_n}{\Omega} = \left(\frac{F_{cr}}{\Omega_c}\right) A_g$ $= 18.0$ ksi$(36.9$ in.$)($⅝ in.$)$ $= 415$ kips 415 kips > 1.1 $(311$ kips$) = 342$ kips **o.k.**

Therefore, the ⅝-in.-thick gusset plate is acceptable.

5.3 SPECIAL CONCENTRICALLY BRACED FRAMES

Check block shear rupture of the gusset plate

$$R_n = 0.60F_u A_{nv} + U_{bs}F_u A_{nt} \leq 0.60F_y A_{gv} + U_{bs}F_u A_{nt} \qquad (Spec.\ Eq.\ J4\text{-}5)$$

Because the gross shear area, A_{gv}, and the net shear area, A_{nv}, are equal in this case, the shear yielding component, $0.60F_y A_{gv}$, governs over the shear rupture component, $0.60F_u A_{nv}$.

$$R_n = 0.60F_y(2)l_w t_p + U_{bs}F_u D_{brace}t_p$$
$$= 0.60(36\ \text{ksi})(2)(26.0\ \text{in.})t_p + 1.0(58\ \text{ksi})(6.875\ \text{in.})t_p$$
$$= 1{,}520\ \text{kip/in.}(t_p)$$

LRFD	ASD
$t_p \geq \dfrac{P_u}{\phi(1{,}520\ \text{kip/in.})}$	$t_p \geq \dfrac{\Omega P_a}{1{,}520\ \text{kip/in.}}$
$\geq \dfrac{550\ \text{kips}}{0.75(1{,}520\ \text{kip/in.})}$	$\geq \dfrac{2.00(385\ \text{kips})}{1{,}520\ \text{kip/in.}}$
$= 0.482\ \text{in.}$	$= 0.507\ \text{in.}$

Use a ⅝-in.-thick gusset plate.

Gusset Analysis

In order to perform the gusset plate checks at vertical and horizontal sections at the interfaces with the beam and column and to perform checks of local limit states within the beam and column, it is necessary to obtain design forces by performing an analysis of the gusset.

For the design method illustrated in this example, the checks of the gusset plate at these vertical and horizontal sections will necessarily be satisfied as a consequence of satisfying the check of yielding of the Whitmore section and of designing the fillet welds at the gusset-beam and gusset-column interfaces to be stronger than the gusset plate. Nevertheless, it is necessary to derive the forces on these interfaces in order to obtain forces for the web local yielding and web local crippling checks on the beam and column.

In this example, the Parallel Force Method (also known as the Ricker method) will be used for simplicity (Thornton, 1991).

Note: Alternatively, the Uniform Force Method is also applicable to this connection. Because of the proportioning of the gusset plate in this example, the Uniform Force Method will result in moments being assigned to the vertical and horizontal interfaces. The forces used to evaluate the limit states of web local yielding and web local crippling would then be adjusted to include these moments as illustrated in Example 5.3.10.

The Parallel Force Method has a disadvantage relative to the Uniform Force Method in that minor moments result at the column face. However, the use of a rigid beam-to-column connection is generally sufficient to resist such moments and they may be disregarded under these conditions.

In the Parallel Force Method, eccentricities are calculated from the brace centerline to the centroids of the gusset plate welds at the beam and column faces. The gusset-to-beam connection is designed for the required shear force, H_b, and the required normal force, V_b. The gusset-to-column connection is designed for the required shear force, V_c, and the required normal force, H_c. As shown in Figure 5-47, a line perpendicular to the brace axis which passes through the centroid of the gusset-to-column-flange interface may be used to find the eccentricity. (This is also done for the gusset-to-beam flange interface.) As discussed previously, total gusset lengths are used for evaluating yielding limit states; local effects due to the corner clip are considered only for rupture limit states.

At the column flange, the gusset-to-column flange centroid is located at this point, relative to the working point:

$$(x_c, y_c) = \left(\frac{d_c}{2}, \frac{d_b}{2} + \frac{b}{2}\right)$$

$$= \left(\frac{12.7 \text{ in.}}{2}, \frac{23.7 \text{ in.}}{2} + \frac{34.5 \text{ in.}}{2}\right)$$

$$= (6.35 \text{ in.}, 29.1 \text{ in.})$$

The point on the brace centerline that is the intersection of a line through this point perpendicular to the brace centerline is given by these equations (as shown in Figure 5-47), with the working point taken as the origin:

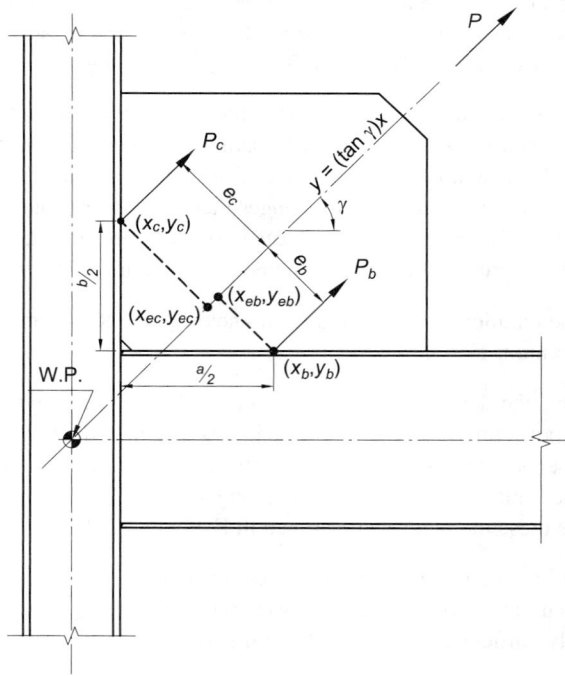

Fig. 5-47. Geometric method of establishing eccentricity from brace centerline.

5.3 SPECIAL CONCENTRICALLY BRACED FRAMES

$$x_{ec} = \frac{(\tan\gamma)y_c + x_c}{\tan^2\gamma + 1}$$

$$= \frac{(\tan 45°)(29.1 \text{ in.}) + 6.35 \text{ in.}}{(\tan 45°)^2 + 1}$$

$$= 17.7 \text{ in.}$$

$$y_{ec} = (\tan\gamma)x_{ec}$$

$$= (\tan 45°)17.7 \text{ in.}$$

$$= 17.7 \text{ in.}$$

The eccentricity between the centroid of the gusset-to-column interface and the brace centerline is therefore:

$$e_c = \sqrt{(x_{ec} - x_c)^2 + (y_{ec} - y_c)^2}$$

$$= \sqrt{(17.7 \text{ in.} - 6.35 \text{ in.})^2 + (17.7 \text{ in.} - 29.1 \text{ in.})^2}$$

$$= 16.1 \text{ in.}$$

At the beam flange, the gusset-to-beam flange centroid is located at this point, relative to the working point:

$$(x_b, y_b) = \left(\frac{d_c}{2} + \frac{a}{2}, \frac{d_b}{2}\right)$$

$$= \left(\frac{12.7 \text{ in.}}{2} + \frac{40.0 \text{ in.}}{2}, \frac{23.7 \text{ in.}}{2}\right)$$

$$= (26.4 \text{ in.}, 11.9 \text{ in.})$$

The point on the brace centerline that is the intersection of a line through this point perpendicular to the brace centerline is given by these equations (see Figure 5-47), relative to the working point:

$$x_{eb} = \frac{(\tan\gamma)y_b + x_b}{\tan^2\gamma + 1}$$

$$= \frac{(\tan 45°)11.9 \text{ in.} + 26.4 \text{ in.}}{(\tan 45°)^2 + 1}$$

$$= 19.2 \text{ in.}$$

$$y_{eb} = (\tan\gamma)x_{eb}$$

$$= (\tan 45°)19.2 \text{ in.}$$

$$= 19.2 \text{ in.}$$

The eccentricity between the centroid of the gusset-to-beam interface and the centerline of the brace is therefore:

$$e_b = \sqrt{(x_{eb} - x_b)^2 + (y_{eb} - y_b)^2}$$

$$\sqrt{(19.2 \text{ in.} - 26.4 \text{ in.})^2 + (19.2 \text{ in.} - 11.9 \text{ in.})^2}$$

$$= 10.3 \text{ in.}$$

Taking moments about point (x_b, y_b), the diagonal force, parallel to the brace force, at the column flange corresponding to the expected strength of the brace in tension is:

LRFD	ASD
$P_{uc} = \dfrac{P_u e_b}{(e_c + e_b)}$ $= \dfrac{(550 \text{ kips})(10.3 \text{ in.})}{(16.1 \text{ in.} + 10.3 \text{ in.})}$ $= 215 \text{ kips}$	$P_{ac} = \dfrac{P_a e_b}{(e_c + e_b)}$ $= \dfrac{(385 \text{ kips})(10.3 \text{ in.})}{(16.1 \text{ in.} + 10.3 \text{ in.})}$ $= 150 \text{ kips}$

Note that summing moments as described will result in a P_c force in the opposite direction to the column flange force as shown in Figure 5-47. Because Figure 5-47 is not actually a free-body diagram of the gusset, forces P_c and P_b are shown as they act on the beam and column. When resolving these forces into components, forces denoted H act in the horizontal direction and forces V act in the vertical direction. Depending on whether the interface is a beam or a column, H or V might be either a shear or a normal force.

The corresponding shear on the column face is:

LRFD	ASD
$V_{uc} = P_{uc} \sin \gamma$ $= 215 \text{ kips}(\sin 45°)$ $= 152 \text{ kips}$	$V_{ac} = P_{ac} \sin \gamma$ $= 150 \text{ kips}(\sin 45°)$ $= 106 \text{ kips}$

The corresponding normal force on the column face is:

LRFD	ASD
$H_{uc} = P_{uc} \cos \gamma$ $= 215 \text{ kips}(\cos 45°)$ $= 152 \text{ kips}$	$H_{ac} = P_{ac} \cos \gamma$ $= 150 \text{ kips}(\cos 45°)$ $= 106 \text{ kips}$

Taking moments about point (x_c, y_c), the diagonal force at the beam flange corresponding to the expected strength of the brace in tension is:

5.3 SPECIAL CONCENTRICALLY BRACED FRAMES

LRFD	ASD
$P_{ub} = \dfrac{P_u e_c}{(e_c + e_b)}$ $= \dfrac{(550 \text{ kips})(16.1 \text{ in.})}{(16.1 \text{ in.} + 10.3 \text{ in.})}$ $= 335 \text{ kips}$	$P_{ab} = \dfrac{P_a e_c}{(e_c + e_b)}$ $= \dfrac{(385 \text{ kips})(16.1 \text{ in.})}{(16.1 \text{ in.} + 10.3 \text{ in.})}$ $= 235 \text{ kips}$

The corresponding shear on the beam face is:

LRFD	ASD
$H_{ub} = P_{ub} \cos\gamma$ $= 335 \text{ kips}(\cos 45°)$ $= 237 \text{ kips}$	$H_{ab} = P_{ab} \cos\gamma$ $= 235 \text{ kips}(\cos 45°)$ $= 166 \text{ kips}$

The corresponding normal force on the beam face is:

LRFD	ASD
$V_{ub} = P_{ub} \sin\gamma$ $= 335 \text{ kips}(\sin 45°)$ $= 237 \text{ kips}$	$V_{ab} = P_{ab} \sin\gamma$ $= 235 \text{ kips}(\sin 45°)$ $= 166 \text{ kips}$

The beam-to-column connection is designed for a moment based on the normal and shear forces at the gusset-to-beam interface. Taking moments about the work point, the resulting moment at the beam-to-column connection due to the brace force is:

LRFD	ASD
$M_u = \left\| H_{ub}\left(\dfrac{d_b}{2}\right) - V_{ub}\left(\dfrac{a}{2}\right) \right\|$ $= \left\| \begin{matrix} 237 \text{ kips}\left(\dfrac{23.7 \text{ in.}}{2}\right) \\ -237 \text{ kips}\left(\dfrac{40.0 \text{ in.}}{2}\right) \end{matrix} \right\|$ $= 1{,}930 \text{ kip-in.}$	$M_a = \left\| H_{ab}\left(\dfrac{d_b}{2}\right) - V_{ab}\left(\dfrac{a}{2}\right) \right\|$ $= \left\| \begin{matrix} 166 \text{ kips}\left(\dfrac{23.7 \text{ in.}}{2}\right) \\ -166 \text{ kips}\left(\dfrac{40.0 \text{ in.}}{2}\right) \end{matrix} \right\|$ $= 1{,}350 \text{ kip-in.}$

The horizontal force at the connection of the beam to the column is affected by both the force entering the frame (defined by the mechanism analysis provisions of AISC *Seismic Provisions* Section F2.3) and the horizontal force transferred from the gusset to the column (H_{uc} or H_{ac}). The total force entering the frame can be computed based on the difference between the total expected frame shear strength above and below the beam, as explained in

Example 5.3.5. These shear strengths are calculated based on the horizontal components of the brace expected strengths. The total force entering the frame is the difference between the expected strengths of the braces above the third level and the braces below the third level:

$$P_x = \begin{bmatrix} \Sigma(\text{Brace expected strengths below beam})\cos\gamma \\ -\Sigma(\text{Brace expected strengths above beam})\cos\gamma \end{bmatrix}$$

LRFD	ASD
$P_x = \cos 45° \begin{bmatrix} (606 \text{ kips} + 518 \text{ kips}) \\ -(550 \text{ kips} + 444 \text{ kips}) \end{bmatrix}$ $= 91.9 \text{ kips}$	$P_x = \cos 45° \begin{bmatrix} (424 \text{ kips} + 363 \text{ kips}) \\ -(385 \text{ kips} + 311 \text{ kips}) \end{bmatrix}$ $= 64.3 \text{ kips}$

Similar to what was illustrated in Example 5.3.5, the mechanism analysis with the compression braces at their post-buckling strengths will not result in a higher force entering the frame in this case.

Since the braced frame is in the middle bay of a three-bay building, the collector force (half of this story force) can be considered to enter the braced frame from each side. These forces are shown in Figures 5-48a and 5-48b.

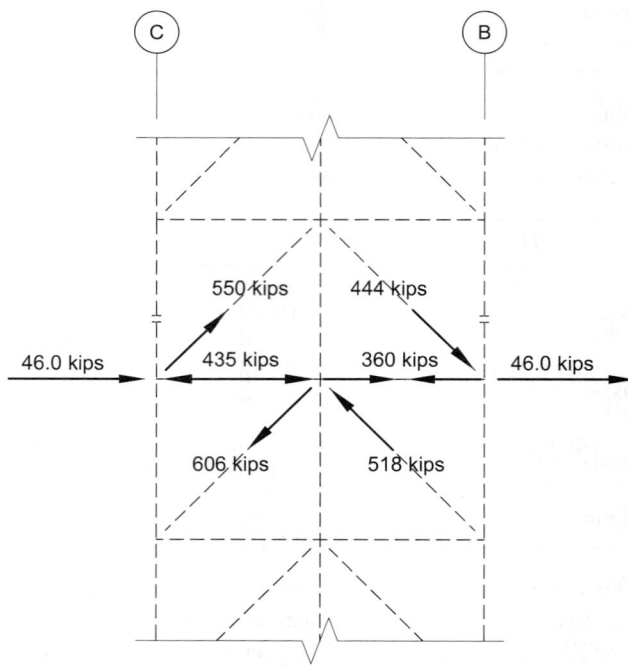

Fig. 5-48a. Collector and frame forces for the third level (LRFD).

5.3 SPECIAL CONCENTRICALLY BRACED FRAMES

LRFD	ASD
$H_{u,\,collector} = \frac{1}{2}(91.9\text{ kips})$ $= 46.0\text{ kips}$	$H_{a,\,collector} = \frac{1}{2}(64.3\text{ kips})$ $= 32.2\text{ kips}$

The force at the beam-to-column connection within the frame must also include H_{uc} (LRFD) and H_{ac} (ASD):

LRFD	ASD
$H_{u,\,connection} = H_{u,\,collector} + H_{uc}$ $= 46.0\text{ kips} + 152\text{ kips}$ $= 198\text{ kips}$	$H_{a,\,connection} = H_{a,\,collector} + H_{ac}$ $= 32.2\text{ kips} + 106\text{ kips}$ $= 138\text{ kips}$

This force may be resisted in the beam flange-to-column welds, the beam web-to-column weld, or shared between the two. In this example the available strength of the flanges will be calculated, and any excess demand will be assigned to the web. For this comparison the required strength of each beam flange is taken as:

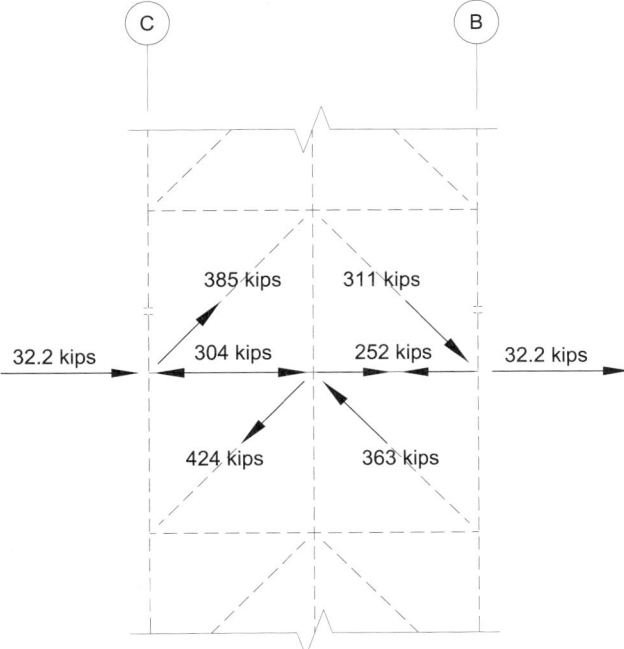

Fig. 5-48b. Collector and frame forces for the third level (ASD).

LRFD	ASD
$R_u = \dfrac{M_u}{d-t_f} + \dfrac{H_{u,\,connection}}{2}$	$R_a = \dfrac{M_a}{d-t_f} + \dfrac{H_{a,\,connection}}{2}$
$= \dfrac{1{,}930 \text{ kip-in.}}{23.7 \text{ in.} - 0.585 \text{ in.}} + \dfrac{198 \text{ kips}}{2}$	$= \dfrac{1{,}350 \text{ kip-in.}}{23.7 \text{ in.} - 0.585 \text{ in.}} + \dfrac{138 \text{ kips}}{2}$
$= 182$ kips	$= 127$ kips

The available strength of each beam flange for the limit state of tensile yielding is calculated as:

$R_n = F_y A_g$ \hfill (*Spec.* Eq. J4-1)
$ = F_y b_f t_f$
$ = 50 \text{ ksi}(8.97 \text{ in.})(0.585 \text{ in.})$
$ = 262$ kips

LRFD	ASD
$\phi R_n = 0.90(262 \text{ kips})$ $= 236$ kips > 182 kips **o.k.**	$\dfrac{R_n}{\Omega} = \dfrac{262 \text{ kips}}{1.67}$ $= 157$ kips > 127 kips **o.k.**

Thus, the entire force can be assigned to the beam flanges, and none need be assigned to the beam web in this case.

Gusset Plate at Column Flange

The combined effects of shear and tension at the gusset-to-column interface may be calculated using von Mises interaction (as shown in terms of stresses in AISC *Manual* Equation 9-1):

LRFD	ASD
$\sqrt{\left(\dfrac{H_{uc}}{t_p b}\right)^2 + 3\left(\dfrac{V_{uc}}{t_p b}\right)^2}$	$\sqrt{\left(\dfrac{H_{ac}}{t_p b}\right)^2 + 3\left(\dfrac{V_{ac}}{t_p b}\right)^2}$
$= \sqrt{\left[\dfrac{152 \text{ kips}}{(\tfrac{5}{8} \text{ in.})(34.5 \text{ in.})}\right]^2 + 3\left[\dfrac{152 \text{ kips}}{(\tfrac{5}{8} \text{ in.})(34.5 \text{ in.})}\right]^2}$	$= \sqrt{\left[\dfrac{106 \text{ kips}}{(\tfrac{5}{8} \text{ in.})(34.5 \text{ in.})}\right]^2 + 3\left[\dfrac{106 \text{ kips}}{(\tfrac{5}{8} \text{ in.})(34.5 \text{ in.})}\right]^2}$
$= 14.1$ ksi	$= 9.83$ ksi

5.3 SPECIAL CONCENTRICALLY BRACED FRAMES

The stress from the von Mises interaction is compared to the strength of the gusset plate, which is taken as ϕF_y (LRFD) and F_y/Ω (ASD) for the limit state of tensile yielding from AISC *Specification* Equation J4-1.

LRFD	ASD
$\phi F_y = 0.90(36 \text{ ksi})$ $= 32.4 \text{ ksi} > 14.1 \text{ ksi}$ **o.k.**	$\dfrac{F_y}{\Omega} = \dfrac{36 \text{ ksi}}{1.67}$ $= 21.6 \text{ ksi} > 9.83 \text{ ksi}$ **o.k.**

A similar check for rupture on this plane is made, considering the 1½-in. corner clip:

LRFD	ASD
$=\sqrt{\left[\dfrac{H_{uc}}{t_p(b-clip)}\right]^2 + 3\left[\dfrac{V_{uc}}{t_p(b-clip)}\right]^2}$ $=\sqrt{\left[\dfrac{152 \text{ kips}}{(\text{⅝ in.})(34.5 \text{ in.} - 1½ \text{ in.})}\right]^2 + 3\left[\dfrac{152 \text{ kips}}{(\text{⅝ in.})(34.5 \text{ in.} - 1½ \text{ in.})}\right]^2}$ $= 14.7$ ksi	$=\sqrt{\left[\dfrac{H_{ac}}{t_p(b-clip)}\right]^2 + 3\left[\dfrac{V_{ac}}{t_p(b-clip)}\right]^2}$ $=\sqrt{\left[\dfrac{106 \text{ kips}}{(\text{⅝ in.})(34.5 \text{ in.} - 1½ \text{ in.})}\right]^2 + 3\left[\dfrac{106 \text{ kips}}{(\text{⅝ in.})(34.5 \text{ in.} - 1½ \text{ in.})}\right]^2}$ $= 10.3$ ksi

The stress from the von Mises interaction is compared to the strength of the gusset plate, which is taken as ϕF_u (LRFD) and F_u/Ω (ASD) for the limit state of tensile rupture, from AISC *Specification* Equation J4-2.

LRFD	ASD
$\phi F_u = 0.75(58 \text{ ksi})$ $= 43.5 \text{ ksi} > 14.7 \text{ ksi}$ **o.k.**	$\dfrac{F_u}{\Omega} = \dfrac{58 \text{ ksi}}{2.00}$ $= 29.0 \text{ ksi} > 10.3 \text{ ksi}$ **o.k.**

Gusset Plate at Beam Flange
Similar to the gusset at the column face, the combined effects of shear and tension on the gusset at the beam flange may be calculated using von Mises interaction:

LRFD	ASD
$\sqrt{\left(\dfrac{V_{ub}}{t_p a}\right)^2 + 3\left(\dfrac{H_{ub}}{t_p a}\right)^2}$	$\sqrt{\left(\dfrac{V_{ab}}{t_p a}\right)^2 + 3\left(\dfrac{H_{ab}}{t_p a}\right)^2}$
$= \sqrt{\left[\dfrac{237 \text{ kips}}{(\text{⅝ in.})(40.0 \text{ in.})}\right]^2 + 3\left[\dfrac{237 \text{ kips}}{(\text{⅝ in.})(40.0 \text{ in.})}\right]^2}$	$= \sqrt{\left[\dfrac{166 \text{ kips}}{(\text{⅝ in.})(40.0 \text{ in.})}\right]^2 + 3\left[\dfrac{166 \text{ kips}}{(\text{⅝ in.})(40.0 \text{ in.})}\right]^2}$
$= 19.0 \text{ ksi} < \phi F_y = 32.4 \text{ ksi}$ **o.k.**	$= 13.3 \text{ ksi} < F_y/\Omega = 21.6 \text{ ksi}$ **o.k.**

A similar check for rupture on this plane is made, considering the 1½-in. corner clip:

LRFD	ASD
$\sqrt{\left[\dfrac{V_{ub}}{t_p(a-clip)}\right]^2 + 3\left[\dfrac{H_{ub}}{t_p(a-clip)}\right]^2}$	$\sqrt{\left(\dfrac{V_{ab}}{t_p(a-clip)}\right)^2 + 3\left(\dfrac{H_{ab}}{t_p(a-clip)}\right)^2}$
$= \sqrt{\left[\dfrac{237 \text{ kips}}{(\text{⅝ in.})(40.0 \text{ in.} - 1\text{½ in.})}\right]^2 + 3\left[\dfrac{237 \text{ kips}}{(\text{⅝ in.})(40.0 \text{ in.} - 1\text{½ in.})}\right]^2}$	$= \sqrt{\left(\dfrac{166 \text{ kips}}{(\text{⅝ in.})(40.0 \text{ in.} - 1\text{½ in.})}\right)^2 + 3\left(\dfrac{166 \text{ kips}}{(\text{⅝ in.})(40.0 \text{ in.} - 1\text{½ in.})}\right)^2}$
$= 19.7 \text{ ksi} < \phi F_u = 43.5 \text{ ksi}$ **o.k.**	$= 13.8 \text{ ksi} < F_u/\Omega = 29.0 \text{ ksi}$ **o.k.**

Column Web at Gusset-to-Column Interface

At the gusset-to-column interface and gusset-to-beam interface, the column and beam webs, respectively, must be checked for the limit states of web local yielding and web local crippling. The length of bearing, l_b, is taken as the height of the gusset plate, b.

Check column web local yielding

For a force applied at a distance less than the depth of the member from the member end, the available web local yielding strength is determined as follows:

$$R_n = F_{yw} t_w (5k_{des} + l_b) \quad \text{(Spec. Eq. J10-2)}$$
$$= (50 \text{ ksi})(0.550 \text{ in.})[5(1.50 \text{ in.}) + 34.5 \text{ in.}]$$
$$= 1{,}160 \text{ kips}$$

LRFD	ASD
$\phi R_n = 1.00(1{,}160 \text{ kips})$ $= 1{,}160 \text{ kips} > H_{uc} = 152 \text{ kips}$ **o.k.**	$\dfrac{R_n}{\Omega} = \dfrac{1{,}160 \text{ kips}}{1.50}$ $= 773 \text{ kips} > H_{ac} = 106 \text{ kips}$ **o.k.**

Check column web local crippling

For a force applied greater than a distance of $d/2$ from the member end:

$$R_n = 0.80 t_w^2 \left[1 + 3\left(\frac{l_b}{d}\right)\left(\frac{t_w}{t_f}\right)^{1.5}\right]\sqrt{\frac{EF_{yw}t_f}{t_w}} \qquad (\textit{Spec. Eq. J10-4})$$

$$= (0.80)(0.550 \text{ in.})^2 \left[1 + 3\left(\frac{34.5 \text{ in.}}{12.7 \text{ in.}}\right)\left(\frac{0.550 \text{ in.}}{0.900 \text{ in.}}\right)^{1.5}\right]$$

$$\times \sqrt{\frac{(29{,}000 \text{ ksi})(50 \text{ ksi})(0.900 \text{ in.})}{0.550 \text{ in.}}}$$

$$= 1{,}820 \text{ kips}$$

This value is not compared to the value of H_{uc} or H_{ac} calculated previously, which is based on tension in the brace, as crippling is a compression limit state. Because the H_{uc} and H_{ac} forces calculated previously are directly proportional to the brace force, they can be scaled down based on the ratio of the brace force in compression to the brace force in tension. The maximum compression force at the gusset-to-column interface is:

LRFD	ASD
$H_{uc}\left(\dfrac{444 \text{ kips}}{550 \text{ kips}}\right) = 152 \text{ kips}\left(\dfrac{444 \text{ kips}}{550 \text{ kips}}\right)$ $= 123 \text{ kips}$	$H_{ac}\left(\dfrac{311 \text{ kips}}{385 \text{ kips}}\right) = 106 \text{ kips}\left(\dfrac{311 \text{ kips}}{385 \text{ kips}}\right)$ $= 85.6 \text{ kips}$

Therefore:

LRFD	ASD
$\phi R_n = 0.75(1{,}820 \text{ kips})$ $= 1{,}370 \text{ kips} > 123 \text{ kips}$ **o.k.**	$\dfrac{R_n}{\Omega} = \dfrac{1{,}820 \text{ kips}}{2.00}$ $= 910 \text{ kips} > 85.6 \text{ kips}$ **o.k.**

Beam Web at Gusset-to-Beam Interface

Check beam web local yielding

Consider that the interface force, V_b, acts at the centroid of the gusset-to-beam interface, a distance of $a/2 = 40.0 \text{ in.}/2 = 20.0 \text{ in.}$ from the face of the column.

For a force applied at a distance greater than the depth of the member from the member end, the available strength is determined as follows:

$$R_n = F_{yw}t_w(2.5k_{des} + l_b) \quad \text{(Spec. Eq. J10-3)}$$
$$= (50 \text{ ksi})(0.415 \text{ in.})[2.5(1.09 \text{ in.}) + 40.0 \text{ in.}]$$
$$= 887 \text{ kips}$$

LRFD	ASD
$\phi R_n = 1.0(887 \text{ kips})$ $= 887 \text{ kips} > V_{ub} = 237 \text{ kips}$ **o.k.**	$\dfrac{R_n}{\Omega} = \dfrac{887 \text{ kips}}{1.50}$ $= 591 \text{ kips} > V_{ab} = 166 \text{ kips}$ **o.k.**

Check beam web local crippling

The resultant force at the centroid of the gusset-to-beam interface is greater than $d/2$ from the member end. Thus, AISC *Specification* Equation J10-4 is applicable.

$$R_n = 0.80t_w^2\left[1 + 3\left(\frac{l_b}{d}\right)\left(\frac{t_w}{t_f}\right)^{1.5}\right]\sqrt{\frac{EF_{yw}t_f}{t_w}} \quad \text{(Spec. Eq. J10-4)}$$

$$= (0.80)(0.415 \text{ in.})^2\left[1 + 3\left(\frac{40.0 \text{ in.}}{23.7 \text{ in.}}\right)\left(\frac{0.415 \text{ in.}}{0.585 \text{ in.}}\right)^{1.5}\right]$$
$$\times \sqrt{\frac{(29,000 \text{ ksi})(50 \text{ ksi})(0.585 \text{ in.})}{0.415 \text{ in.}}}$$
$$= 793 \text{ kips}$$

This value is not compared to the value of V_{ub} or V_{ab} calculated previously, which is based on tension in the brace, as crippling is a compression limit state. Compression in the beam web occurs when the brace is in compression, not when it is in tension, so new V_{ub} and V_{ab} forces need to be determined. Because the V_{ub} and V_{ab} forces calculated previously are directly proportional to the brace force, they can be scaled down based on the ratio of the brace force in compression to the brace force in tension. The maximum compression force at the gusset-to-beam interface is calculated as:

LRFD	ASD
$V_{ub}\left(\dfrac{444 \text{ kips}}{550 \text{ kips}}\right) = 237 \text{ kips}\left(\dfrac{444 \text{ kips}}{550 \text{ kips}}\right)$ $= 191 \text{ kips}$	$V_{ab}\left(\dfrac{311 \text{ kips}}{385 \text{ kips}}\right) = 166 \text{ kips}\left(\dfrac{311 \text{ kips}}{385 \text{ kips}}\right)$ $= 134 \text{ kips}$

Therefore:

5.3 SPECIAL CONCENTRICALLY BRACED FRAMES

LRFD	ASD
$\phi R_n = 0.75(793 \text{ kips})$ $= 595 \text{ kips} > 191 \text{ kips}$ **o.k.**	$\dfrac{R_n}{\Omega} = \dfrac{793 \text{ kips}}{2.00}$ $= 397 \text{ kips} > 134 \text{ kips}$ **o.k.**

Interface Welds

Based on experiments and simulations, Roeder et al. (2011) recommend designing the welds at the gusset-to-beam and gusset-to-column interfaces for the expected tensile strength of the gusset plate in order to increase the deformation and ductility capacity of the system and limit the weld damage. The recommended expression for the size of a pair of fillet welds, where w is the weld size and the 1.5 represents the directional strength increase for transversely loaded fillet welds, is:

$$2(1.5)\beta(0.60)F_{EXX}(0.707)w \geq R_y F_y t_p$$

where $\beta = 0.75$. In order to comply with the AISC *Specification*, use $\phi = 0.75$ instead of $\beta = 0.75$.

This expression, which is based on AISC *Specification* Equations J2-4 and J2-5, may be rearranged to solve for the fillet weld size, w, for the given material strengths (the required strength for ASD is taken to be $R_y F_y/1.5$). From AISC *Seismic Provisions* Table A3.1, for ASTM A36 plate material, $R_y = 1.3$.

LRFD	ASD
$w = \left[\dfrac{R_y F_y}{2(1.5)\phi(0.60)F_{EXX}(0.707)}\right]t_p$ $= \left[\dfrac{(1.3)(36 \text{ ksi})}{2(1.5)(0.75)(0.60)(70 \text{ ksi})(0.707)}\right]t_p$ $= 0.700 t_p$	$w = \left[\dfrac{\Omega R_y F_y}{2(1.5)(1.5)(0.60)F_{EXX}(0.707)}\right]t_p$ $= \left[\dfrac{(2.00)(1.3)(36 \text{ ksi})}{2(1.5)(1.5)(0.60)(70 \text{ ksi})(0.707)}\right]t_p$ $= 0.700 t_p$

For the ⅝-in.-thick gusset plate, the weld size required is:

$w = 0.700(⅝ \text{ in.})$
$= 0.438 \text{ in.}$

Use a ⁷⁄₁₆-in. fillet weld on both sides of the gusset plate to connect the gusset to the beam and column.

Beam-to-Column Connection

The beam-to-column connection must comply with the requirements of AISC *Seismic Provisions* Section F2.6b. For this example, Section F2.6b(b), the moment-resisting beam end connection option, is chosen. This example utilizes a moment connection with CJP groove welds of the beam flanges and web to the column flange, which will be adequate to resist a moment corresponding to the expected beam flexural strength multiplied by 1.1 (LRFD) or 1.1/1.5 (ASD), thereby meeting AISC *Seismic Provisions* Section F2.6b(b)(i). An alternative method of providing a moment connection at the beam-to-column connection and meeting AISC *Seismic Provisions* Section F2.6b(b), which explicitly considers frame rotational forces, is presented in Example 5.3.12. A connection with a simple beam-to-column connection meeting AISC *Seismic Provisions* Section F2.6b(a) was presented in Example 5.3.10. Any of these approaches is satisfactory.

Use CJP groove welds at the beam flanges-to-column and beam web-to-column connections.

To determine whether continuity plates are required, check whether the limit states of web local yielding, web local crippling, and flange local bending of the column are adequate for the required strength. The required strength must be determined. AISC *Seismic Provisions* Section F2.6b(b) requires that the connection be designed to resist a moment equal to the expected beam flexural strength multiplied by 1.1 (LRFD) or 1.1/1.5 (ASD). In this case, the beam web has a CJP groove weld to the column flange and therefore can develop the full expected flexural strength of the beam web. Therefore, for the local column limit states of web local yielding and web local crippling, the demand at the column face will be taken as the expected, strain-hardened strength of the beam flange using a strain-hardening factor of 1.1:

LRFD	ASD
$R_{u\,flange} = 1.1 R_y F_y A_{flange}$ $= 1.1(1.1)(50 \text{ ksi})$ $\times (8.97 \text{ in.})(0.585 \text{ in.})$ $= 317 \text{ kips}$	$R_{a\,flange} = (1.1/1.5) R_y F_y A_{flange}$ $= (1.1/1.5)(1.1)(50 \text{ ksi})$ $\times (8.97 \text{ in.})(0.585 \text{ in.})$ $= 212 \text{ kips}$

Check web local yielding of the column

For a force applied at a distance greater than the depth of the member from the member end, the available web local yielding strength of the column is determined as follows, where the length of bearing, l_b, is taken as the beam flange thickness.

$R_n = F_{yw} t_w (5 k_{des} + l_b)$ (*Spec*. Eq. J10-2)

    $= (50 \text{ ksi})(0.550 \text{ in.})[5(1.50 \text{ in.}) + 0.585 \text{ in.}]$

    $= 222 \text{ kips}$

LRFD	ASD
$\phi R_n = 1.00(222 \text{ kips})$ $= 222 \text{ kips} < 317 \text{ kips}$ **n.g.**	$\dfrac{R_n}{\Omega} = \dfrac{222 \text{ kips}}{1.50}$ $= 148 \text{ kips} < 212 \text{ kips}$ **n.g.**

Check web local crippling of the column

For a force applied greater than a distance of $d/2$ from the member end, the available web local crippling strength of the column is determined as follows, where the length of bearing, l_b, is taken as the beam flange thickness:

$$R_n = 0.80t_w^2\left[1+3\left(\frac{l_b}{d}\right)\left(\frac{t_w}{t_f}\right)^{1.5}\right]\sqrt{\frac{EF_{yw}t_f}{t_w}} \qquad (Spec.\ Eq.\ J10\text{-}4)$$

$$= (0.80)(0.550\ \text{in.})^2\left[1+3\left(\frac{0.585\ \text{in.}}{12.7\ \text{in.}}\right)\left(\frac{0.550\ \text{in.}}{0.900\ \text{in.}}\right)^{1.5}\right]$$

$$\times\sqrt{\frac{(29{,}000\ \text{ksi})(50\ \text{ksi})(0.900\ \text{in.})}{0.550\ \text{in.}}}$$

$$= 397\ \text{kips}$$

LRFD	ASD
$\phi R_n = 0.75(397\ \text{kips})$ $= 298\ \text{kips} < 317\ \text{kips}$ **n.g.**	$\dfrac{R_n}{\Omega} = \dfrac{397\ \text{kips}}{2.00}$ $= 199\ \text{kips} < 212\ \text{kips}$ **n.g.**

Check flange local bending of the column

The available strength of the column due to flange local bending is determined as follows:

$$R_n = 6.25F_{yf}t_f^2 \qquad (Spec.\ Eq.\ J10\text{-}1)$$

$$= 6.25(50\ \text{ksi})(0.900\ \text{in.})^2$$

$$= 253\ \text{kips}$$

LRFD	ASD
$\phi R_n = 0.90(253\ \text{kips})$ $= 228\ \text{kips} < 317\ \text{kips}$ **n.g.**	$\dfrac{R_n}{\Omega} = \dfrac{253\ \text{kips}}{1.67}$ $= 151\ \text{kips} < 212\ \text{kips}$ **n.g.**

Based on the checks of web local yielding, web local crippling, and flange local bending, the column requires continuity plates. The continuity plates must be designed to resist the difference between the flange force, $R_{u\ flange}$ or $R_{a\ flange}$, and the lesser of the column web local yielding, web local crippling, and flange local bending strengths:

LRFD	ASD
$R_u = R_{u\ flange} - \phi R_n$ $= 317\ \text{kips} - 222\ \text{kips}$ $= 95.0\ \text{kips}$	$R_a = R_{a\ flange} - R_n/\Omega$ $= 212\ \text{kips} - 148\ \text{kips}$ $= 64.0\ \text{kips}$

Using a continuity plate width that closely matches the beam flange width:

$$\frac{b_{f\,beam}}{2} - \frac{t_{w\,col}}{2}$$

$$\frac{8.97 \text{ in.}}{2} - \frac{0.550 \text{ in.}}{2} = 4.21 \text{ in.}$$

Select 4.50 in. as the plate width. Make sure that this plate width fits within the column flange:

$$\frac{12.2 \text{ in.}}{2} - \frac{0.550 \text{ in.}}{2} = 5.83 \text{ in.} > 4.50 \text{ in.} \quad \textbf{o.k.}$$

The required thickness for the two continuity plates, based on the limit state of tensile yielding from AISC *Specification* Equation J4-1, is:

LRFD	ASD
$\phi(2 \text{ plates})F_y bt > R_u$	$(2 \text{ plates})F_y bt/\Omega > R_a$
$t > \dfrac{R_u}{\phi(2)F_y b}$	$t > \dfrac{\Omega R_a}{(2)F_y b}$
$> \dfrac{95.0 \text{ kips}}{0.90(2)(50 \text{ ksi})(4.50 \text{ in.})}$	$> \dfrac{1.67(64.0 \text{ kips})}{(2)(50 \text{ ksi})(4.50 \text{ in.})}$
$> 0.235 \text{ in.}$	$> 0.238 \text{ in.}$

Therefore ¼-in.-thick continuity plates will be used.

Design the welds between the continuity plates and column

There are several design considerations that could be used to determine the required weld size. For the welds between the continuity plates and column, the welds will be designed to be at least as strong as the available strength of the contact area of the continuity plate with the flange. This design approach meets the exception in AISC *Seismic Provisions* Section E1.6b(c)(4), although this connection is not required to comply with OMF requirements. Using the expression for the required weld size to develop a plate in tension discussed previously for the gusset plate, the fillet welds at the continuity plate to column flange are sized as follows:

LRFD	ASD
$w = \left[\dfrac{\phi F_y}{2(1.5)\phi(0.60)F_{EXX}(0.707)}\right]t$	$w = \left[\dfrac{\Omega F_y}{2(1.50)\Omega(0.6)F_{EXX}(0.707)}\right]t$
$= \left[\dfrac{0.90(50 \text{ ksi})}{2(1.5)(0.75)(0.60)(70 \text{ ksi})(0.707)}\right]t$	$= \left[\dfrac{2.00(50 \text{ ksi})}{2(1.50)(1.67)(0.60)(70 \text{ ksi})(0.707)}\right]t$
$= 0.674t$	$= 0.672t$

5.3 SPECIAL CONCENTRICALLY BRACED FRAMES

For the ¼-in.-thick continuity plate, the required weld size is:

LRFD	ASD
$w = 0.674t$ $= 0.674(¼ \text{ in.})$ $= 0.169 \text{ in.}$	$w = 0.672t$ $= 0.672(¼ \text{ in.})$ $= 0.168 \text{ in.}$

Use ³⁄₁₆-in. fillet welds between the continuity plate and the column flange (both sides of the plate).

For the welds between the continuity plate and the column web, a weld size will be chosen that is stronger than the available shear strength of the continuity plate contact area with the web. This design approach meets AISC *Seismic Provisions* Section E3.6f(3)(b), although this connection is not required to comply with SMF requirements.

Deriving the weld size as was done previously for the gusset plate in tension:

LRFD	ASD
$w = \left[\dfrac{\phi 0.60 F_y}{2\phi(0.60) F_{EXX}(0.707)}\right] t$ $= \left[\dfrac{1.00(0.60)(50 \text{ ksi})}{2(0.75)(0.60)(70 \text{ ksi})(0.707)}\right] t$ $= 0.674t$	$w = \left[\dfrac{\Omega 0.60 F_y}{2\Omega(0.60) F_{EXX}(0.707)}\right] t$ $= \left[\dfrac{2.00(0.60)(50 \text{ ksi})}{2(1.50)(0.60)(70 \text{ ksi})(0.707)}\right] t$ $= 0.674t$

For the ¼-in.-thick continuity plate, the weld size required is:

$w = 0.674(¼ \text{ in.})$
$= 0.169 \text{ in.}$

Use ³⁄₁₆-in. fillet welds between the continuity plate and the column web (both sides of the plate).

Check beam web-to-column connection

The beam web is subject to gravity forces from beam shear in addition to forces from the brace.

The required shear strength of the beam for the case of tension in the brace is calculated as follows. The gravity shears from the beam act in the opposite direction as the brace force, with $S_{DS} = 1.0$:

LRFD	ASD
LRFD Load Combination 7 from ASCE/SEI 7 Section 12.4.2.3 $V_u = (0.9 - 0.2S_{DS})V_D + V_{Emh}$ $V_u = 0.7(-4.50 \text{ kips}) + 237 \text{ kips}$ $= 234 \text{ kips}$	ASD Load Combination 8 from ASCE/SEI 7 Section 12.4.2.3 $V_a = (0.6 - 0.14S_{DS})V_D + 0.7V_{Emh}$ $V_a = 0.46(-4.50 \text{ kips}) + 166 \text{ kips}$ $= 164 \text{ kips}$

The required shear strength of the beam for the case of compression in the brace is based on a brace expected strength of 444 kips (LRFD) and 311 kips (ASD). As calculated previously for V_{ub} (LRFD) and V_{ab} (ASD):

LRFD	ASD
$V_{ub}\left(\dfrac{444 \text{ kips}}{550 \text{ kips}}\right) = 237 \text{ kips}\left(\dfrac{444 \text{ kips}}{550 \text{ kips}}\right)$ $= 191 \text{ kips}$	$V_{ab}\left(\dfrac{311 \text{ kips}}{385 \text{ kips}}\right) = 166 \text{ kips}\left(\dfrac{311 \text{ kips}}{385 \text{ kips}}\right)$ $= 134 \text{ kips}$

LRFD	ASD
LRFD Load Combination 5 from ASCE/SEI 7 Section 12.4.3.2 $V_u = (1.2 + 0.2S_{DS})V_D + V_{Emh} + 0.5V_L$ $V_u = 1.4(4.50 \text{ kips}) + 191 \text{ kips}$ $\quad + 0.5(3.00 \text{ kips})$ $= 199 \text{ kips}$	ASD Load Combination 5 from ASCE/SEI 7 Section 12.4.3.2 $V_a = (1.0 + 0.14S_{DS})V_D + 0.7V_{Emh}$ $V_a = 1.14(4.50 \text{ kips}) + 134 \text{ kips}$ $= 139 \text{ kips}$

The strength of the beam in shear is, from AISC *Manual* Table 3-6:

LRFD	ASD
$\phi_v V_n = 295 \text{ kips} > 234 \text{ kips}$ **o.k.**	$\dfrac{V_n}{\Omega_v} = 197 \text{ kips} > 164 \text{ kips}$ **o.k.**

At the column face, the available shear strength is reduced by the material removed for the weld access holes. From Table 1-1 and Table 1-3, weld access hole type B applies to the W24×68 and the (3) and (4) dimensions are ¾ in. and ½ in., respectively. The available shear strength is determined from AISC *Specification* Section G2.

$V_n = \{(0.6)(50 \text{ ksi})[23.7 \text{ in.} - 2(0.585 \text{ in.} + \text{¾ in.} + \text{½ in.})]\}(0.415 \text{ in.})$
$\quad = 249 \text{ kips}$

5.3 SPECIAL CONCENTRICALLY BRACED FRAMES

LRFD	ASD
$\phi_v V_n = 1.00(249 \text{ kips})$ $= 249 \text{ kips} > 234 \text{ kips}$ **o.k.**	$\dfrac{V_n}{\Omega_v} = \dfrac{249 \text{ kips}}{1.50}$ $= 166 \text{ kips} > 164 \text{ kips}$ **o.k.**

The final design is shown in Figure 5-45.

Example 5.3.12. SCBF Brace-to-Beam/Column Connection Design—In-Plane Brace Buckling

Given:

Refer to Figure 5-49. Design the brace-to-beam connection at Joint JT-1 shown schematically in Figure 5-49. The brace orientation, connection type, transfer force, and beam shear due to gravity loads are shown in Figure 5-50. The connection configuration shown in Figure 5-51, which makes use of a "hinge plate," allows large inelastic rotations for in-plane brace buckling with small flexural demand on the connection and supporting members. In this configuration, large inelastic rotations are accommodated with the advantage of having a compact connection (Thornton and Fortney, 2012). This is different from the approach shown in Examples 5.3.10 and 5.3.11, where the brace is expected to buckle out of the plane of the frame. The round HSS brace is ASTM A500 Grade B and the beam and column are ASTM A992. Use ASTM A572 Grade 50 plate material. The bolts are ASTM A490-X.

The completed design shown in Figure 5-51 will be verified in this example.

From Example 5.3.11, the expected strengths of the HSS6.875×0.500 brace are:

$P_{tension}$ = 550 kips
$P_{compression}$ = 444 kips
$0.3 P_{compression}$ = 133 kips (post-buckling strength from AISC *Seismic Provisions* Section F2.3ii)

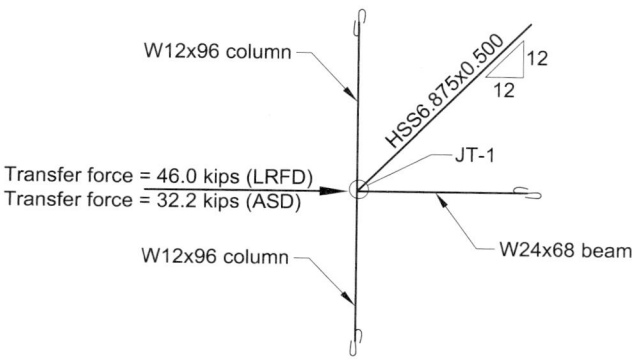

Fig. 5-49. Beam-column joint for Example 5.3.12.

The required strength of the connection from ASCE/SEI 7 Section 12.4.3.2 Load Combination 5 (LRFD) and Load Combination 5 (ASD) is based on the horizontal seismic effect including the overstrength factor, $E_{mh} = \Omega_o Q_E$ (ASCE/SEI 7 Section 12.4.3.1). In this case, E_{mh} is the expected strength given previously for the brace as stipulated in AISC *Seismic Provisions* Section F2.3. The required strength of the connection when the brace is in tension is:

LRFD	ASD
$P_u = 1.0 E_{mh}$ $= 1.0(550 \text{ kips})$ $= 550 \text{ kips}$	$P_a = 0.7 E_{mh}$ $= 0.7(550 \text{ kips})$ $= 385 \text{ kips}$

The required strength of the bracing connection when the brace is in compression is:

LRFD	ASD
$P_u = 1.0 E_{mh}$ $= 1.0(444 \text{ kips})$ $= 444 \text{ kips}$	$P_a = 0.7 E_{mh}$ $= 0.7(444 \text{ kips})$ $= 311 \text{ kips}$

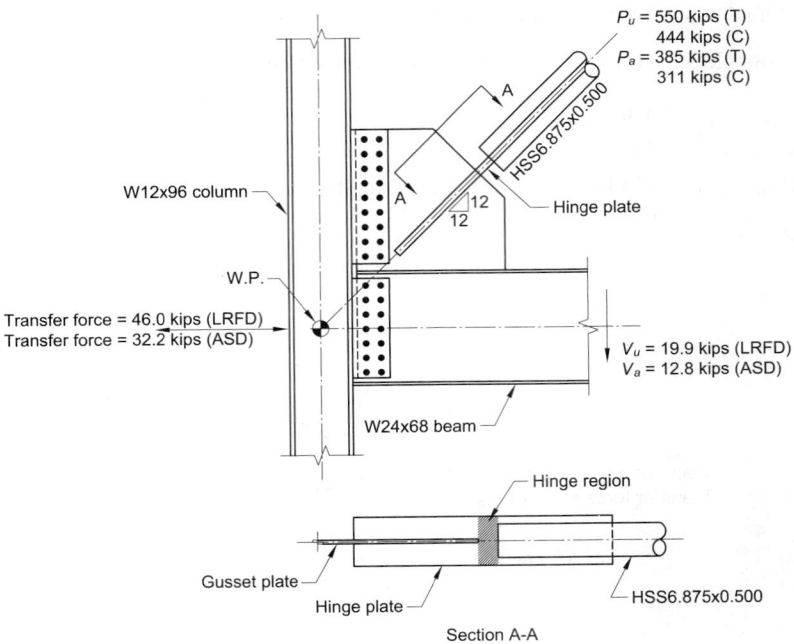

Fig. 5-50. Brace connection to be designed for Example 5.3.12.

5.3 SPECIAL CONCENTRICALLY BRACED FRAMES

The required strength of the bracing connection when the brace is in compression at its post-buckling strength is:

LRFD	ASD
$P_u = 1.0 E_{mh}$	$P_a = 0.7 E_{mh}$
$= 1.0(133 \text{ kips})$	$= 0.7(133 \text{ kips})$
$= 133 \text{ kips}$	$= 93.1 \text{ kips}$

Solution:

From AISC *Manual* Tables 2-4 and 2-5, the material properties are as follows:

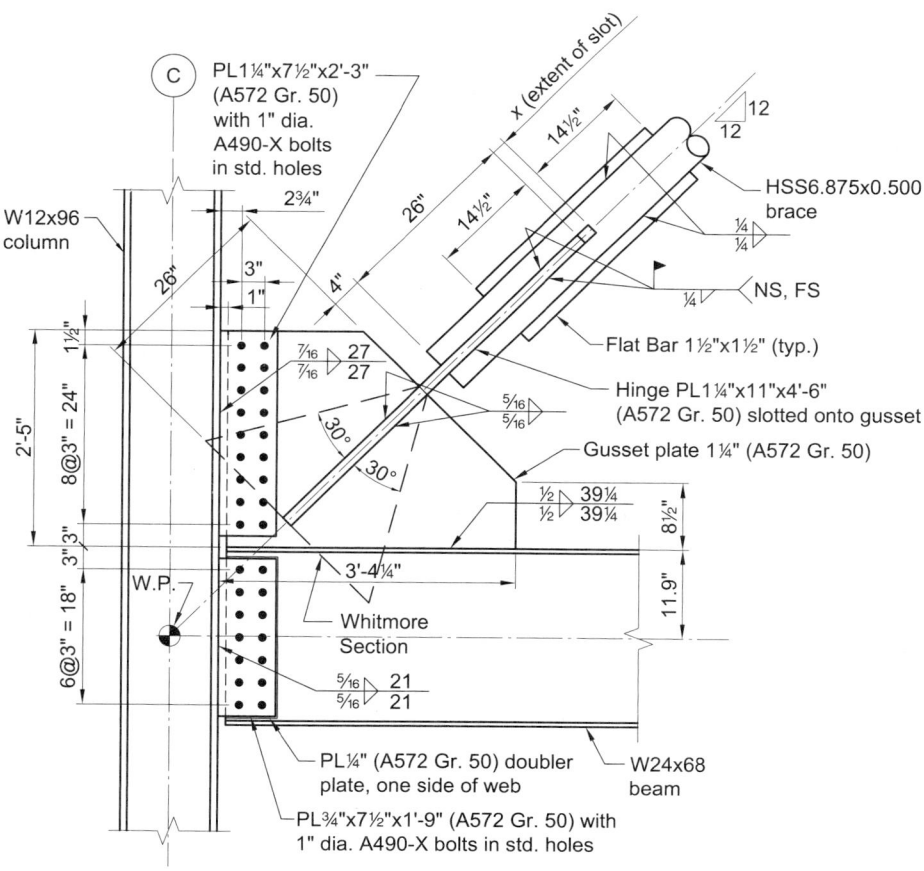

Note: A490-X bolts are to be pretensioned with slip-critical faying surfaces.

Fig. 5-51. Completed connection design for Example 5.3.12.

ASTM A500 Grade B
$F_y = 42$ ksi
$F_u = 58$ ksi

ASTM A572 Grade 50
$F_y = 50$ ksi
$F_u = 65$ ksi

ASTM A992
$F_y = 50$ ksi
$F_u = 65$ ksi

From AISC *Manual* Tables 1-1 and 1-13, the geometric properties are as follows:

Brace
HSS 6.875×0.500
$D = 6.875$ in. $t_{nom} = 0.500$ in. $t_{des} = 0.465$ in. $A = 9.36$ in.^{2}
$r = 2.27$ in.

Beam
W24×68
$A = 20.1$ in.^{2} $d = 23.7$ in. $t_w = 0.415$ in. $b_f = 8.97$ in.
$t_f = 0.585$ in. $k_{des} = 1.09$ in. $Z_x = 177$ in.^{3}

Column
W12×96
$d = 12.7$ in. $t_w = 0.550$ in. $b_f = 12.2$ in. $t_f = 0.900$ in.
$k_{des} = 1.50$ in. $k_{det} = 1^{13}/_{16}$ in. $Z_x = 147$ in.^{3}

AISC *Seismic Provisions* Sections F2.3(i) and F2.3(ii) define the two mechanism analyses that must be considered in determining the required strength of beams, columns and connections. AISC *Seismic Provisions* Section F2.6c specifies the required strength of bracing connections. For the required compressive strength based on buckling limit states, Section F2.6c(2) has an additional 1.1 factor (relative to the requirements of Section F2.3) applied to the required strength of the connection.

For these SCBF connection examples, the requirements of AISC *Seismic Provisions* Section F2.3 will be used for both LRFD and ASD, except for the limit state of compression buckling on the Whitmore section, which will use the 1.1 factor specified in AISC *Seismic Provisions* Section F2.6c(2).

Brace-to-Hinge Plate Connection Design

Example 5.3.10 showed the full brace-to-gusset connection design for the same size brace as used in this example. The calculations for the brace side of the brace-to-gusset connection are not repeated here.

5.3 SPECIAL CONCENTRICALLY BRACED FRAMES

Hinge Plate

Assume the width of the hinge plate is limited by the column flange width of 12.2 in. This limit is an architectural consideration to ensure that the connection does not affect the façade or internal partition width. It is not an engineering requirement.

Choose a hinge plate width, b_p, of 11.0 in. This protrudes beyond the beam flange width, but is less than the column flange width and is sufficient to accommodate the 6.875 in. diameter HSS brace.

To size the hinge plate for the limit state of tension yielding, where t_p is the thickness of the hinge plate:

$$R_n = F_y A_g \qquad \text{(Spec. Eq. J4-1)}$$
$$= F_y t_p b_p$$

The 11.0-in.-wide hinge plate is well within the maximum allowable Whitmore section according to AISC *Manual* Part 9, and therefore the entire hinge plate width can be considered effective in this limit state.

LRFD	ASD
$\phi R_n = \phi F_y t_p b_p \geq P_u$	$\dfrac{R_n}{\Omega} = \dfrac{F_y t_p b_p}{\Omega} \geq P_a$
$t_p \geq \dfrac{P_u}{\phi F_y b_p}$	$t_p \geq \dfrac{\Omega P_a}{F_y b_p}$
$\geq \dfrac{550 \text{ kips}}{0.90(50 \text{ ksi})(11.0 \text{ in.})}$	$\geq \dfrac{1.67(385 \text{ kips})}{(50 \text{ ksi})(11.0 \text{ in.})}$
≥ 1.11 in.	≥ 1.17 in.

Use a 1¼-in.-thick hinge plate.

Check hinge plate net section for tensile rupture strength

Assume the gusset plate thickness is 1¼ in. and verify this assumption later. The hinge plate is slotted over the gusset plate with an additional ¹⁄₁₆ in. increase in slot width on either side of the gusset plate. For the hinge plate:

$$A_n = [11.0 \text{ in.} - 1\tfrac{1}{4} \text{ in.} - 2(\tfrac{1}{16} \text{ in.})](1\tfrac{1}{4} \text{ in.})$$
$$= 12.0 \text{ in.}^2$$

According to AISC *Specification* Table D3.1 Case 1, $U = 1.0$ because the tension load is transmitted directly to the cross-section element. From AISC *Specification* Equation J4-2 with $A_e = A_n$:

LRFD	ASD
$\phi R_n = \phi F_u A_e$ $= 0.75(65 \text{ ksi})(12.0 \text{ in.}^2)$ $= 585 \text{ kips} > 550 \text{ kips}$ **o.k.**	$\dfrac{R_n}{\Omega} = \dfrac{F_u A_e}{\Omega}$ $= \dfrac{65 \text{ ksi}(12.0 \text{ in.}^2)}{2.00}$ $= 390 \text{ kips} > 385 \text{ kips}$ **o.k.**

Hinge Plate-to-HSS Brace Connection Design

The limit state of shear rupture in the brace wall was used in Example 5.3.10 to determine the length of the brace-to-gusset plate connection. Because the brace size in this example is the same as that used in Example 5.3.10, determination of the weld size and length between the brace and the hinge plate are not repeated here. Similarly, the flat bar reinforcement on the brace is kept the same as Example 5.3.10. For the limit state of block shear rupture on the hinge plate, the hinge plate in this example is thicker (1¼ in.) than the gusset plate (¾ in.) in Example 5.3.10 and is also a material with a higher yield strength. Therefore, from Example 5.3.10, block shear on the hinge plate will be adequate and need not be checked.

Check hinge plate for compression buckling

The minimum recommended hinge length for this connection configuration measured between the end of the brace and the gusset, is $3t_p$. Refer to Thornton and Fortney (2012) for discussion on the recommended $3t_p$ hinge length.

$$3t_p = 3(1\tfrac{1}{4} \text{ in.})$$
$$= 3.75 \text{ in.}$$

Use 4.00 in. for the hinge length.

Modeling the hinge plate as fixed at one end and free to rotate at the other end, the effective length factor from AISC *Specification* Commentary Table C-A-7.1 is 1.2. The effective length of the hinge plate is:

$$KL = 1.2(4.00 \text{ in.})$$
$$= 4.80 \text{ in.}$$

The radius of gyration of the hinge plate is:

$$r = \dfrac{t_p}{\sqrt{12}}$$
$$= \dfrac{1\tfrac{1}{4} \text{ in.}}{\sqrt{12}}$$
$$= 0.361 \text{ in.}$$

$$\dfrac{KL}{r} = \dfrac{4.80 \text{ in.}}{0.361 \text{ in.}}$$
$$= 13.3$$

5.3 SPECIAL CONCENTRICALLY BRACED FRAMES

Because $KL/r < 25$, buckling is not considered according to AISC *Specification* Section J4.4. However, the use of AISC *Manual* Table 4-22 will be demonstrated here because the buckling strength of the hinge plate is required to check the requirements of *Seismic Provisions* Section F2.6c(2).

From AISC *Manual* Table 4-22:

LRFD	ASD
$\phi_c F_{cr} = 44.4$ ksi	$\dfrac{F_{cr}}{\Omega_c} = 29.6$ ksi

The available compressive strength of the hinge plate is:

LRFD	ASD
$\phi P_{cr} = \phi F_{cr} A_g$ $= 44.4$ ksi$(11.0$ in.$)(1\frac{1}{4}$ in.$)$ $= 611$ kips 611 kips $> 1.1(444$ kips$) = 488$ kips **o.k.**	$\dfrac{P_{cr}}{\Omega} = \dfrac{F_{cr} A_g}{\Omega}$ $= 29.6$ ksi$(11.0$ in.$)(1\frac{1}{4}$ in.$)$ $= 407$ kips 407 kips $> 1.1(311$ kips$) = 342$ kips **o.k.**

AISC *Seismic Provisions* Section F2.6c(3) requires that the brace connection accommodate the flexural forces or rotation imposed by brace buckling. This can be achieved either by option (a) designing the connection to have an available flexural strength of the expected brace flexural strength, $R_y M_p$, multiplied by 1.1 (LRFD) or by 1.1/1.5 (ASD) or option (b) providing rotation capacity to accommodate the required rotation. Examples 5.3.8 through 5.3.11 used option (b) to satisfy this requirement. This brace configuration also satisfies option (b) because the $3t_p$ length of the hinge plate provides rotation capacity (Thornton and Fortney, 2012).

The hinge plate allows the brace to buckle in the plane of the gusset plate by means of introducing a perpendicular hinge plate. The connection thus accommodates brace rotation according to AISC *Seismic Provisions* Section F2.6c(3)(b); the requirement to withstand flexural forces imposed by brace buckling according to Section F2.6c(3)(a) is not applicable. Note that the commentary to this section implies that buckling in the plane of the gusset is fixed-end buckling (thus requiring application of Section F2.6c(3)(a)); in the context of this connection, the hinge plate takes the place of the gusset for purposes of determining end fixity.

To ensure that rotation of the hinge plate can occur without damage to other parts of the assembly, in this example the expected flexural strength of the hinge plate is used to determine maximum forces on the hinge-plate welds. This ensures that the hinge plate-to-gusset welds are sufficient to allow the hinge plate to achieve its expected flexural strength multiplied by 1.1.

Determine the expected flexural strength of the hinge plate (multiplied by 1.1):

$$M_{hinge} = 1.1 R_y F_y Z_h$$

where

$R_y = 1.1$ from AISC *Seismic Provisions* Table A3.1
Z_h = plastic section modulus of the hinge plate about the weak axis

$$= \frac{b_p t_p^2}{4}$$

$$= \frac{11.0 \text{ in.}(1\frac{1}{4} \text{ in.})^2}{4}$$

$$= 4.30 \text{ in.}^3$$

$$M_{hinge} = 1.1(1.1)(50 \text{ ksi})(4.30 \text{ in.}^3)$$
$$= 260 \text{ kip-in.}$$

This moment can be replaced by two equal and opposite forces, F, acting on the welds between the hinge plate and the brace.

$$F = \frac{M_{hinge}}{t_p}$$

$$= \frac{260 \text{ kip-in.}}{1\frac{1}{4} \text{ in.}}$$

$$= 208 \text{ kips}$$

The weld required to carry the force, F, from AISC *Manual* Equations 8-2a and 8-2b is:

LRFD	ASD
$D = \dfrac{F}{2(1.392 \text{ kip/in.})l}$	$D = \dfrac{F/1.5}{2(0.928 \text{ kip/in.})l}$
$= \dfrac{208 \text{ kips}}{2(1.392 \text{ kip/in.})(26.0 \text{ in.})}$	$= \dfrac{(208 \text{ kips}/1.5)}{2(0.928 \text{ kip/in.})(26.0 \text{ in.})}$
$= 2.87$ sixteenths < 4 sixteenths **o.k.**	$= 2.87$ sixteenths < 4 sixteenths **o.k.**

Hinge Plate-to-Gusset Connection Design

As shown in Figure 5-50, the hinge plate is slotted over the gusset plate. The hinge plate-to-gusset contact length is the same as the hinge plate-to-brace contact length (26.0 in.); therefore, the ¼-in. fillet welds would be appropriate. However, according to AISC *Specification* Table J2.4, the minimum required weld size is ⁵⁄₁₆ in. based on the 1¼ in. thickness of the hinge plate and gusset plate.

Use (4) 26-in.-long, ⁵⁄₁₆-in. fillet welds at the hinge plate-to-gusset connection.

Check tensile yielding of the gusset plate on the Whitmore section

Tension yielding is checked on a section of the gusset plate commonly referred to as the "Whitmore section." This section is explained in AISC *Manual* Part 9 (Figure 9-1) and in Thornton and Lini (2011).

5.3 SPECIAL CONCENTRICALLY BRACED FRAMES

The width of the maximum Whitmore section on the gusset plate at 30° is:

$$l_w = 2(26.0 \text{ in.})(\tan 30°) + 1\tfrac{1}{4} \text{ in.}$$
$$= 31.3 \text{ in.}$$

Part of this Whitmore section lies outside of the gusset plate. Approximately 12.0 in. of this width remains in the gusset at the gusset-to-column interface. In order to avoid accounting for Whitmore width within the bolted joint, a 12.0 in. width will be used on the column side and 15.0 in. on the beam side. On the beam side, approximately 5.00 in. are in the gusset and 10.0 in. are in the beam web (the 10.0 in. within the beam web is included in the Whitmore section area). Try a 1¼-in.-thick gusset. The Whitmore area is:

$$A_w = (12.0 \text{ in.} + 5.00 \text{ in.})(1\tfrac{1}{4} \text{ in.}) + (10.0 \text{ in.})(0.415 \text{ in.})$$
$$= 25.4 \text{ in.}^2$$

From AISC *Specification* Equation J4-1, the available tensile strength is:

LRFD	ASD
$\phi P_n = \phi F_y A_w$ $= 0.90(50 \text{ ksi})(25.4 \text{ in.}^2)$ $= 1{,}140 \text{ kips} > 550 \text{ kips}$ **o.k.**	$\dfrac{P_n}{\Omega} = \dfrac{F_y A_w}{\Omega}$ $= \dfrac{50 \text{ ksi}(25.4 \text{ in.}^2)}{1.67}$ $= 760 \text{ kips} > 385 \text{ kips}$ **o.k.**

Check shear yielding on the gusset plate

From AISC *Specification* Equation J4-3, the available shear strength due to yielding on the gusset plate is:

LRFD	ASD
$\phi R_n = \phi 0.60 F_y l t_p (2)$ $= 1.00(0.60)(50 \text{ ksi})(26.0 \text{ in.})$ $\times (1\tfrac{1}{4} \text{ in.})(2)$ $= 1{,}950 \text{ kips} > 550 \text{ kips}$ **o.k.**	$\dfrac{R_n}{\Omega} = \dfrac{0.60 F_y l t_p (2)}{\Omega}$ $= \dfrac{0.60(50 \text{ ksi})(26.0 \text{ in.})(1\tfrac{1}{4} \text{ in.})(2)}{1.50}$ $= 1{,}300 \text{ kips} > 385 \text{ kips}$ **o.k.**

Therefore, a 1¼-in.-thick gusset plate is adequate.

Check buckling of the gusset plate

The gusset buckling length is 5.00 in., and by inspection buckling will not control.

Gusset Interface Forces

Use the Uniform Force Method presented in AISC *Manual* Part 13. From the geometry of Figure 5-51:

$$e_c = \frac{d_c}{2}$$
$$= \frac{12.7 \text{ in.}}{2}$$
$$= 6.35 \text{ in.}$$

$$e_b = \frac{d_b}{2}$$
$$= \frac{23.7 \text{ in.}}{2}$$
$$= 11.9 \text{ in.}$$

$$\beta = 3.00 \text{ in.} + \frac{24.0 \text{ in.}}{2}$$
$$= 15.0 \text{ in.}$$

$$\theta = 45°$$

For the force distribution to remain free of moments on the connection interfaces, choose a value of α to satisfy the following expression.

$$\alpha - \beta \tan \theta = e_b \tan \theta - e_c \quad \text{(*Manual* Eq. 13-1)}$$

$$\alpha = (\beta + e_b)(\tan 45°) - e_c$$
$$= (15.0 \text{ in.} + 11.9 \text{ in.})(1) - 6.35 \text{ in.}$$
$$= 20.6 \text{ in.}$$

The required axial and shear forces on the connection due to the tensile load on the brace are determined from AISC *Manual* Equations 13-2 through 13-5, where:

$$r = \sqrt{(\alpha + e_c)^2 + (\beta + e_b)^2} \quad \text{(*Manual* Eq. 13-6)}$$
$$= \sqrt{(20.6 \text{ in.} + 6.35 \text{ in.})^2 + (15.0 \text{ in.} + 11.9 \text{ in.})^2}$$
$$= 38.1 \text{ in.}$$

5.3 SPECIAL CONCENTRICALLY BRACED FRAMES

LRFD	ASD
From AISC *Manual* Equation 13-3:	From AISC *Manual* Equation 13-3:
$H_{uc} = \dfrac{e_c}{r} P_u$ $= \dfrac{6.35 \text{ in.}}{38.1 \text{ in.}}(550 \text{ kips})$ $= 91.7 \text{ kips}$	$H_{ac} = \dfrac{e_c}{r} P_a$ $= \dfrac{6.35 \text{ in.}}{38.1 \text{ in.}}(385 \text{ kips})$ $= 64.2 \text{ kips}$
From AISC *Manual* Equation 13-5:	From AISC *Manual* Equation 13-5:
$H_{ub} = \dfrac{\alpha}{r} P_u$ $= \dfrac{20.6 \text{ in.}}{38.1 \text{ in.}}(550 \text{ kips})$ $= 297 \text{ kips}$	$H_{ab} = \dfrac{\alpha}{r} P_a$ $= \dfrac{20.6 \text{ in.}}{38.1 \text{ in.}}(385 \text{ kips})$ $= 208 \text{ kips}$
From AISC *Manual* Equation 13-2:	From AISC *Manual* Equation 13-2:
$V_{uc} = \dfrac{\beta}{r} P_u$ $= \dfrac{15.0 \text{ in.}}{38.1 \text{ in.}}(550 \text{ kips})$ $= 217 \text{ kips}$	$V_{ac} = \dfrac{\beta}{r} P_a$ $= \dfrac{15.0 \text{ in.}}{38.1 \text{ in.}}(385 \text{ kips})$ $= 152 \text{ kips}$
From AISC *Manual* Equation 13-4:	From AISC *Manual* Equation 13-4:
$V_{ub} = \dfrac{e_b}{r} P_u$ $= \dfrac{11.9 \text{ in.}}{38.1 \text{ in.}}(550 \text{ kips})$ $= 172 \text{ kips}$	$V_{ab} = \dfrac{e_b}{r} P_a$ $= \dfrac{11.9 \text{ in.}}{38.1 \text{ in.}}(385 \text{ kips})$ $= 120 \text{ kips}$

These forces are shown in Figures 5-52a and 5-52b.

Beam-to-Column Connection

The beam-to-column connection will be designed to satisfy the requirements of AISC *Seismic Provisions* Section F2.6b(b). The following exemplifies the determination of the required moment and forces on the connection.

In this example, the required flexural strength is resisted through the entire connection, including the gusset plate. The moment resistance is not confined to the beam-to-column portion of the connection. Alternatively, as shown in Example 5.3.11, AISC *Seismic Provisions* Section F2.6b(b) could also be satisfied by providing a fixed beam-to-column connection.

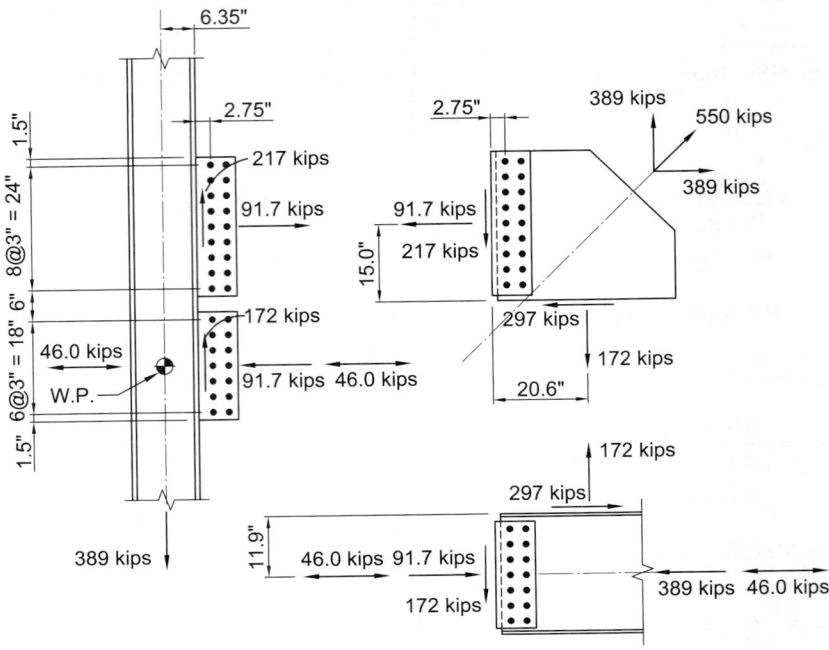

Fig. 5-52a. Gusset interface forces due to brace expected strength (LRFD).

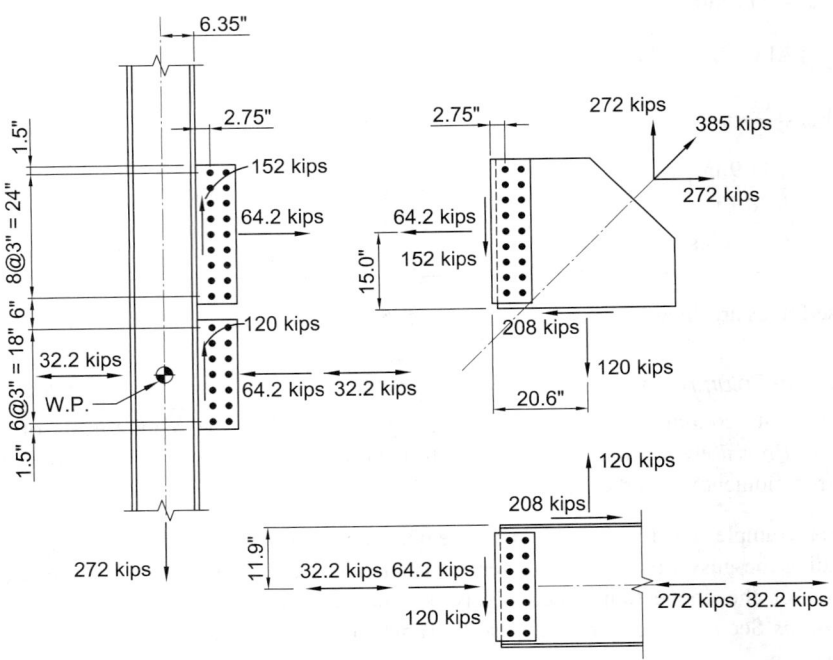

Fig. 5-52b. Gusset interface forces due to brace expected strength (ASD).

5.3 SPECIAL CONCENTRICALLY BRACED FRAMES

The required flexural strength is based on the lesser of the expected flexural strengths of the column and beam multiplied by 1.1 (LRFD) or 1.1/1.5 (ASD) as required by AISC *Seismic Provisions* Section F2.6b(b):

LRFD	ASD
$M_{col} = \Sigma 1.1 R_y F_y Z_x$ $= 2(1.1)(1.1)(50 \text{ ksi})(147 \text{ in.}^3)$ $= 17,800$ kip-in. $M_{beam} = 1.1 R_y F_y Z_x$ $= 1.1(1.1)(50 \text{ ksi})(177 \text{ in.}^3)$ $= 10,700$ kip-in.	$M_{col} = \Sigma(1.1/1.5) R_y F_y Z_x$ $= 2(1.1/1.5)(1.1)(50 \text{ ksi})(147 \text{ in.}^3)$ $= 11,900$ kip-in. $M_{beam} = (1.1/1.5) R_y F_y Z_x$ $= (1.1/1.5)(1.1)(50 \text{ ksi})(177 \text{ in.}^3)$ $= 7,140$ kip-in.

The lesser of these expected flexural strengths is $M_R = M_{beam} = 10,700$ kip-in. (LRFD) and 7,140 kip-in. (ASD). The subscript R is used to denote "rotational" forces and moments because this moment is due to frame action. Refer to Thornton and Muir (2009) for more discussion.

From Figures 5-53a and 5-53b:

LRFD	ASD
$H_R = \dfrac{M_R}{\beta + e_b}$ $= \dfrac{10,700 \text{ kip-in.}}{15.0 \text{ in.} + 11.9 \text{ in.}}$ $= 398$ kips $V_R = \dfrac{H_R \beta}{\alpha}$ $= \dfrac{398 \text{ kips}(15.0 \text{ in.})}{20.6 \text{ in.}}$ $= 290$ kips	$H_R = \dfrac{M_R}{\beta + e_b}$ $= \dfrac{7,140 \text{ kip-in.}}{15.0 \text{ in.} + 11.9 \text{ in.}}$ $= 265$ kips $V_R = \dfrac{H_R \beta}{\alpha}$ $= \dfrac{265 \text{ kips}(15.0 \text{ in.})}{20.6 \text{ in.}}$ $= 193$ kips

These rotational forces due to frame action are shown in Figures 5-54a and 5-54b. Application of moment in the figure is consistent with the angle between the beam and column closing as the brace goes into tension. In addition to the admissible force distribution due to the brace expected strength shown in Figures 5-52a and 5-52b, and the admissible force distribution due to frame action shown in Figures 5-54a and 5-54b, an admissible gravity force distribution must also be determined.

Note that the gravity forces always exist and therefore, must be added to the brace expected strength shown in Figures 5-52a and 5-52b and the rotational forces shown in Figures 5-54a and 5-54b.

AISC *Seismic Provisions* Section F2.6b requires that the rotational forces calculated from the lesser of the column moment strength or the beam moment strength be "considered in combination with the required strength of the brace connection and beam connection, including amplified diaphragm collector forces."

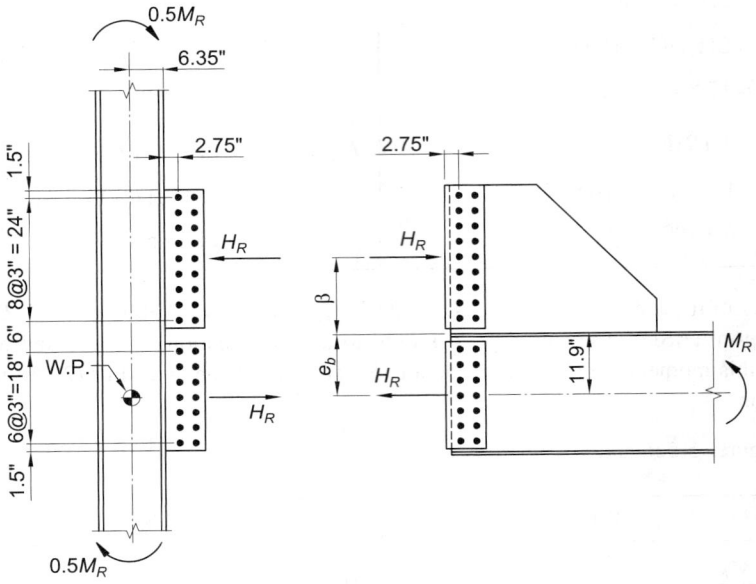

Fig. 5-53a. Rotational forces due to frame action, M_R.

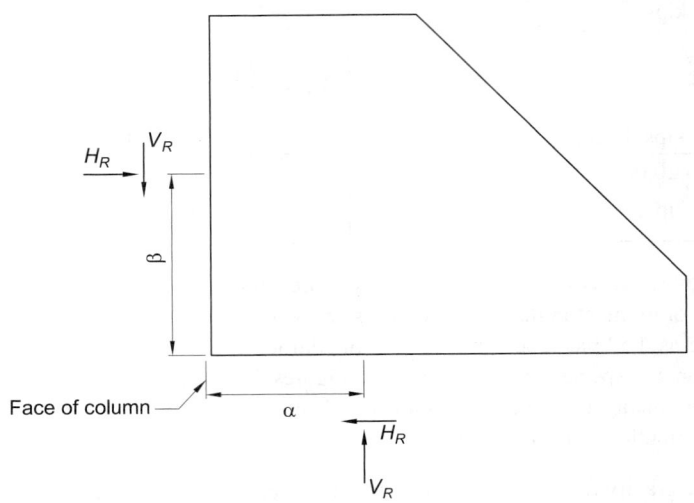

Fig. 5-53b. Gusset plate free body diagram due to rotational forces.

5.3 SPECIAL CONCENTRICALLY BRACED FRAMES

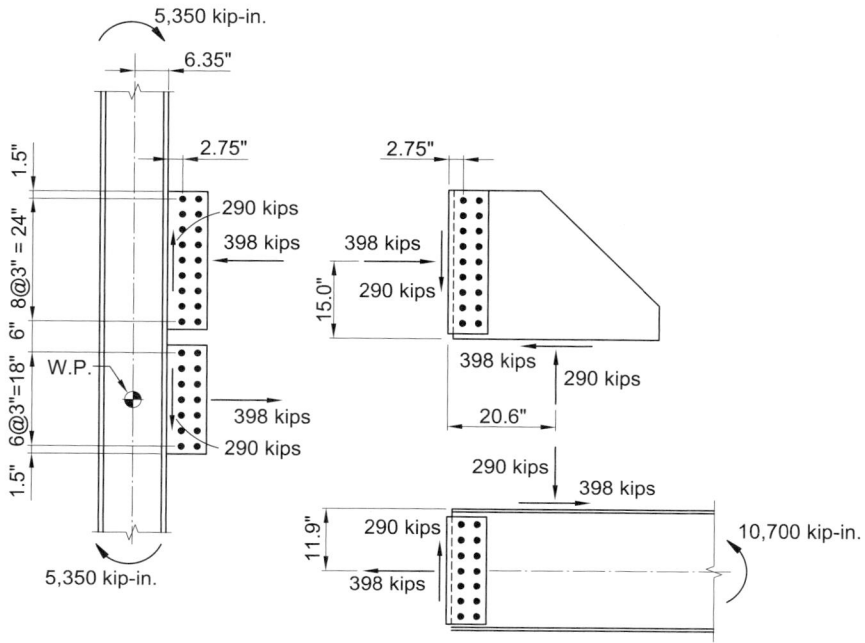

Fig. 5-54a. Rotational force distribution due to frame action (LRFD).

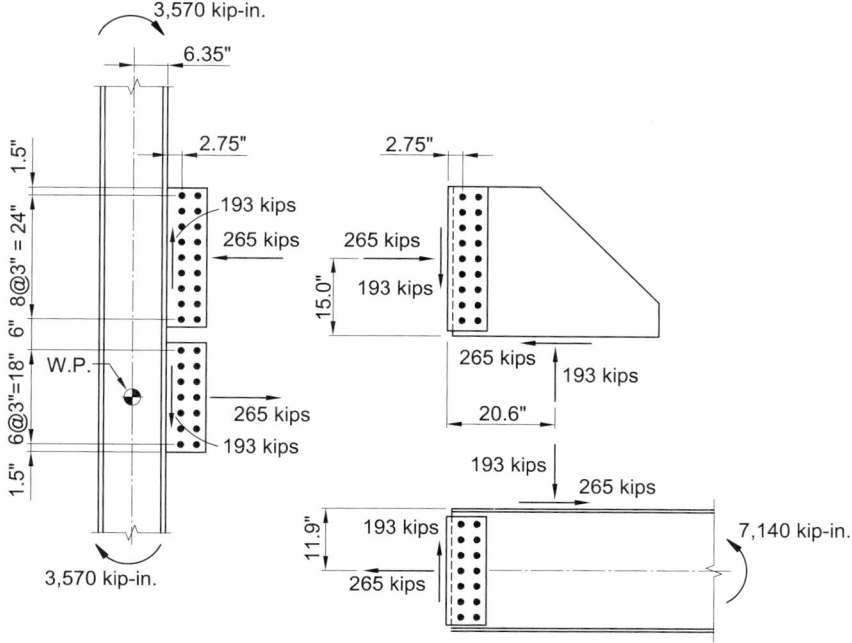

Fig. 5-54b. Rotational force distribution due to frame action (ASD).

Figures 5-55a and 5-55b show the combined brace, rotational, and gravity interface forces as required by AISC *Seismic Provisions* Section F2.6b.

Gusset-to-Column Single-Plate Connection Design

Figures 5-55a and 5-55b show the interface forces for this connection. Note that shear forces from the brace expected strength are additive with shears from the rotational forces, but normal forces from the brace expected strength are counteracted by rotational forces. This figure also shows the total axial load on the column, including the axial load above the column, P_u and P_a.

From Figures 5-55a and 5-55b, the total required strengths are:

LRFD	ASD
$V_u = 507$ kips $N_u = 306$ kips $R_u = \sqrt{V_u^2 + N_u^2}$ $ = \sqrt{(507 \text{ kips})^2 + (306 \text{ kips})^2}$ $ = 592$ kips	$V_a = 345$ kips $N_a = 201$ kips $R_a = \sqrt{V_a^2 + N_a^2}$ $ = \sqrt{(345 \text{ kips})^2 + (201 \text{ kips})^2}$ $ = 399$ kips

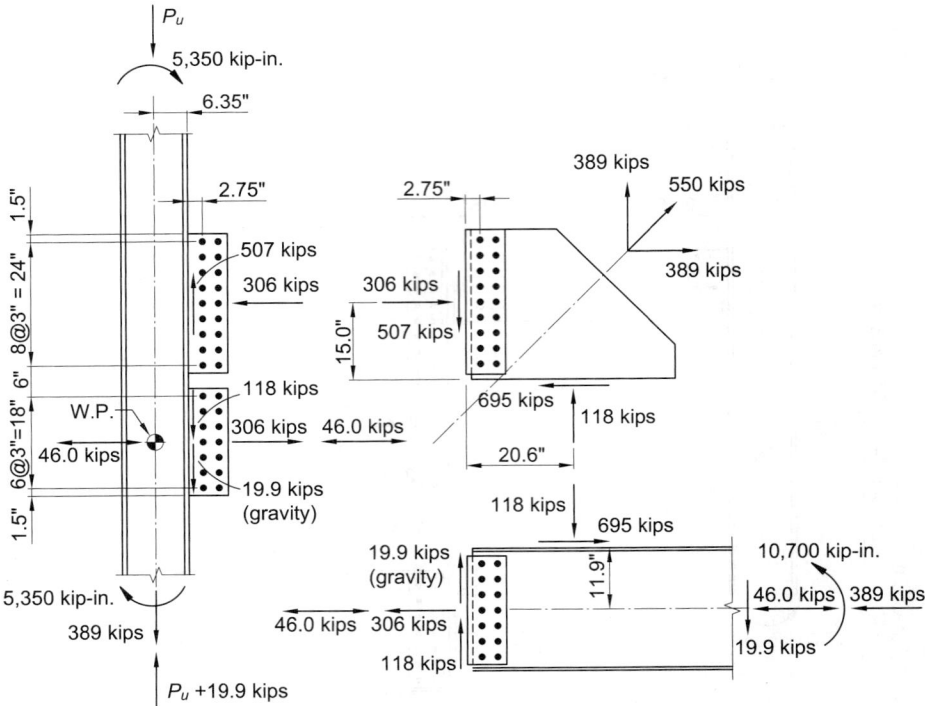

Fig. 5-55a. Combined brace, rotational and gravity forces (LRFD).

5.3 SPECIAL CONCENTRICALLY BRACED FRAMES

Use 1-in.-diameter bolts.

From AISC *Manual* Table 7-1 the available shear strength of 1-in.-diameter ASTM A490-X (Group B) bolts in standard holes is:

LRFD	ASD
$\phi r_n = 49.5$ kips/bolt	$\dfrac{r_n}{\Omega} = 33.0$ kips/bolt

The angle from the vertical is:

LRFD	ASD
$\theta = \tan^{-1}\left(\dfrac{306 \text{ kips}}{507 \text{ kips}}\right)$ $= 31.1°$	$\theta = \tan^{-1}\left(\dfrac{201 \text{ kips}}{345 \text{ kips}}\right)$ $= 30.2°$

The eccentricity from the centerline of the two rows of bolts to the column face is:

$$2.75 \text{ in.} + \frac{3.00 \text{ in.}}{2} = 4.25 \text{ in.}$$

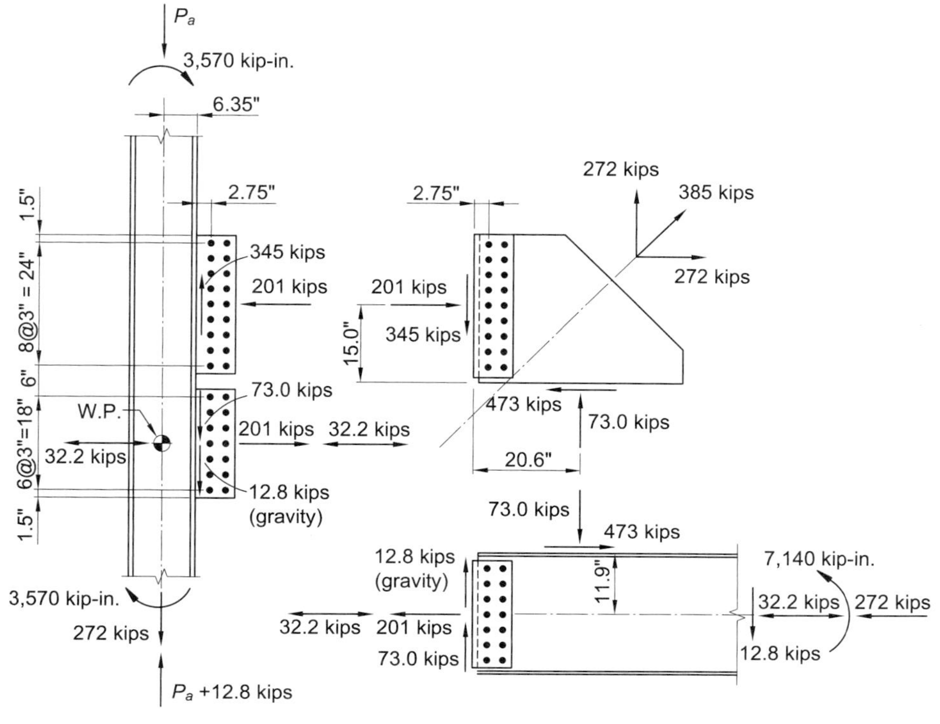

Fig. 5-55b. Combined brace, rotational and gravity forces (ASD).

Using AISC *Manual* Table 7-7 for an angle of 30° with $e_x = 4.25$ in., $n = 9$, and $s = 3$ in.:

$C = 14.9$

The available shear strength of the bolt group is:

LRFD	ASD
$\phi R_n = C\phi r_n$ $= 14.9(49.5 \text{ kips/bolt})$ $= 738 \text{ kips} > 592 \text{ kips}$ **o.k.**	$\dfrac{R_n}{\Omega} = C\left(\dfrac{r_n}{\Omega}\right)$ $= 14.9(33.0 \text{ kips/bolt})$ $= 492 \text{ kips} > 399 \text{ kips}$ **o.k.**

Check gusset gross section for shear yielding strength

From the geometry and edge distances shown in Figure 5-51:

$A_g = 29.0 \text{ in.}(1\tfrac{1}{4} \text{ in.})$

$= 36.3 \text{ in.}^2$

From AISC *Specification* Equation J4-3, the available shear yielding strength of the gusset plate is:

LRFD	ASD
$\phi R_n = \phi 0.60 F_y A_g$ $= 1.00(0.60)(50 \text{ ksi})(36.3 \text{ in.}^2)$ $= 1,090 \text{ kips} > 507 \text{ kips}$ **o.k.**	$\dfrac{R_n}{\Omega} = \dfrac{0.60 F_y A_g}{\Omega}$ $= \dfrac{0.60(50 \text{ ksi})(36.3 \text{ in.}^2)}{1.50}$ $= 726 \text{ kips} > 345 \text{ kips}$ **o.k.**

Check gusset gross section for tensile yielding strength

From AISC *Specification* Equation J4-1, the available tensile yielding strength of the gusset plate is:

LRFD	ASD
$\phi R_n = \phi F_y A_g$ $= 0.90(50 \text{ ksi})(36.3 \text{ in.}^2)$ $= 1,630 \text{ kips} > 306 \text{ kips}$ **o.k.**	$\dfrac{R_n}{\Omega} = \dfrac{F_y A_g}{\Omega}$ $= \dfrac{50 \text{ ksi}(36.3 \text{ in.}^2)}{1.67}$ $= 1,090 \text{ kips} > 201 \text{ kips}$ **o.k.**

Check gusset net section for shear rupture strength

Based on the required hole size for a 1-in.-diameter bolt in standard holes from AISC *Specification* Table J3.3 and the ¹⁄₁₆ in. increase required from AISC *Specification* Section B4.3b, the net area is:

$$A_{nv} = [29.0 \text{ in.} - 9(1\tfrac{1}{16} \text{ in.} + \tfrac{1}{16} \text{ in.})](1\tfrac{1}{4} \text{ in.})$$
$$= 23.6 \text{ in.}^2$$

From AISC *Specification* Equation J4-4, the available shear rupture strength of the gusset plate is:

LRFD	ASD
$\phi R_n = \phi 0.60 F_u A_{nv}$ $= 0.75(0.60)(65 \text{ ksi})(23.6 \text{ in.}^2)$ $= 690 \text{ kips} > 507 \text{ kips}$ **o.k.**	$\dfrac{R_n}{\Omega} = \dfrac{0.60 F_u A_{nv}}{\Omega}$ $= \dfrac{0.60(65 \text{ ksi})(23.6 \text{ in.}^2)}{2.00}$ $= 460 \text{ kips} > 345 \text{ kips}$ **o.k.**

Check gusset net section for tensile rupture strength

The net tension area is:

$$A_{nt} = A_{nv}$$

From AISC *Specification* Equation J4-2, with $A_e = A_{nt}$, the available tensile rupture strength is:

LRFD	ASD
$\phi R_{nt} = \phi F_u A_e$ $= 0.75(65 \text{ ksi})(23.6 \text{ in.}^2)$ $= 1{,}150 \text{ kips} > 306 \text{ kips}$ **o.k.**	$\dfrac{R_{nt}}{\Omega} = \dfrac{F_u A_e}{\Omega}$ $= \dfrac{65 \text{ ksi}(23.6 \text{ in.}^2)}{2.00}$ $= 767 \text{ kips} > 201 \text{ kips}$ **o.k.**

Check net tension and shear rupture interaction

LRFD	ASD
$\left(\dfrac{306 \text{ kips}}{1{,}150 \text{ kips}}\right)^2 + \left(\dfrac{507 \text{ kips}}{690 \text{ kips}}\right)^2$ $= 0.611 < 1.0$ **o.k.**	$\left(\dfrac{201 \text{ kips}}{767 \text{ kips}}\right)^2 + \left(\dfrac{345 \text{ kips}}{460 \text{ kips}}\right)^2$ $= 0.631 < 1.0$ **o.k.**

Check block shear rupture on gusset at gusset-to column interface

The failure path shown in Figure 5-56 controls the block shear rupture strength on the gusset plate relative to the shear force. Because the tension stress is nonuniform, similar to AISC *Specification* Commentary Figure C-J4.2(b), $U_{bs} = 0.5$. From AISC *Specification* Section J4.3:

$$U_{bs} = 0.5$$

$$A_{nt} = [4.75 \text{ in.} - 1.5(1\tfrac{1}{16} \text{ in.} + \tfrac{1}{16} \text{ in.})](1\tfrac{1}{4} \text{ in.})$$
$$= 3.83 \text{ in.}^2$$

$$A_{gv} = 26.0 \text{ in.}(1\tfrac{1}{4} \text{ in.})$$
$$= 32.5 \text{ in.}^2$$

$$A_{nv} = [26.0 \text{ in.} - 8.5(1\tfrac{1}{16} \text{ in.} + \tfrac{1}{16} \text{ in.})](1\tfrac{1}{4} \text{ in.})$$
$$= 20.5 \text{ in.}^2$$

The available strength for the limit state of block shear rupture relative to the shear force on the gusset plate is:

$$R_n = 0.60 F_u A_{nv} + U_{bs} F_u A_{nt} \leq 0.60 F_y A_{gv} + U_{bs} F_u A_{nt} \qquad (Spec. \text{ Eq. J4-5})$$

$$0.60 F_u A_{nv} + U_{bs} F_u A_{nt} = 0.60(65 \text{ ksi})(20.5 \text{ in.}^2) + 0.5(65 \text{ ksi})(3.83 \text{ in.}^2)$$
$$= 924 \text{ kips}$$

$$0.60 F_y A_{gv} + U_{bs} F_u A_{nt} = 0.60(50 \text{ ksi})(32.5 \text{ in.}^2) + 0.5(65 \text{ ksi})(3.83 \text{ in.}^2)$$
$$= 1{,}100 \text{ kips}$$

Therefore, the nominal strength for the limit state of block shear rupture is $R_n = 924$ kips. The available strength for the limit state of block shear rupture on the gusset plate is:

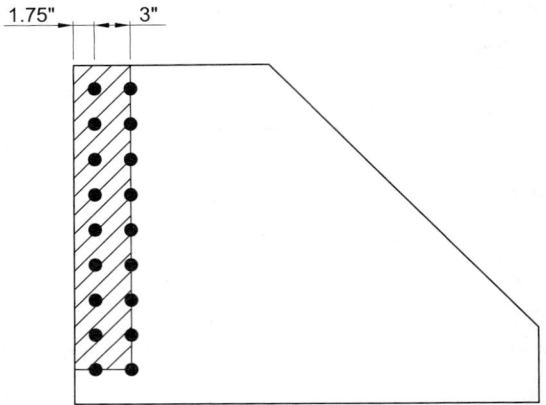

Fig. 5-56. Controlling block shear failure path in gusset plate.

5.3 SPECIAL CONCENTRICALLY BRACED FRAMES

LRFD	ASD
$\phi R_n = 0.75(924 \text{ kips})$ $= 693 \text{ kips} > 507 \text{ kips}$ **o.k.**	$\dfrac{R_n}{\Omega} = \dfrac{924 \text{ kips}}{2.00}$ $= 462 \text{ kips} > 345 \text{ kips}$ **o.k.**

Check block shear relative to the normal force on the gusset plate using the failure path shown in Figure 5-56:

$U_{bs} = 1.0$

$A_{nt} = [26.0 \text{ in.} - 8.5(1\frac{1}{16} \text{ in.} + \frac{1}{16} \text{ in.})](1\frac{1}{4} \text{ in.})$
$\quad = 20.5 \text{ in.}^2$

$A_{gv} = 4.75 \text{ in.}(1\frac{1}{4} \text{ in.})$
$\quad = 5.94 \text{ in.}^2$

$A_{nv} = [4.75 \text{ in.} - 1.5(1\frac{1}{16} \text{ in.} + \frac{1}{16} \text{ in.})](1\frac{1}{4} \text{ in.})$
$\quad = 3.83 \text{ in.}^2$

$R_n = 0.60 F_u A_{nv} + U_{bs} F_u A_{nt} \leq 0.60 F_y A_{gv} + U_{bs} F_u A_{nt}$ *(Spec. Eq. J4-5)*

$0.60 F_u A_{nv} + U_{bs} F_u A_{nt} = 0.60(65 \text{ ksi})(3.83 \text{ in.}^2) + 1.0(65 \text{ ksi})(20.5 \text{ in.}^2)$
$\quad = 1{,}480 \text{ kips}$

$0.60 F_y A_{gv} + U_{bs} F_u A_{nt} = 0.60(50 \text{ ksi})(5.94 \text{ in.}^2) + 1.0(65 \text{ ksi})(20.5 \text{ in.}^2)$
$\quad = 1{,}510 \text{ kips}$

Therefore, the nominal strength for the limit state of block shear rupture is $R_n = 1{,}480$ kips. The available strength for the limit state of block shear rupture on the gusset plate is:

LRFD	ASD
$\phi R_n = 0.75(1{,}480 \text{ kips})$ $= 1{,}110 \text{ kips} > 306 \text{ kips}$ **o.k.**	$\dfrac{R_n}{\Omega} = \dfrac{1{,}480 \text{ kips}}{2.00}$ $= 740 \text{ kips} > 201 \text{ kips}$ **o.k.**

Check shear and tension interaction due to block shear

LRFD	ASD
$\left(\dfrac{306 \text{ kips}}{1{,}110 \text{ kips}}\right)^2 + \left(\dfrac{507 \text{ kips}}{693 \text{ kips}}\right)^2$ $= 0.611 < 1.0$ **o.k.**	$\left(\dfrac{201 \text{ kips}}{740 \text{ kips}}\right)^2 + \left(\dfrac{345 \text{ kips}}{462 \text{ kips}}\right)^2$ $= 0.631 < 1.0$ **o.k.**

Check bolt bearing and tearout on the gusset plate

The gusset vertical edge distance to the end bolt is 2.00 in. at the top and 3.00 in. at the bottom. The gusset horizontal edge dimension is 1.75 in. The resultant force per bolt, based on the C-value taken from AISC *Manual* Table 7-7 previously, is:

LRFD	ASD
$r_u = \dfrac{592 \text{ kips}}{14.9 \text{ bolts}}$ $= 39.7$ kips/bolt	$r_a = \dfrac{399 \text{ kips}}{14.9 \text{ bolts}}$ $= 26.8$ kips/bolt

The edge distance along the line of action of the bolt force may be calculated from the line of action of the given shear and tension. For simplicity, use a conservative value for the bolt edge distance of 1.75 in. If this conservative assumption requires a thicker gusset plate, the aforementioned line of action method will be used.

The bearing and tearout strength limit states are separated for clarity in the following. For SCBF systems, where large deformations are anticipated, it is appropriate to assume that deformation at the bolt hole is not a design consideration for the seismic loading, and the tearout strength is checked at the end bolt with the 1.75 in. edge distance using AISC *Specification* Equation J3-6b. AISC *Seismic Provisions* Section D2.2(1) limits the nominal bearing strength of the gusset plate to $2.4dtF_u$, therefore the available bearing strength is:

LRFD	ASD
$\phi r_n = \phi 2.4 dt_p F_u$ $= 0.75(2.4)(1.00 \text{ in.})(1\frac{1}{4} \text{ in.})(65 \text{ ksi})$ $= 146$ kips/bolt > 39.7 kips/bolt **o.k.**	$\dfrac{r_n}{\Omega} = \dfrac{2.4 dt_p F_u}{\Omega}$ $= \dfrac{2.4(1.00 \text{ in.})(1\frac{1}{4} \text{ in.})(65 \text{ ksi})}{2.00}$ $= 97.5$ kips/bolt > 26.8 kips/bolt **o.k.**

From the lefthand side of AISC *Specification* Equation J3-6b, the available tearout strength is:

LRFD	ASD
$\phi r_n = \phi 1.5 l_c t_p F_u$ $= 0.75(1.5)[1.75 \text{ in.} - 0.5(1\frac{1}{16} \text{ in.})]$ $\times (1\frac{1}{4} \text{ in.})(65 \text{ ksi})$ $= 111$ kips/bolt > 39.7 kips/bolt **o.k.**	$\dfrac{r_n}{\Omega} = \dfrac{1.5 l_c t_p F_u}{\Omega}$ $= \dfrac{\begin{Bmatrix} 1.5[1.75 \text{ in.} - 0.5(1\frac{1}{16} \text{ in.})] \\ \times (1\frac{1}{4} \text{ in.})(65 \text{ ksi}) \end{Bmatrix}}{2.00}$ $= 74.3$ kips/bolt > 26.8 kips/bolt **o.k.**

5.3 SPECIAL CONCENTRICALLY BRACED FRAMES

Check bolt bearing and tearout on the single plate

Assume the single plate is 1¼ in. thick—the same as the gusset plate.

The bearing and tearout strength limit states are separated for clarity in the following. For SCBF systems, where large deformations are anticipated, it is appropriate to assume that deformation at the bolt hole is not a design consideration for the seismic loading, and the tearout strength is checked at the end bolt using AISC *Specification* Equation J3-6b. AISC *Seismic Provisions* Section D2.2(1) limits the nominal bearing strength to $2.4dtF_u$, therefore the available bearing strength of the single plate is:

LRFD	ASD
$\phi r_n = \phi 2.4 dt_p F_u$ $= 0.75(2.4)(1.00 \text{ in.})(1\frac{1}{4} \text{ in.})(65 \text{ ksi})$ $= 146 \text{ kips/bolt} > 39.7 \text{ kips/bolt}$ **o.k.**	$\dfrac{r_n}{\Omega} = \dfrac{2.4 dt_p F_u}{\Omega}$ $= \dfrac{2.4(1.00 \text{ in.})(1\frac{1}{4} \text{ in.})(65 \text{ ksi})}{2.00}$ $= 97.5 \text{ kips/bolt} > 26.8 \text{ kips/bolt}$ **o.k.**

The single plate has top and bottom edge distances of 1.50 in. and a horizontal edge distance of 1.75 in.

From AISC *Specification* Equation J3-6b, the available tearout strength of the single plate is:

LRFD	ASD
$\phi r_n = \phi 1.5 l_c t_p F_u$ $= 0.75(1.5)[1.50 \text{ in.} - 0.5(1\frac{1}{16} \text{ in.})]$ $\times (1\frac{1}{4} \text{ in.})(65 \text{ ksi})$ $= 88.5 \text{ kips/bolt} > 39.7 \text{ kips/bolt}$ **o.k.**	$\dfrac{r_n}{\Omega} = \dfrac{1.5 l_c t_p F_u}{\Omega}$ $= \dfrac{\begin{cases}1.5[1.50 \text{ in.} - 0.5(1\frac{1}{16} \text{ in.})] \\ \times (1\frac{1}{4} \text{ in.})(65 \text{ ksi})\end{cases}}{2.00}$ $= 59.0 \text{ kips/bolt} > 26.8 \text{ kips/bolt}$ **o.k.**

Check gross and net shear and tension on the single plate

From Figure 5-51, the single plate is 27.0 in. long. From AISC *Specification* Equation J4-3, the available shear yielding strength of the single plate is:

LRFD	ASD
$\phi R_n = \phi 0.60 F_y A_{gv}$ $= 1.00(0.60)(50 \text{ ksi})(27.0 \text{ in.})$ $\times (1\frac{1}{4} \text{ in.})$ $= 1,010 \text{ kips} > 507 \text{ kips}$ **o.k.**	$\dfrac{R_n}{\Omega} = \dfrac{0.60 F_y A_{gv}}{\Omega}$ $= \dfrac{0.60(50 \text{ ksi})(27.0 \text{ in.})(1\frac{1}{4} \text{ in.})}{1.50}$ $= 675 \text{ kips} > 345 \text{ kips}$ **o.k.**

From AISC *Specification* Equation J4-1, the available tensile yielding strength of the single plate is:

LRFD	ASD
$\phi R_n = \phi F_y A_g$ $= 0.90(50 \text{ ksi})(27.0 \text{ in.})(1\frac{1}{4} \text{ in.})$ $= 1{,}520 \text{ kips} > 306 \text{ kips}$ **o.k.**	$\dfrac{R_n}{\Omega} = \dfrac{F_y A_g}{\Omega}$ $= \dfrac{50 \text{ ksi}(27.0 \text{ in.})(1\frac{1}{4} \text{ in.})}{1.67}$ $= 1{,}010 \text{ kips} > 201 \text{ kips}$ **o.k.**

The available shear rupture strength of the single plate is determined from AISC *Specification* Equation J4-4, where:

$$A_{nv} = \left[27.0 \text{ in.} - 9\left(1\frac{1}{16} \text{ in.} + \frac{1}{16} \text{ in.}\right)\right]\left(1\frac{1}{4} \text{ in.}\right)$$
$$= 21.1 \text{ in.}^2$$

LRFD	ASD
$\phi R_n = \phi 0.60 F_u A_{nv}$ $= 0.75(0.60)(65 \text{ ksi})\left(21.1 \text{ in.}^2\right)$ $= 617 \text{ kips} > 507 \text{ kips}$ **o.k.**	$\dfrac{R_n}{\Omega} = \dfrac{0.60 F_u A_{nv}}{\Omega}$ $= \dfrac{0.60(65 \text{ ksi})\left(21.1 \text{ in.}^2\right)}{2.00}$ $= 411 \text{ kips} > 345 \text{ kips}$ **o.k.**

The available tensile rupture strength on the single plate is determined from AISC *Specification* Equation J4-2, with $A_e = A_{nt}$, where:

$$A_{nt} = A_{nv}$$
$$= 21.1 \text{ in.}^2$$

LRFD	ASD
$\phi R_{nt} = \phi F_u A_{nt}$ $= 0.75(65 \text{ ksi})\left(21.1 \text{ in.}^2\right)$ $= 1{,}030 \text{ kips} > 306 \text{ kips}$ **o.k.**	$\dfrac{R_{nt}}{\Omega} = \dfrac{F_u A_{nt}}{\Omega}$ $= \dfrac{65 \text{ ksi}\left(21.1 \text{ in.}^2\right)}{2.00}$ $= 686 \text{ kips} > 201 \text{ kips}$ **o.k.**

5.3 SPECIAL CONCENTRICALLY BRACED FRAMES

Check net tension and shear rupture interaction

LRFD	ASD
$\left(\dfrac{306 \text{ kips}}{1{,}030 \text{ kips}}\right)^2 + \left(\dfrac{507 \text{ kips}}{617 \text{ kips}}\right)^2$ $= 0.763 < 1.0$ **o.k.**	$\left(\dfrac{201 \text{ kips}}{686 \text{ kips}}\right)^2 + \left(\dfrac{345 \text{ kips}}{411 \text{ kips}}\right)^2$ $= 0.790 < 1.0$ **o.k.**

Single plate-to-column flange weld

From AISC *Manual* Equations 8-2a and 8-2b, including the increased strength due to the load angle given by AISC *Specification* Equation J2-5, the required single plate-to-column flange weld is:

LRFD	ASD
Load angle $\theta = \tan^{-1}\left(\dfrac{306 \text{ kips}}{507 \text{ kips}}\right)$ $= 31.1°$ $D_{req} = \dfrac{R_u}{2(1.392 \text{ kip/in.})\left(1 + 0.50\sin^{1.5}\theta\right)(l)}$ $= \dfrac{592 \text{ kips}}{2(1.392 \text{ kip/in.})(1.19)(27.0 \text{ in.})}$ $= 6.62$ sixteenths	Load angle $\theta = \tan^{-1}\left(\dfrac{201 \text{ kips}}{345 \text{ kips}}\right)$ $= 30.2°$ $D_{req} = \dfrac{R_a}{2(0.928 \text{ kip/in.})\left(1 + 0.50\sin^{1.5}\theta\right)(l)}$ $= \dfrac{399 \text{ kips}}{2(0.928 \text{ kip/in.})(1.18)(27.0 \text{ in.})}$ $= 6.75$ sixteenths

Use a $\frac{7}{16}$-in. fillet weld.

Gusset-to-Beam Interface

The length of the weld is:

$$l_b = (\alpha - 1.00)(2)$$
$$= (20.6 \text{ in.} - 1.00 \text{ in.})(2)$$
$$= 39.2 \text{ in.}$$

The required strengths at the gusset-to-beam interface from Figures 5-55a and 5-55b are:

LRFD	ASD
$V_u = 695$ kips $N_u = 118$ kips	$V_a = 473$ kips $N_a = 73.0$ kips

Gusset-to-beam weld

From AISC *Manual* Equations 8-2a and 8-2b, including the increased strength due to the load angle, and the 1.25 weld ductility factor discussed in Part 13 of the AISC *Manual*, the required gusset plate-to-beam flange weld is:

LRFD	ASD
Load angle	Load angle
$\theta = \tan^{-1}\left(\dfrac{118 \text{ kips}}{695 \text{ kips}}\right)$	$\theta = \tan^{-1}\left(\dfrac{73.0 \text{ kips}}{473 \text{ kips}}\right)$
$= 9.64°$	$= 8.77°$
$D_{req} = \dfrac{1.25\sqrt{V_u^2 + N_u^2}}{2(1.392 \text{ kip/in.})}$	$D_{req} = \dfrac{1.25\sqrt{V_a^2 + N_a^2}}{2(0.928 \text{ kip/in.})}$
$\times \dfrac{1}{(1.0 + 0.50\sin^{1.5}\theta)(l)}$	$\times \dfrac{1}{(1.0 + 0.50\sin^{1.5}\theta)(l)}$
$= \dfrac{1.25\sqrt{(695 \text{ kips})^2 + (118 \text{ kips})^2}}{2(1.392 \text{ kip/in.})(1.03)(39.2 \text{ in.})}$	$= \dfrac{1.25\sqrt{(473 \text{ kips})^2 + (73.0 \text{ kips})^2}}{2(0.928 \text{ kip/in.})(1.03)(39.2 \text{ in.})}$
$= 7.84$ sixteenths	$= 7.98$ sixteenths

Use a ½-in. fillet weld, 39¼ in. long.

Check gusset plate for shear yielding and tension yielding

LRFD	ASD
Shear yielding on gross section, from AISC *Specification* Equation J4-3:	Shear yielding on gross section, from AISC *Specification* Equation J4-3:
$\phi V_n = \phi 0.60 F_y A_{gv}$	$\dfrac{V_n}{\Omega} = \dfrac{0.60 F_y A_{gv}}{\Omega}$
$= 1.00(0.60)(50 \text{ ksi})(1¼ \text{ in.})$	
$\times (39¼ \text{ in.})$	$= \dfrac{0.60(50 \text{ ksi})(1¼ \text{ in.})(39¼ \text{ in.})}{1.50}$
$= 1,470$ kips > 695 kips **o.k.**	$= 981$ kips > 473 kips **o.k.**
Tension yielding on gross section, from AISC *Specification* Equation J4-1:	Tension yielding on gross section, from AISC *Specification* Equation J4-1:
$\phi R_n = \phi F_y A_g$	$\dfrac{R_n}{\Omega} = \dfrac{F_y A_g}{\Omega}$
$= 0.90(50 \text{ ksi})(1¼ \text{ in.})(39¼ \text{ in.})$	$= \dfrac{50 \text{ ksi}(1¼ \text{ in.})(39¼ \text{ in.})}{1.67}$
$= 2,210$ kips > 118 kips **o.k.**	$= 1,470$ kips > 73.0 kips **o.k.**

5.3 SPECIAL CONCENTRICALLY BRACED FRAMES

Check beam web local yielding

For the W24×68, the available web local yielding strength is determined from AISC *Specification* Equation J10-3 for a force applied from the member end that is less than the member depth as follows:

LRFD	ASD
$\phi R_n = \phi F_y t_w (2.5 k_{des} + l_b)$ $= 1.00(50 \text{ ksi})(0.415 \text{ in.})$ $\times [2.5(1.09 \text{ in.}) + 39\frac{1}{4} \text{ in.}]$ $= 871 \text{ kips} > 118 \text{ kips}$ **o.k.**	$\dfrac{R_n}{\Omega} = \dfrac{F_y t_w (2.5 k_{des} + l_b)}{\Omega}$ $= (1/1.50)(50 \text{ ksi})(0.415 \text{ in.})$ $\times [2.5(1.09 \text{ in.}) + 39\frac{1}{4} \text{ in.}]$ $= 581 \text{ kips} > 73.0 \text{ kips}$ **o.k.**

Check beam web local crippling

The resultant load on the beam from the gusset plate is applied at 20.6 in. from the column face, which is greater than $d/2$; therefore, use the following equation to determine the nominal strength due to web local crippling.

$$R_n = 0.80 t_w^2 \left[1 + 3 \left(\frac{l_b}{d} \right) \left(\frac{t_w}{t_f} \right)^{1.5} \right] \sqrt{\frac{E F_{yf} t_f}{t_w}} \qquad \text{(Spec. Eq. J10-4)}$$

The available strength due to web local crippling is:

LRFD	ASD
$\phi R_n = 0.75(0.80)(0.415 \text{ in.})^2$ $\times \left[1 + 3 \left(\dfrac{39\frac{1}{4} \text{ in.}}{23.7 \text{ in.}} \right) \left(\dfrac{0.415 \text{ in.}}{0.585 \text{ in.}} \right)^{1.5} \right]$ $\times \sqrt{\dfrac{29{,}000 \text{ ksi}(50 \text{ ksi})(0.585 \text{ in.})}{0.415 \text{ in.}}}$ $= 586 \text{ kips} > 118 \text{ kips}$ **o.k.**	$\dfrac{R_n}{\Omega} = (1/2.00)(0.80)(0.415 \text{ in.})^2$ $\times \left[1 + 3 \left(\dfrac{39\frac{1}{4} \text{ in.}}{23.7 \text{ in.}} \right) \left(\dfrac{0.415 \text{ in.}}{0.585 \text{ in.}} \right)^{1.5} \right]$ $\times \sqrt{\dfrac{29{,}000 \text{ ksi}(50 \text{ ksi})(0.585 \text{ in.})}{0.415 \text{ in.}}}$ $= 391 \text{ kips} > 73.0 \text{ kips}$ **o.k.**

Beam-to-Column Connection

The required strengths from Figures 5-55a and 5-55b are:

LRFD	ASD
$V_u = 118 \text{ kips} + 19.9 \text{ kips}$ $\quad = 138 \text{ kips}$ $N_u = 306 \text{ kips} + 46.0 \text{ kips}$ $\quad = 352 \text{ kips}$	$V_a = 73.0 \text{ kips} + 12.8 \text{ kips}$ $\quad = 85.8 \text{ kips}$ $N_a = 201 \text{ kips} + 32.2 \text{ kips}$ $\quad = 233 \text{ kips}$

Check bolt strength

The required bolt strength due to the resultant loading is:

LRFD	ASD
$R_u = \sqrt{V_u^2 + N_u^2}$ $\quad = \sqrt{(138 \text{ kips})^2 + (352 \text{ kips})^2}$ $\quad = 378 \text{ kips}$	$R_a = \sqrt{V_a^2 + N_a^2}$ $\quad = \sqrt{(85.8 \text{ kips})^2 + (233 \text{ kips})^2}$ $\quad = 248 \text{ kips}$

There are 14 ASTM A490-X bolts in standard holes as shown in Figure 5-51. From AISC *Manual* Table 7-1, the available shear strength per bolt is:

LRFD	ASD
$\phi r_n = 49.5 \text{ kips/bolt}$	$\dfrac{r_n}{\Omega} = 33.0 \text{ kips/bolt}$

The angle of the resultant with respect to the vertical is:

LRFD	ASD
$\theta = \tan^{-1}\left(\dfrac{352 \text{ kips}}{138 \text{ kips}}\right)$ $\quad = 68.6°$	$\theta = \tan^{-1}\left(\dfrac{233 \text{ kips}}{85.8 \text{ kips}}\right)$ $\quad = 69.8°$

Using AISC *Manual* Table 7-7 with 60°, $n = 7$, $e_x = 4.25$ in., and $s = 3$ in.:

$\quad C = 11.4$

LRFD	ASD
$\phi R_n = C\phi r_n$ $\quad = 11.4(49.5 \text{ kips/bolt})$ $\quad = 564 \text{ kips} > 378 \text{ kips}$ **o.k.**	$\dfrac{R_n}{\Omega} = C\dfrac{r_n}{\Omega}$ $\quad = 11.4(33.0 \text{ kips/bolt})$ $\quad = 376 \text{ kips} > 248 \text{ kips}$ **o.k.**

5.3 SPECIAL CONCENTRICALLY BRACED FRAMES

Check beam shear strength
From AISC *Manual* Table 3-6, the x-x axis available shear strength of the beam due to shear yielding and shear buckling is:

LRFD	ASD
$\phi V_n = 295$ kips > 138 kips **o.k.**	$\dfrac{V_n}{\Omega} = 197$ kips > 85.8 kips **o.k.**

Check beam tensile yielding strength
From AISC *Specification* Equation D2-1, the available tensile strength due to yielding is:

LRFD	ASD
$\phi P_n = \phi F_y A_g$ $= 0.90(50 \text{ ksi})(20.1 \text{ in.}^2)$ $= 905$ kips > 352 kips **o.k.**	$\dfrac{P_n}{\Omega} = \dfrac{F_y A_g}{\Omega}$ $= \dfrac{50 \text{ ksi}(20.1 \text{ in.}^2)}{1.67}$ $= 602$ kips > 233 kips **o.k.**

Check block shear rupture on beam web
The limit state of block shear rupture due to the shear load on the beam web is not applicable because the remaining beam flange will prevent net section rupture. However, the limit state of block shear rupture must be checked for the tension load on the beam web as follows.

$U_{bs} = 1.0$

$A_{nt} = [18.0 \text{ in.} - 6(1\text{\textonesixteenth} \text{ in.} + \text{\textonesixteenth} \text{ in.})](0.415 \text{ in.})$
$\quad = 4.67 \text{ in.}^2$

$A_{gv} = 4.75 \text{ in.}(0.415 \text{ in.})(2)$
$\quad = 3.94 \text{ in.}^2$

$A_{nv} = [4.75 \text{ in.} - 1.5(1\text{\textonesixteenth} \text{ in.} + \text{\textonesixteenth} \text{ in.})](0.415 \text{ in.})(2)$
$\quad = 2.54 \text{ in.}^2$

$R_n = 0.60 F_u A_{nv} + U_{bs} F_u A_{nt} \le 0.60 F_y A_{gv} + U_{bs} F_u A_{nt}$ $\qquad$ (*Spec.* Eq. J4-5)

$0.60 F_u A_{nv} + U_{bs} F_u A_{nt} = 0.60(65 \text{ ksi})(2.54 \text{ in.}^2) + 1.0(65 \text{ ksi})(4.67 \text{ in.}^2)$
$\qquad\qquad\qquad\qquad\quad = 403$ kips

$0.60 F_y A_{gv} + U_{bs} F_u A_{nt} = 0.60(50 \text{ ksi})(3.94 \text{ in.}^2) + 1.0(65 \text{ ksi})(4.67 \text{ in.}^2)$
$\qquad\qquad\qquad\qquad\quad = 422$ kips

Therefore, the nominal block shear rupture strength is 403 kips and the available block shear rupture strength is:

LRFD	ASD
$\phi R_n = 0.75(403 \text{ kips})$ $= 302 \text{ kips} < 352 \text{ kips}$ **n.g.**	$\dfrac{R_n}{\Omega} = \dfrac{403 \text{ kips}}{2.00}$ $= 202 \text{ kips} < 233 \text{ kips}$ **n.g.**

Therefore, a web doubler plate is required. The required thickness of the doubler plate is:

LRFD	ASD
$t = \left(\dfrac{352 \text{ kips}}{302 \text{ kips}}\right)(0.415 \text{ in.}) - 0.415 \text{ in.}$ $= 0.0687 \text{ in.}$	$t = \left(\dfrac{233 \text{ kips}}{202 \text{ kips}}\right)(0.415 \text{ in.}) - 0.415 \text{ in.}$ $= 0.0637 \text{ in.}$

Use a ¼-in.-thick doubler plate with ¼-in. fillet welds.

Check bolt bearing and tearout on the beam

The resultant load per bolt based on the C-value taken from AISC *Manual* Table 7-7 previously, is:

LRFD	ASD
$r_u = \dfrac{378 \text{ kips}}{11.4 \text{ bolts}}$ $= 33.2 \text{ kips/bolt}$	$r_a = \dfrac{248 \text{ kips}}{11.4 \text{ bolts}}$ $= 21.8 \text{ kips/bolt}$

The bearing and tearout strength limit states are separated for clarity in the following. For SCBF systems, where large deformations are anticipated, it is appropriate to assume that deformation at the bolt hole is not a design consideration, and the tearout strength is checked at the end bolt with the 1.75 in. edge distance using AISC *Specification* Equation J3-6b. AISC *Seismic Provisions* Section D2.2(1) limits the nominal bearing strength to $2.4dtF_u$; therefore the available bearing strength of the beam and doubler plate is:

LRFD	ASD
$\phi r_n = \phi 2.4 dt F_u$ $= 0.75(2.4)(1.00 \text{ in.})(0.415 \text{ in.} + ¼ \text{ in.})$ $\times (65 \text{ ksi})$ $= 77.8 \text{ kips/bolt} > 33.2 \text{ kips/bolt}$ **o.k.**	$\dfrac{r_n}{\Omega} = \dfrac{2.4 dt F_u}{\Omega}$ $= \dfrac{\left[\begin{array}{l}2.4(1.00 \text{ in.})(0.415 \text{ in.} + ¼ \text{ in.})\\ \times(65 \text{ ksi})\end{array}\right]}{2.00}$ $= 51.9 \text{ kips/bolt} > 21.8 \text{ kips/bolt}$ **o.k.**

5.3 SPECIAL CONCENTRICALLY BRACED FRAMES

Assuming that deformation at the bolt hole is not a design consideration, the tearout strength is checked at the end bolt with the 1.75 in. edge distance. The available tearout strength is:

LRFD	ASD
$\phi r_n = \phi 1.5 F_u l_c t$ $= 0.75(1.5)(65 \text{ ksi})$ $\times [1.75 \text{ in.} - 0.5(1\frac{1}{16} \text{ in.})]$ $\times (0.415 \text{ in.} + \frac{1}{4} \text{ in.})$ $= 59.3 \text{ kips/bolt} > 33.2 \text{ kips/bolt}$ **o.k.**	$\dfrac{r_n}{\Omega} = \dfrac{1.5 F_u l_c t}{\Omega}$ $= (1/2.00)(1.5)(65 \text{ ksi})$ $\times [1.75 \text{ in.} - 0.5(1\frac{1}{16} \text{ in.})]$ $\times (0.415 \text{ in.} + \frac{1}{4} \text{ in.})$ $= 39.5 \text{ kips/bolt} > 21.8 \text{ kips/bolt}$ **o.k.**

As previously discussed, this is a conservative treatment of tearout. If the check failed, the edge distance along the line of action of the bolt force would be evaluated before declaring the design inadequate.

Beam-to-column single-plate connection

Determine the required thickness of the 7.50 in. × 21.0 in. single plate connecting the beam web to the column flange. Try a ⅝-in.-thick plate.

From AISC *Specification* Equation J4-3, the available shear yielding strength of the plate is:

LRFD	ASD
$\phi R_n = \phi 0.60 F_y A_{gv}$ $= 1.00(0.60)(50 \text{ ksi})(21.0 \text{ in.})$ $\times (\frac{5}{8} \text{ in.})$ $= 394 \text{ kips} > 138 \text{ kips}$ **o.k.**	$\dfrac{R_n}{\Omega} = \dfrac{0.60 F_y A_{gv}}{\Omega}$ $= \dfrac{0.60(50 \text{ ksi})(21.0 \text{ in.})(\frac{5}{8} \text{ in.})}{1.50}$ $= 263 \text{ kips} > 85.8 \text{ kips}$ **o.k.**

From AISC *Specification* Equation J4-1, the available tensile yielding strength of the plate is:

LRFD	ASD
$\phi R_n = \phi F_y A_g$ $= 0.90(50 \text{ ksi})(21.0 \text{ in.})(\frac{5}{8} \text{ in.})$ $= 591 \text{ kips} > 352 \text{ kips}$ **o.k.**	$\dfrac{R_n}{\Omega} = \dfrac{F_y A_g}{\Omega}$ $= \dfrac{50 \text{ ksi}(21.0 \text{ in.})(\frac{5}{8} \text{ in.})}{1.67}$ $= 393 \text{ kips} > 233 \text{ kips}$ **o.k.**

Single plate-to-column flange weld

Determine the fillet weld size required to connect the single plate on the beam to the column flange. Using AISC *Manual* Equations 8-2a and 8-2b, including the increased strength due to the load angle given by AISC *Specification* Equation J2-5, the required single plate-to-column flange weld is determined as follows:

LRFD	ASD
Resultant load $R_u = 378$ kips Load angle $\theta = \tan^{-1}\left(\dfrac{352 \text{ kips}}{138 \text{ kips}}\right)$ $= 68.6°$ Directional strength increase $\left[1.0 + 0.50\sin^{1.5}\theta\right]$ $= \left[1.0 + 0.50\sin^{1.5} 68.6°\right]$ $= 1.45$	Resultant load $R_a = 248$ kips Load angle $\theta = \tan^{-1}\left(\dfrac{233 \text{ kips}}{85.8 \text{ kips}}\right)$ $= 69.8°$ Directional strength increase $\left[1.0 + 0.50\sin^{1.5}\theta\right]$ $= \left[1.0 + 0.50\sin^{1.5} 69.8°\right]$ $= 1.45$

LRFD	ASD
$D_{req} = \dfrac{378 \text{ kips}}{2(21.0 \text{ in.})(1.392 \text{ kip/in.})(1.45)}$ $= 4.46$ sixteenths	$D_{req} = \dfrac{248 \text{ kips}}{2(21.0 \text{ in.})(0.928 \text{ kip/in.})(1.45)}$ $= 4.39$ sixteenths

Use a 5⁄16-in. fillet weld.

Check bolt bearing and tearout on the single plate

The resultant load per bolt determined previously is:

LRFD	ASD
$r_u = 33.2$ kips/bolt	$r_a = 21.8$ kips/bolt

The bearing and tearout strength limit states are separated for clarity in the following. For SCBF systems, where large deformations are anticipated, it is appropriate to assume that deformation at the bolt hole is not a design consideration, and the tearout strength is checked at the end bolt using AISC *Specification* Equation J3-6b. AISC *Seismic Provisions* Section D2.2(1) limits the nominal bearing strength to $2.4dtF_u$; therefore the available bearing strength of the gusset plate is:

5.3 SPECIAL CONCENTRICALLY BRACED FRAMES

LRFD	ASD
$\phi r_n = \phi 2.4 dt F_u$ $= 0.75(2.4)(1.00 \text{ in.})(\text{⅝ in.})(65 \text{ ksi})$ $= 73.1 \text{ kips/bolt} > 33.2 \text{ kips/bolt}$ **o.k.**	$\dfrac{r_n}{\Omega} = \dfrac{2.4 dt F_u}{\Omega}$ $= \dfrac{2.4(1.00 \text{ in.})(\text{⅝ in.})(65 \text{ ksi})}{2.00}$ $= 48.8 \text{ kips/bolt} > 21.8 \text{ kips/bolt}$ **o.k.**

The tearout strength is checked at the end bolt with the 1.75 in. edge distance using the left side of AISC *Specification* Equation J3-6b. The available tearout strength is:

LRFD	ASD
$\phi r_n = \phi 1.5 l_c t F_u$ $= 0.75(1.5)[1.75 \text{ in.} - 0.5(1\text{1/16 in.})]$ $\times (\text{⅝ in.})(65 \text{ ksi})$ $= 55.7 \text{ kips/bolt} > 33.2 \text{ kips/bolt}$ **o.k.**	$\dfrac{r_n}{\Omega} = \dfrac{1.5 l_c t F_u}{\Omega}$ $= \dfrac{\begin{bmatrix}1.5[1.75 \text{ in.} - 0.5(1\text{1/16 in.})]\\ \times (\text{⅝ in.})(65 \text{ ksi})\end{bmatrix}}{2.00}$ $= 37.1 \text{ kips/bolt} > 21.8 \text{ kips/bolt}$ **o.k.**

Check block shear rupture on single plate at beam-to-column interface

For the shear force

The nominal block shear rupture strength due to shear on the single plate is determined as follows:

$U_{bs} = 0.5$

$A_{nt} = [4.75 \text{ in.} - 1.5(1\text{1/16 in.} + \text{1/16 in.})](\text{⅝ in.})$
$= 1.91 \text{ in.}^2$

$A_{gv} = (19.5 \text{ in.})(\text{⅝ in.})$
$= 12.2 \text{ in.}^2$

$A_{nv} = [19.5 \text{ in.} - 6.5(1\text{1/16 in.} + \text{1/16 in.})](\text{⅝ in.})$
$= 7.62 \text{ in.}^2$

$R_n = 0.60 F_u A_{nv} + U_{bs} F_u A_{nt} \leq 0.60 F_y A_{gv} + U_{bs} F_u A_{nt}$ (Spec. Eq. J4-5)

$0.60 F_u A_{nv} + U_{bs} F_u A_{nt} = 0.60(65 \text{ ksi})(7.62 \text{ in.}^2) + 0.5(65 \text{ ksi})(1.91 \text{ in.}^2)$
$= 359 \text{ kips}$

$0.60 F_y A_{gv} + U_{bs} F_u A_{nt} = 0.60(50 \text{ ksi})(12.2 \text{ in.}^2) + 0.5(65 \text{ ksi})(1.91 \text{ in.}^2)$
$= 428 \text{ kips}$

Therefore, the nominal block shear rupture strength is 359 kips and the available block shear rupture strength is:

LRFD	ASD
$\phi R_n = 0.75(359 \text{ kips})$ $= 269 \text{ kips} > 138 \text{ kips}$ **o.k.**	$\dfrac{R_n}{\Omega} = \dfrac{359 \text{ kips}}{2.00}$ $= 180 \text{ kips} > 85.8 \text{ kips}$ **o.k.**

For the tension force

The nominal block shear rupture strength due to the tension force on the single plate is:

$U_{bs} = 1.0$

$A_{nt} = [19.5 \text{ in.} - 6.5(1\tfrac{1}{16} \text{ in.} + \tfrac{1}{16} \text{ in.})](\tfrac{5}{8} \text{ in.})$
$= 7.62 \text{ in.}^2$

$A_{gv} = 4.75 \text{ in.}(\tfrac{5}{8} \text{ in.})$
$= 2.97 \text{ in.}^2$

$A_{nv} = [4.75 \text{ in.} - 1.5(1\tfrac{1}{16} \text{ in.} + \tfrac{1}{16} \text{ in.})](\tfrac{5}{8} \text{ in.})$
$= 1.91 \text{ in.}^2$

$0.60 F_u A_{nv} + U_{bs} F_u A_{nt} = 0.60(65 \text{ ksi})(1.91 \text{ in.}^2) + 1.0(65 \text{ ksi})(7.62 \text{ in.}^2)$
$= 570 \text{ kips}$

$0.60 F_y A_{gv} + U_{bs} F_u A_{nt} = 0.60(50 \text{ ksi})(2.97 \text{ in.}^2) + 1.0(65 \text{ ksi})(7.62 \text{ in.}^2)$
$= 584 \text{ kips}$

Therefore, the nominal block shear rupture strength is 570 kips and the available block shear rupture strength is:

LRFD	ASD
$\phi R_n = 0.75(570 \text{ kips})$ $= 428 \text{ kips} > 352 \text{ kips}$ **o.k.**	$\dfrac{R_n}{\Omega} = \dfrac{570 \text{ kips}}{2.00}$ $= 285 \text{ kips} > 233 \text{ kips}$ **o.k.**

Check tension-shear interaction (block shear rupture)

The interaction of tension and shear based on the block shear rupture limit state is checked as follows:

5.3 SPECIAL CONCENTRICALLY BRACED FRAMES

LRFD	ASD
$\left(\dfrac{352 \text{ kips}}{428 \text{ kips}}\right)^2 + \left(\dfrac{138 \text{ kips}}{269 \text{ kips}}\right)^2$ $= 0.940 < 1.0$ **o.k.**	$\left(\dfrac{233 \text{ kips}}{285 \text{ kips}}\right)^2 + \left(\dfrac{85.8 \text{ kips}}{180 \text{ kips}}\right)^2$ $= 0.896 < 1.0$ **o.k.**

Use a ⅝-in.-thick plate.

Check shear rupture on the single plate

From AISC *Specification* Equation J4-4, the available shear rupture strength of the single plate is:

LRFD	ASD
$\phi R_n = 0.75(0.60)(65 \text{ ksi})$ $\times [21.0 \text{ in.} - 7(1\frac{1}{16} \text{ in.} + \frac{1}{16} \text{ in.})]$ $\times (\frac{5}{8} \text{ in.})$ $= 240 \text{ kips} > 138 \text{ kips}$ **o.k.**	$\dfrac{R_n}{\Omega} = (1/2.00)(0.60)(65 \text{ ksi})$ $\times [21.0 \text{ in.} - 7(1\frac{1}{16} \text{ in.} + \frac{1}{16} \text{ in.})]$ $\times (\frac{5}{8} \text{ in.})$ $= 160 \text{ kips} > 85.8 \text{ kips}$ **o.k.**

Check tensile rupture on the single plate

From AISC *Specification* Equation J4-2, the available tensile rupture strength of the single plate is:

LRFD	ASD
$\phi R_n = 0.75(65 \text{ ksi})$ $\times [21.0 \text{ in.} - 7(1\frac{1}{16} \text{ in.} + \frac{1}{16} \text{ in.})]$ $\times (\frac{5}{8} \text{ in.})$ $= 400 \text{ kips} > 352 \text{ kips}$ **o.k.**	$\dfrac{R_{nt}}{\Omega} = 65 \text{ ksi}$ $\times [21.0 \text{ in.} - 7(1\frac{1}{16} \text{ in.} + \frac{1}{16} \text{ in.})]$ $\times (\frac{5}{8} \text{ in.})(1/2.00)$ $= 267 \text{ kips} > 233 \text{ kips}$ **o.k.**

Check tension-shear interaction (tensile and shear rupture)

LRFD	ASD
$\left(\dfrac{352 \text{ kips}}{400 \text{ kips}}\right)^2 + \left(\dfrac{138 \text{ kips}}{240 \text{ kips}}\right)^2$ $= 1.11 > 1.0$ **n.g.**	$\left(\dfrac{233 \text{ kips}}{267 \text{ kips}}\right)^2 + \left(\dfrac{85.8 \text{ kips}}{160 \text{ kips}}\right)^2$ $= 1.05 > 1.0$ **n.g.**

With a ¾-in.-thick plate:

LRFD	ASD
$1.11\left(\dfrac{\text{⅝ in.}}{\text{¾ in.}}\right)^2 = 0.771 < 1.0$ **o.k.**	$1.05\left(\dfrac{\text{⅝ in.}}{\text{¾ in.}}\right)^2 = 0.729 < 1.0$ **o.k.**

Use a ¾-in.-thick plate.

Note: Shear yielding and tensile yielding limit states should also be checked, but were assumed to not control this design.

The final connection design is shown in Figure 5-51.

5.4 ECCENTRICALLY BRACED FRAMES (EBF)

In eccentrically braced frame (EBF) systems, lateral forces are resisted by a combination of flexure, shear and axial forces in the framing members. An EBF is essentially a hybrid system, offering lateral stiffness approaching that of a concentrically braced frame system and ductility approaching that of a moment frame system. The design provisions for EBF systems are given in AISC *Seismic Provisions* Section F3 and typical configurations are shown in AISC *Seismic Provisions* Figure C-F3.1. Section F3.1 describes EBF systems as "braced frames for which one end of each brace intersects a beam at an eccentricity from the intersection of the centerlines of the beam and an adjacent brace or column, forming a link that is subject to shear and flexure." The link becomes the focal point in the design and detailing of an EBF system, as it is intended to be the primary location for the inelastic behavior in the frame. The remainder of the members and connections are intended to remain essentially elastic and are required to have sufficient strength to withstand forces corresponding to the expected strength of the link, including strain hardening.

Designers can often fit eccentrically braced frames in locations within the architectural floor plan where concentrically braced frames cannot be located, due to the space limitations presented by doors and windows. Additionally, the system is generally considered to be stiff enough to efficiently limit nonstructural, drift-related damage, as compared to the relatively flexible nature of moment frames. An EBF system may be more complex for design and construction than other common systems. As with all systems, the choice of an EBF as the lateral system requires balancing the needs of the building owner and architect with the project budget. Consideration should be given to "first-costs" of the project versus the life-cycle costs and potential repair costs following a major earthquake. First-cost benefits of using an EBF system include a reduction in the seismic base shear force due to the higher R factor than other braced frame systems, which may result in savings in the construction of the diaphragm and foundation.

While EBF systems combine many concepts of both concentrically braced frames and moment frames, the technology of eccentrically braced frames is relatively new. The system was first developed in Japan in the early 1970s. Research and development in the United States followed later that decade, continuing through the 1980s, with the first codified design procedure appearing in the 1988 Uniform Building Code (UBC). As noted previously, the focal point of the design of an EBF system is the link. The link design procedures put forth in the AISC *Seismic Provisions* are quite extensive and are intended to provide reliable and ductile performance of the link under seismic loading. The first of these provisions relates to width-to-thickness limits in Section D1.1 of the AISC *Seismic Provisions*. For EBF systems, the link must satisfy the width-to-thickness requirements for highly ductile members. There is an exception for the flanges of short, shear dominated links with I-shaped sections. For link lengths less than $1.6M_p/V_p$, the flanges need only satisfy the width-to-thickness requirements for moderately ductile members. Additional limitations on the web include a maximum specified yield stress of 50 ksi, and a requirement that the web be a single thickness of material. Thus, doubler plates and penetrations are not permitted in the link zone. The 2010 AISC *Seismic Provisions* introduce an allowance for the use of built-up box section links; however, the use of HSS links is not allowed.

The nominal shear strength of the link, V_n, is calculated as the lesser of the shear yielding strength of the link, V_p, and the shear associated with the flexural yielding strength of the link, $2M_p/e$. Additional link requirements apply when the required axial strength in the link exceeds $0.15P_y$. These requirements limit the nominal shear strength and the link length in order to provide for more stable inelastic behavior within the link when axial forces become large enough to have a significant effect. For specific requirements, the AISC *Seismic Provisions* should be consulted.

Another consideration in the design of the link is the link length, e. When related to the length of the frame, L, it can be shown that as e/L approaches zero, an EBF system reaches the stiffness of a concentrically braced frame, while values of e/L approaching 1.0 indicate behavior consistent with moment frames. This concept is illustrated in Figure 5-57. Further consideration of link length relates to the behavior of the link itself in the inelastic range. From simple mechanics, it can be demonstrated that when $e = 2.0M_p/V_p$, the yield condition is balanced between shear and flexure. For values less than $1.6M_p/V_p$, the link behavior is generally controlled by shear, whereas for values greater than $2.6M_p/V_p$ it is controlled by flexure. For link lengths between $1.6M_p/V_p$ and $2.6M_p/V_p$, a combination of shear and flexural yielding occurs. Because shear yielding is much more reliable than flexural yielding, it is generally considered advantageous to keep link lengths short enough to be controlled by shear. With this in mind, a target value of $1.6M_p/V_p$ is used for the link length, e. To achieve this, many designers will start the design of the link using a value of $1.3M_p/V_p$. This allows some flexibility in changing the link beam size and frame geometry while still maintaining a final link length consistent with the $1.6M_p/V_p$ goal.

The AISC *Seismic Provisions* address the ratio of M_p/V_p in relation to the overall ductility of the frame by relating the link rotation angle, γ_p, to the value of M_p/V_p in a given frame. Link rotation angle is illustrated in AISC *Seismic Provisions* Figure C-F3.4. The AISC *Seismic Provisions* note that for $e < 1.6M_p/V_p$ the link rotation angle is limited to 0.08 rad, and for $e > 2.6M_p/V_p$ the link rotation angle is limited to 0.02 rad. For values between these limits, the link rotation angle should be interpolated. This is illustrated in Figure 5-58. Additional link design considerations apply when providing stiffener plates in the link zone. The AISC *Seismic Provisions* specify that links of all lengths require stiffeners at each end. Additionally, spacing of intermediate stiffeners varies with link length. Note that when $e > 5.0M_p/V_p$, no intermediate web stiffeners are required.

When the frame is configured such that the link is directly adjacent to a column, there are special requirements for the connection between the link and the column as required by AISC *Seismic Provisions* Section F3.6e. The link-to-column connection must be capable

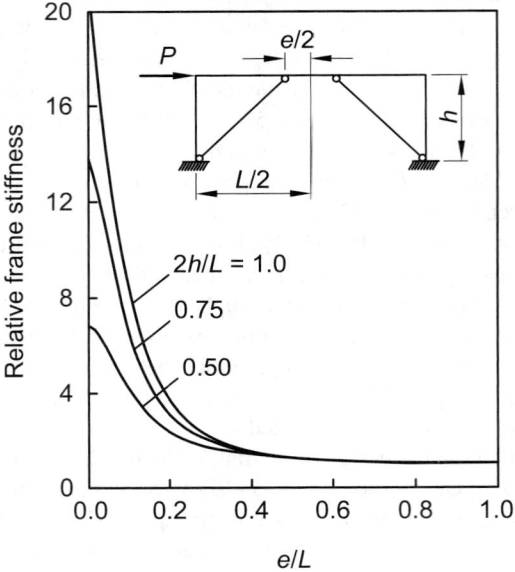

Fig. 5-57. *Frame stiffness versus link length (Engelhardt and Popov, 1989).*

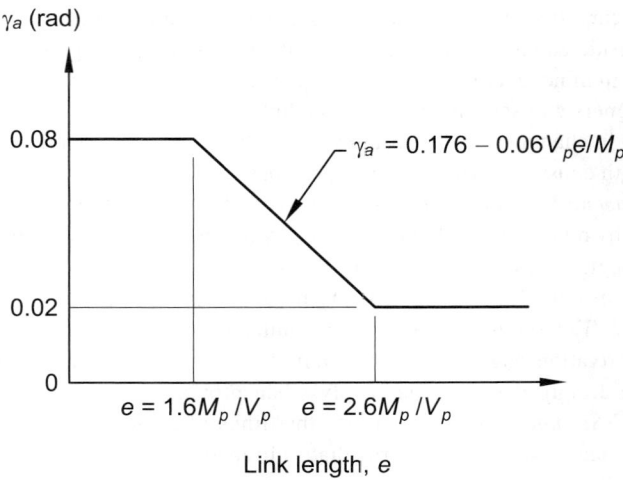

Fig. 5-58. *Maximum allowed link rotation angle versus link length.*

of sustaining the link rotation angle as prescribed by the AISC *Seismic Provisions* based on link length. Additionally, the connection must be able to develop the full value of the expected link shear strength, $R_y V_n$, at such a rotation angle. Furthermore, the link-to-column connection must meet the requirements of moment connections consistent with those as prescribed under the provisions for special moment frames; in other words, a qualified or prequalified connection must be used. The exception to this occurs when the connections are adequately reinforced such that beam yielding is forced to a location away from the face of the column, and when the length of the link is less than $1.6 M_p/V_p$. In this case, the link length is defined as the length from the end of the beam reinforcing to the end of the brace member. If the link-to-column connection meets these requirements, prequalification or qualification of the connection is not required. Full-depth stiffener plates are still required at the end of the link adjacent to the reinforced beam section, however.

AISC *Seismic Provisions* Section F3.4b requires lateral bracing of both the top and bottom flanges at the ends of I-shaped links. These braces must be designed to satisfy the strength and stiffness requirements of AISC *Seismic Provisions* Section D1.2c for special braces at plastic hinge locations.

Once the design of the link is complete, the remaining requirements address the design of the diagonal brace and beam segments away from the link, the connections of the beams to the columns, and the strength of the columns and the column base attachment to the foundation. Due to the nature of EBF systems, the brace members may be subject to large axial and flexural forces resulting from the rotations anticipated in the link segment. Therefore, the diagonal brace is required to have a combined axial and flexural strength due to seismic loading equal to the forces generated by the adjusted link shear strength. The adjusted link shear strength is defined as the expected shear strength of the link, $R_y V_n$, multiplied by a factor to account for strain hardening. This strain hardening factor is equal to 1.25 for I-shaped links and 1.4 for built-up box links. Braces must also satisfy the width-to-thickness requirements of AISC *Seismic Provisions* Section D1.1 for moderately ductile members.

The design of the beam outside of the link is similar, but differs slightly from the design requirements for braces. It is also designed for the forces due to the adjusted shear strength of the link. However, the adjusted shear strength of the link is allowed to be taken equal to 0.88 times the value used in the design of the braces. This accounts for the increased member strength realized by having a concrete slab composite with the beam outside of the link and recognizes the fact that limited yielding in the beam is not likely to be detrimental to EBF performance, as long as stability of the beam is assured. If there is not a concrete slab composite with the beam outside of the link, a strain hardening factor of 1.25 should be used (for additional information see the Commentary to AISC *Seismic Provisions* Section F3.3). Additional lateral bracing along the length of the beam, if required, is designed per AISC *Specification* Appendix 6. If the beam outside of the link is a different section than the link, then it must also satisfy the width-to-thickness requirements of AISC *Seismic Provisions* Section D1.1 for moderately ductile members.

The connection of the brace to the beam is required to meet the same strength requirements as the brace member. The AISC *Seismic Provisions* require this connection to be considered fully restrained (FR) if the connection is detailed such that the brace resists any portion of the link end moment. Because it was considered to be overly conservative in previous editions, the 2010 AISC *Seismic Provisions* no longer require that the connection also be designed for $1.1 R_y P_n$ of the brace and no longer prohibit the brace connection from extending into the link zone. There is a discussion of these changes in the Commentary to Sections F3.6c and F3.5b, respectively.

The beam-to-column connection where a brace connects to both members has design and detailing considerations in addition to the preceding requirements for the brace-to-beam connection. AISC *Seismic Provisions* Section F3.6b requires that these connections either be a simple connection meeting the requirements of AISC *Specification* Section B3.6a with a required rotation of 0.025 rad, or they must be designed as a moment connection. If the latter is chosen, the required strength of the connection is equal to the lesser of the expected beam flexural strength and the sum of the expected flexural strengths of the column above and below the joint.

The columns of the EBF system must satisfy the width-to-thickness requirements of AISC *Seismic Provisions* Section D1.1 for highly ductile members. Additionally, the columns must be designed to resist the forces due to the adjusted shear strengths of all links above the level of the column (as discussed previously for brace design). For columns in frames with three or more stories of bracing, the adjusted shear strength of the link is allowed to be taken equal to 0.88 times the value used in the design of the braces, which recognizes that it is unlikely for all links to be fully strain-hardened at the same time.

EBF Design Example Plan and Elevation

The following section consists of seven design examples for an EBF system. See Figure 5-59 for the elevation of the EBF. Example 5.4.1 checks story drift. Examples 5.4.2 through 5.4.5 illustrate a link design, a beam outside of the link design, a brace design, and a column design, respectively. Examples 5.4.6 and 5.4.7 show the design of a brace-to-link connection and a brace-to-beam/column connection.

From ASCE/SEI 7, the following parameters apply: Seismic Design Category D, $R = 8$, $\Omega_o = 2$, $C_d = 4$, $I_e = 1.0$, $S_{DS} = 1.0$, and $\rho = 1.3$.

The total floor area is 9,000 ft^2, the perimeter is 390 ft, and the code-specified gravity loading is as follows:

$D_{floor} = 85$ psf
$D_{roof} = 68$ psf
$L_{floor} = 80$ psf (50 psf reduced)
$S \quad = 20$ psf
Curtain wall = 175 lb/ft along building perimeter at every level

The loads given in each design example are from a first-order analysis. Assume the effective length method of AISC *Specification* Appendix 7 is used for the stability design.

When designing EBF systems, several design iterations are usually required to obtain the best combination of compatible frame-member sizes. Optimized designs are often difficult to obtain, due to member local buckling requirements, geometric constraints, the resistance of the beam outside of the link to flexure combined with axial effects, and architectural constraints that commonly occur throughout the design process. Nonetheless, EBF systems can be used to provide ductile and cost-effective solutions for seismic load resistance.

Example 5.4.1. EBF Story Drift Check

Given:

Refer to the EBF elevation shown in Figure 5-59. The applicable building code specifies the use of ASCE/SEI 7 for drift requirements. Determine if the third level of the frame satisfies the drift requirements.

From an elastic analysis of the structure using an equivalent lateral force analysis, the story drift between the second and third levels is:

$\delta_{xe} = 0.175$ in.

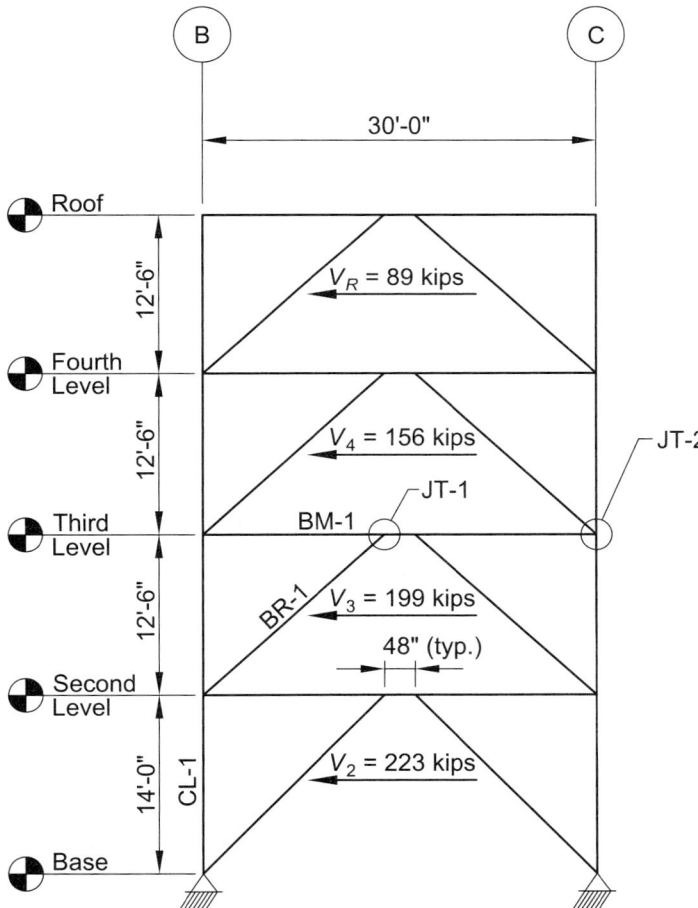

Fig. 5-59. EBF elevation.

Solution:

According to AISC *Seismic Provisions* Section B1, the design story drift and the story drift limits are those stipulated by the applicable building code. From ASCE/SEI 7 Table 12.12-1, the allowable story drift, Δ_a, is $0.025h_{sx}$, where h_{sx} is the story height below level x.

$$\Delta_a = 0.025 h_{sx}$$
$$= 0.025(12.5 \text{ ft})(12 \text{ in./ft})$$
$$= 3.75 \text{ in.}$$

ASCE/SEI 7 defines the design story drift as Δ, the difference of the deflections at level 2 and level 3 at the centers of mass. The deflection at level x, δ_x, is:

$$\delta_x = \frac{C_d \delta_{xe}}{I_e} \qquad \text{(ASCE/SEI 7 Eq. 12.8-15)}$$

Therefore, the design story drift at level 3 is:

$$\delta_x = \Delta_3$$
$$= \frac{C_d \delta_{x3}}{I_e} - \frac{C_d \delta_{x2}}{I_e}$$
$$= \frac{C_d (\delta_{x3} - \delta_{x2})}{I_e}$$
$$= \frac{4(0.175 \text{ in.})}{1.0}$$
$$= 0.700 \text{ in.} < 3.75 \text{ in.} \qquad \text{o.k.}$$

Example 5.4.2 EBF Link Design

Given:

Refer to Beam BM-1 in Figure 5-59. Determine the adequacy of an ASTM A992 W16×77 as the link segment for the following loading. The stiffener material is ASTM A36 plate. The applicable building code specifies the use of ASCE/SEI 7 for calculation of loads. From a first-order analysis:

$P_D = 7.40$ kips $\qquad P_L = 5.30$ kips $\qquad P_{Q_E} = 5.50$ kips
$V_D = 1.80$ kips $\qquad V_L = 1.30$ kips $\qquad V_{Q_E} = 84.0$ kips
$M_D = 14.4$ kip-ft $\qquad M_L = 9.60$ kip-ft $\qquad M_{Q_E} = 168$ kip-ft

Assume the brace-to-beam connection will be that shown in AISC *Seismic Provisions* Figure C-F3.7. The brace will be detailed as fixed to the link in order to decrease the flexural demand on the beam outside of the link. Assume the brace will be an ASTM A992 W10×112.

Solution:

From AISC *Manual* Tables 2-4 and 2-5, and AISC *Seismic Provisions* Table A3.1, the material properties are as follows:

5.4 ECCENTRICALLY BRACED FRAMES

ASTM A36
$F_y = 36$ ksi
$F_u = 58$ ksi

ASTM A992
$F_y = 50$ ksi
$F_u = 65$ ksi
$R_y = 1.1$

From AISC *Manual* Table 1-1, the geometric properties are as follows:

W16×77

$A = 22.6$ in.^{2}	$d = 16.5$ in.	$t_w = 0.455$ in.	$b_f = 10.3$ in.
$t_f = 0.760$ in.	$k_{det} = 1\%$ in.	$k_1 = 1\frac{1}{16}$ in.	$b_f/2t_f = 6.77$
$h/t_w = 31.2$	$I_x = 1,110$ in.^{4}	$Z_x = 150$ in.^{3}	$h_o = 15.7$ in.

Required Strength

Considering the load combinations given in ASCE/SEI 7 that include seismic effects, it was determined that the governing load combination for the link is:

LRFD	ASD
LRFD Load Combination 5 from ASCE/SEI 7 Section 12.4.2.3 (including the 0.5 factor on *L* permitted in Section 12.4.2.3) $(1.2 + 0.2S_{DS})D + \rho Q_E + 0.5L + 0.2S$	ASD Load Combination 5 from ASCE/SEI 7 Section 12.4.2.3 $(1.0 + 0.14S_{DS})D + H + F + 0.7\rho Q_E$

Determine the required shear strength of the link

The required shear strength of the link is:

LRFD	ASD
$V_u = (1.2 + 0.2S_{DS})V_D + \rho V_{Q_E}$ $\quad + 0.5V_L + 0.2V_S$ $= [1.2 + 0.2(1.0)](1.80 \text{ kips})$ $\quad + 1.3(84.0 \text{ kips}) + 0.5(1.30 \text{ kips})$ $\quad + 0.2(0 \text{ kips})$ $= 112 \text{ kips}$	$V_a = (1.0 + 0.14S_{DS})V_D + V_H + V_F$ $\quad + 0.7\rho V_{Q_E}$ $= [1.0 + 0.14(1.0)](1.80 \text{ kips})$ $\quad + 0 \text{ kips} + 0 \text{ kips}$ $\quad + 0.7(1.3)(84.0 \text{ kips})$ $= 78.5 \text{ kips}$

Determine the required axial and flexural strengths of the link

Consider second-order effects
Second-order effects are addressed using AISC *Specification* Appendix 8 as follows:

$$M_r = B_1 M_{nt} + B_2 M_{lt} \qquad \text{(Spec. Eq. A-8-1)}$$

$$P_r = P_{nt} + B_2 P_{lt} \qquad \text{(Spec. Eq. A-8-2)}$$

$$B_1 = \frac{C_m}{1 - \frac{\alpha P_r}{P_{e1}}} \geq 1 \qquad \text{(Spec. Eq. A-8-3)}$$

Since the calculation of B_1 requires P_r, B_2 will be calculated first, although AISC *Specification* Appendix 8, Section 8.1 permits the use of a first-order estimate of P_r.

$$B_2 = \frac{1}{1 - \frac{\alpha P_{story}}{P_{e\,story}}} \geq 1 \qquad \text{(Spec. Eq. A-8-6)}$$

Calculate P_{story}
From the given loading, the total vertical load at the third level is:

LRFD	ASD
$P_{story} = 9{,}000 \text{ ft}^2 \begin{Bmatrix} [1.2 + 0.2(1.0)] \\ \times [68 \text{ psf} + 2(85 \text{ psf})] \\ + 0 \text{ psf} \\ + 0.5(2)(50 \text{ psf}) \\ + 0.2(20 \text{ psf}) \end{Bmatrix}$ $\times (1 \text{ kip}/1{,}000 \text{ lb})$ $+ \begin{Bmatrix} [1.2 + 0.2(1.0)] \\ [175 \text{ lb/ft}(2)(390 \text{ ft})] \end{Bmatrix}$ $\times (1 \text{ kip}/1{,}000 \text{ lb})$ $= 3{,}680 \text{ kips}$	$P_{story} = 9{,}000 \text{ ft}^2 \begin{Bmatrix} [1.0 + 0.14(1.0)] \\ \times [68 \text{ psf} + 2(85 \text{ psf})] \\ + 0 \text{ psf} + 0 \text{ psf} + 0 \text{ psf} \end{Bmatrix}$ $\times (1 \text{ kip}/1{,}000 \text{ lb})$ $+ \begin{Bmatrix} [1.0 + 0.14(1.0)] \\ [175 \text{ lb/ft}(2)(390 \text{ ft})] \end{Bmatrix}$ $\times (1 \text{ kip}/1{,}000 \text{ lb})$ $= 2{,}600 \text{ kips}$

The total story shear, H, is shown in Figure 5-59 as $V_3 = 199$ kips. From Example 5.4.1, an elastic analysis determined that the first-order interstory drift is $\Delta_H = 0.175$ in.

$L = 12.5 \text{ ft}(12.0 \text{ in./ft})$
$\quad = 150 \text{ in.}$

$R_M = 1$ for braced frame systems

5.4 ECCENTRICALLY BRACED FRAMES

$$P_{e\,story} = R_M \frac{HL}{\Delta_H} \quad \textit{(Spec. Eq. A-8-7)}$$

$$= 1.0 \frac{199 \text{ kips}(150 \text{ in.})}{0.175 \text{ in.}}$$

$$= 171{,}000 \text{ kips}$$

Using AISC *Specification* Equation A-8-6:

LRFD	ASD
$\alpha = 1.00$	$\alpha = 1.60$
$B_2 = \dfrac{1}{1 - \dfrac{1.00(3{,}680 \text{ kips})}{171{,}000 \text{ kips}}}$	$B_2 = \dfrac{1}{1 - \dfrac{1.60(2{,}600 \text{ kips})}{171{,}000 \text{ kips}}}$
$= 1.02$	$= 1.02$

P-Δ effects, approximated through the B_2 factor, apply only to axial forces and moments due to lateral translation. Thus, the required axial strength of the link including second-order effects is:

LRFD	ASD
$P_u = (1.2 + 0.2S_{DS})P_D + B_2\rho P_{Q_E} + 0.5P_L$ $\quad + 0.2P_S$	$P_a = (1.0 + 0.14S_{DS})P_D + P_H + P_F$ $\quad + 0.7B_2\rho P_{Q_E}$
$= [1.2 + 0.2(1.0)](7.40 \text{ kips})$ $\quad + 1.02(1.3)(5.50 \text{ kips})$ $\quad + 0.5(5.30 \text{ kips}) + 0.2(0 \text{ kips})$ $= 20.3 \text{ kips}$	$= [1.0 + 0.14(1.0)](7.40 \text{ kips})$ $\quad + 0 \text{ kips} + 0 \text{ kips}$ $\quad + 0.7(1.02)(1.3)(5.50 \text{ kips})$ $= 13.5 \text{ kips}$

Calculate B_1

Conservatively assume $C_m = 1.0$ and the effective length method is used for stability design. From Figure 5-59, the link length is 48.0 in.

$$B_1 = \frac{C_m}{1 - (\alpha P_r / P_{e1})} \geq 1 \quad \textit{(Spec. Eq. A-8-3)}$$

$$P_{e1} = \frac{\pi^2 EI^*}{(K_1 L)^2} \quad \textit{(Spec. Eq. A-8-5)}$$

$$= \frac{\pi^2 (29{,}000 \text{ ksi})(1{,}110 \text{ in.}^4)}{[1.0(48.0 \text{ in.})]^2}$$

$$= 138{,}000 \text{ kips}$$

LRFD	ASD
$\alpha = 1.00$	$\alpha = 1.60$
$B_1 = \dfrac{1.0}{1 - \dfrac{1.00(20.3 \text{ kips})}{138{,}000 \text{ kips}}} \geq 1$	$B_1 = \dfrac{1.0}{1 - \dfrac{1.60(13.5 \text{ kips})}{138{,}000 \text{ kips}}} \geq 1$
$= 1.00$	$= 1.00$

Since $B_1 = 1.00$, the required flexural strength need not be amplified to account for $P\text{-}\delta$ effects.

The required flexural strength of the link including second-order effects is:

LRFD	ASD
$M_u = (1.2 + 0.2S_{DS})M_D + B_2\rho M_{Q_E}$ $\quad + 0.5M_L + 0.2M_S$ $= [1.2 + 0.2(1.0)](14.4 \text{ kip-ft})$ $\quad + 1.02(1.3)(168 \text{ kip-ft})$ $\quad + 0.5(9.60 \text{ kip-ft}) + 0.2(0 \text{ kip-ft})$ $= 248 \text{ kip-ft}$	$M_a = (1.0 + 0.14S_{DS})M_D + M_H + M_F$ $\quad + 0.7B_2\rho M_{Q_E}$ $= [1.0 + 0.14(1.0)](14.4 \text{ kip-ft})$ $\quad + 0 \text{ kips} + 0 \text{ kips}$ $\quad + 0.7(1.02)(1.3)(168 \text{ kip-ft})$ $= 172 \text{ kip-ft}$

Second-order effects are not required to be applied to the required shear strength.

Width-to-Thickness Limitations

According to AISC *Seismic Provisions* Section F3.5b(1), the stiffened and unstiffened elements of links shall comply with AISC *Seismic Provisions* Section D1.1 for highly ductile members. There is an exception given in AISC *Seismic Provisions* Section F3.5b(1) that allows flanges of I-shaped links with length $e \leq 1.6M_p/V_p$ to satisfy the requirements of moderately ductile members. Determine whether the link length satisfies this limit.

$P_y = F_y A_g$ \hfill (*Provisions* Eq. F3-6)

$\quad = 50 \text{ ksi}(22.6 \text{ in.}^2)$

$\quad = 1{,}130 \text{ kips}$

From AISC *Seismic Provisions* Section F3.5b(2):

LRFD	ASD
$P_c = P_y$ $= 1{,}130$ kips	$P_c = \dfrac{P_y}{1.5}$ $= \dfrac{1{,}130 \text{ kips}}{1.5}$ $= 753$ kips
$\dfrac{P_r}{P_c} = \dfrac{20.3 \text{ kips}}{1{,}130 \text{ kips}}$ $= 0.0180$	$\dfrac{P_r}{P_c} = \dfrac{13.5 \text{ kips}}{753 \text{ kips}}$ $= 0.0179$

With $P_r/P_c \leq 0.15$, the AISC *Seismic Provisions* allows the effect of axial force on the link shear strength to be neglected.

$$V_p = 0.6 F_y A_{lw} \qquad \text{(Provisions Eq. F3-2)}$$

Where A_{lw} for I-shaped link sections is defined as:

$$A_{lw} = (d - 2t_f)t_w \qquad \text{(Provisions Eq. F3-4)}$$
$$= [16.5 \text{ in.} - 2(0.760 \text{ in.})](0.455 \text{ in.})$$
$$= 6.82 \text{ in.}^2$$

The link shear strength is:

$$V_p = 0.6(50 \text{ ksi})(6.82 \text{ in.}^2)$$
$$= 205 \text{ kips}$$

With $P_r/P_c \leq 0.15$:

$$M_p = F_y Z \qquad \text{(Provisions Eq. F3-8)}$$
$$= 50 \text{ ksi}(150 \text{ in.}^3)$$
$$= 7{,}500 \text{ kip-in.}$$

$$\dfrac{1.6 M_p}{V_p} = \dfrac{1.6(7{,}500 \text{ kip-in.})}{205 \text{ kips}}$$
$$= 58.5 \text{ in.}$$

Because $e = 48.0$ in. < 58.5 in., link flanges are permitted to comply with the requirements for moderately ductile members. From Table 1-3 of this Manual, a W16×77 satisfies the requirements for moderately ductile link beam flanges.

Table 1-3 of this Manual also shows that a W16×77 satisfies the requirements for a highly ductile link beam web.

Available Shear Strength

AISC *Seismic Provisions* Section F3.5b(2) defines the shear strength of the link as the lesser of that determined based on the limit states of flexural yielding and shear yielding.

For the limit state of shear yielding AISC *Seismic Provisions* Equation F3-1 defines the shear strength as follows, where V_p was previously calculated:

$$V_n = V_p$$
$$= 205 \text{ kips}$$

For the limit state of flexural yielding AISC *Seismic Provisions* Equation F3-7 defines the shear strength as follows, where M_p was previously calculated:

$$V_n = \frac{2M_p}{e}$$
$$= \frac{2(7,500 \text{ kip-in.})}{48.0 \text{ in.}}$$
$$= 313 \text{ kips}$$

Because 205 kips < 313 kips, the limit state of shear yielding from AISC *Seismic Provisions* Equation F3-2 controls:

LRFD	ASD
$\phi_v V_n = 0.90(205 \text{ kips})$ $= 185 \text{ kips} > V_u = 112 \text{ kips}$ **o.k.**	$\dfrac{V_n}{\Omega_v} = 205 \text{ kips}/1.67$ $= 123 \text{ kips} > V_a = 78.5 \text{ kips}$ **o.k.**

Link Rotation Angle

AISC *Seismic Provisions* Section F3.4a specifies a maximum link rotation angle based on the expected behavior of the link. The expected link behavior is determined by solving for the coefficient in front of M_p/V_p based on the given link length.

$$e = X \frac{M_p}{V_p}$$

Solving for the coefficient X:

$$\frac{V_p e}{M_p} = X$$

For the link being investigated:

$$\frac{V_p e}{M_p} = \frac{205 \text{ kips}(48.0 \text{ in.})}{7,500 \text{ kip-in.}}$$
$$= 1.31 < 1.6$$

A value of the ratio, $V_p e/M_p$, less than 1.6 indicates that the link behavior will be dominated by shear yielding. The corresponding limit on the link rotation angle for this type of expected link behavior is 0.08 rad according to AISC *Seismic Provisions* Section F3.4a. AISC *Seismic Provisions* Figure C-F3.4 defines the link rotation angle for this configuration as:

$$\gamma_p = \frac{L}{e}\theta_p$$

where

$$\theta_p = \frac{\Delta_p}{h}$$

AISC *Seismic Provisions* Section F3.3 requires that the inelastic link rotation angle be determined from the inelastic portion of the design story drift. From Example 5.4.1, the inelastic portion of the story drift is:

$$\Delta_p = \delta_x - \delta_{xe}$$
$$= 0.700 \text{ in.} - 0.175 \text{ in.}$$
$$= 0.525 \text{ in.}$$

$$\theta_p = \frac{0.525 \text{ in.}}{12.5 \text{ ft}(12 \text{ in./ft})}$$
$$= 0.00350 \text{ rad}$$

$$\gamma_p = \frac{30.0 \text{ ft}(12 \text{ in./ft})}{48.0 \text{ in.}}(0.00350 \text{ rad})$$
$$= 0.0263 \text{ rad}$$

$\gamma_p < 0.08$ rad **o.k.**

Note that the plastic story drift could have been conservatively assumed to equal the design story drift (0.700 in.). Using the design story drift determined in Example 5.4.1, $\gamma_p = 0.0350$ rad.

Available Compressive Strength

Use $K = 1.0$ for both the *x-x* and *y-y* axis. Use AISC *Manual* Table 6-1, where interpolating between values is approximate because the available compressive strength does not vary linearly with KL. The available strength in axial compression for a **W16×77** with $KL = 4$ ft:

LRFD	ASD
$p = \dfrac{1}{\phi_c P_n}$ $= 1.03 \times 10^{-3}$ $\phi_c P_n = 971 \text{ kips} > 20.3 \text{ kips}$ **o.k.**	$p = \dfrac{\Omega_c}{P_n}$ $= 1.54 \times 10^{-3}$ $\dfrac{P_n}{\Omega_c} = 649 \text{ kips} > 13.5 \text{ kips}$ **o.k.**

Available Flexural Strength

From AISC *Manual* Table 3-2, L_p = 8.72 ft. Since $L_b \le L_p$, the limit state of lateral-torsional buckling does not apply according to AISC *Specification* Section F2.2. Therefore, the available flexural strength is, from AISC *Manual* Table 3-2:

LRFD	ASD
$\phi_b M_p = 563 \text{ kip-ft} > 248 \text{ kip-ft}$ **o.k.**	$\dfrac{M_p}{\Omega_b} = 374 \text{ kip-ft} > 172 \text{ kip-ft}$ **o.k.**

Combined Loading

LRFD	ASD
$\dfrac{P_r}{P_c} = \dfrac{20.3 \text{ kips}}{971 \text{ kips}}$ $= 0.0209$	$\dfrac{P_r}{P_c} = \dfrac{13.5 \text{ kips}}{649 \text{ kips}}$ $= 0.0208$

Because $P_r/P_c < 0.2$, the beam-column design is controlled by the equation:

$$\dfrac{P_r}{2P_c} + \left(\dfrac{M_{rx}}{M_{cx}} + \dfrac{M_{ry}}{M_{cy}}\right) \le 1.0 \qquad \text{(\textit{Spec.} Eq. H1-1b)}$$

LRFD	ASD
$\dfrac{0.0209}{2} + \dfrac{248 \text{ kip-ft}}{563 \text{ kip-ft}} + 0 = 0.451$ $0.451 < 1.0$ **o.k.**	$\dfrac{0.0208}{2} + \dfrac{172 \text{ kip-ft}}{374 \text{ kip-ft}} + 0 = 0.470$ $0.470 < 1.0$ **o.k.**

The **W16×77** is adequate to resist the loads given for the link segment of Beam BM-1.

Lateral Bracing Requirements

AISC *Seismic Provisions* Section F3.4b requires that both flanges at each end of the link be braced. Bracing is required to have strength and stiffness as specified by AISC *Seismic Provisions* Section D1.2c for expected plastic hinge locations. This design uses lateral bracing of the flanges. From AISC *Seismic Provisions* Equations D1-4a and D1-4b, the required lateral brace strength, with $R_y = 1.1$ from AISC *Seismic Provisions* Table A3.1, is:

LRFD	ASD
$P_u = \dfrac{0.06 R_y F_y Z}{h_o}$ $= \dfrac{0.06(1.1)(50 \text{ ksi})(150 \text{ in.}^3)}{15.7 \text{ in.}}$ $= 31.5 \text{ kips}$	$P_a = \dfrac{(0.06/1.5) R_y F_y Z}{h_o}$ $= \dfrac{(0.06/1.5)(1.1)(50 \text{ ksi})(150 \text{ in.}^3)}{15.7 \text{ in.}}$ $= 21.0 \text{ kips}$

The required brace stiffness according to AISC *Seismic Provisions* Section D1.2c(3) is calculated in accordance with AISC *Specification* Appendix 6 with $C_d = 1.0$ and with the value of M_r specified in AISC *Seismic Provisions* Equation D1-6 as:

LRFD	ASD
$M_r = R_y F_y Z$ $= 1.1(50 \text{ ksi})(150 \text{ in.}^3)$ $= 8{,}250 \text{ kip-in.}$	$M_r = R_y F_y Z / 1.5$ $= 1.1(50 \text{ ksi})(150 \text{ in.}^3)/1.5$ $= 5{,}500 \text{ kip-in.}$

This required flexural strength can be used in AISC *Specification* Appendix 6 to determine the required stiffness of the lateral bracing. Use nodal bracing and Equation A-6-8 to calculate the required stiffness, where L_b is the length of the link, 48 in., as:

LRFD	ASD
$\beta_{br} = \dfrac{1}{\phi}\left(\dfrac{10 M_r C_d}{L_b h_o}\right)$ $= \dfrac{1}{0.75}\left(\dfrac{10(8{,}250 \text{ kip-in.})(1.0)}{(48 \text{ in.})(15.7 \text{ in.})}\right)$ $= 146 \text{ kip/in.}$	$\beta_{br} = \Omega\left(\dfrac{10 M_r C_d}{L_b h_o}\right)$ $= 2.00\left(\dfrac{10(5{,}500 \text{ kip-in.})(1.0)}{(48 \text{ in.})(15.7 \text{ in.})}\right)$ $= 146 \text{ kip/in.}$

Top and bottom flange bracing will be provided in accordance with AISC *Specification* Appendix 6 with the strength and stiffness required by these calculations.

Stiffener Requirements

AISC *Seismic Provisions* Section F3.5b(4) requires double-sided, full-depth web stiffeners at each end of the link. The minimum required combined width of the stiffeners is $(b_f - 2t_w)$. Thus, the minimum width of each stiffener is:

$$w_{min} = \frac{b_f - 2t_w}{2}$$

$$= \frac{10.3 \text{ in.} - 2(0.455 \text{ in.})}{2}$$

$$= 4.70 \text{ in.}$$

The minimum required thickness is the larger of $0.75t_w$ and $\frac{3}{8}$ in.:

$$t_{min} = 0.75t_w$$

$$= 0.75(0.455 \text{ in.})$$

$$= 0.341 \text{ in.} < \tfrac{3}{8} \text{ in.}$$

Therefore, $t_{min} = \frac{3}{8}$ in.

Full depth $\frac{3}{8}$ in. × $4\frac{3}{4}$ in. stiffeners will be provided on both sides of the web at each end of the link segment.

AISC *Seismic Provisions* Section F3.5b(4) also requires full depth intermediate web stiffeners (intermediate stiffeners are stiffeners within the link segment). Because the length of the link is less than $1.6M_p/V_p$, the spacing requirements for intermediate web stiffeners are determined based on the link rotation angle.

For a link rotation angle equal to 0.08 rad, the required spacing is:

$$30t_w - \frac{d}{5} = 30(0.455 \text{ in.}) - \left(\frac{16.5 \text{ in.}}{5}\right)$$

$$= 10.4 \text{ in.}$$

For a link rotation angle equal to 0.02 rad or less, the required spacing is:

$$52t_w - \frac{d}{5} = 52(0.455 \text{ in.}) - \left(\frac{16.5 \text{ in.}}{5}\right)$$

$$= 20.4 \text{ in.}$$

Interpolating between these limits using the calculated link rotation angle of $\gamma_p = 0.0263$ rad, the maximum spacing between web stiffeners is 19.4 in.

From AISC *Seismic Provisions* Section F3.5b(4), with a link depth less than 25 in., the intermediate stiffeners are required on one side of the web only. Also, the minimum required thickness of the intermediate web stiffeners on one side only is the larger of t_w and $\frac{3}{8}$ in.

$t_{min} = t_w$
$= 0.455$ in. $\geq \frac{3}{8}$ in.

Therefore, $t_{min} = 0.455$ in.

The required width of intermediate stiffeners on one side only is:

$w_{min} = \dfrac{b_f}{2} - t_w$

$= \dfrac{10.3 \text{ in.}}{2} - 0.455$ in.

$= 4.70$ in.

Full depth ½ in. × 4¾ in. intermediate web stiffeners will be provided within the link segment, on one side of the web only and at a maximum spacing of 19.4 in. With the link length of 48 in. given in Figure 5-59, choose to use two intermediate link stiffeners with a spacing of 16.0 in. on center.

Note that it may be beneficial to also use ½-in.-thick material for the link end stiffeners in order to simplify the detailing and fabrication of the link. This simplification will be made in this example.

AISC *Seismic Provisions* Section F3.5b(4) also specifies that the required strength of the fillet welds connecting the link stiffeners to the link web is $F_y A_{st}$ (LRFD) or $F_y A_{st}/1.5$ (ASD) and of the welds connecting the link stiffeners to the link flanges is $F_y A_{st}/4$ (LRFD) or $F_y A_{st}/4(1.5)$ (ASD), where A_{st} is the horizontal cross-sectional area of the stiffener. For the ½-in.-thick stiffener, the cross-sectional area of the stiffener is:

$A_{st} = \frac{1}{2}$ in.$(4\frac{3}{4}$ in.$)$

$= 2.38$ in.^{2}

The Commentary to AISC *Seismic Provisions* Section F3.5b(4) suggests that welding in the k-area of the beam be avoided. To accomplish this, the stiffener clips will be sized to comply with the requirements of AWS D1.8 clause 4.1.1 Based on AWS D1.8 clause 4.1.1, the clip along the web must extend at least 1½ in. beyond the published k_{det} dimension for the rolled shape. This corresponds to a clip length measured from the edge of the stiffener of at least:

$1\frac{1}{2}$ in. $+ k_{det} - t_f = 1\frac{1}{2}$ in. $+ 1\frac{5}{8}$ in. $- 0.760$ in.

$= 2.37$ in.

Use a clip length of 2⅜ in. along the web. The length of the stiffener along the web is thus:

$L_{st} = d - 2t_f - 2(2\frac{3}{8}$ in.$)$

$= 16.5$ in. $- 2(0.760$ in.$) - 2(2\frac{3}{8}$ in.$)$

$= 10.2$ in.

From AISC *Manual* Equations 8-2a and 8-2b, the double-sided fillet weld required to connect the link stiffeners to the link web is:

LRFD	ASD
$D = \dfrac{F_y A_{st}}{2(1.392 \text{ kip/in.})(L_{st})}$	$D = \dfrac{(F_y A_{st})/1.5}{2(0.928 \text{ kip/in.})(L_{st})}$
$= \dfrac{36 \text{ ksi}(2.38 \text{ in.}^2)}{2(1.392 \text{ kip/in.})(10.2 \text{ in.})}$	$= \dfrac{[36 \text{ ksi}(2.38 \text{ in.}^2)]/1.5}{2(0.928 \text{ kip/in.})(10.2 \text{ in.})}$
$= 3.02$ sixteenths	$= 3.02$ sixteenths

Checking AISC *Specification* Table J2.4, with the 0.455 in. link web thickness, the minimum fillet weld size is $\frac{3}{16}$ in.

Use double-sided ¼-in. fillet welds to connect the link stiffeners to the link web.

Based on AWS D1.8 clause 4.1.2, the clip along the flanges must not exceed a distance of ½ in. beyond the published k_1 detail dimension for the rolled shape. The maximum clip length measured from the edge of the plate is therefore:

$$k_1 - \frac{t_w}{2} + \tfrac{1}{2} \text{ in.} = 1\tfrac{1}{16} \text{ in.} - \frac{0.455 \text{ in.}}{2} + \tfrac{1}{2} \text{ in.}$$

$$= 1.34 \text{ in.}$$

Use a 1.00-in. clip along the flange to allow the stiffeners to clear the fillets. The width of the stiffener along the flange is:

$$w_{st} = \min\left(\frac{b_f - t_w}{2} - 1.00 \text{ in.},\ 4\tfrac{3}{4} \text{ in.} - 1.00 \text{ in.}\right)$$

$$= \min\left(\frac{10.3 \text{ in.} - 0.455 \text{ in.}}{2} - 1.00 \text{ in.},\ 4\tfrac{3}{4} \text{ in.} - 1.00 \text{ in.}\right)$$

$$= \min(3.92 \text{ in.},\ 3\tfrac{3}{4} \text{ in.})$$

$$= 3\tfrac{3}{4} \text{ in.}$$

From AISC *Manual* Equations 8-2a and 8-2b, the double-sided fillet weld size required to connect the link stiffeners to the link flanges is:

LRFD	ASD
$D = \dfrac{F_y A_{st}/4}{2(1.392 \text{ kip/in.})(w_{st})}$	$D = \dfrac{(F_y A_{st})/[1.5(4)]}{2(0.928 \text{ kip/in.})(w_{st})}$
$= \dfrac{[36 \text{ ksi}(2.38 \text{ in.}^2)]/4}{2(1.392 \text{ kip/in.})(3\tfrac{3}{4} \text{ in.})}$	$= \dfrac{[36 \text{ ksi}(2.38 \text{ in.}^2)]/[1.5(4)]}{2(0.928 \text{ kip/in.})(3\tfrac{3}{4} \text{ in.})}$
$= 2.05$ sixteenths	$= 2.05$ sixteenths

Checking AISC *Specification* Table J2.4, with the ½-in. stiffener plate thickness, the minimum fillet weld size is 3/16 in.

Use double-sided 3/16-in. fillet welds to connect the link stiffeners to the link flanges.

Note that it may be beneficial to also use double-sided ¼-in. fillet welds to connect the link stiffeners to the link flanges in order to simplify the detailing and fabrication of the link.

Example 5.4.3. EBF Beam Outside of the Link Design

Given:

Refer to Beam BM-1 in Figure 5-59. Determine the adequacy of the ASTM A992 W16×77 link segment selected in Example 5.4.2 as the beam outside of the link for the following loading. The applicable building code specifies the use of ASCE/SEI 7 for calculation of loads. From a first-order analysis:

$P_D = 1.00$ kips $\qquad P_L = 0.700$ kips $\qquad P_{Q_E} = 105$ kips
$V_D = 6.80$ kips $\qquad V_L = 4.80$ kips $\qquad V_{Q_E} = 8.70$ kips
$M_D = 17.0$ kip-ft $\qquad M_L = 11.3$ kip-ft $\qquad M_{Q_E} = 113$ kip-ft

Relevant seismic parameters are given in the EBF Design Example Plan and Elevation section.

Assume the braces are ASTM A992 W10×112, the columns are W12 wide-flange sections and that the flanges of Beam BM-1 are braced at the columns.

Solution:

From AISC *Manual* Table 2-4 and AISC *Seismic Provisions* Table A3.1, the material properties are as follows:

ASTM A992
$F_y = 50$ ksi
$F_u = 65$ ksi
$R_y = 1.1$

From AISC *Manual* Table 1-1, the geometric properties are as follows:

Beam
W16×77
$A = 22.6$ in.^{2} $\qquad I_x = 1,110$ in.^{4} $\qquad r_y = 2.47$ in.

Brace
W10×112
$I_x = 716$ in.^{4}

Required Strength

According to AISC *Seismic Provisions* Section F3.3, the required strength of the beam outside of the link is a combination of the factored gravity forces plus the forces generated by the adjusted link shear strength. From Example 5.4.2, the nominal shear strength of the link, V_n, was determined to be 205 kips. According to AISC *Seismic Provisions* Section F3.3, the adjusted link shear strength for an I-shaped section (using Exception (1)(a) from Section F3.3) is:

$$0.88(1.25)R_y V_n = 0.88(1.25)(1.1)(205 \text{ kips})$$
$$= 248 \text{ kips}$$

The geometry of the column, brace, half-beam and half-link is shown in Figure 5-60. The axial force in the beam outside of the link based on the adjusted shear strength of the link is:

$$P_{E_{mh}} = \frac{0.88(1.25)R_y V_n L}{2H}$$
$$= \frac{248 \text{ kips}(30.0 \text{ ft})}{2(12.5 \text{ ft})}$$
$$= 298 \text{ kips}$$

The resulting link end moment based on the adjusted shear strength of the link is:

$$M_{link} = \frac{0.88(1.25)R_y V_n e}{2}$$
$$= \frac{248 \text{ kips}(48.0 \text{ in.})}{2}$$
$$= 5,950 \text{ kip-in.}$$

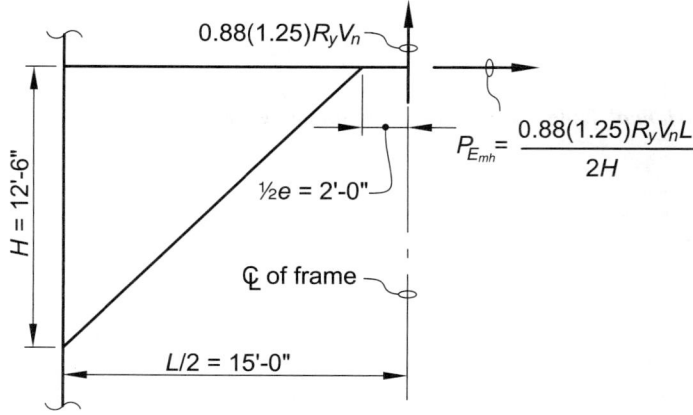

Fig. 5-60. Diagram for Example 5.4.3.

5.4 ECCENTRICALLY BRACED FRAMES

As given in Example 5.4.2, the brace-to-beam connection will be detailed as a fixed connection; therefore, the moment at the end of the link will be distributed between the brace and the beam outside of the link. One way to determine the portion of this moment resisted by the beam outside of the link is based on relative member stiffness. Since the modulus of elasticity is the same for both members, it can be neglected in the stiffness calculation. Using relative member stiffness to distribute the link end moment, the portion of the moment taken by the beam outside of the link (*bol*) is:

$$M_{bol} = \left(\frac{\frac{I_{bol}}{L_{bol}}}{\frac{I_{bol}}{L_{bol}} + \frac{I_{br}}{L_{br}}} \right) M_{link}$$

$$L_{bol} = \frac{30.0 \text{ ft} - 4.00 \text{ ft}}{2}$$
$$= 13.0 \text{ ft}$$

$$L_{br} = \sqrt{(12.5 \text{ ft})^2 + (13.0 \text{ ft})^2}$$
$$= 18.0 \text{ ft}$$

$$\frac{I_{bol}}{L_{bol}} = \frac{1{,}110 \text{ in.}^4}{13.0 \text{ ft}}$$
$$= 85.4 \text{ in.}^4/\text{ft}$$

$$\frac{I_{br}}{L_{br}} = \frac{716 \text{ in.}^4}{18.0 \text{ ft}}$$
$$= 39.8 \text{ in.}^4/\text{ft}$$

$$\frac{\frac{I_{bol}}{L_{bol}}}{\frac{I_{bol}}{L_{bol}} + \frac{I_{br}}{L_{br}}} = \frac{85.4 \text{ in.}^4/\text{ft}}{85.4 \text{ in.}^4/\text{ft} + 39.8 \text{ in.}^4/\text{ft}}$$
$$= 0.682$$

Using this method, the beam outside of the link is assumed to take 68.2% of the link end moment. The moment in the beam outside of the link is then:

$$M_{E_{mh}} = M_{bol}$$
$$= 0.682 M_{link}$$
$$= 0.682(5{,}950 \text{ kip-in.})/(12 \text{ in./ft})$$
$$= 338 \text{ kip-ft}$$

Alternatively, a method based on the calculation of an amplification factor can be used. In this method, the adjusted link shear strength is divided by the link shear generated by the code-specified earthquake forces. The resulting amplification factor is used to amplify the remaining member end forces generated by the analysis using the code-specified earthquake loading. From Example 5.4.2, the link shear force obtained from a computer analysis using the code-specified seismic forces was given as:

$V_{Q_E} = 84.0$ kips

The resulting overstrength factor is:

$$\frac{0.88(1.25)R_y V_n}{V_{Q_E}} = \frac{248 \text{ kips}}{84.0 \text{ kips}}$$
$$= 2.95$$

The moment in the beam outside of the link due to the link mechanism based on the expected shear strength of the link is:

$$M_{E_{mh}} = 2.95 M_{Q_E}$$
$$= 2.95(113 \text{ kip-ft})$$
$$= 333 \text{ kip-ft}$$

The axial force in the beam outside of the link due to the link mechanism based on the expected shear strength of the link is:

$$P_{E_{mh}} = 2.95 P_{Q_E}$$
$$= 2.95(105 \text{ kips})$$
$$= 310 \text{ kips}$$

The shear in the beam outside of the link due to the link mechanism based on the expected shear strength of the link is:

$$V_{E_{mh}} = 2.95 V_{Q_E}$$
$$= 2.95(8.70 \text{ kips})$$
$$= 25.7 \text{ kips}$$

Note that the moments generated by the two methods are very similar. Since the beam outside of the link shear has already been determined, the forces generated using the amplification factor method will be used in the calculation of the required strengths.

Considering the load combinations given in ASCE/SEI 7 that include the amplified seismic loads, it was determined that the governing load combination for the beam outside the link, with $\Omega_o Q_E = E_{mh}$, is:

5.4 ECCENTRICALLY BRACED FRAMES

LRFD	ASD
LRFD Load Combination 5 from ASCE/SEI 7 Section 12.4.3.2 (including the 0.5 factor on L permitted in Section 12.4.3.2)	ASD Load Combination 5 from ASCE/SEI 7 Section 12.4.3.2
$(1.2 + 0.2S_{DS})D + E_{mh} + 0.5L + 0.2S$	$(1.0 + 0.14S_{DS})D + H + F + 0.7E_{mh}$

The required axial strength of the beam outside the link is:

LRFD	ASD
$P_u = (1.2 + 0.2S_{DS})P_D + P_{E_{mh}} + 0.5P_L$ $\quad + 0.2P_S$ $= [1.2 + 0.2(1.0)](1.00 \text{ kips})$ $\quad + 310 \text{ kips} + 0.5(0.700 \text{ kips})$ $\quad + 0.2(0 \text{ kips})$ $= 312 \text{ kips}$	$P_a = (1.0 + 0.14S_{DS})P_D + P_H + P_F$ $\quad + 0.7P_{E_{mh}}$ $= [1.0 + 0.14(1.0)](1.00 \text{ kips})$ $\quad + 0 \text{ kips} + 0 \text{ kips}$ $\quad + 0.7(310 \text{ kips})$ $= 218 \text{ kips}$

The required flexural strength of the beam outside the link is:

LRFD	ASD
$M_u = (1.2 + 0.2S_{DS})M_D + M_{E_{mh}}$ $\quad + 0.5M_L + 0.2M_S$ $= [1.2 + 0.2(1.0)](17.0 \text{ kip-ft})$ $\quad + 333 \text{ kip-ft} + 0.5(11.3 \text{ kip-ft})$ $\quad + 0.2(0 \text{ kip-ft})$ $= 362 \text{ kip-ft}$	$M_a = (1.0 + 0.14S_{DS})M_D + M_H + M_F$ $\quad + 0.7M_{E_{mh}}$ $= [1.0 + 0.14(1.0)](17.0 \text{ kip-ft})$ $\quad + 0 \text{ kip-ft} + 0 \text{ kip-ft}$ $\quad + 0.7(333 \text{ kip-ft})$ $= 252 \text{ kip-ft}$

The required shear strength of the beam outside the link is:

LRFD	ASD
$V_u = (1.2 + 0.2S_{DS})V_D + V_{E_{mh}} + 0.5V_L$ $\quad + 0.2V_S$ $= [1.2 + 0.2(1.0)](6.80 \text{ kips})$ $\quad + 25.7 \text{ kips} + 0.5(4.80 \text{ kips})$ $\quad + 0.2(0 \text{ kips})$ $= 37.6 \text{ kips}$	$V_a = (1.0 + 0.14S_{DS})V_D + V_H + V_F$ $\quad + 0.7V_{E_{mh}}$ $= [1.0 + 0.14(1.0)](6.80 \text{ kips})$ $\quad + 0 \text{ kips} + 0 \text{ kips}$ $\quad + 0.7(25.7 \text{ kips})$ $= 25.7 \text{ kips}$

Width-to-Thickness Limitations

Since the beam outside of the link is the same section as the link, no additional local buckling checks are required.

Unbraced Length

As established in Example 5.4.2, each end of the link will be braced. A nominal column depth of 12 in. will be assumed. Therefore, the unbraced length of the beam outside of the link to the face of the column is:

$$L_b = \frac{L - e - 2\left(\frac{d_c}{2}\right)}{2}$$

$$= \frac{30.0 \text{ ft}(12 \text{ in./ft}) - 48.0 \text{ in.} - 12.0 \text{ in.}}{2}$$

$$= 150 \text{ in.}$$

or

$$L_b = 150 \text{ in.}/(12 \text{ in./ft})$$
$$= 12.5 \text{ ft}$$

Second-Order Effects

From AISC *Specification* Appendix 8, the required flexural and axial strength including second-order effects are determined as follows:

$$M_r = B_1 M_{nt} + B_2 M_{lt} \qquad \text{(\textit{Spec.} Eq. A-8-1)}$$

$$P_r = P_{nt} + B_2 P_{lt} \qquad \text{(\textit{Spec.} Eq. A-8-2)}$$

The multiplier that accounts for P-Δ effects, B_2, is 1.0 because the lateral load effect is based on the adjusted link shear strength. P-Δ effects do not increase the forces corresponding to the fully-yielded, strain-hardened link; instead they may be thought of as contributing to the system reaching that state.

Because $B_2 = 1.0$, the required compression and flexural strengths will not be amplified to account for P-Δ effects. Conservatively use $C_m = 1.0$ and the effective length method is used for stability design.

Calculate B_1

$$B_1 = \frac{C_m}{1 - \alpha P_r / P_{e1}} \geq 1 \qquad \text{(\textit{Spec.} Eq. A-8-3)}$$

where

$$P_{e1} = \frac{\pi^2 EI^*}{(K_1 L)^2} \quad \text{(Spec. Eq. A-8-5)}$$

$$= \frac{\pi^2 (29{,}000 \text{ ksi})(1{,}110 \text{ in.}^4)}{[1.0(150 \text{ in.})]^2}$$

$$= 14{,}100 \text{ kips}$$

From AISC *Specification* Equation A-8-3:

LRFD	ASD
$\alpha = 1.00$	$\alpha = 1.60$
$B_1 = \dfrac{1.0}{1 - \left[\dfrac{1.00(312 \text{ kips})}{14{,}100 \text{ kips}}\right]}$	$B_1 = \dfrac{1.0}{1 - \left[\dfrac{1.60(218 \text{ kips})}{14{,}100 \text{ kips}}\right]}$
$= 1.02$	$= 1.03$

According to AISC *Specification* Equation A-8-1, the B_1 factor (P-δ effect) need only be applied to the first-order moment with the structure restrained against translation.

LRFD	ASD
$M_u = B_1(1.2 + 0.2S_{DS})M_D + M_{E_{mh}}$ $\quad + B_1(0.5M_L) + B_1(0.2M_S)$ $= 1.02[1.2 + 0.2(1.0)](17.0 \text{ kip-ft})$ $\quad + 333 \text{ kip-ft}$ $\quad + 1.02(0.5)(11.3 \text{ kip-ft})$ $\quad + 1.02(0.2)(0 \text{ kip-ft})$ $= 363 \text{ kip-ft}$	$M_a = B_1(1.0 + 0.14S_{DS})M_D + M_H + M_F$ $\quad + 0.7M_{E_{mh}}$ $= 1.03[1.0 + 0.14(1.0)](17.0 \text{ kip-ft})$ $\quad + 0 \text{ kip-ft} + 0 \text{ kip-ft}$ $\quad + 0.7(333 \text{ kip-ft})$ $= 253 \text{ kip-ft}$

Combined Loading

Because the beam outside of the link is the same member as the link, AISC *Seismic Provisions* Section A3.2 permits the use of $R_y F_y$ in lieu of F_y when determining the available strengths of the beam outside of the link.

Determine available compressive strength of the W16×77

Use AISC *Specification* Section E3 to determine the available compressive strength. Note that using AISC *Manual* tables to determine the available compressive strength and multiplying this strength by R_y may not give accurate values, as the compressive strength does not vary linearly with F_y. The applicable critical stress equation can be determined by the ratio of $R_y F_y / F_e$. The elastic buckling stress, F_e, is:

$$F_e = \frac{\pi^2 E}{\left(\dfrac{L_b}{r_y}\right)^2} \quad \text{(Spec. Eq. E3-4)}$$

$$= \frac{\pi^2 (29{,}000 \text{ ksi})}{\left(\dfrac{150 \text{ in.}}{2.47 \text{ in.}}\right)^2}$$

$$= 77.6 \text{ ksi}$$

$$\frac{R_y F_y}{F_e} = \frac{1.1(50 \text{ ksi})}{77.6 \text{ ksi}}$$

$$= 0.709$$

Because $R_y F_y / F_e \leq 2.25$, the critical stress, F_{cr}, is:

$$F_{cr} = \left(0.658^{\frac{R_y F_y}{F_e}}\right) R_y F_y \quad \text{(Spec. Eq. E3-2)}$$

$$= \left[0.658^{\frac{1.1(50 \text{ ksi})}{77.6 \text{ ksi}}}\right](1.1)(50 \text{ ksi})$$

$$= 40.9 \text{ ksi}$$

The available compressive strength is then determined from AISC *Specification* Equation E3-1:

LRFD	ASD
$\phi_c P_n = \phi_c F_{cr} A_g$ $= 0.90(40.9 \text{ ksi})(22.6 \text{ in.}^2)$ $= 832 \text{ kips}$	$\dfrac{P_n}{\Omega_c} = \dfrac{F_{cr} A_g}{\Omega_c}$ $= \dfrac{(40.9 \text{ ksi})(22.6 \text{ in.}^2)}{1.67}$ $= 553 \text{ kips}$

Determine available flexural strength of the W16×77

From AISC *Manual* Table 3-2, for a W16×77, $L_p = 8.72$ ft and $L_r = 27.8$ ft. Therefore, with $C_b = 1.0$ and $L_b = 12.5$ ft ($L_p < L_b \leq L_r$), lateral-torsional buckling applies. Using AISC *Manual* Table 3-10 and adjusting by R_y, the available flexural strength is:

LRFD	ASD
$\phi_b M_n = 1.1(521 \text{ kip-ft})$ $= 573 \text{ kip-ft}$	$\dfrac{M_n}{\Omega_b} = 1.1(347 \text{ kip-ft})$ $= 382 \text{ kip-ft}$

Check combined flexure and compression of the W16×77

LRFD	ASD
$\dfrac{P_r}{P_c} = \dfrac{P_r}{\phi_c P_n}$ $= \dfrac{312 \text{ kips}}{832 \text{ kips}}$ $= 0.375$	$\dfrac{P_r}{P_c} = \dfrac{\Omega_c P_r}{P_n}$ $= \dfrac{218 \text{ kips}}{553 \text{ kips}}$ $= 0.394$

Because $\dfrac{P_r}{P_c} \geq 0.2$, AISC *Specification* Equation H1-1a applies.

$$\dfrac{P_r}{P_c} + \dfrac{8}{9}\left(\dfrac{M_{rx}}{M_{cx}} + \dfrac{M_{ry}}{M_{cy}}\right) < 1.0 \qquad (Spec.\ Eq.\ H1\text{-}1a)$$

LRFD	ASD
$\dfrac{P_r}{\phi_c P_n} + \dfrac{8}{9}\left(\dfrac{M_{rx}}{\phi_b M_{nx}} + \dfrac{M_{ry}}{\phi_b M_{ny}}\right)$ $0.375 + \dfrac{8}{9}\left(\dfrac{363 \text{ kip-ft}}{573 \text{ kip-ft}} + 0\right)$ $= 0.938 < 1.0 \quad \textbf{o.k.}$	$\dfrac{\Omega_c P_r}{P_n} + \dfrac{8}{9}\left(\dfrac{\Omega_b M_{rx}}{M_{nx}} + \dfrac{\Omega_b M_{ry}}{M_{ny}}\right)$ $0.394 + \dfrac{8}{9}\left(\dfrac{253 \text{ kip-ft}}{382 \text{ kip-ft}} + 0\right)$ $= 0.983 < 1.0 \quad \textbf{o.k.}$

Available Shear Strength

From AISC *Manual* Table 3-6, the available shear strength is:

LRFD	ASD
$\phi_v V_n = 225 \text{ kips} > 37.6 \text{ kips} \quad \textbf{o.k.}$	$\dfrac{V_n}{\Omega_v} = 150 \text{ kips} > 25.7 \text{ kips} \quad \textbf{o.k.}$

The W16×77 is adequate to resist the loads given for the beam outside of the link segments of Beam BM-1. Additional flange bracing is not required.

Example 5.4.4. EBF Brace Design

Given:
Refer to Brace BR-1 in Figure 5-59. Select an ASTM A992 wide-flange section to resist the following loads. The applicable building code specifies the use of ASCE/SEI 7 for calculation of loads. From a first-order analysis:

$P_D = 11.8$ kips $P_L = 8.30$ kips $P_{Q_E} = 136$ kips
$V_D = 0.200$ kips $V_L = 0.120$ kips $V_{Q_E} = 3.02$ kips
$M_D = 3.20$ kip-ft $M_L = 2.20$ kip-ft $M_{Q_E} = 54.5$ kip-ft

Relevant seismic parameters are given in the EBF Design Example Plan and Elevation section.

Assume that the link segment and beam outside of the link segments are those selected in Examples 5.4.2 and 5.4.3, and that the column-end of the brace is pinned and braced against translation for both the *x-x* and *y-y* axes.

Solution:
From AISC *Manual* Table 2-4 and AISC *Seismic Provisions* Table A3.1, the material properties are as follows:

ASTM A992
$F_y = 50$ ksi
$F_u = 65$ ksi
$R_y = 1.1$

Required Strengths
According to AISC *Seismic Provisions* Section F3.3, the required strength of the brace is a combination of the factored gravity forces plus the forces generated by the adjusted link shear strength, using the load combinations that include the amplified seismic load. From Example 5.4.2, the nominal shear strength of the link, V_n, is 205 kips.

$$1.25 R_y V_n = 1.25(1.1)(205 \text{ kips})$$
$$= 282 \text{ kips}$$

Using the overstrength factor method described in Example 5.4.3 with the link shear force, V_{Q_E}, given in Example 5.4.2, the overstrength factor is:

$$\frac{1.25 R_y V_n}{V_{Q_E}} = \frac{282 \text{ kips}}{84.0 \text{ kips}}$$
$$= 3.36$$

The moment in the brace due to the link mechanism is:

5.4 ECCENTRICALLY BRACED FRAMES

$$M_{E_{mh}} = 3.36 M_{Q_E}$$
$$= 3.36(54.5 \text{ kip-ft})$$
$$= 183 \text{ kip-ft}$$

The axial force in the brace due to the link mechanism is:

$$P_{E_{mh}} = 3.36 P_{Q_E}$$
$$= 3.36(136 \text{ kips})$$
$$= 457 \text{ kips}$$

The shear in the brace due to the link mechanism is:

$$V_{E_{mh}} = 3.36 V_{Q_E}$$
$$= 3.36(3.02 \text{ kips})$$
$$= 10.1 \text{ kips}$$

Considering the load combinations given in ASCE/SEI 7 that include the amplified seismic load, with $\Omega_o Q_E = E_{mh}$, it was determined that the governing load combination for the brace is:

LRFD	ASD
LRFD Load Combination 5 from ASCE/SEI 7 Section 12.4.3.2 (including the 0.5 factor on L permitted in Section 12.4.3.2) $(1.2 + 0.2 S_{DS})D + E_{mh} + 0.5L + 0.2S$	ASD Load Combination 5 from ASCE/SEI 7 Section 12.4.3.2 $(1.0 + 0.14 S_{DS})D + H + F + 0.7 E_{mh}$

The required axial strength of the brace is:

LRFD	ASD
$P_u = (1.2 + 0.2 S_{DS}) P_D + P_{E_{mh}} + 0.5 P_L$ $+ 0.2 P_S$ $= [1.2 + 0.2(1.0)](11.8 \text{ kips})$ $+ 457 \text{ kips} + 0.5(8.30 \text{ kips})$ $+ 0.2(0 \text{ kips})$ $= 478 \text{ kips}$	$P_a = (1.0 + 0.14 S_{DS}) P_D + P_H + P_F$ $+ 0.7 P_{E_{mh}}$ $= [1.0 + 0.14(1.0)](11.8 \text{ kips})$ $+ 0 \text{ kips} + 0 \text{ kips}$ $+ 0.7(457 \text{ kips})$ $= 333 \text{ kips}$

The required flexural strength of the brace is:

LRFD	ASD
$M_u = (1.2 + 0.2S_{DS})M_D + M_{E_{mh}} + 0.5M_L$ $\quad + 0.2M_S$ $= [1.2 + 0.2(1.0)](3.20 \text{ kip-ft})$ $\quad + 183 \text{ kip-ft} + 0.5(2.20 \text{ kip-ft})$ $\quad + 0.2(0 \text{ kip-ft})$ $= 189 \text{ kip-ft}$	$M_a = (1.0 + 0.14S_{DS})M_D + M_H + M_F$ $\quad + 0.7M_{E_{mh}}$ $= [1.0 + 0.14(1.0)](3.20 \text{ kip-ft})$ $\quad + 0 \text{ kip-ft} + 0 \text{ kip-ft}$ $\quad + 0.7(183 \text{ kip-ft})$ $= 132 \text{ kip-ft}$

The required shear strength of the brace is:

LRFD	ASD
$V_u = (1.2 + 0.2S_{DS})V_D + V_{E_{mh}} + 0.5V_L$ $\quad + 0.2V_S$ $= [1.2 + 0.2(1.0)](0.200 \text{ kips})$ $\quad + 10.1 \text{ kips} + 0.5(0.120 \text{ kips})$ $\quad + 0.2(0 \text{ kips})$ $= 10.4 \text{ kips}$	$V_a = (1.0 + 0.14S_{DS})V_D + V_H + V_F$ $\quad + 0.7V_{E_{mh}}$ $= [1.0 + 0.14(1.0)](0.200 \text{ kips})$ $\quad + 0 \text{ kips} + 0 \text{ kips}$ $\quad + 0.7(10.1 \text{ kips})$ $= 7.30 \text{ kips}$

As assumed in Examples 5.4.2 and 5.4.3, try a **W10×112** for the brace.

From AISC *Manual* Table 1-1, the geometric properties are as follows:

$A = 32.9 \text{ in.}^2 \qquad d = 11.4 \text{ in.} \qquad t_w = 0.755 \text{ in.}$
$b_f = 10.4 \text{ in.} \qquad t_f = 1.25 \text{ in.} \qquad I_x = 716 \text{ in.}^4$

Width-to-Thickness Limitations

According to AISC *Seismic Provisions* Section F3.5a, the stiffened and unstiffened elements of EBF braces shall comply with the requirements of Section D1.1 for moderately ductile members. From Table 1-3 of this Manual, the W10×112 satisfies these limits for EBF braces.

Determine unbraced length

$L_b = \sqrt{(12.5 \text{ ft})^2 + (13.0 \text{ ft})^2}$
$\quad = 18.0 \text{ ft}$

or

$L_b = (18.0 \text{ ft})(12 \text{ in./ft})$
$\quad = 216 \text{ in.}$

5.4 ECCENTRICALLY BRACED FRAMES

Note that the unbraced length is based on the work point-to-work point distance. Shorter lengths may be used provided the lateral support is adequate at each end of the assumed unbraced length.

Second-Order Effects

Second-order effects are addressed using AISC *Specification* Appendix 8. Since the lateral load effect is based on the adjusted link shear strength, $B_2 = 1.0$. P-Δ effects do not increase the forces corresponding to the fully yielded, strain-hardened link; instead they may be thought of as contributing to the system reaching that state.

Because $B_2 = 1.0$, the required compressive and flexural strengths will not be amplified to account for P-Δ effects. The effective length method is used for stability design.

Calculate B_1

$$B_1 = \frac{C_m}{1 - \alpha P_r/P_{e1}} \geq 1 \qquad (Spec.\ Eq.\ A\text{-}8\text{-}3)$$

where

$$P_{e1} = \frac{\pi^2 EI^*}{(K_1 L)^2} \qquad (Spec.\ Eq.\ A\text{-}8\text{-}5)$$

$$= \frac{\pi^2 (29{,}000\ \text{ksi})(716\ \text{in.}^4)}{[1.0(216\ \text{in.})]^2}$$

$$= 4{,}390\ \text{kips}$$

Where there is no transverse loading on the brace, C_m is determined from AISC *Specification* Equation A-8-4. For both LRFD and ASD:

$$C_m = 0.6 - 0.4(M_1/M_2) \qquad (Spec.\ Eq.\ A\text{-}8\text{-}4)$$

$$= 0.6 - 0.4(0)$$

$$= 0.6$$

Therefore:

LRFD	ASD
$\alpha = 1.00$	$\alpha = 1.60$
$B_1 = \dfrac{0.6}{1 - \left[\dfrac{1.00(478\ \text{kips})}{4{,}390\ \text{kips}}\right]} \geq 1.0$	$B_1 = \dfrac{0.6}{1 - \left[\dfrac{1.60(333\ \text{kips})}{4{,}390\ \text{kips}}\right]} \geq 1.0$
$= 0.673$	$= 0.683$

Because $B_1 < 1$, use $B_1 = 1.0$.

Since $B_1 = B_2 = 1.0$, the required flexural strength calculated previously need not be amplified to account for P-δ or P-Δ effects.

Combined Loading
Using AISC *Manual* Table 6-1 for combined loading with $KL_y = L_{bx} = 18.0$ ft:

LRFD	ASD
$p = 1.09 \times 10^{-3}$ kips^{-1}	$p = 1.63 \times 10^{-3}$ kips^{-1}
$b_x = 1.72 \times 10^{-3}$ (kip-ft)$^{-1}$	$b_x = 2.59 \times 10^{-3}$ (kip-ft)$^{-1}$

LRFD	ASD
$\dfrac{P_r}{P_c} = pP_r$	$\dfrac{P_r}{P_c} = pP_r$
$= 1.09 \times 10^{-3} (478 \text{ kips})$	$= 1.63 \times 10^{-3} (333 \text{ kips})$
$= 0.521$	$= 0.543$

Because $\dfrac{P_r}{P_c} \geq 0.2$, AISC *Specification* Equation H1-1a applies:

$$\frac{P_r}{P_c} + \frac{8}{9}\left(\frac{M_{rx}}{M_{cx}} + \frac{M_{ry}}{M_{cy}}\right) < 1.0 \qquad \text{(Spec. Eq. H1-1a)}$$

Or, in terms of AISC *Manual* Equation 6-1:

LRFD	ASD
$pP_r + b_x M_{rx} + b_y M_{ry}$	$pP_r + b_x M_{rx} + b_y M_{ry}$
$0.521 + 1.72 \times 10^{-3} \text{(kip-ft)}^{-1} (189 \text{ kip-ft})$	$0.543 + 2.59 \times 10^{-3} \text{(kip-ft)}^{-1} (132 \text{ kip-ft})$
$+ 0$	$+ 0$
$= 0.846 \leq 1.0$ **o.k.**	$= 0.885 \leq 1.0$ **o.k.**

Available Shear Strength
From AISC *Manual* Table 3-6:

LRFD	ASD
$\phi_v V_n = 258$ kips > 10.4 kips **o.k.**	$\dfrac{V_n}{\Omega_v} = 172$ kips > 7.30 kips **o.k.**

The W10×112 is adequate to resist the loads given for Brace BR-1.

Example 5.4.5. EBF Column Design

Given:
Refer to Column CL-1 in Figure 5-59. Select an ASTM A992 wide-flange section to resist the following loading between the base and second level. The applicable building code specifies the use of ASCE/SEI 7 for calculation of loads. From a first-order analysis:

$P_D = 151$ kips $P_L = 46.0$ kips $P_{Q_E} = 172$ kips
$M_{Dx} = 15.0$ kip-ft $M_{Lx} = 9.00$ kip-ft $M_{E_{mhx}} = 0$ kip-ft
$M_{Dy} = 10.0$ kip-ft $M_{Ly} = 6.00$ kip-ft $M_{E_{mhy}} = 0$ kip-ft

Relevant seismic parameters are given in the EBF Design Example Plan and Elevation section.

Assume that the ends of the column are pinned and braced against translation for both the x-x and y-y axes and that the beam at the third level and brace between the second and third levels are as designed in Examples 5.4.2, 5.4.3 and 5.4.4.

Solution:
From AISC *Manual* Table 2-4 and AISC *Seismic Provisions* Table A3.1, the material properties are as follows:

ASTM A992
$F_y = 50$ ksi
$F_u = 65$ ksi
$R_y = 1.1$

Required Strength
Using the load combinations in ASCE/SEI 7 that include the amplified seismic load, with $\Omega_o Q_E = E_{mh}$, it was determined that the governing load combination for the column in compression is:

LRFD	ASD
LRFD Load Combination 5 from ASCE/SEI 7 Section 12.4.3.2 (including the 0.5 factor on *L* permitted in Section 12.4.3.2)	ASD Load Combination 5 from ASCE/SEI 7 Section 12.4.3.2
$(1.2 + 0.2S_{DS})D + E_{mh} + 0.5L + 0.2S$	$(1.0 + 0.14S_{DS})D + H + F + 0.7E_{mh}$

And the governing load combination for the column in tension is:

LRFD	ASD
LRFD Load Combination 7 from ASCE/SEI 7 Section 12.4.3.2 $(0.9 - 0.2S_{DS})D + E_{mh} + 1.6H$	ASD Load Combination 8 from ASCE/SEI 7 Section 12.4.3.2 $(0.6 - 0.14S_{DS})D + 0.7E_{mh} + H$

AISC *Seismic Provisions* Section F3.3 requires the column to have the strength to resist the forces generated by the sum of the adjusted link shear strengths of the links above the level of the column top in addition to the factored gravity forces. From Example 5.4.2, the nominal shear strength of the link at the third level is 205 kips. By calculations not shown here, it was determined that the sum of the nominal shear strengths of the links at the fourth level and the roof is 318 kips. There is also a small axial load due to the shear from the beam outside of the link at level 2. It is neglected in the following calculation due to its negligible effect on the result. Therefore, accounting for the exception allowed in AISC *Seismic Provisions* Section F3.3(1)(b), the sum of the adjusted link yield strengths of the links at the third level, fourth level and roof is:

$$0.88(1.25)R_y\Sigma V_n = 0.88(1.25)(1.1)(318 \text{ kips} + 205 \text{ kips})$$
$$= 633 \text{ kips}$$

Using the governing load combination for the column in compression, the required axial compressive strength of the column is:

LRFD	ASD
$P_u = (1.2 + 0.2S_{DS})P_D + P_{E_{mh}} + 0.5P_L$ $+ 0.2P_S$ $= [1.2 + 0.2(1.0)](151 \text{ kips})$ $+ 633 \text{ kips} + 0.5(46.0 \text{ kips})$ $+ 0.2(0 \text{ kips})$ $= 867 \text{ kips}$	$P_a = (1.0 + 0.14S_{DS})P_D + P_H + P_F$ $+ 0.7P_{E_{mh}}$ $= [1.0 + 0.14(1.0)](151 \text{ kips})$ $+ 0 \text{ kips} + 0 \text{ kips}$ $+ 0.7(633 \text{ kips})$ $= 615 \text{ kips}$

The required flexural strength of the column simultaneous with the axial compression is:

5.4 ECCENTRICALLY BRACED FRAMES

LRFD	ASD
$M_{ux} = (1.2 + 0.2S_{DS})M_D + M_{E_{mhx}}$ $\quad + 0.5M_{Lx} + 0.2M_S$ $\quad = [1.2 + 0.2(1.0)](15.0 \text{ kip-ft})$ $\quad + 0 \text{ kip-ft} + 0.5(9.00 \text{ kip-ft})$ $\quad + 0.2(0 \text{ kip-ft})$ $\quad = 25.5 \text{ kip-ft}$	$M_{ax} = (1.0 + 0.14S_{DS})M_D + M_H + M_F$ $\quad + 0.7M_{E_{mhx}}$ $\quad = [1.0 + 0.14(1.0)](15.0 \text{ kip-ft})$ $\quad + 0 \text{ kip-ft} + 0 \text{ kip-ft}$ $\quad + 0.7(0 \text{ kip-ft})$ $\quad = 17.1 \text{ kip-ft}$
$M_{uy} = (1.2 + 0.2S_{DS})M_D + M_{E_{mhy}}$ $\quad + 0.5M_{Ly} + 0.2M_S$ $\quad = [1.2 + 0.2(1.0)](10.0 \text{ kip-ft})$ $\quad + 0 \text{ kip-ft} + 0.5(6.0 \text{ kip-ft})$ $\quad + 0.2(0 \text{ kip-ft})$ $\quad = 17.0 \text{ kip-ft}$	$M_{ay} = (1.0 + 0.14S_{DS})M_D + M_H + M_F$ $\quad + 0.7M_{E_{mhy}}$ $\quad = [1.0 + 0.14(1.0)](10.0 \text{ kip-ft})$ $\quad + 0 \text{ kip-ft} + 0 \text{ kip-ft}$ $\quad + 0.7(0 \text{ kip-ft})$ $\quad = 11.4 \text{ kip-ft}$

Using the governing load combination for the column in tension, the required axial tensile strength of the column is:

LRFD	ASD
$P_u = (0.9 - 0.2S_{DS})P_D + P_{E_{mh}} + 1.6P_H$ $\quad = [0.9 - 0.2(1.0)](151 \text{ kips})$ $\quad + (-633 \text{ kips}) + 1.6(0 \text{ kips})$ $\quad = -527 \text{ kips}$	$P_a = (0.6 - 0.14S_{DS})P_D + 0.7P_{E_{mh}} + P_H$ $\quad = [0.6 - 0.14(1.0)](151 \text{ kips})$ $\quad + 0.7(-633 \text{ kips}) + 0 \text{ kips}$ $\quad = -374 \text{ kips}$

The required flexural strength of the column simultaneous with the axial tension is:

LRFD	ASD
$M_{ux} = (0.9 - 0.2S_{DS})M_{Dx} + M_{E_{mhx}}$ $\quad + 1.6M_H$ $\quad = [0.9 - 0.2(1.0)](15.0 \text{ kip-ft})$ $\quad + 0 \text{ kip-ft} + 1.6(0 \text{ kip-ft})$ $\quad = 10.5 \text{ kip-ft}$	$M_{ax} = (0.6 - 0.14S_{DS})M_{Dx} + 0.7M_{E_{mhx}}$ $\quad + M_H$ $\quad = [0.6 - 0.14(1.0)](15.0 \text{ kip-ft})$ $\quad + 0.7(0 \text{ kip-ft}) + 0 \text{ kip-ft}$ $\quad = 6.90 \text{ kip-ft}$

LRFD	ASD
$M_{uy} = (0.9 - 0.2S_{DS})M_{Dy} + M_{E_{mhy}}$ $\quad + 1.6M_H$ $\quad = [0.9 - 0.2(1.0)](10.0 \text{ kip-ft})$ $\quad + 0 \text{ kip-ft} + 1.6(0 \text{ kip-ft})$ $\quad = 7.00 \text{ kip-ft}$	$M_{ay} = (0.6 - 0.14S_{DS})M_{Dy} + 0.7M_{E_{mhy}}$ $\quad + M_H$ $\quad = [0.6 - 0.14(1.0)](10.0 \text{ kip-ft})$ $\quad + 0.7(0 \text{ kip-ft}) + 0 \text{ kip-ft}$ $\quad = 4.60 \text{ kip-ft}$

The load combination that will govern the design of the column is that for compression. The resulting required strengths are:

LRFD	ASD
$P_u = 867$ kips	$P_a = 615$ kips
$M_{ux} = 25.5$ kip-ft	$M_{ax} = 17.1$ kip-ft
$M_{uy} = 17.0$ kip-ft	$M_{ay} = 11.4$ kip-ft

Try a **W12×96**.

From AISC *Manual* Table 1-1, the geometric properties are as follows:

$A = 28.2 \text{ in.}^2 \quad d = 12.7 \text{ in.} \quad t_w = 0.550 \text{ in.} \quad t_f = 0.900 \text{ in.}$
$I_x = 833 \text{ in.}^4 \quad I_y = 270 \text{ in.}^4 \quad b_f/2t_f = 6.76 \quad h/t_w = 17.7$

Width-to-Thickness Limitations

According to AISC *Seismic Provisions* Section F3.5a, the column must comply with the requirements of Section D1.1 for highly ductile members. From Table 1-3 of this Manual, these requirements are satisfied for a **W12×96** column (both flanges and web).

Consider second-order effects

From AISC *Specification* Appendix 8, the required flexural and axial strength including second-order effects are determined as follows:

$$M_r = B_1 M_{nt} + B_2 M_{lt} \quad \text{(Spec. Eq. A-8-1)}$$

$$P_r = P_{nt} + B_2 P_{lt} \quad \text{(Spec. Eq. A-8-2)}$$

Since the lateral load effect is based on the adjusted link shear strength, $B_2 = 1.0$. P-Δ effects do not increase the forces corresponding to the fully yielded, strain-hardened link; instead they may be thought of as contributing to the system reaching that state.

Because $B_2 = 1.0$, the required compressive and flexural strengths will not be amplified to account for P-Δ effects. Determine B_1 as follows from AISC *Specification* Appendix 8. The effective length method is used for stability design.

5.4 ECCENTRICALLY BRACED FRAMES

$$P_{elx} = \frac{\pi^2 EI_x^*}{(K_1 L)^2} \qquad \text{(Spec. Eq. A-8-5)}$$

$$= \frac{\pi^2 (29{,}000 \text{ ksi})(833 \text{ in.}^4)}{[1.0(14.0 \text{ ft})(12 \text{ in./ft})]^2}$$

$$= 8{,}450 \text{ kips}$$

$$P_{ely} = P_{elx} \frac{I_y}{I_x}$$

$$= 8{,}450 \text{ kips} \left(\frac{270 \text{ in.}^4}{833 \text{ in.}^4} \right)$$

$$= 2{,}740 \text{ kips}$$

The columns are assumed to be pinned at the base, so M_1 in AISC *Specification* Equation A-8-4 is zero. Because the column is not subject to transverse (perpendicular to the axis of the member) loading, C_m is determined for both LRFD and ASD as follows:

$$C_m = 0.6 - 0.4(M_1/M_2) \qquad \text{(Spec. Eq. A-8-4)}$$
$$= 0.6 - 0.4(0)$$
$$= 0.6$$

$$C_{mx} = C_{my} = 0.6.$$

$$B_1 = \frac{C_m}{1 - \alpha P_r / P_{e1}} \geq 1 \qquad \text{(Spec. Eq. A-8-3)}$$

Therefore:

LRFD	ASD
$\alpha = 1.00$	$\alpha = 1.60$
$B_{1x} = \dfrac{0.6}{1 - \dfrac{1.00(867 \text{ kips})}{8{,}450 \text{ kips}}} \geq 1.0$	$B_{1x} = \dfrac{0.6}{1 - \dfrac{1.60(615 \text{ kips})}{8{,}450 \text{ kips}}} \geq 1.0$
$= 0.669$	$= 0.679$
$B_{1y} = \dfrac{0.6}{1 - \dfrac{1.00(867 \text{ kips})}{2{,}740 \text{ kips}}} \geq 1.0$	$B_{1y} = \dfrac{0.6}{1 - \dfrac{1.60(615 \text{ kips})}{2{,}740 \text{ kips}}} \geq 1.0$
$= 0.878$	$= 0.936$

Because the calculated B_{1x} and B_{1y} are less than 1.0, $B_{1x} = B_{1y} = 1.0$, and there is no need to amplify the required flexural strengths.

Combined Loading

Using AISC *Manual* Table 6-1 for combined loading with $L_{by} = L_{bx} = 14$ ft:

LRFD	ASD
$p = 0.978 \times 10^{-3}$ kips^{-1}	$p = 1.47 \times 10^{-3}$ kips^{-1}
$b_x = 1.67 \times 10^{-3}$ (kip-ft)$^{-1}$	$b_x = 2.50 \times 10^{-3}$ (kip-ft)$^{-1}$
$b_y = 3.51 \times 10^{-3}$ (kip-ft)$^{-1}$	$b_y = 5.28 \times 10^{-3}$ (kip-ft)$^{-1}$

LRFD	ASD
$\dfrac{P_r}{P_c} = pP_r$	$\dfrac{P_r}{P_c} = pP_r$
$= 0.978 \times 10^{-3} (867 \text{ kips})$	$= 1.47 \times 10^{-3} (615 \text{ kips})$
$= 0.848$	$= 0.904$

Because $\dfrac{P_r}{P_c} \geq 0.2$, AISC *Specification* Equation H1-1a applies:

$$\frac{P_r}{P_c} + \frac{8}{9}\left(\frac{M_{rx}}{M_{cx}} + \frac{M_{ry}}{M_{cy}}\right) < 1.0 \qquad \text{(Spec. Eq. H1-1a)}$$

Or, in terms of AISC *Manual* Equation 6-1:

LRFD	ASD
$pP_r + b_x M_{rx} + b_y M_{ry} \leq 1.0$	$pP_r + b_x M_{rx} + b_y M_{ry} \leq 1.0$
$0.848 + 1.67 \times 10^{-3}$ (kip-ft)$^{-1}$ (25.5 kip-ft)	$0.904 + 2.50 \times 10^{-3}$ (kip-ft)$^{-1}$ (17.1 kip-ft)
$+ 3.51 \times 10^{-3}$ (kip-ft)$^{-1}$ (17.0 kip-ft)	$+ 5.28 \times 10^{-3}$ (kip-ft)$^{-1}$ (11.4 kip-ft)
$= 0.950 < 1.0$ **o.k.**	$= 1.00 \leq 1.0$ **o.k.**

The W12×96 is adequate to resist the loads given for Column CL-1 between the base and second level.

Example 5.4.6. EBF Brace-to-Link Connection Design

Given:

Refer to Joint JT-1 in Figure 5-59. Design the connection between Brace BR-1 and Beam BM-1 assuming the brace is oriented with the web in the plane of the frame. Use ASTM A36 material for all plate material and 70-ksi electrodes for all welds. Assume the link, beam outside of the link, and brace are as designed in Examples 5.4.2, 5.4.3 and 5.4.4, respectively.

5.4 ECCENTRICALLY BRACED FRAMES

Solution:

From AISC *Manual* Table 2-5, the material properties are as follows:

ASTM A36
$F_y = 36$ ksi
$F_u = 58$ ksi

From AISC *Manual* Table 1-1 the geometric properties are as follows:

Brace
W10×112
$A = 32.9$ in.^{2} $\quad d = 11.4$ in. $\quad b_f = 10.4$ in. $\quad t_f = 1.25$ in.
$I_x = 716$ in.^{4} $\quad k_{des} = 1.75$ in.

Beam
W16×77
$d = 16.5$ in. $\quad t_w = 0.455$ in. $\quad b_f = 10.3$ in. $\quad t_f = 0.760$ in.
$k_{des} = 1.16$ in.

Determine the brace connection forces

According to AISC *Seismic Provisions* Section F3.3, brace connections must consider the forces generated by the adjusted link shear strength. From Example 5.4.4 for the design of the brace, the required strengths of the brace based on the adjusted link shear strength are:

LRFD	ASD
$P_u = 478$ kips	$P_a = 333$ kips
$V_u = 10.4$ kips	$V_a = 7.30$ kips
$M_u = 189$ kip-ft	$M_a = 132$ kip-ft

Determine the brace flange force

Assuming the axial force is resisted entirely by the flanges, the force in each flange due to axial load is:

LRFD	ASD
$P_{fa} = \dfrac{P_u}{2}$ $= \dfrac{478 \text{ kips}}{2}$ $= 239$ kips	$P_{fa} = \dfrac{P_a}{2}$ $= \dfrac{333 \text{ kips}}{2}$ $= 167$ kips

Assuming the entire moment will be taken by the flanges, the force in each flange due to the moment is:

LRFD	ASD
$P_{ff} = \dfrac{M_u}{d - t_f}$ $= \dfrac{189 \text{ kip-ft}(12 \text{ in./ft})}{11.4 \text{ in.} - 1.25 \text{ in.}}$ $= 223 \text{ kips}$	$P_{ff} = \dfrac{M_a}{d - t_f}$ $= \dfrac{132 \text{ kip-ft}(12 \text{ in./ft})}{11.4 \text{ in.} - 1.25 \text{ in.}}$ $= 156 \text{ kips}$

The maximum resultant force in each flange is:

LRFD	ASD
$P_{uf} = P_{fa} + P_{ff}$ $= 239 \text{ kips} + 223 \text{ kips}$ $= 462 \text{ kips}$	$P_{af} = P_{fa} + P_{ff}$ $= 167 \text{ kips} + 156 \text{ kips}$ $= 323 \text{ kips}$

Determine the brace web force
It will be assumed that the entire shear force will be taken by the web.

LRFD	ASD
$V_w = V_u$ $= 10.4 \text{ kips}$	$V_w = V_a$ $= 7.30 \text{ kips}$

Brace Flange Connection

From Example 5.4.2, because the brace was designed to resist a portion of the link end moment, AISC *Seismic Provisions* Section F3.6c requires that this connection be designed as fully restrained. Use a fully welded connection.

Try a complete-joint-penetration (CJP) groove weld to connect the brace flanges to the beam flange.

From AISC *Specification* Table J2.5, the strength of the CJP groove weld in tension is based on the strength of the base material. The tensile rupture strength of each brace flange, with $A_e = A_g$, is:

$R_n = F_u A_e$ (Spec. Eq. J4-2)
$\quad = F_u b_f t_f$
$\quad = 65 \text{ ksi}(10.4 \text{ in.})(1.25 \text{ in.})$
$\quad = 845 \text{ kips}$

5.4 ECCENTRICALLY BRACED FRAMES

LRFD	ASD
$\phi R_n = 0.75(845 \text{ kips})$ $= 634 \text{ kips} > P_{uf} = 462 \text{ kips}$ **o.k.**	$\dfrac{R_n}{\Omega} = \dfrac{845 \text{ kips}}{2.00}$ $= 423 \text{ kips} > P_{af} = 323 \text{ kips}$ **o.k.**

Check concentrated forces at brace flange connection

The vertical component of the flange force is:

LRFD	ASD
$V_{uf} = P_{uf}\left(\dfrac{12.5 \text{ ft}}{18.0 \text{ ft}}\right)$ $= 462 \text{ kips}\left(\dfrac{12.5 \text{ ft}}{18.0 \text{ ft}}\right)$ $= 321 \text{ kips}$	$V_{af} = P_{af}\left(\dfrac{12.5 \text{ ft}}{18.0 \text{ ft}}\right)$ $= 323 \text{ kips}\left(\dfrac{12.5 \text{ ft}}{18.0 \text{ ft}}\right)$ $= 224 \text{ kips}$

Because the concentrated force is applied at a distance greater than d, the beam depth, from the beam end, the beam web local yielding strength at the brace flange connection is:

$R_n = F_{yw}t_w(5k_{des} + l_b)$ *(Spec. Eq. J10-2)*

$= (50 \text{ ksi})(0.455 \text{ in.})[5(1.16 \text{ in.}) + 1.25 \text{ in.}]$

$= 160 \text{ kips}$

LRFD	ASD
$\phi R_n = 1.00(160 \text{ kips})$ $= 160 \text{ kips} < V_{uf} = 321 \text{ kips}$ **n.g.**	$\dfrac{R_n}{\Omega} = \dfrac{160 \text{ kips}}{1.50}$ $= 107 \text{ kips} < V_{af} = 224 \text{ kips}$ **n.g.**

Because the concentrated force is appplied at a distance greater than or equal to $d/2$ from the beam end, the beam web local crippling strength at the brace flange connection is:

$R_n = 0.80 t_w^2 \left[1 + 3\left(\dfrac{l_b}{d}\right)\left(\dfrac{t_w}{t_f}\right)^{1.5}\right]\sqrt{\dfrac{EF_{yw}t_f}{t_w}}$ *(Spec. Eq. J10-4)*

$= (0.80)(0.455 \text{ in.})^2 \left[1 + 3\left(\dfrac{1.25 \text{ in.}}{16.5 \text{ in.}}\right)\left(\dfrac{0.455 \text{ in.}}{0.760 \text{ in.}}\right)^{1.5}\right]$

$\quad \times \sqrt{\dfrac{29{,}000 \text{ ksi}(50 \text{ ksi})(0.760 \text{ in.})}{0.455 \text{ in.}}}$

$= 285 \text{ kips}$

LRFD	ASD
$\phi R_n = 0.75(285 \text{ kips})$ $= 214 \text{ kips} < V_{uf} = 321 \text{ kips}$ **n.g.**	$\dfrac{R_n}{\Omega} = \dfrac{285 \text{ kips}}{2.00}$ $= 143 \text{ kips} < V_{af} = 224 \text{ kips}$ **n.g.**

The flange local bending strength is:

$$R_n = 6.25 F_{yf} t_f^2 \hspace{4em} (Spec. \text{ Eq. J10-1})$$
$$= 6.25(50 \text{ ksi})(0.760 \text{ in.})^2$$
$$= 181 \text{ kips}$$

LRFD	ASD
$\phi R_n = 0.90(181 \text{ kips})$ $= 163 \text{ kips} < V_{uf} = 321 \text{ kips}$ **n.g.**	$\dfrac{R_n}{\Omega} = \dfrac{181 \text{ kips}}{1.67}$ $= 108 \text{ kips} < V_{af} = 224 \text{ kips}$ **n.g.**

Beam web stiffeners are required adjacent to the brace flanges as shown in Figure 5-61. The controlling limit state for concentrated loading is beam web local yielding, and the required

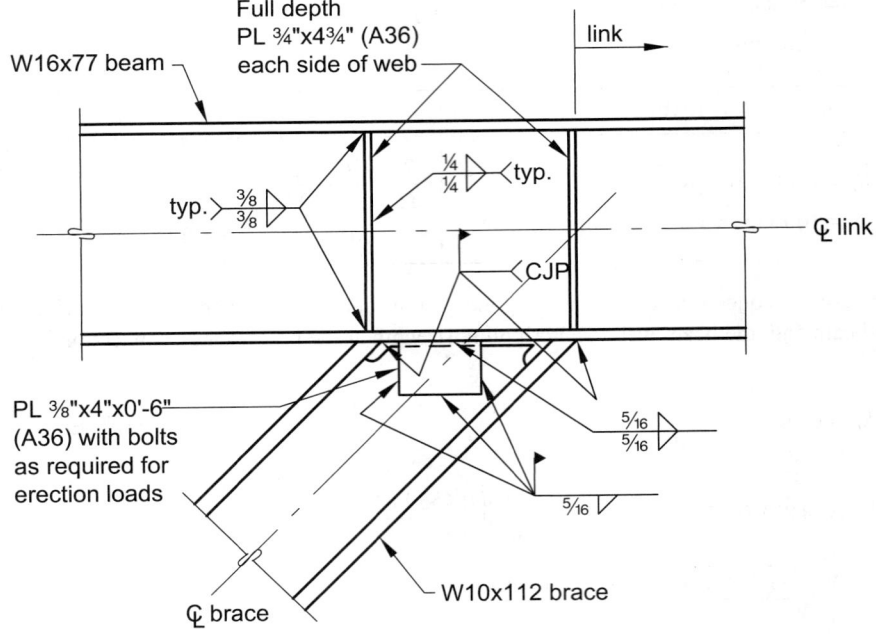

Fig. 5-61. Connection as designed in Example 5.4.6.

5.4 ECCENTRICALLY BRACED FRAMES

strength of the stiffeners is the difference between the vertical component of the flange force, V_{uf} or V_{af}, and the available strength of the beam web due to web local yielding.

Size beam web stiffeners

Using one stiffener on each side of the beam web, the portion of the vertical component of the brace flange force to be resisted by each stiffener is:

LRFD	ASD
$P_s = \dfrac{V_{uf} - \phi R_n}{2}$ $= \dfrac{321 \text{ kips} - 160 \text{ kips}}{2}$ $= 80.5 \text{ kips}$	$P_s = \dfrac{V_{af} - \left(\dfrac{R_n}{\Omega}\right)}{2}$ $= \dfrac{224 \text{ kips} - 107 \text{ kips}}{2}$ $= 58.5 \text{ kips}$

For convenience, use the same stiffener geometry as used in Example 5.4.2 for the link stiffeners. Try a 4¾-in. stiffener width with 1 in. × 2⅜ in. corner clips. From Example 5.4.2, accounting for the corner clips, the length of stiffener in contact with the flange is $w_{st} = 3\frac{3}{4}$ in. and the length of stiffener in contact with the web is $L_{st} = 10.2$ in. The stiffener thickness necessary to develop the required strength, based on the limit state of tensile yielding from AISC *Specification* Equation J4-1, is:

LRFD	ASD
$\phi R_n \geq P_s$ $\phi F_y w_{st} t_{min} \geq P_s$ $t_{min} \geq \dfrac{P_s}{\phi F_y w_{st}}$ $\geq \dfrac{80.5 \text{ kips}}{0.90(36 \text{ ksi})(3\frac{3}{4} \text{ in.})}$ $\geq 0.663 \text{ in.}$	$\dfrac{R_n}{\Omega} \geq P_s$ $\dfrac{F_y w_{st} t_{min}}{\Omega} \geq P_s$ $t_{min} \geq \dfrac{\Omega P_s}{F_y w_{st}}$ $\geq \dfrac{1.67(58.5 \text{ kips})}{36 \text{ ksi}(3\frac{3}{4} \text{ in.})}$ $\geq 0.724 \text{ in.}$

Note that one flange of each brace frames into the beam at the end of the link segment. In Example 5.4.2, the AISC *Seismic Provisions* requirements resulted in a ⅜ in. minimum thickness for the stiffeners at the end of the link.

Use ¾ in. × 4¾ in. full-depth stiffeners on each side of the beam at the locations where a brace flange intersects the beam flange. These will replace the link end stiffeners designed in Example 5.4.2.

Design stiffener welds

Using the increased strength allowed for transversely loaded fillet welds according to AISC *Specification* Equation J2-5, $(1.0 + 0.50 \sin^{1.5} \theta) = (1.0 + 0.50 \sin^{1.5} 90°) = 1.5$, the minimum double-sided fillet weld size required to transfer the required stiffener load from the beam flange to the stiffener is:

LRFD	ASD
$D_{min} = \dfrac{P_s}{2(1.5)(1.392 \text{ kip/in.})w_{st}}$ $= \dfrac{80.5 \text{ kips}}{2(1.5)(1.392 \text{ kip/in.})(3¾ \text{ in.})}$ $= 5.14 \text{ sixteenths}$	$D_{min} = \dfrac{P_s}{2(1.5)(0.928 \text{ kip/in.})w_{st}}$ $= \dfrac{58.5 \text{ kips}}{2(1.5)(0.928 \text{ kip/in.})(3¾ \text{ in.})}$ $= 5.60 \text{ sixteenths}$

The minimum double-sided fillet weld size required to transfer the stiffener force to the web is:

LRFD	ASD
$D_{min} = \dfrac{P_s}{2(1.392 \text{ kip/in.})L_{st}}$ $= \dfrac{80.5 \text{ kips}}{2(1.392 \text{ kip/in.})(10.2 \text{ in.})}$ $= 2.83 \text{ sixteenths}$	$D_{min} = \dfrac{P_s}{2(0.928 \text{ kip/in.})L_{st}}$ $= \dfrac{58.5 \text{ kips}}{2(0.928 \text{ kip/in.})(10.2 \text{ in.})}$ $= 3.09 \text{ sixteenths}$

Note that per AISC *Specification* Table J2.4, the minimum stiffener-to-web weld is ³⁄₁₆ in. based on the thinner part joined, $t_w = 0.455$ in. The minimum stiffener-to-flange weld is ¼ in. based on the ¾-in. stiffener, which is the thinner part joined.

Use double-sided ⅜-in. fillet welds to connect the stiffener to the beam flanges and double-sided ¼-in. fillet welds to connect the stiffener to the beam web.

Design the brace web connection

Use a ⅜ in. × 4 in. × 0 ft 6 in. single-plate connection with ⁵⁄₁₆-in. fillet welds to connect the brace to the beam. This connection will be adequate for the small required shear strength.

The final connection design and geometry is shown in Figure 5-61.

Example 5.4.7. EBF Brace-to-Beam/Column Connection Design

Given:

Refer to Joint JT-2 in Figure 5-59. Design the connection between brace, beam and column. Use ASTM A572 Grade 50 for all plate material and 70-ksi electrodes for all welds. Use ASTM A325-N bolts. Assume that the beam is as designed in Example 5.4.3, the brace size is the same as that determined in Example 5.4.4 and the column is as designed in Example 5.4.5. The applicable building code specifies the use of ASCE/SEI 7 for calculation of loads.

Relevant seismic parameters are given in the EBF Design Example Plan and Elevation section.

The brace will be connected to the beam-to-column joint through a gusset plate. The connection of the brace to the gusset plate will consist of WT sections with flanges bolted to each side of the brace web and gusset plate. The gusset plate and beam will be connected to the column using a bolted end plate. Figure 5-62 is a schematic drawing showing the relevant forces on the connection. This is not a realistic drawing of the connection and shows only minimal detail.

Solution:

From AISC *Manual* Tables 2-4 and 2-5, the material properties are as follows:

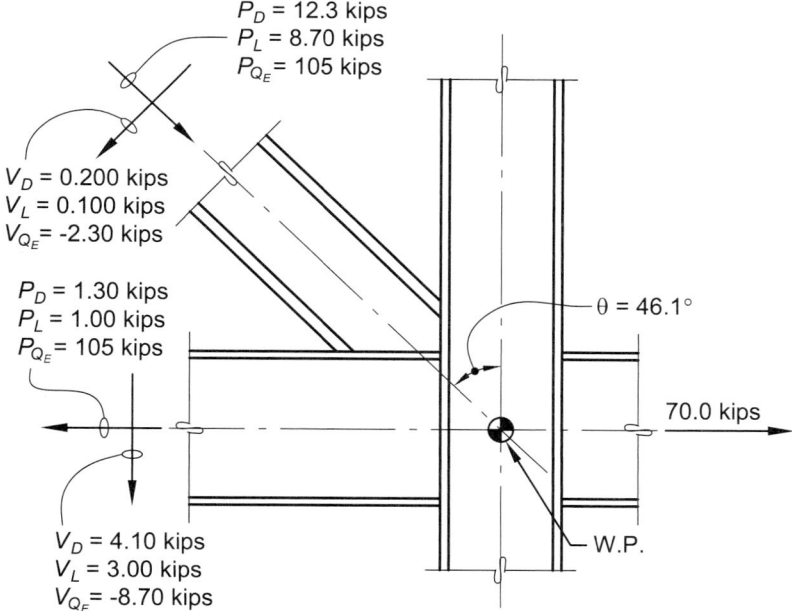

Fig. 5-62. Connection forces for Example 5.4.7.

ASTM A572 Grade 50
$F_y = 50$ ksi
$F_u = 65$ ksi

ASTM A992
$F_y = 50$ ksi
$F_u = 65$ ksi

From AISC *Manual* Table 1-1, the geometric properties are as follows:

Beam outside of the link
W16×77
$A = 22.6$ in.²	$d = 16.5$ in.	$t_w = 0.455$ in.	$b_f = 10.3$ in.
$t_f = 0.760$ in.	$k_{des} = 1.16$ in.	$T = 13¼$ in.	

Brace
W10×112
$A = 32.9$ in.²	$d = 11.4$ in.	$b_f = 10.4$ in.	$t_w = 0.755$ in.
$t_f = 1.25$ in.	$T = 7½$ in.		

Column
W12×96
$A = 28.2$ in.²	$d = 12.7$ in.	$t_w = 0.550$ in.	$t_f = 0.900$ in.
$k_{des} = 1.50$ in.			

In order to envelope the design, two conditions should be examined. Forces from both conditions are shown in Figure 5-62.

Condition 1: The brace force required to develop the adjusted link yield strength at the fourth level must be transferred through the connection and into the column and beam outside of the link. The additional collector force required to develop the adjusted link yield strength at the third level must be transferred from the collector element through the beam-to-column connection. This collector force need not exceed that determined using the amplified seismic load. The shear in the beam outside of the link must be transferred into the column.

Condition 2: The amplified collector force must be transferred into the beam outside of the link. The additional brace force required to develop the adjusted link yield strength at the third level must be transferred through the connection and into the column and beam outside of the link. The brace force need not exceed that required to develop the adjusted link yield strength at the fourth level. The shear in the beam outside of the link must be transferred into the column.

Required Strength
The governing load combination, with $\Omega_o Q_E = E_{mh}$, is:

5.4 ECCENTRICALLY BRACED FRAMES

LRFD	ASD
LRFD Load Combination 5 from ASCE/SEI 7 Section 12.4.3.2 (including the 0.5 factor on L permitted in Section 12.4.3.2) $(1.2 + 0.2S_{DS})D + E_{mh} + 0.5L + 0.2S$	ASD Load Combination 5 from ASCE/SEI 7 Section 12.4.3.2 $(1.0 + 0.14S_{DS})D + H + F + 0.7E_{mh}$

The governing seismic load case causes compression in the brace. Assume the connection forces are as shown in Figure 5-62.

Determine the load from the beam outside of the link
(considered in both Conditions 1 and 2)

The adjusted link yield strength used in the design of the beam outside of the link was allowed to be reduced by 0.88 according to Exception (1)(a) in AISC *Seismic Provisions* Section F3.3. This reduction is not allowed for connections. From Example 5.4.4, the overstrength factor for the link at the third level is 3.36. The factored forces at the connection due to the beam outside of the link are:

LRFD	ASD
$P_u = (1.2 + 0.2S_{DS})P_D + P_{E_{mh}} + 0.5P_L$ $\quad + 0.2P_S$ $= [1.2 + 0.2(1.0)](1.30 \text{ kips})$ $\quad + 3.36(105 \text{ kips}) + 0.5(1.00 \text{ kips})$ $\quad + 0.2(0 \text{ kips})$ $= 355 \text{ kips}$	$P_a = (1.0 + 0.14S_{DS})P_D + P_H + P_F$ $\quad + 0.7P_{E_{mh}}$ $= [1.0 + 0.14(1.0)](1.30 \text{ kips})$ $\quad + 0 \text{ kips} + 0 \text{ kips}$ $\quad + 0.7(3.36)(105 \text{ kips})$ $= 248 \text{ kips}$

LRFD	ASD
$V_u = (1.2 + 0.2S_{DS})V_D + V_{E_{mh}} + 0.5V_L$ $\quad + 0.2V_S$ $= [1.2 + 0.2(1.0)](4.10 \text{ kips})$ $\quad + 3.36(8.70 \text{ kips}) + 0.5(3.00 \text{ kips})$ $\quad + 0.2(0 \text{ kips})$ $= 36.5 \text{ kips}$	$V_a = (1.0 + 0.14S_{DS})V_D + V_H + V_F$ $\quad + 0.7V_{E_{mh}}$ $= [1.0 + 0.14(1.0)](4.10 \text{ kips})$ $\quad + 0 \text{ kips} + 0 \text{ kips}$ $\quad + 0.7(3.36)(8.70 \text{ kips})$ $= 25.1 \text{ kips}$

Determine the load from the brace (Condition 1)

AISC *Seismic Provisions* Section F3.3 requires that the brace connections have sufficient strength to develop the adjusted link yield strength. Use the overstrength factor method described in Example 5.4.3 and assume that the overstrength factor is 3.36, the same as that used in Example 5.4.4 for the design of the brace. The required strengths of the connection from the brace, based on the forces shown in Figure 5-62, are:

LRFD	ASD
$P_u = (1.2 + 0.2S_{DS})P_D + P_{E_{mh}} + 0.5P_L$ $\quad + 0.2P_S$ $= [1.2 + 0.2(1.0)](12.3 \text{ kips})$ $\quad + 3.36(105 \text{ kips}) + 0.5(8.70 \text{ kips})$ $\quad + 0.2(0 \text{ kips})$ $= 374 \text{ kips}$	$P_a = (1.0 + 0.14S_{DS})P_D + P_H + P_F$ $\quad + 0.7P_{E_{mh}}$ $= [1.0 + 0.14(1.0)](12.3 \text{ kips})$ $\quad + 0 \text{ kips} + 0 \text{ kips}$ $\quad + 0.7(3.36)(105 \text{ kips})$ $= 261 \text{ kips}$

LRFD	ASD
$V_u = (1.2 + 0.2S_{DS})V_D + V_{E_{mh}} + 0.5V_L$ $\quad + 0.2V_S$ $= [1.2 + 0.2(1.0)](0.200 \text{ kips})$ $\quad + 3.36(2.30 \text{ kips}) + 0.5(0.100 \text{ kips})$ $\quad + 0.2(0 \text{ kips})$ $= 8.06 \text{ kips}$	$V_a = (1.0 + 0.14S_{DS})V_D + V_H + V_F$ $\quad + 0.7V_{E_{mh}}$ $= [1.0 + 0.14(1.0)](0.200 \text{ kips})$ $\quad + 0 \text{ kips} + 0 \text{ kips}$ $\quad + 0.7(3.36)(2.30 \text{ kips})$ $= 5.64 \text{ kips}$

The resulting collector force in Condition 1 is what is needed to achieve horizontal equilibrium. Ignoring the small contribution to horizontal forces from the brace shear, the collector force in Condition 1 is:

LRFD	ASD
$374 \text{ kips}\left(\dfrac{13 \text{ ft}}{\sqrt{(13 \text{ ft})^2 + (12.5 \text{ ft})^2}}\right)$ $\quad - 355 \text{ kips} + P_{drag} = 0$ Therefore: $P_{drag} = 355 \text{ kips} - 374 \text{ kips}$ $\quad \times \left[\dfrac{13 \text{ ft}}{\sqrt{(13 \text{ ft})^2 + (12.5 \text{ ft})^2}}\right]$ $= 85.4 \text{ kips}$	$261 \text{ kips}\left(\dfrac{13 \text{ ft}}{\sqrt{(13 \text{ ft})^2 + (12.5 \text{ ft})^2}}\right)$ $\quad - 248 \text{ kips} + P_{drag} = 0$ Therefore: $P_{drag} = 248 \text{ kips} - 261 \text{ kips}$ $\quad \times \left[\dfrac{13 \text{ ft}}{\sqrt{(13 \text{ ft})^2 + (12.5 \text{ ft})^2}}\right]$ $= 59.9 \text{ kips}$

Determine the load from the brace (Condition 2)

Determine the collector force based on the amplified seismic load. The amplified collector force is:

5.4 ECCENTRICALLY BRACED FRAMES

LRFD	ASD
$\Omega_o P_{QE} = 2.0(70.0 \text{ kips})$ $= 140 \text{ kips}$	$0.7\Omega_o P_{QE} = 0.7(2.0)(70.0 \text{ kips})$ $= 98.0 \text{ kips}$

To achieve equilibrium at the joint, the force from the brace must be adjusted accordingly. The net horizontal force due to the collector force and the axial force in the beam outside of the link is:

LRFD	ASD
$F_h = 140 \text{ kips} - 355 \text{ kips}$ $= -215 \text{ kips}$	$F_h = 98.0 - 248 \text{ kips}$ $= -150 \text{ kips}$

Thus, the force from the brace to achieve equilibrium is:

LRFD	ASD
$P_u = 215 \text{ kips} \left(\dfrac{\sqrt{(13 \text{ ft})^2 + (12.5 \text{ ft})^2}}{13 \text{ ft}} \right)$ $= 298 \text{ kips}$	$P_a = 150 \text{ kips} \left(\dfrac{\sqrt{(13 \text{ ft})^2 + (12.5 \text{ ft})^2}}{13 \text{ ft}} \right)$ $= 208 \text{ kips}$

Force diagrams for Conditions 1 and 2 are shown in Figure 5-63. For the purposes of this example, these forces will be assumed to be equal but opposite for the condition of the brace in tension. This is a conservative assumption for the connection being designed in this example. However, this may not be a conservative assumption for all connection geometries and loading conditions.

Determine the required strength of the brace-to-gusset connection (Condition 1)

Using the required strength of the brace (Condition 1), the resultant force on the connection is:

LRFD	ASD
$R_u = \sqrt{P_u^2 + V_u^2}$ $= \sqrt{(374 \text{ kips})^2 + (8.06 \text{ kips})^2}$ $= 374 \text{ kips}$	$R_a = \sqrt{P_a^2 + V_a^2}$ $= \sqrt{(261 \text{ kips})^2 + (5.64 \text{ kips})^2}$ $= 261 \text{ kips}$

Because this is greater than $P_u = 298$ kips (LRFD) and $P_a = 208$ kips (ASD) calculated previously for Condition 2, use Condition 1 values.

AMERICAN INSTITUTE OF STEEL CONSTRUCTION

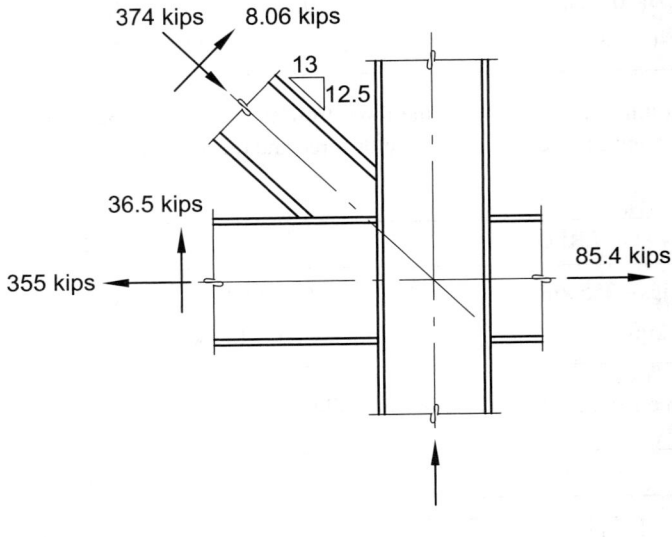

Condition 1

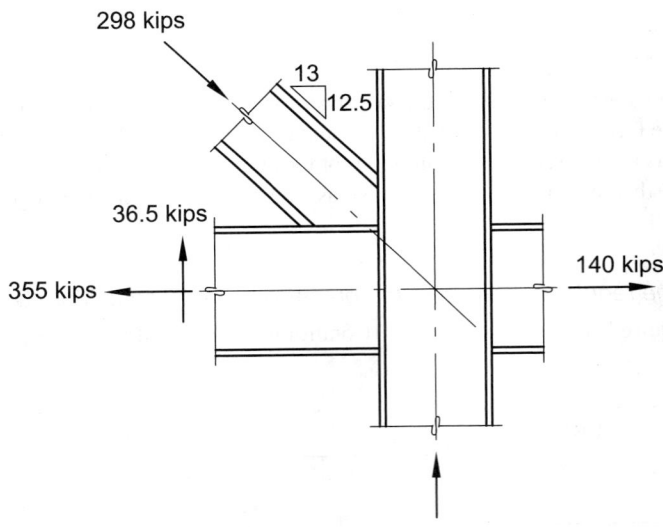

Condition 2

(a) Schematic of LRFD force diagram for Example 5.4.7

Fig. 5-63. Schematic force diagrams for Example 5.4.7.

5.4 ECCENTRICALLY BRACED FRAMES

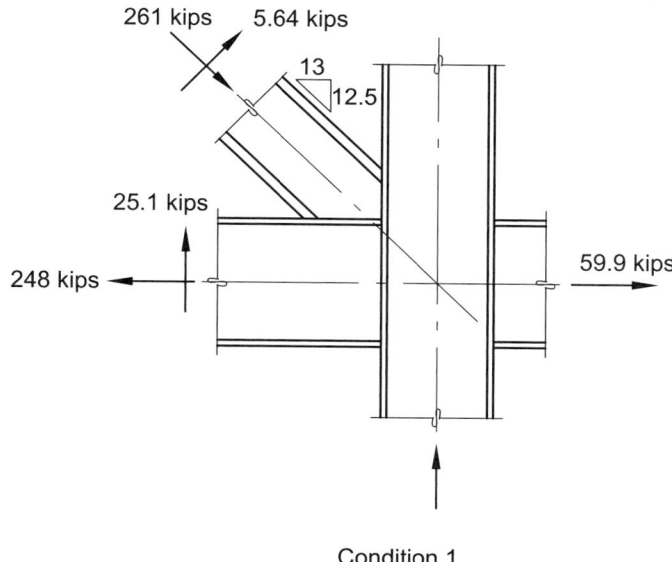

Condition 1

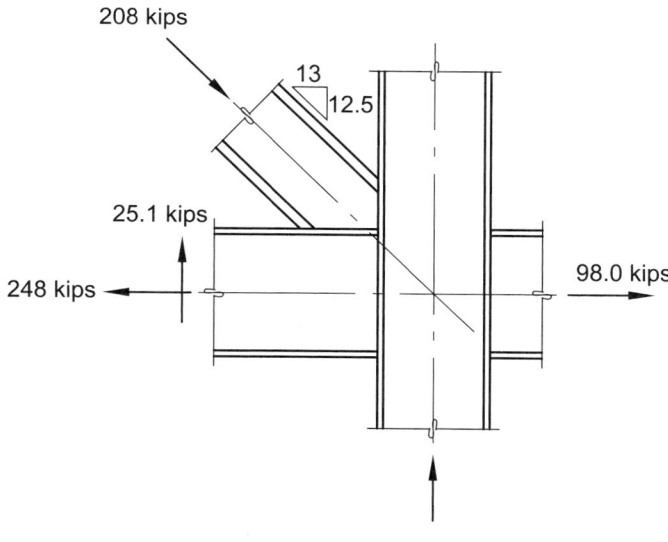

Condition 2

(b) Schematic of ASD force diagram for Example 5.4.7

Fig. 5-63. Schematic force diagrams for Example 5.4.7.(continued)

Connection Design

Determine the required number of bolts

Using AISC *Manual* Table 7-1, the minimum number of 1-in.-diameter ASTM A325-N bolts in double shear required to develop the required strength is:

LRFD	ASD
$n_{min} = \dfrac{R_u}{\phi r_n}$ $= \dfrac{374 \text{ kips}}{63.6 \text{ kips/bolt}}$ $= 5.88 \text{ bolts}$	$n_{min} = \dfrac{R_a}{r_n / \Omega}$ $= \dfrac{261 \text{ kips}}{42.4 \text{ kips/bolt}}$ $= 6.16 \text{ bolts}$

Try eight bolts in standard holes with 3-in. spacing and 2-in. edge distance as shown in Figure 5-64.

Check bearing strength of gusset plate

Try an initial gusset plate thickness of ¾ in. Using AISC *Manual* Table 7-4 for 1-in.-diameter bolts in standard holes and ASTM A572 Grade 50 plate material, the available bearing strength of the plate at each of the interior bolts is:

LRFD	ASD
$\phi r_n = 113 \text{ kip/in.}(\text{¾ in.})$ $= 84.8 \text{ kips}$	$\dfrac{r_n}{\Omega} = 75.6 \text{ kip/in.}(\text{¾ in.})$ $= 56.7 \text{ kips}$

Using AISC *Manual* Table 7-5 for 1-in.-diameter bolts in standard holes with 2-in. edge distance, the available bearing strength of the plate at each of the edge bolts is:

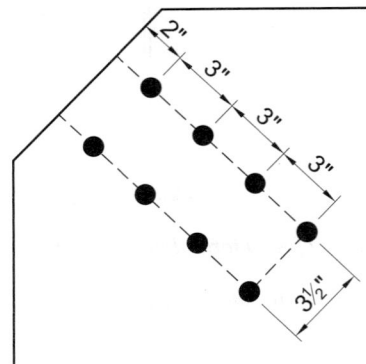

Fig. 5-64. Initial bolt configuration.

5.4 ECCENTRICALLY BRACED FRAMES

LRFD	ASD
$\phi r_n = 85.9$ kips/in.(¾ in.) $= 64.4$ kips	$\dfrac{r_n}{\Omega} = 57.3$ kip/in.(¾ in.) $= 43.0$ kips

Therefore, the total available bearing strength of the gusset plate is:

LRFD	ASD
$\phi R_n = 6(84.8 \text{ kips}) + 2(64.4 \text{ kips})$ $= 638$ kips > 374 kips **o.k.**	$\dfrac{R_n}{\Omega} = 6(56.7 \text{ kips}) + 2(43.0 \text{ kips})$ $= 426$ kips > 261 kips **o.k.**

Check block shear strength of gusset plate

Assume that the brace force P_u (LRFD) or P_a (ASD) can act as a tensile force and check the block shear rupture strength using AISC *Specification* Equation J4-5. As assumed previously, use bolt spacing of 3.00 in. and edge distance of 2.00 in. The gage is equal to 3½ in. and from AISC *Specification* Table J3.3, the bolt hole is 1¹⁄₁₆ in.

$U_{bs} = 1.0$

$A_{gv} = 2[2.00 \text{ in.} + 3(3.00 \text{ in.})](¾ \text{ in.})$
$\phantom{A_{gv}} = 16.5 \text{ in.}^2$

$A_{nv} = 16.5 \text{ in.}^2 - 2(3½)(1¹⁄₁₆ \text{ in.} + ¹⁄₁₆ \text{ in.})(¾ \text{ in.})$
$\phantom{A_{nv}} = 10.6 \text{ in.}^2$

$A_{gt} = 3½ \text{ in.}(¾ \text{ in.})$
$\phantom{A_{gt}} = 2.63 \text{ in.}^2$

$A_{nt} = 2.63 \text{ in.}^2 - (1¹⁄₁₆ \text{ in.} + ¹⁄₁₆ \text{ in.})(¾ \text{ in.})$
$\phantom{A_{nt}} = 1.79 \text{ in.}^2$

$F_u A_{nt} = 65 \text{ ksi}(1.79 \text{ in.}^2)$
$\phantom{F_u A_{nt}} = 116 \text{ kips}$

$0.60 F_u A_{nv} = 0.60(65 \text{ ksi})(10.6 \text{ in.}^2)$
$\phantom{0.60 F_u A_{nv}} = 413 \text{ kips}$

$0.60 F_y A_{gv} = 0.60(50 \text{ ksi})(16.5 \text{ in.}^2)$
$\phantom{0.60 F_y A_{gv}} = 495 \text{ kips}$

Because $0.60F_uA_{nv} < 0.60F_yA_{gv}$, from AISC *Specification* Equation J4-5:

LRFD	ASD
$\phi R_n = \phi(0.60F_uA_{nv} + U_{bs}F_uA_{nt})$ $= 0.75[413 \text{ kips} + (1.0)(116 \text{ kips})]$ $= 397 \text{ kips} > 374 \text{ kips}$ **o.k.**	$\dfrac{R_n}{\Omega} = \dfrac{(0.6F_uA_{nv} + U_{bs}F_uA_{nt})}{\Omega}$ $= \dfrac{[413 \text{ kips} + 1.0(116 \text{ kips})]}{2.00}$ $= 265 \text{ kips} > 261 \text{ kips}$ **o.k.**

See Figure 5-65 for initial connection geometry.

Check compression buckling strength of the gusset

As can be seen in Figure 5-65, the width of the Whitmore section is:

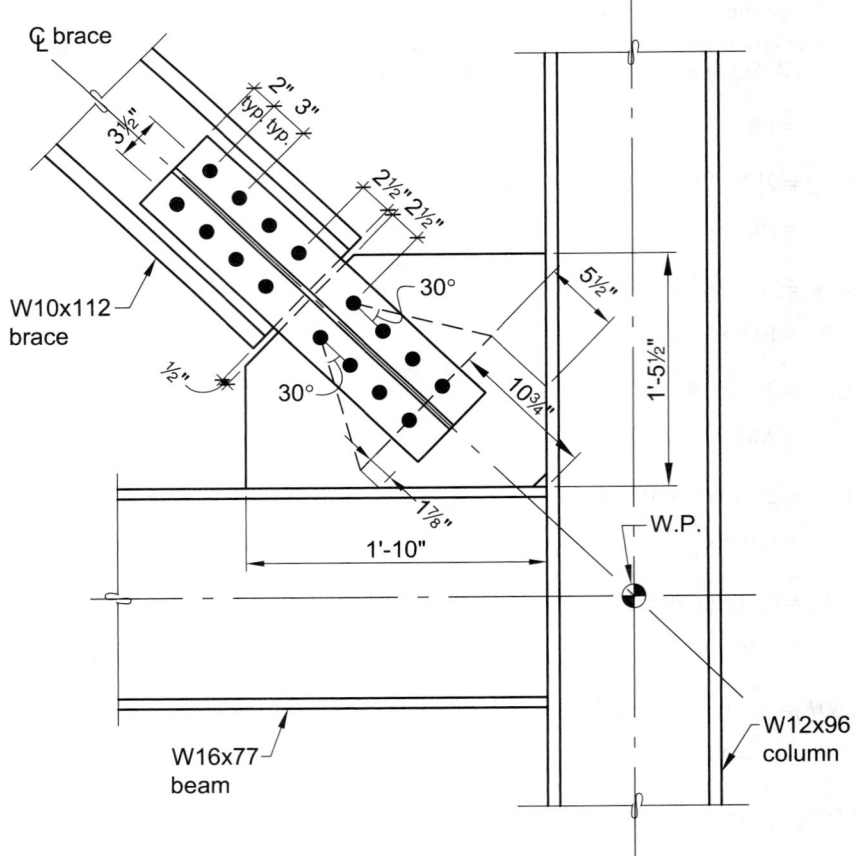

Fig. 5-65. Initial connection geometry for Example 5.4.7.

5.4 ECCENTRICALLY BRACED FRAMES

$L_w = 3\frac{1}{2}$ in. $+ 2(3)(3$ in.$)\tan 30°$
$= 13.9$ in.

The average unbraced length of the gusset plate, using the dimensions given in Figure 5-65, is:

$$L = \frac{10\frac{3}{4} \text{ in.} + 5\frac{1}{2} \text{ in.} + 1\frac{7}{8} \text{ in.}}{3}$$
$= 6.04$ in.

Continuing with the assumed ¾ in. thickness, the radius of gyration of the gusset plate is:

$$r = \frac{t}{\sqrt{12}}$$
$$= \frac{¾ \text{ in.}}{\sqrt{12}}$$
$= 0.217$ in.

Using a column effective length factor of 0.65 from AISC *Specification* Commentary Table C-A-7.1:

$$\frac{KL}{r} = \frac{0.65(6.04 \text{ in.})}{0.217 \text{ in.}}$$
$= 18.1$

With $KL/r \leq 25$, AISC *Specification* Section J4.4(a) applies and $F_{cr} = F_y$. The design strength of the gusset is determined as follows:

$P_n = F_{cr} A_g$ *(Spec. Eq. J4-6)*
$= 50$ ksi$(13.9$ in.$)(¾$ in.$)$
$= 521$ kips

LRFD	ASD
$\phi P_n = 0.90(521 \text{ kips})$ $= 469$ kips > 374 kips **o.k.**	$\dfrac{R_n}{\Omega} = \dfrac{521 \text{ kips}}{1.67}$ $= 312$ kips > 261 kips **o.k.**

Use a ¾-in.-thick gusset plate.

Select trial connection between gusset and brace

Use a pair of bolted WT-sections to connect the brace to the gusset plate. The flange width of the WT-sections must be less than or equal to the *T*-dimension of the W10×112 brace ($T = 7½$ in.). Try (2) **WT8×28.5**.

From AISC *Manual* Table 1-8, the geometric properties of a **WT8×28.5** are:

$A = 8.39$ in.^{2} $d = 8.22$ in. $b_f = 7.12$ in. $t_f = 0.715$ in.
$t_w = 0.430$ in. $r_y = 1.60$ in. $\bar{y} = 1.94$ in.

$b_f = 7.12$ in. $< T_{brace} = 7\frac{1}{2}$ in.   **o.k.**

Check tensile yielding strength of WT-sections
(for the required strength of the brace considered as a tension force)

From AISC *Specification* Equation J4-1, the tensile yielding strength of the two WT-sections is:

$$R_n = F_y A_g \qquad \text{(Spec. Eq. J4-1)}$$
$$= 50 \text{ ksi}(2)(8.39 \text{ in.}^2)$$
$$= 839 \text{ kips}$$

LRFD	ASD
$\phi R_n = 0.90(839 \text{ kips})$ $= 755 \text{ kips} > 374 \text{ kips}$ **o.k.**	$\dfrac{R_n}{\Omega} = \dfrac{839 \text{ kips}}{1.67}$ $= 502 \text{ kips} > 261 \text{ kips}$ **o.k.**

Check tensile rupture strength of the WT-sections

Assume that all bolts will be 1-in.-diameter A325-N bolts. The net area of the two WT-sections is:

$$A_n = 2(A_g - 2d_h t_f)$$
$$= 2\left[8.39 \text{ in.}^2 - 2(1\tfrac{1}{16} \text{ in.} + \tfrac{1}{16} \text{ in.})(0.715 \text{ in.})\right]$$
$$= 13.6 \text{ in.}^2$$

Because the WT webs are not connected to the brace, an effective area of the WT-sections needs to be determined. From AISC *Specification* Table D3.1 with $\bar{x} = \bar{y}$ for the WT-section, the shear lag factor is:

$$U = 1 - \frac{\bar{x}}{l}$$
$$= 1 - \frac{1.94 \text{ in.}}{3(3 \text{ in.})}$$
$$= 0.784$$

$A_e = UA_n \le 0.85 A_g$

$= 0.784(13.6 \text{ in.}^2) \le 0.85(2)(8.39 \text{ in.}^2)$

$= 10.7 \text{ in.}^2 \le 14.3 \text{ in.}^2$

$= 10.7 \text{ in.}^2$

The tensile rupture strength of the two WT-sections is:

$R_n = F_u A_e$ (Spec. Eq. J4-2)

$= 65 \text{ ksi}(10.7 \text{ in.}^2)$

$= 696 \text{ kips}$

LRFD	ASD
$\phi R_n = 0.75(696 \text{ kips})$ $= 522 \text{ kips} > 374 \text{ kips}$ **o.k.**	$\dfrac{R_n}{\Omega} = \dfrac{696 \text{ kips}}{2.00}$ $= 348 \text{ kips} > 261 \text{ kips}$ **o.k.**

Check compressive strength of the WT-sections

The unbraced length of each WT is 5½ in., measured from the last bolt on the brace to the first bolt on the gusset plate, as shown in Figure 5-65. The effective slenderness ratio is:

$\dfrac{KL}{r_y} = \dfrac{0.65(5½ \text{ in.})}{1.60 \text{ in.}}$

$= 2.23$

With $KL/r \le 25$, AISC *Specification* Section J4.4(a) applies and $F_{cr} = F_y$. The nominal compressive strength of the two WT-sections is:

$P_n = F_y A_g$ (Spec. Eq. J4-6)

$= 50 \text{ ksi}(2)(8.39 \text{ in.}^2)$

$= 839 \text{ kips}$

LRFD	ASD
$\phi P_n = 0.90(839 \text{ kips})$ $= 755 \text{ kips} > 374 \text{ kips}$ **o.k.**	$\dfrac{P_n}{\Omega} = \dfrac{839 \text{ kips}}{1.67}$ $= 502 \text{ kips} > 261 \text{ kips}$ **o.k.**

Check bearing strength of the WT-sections

Because the specified minimum tensile strength of the WT-sections is equal to the specified minimum tensile strength of the gusset plate and the sum of the WT flange thicknesses is greater than the gusset plate thickness, the bearing strength of the WT-sections is adequate.

Check block shear rupture strength of the WT-sections

Because the specified minimum tensile strength of the WT-sections is equal to the specified minimum tensile strength of the gusset plate and the shear and tensile areas of the WT flanges in block shear are each greater than the corresponding gusset areas, the block shear rupture strength of the WT-sections is adequate.

Use (2) **WT8×28.5** to connect the brace web to the gusset plate.

Use (8) 1-in.-diameter ASTM A325-N bolts in standard holes to connect the WT-sections to the gusset plate. Use a 3-in. spacing, 2-in. edge distance and 3½-in. gage for the bolts.

Check bearing strength at bolt holes in the brace web

Because the specified minimum tensile strength of the brace is equal to the specified minimum tensile strength of the gusset plate and the brace web thickness is greater than the gusset plate thickness, the bearing strength of the brace web is adequate.

Check block shear rupture strength of the brace web

Because the material strength of the brace is equal to the material strength of the gusset plate and the brace web thickness is greater than the gusset plate thickness, the block shear rupture strength of the brace web is adequate.

Check tensile rupture strength of the brace

The net area of the brace is:

$$A_n = A_g - 2(d_h + \tfrac{1}{16} \text{ in.})t_w$$
$$= 32.9 \text{ in.}^2 - 2(1\tfrac{1}{16} \text{ in.} + \tfrac{1}{16} \text{ in.})(0.755 \text{ in.})$$
$$= 31.2 \text{ in.}^2$$

To determine the connection eccentricity for shear lag in a W-shape connected by the web only, AISC *Specification* Commentary Figure C-D3.1 recommends that half of the flange and a portion of the web be treated as an angle.

Conservatively ignoring the fillets, the distance from the web centerline to the centroid of the effective section is:

$$\bar{x} = \frac{t_f\left(\dfrac{b_f}{2}\right)\left(\dfrac{b_f}{4}\right) + \dfrac{t_w}{2}\left(\dfrac{d}{2} - t_f\right)\left(\dfrac{t_w}{4}\right)}{t_f\left(\dfrac{b_f}{2}\right) + \dfrac{t_w}{2}\left(\dfrac{d}{2} - t_f\right)}$$

$$= \frac{1.25 \text{ in.}\left(\dfrac{10.4 \text{ in.}}{2}\right)\left(\dfrac{10.4 \text{ in}}{4}\right) + \dfrac{0.755 \text{ in.}}{2}\left(\dfrac{11.4 \text{ in.}}{2} - 1.25 \text{ in.}\right)\left(\dfrac{0.755 \text{ in.}}{4}\right)}{1.25 \text{ in.}\left(\dfrac{10.4 \text{ in.}}{2}\right) + \dfrac{0.755 \text{ in.}}{2}\left(\dfrac{11.4 \text{ in.}}{2} - 1.25 \text{ in.}\right)}$$

$$= 2.10 \text{ in.}$$

5.4 ECCENTRICALLY BRACED FRAMES

From AISC *Specification* Table D3.1, Case 2:

$$U = 1 - \frac{\bar{x}}{l}$$
$$= 1 - \frac{2.10 \text{ in.}}{9.00 \text{ in.}}$$
$$= 0.767$$

$$A_e = UA_n$$
$$= 0.767(31.2 \text{ in.}^2)$$
$$= 23.9 \text{ in.}^2$$

$$P_n = F_u A_e \qquad \qquad (\textit{Spec.} \text{ Eq. D2-2})$$
$$= 65 \text{ ksi}(23.9 \text{ in.}^2)$$
$$= 1{,}550 \text{ kips}$$

LRFD	ASD
$\phi P_n = 0.75(1{,}550 \text{ kips})$ $= 1{,}160 \text{ kips} > 374 \text{ kips}$ **o.k.**	$\dfrac{P_n}{\Omega} = \dfrac{1{,}550 \text{ kips}}{2.00}$ $= 775 \text{ kips} > 261 \text{ kips}$ **o.k.**

Use (8) 1-in.-diameter ASTM A325-N bolts in standard holes to connect the WT-sections to the brace web. Use a 3-in. spacing, 2-in. edge distance and 3½-in. gage for the bolts.

Determine gusset-to-beam and column connection interface forces

The forces at the gusset-to-beam and gusset-to-column interfaces are determined using the geometry shown in Figure 5-65 and the Uniform Force Method. It will be assumed that a 1-in. clip in the corner of the gusset will be necessary to clear a fillet weld on the top flange of the beam, and a ⅝-in.-thick bolted end-plate will be used to connect the gusset and beam to the column.

$e_b = 8.25 \text{ in.} \qquad e_c = 6.35 \text{ in.} \qquad \theta = 46.1°$

$$\bar{\alpha} = \tfrac{1}{2}(22.0 \text{ in.} - 1 \text{ in.} - \tfrac{5}{8} \text{ in.}) + 1 \text{ in.} + \tfrac{5}{8} \text{ in.}$$
$$= 11.8 \text{ in.}$$

$$\bar{\beta} = \tfrac{1}{2}(17.5 \text{ in.} - 1 \text{ in.}) + 1 \text{ in.}$$
$$= 9.25 \text{ in.}$$

Using $\beta = \bar{\beta}$,

$$\alpha = (e_b + \beta)\tan\theta - e_c \qquad \qquad \text{(from \textit{Manual} Eq. 13-1)}$$
$$= (8.25 \text{ in.} + 9.25 \text{ in.})\tan 46.1° - 6.35 \text{ in.}$$
$$= 11.8 \text{ in.}$$

Because $\bar{\alpha} = \alpha$, there is no moment at the beam or column interface.

$$r = \sqrt{(\alpha + e_c)^2 + (\beta + e_b)^2} \qquad \text{(Manual Eq. 13-6)}$$
$$= \sqrt{(11.8 \text{ in.} + 6.35 \text{ in.})^2 + (9.25 \text{ in.} + 8.25 \text{ in.})^2}$$
$$= 25.2 \text{ in.}$$

The forces on the gusset-to-beam and gusset-to-column interface are:

LRFD	ASD
From AISC *Manual* Equation 13-4: $V_{ub} = \dfrac{e_b}{r} P_u$ $= \dfrac{8.25 \text{ in.}}{25.2 \text{ in.}}(374 \text{ kips})$ $= 122 \text{ kips}$	From AISC *Manual* Equation 13-4: $V_{ab} = \dfrac{e_b}{r} P_a$ $= \dfrac{8.25 \text{ in.}}{25.2 \text{ in.}}(261 \text{ kips})$ $= 85.4 \text{ kips}$
From AISC *Manual* Equation 13-2: $V_{uc} = \dfrac{\beta}{r} P_u$ $= \dfrac{9.25 \text{ in.}}{25.2 \text{ in.}}(374 \text{ kips})$ $= 137 \text{ kips}$	From AISC *Manual* Equation 13-2: $V_{ac} = \dfrac{\beta}{r} P_a$ $= \dfrac{9.25 \text{ in.}}{25.2 \text{ in.}}(261 \text{ kips})$ $= 95.8 \text{ kips}$
From AISC *Manual* Equation 13-5: $H_{ub} = \dfrac{\alpha}{r} P_u$ $= \dfrac{11.8 \text{ in.}}{25.2 \text{ in.}}(374 \text{ kips})$ $= 175 \text{ kips}$	From AISC *Manual* Equation 13-5: $H_{ab} = \dfrac{\alpha}{r} P_a$ $= \dfrac{11.8 \text{ in.}}{25.2 \text{ in.}}(261 \text{ kips})$ $= 122 \text{ kips}$
From AISC *Manual* Equation 13-3: $H_{uc} = \dfrac{e_c}{r} P_u$ $= \dfrac{6.35 \text{ in.}}{25.2 \text{ in.}}(374 \text{ kips})$ $= 94.2 \text{ kips}$	From AISC *Manual* Equation 13-3: $H_{ac} = \dfrac{e_c}{r} P_a$ $= \dfrac{6.35 \text{ in.}}{25.2 \text{ in.}}(261 \text{ kips})$ $= 65.8 \text{ kips}$

The connection interface forces are shown in Figure 5-66. It should be noted that the forces are for the brace in compression. For the purposes of this example, equal and opposite forces have been assumed for the brace in tension.

5.4 ECCENTRICALLY BRACED FRAMES

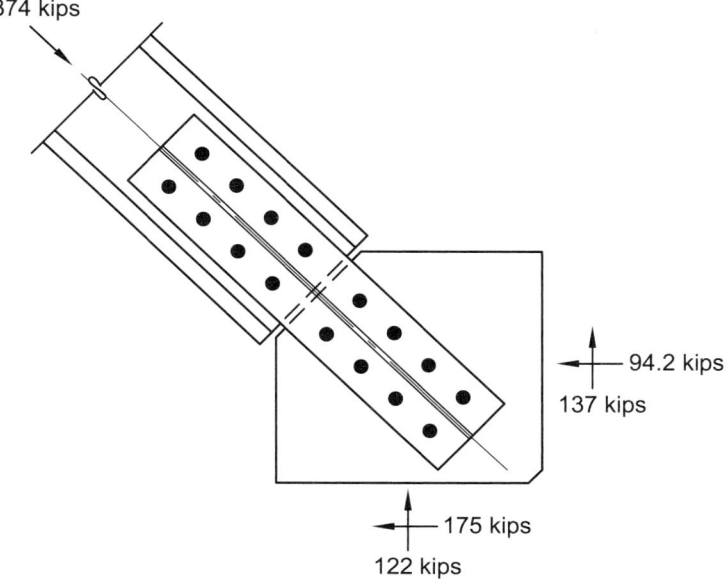

(a) Connection interface forces for Example 5.4.7 (LRFD)

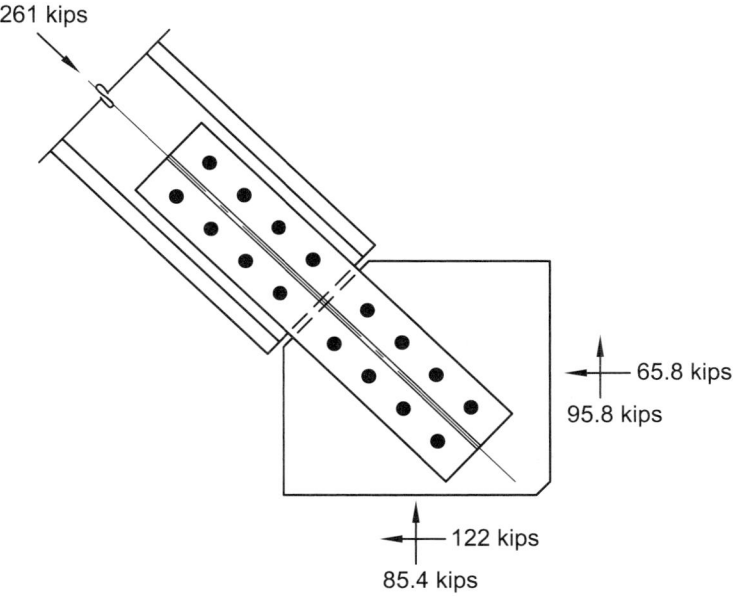

(b) Connection interface forces for Example 5.4.7 (ASD)

Fig. 5-66. Connection interface forces for Example 5.4.7.

Design the weld at the gusset-to-beam interface

Assuming a ⅝-in.-thick end-plate and 1-in. corner clip, the length of the weld connecting the gusset plate to the beam flange is:

$l_w = 22.0$ in. $- 1$ in. $- ⅝$ in.
$ = 20.4$ in.

The stresses at the gusset-to-beam interface are:

LRFD	ASD
$f_{uv} = \dfrac{H_{ub}}{l_w}$	$f_{av} = \dfrac{H_{ab}}{l_w}$
$= \dfrac{175 \text{ kips}}{20.4 \text{ in.}}$	$= \dfrac{122 \text{ kips}}{20.4 \text{ in.}}$
$= 8.58$ kip/in.	$= 5.98$ kip/in.
$f_{ua} = \dfrac{V_{ub}}{l_w}$	$f_{aa} = \dfrac{V_{ab}}{l_w}$
$= \dfrac{122 \text{ kips}}{20.4 \text{ in.}}$	$= \dfrac{85.4 \text{ kips}}{20.4 \text{ in.}}$
$= 5.98$ kip/in.	$= 4.19$ kip/in.
$f_{ur} = \sqrt{f_{uv}^2 + f_{ua}^2}$	$f_{ar} = \sqrt{f_{av}^2 + f_{aa}^2}$
$= \sqrt{(8.58 \text{ kip/in.})^2 + (5.98 \text{ kip/in.})^2}$	$= \sqrt{(5.98 \text{ kip/in.})^2 + (4.19 \text{ kip/in.})^2}$
$= 10.5$ kip/in.	$= 7.30$ kip/in.

Multiplying by the weld ductility factor of 1.25 discussed in Part 13 of the AISC *Manual*, the required strength per inch of weld is:

LRFD	ASD
$f_{ur} = 1.25(10.5 \text{ kip/in.})$	$f_{ar} = 1.25(7.30 \text{ kip/in.})$
$= 13.1$ kip/in.	$= 9.13$ kip/in.

The resultant load angle with respect to the longitudinal axis of the weld group is:

5.4 ECCENTRICALLY BRACED FRAMES

LRFD	ASD
$\theta = \tan^{-1}\left(\dfrac{V_{ub}}{H_{ub}}\right)$ $= \tan^{-1}\left(\dfrac{122 \text{ kips}}{175 \text{ kips}}\right)$ $= 34.9°$	$\theta = \tan^{-1}\left(\dfrac{V_{ab}}{H_{ab}}\right)$ $= \tan^{-1}\left(\dfrac{85.4 \text{ kips}}{122 \text{ kips}}\right)$ $= 35.0°$

AISC *Specification* Section J2.4 allows an increase in the available strength of fillet welds when the angle of loading is not along the weld longitudinal axis. Using AISC *Manual* Equations 8-2a and 8-2b in conjunction with AISC *Specification* Equation J2-5, the required fillet weld size for two lines of weld is:

LRFD	ASD
$D_{min} = \dfrac{13.1 \text{ kip/in.}}{\left[2(1.392 \text{ kip/in.})\right. \\ \left. \times\left(1.0+0.50\sin^{1.5} 34.9°\right)\right]}$ $= 3.87$ sixteenths	$D_{min} = \dfrac{9.13 \text{ kip/in.}}{\left[2(0.928 \text{ kip/in.})\right. \\ \left. \times\left(1.0+0.50\sin^{1.5} 35.0°\right)\right]}$ $= 4.04$ sixteenths

From AISC *Specification* Table J2.4, the minimum weld size is ¼ in.

Use double-sided 5⁄16-in. fillet welds to connect the gusset plate to the beam.

Check gusset rupture at weld

The shear rupture strength of the gusset is:

$R_n = 0.60 F_u A_{nv}$ *(Spec. Eq. J4-4)*
$= 0.60(65 \text{ ksi})(¾ \text{ in.})$
$= 29.3$ kip/in.

LRFD	ASD
$\phi R_n = 0.75(29.3 \text{ kip/in.})$ $= 22.0$ kip/in. > 13.1 kip/in. **o.k.**	$\dfrac{R_n}{\Omega} = \dfrac{29.3 \text{ kip/in.}}{2.00}$ $= 14.7$ kip/in. > 9.13 kip/in. **o.k.**

Check yielding of the gusset

The shear yielding strength of the gusset plate is:

$R_n = 0.6 F_y A_{gv}$ *(Spec. Eq. J4-3)*
$ = 0.6 F_y t l_w$

$R_n/l_w = 0.6(50 \text{ ksi})(¾ \text{ in.})$
$= 22.5 \text{ kip/in.}$

LRFD	ASD
$\phi R_n = 1.00(22.5 \text{ kip/in.})$ $= 22.5 \text{ kip/in.} > 13.1 \text{ kip/in.}$ **o.k.**	$\dfrac{R_n}{\Omega} = \dfrac{22.5 \text{ kip/in.}}{1.50}$ $= 15.0 \text{ kip/in.} > 9.13 \text{ kip/in.}$ **o.k.**

Check beam web local yielding

With the centroid of the compressive force applied less than d (the beam depth) from the member end, and l_b is the length of bearing, the web local yielding available strength is determined as follows:

$R_n = F_{yw}t_w(2.5k_{des} + l_b)$ (Spec. Eq. J10-3)
$= 50 \text{ ksi}(0.455 \text{ in.})[2.5(1.16 \text{ in.}) + 20.4 \text{ in.}]$
$= 530 \text{ kips}$

LRFD	ASD
$\phi R_n = 1.00(530 \text{ kips})$ $= 530 \text{ kips} > V_{ub} = 122 \text{ kips}$ **o.k.**	$\dfrac{R_n}{\Omega} = \dfrac{530 \text{ kips}}{1.50}$ $= 353 \text{ kips} > V_{ab} = 85.4 \text{ kips}$ **o.k.**

Check beam web local crippling

With the centroid of the compressive force applied greater than $d/2$ from the beam end, the web local crippling available strength is determined as follows:

$R_n = 0.80 t_w^2 \left[1 + 3\left(\dfrac{l_b}{d}\right)\left(\dfrac{t_w}{t_f}\right)^{1.5}\right]\sqrt{\dfrac{EF_{yw}t_f}{t_w}}$ (Spec. Eq. J10-4)

$= (0.80)(0.455 \text{ in.})^2 \left[1 + 3\left(\dfrac{20.4 \text{ in.}}{16.5 \text{ in.}}\right)\left(\dfrac{0.455 \text{ in.}}{0.760 \text{ in.}}\right)^{1.5}\right]\sqrt{\dfrac{29{,}000(50 \text{ ksi})(0.760 \text{ in.})}{0.455 \text{ in.}}}$

$= 701 \text{ kips}$

LRFD	ASD
$\phi R_n = 0.75(701 \text{ kips})$ $= 526 \text{ kips} > V_{ub} = 122 \text{ kips}$ **o.k.**	$\dfrac{R_n}{\Omega} = \dfrac{701 \text{ kips}}{2.00}$ $= 351 \text{ kips} > V_{ab} = 85.4 \text{ kips}$ **o.k.**

5.4 ECCENTRICALLY BRACED FRAMES

Design the weld between the gusset and the end plate

From Figure 5-65, the length of weld is 17.5 in. Subtracting the 1-in. clip in the gusset plate, the length of weld is 16.5 in. The forces on the gusset per unit length are:

LRFD	ASD
$f_{uv} = \dfrac{V_{uc}}{l_w}$ $= \dfrac{137 \text{ kips}}{16.5 \text{ in.}}$ $= 8.30 \text{ kip/in.}$	$f_{av} = \dfrac{V_{ac}}{l_w}$ $= \dfrac{95.8 \text{ kips}}{16.5 \text{ in.}}$ $= 5.81 \text{ kip/in.}$
$f_{ua} = \dfrac{H_{uc}}{l_w}$ $= \dfrac{94.2 \text{ kips}}{16.5 \text{ in.}}$ $= 5.71 \text{ kip/in.}$	$f_{aa} = \dfrac{H_{ac}}{l_w}$ $= \dfrac{65.8 \text{ kips}}{16.5 \text{ in.}}$ $= 3.99 \text{ kip/in.}$
$f_{ur} = \sqrt{f_{uv}^2 + f_{ua}^2}$ $= \sqrt{(8.30 \text{ kip/in.})^2 + (5.71 \text{ kip/in.})^2}$ $= 10.1 \text{ kip/in.}$	$f_{ar} = \sqrt{f_{av}^2 + f_{aa}^2}$ $= \sqrt{(5.81 \text{ kip/in.})^2 + (3.99 \text{ kip/in.})^2}$ $= 7.05 \text{ kip/in.}$

Multiplying by the weld ductility factor of 1.25 discussed in Part 13 of the AISC *Manual*, the required strength per inch of weld is:

LRFD	ASD
$f_{ur} = 1.25(10.1 \text{ kip/in.})$ $= 12.6 \text{ kip/in.}$	$f_{ar} = 1.25(7.05 \text{ kip/in.})$ $= 8.81 \text{ kip/in.}$

The load angle with respect to the longitudinal axis of the weld group is:

LRFD	ASD
$\theta = \tan^{-1}\left(\dfrac{H_{uc}}{V_{uc}}\right)$ $= \tan^{-1}\left(\dfrac{94.2 \text{ kips}}{137 \text{ kips}}\right)$ $= 34.5°$	$\theta = \tan^{-1}\left(\dfrac{H_{ac}}{V_{ac}}\right)$ $= \tan^{-1}\left(\dfrac{65.8 \text{ kips}}{95.8 \text{ kips}}\right)$ $= 34.5°$

AISC *Specification* Section J2.4 allows an increase in the available strength of fillet welds when the angle of loading is not along the weld longitudinal axis. Using AISC *Manual*

Equations 8-2a and 8-2b in conjunction with AISC *Specification* Equation J2-5, the required fillet weld size for two lines of weld is:

LRFD	ASD
$D_{min} = \dfrac{12.6 \text{ kip/in.}}{\left[2(1.392 \text{ kip/in.})\right] \times \left(1.0 + 0.50\sin^{1.5} 34.5°\right)}$	$D_{min} = \dfrac{8.81 \text{ kip/in.}}{\left[2(0.928 \text{ kip/in.})\right] \times \left(1.0 + 0.50\sin^{1.5} 34.5°\right)}$
$= 3.73$ sixteenths	$= 3.91$ sixteenths

From AISC *Specification* Table J2.4, the minimum weld size is ¼ in. Therefore, a double-sided ¼-in. fillet weld is required at the gusset-to-end plate connection.

For ease of fabrication, use the maximum required weld size of the gusset-to-end plate connection and the beam-to-end plate connection.

Check gusset rupture at gusset-to-end plate weld

Use the gusset shear rupture strength previously determined for the gusset-to-beam interface.

LRFD	ASD
$\phi R_n = 22.5$ kip/in. > 12.6 kip/in. **o.k.**	$\dfrac{R_n}{\Omega} = 14.7$ kip/in. > 8.81 kip/in. **o.k.**

Check yielding of the gusset at gusset-to-end plate

Use the gusset shear yielding strength previously determined for the gusset-to-beam interface.

LRFD	ASD
$\phi R_n = 22.5$ kip/in. > 12.6 kip/in. **o.k.**	$\dfrac{R_n}{\Omega} = 15.0$ kip/in. > 8.81 kip/in. **o.k.**

Design the weld between the beam and the end plate

From Figures 5-63 and 5-66, the vertical force component at the beam-to-end plate interface is:

LRFD	ASD
$V_{ub} + V_{ubeam} = 122$ kips $+ 36.5$ kips $= 159$ kips	$V_{ab} + V_{abeam} = 85.4$ kips $+ 25.1$ kips $= 111$ kips

5.4 ECCENTRICALLY BRACED FRAMES

The minimum double-sided fillet weld size required to develop the vertical force through the beam web T-dimension is:

LRFD	ASD
$D \geq \dfrac{159 \text{ kips}}{2(1.392 \text{ kip/in.})(13\frac{1}{4} \text{ in.})}$ = 4.31 sixteenths	$D \geq \dfrac{111 \text{ kips}}{2(0.928 \text{ kip/in.})(13\frac{1}{4} \text{ in.})}$ = 4.51 sixteenths

A $\frac{3}{16}$-in. weld size is the minimum required by AISC *Specification* Table J2.4 for the W16×77 web and $\frac{3}{4}$-in.-thick gusset plate.

Use a $\frac{5}{16}$-in. double-sided fillet weld to connect the beam web to the end plate. Also use a $\frac{5}{16}$-in. double-sided fillet weld to connect the gusset to the end plate.

Check beam web rupture strength at weld
The shear rupture strength of the beam web is:

$R_n = 0.60 F_u A_{nv}$ (*Spec.* Eq. J4-4)
$= 0.60(65 \text{ ksi})(0.455 \text{ in.})(13\frac{1}{4} \text{ in.})$
$= 235 \text{ kips}$

LRFD	ASD
$\phi R_n = 0.75(235 \text{ kips})$ = 176 kips > 159 kips **o.k.**	$\dfrac{R_n}{\Omega} = \dfrac{235 \text{ kips}}{2.00}$ = 118 kips > 111 kips **o.k.**

Design the weld between the beam flanges and the end plate
The horizontal force component is the maximum of the following three load conditions:

1. The amplified collector force from Figure 5-63 (Condition 1).

LRFD	ASD
$H_u = 85.4 \text{ kips}$	$H_a = 59.9 \text{ kips}$

2. The axial force in the beam outside the link corresponding to $1.25 R_y V_n - H_b$, where the force in the beam outside the link corresponding to $1.25 R_y V_n$ is shown in Figure 5-63.

LRFD	ASD
$H_u = 355 \text{ kips} - 175 \text{ kips}$ = 180 kips	$H_a = 248 \text{ kips} - 122 \text{ kips}$ = 126 kips

3. H_{uc} or H_{ac}: The horizontal component at the gusset-to-column interface from the Uniform Force Method, calculated previously for Condition 1.

LRFD	ASD
$H_u = 94.2$ kips	$H_a = 65.8$ kips

Therefore, the required horizontal strength of the beam-to-column connection is $H_u = 180$ kips and $H_a = 126$ kips, as provided by the second condition. Assuming that the horizontal force is transferred by the beam flanges, the force in each flange is:

LRFD	ASD
$R_{uf} = \dfrac{180 \text{ kips}}{2}$ $= 90.0$ kips	$R_{af} = \dfrac{126 \text{ kips}}{2}$ $= 63.0$ kips

Using the full beam flange width and the directional strength increase for a transversely loaded fillet weld, the minimum required single-sided fillet weld size to develop the flange force is:

LRFD	ASD
$D \geq \dfrac{90.0 \text{ kips}}{1.5(1.392 \text{ kip/in.})(10.3 \text{ in.})}$ $= 4.18$ sixteenths	$D \geq \dfrac{63.0 \text{ kips}}{1.5(0.928 \text{ kip/in.})(10.3 \text{ in.})}$ $= 4.39$ sixteenths

A ¼-in. weld size is the minimum required by AISC *Specification* Table J2.4 for the thinner part joined—the ⅝-in. end plate.

Use single-sided 5⁄16-in. fillet welds to connect the beam flanges to the end plate.

Check beam flange rupture at weld

The tension rupture strength of the beam flange is:

$R_n = F_u A_e$ (*Spec.* Eq. J4-2)
$\quad = F_u b_f t_f$
$\quad = 65 \text{ ksi}(0.760 \text{ in.})(10.3 \text{ in.})$
$\quad = 509$ kips

LRFD	ASD
$\phi R_n = 0.75(509 \text{ kips})$ $= 382$ kips > 90.0 kips **o.k.**	$\dfrac{R_n}{\Omega} = \dfrac{509 \text{ kips}}{2.00}$ $= 255$ kips > 63.0 kips **o.k.**

5.4 ECCENTRICALLY BRACED FRAMES

Design end-plate bolts

Try seven rows of two 1-in.-diameter ASTM A325-N bolts at a 5½-in. gage. Use four bolts adjacent to each beam flange and an additional three bolts on each side of the gusset plate as shown in Figure 5-67. Assuming the total shear is shared equally among all bolts (refer to Figure 5-63), the required shear force per bolt is:

LRFD	ASD
$r_{uv} = \dfrac{V_{uc} + V_{ub} - V_{ubeam}}{n_b}$ $= \dfrac{137 \text{ kips} + 122 \text{ kips} - 36.5 \text{ kips}}{14 \text{ bolts}}$ $= 15.9 \text{ kips/bolt}$	$r_{av} = \dfrac{V_{ac} + V_{ab} - V_{abeam}}{n_b}$ $= \dfrac{95.8 \text{ kips} + 85.4 \text{ kips} - 25.1 \text{ kips}}{14 \text{ bolts}}$ $= 11.2 \text{ kips/bolt}$

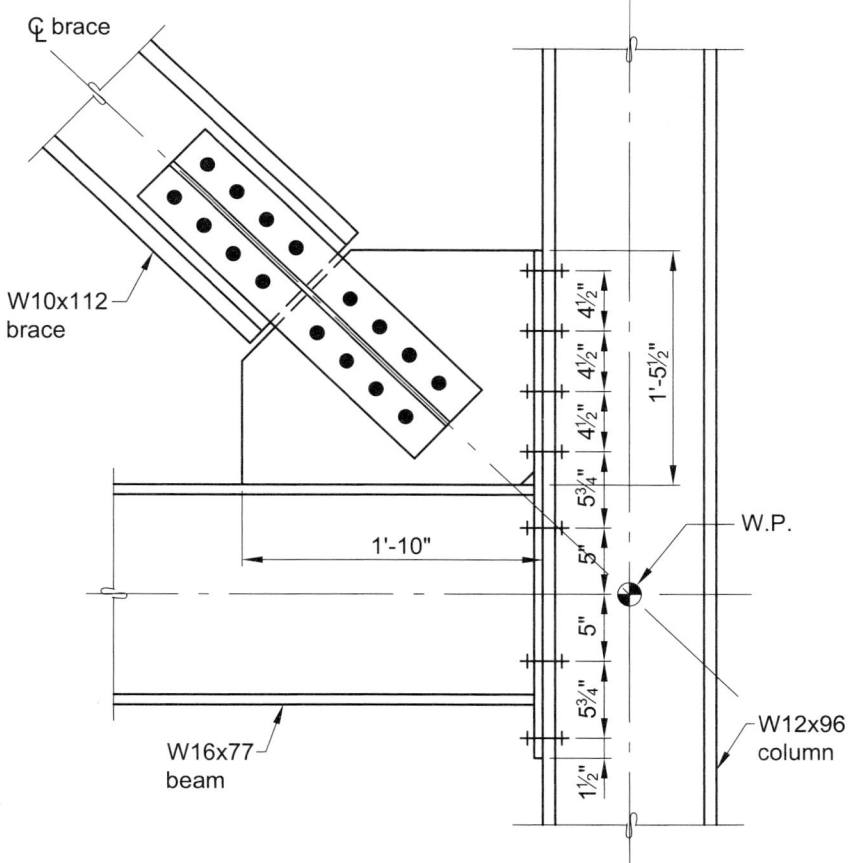

Fig. 5-67. End-plate geometry for Example 5.4.7.

From AISC *Specification* Table J3.2 for Group A bolts with the threads not excluded, F_{nt} = 90 ksi and F_{nv} = 54 ksi. From AISC *Manual* Table 7-1, the area of a 1-in.-diameter bolt is 0.785 in.² Based on the required shear force per bolt, the nominal tensile strength of each bolt subject to combined tension and shear rupture, from AISC *Specification* Equation J3-3, is:

LRFD	ASD
$F'_{nt} = 1.3 F_{nt} - \dfrac{F_{nt}}{\phi F_{nv}} f_{rv} \leq F_{nt}$	$F'_{nt} = 1.3 F_{nt} - \dfrac{\Omega F_{nt}}{F_{nv}} f_{rv} \leq F_{nt}$
$f_{rv} = \dfrac{r_{uv}}{A_b}$ $= \dfrac{15.9 \text{ kips/bolt}}{0.785 \text{ in.}^2}$ $= 20.3 \text{ ksi}$	$f_{rv} = \dfrac{r_{av}}{A_b}$ $= \dfrac{11.2 \text{ kips/bolt}}{0.785 \text{ in.}^2}$ $= 14.3 \text{ ksi}$
$F'_{nt} = 1.3(90 \text{ ksi})$ $\quad - \dfrac{90 \text{ ksi}}{0.75(54 \text{ ksi})}(20.3 \text{ ksi})$ $= 71.9 \text{ ksi} \leq 90 \text{ ksi}$ Use $F'_{nt} = 71.9$ ksi.	$F'_{nt} = 1.3(90 \text{ ksi})$ $\quad - \dfrac{2.00(90 \text{ ksi})}{54 \text{ ksi}}(14.3 \text{ ksi})$ $= 69.3 \text{ ksi} \leq 90 \text{ ksi}$ Use $F'_{nt} = 69.3$ ksi.

The available tensile strength of each bolt is, from AISC *Specification* Equation J3-2:

LRFD	ASD
$\phi r_{nt} = \phi F'_{nt} A_b$ $= 0.75(71.9 \text{ ksi})(0.785 \text{ in.}^2)$ $= 42.3 \text{ kips}$	$\dfrac{r'_{nt}}{\Omega} = \dfrac{F'_{nt} A_b}{\Omega}$ $= \dfrac{(69.3 \text{ ksi})(0.785 \text{ in.}^2)}{2.00}$ $= 27.2 \text{ kips}$

When the brace is in compression, a tensile force is transmitted across the beam-to-column interface. Assuming the four bolts adjacent to each beam flange transfer the tensile load, the required tensile force per bolt is:

5.4 ECCENTRICALLY BRACED FRAMES

LRFD	ASD
$r_{ut} = \dfrac{180 \text{ kips}}{8 \text{ bolts}}$ $= 22.5 \text{ kips/bolt}$ $< \phi r'_{nt} = 42.3 \text{ kips}$ **o.k.**	$r_{at} = \dfrac{126 \text{ kips}}{8 \text{ bolts}}$ $= 15.8 \text{ kips/bolt}$ $< \dfrac{r'_{nt}}{\Omega} = 27.2 \text{ kips}$ **o.k.**

When the brace is in tension, a tensile force is transmitted across the gusset-to-column interface. Assuming the four rows of bolts adjacent to the gusset plate transfer the tensile load, the required tensile force per bolt is:

LRFD	ASD
$r_{ut} = \dfrac{H_{uc}}{n}$ $= \dfrac{94.2 \text{ kips}}{8 \text{ bolts}}$ $= 11.8 \text{ kips/bolt}$ $< \phi r'_{nt} = 42.3 \text{ kips}$ **o.k.**	$r_{at} = \dfrac{H_{ac}}{n}$ $= \dfrac{65.8 \text{ kips}}{8 \text{ bolts}}$ $= 8.23 \text{ kips/bolt}$ $< r'_{nt}/\Omega = 27.2 \text{ kips}$ **o.k.**

Select end-plate thickness

Part 9 of the AISC *Manual* will be used to account for the effects of prying action on the bolts. Since the bolts are used to resist combined shear and tension, the available tensile strength per bolt used in the prying action calculations will be taken as calculated previously, with a reduction to include the effects of shear stress.

The two locations that need to be investigated for prying action are at the bolts adjacent to the gusset plate and the bolts adjacent to each beam flange. The controlling condition for prying action in this case is for the bolts adjacent to the beam flanges when the brace is in compression. Using the dimensions shown in Figure 5-67, an 11-in. end-plate width, and standard holes in the end plate, determine the applicable parameters for the bolts through the end plate:

$$b = \dfrac{5\tfrac{1}{2} \text{ in.} - \tfrac{3}{4} \text{ in.}}{2}$$
$$= 2.38 \text{ in.}$$

$$b' = b - \dfrac{d_b}{2}$$
$$= 2.38 \text{ in.} - \dfrac{1.00 \text{ in.}}{2}$$
$$= 1.88 \text{ in.}$$

$$a = \frac{11.0 \text{ in.} - 5\frac{1}{2} \text{ in.}}{2}$$
$$= 2.75 \text{ in.}$$

$$a' = a + \frac{d_b}{2} \leq \left(1.25b + \frac{d_b}{2}\right) \quad \text{(\textit{Maual} Eq. 9-27)}$$
$$= 2.75 \text{ in.} + \frac{1.00 \text{ in.}}{2} \leq 1.25(2.38 \text{ in.}) + \frac{1.00 \text{ in.}}{2}$$
$$= 3.25 \text{ in.} \leq 3.48$$

Use $a = 3.25$ in.

The tributary length, p, as shown in AISC *Manual* Figure 9-4, for the bolts adjacent to the flanges will be limited by b. For the lower flange, edge distance will also affect the tributary length for the bottom bolts. The average value for p at the lower flange is:

$$p = \frac{2(2.38 \text{ in.}) + 2.38 \text{ in.} + 1.5 \text{ in.}}{2}$$
$$= 4.32 \text{ in.}$$

$d' = 1\frac{1}{16}$ in.

$$\delta = 1 - \frac{d'}{p} \quad \text{(\textit{Maual} Eq. 9-24)}$$
$$= 1 - \frac{1\frac{1}{16} \text{ in.}}{4.32 \text{ in.}}$$
$$= 0.754$$

$$\rho = \frac{b'}{a'} \quad \text{(\textit{Maual} Eq. 9-26)}$$
$$= \frac{1.88 \text{ in.}}{3.25 \text{ in.}}$$
$$= 0.578$$

From AISC *Manual* Equation 9-25:

LRFD	ASD
$\beta = \frac{1}{\rho}\left(\frac{\phi r'_{nt}}{r_{ut}} - 1\right)$	$\beta = \frac{1}{\rho}\left(\frac{r'_{nt}}{\Omega r_{at}} - 1\right)$
$= \frac{1}{0.578}\left(\frac{42.3 \text{ kips}}{17.8 \text{ kips}} - 1\right)$	$= \frac{1}{0.578}\left(\frac{27.2 \text{ kips}}{12.5 \text{ kips}} - 1\right)$
$= 2.38$	$= 2.03$

Because $\beta > 1$, $\alpha' = 1.0$. The minimum required end-plate thickness is, from AISC *Manual* Equation 9-23:

LRFD	ASD
$t_{min} = \sqrt{\dfrac{4r_{ut}b'}{\phi p F_u (1+\delta\alpha')}}$ $= \sqrt{\dfrac{4(17.8 \text{ kips})(1.88 \text{ in.})}{\begin{bmatrix}0.90(4.32 \text{ in.})(65 \text{ ksi})\\ \times[1+0.754(1.0)]\end{bmatrix}}}$ $= 0.550$ in.	$t_{min} = \sqrt{\dfrac{\Omega 4r_{at}b'}{p F_u (1+\delta\alpha')}}$ $= \sqrt{\dfrac{1.67(4)(12.5 \text{ kips})(1.88 \text{ in.})}{4.32 \text{ in.}(65 \text{ ksi})[1+0.754(1.0)]}}$ $= 0.565$ in.

Try a ⅝-in.-thick end plate.

Check bearing strength of end plate

From AISC *Manual* Table 7-4, the minimum spacing required to achieve full bearing strength for 1-in.-diameter bolts is 3¹/₁₆ in. Using the smallest bolt spacing on the end plate (4½ in.) and ASTM A572 Grade 50 plate, the available bearing strength at each interior bolt is (given in the row noted as $s \geq s_{full}$):

LRFD	ASD
$\phi r_n = 117$ kip/in.(⅝ in.) $= 73.1$ kips > 15.9 kips/bolt	$\dfrac{r_n}{\Omega} = 78.0$ kip/in.(⅝ in.) $= 48.8$ kips > 11.2 kips/bolt

Conservatively using AISC *Manual* Table 7-5 with an edge distance equal to 1¼ in., the available bearing strength at each edge bolt is:

LRFD	ASD
$\phi r_n = 42.0$ kip/in.(⅝ in.) $= 26.3$ kips > 15.9 kips/bolt	$\dfrac{r_n}{\Omega} = 28.0$ kip/in.(⅝ in.) $= 17.5$ kips > 11.2 kips/bolt

Because the available strength of each bolt is greater than the required shear strength per bolt, the bearing strength of the end plate is adequate.

Check bearing strength of column flange

Since the column flange thickness is greater than the end-plate thickness and the end plate and column have the same specified minimum tensile strength, the bearing strength of the column flange is adequate.

Use seven rows of two 1-in.-diameter ASTM A325-N bolts at a 5½ in. gage. Use four bolts adjacent to each beam flange and an additional three bolts on each side of the gusset plate as shown in Figure 5-68.

Check shear yielding strength of the end plate

The available shear yielding strength of the end plate is determined as follows:

$$R_n = 2(0.60)F_y A_{gv}$$ (from *Spec.* Eq. J4-3)

$$R_n/l = 2(0.60)F_y t$$
$$= 2(0.60)(50 \text{ ksi})(\tfrac{5}{8} \text{ in.})$$
$$= 37.5 \text{ kip/in.}$$

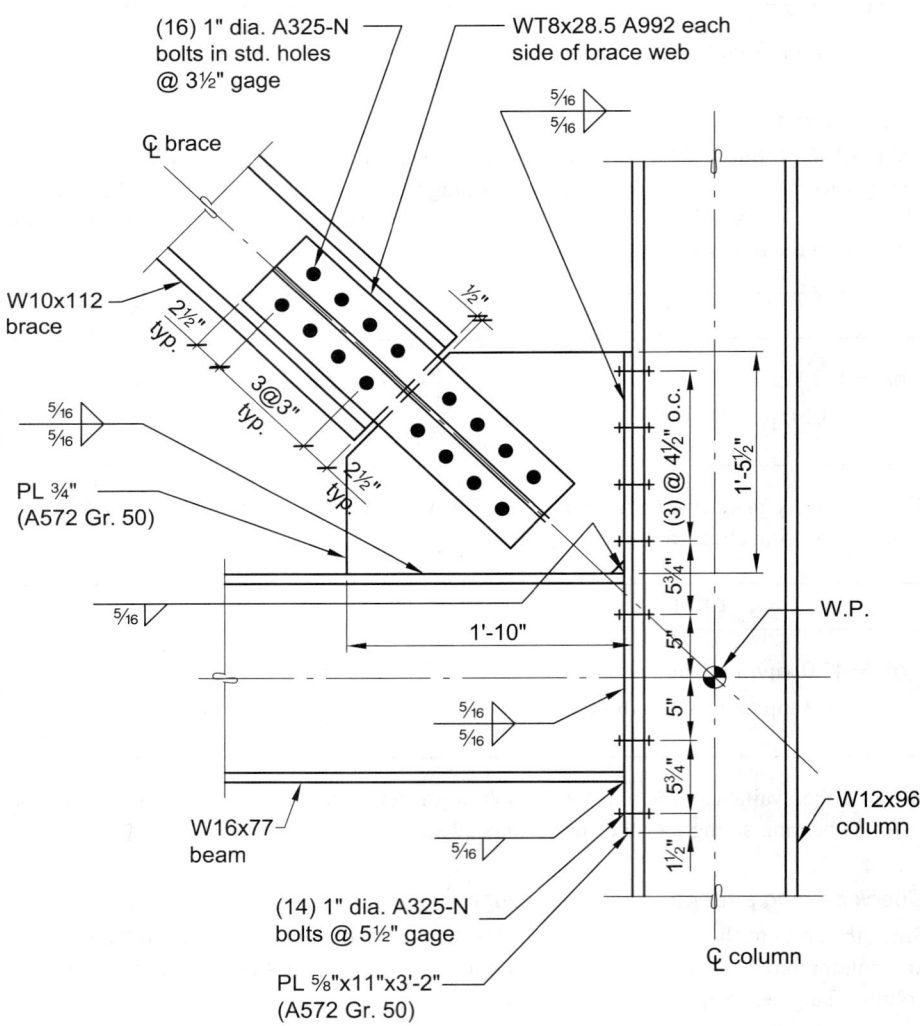

Fig. 5-68. Connection designed in Example 5.4.7.

5.4 ECCENTRICALLY BRACED FRAMES

LRFD	ASD
$\phi R_n = 1.00(37.5 \text{ kip/in.})$ $= 37.5$ kip/in.	$\dfrac{R_n}{\Omega} = \dfrac{37.5 \text{ kip/in.}}{1.50}$ $= 25.0$ kip/in.

This is greater than the required strength at both the beam and gusset connections to the end plate. Therefore, the available end plate shear yielding strength is adequate.

Check end plate rupture at beam web weld

The available shear rupture strength of the end plate at the beam web weld is determined as follows:

$R_n = 2(0.60)F_u A_{nv}$ (from *Spec.* Eq. J4-4)
$= 2(0.60)F_u T_{beam} t$
$= 2(0.60)(65 \text{ ksi})(13\frac{1}{4} \text{ in.})(\frac{5}{8} \text{ in.})$
$= 646$ kips

LRFD	ASD
$\phi R_n = 0.75(646 \text{ kips})$ $= 485$ kips $> V_{ub} + V_{ubeam} = 159$ kips **o.k.**	$\dfrac{R_n}{\Omega} = \dfrac{646 \text{ kips}}{2.00}$ $= 323$ kips $> V_{ab} + V_{abeam} = 111$ kips **o.k.**

Check end-plate rupture at beam flange weld

The tensile rupture strength of the end plate at each beam flange weld is:

$R_n = F_u A_e$ (from *Spec.* Eq. J4-2)
$= F_u t b_f$
$= 65 \text{ ksi}(\frac{5}{8} \text{ in.})(10.3 \text{ in.})$
$= 418$ kips

LRFD	ASD
$\phi R_n = 0.75(418 \text{ kips})$ $= 314$ kips > 90.0 kips **o.k.**	$\dfrac{R_n}{\Omega} = \dfrac{418 \text{ kips}}{2.00}$ $= 209$ kips > 63.0 kips **o.k.**

Check end-plate shear rupture at bolt line

The total height of the end plate is 38.0 in., as shown in Figure 5-68. The available shear rupture strength of the end plate at the bolt line is determined as follows:

$$A_n = 2(\tfrac{5}{8} \text{ in.})\left[38.0 \text{ in.} - 7(1\tfrac{1}{16} \text{ in.} + \tfrac{1}{16} \text{ in.})\right]$$
$$= 37.7 \text{ in.}^2$$

$$R_n = 0.60 F_u A_n \hspace{2cm} (Spec.\ Eq.\ J4\text{-}4)$$
$$= 0.60(65 \text{ ksi})(37.7 \text{ in.}^2)$$
$$= 1{,}470 \text{ kips}$$

LRFD	ASD
$\phi R_n = 0.75(1{,}470 \text{ kips})$ $= 1{,}100 \text{ kips}$	$\dfrac{R_n}{\Omega} = \dfrac{1{,}470 \text{ kips}}{2.00}$ $= 735 \text{ kips}$

The total required shear strength of the end plate is:

LRFD	ASD
$V_u = V_{uc} + V_{ub} - V_{ubeam}$ $= 137 \text{ kips} + 122 \text{ kips} - 36.5 \text{ kips}$ $= 223 \text{ kips}$ 1,100 kips > 223 kips **o.k.**	$V_a = V_{ac} + V_{ab} - V_{abeam}$ $= 95.8 \text{ kips} + 85.4 \text{ kips} - 25.1 \text{ kips}$ $= 156 \text{ kips}$ 735 kips > 156 kips **o.k.**

Use a ⅝ in. × 11 in. end plate.

Check column web local yielding

The centroid of the compressive force is applied at a distance greater than the column depth, d. Therefore, adjacent to each beam flange, the column web local yielding available strength, with l_b taken as the beam flange thickness, is determined as follows:

$$R_n = F_{yw} t_w (5 k_{des} + l_b) \hspace{2cm} (Spec.\ Eq.\ J10\text{-}2)$$
$$= (50 \text{ ksi})(0.550 \text{ in.})\left[5(1.50 \text{ in.}) + 0.760 \text{ in.}\right]$$
$$= 227 \text{ kips}$$

LRFD	ASD
$\phi R_n = 1.00(227 \text{ kips})$ $= 227 \text{ kips} > 90.0 \text{ kips}$ **o.k.**	$\dfrac{R_n}{\Omega} = \dfrac{227 \text{ kips}}{1.50}$ $= 151 \text{ kips} > 63.0 \text{ kips}$ **o.k.**

This available strength can conservatively be applied to check concentrated forces from the gusset plate, since this gusset has a longer bearing length.

LRFD	ASD
$\phi R_n > H_{uc}$ o.k.	$\dfrac{R_n}{\Omega} > H_{ac}$ o.k.

Check column web local crippling

With the centroid of the compressive force applied greater than $d/2$ from the column end, where d is the column depth, the column web local crippling available strength adjacent to each beam flange is determined as follows:

$$R_n = 0.80 t_w^2 \left[1 + 3\left(\dfrac{l_b}{d}\right)\left(\dfrac{t_w}{t_f}\right)^{1.5}\right]\sqrt{\dfrac{EF_{yw}t_f}{t_w}} \qquad \text{(Spec. Eq. J10-4)}$$

$$= 0.80(0.550 \text{ in.})^2 \left[1 + 3\left(\dfrac{0.760 \text{ in.}}{12.7 \text{ in.}}\right)\left(\dfrac{0.550 \text{ in.}}{0.900 \text{ in.}}\right)^{1.5}\right]\sqrt{\dfrac{29{,}000 \text{ ksi}(50)(0.900 \text{ in.})}{0.550 \text{ in.}}}$$

$$= 405 \text{ kips}$$

LRFD	ASD
$\phi R_n = 0.75(405 \text{ kips})$ $= 304 \text{ kips} > 90.0 \text{ kips}$ o.k.	$\dfrac{R_n}{\Omega} = \dfrac{405 \text{ kips}}{2.00}$ $= 203 \text{ kips} > 63.0 \text{ kips}$ o.k.

This available strength can conservatively be applied to check concentrated forces from the gusset plate, since this gusset has a longer bearing length.

LRFD	ASD
$\phi R_n > H_{uc}$ o.k.	$\dfrac{R_n}{\Omega} > H_{ac}$ o.k.

Check prying action on column flange

The prying action model found in the AISC *Manual* can be used to determine the minimum column flange thickness required to prevent flexural yielding of the flange. This flange is thicker than the end plate, which was previously determined to have adequate thickness.

Therefore:

$t_f = 0.900$ in.   o.k.

Check column web panel zone shear

The maximum shear in the column is equal to the gusset-to-column force, H_{uc} (LRFD) or H_{ac} (ASD). Using the required axial compressive strength of the column based on the sum of the strain-hardened expected yield strengths of the links at the third and fourth levels as determined in Example 5.4.5, $P_r = 867$ kips (LRFD) or $P_r = 615$ kips (ASD).

LRFD	ASD
$\dfrac{P_r}{P_c} = \dfrac{867 \text{ kips}}{50 \text{ ksi}(28.2 \text{ in.}^2)}$ $= 0.615$	$\dfrac{P_r}{P_c} = \dfrac{615 \text{ kips}}{0.60(50 \text{ ksi})(28.2 \text{ in.}^2)}$ $= 0.727$

From AISC *Specification* Section J10.6 with $\dfrac{P_r}{P_c} > 0.4$:

$$R_n = 0.60 F_y d_c t_w \left(1.4 - \dfrac{P_r}{P_c}\right) \qquad \text{(Spec. Eq. J10-10)}$$

LRFD	ASD
$\phi R_n = 0.90(0.60)(50 \text{ ksi})(12.7 \text{ in.})$ $\times (0.550 \text{ in.})(1.4 - 0.615)$ $= 148 \text{ kips} > 94.2 \text{ kips} \quad \textbf{o.k.}$	$\dfrac{R_n}{\Omega} = 0.60(50 \text{ ksi})(12.7 \text{ in.})$ $\times (0.550 \text{ in.})(1.4 - 0.727)/1.67$ $= 84.4 \text{ kips} > 65.8 \text{ kips} \quad \textbf{o.k.}$

Check rotational ductility of the beam-to-column connection

AISC *Seismic Provisions* Section F3.6b includes requirements for beam-to-column connections at the location of a brace connection. This example uses option (a), a simple connection which is capable of providing the required rotation. The method for determining rotational ductility of a tee stub connection presented by Thornton (1997) will be used. This is a generalized form of the rotational ductility check for a tee stub connection found in Part 9 of the AISC *Manual*. Thornton (1997) presents the minimum bolt diameter, d_b, required to develop the simple beam end rotation as:

$$d_b = 0.892 t \sqrt{\dfrac{F_y s}{F_t b}\left(\dfrac{b^2}{L^2} + 2\right)}$$

where

- t = end-plate thickness = ⅝ in.
- F_y = specified minimum yield stress of the end plate = 50 ksi
- F_t = tensile strength of the bolt = 120 ksi
- s = bolt spacing = 38 in./7 rows = 5.43 in. (average)
- b = 2.38 in., as previously determined for prying action
- L = depth of connection element = 38 in.

$$d_b = 0.892(\text{\textfrac{5}{8} in.})\sqrt{\frac{50 \text{ ksi}(5.43 \text{ in.})}{120 \text{ ksi}(2.38 \text{ in.})}\left[\frac{(2.38 \text{ in.})^2}{(38.0 \text{ in})^2}+2\right]}$$
$$= 0.769 \text{ in.}$$

The 1-in.-diameter bolts used satisfy this minimum bolt diameter.

The final connection design and geometry is shown in Figure 5-68.

5.5 BUCKLING-RESTRAINED BRACED FRAMES (BRBF)

Buckling-restrained braced frame (BRBF) systems are a special class of concentrically braced frames addressed in AISC *Seismic Provisions* Section F4. Like other concentrically braced frames, BRBF systems resist lateral forces and displacements primarily through the axial strength and stiffness of the brace members. The centerlines of BRBF framing members which meet at a common joint (braces, columns and beams) coincide or nearly coincide, forming a vertical truss capable of being detailed to minimize the effects of flexure. BRBF systems have more ductility and energy dissipation capability than other types of concentrically braced frames because overall buckling of the buckling-restrained brace (BRB) is precluded at forces and deformations corresponding to the design story drift.

Buckling-restrained braces are characterized by their ability to yield in compression as well as in tension. This is accomplished by separating the actions of resisting axial loads and resisting global buckling. AISC *Seismic Provisions* Commentary Figure C-F4.1 illustrates the components of a BRB. Global buckling of the brace is resisted by the BRB casing, which is typically a square or round HSS section, and can be sized as needed for this requirement. Axial tension and compression loads in the brace are resisted by the BRB core which consists of a shaped plate, in either a flat or cruciform section, sized as required by the AISC *Seismic Provisions*. Because both the casing and the core are sized independently, the brace strength can be fine-tuned, eliminating much of the overstrength that other braced frame systems impart to the structure.

Because buckling of the BRB core is restrained to very small amplitudes, the core achieves the same, or greater, strength in compression as in tension. This behavior is repeatable throughout multiple loading cycles without the occurrence of brace buckling (and the consequent degradation associated with it), dissipating high levels of energy and resulting in a highly ductile system. This uniform, predictable behavior eliminates the lateral-force-distribution requirement that exists for SCBF systems (AISC *Seismic Provisions* Section F2.4a) where the percentage of braces in a given line that may be in tension at one time is limited.

Buckling-restrained braced frame systems tend to be cost-competitive and often more economical than SMF, EBF and SCBF systems in terms of material, fabrication and erection costs. Similar to SCBF systems, BRBF systems may have reduced flexibility in floor-plan layout, space planning, and electrical and mechanical routing as a result of the presence of braces. In certain circumstances, however, the frames are exposed and featured in the architecture of the building.

AISC *Seismic Provisions* Section F4.5a requires that beam and column members in BRBF systems satisfy the requirements for highly ductile members. Section F4.6c also requires that BRB connections be designed for 1.1 times the adjusted brace strength in compression to ensure that they can withstand the maximum forces that the braces can deliver, including the effects of strain-hardening beyond the degree corresponding to the calculated drift. These requirements are intended to result in a system with braces that maintain a high level of ductility and hysteretic damping when subjected to severe seismic forces while ensuring that the connecting elements remain essentially elastic allowing the BRB to be the energy dissipating member in the system.

V-type and inverted V-type BRBF systems are required to meet the additional criteria given in AISC *Seismic Provisions* Section F4.4a. These requirements include:

(1) Beams, connections and their supporting members must be designed for gravity dead and live loads, assuming the bracing provides no support.
(2) Beams intersected by braces must be designed for the vertical and horizontal unbalanced loads resulting from the effects of adjusted brace strengths in compression and tension.
(3) Beams must be continuous between columns.
(4) Beams must be braced to satisfy the requirements for moderately ductile members in accordance with AISC *Seismic Provisions* Section D1.2(a).

Because the adjusted brace compression and tension forces are nearly equal, the vertical unbalanced load on the beam is minimal. The available compressive strength of the BRB is greater than the available tensile strength by an amount equal to $(\beta - 1)$ times the adjusted brace strength, where β is the compression strength adjustment factor discussed in AISC *Seismic Provisions* Section F4.2. The vertical component of this difference in force is the unbalanced load that will be developed. Brace configurations that utilize a two-story X-configuration may have even lower unbalanced forces at the beam.

Columns in BRBF systems, like beams, are required to meet the requirements for highly ductile members. According to AISC *Seismic Provisions* Section F4.6d, column splices are required to develop at least 50% of the lesser available flexural strength of the connected members and to have a required shear strength equal to $\Sigma M_{pc}/H_c$ (LRFD) or $\Sigma M_{pc}/(1.5H_c)$ (ASD). This requirement is identical to that for SCBF systems and is intended to account for the possibility of the columns sharing some of the lateral force demand through frame action as the brace elements deform inelastically, deflecting the frames beyond what elastic calculations might predict.

Buckling-restrained braces are required to be designed based upon results from qualifying cyclic tests in accordance with the requirements of AISC *Seismic Provisions* Section K3. Qualifying tests must consist of at least two successful cyclic tests. One of these tests must be a subassemblage test that includes rotational demands at the ends of the BRB. The second test may be either a uniaxial test or a subassemblage test. Qualifying tests may be done specifically for a project or may consist of previous tests documented elsewhere. Contract documents should include requirements for testing of the braces conforming to the AISC *Seismic Provisions*. This requirement demonstrates to the contractor that this is a specialty item and cannot simply be fabricated by a typical steel contractor, but must be procured by a company that has conducted the necessary testing to qualify the braces. Testing of each brace type is required to confirm that the brace design concept meets the requirements to be considered a buckling-restrained brace. It is also performed to determine the load ranges acceptable for a given brace design.

In most systems, member sizes are selected from a table of discrete values. In this way the yielding members are selected to meet the minimum strength requirements, and material variability is addressed through use of the R_y factor for the design of connections and adjacent members.

Buckling-restrained braces, however, are manufactured to match the project requirements and the yielding area can be precisely defined. The details of the brace design, such as the area and length of the yielding zone, can be tuned considering the yield stress of the core material. However, during the design phase the yield stress of the core material is not known precisely, although an acceptable range may be specified. This range should be sufficiently wide to permit a reasonable procurement process for brace manufacturers. The range of 38 to 46 ksi is the de facto industry standard; typically the engineer defines this range by specifying the minimum and maximum core material yield stress ($F_{y\,min}$ and $F_{y\,max}$). Compliance with these limits is verified by the brace manufacturer through coupon tests of the material to be used in the fabrication of the brace. The engineer may account for this material variability in one of two ways: the area-based or the strength-based approach.

The area-based approach is the more common approach for designing BRBF systems. In the area-based approach, the engineer defines the core area; the brace strength is defined by the core area and yield stress. The engineer uses the lower bound yield stress, $F_{y\,min}$, for choosing the core area of the brace. Once this area is established, the upper bound yield stress, $F_{y\,max}$, is used to determine the adjusted brace strength for design of connections and adjacent frame members (factoring up as required to address strain hardening and compression overstrength). Brace core areas may be defined precisely, although there is typically little benefit in precision beyond the nearest ½ in.^{2} In the area-based approach, the brace stiffness used in the analysis is established by the engineer based on the area determined from $F_{y\,min}$ and adjusted upward by the applicable factor (which accounts for the nonprismatic configuration of the brace and is normally supplied by the brace manufacturer). With an area-based approach, stiffness can be specified on the design drawings in terms of brace core area and adjustment factor. In the area-based approach there is a necessary variability in strength resulting from the range of core material yield strength allowed. While it is theoretically possible to specify precisely the stiffness required, it is nonetheless preferable to allow a reasonable tolerance (typically 10% or less) in order to permit the manufacturer to adjust the details of the brace to control brace core strain, provide optimal brace-end conditions, and to allow for the differing details and proportioning used by different manufacturers.

The second option for accounting for material variability is the strength-based approach. In this method the engineer defines the required strength of the brace. (Engineers using this approach should be explicit as to whether they are defining the available strength, ϕP_{ysc} or P_{ysc}/Ω, or the nominal strength, P_{ysc}.) The engineer should specify an acceptable range (e.g., for LRFD, ϕP_{ysc} = 500 kips + 25 kips/− 0 kips) and should use the upper bound for design of connections and adjacent members (factoring up as described previously to account for strain hardening and compression overstrength). This 25 kips tolerance is roughly equivalent to the ½ in.^{2} tolerance recommended for the area-based approach. This method allows the manufacturer to set the brace area provided to adjust for the measured yield stress of the core material so that P_{ysc} is obtained as the yield capacity of the brace. If P_{ysc} is established using yield stress determined from a coupon test, the R_y factor is not applied. Brace stiffness is estimated in the design based on the area from an assumed yield stress in the middle of the specified range and the applicable adjustment factor as described previously. With a strength-based approach, stiffness must be specified explicitly (in kip/in.) on the design drawings with a specified tolerance. This tolerance is typically ± 10%. That range, however, is insufficient to cover both the adjustments implied by the range of core yield stress (± 10%) and the differing details and proportioning used by different manufacturers. Thus for material stress at the extremes of the permitted range, the manufacturer may need to make adjustments in the core length (along with other details such as the core area outside the yielding zone) in order to maintain stiffness in the specified range of ± 10%. The details so configured must comply with the range of the brace tests. (For example, core strains must be calculated using the detailed core length and compared to values in the tests.) Such adjustments in the brace details would not be necessary if the specified range were increased to ± 20%, but this is not typical practice. In the strength-based approach, therefore, the material variability becomes a variability in stiffness and may also limit the applicability of tests to a smaller range. This is a consequence of the overall variability implied by the tolerances commonly used in this method. While it is theoretically possible to specify precisely the strength required in the strength-based approach, it is nonetheless preferable to allow a reasonable tolerance (typically 25 kips) in order to permit the manufacturer to use ½-in. dimensions; precision beyond this is not warranted given the methods used to establish yield stress.

Table 5-4 summarizes how the area-based and strength-based approaches address the effect of material variability on the strength and stiffness of buckling-restrained braces.

It should be noted that both the elastic stiffness and the first yield strength (both necessary properties for use in code-based seismic design) are transient properties in the actual seismic response of systems; these properties change significantly as drifts exceed the drift corresponding to first yield. Designers should not perform bounding analyses or otherwise place undue emphasis on the effects of variability beyond accounting for maximum brace forces in the design of connections, beams and columns. Such variability in stiffness is routinely (and justly) neglected in the seismic design of many systems and is minimal in the context of the use of elastic methods to represent inelastic response.

Brace strength is controlled by brace core area, but the use of this core area in the structural model without any adjustment will not correctly capture the stiffness of the brace. Overall brace stiffness includes contributions from not only the yielding core, but also from the nonyielding portions of the brace and connection materials. This stiffness is usually captured in the model through the use of a stiffness modification factor, KF. The modeled brace stiffness would then be represented by the following equation:

$$K_{model} = \frac{KF(A_{sc})E}{L_{wp-wp}} \qquad (5\text{-}1)$$

where A_{sc} is the steel core area, E is the modulus of elasticity, and $L_{wp\text{-}wp}$ is the work point-to-work point distance along the axis of the brace. The modeled brace stiffness can also be represented as a spring with a defined stiffness, K_{model}.

The stiffness factor or modeled brace stiffness is unique to each brace manufacturers' design, although it may be similar between manufacturers. It is also dependent on brace strength, bay geometry and connection details. The design engineer will need to assume an initial value for this factor for early estimation of required brace strength and preliminary beam and column sizes and will send this information to a brace manufacturer for early coordination to obtain the recommended stiffness modification factors for the braces. If brace strengths are adjusted, final values should also be confirmed with the manufacturer prior to finalizing contract documents.

Table 5-4
Summary of Variability in the Area-Based and Strength-Based Approaches

Method	Strength Variability	Stiffness variability
Area-based approach	Implicit, 46 ksi/38 ksi ~ 1.2 (+20%/−0%)	Engineer-specified, typically ±10%
Strength-based approach	Engineer-specified, typically ≤ 5% (+5%/−0%)	Engineer-specified, typically ±10%

As buckling-restrained braces are typically provided by a specialty manufacturer who designs the details of the brace (such as sizing the casing, determining the details of the transitions between yielding and nonyielding zones, etc.), the design process may be slightly different from that of other systems in that it ideally involves input from brace manufacturers during the design process.

The manufacturer typically proposes certain details of the brace connection. It is often convenient for the engineer to delegate certain parts of the connection design to the manufacturer, such as the connection of the brace to the gusset plate. The engineer must explicitly identify any such delegated design, and review the corresponding calculations with the brace submittal.

The following design process illustrates the interaction between the engineer of record (EOR) and the manufacturer.

1. Preliminary design phase.
   (a) EOR determines base shear, frame layout, etc.
   (b) EOR sizes braces (required core area or required strength).
   (c) EOR assumes brace stiffness factors, KF, and overstrength factors, β and ω. (A preliminary consultation with the manufacturer may be helpful at this stage.)
   (d) EOR sizes beams and columns.
   (e) EOR checks drift.
   (f) EOR estimates brace deformations.

2. Consultation with manufacturer(s). EOR consults with manufacturer for:
   (a) Sufficient applicable testing for the brace sizes proposed
   (b) Stiffness of braces or stiffness factors, KF, used in the EOR's analysis
   (c) Overstrength factors β and ω used in the EOR's design of beams and columns.

3. Design iteration. EOR reanalyzes (as required by change in member size or change in stiffness factors, etc.).
   (a) EOR finalizes brace sizes, beam and column sizes, brace stiffness factors and brace deformations.
   (b) EOR consults with the brace manufacturer if the brace sizes or deformations are substantially different than the preliminary design.

4. Specification. EOR specifies:
   (a) Required brace sizes (core area or required strength), with tolerance
   (b) Minimum and maximum core material yield stress
   (c) Overstrength factors β and ω.
   (d) Brace stiffness (or stiffness factors), with tolerance
   (e) Required brace axial deformation and connection rotation
   (f) Testing per the AISC *Seismic Provisions*
   (g) Connection design or portions thereof delegated to the brace manufacturer
5. Brace submittal.
   (a) Manufacturer submits:
      i. Brace shop drawings
      ii. Supporting documentation
         (a) Justifying applicable tests (in terms of brace size, strain at the specified displacements)
         (b) Overstrength factors, β and ω (based on specified displacements)
         (c) Brace stiffness calculations
         (d) Connection design, where delegated to brace manufacturer
   (b) Test reports for submitted brace types and sizes
   (c) EOR verifies compliance with specification

BRBF Design Example Plan and Elevation

The following examples illustrate the design of a BRBF based on AISC *Seismic Provisions* Section F4. The plan and elevation are shown in Figure 5-69 and Figure 5-70.

The lateral forces shown in Figure 5-70 are the seismic forces from an equivalent lateral force procedure of ASCE/SEI 7 Section 12.8 and apply to the entire frame.

The code-specified gravity loading is as follows:

D_{floor} = 85 psf
D_{roof} = 68 psf
L_{floor} = 80 psf (50 psf reduced)
S = 20 psf
Curtain wall = 175 lb/ft along building perimeter at every level

The applicable building code specifies the use of ASCE/SEI 7 for calculation of loads. From ASCE/SEI 7, the Seismic Design Category is D, ρ = 1.3, I_e = 1.0, S_{DS} = 1.0, and C_d = 5.0.

The vertical seismic load effect, E_v, according to ASCE/SEI 7 is:

$0.2 S_{DS} D = 0.2(1.0) D$ (ASCE/SEI 7 Eq. 12.4-4)
$\quad\quad\quad\quad = 0.2 D$

Assume that the ends of the diagonal braces are pinned and braced against translation for both the *x-x* and *y-y* axes. The loads given for each example are from a first-order analysis. Assume that the effective length method of AISC *Specification* Appendix 7 is used for the

5.5 BUCKLING-RESTRAINED BRACED FRAMES

stability design. AISC *Specification* Appendix 8 will be applied to approximate a second-order analysis.

Example 5.5.1. BRBF Brace Design

Given:

Refer to Brace BRB-1 in Figure 5-70. Frame configurations and preliminary loads have been sent to a BRB manufacturer and the elastic stiffness of the braces have been found to be 1.5 times higher than the stiffness of the yielding core area alone, if it were extended from work point to work point ($KF = K_{actual}/K_{core} = 1.28$). These stiffness factors may be used to determine the horizontal load distribution on each story. Design a buckling-restrained brace to resist the resulting axial loading, $P_{Q_E} = 113$ kips. The applicable building code specifies the use of ASCE/SEI 7 for calculation of loads. According to AISC *Seismic Provisions* Section F4.3, buckling-restrained braces should not be considered as resisting gravity forces.

Using the area-based approach described previously, allow for material variability of 42 ksi ± 4 ksi.

$F_{ysc\ min} = 38$ ksi $\qquad F_{ysc\ max} = 46$ ksi

From an elastic analysis, the first-order interstory drift is $\Delta_H = 0.223$ in.

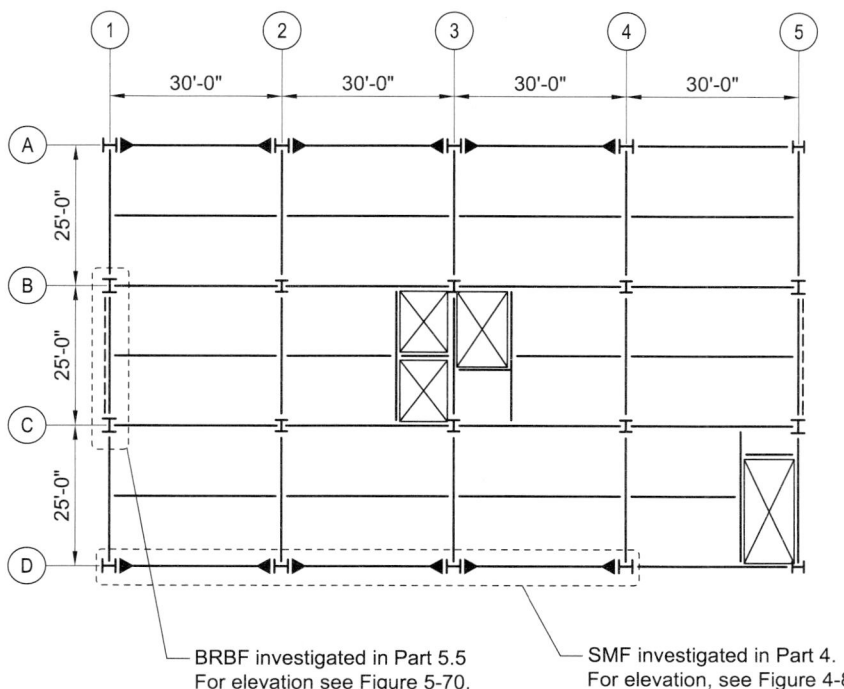

Fig. 5-69. Floor plan for BRBF Examples.

Assume that the ends of the brace are pinned and braced against translation for both the x-x and y-y axes.

Solution:

The governing load combinations in ASCE/SEI 7 including seismic effects are:

LRFD	ASD
LRFD Load Combinations 5 and 6 from ASCE/SEI 7 Section 12.4.2.3 (including the 0.5 factor on L permitted in Section 12.4.2.3)	ASD Load Combinations 5 and 8 from ASCE/SEI 7 Section 12.4.2.3
$(1.2 + 0.2S_{DS})D + \rho Q_E + 0.5L + 0.2S$	$(1.0 + 0.14S_{DS})D + H + F + 0.7\rho Q_E$
$(0.9 - 0.2S_{DS})D + \rho Q_E + 1.6H$	$(0.6 - 0.14S_{DS})D + 0.7\rho Q_E + H$

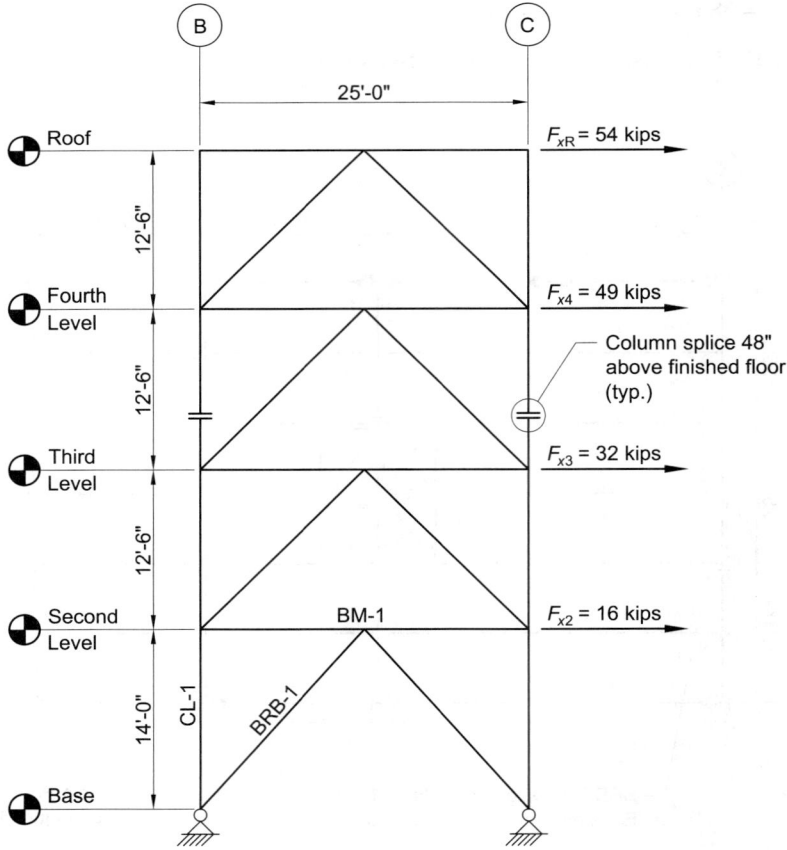

Fig. 5-70. Frame elevation for BRBF examples.

The required compressive and tensile strengths of the brace are:

LRFD	ASD
$P_u = T_u$ $= \rho P_{QE}$ $= 1.3(113 \text{ kips})$ $= 147 \text{ kips}$	$P_a = T_a$ $= 0.7\rho P_{QE}$ $= 0.7(1.3)(113 \text{ kips})$ $= 103 \text{ kips}$

Required Strength

Consider second-order effects

AISC *Specification* Appendix 8 is used to address second-order effects. The required second-order axial strength is:

$$P_r = P_{nt} + B_2 P_{lt} \qquad \text{(Spec. Eq. A-8-2)}$$

For the calculation of B_2:

$$B_2 = \frac{1}{1 - \frac{\alpha P_{story}}{P_{e\,story}}} \geq 1 \qquad \text{(Spec. Eq. A-8-6)}$$

To determine P_{story}, use an area of 9,000 ft² on each floor and the surface gravity loads given in the BRBF Design Example Plan and Elevation section. Use load combinations that include seismic effects.

LRFD	ASD
$P_{story} = 9{,}000 \text{ ft}^2 \begin{Bmatrix} [1.2 + 0.2(1.0)] \\ \times [68 \text{ psf} + 3(85 \text{ psf})] \\ + 0 \text{ psf} \\ + 0.5(3)(50 \text{ psf}) \\ + 0.2(20 \text{ psf}) \end{Bmatrix}$ $\times (1 \text{ kip}/1{,}000 \text{ lb})$ $+ \begin{Bmatrix} [1.2 + 0.2(1.0)] \\ \times [175 \text{ lb/ft}(4)(390 \text{ ft})] \\ \times (1 \text{ kip}/1{,}000 \text{ lb}) \end{Bmatrix}$ $= 5{,}160 \text{ kips}$	$P_{story} = 9{,}000 \text{ ft}^2 \begin{Bmatrix} [1.0 + 0.14(1.0)] \\ \times [68 \text{ psf} + 3(85 \text{ psf})] \\ + 0 \text{ psf} + 0 \text{ psf} + 0 \text{ psf} \end{Bmatrix}$ $\times (1 \text{ kip}/1{,}000 \text{ lb})$ $+ \begin{Bmatrix} [1.0 + 0.14(1.0)] \\ \times [175 \text{ lb/ft}(4)(390 \text{ ft})] \\ \times (1 \text{ kip}/1{,}000 \text{ lb}) \end{Bmatrix}$ $= 3{,}630 \text{ kips}$

The total story shear, H, with two bays of bracing in the direction under consideration where each braced frame is designed to resist the seismic loads shown in Figure 5-70, is determined in the following. From an elastic analysis, the first-order interstory drift is $\Delta_H = 0.223$ in.

$H = 2(54.0 \text{ kips} + 49.0 \text{ kips} + 32.0 \text{ kips} + 16.0 \text{ kips})$
$= 302 \text{ kips}$

$L = 14.0 \text{ ft}$

$R_M = 1.0$ for braced frames

$P_{e\ story} = R_M \dfrac{HL}{\Delta_H}$ (*Spec.* Eq. A-8-7)

$= 1.0 \dfrac{302 \text{ kips}(14.0 \text{ ft})}{0.223 \text{ in.}(1 \text{ ft}/12 \text{ in.})}$

$= 228{,}000 \text{ kips}$

Using AISC *Specification* Equation A-8-6:

LRFD	ASD
$\alpha = 1.00$	$\alpha = 1.60$
$B_2 = \dfrac{1}{1 - \dfrac{\alpha P_{story}}{P_{e\ story}}} \geq 1$	$B_2 = \dfrac{1}{1 - \dfrac{\alpha P_{story}}{P_{e\ story}}} \geq 1$
$= \dfrac{1}{1 - \dfrac{1.00(5{,}160 \text{ kips})}{228{,}000 \text{ kips}}}$	$= \dfrac{1}{1 - \dfrac{1.60(3{,}630 \text{ kips})}{228{,}000 \text{ kips}}}$
$= 1.02$	$= 1.03$

Considering second-order effects, the required compressive and tensile strengths of the brace are:

LRFD	ASD
$P_u = T_u$	$P_a = T_a$
$= 1.02(147 \text{ kips})$	$= 1.03(103 \text{ kips})$
$= 150 \text{ kips}$	$= 106 \text{ kips}$

Determination of the brace area required to resist the required brace strength must use the minimum yield of the core material, $F_{ysc\ min}$. For the limit state of tensile or compressive yielding, set the required strength equal to AISC *Seismic Provisions* Equation F4-1 and solve for $A_{sc\ min}$:

5.5 BUCKLING-RESTRAINED BRACED FRAMES

LRFD	ASD
$A_{sc\ min} = \dfrac{P_u}{\phi F_{ysc\ min}}$ $= \dfrac{150 \text{ kips}}{0.90(38 \text{ ksi})}$ $= 4.39 \text{ in.}^2$	$A_{sc\ min} = \dfrac{\Omega P_a}{F_{ysc\ min}}$ $= \dfrac{1.67(106 \text{ kips})}{38 \text{ ksi}}$ $= 4.66 \text{ in.}^2$

In design practice, either LRFD or ASD design should be used consistently. The two methods give slightly different results here. In order not to show two separate designs, the LRFD result will be used.

Try a BRB with a core area, A_{sc}, of 4.50 in.^{2}

Note that while BRB manufacturers can fabricate a BRB with the accuracy to which the core can be cut (generally ± 1/8 in. in width) it is common to round the required core area up to standard increments. Generally, it is good practice to specify core areas in 0.25 in.^{2} increments for 0 in.^{2} < A_{sc} ≤ 5.00 in.^{2}, in 0.50 in.^{2} increments for 5.00 in.^{2} < A_{sc} ≤ 10.0 in.^{2}, in increments of 1.00 in.^{2} for 10.0 in.^{2} < A_{sc} ≤ 20.0 in.^{2}, and in 2.00 in.^{2} increments for A_{sc} > 20.0 in.^{2} (or maintaining increment amounts in the range of 5% to 10% of the total amount). When specifying BRB area greater than required, the EOR must account for the increased demand that the specified area will place on the structure, because the beams and columns are designed to be stronger than the adjusted brace strength.

For LRFD, the available axial strength for the limit state of tensile or compression yielding is:

$\phi P_{n\ min} = \phi F_{ysc\ min} A_{sc}$ (Spec. Eq. D2-1)

$= 0.90(38 \text{ ksi})(4.50 \text{ in.}^2)$

$= 154 \text{ kips} < 150 \text{ kips}$ **o.k.**

Verify with the brace manufacturer that the stiffness factor $KF = 1.28$ is acceptable for a 4.50 in.^{2} brace of this length. The remainder of the brace design is performed by the BRB manufacturer. Overstrength factors, β and ω, along with the maximum deformation capability of the brace, must be provided by the brace manufacturer in order to design the columns and beams of the BRBF and to determine the BRB applicability to the design.

The final part of the brace design is establishing the expected deformation of the brace and using this deformation to determine forces that the brace imposes on the columns, beams and connections. AISC *Seismic Provisions* Section F4.2 requires consideration of deformations at the greater of 2% drift or two times the design story drift.

The design story drift is defined in the AISC *Seismic Provisions* Glossary as the calculated story drift including the effect of expected inelastic action. As given, the first-order interstory drift is $\Delta_H = 0.223$ in. This drift does not include the redundancy factor, ρ. Note that

ASCE/SEI 7 Section 12.3.4.1 permits ρ to be taken equal to 1 for drift calculations. The design story drift including inelastic action is:

$$\Delta = \frac{C_d \Delta_H}{I_e} \quad \text{(ASCE/SEI 7 Eq. 12.8-15)}$$

$$= \frac{5.0(0.223 \text{ in.})}{1.0}$$

$$= 1.12 \text{ in.}$$

Twice the story drift including inelastic action is:

$$2\Delta = 2(1.12 \text{ in.})$$
$$= 2.24 \text{ in.}$$

2% drift corresponds to a deflection of:

$$\Delta = 0.02H$$
$$= 0.02(14.0 \text{ ft})$$
$$= 0.280 \text{ ft}$$

$$\Delta = 0.280 \text{ ft}(12 \text{ in./1 ft})$$
$$= 3.36 \text{ in.}$$

In this case, 2% drift governs. The brace spans 14.0 ft vertically and 12.5 ft horizontally. The brace deformation can be calculated to be:

$$\Delta_{br} = \left[\sqrt{(14.0 \text{ ft})^2 + (12.5 \text{ ft} + 0.280 \text{ ft})^2} - \sqrt{(14.0 \text{ ft})^2 + (12.5 \text{ ft})^2}\right](12 \text{ in./1 ft})$$
$$= 2.25 \text{ in.}$$

Consulting with the brace manufacturer, the yield length for this brace is determined to be 70% of the work-point length.

$$L_y \geq 0.7L$$
$$= 0.7\sqrt{(14.0 \text{ ft})^2 + (12.5 \text{ ft})^2}(12 \text{ in./1 ft})$$
$$= 158 \text{ in.}$$

The strain is therefore:

$$\varepsilon = \frac{\Delta_{br}}{L_y}$$
$$= \frac{2.25 \text{ in.}}{158 \text{ in.}}$$
$$= 1.42\%$$

5.5 BUCKLING-RESTRAINED BRACED FRAMES

Determination of the strain and the yield length is typically performed by the brace manufacturer and is shown here for illustrative purposes only.

Consulting with the brace manufacturer, the β and ω factors corresponding to this level of strain are determined to be:

$\omega = 1.36$

$\beta = 1.1$

Alternatively, according to AISC *Seismic Provisions* Section F4.3 and ASCE/SEI 7 Chapter 16, brace deformation is permitted to be determined from a nonlinear analysis in lieu of the expected deformation requirements in AISC *Seismic Provisions* Section F4.2 illustrated here.

Example 5.5.2. BRBF Column Design

Given:
Refer to Column CL-1 in the frame shown in Figure 5-70. Select an ASTM A992 wide-flange section to resist the following axial loading between the base and the second level. The applicable building code specifies the use of ASCE/SEI 7 for calculation of loads.

$P_D = 147$ kips $P_L = 60.0$ kips $P_S = 7.00$ kips

Relevant seismic parameters are given in the BRBF Design Example Plan and Elevation Section.

The brace core areas are as indicated in Figure 5-71 (BRB X.X indicates a brace with a core area of X.X in.^{2}) Allow for BRB core material variability of 42 ksi ± 4 ksi ($F_{ysc\ min} = 38$ ksi, $F_{ysc\ max} = 46$ ksi). The brace manufacturer has provided the given overstrength factors. From AISC *Seismic Provisions* Section F4.2a, the factor R_y need not be applied if $P_{ysc}(= F_{ysc})$ is determined from a coupon test, as is the case here. Therefore, R_y will not be shown in the examples in Section 5.5.

$\omega = 1.36$ $\beta = 1.1$ $\beta\omega = 1.5$

Solution:
From AISC *Manual* Table 2-4, the material properties are as follows:

ASTM A992
$F_y = 50$ ksi
$F_u = 65$ ksi

According to AISC *Seismic Provisions* Section F4.3, the required strength of columns due to the applied seismic load, $P_{E_{mh}}$, is based on the adjusted strengths of the braces in the frame, where adjusted strength is defined in AISC *Seismic Provisions* Section F4.2a. Use the specified A_{sc} and $F_{y\ max}$ to determine the brace forces in the design of the column to account for material variability. Starting at the lower braces, the adjusted brace strengths for the braces contributing to the load on Column CL-1 in compression are:

$$\beta\omega P_{ysc\,max2} = \beta\omega A_{sc2} F_{ysc\,max}$$
$$= 1.1(1.36)(3.75 \text{ in.}^2)(46 \text{ ksi})$$
$$= 258 \text{ kips}$$

$$\beta\omega P_{ysc\,max3} = \beta\omega A_{sc3} F_{ysc\,max}$$
$$= 1.1(1.36)(3.00 \text{ in.}^2)(46 \text{ ksi})$$
$$= 206 \text{ kips}$$

$$\beta\omega P_{ysc\,max4} = \beta\omega A_{sc4} F_{ysc\,max}$$
$$= 1.1(1.36)(1.50 \text{ in.}^2)(46 \text{ ksi})$$
$$= 103 \text{ kips}$$

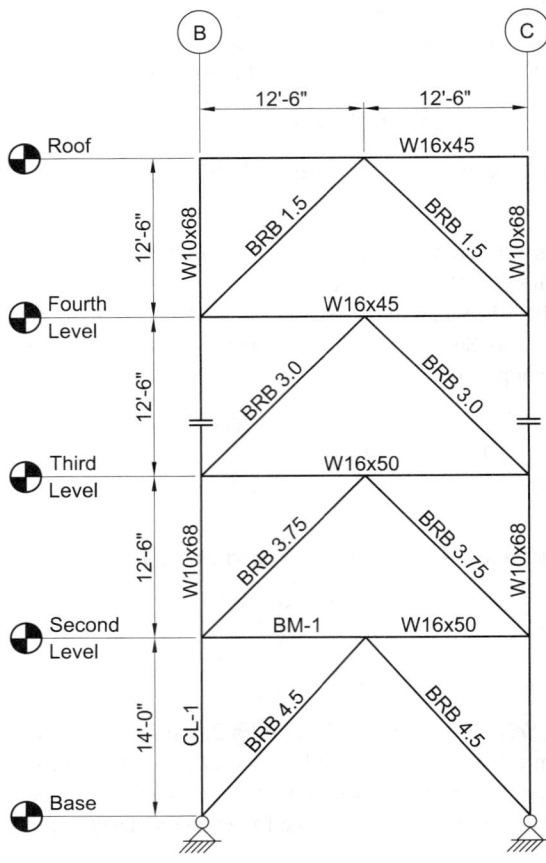

Fig. 5-71. Frame elevation for BRBF examples with member sizes.

5.5 BUCKLING-RESTRAINED BRACED FRAMES

The axial compressive force, $P_{E_{mh}}$, is then determined from the force diagram of the column, as shown in Figure 5-72.

The vertical force on the column from the braces' adjusted strength is:

$$\sum \beta \omega P_{ysc\ max} \sin \theta = (258 \text{ kips} + 206 \text{ kips} + 103 \text{ kips})(\sin 45°)$$
$$= 401 \text{ kips}$$

The vertical component of the force from the tension brace on the beam will be $\omega P_{ysc\ max} \sin \theta$ and the vertical component of the force from the compression brace on the beam will be $\omega \beta P_{ysc\ max} \sin \theta$. The net sum of these forces, which act in opposite directions, is $\omega \beta P_{ysc\ max} \sin \theta - \omega P_{ysc\ max} \sin \theta = (\beta - 1) \omega P_{ysc\ max} \sin \theta$, with half of this force reacting at each end of the beam. Thus the force due to beam shears resulting from unbalanced brace-induced vertical forces is:

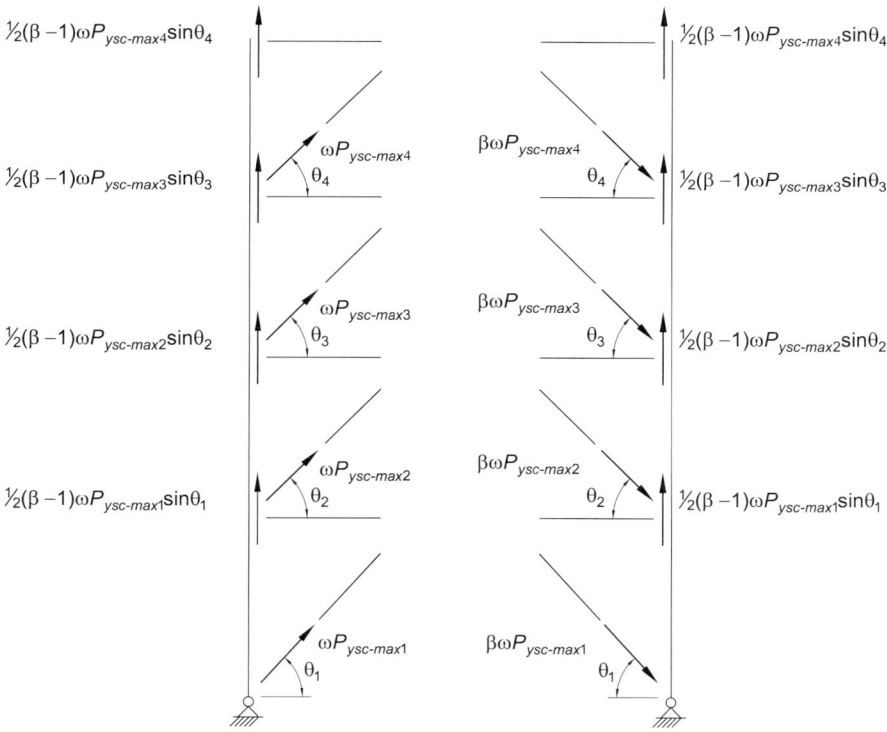

Fig. 5-72. BRBF column forces for Example 5.5.2.

$$\sum \tfrac{1}{2}(\beta-1)\omega P_{ysc\ max}\sin\theta = \tfrac{1}{2}(\beta-1)\omega F_{ysc\ max}\sum A_{sc}\sin\theta$$
$$= \tfrac{1}{2}(1.1-1)(1.36)(46\text{ ksi})$$
$$\times \begin{bmatrix} (1.50\text{ in.}^2)(\sin 45.0°) \\ +(3.00\text{ in.}^2)(\sin 45.0°) \\ +(3.75\text{ in.}^2)(\sin 45.0°) \\ +(4.50\text{ in.}^2)(\sin 48.2°) \end{bmatrix}$$
$$= 28.7\text{ kips}$$

See Example 5.5.3 for calculation of vertical unbalanced forces from the braces on the beam.

The total axial compression in the column due to the braces is:

$$P_{E_{mh}} = 401\text{ kips} - 28.7\text{ kips}$$
$$= 372\text{ kips}$$

Starting at the lower braces, the adjusted brace strengths for the braces contributing to the load on Column CL-1 in tension are:

$$\omega P_{ysc\ max2} = \omega A_{sc2} F_{ysc\ max}$$
$$= 1.36(3.75\text{ in.}^2)(46\text{ ksi})$$
$$= 235\text{ kips}$$

$$\omega P_{ysc\ max3} = \omega A_{sc3} F_{ysc\ max}$$
$$= 1.36(3.00\text{ in.}^2)(46\text{ ksi})$$
$$= 188\text{ kips}$$

$$\omega P_{ysc\ max4} = \omega A_{sc4} F_{ysc\ max}$$
$$= 1.36(1.50\text{ in.}^2)(46\text{ ksi})$$
$$= 93.8\text{ kips}$$

The force due to the adjusted brace strength is:

$$\sum \omega P_{ysc\ max}\sin\theta = (235\text{ kips} + 188\text{ kips} + 93.8\text{ kips})(\sin 45°)$$
$$= 365\text{ kips}$$

From the calculation for the column in compression, the force on the column due to brace-induced beam shear is 28.7 kips.

The total axial tension in the column due to the braces is:

$$T_{E_{mh}} = 365\text{ kips} + 28.7\text{ kips}$$
$$= 394\text{ kips}$$

5.5 BUCKLING-RESTRAINED BRACED FRAMES

Using the load combinations in ASCE/SEI 7 that include the amplified seismic load, where the amplified seismic load is substituted with the analysis described in AISC *Seismic Provisions* Section F4.3, the required compressive force in the column is:

LRFD	ASD
LRFD Load Combination 5 from ASCE/SEI 7 Section 12.4.3.2 (including the 0.5 factor on L permitted in Section 12.4.3.2)	ASD Load Combination 5 from ASCE/SEI 7 Section 12.4.3.2
$P_u = (1.2 + 0.2S_{DS})P_D + P_{E_{mh}} + 0.5P_L$ $+ 0.2P_S$ $= [1.2 + 0.2(1.0)](147 \text{ kips}) + 372 \text{ kips}$ $+ 0.5(60.0 \text{ kips}) + 0.2(7.00 \text{ kips})$ $= 609 \text{ kips}$	$P_a = (1.0 + 0.14S_{DS})P_D + P_H + P_F$ $+ 0.7P_{E_{mh}}$ $= 1.14(147 \text{ kips}) + 0 \text{ kips} + 0 \text{ kips}$ $+ 0.7(372 \text{ kips})$ $= 428 \text{ kips}$

The required axial tensile strength of the column is:

LRFD	ASD
LRFD Load Combination 7 from ASCE/SEI 7 Section 12.4.3.2	ASD Load Combination 8 from ASCE/SEI 7 Section 12.4.3.2
$P_u = (0.9 - 0.2S_{DS})P_D + P_{E_{mh}} + 1.6P_H$ $= [0.9 - 0.2(1.0)](147 \text{ kips})$ $+ (-394 \text{ kips}) + 1.6(0 \text{ kips})$ $= -291 \text{ kips}$	$P_a = (0.6 - 0.14S_{DS})P_D + 0.7P_{E_{mh}} + P_H$ $= [0.6 - 0.14(1.0)](147 \text{ kips})$ $+ 0.7(-394 \text{ kips}) + 0 \text{ kips}$ $= -208 \text{ kips}$

Consider second-order effects

Because the seismic component of the beam required strength comes from the mechanism analysis of AISC *Seismic Provisions* Section F4.3 and is based on the expected strengths of the braces, P-Δ effects need not be considered and B_2 from AISC *Specification* Appendix 8 need not be applied. P-Δ effects do not increase the forces corresponding to the expected brace strengths in compression and tension; instead, they may be thought of as contributing to the system reaching that state. P-δ effects do apply, but because the column does not have moments, there is no need to calculate B_1 factors.

Try a W10×68.

From AISC *Manual* Table 1-1, the geometric properties are as follows:

$A = 19.9$ in.^{2} $d = 10.4$ in. $t_w = 0.470$ in. $b_f = 10.1$ in.
$t_f = 0.770$ in. $r_x = 4.44$ in. $r_y = 2.59$ in. $Z_x = 85.3$ in.^{3}
$S_x = 75.7$ in.^{3} $h/t_w = 16.7$ $b_f/2t_f = 6.58$

Width-to-Thickness Limitations

According to AISC *Seismic Provisions* Section F4.5a, the stiffened and unstiffened elements of columns must comply with the width-to-thickness limits for highly ductile members given in AISC *Seismic Provisions* Table D1.1.

From Table 1-3 in this Manual, the W10×68 satisfies the width-to-thickness requirements for a BRBF column.

Available Compressive Strength

From AISC *Specification* Section C1.2 and Appendix 7, Section 7.2.1, the effective length method is limited to conditions in which the structure supports gravity loads primarily through nominally vertical columns, walls or frames and the ratio of maximum second-order drift to maximum first-order drift in all stories is equal to or less than 1.5. Assume both conditions are met for BRBF systems. $K = 1.0$ for both the x-x and y-y axes. Since the unbraced length is the same for both axes, use the least radius of gyration. From AISC *Manual* Table 4-1, the available compressive strength for a W10×68 with $KL = 14$ ft is:

LRFD	ASD
$\phi_c P_n = 658$ kips > 609 kips **o.k.**	$\dfrac{P_n}{\Omega_c} = 438$ kips > 428 kips **o.k.**

Available Tensile Strength

From AISC *Manual* Table 5-1, the available strength of the W10×68 column in axial tension for yielding on the gross section is:

LRFD	ASD
$\phi_t P_n = 896$ kips > 291 kips **o.k.**	$\dfrac{P_n}{\Omega_t} = 596$ kips > 208 kips **o.k.**

Use a W10×68 for Column CL-1. Verify with the BRB manufacturer that the stiffness and overstrength factors are still applicable for the final bay geometry. Verify that the ratio of second-order drift to first-order drift is less than or equal to 1.5.

Example 5.5.3. BRBF Beam Design

Given:

Refer to Beam BM-1 in Figure 5-70. Select a noncomposite ASTM A992 wide-flange section (the beam may be constructed as a composite member, but for simplicity it is designed as a noncomposite beam). Assume that the bottom flange of the beam is laterally braced at its quarter points. The applicable building code specifies the use of ASCE/SEI 7 for calculation of loads. Relevant seismic parameters were given in the BRBF Design Example Plan and Elevation section. The gravity shears and moments on the beam are:

5.5 BUCKLING-RESTRAINED BRACED FRAMES

$V_D = 11.2$ kips $V_L = 8.50$ kips $M_D = 120$ kip-ft $M_L = 100$ kip-ft

The brace core areas are as indicated on Figure 5-71. (BRB X.X indicates a brace with a core area of X.X in.^{2}) Allow for BRB core material variability of 42 ksi ± 4 ksi ($F_{ysc\ min} = 38$ ksi, $F_{ysc\ max} = 46$ ksi). The brace manufacturer has provided the given overstrength factors:

$\omega = 1.36$ $\beta = 1.1$

Solution:

From AISC *Manual* Table 2-4, the material properties are:

ASTM A992
$F_y = 50$ ksi
$F_u = 65$ ksi

Use the specified A_{sc} and $F_{ysc\ max}$ to determine the brace forces in the design of the beam to account for material variability.

Required Strength

Determine the adjusted brace strength of the tension Brace BRB-1

$$\omega A_{sc} F_{ysc\ max} = 1.36(4.50 \text{ in.}^2)(46 \text{ ksi})$$
$$= 282 \text{ kips}$$

Determine the adjusted brace strength of the compression Brace BRB-1

$$\beta \omega A_{sc} F_{ysc\ max} = 1.1(1.36)(4.50 \text{ in.}^2)(46 \text{ ksi})$$
$$= 310 \text{ kips}$$

Determine the unbalanced vertical load on the beam

The difference between the vertical components of the brace forces is:

$$P_y = (310 \text{ kips} - 282 \text{ kips})(\sin 48.2°)$$
$$= 20.9 \text{ kips}$$

Consequently, there is a 20.9-kip force acting upward on the beam.

Determine the shear and moment in the beam due to the brace analysis

Assuming a simply supported beam:

$$V_{E_{mh}} = \frac{-P_y}{2}$$
$$= \frac{-20.9 \text{ kips}}{2}$$
$$= -10.5 \text{ kips}$$

$$M_{E_{mh}} = \frac{-P_y L}{4}$$
$$= \frac{-20.9 \text{ kips}(25.0 \text{ ft})}{4}$$
$$= -131 \text{ kip-ft}$$

Determine the axial force in the beam

The horizontal components of the brace forces are:

$$P_{tx} = 282 \text{ kips}(\cos 48.2°)$$
$$= 188 \text{ kips}$$

$$P_{cx} = 310 \text{ kips}(\cos 48.2°)$$
$$= 207 \text{ kips}$$

These forces are delivered to the brace through axial forces in the beam—tension in the segment of the beam on one side of the midspan connection (braces-to-beam) and compression in the other segment. The distribution of the total horizontal force between tension and compression depends on the load path and tributary mass. Forces from collectors on each side of the frame may differ significantly based on the collector length and tributary width. The method presented for SCBF provides guidance.

In this example, with a symmetrical brace configuration and symmetrical collector conditions, the horizontal force may be assumed to be distributed evenly between the two segments of the beam.

$$P_{E_{mh}} = \frac{P_{tx} + P_{cx}}{2}$$
$$= \frac{188 \text{ kips} + 207 \text{ kips}}{2}$$
$$= 198 \text{ kips}$$

BRBF beams and columns must be designed to resist all gravity loads. Using the load combinations in ASCE/SEI 7 that include the amplified seismic load, where the amplified seismic load is substituted with the analysis described in AISC *Seismic Provisions* Section F4.3, the required flexural strength of Beam BM-1 is:

5.5 BUCKLING-RESTRAINED BRACED FRAMES

LRFD	ASD
LRFD Load Combination 5 from ASCE/SEI 7 Section 12.4.3.2 $$M_u = (1.2 + 0.2S_{DS})M_D + M_{E_{mh}} + 0.5M_L$$ $$+ 0.2M_S$$ $$= 1.4(120 \text{ kip-ft}) + (-131 \text{ kip-ft})$$ $$+ 0.5(100 \text{ kip-ft}) + 0.2(0 \text{ kip-ft})$$ $$= 87.0 \text{ kip-ft}$$	ASD Load Combination 5 from ASCE/SEI 7 Section 12.4.3.2 $$M_a = (1.0 + 0.14S_{DS})M_D + M_H + M_F$$ $$+ 0.7M_{E_{mh}}$$ $$= 1.14(120 \text{ kip-ft}) + 0 \text{ kip-ft}$$ $$+ 0 \text{ kip-ft} + 0.7(-131 \text{ kip-ft})$$ $$= 45.1 \text{ kip-ft}$$

This required flexural strength is concurrent with the following required axial strength:

LRFD	ASD
LRFD Load Combination 5 from ASCE/SEI 7 Section 12.4.3.2 $$P_u = (1.2 + 0.2S_{DS})P_D + P_{E_{mh}} + 0.5P_L$$ $$+ 0.2P_S$$ $$= 1.4(0 \text{ kips}) + 198 \text{ kips}$$ $$+ 0.5(0 \text{ kips}) + 0.2(0 \text{ kips})$$ $$= 198 \text{ kips}$$	ASD Load Combination 5 from ASCE/SEI 7 Section 12.4.3.2 $$P_a = (1.0 + 0.14S_{DS})P_D + P_H + P_F$$ $$+ 0.7P_{E_{mh}}$$ $$= 1.14(0 \text{ kips}) + 0 \text{ kips} + 0 \text{ kips}$$ $$+ 0.7(198 \text{ kips})$$ $$= 139 \text{ kips}$$

The required flexural strength according to the analysis requirements of AISC *Seismic Provisions* Section F4.3 is also determined as follows:

LRFD	ASD
LRFD Load Combination 7 from ASCE/SEI 7 Section 12.4.3.2 $$M_u = (0.9 - 0.2S_{DS})M_D + M_{E_{mh}}$$ $$+ 1.6M_H$$ $$= 0.7(120 \text{ kip-ft}) + (-131 \text{ kip-ft}) + 0$$ $$= -47.0 \text{ kip-ft}$$	ASD Load Combination 8 from ASCE/SEI 7 Section 12.4.3.2 $$M_a = (0.6 - 0.14S_{DS})M_D + 0.7M_{E_{mh}}$$ $$+ M_H$$ $$= 0.46(120 \text{ kip-ft})$$ $$+ 0.7(-131 \text{ kip-ft}) + 0$$ $$= -36.5 \text{ kip-ft}$$

This required flexural strength is concurrent with the following required axial strength:

LRFD	ASD
LRFD Load Combination 7 from ASCE/SEI 7 Section 12.4.3.2 $$P_u = (0.9 - 0.2S_{DS})P_D + P_{E_{mh}} + 1.6P_H$$ $$= 0.7(0 \text{ kips}) + 198 \text{ kips} + 0 \text{ kips}$$ $$= 198 \text{ kips}$$	ASD Load Combination 8 from ASCE/SEI 7 Section 12.4.3.2 $$P_a = (0.6 - 0.14S_{DS})P_D + 0.7P_{E_{mh}} + P_H$$ $$= 0.46(0 \text{ kips}) + 0.7(198 \text{ kips})$$ $$+ 0 \text{ kips}$$ $$= 139 \text{ kips}$$

It is worth noting that the unbalanced load resulting from the adjusted brace strength in tension and compression imparts an upward point load on the beam, acting in opposition to gravity forces. This is true regardless of the direction of earthquake loading for an inverted-V brace configuration, because the adjusted brace strength in compression is higher than the adjusted brace strength in tension.

Because the moment due to seismic forces counteracts the moment due to gravity load, it is important to also consider load combinations that do not include the seismic load. The required flexural strength of the beam for the governing load combination that does not include seismic load is:

LRFD	ASD
LRFD Load Combination 2 from ASCE/SEI 7 Section 2.3.2 $$M_u = 1.2M_D + 1.6M_L$$ $$+ 0.5(M_{L_r} \text{ or } M_S \text{ or } M_R)$$ $$= 1.2(120 \text{ kip-ft}) + 1.6(100 \text{ kip-ft})$$ $$+ 0.5(0 \text{ kip-ft})$$ $$= 304 \text{ kip-ft}$$	ASD Load Combination 2 from ASCE/SEI 7 Section 2.4.1 $$M_a = M_D + M_L$$ $$= 120 \text{ kip-ft} + 100 \text{ kip-ft}$$ $$= 220 \text{ kip-ft}$$

Because dead and live loads do not result in axial forces in the beam, there is no axial load acting concurrently with this moment.

The required shear strength of the beam is shown in the following calculation. By inspection, because the unbalanced load from the braces always acts upward on the beam, the seismic component of the required shear strength will always counteract the gravity shears. Therefore, the governing load combination is one that does not include seismic effects:

5.5 BUCKLING-RESTRAINED BRACED FRAMES

LRFD	ASD
LRFD Load Combination 2 from ASCE/SEI 7 Section 2.3.2 $V_u = 1.2V_D + 1.6V_L + 0.5(V_{L_r} \text{ or } V_S \text{ or } V_R)$ $= 1.2(11.2 \text{ kips}) + 1.6(8.50 \text{ kips})$ $+ (0 \text{ kip-ft})$ $= 27.0 \text{ kips}$	ASD Load Combination 2 from ASCE/SEI 7 Section 2.4.1 $V_a = V_D + V_L$ $= 11.2 \text{ kips} + 8.50 \text{ kips}$ $= 19.7 \text{ kips}$

Try a W16×50.

From AISC *Manual* Table 1-1, the geometric properties are as follows:

$A = 14.7 \text{ in.}^2$ $d = 16.3 \text{ in.}$ $t_w = 0.380 \text{ in.}$ $b_f = 7.07 \text{ in.}$
$t_f = 0.630 \text{ in.}$ $k_{des} = 1.03 \text{ in.}$ $b_f/2t_f = 5.61$ $h/t_w = 37.4$
$I_x = 659 \text{ in.}^4$ $S_x = 81.0 \text{ in.}^3$ $r_x = 6.68 \text{ in.}$ $Z_x = 92.0 \text{ in.}^3$
$I_y = 37.2 \text{ in.}^4$ $r_y = 1.59 \text{ in.}$ $h_o = 15.7 \text{ in.}$ $J = 1.52 \text{ in.}^4$
$C_w = 2,270 \text{ in.}^6$

Width-to-Thickness Limitations

According to AISC *Seismic Provisions* Section F4.5a, beam members must satisfy the requirements for highly ductile members stipulated in AISC *Seismic Provisions* Table D1.1.

From Table 1-3 of this Manual, a W16×50 satisfies the ductility requirements for beams in BRBF systems.

Available Flexural Strength (negative flexure)

For negative flexure (bottom flange in compression), consider the bottom flange of the beam to be laterally braced at quarter points.

$$L_b = \frac{25.0 \text{ ft}}{4}$$
$$= 6.25 \text{ ft}$$

From AISC *Manual* Table 3-2, for a W16×50:

$L_p = 5.62 \text{ ft}$ $L_r = 17.2 \text{ ft}$

Therefore, $L_p < L_b \le L_r$, and lateral-torsional buckling applies. Conservatively use $C_b = 1.0$. From AISC *Manual* Table 3-10, the available flexural strength (negative flexure) is:

LRFD	ASD
$\phi_b M_n = 338 \text{ kip-ft} > \|-47.0 \text{ kip-ft}\|$ **o.k.**	$\dfrac{M_n}{\Omega_b} = 225 \text{ kip-ft} > \|-36.5 \text{ kip-ft}\|$ **o.k.**

Available Flexural Strength (positive flexure)

For positive flexure (top flange in compression), the beam can be considered fully braced by the slab, and therefore the limit state of lateral-torsional buckling does not apply. From AISC *Manual* Table 3-6, the available flexural strength of a W16×50 is:

LRFD	ASD
$\phi_b M_n = \phi_b M_p$ $= 345$ kip-ft > 304 kip-ft **o.k.**	$\dfrac{M_n}{\Omega_b} = \dfrac{M_p}{\Omega_b}$ $= 230$ kip-ft > 220 kip-ft **o.k.**

Available Compressive Strength

As explained in Part 8 for collectors, torsional buckling is considered because the torsional unbraced length is not the same as the minor-axis flexural buckling unbraced length. Because the top flange is constrained by the slab, the applicable torsional limit state is constrained-axis flexural-torsional buckling, as discussed in Part 8 of this Manual. The design compressive strength of the beam is the lowest value obtained based on the limit states of flexural buckling and torsional buckling.

To determine the unbraced length for flexural buckling about the *x*-*x* axis, it is necessary to verify whether the BRB provide a braced point for the beam at midspan.

For this purpose, assume first that the beam buckles and its midpoint moves upwards. Displacement compatibility will cause an increase in the demand on the BRB resisting tension due to lateral forces, while the load in the BRB resisting compression will be reduced. Conversely, if the beam buckles and the midpoint moves downwards, the BRB in compression will experience a load increase while the BRB in tension will be relieved of some of its original load. In both cases the BRB that is relieved of load will rebound along the path of its elastic stiffness.

If the strength and stiffness of the unloaded BRB in each of the aforementioned cases meet the requirements of AISC *Specification* Appendix 6, Section 6.2.2, then the beam can be considered to be braced by the BRB at midspan. From AISC *Specification* Equation A-6-3, the required strength of nodal bracing is:

LRFD	ASD
$P_{rb} = 0.01 P_u$ $= 0.01(198$ kips$)$ $= 1.98$ kips	$P_{rb} = 0.01 P_a$ $= 0.01(139$ kips$)$ $= 1.39$ kips

If the braces act as a braced point, the required stiffness of nodal bracing, from AISC *Specification* Equation A-6-4 with $L_b = 12.5$ ft, is:

5.5 BUCKLING-RESTRAINED BRACED FRAMES

LRFD	ASD
$\beta_{br} = \dfrac{1}{\phi}\left(\dfrac{8P_u}{L_b}\right)$	$\beta_{br} = \Omega\left(\dfrac{8P_a}{L_b}\right)$
$= \dfrac{1}{0.75}\left[\dfrac{8(198 \text{ kips})}{(12.5 \text{ ft})(12 \text{ in./ft})}\right]$	$= 2.00\left[\dfrac{8(139 \text{ kips})}{(12.5 \text{ ft})(12 \text{ in./ft})}\right]$
$= 14.1$ kip/in.	$= 14.8$ kip/in.

The minimum BRB axial strength is:

LRFD	ASD
$\phi P_n = \phi F_{ysc\ min} A_{sc}$	$\dfrac{P_n}{\Omega} = \dfrac{F_{ysc\ min} A_{sc}}{\Omega}$
$= 0.90(38 \text{ ksi})(4.50 \text{ in.}^2)$	$= \dfrac{38 \text{ ksi}(4.50 \text{ in.}^2)}{1.67}$
$= 154$ kips > 1.98 kips	$= 102$ kips > 1.39 kips

The elastic stiffness of the BRB adjusted by the angle of inclination, where $KF = K_{actual}/K_{core} = 1.28$ as given in Example 5.5.1, is:

$$\beta_{act} = (KF)\left(\dfrac{EA_{sc}}{L}\right)\sin\theta$$

$$= 1.28\left[\dfrac{29,000 \text{ ksi}(4.50 \text{ in.}^2)}{18.8 \text{ ft}(12 \text{ in./ft})}\right](\sin 48.2°)$$

$$= 552 \text{ kip/in.}$$

Nodal bracing requirements are met; therefore, for the beam, $KL_x = 12.5$ ft.

For flexural buckling about the y-y axis, the slab braces the beam continuously, so $KL_y = 0$ ft.

For constrained-axis flexural-torsional buckling, the unbraced length is the distance between bottom-flange braces, i.e., $KL_z = 6.25$ ft.

$$\dfrac{KL_x}{r_x} = \dfrac{1.0(12.5 \text{ ft})(12 \text{ in./ft})}{6.68 \text{ in.}}$$

$$= 22.5$$

$$\dfrac{KL_y}{r_y} = \dfrac{1.0(0 \text{ ft})(12 \text{ in./ft})}{1.59 \text{ in.}}$$

$$= 0$$

$$\frac{KL_z}{r_y} = \frac{1.0(6.25 \text{ ft})(12 \text{ in./ft})}{1.59 \text{ in.}}$$
$$= 47.2$$

From AISC *Manual* Table 1-1 and AISC *Specification* Table B4.1, the web is slender for compression with $F_y = 50$ ksi. Therefore, the reduction factor for slender stiffened elements, Q_a, must be determined. First, determine the governing limit state.

Determine the critical buckling strength for flexural buckling about the x-x axis, assuming Q = 1

The elastic buckling stress is:

$$F_e = \frac{\pi^2 E}{\left(\frac{KL}{r}\right)^2} \qquad \text{(Spec. Eq. E3-4)}$$

$$= \frac{\pi^2(29{,}000 \text{ ksi})}{(22.5)^2}$$

$$= 565 \text{ ksi}$$

The value of F_{cr} before local buckling effects are considered is determined as follows:

$$\frac{F_y}{F_e} = \frac{50 \text{ ksi}}{565 \text{ ksi}}$$

$$= 0.0885$$

Because $0.0885 < 2.25$, use AISC *Specification* Equation E3-2 to determine the critical buckling stress.

$$F_{cr} = \left(0.658^{\frac{F_y}{F_e}}\right) F_y \qquad \text{(Spec. Eq. E3-2)}$$

$$= \left(0.658^{\frac{50 \text{ ksi}}{565 \text{ ksi}}}\right) 50 \text{ ksi}$$

$$= 48.2 \text{ ksi}$$

Determine the critical buckling strength for constrained-axis flexural-torsional buckling, assuming Q = 1

For the limit state of constrained-axis flexural-torsional buckling, the unbraced length is 6.25 ft and the top flange of the beam is considered continuously braced by the slab as described in Part 8 of this Manual.

$$F_e = \left\{\frac{\pi^2 E\left[C_w + I_y(d/2)^2\right]}{(K_z L)^2} + GJ\right\}\left[\frac{1}{I_x + I_y + (d/2)^2 A_g}\right] \quad (8\text{-}3)$$

$$= \left\{\frac{\pi^2(29{,}000 \text{ ksi})\left[2{,}270 \text{ in.}^6 + 37.2 \text{ in.}^4 (16.3 \text{ in.}/2)^2\right]}{[1.0(6.25 \text{ ft})(12 \text{ in./ft})]^2} + 11{,}200 \text{ ksi}\left(1.52 \text{ in.}^4\right)\right\}$$

$$\times \left[\frac{1}{659 \text{ in.}^4 + 37.2 \text{ in.}^4 + (16.3 \text{ in.}/2)^2 \left(14.7 \text{ in.}^2\right)}\right]$$

$$= 154 \text{ ksi}$$

The value of F_{cr} before local buckling effects are considered is determined as follows:

$$\frac{F_y}{F_e} = \frac{50 \text{ ksi}}{154 \text{ ksi}}$$

$$= 0.325$$

Because $0.325 < 2.25$, use AISC *Specification* Equation E3-2 to determine the critical buckling stress.

$$F_{cr} = \left(0.658^{\frac{F_y}{F_e}}\right) F_y \quad (\textit{Spec. Eq. E3-2})$$

$$= \left(0.658^{\frac{50 \text{ ksi}}{154 \text{ ksi}}}\right) 50 \text{ ksi}$$

$$= 43.6 \text{ ksi}$$

Because F_{cr} is lower for constrained-axis flexural-torsional buckling, this limit state governs over major axis flexural buckling.

Determine the reduction factor, Q, for slender elements

To determine the reduction factor, Q, use AISC *Specification* Section E7.2, with $f = F_{cr}$ and the minimum F_{cr} from the two preceding limit states. The reduced effective width of the slender web is determined as follows:

$$b = h$$
$$= d - 2k_{des}$$
$$= 16.3 \text{ in.} - 2(1.03 \text{ in.})$$
$$= 14.2 \text{ in.}$$

$f = F_{cr}$
  $= 43.6$ ksi

$$b_e = 1.92t\sqrt{\frac{E}{f}}\left[1 - \frac{0.34}{(b/t)}\sqrt{\frac{E}{f}}\right] \le b \qquad \textit{(Spec. Eq. E7-17)}$$

$\quad = 1.92(0.380 \text{ in.})\sqrt{\dfrac{29{,}000 \text{ ksi}}{43.6 \text{ ksi}}}\left[1 - \dfrac{0.34}{37.4}\sqrt{\dfrac{29{,}000 \text{ ksi}}{43.6 \text{ ksi}}}\right]$

$\quad = 14.4$ in.

Because $b_e > b$, use $b_e = 14.2$ in.

Since $b_e = h$, there is no need to calculate the reduction factor and F_{cr} is the minimum calculated for the two preceding limit states. Constrained-axis flexural-torsional buckling governs.

Therefore, from AISC *Specification* Equation E3-1, the available compressive strength of a W16×50 is:

LRFD	ASD
$\phi_c P_n = 0.90(43.6 \text{ ksi})(14.7 \text{ in.}^2)$ $\quad = 577$ kips	$\dfrac{P_n}{\Omega_c} = \dfrac{43.6 \text{ ksi}(14.7 \text{ in.}^2)}{1.67}$ $\quad = 384$ kips

Combined Loading

LRFD	ASD
$\dfrac{P_r}{P_c} = \dfrac{198 \text{ kips}}{577 \text{ kips}}$ $\quad = 0.343$	$\dfrac{P_r}{P_c} = \dfrac{139 \text{ kips}}{384 \text{ kips}}$ $\quad = 0.362$

Because $P_r/P_c \ge 0.2$, the beam-column design is controlled by the equation:

$$\frac{P_r}{P_c} + \frac{8}{9}\left(\frac{M_{rx}}{M_{cx}} + \frac{M_{ry}}{M_{cy}}\right) \le 1.0 \qquad \textit{(Spec. Eq. H1-1a)}$$

Note that the maximum moment results from a load combination that does not include seismic effects. This moment is not concurrent with axial force in the beam, because the axial force is from seismic effects. Therefore, the maximum moment need not be considered in the combined loading check.

Determine the moment ratio. For positive moments (top flange in compression) due to seismic effects:

LRFD	ASD
$\dfrac{M_{rx}}{M_{cx}} = \dfrac{87.0 \text{ kip-ft}}{345 \text{ kip-ft}}$ $= 0.252$	$\dfrac{M_{rx}}{M_{cx}} = \dfrac{45.1 \text{ kip-ft}}{230 \text{ kip-ft}}$ $= 0.196$

For negative moments (bottom flange in compression) due to seismic effects:

LRFD	ASD
$\dfrac{M_{rx}}{M_{cx}} = \dfrac{47.0 \text{ kip-ft}}{338 \text{ kip-ft}}$ $= 0.139$	$\dfrac{M_{rx}}{M_{cx}} = \dfrac{36.5 \text{ kip-ft}}{225 \text{ kip-ft}}$ $= 0.162$

Use the positive flexure values for interaction.

LRFD	ASD
$0.343 + \dfrac{8}{9}(0.252 + 0) = 0.567$ $0.567 < 1.0$ **o.k.**	$0.362 + \dfrac{8}{9}(0.196 + 0) = 0.536$ $0.536 < 1.0$ **o.k.**

Available Shear Strength

From AISC *Manual* Table 3-2, the available shear strength of the **W16×50** is:

LRFD	ASD
$\phi_v V_n = 186$ kips > 27.0 kips **o.k.**	$\dfrac{V_n}{\Omega_v} = 124$ kips > 19.7 kips **o.k.**

Use a **W16×50** for Beam BM-1. Verify with the BRB manufacturer that the stiffness and overstrength factors are still valid with the final bay geometry.

Beam Bracing Requirements

From AISC *Seismic Provisions* Section F4.4a(2), beams in V- and inverted V-braced frames should be braced to satisfy the requirements for moderately ductile members in Section D1.2(a). AISC *Seismic Provisions* Section D1.2a(3) requires that beam bracing in moderately ductile members have a maximum spacing of:

$L_b = 0.17 r_y E/F_y$ (*Provisions* Eq. D1-2)

$= 0.17(1.59 \text{ in.})(29{,}000 \text{ ksi}/50 \text{ ksi})$

$= 157$ in.

$L_b = 157$ in.(1 ft/12 in.)
     $= 13.1$ ft

The bracing of the bottom flange at the quarter points of the beam (6.25 ft) satisfies this requirement.

Beam bracing requirements are given in AISC *Specification* Appendix 6. The required strength of lateral nodal bracing is:

$$P_{rb} = 0.02 M_r C_d / h_o \qquad \text{(Spec. Eq. A-6-7)}$$

where
$h_o = 15.7$ in.
$C_d = 1.0$

From AISC *Seismic Provisions* Equation D1-1, the required flexural strength to be used in AISC *Specification* Appendix 6 equations is:

LRFD	ASD
$M_r = R_y F_y Z$	$M_r = R_y F_y Z / 1.5$
$= 1.1(50 \text{ ksi})(92.0 \text{ in.}^3)$	$= 1.1(50 \text{ ksi})(92.0 \text{ in.}^3)/1.5$
$= 5{,}060$ kip-in.	$= 3{,}370$ kip-in.

From AISC *Specification* Equation A-6-7, the required brace strength is:

LRFD	ASD
$P_{rb} = 0.02 M_r C_d / h_o$	$P_{rb} = 0.02 M_r C_d / h_o$
$= 0.02(5{,}060 \text{ kip-in.})(1.0)/15.7 \text{ in.}$	$= 0.02(3{,}370 \text{ kip-in.})(1.0)/15.7 \text{ in.}$
$= 6.45$ kips	$= 4.29$ kips

From AISC *Specification* Equation A-6-8, the required brace stiffness is:

LRFD	ASD
$\beta_{br} = \dfrac{1}{\phi}\left(\dfrac{10 M_r C_d}{L_b h_o}\right)$	$\beta_{br} = \left(\dfrac{10 M_r C_d}{L_b h_o}\right)$
$= \dfrac{1}{0.75}\left[\dfrac{10(5{,}060 \text{ kip-in.})(1.0)}{(6.25 \text{ ft})(12 \text{ in./ft})(15.7 \text{ in.})}\right]$	$= 2.00\left[\dfrac{10(3{,}370 \text{ kip-in.})(1.0)}{(6.25 \text{ ft})(12 \text{ in./ft})(15.7 \text{ in.})}\right]$
$= 57.3$ kip/in.	$= 57.2$ kip/in.

Provide top and bottom flange beam bracing with these minimum strengths and stiffnesses at quarter points of the beam.

5.6 NONBUILDING STRUCTURES: A SPECIAL CASE

Typical building structures have both in-plane and out-of-plane support at the location of bracing connections to columns. In nonbuilding structures, however, it is common to have multistory planar braced frames with in-plane support but no out-of-plane support at intermediate locations where braces connect to the column. The AISC *Seismic Provisions* classifies this configuration as K-bracing due to the lack of out-of-plane support. Although K-bracing is not permitted for systems in the AISC *Seismic Provisions*, the steel ordinary concentrically braced frame designed to satisfy the AISC *Specification* (using $R = 1\frac{1}{2}$ per ASCE/SEI 7 Table 15.4-1) is permitted.

PART 5 REFERENCES

Ashaneh-Asl, A. (1998), "Seismic Behavior and Design of Gusset Plates for Braced Frames," *Steel Tips*, Structural Steel Education Council, Moraga, CA.

AISC (2011), *Design Examples*, V14.0, American Institute of Steel Construction, Chicago, IL, **www.aisc.org.**

Dowswell, B. (2006), "Effective Length Factors for Gusset Plate Buckling," *Engineering Journal*, AISC, Vol. 43, No. 2, 2nd Quarter, pp. 91–101.

Dowswell, B. (2012), "Effective Length Factors for Gusset Plates in Chevron Braced Frames," *Engineering Journal*, AISC, Vol. 49, No. 3, 3rd Quarter, pp. 115–117.

El-Tayem, A. and Goel, S. (1986), "Effective Length Factor for the Design of X-bracing Systems," *Engineering Journal*, AISC, Vol. 23, No.1, 1st Quarter, pp. 41–45.

Engelhardt, M.D. and Popov, E.P. (1989), "Behavior of Long Links in Eccentrically Braced Frames," *UCB/EERC - 89/01*, Earthquake Engineering Research Center, University of California, Berkeley, CA.

Gross, J.L. (1990), "Experimental Study of Gusseted Connections," *Engineering Journal*, AISC, Vol. 27, No. 3, 3rd Quarter, pp. 89–97.

Hewitt, C.M. and Thornton, W.A. (2004), "Rationale Behind and Proper Application of the Ductility Factor for Bracing Connections Subjected to Shear and Transverse Loading," *Engineering Journal*, AISC, Vol. 41, No. 1, 1st Quarter, pp. 3–6.

Kotulka, B.A. (2007), "Analysis for a Design Guide on Gusset Plates Used in Special Concentrically Braced Frames," M.S. Thesis, Department of Civil Engineering, University of Washington, Seattle, WA.

Lehman, D.E, Roeder, C.W., Herman, D., Johnson S. and Kotulka, B. (2008), "Improved Seismic Performance of Gusset Plate Connections," *Journal of Structural Engineering*, Vol. 134, No. 6, pp. 890–901.

Nair, S. (1997), "Practical Application of Energy Methods to Structural Stability Problems," *Engineering Journal*, AISC, Vol. 34, No. 4, 4th Quarter, pp. 126–134.

Picard, A. and Beaulieu, D. (1987), "Design of Diagonal Cross Bracings Part 1: Theoretical Study," *Engineering Journal*, AISC, Vol. 24, No. 3, 3rd Quarter, pp. 122–126.

Richard, R.M. (1986), "Analysis of Large Bracing Connection Designs for Heavy Construction," *National Steel Construction Conference Proceedings*, AISC, pp. 31.1–31.24, Chicago, IL.

Roeder, C.W., Lumpkin, E.J. and Lehman, D.E. (2011), "A Balanced Design Procedure for Special Concentrically Braced Frame Connections," *Journal of Constructional Steel Research*, Vol. 67, pp. 1,760–1,772.

SEAOC (2006), *IBC Structural/Seismic Design Manual*, Structural Engineers Association of California, Sacramento, CA.

Thornton, W.A. (1991), "On the Analysis and Design of Bracing Connections," *National Steel Construction Conference Proceedings*, AISC, pp. 26.1–26.33, Chicago, IL.

Thornton, W.A. (1996), "The Effect of Eccentricity on Brace-to-Gusset Angles," *Engineering Journal*, AISC, Vol. 33, No. 4, 4th Quarter, pp. 123–128.

Thornton, W.A. (1997), "Strength and Ductility Requirements for Simple Shear Connections With Shear and Axial Load," *National Steel Construction Conference Proceedings*, AISC, Chicago, IL.

Thornton, W.A. and Muir, L.S. (2009), "Design of Vertical Bracing Connections for High-Seismic Drift," *National Steel Construction Conference Proceedings*, AISC, Chicago, IL.

Thornton, W.A. and Fortney, P. (2012), "Satisfying Inelastic Rotation Requirements for In-Plane Critical Axis Brace Buckling for High-Seismic Design," *Engineering Journal*, AISC, Vol. 49, No. 3, 3rd Quarter, pp. 99–108.

Thornton, W.A. and Lini, C. (2011), "How to Use the Whitmore Method for Tension and Compression Strength Checks," *Modern Steel Construction*, July.

PART 6
COMPOSITE MOMENT FRAMES

6.1 SCOPE 6–2
6.2 COMPOSITE ORDINARY MOMENT FRAMES (C-OMF) 6–2
    Overview of Applicable Design Provisions 6–2
6.3 COMPOSITE INTERMEDIATE MOMENT FRAMES (C-IMF) 6–4
    Overview of Applicable Design Provisions 6–4
6.4 COMPOSITE SPECIAL MOMENT FRAMES (C-SMF) 6–7
    Overview of Applicable Design Provisions 6–8
6.5 COMPOSITE PARTIALLY RESTRAINED MOMENT FRAMES
    (C-PRMF) 6–11
    Overview of Applicable Design Provisions 6–12
6.6 CONNECTION DESIGN 6–14
    Reinforced Concrete Column-to-Steel Beam Connections 6–15
    Round Filled Composite Column-to-Steel Beam Connections 6–19
    Rectangular Filled Composite Column-to-Steel Beam Connections 6–21
PART 6 REFERENCES 6–26

6.1 SCOPE

The following types of composite moment frames are addressed in this Part: composite ordinary moment frames, composite intermediate moment frames, composite special moment frames, and composite partially restrained moment frames. The AISC *Seismic Provisions* and other design considerations summarized in this Part apply to the design of the members and connections in composite moment frames that require seismic detailing. AISC *Seismic Provisions* Sections A1 and B2 state that systems with reinforced concrete elements that must be designed according to ACI 318 should be designed only by the load and resistance factor design (LRFD) method because ACI 318 does not address allowable strength design (ASD).

6.2 COMPOSITE ORDINARY MOMENT FRAMES (C-OMF)

Composite ordinary moment frame (C-OMF) systems consist of: (i) composite or reinforced concrete columns; (ii) structural steel, concrete-encased composite, or composite beams; and (iii) fully restrained connections. C-OMF systems are designed and detailed according to AISC *Seismic Provisions* Section G1. They are expected to provide minimal inelastic deformation capacity in their members and connections.

ASCE/SEI 7 permits the use of C-OMF systems in Seismic Design Categories A and B only, as is the case for ordinary reinforced concrete moment frames. The use of C-OMF systems is limited because they can potentially involve the use of reinforced concrete columns or beams that are not designed or detailed to meet the seismic requirements of ACI 318 Chapter 21.

Because C-OMF systems are limited to Seismic Design Categories A and B, they are expected to withstand minimal inelastic drift through inelastic behavior of the composite beams, columns and panel zones. As a result, there are no requirements for: (i) structural analysis; (ii) system configuration; and (iii) designing steel or composite members other than those given in the AISC *Specification* and the applicable building code. There are no additional requirements for designing reinforced concrete members besides those provided in ACI 318, excluding Chapter 21.

Overview of Applicable Design Provisions

An overview of the applicable provisions of the AISC *Seismic Provisions* for the design of C-OMF systems follows and is presented in a simplified format in Table 6-1. All requirements of the AISC *Specification* apply, unless stated otherwise in the AISC *Seismic Provisions*.

Note 1. The structural steel material used for C-OMF systems is limited by the requirements of AISC *Seismic Provisions* Section A3.1 with the exception that the specified minimum yield stress of the steel for members in which inelastic behavior is expected is not to exceed 55 ksi. This specified minimum yield stress can be exceeded when the suitability of the material is determined by testing or other rational criteria. The concrete and steel reinforcement is selected to satisfy the requirements of AISC *Seismic Provisions* Section A3.5. The weld filler metal

Table 6-1
Simplified Overview of Provisions for C-OMF Systems

Note in Overview	Item	Referenced Standards[a]
1	Materials	*Seismic Prov.* Sects. A3.1, A3.4 & A3.5
2	Structural design drawings and specifications	*Seismic Prov.* Sects. A4.1, A4.2 & A4.3
3	Loads and load combinations	*Seismic Prov.* Sect. B2
4	Required strength for structural members and connections	*Seismic Prov.* Sect. B3.1
5	Structural analysis	*Seismic Prov.* Ch. C
(a)	Elastic stiffness of concrete/composite members	Commentary to *Seismic Prov.* Ch. C
6	Column members	ACI 318 (excl. Ch. 21)
7	Beam members	—
8	Beam-to-column connections	*Seismic Prov.* Sect. D2
9	Column splices	*Seismic Prov.* Sect. D2.5
10	Column bases	*Seismic Prov.* Sect. D2.6
11	Steel headed stud anchors or welded reinforcing bar anchors	*Seismic Prov.* Sect. D2.8
12	Composite slab diaphragms	*Seismic Prov.* Sect. D1.5

[a] The referenced standards listed are in addition to the AISC *Specification*.

used in the members and connections of seismic force resisting systems is selected to meet the requirements of AISC *Seismic Provisions* Section A3.4a. For C-OMF systems, there are no welds designated as demand critical welds.

Note 2. The structural design drawings and specifications for C-OMF systems are to meet the requirements of AISC *Seismic Provisions* Sections A4.1, A4.2 and A4.3.

Note 3. Loads and load combinations as defined by the applicable building code are to be followed as indicated in AISC *Seismic Provisions* Section B2. C-OMF systems including reinforced concrete components are to be designed using load and resistance factor design (LRFD) because allowable strength design (ASD) is not addressed in ACI 318.

Note 4. The required strength for structural members and connections is determined according to AISC *Seismic Provisions* Section B3.1.

Note 5. Structural analysis for the appropriate load combinations is to be performed in accordance with the requirements of AISC *Seismic Provisions* Chapter C.

(a) For elastic analysis, the stiffness of composite members includes the effects of cracked sections. Additional guidelines for estimating the stiffness of concrete beam and column members, concrete-encased and concrete-filled members, and steel beams with composite slabs are provided in the Commentary to the AISC *Seismic Provisions* Chapter C. These concrete and composite member properties reflect the effective stiffness at the onset of significant yielding in the members.

Note 6. Columns of C-OMF systems are designed in accordance with AISC *Specification* Chapter I or ACI 318 (excluding Chapter 21).

Note 7. Beams of C-OMF systems are designed in accordance with the AISC *Specification*.

Note 8. The beam-to-column connections are designed in accordance with the AISC *Specification* and AISC *Seismic Provisions* Section D2.

Note 9. Column splices are designed in accordance with the AISC *Specification* and AISC *Seismic Provisions* Section D2.5.

Note 10. Column bases are designed in accordance with the AISC *Specification* and AISC *Seismic Provisions* Section D2.6.

Note 11. Steel headed stud anchors and welded reinforcing bar anchors are designed in accordance with the AISC *Specification* and AISC *Seismic Provisions* Section D2.8.

Note 12. Composite slab diaphragms are to satisfy the requirements of AISC *Seismic Provisions* Section D1.5.

6.3 COMPOSITE INTERMEDIATE MOMENT FRAMES (C-IMF)

Composite intermediate moment frame (C-IMF) systems consist of: (i) composite or reinforced concrete columns; (ii) structural steel, concrete-encased or composite beams; and (iii) fully restrained connections. C-IMF systems are designed and detailed according to AISC *Seismic Provisions* Section G2. ASCE/SEI 7 limits the use of C-IMF systems to Seismic Design Categories A, B and C. The provisions for C-IMF systems as well as the associated R and C_d values in ASCE/SEI 7 are comparable to those required for reinforced concrete IMF systems.

C-IMF systems are expected to provide limited inelastic deformation capacity through flexural yielding of the C-IMF beams and columns and shear yielding of the column panel zones. The inelastic drift capability of C-IMF systems is permitted to be obtained from inelastic deformations of beams, columns and panel zones. The C-IMF system connection is based on a tested design with a qualifying story drift angle of 0.02 rad.

Overview of Applicable Design Provisions

An overview of the applicable provisions of the AISC *Seismic Provisions* for the design of C-IMF systems follows and is presented in a simplified format in Table 6-2. All requirements of the AISC *Specification* apply, unless stated otherwise in the AISC *Seismic Provisions*.

6.3 COMPOSITE INTERMEDIATE MOMENT FRAMES (C-IMF)

Note 1. The structural steel material used for C-IMF systems is limited by the requirements of AISC *Seismic Provisions* Section A3.1 with the exception that the specified minimum yield stress of the steel for members in which inelastic behavior is expected is not to exceed 50 ksi. This specified minimum yield stress can be exceeded when the suitability of the material is determined by testing or other rational criteria. Expected material strength is discussed in AISC *Seismic Provisions* Section A3.2 and values of R_y and R_t required to calculate the expected yield and tensile strength of steel are provided in AISC *Seismic Provisions* Table A3.1. The concrete and steel reinforcement is selected to satisfy the requirements of AISC *Seismic Provisions* Section A3.5. The weld filler metal used in the members and connections of seismic force resisting systems is selected to meet the requirements of AISC *Seismic Provisions* Section A3.4a. For C-IMF systems, there are no welds designated as demand critical welds.

Note 2. The structural design drawings and specifications for C-IMF systems are to meet the requirements of AISC *Seismic Provisions* Sections A4.1, A4.2 and A4.3.

Note 3. Loads and load combinations as defined by the applicable building code are to be followed as indicated in AISC *Seismic Provisions* Section B2. C-IMF systems including reinforced concrete components are to be designed using load and resistance factor design (LRFD) because allowable strength design (ASD) is not addressed in ACI 318.

Note 4. The general provisions for the required strength for structural members and connections is determined according to AISC *Seismic Provisions* Section B3.1. The required strength of columns is determined according to AISC *Seismic Provisions* Section D1.4a.

Note 5. Structural analysis for the appropriate load combinations is to be performed in accordance with the requirements of AISC *Seismic Provisions* Chapter C.

  (a) For elastic analysis, the stiffness of composite members includes the effects of cracked sections. Additional guidelines for estimating the stiffness of concrete beam and column members, concrete-encased and concrete-filled members, and steel beams with composite slabs are provided in the Commentary to AISC *Seismic Provisions* Chapter C. These concrete and composite member properties reflect the effective stiffness at the onset of significant yielding in the members.

Note 6. As stipulated in AISC *Seismic Provisions* Section G2.5a, steel and composite columns of C-IMF systems are required to meet the moderately ductile member requirements of AISC *Seismic Provisions* Section D1.1.

  (a) Encased composite columns must satisfy the requirements of AISC *Seismic Provisions* Section D1.4b(1). The width-to-thickness ratios of steel compression elements must not exceed the limiting width-to-thickness ratios, λ_{md}, from AISC *Seismic Provisions* Table D1.1.

  (b) Filled composite columns must satisfy the requirements of AISC *Seismic Provisions* Section D1.4c. The width-to-thickness ratios of steel compression elements must not exceed the limiting width-to-thickness ratios, λ_{md}, from AISC *Seismic Provisions* Table D1.1.

  (c) Concrete columns must satisfy the requirements of ACI 318 Section 21.3.

Table 6-2
Simplified Overview of Provisions for C-IMF Systems

Note in Overview	Item	Referenced Standards[a]
1	Steel and concrete materials	*Seismic Prov.* Sects. A3.1, A3.2, A3.4 & A3.5
2	Structural design drawings and specification	*Seismic Prov.* Sects. A4.1, A4.2 & A4.3
3	Loads and load combinations	*Seismic Prov.* Sect. B2
4	Required strength for structural members and connections	*Seismic Prov.* Sects. B3.1 & D1.4a
5	Structural analysis	*Seismic Prov.* Ch. C
(a)	Elastic stiffness of concrete/composite members	Commentary to *Seismic Prov.* Ch. C
6	Moderately ductile column members	*Seismic Prov.* Sects. D1.1 & G2.5a
(a)	Encased composite columns	*Seismic Prov.* Sect. D1.4b(1) & Table D1.1
(b)	Filled composite columns	*Seismic Prov.* Sect. D1.4c & Table D1.1
(c)	Reinforced concrete columns	ACI 318 Sect. 21.3
7	Moderately ductile beam members	*Seismic Prov.* Sects. D1.1 & G2.5a
(a)	Limiting width-to-thickness ratios	*Seismic Prov.* Table D1.1
(b)	Lateral bracing of moderately ductile beam members	*Seismic Prov.* Sect. D1.2a
(c)	Lateral bracing at plastic hinge locations	*Seismic Prov.* Sects. D1.2c & G2.4a
8	Beam-to-column connections	*Seismic Prov.* Sects. D2 & G2.6
(a)	Beam-to-column connection performance requirements	*Seismic Prov.* Sect. G2.6b
(b)	Beam-to-column conformance demonstration	*Seismic Prov.* Sect. G2.6c
(c)	Beam-to-column required shear strength	*Seismic Prov.* Sect. G2.6d
9	Connection diaphragm plates and continuity plates	*Seismic Prov.* Sect. G2.6e
10	Column splices	*Seismic Prov.* Sects. D2.5 & G2.6f
11	Column bases	*Seismic Prov.* Sect. D2.6
12	Steel headed stud anchors or welded reinforcing bar anchor	*Seismic Prov.* Sect. D2.8
13	Composite slab diaphragms	*Seismic Prov.* Sect. D1.5

[a] The referenced standards listed are in addition to the AISC *Specification*.

Note 7. As stipulated in AISC *Seismic Provisions* Section G2.5a, steel and composite beams of C-IMF systems are required to meet the moderately ductile member requirements of AISC *Seismic Provisions* Section D1.1.

   (a) The width-to-thickness ratios of steel compression elements are not to exceed the limiting width-to-thickness ratios, λ_{md}, from AISC *Seismic Provisions* Table D1.1.

   (b) The lateral bracing for beams is to be designed according to the requirements of AISC *Seismic Provisions* Section D1.2a.

   (c) Special bracing at plastic hinge locations required by AISC *Seismic Provisions* Section G2.4a must meet the requirements of AISC *Seismic Provisions* Section D1.2c.

Note 8. Beam-to-column connections are to be designed according to AISC *Seismic Provisions* Sections D2 and G2.6.

   (a) The performance requirements for beam-to-column connections are given in AISC *Seismic Provisions* Section G2.6b.

   (b) The methodology for conformance demonstration is given in AISC *Seismic Provisions* Section G2.6c.

   (c) The required shear strength for connections is based on AISC *Seismic Provisions* Section G2.6d.

Note 9. Connection diaphragm plates and continuity plates are designed according to the requirements of AISC *Seismic Provisions* Section G2.6e.

Note 10. Column splices are designed according to the requirements of AISC *Seismic Provisions* Sections D2.5 and G2.6f.

Note 11. Column bases are to satisfy the requirements of AISC *Seismic Provisions* Section D2.6.

Note 12. Steel headed stud anchors or welded reinforcing bar anchors are designed to meet the requirements of AISC *Seismic Provisions* Section D2.8.

Note 13. Composite slab diaphragms are to satisfy the requirements of AISC *Seismic Provisions* Section D1.5.

6.4 COMPOSITE SPECIAL MOMENT FRAMES (C-SMF)

Composite special moment frame (C-SMF) systems consist of: (i) composite or reinforced concrete columns; (ii) structural steel, concrete-encased composite, or composite beams; and (iii) fully restrained connections. C-SMF systems are designed and detailed according to AISC *Seismic Provisions* Section G3. ASCE/SEI 7 permits C-SMF systems in any seismic design category but they are primarily intended for use in Seismic Design Categories D, E and F. Design and detailing provisions for C-SMF systems are comparable to those required for steel and reinforced concrete SMF systems.

C-SMF systems are generally expected to experience significant inelastic deformations during a large seismic event. It is expected that most of the inelastic deformation will take place as rotation in beam "hinges" with limited inelastic deformation in the panel zone of the column. The beam-to-column connections for these systems are required to be qualified based on tests that demonstrate that the connection can sustain a story drift angle of at least 0.04 rad based on the loading protocol specified in AISC *Seismic Provisions* Chapter K. The connection configuration and design procedures are based on the results of these qualifying tests.

Other provisions are intended to limit or prevent excessive panel zone distortion, failure of connectivity plates or diaphragms, column hinging, and local buckling that may lead to inadequate system performance in spite of good connection performance.

Overview of Applicable Design Provisions

An overview of the AISC *Seismic Provisions* for the design of C-SMF systems follows and is presented in a simplified format in Table 6-3. All requirements of the AISC *Specification* apply, unless stated otherwise in the AISC *Seismic Provisions*.

Note 1. The structural steel material used for the C-SMF systems is limited by the requirements of AISC *Specification* Section A3.1 and AISC *Seismic Provisions* Section A3.1 with the exception that the specified minimum yield stress of the steel for members in which inelastic behavior is expected is not to exceed 50 ksi, as stipulated in AISC *Seismic Provisions* Section A3.1. This specified minimum yield stress can be exceeded when the suitability of the material is determined by testing or other rational criteria. For columns in C-SMF systems, the specified minimum yield stress is not to exceed 65 ksi. Expected material strength is discussed in AISC *Seismic Provisions* Section A3.2 and values of R_y and R_t required to calculate the expected yield and tensile strength of steel are provided in AISC *Seismic Provisions* Table A3.1. This specified minimum yield stress can be exceeded when the suitability of the material is determined by testing or other rational criteria. The concrete and steel reinforcement is selected to meet the requirements of AISC *Seismic Provisions* Section A3.5. The weld filler metal used in the members and connections of seismic force resisting systems is selected to meet the requirements of AISC *Seismic Provisions* Section A3.4a. Filler metals used in welds designated as demand critical welds in AISC *Seismic Provisions* Section G3.6a are expected to meet the requirements of AISC *Seismic Provisions* Section A3.4b.

Note 2. The structural design drawings and specifications for C-SMF systems are to meet the requirements of AISC *Seismic Provisions* Sections A4.1, A4.2 and A4.3.

Note 3. Loads and load combinations as defined by the applicable building code are to be followed as indicated in AISC *Seismic Provisions* Section B2. C-SMF systems including reinforced concrete components must be designed using load and resistance factor design (LRFD) because allowable strength design (ASD) is not addressed in ACI 318.

Note 4. The required strength for structural members and connections is determined according to AISC *Seismic Provisions* Section B3.1. The required strength of columns is determined according to AISC *Seismic Provisions* Section D1.4a.

Note 5. Structural analysis for the appropriate load combinations is to be performed in accordance with the requirements of AISC *Seismic Provisions* Chapter C.

6.4 COMPOSITE SPECIAL MOMENT FRAMES (C-SMF)

Table 6-3
Simplified Overview of Provisions for C-SMF Systems

Note in Overview	Item	Referenced Standards[a]
1	Materials	*Seismic Prov.* Sects. A3.1, A3.2, A3.4 & A3.5
2	Structural design drawings and specification	*Seismic Prov.* Sects. A4.1, A4.2 & A4.3
3	Loads and load combinations	*Seismic Prov.* Sect. B2
4	Required strength for structural members and connections	*Seismic Prov.* Sects. B3.1 & D1.4a
5	Structural analysis	*Seismic Prov.* Ch. C
(a)	Elastic stiffness of concrete/composite members	Commentary to *Seismic Prov.* Ch. C
6	Highly ductile column members	*Seismic Prov.* Sects. D1.1 & G3.5a
(a)	Encased composite columns	*Seismic Prov.* Sect. D1.4b(2) & Table D1.1
(b)	Filled composite columns	*Seismic Prov.* Sect. D1.4c & Table D1.1
(c)	Reinforced concrete columns	ACI 318 Sect. 21.6
7	Highly ductile beam members	*Seismic Prov.* Sects. D1.1 & G3.5a
(a)	Width-to-thickness ratios of highly ductile members	*Seismic Prov.* Table D1.1
(b)	Lateral bracing of highly ductile beam members	*Seismic Prov.* Sect. D1.2b
(c)	Lateral bracing at plastic hinge locations	*Seismic Prov.* Sects. D1.2c & G3.4b
8	Proportioning of columns and beams at joints	*Seismic Prov.* Sect. G3.4a
9	Beam-to-column connections	*Seismic Prov.* Sect. G3.6
(a)	Demand critical welds	*Seismic Prov.* Sect. G3.6a
(b)	Beam-to-column connection performance requirements	*Seismic Prov.* Sect. G3.6b
(c)	Beam-to-column conformance demonstration	*Seismic Prov.* Sect. G3.6c
(d)	Beam-to-column required shear strength	*Seismic Prov.* Sect. G3.6d
10	Connection diaphragm plates and continuity plates	*Seismic Prov.* Sect. G3.6e
11	Column Splices	*Seismic Prov.* Sects. D2.5 & G3.6f
12	Column bases	*Seismic Prov.* Sect. D2.6
13	Steel headed stud anchors or welded reinforcing bar anchor	*Seismic Prov.* Sect. D2.8
14	Composite slab diaphragms	*Seismic Prov.* Sect. D1.5

[a] The referenced standards listed are in addition to the AISC *Specification*.

(a) For elastic analysis, the stiffness of composite members includes the effects of cracked sections. Additional guidelines for estimating the stiffness of concrete beam and column members, concrete-encased and concrete-filled members, and steel beams with composite slabs are provided in the Commentary to AISC *Seismic Provisions* Chapter C. These concrete and composite member properties reflect the effective stiffness at the onset of significant yielding in the members.

Note 6. As stipulated in AISC *Seismic Provisions* Section G3.5a, composite columns of C-SMF systems are required to meet the highly ductile member requirements of AISC *Seismic Provisions* Section D1.1.

(a) Encased composite columns must satisfy the requirements of AISC *Seismic Provisions* Section D1.4b(2). The width-to-thickness ratios of steel compression elements must not exceed the limiting width-to-thickness ratios, λ_{hd}, from AISC *Seismic Provisions* Table D1.1.

(b) Filled composite columns must satisfy the requirements of AISC *Seismic Provisions* Section D1.4c. The width-to-thickness ratios of steel compression elements must not exceed the limiting width-to-thickness ratios, λ_{hd}, from AISC *Seismic Provisions* Table D1.1.

(c) Concrete columns must satisfy the requirements of ACI 318 Section 21.6.

Note 7. As stipulated in AISC *Seismic Provisions* Section G3.5a, beams of C-SMF systems are required to meet the highly ductile member requirements of AISC *Seismic Provisions* Section D1.1.

(a) The width-to-thickness ratios of steel compression elements must not exceed the limiting width-to-thickness ratios, λ_{hd}, from AISC *Seismic Provisions* Table D1.1.

(b) The lateral bracing for beams is designed according to the requirements of AISC *Seismic Provisions* Section D1.2b.

(c) Special bracing at plastic hinge locations required by AISC *Seismic Provisions* Section G3.4b must meet the requirements of AISC *Seismic Provisions* Section D1.2c.

Note 8. Columns and beams of C-SMF systems are proportioned to meet the strong-column weak-beam requirements of AISC *Seismic Provisions* Section G3.4a.

Note 9. Beam-to-column connections are designed according to AISC *Seismic Provisions* Section G3.6.

(a) Welds designated as demand critical are stipulated in AISC *Seismic Provisions* Section G3.6a.

(b) The performance requirements for beam-to-column connections are given in AISC *Seismic Provisions* Section G3.6b.

(c) The methodology for conformance demonstration is based on AISC *Seismic Provisions* Section G3.6c.

(d) The required shear strength for connections is based on AISC *Seismic Provisions* Section G3.6d.

Note 10. Connection diaphragm plates and continuity plates are designed according to the requirements of AISC *Seismic Provisions* Section G3.6e.

Note 11. Column splices are designed according to the requirements of AISC *Seismic Provisions* Sections D2.5 and G3.6f.

Note 12. Column bases are to satisfy the requirements of AISC *Seismic Provisions* Section D2.6.

Note 13. Steel headed stud anchors or welded reinforcing bar anchors are designed to meet the requirements of AISC *Seismic Provisions* Section D2.8.

Note 14. Composite slab diaphragms are to satisfy the requirements of AISC *Seismic Provisions* Section D1.5.

6.5 COMPOSITE PARTIALLY RESTRAINED MOMENT FRAMES (C-PRMF)

Composite partially restrained moment frame (C-PRMF) systems consist of structural steel columns and composite beams that are connected with partially restrained moment connections. C-PRMF systems are designed and detailed according to AISC *Seismic Provisions* Section G4.

C-PRMF systems resist lateral forces and displacements through the flexural and shear strengths of the beams and columns similar to other moment frame systems. The primary difference between C-PRMF systems and the other moment frame systems is that the beam-to-column connections in C-PRMF are not designed for the full flexural strength of the beam. Consequently, hinging is forced to occur in the partially restrained composite connections (PRCC) rather than the beam ends and column panel zone. The beams and columns in a properly designed C-PRMF will typically remain elastic with low ductility demands with the exception of expected hinging at the base of the columns.

The design of a C-PRMF is different from the design of a more traditional steel moment frame in three important ways. First, PRCC are not designed to be stronger than the beam it is connecting. Consequently, the lateral system typically will hinge within the connections and not within the associated beams or columns. Second, because the connections are neither pinned nor fixed, their stiffness must be accounted for in the frame analysis. Third, because the connections are weaker than fully restrained moment connections, the lateral force resisting system requires more frames with more connections, resulting in a highly redundant system.

The work that forms the basis of many of the recommendations for the C-PRMF has been summarized in *Partially Restrained Composite Connections,* Design Guide 8 (Leon et al., 1996) and ASCE (1998). The type of C-PRMF system envisioned under the current AISC *Seismic Provisions* is one using bare steel W-shape columns and composite steel beam framing. Most research addressing C-PRMF systems has investigated systems with a reinforced composite slab, a double-angle bolted web connection, and a bolted seat angle.

With the magnitude of the seismic response modification coefficient, R, between those of the IMF and SMF systems, the C-PRMF system is expected to experience significant inelastic behavior during a seismic event and the PRCC must be capable of providing stable moment-rotation behavior up to 0.02 rad. The PRCC must also exhibit a moment strength of at least 50% of the nominal flexural strength of the steel beam at a connection rotation of 0.02 rad. The AISC *Seismic Provisions* do not provide an upper bound on the characteristic connection moment strength; however, 100% of the nominal plastic flaxural strength of the bare steel beam is recommended.

The design concept of "strong column-weak beam" is not specifically required by the AISC *Seismic Provisions* for C-PRMF systems; however, it is recommended for C-PRMF systems in *Seismic Provisions* Commentary Section G4.4. Similar to the special moment frame, this provision is not intended to eliminate all yielding in the columns. Rather, it is intended to result in framing systems that have distributed inelasticity in large seismic events and discourage story mechanisms.

Overview of Applicable Design Provisions

An overview of the AISC *Seismic Provisions* requirements for the design of C-PRMF systems follows and is presented graphically in Figure 6-1 and in a simplified format in Table 6-4. All requirements of the AISC *Specification* apply, unless stated otherwise in the AISC *Seismic Provisions*.

Note 1. The structural steel material used for the C-PRMF systems is limited by the requirements of AISC *Specification* Section A3.1 and AISC *Seismic Provisions* Section A3.1 with the exception that the specified minimum yield stress of the steel for members in which inelastic behavior is expected is not to exceed 50 ksi, as stipulated in AISC *Seismic Provisions* Section A3.1. This specified minimum yield stress can be exceeded when the suitability of the material is determined by testing or other rational criteria. Expected material strength is discussed in AISC *Seismic Provisions* Section A3.2 and values of R_y and R_t required to calculate the expected yield and tensile strength of steel are provided in AISC *Seismic Provisions* Table A3.1. The concrete and steel reinforcement is selected to meet the requirements of AISC *Seismic Provisions* Section A3.5. The weld filler metal used in the members and connections of seismic force resisting systems is selected to meet the requirements of AISC *Seismic Provisions* Section A3.4a. Welds designated as demand critical welds in AISC *Seismic Provisions* Section G4.6a are expected to meet the requirements of AISC *Seismic Provisions* Section A3.4b.

Note 2. The structural design drawings and specifications for C-PRMF systems are to meet the requirements of AISC *Seismic Provisions* Sections A4.1, A4.2 and A4.3.

Note 3. Loads and load combinations as defined by the applicable building code are to be followed as indicated in AISC *Seismic Provisions* Section B2.

Note 4. The required strength for structural members and connections is determined according to AISC *Seismic Provisions* Section B3.1. The required strength of columns is determined according to AISC *Seismic Provisions* Section D1.4a.

Note 5. Structural analysis for the appropriate load combinations is to be performed in accordance with the requirements of AISC *Seismic Provisions* Chapter C and Section G4.3.

(a) For elastic analysis, the stiffness of composite members includes the effects of cracked sections. Additional guidelines for estimating the stiffness of steel beams with composite slabs are provided in the commentary to AISC *Seismic*

6.5 COMPOSITE PARTIALLY RESTRAINED MOMENT FRAMES (C-PRMF)

Provisions Chapter C. These composite member properties reflect the effective stiffness at the onset of significant yielding in the members.

Note 6. As stipulated in AISC *Seismic Provisions* Section G4.5a, columns of C-PRMF systems are required to meet the highly ductile member requirements of AISC *Seismic Provisions* Section D1.1.

Note 7. As stipulated in AISC *Seismic Provisions* Section G4.5b, beams of C-PRMF systems are required to meet the highly ductile member requirements of AISC *Seismic Provisions* Section D1.1.

(a) The width-to-thickness ratios of steel compression elements must not exceed the limiting width-to-thickness ratios, λ_{hd}, from AISC *Seismic Provisions* Table D1.1.

(b) The lateral bracing for beams is designed according to the requirements of AISC *Seismic Provisions* Section D1.2b.

(c) A solid slab is to be provided as stipulated in AISC *Seismic Provisions* Section G4.5b.

Note 8. Beam-to-column connections are designed according to AISC *Seismic Provisions* Sections D2 and G4.6. Specifically, steel reinforcement must be designed to satisfy the requirements of AISC *Seismic Provisions* Section D2.7(5).

Note 9. Column splices are to satisfy the requirements of AISC *Seismic Provisions* Sections D2.5, G4.6a(1) and G4.6f.

Note 10. Column bases are to satisfy the requirements of AISC *Seismic Provisions* Section D2.6.

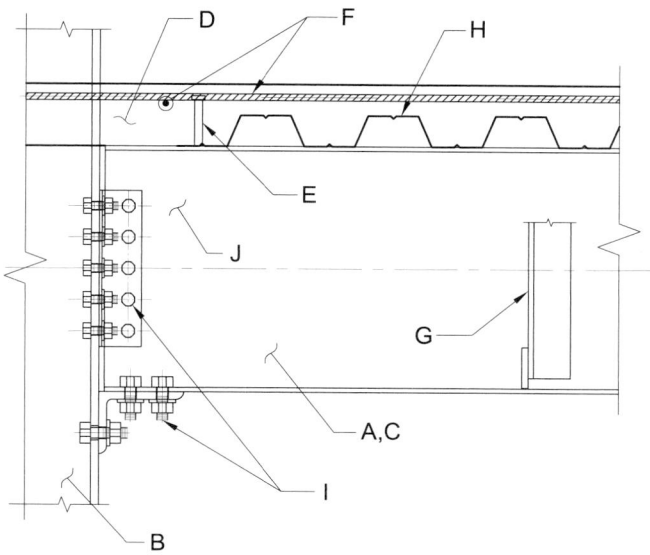

Fig. 6-1. Notes key for Table 6-4 for C-PRMF systems.

Table 6-4
Notes to Figure 6-1: C-PRMF Systems

Note in Fig. 6-1	Note in Overview	Item	AISC *Seismic Provisions* Reference
–	1	Materials	Sects. A3.1, A3.2, A3.4 & A3.5
–	2	Structural design drawings and specification	Sects. A4.1, A4.2 & A4.3
–	3	Loads and load combinations	Sect. B2
–	4	Required and available strength for structural members and connections	Sects. B3.1 & D1.4a
–	5	Structural analysis	Chapter C, Sect. G4.3
A	(a)	Composite member stiffness	Comm. to Ch. C
B	6	Highly ductile column members	Sects. D1.1 & G4.5a
C	7	Beam members	Sects. D1.1 & G4.5b
	(a)	Width-to-thickness ratios of highly ductile members	Table D1.1
G	(b)	Lateral bracing of highly ductile beam members	Sect. D1.2b
D	(c)	Solid slab zone	Sect. G4.5b
F, I, J	8	Beam-to-column connections (including composite partially restrained connections)	Sects. D2 & G4.6
B	9	Column splices	Sects. D2.5, G4.6a(1) & G4.6f
B	10	Column bases	Sect. D2.6
E	11	Steel headed stud anchors	Sect. D2.8
H	12	Composite slab diaphragm	Sect. D1.5

Note 11. Steel headed stud anchors are to satisfy the requirements of AISC *Seismic Provisions* Section D2.8.

Note 12. Composite slab diaphragms are to satisfy the requirements of AISC *Seismic Provisions* Section D1.5.

6.6 CONNECTION DESIGN

Unlike steel moment resisting frames, there currently are no prequalified connections available for use in composite moment resisting frames. Therefore, the following summarizes the results of testing and evaluation of selected types of connections for composite moment resisting frames. The discussion focuses on reinforced concrete column-to-steel beam connections (RCS), round filled composite column-to-steel beam connections, and rectangular filled composite column-to-steel beam connections.

Reinforced Concrete Column-to-Steel Beam Connections

During the 1980s and 1990s, more than 400 RCS connections were tested in Japan and 36 in the United States (Deierlein and Noguchi, 2004). Through the U.S.-Japan Cooperative Research Program, 56 more connection subassemblies were tested to fill knowledge gaps for certain connection configurations and force-transfer mechanisms (U.S.-Japan, 1983). Examples of the wide variety of RCS connection details tested in Japan and the U.S. are shown in Figures 6-2a and 6-2b, which are taken from Deierlein and Noguchi (2004).

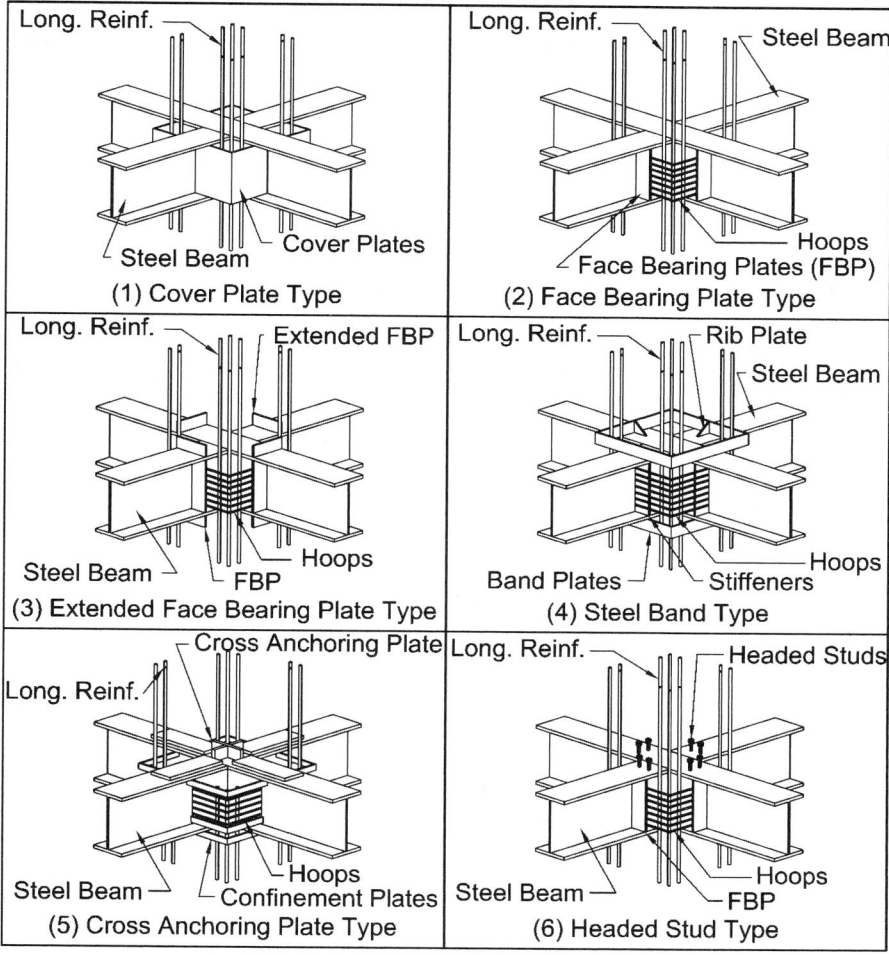

Fig. 6-2a. Example details 1 through 6 of reinforced concrete column-to-steel beam connections tested in the U.S. and Japan (from Deierlein and Noguchi, 2004). Reprinted with permission from ASCE.

In Figures 6-2a and 6-2b, details 1 through 7 are through-beam type connections where the beam is continuous through the joint. By not interrupting (splicing) the beam at the point of maximum moment at the column face, the through-beam details provide the ductility that is generally desirable in conventional steel construction. Details 8 through 11 are through-column type connections where the beam flanges are interrupted to minimize the impact on the column reinforcing bar arrangement and to facilitate concrete placement in the joint. Detail 12 is an example of a hybrid detail, combining conventional reinforced concrete concepts by encasing the steel beam ends connecting to the concrete column.

Through-beam-type connections have been the preferred detail in the U.S., however, both types have been used in Japan. The primary differences between the details in Figures 6-2a

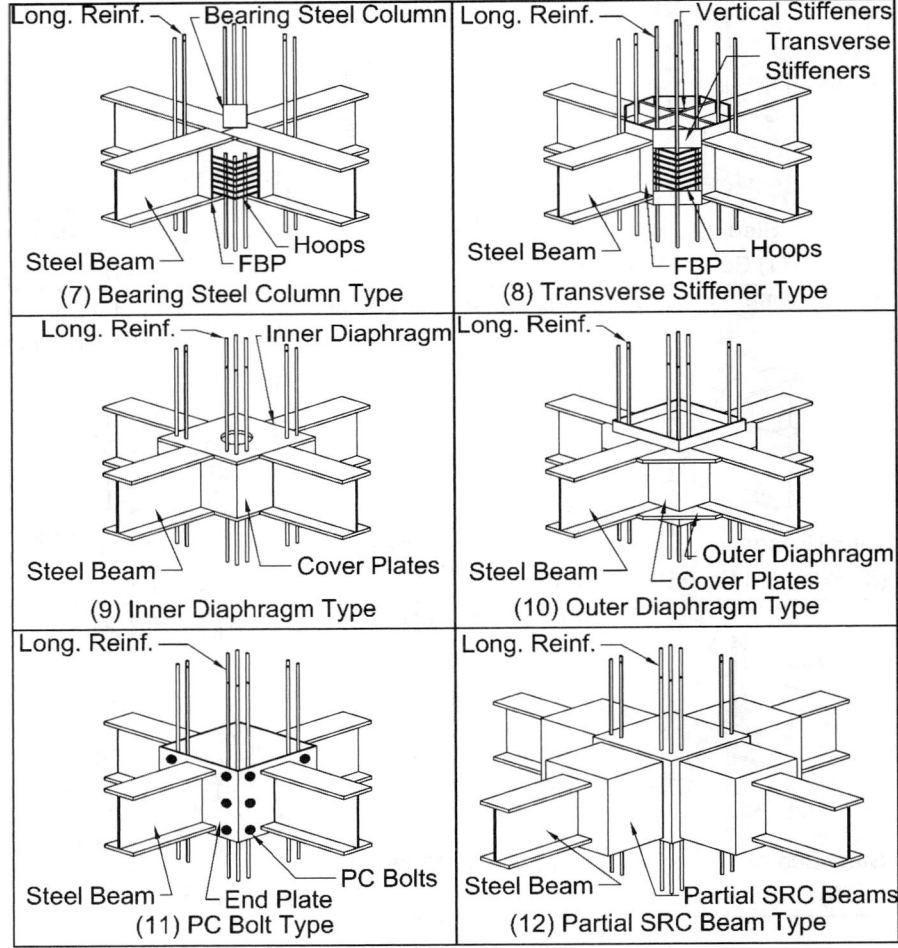

Fig. 6-2b. Example details 7 through 12 of reinforced concrete column-to-steel beam connections tested in the U.S. and Japan (from Deierlein and Noguchi, 2004). Reprinted with permission from ASCE.

Table 6-5
Summary of Reinforced Concrete Column-to-Steel Beam Connection Tests

Organization	Test Description	References
Building Research Inst.	10 planar, through-column joints	Kuramoto and Noguchi (1997)
Building Contractors Society	six three-dimensional, through-column joints	Nishiyama et al. (1998, 2000) Kuramoto and Nishiyama (2004)
Chiba Univ.	six planar through-beam joints	Kuramoto and Noguchi (1997)
	five planar through-column joints	Noguchi and Kim (1997, 1998)
Osaka Inst. of Technology	six planar through-beam joints, investigation of specific internal force transfer mechanisms	Baba and Nishimura (2000)
Univ. of Michigan	15 through-beam joints (nine exterior configurations, four with composite slab, and two post-earthquake repairs)	Parra-Montesinos et al. (2000a, 2000b, 2001a, 2001b, 2003) Liang and Parra-Montesinos (2004)
Texas A & M	six three-dimensional through-beam joints, with composite slab	Bugeja et al. (1999, 2000) Bracci et al. (1999) Esche et al. (1999)
U.C. San Diego	two planar tests of steel beams to composite column with reduced beam sections	Chou and Uang (2002)
Cornell Univ.	19 through-beam joints	Kanno and Deierlein (1993, 1997)
Univ. of Texas	17 through-beam joints	Sheikh et al. (1989) Deierlein et al. (1989)

From Deierlein and Noguchi (2004). Reprinted with permission from ASCE.

and 6-2b lie in attachments of various stiffener plates, cover plates and bearing plates, which act together with reinforcing bars to effect force transfer between the steel and concrete.

A summary of test results available in the literature is presented in Table 6-5. The test specimens were approximately one-half to two-thirds of full scale, with typical reinforced concrete column sizes ranging from 10 to 18 in. deep. The tests were generally conducted under cyclic loading, and several of the tests included axial loading of the reinforced concrete columns to represent gravity loading and earthquake-induced overturning. The typical yield strength of the steel beams was 50 ksi, and the concrete compressive strength was 4 ksi minimum. Most connection test assemblies were designed to fail in the joint to allow study of the internal force transfer mechanisms. This is counter to design practice, where the joints are typically designed to be stronger than the beams. This should be kept in mind when reviewing test results from literature.

Overall, the tests show that, when properly detailed to mobilize internal force transfer mechanisms, RCS connections provide reliable strength and ductility for seismic design. A limited suite of details (face bearing plates, vertical joint reinforcement, web doubler plates, etc.) have been tested and shown to enhance stiffness and strength of the connection. Other details adjusted to suit design and fabrication that provide similar levels of confinement and force transfer may be suitable but would need engineering evaluation.

Models to calculate the stiffness and strength of RCS joints have been synthesized into guidelines (ASCE, 1994). The ASCE guidelines have been validated for seismic design using the tests noted in Table 6-5. Several proposals have been made to improve them (e.g., Parra-Montesinos and Wight, 2001a; Parra-Montesinos et al., 2003; Kuramoto and Nishiyama, 2004). In particular, through-beam type connections eliminate the need for field welding of the beam flanges and are generally not susceptible to rupture behavior. Tests have shown that, of the many possible ways of strengthening the joint, face bearing plates and steel band plates attached to the beam are very effective for both mobilizing the joint shear strength of reinforced concrete and providing confinement to the concrete. Further information on design methods and equations for these composite connections is available in published guidelines, e.g., Nishiyama et al. (1990) and Parra-Montesinos and Wight (2001a).

Liang and Parra-Montesinos (2004) have demonstrated the experimental behavior of these connections by testing two interior and two exterior RCS subassemblies under cyclic load reversals. The test specimens included reinforced concrete columns or RCS columns and composite beams with the steel beam running continuously through the columns and a reinforced concrete slab cast upon metal decks supported by the steel beams. Strong column-weak beam design philosophy was implemented by designing the interior specimens to have a column-to-beam moment strength ratio of 1.3, and the exterior specimens to have a ratio of 2.2. Figure 6-3 shows the two types of composite joint details used for the interior and exterior RCS connection subassemblies.

As shown in Figure 6-3(a), one of the details consisted of overlapping U-shaped stirrups passing through holes drilled in the beam web. For this detail, the transverse beam was

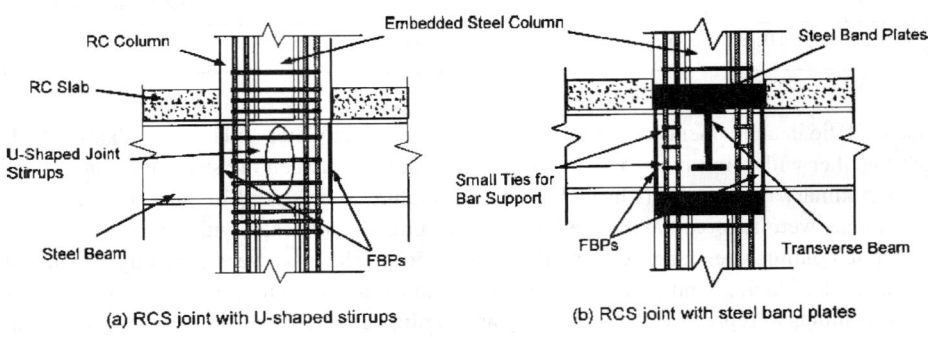

Fig. 6-3. RCS connections tested by Liang and Parra-Montesinos (2004) and demonstrated to achieve 0.04 rad interstory drift.

assumed to frame in to the main beam some distance away from the connection. Also closely spaced stirrups were placed in the column regions directly above and below the steel beams to provide confinement to regions susceptible to bearing failure and to mobilize concrete regions outside the width of the steel beam flanges.

The second detail shown in Figure 6-3(b) features steel band plates wrapping around the column regions just above and below the steel beams. The U-shaped stirrups that pass through the steel web panel were eliminated because of the confinement provided by the steel band plates. This further allows transverse beams to frame into the main beam at the connection region. In order to prevent outward buckling of longitudinal bars through the joint region, small ties that do not penetrate the steel web panel were provided over the joint depth.

Experimental results indicated excellent performance and only moderate damage in the connections. Plastic hinges formed in the beam regions adjacent to the connections and dissipated energy under cyclic loading to achieve story drift angles greater than 0.04 rad, which is required for C-SMF systems.

Round Filled Composite Column-to-Steel Beam Connections

The behavior of different types of round filled column-to-steel beam connections for composite frames has been investigated in the U.S. (Azizinamini and Schneider, 2004). Six different types of composite connections were tested; these test configurations are presented in Figure 6-4.

Each tested connection consisted of a round filled composite column connected to an ASTM A992 W14×38 beam. The composite column was an ASTM A500 Grade B round hollow structural section (HSS) that was 14 in. in diameter, 0.25 in. thick, and filled with f'_c = 5 ksi concrete. The test setup consisted of an exterior subassembly (girder on only one side of the column) that was subjected to cyclic deformations on the tip of the cantilever girder at a distance of 9 ft from the face of the column. The cyclic deformation history followed ATC-24 (ATC, 1992) guidelines.

When the steel beam is welded directly to the round HSS of the composite column, as shown in connection type I, large distortions of the HSS walls occurred, and the connection was susceptible to weld, flange or HSS wall rupture. This type of connection had a rotation capacity less than 0.02 rad and is acceptable only for C-OMF systems. Connection types II and III, with external diaphragm and continuous web details, respectively, had better inelastic behavior, but the flexural strength of these connections deteriorated early in the imposed deformation history after reaching a rotation capacity of 0.02 rad, which is not acceptable for C-SMF systems.

Connection type IV was similar to type I with the addition of four No. 6 rebars that were welded to the girder flanges and anchored into the concrete infill of the composite column. The behavior of this connection was better, but there was some local tearing of the steel HSS at a rotation of 0.03 rad, and rupture of the deformed bars at a rotation of 0.0375 rad. As a result, this connection type is also not acceptable for C-SMF systems. Connection type V with the girder flange through the composite column had rupture failures at the flange welds and is not recommended for any of the systems.

Connection type VI, the through-beam-type design, had excellent cyclic behavior and developed 0.04 rad rotation. This is the only connection type acceptable for C-SMF systems.

Connection types II, III and IV are acceptable for C-IMF systems where the required rotation capacity is 0.02 rad.

Elremaily and Azizinamini (2000) conducted additional research to develop design guidelines for through-beam-type connections for systems with round filled composite columns. They conducted seven two-thirds scale tests on connection systems consisting of a round composite column and a steel beam passing through the column representing an interior

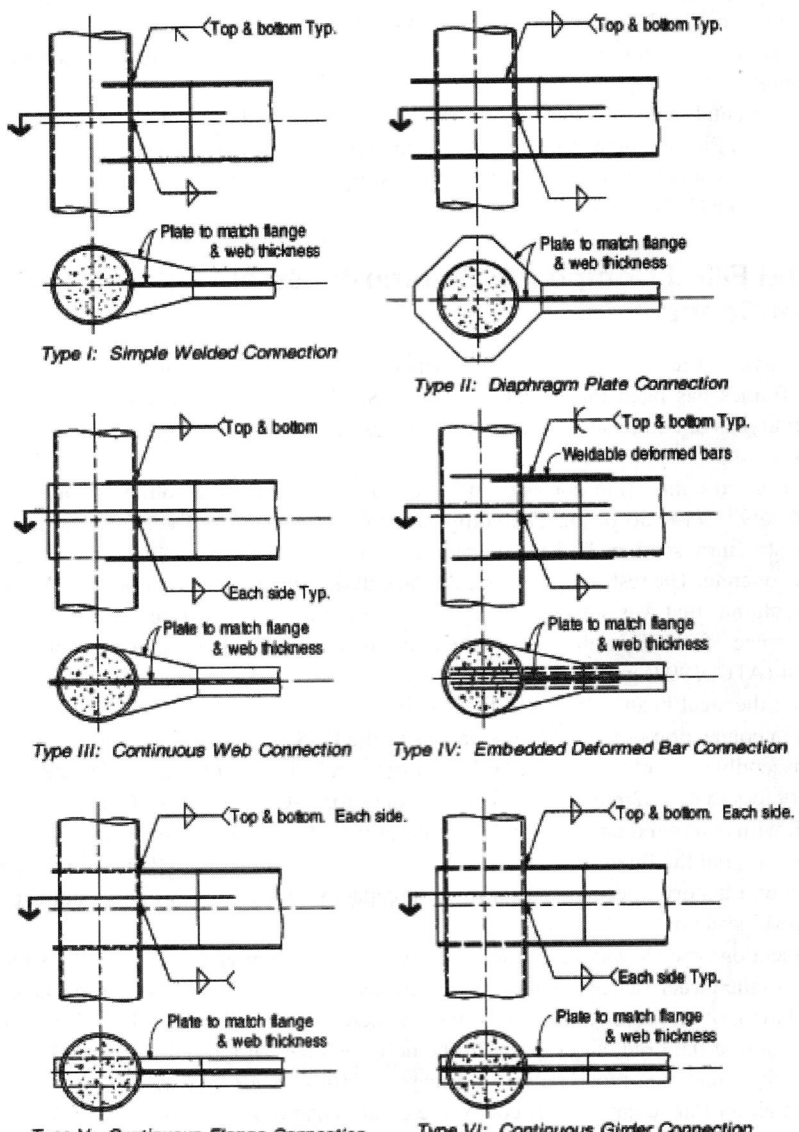

Fig. 6-4. Round filled composite column-to-steel beam connection test configurations (Azizinamini and Schneider, 2004). Reprinted with permission from ASCE.

Table 6-6
Matrix of Specimens Tested (Ricles et al., 2004)

Specimen	Connection Detail
1	Interior diaphragms (four-sided CJP weld), weak beam
2	Interior diaphragms (three-sided CJP weld), weak beam
1R	Interior diaphragms (four-sided CJP weld), weak panel zone
2R	Interior diaphragms (three-sided CJP weld), weak panel zone
3	Extended tee, weak beam
3R	Extended tee with taper, weak beam
4	Bolted split-tee connection with shear tab, weak beam
5	Bolted split-tee connection without shear tab, weak beam
6	Welded split-tee connection without shear tab, weak beam
7	Welded split-tee connection without shear tab, weak beam

Reprinted with permission from ASCE.

subassembly. The specimens were designed to investigate different possible failure modes and develop connection strength equations. The main test variables were the column-to-beam flexural strength ratio (moment ratio) and the type of weld used to attach the beam to the HSS. The ASTM A500 Grade B steel HSS varied from 12 to 16 in. in diameter with 0.25 in. wall thickness, and the ASTM A992 steel beams varied from W16×31 to W18×50.

Rectangular Filled Composite Column-to-Steel Beam Connections

Extensive research has been conducted in Japan to study the behavior of moment connections between filled composite columns and wide flange beams under seismic loading conditions (Ricles et al., 2004). Research on welded beam-to-filled composite column connections having interior or exterior diaphragms has shown that these elements are susceptible to buckling or shear yielding of the steel HSS within the panel zone of the connection.

Ricles et al. (2004) conducted full-scale tests representative of the interior subassemblies in the middle to upper floors of moment frames with six to 12 stories. A total of 10 full-scale tests were conducted. Each test specimen consisted of two W24×62 beam sections made from ASTM A36 steel attached to an HSS16×16×1/2 rectangular filled column made from ASTM A500 Grade B steel and filled with concrete with a measured compressive strength of 7 to 8.5 ksi. The test specimens are as defined in Table 6-6.

Specimens 1, 2, 3, 4, 5, 6 and 7 in this test were designed using the strong column-weak beam principle, where the connection elements were designed to resist 1.50 times the nominal plastic moment strength of the beam. The details of these connections are shown in Figure 6-5 and described as follows:

- Figure 6-5(a) shows the detail of connection specimens 1 and 2 that consisted of interior diaphragms and welded details for the filled composite column-to-steel beam connection. The only difference between the two specimens was that the interior diaphragms of specimen 2 were welded on only three sides. The complete-joint-

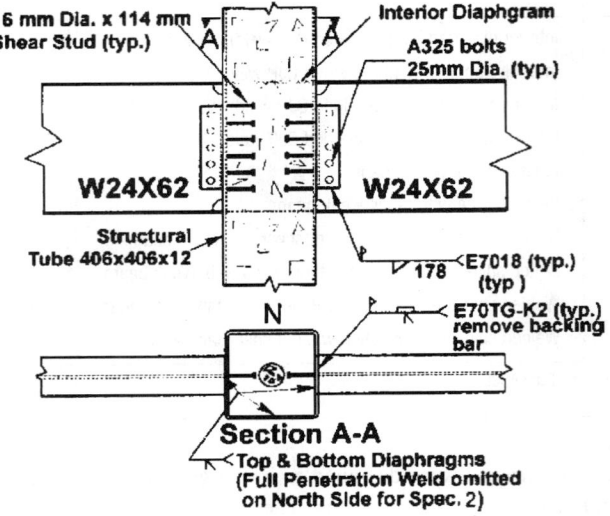

(a) Details of specimens 1 and 2 in Table 6-6

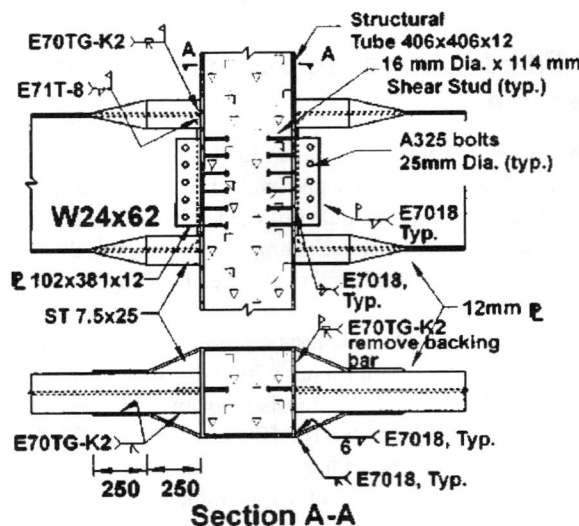

(b) Details of specimen 3 in Table 6-6

Fig. 6-5. Details of rectangular filled composite column-to-steel beam connection (Ricles et al., 2004). Reprinted with permission from ASCE.

penetration groove weld on the north side adjacent to the panel zone (i.e., web of HSS) was omitted.
- Specimen 3 had an extended-tee moment connection detail as shown in Figure 6-5(b). As shown, the extended tee was an ST7.5×25 section that was attached to the beam flanges and column by complete-joint-penetration groove welds.

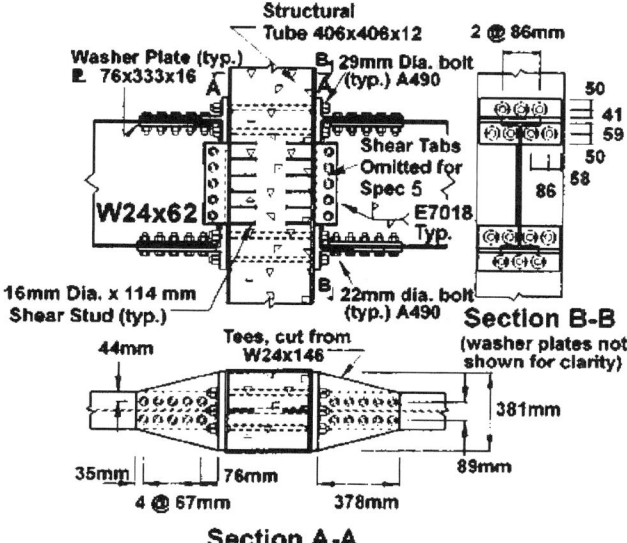

(c) Details of specimens 4 and 5 in Table 6-6

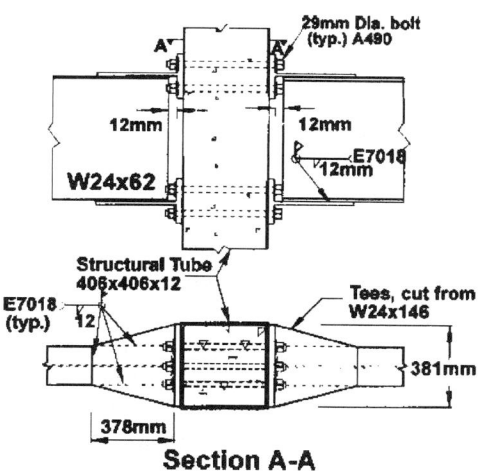

(d) Details of specimens 6 and 7 in Table 6-6

Fig. 6-5 (cont'd). Details of rectangular filled composite column-to-steel beam connection tested by Ricles et al. (2004). Reprinted with permission from ASCE.

- Specimens 4, 5, 6 and 7 had split-tee moment connection details as shown in Figure 6-5(c) and (d). The split-tee connections were designed to activate a diagonal concrete compression strut within the connection's panel zone under the action of overturning moment. This was achieved by the use of ASTM A490 bolts to develop a horizontal tension force through the joint. These bolts were passed through the column with the use of PVC conduits placed prior to casting concrete and tensioned after curing of the concrete. The split-tee detail was designed to avoid prying action in the ASTM A490 bolts.
- In specimens 4 and 5 the stem of the tees were attached to the beam flanges using $\frac{7}{8}$-in.-diameter ASTM A325 bolts with $\frac{1}{8}$-in. oversized bolt holes, whereas in specimens 6 and 7 a $\frac{1}{2}$-in. fillet weld was used. The structural tees in all specimens were cut from a W24×146 section of ASTM A572 Grade 50 steel that had a measured yield strength of 49.6 ksi.

Specimen 1 achieved a maximum story drift of about 0.04 rad when a rupture initiated at the fusion line of the beam flange and the weld. Prior to rupture, the beams developed appreciable yielding in the flanges and web. This type of connection detail is not acceptable for C-SMF systems, but it can be used for C-IMF systems that require only 0.02 rad story drift angle.

Specimen 3 developed a rupture in the beam tension flange, adjacent to an extended tee at the end of the connection during the first half cycle of 0.03 rad story drift. An examination of the beam flange in the ruptured area revealed that the material had necked at the crack, indicating that a significant amount of strain had developed. This type of connection detail is not acceptable for C-SMF systems, but it can be used for C-IMF systems that require only 0.02 rad of interstory drift.

Specimens 4, 5, 6 and 7 developed significant yielding at the base of tee stem during the inelastic displacement cycles during the test. These specimens also developed full plastic flexural hinges in the beams at the end of the connection, where pronounced flange and web yielding occurred and was followed by local flange and web buckling. Each test was stopped after a story drift of 0.06 rad was imposed to the top of the column of these specimens. Figure 6-6 shows the moment-plastic rotation behavior of these connections. All of these connection details are acceptable for C-SMF systems that require 0.04 rad of interstory drift.

As shown in Figure 6-6, pinching occurred in the cyclic behavior of specimen 4 due to the bolt hole elongation and resulting slippage between the beam and the connection under cyclic loading. At the end of the test a net section rupture occurred in the flange bolt line, leading to deterioration in strength. The welding of the washer plates in specimen 5 and tee stems to the beam flanges in specimens 6 and 7 served to reinforce the bearing strength and increase the net area in the beam flanges. This avoided hole elongation and subsequent problems from developing.

Figure 6-7 shows the inelastic story drift capacity of tested specimens, and the required inelastic story drift for design basis and maximum considered earthquakes. As shown, the split-tee moment connections (specimens 4 through 7) have acceptable behavior for use with C-SMF systems. The required story drifts were calculated by conducting nonlinear time history analysis of several CFT moment resisting frames, subjecting them each to several ground motion records.

The panel zone shear strength of the split-tee moment connections can be estimated using the Kanatani et al. (1991) model developed based on the Japanese test results. Detailed seismic design guidelines for the split-tee moment connections are included in Peng (2001).

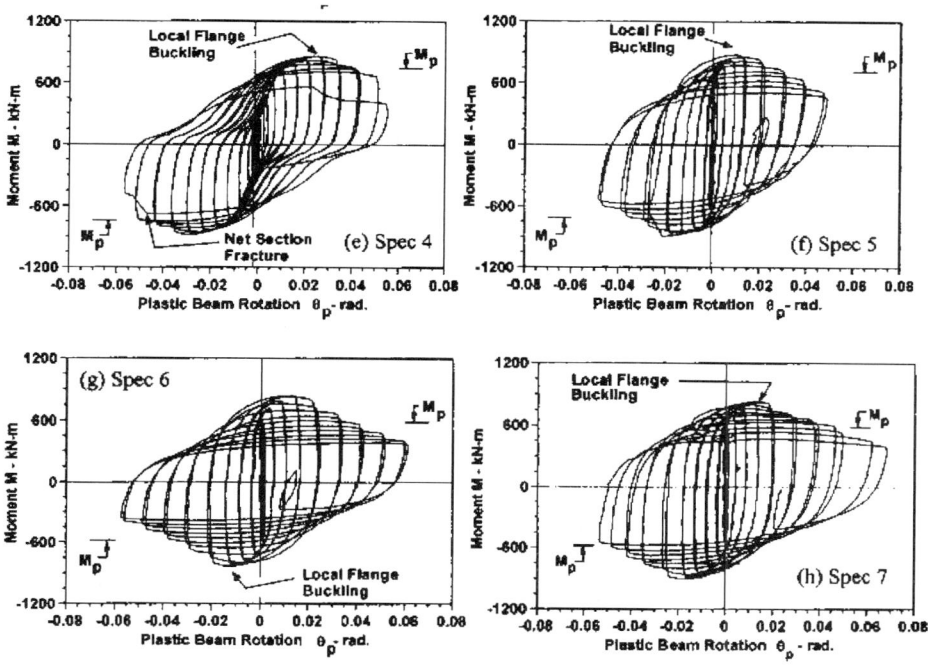

Fig. 6-6. Moment-plastic beam rotation behavior of tested split-tee moment connections (Ricles et al., 2004). Reprinted with permission from ASCE.

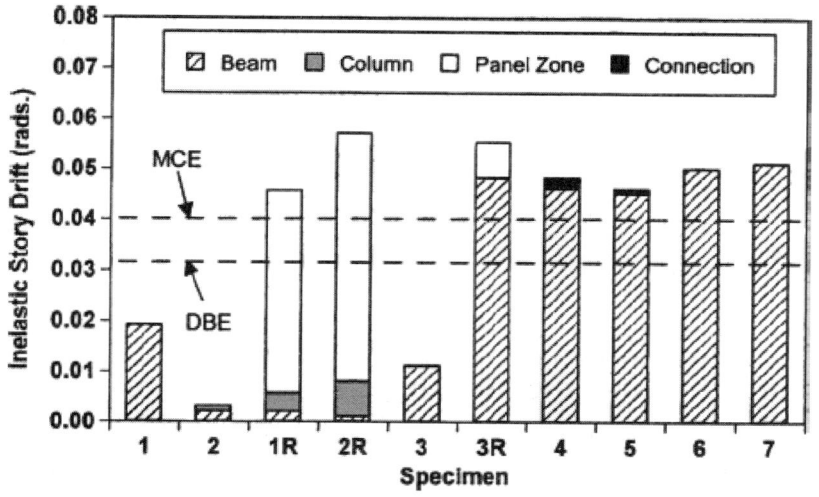

Fig. 6-7. Inelastic story drift capacity of connection test specimens (Ricles et al., 2004). Reprinted with permission from ASCE.

PART 6 REFERENCES

ASCE (1994), "Guidelines for Design of Joints between Steel Beams and Reinforced Concrete Columns," ASCE Task Committee on Design Criteria for Composite Structures in Steel and Concrete, *Journal of Structural Engineering*, ASCE, Vol. 20, No. 8, pp. 2,330–2,357.

ASCE (1998), "Design Guide for Partially Restrained Composite Connections," ASCE Task Committee on Design Criteria for Composite Structures in Steel and Concrete, *Journal of Structural Engineering*, ASCE, Vol. 124, No. 10, pp. 1,099–1,114.

ATC (1992), "Guidelines for Cyclic Seismic Testing of Components of Steel Structures," ATC-24, Applied Technology Council, Redwood City, CA.

Azizinamini, A. and Schneider, S.P. (2004), "Moment Connections to Circular Concrete-Filled Steel Tube Columns," *Journal of Structural Engineering*, ASCE, Vol. 130, No. 2, pp. 213–222.

Baba, N. and Nishimura, Y. (2000), "Stress Transfer on Through Beam Type Steel Beam-Reinforced Concrete Column Joints," *Proceedings of the 6th ASCCS International Conference on Steel-Concrete Composite Structures,* Y. Xiao and S. Mahin, eds., pp. 753–760.

Bracci, J.M., Moore, W.P., Jr. and Bugeja, M.N. (1999), "Seismic Design and Constructability of RCS Special Moment Frames," *Journal of Structural Engineering*, Vol. 25, No. 4, pp. 385–392.

Bugeja, M., Bracci, J.M. and Moore, W.P. (1999), "Seismic Behavior of Composite Moment Resisting Frame Systems," *Technical Report No. CBDC-99-01,* Dept. of Civil Engineering, Texas A&M University.

Bugeja, M., Bracci, J.M. and Moore, W.P. (2000), "Seismic Behavior of Composite RCS Frame Systems," *Journal of Structural Engineering*, Vol. 126, No. 4, pp. 429–436.

Chou, C.C. and Uang, C.M. (2002), "Cyclic Performance of a Type of Steel Beam to Steel-Encased Reinforced Concrete Column Moment Connection," *Journal of Constructional Steel Research,* Vol. 58, No. 5–8, pp. 637–663.

Deierlein, G.G., Sheikh, T.M., Yura, J.A. and Jirsa, J.O. (1989), "Beam-Column Moment Connections for Composite Frames: Part 2," *Journal of Structural Engineering*, Vol. 115, No. 11, pp. 2,877–2,896.

Deierlein, G.G. and Noguchi, H. (2004), "Overview of U.S.-Japan Research on the Seismic Design of Composite Reinforced Concrete and Steel Moment Frame Structures," *Journal of Structural Engineering*, ASCE, Vol. 130, No. 2, February, pp. 361–367.

Elremaily, A. and Azizinamini, A. (2000), "Experimental Behavior of Steel Beam to CFT Column Connections," *Journal of Constructional Steel Research*, Vol. 57, No. 10, pp. 1,099–1,119.

Esche, C.D., Bracci, J.M. and Moore, W.P. (1999), "Joint Strength in RCS Frames," *Technical Report No. CBDC-99-02,* Department of Civil Engineering, Texas A&M University.

Kanatani, H., Tabuchi, M., Kamba, T., Hsiaolien, J. and Ishikawa, M. (1991), "A Study on Concrete Filled RHS Column to H-Beam Connections Fabricated with HT Bolts in Rigid Frames," *Proceedings, Composite Construction in Steel and Concrete Conference*, Henniker, NH, pp. 614–635.

Kanno, R. and Deierlein, G.G. (1993), "Strength, Deformation, Seismic Resistance of Joints Between Steel Beams and Reinforced Concrete Columns," *Structural Engineering Report No. 93-6*, Cornell University, Ithaca, NY.

Kanno, R. and Deierlein, G.G. (1997), Seismic Behavior of Composite (RCS) Beam-Column Joint Subassemblies," *Composite Construction in Steel and Concrete III*, ASCE, New York, pp. 236–249.

Kuramoto, H. and Nishiyama, I. (2004), "Seismic Performance and Stress Transferring Mechanism of Through-Column-Type Joints for Composite Reinforced Concrete and Steel Frames," *SP 196-6*, ACI, Farmington Hills, MI, pp. 109–123.

Kuramoto, H. and Noguchi, H. (1997), "An Overview of Japanese Research on RCS Systems," *Proceedings, ASCE Structures Congress XV*, ASCE, Reston, VA, pp. 716–720.

Leon, R., Hoffman, J. and Staeger, T. (1996), *Partially Restrained Composite Connections*, Design Guide 8, AISC, Chicago, IL.

Liang, X. and Parra-Montesinos, G. (2004), "Seismic Behavior of RCS Beam-Column-Slab Subassemblies and Frame Systems," *Journal of Structural Engineering*, ASCE, Vol. 130, No. 2, pp. 310–319.

Nishiyama, I., Hasegawa, T. and Yamanouchi, H. (1990), "Strength and Deformation Capacity of Reinforced Concrete Column to Steel Beam Joint Panels," *Building Research Institute Report 71*, Ministry of Construction, Tsukuba, Japan.

Nishiyama, I. Itadani, H. and Suginiro, K. (1998), "Bidirectional Seismic Response of Reinforced Concrete Column and Structural Steel Beam Subassemblies," *Proceedings, Structural Engineers World Congress*, ASCE, *Paper Ref. T177-2*, Reston, VA.

Nishiyama, I., Kuramoto, H., Itadani, H. and Sugihiro, K. (2000), "Bidirectional Behavior of Interior, Exterior and Corner Joints of RCS System," *Proceedings, 12 WCEE*, Paper No. 1911/6/A.

Noguchi, H. and Kim, K. (1997), "Analysis of Beam-Column Joints in Hybrid Structures," *Proceedings, ASCE Structures Congress XV*, ASCE, Reston, VA, pp. 726–730.

Noguchi, H. and Kim, K. (1998), "Shear Strength of Beam-to-Column Connections in RCS System," *Proceedings, Structural Engineers World Congress*, ASCE, Reston, VA, Paper Ref. T177-3.

Parra-Montesinos, G. and Wight, J.K. (2000a), "Seismic Behavior, Strength, and Retrofit of Exterior RC Column-to Steel Beam Connections," *UMCEE 00-09*, Department of Civil and Environmental Engineering, University of Michigan.

Parra-Montesinos, G. and Wight, J.K. (2000b), "Seismic Response of Exterior RC Column-to-Steel Beam Connections," *Journal of Structural Engineering*, Vol. 126, No. 10, pp. 1,113–1,121.

Parra-Montesinos, G. and Wight, J.K. (2001a), "Modeling Shear Behavior of Hybrid RCS Beam-Column Connections," *Journal of Structural Engineering*, Vol. 127, No. 1, pp. 3–11.

Parra-Montesinos, G. and Wight, J.K. (2001b), "Seismic Repair of Hybrid RCS Beam-Column Connections," *ACI Structural Journal,* Vol. 98, No. 5, pp. 762–770.

Parra-Montesinos, G., Liang, X. and Wight, J.K. (2003), "Towards Deformation-Based Capacity Design of RCS Beam-Column Connections," *Engineering Structures*, Vol. 25, No. 5, pp. 681–690.

Peng, S.W. (2001), "Full Scale Testing of Seismically Resistant Moment Connections for Concrete Filled Tube Column to WF Beam Hybrid Systems," *Composite and Hybrid Structures: Proceedings of the Sixth ASCCS International Conference on Steel-Concrete Composite Structures*, Vol. 1.

Ricles J., Peng, S. and Lu, L. (2004), "Seismic Behavior of Composite Concrete Filled Steel Tube Column-Wide Flange Beam Moment Connections," *Journal of Structural Engineering*, ASCE, Vol. 130, No. 2, pp. 223–243.

Sheikh, T.M., Deierlein, G.G., Yura, J.A. and Jirsa, J.O. (1989), "Beam-Column Moment Connections for Composite Frames: Part 1," *Journal of Structural Engineering*, ASCE, Vol. 115, No. 11, pp. 2,858–2,876.

U.S.-Japan (1983), "U.S.-Japan Cooperative Research Program: Construction of the Full Scale Reinforced Concrete Test Structure," *Technical Report UMEE 83R2*, Department of Engineering, University of Michigan, August.

PART 7

COMPOSITE BRACED FRAMES AND SHEAR WALLS

7.1 SCOPE 7–3
7.2 COMPOSITE ORDINARY BRACED FRAMES (C-OBF) 7–3
    Overview of Applicable Design Provisions 7–3
7.3 COMPOSITE SPECIAL CONCENTRICALLY BRACED FRAMES
    (C-SCBF) 7–5
    Overview of Applicable Design Provisions 7–5
7.4 COMPOSITE ECCENTRICALLY BRACED FRAMES (C-EBF) 7–7
    Overview of Applicable Design Provisions 7–7
7.5 COMPOSITE SHEAR WALLS 7–10
    General System Behavior 7–10
        Shear Wall Coupling 7–10
        Degree of Coupling 7–11
        Steel Coupling Beam Design 7–11
        Beam Embedment Length (Connection) 7–14
        Detailing Requirements in the Embedded Region 7–14
        Wall Overstrength 7–15
    Composite Ordinary Shear Walls (C-OSW) 7–16
        Overview of Applicable Design Provisions 7–16
        Steel Coupling Beam Design 7–19
        Beam Embedment Length 7–19
        Composite Coupling Beams 7–19
            Expected Plastic Moment 7–19
            Shear Strength 7–21
            Embedment Length 7–21
        Example 7.5.1. C-OSW Steel Coupling Beam Design 7–21
        Example 7.5.2. C-OSW Composite Coupling Beam Design 7–30
    Composite Special Shear Walls (C-SSW) 7–36
        Overview of Applicable Design Provisions 7–36
        Steel Coupling Beam Design 7–39

	Wall Overstrength 7–40
	Beam Embedment Length 7–40
	Intermediate Web Stiffeners 7–40
	Face Bearing Plates 7–41
	Stiffeners within the Embedded Region 7–41
	Vertical Transfer Bars 7–41
	Example 7.5.3. C-SSW Steel Coupling Beam Design 7–41
7.6	DESIGN TABLE DISCUSSION 7–51
	DESIGN TABLES 7–52
	Table 7-10A. Plastic Capacities for Rectangular Encased W-Shapes Bent About the X-X Axis 7–52
	Table 7-10B. Plastic Capacities for Rectangular Encased W-Shapes Bent About the Y-Y Axis 7–53
	Table 7-11A. Plastic Capacities for Composite Filled HSS Bent About Either Axis 7–54
	Table 7-11B. Plastic Capacities for Composite, Filled Round HSS Bent About Any Axis 7–55
	PART 7 REFERENCES 7–56

7.1 SCOPE

The following types of composite braced frame and shear wall systems are addressed in this Part: composite ordinary braced frames, composite special concentrically braced frames, composite eccentrically braced frames, composite ordinary shear walls, and composite special shear walls. The AISC *Seismic Provisions* and other design considerations summarized in this Part apply to the design of the members and connections in composite braced frame and shear wall systems that require seismic detailing. Where these systems utilize reinforced concrete elements, these elements are to be designed in accordance with ACI 318. Reinforced concrete elements are permitted to be used in Section H1 (C-OBF), Section H4 (C-OSW), and Section H5 (C-SSW). However, the requirements of ACI 318 Chapter 21 are applicable only in the design of the reinforced concrete walls used in Section H5 (C-SSW). AISC *Seismic Provisions* Sections A1 and B2 state that systems with reinforced concrete elements that must be designed according to ACI 318 should be designed only by the load and resistance factor design (LRFD) method because ACI 318 does not address allowable strength design (ASD). The design examples in this Part are limited to the LRFD method since in each example there is a concrete element that must be designed according to ACI 318.

7.2 COMPOSITE ORDINARY BRACED FRAMES (C-OBF)

Composite ordinary braced frame (C-OBF) systems consist of structural steel, composite or reinforced concrete columns; structural steel or composite beams; and structural steel or filled composite brace members, provided at least one element is either composite or reinforced concrete. Concentrically connected members are required; however, eccentricities less than the beam depth are permitted if accounted for in the member design. C-OBF systems are designed and detailed according to AISC *Seismic Provisions* Section H1. They are expected to provide minimal inelastic deformation capacity in the members and connections.

Overview of Applicable Design Provisions

An overview of the AISC *Seismic Provisions* applicable for design of C-OBF systems follows and is presented in a simplified format in Table 7-1.

Note 1. The structural steel material used for C-OBF systems is limited by the requirements of AISC *Seismic Provisions* Section A3.1, where the specified minimum yield stress is not to exceed 55 ksi for members in which inelastic behavior is expected. These specified minimum yield stresses can be exceeded when the suitability of the material is determined by testing or other rational criteria. For columns in C-OBF systems, the specified minimum yield stress is not to exceed 65 ksi. The concrete and steel reinforcing materials used in composite components are to satisfy the requirements of AISC *Seismic Provisions* Section A3.5. The weld filler metal used in the members and connections of seismic force resisting systems is selected to meet the requirements of AISC *Seismic Provisions* Section A3.4a.

Note 2. The structural design drawings and specifications for C-OBF systems are to meet the requirements of AISC *Seismic Provisions* Sections A4.1, A4.2 and A4.3.

Table 7-1
Simplified Overview of Provisions for C-OBF Systems

Note	Item	Referenced Standard*
1	Steel and concrete materials	*Seismic Prov.* Sects. A3.1, A3.4 & A3.5
2	Design drawings and specifications	*Seismic Prov.* Sects. A4.1, A4.2 & A4.3
3	Loads and load combinations	*Seismic Prov.* Sect. B2
4	Required strength for members and connections	*Seismic Prov.* Sect. B3.1
5	Structural analysis	*Seismic Prov.* Ch. C
5(a)	Elastic stiffness of concrete/composite members	*Seismic Prov.* Ch. C. See *Seismic Prov.* Commentary for discussion.
6	Column members	ACI 318 (excl. Ch. 21)
7	Beam members	None
8	Brace members	None
9	Connections	*Seismic Prov.* Sect. D2.7
10	Column splices	*Seismic Prov.* Sect. D2.5
11	Column bases	*Seismic Prov.* Sect. D2.6

*The referenced standards are in addition to the requirements of the AISC *Specification*.

Note 3. Loads and load combinations as defined by the applicable building code are to be followed as indicated in AISC *Seismic Provisions* Section B2. C-OBF systems including reinforced concrete components are to be designed using LRFD because ASD is not addressed in ACI 318.

Note 4. The required strength for structural members and connections is determined according to AISC *Seismic Provisions* Section B3.1.

Note 5. Structural analysis for the appropriate load combinations is to be performed in accordance with the requirements of AISC *Seismic Provisions* Chapter C.

(a) For elastic analysis, the stiffness of composite members includes the effects of cracked sections. Additional guidelines for estimating the stiffness of concrete beam and column members, concrete-encased and concrete-filled members, and steel beams with composite slabs are provided in the Commentary to the AISC *Seismic Provisions* Chapter C. These concrete and composite member properties reflect the effective stiffness at the onset of significant yielding in the members.

Note 6. Columns of C-OBF systems are designed in accordance with the AISC *Specification* or ACI 318 (excluding Chapter 21).
Note 7. Beams of C-OBF systems are designed in accordance with the AISC *Specification*.
Note 8. Diagonal braces of C-OBF systems are designed in accordance with the AISC *Specification*.
Note 9. Connections are designed in accordance with the AISC *Specification* and AISC *Seismic Provisions* Section D2.7.
Note 10. Splices in structural steel or composite columns are designed in accordance with the AISC *Seismic Provisions* Section D2.5.
Note 11. Column bases are designed in accordance with the AISC *Specification* and AISC *Seismic Provisions* Section D2.6.

Discussion

ASCE/SEI 7 permits the use of C-OBF systems in Seismic Design Categories A, B and C only. This is in contrast to steel ordinary concentrically braced frame (OCBF) systems that are also permitted in Seismic Design Categories D, E and F with height limitations and roof load restrictions for Seismic Design Category F.

Because C-OBF systems are limited to Seismic Design Categories A, B and C, they are expected to withstand minimal inelastic drift through inelastic behavior of composite beams, columns or braces. There are no additional requirements for designing reinforced concrete columns beyond those provided in ACI 318, excluding Chapter 21.

7.3 COMPOSITE SPECIAL CONCENTRICALLY BRACED FRAMES (C-SCBF)

Composite special concentrically braced frame (C-SCBF) systems consist of either encased or filled composite columns; structural steel or composite beams; and structural steel or filled composite brace members. Concentrically connected members are required; however, members connected with an eccentricity less than the depth of the beam are permitted provided the eccentricity is included in the analysis. C-SCBF systems are designed and detailed according to AISC *Seismic Provisions* Section H2. They are expected to provide significant inelastic deformation capacity primarily through brace buckling in compression and yielding in tension.

Overview of Applicable Design Provisions

An overview of the AISC *Seismic Provisions* requirements applicable for design of C-SCBF systems follows and is presented in a simplified format in Table 7-2.

Note 1. The structural steel material used for C-SCBF systems is limited by the requirements of the AISC *Seismic Provisions* Section A3.1, where the specified minimum yield stress is not to exceed 50 ksi for members in which inelastic behavior is expected. This specified minimum yield stress can be exceeded when the suitability of the material is determined by testing or other rational criteria. For columns in C-SCBF systems, the specified minimum yield stress is not to exceed 65 ksi.

Table 7-2
Simplified Overview of Provisions for C-SCBF Systems

Note	Item	AISC *Seismic Provisions* Reference*
1	Steel and concrete materials	Sects. A3.1 & A3.5
2	Design drawings and specifications	Sects. A4.1, A4.2 & A4.3
3	Loads and load combinations	Sect. B2
4	Required strength for members Required strength for connections	Sects. B3.1, H2.3 & H2.5 Sects. H2.3 & H2.6
5	Structural analysis	Ch. C & Sect. H2.3
5(a)	Elastic stiffness of concrete/composite members	Commentary to *Seismic Prov.* Ch. C
6	Column members	Sects. D1.1 & H2.5a
7	Beam members	Sects. D1.1 & H2.5a
8	Brace members	Sects. H2.5a & H2.5b
9	Connections	Sects. D2 & H2.6
10	Column splices	Sects. G2.6f & H2.6d
11	Column bases	Sect. D2.6
12	Demand critical welds	Sects. A3.4b, H2.6a & I2.3

*The referenced standards are in addition to the requirements of the AISC *Specification*.

The concrete and steel reinforcing materials used in composite components are to satisfy the requirements of AISC *Seismic Provisions* Section A3.5. The weld filler metal used in the members and connections of seismic force resisting systems is selected to meet the requirements of AISC *Seismic Provisions* Section A3.4a.

Note 2. The structural design drawings and specifications for C-SCBF systems are to satisfy the requirements of AISC *Seismic Provisions* Sections A4.1, A4.2 and A4.3.

Note 3. Loads and load combinations as defined by the applicable building code are to be followed as indicated in AISC *Seismic Provisions* Section B2.

Note 4. The required strength for structural members and connections is determined according to AISC *Seismic Provisions* Sections B3.1, H2.3, H2.5 and H2.6.

Note 5. Structural analysis for the appropriate load combinations is to be performed in accordance with the requirements in AISC *Seismic Provisions* Chapter C and Section H2.3.

(a) For elastic analysis, the stiffness of composite members includes the effects of cracked sections. Additional guidelines for estimating the stiffness of concrete

7.4 COMPOSITE ECCENTRICALLY BRACED FRAMES 7–7

beam and column members, concrete-encased and concrete-filled members, and steel beams with composite slabs are provided in the Commentary to the AISC *Seismic Provisions* Chapter C. These concrete and composite member properties reflect the effective stiffness at the onset of significant yielding in the members.

Note 6. Columns of C-SCBF systems are designed in accordance with AISC *Specification* Chapter I and the requirements of AISC *Seismic Provisions* Section H2.5a. Composite columns are required to satisfy the highly ductile member requirements of AISC *Seismic Provisions* Section D1.1.

Note 7. Beams of C-SCBF systems are designed in accordance with the AISC *Specification* and the requirements of AISC *Seismic Provisions* Section H2.5a. Composite beams are required to satisfy the moderately ductile member requirements of AISC *Seismic Provisions* Section D1.1.

Note 8. Diagonal braces of C-SCBF systems are designed in accordance with AISC *Seismic Provisions* Section H2.5a and b. The radius of gyration for filled composite braces is taken as that of the steel section alone.

Note 9. Connections are designed in accordance with the AISC *Specification* and AISC *Seismic Provisions* Sections D2 and H2.6.

Note 10. Column splices are designed in accordance with AISC *Seismic Provisions* Sections G2.6f and H2.6d.

Note 11. Column bases are designed in accordance with the AISC *Specification* and AISC *Seismic Provisions* Section D2.6.

Note 12. Demand critical welds are designed in accordance with AISC *Seismic Provisions* Sections A3.4b, H2.6a and I2.3.

Discussion

ASCE/SEI 7 permits the use of C-SCBF systems in Seismic Design Categories A, B and C without height limitations and in Seismic Design Categories D, E and F with height limitations. These limitations are the same as those applied to steel concentrically braced frame systems. This system is expected to resist inelastic drift through inelastic behavior of composite beams, columns and braces.

7.4. COMPOSITE ECCENTRICALLY BRACED FRAMES (C-EBF)

Composite eccentrically braced frame (C-EBF) systems consist of encased or filled composite columns; structural steel or composite beams; structural steel links; and structural steel or filled composite braces. C-EBF systems are designed and detailed in accordance with AISC *Seismic Provisions* Section H3. They are expected to provide significant inelastic deformation capacity primarily through shear or flexural yielding in the links.

Overview of Applicable Design Provisions

An overview of the AISC *Seismic Provisions* requirements applicable for design of C-EBF systems follows and is presented in simplified format in Table 7-3.

Table 7-3
Simplified Overview of Provisions for C-EBF Systems

Note	Item	AISC *Seismic Provisions* Reference*
1	Steel and concrete materials	Sects. A3.1 & A3.5
2	Design drawings and specifications	Sects. A4.1, A4.2 & A4.3
3	Loads and load combinations	Sect. B2
4	Required strength for members Required strength for connections	Sects. B3.1 & H3.5 Sects. B3.1 & H3.6
5	Structural analysis	Ch. C
5(a)	Elastic stiffness of concrete/composite members	Commentary to *Seismic Prov.* Ch. C
6	Additional analysis and system requirements	Sect. H3.4
7	Column members	Sect. H3.5
8	Beam members	Sect. H3.5
9	Braces	Sect. H3.5
10	Connections	Sect. H3.6
11	Column splices	Sects. D2.5 & H3.6
12	Column bases	Sect. D2.6
13	Protected zones	Sects. D1.3 & H3.5
14	Demand critical welds	Sects. A3.4b, H3.6 & I2.3

*The referenced standards are in addition to the requirements of the AISC *Specification*.

Note 1. The structural steel material used for C-EBF systems is limited by the requirements of AISC *Seismic Provisions* Section A3.1 where the specified minimum yield stress of the steel is not to exceed 50 ksi for members in which inelastic behavior is expected. This specified minimum yield stress can be exceeded when the suitability of the material is determined by testing or other rational criteria. For columns of C-EBF, the specified minimum yield stress is not to exceed 65 ksi. The concrete and steel reinforcing materials used in composite components should satisfy the requirements of AISC *Seismic Provisions* Section A3.5. The weld filler metal used in the members and connections of seismic force resisting systems is selected to meet the requirements of AISC *Seismic Provisions* Section A3.4a.

Note 2. The structural design drawings and specifications for C-EBF systems are to satisfy the requirements of AISC *Seismic Provisions* Sections A4.1, A4.2 and A4.3.

Note 3. Loads and load combinations as defined by the applicable building code are to be followed as indicated in AISC *Seismic Provisions* Section B2.

7.4 COMPOSITE ECCENTRICALLY BRACED FRAMES

Note 4. The required strength for structural members and connections is determined according to AISC *Seismic Provisions* Sections B3.1, H3.5 and H3.6.

Note 5. Structural analysis for the appropriate load combinations is to be performed in accordance with the requirements of AISC *Seismic Provisions* Chapter C.

  (a) For elastic analysis, the stiffness of composite members includes the effects of cracked sections. Additional guidelines for estimating the stiffness of concrete beam and column members, concrete-encased and concrete-filled members, and steel beams with composite slabs are provided in the Commentary to the AISC *Seismic Provisions* Chapter C. These concrete and composite member properties reflect the effective stiffness at the onset of significant yielding in the members.

Note 6. System requirements are as given in AISC *Seismic Provisions* Section H3.4.

Note 7. Columns of C-EBF systems are designed in accordance with AISC *Specification* Chapter I and AISC *Seismic Provisions* Section H3.5. By reference to AISC *Seismic Provisions* Section F3.5, the composite member must satisfy the requirements for highly ductile members.

Note 8. Beams of C-EBF systems are designed in accordance with the AISC *Specification* and AISC *Seismic Provisions* Section H3.5. Links must satisfy the requirement for highly ductile members with the exception that if the beam outside the link is a different section than that of the link, the beam outside the link need only satisfy the requirements for moderately ductile members. Additionally, flanges of I-shaped links of a certain length defined in Section F3.5b(1), may also satisfy the moderately ductile member requirements.

Note 9. Braces of C-EBF systems are designed in accordance with the AISC *Specification* and AISC *Seismic Provisions* Section H3.5. Braces must satisfy the requirements for moderately ductile members.

Note 10. Connections are designed in accordance with the AISC *Specification* and AISC *Seismic Provisions* Section H3.6.

Note 11. Column splices are designed in accordance with the requirements of AISC *Seismic Provisions* Sections D2.5 and H3.6.

Note 12. Column bases are designed in accordance with the AISC *Specification*, with additional requirements for groove-welded bases and concrete elements as given in AISC *Seismic Provisions* Section D2.6.

Note 13. Links in C-EBF systems are protected zones in accordance with AISC *Seismic Provisions* Section H3.5 and must satisfy the requirements of AISC *Seismic Provisions* Section D1.3.

Note 14. Demand critical welds are evaluated using the requirements of AISC *Seismic Provisions* Sections A3.4b, H3.6 and I2.3.

Discussion

ASCE/SEI 7 permits the use of composite eccentrically braced frame systems in Seismic Design Categories A, B and C without height limitations and Seismic Design Categories D, E and F with height limitations. This system is expected to resist inelastic drift through inelastic behavior of structural steel links.

7.5. COMPOSITE SHEAR WALLS

General System Behavior

Composite shear wall systems are addressed in AISC *Seismic Provisions* Sections H4, H5 and H6. The composite shear walls addressed are those that include steel or composite boundary elements and/or steel or composite coupling beams and walls consisting of steel plate encased in concrete. Since all composite shear wall systems incorporate a reinforced concrete wall that will be designed according to ACI 318, the discussion and examples that follow will only consider design by LRFD.

Shear Wall Coupling

The benefits of coupling shear walls are well recognized and understood. The coupling beams provide transfer of vertical forces between adjacent walls, which create a frame-like coupling action that resists a portion of the total overturning moment induced by the seismic action. Figure 7-1 shows the overturning resisting mechanisms formed in a coupled system. The total overturning resistance is a combination of the flexural resistance of the individual wall piers (M_1 and M_2), and the resistance provided by the coupling action ($M_{cpl} = TL$ or CL).

The coupling beam action has three desirable effects: (1) the required flexural strength of the wall piers is reduced; (2) steel and composite coupling beams dissipate energy; and (3) lateral stiffness of the coupled system is greater than the sum of the individual wall piers.

When the beams are proportioned properly, beam yielding over the height of the building can occur, providing a desirable distribution of energy dissipation over the height of the building. A comprehensive discussion of coupled wall system behavior is presented in *Recommendation for Seismic Design of Hybrid Coupled Wall Systems* (El-Tawil et al., 2009).

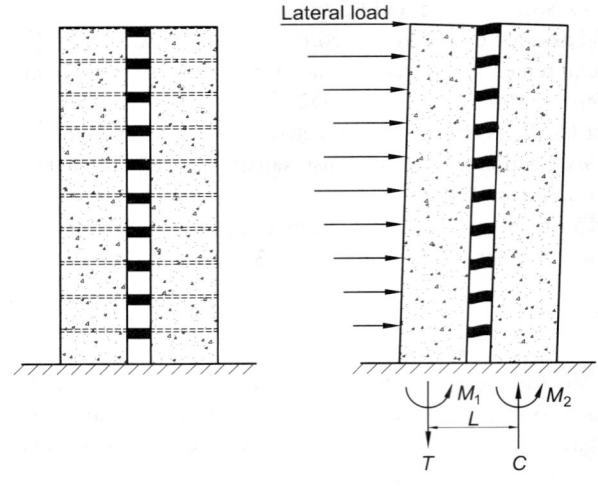

Fig. 7-1. Overturning mechanisms in a coupled wall system.

Degree of Coupling

The efficiency of coupled wall systems is generally measured by the degree to which the coupling action participates in the overall overturning resistance to lateral loads. This measurement is referred to as the degree of coupling. Consider a wall system similar to that shown in Figure 7-1. If the system has no coupling beams, the degree of coupling would be zero. As the flexural and shear stiffness of coupling beams increase, the degree of coupling increases. The degree of coupling is measured as the ratio of overturning resistance due to the coupling effect to the total overturning resistance, as shown in Equation 7-1. In Figure 7-1, C and T are the accumulation of beam shears over the height of the building in the compression and tension walls, respectively.

$$\text{Degree of coupling} = \frac{M_{cpl}}{M_{cpl} + (M_1 + M_2)} \qquad (7\text{-}1)$$

The degree of coupling can be measured at any stage of loading and at any floor level. However, it is generally measured at the base of the building and at the stage of loading where mechanisms have formed in the coupling beams. Designers should be aware that the degree of coupling has an impact on the total lateral stiffness, wall pier required strength, and economy of construction among other things. Compromises between building performance and construction costs must be made.

The degree of coupling also has an impact on the total wall pier axial forces. As the degree of coupling increases, the wall pier required moment strength decreases. However, the wall pier axial forces simultaneously increase as the degree of coupling increases. Most model codes have upper limits on the required axial strength of reinforced concrete wall piers. When the wall pier required axial strength exceeds prescribed limits, reducing the degree of coupling can help to reduce the required axial strength. It is worth noting that most model codes limit the wall pier required axial strength to some percentage of the nominal axial strength of the wall pier. This limit is intended to keep the required axial strength at or below the balanced point of the axial-moment interaction surface of the wall pier. Considering that the balanced point location is sensitive to wall pier cross-sectional geometry and reinforcing layout and ratio, the required axial strength should be evaluated against the axial load component of the balanced point in addition to some percentage of the axial load strength. Further discussion of axial strength limits on wall piers in coupled systems is presented in *Recommendation for Seismic Design of Hybrid Coupled Wall Systems* (El-Tawil et al., 2009).

Steel Coupling Beam Design

Steel coupling beams are designed in a manner similar to shear links in an eccentrically braced frame. Cross-sectional proportioning is dependent on the desired performance of the beams. Flexure-critical beams will have cross-sectional properties that ensure inelastic deformations are resisted through flexural yielding. Shear-critical beams will have cross-sectional properties that ensure inelastic deformations are resisted through shear yielding.

AISC *Seismic Provisions* Sections H4 and H5 permit either shear-critical or flexure-critical coupling beams. Shear-critical and flexure-critical coupling beams have lengths less than or equal to $1.6M_p/V_p$ or greater than or equal to $2.6M_p/V_p$, respectively. Coupling beams with lengths between these two lengths are considered to yield in shear and flexure (refer to the commentary to Section F3.5b(4) of the AISC *Seismic Provisions* for further discussion). With these relationships between coupling beam length and M_p/V_p, the cross-sectional properties of the beam can be determined and evaluated depending on the type of yielding (shear or flexure) desired by the designer.

Built-up I-shapes or W-shapes may be used to achieve the desired cross-sectional properties. Although rolled shapes are generally more economical, built-up shapes provide more flexibility for proportioning cross sections to satisfy design requirements. Flanges and webs of the beams must satisfy seismic ductility requirements regardless of whether built-up or rolled shapes are used. Coupling beams in ordinary systems are required to be moderately ductile. Coupling beams in special systems are required to be highly ductile.

The required shear strength of a coupling beam is the shear corresponding to the required flexural strength, assuming the required flexural strength acts as equal moments at the ends of the beam bending in reverse curvature, as shown in Equation 7-2.

$$V_u = \frac{2M_u}{L} \quad (7\text{-}2)$$

where
L = length of beam
M_u = required flexural strength
V_u = required shear strength

As discussed previously, steel coupling beams are treated similar to steel links in eccentrically braced frame systems. For a given beam, a relationship between the beam length, plastic section modulus, and web area can be written using the relationship between length, plastic flexural strength, and shear strength. Thus, for a shear-critical coupling beam:

$$L \leq \frac{1.6 M_p}{V_p} \quad (7\text{-}3)$$

and for a flexure-critical coupling beam:

$$L \geq \frac{2.6 M_p}{V_p} \quad (7\text{-}4)$$

For an ordinary system, according to AISC *Seismic Provisions* Section H4.5b, the shear strength of the beam is calculated using AISC *Specification* Chapter G where the area of the web is calculated as dt_w. For a special system, according to AISC *Seismic Provisions* Section H5.5c, the shear strength of the beam is calculated according to AISC *Seismic Provisions* Equation F3-2. Note that there is typically little, if any, axial load demand on a coupling beam. Therefore, the area of the web is calculated as $(d-2t_f)t_w$. The plastic flexural strength and shear strength of the beam are given as:

$$M_p = F_y Z_x \quad (7\text{-}5)$$

$$V_p = 0.6 F_y A_w \quad (7\text{-}6)$$

7.5 COMPOSITE SHEAR WALLS

Using Equation 7-3 for a shear-critical coupling beam and substituting Equations 7-5 and 7-6 gives the relationship between the required plastic section modulus, Z_x, and length of the beam, L, assuming a homogeneous member, as follows:

$$L \leq \frac{1.6M_p}{V_p} = \frac{1.6F_yZ_x}{0.6F_yA_w} = \frac{1.6Z_x}{0.6A_w} \tag{7-7}$$

Solving Equation 7-7 for the required plastic section modulus yields, for a shear-critical coupling beam:

$$Z_x \geq \frac{LA_w}{2.67} \tag{7-8}$$

For an ordinary system, the area of the web is calculated as $A_w = dt_w$. Therefore, for a shear-critical beam in an ordinary system:

$$Z_x \geq \frac{Ldt_w}{2.67} \tag{7-9}$$

where
d = depth of beam
t_w = thickness of beam web

For a special system, the area of the web is calculated as $A_{lw} = (d-2t_f)t_w$. Therefore, for a shear-critical beam in a special system:

$$Z_x \geq \frac{L(d-2t_f)t_w}{2.67} \tag{7-10}$$

where
t_f = thickness of beam flange

Using Equation 7-4 for a flexure-critical coupling beam and substituting Equations 7-5 and 7-6 gives the relationship between the required plastic section modulus and length of the beam, assuming a homogeneous member, as follows:

$$L \geq \frac{2.6M_p}{V_p} = \frac{2.6F_yZ_x}{0.6F_yA_w} = \frac{2.6Z_x}{0.6A_w} \tag{7-11}$$

Solving Equation 7-11 for the required plastic section modulus yields, for a flexure-critical coupling beam:

$$Z_x \leq \frac{LA_w}{4.33} \tag{7-12}$$

For an ordinary system, the area of the web is calculated as $A_w = dt_w$. Therefore, for a flexure-critical beam in an ordinary system:

$$Z_x \leq \frac{Ldt_w}{4.33} \tag{7-13}$$

For a special system, the area of the web is calculated as $A_{lw} = (d-2t_f)t_w$. Therefore, for a flexure-critical beam in a special system:

$$Z_x \leq \frac{L(d-2t_f)t_w}{4.33} \qquad (7\text{-}14)$$

Using a model of three rectangles for an I-shaped coupling beam, the plastic section modulus can be taken as:

$$Z_{req} = b_f t_f (d-t_f) + \frac{t_w(d-2t_f)^2}{4} \qquad (7\text{-}15)$$

where
b_f = width of beam flange

Equations 7-9, 7-10, 7-13, 7-14 and 7-15 can then be used to establish the cross-sectional dimensions of the beam.

Making further assumptions regarding beam depth, flange size or web size, the remaining cross-sectional dimensions can be calculated. For example, assuming a beam depth, d, a flange thickness, t_f, and web thickness, t_w, the required flange width, b_f, can be calculated. There are other considerations to address in the detailing of the coupling beam. The embedment length into the wall pier, intermediate web stiffeners, face bearing plates, and connection detailing all also need to be determined. These considerations vary depending on the type of system (i.e., ordinary or special systems), and are discussed in separate sections of this Part of the Manual.

Beam Embedment Length (Connection)

The required beam embedment length is determined based on the expected shear strength of the coupling beams. The AISC *Seismic Provisions* require the embedded length to be measured from the location of the first reinforcement layer of the confining reinforcement steel in the boundary element of the wall. The expected shear strength of the beam is determined using a form of Equation 7-2, where the expected plastic flexural strength of the beam, $R_y M_p$, is substituted for M_u. This is the static shear associated with a required flexural strength equal to the expected plastic flexural strength of the beam. For special systems, the shear calculated based on the expected plastic flexural strength must be amplified by a factor of 1.1 when computing embedment length, as required by AISC *Seismic Provisions* Section H5.5c. This is to account for strain hardening.

The embedment length can be determined through AISC *Seismic Provisions* Equation H4-2 or H4-2M. The coupling beam clear span is identified as g in Equation H4-2.

Detailing Requirements in the Embedded Region

For ordinary seismic force resisting systems, there are no special detailing requirements in the embedded region. For special systems, the embedded region must be detailed to provide resistance against connection strength and stiffness degradation, and to ensure proper distribution of bearing stresses within the embedded region. Face bearing plates, web stiffeners, and vertical transfer bars are required. In addition to the discussion presented in this Part of

7.5 COMPOSITE SHEAR WALLS

the Manual, further discussion of detailing requirements is provided in the Commentary to AISC *Seismic Provisions* Sections H4 and H5.

Face bearing plates (link stiffeners) are provided on both sides of the beam web, and located at the face of the wall pier. These plates should meet the requirements of stiffeners in links at the diagonal brace ends in an EBF as required in AISC *Seismic Provisions* Section F3.5b(4).

The web of the beam, over the clear span, must be supported with web stiffeners meeting the requirements for intermediate link stiffeners in Section F3.5b(4). From AISC *Seismic Provisions* Section H5.5c, the beams in special systems are to have inelastic deformation capacities equal to 0.08 rad. Smaller rotations are permitted if justified by a rational analysis of the inelastic deformations expected under design story drift.

Wall Overstrength

The beam required shear and flexural strengths delivered to the wall piers as an axial force and moment, respectively, must be accounted for in the design axial and flexural demands on the wall piers. In ordinary systems, the required axial strength from the coupling action is based on an accumulation of expected beam shear strengths (i.e., ΣV_n). In special systems, the expected beam shear strength must be amplified by a factor of 1.1 to account for strain hardening $[\text{i.e., } \Sigma(1.1V_n)]$. These amplified axial loads are generally referred to as wall overstrength.

The proportioning of beam shear strengths can have a significant impact on the required axial strength of the wall piers. To minimize the wall overstrength, beam sizes can be grouped over the height of the building to minimize the ratio of nominal beam shear strength to required beam shear strength at each floor level. Figure 7-2(a) shows a representative plot of the beam required shear strengths over the height of the building when the same coupling beam size is used over the entire height of the structure. When the same beam size is used, the beam size is proportioned based on the maximum required beam shear strength. Beam

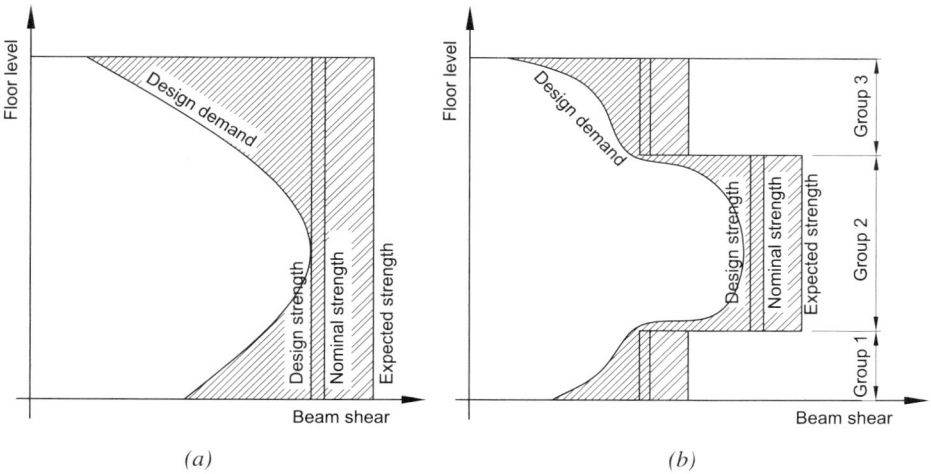

Fig. 7-2. Beam shears for (a) same size beam over the entire height and (b) three groups of different sizes.

strengths at all other floors, other than the floor corresponding to maximum required strength, will be relatively stronger than required. This effect is amplified in the upper and lower floors. In Figure 7-2(a), the hatched region bounded by the nominal strength and required strength is a graphical representation of the magnitude of the wall overstrength required. At the floor level of maximum required strength, the design strength-to-required strength ratio approaches one, with that ratio increasing at floors above or below that floor. Even at the floor level of maximum required strength, some degree of overstrength will exist when the shear strength resistance factor, ϕ_v, is less than 1.00.

When wall overstrength is based on expected beam shear strengths, wall overstrength requirements increase further. Although proportioning a system in this manner is advantageous for drift-controlled systems, it represents the worst case for wall overstrength requirements. Considering that coupled systems rarely are drift-controlled, proportioning beam sizes by groups over the height of the building will reduce the wall overstrength required for the wall piers without compromising drift limits.

Figure 7-2(b) represents beam required shear strengths and available strengths varied over the height of the building. In the representation shown in Figure 7-2(b), three groups of different size beams are used. This type of proportioning alters the distribution of required shear strength over the height as a result of the varying beam stiffness and reduces the design strength-to-required strength ratios at each of the floor levels relative to the case where the same beam size is used over the entire height. This type of proportioning is referred to as tuning the beam shear strengths. The extent of tuning performed is up to the designer based on the level of efficiency desired. Further information regarding tuning and wall overstrength can be found in Fortney et al. (2008) and Harries and McNeice (2006).

Composite Ordinary Shear Walls (C-OSW)

Composite ordinary shear wall (C-OSW) systems are designed in accordance with AISC *Seismic Provisions* Section H4.

Overview of Applicable Design Provisions

An overview of the AISC *Seismic Provisions* requirements for the design of C-OSW systems follows. Figure 7-3 illustrates an embedded steel coupling beam in an ordinary system. Areas of the figure are labeled to identify pertinent design considerations that correspond to the "Notes in Figure 7-3" listed in Table 7-4. Table 7-4 also provides a simplified overview of the design requirements that follow.

Note 1. The structural steel material used for C-OSW systems is limited by the requirements of AISC *Seismic Provisions* Section A3.1, where the specified minimum yield stress of the steel for members in which inelastic behavior is expected is not to exceed 55 ksi. These specified minimum yield stresses can be exceeded when the suitability of the material is determined by testing or other rational criteria. The weld filler metal used in the members and connections of the seismic force resisting system is selected to meet the requirements of AISC *Seismic Provisions* Section A3.4a. The concrete and steel reinforcing materials used in composite components should satisfy the requirements of AISC *Seismic Provisions* Section A3.5.

Note 2. The structural design drawings and specifications for C-OSW systems are to meet the requirements of AISC *Seismic Provisions* Sections A4.1, A4.2 and A4.3.

7.5 COMPOSITE SHEAR WALLS

Note 3. Loads and load combinations as defined by the applicable building code are to be followed as indicated in AISC *Seismic Provisions* Section B2.

Note 4. The required strength for structural members and connections is determined according to AISC *Seismic Provisions* Sections B3.1 and H4.5.

Note 5. Structural analysis for the appropriate load combinations is to be performed in accordance with the requirements of AISC *Seismic Provisions* Chapter C and Section H4.3.

    (a) For elastic analysis, the stiffness of composite members includes the effects of cracked sections. Additional guidelines for estimating the stiffness of concrete beam and column members, concrete-encased and concrete-filled members, and steel beams with composite slabs are provided in the Commentary to the AISC *Seismic Provisions* Chapter C. These concrete and composite member properties reflect the effective stiffness at the onset of significant yielding in the members.

Note 6. System requirements are as given in AISC *Seismic Provisions* Section H4.4.

Note 7. Boundary members of C-OSW systems are designed in accordance with the AISC *Specification* and AISC *Seismic Provisions* Section H4.5a.

Note 8. Coupling beams of C-OSW systems are designed in accordance with the AISC *Specification* and AISC *Seismic Provisions* Section H4.5b. Structural steel links must satisfy the requirement for moderately ductile members.

    (a) The beam flange must meet the width-to-thickness requirements for a moderately ductile element given in AISC *Seismic Provisions* Table D1.1.

    (b) The beam web must meet the width-to-thickness requirements for a moderately ductile element given in AISC *Seismic Provisions* Table D1.1.

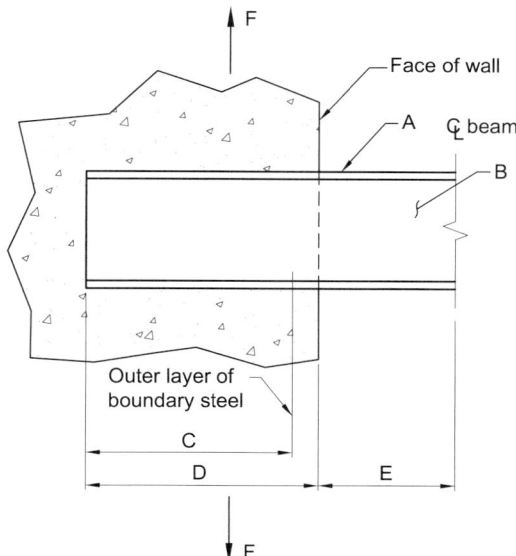

Fig. 7-3. Illustration of an embedded steel coupling beam for a C-OSW system. Notes are keyed to Table 7-4.

Table 7-4
Notes in Figure 7-3 and Simplified Overview of C-OSW Requirements

Note in Fig. 7-3	Note in Overview	Item	Referenced Standard*
–	1	Materials	*Seismic Prov.* Sects. A3.1, A3.2, A3.4a & A3.5
–	2	Structural design drawings and specifications	*Seismic Prov.* Sects. A4.1, A4.2 & A4.3
–	3	Loads and load combinations	*Seismic Prov.* Sect. B2
–	4	Required strength for structural members and connections	*Seismic Prov.* Sects. B3.1 & H4.5
–	5	Structural analysis	*Seismic Prov.* Ch. C & Sect. H4.3
–	5(a)	Composite member stiffness	*Seismic Prov.* Comm. to Ch. C
–	6	System requirements	*Seismic Prov.* Sect. H4.4
–	7	Boundary members	*Seismic Prov.* Sect. H4.5a
–	8	Coupling beams	*Seismic Prov.* Sect. H4.5b
A	8(a)	Beam flange local buckling	*Seismic Prov.* Sect. H4.5b(1)(1) & Table D1.1
B	8(b)	Web local buckling	*Seismic Prov.* Sect. H4.5b(1)(1) & Table D1.1
C	8(c)	Calculated embedment length	*Seismic Prov.* Sect. H4.5b(1)(3)
D		Total embedment length	
E		Beam clear span, for calculation of embedment (definition of g)	
–	9	Reinforced concrete walls	ACI 318 Sect. 11.9 & Ch. 14
F	–	Wall pier axial load due to coupling action	Undefined in the *Seismic Provisions*. See this Part of the Manual for guidance.

*The referenced standards are in addition to the requirements of the AISC *Specification*.

  (c) The embedment length is determined from AISC *Seismic Provisions* Equation H4-2 or H4-2M, and is considered to begin inside the outer layer of confining reinforcement in the wall boundary layer.

Note 9. Reinforced concrete walls are designed in accordance with ACI 318 Section 11.9 and Chapter 14.

7.5 COMPOSITE SHEAR WALLS

Steel Coupling Beam Design

The steel coupling beams used in C-OSW systems do not require special detailing. The proportioning of the beam cross sections over the height of the building need only satisfy the required shear and moment strengths determined from a linear elastic analysis (e.g., equivalent lateral force analysis). Flexural and shear strengths are determined using AISC *Specification* Chapters F and G. However, AISC *Seismic Provisions* Section H4 requires the beam-to-wall connection to transfer the expected beam shear strength. Therefore, it is advantageous to consider grouping beam strengths over the height of the building in an effort to minimize the required embedment lengths at each level.

Flanges and webs must satisfy the ductility requirements of AISC *Seismic Provisions* Section D1.1 for moderately ductile members. For these checks, the value of C_a in Table D1.1 can be taken as zero because these coupling beams have little or no required axial strength.

Beam Embedment Length

The required length of the beam embedded into the wall pier is determined using AISC *Seismic Provisions* Equation H4-2 or H4-2M. The V_n term in this equation is the expected beam shear strength. In this equation, the term g is the clear span of the coupling beam. However, the actual embedment length is measured from the first line of boundary element wall reinforcement. So, the actual embedment length of the beam, from the face of the wall, is the length calculated using Equation H4-2 plus the concrete cover of the boundary element reinforcement as illustrated in Figure 7-4.

Composite Coupling Beams

Expected Plastic Moment

AISC *Specification* Section I3.3(c) permits the use of the plastic stress distribution or strain-compatibility methods for determination of flexural strength on the composite section when

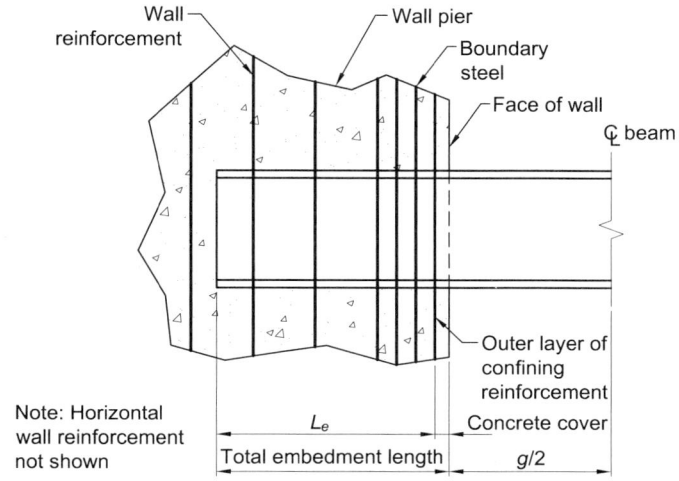

Fig. 7-4. Total embedment length of beam.

steel anchors are provided. At the expected plastic moment strength of a composite coupling beam, it is reasonable to assume that the concrete in tension has cracked. To calculate plastic moment strength, the location of the plastic neutral axis of the cracked section must be determined. Depending on the position within the cross section of the constituent elements there are many different locations of the plastic neutral axis that can be conceived when determining the internal forces acting on the section. For example, assume that the structural steel section does not extend up to the top layer or below the bottom layer of longitudinal reinforcement as shown in Figure 7-5. If the centroids of the top and bottom flange

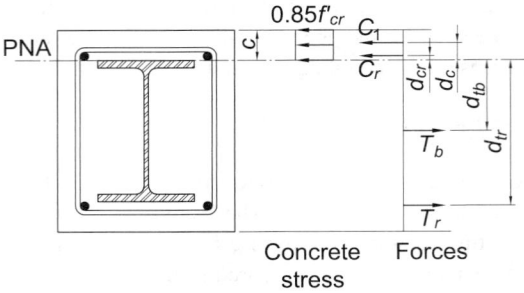

Case 1: Plastic Neutral Axis above Steel Shape

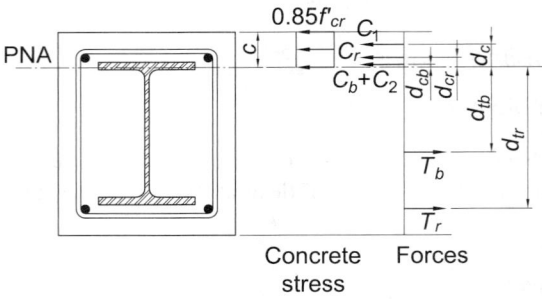

Case 2: Plastic Neutral Axis in Flange

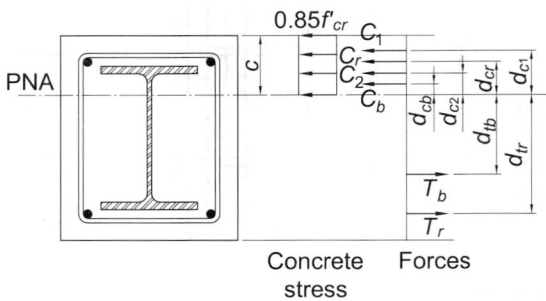

Case 3: Plastic Neutral Axis in Web

Fig. 7-5. Possible internal forces based on the plastic stress distribution.

7.5 COMPOSITE SHEAR WALLS

of the structural steel coincide with the elevation of the upper and lower reinforcement, respectively, an entirely new set of geometries exists. Additional configurations are possible, depending on the placement of the steel member and the reinforcing.

Figure 7-5 shows three possibilities for the location of the plastic neutral axis in a cross section where the structural steel does not extend into the elevations of the steel reinforcement. For Case 1, the plastic neutral axis is above the top of the steel shape. Although it is possible, it is unlikely that the reinforcing steel in this region would be below the plastic neutral axis, so that possible arrangement is not illustrated. For Cases 2 and 3, the plastic neutral axis extends into the structural section.

Regardless of the position of the elements, the plastic moment strength can be determined using either the plastic stress distribution method or the strain compatibility method of AISC *Specification* Section I3, as appropriate. The challenge is in determining the location of the plastic neutral axis. One approach would be to use the equations for pure bending given for composite beam-columns in Tables 7-10A, 7-10B, 7-11A and 7-11B in this Part of the Manual.

Shear Strength

The limiting expected shear strength of a composite coupling beam is calculated using AISC *Seismic Provisions* Equations H4-4 or H4-4M and is based on the expected strength of the steel shape. The total expected shear strength is the sum of the resistances provided by the structural steel section, the concrete, and the transverse reinforcement. Of the three material contributions to the total shear resistance of the beam, expected material strength is considered only for the structural steel.

From the ACI 318 requirements for shear reinforcement, the size and spacing of transverse reinforcement depends on the magnitude of shear stress being resisted. However, regardless of the magnitude of shear stress, at least the minimum shear reinforcement requirements must be provided. The Commentary to AISC *Seismic Provisions* Sections H4.2 and H5.2 provides further discussion on this topic.

Embedment Length

As with steel coupling beams, the embedment length of the steel section of the composite beam is considered to begin within the outer layer of confining reinforcement. Similar to steel coupling beams, the embedment length is determined through AISC *Seismic Provisions* Equation H4-2, but is based on the nominal shear strength defined by Equation H4-3. As discussed previously, it is permitted to have a composite beam detailed to be shear- or flexure-critical.

Example 7.5.1. C-OSW Steel Coupling Beam Design

Given:

The sixth floor core plan of a 15-story core wall system is shown in Figure 7-6. The composite ordinary shear wall system is coupled with steel coupling beams. Coupling beam sizes are grouped such that different beam sizes are used at floor levels 1–5, 6–10, and 11–15. Table 7-5 tabulates the maximum LRFD beam required shear strength for each group of beams. A modal response spectrum analysis procedure according to ASCE/SEI 7 was used to determine the seismic loading, which was then combined with gravity loads using the basic seismic load combinations of ASCE/SEI 7 Section 12.4.2.3 (not using the amplified seismic load). The analysis meets AISC *Seismic Provisions* Section H4.3 requirements: 1) uncracked effective stiffness values were used, and 2) flexibility of the connection between coupling beams and wall piers was taken into account. Second order effects were also considered in the analysis.

From ASCE/SEI 7, the following parameters apply: $R = 5$, $C_d = 4.5$, $\Omega_o = 2.5$, Seismic Design Category C, and $I_e = 1.0$.

The compressive strength of the wall pier concrete is 8 ksi, the steel reinforcement is ASTM Grade 60, and the steel beams are built-up I-shapes of ASTM A572 Grade 50 plate material. The clear cover from face of wall to boundary reinforcement is 0.750 in. The maximum beam depth permitted is 30.0 in. Assume a maximum rotation of 0.08 rad and no axial load in the coupling beam.

Perform the following:

1. Specify the cross-sectional dimensions of the coupling beams on levels 6–10 assuming that the beam will be shear-critical.
2. For the beam sized in Part 1 of this problem, compute the required embedment length of the beam into the wall pier.
3. Given the LRFD beam required shear strengths over the height of the building provided in Table 7-5, determine the LRFD required axial strength at the base of the wall piers due to coupling action only.

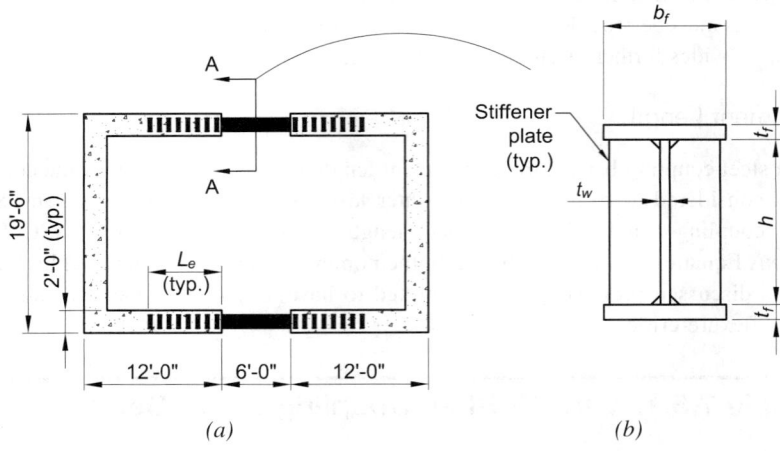

Fig. 7-6. (a) Core plan at sixth floor and (b) Section A-A—beam cross section.

Table 7-5
LRFD Beam Required Shear Strength

Floor Level	V_u, kips
11–15	295
6–10	486
1–5	380

Table 7-6
Expected Beam Shear Strengths, V_n

Floor Level	V_n, kips
11–15	325
6–10	To be determined
1–5	429

Solution:

Part 1: Coupling Beam Design

From AISC *Manual* Table 2-4, the material properties are as follows:

ASTM A572 Grade 50
$F_y = 50$ ksi
$F_u = 65$ ksi

Local Buckling

For a C-OSW system, the flanges and webs of the steel coupling beam are to satisfy the requirements for moderately ductile members. Referring to Table D1.1 of the AISC *Seismic Provisions*, the limiting width-to-thickness ratio for a built-up member is given.

From AISC *Seismic Provisions* Table D1.1, the limiting width-to-thickness ratio for the flanges of the beam is:

$$\frac{b}{t} \leq \lambda_{md}$$

$$\lambda_{md} = 0.38\sqrt{\frac{E}{F_y}}$$

$$= 0.38\sqrt{\frac{29{,}000 \text{ ksi}}{50 \text{ ksi}}}$$

$$= 9.15$$

The limiting width-to-thickness ratio for the beam web is also determined from Table D1.1 as follows:

$$\frac{h}{t} \leq \lambda_{md}$$

Because there is no axial load on the beam, $P_u = 0$; therefore, $C_a = 0$.

$$\lambda_{md} = 3.76\sqrt{\frac{E}{F_y}}(1-2.75C_a)$$

$$= 3.76\sqrt{\frac{29{,}000 \text{ ksi}}{50 \text{ ksi}}}(1-0)$$

$$= 90.6$$

Link Length Requirement

The problem statement requires a chord rotation of 0.08 rad and a shear-critical coupling beam. Using Equations 7-9 and 7-15:

$$Z_{req} \geq \frac{Ldt_w}{2.67}$$

$$b_f t_f (d-t_f) + \frac{t_w(d-2t_f)^2}{4} \geq \frac{Ldt_w}{2.67}$$

Assume the beam depth, d, is the maximum permitted of 30.0 in. The maximum flange width that can fit within the steel reinforcement in the wall piers is (assuming No. 8 horizontal and vertical reinforcing bars):

$$b_{f,\,max} = t_{wall} - 2C_c - 2d_{b,v} - 2d_{b,h}$$

$$= 24.0 \text{ in.} - 2(0.75 \text{ in.}) - 2(1.00 \text{ in.}) - 2(1.00 \text{ in.})$$

$$= 18.5 \text{ in.}$$

where
$\quad C_c$ = concrete cover
$\quad d_{b,v}$ = diameter of vertical reinforcement bar
$\quad d_{b,h}$ = diameter of horizontal reinforcement bar

Use $b_f = 15.0$ in.

7.5 COMPOSITE SHEAR WALLS

Determine the required flange thickness to satisfy flange local buckling based on the limiting width-to-thickness ratio determined previously.

$$\frac{b}{t} \leq 9.15$$

$$\frac{b}{t} = \frac{b_f}{2t_f}$$

$$= 9.15$$

$$t_f \geq \frac{15.0 \text{ in.}}{2(9.15)} = 0.820 \text{ in.}$$

Assume $t_f = 1.00$ in.

Knowing the depth ($d = 30.0$ in.) and flange thickness ($t_f = 1.00$ in.) of the cross section, the value, $d - 2t_f$, can be calculated.

$$d - 2t_f = 30.0 \text{ in.} - 2(1.00 \text{ in.})$$

$$= 28.0 \text{ in.}$$

Using the trial values for the flange and web dimensions, the thickness of the web can be determined. The thickness required to satisfy the link length equation is used to compute the required thickness and $A_w = dt_w$ according to AISC *Specification* Section G2. Solving for t_w:

$$b_f t_f (d - t_f) + \frac{t_w (d - 2t_f)^2}{4} = \frac{Ldt_w}{2.67}$$

$$(15.0 \text{ in.})(1.00 \text{ in.})(30.0 \text{ in.} - 1.00 \text{ in.}) + \frac{t_w (28.0 \text{ in.})^2}{4} = \frac{72.0 \text{ in.}(30.0 \text{ in.})t_w}{2.67}$$

$$t_w = 0.710 \text{ in.}$$

Use $t_w = 0.625$ in.

The following trial section dimensions can be used to check the requirements for the design of the beam.

$b_f = 15.0$ in. $t_f = 1.00$ in. $A = 47.5$ in.²
$d = 30.0$ in. $t_w = 0.625$ in. $I_y = 563$ in.⁴
$r_y = 3.44$ in.

Check web local buckling

$$\frac{h}{t} \leq 90.6$$

$$\frac{h}{t} = \frac{28.0 \text{ in.}}{0.625 \text{ in.}}$$

$$= 44.8 < 90.6 \quad \textbf{o.k.}$$

Check flange local buckling

$$\frac{b}{t} \leq 9.15$$

$$\frac{b}{t} = \frac{b_f}{2t_f}$$

$$= \frac{15.0 \text{ in.}}{2(1.00 \text{ in.})}$$

$$= 7.50 < 9.15 \quad \textbf{o.k.}$$

Maximum Link Length

For a shear-critical beam in an ordinary system, the maximum link length is determined from Equation 7-3 as follows:

$$L \leq \frac{1.6M_p}{V_p} \qquad (7\text{-}3)$$

$$\frac{1.6M_p}{V_p} = \frac{1.6F_y Z_x}{0.6F_y A_{lw}}$$

where

$$Z_x = b_f t_f (d - t_f) + \frac{t_w (d - 2t_f)^2}{4} \qquad (7\text{-}15)$$

$$= (15.0 \text{ in.})(1.00 \text{ in.})(30.0 \text{ in.} - 1.00 \text{ in.}) + \frac{0.625 \text{ in.}(28.0 \text{ in.})^2}{4}$$

$$= 558 \text{ in.}^3$$

$$M_p = F_y Z_x \qquad (7\text{-}5)$$

$$= 50 \text{ ksi}(558 \text{ in.}^3)$$

$$= 27{,}900 \text{ kip-in.}$$

$$V_p = 0.6 F_y A_{lw} \qquad (7\text{-}6)$$

$$= 0.6(50 \text{ ksi})(30.0 \text{ in.})(0.625 \text{ in.})$$

$$= 563 \text{ kips}$$

Therefore:

$$L \leq \frac{1.6(27{,}900 \text{ kip-in.})}{563 \text{ kips}}$$

$$\leq 79.3 \text{ in.}$$

$$72.0 \text{ in.} \leq 79.3 \text{ in.} \quad \textbf{o.k.}$$

The given link length of 6 ft meets this requirement.

7.5 COMPOSITE SHEAR WALLS

Flexural Strength

The required flexural strength at levels 6–10, from Equation 7-2 with V_u from Table 7-5, is:

$$M_u = \frac{V_u L}{2}$$
$$= \frac{486 \text{ kips}(72.0 \text{ in.})}{2}$$
$$= 17,500 \text{ kip-in.}$$

Because the shape is compact according to the AISC *Specification*, determine whether the limit state of lateral-torsional buckling applies.

$$L_p = 1.76 r_y \sqrt{\frac{E}{F_y}} \qquad \textit{(Spec. Eq. F2-5)}$$
$$= 1.76(3.44 \text{ in.})\sqrt{\frac{29,000 \text{ ksi}}{50 \text{ ksi}}}$$
$$= 146 \text{ in.}$$

Because 72 in. < 146 in., lateral-torsional buckling does not apply, and the design flexural strength based on flexural yielding is determined in accordance with AISC *Specification* Section F2.1:

$$\phi M_n = \phi M_p$$
$$= 0.90(27,900 \text{ kip-in.})$$
$$= 25,100 \text{ kip-in.}$$
$$25,100 \text{ kip-in.} > 17,500 \text{ kip-in.} \qquad \textbf{o.k.}$$

Shear Strength

The available shear strength from AISC *Specification* Section G2 is determined as follows.

For a built-up member:

$$\phi = 0.90$$

For a beam without stiffeners, $k_v = 5$, and

$$1.10\sqrt{\frac{k_v E}{F_y}} = 1.10\sqrt{\frac{5(29,000 \text{ ksi})}{50 \text{ ksi}}}$$
$$= 59.2 > 44.8$$

Therefore:

$$C_v = 1.0 \qquad \textit{(Spec. Eq. G2-3)}$$

The available shear strength is:

$$\phi_v V_n = \phi_v 0.6 F_y A_w C_v \qquad \text{(from \textit{Spec.} Eq. G2-1)}$$
$$= 0.90(0.6)(50 \text{ ksi})(0.625 \text{ in.})(30.0 \text{ in.})(1.0)$$
$$= 506 \text{ kips} > 486 \text{ kips} \quad \textbf{o.k.}$$

Part 2: Beam Embedment Length

From Part 1 of this solution, $V_n = 563$ kips and $M_n = 27{,}900$ kip-in.

From AISC *Seismic Provisions* Section H4.5b(1)(3):

$$V_n = 1.54\sqrt{f'_c}\left(\frac{b_w}{b_f}\right)^{0.66} \beta_1 b_f L_e \left(\frac{0.58 - 0.22\beta_1}{0.88 + \dfrac{g}{2L_e}}\right) \qquad \text{(\textit{Provisions} Eq. H4-2)}$$

where
- $b_w = 24$ in.
- $f'_c = 8$ ksi
- $\beta_1 = 0.65$ from ACI 318 Section 10.2.7.3
- $g = L = 72.0$ in.

In AISC *Seismic Provisions* Equation H4-2, g could be replaced with g_{eff}, an effective clear span to account for spalling at the face of the wall, but this is not required by the AISC *Seismic Provisions* and will have very little impact on the final design. This would require the clear span to be increased by the concrete cover over the first reinforcing bar at each side.

The expected shear strength of the steel coupling beam is:

$$V_n = \frac{2R_y M_p}{g} \le R_y V_p \qquad \text{(\textit{Provisions} Eq. H4-1)}$$

From AISC *Seismic Provisions* Table A3.1, for ASTM A572 Grade 50 material:

$R_y = 1.1$

$$V_n = \frac{2R_y M_p}{g}$$
$$= \frac{2(1.1)(27{,}900 \text{ kip-in.})}{72.0 \text{ in.}}$$
$$= 853 \text{ kips}$$

$$V_n = R_y V_p$$
$$= 1.1(563 \text{ kips})$$
$$= 619 \text{ kips}$$

7.5 COMPOSITE SHEAR WALLS

Use $V_n = 619$ kips.

Determine the embedment length by solving for L_e in Equation H4-2:

$$619 = 1.54\sqrt{8 \text{ ksi}} \left(\frac{24.0 \text{ in.}}{15.0 \text{ in.}}\right)^{0.66} (0.65)(15.0 \text{ in.})L_e \left[\frac{0.58 - (0.22)(0.65)}{0.88 + \frac{72.0}{2L_e}}\right]$$

$$619 = \frac{25.3 L_e}{0.88 + \frac{36.0}{L_e}}$$

$L_e = 42.3$ in.

The total embedment length, from the face of the wall, is:

$L_e + C_c = 42.3$ in. $+ 0.750$ in.
    $= 43.1$ in.

Therefore, each end of the beam will be embedded a minimum of 44.0 in. from the face of the wall.

Part 3: Wall Pier Required Axial Strength

AISC *Seismic Provisions* Section H4.5b(1)(4) requires that vertical wall reinforcement with a nominal strength equal to the expected shear strength, V_n, of the steel coupling beam be provided over the embedded length of the coupling beam. The axial load resulting from the coupling action on the base wall piers is the accumulation of the expected beam shear strengths over the height of the building.

For the given core wall system, two coupling beams frame into each shear wall.

Based on the nominal shear strengths given in Table 7-6, the required wall pier axial strength is determined as follows.

Expected beam shear strength at levels 11–15, from Table 7-6:

$(V_n)_{11-15} = (325 \text{ kips/floor/beam})$
$(\Sigma V_n)_{11-15} = (5 \text{ floors})(325 \text{ kips/floor/beam})(2 \text{ beams})$
    $= 3{,}250$ kips

Expected beam shear strengths at levels 6–10 from previous calculations:

$(V_n)_{6-10} = (619 \text{ kips/floor/beam})$
$(\Sigma V_n)_{6-10} = (5 \text{ floors})(619 \text{ kips/floor/beam})(2 \text{ beams})$
    $= 6{,}190$ kips

Expected beam shear strength at levels 1–5, from Table 7-6:

$(V_n)_{1-5} = (429 \text{ kips/floor/beam})$

$(\Sigma V_n)_{1-5} = (5 \text{ floors})(429 \text{ kips/floor/beam})(2 \text{ beams})$

$\phantom{(\Sigma V_n)_{1-5}} = 4{,}290 \text{ kips}$

The total axial load effect due to coupling below level 1 is:

$P_{u,\,wall,\,coupling} = 3{,}250 \text{ kips} + 6{,}190 \text{ kips} + 4{,}290 \text{ kips}$

$\phantom{P_{u,\,wall,\,coupling}} = 13{,}700 \text{ kips}$

The required axial strength of the wall piers in upper floors is calculated in a similar manner. In such a case, the total axial load due to coupling, at a given floor level, is an accumulation of the beam shear strengths for the beam at the floor being considered and beams above.

Example 7.5.2. C-OSW Composite Coupling Beam Design

Given:

A composite coupling beam is used to couple the 16-in.-thick shear walls of a composite ordinary shear wall system. A cross section of the coupling beam is shown in Figure 7-7. A modal response spectrum analysis procedure according to ASCE/SEI 7 was used to determine the seismic loading, which was then combined with gravity loads using the basic seismic load combinations of ASCE/SEI 7 Section 12.4.2.3 (not using the amplified seismic load). The analysis meets AISC *Seismic Provisions* Section H4.3 requirements: 1) uncracked effective stiffness values were used, and 2) flexibility of the connection between coupling beams and wall piers was taken into account. Second-order effects were also considered in the analysis. The LRFD required shear and flexural strengths are:

$V_u = 232 \text{ kips}$
$M_u = 6{,}960 \text{ kip-in.}$

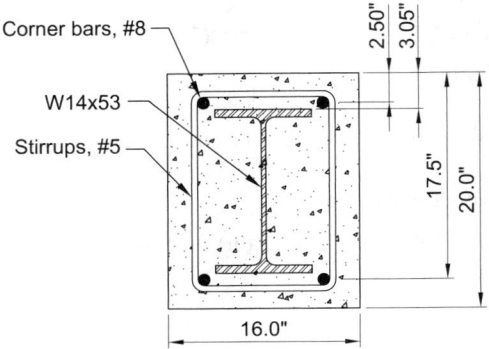

Fig. 7-7. Composite coupling beam section.

7.5 COMPOSITE SHEAR WALLS

ASTM A992 material is used for the structural steel, ASTM A615 Grade 60 material is used for all steel reinforcement, and the concrete compressive strength is 4 ksi.

1. Specify the required spacing of the transverse reinforcement.
2. Calculate the expected plastic moment strength of the composite beam.
3. Calculate the required embedment length of the structural steel section into the wall pier. Assume the span length of the beam is 5.00 ft between wall faces.

Solution:

From AISC *Manual* Table 2-4, the material properties are as follows:

ASTM A992
$F_y = 50$ ksi
$F_u = 65$ ksi

From AISC *Manual* Table 1-1, the geometric properties are as follows:

W14×53
$d = 13.9$ in. $t_w = 0.370$ in. $b_f = 8.06$ in. $t_f = 0.660$ in.
$Z = 87.1$ in.^{3}

Part 1: Specify Transverse Reinforcement Spacing

According to AISC *Specification* Section I4.1, the shear strength of the composite beam can be taken as the sum of that contributed by the steel shape and that contributed by the reinforcing, with $\phi = 0.75$. With the required shear strength given as 232 kips, the nominal shear strength of the composite coupling beam must be at least the following:

$$V_n \geq \frac{232}{0.75} = 309 \text{ kips}$$

For the rolled steel shape with $h/t_w \leq 2.24\sqrt{E/F_y}$, $C_v = 1.0$; therefore, the nominal shear strength of the steel is:

$$V_{n,\,steel} = 0.6 F_y A_w C_v \quad\quad\quad\quad (Spec.\ Eq.\ G2\text{-}1)$$
$$= 0.6(50 \text{ ksi})(13.9 \text{ in.})(0.370 \text{ in.})(1.0)$$
$$= 154 \text{ kips}$$

The reinforcing steel must provide the following nominal shear strength:

$$V_{n,\,sr} \geq 309 \text{ kips} - 154 \text{ kips} = 155 \text{ kips}$$

For No. 5 stirrups, the maximum spacing can be determined using:

$$V_{n,\,sr} = \frac{A_v f_y d}{s} \quad\quad\quad\quad (ACI\ 318\ Eq.\ 11\text{-}15)$$

Therefore, solving for s, the required stirrup spacing is:

$$s = \frac{A_v f_y d}{V_{n,sr}}$$

$$= \frac{2(0.31 \text{ in.}^2)(60 \text{ ksi})(17.5 \text{ in.})}{154 \text{ kips}}$$

$$= 4.23 \text{ in.}$$

Maximum Transverse Spacing Requirements

The maximum spacing of the transverse reinforcement is determined from ACI 318 Section 11.4.5.1 and solving for s in the equations for $A_{v,\,min}$ given in Section 11.4.6.3. With the concrete strength taken in psi for use in ACI 318 equations:

$$s_{max} \leq \min \begin{cases} \dfrac{d}{2} = \dfrac{17.5 \text{ in.}}{2} = 8.75 \text{ in.} \\ 24 \text{ in.} \\ \dfrac{A_v F_y}{50 b_w} = \dfrac{2(0.31 \text{ in.}^2)(60{,}000 \text{ psi})}{50(16.0 \text{ in.})} = 46.5 \text{ in.} \\ \dfrac{A_v F_y}{0.75\sqrt{f_c'} b_w} = \dfrac{2(0.31 \text{ in.}^2)(60{,}000 \text{ psi})}{0.75\sqrt{4{,}000 \text{ psi}}(16.0 \text{ in.})} = 49.0 \text{ in.} \end{cases}$$

Therefore, the maximum spacing is 8.75 in.

The spacing requirements based on shear strength were determined as 4.23 in. Thus, use No. 5 closed stirrups at 4.00 in. on-center spacing.

Part 2: Expected Flexural Strength of Composite Beam

It may take several iterations to identify the case (see Figure 7-5) that applies to the cross section in any given problem. For this problem it is assumed that the plastic neutral axis is in the flange of the steel shape—Case 2 in Figure 7-5. The plastic flexural strength may be determined by application of equilibrium principles or by the equations provided for pure bending of encased composite beam-columns given in Table 7-10A of this Part of the Manual (Geschwindner, 2010). Because these equations are somewhat more straightforward, they will be illustrated here.

AISC *Seismic Provisions* Section H4.5b(2) requires the use of the expected strength of the steel shape, the reinforcing steel, and the concrete. As for other expected strength calculations, AISC *Seismic Provisions* Table A3.1 gives $R_y = 1.1$ for ASTM A992 steel and $R_y = 1.25$ for ASTM A615 reinforcing steel.

The User Note in AISC *Seismic Provisions* Section A3.2 states that to determine the expected strength of reinforced concrete, reference should be made to *Seismic Rehabilitation of Existing Buildings,* ASCE/SEI 41-06 (ASCE, 2006). The expected strength can be determined by applying the factor provided in ASCE/SEI 41-06 Table 6-4 for concrete compressive strength, where a value of 1.50 is given. Thus, the expected yield strengths used in determining the expected plastic flexural strength are:

7.5 COMPOSITE SHEAR WALLS

ASTM A992: $R_y F_y = 1.1(50 \text{ ksi}) = 55 \text{ ksi}$
ASTM A615: $R_y F_y = 1.25(60 \text{ ksi}) = 75 \text{ ksi}$
4-ksi concrete: $1.50 f_c' = 1.50(4 \text{ ksi}) = 6 \text{ ksi}$

From Table 7-10A, assuming that the plastic neutral axis is in the flange, the variable h_n can be determined. This is the distance from the centroid of the section to the location of the plastic neutral axis. Thus:

$$h_n = \frac{0.85 f_c'(A_c + A_s - db_f) - 2F_y(A_s - db_f)}{2\left[0.85 f_c'(h_1 - b_f) + 2F_y b_f\right]}$$

Note there is no term for the reinforcing steel because there is no reinforcing steel at the midpoint of the section. Determine the area of the steel shape using the same geometry used in the derivation of the equations, which is the model of three rectangles.

$$A_s = 2(8.06 \text{ in.})(0.660 \text{ in.}) + [13.9 \text{ in.} - 2(0.660 \text{ in.})](0.370 \text{ in.})$$
$$= 15.3 \text{ in.}^2$$

The area of the concrete is:

$$A_c = 16.0 \text{ in.}(20.0 \text{ in.}) - 15.3 \text{ in.}^2 - 4(0.79 \text{ in.}^2)$$
$$= 302 \text{ in.}^2$$

Thus:

$$h_n = \frac{0.85(6 \text{ ksi})\left[302 \text{ in.}^2 + 15.3 \text{ in.}^2 - 13.9 \text{ in.}(8.06 \text{ in.})\right] - 2(55 \text{ ksi}) \times \left[15.3 \text{ in.}^2 - 13.9 \text{ in.}(8.06 \text{ in.})\right]}{2\left[0.85(6 \text{ ksi})(16.0 \text{ in.} - 8.06 \text{ in.}) + 2(55 \text{ ksi})(8.06 \text{ in.})\right]}$$
$$= 6.30 \text{ in.}$$

Because this dimension is greater than the distance from the center of the section to the underside of the flange and less than half of the steel beam depth, the assumption that the plastic neutral axis is in the flange is correct.

Using the series of equations given in Table 7-10A in this Part of the Manual for Point B, for h_n within the flange, and Point D:

$$Z_{sn} = Z_s - b_f\left(\frac{d}{2} - h_n\right)\left(\frac{d}{2} + h_n\right)$$
$$= 87.1 \text{ in.}^3 - 8.06 \text{ in.}\left(\frac{13.9 \text{ in.}}{2} - 6.30 \text{ in.}\right)\left(\frac{13.9 \text{ in.}}{2} + 6.30 \text{ in.}\right)$$
$$= 17.7 \text{ in.}^3$$

$$Z_{cn} = h_1 h_n^2 - Z_{sn}$$
$$= 16.0 \text{ in.}(6.30 \text{ in.})^2 - 17.7 \text{ in.}^3$$
$$= 617 \text{ in.}^3$$

$$Z_r = (A_{sr} - A_{srs})\left(\frac{h_2}{2} - c\right)$$
$$= \left[4(0.79 \text{ in.}^2) - 0 \text{ in.}^2\right]\left(\frac{20.0 \text{ in.}}{2} - 2.5 \text{ in.}\right)$$
$$= 23.7 \text{ in.}^3$$

$$Z_c = \frac{h_1 h_2^2}{4} - Z_s - Z_r$$
$$= \frac{16.0 \text{ in.}(20.0 \text{ in.})^2}{4} - 87.1 \text{ in.}^3 - 23.7 \text{ in.}^3$$
$$= 1,490 \text{ in.}^3$$

$$M_D = Z_s F_y + Z_r F_{yr} + \frac{Z_c}{2}(0.85 f_c')$$
$$= 87.1 \text{ in.}^3 (55 \text{ ksi}) + 23.7 \text{ in.}^3 (75 \text{ ksi}) + \frac{1,490 \text{ in.}^3}{2}[0.85(6 \text{ ksi})]$$
$$= 10,400 \text{ kip-in.}$$

$$M_B = M_D - Z_{sn} F_y - \frac{Z_{cn}}{2}(0.85 f_c')$$
$$= 10,400 \text{ kip-in.} - 17.7 \text{ in.}(55 \text{ ksi}) - \frac{617 \text{ in.}^3}{2}[0.85(6 \text{ ksi})]$$
$$= 7,850 \text{ kip-in.}$$

Thus:

$$M_{p,exp} = 7,850 \text{ kip-in.}/(12 \text{ in./ft})$$
$$= 654 \text{ kip-ft}$$

Part 3: Beam Embedment Length

From AISC *Seismic Provisions* Section H4.5b(1)(3), modified according to Section H4.5b(2)(2):

$$V_{n,comp} = 1.54\sqrt{f_c'}\left(\frac{b_w}{b_f}\right)^{0.66} \beta_1 b_f L_e \left[\frac{0.58 - 0.22\beta_1}{0.88 + \frac{g}{2L_e}}\right] \quad \text{(\textit{Provisions} Eq. H4-2)}$$

$$f_c' = 4 \text{ ksi}$$
$$\beta_1 = 0.85$$

7.5 COMPOSITE SHEAR WALLS

The clear span, g, for this composite coupling beam is 5.00 ft.

The expected shear strength, $V_{n,comp}$, is determined from AISC *Seismic Provisions* Equation H4-3 as follows:

$$V_{n,\,comp} = \frac{2M_{p,\,exp}}{g} \leq V_{comp} \qquad \text{(\textit{Provisions} Eq. H4-3)}$$

$$= \frac{2(654 \text{ kip-ft})}{5.00 \text{ ft}}$$

$$= 262 \text{ kips} < V_{comp}$$

where

$$V_{comp} = R_y V_p + \left(0.0632\sqrt{f'_c}\, b_{wc} d_c + \frac{A_s F_{ysr} d_c}{s}\right) \qquad \text{(from \textit{Provisions} Eq. H4-4)}$$

$$= 1.1(0.6)(50 \text{ ksi})(13.9 \text{ in.})(0.370 \text{ in.})$$

$$+ 0.0632\sqrt{4 \text{ ksi}}\,(16.0 \text{ in.})(17.5 \text{ in.})$$

$$+ \frac{2(0.31 \text{ in.}^2)(60 \text{ ksi})(17.5 \text{ in.})}{4.00 \text{ in.}}$$

$$= 368 \text{ kips}$$

Therefore, $V_{n,\,comp} = 262$ kips.

Note that the equation used here for V_{comp} includes a conversion factor so that f'_c can be applied in ksi units.

The embedment length can then be determined from AISC *Seismic Provisions* Equation H4-2 as follows:

$$262 = 1.54\sqrt{4 \text{ ksi}} \left(\frac{16.0 \text{ in.}}{8.06 \text{ in.}}\right)^{0.66} (0.85)(8.06 \text{ in.})$$

$$\times L_e \left[\frac{0.58 - 0.22(0.85)}{0.88 + \frac{60.0 \text{ in.}}{2L_e}}\right]$$

$$262 = \frac{13.0 L_e}{0.88 + \frac{30.0}{L_e}}$$

$$L_e = 35.0 \text{ in.}$$

The total embedded length from the face of the wall is:

$$L_e + C_c = 35.0 \text{ in.} + 0.75 \text{ in.}$$

$$= 35.8 \text{ in.}$$

Each end of the beam will be embedded a minimum of 36 in. beyond the face of the wall.

Composite Special Shear Walls (C-SSW)

Composite special shear wall (C-SSW) systems are designed in accordance with AISC *Seismic Provisions* H5. C-SSW systems are reinforced concrete walls composite with structural steel, including steel or composite boundary members and steel or composite coupling beams.

Overview of Applicable Design Provisions

An overview of the AISC *Seismic Provisions* and ACI 318 requirements for the design of C-SSW systems follows. Figure 7-8 illustrates an embedded steel coupling beam in a composite special shear wall system. Areas of the figure are labeled to identify pertinent design considerations. Table 7-7 identifies specific requirements of the AISC *Seismic Provisions* that correspond to the areas labeled in the figure and provides a simplified summary of the design requirements.

Note 1. The structural steel material used for C-SSW systems is limited by the requirements of AISC *Seismic Provisions* Section A3.1, where the specified minimum yield stress of the steel is not to exceed 50 ksi for members in which inelastic behavior is expected. This specified minimum yield stress can be exceeded when the suitability of the material is determined by testing or other rational criteria. The weld filler metal used in the members and connections of the seismic force resisting system is selected to meet the requirements of AISC *Seismic Provisions* Section A3.4a. The concrete and steel reinforcing materials used in composite components should satisfy the requirements of AISC *Seismic Provisions* Section A3.5.

Note 2. The structural design drawings and specifications for C-SSW systems are to meet the requirements of AISC *Seismic Provisions* Sections A4.1, A4.2 and A4.3.

Note 3. Loads and load combinations as defined by the applicable building code are to be followed as indicated in AISC *Seismic Provisions* Section B2.

Note 4. The required strength for structural members and connections is determined according to AISC *Seismic Provisions* Sections B3.1 and H5.5.

Note 5. Structural analysis for the appropriate load combinations is to be performed in accordance with the requirements of AISC *Seismic Provisions* Chapter C and Section H5.3.

(a) For elastic analysis, the stiffness of composite members shall include the effects of cracked sections. Additional guidelines for estimating the stiffness of concrete beam and column members, concrete-encased and concrete-filled members, and steel beams with composite slabs are provided in the Commentary to the AISC *Seismic Provisions* Chapter C. These concrete and composite member properties reflect the effective stiffness at the onset of significant yielding in the members.

Note 6. System requirements are as given in AISC *Seismic Provisions* Section H5.4.

Note 7. Boundary members of C-SSW systems are designed in accordance with the AISC *Specification* and AISC *Seismic Provisions* Section H5.5b.

Note 8. Steel coupling beams of C-SSW systems are designed in accordance with the AISC *Specification* and AISC *Seismic Provisions* Sections H5.5a, H5.5c and H5.5d. Structural steel links must satisfy the requirements for highly ductile members.

7.5 COMPOSITE SHEAR WALLS

(a) As stipulated in AISC *Seismic Provisions* Section H5.5c by reference to Section F3.5b, for I-shaped beams, with link lengths $\leq 1.6M_p/V_p$, the steel beam flange may meet the width-to-thickness requirements for a moderately ductile element given in AISC *Seismic Provisions* Table D1.1; otherwise the requirements for highly ductile elements must be met.

(b) As stipulated in AISC *Seismic Provisions* Section H5.5c, the steel beam web must meet the width-to-thickness requirements for a highly ductile element given in AISC *Seismic Provisions* Table D1.1.

(c) As stipulated in AISC *Seismic Provisions* Section H5.5c by reference to Section F3.5b, for links made of built-up cross sections, complete-joint-penetration groove welds are used to connect the web to the flanges.

(d) Intermediate web stiffeners are designed in accordance with AISC *Seismic Provisions* Sections H5.5a by reference to Section F3.5b(4), and Section H5.5c, which also references Section H4.5b.

(e) Face bearing plates are designed in accordance with AISC *Seismic Provisions* Sections H5.5c and F3.5b by reference.

(f) Vertical transfer bars are designed in accordance with AISC *Seismic Provisions* Section H5.5c.

(g) The embedment length is determined from AISC *Seismic Provisions* Equation H4-2 or H4-2M with modifications given in Section H5.5c and is considered to begin inside the first layer of confining reinforcement in the wall boundary layer.

(h) The link length is determined in accordance with AISC *Seismic Provisions* Section H5.5c, by reference to Sections F3.5b and H4.5b.

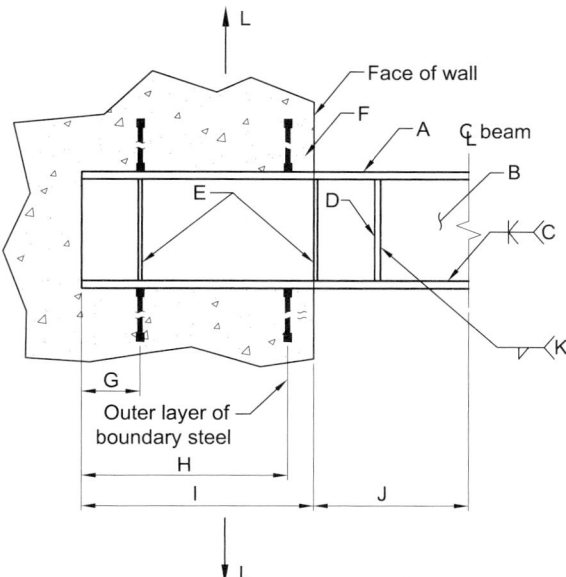

Fig. 7-8. Notes key for AISC Seismic Provisions design requirements for a coupling beam in a C-SSW system (see Table 7-7).

Table 7-7
Notes in Figure 7-8 and Overview of Requirements for C-SSW Systems

Note in Fig. 7-8	Note in Summary	Item	Referenced Standard*
–	1	Materials	Seismic Prov. Sects. A3.1, A3.4 & A3.5
–	2	Structural design drawings and specifications	Seismic Prov. Sects. A4.1, A4.2 & A4.3
–	3	Load and load combinations	Seismic Prov. Sect. B2
–	4	Required strength for structural members and connections	Seismic Prov. Sects. B3.1 & H5.5
–	5	Structural analysis	Seismic Prov. Sect. H5.3
–	5(a)	Elastic stiffness of concrete/composite members	Seismic Prov. Comm. to Ch. C
–	6	System requirements	Seismic Prov. Sect. H5.4
–	7	Boundary members	Seismic Prov. Sect. H5.5b
–	8	Steel coupling beams	Seismic Prov. Sects. H5.5a & H5.5c
A	8(a)	Beam flange local buckling	Seismic Prov. Sects. H5.5c, F3.5b(1) (Exception) & Table D1.1
B	8(b)	Web local buckling	Seismic Prov. Sect. H5.5c & Table D1.1
C	8(c)	Flange-web weld (built-up I-shape)	Seismic Prov. Sect. F3.5b(1)
D	8(d)	Intermediate web stiffeners	Seismic Prov. Sects. F3.5b(4), H4.5b & H5.5c
E	8(e)	Face bearing plates	Seismic Prov. Sects. F3.5b(4) & H5.5c
F	8(f)	Vertical transfer bars	Seismic Prov. Sect. H5.5c
G		Location of end vertical transfer bar and stiffener	Seismic Prov. Sect. H5.5c
H	8(g)	Embedment length: Calculated embed length	Seismic Prov. Sect. H4.5b(1)(3)
		Magnitude of shear used to calculate embed length	Seismic Prov. Sect. H5.5c
I		Total embedment length	Seismic Prov. Sect. H4.5b(1)(3)

Table 7-7 (continued)
Notes in Figure 7-8 and Overview of Requirements for C-SSW Systems

Note in Fig. 7-8	Note in Summary	Item	Referenced Standard*
J	8(h)	Clear span of beam: Link length For calculation of embedment (definition of g)	*Seismic Prov.* Sect. F3.5b(3) *Seismic Prov.* Sect. H4.5b(1)
K	–	Stiffener welds	*Seismic Prov.* Sect. F3.5b(4)
L	–	Wall pier axial load due to coupling action	*Seismic Prov.* Sect. H5.5c
–	9	Composite coupling beams	*Seismic Prov.* Sect. H5.5d
–	10	Demand critical welds	*Seismic Prov.* Sect. H5.6a
–	11	Column splices	*Seismic Prov.* Sect. H5.6b
–	12	Reinforced concrete walls	ACI 318 Sect. 11.9, Ch. 14 & Sect. 21.9

*The referenced standards are in addition to the AISC *Specification*.

Note 9. Composite coupling beams are designed in accordance with AISC *Seismic Provisions* Section H5.5d.

Note 10. Demand critical welds are required as defined in AISC *Seismic Provisions* Section H5.6a.

Note 11. Column splices are designed in accordance with the AISC *Specification* and AISC *Seismic Provisions* Section H5.6b.

Note 12. Reinforced concrete walls are designed in accordance with ACI 318 Section 11.9 and Chapter 14, in addition to Section 21.9.

Steel Coupling Beam Design

The steel coupling beams used in C-SSW systems require special detailing as outlined in AISC *Seismic Provisions* Section H5.5c. The proportions of the beam cross sections must meet the requirements of Sections H4.5b and F3.5 where the coupling beam is treated as a link in an eccentrically braced frame. The anticipated rotational demand of the beams in special systems is equal to or larger than 0.08 rad.

Moment and shear strength is determined using AISC *Specification* Chapters F and G. The requirements of AISC *Seismic Provisions* Section H5 stipulate the transfer of the expected beam shear strength, amplified by a factor of 1.1, to the wall piers. Therefore, it is advantageous to consider grouping beam strengths over the height of the building in an effort to reduce the wall overstrength requirement as was discussed for C-OSW systems.

Flange and web width-to-thickness ratios must satisfy the requirements of AISC *Seismic Provisions* Section D1.1 for highly ductile members.

Wall Overstrength

From AISC *Seismic Provisions* H5.4(2), the expected shear strength of the coupling beams, amplified by a factor of 1.1, must be considered as the shear required to be transferred to the wall piers. In addition, when computing the required embedment length for steel coupling beams, the V_n calculated using Equation H4-1 must be amplified by 1.1 as indicated in AISC *Seismic Provisions* Section H5.5c.

Beam Embedment Length

The length of the steel coupling beam embedded into the wall pier is computed using AISC *Seismic Provisions* Equation H4-2. The V_n term in this equation is the same expected beam shear strength used to determine wall overstrength for the C-SSW system. In this equation, the term g is the clear span of the beam. However, the embedment length, L_e, is measured from the outer layer of boundary element wall reinforcement. Thus, the embedment length of the beam, from the face of the wall, is the length calculated using Equation H4-2 plus the concrete cover on the boundary element reinforcement as illustrated in Figure 7-9.

Intermediate Web Stiffeners

From AISC *Seismic Provisions* Section H5.5c, web stiffeners must meet the requirements for intermediate link stiffeners given in AISC *Seismic Provisions* Section F3.5b(4).

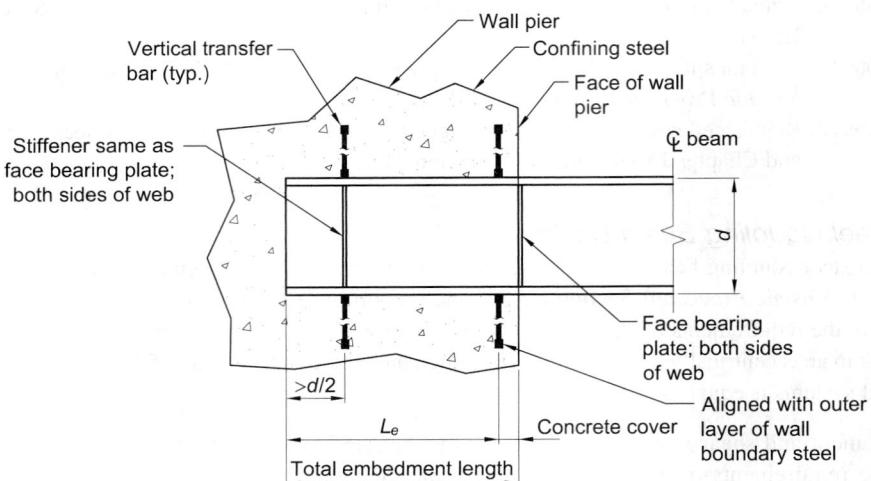

Fig. 7-9. Placement of vertical transfer bars and face bearing plates.

7.5 COMPOSITE SHEAR WALLS

Face Bearing Plates
From AISC *Seismic Provisions* Section H5.5c, face bearing plates (link stiffeners) are provided on both sides of the beam web and located at the face of the wall pier. These plates should meet the requirements of stiffeners in links at "the diagonal brace ends" as required in AISC *Seismic Provisions* Section F3.5b(4). Figure 7-9 illustrates the placement of face bearing plates.

Stiffeners within the Embedded Region
Although not specifically required by the AISC *Seismic Provisions*, stiffeners on both sides of the web, aligned with the outermost pair of vertical transfer bars, provide significantly higher connection ductility than when these stiffeners are not present. The same size stiffener specified for the face bearing plate should be used and placed as shown in Figure 7-9. The AISC *Seismic Provisions* Commentary Section H5 discusses this further.

Vertical Transfer Bars
In C-SSW systems, reinforcing bars are attached to the flanges within the embedded region to improve the ductility and general hysteretic behavior of the connection region. The requirements for size and development of the transfer bars are specified in AISC *Seismic Provisions* Section H5.5c. Figure 7-9 illustrates the placement of these bars. A minimum of two bars are required on each flange in each embed region. At a minimum, one pair is placed near the face of the wall to coincide with the wall boundary steel, and one pair is placed near the end of the embed region no less than one-half the depth of the beam from the end.

The AISC *Seismic Provisions* permit the attachment and development of these bars to be done mechanically. When mechanical devices are not used, weldable grade reinforcing bars (e.g., ASTM A706) may be welded directly to the flanges of the beam, and the development length is computed using the provisions of ACI 318 for the development length of straight reinforcement bars in tension. It should be noted that, depending on the diameter of the reinforcing bar used, the development length might be significant. Where geometry is tight, mechanical anchorage will reduce the space required for these bars.

Example 7.5.3. C-SSW Steel Coupling Beam Design

Given:
The sixth floor core plan of a 15-story core wall system is shown in Figure 7-10. The composite special shear wall system includes steel coupling beams. Table 7-8 tabulates the LRFD required shear strengths and Table 7-9 tabulates the nominal beam shear strengths over the height of the building. At the sixth floor level, the LRFD required shear and moment strengths (determined using the equivalent lateral force procedure) on the coupling beams are 795 kips and 2,390 kip-ft, respectively. There is no axial load on the beams. The applicable ASCE/SEI 7 parameters are: $R = 6$, $C_d = 5$, $\Omega_o = 2.5$, Seismic Design Category C, and $I_e = 1.0$.

The compressive strength of the wall pier concrete is 8 ksi, the steel reinforcement is ASTM A615 Grade 60, and the steel beams are built-up I-shapes of ASTM A572 Grade 50 plate material. The stiffener material is also ASTM A572 Grade 50 plate. The beam chord rotation demands are expected to be equal to or greater than 0.08 rad. The clear cover from the face of the wall to the reinforcement is 0.75 in. The coupling beam dimensions are given in Figure 7-10(b).

Perform the following for the coupling beam at the sixth floor:

1. Check the width-to-thickness requirements for the flanges and web of the coupling beam given in Figure 7-10(b).
2. Determine if the clear span length of the beam is sufficient given the expected chord rotation demands.
3. Determine the size and spacing of the web stiffeners over the clear span region of the beam.
4. Compute the required embedment length of the beam into the wall pier.
5. Specify the diameter, quantity and location of vertical transfer bars needed at the flanges within the embedded regions of the beam.
6. Detail the face bearing plates required at the face of the wall and stiffener near the end of the embedded region.
7. Given the LRFD beam required shear strengths over the height of the building provided in Table 7-8, determine the LRFD required axial strength at the base of the wall piers due to coupling action.

Solution:

From AISC *Manual* Table 2-4, the material properties are as follows:

ASTM A572 Grade 50
$F_y = 50$ ksi
$F_u = 65$ ksi

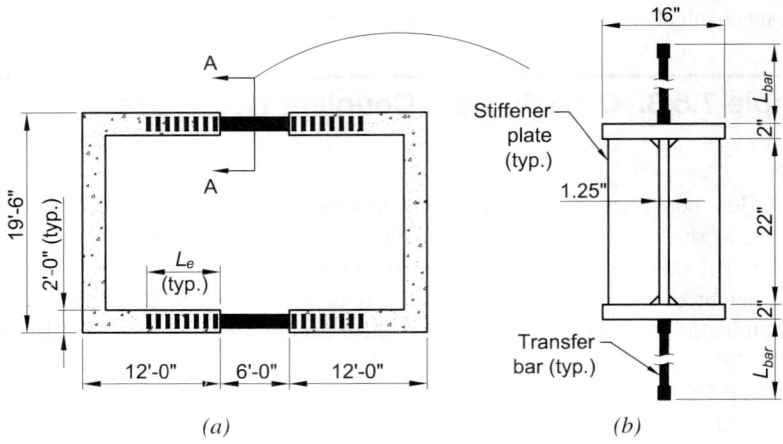

Fig. 7-10. (a) Core plan and (b) Section A-A—beam cross section.

7.5 COMPOSITE SHEAR WALLS

Table 7-8
LRFD Beam Required Shear Strengths

Floor Level	V_u, kips
11–15	340
6–10	795
1–5	318

Table 7-9
Nominal Beam Shear Strengths, V_n

Floor Level	V_u, kips
11–15	404
6–10	To be determined
1–5	462

The geometric properties of the built-up section are:

$A = 91.5$ in.² $d = 26$ in. $t_w = 1.25$ in. $b_f = 16.0$ in.
$t_f = 2.00$ in. $r_y = 3.87$ in. $Z_x = 919$ in.³

Part 1: Local Buckling

Check member ductility

From AISC *Seismic Provisions* Table D1.1, the limiting width-to-thickness ratio for the flanges of a highly ductile member is:

$$\frac{b}{t} \leq \lambda_{hd}$$

$$= 0.30\sqrt{\frac{E}{F_y}}$$

$$= 0.30\sqrt{\frac{29{,}000 \text{ ksi}}{50 \text{ ksi}}}$$

$$= 7.22$$

$$\frac{b}{t} = \frac{b_f}{2t_f}$$

$$= \frac{16.0 \text{ in.}}{2(2.00 \text{ in.})}$$

$$= 4.00 < 7.22 \quad \textbf{o.k.}$$

From AISC *Seismic Provisions* Table D1.1, the limiting width-to-thickness ratio for the web of a highly ductile member is:

$$\frac{h}{t} \leq 2.45 \sqrt{\frac{E}{F_y}} (1 - 0.93 C_a)$$

$$= 2.45 \sqrt{\frac{29,000 \text{ ksi}}{50 \text{ ksi}}} (1 - 0)$$

$$= 59.0$$

$$\frac{h}{t} = \frac{22.0 \text{ in.}}{1.25 \text{ in.}}$$

$$= 17.6 < 59.0 \quad \textbf{o.k.}$$

This member also meets the compact limits according to AISC *Specification* Table B4.1b.

Part 2: Beam Length

Determine whether the limit state of lateral-torsional buckling applies. Because the beam is compact, according to AISC *Specification* Section F2, Equation F2-5 gives the maximum unbraced length permitted for the beam to reach the plastic moment. Thus:

$$L_p = 1.76 r_y \sqrt{\frac{E}{F_y}}$$

$$= (1 \text{ ft}/12 \text{ in.})(1.76)(3.87 \text{ in.}) \sqrt{\frac{29,000 \text{ ksi}}{50 \text{ ksi}}}$$

$$= 13.7 \text{ ft}$$

$$L_b = 6.00 \text{ ft}$$

6.00 ft < 13.7 ft; therefore, the limit state of lateral-torsional buckling does not apply and yielding controls.

From AISC *Specification* Section F2.1, the nominal flexural strength is:

$$M_n = M_p = F_y Z_x \quad \text{(Spec. Eq. F2-1)}$$

$$= \frac{(50 \text{ ksi})(919 \text{ in.}^3)}{12 \text{ in./ft}}$$

$$= 3,830 \text{ kip-ft}$$

7.5 COMPOSITE SHEAR WALLS

From AISC Seismic Provisions Section F3.5b(2), noting that the area of the web in the beam of a special system is calculated as $A_{lw} = (d - 2t_f)t_w$:

$V_p = 0.6 F_y A_{lw}$ (*Provisions* Eq. F3-2)
$= 0.6(50 \text{ ksi})(22.0 \text{ in.})(1.25 \text{ in.})$
$= 825 \text{ kips}$

For a shear-critical beam in a special system, Equation 7-3 can be used to check the length of the beam:

$L \leq \dfrac{1.6 M_p}{V_p}$

$= \dfrac{1.6(3,830 \text{ kip-ft})}{825 \text{ kips}}$

$= 7.43 \text{ ft}$

$L = 6.00 \text{ ft} < 7.43 \text{ ft}$ **o.k.**

Part 3: Size and Spacing of Web Stiffeners

AISC *Seismic Provisions* Section F3.5b(4) addresses provisions for stiffener thickness and spacing requirements as well as requirements for one- or two-sided stiffeners. Since the length of the beam is less than $1.6 M_p / V_p$, and the expected chord rotation is greater than or equal to 0.08 rad, part (a) of AISC *Seismic Provisions* Section F3.5b(4) is used to determine the stiffener requirements.

Stiffener Spacing

$s \leq 30 t_w - \dfrac{d}{5}$

$= 30(1.25 \text{ in.}) - \dfrac{26.0 \text{ in.}}{5}$

$= 32.3 \text{ in.}$

Because the depth of the beam is 26 in. (≥ 25.0 in.), stiffeners are required on both sides of the web.

Stiffener Thickness

$t_s \geq \max \begin{cases} t_w = 1.25 \text{ in.} \\ \frac{3}{8} \text{ in.} \end{cases}$

Use $t_s = 1.25$ in.

Single Stiffener Width

$$b_s \geq \frac{b_f}{2} - t_w$$

$$= \frac{16.0 \text{ in.}}{2} - 1.25 \text{ in.}$$

$$= 6.75 \text{ in.}$$

Stiffener-to-Flange Weld

According to AISC *Seismic Provisions* Section F3.5b(4), the required strength of the stiffener-to-flange weld is determined as follows:

$$R_{uw} \geq \frac{F_y A_{st}}{4}$$

$$= \frac{(50 \text{ ksi})(1.25 \text{ in.})(6.75 \text{ in.})}{4}$$

$$= 105 \text{ kips}$$

Assuming a 1-in. corner clip, the weld size is determined from AISC *Manual* Equation 8-2a as follows:

R_{uw} = (1.392 kip/in.)DL

105 kips = (1.392 kip/in.)D(5.75 in.)(2)

D = 6.56 sixteenths

Use 7/16-in. fillet weld on both sides of the stiffener.

Stiffener-to-Web Weld

According to AISC *Seismic Provisions* Section F3.5b(4), the required strength of the stiffener-to-web weld is determined as follows:

$$R_{uw} \geq F_y A_{st}$$

$$= (50 \text{ ksi})(1.25 \text{ in.})(6.75 \text{ in.})$$

$$= 422 \text{ kips}$$

Assuming a 1-in. × 1-in. corner clip at the flange-to-web corner of the stiffener, the weld size is determined from AISC *Manual* Equation 8-2a as follows:

R_{uw} = (1.392 kip/in.)DL

422 kips = (1.392 kip/in.)D(20.0 in.)(2)

D = 7.58 sixteenths

Use 1/2-in. fillet welds on both sides of the stiffener.

7.5 COMPOSITE SHEAR WALLS

Provide 1.25-in. × 6.75-in. full depth stiffeners on each side of the web spaced no farther apart than 32 in. See Figure 7-11 for final beam detailing.

Part 4: Beam Embedment Length

From AISC *Seismic Provisions* Section H4.5b(1)(3), L_e is determined from:

$$V_n = 1.54\sqrt{f_c'}\left(\frac{b_w}{b_f}\right)^{0.66} \beta_1 b_f L_e \left[\frac{0.58 - 0.22\beta_1}{0.88 + \frac{g}{2L_e}}\right] \qquad (\text{\textit{Provisions} Eq. H4-2})$$

where

$f_c' = 8$ ksi
$\beta_1 = 0.65$ from ACI 318 Section 10.2.7.3
$g = L = 72.0$ in.

The expected shear strength of the steel coupling beam is:

$$V_n = \frac{2R_y M_p}{g} \leq R_y V_p \qquad (\text{\textit{Provisions} Eq. H4-1})$$

$R_y = 1.1$ (from AISC *Seismic Provisions* Table A3.1)

$$V_n = \frac{2(1.1)(3{,}830 \text{ kip-ft})(12 \text{ in./ft})}{72.0 \text{ in.}}$$
$= 1{,}400$ kips

$V_n = R_y V_p$
$= 1.1(0.6)(50 \text{ ksi})(22.0 \text{ in.})(1.25 \text{ in.})$
$= 1.1(825 \text{ kips})$
$= 908$ kips

Use $V_n = 908$ kips

As a C-SSW system, in accordance with AISC *Seismic Provisions* Section H5.5c, the expected shear strength of the beam for which embedment length is calculated must be increased by a factor of 1.1 to account for strain hardening. Therefore:

$V_n = 1.1(908 \text{ kips})$
$= 999$ kips

and AISC *Seismic Provisions* Equation H4-2 gives:

$$999 \text{ kips} = 1.54\sqrt{8 \text{ ksi}} \left(\frac{24.0 \text{ in.}}{16.0 \text{ in.}}\right)^{0.66} (0.65)(16.0 \text{ in.}) L_e \left[\frac{0.58 - 0.22(0.65)}{0.88 + \frac{72.0 \text{ in.}}{2L_e}}\right]$$

$$= \frac{25.9 L_e}{0.88 + \frac{36.0}{L_e}}$$

$L_e = 57.9$ in.

The total embedded length from the face of the wall is:

$L_e + C_c = 57.9$ in. $+ 0.75$ in.
$= 58.7$ in.

Each end of the beam will be embedded a minimum of 60 in. beyond the face of the wall.

Part 5: Vertical Transfer Bars

From AISC *Seismic Provisions* Section H5.5c, the required cross-sectional area of vertical transfer reinforcement, attached to the top and bottom flanges of the beam is determined using Equation H5-1. As calculated previously, the embedment length is 60.0 in. less the cover on the reinforcing. Here the embedment length will conservatively be taken as 60.0 in.

$$A_{tb} \geq \frac{0.03 f'_c L_e b_f}{F_{ysr}} \qquad (Provisions \text{ Eq. H5-1})$$

$$A_{tb} \geq \frac{0.03(8 \text{ ksi})(60.0 \text{ in.})(16.0 \text{ in.})}{60 \text{ ksi}}$$

$= 3.84$ in.^{2}

From requirements of AISC *Seismic Provisions* Section H5.5c, A_{tb} is the area of vertical transfer reinforcement required at the top and bottom flanges in each region of the embedded length. Assuming four bars will be used in each of the four required locations,

$$A_{req} = \frac{3.84 \text{ in.}^2}{4}$$

$= 0.960$ in.^{2} / location

Thus, provide (4) No. 9 vertical transfer bars on each flange at each of the four regions of the embedment length. See Figure 7-11 for details of transfer bar arrangements.

Another option is to use alternating U-shaped hairpins (see AISC *Seismic Provisions* Figure C-H5-3). The hairpins extending above (or below) the flange need to provide $A_{tb} = 3.84$ in.^{2} Limit the bar size to No. 5 in order to ensure reasonable bend radii. U-shaped hairpins have two legs, and the area of a No. 5 bar is 0.31 in.^{2} Therefore:

N (number of No. 5 hair pins) = (3.84 in.^{2}/2)/0.31 in.^{2} = 6.19; therefore use 7

7.5 COMPOSITE SHEAR WALLS

Note that the transfer bars must be developed in a manner consistent with AISC *Seismic Provisions* Section H5.5c. Also note that AISC *Seismic Provisions* Equation H5-2 provides an upper limit on A_{tb}. Assume the longitudinal wall reinforcement ratio is 0.0025. Therefore, the area of longitudinal wall reinforcement along the embedment length is:

$$A_s = 0.0025 b_{wc} L_e$$
$$= 0.0025(24.0 \text{ in.})(60.0 \text{ in.})$$
$$= 3.60 \text{ in.}^2$$

Use AISC *Seismic Provisions* Equation H5-2 to check the limit on A_{tb}.

$$0.08 L_e b_w - A_s = 0.08(60.0 \text{ in.})(24.0 \text{ in.}) - 3.60 \text{ in.}^2$$
$$= 112 \text{ in.}^2$$

The provided A_{tb} is 4.00 in.^{2} [(4) No. 9 bars mechanically attached to the flanges] or 4.34 in.^{2} if (7) No. 5 U-shaped hairpins are provided. Either of these values is well below the limit of 112 in.^{2}

Part 6: Face Bearing Plates

AISC *Seismic Provisions* Section H5.5c, requires bearing plates at the face of the wall pier. These face bearing plates must meet the detailing requirements of AISC *Seismic Provisions* Section F3.5b(4), and must be placed on both sides of the web regardless of beam depth according to Section H5.5c.

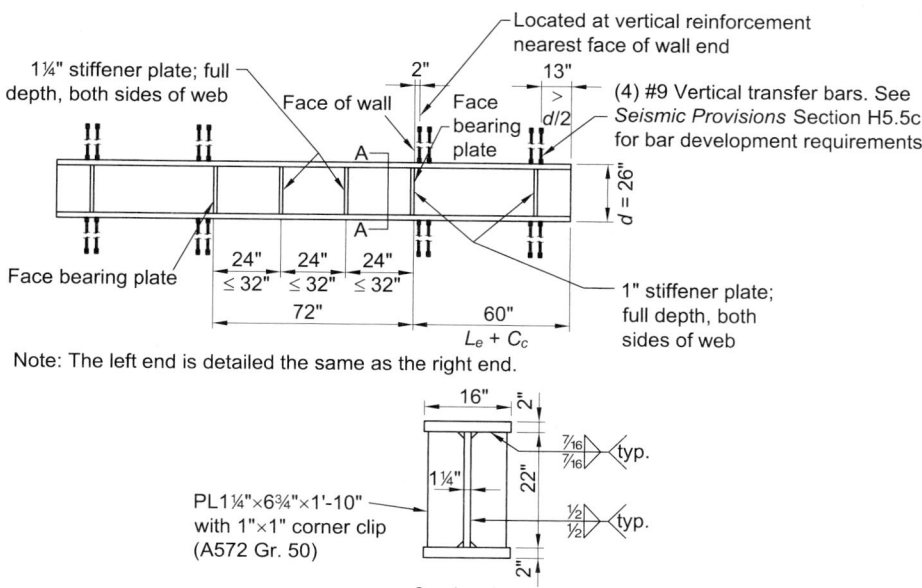

Fig. 7-11. Steel coupling beam detail for Example 7.5.3.

The face bearing plate is located at the beam-wall interface, and therefore should satisfy the requirements for the "end of a link."

Stiffeners are required at two locations: one pair at the beam-wall interface and one pair at the location of vertical transfer bars nearest the end of the embedded region.

Stiffeners in the embedded region and at the beam-wall interface must be two-sided stiffeners.

Stiffener Thickness

$$t_s \geq \max \begin{cases} 0.75t_w = (0.75)(1.25 \text{ in.}) = 0.938 \text{ in.} \\ \frac{3}{8} \text{ in.} \end{cases}$$

Stiffener Width

$$\begin{aligned} b_{s,\,combined} &\geq b_f - 2t_w \\ &= 16.0 \text{ in.} - 2(1.25 \text{ in.}) \\ &= 13.5 \text{ in.} \end{aligned}$$

Provide 1 in. × 6.75 in. full depth stiffeners on each side of the web at the beam-wall interface and at the location of the vertical transfer bars nearest the end of the embedded region. See Figure 7-11 for final beam detailing.

Part 7: Wall Pier Axial Load

As discussed in AISC *Seismic Provisions* Section H5.5c, the embedded regions of the beams must transfer $1.1V_n$ of beam shear strength in a composite special shear wall system. This expected shear strength, increased to account for strain hardening, must be accounted for in the calculated wall pier required axial strengths. The required axial strength resulting from the coupling action on the base wall piers is the accumulation of the amplified shear strengths over the height of the building.

For the given core wall system, two coupling beams frame into each shear wall. Based on the nominal beam shear strengths provided in Table 7-9, the total wall pier required axial strength is determined as follows.

Amplified shears at levels 11–15:

$$\begin{aligned} (1.1V_n)_{11-15} &= 1.1(404 \text{ kips}) \\ &= 444 \text{ kips/floor/beam} \\ \Sigma(1.1V_n)_{11-15} &= (5 \text{ floors})(444 \text{ kips/floor/beam})(2 \text{ beams}) \\ &= 4{,}440 \text{ kips} \end{aligned}$$

Amplified shears at levels 6–10:

$(1.1V_n)_{6-10}$ = 999 kips/floor/beam

$\Sigma(1.1R_y V_n)_{6-10}$ = (5 floors)(999 kips/floor/beam)(2 beams)
= 9,990 kips

Amplified shears at levels 1–5:

$(1.1V_n)_{1-5}$ = 1.1(462 kips)
= 508 kips/floor/beam

$\Sigma(1.1V_n)_{1-5}$ = (5 floors)(508 kips/floor/beam)(2 beams)
= 5,080 kips

The total axial load effect due to coupling is:

$P_{u,\,wall,\,coupling}$ = 4,440 kips + 9,990 kips + 5,080 kips
= 19,500 kips

Thus, the 19,500 kips will be added to the force in the wall due to other loads.

The final coupling beam with transfer bars and stiffeners is shown in Figure 7-11.

7.6 DESIGN TABLE DISCUSSION

Design Tables 7-10A, 7-10B, 7-11A and 7-11B present equations applicable to the design of members subject to combined compression and bending (Geschwindner, 2010). The nominal axial and flexural strengths are given for rectangular encased W-shapes bent about the x-x axis and y-y axis in Tables 7-10A and 7-10B, respectively, depending on where the plastic neutral axis is located in the member. Equations for the pertinent properties are also included. Tables 7-11A and 7-11B provide similar equations for composite filled HSS bent about either principal axis and composite filled round HSS bent about any axis, respectively. The given equations may be used with the simplified interaction diagram discussed in AISC *Specification* Commentary Section I5 (see Method 2).

Table 7-10A
Plastic Capacities for Rectangular Encased W-Shapes Bent About the X-X Axis

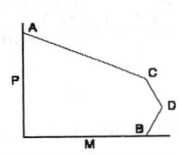

Section	Stress Distribution	Pt.	Defining Equations
A		A	$P_A = A_s F_y + A_{sr} F_{yr} + 0.85 f'_c A_c$ $M_A = 0$ A_s = area of steel shape A_{sr} = area of all continuous reinforcing bars $A_c = h_1 h_2 - A_s - A_{sr}$
		C	$P_C = 0.85 f'_c A_c$ $M_C = M_B$
C		D	$P_D = \dfrac{0.85 f'_c A_c}{2}$ $M_D = Z_s F_y + Z_r F_{yr} + \dfrac{Z_c}{2}(0.85 f'_c)$ Z_s = full x-axis plastic section modulus of steel shape A_{srs} = area of continuous reinforcing bars at the centerline $Z_r = (A_{sr} - A_{srs})\left(\dfrac{h_2}{2} - c\right)$ $Z_c = \dfrac{h_1 h_2^2}{4} - Z_s - Z_r$
D		B	$P_B = 0$ $M_B = M_D - Z_{sn} F_y - \dfrac{Z_{cn}}{2}(0.85 f'_c)$ $Z_{cn} = h_1 h_n^2 - Z_{sn}$ For h_n below the flange $\left(h_n \leq \dfrac{d}{2} - t_f\right)$ $h_n = \dfrac{0.85 f'_c (A_c + A_{srs}) - 2 F_{yr} A_{srs}}{2\left[0.85 f'_c (h_1 - t_w) + 2 F_y t_w\right]}$ $Z_{sn} = t_w h_n^2$ For h_n within the flange $\left(\dfrac{d}{2} - t_f < h_n \leq \dfrac{d}{2}\right)$ $h_n = \dfrac{0.85 f'_c (A_c + A_s - d b_f + A_{srs}) - 2 F_y (A_s - d b_f) - 2 F_{yr} A_{srs}}{2\left[0.85 f'_c (h_1 - b_f) + 2 F_y b_f\right]}$ $Z_{sn} = Z_s - b_f\left(\dfrac{d}{2} - h_n\right)\left(\dfrac{d}{2} + h_n\right)$ For h_n above the flange $\left(h_n > \dfrac{d}{2}\right)$ $h_n = \dfrac{0.85 f'_c (A_c + A_s + A_{srs}) - 2 F_y A_s - 2 F_{yr} A_{srs}}{2(0.85 f'_c h_1)}$ $Z_{sn} = Z_s$

Table 7-10B
Plastic Capacities for Rectangular Encased W-Shapes Bent About the Y-Y Axis

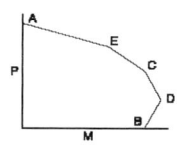

Section	Stress Distribution	Pt.	Defining Equations
A	0.85f'_c, F_y, F_yr	A	$P_A = A_s F_y + A_{sr} F_{yr} + 0.85 f'_c A_c$ $M_A = 0$ A_s = area of steel shape A_{sr} = area of continuous reinforcing bars $A_c = h_1 h_2 - A_s - A_{sr}$
E		E	$P_E = A_s F_y + (0.85 f'_c)\left[A_c - \dfrac{h_1}{2}(h_2 - b_f) + \dfrac{A_{sr}}{2}\right]$ $M_E = M_D - Z_{sE} F_y - \dfrac{Z_{cE}}{2}(0.85 f'_c)$ $Z_{sE} = Z_s$ = full y-axis plastic section modulus of steel shape $Z_{cE} = \dfrac{h_1 b_f^2}{4} - Z_{sE}$
C		C	$P_C = 0.85 f'_c A_c$ $M_C = M_B$
D		D	$P_D = \dfrac{0.85 f'_c A_c}{2}$ $M_D = Z_s F_y + Z_r F_{sr} + \dfrac{Z_c}{2}(0.85 f'_c)$ $Z_r = A_{sr}\left(\dfrac{h_2}{2} - c\right)$ $Z_c = \dfrac{h_1 h_2^2}{4} - Z_s - Z_r$
B		B	$P_B = 0$ $M_B = M_D - Z_{sn} F_y - \dfrac{Z_{cn}}{2}(0.85 f'_c)$ $Z_{cn} = h_1 h_n^2 - Z_{sn}$ For h_n below the flange $\left(\dfrac{t_w}{2} < h_n \leq \dfrac{b_f}{2}\right)$ $h_n = \dfrac{0.85 f'_c (A_c + A_s - 2 t_f b_f) - 2 F_y (A_s - 2 t_f b_f)}{2[4 t_f F_y + (h_1 - 2 t_f) 0.85 f'_c]}$ $Z_{sn} = Z_s - 2 t_f \left(\dfrac{b_f}{2} + h_n\right)\left(\dfrac{b_f}{2} - h_n\right)$ For h_n above the flange $\left(h_n > \dfrac{b_f}{2}\right)$ $h_n = \dfrac{0.85 f'_c (A_c + A_s) - 2 F_y A_s}{2[0.85 f'_c h_1]}$ $Z_{sn} = Z_s$

AMERICAN INSTITUTE OF STEEL CONSTRUCTION

Table 7-11A
Plastic Capacities for Composite Filled HSS Bent About Either Principal Axis

Section	Stress Distribution	Pt.	Defining Equations
A	$0.85f'_c \quad F_y$	A	$P_A = F_y A_s + 0.85 f'_c A_c$ $M_A = 0$ A_s = area of steel shape $A_c = b_i h_i - 0.858 r_i^2$ $b_i = B - 2t$ $h_i = H - 2t$ $r_i = t$
E		E	$P_E = \dfrac{0.85 f'_c A_c}{2} + 0.85 f'_c b_i h_E + 4 F_y t h_E$ $M_E = M_D - F_y Z_{sE} - \dfrac{0.85 f'_c Z_{cE}}{2}$ $Z_{cE} = b_i h_E^2$ $Z_{sE} = 2 t h_E^2$ $h_E = \dfrac{h_n}{2} + \dfrac{H}{4}$
C		C	$P_C = 0.85 f'_c A_c$ $M_C = M_B$
D		D	$P_D = \dfrac{0.85 f'_c A_c}{2}$ $M_D = F_y Z_s + \dfrac{0.85 f'_c Z_c}{2}$ Z_s = full x-axis plastic section modulus of HSS $Z_c = \dfrac{b_i h_i^2}{4} - 0.192 r_i^3$
B		B	$P_B = 0$ $M_B = M_D - F_y Z_{sn} - \dfrac{0.85 f'_c Z_{cn}}{2}$ $Z_{sn} = 2 t h_n^2$ $Z_{cn} = b_i h_n^2$ $h_n = \dfrac{0.85 f'_c A_c}{2 \left[0.85 f'_c b_i + 4 t F_y \right]} \leq \dfrac{h_i}{2}$

Note: Equations in this table are applicable to single-axis bending of the shape about its x-x axis (when $H \geq B$) or about its y-y axis (when $B > H$).

Table 7-11B
Plastic Capacities for Composite Filled Round HSS Bent About Any Axis

Section	Stress Distribution	Pt.	Defining Equations
A	$0.95f'_c$ F_y	A	$P_A = F_y A_s + 0.95f'_c A_c$ $M_A = 0$ $A_s = \pi(dt - t^2)$ $A_c = \dfrac{\pi h^2}{4}$
E		E	$P_E = P_A - \tfrac{1}{4}\left[F_y(d^2 - h^2) + \dfrac{0.95f'_c}{2} h^2\right](\theta_2 - \sin\theta_2)$ $M_E = F_y Z_{sE} + \dfrac{0.95f'_c Z_{cE}}{2}$ $Z_{cE} = \dfrac{h^3}{6}\sin^3\left(\dfrac{\theta_2}{2}\right)$ $Z_{sE} = \dfrac{(d^3 - h^3)}{6}\sin\left(\dfrac{\theta_2}{2}\right)$ $h_E = \dfrac{h_n}{2} + \dfrac{h}{4}$ $\theta_2 = \pi - 2\arcsin\left(\dfrac{2h_E}{h}\right)$
C		C	$P_C = 0.95f'_c A_c$ $M_C = M_B$
D		D	$P_D = \dfrac{0.95f'_c A_c}{2}$ $M_D = F_y Z_s + \dfrac{0.95f'_c Z_c}{2}$ Z_s = plastic section modulus of steel shape $= \dfrac{d^3}{6} - Z_c$ $Z_c = \dfrac{h^3}{6}$
B		B	$P_B = 0$ $M_B = F_y Z_{sB} + \dfrac{0.95f'_c Z_{cB}}{2}$ $Z_{sB} = \dfrac{(d^3 - h^3)}{6}\sin\left(\dfrac{\theta}{2}\right)$ $Z_{cB} = \dfrac{h^3 \sin^3\left(\dfrac{\theta}{2}\right)}{6}$ $\theta = \dfrac{0.0260K_c - 2K_s}{0.0848K_c}$ $\quad + \dfrac{\sqrt{(0.0260K_c + 2K_s)^2 + 0.857K_c K_s}}{0.0848K_c}$ (rad) $K_c = f'_c h^2$ $K_s = F_y\left(\dfrac{d-t}{2}\right)t$ ("thin" HSS wall assumed) $h_n = \dfrac{h}{2}\sin\left(\dfrac{\pi - \theta}{2}\right) \leq \dfrac{h}{2}$

PART 7 REFERENCES

ASCE (2006), *Seismic Rehabilitation of Existing Buildings*, ASCE/SEI 41-06, American Society of Civil Engineers, Reston, VA.

El-Tawil, S., Harries, K.A., Fortney, P.J., Shahrooz, B.M., Kurama, Y., Hassan, M. and Tong, X. (2009), *Recommendation for Seismic Design of Hybrid Coupled Wall Systems*, ASCE Special Publication, American Society of Civil Engineers, Reston, VA.

Fortney, P.J., Shahrooz, B.M. and Rassati, G.A. (2008), "Seismic Performance Evaluation of Coupled Core Walls with Concrete and Steel Coupling Beams," *Steel and Composite Structures Journal*, Vol. 7, No. 4, pp. 279-301.

Geschwindner, L.F. (2010), "Discussion of Limit State Response of Composite Columns and Beam-Columns Part II: Application of Design Provisions for the 2005 AISC Specification," *Engineering Journal*, AISC, Vol. 47, No.2, 2nd Quarter, pp.131-140.

Harries, K.A. and McNeice, D.S. (2006), "Performance-Based Design of High-Rise Coupled Wall Systems," *The Structural Design of Tall and Special Structures*, Vol. 15, No. 3, pp. 289-306.

PART 8
DIAPHRAGMS, COLLECTORS AND CHORDS

8.1 SCOPE 8–2
8.2 GENERAL DISCUSSION 8–2
8.3 FLEXURAL AND TORSIONAL BUCKLING OF
    COLLECTOR ELEMENTS 8–4
    Bracing and Compressive Strength of Collectors 8–4
    Major Axis Buckling 8–5
    Minor Axis and Torsional Buckling 8–5
8.4 DESIGN EXAMPLES 8–8
    Example 8.4.1. Diaphragm Chord and Collector Design 8–8
    Example 8.4.2. Collector Connection Design 8–22
PART 8 REFERENCES 8–37

8.1 SCOPE

The requirements and other design considerations summarized in this Part apply to elements and connections of buildings and of frames that are specifically detailed for seismic resistance (or other lateral loads) but are not covered in Parts 4, 5, 6 or 7.

8.2 GENERAL DISCUSSION

Seismic design requires that components of the structure be connected or tied together in such a manner that they behave as a unit. Diaphragms are an important structural element for creating this interconnection. Diaphragm elements:

- connect the distributed mass of the building to the vertical elements of the lateral force resisting system (braced frames, moment frames or shear walls);
- interconnect the vertical elements of the lateral force resisting system, thus completing the system for resistance to building torsion;
- provide lateral stability to columns and beams including nonlateral force resisting system columns and beams; and
- provide out-of-plane support for walls and cladding.

The elements that make up a diaphragm are generally already present in a building to carry other loads, such as gravity loads.

Floors, roofs, and other membrane or bracing systems are generally used as diaphragm elements. Diaphragms are typically horizontally spanning members, analogous to deep beams, which distribute the seismic loads from their origin to the vertically oriented lateral force resisting frames (braced frames, moment frames, etc.). Diaphragms are idealized as simple-span or continuous horizontally spanning deep beams, and hence are subject to shear, moment and axial forces, and the associated deformations. Figure 8-1 shows typical loading, shear and moment diagrams for the analysis and design of a diaphragm. The floor- or roof-deck system is usually designed as the shear-resistant element (analogous to the web of a beam) and the beams or supplemental deck reinforcing at the boundaries of the diaphragm are designed to resist axial force (analogous to the flanges of a beam).

Diaphragms act as beams on elastic supports, with the diaphragm acting as the beam and the vertical elements of the lateral force resisting system acting as the supports. The relative rigidity of the diaphragm and the vertical elements is used to classify diaphragms into one of three categories: rigid, flexible or semi-rigid. Rigid diaphragms are those in which the flexibility of the supports is far greater than the in-plane flexibility of the diaphragm. They also possess the strength and stiffness to distribute the lateral forces to the lateral force resisting frames in proportion to the relative stiffness of the individual frames, without significant deformation in the diaphragm. Where the in-plane flexibility of the diaphragm is far greater than that of the vertical elements, the diaphragm is classified as flexible. A flexible diaphragm distributes the lateral forces to the lateral force resisting frames in a manner analogous to a simple beam spanning between the lateral force resisting elements. The distribution of the lateral forces through a flexible diaphragm is independent of the relative stiffness of the lateral force resisting frames. Where the flexibility of the diaphragm and its supports (the vertical elements) is similar (or where the diaphragm cannot be uniformly categorized as either rigid or flexible in all spans in each direction) the diaphragm is considered

8.2 GENERAL DISCUSSION

semi-rigid. A semi-rigid diaphragm distributes lateral forces in proportion to the stiffness of the diaphragm and the relative stiffness of the lateral force resisting frames. Semi-rigid diaphragms are often analyzed using the analogy of a beam on elastic supports, where the beam represents the stiffness of the diaphragm and the elastic supports represent the stiffness of the lateral force resisting frames.

In a building with flexible diaphragms, the diaphragm is analyzed first (for diaphragm forces); the effect of the reactions on the supports is used in the design of the vertical elements of the lateral force resisting system. These reactions may need to be adjusted to be consistent with the base shear. In buildings with rigid or semi-rigid diaphragms, a full building analysis is done (for seismic lateral forces), and the diaphragm is designed based on the forces from that analysis. These reactions are adjusted to be consistent with the required diaphragm forces. For more information, see Sabelli et al. (2011).

Because many buildings have lateral force resisting frames that are not uniformly spaced and continuous around the diaphragm boundaries, collector elements are utilized. Collector elements are tension and compression members that deliver the diaphragm forces to the

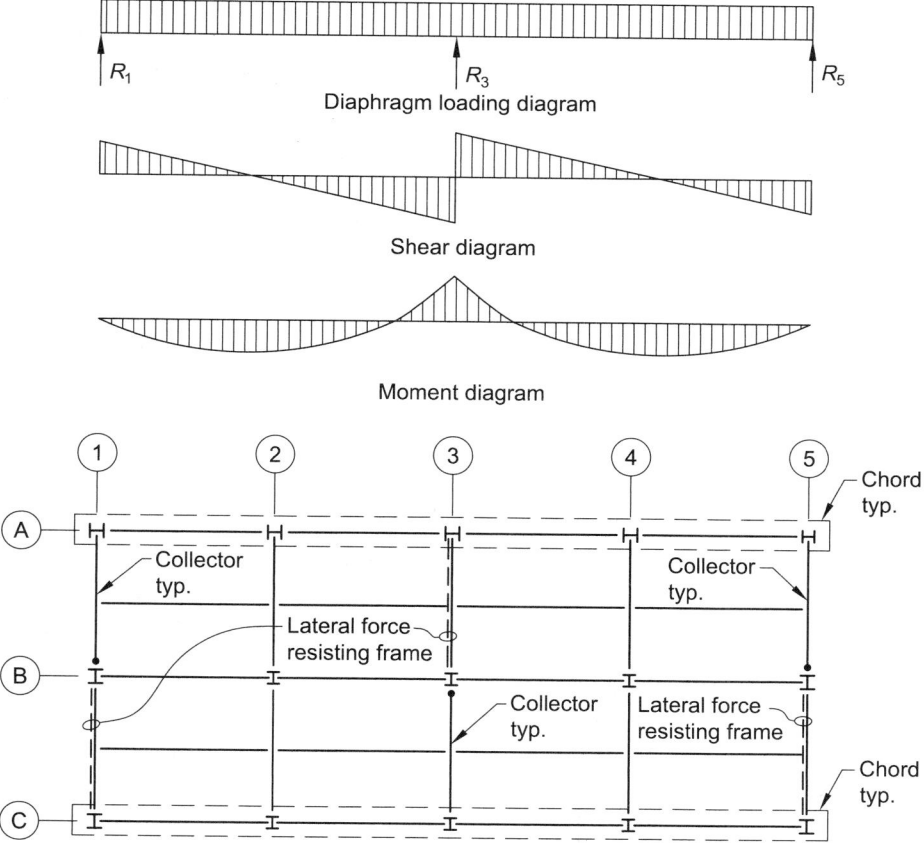

Fig. 8-1. Diaphragm force distribution.

lateral force resisting frames. A redistribution of collector forces can occur as ductile design mechanisms form in the lateral force resisting frames.

When horizontal truss bracing is used as a diaphragm, the chords should be regarded and designed as collectors using the appropriate load combinations. The diagonal and cross brace members can also be regarded and designed as collectors to ensure that they will not buckle or hinge before they deliver forces to the vertical lateral force resisting frame. Alternatively, diagonal diaphragm braces can be allowed to buckle or hinge and be a source of additional energy absorption. Neither ASCE/SEI 7 nor the AISC *Seismic Provisions* provide prescriptive direction on how to consider horizontal truss bracing. For recommendations on the design of diaphragms, see Sabelli et al. (2011).

8.3 FLEXURAL AND TORSIONAL BUCKLING OF COLLECTOR ELEMENTS

Bracing and Compressive Strength of Collectors

In buildings, collectors are typically floor or roof framing members that transfer loads to the seismic force resisting system. In nonbuilding structures, collectors may be connected to horizontal bracing. In many of these conditions the effective lengths may be different for major axis flexural buckling, minor axis flexural buckling, and torsional buckling. Additionally, the torsional buckling strength determined in AISC *Specification* Section E4 is not applicable to members constrained to twist about an axis other than the centroidal axis. This is the case for beams with the top flange laterally braced continuously at the top flange by the deck or slab and the bottom flange unbraced between lateral brace points. This condition is termed constrained axis flexural-torsional buckling.

For W-shapes the constrained axis flexural-torsional buckling strength will be greater than the flexural buckling strength if the unbraced lengths are equal. The constrained axis flexural-torsional buckling length is taken as the bottom-flange unbraced length. This buckling strength will be less than the torsional buckling strength for the same unbraced length. Thus, when constrained axis flexural-torsional buckling is an applicable limit state in wide-flange members, torsional buckling typically need not be evaluated.

Designers often simplify the determination of the compressive strength of collectors with conservative assumptions and methods, such as neglecting the continuous bracing of the top flange and taking the minor axis unbraced length as the distance between bottom-flange lateral supports so that torsional and constrained axis flexural-torsional buckling may be neglected. While such approaches are acceptable, they often indicate the need for additional braces or increases in beam size well beyond what is actually required. Note that Appendix 6 of the AISC *Specification* does not provide requirements for torsional bracing of compressive members. Criteria for torsional bracing of columns can be found in Helwig and Yura (1999). The following discussion provides guidance for a more explicit determination of the governing limit states and a more efficient design approach.

Once the available axial compressive strength of the collector is determined, the combined effects of flexural and axial forces are evaluated per Chapter H of the AISC *Specification*. In many cases, a more detailed stability analysis than the following will permit even greater efficiency. Such approaches can include explicit consideration of the torsional bracing provided by the steel or composite deck, or a beam-column stability analysis considering both flexure and axial forces simultaneously in lieu of the Chapter H interaction method.

Major Axis Buckling

For collectors, the major axis flexural buckling length is typically taken as the full member length as described in AISC *Specification* Commentary Section I7, assuming webs are oriented vertically. Exceptions to this include certain cases in which braces may be considered to provide in-plane bracing under design conditions. For seismic loads, such cases include beams in eccentrically braced frames and beams in V- and inverted V-configuration braced frames not specifically detailed for seismic resistance; the diagonal braces in these systems provide a braced point.

Minor Axis and Torsional Buckling

Steel deck with ribs parallel to the beam is generally assumed not to provide lateral bracing. Lateral and torsional bracing may be provided by transverse members at points along the length of the beam because typically the connection to the transverse member is designed to provide both torsional and lateral restraint. For this case the minor axis flexural buckling lengths and torsional buckling lengths are the same and equal to the distance between these bracing points; thus, the minor axis flexural buckling strength will be lower than the torsional buckling strength.

Steel deck with ribs perpendicular to the beam is generally assumed to provide continuous lateral bracing but not torsional bracing. Torsional bracing may be provided at points along the beam length. For this case the compression strength may be governed by constrained axis flexural-torsional buckling. Figure 8-2 shows minor axis flexural buckling, torsional buckling, and constrained axis flexural-torsional buckling about the top flange.

Collector beams with composite deck or slabs are likewise continuously braced for minor axis flexural buckling as noted in AISC *Specification* Commentary Section I7. The composite deck or slab also provides significant continuous torsional bracing. This continuous torsional bracing is often sufficient to preclude torsional buckling altogether. This can be

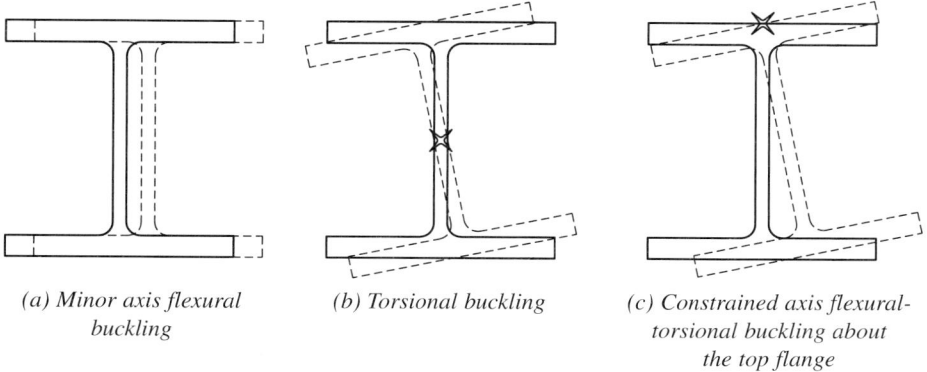

(a) Minor axis flexural buckling

(b) Torsional buckling

(c) Constrained axis flexural-torsional buckling about the top flange

Fig. 8-2. Types of collector buckling.

verified using methods developed by Helwig and Yura (1999). For simplicity, designers can conservatively compute the constrained axis flexural-torsional buckling strength about the top flange and neglect the effect of the continuous torsional bracing.

For collectors in diaphragms with horizontal diagonal bracing, if the brace connections provide torsional bracing such that both minor axis flexural buckling and torsional buckling lengths are equal, the compressive strength is likely governed by flexural buckling. Where the torsional and minor axis flexural buckling lengths are equal, the torsional buckling strength will exceed the minor axis flexural buckling strength for doubly symmetric I-shaped members. If the brace connections do not provide torsional bracing (for example, bracing only one flange), the minor axis flexural buckling and torsional buckling lengths are not equal and both limit states, in addition to major axis buckling, must be considered.

Methods for computing the compressive strength of members governed by torsional buckling about the centroidal axis are presented in the AISC *Specification* Section E4. For constrained axis flexural-torsional buckling, as shown in Figure 8-2(c), Helwig and Yura (1999) give the following expression:

$$P_T = \frac{P_{ey}\left(\dfrac{d^2}{4} + a^2\right) + GJ}{r_x^2 + r_y^2 + a^2} \tag{8-1}$$

where

$P_{ey} = \dfrac{\pi^2 \tau E I_y}{L_T^2}$

E = modulus of elasticity, ksi
G = shear modulus, ksi
I_y = minor axis moment of inertia, in.^{4}
J = torsional constant, in.^{4}
L_T = torsional unbraced length, in.
P_T = torsional buckling strength, kips
a = distance from centroid to lateral restraint on the member minor axis, in.
d = member depth, in.
r_x = radius of gyration about the major axis, in.
r_y = radius of gyration about the minor axis, in.
τ = tanget modulus stiffness reduction factor

For use with AISC *Specification* equations, this can be expressed in a form similar to AISC *Specification* Equation E4-4, in which the elastic torsional buckling stress, F_e, is computed as follows:

$$F_e = \left[\frac{\pi^2 E\left(C_w + I_y a^2\right)}{(K_z L)^2} + GJ\right]\frac{1}{I_x + I_y + a^2 A_g} \tag{8-2}$$

Table 8-1
Summary of Unbraced Lengths and Restraint Conditions for Collector Beams
(Compressive Strength)

Condition		Major Axis Flexural Buckling Length	Minor Axis Flexural Buckling Length	Constrained-Axis Flexural-Torsional Buckling Length	Torsional Buckling Length
Steel deck	Ribs parallel to beam	Full length	Between lateral brace points	Not applicable	Between torsional brace points
	Ribs perpendicular to beam	Full length	Not applicable (continuously braced)	Between torsional brace points	Not applicable
Composite deck or slab		Full length	Not applicable (continuously braced)	Between torsional brace points[1]	Not applicable[1]
Horizontal diagonal bracing		Full length	Between lateral brace points	Not applicable (if braced at centroid)	Between torsional brace points

[1] The composite deck or slab provides some continuous torsional bracing. In some cases this torsional bracing is sufficient to preclude flexural-torsional buckling. Methods for determining adequacy of such bracing are not presented in this Manual and for simplicity these effects are not considered. See Helwig and Yura (1999) for guidance on evaluating continuous torsional bracing.

where

A_g = gross cross-sectional area of member, in.^{2}
I_x = major axis moment of inertia, in.^{4}
C_w = warping constant, in.^{6}
K_z = effective length factor for torsional buckling

For $a = d/2$, the case for restraint at the top flange, Equation 8-2 simplifies to:

$$F_e = \left\{ \frac{\pi^2 E\left[C_w + I_y (d/2)^2\right]}{(K_z L)^2} + GJ \right\} \frac{1}{I_x + I_y + (d/2)^2 A_g} \quad (8\text{-}3)$$

The value of F_e is used in AISC *Specification* Equation E3-2 for $F_y/F_e \leq 2.25$ and AISC *Specification* Equation E3-3 for $F_y/F_e > 2.25$.

A summary of the buckling lengths and this discussion is provided in Table 8-1.

8.4 DESIGN EXAMPLES

Example 8.4.1. Diaphragm Chord and Collector Design

Given:
Refer to Figure 8-3a for the plan and Figure 8-3b for the braced frame elevations called out on the plan. Assume the braced frames are special concentrically braced frames (SCBF). Based on the following information given for a north-south motion, determine the required strengths of a collector and a chord at the third level and design the chord. (A similar calculation must be performed for east-west loading; this is not illustrated here.) Design the collector on grid 1 between grids C and D using ASTM A992 material. The diaphragm consists of 2-in. metal deck with 2½-in. normal weight concrete topping (total slab thickness = 4½ in.) with ¾-in.-diameter steel headed stud anchors spaced every 12 in. along the beam. The specified compressive strength of the concrete is 4,000 psi and the metal-deck span is north-south. The applicable building code specifies the use of ASCE/SEI 7 for calculation of loads. Assume surface loads of $D = 85$ psf and $L = 80$ psf ($L_{reduced} = 50$ psf) on typical levels, and $D_r = 85$ psf and $L_r = 20$ psf on the roof. Due to seismic forces from an equivalent lateral force analysis (ASCE/SEI 7 Section 12.8), the first-order interstory drift at level 3, Δ_H, is 0.375 in.

For the collector beam at the third level along gridline 1 and between gridlines C and D, the gravity moments are:

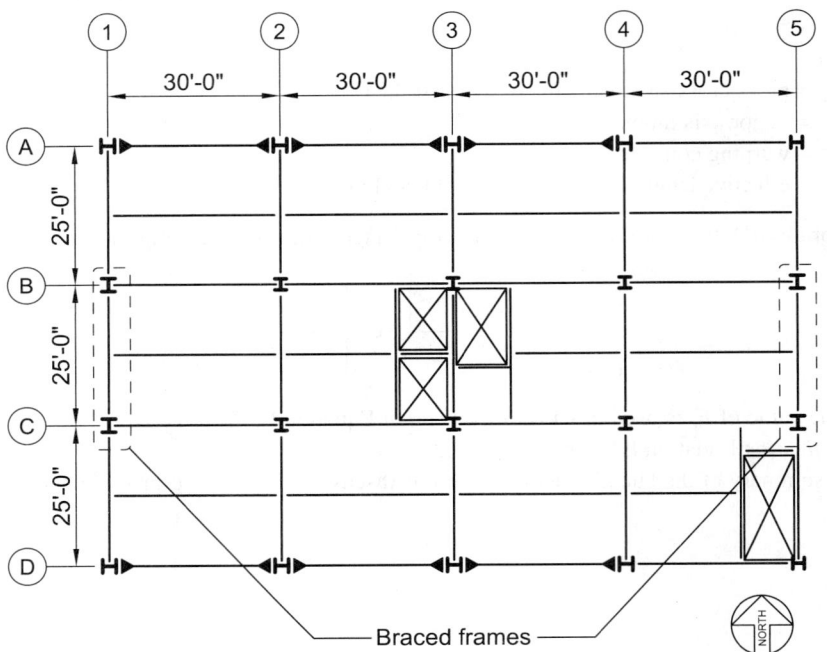

Fig. 8-3a. Floor plan for Example 8.4.1.

$M_D = 123$ kip-ft
$M_L = 96.2$ kip-ft

The gravity shear loads are:

$V_D = 11.8$ kips
$V_L = 8.29$ kips

From ASCE/SEI 7, this structure is assigned to Seismic Design Category D, $\Omega_o = 2.0$, $\rho = 1.3$, $I_e = 1.0$, $R = 6$, $S_{DS} = 1.0$, $k = 1.0$ and $C_s = 0.167$. The seismic base shear is:

$V = C_s W$ (ASCE/SEI 7 Eq. 12.8-1)
$= 0.167(4)(765 \text{ kips})$
$= 511$ kips

where W is the effective seismic weight including the total dead load of the building as required by ASCE/SEI 7 Section 12.7.2 (assuming no other loading applies). The seismic

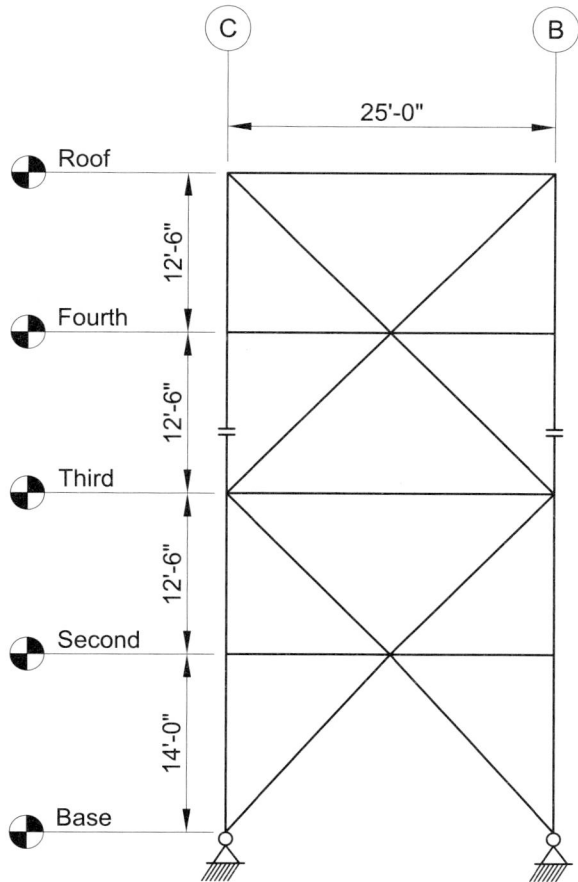

Fig. 8-3b. SCBF elevation.

forces in the north-south direction using the equivalent lateral force procedure of ASCE/SEI 7 are:

Level	Story Height H, ft	Seismic Weight w_i, kips	Force F_i, kips
Roof	12.5	765	201
4	12.5	765	152
3	12.5	765	103
2	14.0	765	55

Solution:

The diaphragm force is:

$$F_{px} = \frac{\sum_{i=x}^{n} F_i}{\sum_{i=x}^{n} w_i} w_{px} \qquad \text{(ASCE/SEI 7 Eq. 12.10-1)}$$

However, ASCE/SEI 7 requires that this force must be greater than or equal to $0.2S_{DS}I_e w_{px}$, but need not exceed $0.4S_{DS}I_e w_{px}$. Values of F_{px} are calculated in the table below. Shaded values indicate the governing force, not including Ω_o.

Level	$w_i = w_{px}$ kips	Σw_i kips	F_i kips	ΣF_i kips	F_{px} kips	$\Omega_o F_{px}$ kips	$0.2S_{DS}I_e w_{px}$ kips	$0.4S_{DS}I_e w_{px}$ kips
Roof	765	765	201	201	201	402	153	306
4	765	1,530	152	353	177	354	153	306
3	765	2,295	103	456	152	306	153	306
2	765	3,060	55	511	128	256	153	306

Chord Force at the Third Level

The governing required strength for the diaphragm at the third level is 153 kips. Analyze the diaphragm as a uniformly loaded beam with a length, L, equal to 120 ft (this is the distance between the braced frame along grid 1 and the braced frame along grid 5). The distributed load is equal to the diaphragm force, F_p, divided by the diaphragm length, as shown.

$$w = \frac{F_p}{L}$$
$$= \frac{153 \text{ kips}}{120 \text{ ft}}$$
$$= 1.28 \text{ kip/ft}$$

As shown in Figure 8-4, the maximum moment in the diaphragm at the third level is:

$$M = \frac{wL^2}{8}$$
$$= \frac{1.28 \text{ kip/ft}(120 \text{ ft})^2}{8}$$
$$= 2{,}300 \text{ kip-ft}$$

The reactions at the braced frames should be consistent with the force distribution from the lateral analysis. In this case, due to symmetry, the maximum shear reactions may be taken as:

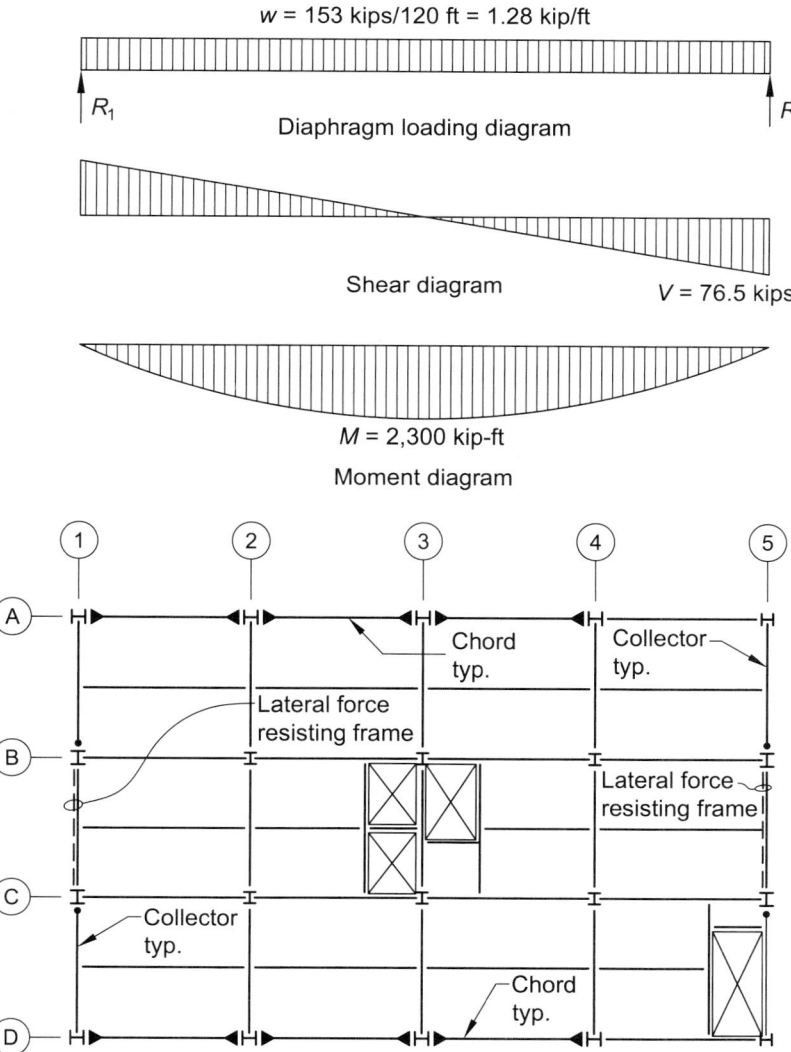

Fig. 8-4. Diaphragm load, shear and moment diagram at the third level.

$$V = \frac{F_p}{2}$$
$$= \frac{153 \text{ kips}}{2}$$
$$= 76.5 \text{ kips}$$

For unsymmetric cases with rigid or semi-rigid diaphragms the distribution should be determined from the lateral analysis. For more information, see Sabelli (2011).

Assuming the diaphragm depth, d, is equal to 75.0 ft (the distance between grids A and D) and the moment is resisted by chord members along grid lines A and D, the maximum tension and compression force in the chords along gridlines A and D is:

$$T = C$$
$$= \frac{M}{d}$$
$$= \frac{2{,}300 \text{ kip-ft}}{75.0 \text{ ft}}$$
$$= 30.7 \text{ kips}$$

A chord member with adequate tensile strength to resist this force can be provided by the addition of supplemental slab reinforcement such as ASTM A615 Grade 60 deformed reinforcing bars, or the force could be assumed to be carried by the steel members alone. If the concrete slab is utilized as the collector, the concrete chord must be designed using the strength design provisions of ACI 318, whether the structural steel is designed using LRFD or ASD. The governing load combination is LRFD Load Combination 5 (the load factor on L is permitted to equal 0.5 since the live load is less than 100 psf) from ASCE/SEI 7 Section 12.4.2.3:

$$(1.2 + 0.2S_{DS})D + \rho Q_E + L + 0.2S$$

Therefore, the required tension force in the chord is:

$$T_u = \rho Q_E$$
$$= 1.3(30.7 \text{ kips})$$
$$= 39.9 \text{ kips}$$

The required area of slab reinforcement is:

$$A_{s\ req} = \frac{T_u}{\phi F_y}$$
$$= \frac{39.9 \text{ kips}}{0.90(60 \text{ ksi})}$$
$$= 0.739 \text{ in.}^2$$

Two No. 6 bars ($A_s = 0.88$ in.^{2}) can provide this supplemental slab reinforcement at the chord locations for the tension force in the chord. Per ACI 318 Section 21.11.7.5, additional transverse reinforcement to confine the concrete and reinforcement under compression

8.4 DESIGN EXAMPLES

forces is not required if the extreme compressive fiber stress in the concrete is equal to or less than $0.2f'_c$. Because the deck span is perpendicular to the chord span, assume that only the concrete above the top of the metal deck is effective in resisting the chord force. The plastic section modulus of the diaphragm is:

$$Z = \frac{bd^2}{4}$$

$$= \frac{2\frac{1}{2} \text{ in.}(75.0 \text{ ft})^2}{4(12 \text{ in./ft})}$$

$$= 293 \text{ ft}^3$$

The extreme compressive fiber stress at the chord is:

$$f_c = \frac{M}{Z}$$

$$= \frac{2{,}300 \text{ kip-ft}(1{,}000 \text{ lb/kip})}{293 \text{ ft}^3 (12 \text{ in./ft})^2}$$

$$= 54.5 \text{ psi}$$

$$0.2f'_c = 0.2(4{,}000 \text{ psi})$$
$$= 800 \text{ psi}$$

$f_c < 0.2f'_c$; therefore, additional transverse reinforcing is not required at the diaphragm chord.

Provide two No. 6 continuous reinforcing bars at the edges of the concrete floor. Per Section 12.2.2 of ACI 318, the development length, l_d, is computed as:

$$l_d = 37.9d_b$$
$$= 37.9(0.750 \text{ in.})$$
$$= 28.4 \text{ in.}$$

Per Section 12.15.1 of ACI 318, the minimum lap length for a Class B lap splice is calculated as:

$$\text{lap length} \geq 1.3l_d$$
$$= 1.3(28.4 \text{ in.})$$
$$= 36.9 \text{ in.}$$

Lap all splices a minimum of 37 in.

The maximum shear in the diaphragm occurs at each end; therefore the total shear force along grid 1 is 76.5 kips. This shear is considered to be uniformly distributed along the depth of the diaphragm (grid 1). This assumption is a simple and rational approach to determine the required strength of the collector beam. Other assumptions about shear transfer through the diaphragm may also be valid but are not considered here since the focus of this example is collector beam design. Assuming a uniform distribution of shear along the depth of the diaphragm, the shear demand on the diaphragm is:

$$v = \frac{V}{d}$$
$$= \frac{76.5 \text{ kips}}{75.0 \text{ ft}}$$
$$= 1.02 \text{ kip/ft}$$

As noted previously, ASCE/SEI 7 requires that collector elements in structures assigned to Seismic Design Category C through F be designed to resist the amplified seismic loads (Ω_o-level loads). The required strength per foot is:

$$V_u = 2.0(1.02 \text{ kip/ft})$$
$$= 2.04 \text{ kip/ft}$$

A diaphragm should be selected that has a shear strength greater than 2.04 kip/ft. If steel headed stud anchors are used, they must resist this shear strength. The diaphragm should be attached to the collector in order to transfer this shear. This may be accomplished by using puddle welds between the collector and metal deck or headed shear studs welded to the collector. Gravity loads should also be considered.

Wide-Flange Collector Beam Between Grids C and D Along Gridline 1

The collector axial force diagram is shown in Figure 8-5. ASCE/SEI 7 Section 12.10.2.1 stipulates the load combination to use for collector elements in structures assigned to Seismic Design Category D. In this case, the load combination including the amplified seismic loads (Ω_o-level loads) controls; therefore, from ASCE/SEI 7 Section 12.4.3.2, the

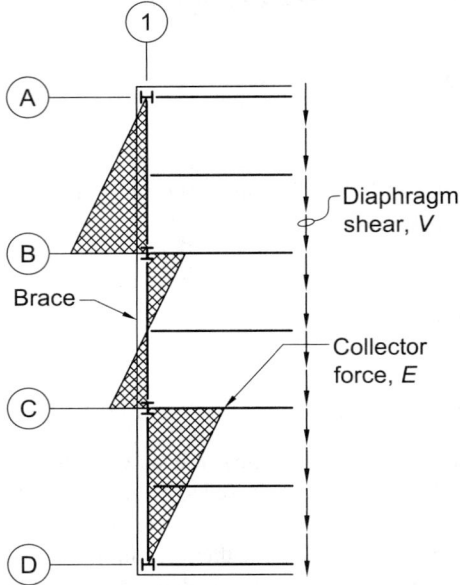

Fig. 8-5. Collector axial load diagram for Example 8.4.1.

governing LRFD load combination is Load Combination 5 (the load factor on L is permitted to equal 0.5 since the live load is less than 100 psf) and the governing ASD load combination is Load Combination 6 as follows:

LRFD	ASD
$(1.2+0.2S_{DS})D+\Omega_o Q_E+0.5L+0.2S$	$(1.0+0.105S_{DS})D+H+F+0.525\Omega_o Q_E$ $+0.75L+0.75S$

The required flexural strength is:

LRFD	ASD
$M_u = 1.4(123 \text{ kip-ft})+2.0(0 \text{ kip-ft})$ $+0.5(96.2 \text{ kip-ft})+0.2(0 \text{ kip-ft})$ $= 220 \text{ kip-ft}$	$M_a = 1.105(123 \text{ kip-ft})+0 \text{ kip-ft}+0 \text{ kip-ft}$ $+0.525(2.0)(0 \text{ kip-ft})$ $+0.75(96.2 \text{ kip-ft})+0.75(0 \text{ kip-ft})$ $= 208 \text{ kip-ft}$

Using the shear strength along grid 1, the axial force in the collector, due to the seismic load, at the intersection of grids C and 1 is:

$P_{Q_E} = 25.0 \text{ ft } (1.02 \text{ kip/ft})$
$= 25.5 \text{ kips (tension or compression)}$

Therefore, from ASCE/SEI 7 Section 12.4.3.2, the governing LRFD load combination is Load Combination 5 and the governing ASD load combination is Load Combination 5, and the required first order axial force in the beam is:

LRFD	ASD
$P_u = (1.2+0.2S_{DS})P_D+\Omega_o P_{Q_E}$ $+0.5P_L+0.2P_S$ $= 1.4(0 \text{ kips})+2.0(25.5 \text{ kips})$ $+0.5(0 \text{ kips})+0.2(0 \text{ kips})$ $= 51.0 \text{ kips (tension or compression)}$	$P_a = (1.0+0.14S_{DS})P_D+P_H+P_F$ $+0.7\Omega_o P_{Q_E}$ $= 1.14(0 \text{ kips})+0 \text{ kips}+0 \text{ kips}$ $+0.7(2.0)(25.5 \text{ kips})$ $= 35.7 \text{ kips (tension or compression)}$

Try a W18×50.

From AISC *Manual* Table 2-4, the geometric properties are as follows:

ASTM A992
$F_y = 50$ ksi
$F_u = 65$ ksi

From AISC *Manual* Tables 1-1 and 6-1, the geometric properties are as follows:

W18×50
$A = 14.7$ in.^{2} $d = 18.0$ in. $t_w = 0.355$ in. $h/t_w = 45.2$ $J = 1.24$ in.^{4}
$r_x = 7.38$ in. $r_y = 1.65$ in. $I_x = 800$ in.^{4} $I_y = 40.1$ in.^{4} $C_w = 3,040$ in.^{6}
$r_x/r_y = 4.47$

Required Second Order Axial Strength

Consider second-order effects with $L = 12.5$ ft using AISC *Specification* Appendix 8.

B_2 is calculated based on an elastic analysis of the structure. Alternatively, a maximum permitted drift can be used to calculate B_2. Note that B_2 and Ω_o apply to the forces derived from the base shear. They do not apply to the minimum diaphragm force from ASCE/SEI 7 Equation 12.10-2.

Calculate B_2 with a first-order interstory drift, Δ_H, of 0.375 in.

H = 201 kips + 152 kips + 103 kips
= 456 kips
L = 12.5 ft
R_M = 1.0 for braced frames

$$P_{e\,story} = R_M \frac{HL}{\Delta_H} \quad \text{(Spec. Eq. A-8-7)}$$

$$= 1.0 \frac{456 \text{ kips}(12.5 \text{ ft})}{(0.375 \text{ in.})(1 \text{ ft}/12 \text{ in.})}$$

= 182,000 kips

Calculate P_{story}, the total vertical load supported by the story. Use a surface area of 9,000 ft^2 on each floor and the following surface loads:

Floor $D = 85$ psf $L_{reduced} \approx 50$ psf
Roof $D_r = 85$ psf $L_r = 20$ psf

Using the ASCE/SEI 7 Section 12.4.2.3, the governing load combinations are as follows:

LRFD	ASD
For LRFD, use Load Combination 5, with the 0.5 factor on L permitted by ASCE/SEI 7 Section 12.4.2.3.	For ASD, use Load Combination 5.
$P_{story} = (1/1,000 \text{ lb/kip})(9,000 \text{ ft}^2)$ $\times\{[1.2+0.2(1.0)]$ $\times[85 \text{ psf} + 2(85 \text{ psf})] + 2.0(0 \text{ psf})$ $+ 0.5[20 \text{ psf} + 2(50 \text{ psf})]$ $+ 0.2(0 \text{ psf})\}$ $= 3,750 \text{ kips}$	$P_{story} = (1/1,000 \text{ lb/kip})(9,000 \text{ ft}^2)$ $\times\{[1.0+0.14(1.0)]$ $\times[85 \text{ psf} + 2(85 \text{ psf})] + 0 \text{ psf}$ $+ 0 \text{ psf}$ $+ 0.7(2.0)(0 \text{ psf})\}$ $= 2,620 \text{ kips}$

8.4 DESIGN EXAMPLES

B_2 is calculated from AISC *Specification* Equation A-8-6:

LRFD	ASD
$\alpha = 1.00$	$\alpha = 1.60$
$B_2 = \dfrac{1}{1 - \dfrac{\alpha P_{story}}{P_{e\,story}}} \geq 1$	$B_2 = \dfrac{1}{1 - \dfrac{\alpha P_{story}}{P_{e\,story}}} \geq 1$
$= \dfrac{1}{1 - \dfrac{1.00(3{,}750 \text{ kips})}{182{,}000 \text{ kips}}}$	$= \dfrac{1}{1 - \dfrac{1.60(2{,}620 \text{ kips})}{182{,}000 \text{ kips}}}$
$= 1.02$	$= 1.02$

Determine the required second-order axial force, P_r, using AISC *Specification* Equation A-8-2, with P_{lt} equal to P_u and P_a for LRFD and ASD, respectively, as determined previously.

LRFD	ASD
$P_r = P_{nt} + B_2 P_{lt}$	$P_r = P_{nt} + B_2 P_{lt}$
$= 0 \text{ kips} + 1.02(51.0 \text{ kips})$	$= 0 \text{ kips} + 1.02(35.7 \text{ kips})$
$= 52.0 \text{ kips}$	$= 36.4 \text{ kips}$

Note: The amplification calculated here is for the lateral system and the force it delivers through the diaphragm to the collector beam.

Compressive Strength of the W18×50

Assume that the W18×50 collector beam has the following unbraced lengths in compression:

$(KL)_x = 25.0$ ft
$(KL)_y = 0$ ft (assume lateral movement is braced by the slab)
$(KL)_z = 12.5$ ft

For the compressive strength based on the limit state of flexural buckling, assume the composite slab fully braces the beam in the weak axis but not in the strong axis. Calculate the strong axis compressive strength using AISC *Manual* Table 6-1. Enter the table using $(KL)_{y\,eq}$:

$$(KL)_{y\,eq} = \frac{(KL)_x}{r_x/r_y} \qquad (Manual \text{ Eq. 4-1})$$

$$= \frac{25.0 \text{ ft}}{4.47}$$

$$= 5.59 \text{ ft}$$

Interpolating p from AISC *Manual* Table 6-1, the compressive strength due to strong axis flexural buckling is:

LRFD	ASD
$\phi_c P_n = \dfrac{1}{p}$ $= \dfrac{1}{0.00180 \text{ kips}^{-1}}$ $= 556 \text{ kips}$	$\dfrac{P_n}{\Omega_c} = \dfrac{1}{p}$ $= \dfrac{1}{0.00270 \text{ kips}^{-1}}$ $= 370 \text{ kips}$

Weak axis compressive strength due to flexural buckling will not govern by inspection ($KL = 0$).

For the limit state of constrained axis torsional buckling, the unbraced length is 12.5 ft. Use Equation 8-3 for F_e.

The W18×50 has a slender web in compression, as indicated in Table 1-1 of the AISC *Manual*. To determine the reduction factor, Q_a, for slender stiffened elements, use AISC *Specification* Section E7.2, with $f = F_{cr}$ computed with $Q = 1.0$.

$$F_e = \left\{ \frac{\pi^2 E \left[C_w + I_y (d/2)^2 \right]}{(K_z L)^2} + GJ \right\} \frac{1}{I_x + I_y + (d/2)^2 A_g} \quad (8\text{-}3)$$

$$= \left\{ \frac{\pi^2 (29{,}000 \text{ ksi}) \left[(3{,}040 \text{ in.}^6) + (40.1 \text{ in.}^4)(18.0 \text{ in.}/2)^2 \right]}{[1.0(12.5 \text{ ft})(12 \text{ in./ft})]^2} + 11{,}200 \text{ ksi} (1.24 \text{ in.}^4) \right\}$$

$$\times \frac{1}{800 \text{ in.}^4 + 40.1 \text{ in.}^4 + (18.0 \text{ in.}/2)^2 (14.7 \text{ in.}^2)}$$

$$= 46.2 \text{ ksi}$$

$$\frac{QF_y}{F_e} = \frac{1.0(50.0 \text{ ksi})}{46.2 \text{ ksi}}$$

$$= 1.08 \leq 2.25$$

Therefore:

$$F_{cr} = Q \left[0.658^{\frac{QF_y}{F_e}} \right] F_y \quad (\textit{Spec.} \text{ Eq. E7-2})$$

$$= 1.0 \left[0.658^{\frac{1.0(50 \text{ ksi})}{46.2 \text{ ksi}}} \right] 50 \text{ ksi}$$

$$= 31.8 \text{ ksi}$$

8.4 DESIGN EXAMPLES

With $f = F_{cr}$, check that $h/t_w \geq 1.49\sqrt{E/f} = 45.0$ as given in AISC *Specification* Section E7.2(a). Because $h/t_w = 45.2 \approx 45.0$, there is negligible reduction in the gross cross section based on AISC *Specification* Section E7.2.

The available axial strength is determined as follows:

$P_n = F_{cr} A_g$ (*Spec.* Eq. E3-1)

   $= 31.8 \text{ ksi}\left(14.7 \text{ in.}^2\right)$

   $= 467$ kips

LRFD	ASD
$\phi_c P_n = 0.90 P_n$ $= 0.90(467 \text{ kips})$ $= 420 \text{ kips}$	$\dfrac{P_n}{\Omega_c} = \dfrac{P_n}{1.67}$ $= \dfrac{467 \text{ kips}}{1.67}$ $= 280 \text{ kips}$

The following is a summary of the limit states in compression on the collector.

Strong Axis Flexural Buckling	Constrained Axis Flexural-Torsional Buckling
$(KL)_x = 25.0$ ft	$(KL)_z = 12.5$ ft
$\phi_c P_n = 556$ kips	$\phi_c P_n = 420$ kips
$\dfrac{P_n}{\Omega_c} = 370$ kips	$\dfrac{P_n}{\Omega_c} = 280$ kips

The available axial compressive strength of the section is governed by constrained axis flexural-torsional buckling.

Required Flexural Strength

Calculate B_1 from AISC *Specification* Appendix 8, Section 8.2.1.

$P_{e1} = \dfrac{\pi^2 EI^*}{(K_1 L)^2}$ (*Spec.* Eq. A-8-5)

$= \dfrac{\pi^2 (29,000 \text{ ksi})\left(800 \text{ in.}^4\right)}{\left[1.0(25.0 \text{ ft})(12 \text{ in./ft})\right]^2}$

$= 2,540$ kips

Because the beam is subject to transverse loading between supports:

   $C_m = 1.0$

LRFD	ASD
$B_1 = \dfrac{C_m}{1 - \dfrac{\alpha P_r}{P_{e1}}} \geq 1$ (*Spec.* Eq. A-8-3)	$B_1 = \dfrac{C_m}{1 - \dfrac{\alpha P_r}{P_{e1}}} \geq 1$ (*Spec.* Eq. A-8-3)
$= \dfrac{1.0}{1 - \dfrac{1.00(52.0 \text{ kips})}{2{,}540 \text{ kips}}}$	$= \dfrac{1.0}{1 - \dfrac{1.60(36.4 \text{ kips})}{2{,}540 \text{ kips}}}$
$= 1.02$	$= 1.02$

From AISC *Specification* Equation A-8-1, the required second-order flexural strength is:

LRFD	ASD
$M_{rx} = B_1 M_{nt} + B_2 M_{lt}$	$M_{rx} = B_1 M_{nt} + B_2 M_{lt}$
$= 1.02(0 \text{ kip-ft}) + 1.02(220 \text{ kips})$	$= 1.02(0 \text{ kip-ft}) + 1.02(208 \text{ kips})$
$= 224 \text{ kip-ft}$	$= 212 \text{ kip-ft}$

Available Flexural Strength of the W18×50 Beam

The composite flexural strength may be used for collectors. The following demonstrates that the noncomposite beam is adequate. Assuming it is fully braced and using AISC *Manual* Table 3-2 for a W18×50, the available flexural strength is:

LRFD	ASD
$\phi M_n = 379 \text{ kip-ft}$	$\dfrac{M_n}{\Omega} = 252 \text{ kip-ft}$

Check combined loading of the W18×50 using AISC *Specification* Section H1.1.

LRFD	ASD
$\dfrac{P_r}{P_c} = \dfrac{52.0 \text{ kips}}{420 \text{ kips}}$	$\dfrac{P_r}{P_c} = \dfrac{36.4 \text{ kips}}{280 \text{ kips}}$
$= 0.124$	$= 0.130$
Because $P_r/P_c < 0.2$, use AISC *Specification* Equation H1-1b.	Because $P_r/P_c < 0.2$, use AISC *Specification* Equation H1-1b.
$\dfrac{0.124}{2} + \dfrac{224 \text{ kip-ft}}{379 \text{ kip-ft}} + 0 = 0.653$	$\dfrac{0.130}{2} + \dfrac{212 \text{ kip-ft}}{252 \text{ kip-ft}} + 0 = 0.906$
$0.653 < 1.0$ **o.k.**	$0.906 < 1.0$ **o.k.**

Because the member does not require composite action, the studs are only required to resist the diaphragm shear transfer. Where composite flexural action is required, the shear studs may be considered fully effective for both flexural shear transfer and diaphragm shear transfer as described in Burmeister and Jacobs (2008).

Use a W18×50 for the collector.

Alternatively, a collector with adequate tensile strength to resist the diaphragm shear can be provided by the addition of supplemental slab reinforcement, such as ASTM A615 Grade 60 deformed reinforcing bars. In this case, the required area of slab reinforcement is:

$$A_{s\ req} = \frac{T_u}{\phi F_y}$$

$$= \frac{52.0 \text{ kips}}{0.90(60 \text{ ksi})}$$

$$= 0.963 \text{ in.}^2$$

Four No. 5 bars ($A_s = 1.24$ in.^{2}) can be used to provide this supplemental slab reinforcement at the collector location. Per ACI 318 Section 21.11.7.5, additional transverse reinforcement is not required if the extreme fiber stress in the concrete is kept below $0.2f_c'$. Because the deck span is parallel to the collector axis, the concrete above and below the top of the metal deck will be effective in resisting the collector force. Assuming the metal deck profile is such that one-half of the area below the top of the metal deck is filled with concrete, the effective thickness of the concrete collector is 3½ in. The minimum width of slab required to resist the collector force is,

$$b_{min} = \frac{P_u}{0.2 f_c' t}$$

$$= \frac{52.0 \text{ kips}}{0.2(4.00 \text{ ksi})(3\frac{1}{2} \text{ in.})}$$

$$= 18.6 \text{ in.}$$

This collector width can be easily accommodated along grid A. Note that a mechanism needs to be provided to transfer the force from the slab reinforcement into the structure.

Using the $0.2f_c'$ compression limitation set forth in ACI 318 Section 21.11.7.5 in conjunction with Ω_o-level forces may be conservative. Alternate approaches can also be used such as limiting compressive strains in the concrete collector to 0.003 (which is analogous to the strain limits for unconfined concrete resisting seismic loads), treating the collector as a short compression member, or any other rational design method that provides a load path between the inertial mass and the seismic force resisting system.

Example 8.4.2. Collector Connection Design

Given:
Refer to Figure 8-6. Check the adequacy of the ASTM A36 single-plate connection shown to resist the collector forces determined in Example 8.4.1, where V_D = 11.8 kips, V_L = 8.29 kips, and P_{Q_E} = 25.5 kips as determined in Example 8.4.1. The beam is ASTM A992 material.

Solution:
From AISC *Manual* Table 2-4, the plate material properties are as follows:

ASTM A36
F_y = 36 ksi
F_u = 58 ksi

From AISC *Manual* Table 1-1, the geometric properties are as follows:

W18×50
t_w = 0.355 in.

As noted in Example 8.4.1, ASCE/SEI 7 requires that collector elements in structures assigned to Seismic Design Categories C through F be designed to resist the amplified seismic loads (Ω_o-level loads).

From ASCE/SEI 7 Section 12.4.3.2, the required shear strength is determined from the governing Load Combination 5 for LRFD and Load Combination 5 for ASD. These load combinations govern based on the resultant required strength determined in the following.

LRFD	ASD
$R_{uv} = (1.2 + 0.2S_{DS})D + \Omega_o Q_E$ $\quad + 0.5L + 0.2S$ $\quad = 1.4(11.8 \text{ kips}) + 2.0(0)$ $\quad + 0.5(8.29 \text{ kips}) + 0.2(0)$ $\quad = 20.7 \text{ kips}$ (includes 0.5 factor on L permitted by ASCE/SEI 7 Section 12.4.3.2)	$R_{av} = (1.0 + 0.14S_{DS})D + H + F$ $\quad + 0.7\Omega_o Q_E$ $\quad = 1.14(11.8 \text{ kips}) + 0 \text{ kips} + 0 \text{ kips}$ $\quad + 0.7(2.0)(0)$ $\quad = 13.5 \text{ kips}$

The required axial strength is:

LRFD	ASD
$R_{ua} = (1.2 + 0.2S_{DS})D + \Omega_o Q_E$ $+ 0.5L + 0.2S$ $= 1.4(0 \text{ kips}) + 2.0(25.5 \text{ kips})$ $+ 0.5(0 \text{ kips}) + 0.2(0 \text{ kips})$ $= 51.0 \text{ kips (tension or compression)}$	$R_{aa} = (1.0 + 0.14S_{DS})D + H + F$ $+ 0.7\Omega_o Q_E$ $= 1.14(0 \text{ kips}) + 0 \text{ kips} + 0 \text{ kips}$ $+ 0.7(2.0)(25.5 \text{ kips})$ $= 35.7 \text{ kips (tension or compression)}$

The resultant required strength is:

LRFD	ASD
$R_u = \sqrt{(20.7 \text{ kips})^2 + (51.0 \text{ kips})^2}$ $= 55.0 \text{ kips}$	$R_a = \sqrt{(13.5 \text{ kips})^2 + (35.7 \text{ kips})^2}$ $= 38.2 \text{ kips}$

Design the connection in accordance with the procedure for extended single-plate shear connections in Part 10 of the AISC *Manual*, modified for the presence of an axial force in the beam.

Available Shear Strength of the Bolt Group

The angle of the resultant load with respect to the longitudinal axis of the bolt group is:

LRFD	ASD
$\theta = \tan^{-1}\left(\dfrac{R_{ua}}{R_{uv}}\right)$ $= \tan^{-1}\left(\dfrac{51.0 \text{ kips}}{20.7 \text{ kips}}\right)$ $= 67.9°$	$\theta = \tan^{-1}\left(\dfrac{R_{aa}}{R_{av}}\right)$ $= \tan^{-1}\left(\dfrac{35.7 \text{ kips}}{13.5 \text{ kips}}\right)$ $= 69.3°$
The calculated load angle, 67.9°, is between the tabulated values for 60° and 75° in AISC *Manual* Table 7-6. The eccentricity of the shear load is determined by the method used for extended single-plate connections.	The calculated load angle, 69.3°, is between the tabulated values for 60° and 75° in AISC *Manual* Table 7-6. The eccentricity of the shear load is determined by the method used for extended single-plate connections.
Interpolating from AISC *Manual* Table 7-6 with $\theta = 67.9°$, $a = e_x = 2\frac{1}{2}$ in., $s = 3$ in. and using $n = 4$ bolts,	Interpolating from AISC *Manual* Table 7-6 with $\theta = 69.3°$, $a = e_x = 2\frac{1}{2}$ in., $s = 3$ in. and using $n = 4$ bolts,
$C = 3.29$	$C = 3.31$

Try a ⅜-in. single plate.

From AISC *Manual* Table 7-1, for a ⅞-in.-diameter ASTM A325-N (Group A) bolt with the threads included in single shear, the available shear strength of one bolt is:

LRFD	ASD
$\phi r_n = 24.3$ kips	$\dfrac{r_n}{\Omega} = 16.2$ kips

The available shear strength of the bolt group is:

LRFD	ASD
$\phi R_n = C \phi r_n$ $= 3.29(24.3 \text{ kips})$ $= 79.9$ kips 55.0 kips < 79.9 kips **o.k.**	$\dfrac{R_n}{\Omega} = C\left(\dfrac{r_n}{\Omega}\right)$ $= 3.31(16.2 \text{ kips})$ $= 53.6$ kips 38.2 kips < 53.6 kips **o.k.**

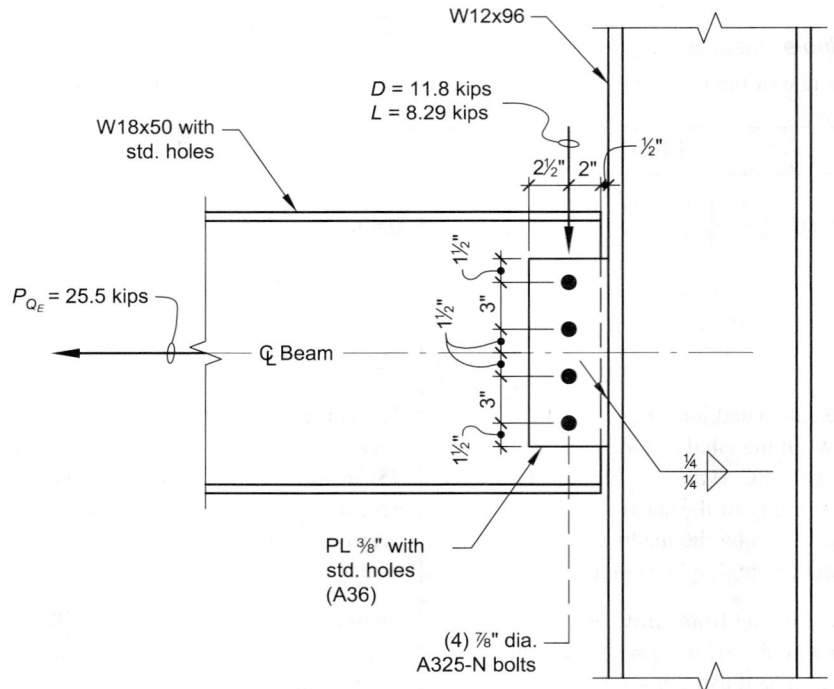

Fig. 8-6. Collector connection investigated in Example 8.4.2.

Bearing on the Plate

At the plate edge:

$$r_n = 1.2l_c tF_u \leq 2.4dtF_u \qquad \text{(Spec. Eq. J3-6a)}$$

where l_c is conservatively taken as the least dimension to the edge in either the horizontal or vertical direction (the vertical edge distance of 1½ in. governs).

$$r_n = 1.2[1\tfrac{1}{2}\text{ in.} - 0.5(^{15}\!/_{16}\text{ in.})](^{3}\!/_{8}\text{ in.})(58\text{ ksi})$$
$$= 26.9 \text{ kips}$$

$$r_n \leq 2.4(^{7}\!/_{8}\text{ in.})(^{3}\!/_{8}\text{ in.})(58\text{ ksi})$$
$$= 45.7 \text{ kips}$$

User $r_n = 26.9$ kips.

LRFD	ASD
$\phi R_n = C\phi r_n$ $= 3.29(0.75)(26.9 \text{ kips})$ $= 66.4$ kips	$\dfrac{R_n}{\Omega} = C\left(\dfrac{r_n}{\Omega}\right)$ $= 3.31\left(\dfrac{26.9 \text{ kips}}{2.00}\right)$ $= 44.5$ kips
55.0 kips < 66.4 kips **o.k.**	38.2 kips < 44.5 kips **o.k.**

Bearing on the Beam web

The beam web is nearly as thick as and stronger than the plate. The beam web has sufficient bearing strength by inspection.

Maximum Permissible Plate Thickness

As discussed in AISC *Manual* Part 10, the connection will be proportioned such that bolt shear failure cannot occur prior to plate flexural yielding. This proportioning ensures sufficient ductility to permit both beam end rotation and rotational deformations corresponding to lateral drift.

In order to achieve this proportioning of limit states, the flexural strength of the bolt group is computed, and the maximum shear plate thickness is determined using the plate yield stress and depth (which is already known).

From AISC *Manual* Table 7-6 (for $\theta = 0°$ in this case), $C' = 11.3$ in. From AISC *Specification* Table J3.2, $F_{nv} = 54$ ksi for A325-N bolts (Group A).

$$M_{max} = \frac{F_y}{0.90}(A_bC') \quad \text{(Manual Eq. 10-4)}$$

$$= \frac{54 \text{ ksi}}{0.90}(0.601 \text{ in.}^2)(11.3 \text{ in.})$$

$$= 407 \text{ kip-in.}$$

$$t_{max} = \frac{6M_{max}}{F_y d^2} \quad \text{(Manual Eq. 10-3)}$$

$$= \frac{6(407 \text{ kip-in.})}{36 \text{ ksi}(12.0 \text{ in.})^2}$$

$$= 0.471 \text{ in.} > 0.375 \text{ in.} \quad \textbf{o.k.}$$

Yielding of the Plate under Combined Shear and Tension

In evaluating this limit state, the minimum plate thickness is calculated independently for each force and the resulting thicknesses are combined for the interaction of forces using an elliptical approximation. The elliptical approximation is not required by the AISC *Specification*, but it is a rational approach for determining the minimum plate thickness.

For shear yielding, the nominal strength is:

$$R_n = 0.60 F_y A_{gv} \quad \text{(Spec. Eq. J4-3)}$$

$$= 0.60 F_y d t_p$$

LRFD	ASD
$t_{pv\,min} = \dfrac{R_{uv}}{\phi 0.60 F_y d}$	$t_{pv\,min} = \dfrac{R_{av}}{(0.60 F_y d / \Omega)}$
$= \dfrac{20.7 \text{ kips}}{1.00(0.60)(36 \text{ ksi})(12.0 \text{ in.})}$	$= \dfrac{13.5 \text{ kips}}{(0.60)(36 \text{ ksi})(12.0 \text{ in.})/1.50}$
$= 0.0799$ in.	$= 0.0781$ in.

For tension yielding, the nominal strength is:

$$R_n = F_y A_g \quad \text{(Spec. Eq. J4-1)}$$

$$= F_y d t_p$$

LRFD	ASD
$t_{pa\,min} = \dfrac{R_{ua}}{\phi F_y d}$	$t_{pa\,min} = \dfrac{R_{aa}}{(F_y d / \Omega)}$
$= \dfrac{51.0 \text{ kips}}{0.90(36 \text{ ksi})(12.0 \text{ in.})}$	$= \dfrac{35.7 \text{ kips}}{36 \text{ ksi}(12.0 \text{ in.})/1.67}$
$= 0.131$ in.	$= 0.138$ in.

8.4 DESIGN EXAMPLES

Therefore, the minimum required plate thickness is:

LRFD	ASD
$t_{p\,min} = \sqrt{(t_{pv\,min})^2 + (t_{pa\,min})^2}$ $= \sqrt{(0.0799\text{ in.})^2 + (0.131\text{ in.})^2}$ $= 0.153\text{ in.}$ 0.375 in. > 0.153 in. **o.k.**	$t_{p\,min} = \sqrt{(t_{pv\,min})^2 + (t_{pa\,min})^2}$ $= \sqrt{(0.0781\text{ in.})^2 + (0.138\text{ in.})^2}$ $= 0.159\text{ in.}$ 0.375 in. > 0.159 in. **o.k.**

Net Section Rupture of the Plate under Combined Shear and Tension

In evaluating this limit state, the minimum plate thickness is calculated independently for each force and the resulting thicknesses are combined for the interaction of forces using an elliptical approximation.

$A_{nv} = A_{nt}$
$\quad = L_{net}t_p$
$\quad = [12.0\text{ in.} - 4(^{15}\!/_{16}\text{ in.} + ^{1}\!/_{16}\text{ in.})]t_p$
$\quad = 8.00t_p\text{ in.}^2$

For shear rupture, the nominal strength is:

$R_n = 0.60F_u A_{nv}$ (Spec. Eq. J4-4)

LRFD	ASD
$t_{pv\,min} = \dfrac{R_{uv}}{\phi 0.60 F_u (8.00\text{ in.})}$ $= \dfrac{20.7\text{ kips}}{0.75(0.60)(58\text{ ksi})(8.00\text{ in.})}$ $= 0.0991\text{ in.}$	$t_{pv\,min} = \dfrac{R_{av}}{0.60 F_u (8.00\text{ in.})/\Omega}$ $= \dfrac{13.5\text{ kips}}{(0.60)(58\text{ ksi})(8.00\text{ in.})/2.00}$ $= 0.0970\text{ in.}$

For tensile rupture, the nominal strength is:

$R_n = F_u A_e$ (Spec. Eq. J4-2)
$\quad = F_u A_{nt}$

LRFD	ASD
$t_{pa\ min} = \dfrac{R_{ua}}{\phi F_u (8.00\ \text{in.})}$ $= \dfrac{51.0\ \text{kips}}{0.75(58\ \text{ksi})(8.00\ \text{in.})}$ $= 0.147\ \text{in.}$	$t_{pa\ min} = \dfrac{R_{aa}}{F_u (8.00\ \text{in.})/\Omega}$ $= \dfrac{35.7\ \text{kips}}{(58\ \text{ksi})(8.00\ \text{in.})/2.00}$ $= 0.154\ \text{in.}$
Therefore, the minimum required plate thickness is:	Therefore, the minimum required plate thickness is:
$t_{p\ min} = \sqrt{(t_{pv\ min})^2 + (t_{pa\ min})^2}$ $= \sqrt{(0.0991\ \text{in.})^2 + (0.147\ \text{in.})^2}$ $= 0.177\ \text{in.} < 0.375\ \text{in.}$ **o.k.**	$t_{p\ min} = \sqrt{(t_{pv\ min})^2 + (t_{pa\ min})^2}$ $= \sqrt{(0.0970\ \text{in.})^2 + (0.154\ \text{in.})^2}$ $= 0.182\ \text{in.} < 0.375\ \text{in.}$ **o.k.**

Block Shear Rupture of the Plate

In evaluating this limit state, the minimum plate thickness is calculated independently for each force, and the resulting thicknesses are combined for the interaction of forces using an elliptical approximation.

Investigate the failure path 1-2-3-4 in Figure 8-7(b) for the shear force.

Tensile Rupture

$A_{gt} = 2.50 t_p\ \text{in.}^2$

$A_{nt} = 2.50 t_p - 0.50(^{15}\!/_{16}\ \text{in.} + {}^1\!/_{16}\ \text{in.}) t_p$
$= 2.50 t_p - 0.500 t_p$
$= 2.00 t_p\ \text{in.}^2$

$U_{bs} = 1.0$

$U_{bs} F_u A_{nt} = 1.0(58\ \text{ksi})(2.00 t_p\ \text{in.}^2)$
$= 116 t_p\ \text{kips}$

Shear Yielding

$A_{gv} = (12.0\ \text{in.} - 1.50\ \text{in.}) t_p\ \text{in.}^2$
$= 10.5 t_p\ \text{in.}^2$

$0.60 F_y A_{gv} = 0.60(36\ \text{ksi})(10.5 t_p\ \text{in.}^2)$
$= 227 t_p\ \text{kips}$

8.4 DESIGN EXAMPLES

Shear Rupture

$$A_{nv} = 10.5t_p - 3.50(^{15}/_{16} \text{ in.} + ^1/_{16} \text{ in.})t_p$$
$$= 7.00t_p \text{ in.}^2$$

$$0.60F_u A_{nv} = 0.60(58 \text{ ksi})(7.00t_p \text{ in.}^2)$$
$$= 244t_p \text{ kips}$$

From AISC *Specification* Equation J4-5:

LRFD	ASD
$\phi R_n = \phi[0.60F_u A_{nv} + U_{bs} F_u A_{nt}]$ $\leq \phi[0.60F_y A_{gv} + U_{bs} F_u A_{nt}]$	$\dfrac{R_n}{\Omega} = \dfrac{0.60F_u A_{nv} + U_{bs} F_u A_{nt}}{\Omega}$ $\leq \dfrac{0.60F_y A_{gv} + U_{bs} F_u A_{nt}}{\Omega}$
$\dfrac{\phi[0.60F_u A_{nv} + U_{bs} F_u A_{nt}]}{t_p}$ $= 0.75(244 \text{ kip/in.} + 116 \text{ kip/in.})$ $= 270 \text{ kip/in.}$	$\dfrac{0.60F_u A_{nv} + U_{bs} F_u A_{nt}}{\Omega t_p}$ $= \dfrac{244 \text{ kip/in.} + 116 \text{ kip/in.}}{2.00}$ $= 180 \text{ kip/in.}$
$\dfrac{\phi[0.60F_y A_{gv} + U_{bs} F_u A_{nt}]}{t_p}$ $= 0.75(227 \text{ kip/in.} + 116 \text{ kip/in.})$ $= 257 \text{ kip/in.}$	$\dfrac{0.60F_y A_{gv} + U_{bs} F_u A_{nt}}{\Omega t_p}$ $= \dfrac{227 \text{ kip/in.} + 116 \text{ kip/in.}}{2.00}$ $= 172 \text{ kip/in.}$
Use $\phi R_n / t_p = 257$ kip/in.	Use $R_n / \Omega t_p = 172$ kip/in.
$t_{pv \text{ min}} = \dfrac{R_{uv}}{\phi R_n / t_p}$ $= \dfrac{20.7 \text{ kips}}{257 \text{ kip/in.}}$ $= 0.0805 \text{ in.} < 0.375 \text{ in.}$ **o.k.**	$t_{pv \text{ min}} = \dfrac{R_{av}}{(R_n / \Omega) / t_p}$ $= \dfrac{13.5 \text{ kips}}{172 \text{ kip/in.}}$ $= 0.0785 \text{ in.} < 0.375 \text{ in.}$ **o.k.**

Investigate the failure path 1-2-3-4 in Figure 8-7(b) for the axial force.

Tensile Rupture

$$A_{gt} = (12.0 \text{ in.} - 1.50 \text{ in.})t_p$$
$$= 10.5t_p \text{ in.}^2$$

$$A_{nt} = 10.5t_p - 3.50(^{15}\!/_{16} \text{ in.} + ^{1}\!/_{16} \text{ in.})t_p$$
$$= 7.00t_p \text{ in.}^2$$

$$U_{bs} = 1.0$$

$$U_{bs}F_u A_{nt} = 1.0(58 \text{ ksi})(7.00t_p \text{ in.})$$
$$= 406t_p \text{ kips}$$

Shear Yielding

$$A_{gv} = 2.50t_p \text{ in.}^2$$

$$0.60F_y A_{gv} = 0.60(36 \text{ ksi})(2.50t_p \text{ in.}^2)$$
$$= 54.0t_p \text{ kips}$$

Shear Rupture

$$A_{nv} = 2.50t_p - 0.50(^{15}\!/_{16} \text{ in.} + ^{1}\!/_{16} \text{ in.})t_p$$
$$= 2.50t_p - 0.500t_p$$
$$= 2.00t_p \text{ in.}^2$$

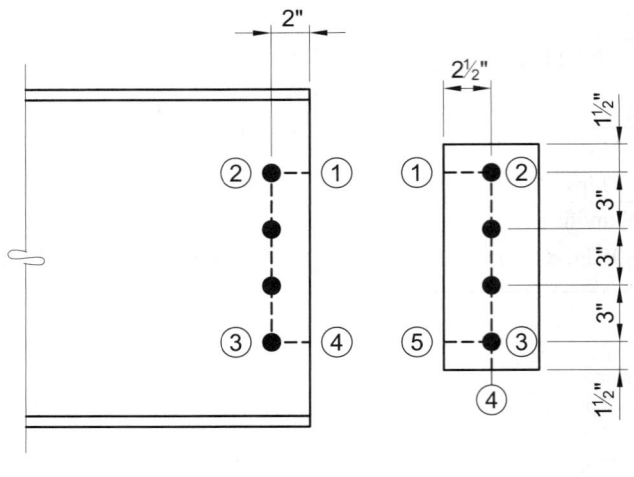

(a) Beam web   *(b) Single plate*

Fig. 8-7. Block shear failure paths for Example 8.4.2.

8.4 DESIGN EXAMPLES

$$0.60F_u A_{nv} = 0.60(58 \text{ ksi})(2.00t_p \text{ in.}^2)$$
$$= 69.6t_p \text{ kips}$$

From AISC *Specification* Equation J4-5:

LRFD	ASD
$\phi R_n = \phi[0.60F_u A_{nv} + U_{bs}F_u A_{nt}]$ $\leq \phi[0.60F_y A_{gv} + U_{bs}F_u A_{nt}]$	$\dfrac{R_n}{\Omega} = \dfrac{0.60F_u A_{nv} + U_{bs}F_u A_{nt}}{\Omega}$ $\leq \dfrac{0.60F_y A_{gv} + U_{bs}F_u A_{nt}}{\Omega}$
$\dfrac{\phi[0.60F_u A_{nv} + U_{bs}F_u A_{nt}]}{t_p}$ $= 0.75(69.6 \text{ kip/in.} + 406 \text{ kip/in.})$ $= 357 \text{ kip/in.}$	$\dfrac{0.60F_u A_{nv} + U_{bs}F_u A_{nt}}{\Omega\, t_p}$ $= \dfrac{69.6 \text{ kip/in.} + 406 \text{ kip/in.}}{2.00}$ $= 238 \text{ kip/in.}$
$\dfrac{\phi[0.60F_y A_{gv} + U_{bs}F_u A_{nt}]}{t_p}$ $= 0.75(54.0 \text{ kip/in.} + 406 \text{ kip/in.})$ $= 345 \text{ kip/in.}$	$\dfrac{0.60F_y A_{gv} + U_{bs}F_u A_{nt}}{\Omega}$ $= \dfrac{54.0 \text{ kip/in.} + 406 \text{ kip/in.}}{2.00}$ $= 230 \text{ kip/in.}$
Therefore, use $\phi R_n/t_p = 345$ kip/in.	Therefore, use $R_n/\Omega t_p = 230$ kip/in.
$t_{pa\,min} = \dfrac{R_{ua}}{\phi R_n / t_p}$ $= \dfrac{51.0 \text{ kips}}{345 \text{ kip/in.}}$ $= 0.148 \text{ in.}$	$t_{pa\,min} = \dfrac{R_{aa}}{R_n / \Omega t_p}$ $= \dfrac{35.7 \text{ kips}}{230 \text{ kip/in.}}$ $= 0.155 \text{ in.}$
The minimum required plate thickness due to both forces is:	The minimum required plate thickness due to both forces is:
$t_{p\,min} = \sqrt{(t_{pv\,min})^2 + (t_{pa\,min})^2}$ $= \sqrt{(0.0805 \text{ in.})^2 + (0.148 \text{ in.})^2}$ $= 0.168 \text{ in.} < 0.375 \text{ in.}$ **o.k.**	$t_{p\,min} = \sqrt{(t_{pv\,min})^2 + (t_{pa\,min})^2}$ $= \sqrt{(0.0785 \text{ in.})^2 + (0.155 \text{ in.})^2}$ $= 0.174 \text{ in.} < 0.375 \text{ in.}$ **o.k.**

Investigate the failure path 1-2-3-5 on Figure 8.7(b) for the shear force.

For the shear force conservatively use the minimum thickness determined from path 1-2-3-4.

LRFD	ASD
$t_{pv\ min} = 0.0805$ in.	$t_{pv\ min} = 0.0785$ in.

Investigate the failure path 1-2-3-5 in Figure 8-7(b) for the axial force.

Tensile Rupture

$$A_{gt} = (12.0 \text{ in.} - 3.00 \text{ in.})t_p$$
$$= 9.00t_p \text{ in.}^2$$

$$A_{nt} = 9.00t_p - 3(^{15}\!/_{16} \text{ in.} + ^1\!/_{16} \text{ in.})t_p$$
$$= 6.00t_p \text{ in.}^2$$

$$U_{bs} = 1.0$$

$$U_{bs}F_u A_{nt} = 1.0(58 \text{ ksi})\left(6.00t_p \text{ in.}^2\right)$$
$$= 348t_p \text{ kips}$$

Shear Yielding

$$A_{gv} = 2(2.50 \text{ in.})t_p$$
$$= 5.00t_p \text{ in.}^2$$

$$0.60F_y A_{gv} = 0.60(36 \text{ ksi})\left(5.00t_p \text{ in.}^2\right)$$
$$= 108t_p \text{ kips}$$

Shear Rupture

$$A_{nv} = 5.00t_p - (^{15}\!/_{16} \text{ in.} + ^1\!/_{16} \text{ in.})t_p$$
$$= 5.00t_p - t_p$$
$$= 4.00t_p \text{ in.}^2$$

$$0.60F_u A_{nv} = 0.60(58 \text{ ksi})\left(4.00t_p \text{ in.}^2\right)$$
$$= 139t_p \text{ kips}$$

From AISC *Specification* Equation J4-5:

LRFD	ASD
$\phi R_n = \phi\left[0.60F_u A_{nv} + U_{bs}F_u A_{nt}\right]$ $\leq \phi\left[0.60F_y A_{gv} + U_{bs}F_u A_{nt}\right]$	$\dfrac{R_n}{\Omega} = \dfrac{0.60F_u A_{nv} + U_{bs}F_u A_{nt}}{\Omega}$ $\leq \dfrac{0.60F_y A_{gv} + U_{bs}F_u A_{nt}}{\Omega}$

LRFD	ASD
$\dfrac{\phi\left[0.60F_u A_{nv}+U_{bs}F_u A_{nt}\right]}{t_p}$	$\dfrac{0.60F_u A_{nv}+U_{bs}F_u A_{nt}}{\Omega\, t_p}$
$= 0.75(139 \text{ kip/in.} + 348 \text{ kip/in.})$	$= \dfrac{139 \text{ kip/in.} + 348 \text{ kip/in.}}{2.00}$
$= 365$ kip/in.	$= 244$ kip/in.
$\dfrac{\phi\left[0.60F_y A_{gv}+U_{bs}F_u A_{nt}\right]}{t_p}$	$\dfrac{0.60F_y A_{gv}+U_{bs}F_u A_{nt}}{\Omega\, t_p}$
$= 0.75(108 \text{ kip/in.} + 348 \text{ kip/in.})$	$= \dfrac{108 \text{ kip/in.} + 348 \text{ kip/in.}}{2.00}$
$= 342$ kip/in.	$= 228$ kip/in.
Therefore, use $\dfrac{\phi R_n}{t_p} = 342$ kip/in.	Therefore, use $\dfrac{R_n}{\Omega\, t_p} = 228$ kip/in.
$t_{pa\,min} = \dfrac{R_{ua}}{\phi R_n / t_p}$	$t_{pa\,min} = \dfrac{R_{aa}}{(R_n/\Omega)/t_p}$
$= \dfrac{51.0 \text{ kips}}{342 \text{ kip/in.}}$	$= \dfrac{35.7 \text{ kips}}{228 \text{ kips}}$
$= 0.149$ in.	$= 0.157$ in.
The minimum required plate thickness due to both forces is:	The minimum required plate thickness due to both forces is:
$t_{p\,min} = \sqrt{(t_{pv\,min})^2 + (t_{pa\,min})^2}$	$t_{p\,min} = \sqrt{(t_{pv\,min})^2 + (t_{pa\,min})^2}$
$= \sqrt{(0.0805 \text{ in.})^2 + (0.149 \text{ in.})^2}$	$= \sqrt{(0.0785 \text{ in.})^2 + (0.157 \text{ in.})^2}$
$= 0.169$ in. < 0.375 in. **o.k.**	$= 0.176$ in. < 0.375 in. **o.k.**

Block Shear Rupture and Shear Rupture of the Beam Web

In evaluating these limit states, the minimum beam web thickness is calculated independently for each force, and the resulting thicknesses are combined for the interaction of forces using an elliptical approximation.

Block shear rupture can occur on the beam web due to the axial load on the beam (block shear rupture cannot occur due to shear on an uncoped beam). For the block shear calculations, investigate the failure path 1-2-3-4 as shown in Figure 8-7(a). A beam underrun of ¼ in. will be assumed.

Tensile Rupture

$A_{gt} = 3(3.00 \text{ in.}) t_w$

$\quad\quad = 9.00 t_w$ in.^{2}

$$A_{nt} = 9.00t_w \text{ in.}^2 - 3(^{15}/_{16} \text{ in.} + ^{1}/_{16} \text{ in.})t_w$$
$$= 6.00t_w \text{ in.}^2$$

$$U_{bs} = 1.0$$

$$F_u A_{nt} = U_{bs}(65 \text{ ksi})(6.00t_w \text{ in.}^2)$$
$$= 390t_w \text{ kips}$$

Shear Yielding

$$A_{gv} = 2(2.00 \text{ in.} - \tfrac{1}{4} \text{ in.})t_w$$
$$= 3.50t_w \text{ in.}^2$$

$$0.60F_y A_{gv} = 0.60(50 \text{ ksi})(3.50t_w \text{ in.}^2)$$
$$= 105t_w \text{ kips}$$

Shear Rupture

$$A_{nv} = 3.50t_w \text{ in.}^2 - 2(0.50)(^{15}/_{16} \text{ in.} + ^{1}/_{16} \text{ in.})t_w$$
$$= 2.50t_w \text{ in.}^2$$

$$0.60F_u A_{nv} = 0.60(65 \text{ ksi})(2.50t_w \text{ in.}^2)$$
$$= 97.5t_w \text{ kips}$$

From AISC *Specification* Equation J4-5:

LRFD	ASD
$\phi R_n = \phi[0.60F_u A_{nv} + U_{bs} F_u A_{nt}]$ $\leq \phi[0.60F_y A_{gv} + U_{bs} F_u A_{nt}]$	$\dfrac{R_n}{\Omega} = \dfrac{0.60F_u A_{nv} + U_{bs} F_u A_{nt}}{\Omega}$ $\leq \dfrac{0.60F_y A_{gv} + U_{bs} F_u A_{nt}}{\Omega}$
$\dfrac{\phi[0.60F_u A_{nv} + U_{bs} F_u A_{nt}]}{t_w}$ $= 0.75(97.5 \text{ kip/in.} + 390 \text{ kip/in.})$ $= 366 \text{ kip/in.}$	$\dfrac{0.60F_u A_{nv} + U_{bs} F_u A_{nt}}{\Omega t_w}$ $= \dfrac{97.5 \text{ kip/in.} + 390 \text{ kip/in.}}{2.00}$ $= 244 \text{ kip/in.}$
$\dfrac{\phi[0.60F_y A_{gv} + U_{bs} F_u A_{nt}]}{t_w}$ $= 0.75(105 \text{ kip/in.} + 390 \text{ kip/in.})$ $= 371 \text{ kip/in.}$	$\dfrac{0.60F_y A_{gv} + U_{bs} F_u A_{nt}}{\Omega t_w}$ $= \dfrac{105 \text{ kip/in.} + 390 \text{ kip/in.}}{2.00}$ $= 248 \text{ kip/in.}$

8.4 DESIGN EXAMPLES

LRFD	ASD
Therefore, use $\phi R_n = 366$ kip/in.	Therefore, use $\dfrac{R_n}{\Omega} = 244$ kip/in.
$t_{wa\ min} = \dfrac{R_{ua}}{\phi R_n / t_p}$	$t_{wa\ min} = \dfrac{R_{aa}}{R_n / \Omega t_p}$
$= \dfrac{51.0 \text{ kips}}{366 \text{ kip/in.}}$	$= \dfrac{35.7 \text{ kips}}{244 \text{ kip/in.}}$
$= 0.139$ in. < 0.355 in. **o.k.**	$= 0.146$ in. < 0.355 in. **o.k.**

Also check shear rupture of the beam web according to AISC *Specification* Section J4.2, as follows:

LRFD	ASD
$t_{wv\ min} = \dfrac{R_{uv}}{\phi R_n / t_w}$	$t_{wv\ min} = \dfrac{R_{av}}{(R_n / \Omega) / t_w}$
$= \dfrac{R_{uv}}{\phi(0.60) F_u d_{net}}$	$= \dfrac{R_{av}}{[(0.60) F_u / \Omega] d_{net}}$
$= \dfrac{20.7 \text{ kips}}{0.75(0.60)(65 \text{ ksi})}$	$= \dfrac{13.5 \text{ kips}}{[(0.60)(65 \text{ ksi})/2.00]}$
$\times \dfrac{1}{[18.0 \text{ in.} - 4(^{15}/_{16} \text{ in.} + ^{1}/_{16} \text{ in.})]}$	$\times \dfrac{1}{[18.0 \text{ in.} - 4(^{15}/_{16} \text{ in.} + ^{1}/_{16} \text{ in.})]}$
$= 0.0505$ in.	$= 0.0495$ in.
$t_{w\ min} = \sqrt{(t_{wv\ min})^2 + (t_{wa\ min})^2}$	$t_{w\ min} = \sqrt{(t_{wv\ min})^2 + (t_{wa\ min})^2}$
$= \sqrt{(0.0505 \text{ in.})^2 + (0.139 \text{ in.})^2}$	$= \sqrt{(0.0495 \text{ in.})^2 + (0.146 \text{ in.})^2}$
$= 0.148$ in. < 0.355 in. **o.k.**	$= 0.154$ in. < 0.355 in. **o.k.**

Plate Required Flexural Strength

LRFD	ASD
$M_u = R_{uv} e$	$M_u = R_{av} e$
$= 20.7 \text{ kips}(2.50 \text{ in.})$	$= 13.5 \text{ kips}(2.50 \text{ in.})$
$= 51.8$ kip-in.	$= 33.8$ kip-in.

Plate Available Flexural Strength

The Von Mises criterion will be used to reduce the permitted bending strength to account for the shear stress present. The axial tension stress present is deducted directly from the bending stress. Rearranging AISC *Manual* Equation 9-1 to solve for f_x, eliminating the perpendicular axial component ($f_t = 0$), setting $f_x = F_{cr}, f_e = F_y$, and $f_{xy} = f_v$, and subtracting the axial tension stress, results in the following equations. A factor, α, has been incorporated into the equation. This is the factor used in the AISC *Specification* when calculating loads or stresses at the ultimate strength level.

LRFD	ASD
$\alpha = 1.00$	$\alpha = 1.60$
$F_{cr} = \sqrt{F_y^2 - 3\alpha f_v^2} - \alpha f_t$	$F_{cr} = \sqrt{F_y^2 - 3\alpha f_v^2} - \alpha f_t$
$= \sqrt{(36 \text{ ksi})^2 - 3(1.00)\left(\dfrac{20.7 \text{ kips}}{12.0 \text{ in.}(\frac{3}{8} \text{ in.})}\right)^2}$	$= \sqrt{(36 \text{ ksi})^2 - 3(1.60)\left(\dfrac{13.5 \text{ kips}}{12.0 \text{ in.}(\frac{3}{8} \text{ in.})}\right)^2}$
$\quad - \dfrac{1.00(51.0 \text{ kips})}{12.0 \text{ in.}(\frac{3}{8} \text{ in.})}$	$\quad - \dfrac{1.60(35.7 \text{ kips})}{12.0 \text{ in.}(\frac{3}{8} \text{ in.})}$
$= 23.8$ ksi	$= 22.7$ ksi

$$Z = \frac{\frac{3}{8} \text{ in.}(12.0 \text{ in.})^2}{4}$$

$$= 13.5 \text{ in.}^3$$

LRFD	ASD
$\phi M_n = \phi F_{cr} Z$	$\dfrac{M_n}{\Omega} = \dfrac{F_{cr} Z}{\Omega}$
$= 0.90(23.8 \text{ ksi})(13.5 \text{ in.}^3)$	
$= 289$ kip-in. > 51.8 kip-in. **o.k.**	$= \dfrac{(22.7 \text{ ksi})(13.5 \text{ in.}^3)}{1.67}$
	$= 184$ kip-in. > 33.8 kip-in. **o.k.**

The selected ⅜-in.-thick plate is acceptable.

Weld Strength

The welds are sized as ⅝t_p to develop the strength of the plate as discussed in Part 10 of the AISC *Manual*.

$⅝ t_p = ⅝$ (⅜ in.)

$\quad = 0.234$ in.

Use a ¼-in. fillet weld to connect the single plate to the column flange.

The single-plate connection shown in Figure 8-6 is adequate to resist the collector forces determined in Example 8.4.1.

PART 8 REFERENCES

Burmeister, S. and Jacobs, W.P. (2008), "Under Foot: Horizontal Floor Diaphragm Load Effects on Composite Beam Design," *Modern Steel Construction*, AISC, December.

Helwig, T.A. and Yura, J.A. (1999), "Torsional Bracing of Columns," *Journal of Structural Engineering*, ASCE, Vol. 125, No. 5, pp. 547–555.

Sabelli, R., Sabol, T.A. and Easterling, S.W. (2011), *NEHRP Seismic Design Technical Brief No. 5 Seismic Design of Composite Steel Deck and Concrete-filled Diaphragms: A Guide for Practicing Engineers,* NIST GCR 11-917-10, NEHRP Consultants Joint Venture, partnership of the Applied Technology Council and the Consortium of Universities for Research in Earthquake Engineering, National Institute of Standards and Technology, Gaithersburg, MD.

PART 9

PROVISIONS AND STANDARDS

9.1 SEISMIC PROVISIONS FOR STRUCTURAL STEEL BUILDINGS,
     JUNE 22, 2010 . . . . 9.1–i
     Preface 9.1–iii
     Table of Contents 9.1–vii
     Cross Reference 9.1–xxix
     Symbols 9.1–xxxiii
     Glossary 9.1–xxxix
     Acronyms 9.1–xlv
     Provisions 9.1–1
     Commentary 9.1–135

9.2 PREQUALIFIED CONNECTIONS FOR SPECIAL AND
     INTERMEDIATE STEEL MOMENT FRAMES FOR
     SEISMIC APPLICATIONS, 2010 . . . . 9.2–i
     Preface 9.2–iii
     Table of Contents 9.2–v
     Symbols 9.2–xiii
     Glossary 9.2–xix
     Standard 9.2–1
     Commentary 9.2–89

ANSI/AISC 341-10
An American National Standard

Seismic Provisions for Structural Steel Buildings

June 22, 2010

Supersedes the
Seismic Provisions for Structural Steel Buildings
dated March 9, 2005,
Supplement No. 1 dated November 16, 2005,
and all previous versions

Approved by the AISC Committee on Specifications

AMERICAN INSTITUTE OF STEEL CONSTRUCTION
One East Wacker Drive, Suite 700
Chicago, Illinois 60601-1802

Copyright © 2010

by

American Institute of Steel Construction

All rights reserved. This book or any part thereof must not be reproduced in any form without the written permission of the publisher.

The AISC logo is a registered trademark of AISC.

The information presented in this publication has been prepared in accordance with recognized engineering principles and is for general information only. While it is believed to be accurate, this information should not be used or relied upon for any specific application without competent professional examination and verification of its accuracy, suitability and applicability by a licensed professional engineer, designer or architect. The publication of the material contained herein is not intended as a representation or warranty on the part of the American Institute of Steel Construction or of any other person named herein, that this information is suitable for any general or particular use or of freedom from infringement of any patent or patents. Anyone making use of this information assumes all liability arising from such use.

Caution must be exercised when relying upon other specifications and codes developed by other bodies and incorporated by reference herein since such material may be modified or amended from time to time subsequent to the printing of this edition. The Institute bears no responsibility for such material other than to refer to it and incorporate it by reference at the time of the initial publication of this edition.

Printed in the United States of America

First Printing: September 2011
Second Printing: January 2012
Third Printing: September 2012

PREFACE

This Preface is not a part of ANSI/AISC 341-10, *Seismic Provisions for Structural Steel Buildings,* but is included for informational purposes only.

The AISC *Specification for Structural Steel Buildings* (ANSI/AISC 360-10) is intended to cover common design criteria. Accordingly, it is not feasible for it to also cover all of the special and unique problems encountered within the full range of structural design practice. This document, the AISC *Seismic Provisions for Structural Steel Buildings* (ANSI/AISC 341-10) (hereafter referred to as the Provisions) is a separate consensus standard that addresses one such topic: the design and construction of structural steel and composite structural steel/reinforced concrete building systems for high-seismic applications.

A list of Symbols and a Glossary are part of this document. Terms that appear in the Glossary are generally italicized where they first appear in a sub-section, throughout these Provisions. A nonmandatory Commentary with background information is also provided. Nonmandatory user notes are interspersed throughout these Provisions to provide guidance on the application of the document.

This edition of the AISC *Seismic Provisions for Structural Steel Buildings* was developed in concert with both ANSI/AISC 360-10 and ASCE/SEI 7-10, *Minimum Design Loads for Buildings and Other Structures.* This will allow these Provisions to be incorporated by reference into the 2012 IBC, which will use ASCE/SEI 7-10 as its basis of design for loadings.

Some of the most significant modifications to this edition of these Provisions are related to format. The organization of the chapters has been changed to be more consistent with that of ANSI/AISC 360-10. In the 2005 edition, these Provisions separated the requirements for structural steel buildings from that of composite structural steel/reinforced concrete construction into two parts. In this edition of the Provisions, Part I and Part II have been combined into one document. In addition, each structural system is presented in a unified manner with parallel headings that will ease comparison of requirements between systems and application of the document. A Cross Reference listing is provided comparing the 2010 to the 2005 version of the Provisions.

A number of significant technical modifications have also been made since the 2005 edition of these Provisions, including the following:

- Clarifying the intended combination of this document with the provisions of ACI 318 for composite construction systems
- Establishing a new chapter on analysis requirements that applies to all systems
- Adding terms to clearly identify the level of ductile response capable of various members in the seismic force resisting system (SFRS)
- Adding language to clarify the design of members and connections that are not part of the SFRS for deformation compatibility
- Including a discussion of the "Basis of Design" that explains the intended seismic response characteristics of each structural system
- Improving the consistency, clarity and completeness of how each structural system treats all aspects of the seismic design and detailing
- Adding requirements for two cantilever column systems to be consistent with other systems in these Provisions and the seismic design parameters ASCE/SEI 7-10

- Adding analysis requirements to address the inelastic response of special concentrically braced frames
- Modifying the connection requirements for braced frame systems to ensure that the expected deformation demands can be accommodated
- Adding requirements for the use of box-shaped link beams in eccentrically braced frames
- Adding requirements for the use of perforated plates in special plate shear walls
- Significantly increasing the detail for the design requirements of composite systems, such that they are consistent with structural steel systems
- Incorporating AWS D1.8/D1.8M by reference for welding related issues

The AISC Committee on Specifications, Task Committee 9—Seismic Design is responsible for the ongoing development of these Provisions. The AISC Committee on Specifications gives final approval of the document through an ANSI-accredited balloting process, and has enhanced these Provisions through careful scrutiny, discussion and suggestions for improvement. The contributions of these two groups, comprising well more than 80 structural engineers with experience from throughout the structural steel industry, is gratefully acknowledged. AISC further acknowledges the significant contributions of several groups to the completion of this document: the Building Seismic Safety Council (BSSC), the Federal Emergency Management Agency (FEMA), the National Science Foundation (NSF), and the Structural Engineers Association of California (SEAOC).

The reader is cautioned that professional judgment must be exercised when data or recommendations in these provisions are applied, as described more fully in the disclaimer notice preceding the Preface.

This specification was approved by the AISC Committee on Specifications:

James M. Fisher, Chairman	Mark V. Holland
Edward E. Garvin, Vice Chairman	Ronald J. Janowiak
Hansraj G. Ashar	Richard C. Kaehler
William F. Baker	Lawrence A. Kloiber
John M. Barsom	Lawrence F. Kruth
William D. Bast	Jay W. Larson
Reidar Bjorhovde	Roberto T. Leon
Roger L. Brockenbrough	James O. Malley
Gregory G. Deierlein	Sanjeev R. Malushte
Bruce R. Ellingwood	David L. McKenzie
Michael D. Engelhardt	Duane K. Miller
Shu-Jin Fang	Larry S. Muir
Steven J. Fenves	Thomas M. Murray
John W. Fisher	R. Shankar Nair
Theodore V. Galambos	Jack E. Petersen
Louis F. Geschwindner	Douglas A. Rees-Evans
Lawrence G. Griffis	Thomas A. Sabol
John L. Gross	Robert E. Shaw, Jr.
Jerome F. Hajjar	Donald R. Sherman
Patrick M. Hassett	W. Lee Shoemaker
Tony C. Hazel	William A. Thornton

Raymond H.R. Tide
Chia-Ming Uang

Donald W. White
Cynthia J. Duncan, Secretary

The Committee gratefully acknowledges the following task committee (TC 9—Seismic Design) for their development of this document.

James O. Malley, Chairman
C. Mark Saunders, Vice Chairman
Michel Bruneau
Gregory G. Deierlein
Richard M. Drake
Michael D. Engelhardt
Timothy P. Fraser
Subhash C. Goel
Jerome F. Hajjar
Ronald O. Hamburger
James R. Harris
Patrick M. Hassett
John D. Hooper
Brian T. Knight
Keith Landwehr

Roberto T. Leon
Sanjeev R. Malushte
Bonnie E. Manley
Clarkson W. Pinkham
John A. Rolfes
Rafael Sabelli
Thomas A. Sabol
Bahram M. Shahrooz
Robert E. Shaw, Jr.
W. Lee Shoemaker
Kurt D. Swensson
Robert Tremblay
Jamie Winans
Cynthia J. Duncan, Secretary
Leigh Arber, Secretary

TABLE OF CONTENTS

CROSS REFERENCE 9.1–xxix
SYMBOLS 9.1–xxxiii
GLOSSARY 9.1–xxxix
ACRONYMS 9.1–xlv

PROVISIONS

A. GENERAL REQUIREMENTS 9.1–1
   A1. Scope 9.1–1
   A2. Referenced Specifications, Codes and Standards 9.1–2
   A3. Materials 9.1–2
      1. Material Specifications 9.1–2
      2. Expected Material Strength 9.1–3
      3. Heavy Sections 9.1–5
      4. Consumables for Welding 9.1–5
      4a. Seismic Force Resisting System Welds 9.1–5
      4b. Demand Critical Welds 9.1–5
      5. Concrete and Steel Reinforcement 9.1–6
   A4. Structural Design Drawings and Specifications 9.1–6
      1. General 9.1–6
      2. Steel Construction 9.1–6
      3. Composite Construction 9.1–7

B. GENERAL DESIGN REQUIREMENTS 9.1–8
   B1. General Seismic Design Requirements 9.1–8
   B2. Loads and Load Combinations 9.1–8
   B3. Design Basis 9.1–9
      1. Required Strength 9.1–9
      2. Available Strength 9.1–9
   B4. System Type 9.1–9

C. ANALYSIS 9.1–10
   C1. General Requirements 9.1–10
   C2. Additional Requirements 9.1–10
   C3. Nonlinear Analysis 9.1–10

D. GENERAL MEMBER AND CONNECTION DESIGN REQUIREMENTS 9.1–11
   D1. Member Requirements 9.1–11
      1. Classification of Sections for Ductility 9.1–11
      1a. Section Requirements for Ductile Members 9.1–11

		1b.	Width-to-Thickness Limitations of Steel and Composite Sections 9.1–11
		2.	Stability Bracing of Beams 9.1–14
		2a.	Moderately Ductile Members 9.1–14
		2b.	Highly Ductile Members 9.1–15
		2c.	Special Bracing at Plastic Hinge Locations 9.1–15
		3.	Protected Zones 9.1–16
		4.	Columns 9.1–16
		4a.	Required Strength 9.1–16
		4b.	Encased Composite Columns 9.1–17
		4c.	Filled Composite Columns 9.1–19
		5.	Composite Slab Diaphragms 9.1–19
		5a.	Load Transfer 9.1–19
		5b.	Nominal Shear Strength 9.1–19
	D2.	Connections 9.1–20
		1.	General 9.1–20
		2.	Bolted Joints 9.1–20
		3.	Welded Joints 9.1–21
		4.	Continuity Plates and Stiffeners 9.1–21
		5.	Column Splices 9.1–21
		5a.	Location of Splices 9.1–21
		5b.	Required Strength 9.1–21
		5c.	Required Shear Strength 9.1–22
		5d.	Structural Steel Splice Configurations 9.1–22
		5e.	Splices in Encased Composite Columns 9.1–22
		6.	Column Bases 9.1–22
		6a.	Required Axial Strength 9.1–23
		6b.	Required Shear Strength 9.1–23
		6c.	Required Flexural Strength 9.1–24
		7.	Composite Connections 9.1–24
		8.	Steel Anchors 9.1–26
	D3.	Deformation Compatibility of Non-SFRS Members and Connections 9.1–26
	D4.	H-Piles 9.1–26
		1.	Design Requirements 9.1–26
		2.	Battered H-Piles 9.1–26
		3.	Tension 9.1–26
		4.	Protected Zone 9.1–27
E.	**MOMENT-FRAME SYSTEMS** 9.1–28
	E1.	Ordinary Moment Frames (OMF) 9.1–28
		1.	Scope 9.1–28
		2.	Basis of Design 9.1–28
		3.	Analysis 9.1–28

		4.	System Requirements	9.1–28
		5.	Members	9.1–28
		5a.	Basic Requirements	9.1–28
		5b.	Protected Zones	9.1–28
		6.	Connections	9.1–29
		6a.	Demand Critical Welds	9.1–29
		6b.	FR Moment Connections	9.1–29
		6c.	PR Moment Connections	9.1–30
	E2.		Intermediate Moment Frames (IMF)	9.1–30
		1.	Scope	9.1–30
		2.	Basis of Design	9.1–31
		3.	Analysis	9.1–31
		4.	System Requirements	9.1–31
		4a.	Stability Bracing of Beams	9.1–31
		5.	Members	9.1–31
		5a.	Basic Requirements	9.1–31
		5b.	Beam Flanges	9.1–31
		5c.	Protected Zones	9.1–32
		6.	Connections	9.1–32
		6a.	Demand Critical Welds	9.1–32
		6b.	Beam-to-Column Connection Requirements	9.1–32
		6c.	Conformance Demonstration	9.1–33
		6d.	Required Shear Strength	9.1–33
		6e.	Panel Zone	9.1–33
		6f.	Continuity Plates	9.1–34
		6g.	Column Splices	9.1–34
	E3.		Special Moment Frames (SMF)	9.1–34
		1.	Scope	9.1–34
		2.	Basis of Design	9.1–34
		3.	Analysis	9.1–34
		4.	System Requirements	9.1–34
		4a.	Moment Ratio	9.1–34
		4b.	Stability Bracing of Beams	9.1–36
		4c.	Stability Bracing at Beam-to-Column Connections	9.1–37
		5.	Members	9.1–38
		5a.	Basic Requirements	9.1–38
		5b.	Beam Flanges	9.1–38
		5c.	Protected Zones	9.1–38
		6.	Connections	9.1–38
		6a.	Demand Critical Welds	9.1–38
		6b.	Beam-to-Column Connections	9.1–39
		6c.	Conformance Demonstration	9.1–39
		6d.	Required Shear Strength	9.1–40

		6e.	Panel Zone 9.1–40

- 6e. Panel Zone 9.1–40
- 6f. Continuity Plates 9.1–41
- 6g. Column Splices 9.1–43

E4. Special Truss Moment Frames (STMF) 9.1–43
  1. Scope 9.1–43
  2. Basis of Design 9.1–43
  3. Analysis 9.1–43
  3a. Special Segment 9.1–43
  3b. Nonspecial Segment 9.1–44
  4. System Requirements 9.1–44
  4a. Special Segment 9.1–44
  4b. Stability Bracing of Trusses 9.1–44
  4c. Stability Bracing of Truss-to-Column Connections 9.1–45
  4d. Stiffness of Stability Bracing 9.1–45
  5. Members 9.1–45
  5a. Special Segment Members 9.1–45
  5b. Expected Vertical Shear Strength of Special Segment 9.1–45
  5c. Width-to-Thickness Limitations 9.1–46
  5d. Built-Up Chord Members 9.1–46
  5e. Protected Zones 9.1–46
  6. Connections 9.1–46
  6a. Demand Critical Welds 9.1–46
  6b. Connections of Diagonal Web Members in the
       Special Segment 9.1–47
  6c. Column Splices 9.1–47

E5. Ordinary Cantilever Column Systems (OCCS) 9.1–47
  1. Scope 9.1–47
  2. Basis of Design 9.1–47
  3. Analysis 9.1–47
  4. System Requirements 9.1–47
  4a. Columns 9.1–47
  4b. Stability Bracing of Columns 9.1–47
  5. Members 9.1–47
  5a. Basic Requirements 9.1–47
  5b. Column Flanges 9.1–48
  5c. Protected Zones 9.1–48
  6. Connections 9.1–48
  6a. Demand Critical Welds 9.1–48
  6b. Column Bases 9.1–48

E6. Special Cantilever Column Systems (SCCS) 9.1–48
  1. Scope 9.1–48
  2. Basis of Design 9.1–48
  3. Analysis 9.1–48

		4. System Requirements	9.1–48
		4a. Columns	9.1–48
		4b. Stability Bracing of Columns	9.1–48
		5. Members	9.1–48
		5a. Basic Requirements	9.1–48
		5b. Column Flanges	9.1–48
		5c. Protected Zones	9.1–49
		6. Connections	9.1–49
		6a. Demand Critical Welds	9.1–49
		6b. Column Bases	9.1–49
F.	**BRACED-FRAME AND SHEAR-WALL SYSTEMS**		9.1–50
	F1.	Ordinary Concentrically Braced Frames (OCBF)	9.1–50
		1. Scope	9.1–50
		2. Basis of Design	9.1–50
		3. Analysis	9.1–50
		4. System Requirements	9.1–50
		4a. V-Braced and Inverted V-Braced Frames	9.1–50
		4b. K-Braced Frames	9.1–51
		5. Members	9.1–51
		5a. Basic Requirements	9.1–51
		5b. Slenderness	9.1–51
		6. Connections	9.1–51
		6a. Diagonal Brace Connections	9.1–51
		7. Ordinary Concentrically Braced Frames above Seismic Isolation Systems	9.1–52
		7a. System Requirements	9.1–52
		7b. Members	9.1–52
	F2.	Special Concentrically Braced Frames (SCBF)	9.1–52
		1. Scope	9.1–52
		2. Basis of Design	9.1–52
		3. Analysis	9.1–52
		4. System Requirements	9.1–53
		4a. Lateral Force Distribution	9.1–53
		4b. V- and Inverted V-Braced Frames	9.1–53
		4c. K-Braced Frames	9.1–54
		4d. Tension-Only Frames	9.1–54
		5. Members	9.1–54
		5a. Basic Requirements	9.1–54
		5b. Diagonal Braces	9.1–54
		5c. Protected Zones	9.1–55
		6. Connections	9.1–55
		6a. Demand Critical Welds	9.1–55
		6b. Beam-to-Column Connections	9.1–55

		6c.	Required Strength of Brace Connections	9.1–56
		6d.	Column Splices	9.1–57
	F3.		Eccentrically Braced Frames (EBF)	9.1–57
		1.	Scope	9.1–57
		2.	Basis of Design	9.1–57
		3.	Analysis	9.1–58
		4.	System Requirements	9.1–58
		4a.	Link Rotation Angle	9.1–58
		4b.	Bracing of Link	9.1–59
		5.	Members	9.1–59
		5a.	Basic Requirements	9.1–59
		5b.	Links	9.1–59
		5c.	Protected Zones	9.1–62
		6.	Connections	9.1–62
		6a.	Demand Critical Welds	9.1–62
		6b.	Beam-to-Column Connections	9.1–63
		6c.	Diagonal Brace Connections	9.1–63
		6d.	Column Splices	9.1–63
		6e.	Link-to-Column Connections	9.1–64
	F4.		Buckling-Restrained Braced Frames (BRBF)	9.1–65
		1.	Scope	9.1–65
		2.	Basis of Design	9.1–65
		2a.	Brace Strength	9.1–65
		3.	Analysis	9.1–66
		4.	System Requirements	9.1–67
		4a.	V- and Inverted V-Braced Frames	9.1–67
		4b.	K-Braced Frames	9.1–67
		5.	Members	9.1–67
		5a.	Basic Requirements	9.1–67
		5b.	Diagonal Braces	9.1–67
		5c.	Protected Zones	9.1–69
		6.	Connections	9.1–69
		6a.	Demand Critical Welds	9.1–69
		6b.	Beam-to-Column Connections	9.1–69
		6c.	Diagonal Brace Connections	9.1–69
		6d.	Column Splices	9.1–70
	F5.		Special Plate Shear Walls (SPSW)	9.1–70
		1.	Scope	9.1–70
		2.	Basis of Design	9.1–71
		3.	Analysis	9.1–71
		4.	System Requirements	9.1–71
		4a.	Stiffness of Boundary Elements	9.1–71
		4b.	HBE-to-VBE Connection Moment Ratio	9.1–72

		4c.	Bracing 9.1–72
		4d.	Openings in Webs 9.1–72
		5.	Members 9.1–72
		5a.	Basic Requirements 9.1–72
		5b.	Webs 9.1–72
		5c.	Protected Zone 9.1–73
		6.	Connections 9.1–73
		6a.	Demand Critical Welds 9.1–73
		6b.	HBE-to-VBE Connections 9.1–73
		6c.	Connections of Webs to Boundary Elements 9.1–73
		6d.	Column Splices 9.1–73
		7.	Perforated Webs 9.1–74
		7a.	Regular Layout of Circular Perforations 9.1–74
		7b.	Reinforced Corner Cut-Out 9.1–75

G. COMPOSITE MOMENT-FRAME SYSTEMS 9.1–77

	G1.	Composite Ordinary Moment Frames (C-OMF) 9.1–77
		1. Scope 9.1–77
		2. Basis of Design 9.1–77
		3. Analysis 9.1–77
		4. System Requirements 9.1–77
		5. Members 9.1–78
		5a. Protected Zones 9.1–78
		6. Connections 9.1–78
		6a. Demand Critical Welds 9.1–78
	G2.	Composite Intermediate Moment Frames (C-IMF) 9.1–78
		1. Scope 9.1–78
		2. Basis of Design 9.1–78
		3. Analysis 9.1–78
		4. System Requirements 9.1–79
		4a. Stability Bracing of Beams 9.1–79
		5. Members 9.1–79
		5a. Basic Requirements 9.1–79
		5b. Beam Flanges 9.1–79
		5c. Protected Zones 9.1–79
		6. Connections 9.1–79
		6a. Demand Critical Welds 9.1–79
		6b. Beam-to-Column Connections 9.1–79
		6c. Conformance Demonstration 9.1–80
		6d. Required Shear Strength 9.1–80
		6e. Connection Diaphragm Plates 9.1–80
		6f. Column Splices 9.1–81
	G3.	Composite Special Moment Frames (C-SMF) 9.1–81
		1. Scope 9.1–81

	2.	Basis of Design	9.1–81
	3.	Analysis	9.1–81
	4.	System Requirements	9.1–81
	4a.	Moment Ratio	9.1–81
	4b.	Stability Bracing of Beams	9.1–82
	4c.	Stability Bracing at Beam-to-Column Connections	9.1–82
	5.	Members	9.1–82
	5a.	Basic Requirements	9.1–82
	5b.	Beam Flanges	9.1–83
	5c.	Protected Zones	9.1–83
	6.	Connections	9.1–83
	6a.	Demand Critical Welds	9.1–83
	6b.	Beam-to-Column Connections	9.1–84
	6c.	Conformance Demonstration	9.1–84
	6d.	Required Shear Strength	9.1–84
	6e.	Connection Diaphragm Plates	9.1–85
	6f.	Column Splices	9.1–85
G4.		Composite Partially Restrained Moment Frames (C-PRMF)	9.1–85
	1.	Scope	9.1–85
	2.	Basis of Design	9.1–85
	3.	Analysis	9.1–85
	4.	System Requirements	9.1–85
	5.	Members	9.1–85
	5a.	Columns	9.1–85
	5b.	Beams	9.1–86
	5c.	Protected Zones	9.1–86
	6.	Connections	9.1–86
	6a.	Demand Critical Welds	9.1–86
	6b.	Required Strength	9.1–86
	6c.	Beam-to-Column Connections	9.1–86
	6d.	Conformance Demonstration	9.1–86
	6e.	Column Splices	9.1–87
H.	**COMPOSITE BRACED-FRAME AND SHEAR-WALL SYSTEMS**		**9.1–88**
H1.		Composite Ordinary Braced Frames (C-OBF)	9.1–88
	1.	Scope	9.1–88
	2.	Basis of Design	9.1–88
	3.	Analysis	9.1–89
	4.	System Requirements	9.1–89
	5.	Members	9.1–89
	5a.	Basic Requirements	9.1–89
	5b.	Columns	9.1–89
	5c.	Braces	9.1–89
	5d.	Protected Zones	9.1–89

		6.	Connections	9.1–89
		6a.	Demand Critical Welds	9.1–89
H2.	Composite Special Concentrically Braced Frames (C-SCBF)			9.1–89
		1.	Scope	9.1–89
		2.	Basis of Design	9.1–90
		3.	Analysis	9.1–90
		4.	System Requirements	9.1–90
		5.	Members	9.1–90
		5a.	Basic Requirements	9.1–90
		5b.	Diagonal Braces	9.1–90
		5c.	Protected Zones	9.1–90
		6.	Connections	9.1–90
		6a.	Demand Critical Welds	9.1–90
		6b.	Beam-to-Column Connections	9.1–91
		6c.	Required Strength of Brace Connections	9.1–91
		6d.	Column Splices	9.1–91
H3.	Composite Eccentrically Braced Frames (C-EBF)			9.1–91
		1.	Scope	9.1–91
		2.	Basis of Design	9.1–91
		3.	Analysis	9.1–92
		4.	System Requirements	9.1–92
		5.	Members	9.1–92
		6.	Connections	9.1–92
		6a.	Beam-to-Column Connections	9.1–92
H4.	Composite Ordinary Shear Walls (C-OSW)			9.1–92
		1.	Scope	9.1–92
		2.	Basis of Design	9.1–93
		3.	Analysis	9.1–93
		4.	System Requirements	9.1–93
		5.	Members	9.1–93
		5a.	Boundary Members	9.1–93
		5b.	Coupling Beams	9.1–94
		5c.	Protected Zones	9.1–96
		6.	Connections	9.1–96
		6a.	Demand Critical Welds	9.1–96
H5.	Composite Special Shear Walls (C-SSW)			9.1–96
		1.	Scope	9.1–96
		2.	Basis of Design	9.1–96
		3.	Analysis	9.1–96
		4.	System Requirements	9.1–96
		5.	Members	9.1–97
		5a.	Ductile Elements	9.1–97
		5b.	Boundary Members	9.1–97

		5c.	Steel Coupling Beams 9.1–97

            5c.   Steel Coupling Beams 9.1–97
            5d. Composite Coupling Beams 9.1–98
            5e.   Protected Zones 9.1–99
            6.   Connections 9.1–99
            6a. Demand Critical Welds 9.1–99
            6b. Column Splices 9.1–99
     H6. Composite Plate Shear Walls (C-PSW) 9.1–99
            1.   Scope 9.1–99
            2.   Basis of Design 9.1–99
            3.   Analysis 9.1–99
            3a. Webs 9.1–99
            3b. Other Members and Connections 9.1–100
            4.   System Requirements 9.1–100
            4a. Steel Plate Thickness 9.1–100
            4b. Stiffness of Vertical Boundary Elements 9.1–100
            4c. HBE-to-VBE Connection Moment Ratio 9.1–100
            4d. Bracing 9.1–100
            4e. Openings in Webs 9.1–100
            5.   Members 9.1–100
            5a. Basic Requirements 9.1–100
            5b. Webs 9.1–100
            5c. Concrete Stiffening Elements 9.1–101
            5d. Boundary Members 9.1–101
            5e. Protected Zones 9.1–101
            6.   Connections 9.1–101
            6a. Demand Critical Welds 9.1–101
            6b. HBE-to-VBE Connections 9.1–101
            6c. Connections of Steel Plate to Boundary Elements 9.1–101
            6d. Connections of Steel Plate to Reinforced Concrete Panel 9.1–102
            6e. Column Splices 9.1–102

I.   **FABRICATION AND ERECTION** 9.1–103
     I1.   Shop and Erection Drawings 9.1–103
            1.   Shop Drawings for Steel Construction 9.1–103
            2.   Erection Drawings for Steel Construction 9.1–103
            3.   Shop and Erection Drawings for Composite Construction 9.1–103
     I2.   Fabrication and Erection 9.1–104
            1.   Protected Zone 9.1–104
            2.   Bolted Joints 9.1–104
            3.   Welded Joints 9.1–104
            4.   Continuity Plates and Stiffeners 9.1–105

J.   **QUALITY CONTROL AND QUALITY ASSURANCE** 9.1–106
     J1.   Scope 9.1–106

	J2.	Fabricator and Erector Documents	9.1–107
		1. Documents to be Submitted for Steel Construction	9.1–107
		2. Documents to be Available for Review for Steel Construction	9.1–107
		3. Documents to be Submitted for Composite Construction	9.1–107
		4. Documents to be Available for Review for Composite Construction	9.1–107
	J3.	Quality Assurance Agency Documents	9.1–108
	J4.	Inspection and Nondestructive Testing Personnel	9.1–108
	J5.	Inspection Tasks	9.1–108
		1. Observe	9.1–109
		2. Perform	9.1–109
		3. Document	9.1–109
		4. Coordinated Inspection	9.1–109
	J6.	Welding Inspection and Nondestructive Testing	9.1–109
		1. Visual Welding Inspection	9.1–109
		2. NDT of Welded Joints	9.1–112
		2a. k-Area NDT	9.1–112
		2b. CJP Groove Weld NDT	9.1–112
		2c. Base Metal NDT for Lamellar Tearing and Laminations	9.1–113
		2d. Beam Cope and Access Hole NDT	9.1–113
		2e. Reduced Beam Section Repair NDT	9.1–113
		2f. Weld Tab Removal Sites	9.1–113
		2g. Reduction of Percentage of Ultrasonic Testing	9.1–113
		2h. Reduction of Percentage of Magnetic Particle Testing	9.1–113
	J7.	Inspection of High-Strength Bolting	9.1–113
	J8.	Other Steel Structure Inspections	9.1–115
	J9.	Inspection of Composite Structures	9.1–115
	J10.	Inspection of H-Piles	9.1–117
K.	**PREQUALIFICATION AND CYCLIC QUALIFICATION TESTING PROVISIONS**		**9.1–118**
	K1.	Prequalification of Beam-to-Column and Link-to-Column Connections	9.1–118
		1. Scope	9.1–118
		2. General Requirements	9.1–118
		2a. Basis for Prequalification	9.1–118
		2b. Authority for Prequalification	9.1–118
		3. Testing Requirements	9.1–119
		4. Prequalification Variables	9.1–119
		4a. Beam or Link Parameters	9.1–119
		4b. Column Parameters	9.1–119
		4c. Beam-to-Column or Link-to-Column Relations	9.1–120
		4d. Continuity Plates	9.1–120
		4e. Welds	9.1–120

	4f.	Bolts 9.1–120
	4g.	Workmanship 9.1–120
	4h.	Additional Connection Details 9.1–120
	5.	Design Procedure 9.1–121
	6.	Prequalification Record 9.1–121
K2.	Cyclic Tests for Qualification of Beam-to-Column and Link-to-Column Connections 9.1–121
	1.	Scope 9.1–121
	2.	Test Subassemblage Requirements 9.1–121
	3.	Essential Test Variables 9.1–122
	3a.	Sources of Inelastic Rotation 9.1–122
	3b.	Size of Members 9.1–122
	3c.	Connection Details 9.1–123
	3d.	Continuity Plates 9.1–123
	3e.	Steel Strength 9.1–123
	3f.	Welded Joints 9.1–124
	3g.	Bolted Joints 9.1–125
	4.	Loading History 9.1–125
	4a.	General Requirements 9.1–125
	4b.	Loading Sequence for Beam-to-Column Moment Connections 9.1–126
	4c.	Loading Sequence for Link-to-Column Connections 9.1–126
	5.	Instrumentation 9.1–126
	6.	Testing Requirements for Material Specimens 9.1–127
	6a.	Tension Testing Requirements for Structural Steel Material Specimens 9.1–127
	6b.	Methods of Tension Testing for Structural Steel Material Specimens 9.1–127
	6c.	Testing Requirements for Weld Metal Material Specimens 9.1–127
	7.	Test Reporting Requirements 9.1–127
	8.	Acceptance Criteria 9.1–128
K3.	Cyclic Tests for Qualification of Buckling-Restrained Braces 9.1–129
	1.	Scope 9.1–129
	2.	Subassemblage Test Specimen 9.1–129
	3.	Brace Test Specimen 9.1–130
	3a.	Design of Brace Test Specimen 9.1–130
	3b.	Manufacture of Brace Test Specimen 9.1–130
	3c.	Similarity of Brace Test Specimen and Prototype 9.1–130
	3d.	Connection Details 9.1–130
	3e.	Materials 9.1–131
	3f.	Connections 9.1–131
	4.	Loading History 9.1–131
	4a.	General Requirements 9.1–131
	4b.	Test Control 9.1–131

	4c.	Loading Sequence	9.1–131
	5.	Instrumentation	9.1–132
	6.	Materials Testing Requirements	9.1–132
	6a.	Tension Testing Requirements	9.1–132
	6b.	Methods of Tension Testing	9.1–132
	7.	Test Reporting Requirements	9.1–132
	8.	Acceptance Criteria	9.1–133

COMMENTARY ... 9.1–135

A. GENERAL REQUIREMENTS ... 9.1–139
- A1. Scope ... 9.1–139
- A2. Referenced Specifications, Codes and Standards ... 9.1–141
- A3. Materials ... 9.1–141
  1. Material Specifications ... 9.1–141
  2. Expected Material Strength ... 9.1–142
  3. Heavy Sections ... 9.1–144
  4. Consumables for Welding ... 9.1–146
  5. Concrete and Steel Reinforcement ... 9.1–147
- A4. Structural Design Drawings and Specifications ... 9.1–148
  1. General ... 9.1–148
  2. Steel Construction ... 9.1–148
  3. Composite Construction ... 9.1–150

B. GENERAL DESIGN REQUIREMENTS ... 9.1–151
- B1. General Seismic Design Requirements ... 9.1–151
- B2. Loads and Load Combinations ... 9.1–151
- B3. Design Basis ... 9.1–153
  2. Available Strength ... 9.1–153

C. ANALYSIS ... 9.1–154
- C1. General Requirements ... 9.1–154
- C2. Additional Requirements ... 9.1–156
- C3. Nonlinear Analysis ... 9.1–158

D. GENERAL MEMBER AND CONNECTION DESIGN REQUIREMENTS ... 9.1–159
- D1. Member Requirements ... 9.1–159
  1. Classification of Sections for Ductility ... 9.1–159
  1a. Section Requirements for Ductile Members ... 9.1–159
  1b. Width-to-Thickness Limitations of Steel and Composite Sections ... 9.1–159
  2. Stability Bracing of Beams ... 9.1–161
  2a. Moderately Ductile Members ... 9.1–161
  2b. Highly Ductile Members ... 9.1–162
  2c. Special Bracing at Plastic Hinge Locations ... 9.1–162

		3.	Protected Zones	9.1–163
		4.	Columns	9.1–163
		4a.	Required Strength	9.1–163
		4b.	Encased Composite Columns	9.1–164
		4c.	Filled Composite Columns	9.1–167
		5.	Composite Slab Diaphragms	9.1–167
	D2.	Connections		9.1–168
		1.	General	9.1–168
		2.	Bolted Joints	9.1–169
		3.	Welded Joints	9.1–172
		4.	Continuity Plates and Stiffeners	9.1–172
		5.	Column Splices	9.1–173
		5a.	Location of Splices	9.1–173
		5b.	Required Strength	9.1–173
		5c.	Required Shear Strength	9.1–174
		5d.	Structural Steel Splice Configurations	9.1–175
		6.	Column Bases	9.1–175
		6a.	Required Axial Strength	9.1–176
		6b.	Required Shear Strength	9.1–176
		6c.	Required Flexural Strength	9.1–178
		7.	Composite Connections	9.1–180
		8.	Steel Anchors	9.1–183
	D3.	Deformation Compatibility of Non-SFRS Members and Connections		9.1–184
	D4.	H-Piles		9.1–186
		1.	Design Requirements	9.1–187
		2.	Battered H-Piles	9.1–187
		3.	Tension	9.1–188
		4.	Protected Zone	9.1–188
E.	**MOMENT-FRAME SYSTEMS**			9.1–189
	E1.	Ordinary Moment Frames (OMF)		9.1–189
		1.	Scope	9.1–189
		2.	Basis of Design	9.1–189
		4.	System Requirements	9.1–191
		5.	Members	9.1–191
		6.	Connections	9.1–191
		6b.	FR Moment Connections	9.1–191
		6c.	PR Moment Connections	9.1–193
	E2.	Intermediate Moment Frames (IMF)		9.1–193
		1.	Scope	9.1–193
		2.	Basis of Design	9.1–193
		4.	System Requirements	9.1–193
		4a.	Stability Bracing of Beams	9.1–193

	5.	Members	9.1–193
	5a.	Basic Requirements	9.1–193
	5b.	Beam Flanges	9.1–194
	5c.	Protected Zones	9.1–194
	6.	Connections	9.1–194
	6a.	Demand Critical Welds	9.1–194
	6b.	Beam-to-Column Connection Requirements	9.1–194
	6c.	Conformance Demonstration	9.1–194
	6d.	Required Shear Strength	9.1–194
	6e.	Panel Zone	9.1–195
	6f.	Continuity Plates	9.1–195
	6g.	Column Splices	9.1–195
E3.		Special Moment Frames (SMF)	9.1–195
	1.	Scope	9.1–195
	2.	Basis of Design	9.1–195
	4.	System Requirements	9.1–196
	4a.	Moment Ratio	9.1–196
	4b.	Stability Bracing of Beams	9.1–197
	4c.	Stability Bracing at Beam-to-Column Connections	9.1–197
	5.	Members	9.1–198
	5a.	Basic Requirements	9.1–198
	5b.	Beam Flanges	9.1–198
	5c.	Protected Zones	9.1–199
	6.	Connections	9.1–199
	6a.	Demand Critical Welds	9.1–199
	6b.	Beam-to-Column Connections	9.1–199
	6c.	Conformance Demonstration	9.1–201
	6d.	Required Shear Strength	9.1–201
	6e.	Panel Zone	9.1–201
	6f.	Continuity Plates	9.1–205
	6g.	Column Splices	9.1–207
E4.		Special Truss Moment Frames (STMF)	9.1–207
	1.	Scope	9.1–207
	2.	Basis of Design	9.1–208
	3.	Analysis	9.1–208
	3a.	Special Segment	9.1–208
	3b.	Nonspecial Segment	9.1–208
	4.	System Requirements	9.1–208
	4a.	Special Segment	9.1–208
	4b.	Stability Bracing of Trusses	9.1–209
	4c.	Stability Bracing of Truss-to-Column Connections	9.1–209
	5.	Members	9.1–209
	5a.	Special Segment Members	9.1–209

		5b.	Expected Vertical Shear Strength of Special Segment	9.1–209
		5c.	Width-to-Thickness Limitations	9.1–210
		5d.	Built-Up Chord Members	9.1–210
		5e.	Protected Zones	9.1–210
		6.	Connections	9.1–211
		6a.	Demand Critical Welds	9.1–212
		6b.	Connections of Diagonal Web Members in the Special Segment	9.1–212
		6c.	Column Splices	9.1–212
	E5.		Ordinary Cantilever Column Systems (OCCS)	9.1–212
		1.	Scope	9.1–212
		2.	Basis of Design	9.1–212
		4.	System Requirements	9.1–213
		4a.	Columns	9.1–213
	E6.		Special Cantilever Column Systems (SCCS)	9.1–213
		1.	Scope	9.1–213
		2.	Basis of Design	9.1–213
		4.	System Requirements	9.1–213
		4a.	Columns	9.1–213
		5.	Members	9.1–213
		5a.	Basic Requirements	9.1–213
		5b.	Column Flanges	9.1–214
		5c.	Protected Zones	9.1–214
		6.	Connections	9.1–214
		6a.	Demand Critical Welds	9.1–214
		6b.	Column Bases	9.1–214
F.	**BRACED-FRAME AND SHEAR-WALL SYSTEMS**			9.1–215
	F1.		Ordinary Concentrically Braced Frames (OCBF)	9.1–215
		1.	Scope	9.1–215
		2.	Basis of Design	9.1–215
		3.	Analysis	9.1–215
		4.	System Requirements	9.1–215
		4a.	V-Braced and Inverted V-Braced Frames	9.1–215
		4b.	K-Braced Frames	9.1–216
		5.	Members	9.1–216
		5a.	Basic Requirements	9.1–216
		5b.	Slenderness	9.1–216
		6.	Connections	9.1–216
		6a.	Diagonal Brace Connections	9.1–216
		7.	Ordinary Concentrically Braced Frames above Seismic Isolation Systems	9.1–216
	F2.		Special Concentrically Braced Frames (SCBF)	9.1–217
		1.	Scope	9.1–217

	2. Basis of Design	9.1–217
	3. Analysis	9.1–219
	4. System Requirements	9.1–222
	4a. Lateral Force Distribution	9.1–222
	4b. V- and Inverted V-Braced Frames	9.1–222
	4c. K-Braced Frames	9.1–223
	4d. Tension-Only Frames	9.1–223
	5. Members	9.1–223
	5a. Basic Requirements	9.1–223
	5b. Diagonal Braces	9.1–224
	5c. Protected Zones	9.1–225
	6. Connections	9.1–225
	6a. Demand Critical Welds	9.1–225
	6b. Beam-to-Column Connections	9.1–227
	6c. Required Strength of Brace Connections	9.1–228
	6d. Column Splices	9.1–230
F3.	Eccentrically Braced Frames (EBF)	9.1–232
	1. Scope	9.1–232
	2. Basis of Design	9.1–233
	3. Analysis	9.1–235
	4. System Requirements	9.1–239
	4a. Link Rotation Angle	9.1–239
	4b. Bracing of Link	9.1–241
	5. Members	9.1–241
	5a. Basic Requirements	9.1–241
	5b. Links	9.1–242
	5c. Protected Zones	9.1–247
	6. Connections	9.1–247
	6a. Demand Critical Welds	9.1–247
	6b. Beam-to-Column Connections	9.1–247
	6c. Diagonal Brace Connections	9.1–247
	6d. Column Splices	9.1–248
	6e. Link-to-Column Connections	9.1–248
F4.	Buckling-Restrained Braced Frames (BRBF)	9.1–249
	1. Scope	9.1–249
	2. Basis of Design	9.1–250
	2a. Brace Strength	9.1–252
	3. Analysis	9.1–253
	4. System Requirements	9.1–254
	4a. V- and Inverted V-Braced Frames	9.1–254
	4b. K-Braced Frames	9.1–254
	5. Members	9.1–255
	5a. Basic Requirements	9.1–255

	5b.	Diagonal Braces	9.1–255
	5c.	Protected Zones	9.1–256
	6.	Connections	9.1–256
	6a.	Demand Critical Welds	9.1–256
	6b.	Beam-to-Column Connections	9.1–256
	6c.	Diagonal Brace Connections	9.1–257
	6d.	Column Splices	9.1–258
F5.		Special Plate Shear Walls (SPSW)	9.1–258
	1.	Scope	9.1–258
	2.	Basis of Design	9.1–259
	3.	Analysis	9.1–260
	4.	System Requirements	9.1–266
	4a.	Stiffness of Boundary Elements	9.1–266
	4c.	Bracing	9.1–267
	4d.	Openings in Webs	9.1–267
	5.	Members	9.1–267
	5a.	Basic Requirements	9.1–267
	5b.	Webs	9.1–267
	5c.	Protected Zone	9.1–269
	6.	Connections	9.1–269
	6a.	Demand Critical Welds	9.1–269
	6b.	HBE-to-VBE Connections	9.1–269
	6c.	Connections of Webs to Boundary Elements	9.1–269
	6d.	Column Splices	9.1–270
	7.	Perforated Webs	9.1–270
	7a.	Regular Layout of Circular Perforations	9.1–270
	7b.	Reinforced Corner Cut-Out	9.1–271

G. COMPOSITE MOMENT-FRAME SYSTEMS 9.1–273

G1.		Composite Ordinary Moment Frames (C-OMF)	9.1–273
	2.	Basis of Design	9.1–273
G2.		Composite Intermediate Moment Frames (C-IMF)	9.1–273
	2.	Basis of Design	9.1–273
	4.	System Requirements	9.1–273
	4a.	Stability Bracing of Beams	9.1–273
	5.	Members	9.1–273
	5a.	Basic Requirements	9.1–273
	5b.	Beam Flanges	9.1–274
	5c.	Protected Zones	9.1–274
	6.	Connections	9.1–274
	6a.	Demand Critical Welds	9.1–274
	6b.	Beam-to-Column Connections	9.1–274
	6c.	Conformance Demonstration	9.1–274
	6d.	Required Shear Strength	9.1–274

	6e. Connection Diaphragm Plates	9.1–274
	6f. Column Splices	9.1–274
G3.	Composite Special Moment Frames (C-SMF)	9.1–275
	1. Scope	9.1–275
	2. Basis of Design	9.1–275
	4. System Requirements	9.1–275
	4a. Moment Ratio	9.1–275
	4b. Stability Bracing of Beams	9.1–276
	4c. Stability Bracing at Beam-to-Column Connections	9.1–276
	5. Members	9.1–276
	5a. Basic Requirements	9.1–276
	5b. Beam Flanges	9.1–276
	5c. Protected Zones	9.1–276
	6. Connections	9.1–276
	6a. Demand Critical Welds	9.1–277
	6b. Beam-to-Column Connections	9.1–277
	6c. Conformance Demonstration	9.1–279
	6d. Required Shear Strength	9.1–280
	6e. Connection Diaphragm Plates	9.1–280
	6f. Column Splices	9.1–280
G4.	Composite Partially Restrained Moment Frames (C-PRMF)	9.1–280
	1. Scope	9.1–280
	2. Basis of Design	9.1–281
	3. Analysis	9.1–281
	4. System Requirements	9.1–282
	5. Members	9.1–282
	5a. Columns	9.1–282
	5b. Beams	9.1–282
	6. Connections	9.1–282
	6c. Beam-to-Column Connections	9.1–283
	6d. Conformance Demonstration	9.1–284
H.	**COMPOSITE BRACED-FRAME AND SHEAR-WALL SYSTEMS**	9.1–285
H1.	Composite Ordinary Braced Frames (C-OBF)	9.1–285
	1. Scope	9.1–285
	6. Connections	9.1–285
H2.	Composite Special Concentrically Braced Frames (C-SCBF)	9.1–287
	1. Scope	9.1–287
	2. Basis of Design	9.1–287
	5. Members	9.1–287
	5b. Diagonal Braces	9.1–287
	6. Connections	9.1–287

	6a. Demand Critical Welds	9.1–288
	6b. Beam-to-Column Connections	9.1–288
	6d. Column Splices	9.1–288
H3.	Composite Eccentrically Braced Frames (C-EBF)	9.1–288
	1. Scope	9.1–288
	2. Basis of Design	9.1–288
	3. Analysis	9.1–289
	6. Connections	9.1–289
	6a. Beam-to-Column Connections	9.1–289
H4.	Composite Ordinary Shear Walls (C-OSW)	9.1–290
	1. Scope	9.1–290
	2. Basis of Design	9.1–292
	3. Analysis	9.1–293
	4. System Requirements	9.1–294
	5. Members	9.1–294
	5b. Coupling Beams	9.1–294
H5.	Composite Special Shear Walls (C-SSW)	9.1–296
	1. Scope	9.1–296
	2. Basis of Design	9.1–296
	3. Analysis	9.1–296
	4. System Requirements	9.1–296
	5. Members	9.1–297
	5a. Ductile Elements	9.1–297
	5b. Boundary Members	9.1–297
	5c. Steel Coupling Beams	9.1–298
	5d. Composite Coupling Beams	9.1–300
	6. Connections	9.1–300
H6.	Composite Plate Shear Walls (C-PSW)	9.1–300
	1. Scope	9.1–300
	3. Analysis	9.1–300
	3a. Webs	9.1–300
	3b. Other Members and Connections	9.1–300
	4. System Requirements	9.1–300
	4e. Openings in Webs	9.1–301
	5. Members	9.1–301
	5b. Webs	9.1–301
	5c. Concrete Stiffening Elements	9.1–301
	5d. Boundary Members	9.1–301
	6. Connections	9.1–301
	6a. Demand Critical Welds	9.1–301
	6b. HBE-to-VBE Connections	9.1–301
	6c. Connections of Steel Plate to Boundary Elements	9.1–303
	6d. Connections of Steel Plate to Reinforced Concrete Panel	9.1–303

I. FABRICATION AND ERECTION ... 9.1–304
  I1. Shop and Erection Drawings ... 9.1–304
      3. Shop and Erection Drawings for Composite Construction ... 9.1–304
  I2. Fabrication and Erection ... 9.1–304
      1. Protected Zone ... 9.1–304
      2. Bolted Joints ... 9.1–304
      3. Welded Joints ... 9.1–305

J. QUALITY CONTROL AND QUALITY ASSURANCE ... 9.1–306
  J1. Scope ... 9.1–306
  J2. Fabricator and Erector Documents ... 9.1–307
      1. Documents to be Submitted for Steel Construction ... 9.1–307
      2. Documents to be Available for Review for Steel Construction ... 9.1–307
      3. Documents to be Submitted for Composite Construction ... 9.1–308
      4. Documents to be Available for Review for Composite Construction ... 9.1–308
  J3. Quality Assurance Agency Documents ... 9.1–308
  J4. Inspection and Nondestructive Testing Personnel ... 9.1–308
  J5. Inspection Tasks ... 9.1–309
      1. Observe ... 9.1–309
      2. Perform ... 9.1–309
      3. Document ... 9.1–309
  J6. Welding Inspection and Nondestructive Testing ... 9.1–309
      1. Visual Welding Inspection ... 9.1–309
      2. NDT of Welded Joints ... 9.1–310
      2a. k-Area NDT ... 9.1–310
      2b. CJP Groove Weld NDT ... 9.1–310
      2c. Base Metal NDT for Lamellar Tearing and Laminations ... 9.1–310
      2d. Beam Cope and Access Hole NDT ... 9.1–311
      2e. Reduced Beam Section Repair NDT ... 9.1–311
      2f. Weld Tab Removal Sites ... 9.1–311
  J7. Inspection of High-Strength Bolting ... 9.1–311
  J8. Other Steel Structure Inspections ... 9.1–311
  J9. Inspection of Composite Structures ... 9.1–312
  J10. Inspection of H-Piles ... 9.1–312

K. PREQUALIFICATION AND CYCLIC QUALIFICATION TESTING PROVISIONS ... 9.1–313
  K1. Prequalification of Beam-to-Column and Link-to-Column Connections ... 9.1–313
      1. Scope ... 9.1–313
      2. General Requirements ... 9.1–314
      2a. Basis for Prequalification ... 9.1–314
      2b. Authority for Prequalification ... 9.1–314

	3. Testing Requirements	9.1–315
	4. Prequalification Variables	9.1–316
	5. Design Procedure	9.1–317
	6. Prequalification Record	9.1–317
K2.	Cyclic Tests for Qualification of Beam-to-Column and Link-to-Column Connections	9.1–317
	1. Scope	9.1–317
	2. Test Subassemblage Requirements	9.1–318
	3. Essential Test Variables	9.1–319
	3a. Sources of Inelastic Rotation	9.1–319
	3b. Size of Members	9.1–320
	3e. Steel Strength	9.1–322
	3f. Welded Joints	9.1–322
	4. Loading History	9.1–322
	6. Testing Requirements for Material Specimens	9.1–324
	8. Acceptance Criteria	9.1–325
K3.	Cyclic Tests for Qualification of Buckling-Restrained Braces	9.1–325
	1. Scope	9.1–326
	2. Subassemblage Test Specimen	9.1–327
	3. Brace Test Specimen	9.1–329
	5. Instrumentation	9.1–332
	6. Materials Testing Requirements	9.1–332
	7. Test Reporting Requirements	9.1–333
	8. Acceptance Criteria	9.1–333
REFERENCES		9.1–335

This table is not part of the Provisions. It provides a cross reference of the 2010 Provisions to the 2005 Provisions. A "—" indicates there was no section in the 2005 Provisions corresponding to the 2010 Provisions.

CROSS REFERENCE
2010 Seismic Provisions – 2005 Seismic Provisions

2010	2005
A. General Requirements	
A1. Scope	I-1, II-1
A2. Referenced Specifications, Codes and Standards	I-2, II-2
A3. Materials	I-6, I-7.3, II-5
A4. Structural Design Drawings and Specifications	I-5, App. W2.1, II-18
B. General Design Requirements	
B1. General Seismic Design Requirements	I-3
B2. Loads and Load Combinations	I-4.1, II-4.1
B3. Design Basis	I-3
B4. System Type	—
C. Analysis	
C1. General Requirements	—
C2. Additional Requirements	—
C3. Nonlinear Analysis	—
D. General Member and Connection Design Requirements	
D1. Member Requirements	I-7.4, I-8.2, I-8.3, I-9.8, II-6.2, II-6.4, II-6.5
D2. Connections	I-7.1, I-7.2, I-7.3, I-7.5, I-8.4, I-8.5, II-7
D3. Deformation Compatibility of Non-SFRS Members and Connections	—
D4. H-Piles	I-8.6
E. Moment-Frame Systems	
E1. Ordinary Moment Frames (OMF)	I-11
E2. Intermediate Moment Frames (IMF)	I-10
E3. Special Moment Frames (SMF)	I-9
E4. Special Truss Moment Frames (STMF)	I-12
E5. Ordinary Cantilever Column Systems (OCCS)	—
E6. Special Cantilever Column Systems (SCCS)	—

CROSS REFERENCE
2010 Seismic Provisions – 2005 Seismic Provisions

2010	2005
F. Braced-Frame and Shear-Wall Systems	
F1. Ordinary Concentrically Braced Frames (OCBF)	I-14
F2. Special Concentrically Braced Frames (SCBF)	I-13
F3. Eccentrically Braced Frames (EBF)	I-15
F4. Buckling-Restrained Braced Frames (BRBF)	I-16
F5. Special Plate Shear Walls (SPSW)	I-17
G. Composite Moment-Frame Systems	
G1. Composite Ordinary Moment Frames (C-OMF)	II-11
G2. Composite Intermediate Moment Frames (C-IMF)	II-10
G3. Composite Special Moment Frames (C-SMF)	II-9
G4. Composite Partially Restrained Moment Frames (C-PRMF)	II-8
H. Composite Braced-Frame and Shear-Wall Systems	
H1. Composite Ordinary Braced Frames (C-OBF)	II-13
H2. Composite Special Concentrically Braced Frames (C-SCBF)	II-12
H3. Composite Eccentrically Braced Frames (C-EBF)	II-14
H4. Composite Ordinary Shear Walls (C-OSW)	II-15
H5. Composite Special Shear Walls (C-SSW)	II-16
H6. Composite Plate Shear Walls (C-PSW)	II-17
I. Fabrication and Erection	
I1. Shop and Erection Drawings	I-5.2, I-5.3, App. W2.2, App. W2.3, II-18
I2. Fabrication and Erection	I-7.2, I-7.3, I-7.4, I-7.5
J. Quality Control and Quality Assurance	
J1. Scope	I-18, App. Q1, II-19
J2. Fabricator and Erector Documents	App. Q3
J3. Quality Assurance Agency Documents	App. Q4
J4. Inspection and Nondestructive Testing Personnel	App. Q2
J5. Inspection Tasks	App. Q5
J6. Welding Inspection and Nondestructive Testing	App. Q5.1, App. Q5.2
J7. Inspection of High-Strength Bolting	App. Q5.3

CROSS REFERENCE
2010 Seismic Provisions – 2005 Seismic Provisions

2010	2005
J8. Other Steel Structure Inspections	App. Q5.4
J9. Inspection of Composite Structures	—
J10. Inspection of H-Piles	—
K. Prequalification and Cycle Qualification Testing Provisions	
K1. Prequalification of Beam-to-Column and Link-to-Column Connections	App. P
K2. Cyclic Tests for Qualification of Beam-to-Column and Link-to-Column Connections	App. S
K3. Cyclic Tests for Qualification of Buckling-Restrained Braces	App. T

SYMBOLS

The symbols listed below are to be used in addition to or replacements for those in the AISC *Specification for Structural Steel Buildings*. Where there is a duplication of the use of a symbol between the Provisions and the AISC *Specification for Structural Steel Buildings*, the symbol listed herein takes precedence. The section or table number in the right-hand column refers to where the symbol is first used.

Symbol	Definition	Reference
A_b	Cross-sectional area of a horizontal boundary element, in.^{2} (mm^2)	F5.5b
A_c	Cross-sectional area of a vertical boundary element, in.^{2} (mm^2)	F5.5b
A_f	Gross area of the flange, in.^{2} (mm^2)	E4.4b
A_g	Gross area, in.^{2} (mm^2)	E3.4a
A_{lw}	Link web area (excluding flanges), in.^{2} (mm^2)	F3.5a
A_s	Cross-sectional area of the structural steel core, in.^{2} (mm^2)	D1.4b
A_s	Area of transverse reinforcement in coupling beam, in.^{2} (mm^2)	H4.5b
A_s	Area of longitudinal wall reinforcement provided over the embedment length, L_e, in.^{2} (mm^2)	H5.5c
A_{sc}	Cross-sectional area of the yielding segment of steel core, in.^{2} (mm^2)	F4.5b
A_{sh}	Minimum area of tie reinforcement, in.^{2} (mm^2)	D1.4b
A_{sp}	Horizontal area of stiffened steel plate in composite plate shear wall, in.^{2} (mm^2)	H6.3b
A_{st}	Horizontal cross-sectional area of the *link* stiffener, in.^{2} (mm^2)	F3.5b
A_{tb}	Area of transfer reinforcement required in each of the first and second regions attached to each of the top and bottom flanges, in.^{2} (mm^2)	H5.5c
A_{tw}	Area of steel beam web, in.^{2} (mm^2)	H4.5b
C_a	Ratio of required strength to available strength	Table D1.1
C_d	Coefficient relating relative brace stiffness and curvature	D1.2a
D	Dead load due to the weight of the structural elements and permanent features on the building, kips (N)	D1.4b
D	Outside diameter, in. (mm)	Table D1.1
D	Diameter of the holes, in. (mm)	F5.7a
E	Seismic load effect, kips (N)	F1.4a
E	Modulus of elasticity of steel, E = 29,000 ksi (200 000 MPa)	Table D1.1
E_{mh}	Horizontal seismic load effect including overstrength factor, kips (N)	B2
F_{cr}	Critical stress, ksi (MPa)	F1.6a

Symbol	Definition	Reference
F_{cre}	Critical stress calculated from *Specification* Chapter E using expected yield stress, ksi (MPa)	F1.6a
F_y	Specified minimum yield stress of the type of steel to be used, ksi (MPa). As used in the *Specification*, "yield stress" denotes either the minimum specified yield point (for those steels that have a yield point) or the specified yield strength (for those steels that do not have a yield point).	A3.2
F_{yb}	Specified minimum yield stress of a beam, ksi (MPa)	E3.4a
F_{yc}	Specified minimum yield stress of a column, ksi (MPa)	E3.4a
F_{ysc}	Specified minimum yield stress of the steel core, or actual yield stress of the steel core as determined from a coupon test, ksi (MPa)	F4.5b
F_{ysr}	Specified minimum yield stress of the ties, ksi (MPa)	D1.4b
F_{ysr}	Specified minimum yield stress of transverse reinforcement, ksi (MPa)	H4.5b
F_{ysr}	Specified minimum yield stress of transfer reinforcement, ksi (MPa)	H5.5c
F_u	Specified minimum tensile strength, ksi (MPa)	A3.2
H	Height of story, which is permitted to be taken as the distance between the centerline of floor framing at each of the levels above and below, or the distance between the top of floor slabs at each of the levels above and below, in. (mm)	D2.5c
H_c	Clear height of the column between beam connections, including a structural slab, if present, in. (mm)	F2.6d
I	Moment of inertia, in.^{4} (mm^4)	E4.5b
I_b	Moment of inertia of a horizontal boundary element taken perpendicular to the direction of the web plate line, in.^{4} (mm^4)	F5.4a
I_c	Moment of inertia of a vertical boundary element taken perpendicular to the direction of the web plate line, in.^{4} (mm^4)	F5.4a
I_y	Moment of inertia about an axis in the plane of the EBF in.^{4} (mm^4)	F3.5b
I_y	Moment of inertia of the plate, in.^{4} (mm^4)	F5.7b
K	Effective length factor for prismatic member	F1.5b
L	Live load due to occupancy and moveable equipment, kips (N)	D1.4b
L	Length of column, in. (mm)	E3.4c
L	Span length of the truss, in. (mm)	E4.5b
L	Length of brace, in. (mm)	F1.5b
L	Distance between vertical boundary element centerlines, in. (mm)	F5.4a
L_b	Length between points which are either braced against lateral displacement of compression flange or braced against twist of the cross section, in. (mm)	D1.2a
L_{cf}	Clear length of beam, in. (mm)	E1.6b
L_{cf}	Clear distance between column flanges, in. (mm)	F5.5b
L_e	Embedment length of coupling beam, in. (mm)	H4.5b
L_h	Distance between plastic hinge locations, as defined within the test report or ANSI/AISC 358, in. (mm)	E2.6d

SYMBOLS

Symbol	Definition	Reference
L_s	Length of the special segment, in. (mm)	E4.5b
M_a	Required flexural strength, using ASD load combinations, kip-in. (N-mm)	D1.2c
M_{av}	Additional moment due to shear amplification from the location of the plastic hinge to the column centerline based on ASD load combinations, kip-in. (N-mm)	E3.4a
M_{nc}	Nominal flexural strength of the chord member of the special segment, kip-in. (N-mm)	E4.5b
$M_{n,PR}$	Nominal flexural strength of PR connection at a rotation of 0.02 rad, kip-in. (N-mm)	E1.6c
M_p	Nominal plastic flexural strength, kip-in. (N-mm)	E1.6b
M_{pc}	Nominal plastic flexural strength of the column, kip-in. (N-mm)	D2.5c
M_{pcc}	Nominal flexural strength of a composite column, kip-in. (N-mm	G2.6f
$M_{p,exp}$	Expected flexural strength, kip-in. (N-mm)	D1.2c
M_r	Required flexural strength, kip-in. (N-mm)	D1.2a
M_u	Required flexural strength, using LRFD load combinations, kip-in. (N-mm)	D1.2c
M_{uv}	Additional moment due to shear amplification from the location of the plastic hinge to the column centerline based on LRFD load combinations, kip-in. (N-mm)	E3.4a
M_{uv}	Moment due to shear amplification from the location of the plastic hinge to the column centerline, kip-in. (N-mm)	G3.4a
M^*_{pb}	Moment at the intersection of the beam and column centerlines determined by projecting the beam maximum developed moments from the column face, kip-in. (N-mm)	E3.4a
M^*_{pc}	Moment at beam and column centerline determined by projecting the sum of the nominal column plastic moment strength, reduced by the axial stress P_{uc}/A_g, from the top and bottom of the beam moment connection, kip-in. (N-mm)	E3.4a
M^*_{pcc}	Moment in the column above or below the joint at the intersection of the beam and column centerlines, kip-in. (N-mm)	G3.4a
$M^*_{p,exp}$	Moment in the steel beam or concrete-encased composite beam at the intersection of the beam and column centerlines, kip-in. (N-mm)	G3.4a
N_r	Number of horizontal rows of perforations	F5.7a
P_a	Required axial strength of a column using ASD load combinations, kips (N)	Table D1.1
P_{ac}	Required compressive strength using ASD load combinations, kips (N)	E3.4a
P_b	Axial design strength of wall at balanced condition, kips (N)	H5.4
P_c	Available axial strength of a column, kips (N)	E3.4a
P_n	Nominal axial strength of a column, kips (N)	E4.5a
P_n	Nominal compressive strength of the composite column calculated in accordance with the *Specification*, kips (N)	D1.4b
P_{nc}	Nominal compressive strength of the chord member at the ends, kips (N)	E4.4c

Symbol	Definition	Reference
P_{nt}	Nominal axial tensile strength of diagonal members of the special segment, kips (N)	E4.5b
P_r	Required compressive strength, kips (N)	E4.4d
P_{rc}	Required compressive strength of columns using ASD or LRFD load combinations, kips (N)	E3.4a
P_u	Required axial strength using LRFD load combinations, kips (N)	Table D1.1
P_{uc}	Required compressive strength using LRFD load combinations, kips (N)	E3.4a
P_y	Nominal axial yield strength of a member, equal to $F_y A_g$, kips (N)	Table D1.1
P_{ysc}	Axial yield strength of steel core, kips (N)	F4.2a
R	Seismic response modification coefficient	A1
R	Radius of the cut-out, in. (mm)	F5.7b
R_n	Nominal strength, kips (N)	A3.2
R_t	Ratio of the expected tensile strength to the specified minimum tensile strength F_u, as related to overstrength in material yield stress, R_y	A3.2
R_y	Ratio of the expected yield stress to the specified minimum yield stress, F_y	A3.2
R_{yb}	Ratio of the expected yield stress of the beam material to the specified minimum yield stress	E3.6f
R_{yc}	Ratio of the expected yield stress of the column material to the specified minimum yield stress	E3.6f
S_{diag}	Shortest center-to-center distance between holes, in. (mm)	F5.7a
V_a	Required shear strength using ASD load combinations, kips (N)	E1.6b
V_c	V_y (LRFD) or $V_y/1.5$ (ASD) as appropriate, kips (N)	F3.5b
V_{comp}	Limiting expected shear strength of an encased composite coupling beam, kips (N)	H4.5b
V_n	Link nominal shear strength, kips (N)	F3.3
V_n	Expected shear strength of a steel coupling beam, kips (N)	H4.5b
$V_{n,comp}$	Expected shear strength of an encased composite coupling beam, kips (N)	H4.5b
V_{ne}	Expected vertical shear strength of the special segment, kips (N)	E4.5b
V_{ns}	Nominal shear strength of the steel plate in a composite plate shear wall, kips (N)	H6.5c
V_r	V_u (LRFD) or V_a (ASD) as appropriate, kips (N)	F3.5b
V_p	Nominal shear strength of an active link, kips (N)	F3.4a
V_u	Required shear strength using LRFD load combinations, kips (N)	E1.6b
V_y	Nominal shear yield strength, kips (N)	F3.5b
Y_{con}	Distance from the top of the steel beam to the top of concrete slab or encasement, in. (mm)	G3.5a
Y_{PNA}	Maximum distance from the maximum concrete compression fiber to the plastic neutral axis, in. (mm)	G3.5a
Z	Plastic section modulus of a member, in.^{3} (mm^3)	D1.2a

SYMBOLS

Symbol	Definition	Reference
Z_b	Plastic section modulus of the beam, in.³ (mm³)	E3.4a
Z_c	Plastic section modulus of the column, in.³ (mm³)	E3.4a
Z_x	Plastic section modulus about x-axis, in.³ (mm³)	E2.6g
Z_{RBS}	Minimum plastic section modulus at the reduced beam section, in.³ (mm³)	E3.4a
a	Distance between connectors, in. (mm)	F2.5b
b	Width of compression element as defined in *Specification* Section B4.1, in. (mm)	Table D1.1
b	Inside width of a box section, in. (mm)	F3.5b
b_{bf}	Flange width of beam, in. (mm)	E3.6f
b_{cf}	Flange width of column, in. (mm)	E3.6f
b_f	Flange width, in. (mm)	D2.5b
b_w	Thickness of wall pier, in. (mm)	H4.5b
b_w	Wall width, in. (mm)	H5.5c
b_{wc}	Width of concrete encasement, in. (mm)	H4.5b
d	Overall beam depth, in. (mm)	Table D1.1
d	Nominal bolt diameter, in. (mm)	D2.2
d	Overall link depth, in. (mm)	F3.5b
d_c	Effective depth of concrete encasement, in. (mm)	H4.5b
d_z	$d-2t_f$ of the deeper beam at the connection, in. (mm)	E3.6e
e	EBF link length, in. (mm)	F3.5b
f'_c	Specified compressive strength of concrete, ksi (MPa)	D1.4b
g	Coupling beam clear span, in. (mm)	H4.5b
h	Clear distance between flanges less the fillet or corner radius for rolled shapes; and for built-up sections, the distance between adjacent lines of fasteners or the clear distance between flanges when welds are used; for tees, the overall depth; and for rectangular HSS, the clear distance between the flanges less the inside corner radius on each side, in. (mm)	Table D1.1
h	Distance between horizontal boundary element centerlines, in. (mm)	F5.4a
h	Overall depth of the boundary member in the plane of the wall, in. (mm)	H5.5b
h_{cc}	Cross-sectional dimension of the confined core region in composite columns measured center-to-center of the transverse reinforcement, in. (mm)	D1.4b
h_o	Distance between flange centroids, in. (mm)	D1.2c
r	Governing radius of gyration, in. (mm)	E3.4c
r_i	Minimum radius of gyration of individual component, in. (mm)	F2.5b
r_y	Radius of gyration about y-axis, in. (mm)	D1.2a
r_y	Radius of gyration of individual components about their weak axis, in. (mm)	E4.5d
s	Spacing of transverse reinforcement, in. (mm)	D1.4b
t	Thickness of element, in. (mm)	Table D1.1
t	Thickness of column web or doubler plate, in. (mm)	E3.6e

Symbol	Definition	Reference
t_{bf}	Thickness of beam flange, in. (mm)	E3.4c
t_{cf}	Minimum required thickness of column flange when no continuity plates are provided, in. (mm)	E3.6f
t_{eff}	Effective web-plate thickness, in. (mm)	F5.7a
t_f	Thickness of the flange, in. (mm)	D2.5b
t_w	Thickness of the web, in. (mm)	F3.5b
t_w	Web-plate thickness, in. (mm)	F5.7a
w_z	Width of panel zone between column flanges, in. (mm)	E3.6e
Δ	Design story drift, in. (mm)	F3.4a
Δ_b	Deformation quantity used to control loading of test specimen (total brace end rotation for the subassemblage test specimen; total brace axial deformation for the brace test specimen), in. (mm)	K3.4c
Δ_{bm}	Value of deformation quantity, Δ_b, corresponding to the design story drift, in. (mm)	K3.4c
Δ_{by}	Value of deformation quantity, Δ_b, at first significant yield of test specimen, in. (mm)	K3.4c
Ω	Safety factor	B3.2
Ω_c	Safety factor for compression	Table D1.1
Ω_o	System overstrength factor	B2
Ω_v	Safety factor for shear strength of panel zone of beam-to-column connections	E3.6e
α	Angle of diagonal members with the horizontal, degrees	E4.5b
α	Angle of web yielding, as measured relative to the vertical, degrees	F5.5b
α	Angle of the shortest center-to-center lines in the opening array to vertical, degrees	F5.7a
β	Compression strength adjustment factor	F4.2a
β_1	Factor relating depth of equivalent rectangular compressive stress block to neutral axis depth, as defined in ACI 318	H4.5b
γ_{total}	Total link rotation angle	K2.4c
θ	Story drift angle, radians	K2.4b
$\lambda_{hd}, \lambda_{md}$	Limiting slenderness parameter for highly and moderately ductile compression elements, respectively	D1.1b
ϕ	Resistance factor	B3.2
ϕ_c	Resistance factor for compression	Table D1.1
ϕ_v	Resistance factor for shear	E3.6e
ω	Strain hardening adjustment factor	F4.2a

GLOSSARY

The terms listed below are to be used in addition to those in the AISC *Specification for Structural Steel Buildings*. Some commonly used terms are repeated here for convenience. Glossary terms are generally *italicized* throughout these Provisions, where they first appear within a section.

Notes:
(1) Terms designated with † are common AISI-AISC terms that are coordinated between the two standards developers.
(2) Terms designated with * are usually qualified by the type of *load effect*, for example, *nominal tensile strength, available compressive strength, design flexural strength.*

Adjusted brace strength. Strength of a brace in a *buckling-restrained braced frame* at deformations corresponding to 2.0 times the *design story drift.*

Allowable strength*†. Nominal strength divided by the *safety factor*, R_n/Ω.

Amplified seismic load. Seismic load effect including overstrength factor.

Applicable building code†. Building code under which the structure is designed.

ASD (allowable strength design)†. Method of proportioning structural components such that the *allowable strength* equals or exceeds the *required strength* of the component under the action of the ASD load combinations.

ASD load combination†. Load combination in the *applicable building code* intended for *allowable strength design* (allowable stress design).

Authority having jurisdiction (AHJ). Organization, political subdivision, office or individual charged with the responsibility of administering and enforcing the provisions of this Standard.

Available strength*†. *Design strength* or *allowable strength*, as appropriate.

Boundary member. Portion along wall or diaphragm edge strengthened with structural steel sections and/or longitudinal steel reinforcement and transverse reinforcement.

Brace test specimen. A single buckling-restrained brace element used for laboratory testing intended to model the brace in the *prototype.*

Braced frame†. An essentially vertical truss system that provides resistance to lateral forces and provides stability for the structural system.

Buckling-restrained brace. A pre-fabricated, or manufactured, brace element consisting of a steel core and a buckling-restraining system as described in Section F4 and qualified by testing as required in Section K3.

Buckling-restrained braced frame (BRBF). A diagonally braced frame employing buckling-restrained braces and meeting the requirements of Section F4.

Buckling-restraining system. System of restraints that limits buckling of the steel core in BRBF. This system includes the casing surrounding the steel core and structural elements adjoining its connections. The buckling-restraining system is intended to permit the transverse expansion and longitudinal contraction of the steel core for deformations corresponding to 2.0 times the *design story drift*.

Casing. Element that resists forces transverse to the axis of the diagonal brace thereby restraining buckling of the core. The casing requires a means of delivering this force to the remainder of the buckling-restraining system. The casing resists little or no force along the axis of the diagonal brace.

Collector. Also known as drag strut, member that serves to transfer loads between floor diaphragms and the members of the *seismic force resisting system*.

Column base. Assemblage of structural shapes, plates, connectors, bolts and rods at the base of a column used to transmit forces between the steel superstructure and the foundation.

Complete loading cycle. A cycle of rotation taken from zero force to zero force, including one positive and one negative peak.

Composite beam. Structural steel beam in contact with and acting compositely with a reinforced concrete slab designed to act compositely for seismic forces.

Composite brace. Concrete-encased structural steel section (rolled or built-up) or concrete-filled steel section used as a diagonal brace.

Composite column. Concrete-encased structural steel section (rolled or built-up) or concrete-filled steel section used as a column.

Composite eccentrically braced frame (C-EBF). Composite braced frame meeting the requirements of Section H3.

Composite intermediate moment frame (C-IMF). Composite moment frame meeting the requirements of Section G2.

Composite ordinary braced frame (C-OBF). Composite braced frame meeting the requirements of Section H1.

Composite ordinary moment frame (C-OMF). Composite moment frame meeting the requirements of Section G1.

Composite ordinary shear wall (C-OSW). Composite shear wall meeting the requirements of Section H4.

Composite partially restrained moment frame (C-PRMF). Composite moment frame meeting the requirements of Section G4.

Composite plate shear wall (C-PSW). Wall consisting of steel plate with reinforced concrete encasement on one or both sides that provides out-of-plane stiffening to prevent buckling of the steel plate and meeting the requirements of Section H6.

Composite shear wall. Steel plate wall panel composite with reinforced concrete wall panel or reinforced concrete wall that has steel or concrete-encased structural steel sections as boundary members.

Composite slab. Reinforced concrete slab supported on and bonded to a formed steel deck that acts as a diaphragm to transfer load to and between elements of the *seismic force resisting system*.

Composite special concentrically braced frame (C-SCBF). Composite braced frame meeting the requirements of Section H2.

Composite special moment frame (C-SMF). Composite moment frame meeting the requirements of Section G3.

Composite special shear wall (C-SSW). Composite shear wall meeting the requirements of Section H5.

Concrete-encased shapes. Structural steel sections encased in concrete.

Continuity plates. Column stiffeners at the top and bottom of the *panel zone*; also known as transverse stiffeners.

Coupling beam. Structural steel or composite beam connecting adjacent reinforced concrete wall elements so that they act together to resist lateral loads.

Demand critical weld. Weld so designated by these Provisions.

Design earthquake. The earthquake represented by the *design response spectrum* as specified in the *applicable building code*.

Design story drift. Calculated story drift, including the effect of expected inelastic action, due to design level earthquake forces as determined by the *applicable building code*.

Design strength†.* *Resistance factor* multiplied by the *nominal strength*, ϕR_n.

Diagonal brace. Inclined structural member carrying primarily axial force in a braced frame.

Eccentrically braced frame (EBF). Diagonally braced frame meeting the requirements of Section F3 that has at least one end of each diagonal brace connected to a beam with a defined eccentricity from another beam-to-brace connection or a beam-to-column connection.

Encased composite beam. Composite beam completely enclosed in reinforced concrete.

Encased composite column. Structural steel column completely encased in reinforced concrete.

Engineer of record. Licensed professional responsible for sealing the contract documents.

Exempted column. Column not meeting the requirements of Equation E3-1 for SMF.

Expected tensile strength.* Tensile strength of a member, equal to the specified minimum tensile strength, F_u, multiplied by R_t.

Expected yield strength. Yield strength in tension of a member, equal to the expected yield stress multiplied by A_g.

Expected yield stress. Yield stress of the material, equal to the specified minimum yield stress, F_y, multiplied by R_y.

Face bearing plates. Stiffeners attached to structural steel beams that are embedded in reinforced concrete walls or columns. The plates are located at the face of the reinforced concrete to provide confinement and to transfer loads to the concrete through direct bearing.

Filled composite column. HSS filled with structural concrete.

Fully composite beam. Composite beam that has a sufficient number of steel headed stud anchors to develop the nominal plastic flexural strength of the composite section.

Highly ductile member. A member expected to undergo significant plastic rotation (more than 0.02 rad) from either flexure or flexural buckling under the *design earthquake*.

Horizontal boundary element (HBE). A beam with a connection to one or more web plates in an SPSW.

Intermediate boundary element (IBE). A member, other than a beam or column, that provides resistance to web plate tension adjacent to an opening in an SPSW.

Intermediate moment frame (IMF). Moment frame system that meets the requirements of Section E2.

Inverted-V-braced frame. See *V-braced frame*.

k-area. The region of the web that extends from the tangent point of the web and the flange-web fillet (AISC "*k*" dimension) a distance of $1\frac{1}{2}$ in. (38 mm) into the web beyond the *k* dimension.

K-braced frame. A braced-frame configuration in which braces connect to a column at a location with no out-of-plane support.

Link. In EBF, the segment of a beam that is located between the ends of the connections of two diagonal braces or between the end of a diagonal brace and a column. The length of the *link* is defined as the clear distance between the ends of two diagonal braces or between the diagonal brace and the column face.

Link intermediate web stiffeners. Vertical web stiffeners placed within the link in EBF.

Link rotation angle. Inelastic angle between the link and the beam outside of the *link* when the total story drift is equal to the *design story drift*.

Link rotation angle, total. The relative displacement of one end of the link with respect to the other end (measured transverse to the longitudinal axis of the undeformed link), divided by the link length. The total link rotation angle includes both elastic and inelastic components of deformation of the link and the members attached to the link ends.

Link design shear strength. Lesser of the available shear strength of the *link* based on the flexural or shear strength of the link member.

Load-carrying reinforcement. Reinforcement in composite members designed and detailed to resist the required loads.

Lowest anticipated service temperature (LAST). The lowest 1-hour average temperature with a 100-year mean recurrence interval.

LRFD (load and resistance factor design)†. Method of proportioning structural components such that the *design strength* equals or exceeds the *required strength* of the component under the action of the *LRFD load combinations*.

LRFD load combination†. Load combination in the *applicable building code* intended for strength design *(load and resistance factor design)*.

Material test plate. A test specimen from which steel samples or weld metal samples are machined for subsequent testing to determine mechanical properties.

Member brace. Member that provides stiffness and strength to control movement of another member out-of-the plane of the frame at the braced points.

Moderately ductile member A member expected to undergo moderate plastic rotation (0.02 rad or less) from either flexure or flexural buckling under the *design earthquake*.

Nominal strength*†. Strength of a structure or component (without the *resistance factor* or *safety factor* applied) to resist load effects, as determined in accordance with the Specification.

Ordinary cantilever column system (OCCS). A seismic force resisting system in which the seismic forces are resisted by one or more columns that are cantilevered from the foundation or from the diaphragm level below and that meets the requirements of Section E5.

Ordinary concentrically braced frame (OCBF). Diagonally braced frame meeting the requirements of Section F1 in which all members of the braced-frame system are subjected primarily to axial forces.

Ordinary moment frame (OMF). Moment frame system that meets the requirements of Section E1.

Partially composite beam. Steel beam with a composite slab with a nominal flexural strength controlled by the strength of the steel headed stud anchors.

Partially restrained composite connection. Partially restrained (PR) connections as defined in the *Specification* that connect partially or *fully composite beams* to steel columns with flexural resistance provided by a force couple achieved with steel reinforcement in the slab and a steel seat angle or similar connection at the bottom flange.

Plastic hinge. Yielded zone that forms in a structural member when the plastic moment is attained. The member is assumed to rotate further as if hinged, except that such rotation is restrained by the plastic moment.

Prequalified connection. Connection that complies with the requirements of Section K1 or ANSI/AISC 358.

Protected zone. Area of members or connections of members in which limitations apply to fabrication and attachments.

Prototype. The connection or diagonal brace that is to be used in the building (SMF, IMF, EBF, BRBF, C-IMF, C-SMF and C-PRMF).

Provisions. Refers to this document, the AISC *Seismic Provisions for Structural Steel Buildings* (ANSI/AISC 341).

Quality assurance plan. Written description of qualifications, procedures, quality inspections, resources and records to be used to provide assurance that the structure complies with the engineer's quality requirements, specifications and contract documents.

Reduced beam section. Reduction in cross section over a discrete length that promotes a zone of inelasticity in the member.

Required strength*. Forces, stresses and deformations acting on a structural component, determined by either structural analysis, for the *LRFD* or *ASD load combinations*, as appropriate, or as specified by the *Specification* and these Provisions.

Resistance factor, ϕ†. Factor that accounts for unavoidable deviations of the *nominal strength* from the actual strength and for the manner and consequences of failure.

Response modification coefficient, R. Factor that reduces seismic load effects to strength level as specified by the *applicable building code*.

Risk category. Classification assigned to a structure based on its use as specified by the *applicable building code*.

Safety factor, Ω†. Factor that accounts for deviations of the actual strength from the nominal strength, deviations of the actual load from the nominal load, uncertainties in the analysis that transforms the load into a load effect, and for the manner and consequences of failure.

Seismic design category. Classification assigned to a building by the *applicable building code* based upon its *risk category* and the design spectral response acceleration coefficients.

Seismic force resisting system (SFRS). That part of the structural system that has been considered in the design to provide the required resistance to the seismic forces prescribed in ASCE/SEI 7.

Special cantilever column system (SCCS). A seismic force resisting system in which the seismic forces are resisted by one or more columns that are cantilevered from the foundation or from the diaphragm level below and that meets the requirements of Section E6.

Special concentrically braced frame (SCBF). Diagonally braced frame meeting the requirements of Section F2 in which all members of the braced-frame system are subjected primarily to axial forces.

Special moment frame (SMF). Moment frame system that meets the requirements of Section E3.

Special plate shear wall (SPSW). Plate shear wall system that meets the requirements of Section F5.

Special truss moment frame (STMF). Truss moment frame system that meets the requirements of Section E4.

Specification. Refers to the AISC *Specification for Structural Steel Buildings* (ANSI/AISC 360).

Steel core. Axial-force-resisting element of a buckling-restrained brace. The steel core contains a yielding segment and connections to transfer its axial force to adjoining elements; it may also contain projections beyond the casing and transition segments between the projections and yielding segment.

Story drift angle. Interstory displacement divided by story height.

Subassemblage test specimen. The combination of members, connections and testing apparatus that replicate as closely as practical the boundary conditions, loading and deformations in the *prototype*.

System overstrength factor, Ω_o. Factor specified by the *applicable building code* in order to determine the amplified seismic load, where required by these Provisions.

Test setup. The supporting fixtures, loading equipment and lateral bracing used to support and load the Test Specimen.

Test specimen. A member, connection or subassemblage test specimen.

Test subassemblage. The combination of the test specimen and pertinent portions of the test setup.

V-braced frame. Concentrically braced frame (SCBF, OCBF, BRBF, C-OBF or C-SCBF) in which a pair of diagonal braces located either above or below a beam is connected to a single point within the clear beam span. Where the diagonal braces are below the beam, the system is also referred to as an *inverted-V-braced frame*.

Vertical boundary element (VBE). A column with a connection to one or more web plates in an SPSW.

X-braced frame. Concentrically braced frame (OCBF, SCBF, C-OBF or C-SCBF) in which a pair of diagonal braces crosses near the mid-length of the diagonal braces.

ACRONYMS

The following acronyms appear in the AISC *Seismic Provisions for Structural Steel Buildings*. The acronyms are written out where they first appear within a Section.

ACI (American Concrete Institute)
ANSI (American National Standards Institute)
ASCE (American Society of Civil Engineers)
ASD (allowable strength design)
ASTM (American Society for Testing of Materials)
AWS (American Welding Society)
BRBF (buckling-restrained braced frame)
CAC-A (air carbon arc cutting)
C-EBF (composite eccentrically braced frame)
C-IMF (composite intermediate moment frame)
CJP (complete joint penetration)
C-OBF (composite ordinary braced frame)
C-OMF (composite ordinary moment frame)
C-OSW (composite ordinary shear wall)
C-PRMF (composite partially restrained moment frame)
CPRP (connection prequalification review panel)
C-PSW (composite plate shear wall)
C-SCBF (composite special concentrically braced frame)
C-SMF (composite special moment frame)
C-SSW (composite special shear wall)
CVN (Charpy V-notch)
EBF (eccentrically braced frame)
FCAW (flux cored arc welding)
FEMA (Federal Emergency Management Agency)
FR (fully restrained)
GMAW (gas metal arc welding)
HBE (horizontal boundary element)
HSS (hollow structural section)
IBE (intermediate boundary element)
IMF (intermediate moment frame)
LAST (lowest anticipated service temperature)
LRFD (load and resistance factor design)
MT (magnetic particle testing)
NDT (nondestructive testing)
OCBF (ordinary concentrically braced frame)
OCCS (ordinary cantilever column system)
OMF (ordinary moment frame)
OVS (oversized)
PJP (partial joint penetration)
PR (partially restrained)

QA (quality assurance)
QC (quality control)
RBS (reduced beam section)
RCSC (Research Council on Structural Connections)
SAW (submerged arc welding)
SCBF (special concentrically braced frame)
SCCS (special cantilever column system)
SDC (seismic design category)
SEI (Structural Engineering Institute)
SFRS (seismic force resisting system)
SMAW (shielded metal arc welding)
SMF (special moment frame)
SPSPW (special perforated steel plate wall)
SPSW (special plate shear wall)
SRC (steel-reinforced concrete)
STMF (special truss moment frame)
UT (ultrasonic testing)
VBE (vertical boundary element)
WPQR (welder performance qualification records)
WPS (welding procedure specification)

CHAPTER A
GENERAL REQUIREMENTS

This chapter states the scope of the Provisions, summarizes referenced specification, code and standard documents, and provides requirements for materials and contract documents.

The chapter is organized as follows:

    A1.   Scope
    A2.   Referenced Specifications, Codes and Standards
    A3.   Materials
    A4.   Structural Design Drawings and Specifications

A1. SCOPE

The *Seismic Provisions for Structural Steel Buildings*, hereafter referred to as these Provisions, shall govern the design, fabrication and erection of structural steel members and connections in the *seismic force resisting systems* (SFRS), and splices and bases of columns in gravity framing systems of buildings and other structures with moment frames, *braced frames* and *shear walls*. Other structures are defined as those structures designed, fabricated and erected in a manner similar to buildings, with building-like vertical and lateral force-resisting-elements. These Provisions shall apply to the design of seismic force resisting systems of structural steel or of structural steel acting compositely with reinforced concrete, unless specifically exempted by the *applicable building code*.

Wherever these Provisions refer to the applicable building code and there is none, the loads, *load combinations*, system limitations, and general design requirements shall be those in ASCE/SEI 7.

User Note: ASCE/SEI 7 (Table 12.2-1, Line H) specifically exempts structural steel systems, but not composite systems, from these Provisions in *seismic design categories* B and C if they are designed in accordance with the *Specification for Structural Steel Buildings* and the seismic loads are computed using a response modification coefficient, R, of 3. For seismic design category A, ASCE/SEI specifies lateral forces to be used as the seismic loads and effects, but these calculations do not involve the use of a *response modification coefficient*. Thus for *seismic design category* A it is not necessary to define a seismic force resisting system that meets any special requirements and these Provisions do not apply.

User Note: ASCE/SEI (Table 15.4-1) permits certain nonbuilding structures to be designed in accordance with the *Specification* in lieu of the Provisions with an appropriately reduced R factor.

> **User Note:** Composite seismic force resisting systems include those systems with members of structural steel acting compositely with reinforced concrete, as well as systems in which structural steel members and reinforced concrete members act together to form a seismic force resisting system.

These Provisions shall be applied in conjunction with the AISC *Specification for Structural Steel Buildings*, hereafter referred to as the *Specification*. All requirements of the *Specification* are applicable unless otherwise stated in these Provisions. Members and connections of the SFRS shall satisfy the requirements of the applicable building code, the *Specification*, and these Provisions.

Building Code Requirements for Structural Concrete (ACI 318), as modified in these Provisions, shall be used for the design and construction of reinforced concrete components in composite construction. For the SFRS in composite construction incorporating reinforced concrete components designed in accordance with ACI 318, the requirements of *Specification* Section B3.3, Design for Strength Using Load and Resistance Factor Design, shall be used.

A2. REFERENCED SPECIFICATIONS, CODES AND STANDARDS

The documents referenced in these Provisions shall include those listed in *Specification* Section A2 with the following additions:

American Institute of Steel Construction (AISC)

ANSI/AISC 360-10 *Specification for Structural Steel Buildings*
ANSI/AISC 358-10 *Prequalified Connections for Special and Intermediate Steel Moment Frames for Seismic Applications*

American Welding Society (AWS)

AWS D1.8/D1.8M:2009 *Structural Welding Code—Seismic Supplement*
AWS B4.0:2007 *Standard Methods for Mechanical Testing of Welds* (U.S. Customary Units)
AWS B4.0M:2000 *Standard Methods for Mechanical Testing of Welds* (Metric Customary Units)
AWS D1.4/D1.4M:2005 *Structural Welding Code—Reinforcing Steel*

A3. MATERIALS

1. Material Specifications

Structural steel used in the *seismic force resisting system* (SFRS) shall satisfy the requirements of *Specification* Section A3.1, except as modified in these Provisions. The specified minimum yield stress of steel to be used for members in which inelastic behavior is expected shall not exceed 50 ksi (345 MPa) for systems defined in Chapters E, F, G and H, except that for systems defined in Sections E1, F1, G1, H1 and H4 this limit shall not exceed 55 ksi (380 MPa). Either of these specified minimum yield stress limits are permitted to be exceeded when the suitability of the material is determined by testing or other rational criteria.

Exception: Specified minimum yield stress of structural steel shall not exceed 65 ksi (450 MPa) for columns in systems defined in Sections E3, E4, G3, H1, H2 and H3, and for columns in all systems in Chapter F.

The structural steel used in the SFRS described in Chapters E, F, G and H shall meet one of the following ASTM Specifications:

(1) A36/A36M

(2) A53/A53M

(3) A500/A500M (Gr. B or C)

(4) A501

(5) A529/A529M

(6) A572/A572M [Gr. 42 (290), 50 (345) or 55 (380)]

(7) A588/A588M

(8) A913/A913M [Gr. 50 (345), 60 (415) or 65 (450)]

(9) A992/A992M

(10) A1011/A1011M HSLAS Gr. 55 (380)

(11) A1043/A1043M

The structural steel used for column base plates shall meet one of the preceding ASTM specifications or ASTM A283/A283M Grade D.

Other steels and nonsteel materials in *buckling-restrained braced frames* are permitted to be used subject to the requirements of Sections F4 and K3.

User Note: This section only covers material properties for structural steel used in the SFRS and included in the definition of structural steel given in Section 2.1 of the AISC *Code of Standard Practice*. Other steel, such as cables for permanent bracing, is not covered. Steel reinforcement used in components in composite SFRS is covered in Section A3.5.

2. Expected Material Strength

When required in these Provisions, the *required strength* of an element (a member or a connection of a member) shall be determined from the *expected yield stress*, $R_y F_y$, of the member or an adjoining member, as applicable, where F_y is the specified minimum yield stress of the steel to be used in the member and R_y is the ratio of the expected yield stress to the specified minimum yield stress, F_y, of that material.

When required to determine the *nominal strength*, R_n, for limit states within the same member from which the required strength is determined, the expected yield stress, $R_y F_y$, and the *expected tensile strength*, $R_t F_u$, are permitted to be used in lieu of F_y and F_u, respectively, where F_u is the specified minimum tensile strength and R_t is the ratio of the expected tensile strength to the specified minimum tensile strength, F_u, of that material.

TABLE A3.1
R_y and R_t Values for Steel and Steel Reinforcement Materials

Application	R_y	R_t
Hot-rolled structural shapes and bars:		
• ASTM A36/A36M	1.5	1.2
• ASTM A1043/1043M Gr. 36 (250)	1.3	1.1
• ASTM A572/572M Gr. 50 (345) or 55 (380), ASTM A913/A913M Gr. 50 (345), 60 (415), or 65 (450), ASTM A588/A588M, ASTM A992/A992M	1.1	1.1
• ASTM A1043/A1043M Gr. 50 (345)	1.2	1.1
• ASTM A529 Gr. 50 (345)	1.2	1.2
• ASTM A529 Gr. 55 (380)	1.1	1.2
Hollow structural sections (HSS):		
• ASTM A500/A500M (Gr. B or C), ASTM A501	1.4	1.3
Pipe:		
• ASTM A53/A53M	1.6	1.2
Plates, Strips and Sheets:		
• ASTM A36/A36M	1.3	1.2
• ASTM A1043/1043M Gr. 36 (250)	1.3	1.1
• A1011/A1011M HSLAS Gr. 55 (380)	1.1	1.1
• ASTM A572/A572M Gr. 42 (290)	1.3	1.0
• ASTM A572/A572M Gr. 50 (345), Gr. 55 (380), ASTM A588/A588M	1.1	1.2
• ASTM 1043/1043M Gr. 50 (345)	1.2	1.1
Steel Reinforcement:		
• ASTM A615, ASTM A706	1.25	1.25

User Note: In several instances a member, or a connection limit state within that member, is required to be designed for forces corresponding to the expected strength of the member itself. Such cases include determination of the nominal strength, R_n, of the beam outside of the *link* in EBF, *diagonal brace* rupture limit states (block shear rupture and net section rupture in the diagonal brace in SCBF), etc. In such cases it is permitted to use the expected material strength in the determination of available member strength. For connecting elements and for other members, specified material strength should be used.

The values of R_y and R_t for various steel and steel reinforcement materials are given in Table A3.1. Other values of R_y and R_t are permitted if the values are determined by testing of specimens, similar in size and source to the materials to be used, conducted in accordance with the testing requirements per the ASTM specifications for the specified grade of steel.

User Note: The expected compressive strength of concrete may be estimated using values from *Seismic Rehabilitation of Existing Buildings*, ASCE/SEI 41-06.

3. Heavy Sections

For structural steel in the SFRS, in addition to the requirements of *Specification* Section A3.1c, hot rolled shapes with flanges $1^1/_2$ in. thick (38 mm) and thicker shall have a minimum Charpy V-notch toughness of 20 ft-lb (27 J) at 70 °F (21 °C), tested in the alternate core location as described in ASTM A6 Supplementary Requirement S30. Plates 2 in. (50 mm) thick and thicker shall have a minimum Charpy V-notch toughness of 20 ft-lb (27 J) at 70 °F (21 °C), measured at any location permitted by ASTM A673, Frequency P, where the plate is used for the following:

(a) Members built up from plate

(b) Connection plates where inelastic strain under seismic loading is expected

(c) The *steel core* of buckling-restrained braces

4. Consumables for Welding

4a. Seismic Force Resisting System Welds

All welds used in members and connections in the SFRS shall be made with filler metals meeting the requirements specified in clause 6.3 of *Structural Welding Code—Seismic Supplement* (AWS D1.8/D1.8M), hereafter referred to as AWS D1.8/D1.8M.

User Note: AWS D1.8/D1.8M subclauses 6.3.5, 6.3.6, 6.3.7 and 6.3.8 apply only to *demand critical welds*.

4b. Demand Critical Welds

Welds designated as demand critical shall be made with filler metals meeting the requirements specified in AWS D1.8/D1.8M clause 6.3.

User Note: AWS D1.8/D1.8M requires that all seismic force resisting system welds are to be made with filler metals classified using AWS A5 standards that achieve the following mechanical properties:

Filler Metal Classification Properties for Seismic Force Resisting System Welds		
Property	Classification	
	70 ksi (480 MPa)	80 ksi (550 MPa)
Yield Strength, ksi (MPa)	58 (400) min.	68 (470) min.
Tensile Strength, ksi (MPa)	70 (480) min.	80 (550) min.
Elongation, %	22 min.	19 min.
CVN Toughness, ft-lb (J)	20 (27) min. @ 0 °F (−18 °C)[a]	

[a] Filler metals classified as meeting 20 ft-lbf (27 J) min. at a temperature lower than 0 °F (−18 °C) also meet this requirement.

In addition to the above requirements, AWS D1.8/D1.8M requires, unless otherwise exempted from testing, that all demand critical welds are to be made with filler metals receiving Heat Input Envelope Testing that achieve the following mechanical properties in the weld metal:

Mechanical Properties for Demand Critical Welds		
Property	Classification	
	70 ksi (480 MPa)	80 ksi (550 MPa)
Yield Strength, ksi (MPa)	58 (400) min.	68 (470) min.
Tensile Strength, ksi (MPa)	70 (480) min.	80 (550) min.
Elongation, %	22 min.	19 min.
CVN Toughness, ft-lb (J)	40 (54) min. @ 70 °F (20 °C)[b, c]	

[b] For LAST of +50 °F (+10 °C). For LAST less than +50 °F (+10 °C), see AWS D1.8/D1.8M subclause 6.3.6.
[c] Tests conducted in accordance with AWS D1.8/D1.8M Annex A meeting 40 ft-lb (54 J) min. at a temperature lower than +70 °F (+20 °C) also meet this requirement.

5. Concrete and Steel Reinforcement

Concrete and steel reinforcement used in composite components in composite intermediate or special SFRS of Sections G2, G3, G4, H2, H3, H5 and H6 shall satisfy the requirements of ACI 318, Chapter 21. Concrete and steel reinforcement used in composite components in composite ordinary SFRS of Sections G1, H1 and H4 shall satisfy the requirements of ACI 318, Section 21.1.1.5.

A4. STRUCTURAL DESIGN DRAWINGS AND SPECIFICATIONS

1. General

Structural design drawings and specifications shall indicate the work to be performed, and include items required by the *Specification*, the AISC *Code of Standard Practice for Steel Buildings and Bridges*, the *applicable building code*, and the following, as applicable:

(1) Designation of the SFRS

(2) Identification of the members and connections that are part of the SFRS

(3) Locations and dimensions of *protected zones*

(4) Connection details between concrete floor diaphragms and the structural steel elements of the SFRS

(5) Shop drawing and erection drawing requirements not addressed in Section I1

2. Steel Construction

In addition to the requirements of Section A4.1, structural design drawings and specifications for steel construction shall indicate the following items, as applicable:

(1) Configuration of the connections

(2) Connection material specifications and sizes

(3) Locations of *demand critical welds*

(4) Locations where gusset plates are to be detailed to accommodate inelastic rotation

(5) Locations of connection plates requiring Charpy V-notch (CVN) toughness in accordance with Section A3.3(b)

(6) *Lowest anticipated service temperature* (LAST) of the steel structure, if the structure is not enclosed and maintained at a temperature of 50 °F (10 °C) or higher

(7) Locations where weld backing is required to be removed

(8) Locations where fillet welds are required when weld backing is permitted to remain

(9) Locations where fillet welds are required to reinforce groove welds or to improve connection geometry

(10) Locations where weld tabs are required to be removed

(11) Splice locations where tapered transitions are required

(12) The shape of weld access holes, if a shape other than those provided for in the *Specification* is required

(13) Joints or groups of joints in which a specific assembly order, welding sequence, welding technique or other special precautions where such items are designated to be submitted to the *engineer of record*

3. Composite Construction

In addition to the requirements of Section A4.1, and the requirements of Section A4.2 as applicable for the steel components of reinforced concrete or composite elements, structural design drawings and specifications for composite construction shall indicate the following items, as applicable:

(1) Bar placement, cutoffs, lap and mechanical splices, hooks and mechanical anchorage, placement of ties and other transverse reinforcement

(2) Requirements for dimensional changes resulting from temperature changes, creep and shrinkage

(3) Location, magnitude and sequencing of any prestressing or post-tensioning present

(4) Location of steel headed stud anchors and welded reinforcing bar anchors

CHAPTER B

GENERAL DESIGN REQUIREMENTS

This chapter addresses the general requirements for the seismic design of steel structures that are applicable to all chapters of the Provisions.

This chapter is organized as follows:

    B1.   General Seismic Design Requirements
    B2.   Loads and Load Combinations
    B3.   Design Basis
    B4.   System Type

B1.   GENERAL SEISMIC DESIGN REQUIREMENTS

The *required strength* and other seismic design requirements for *seismic design categories* (SDCs), *risk categories*, and the limitations on height and irregularity shall be as specified in the *applicable building code*.

The *design story drift* and the limitations on story drift shall be determined as required in the applicable building code.

B2.   LOADS AND LOAD COMBINATIONS

The loads and *load combinations* shall be as stipulated by the *applicable building code*. Unless otherwise defined in these Provisions, where *amplified seismic loads* are required by these Provisions, the seismic load effect including the system overstrength factor shall be applied as prescribed by the applicable building code. Where the effects of horizontal forces including overstrength, E_{mh}, are defined in these Provisions they shall be combined with the vertical seismic load effect as required by the applicable building code and multiplied by 1.0 for use in *LRFD load combinations* and 0.7 for use in *ASD load combinations*.

> **User Note**: The seismic load effect including the system overstrength factor is defined in ASCE/SEI Section 12.4.3. Where E_{mh} is defined in these Provisions it is intended to replace E_{mh} in ASCE/SEI 7 Section 12.4.3.

In composite construction, incorporating reinforced concrete components designed in accordance with the requirements of ACI 318, the requirements of *Specification* Section B3.3, Design for Strength Using Load and Resistance Factor Design, shall be used for the *seismic force resisting system* (SFRS).

> **User Note**: When not defined in the applicable building code, Ω_o should be determined in accordance with ASCE/SEI 7.

B3. DESIGN BASIS

1. Required Strength

The *required strength* of structural members and connections shall be the greater of:

(1) The required strength as determined by structural analysis for the appropriate load combinations, as stipulated in the applicable building code, and in Chapter C.

(2) The required strength given in Chapters D, E, F, G and H.

2. Available Strength

The *available strength* is stipulated as the *design strength*, ϕR_n, for design in accordance with the provisions for *load and resistance factor design (LRFD)* and the *allowable strength*, R_n/Ω, for design in accordance with the provisions for *allowable strength design (ASD)*. The available strength of systems, members and connections shall be determined in accordance with the *Specification*, except as modified throughout these Provisions.

B4. SYSTEM TYPE

The *seismic force resisting system* (SFRS) shall contain one or more moment frame, *braced frame* or shear wall system conforming to the requirements of one of the seismic systems designated in Chapters E, F, G and H.

CHAPTER C

ANALYSIS

This chapter addresses design related analysis requirements. The chapter is organized as follows:

    C1.   General Requirements
    C2.   Additional Requirements
    C3.   Nonlinear Analysis

C1. GENERAL REQUIREMENTS

An analysis conforming to the requirements of the *applicable building code* and the *Specification* shall be performed for design of the system.

When the design is based upon elastic analysis, the stiffness properties of component members of steel systems shall be based on elastic sections and those of composite systems shall include the effects of cracked sections.

C2. ADDITIONAL REQUIREMENTS

Additional analysis shall be performed as specified in Chapters E, F, G and H of these Provisions.

C3. NONLINEAR ANALYSIS

When nonlinear analysis is used to satisfy the requirements of these Provisions, it shall be performed in accordance with Chapter 16 of ASCE/SEI 7.

CHAPTER D

GENERAL MEMBER AND CONNECTION DESIGN REQUIREMENTS

This chapter addresses general requirements for the design of members and connections.

The chapter is organized as follows:

  D1. Member Requirements
  D2. Connections
  D3. Deformation Compatibility of Non-SFRS Members and Connections
  D4. H-Piles

D1. MEMBER REQUIREMENTS

Members of moment frames, *braced frames* and shear walls in the *seismic force resisting system* (SFRS) shall comply with the *Specification* and this section. Certain members of the SFRS that are expected to undergo inelastic deformation under the *design earthquake* are designated in these provisions as *moderately ductile members* or *highly ductile members*.

1. Classification of Sections for Ductility

When required for the systems defined in Chapters E, F, G, H and Section D4, members designated as moderately ductile members or highly ductile members shall comply with this section.

1a. Section Requirements for Ductile Members

Structural steel sections for both moderately ductile members and highly ductile members shall have flanges continuously connected to the web or webs.

Encased composite columns shall comply with the requirements of Section D1.4b(1) for moderately ductile members and Section D1.4b(2) for highly ductile members.

Filled composite columns shall comply with the requirements of Section D1.4c for both moderately and highly ductile members.

Concrete sections shall comply with the requirements of ACI 318 Section 21.3 for moderately ductile members and ACI 318 Section 21.6 for highly ductile members.

1b. Width-to-Thickness Limitations of Steel and Composite Sections

For members designated as moderately ductile members, the width-to-thickness ratios of compression elements shall not exceed the limiting width-to-thickness ratios, λ_{md}, from Table D1.1.

For members designated as highly ductile members, the width-to-thickness ratios of compression elements shall not exceed the limiting width-to-thickness ratios, λ_{hd}, from Table D1.1.

TABLE D1.1
Limiting Width-to-Thickness Ratios for Compression Elements For Moderately Ductile and Highly Ductile Members

	Description of Element	Width-to-Thickness Ratio	Limiting Width-to-Thickness Ratio		Example
			λ_{hd} Highly Ductile Members	λ_{md} Moderately Ductile Members	
Unstiffened Elements	Flanges of rolled or built-up I-shaped sections, channels and tees; legs of single angles or double angle members with separators; outstanding legs of pairs of angles in continuous contact	b/t	$0.30\sqrt{E/F_y}$	$0.38\sqrt{E/F_y}$	
	Flanges of H-pile sections per Section D4	b/t	$0.45\sqrt{E/F_y}$	not applicable	
	Stems of tees	d/t	$0.30\sqrt{E/F_y}$ [a]	$0.38\sqrt{E/F_y}$	
Stiffened Elements	Walls of rectangular HSS	b/t			
	Flanges of boxed I-shaped sections and built-up box sections	b/t	$0.55\sqrt{E/F_y}$ [b]	$0.64\sqrt{E/F_y}$ [c]	
	Side plates of boxed I-shaped sections and walls of built-up box shapes used as diagonal braces	h/t			
	Webs of rolled or built-up I-shaped sections used as diagonal braces	h/t_w	$1.49\sqrt{E/F_y}$	$1.49\sqrt{E/F_y}$	

TABLE D1.1 (CONTINUED)
Limiting Width-to-Thickness Ratios for Compression Elements For Moderately Ductile and Highly Ductile Members

	Description of Element	Width-to-Thickness Ratio	Limiting Width-to-Thickness Ratio		Example
			λ_{hd} Highly Ductile Members	λ_{md} Moderately Ductile Members	
Stiffened Elements	Webs of rolled or built-up I-shaped sections used as beams or columns[d]	h/t_w	For $C_a \leq 0.125$ $2.45\sqrt{E/F_y}(1-0.93C_a)$	For $C_a \leq 0.125$ $3.76\sqrt{E/F_y}(1-2.75C_a)$	
Stiffened Elements	Side plates of boxed I-shaped sections used as beams or columns	h/t	For $C_a > 0.125$ $0.77\sqrt{E/F_y}(2.93-C_a)$ $\geq 1.49\sqrt{E/F_y}$ where $C_a = \dfrac{P_u}{\phi_c P_y}$ (LRFD) $C_a = \dfrac{\Omega_c P_a}{P_y}$ (ASD)	For $C_a > 0.125$ $1.12\sqrt{E/F_y}(2.33-C_a)$ $\geq 1.49\sqrt{E/F_y}$ where $C_a = \dfrac{P_u}{\phi_c P_y}$ (LRFD) $C_a = \dfrac{\Omega_c P_a}{P_y}$ (ASD)	
Stiffened Elements	Webs of built-up box sections used as beams or columns	h/t			
Stiffened Elements	Webs of H-Pile sections	h/t_w	$0.94\sqrt{E/F_y}$	not applicable	
Composite Elements	Walls of round HSS	D/t	$0.038 E/F_y$	$0.044 E/F_y$ [e]	
Composite Elements	Walls of rectangular filled composite members	b/t	$1.4\sqrt{E/F_y}$	$2.26\sqrt{E/F_y}$	
Composite Elements	Walls of round filled composite members	D/t	$0.076 E/F_y$	$0.15 E/F_y$	

[a] For tee shaped compression members, the limiting width-to-thickness ratio for highly ductile members for the stem of the tee can be increased to $0.38\sqrt{E/F_y}$ if either of the following conditions are satisfied:
  (1) Buckling of the compression member occurs about the plane of the stem.
  (2) The axial compression load is transferred at end connections to only the outside face of the flange of the tee resulting in an eccentric connection that reduces the compression stresses at the tip of the stem.

[b] The limiting width-to-thickness ratio of flanges of boxed I-shaped sections and built-up box sections of columns in SMF systems shall not exceed $0.6\sqrt{E/F_y}$.

[c] The limiting width-to-thickness ratio of walls of rectangular *HSS* members, flanges of boxed I-shaped sections and flanges of built-up box sections used as beams or columns shall not exceed $1.12\sqrt{E/F_y}$.

[d] For I-shaped beams in SMF systems, where C_a is less than or equal to 0.125, the limiting ratio h/t_w shall not exceed $2.45\sqrt{E/F_y}$. For I-shaped beams in IMF systems, where C_a is less than or equal to 0.125, the limiting width-to-thickness ratio shall not exceed $3.76\sqrt{E/F_y}$.

[e] The limiting diameter-to-thickness ratio of round *HSS* members used as beams or columns shall not exceed $0.07E/F_y$.

2. Stability Bracing of Beams

When required in Chapters E, F, G and H, stability bracing shall be provided as required in this section to restrain lateral-torsional buckling of structural steel or concrete-encased beams subject to flexure and designated as *moderately ductile members* or *highly ductile members*.

> **User Note:** In addition to the requirements in Chapters E, F, G and H to provide stability bracing for various beam members such as intermediate and *special moment frame* beams, stability bracing is also required for columns in the *special cantilever column system* (SCCS) in Section E6.

2a. Moderately Ductile Members

(a) The bracing of moderately ductile steel beams shall satisfy the following requirements:

   (1) Both flanges of beams shall be laterally braced or the beam cross section shall be torsionally braced.

   (2) Beam bracing shall meet the requirements of Appendix 6 of the *Specification* for lateral or torsional bracing of beams, where the required flexural strength of the member shall be:

$$M_r = R_y F_y Z \text{ (LRFD)} \qquad \text{(D1-1a)}$$

or

$$M_r = R_y F_y Z/1.5 \text{ (ASD)} \qquad \text{(D1-1b)}$$

where
$C_d = 1.0$
R_y = ratio of the *expected yield stress* to the specified minimum yield stress
Z = plastic section modulus, in.^{3} (mm^3)

   (3) Beam bracing shall have a maximum spacing of

$$L_b = 0.17 r_y E/F_y \qquad \text{(D1-2)}$$

(b) The bracing of moderately ductile concrete-*encased composite beams* shall satisfy the following requirements:

   (1) Both flanges of members shall be laterally braced or the beam cross section shall be torsionally braced.

   (2) Lateral bracing shall meet the requirements of Appendix 6 of the *Specification* for lateral or torsional bracing of beams, where $M_r = M_{p,\,exp}$ of the beam as specified in Section G2.6d, and $C_d = 1.0$.

   (3) Member bracing shall have a maximum spacing of

$$L_b = 0.17 r_y E/F_y \qquad \text{(D1-3)}$$

Sect. D1.] MEMBER REQUIREMENTS

using the material properties of the steel section and r_y in the plane of buckling calculated based on the elastic transformed section.

2b. Highly Ductile Members

In addition to the requirements of Sections D1.2a(a)(1) and (2), and D1.2a(b)(1) and (2), the bracing of highly ductile beam members shall have a maximum spacing of $L_b = 0.086 r_y E/F_y$. For concrete-encased composite beams, the material properties of the steel section shall be used and the calculation for r_y in the plane of buckling shall be based on the elastic transformed section.

2c. Special Bracing at Plastic Hinge Locations

Special bracing shall be located adjacent to expected *plastic hinge* locations where required by Chapters E, F, G or H.

(a) For structural steel beams, such bracing shall satisfy the following requirements:

  (1) Both flanges of beams shall be laterally braced or the member cross section shall be torsionally braced.

  (2) The *required strength* of lateral bracing of each flange provided adjacent to plastic hinges shall be:

$$P_u = 0.06 R_y F_y Z/h_o \text{ (LRFD)} \tag{D1-4a}$$

or

$$P_a = (0.06/1.5) R_y F_y Z/h_o \text{ (ASD)} \tag{D1-4b}$$

where
h_o = distance between flange centroids, in. (mm)

The required strength of torsional bracing provided adjacent to plastic hinges shall be:

$$M_u = 0.06 R_y F_y Z \text{ (LRFD)} \tag{D1-5a}$$

or

$$M_a = (0.06/1.5) R_y F_y Z \text{ (ASD)} \tag{D1-5b}$$

  (3) The required bracing stiffness shall satisfy the requirements of Appendix 6 of the *Specification* for lateral or torsional bracing of beams with $C_d = 1.0$ and where the expected flexural strength of the beam shall be:

$$M_r = M_u = R_y F_y Z \text{ (LRFD)} \tag{D1-6a}$$

or

$$M_r = M_a = R_y F_y Z/1.5 \text{ (ASD)} \tag{D1-6b}$$

(b) For concrete-encased composite beams, such bracing shall satisfy the following requirements:

  (1) Both flanges of beams shall be laterally braced or the beam cross section shall be torsionally braced.

(2) The required strength of lateral bracing provided adjacent to plastic hinges shall be

$$P_u = 0.06 M_{p,exp}/h_o \qquad (D1\text{-}7)$$

of the beam, where $M_{p,exp}$ is determined in accordance with Section G2.6d.

The required strength for torsional bracing provided adjacent to plastic hinges shall be $M_u = 0.06 M_{p,exp}$ of the beam.

(3) The required bracing stiffness shall satisfy the requirements of Appendix 6 of the *Specification* for lateral or torsional bracing of beams where $M_r = M_u = M_{p,exp}$ of the beam is determined in accordance with Section G2.6d, and $C_d = 1.0$.

3. Protected Zones

Discontinuities specified in Section I2.1 resulting from fabrication and erection procedures and from other attachments are prohibited in the area of a member or a connection element designated as a *protected zone* by these Provisions or ANSI/AISC 358.

Exception: Welded steel headed stud anchors and other connections are permitted in protected zones when designated in ANSI/AISC 358, or as otherwise determined with a connection prequalification in accordance with Section K1, or as determined in a program of qualification testing in accordance with Sections K2 and K3.

4. Columns

Columns in moment frames, braced frames and shear walls shall satisfy the requirements of this section.

4a. Required Strength

The required strength of columns in the SFRS shall be determined from the following:

(1) The load effect resulting from the analysis requirements for the applicable system per Sections E, F, G and H.

Exception: Section D1.4a need not apply to Sections G1, H1 or H4.

(2) The compressive axial strength and tensile strength as determined using the load combinations stipulated in the *applicable building code* including the *amplified seismic load*. It is permitted to neglect applied moments in this determination unless the moment results from a load applied to the column between points of lateral support. The required axial compressive strength and tensile strength need not exceed either of the following:

(a) The maximum load transferred to the column by the system, including the effects of material overstrength and strain hardening in those members where yielding is expected.

(b) The forces corresponding to the resistance of the foundation to overturning uplift.

4b. Encased Composite Columns

Encased composite columns shall satisfy the requirements of *Specification* Chapter I, in addition to the requirements of this section. Additional requirements, as specified for moderately ductile members and highly ductile members in Sections D1.4b(1) and (2), shall apply as required in the descriptions of the composite seismic systems in Chapters G and H.

(1) Moderately Ductile Members

Encased composite columns used as moderately ductile members shall satisfy the following requirements:

(1) The maximum spacing of transverse reinforcement at the top and bottom shall be the least of the following:

  (i) one-half the least dimension of the section

  (ii) 8 longitudinal bar diameters

  (iii) 24 tie bar diameters

  (iv) 12 in. (300 mm)

(2) This spacing shall be maintained over a vertical distance equal to the greatest of the following lengths, measured from each joint face and on both sides of any section where flexural yielding is expected to occur:

  (i) one-sixth the vertical clear height of the column

  (ii) the maximum cross-sectional dimension

  (iii) 18 in. (450 mm)

(3) Tie spacing over the remaining column length shall not exceed twice the spacing defined in Section D1.4b(1)(1).

(4) Splices and end bearing details for encased composite columns in composite ordinary SFRS of Sections G1, H1 and H4 shall satisfy the requirements of the *Specification* and ACI 318 Section 7.8.2. The design shall comply with ACI 318 Sections 21.1.6 and 21.1.7. The design shall consider any adverse behavioral effects due to abrupt changes in either the member stiffness or the nominal *tensile strength*. Transitions to reinforced concrete sections without embedded structural steel members, transitions to bare structural steel sections, and *column bases* shall be considered abrupt changes.

(5) Welded wire fabric shall be prohibited as transverse reinforcement in moderately ductile members.

(2) Highly Ductile Members

Encased composite columns used as highly ductile members shall satisfy Section D1.4b(1) in addition to the following requirements:

(1) Longitudinal *load-carrying reinforcement* shall satisfy the requirements of ACI 318 Section 21.6.3.

(2) Transverse reinforcement shall be hoop reinforcement as defined in ACI 318 Chapter 21 and shall satisfy the following requirements:

(i) The minimum area of tie reinforcement, A_{sh}, shall be:

$$A_{sh} = 0.09 h_{cc} s \left(1 - \frac{F_y A_s}{P_n}\right)\left(\frac{f'_c}{F_{ysr}}\right) \quad \text{(D1-8)}$$

where
- A_s = cross-sectional area of the structural steel core, in.² (mm²)
- F_y = specified minimum yield stress of the structural *steel core*, ksi (MPa)
- F_{ysr} = specified minimum yield stress of the ties, ksi (MPa)
- P_n = nominal compressive strength of the *composite column* calculated in accordance with the *Specification*, kips (N)
- h_{cc} = cross-sectional dimension of the confined core measured center-to-center of the tie reinforcement, in. (mm)
- f'_c = specified compressive strength of concrete, ksi (MPa)
- s = spacing of transverse reinforcement measured along the longitudinal axis of the structural member, in. (mm)

Equation D1-8 need not be satisfied if the *nominal strength* of the concrete-encased structural steel section alone is greater than the *load effect* from a load combination of $1.0D + 0.5L$.

where
- D = dead load due to the weight of the structural elements and permanent features on the building, kips (N)
- L = live load due to occupancy and moveable equipment, kips (N)

(ii) The maximum spacing of transverse reinforcement along the length of the column shall be the lesser of six longitudinal load-carrying bar diameters or 6 in. (150 mm).

(iii) When specified in Sections D1.4b(1)(2), (3) or (4), the maximum spacing of transverse reinforcement along the member length shall be the lesser of one-fourth the least member dimension or 4 in. (100 mm). Confining reinforcement shall be spaced not more than 14 in. (350 mm) on center in the transverse direction.

(3) Encased composite columns in braced frames with required compressive strengths, without consideration of the amplified seismic loads, greater than $0.2P_n$ shall have transverse reinforcement as specified in Section D1.4b(2)(2)(iii) over the total element length. This requirement need not be satisfied if the nominal strength of the concrete-encased steel section alone is greater than the load effect from a load combination of $1.0D + 0.5L$.

(4) Composite columns supporting reactions from discontinued stiff members, such as walls or braced frames, shall have transverse reinforcement as specified in Section D1.4b(2)(2)(iii) over the full length beneath the level at which the discontinuity occurs if the required compressive strengths, without consideration of the amplified seismic loads, exceeds $0.1P_n$. Transverse

reinforcement shall extend into the discontinued member for at least the length required to develop full yielding in the concrete-encased steel section and longitudinal reinforcement. This requirement need not be satisfied if the nominal strength of the concrete-encased steel section alone is greater than the load effect from a load combination of $1.0D + 0.5L$.

(5) Encased composite columns used in a C-SMF shall satisfy the following requirements:

    (i) Transverse reinforcement shall satisfy the requirements in Section D1.4b(2)(2) at the top and bottom of the column over the region specified in Section D1.4b(1)(2).

    (ii) The strong-column/weak-beam design requirements in Section G3.4a shall be satisfied. Column bases shall be detailed to sustain inelastic flexural hinging.

    (iii) The *required shear strength* of the column shall satisfy the requirements of ACI 318 Section 21.6.5.1.

(6) When the column terminates on a footing or mat foundation, the transverse reinforcement as specified in this section shall extend into the footing or mat at least 12 in. (300 mm). When the column terminates on a wall, the transverse reinforcement shall extend into the wall for at least the length required to develop full yielding in the concrete-encased shape and longitudinal reinforcement.

4c. Filled Composite Columns

This section applies to columns that meet the limitations of *Specification* Section I2.2. Such columns shall be designed to satisfy the requirements of *Specification* Chapter I, except that the *nominal shear strength* of the composite column shall be the nominal shear strength of the structural steel section alone, based on its effective shear area.

5. Composite Slab Diaphragms

The design of composite floor and roof slab diaphragms for seismic effects shall meet the following requirements.

5a. Load Transfer

Details shall be provided to transfer loads between the diaphragm and *boundary members*, *collector* elements, and elements of the horizontal framing system.

5b. Nominal Shear Strength

The nominal in-plane shear strength of composite diaphragms and concrete slab on steel deck diaphragms shall be taken as the nominal shear strength of the reinforced concrete above the top of the steel deck ribs in accordance with ACI 318 excluding Chapter 22. Alternatively, the composite diaphragm nominal shear strength shall be determined by in-plane shear tests of concrete-filled diaphragms.

D2. CONNECTIONS

1. General

Connections, joints and fasteners that are part of the SFRS shall comply with *Specification* Chapter J, and with the additional requirements of this section.

Splices and bases of columns that are not designated as part of the SFRS shall satisfy the requirements of Sections D2.5a, D2.5c and D2.6.

Where *protected zones* are designated in connection elements by these Provisions or ANSI/AISC 358, they shall satisfy the requirements of Sections D1.3 and I2.1.

2. Bolted Joints

Bolted joints shall satisfy the following requirements:

(1) The available shear strength of bolted joints using standard holes shall be calculated as that for bearing-type joints in accordance with *Specification* Sections J3.6 and J3.10. The nominal bearing strength at bolt holes shall not be taken greater than $2.4dtF_u$.

(2) Bolts and welds shall not be designed to share force in a joint or the same force component in a connection.

> **User Note:** A member force, such as a *diagonal brace* axial force, must be resisted at the connection entirely by one type of joint (in other words, either entirely by bolts or entirely by welds). A connection in which bolts resist a force that is normal to the force resisted by welds, such as a moment connection in which welded flanges transmit flexure and a bolted web transmits shear, is not considered to be sharing the force.

(3) Bolt holes shall be standard holes or short-slotted holes perpendicular to the applied load.

Exception: For diagonal braces specified in Sections F1, F2, F3 and F4, oversized holes are permitted in one connection ply only when the connection is designed as a slip-critical joint for the required brace connection strength in Sections F1, F2, F3 and F4.

> **User Note:** Diagonal brace connections with oversized holes must also satisfy other limit states including bolt bearing and bolt shear for the *required strength* of the connection as defined in Sections F1, F2, F3 and F4. Alternative hole types are permitted if designated in ANSI/AISC 358, or if otherwise determined in a connection prequalification in accordance with Section K1, or if determined in a program of qualification testing in accordance with Section K2 or Section K3.

(4) All bolts shall be installed as pretensioned high-strength bolts. Faying surfaces shall satisfy the requirements for slip-critical connections in accordance with *Specification* Section J3.8 with a faying surface with a Class A slip coefficient or higher.

Exceptions: Connection surfaces are permitted to have coatings with a slip coefficient less than that of a Class A faying surface for the following:

(1) End plate moment connections conforming to the requirements of Section E1, or ANSI/AISC 358
(2) Bolted joints where the load effects due to seismic are transferred either by tension in bolts or by compression bearing but not by shear in bolts

3. Welded Joints

Welded joints shall be designed in accordance with Chapter J of the *Specification.*

4. Continuity Plates and Stiffeners

The design of *continuity plates* and stiffeners located in the webs of rolled shapes shall allow for the reduced contact lengths to the member flanges and web based on the corner clip sizes in Section I2.4.

5. Column Splices

5a. Location of Splices

For all building columns, including those not designated as part of the SFRS, column splices shall be located 4 ft (1.2 m) or more away from the beam-to-column flange connections.

Exceptions:

(1) When the column clear height between beam-to-column flange connections is less than 8 ft (2.4 m), splices shall be at half the clear height
(2) Column splices with webs and flanges joined by complete-joint-penetration groove welds are permitted to be located closer to the beam-to-column flange connections, but not less than the depth of the column
(3) Splices in *composite columns*

User Note: Where possible, splices should be located at least 4 ft (1.2 m) above the finished floor elevation to permit installation of perimeter safety cables prior to erection of the next tier and to improve accessibility.

5b. Required Strength

The required strength of column splices in the SFRS shall be the greater of:

(a) The required strength of the columns, including that determined from Chapters E, F, G and H and Section D1.4a; or,
(b) The required strength determined using the *load combinations* stipulated in the *applicable building code* including the *amplified seismic load*. The required strength need not exceed the maximum loads that can be transferred to the splice by the system.

In addition, welded column splices in which any portion of the column is subject to a calculated net tensile load effect determined using the load combinations stipulated in the applicable building code, including the amplified seismic load, shall satisfy all of the following requirements:

(1) The *available strength* of partial-joint-penetration (PJP) groove welded joints, if used, shall be at least equal to 200% of the required strength.

(2) The available strength for each flange splice shall be at least equal to $0.5R_y F_y b_f t_f$ (LRFD) or $(0.5/1.5)R_y F_y b_f t_f$ (ASD), as applicable, where $R_y F_y$ is the *expected yield stress* of the column material and $b_f t_f$ is the area of one flange of the smaller column connected.

(3) Where butt joints in column splices are made with complete-joint-penetration (CJP) groove welds, when tension stress at any location in the smaller flange exceeds $0.30F_y$ (LRFD) or $0.20F_y$ (ASD), tapered transitions are required between flanges of unequal thickness or width. Such transitions shall be in accordance with AWS D1.8/D1.8M clause 4.2.

5c. Required Shear Strength

For all building columns including those not designated as part of the SFRS, the *required shear strength* of column splices with respect to both orthogonal axes of the column shall be M_{pc}/H (LRFD) or $M_{pc}/(1.5H)$ (ASD), as applicable, where M_{pc} is the lesser nominal plastic flexural strength of the column sections for the direction in question, and H is the height of the story.

The required shear strength of splices of columns in the SFRS shall be the greater of the above requirement or the required shear strength determined per Section D2.5b(a) and (b).

5d. Structural Steel Splice Configurations

Structural steel column splices are permitted to be either bolted or welded, or welded to one column and bolted to the other. Splice configurations shall meet all specific requirements in Chapters E, F, G or H.

Splice plates or channels used for making web splices in SFRS columns shall be placed on both sides of the column web.

For welded butt joint splices made with groove welds, weld tabs shall be removed in accordance with AWS D1.8/D1.8M clause 6.11. Steel backing of groove welds need not be removed.

5e. Splices in Encased Composite Columns

For *encased composite columns*, column splices shall conform to Section D1.4b and ACI 318 Section 21.6.3.2.

6. Column Bases

The required strength of *column bases*, including those that are not designated as part of the SFRS, shall be calculated in accordance with this section.

The available strength of steel elements at the column base, including base plates, anchor rods, stiffening plates, and shear lug elements shall be in accordance with the *Specification*.

Where columns are welded to base plates with groove welds, weld tabs and weld backing shall be removed, except that weld backing located on the inside of flanges and weld backing on the web of I-shaped sections need not be removed if backing is attached to the column base plate with a continuous $^5/_{16}$-in. fillet weld. Fillet welds of backing to the inside of column flanges are prohibited.

The available strength of concrete elements at the column base, including anchor rod embedment and reinforcing steel, shall be in accordance with ACI 318 Appendix D.

User Note: When using concrete reinforcing steel as part of the anchorage embedment design, it is important to consider the anchor failure modes and provide reinforcement that is developed on both sides of the expected failure surface. See ACI 318 Appendix D, including Commentary.

6a. Required Axial Strength

The *required axial strength* of column bases that are designated as part of the SFRS, including their attachment to the foundation, shall be the summation of the vertical components of the *required connection strengths* of the steel elements that are connected to the column base, but not less than the greater of:

(a) The column axial load calculated using the load combinations of the applicable building code, including the amplified seismic load

(b) The required axial strength for column splices, as prescribed in Section D2.5

User Note: The vertical components can include both the axial load from columns and the vertical component of the axial load from diagonal members framing into the column base. Section D2.5 includes references to Section D1.4a and Chapters E, F, G and H. Where diagonal braces frame to both sides of a column, the effects of compression brace buckling should be considered in the summation of vertical components. See Section F2.3.

6b. Required Shear Strength

The required shear strength of column bases, including those not designated as part of the SFRS, and their attachments to the foundations, shall be the summation of the horizontal component of the required connection strengths of the steel elements that are connected to the column base as follows:

(a) For diagonal braces, the horizontal component shall be determined from the required strength of diagonal brace connections for the SFRS.

(b) For columns, the horizontal component shall be equal to the required shear strength for column splices prescribed in Section D2.5c.

Exception: Single story columns with simple connections at both ends need not comply with Section D2.6b(b).

> **User Note:** The horizontal components can include the shear load from columns and the horizontal component of the axial load from diagonal members framing into the column base. Section D2.5 includes references to Section D1.4a and Chapters E, F, G and H.

6c. Required Flexural Strength

Where column bases are designed as moment connections to the foundation, the *required flexural strength* of column bases that are designated as part of the SFRS, including their attachment to the foundation, shall be the summation of the required connection strengths of the steel elements that are connected to the column base as follows:

(a) For diagonal braces, the required flexural strength shall be at least equal to the required flexural strength of diagonal brace connections.

(b) For columns, the required flexural strength shall be at least equal to the lesser of the following:

  (i) $1.1 R_y F_y Z$ (LRFD) or $(1.1/1.5) R_y F_y Z$ (ASD), as applicable, of the column, or

  (ii) the moment calculated using the load combinations of the applicable building code, including the amplified seismic load.

> **User Note:** Moments at column to column base connections designed as simple connections may be ignored.

7. Composite Connections

This section applies to connections in buildings that utilize composite steel and concrete systems wherein seismic load is transferred between structural steel and reinforced concrete components. Methods for calculating the connection strength shall satisfy the requirements in this section. Unless the connection strength is determined by analysis or testing, the models used for design of connections shall satisfy the following requirements:

(1) Force shall be transferred between structural steel and reinforced concrete through:

  (a) direct bearing from internal bearing mechanisms;

  (b) shear connection;

  (c) shear friction with the necessary clamping force provided by reinforcement normal to the plane of shear transfer; or

  (d) a combination of these means.

   The contribution of different mechanisms is permitted to be combined only if the stiffness and deformation capacity of the mechanisms are compatible.

Any potential bond strength between structural steel and reinforced concrete shall be ignored for the purpose of the connection force transfer mechanism.

(2) The nominal bearing and shear-friction strengths shall meet the requirements of ACI 318 Chapters 10 and 11. Unless a higher strength is substantiated by cyclic testing, the nominal bearing and shear-friction strengths shall be reduced by 25% for the composite seismic systems described in Sections G3, H2, H3, H5 and H6.

(3) *Face bearing plates* consisting of stiffeners between the flanges of steel beams shall be provided when beams are embedded in reinforced concrete columns or walls.

(4) The *nominal shear strength* of concrete-encased steel panel zones in beam-to-column connections shall be calculated as the sum of the nominal strengths of the structural steel and confined reinforced concrete shear elements as determined in Section E3.6e and ACI 318 Section 21.7, respectively.

(5) Reinforcement shall be provided to resist all tensile forces in reinforced concrete components of the connections. Additionally, the concrete shall be confined with transverse reinforcement. All reinforcement shall be fully developed in tension or compression, as applicable, beyond the point at which it is no longer required to resist the forces. Development lengths shall be determined in accordance with ACI 318 Chapter 12. Additionally, development lengths for the systems described in Sections G3, H2, H3, H5 and H6 shall satisfy the requirements of ACI 318 Section 21.7.5.

(6) Composite connections shall satisfy the following additional requirements:

   (i) When the slab transfers horizontal diaphragm forces, the slab reinforcement shall be designed and anchored to carry the in-plane tensile forces at all critical sections in the slab, including connections to *collector* beams, columns, diagonal braces and walls.

   (ii) For connections between structural steel or *composite beams* and reinforced concrete or encased composite columns, transverse hoop reinforcement shall be provided in the connection region of the column to satisfy the requirements of ACI 318 Section 21.7, except for the following modifications:

   (1) Structural steel sections framing into the connections are considered to provide confinement over a width equal to that of face bearing plates welded to the beams between the flanges.

   (2) Lap splices are permitted for perimeter ties when confinement of the splice is provided by face bearing plates or other means that prevents spalling of the concrete cover in the systems described in Sections G1, G2, H1 and H4.

   (3) The longitudinal bar sizes and layout in reinforced concrete and composite columns shall be detailed to minimize slippage of the bars through the beam-to-column connection due to high force transfer associated with the change in column moments over the height of the connection.

8. Steel Anchors

Where steel headed stud anchors or welded reinforcing bar anchors are part of the intermediate or special SFRS of Sections G2, G3, G4, H2, H3, H5 and H6, their shear and tensile strength shall be reduced by 25% from the specified strengths given in *Specification* Chapter I.

> **User Note:** The 25% reduction is not necessary for gravity and collector components in structures with intermediate or special *seismic force resisting systems* designed for the amplified seismic load.

D3. DEFORMATION COMPATIBILITY OF NON-SFRS MEMBERS AND CONNECTIONS

Where deformation compatibility of members and connections that are not part of the *seismic force resisting system* (SFRS) is required by the *applicable building code*, these elements shall be designed to resist the combination of gravity load effects and the effects of deformations occurring at the *design story drift* calculated in accordance with the applicable building code.

> **User Note:** ASCE/SEI 7 stipulates the above requirement for both structural steel and composite members and connections. Flexible shear connections that allow member end rotations per Section J1.2 of the *Specification* should be considered to meet these requirements. Inelastic deformations are permitted in connections or members provided they are self-limiting and do not create instability in the member. See the Commentary for further discussion.

D4. H-PILES

1. Design Requirements

Design of H-piles shall comply with the requirements of the *Specification* regarding design of members subjected to combined loads. H-piles shall satisfy the requirements for *highly ductile members* of Section D1.1.

2. Battered H-Piles

If battered (sloped) and vertical piles are used in a pile group, the vertical piles shall be designed to support the combined effects of the dead and live loads without the participation of the battered piles.

3. Tension

Tension in each pile shall be transferred to the pile cap by mechanical means such as shear keys, reinforcing bars, or studs welded to the embedded portion of the pile.

4. **Protected Zone**

At each pile, the length equal to the depth of the pile cross section located directly below the bottom of the pile cap shall be designated as a *protected zone* meeting the requirements of Sections D1.3 and I2.1.

CHAPTER E

MOMENT-FRAME SYSTEMS

This chapter provides the basis of design, the requirements for analysis, and the requirements for the system, members and connections for steel moment-frame systems.

The chapter is organized as follows:

- E1. Ordinary Moment Frames (OMF)
- E2. Intermediate Moment Frames (IMF)
- E3. Special Moment Frames (SMF)
- E4. Special Truss Moment Frames (STMF)
- E5. Ordinary Cantilever Column Systems (OCCS)
- E6. Special Cantilever Column Systems (SCCS)

User Note: The requirements of this chapter are in addition to those required by the *Specification* and the *applicable building code*.

E1. ORDINARY MOMENT FRAMES (OMF)

1. Scope

Ordinary moment frames (OMF) of structural steel shall be designed in conformance with this section.

2. Basis of Design

OMF designed in accordance with these provisions are expected to provide minimal inelastic deformation capacity in their members and connections.

3. Analysis

There are no additional analysis requirements.

4. System Requirements

There are no additional system requirements.

5. Members

5a. Basic Requirements

There are no limitations on width-to-thickness ratios of members for OMF, beyond those in the *Specification*. There are no requirements for stability bracing of beams or joints in OMF, beyond those in the *Specification*. Structural steel beams in OMF are permitted to be composite with a reinforced concrete slab to resist gravity loads.

5b. Protected Zones

There are no designated *protected zones* for OMF members.

6. Connections

Beam-to-column connections are permitted to be fully restrained (FR) or partially restrained (PR) moment connections in accordance with this section.

6a. Demand Critical Welds

Complete-joint-penetration (CJP) groove welds of beam flanges to columns are *demand critical welds*, and shall satisfy the requirements of Section A3.4b and I2.3.

6b. FR Moment Connections

FR moment connections that are part of the *seismic force resisting system* (SFRS) shall satisfy at least one of the following requirements:

(a) FR moment connections shall be designed for a *required flexural strength* that is equal to the expected beam flexural strength multiplied by 1.1 (LRFD) or by 1.1/1.5 (ASD), as appropriate. The expected beam flexural strength shall be determined as $R_y M_p$.

The *required shear strength*, V_u or V_a, as appropriate, of the connection shall be based on the load combinations in the *applicable building code* that include the *amplified seismic load*. In determining the amplified seismic load the effect of horizontal forces including overstrength, E_{mh}, shall be taken as:

$$E_{mh} = 2[1.1 R_y M_p]/L_{cf} \qquad \text{(E1-1)}$$

where
L_{cf} = clear length of beam, in. (mm)
$M_p = F_y Z$, kip-in. (N-mm)
R_y = ratio of *expected yield stress* to the specified minimum yield stress, F_y

(b) FR moment connections shall be designed for a required flexural strength and a required shear strength equal to the maximum moment and corresponding shear that can be transferred to the connection by the system, including the effects of material overstrength and strain hardening.

> **User Note:** Factors that may limit the maximum moment and corresponding shear that can be transferred to the connection include:
> (1) the strength of the columns, and
> (2) the resistance of the foundations to uplift.

For options (a) and (b) in Section E1.6b, *continuity plates* should be provided as required by Sections J10.1, J10.2 and J10.3 of the *Specification*. The bending moment used to check for continuity plates should be the same bending moment used to design the beam-to-column connection; in other words, either $1.1 R_y M_p$ (LRFD) or $(1.1/1.5) R_y M_p$ (ASD) or the maximum moment that can be transferred to the connection by the system.

(c) FR moment connections between wide flange beams and the flange of wide flange columns shall either satisfy the requirements of Section E2.6 or E3.6, or shall satisfy the following requirements:

(1) All welds at the beam-to-column connection shall satisfy the requirements of Chapter 3 of ANSI/AISC 358.

(2) Beam flanges shall be connected to column flanges using complete-joint-penetration (CJP) groove welds.

(3) The shape of weld access holes shall be in accordance with subclause 6.10.1.2 of AWS D1.8/D1.8M. Weld access hole quality requirements shall be in accordance with subclause 6.10.2 of AWS D1.8/D1.8M.

(4) Continuity plates shall satisfy the requirements of Section E3.6f.

Exception: The welded joints of the continuity plates to the column flanges are permitted to be complete-joint-penetration groove welds, two-sided partial-joint-penetration groove welds with reinforcement, or two-sided fillet welds. The *required strength* of these joints shall not be less than the *available strength* of the contact area of the plate with the column flange.

(5) The beam web shall be connected to the column flange using either a CJP groove weld extending between weld access holes, or using a bolted single plate shear connection designed for required shear strength per Equation E1-1.

User Note: For FR moment connections, panel zone shear strength should be checked in accordance with Section J10.6 of the *Specification*. The required shear strength of the panel zone should be based on the beam end moments computed from the load combinations stipulated by the applicable building code, not including the amplified seismic load.

6c. PR Moment Connections

PR moment connections shall satisfy the following requirements:

(1) Connections shall be designed for the maximum moment and shear from the applicable load combinations as described in Sections B2 and B3.

(2) The stiffness, strength and deformation capacity of PR moment connections shall be considered in the design, including the effect on overall frame stability.

(3) The nominal flexural strength of the connection, $M_{n,PR}$, shall be no less than 50% of M_p of the connected beam.

Exception: For one-story structures, $M_{n,PR}$ shall be no less than 50% of M_p of the connected column.

(4) V_u or V_a as appropriate, shall be determined per Section E1.6b(a) with M_p in Equation E1-1 taken as $M_{n,PR}$.

E2. INTERMEDIATE MOMENT FRAMES (IMF)

1. Scope

Intermediate moment frames (IMF) of structural steel shall be designed in conformance with this section.

2. Basis of Design

IMF designed in accordance with these provisions are expected to provide limited inelastic deformation capacity through flexural yielding of the IMF beams and columns, and shear yielding of the column panel zones. Design of connections of beams to columns, including panel zones and *continuity plates*, shall be based on connection tests that provide the performance required by Section E2.6b, and demonstrate this conformance as required by Section E2.6c.

3. Analysis

There are no additional analysis requirements.

4. System Requirements

4a. Stability Bracing of Beams

Beams shall be braced to satisfy the requirements for *moderately ductile members* in Section D1.2a.

In addition, unless otherwise indicated by testing, beam braces shall be placed near concentrated forces, changes in cross section, and other locations where analysis indicates that a *plastic hinge* will form during inelastic deformations of the IMF. The placement of stability bracing shall be consistent with that documented for a *prequalified connection* designated in ANSI/AISC 358, or as otherwise determined in a connection prequalification in accordance with Section K1, or in a program of qualification testing in accordance with Section K2.

The *required strength* of lateral bracing provided adjacent to plastic hinges shall be as required by Section D1.2c.

5. Members

5a. Basic Requirements

Beam and column members shall satisfy the requirements of Section D1 for moderately ductile members, unless otherwise qualified by tests.

Structural steel beams in IMF are permitted to be composite with a reinforced concrete slab to resist gravity loads.

5b. Beam Flanges

Abrupt changes in beam flange area shall not be permitted in plastic hinge regions. The drilling of flange holes or trimming of beam flange width shall not be permitted unless testing or qualification demonstrates that the resulting configuration can develop stable plastic hinges to accommodate the required *story drift angle*. The configuration shall be consistent with a prequalified connection designated in ANSI/AISC 358, or as otherwise determined in a connection prequalification in accordance with Section K1, or in a program of qualification testing in accordance with Section K2.

5c. **Protected Zones**

The region at each end of the beam subject to inelastic straining shall be designated as a *protected zone*, and shall satisfy the requirements of Section D1.3. The extent of the protected zone shall be as designated in ANSI/AISC 358, or as otherwise determined in a connection prequalification in accordance with Section K1, or as determined in a program of qualification testing in accordance with Section K2.

> **User Note:** The plastic hinging zones at the ends of IMF beams should be treated as protected zones. The plastic hinging zones should be established as part of a prequalification or qualification program for the connection, per Section E2.6c. In general, for unreinforced connections, the protected zone will extend from the face of the column to one half of the beam depth beyond the plastic hinge point.

6. **Connections**

6a. **Demand Critical Welds**

The following welds are *demand critical welds*, and shall satisfy the requirements of Section A3.4b and I2.3:

(1) Groove welds at column splices.

(2) Welds at column-to-base plate connections.

Exception: Where it can be shown that column hinging at, or near, the base plate is precluded by conditions of restraint, and in the absence of net tension under load combinations including the *amplified seismic load*, demand critical welds are not required.

(3) Complete-joint-penetration groove welds of beam flanges and beam webs to columns, unless otherwise designated by ANSI/AISC 358, or otherwise determined in a connection prequalification in accordance with Section K1, or as determined in a program of qualification testing in accordance with Section K2.

> **User Note:** For the designation of demand critical welds, standards such as ANSI/AISC 358 and tests addressing specific connections and joints should be used in lieu of the more general terms of these Provisions. Where these Provisions indicate that a particular weld is designated demand critical, but the more specific standard or test does not make such a designation, the more specific standard or test should govern. Likewise, these standards and tests may designate welds as demand critical that are not identified as such by these Provisions.

6b. **Beam-to-Column Connection Requirements**

Beam-to-column connections used in the SFRS shall satisfy the following requirements:

(1) The connection shall be capable of accommodating a story drift angle of at least 0.02 rad.

(2) The *measured flexural resistance* of the connection, determined at the column face, shall equal at least $0.80M_p$ of the connected beam at a story drift angle of 0.02 rad.

6c. Conformance Demonstration

Beam-to-column connections used in the SFRS shall satisfy the requirements of Section E2.6b by one of the following:

(a) Use of IMF connections designed in accordance with ANSI/AISC 358.

(b) Use of a connection prequalified for IMF in accordance with Section K1.

(c) Provision of qualifying cyclic test results in accordance with Section K2. Results of at least two cyclic connection tests shall be provided and are permitted to be based on one of the following:

  (i) Tests reported in the research literature or documented tests performed for other projects that represent the project conditions, within the limits specified in Section K2.

  (ii) Tests that are conducted specifically for the project and are representative of project member sizes, material strengths, connection configurations, and matching connection processes, within the limits specified in Section K2.

6d. Required Shear Strength

The *required shear strength* of the connection shall be based on the load combinations in the *applicable building code* that include the amplified seismic load. In determining the amplified seismic load the effect of horizontal forces including overstrength, E_{mh}, shall be taken as:

$$E_{mh} = 2[1.1R_y M_p]/L_h \tag{E2-1}$$

where
  L_h = distance between beam plastic hinge locations as defined within the test report or ANSI/AISC 358, in. (mm)
  $M_p = F_y Z$ = nominal plastic flexural strength, kip-in. (N-mm)
  R_y = ratio of the *expected yield stress* to the specified minimum yield stress, F_y

Exception: In lieu of Equation E2-1, the required shear strength of the connection shall be as specified in ANSI/AISC 358, or as otherwise determined in a connection prequalification in accordance with Section K1, or in a program of qualification testing in accordance with Section K2.

6e. Panel Zone

There are no additional panel zone requirements.

User Note: Panel zone shear strength should be checked in accordance with Section J10.6 of the *Specification*. The required shear strength of the panel zone should be based on the beam end moments computed from the load combinations stipulated by the applicable building code, not including the amplified seismic load.

6f. Continuity Plates

Continuity plates shall be provided in accordance with the provisions of Section E3.6f.

6g. Column Splices

Column splices shall comply with the requirements of Section D2.5. Where welds are used to make the splice, they shall be complete-joint-penetration groove welds.

When bolted column splices are used, they shall have a *required flexural strength* that is at least equal to $R_y F_y Z_x$ (LRFD) or $R_y F_y Z_x/1.5$ (ASD), as appropriate, of the smaller column, where Z_x is the plastic section modulus about the x-axis. The required shear strength of column web splices shall be at least equal to $\Sigma M_{pc}/H$ (LRFD) or $\Sigma M_{pc}/(1.5H)$ (ASD), as appropriate, where ΣM_{pc} is the sum of the nominal plastic flexural strengths of the columns above and below the splice.

Exception: The required strength of the column splice considering appropriate stress concentration factors or fracture mechanics stress intensity factors need not exceed that determined by a nonlinear analysis as specified in Chapter C.

E3. SPECIAL MOMENT FRAMES (SMF)

1. Scope

Special moment frames (SMF) of structural steel shall be designed in conformance with this section.

2. Basis of Design

SMF designed in accordance with these provisions are expected to provide significant inelastic deformation capacity through flexural yielding of the SMF beams and limited yielding of column panel zones. Except where otherwise permitted in this section, columns shall be designed to be stronger than the fully yielded and strain-hardened beams or girders. Flexural yielding of columns at the base is permitted. Design of connections of beams to columns, including panel zones and *continuity plates*, shall be based on connection tests that provide the performance required by Section E3.6b, and demonstrate this conformance as required by Section E3.6c.

3. Analysis

There are no additional analysis requirements.

4. System Requirements

4a. Moment Ratio

The following relationship shall be satisfied at beam-to-column connections:

$$\frac{\Sigma M^*_{pc}}{\Sigma M^*_{pb}} > 1.0 \tag{E3-1}$$

where

ΣM^*_{pc} = the sum of the projections of the nominal flexural strengths of the columns (including haunches where used) above and below the joint to the beam centerline with a reduction for the axial force in the column. It is permitted to determine ΣM^*_{pc} as follows:

$$\Sigma M^*_{pc} = \Sigma Z_c(F_{yc} - P_{uc}/A_g) \text{ (LRFD)} \tag{E3-2a}$$

or

$$\Sigma M^*_{pc} = \Sigma Z_c(F_{yc} - 1.5P_{ac}/A_g) \text{ (ASD)}, \tag{E3-2b}$$

as appropriate.

When the centerlines of opposing beams in the same joint do not coincide, the mid-line between centerlines shall be used.

ΣM^*_{pb} = the sum of the projections of the expected flexural strengths of the beams at the *plastic hinge* locations to the column centerline. It is permitted to determine ΣM^*_{pb} as follows:

$$\Sigma M^*_{pb} = \Sigma(1.1 R_y F_{yb} Z_b + M_{uv}) \text{ (LRFD)} \tag{E3-3a}$$

or

$$\Sigma M^*_{pb} = \Sigma(1.1 R_y F_{yb} Z_b + 1.5 M_{av}) \text{ (ASD)}, \tag{E3-3b}$$

as appropriate.

Alternatively, it is permitted to determine ΣM^*_{pb} consistent with a *prequalified connection* design as designated in ANSI/AISC 358, or as otherwise determined in a connection prequalification in accordance with Section K1, or in a program of qualification testing in accordance with Section K2. When connections with *reduced beam sections* are used, it is permitted to determine ΣM^*_{pb} as follows:

$$\Sigma M^*_{pb} = \Sigma(1.1 R_y F_{yb} Z_{RBS} + M_{uv}) \text{ (LRFD)} \tag{E3-4a}$$

or

$$\Sigma M^*_{pb} = \Sigma(1.1 R_y F_{yb} Z_{RBS} + 1.5 M_{av}) \text{ (ASD)}, \tag{E3-4b}$$

as appropriate.

A_g = gross area of column, in.² (mm²)
F_{yb} = specified minimum yield stress of beam, ksi (MPa)
F_{yc} = specified minimum yield stress of column, ksi (MPa)
M_{av} = additional moment due to shear amplification from the location of the plastic hinge to the column centerline based on *ASD load combinations*, kip-in. (N-mm)
M_{uv} = additional moment due to shear amplification from the location of the plastic hinge to the column centerline based on *LRFD load combinations*, kip-in. (N-mm)
P_{ac} = required compressive strength using ASD load combinations, including the *amplified seismic load*, kips (N)
P_{uc} = required compressive strength using LRFD load combinations, including the amplified seismic load, kips (N)
Z_b = plastic section modulus of the beam, in.³ (mm³)

Z_c = plastic section modulus of the column, in.³ (mm³)
Z_{RBS} = minimum plastic section modulus at the reduced beam section, in.³ (mm³)

Exception: This requirement shall not apply if the following conditions in (a) or (b) are satisfied.

(a) Columns with $P_{rc} < 0.3 P_c$ for all load combinations other than those determined using the *amplified seismic load* that satisfy either of the following:

(i) Columns used in a one-story building or the top story of a multistory building.

(ii) Columns where: (1) the sum of the *available shear strengths* of all *exempted columns* in the story is less than 20% of the sum of the available shear strengths of all moment frame columns in the story acting in the same direction; and (2) the sum of the available shear strengths of all exempted columns on each moment frame column line within that story is less than 33% of the available shear strength of all moment frame columns on that column line. For the purpose of this exception, a column line is defined as a single line of columns or parallel lines of columns located within 10% of the plan dimension perpendicular to the line of columns.

User Note: For purposes of this exception, the available shear strengths of the columns should be calculated as the limit strengths considering the flexural strength at each end as limited by the flexural strength of the attached beams, or the flexural strength of the columns themselves, divided by H, where H is the story height in inches (mm).

The nominal compressive strength, P_c, shall be

$$P_c = F_{yc}A_g \text{ (LRFD)} \qquad \text{(E3-5a)}$$

or

$$P_c = F_{yc}A_g/1.5 \text{ (ASD)} \qquad \text{(E3-5b)}$$

and $P_{rc} = P_{uc}$, (LRFD) or $P_{rc} = P_{ac}$ (ASD), as appropriate.

(b) Columns in any story that has a ratio of available shear strength to *required shear strength* that is 50% greater than the story above.

4b.   Stability Bracing of Beams

Beams shall be braced to satisfy the requirements for *highly ductile members* in Section D1.2b.

In addition, unless otherwise indicated by testing, beam braces shall be placed near concentrated forces, changes in cross section, and other locations where analysis indicates that a plastic hinge will form during inelastic deformations of the SMF. The placement of lateral bracing shall be consistent with that documented for a prequalified connection designated in ANSI/AISC 358, or as otherwise determined in a

Sect. E3.] SPECIAL MOMENT FRAMES (SMF) 9.1–37

connection prequalification in accordance with Section K1, or in a program of qualification testing in accordance with Section K2.

The *required strength* of stability bracing provided adjacent to plastic hinges shall be as required by Section D1.2c.

4c. Stability Bracing at Beam-to-Column Connections

(1) Braced Connections

When the webs of the beams and column are co-planar, and a column is shown to remain elastic outside of the panel zone, column flanges at beam-to-column connections shall require stability bracing only at the level of the top flanges of the beams. It shall be permitted to assume that the column remains elastic when the ratio calculated using Equation E3-1 is greater than 2.0.

When a column cannot be shown to remain elastic outside of the panel zone, the following requirements shall apply:

(1) The column flanges shall be laterally braced at the levels of both the top and bottom beam flanges. Stability bracing is permitted to be either direct or indirect.

> **User Note:** Direct stability bracing of the column flange is achieved through use of *member braces* or other members, deck and slab, attached to the column flange at or near the desired bracing point to resist lateral buckling. Indirect stability bracing refers to bracing that is achieved through the stiffness of members and connections that are not directly attached to the column flanges, but rather act through the column web or stiffener plates.

(2) Each column-flange member brace shall be designed for a required strength that is equal to 2% of the available beam flange strength $F_y b_f t_{bf}$ (LRFD) or $F_y b_f t_{bf}/1.5$ (ASD), as appropriate.

(2) Unbraced Connections

A column containing a beam-to-column connection with no member bracing transverse to the seismic frame at the connection shall be designed using the distance between adjacent member braces as the column height for buckling transverse to the seismic frame and shall conform to *Specification* Chapter H, except that:

(1) The required column strength shall be determined from the load combinations in the *applicable building code* that include the amplified seismic load.

In determining the amplified seismic load the effect of horizontal forces including overstrength, E_{mh}, need not exceed 125% of the frame *available strength* based upon either the beam *available flexural strength* or panel zone *available shear strength*.

(2) The slenderness L/r for the column shall not exceed 60, where

L = length of column, in. (mm)

r = governing radius of gyration, in. (mm)

(3) The column *required flexural strength* transverse to the seismic frame shall include that moment caused by the application of the beam flange force specified in Section E3.4c(1)(2) in addition to the second-order moment due to the resulting column flange lateral displacement.

5. Members

5a. Basic Requirements

Beam and column members shall satisfy the requirements of Section D1.1 for highly ductile members, unless otherwise qualified by tests.

Structural steel beams in SMF are permitted to be composite with a reinforced concrete slab to resist gravity loads.

5b. Beam Flanges

Abrupt changes in beam flange area are prohibited in plastic hinge regions. The drilling of flange holes or trimming of beam flange width shall not be permitted unless testing or qualification demonstrates that the resulting configuration can develop stable plastic hinges to accommodate the required *story drift angle*. The configuration shall be consistent with a prequalified connection designated in ANSI/AISC 358, or as otherwise determined in a connection prequalification in accordance with Section K1, or in a program of qualification testing in accordance with Section K2.

5c. Protected Zones

The region at each end of the beam subject to inelastic straining shall be designated as a *protected zone*, and shall satisfy the requirements of Section D1.3. The extent of the protected zone shall be as designated in ANSI/AISC 358, or as otherwise determined in a connection prequalification in accordance with Section K1, or as determined in a program of qualification testing in accordance with Section K2.

> **User Note:** The plastic hinging zones at the ends of SMF beams should be treated as protected zones. The plastic hinging zones should be established as part of a prequalification or qualification program for the connection, per Section E3.6c. In general, for unreinforced connections, the protected zone will extend from the face of the column to one half of the beam depth beyond the plastic hinge point.

6. Connections

6a. Demand Critical Welds

The following welds are *demand critical welds*, and shall satisfy the requirements of Section A3.4b and I2.3:

(1) Groove welds at column splices
(2) Welds at column-to-base plate connections

   Exception: Where it can be shown that column hinging at, or near, the base plate is precluded by conditions of restraint, and in the absence of net tension under load combinations including the amplified seismic load, demand critical welds are not required.

(3) Complete-joint-penetration groove welds of beam flanges and beam webs to columns, unless otherwise designated by ANSI/AISC 358, or otherwise determined in a connection prequalification in accordance with Section K1, or as determined in a program of qualification testing in accordance with Section K2.

User Note: For the designation of demand critical welds, standards such as ANSI/AISC 358 and tests addressing specific connections and joints should be used in lieu of the more general terms of these Provisions. Where these Provisions indicate that a particular weld is designated demand critical, but the more specific standard or test does not make such a designation, the more specific standard or test should govern. Likewise, these standards and tests may designate welds as demand critical that are not identified as such by these Provisions.

6b. Beam-to-Column Connections

Beam-to-column connections used in the *seismic force resisting system* (SFRS) shall satisfy the following requirements:

(1) The connection shall be capable of accommodating a story drift angle of at least 0.04 rad.
(2) The *measured flexural resistance* of the connection, determined at the column face, shall equal at least $0.80M_p$ of the connected beam at a story drift angle of 0.04 rad.

6c. Conformance Demonstration

Beam-to-column connections used in the SFRS shall satisfy the requirements of Section E3.6b by one of the following:

(a) Use of SMF connections designed in accordance with ANSI/AISC 358.
(b) Use of a connection prequalified for SMF in accordance with Section K1.
(c) Provision of qualifying cyclic test results in accordance with Section K2. Results of at least two cyclic connection tests shall be provided and shall be based on one of the following:

   (i) Tests reported in the research literature or documented tests performed for other projects that represent the project conditions, within the limits specified in Section K2
   (ii) Tests that are conducted specifically for the project and are representative of project member sizes, material strengths, connection configurations, and matching connection processes, within the limits specified in Section K2

6d. Required Shear Strength

The required shear strength of the connection shall be based on the load combinations in the applicable building code that include the amplified seismic load. In determining the amplified seismic load the effect of horizontal forces including overstrength, E_{mh}, shall be taken as:

$$E_{mh} = 2(1.1 R_y M_p)/L_h \qquad (E3\text{-}6)$$

where
L_h = distance between plastic hinge locations as defined within the test report or ANSI/AISC 358, in. (mm)
M_p = nominal plastic flexural strength, kip-in. (N-mm)
R_y = ratio of the *expected yield stress* to the specified minimum yield stress, F_y

Exception: In lieu of Equation E3-6, the required shear strength of the connection shall be as specified in ANSI/AISC 358, or as otherwise determined in a connection prequalification in accordance with Section K1, or in a program of qualification testing in accordance with Section K2.

6e. Panel Zone

(1) Required Shear Strength

The required shear strength of the panel zone shall be determined from the summation of the moments at the column faces as determined by projecting the expected moments at the plastic hinge points to the column faces. The *design shear strength* shall be $\phi_v R_n$ and the *allowable shear strength* shall be R_n/Ω_v where

$$\phi_v = 1.0 \text{ (LRFD)} \qquad \Omega_v = 1.50 \text{ (ASD)}$$

and the *nominal shear strength*, R_n, in accordance with the limit state of shear yielding, is determined as specified in *Specification* Section J10.6.

Alternatively, the required thickness of the panel zone shall be determined in accordance with the method used in proportioning the panel zone of the tested or prequalified connection.

(2) Panel Zone Thickness

The individual thicknesses, t, of column webs and doubler plates, if used, shall conform to the following requirement:

$$t \geq (d_z + w_z)/90 \qquad (E3\text{-}7)$$

where
$d_z = d - 2t_f$ of the deeper beam at the connection, in. (mm)
t = thickness of column web or doubler plate, in. (mm)
w_z = width of panel zone between column flanges, in. (mm)

Alternatively, when local buckling of the column web and doubler plate is prevented by using plug welds joining them, and dividing the plate to conform with

Equation E3-7, the total panel zone thickness shall satisfy Equation E3-7. When plug welds are required, a minimum of four plug welds shall be provided.

(3) Panel Zone Doubler Plates

Doubler plates shall be applied directly to the column web, when the web is not in compliance with Section E3.6e(2). Otherwise, doubler plates are permitted to be applied directly to the column web, or spaced away from the web.

(1) Doubler plates in contact with the web

Doubler plates shall be welded to the column flanges to develop the available strength of the full doubler plate thickness, using either a complete-joint-penetration groove welded or fillet welded joint. When continuity plates are not used, the doubler plate shall be fillet welded across the top and bottom to develop the proportion of the total force that is transmitted to the doubler plate, unless the doubler plates and the web satisfy Section E3.6e(2).

(2) Spaced doubler plates

Doubler plates shall be welded to the column flanges to develop the available strength of the full doubler plate thickness, using a complete-joint-penetration groove welded joint. Doubler plates shall be placed symmetrically in pairs and located between $^1/_3$ and $^2/_3$ of the distance between the beam flange tip and column centerline.

(3) Doubler plates used with continuity plates

Each doubler plate shall be welded to the continuity plates to develop the proportion of the total force that is transmitted to the doubler plate.

(4) Doubler plates used without continuity plates

When continuity plates are not used, doubler plates shall be extended a minimum of 6 in. (150 mm) above and below the top and bottom of the deeper moment frame beam.

User Note: When a doubler plate interferes with connecting continuity plates directly to the column web, the designer must provide a load path that satisfies the requirements of ANSI/AISC 358 Section 2.4.4b. This may be accomplished by sizing the doubler plate such that it is capable of developing the required strength of the continuity plate to column web connection. Alternatively, the doubler plate can stop inside the continuity plates. A similar load path must be provided when the web plate for a beam perpendicular to the column web connects to a doubler plate.

6f. Continuity Plates

(1) Continuity Plate Requirements

Continuity plates shall be provided with the exception of the following conditions:

(1) When otherwise determined in a connection prequalification in accordance with Section K1, or as determined in a program of qualification testing in accordance with Section K2.

(2) When the beam flange is welded to the flange of a wide-flange or built-up I-shaped column having a thickness that satisfies Equations E3-8 and E3-9, continuity plates need not be provided:

$$t_{cf} \geq 0.4\sqrt{1.8 b_{bf} t_{bf} \frac{R_{yb} F_{yb}}{R_{yc} F_{yc}}} \qquad (E3\text{-}8)$$

$$t_{cf} \geq \frac{b_{bf}}{6} \qquad (E3\text{-}9)$$

where
F_{yb} = specified minimum yield stress of the beam flange, ksi (MPa)
F_{yc} = specified minimum yield stress of the column flange, ksi (MPa)
R_{yb} = ratio of the expected yield stress of the beam material to the specified minimum yield stress
R_{yc} = ratio of the expected yield stress of the column material to the specified minimum yield stress
b_{bf} = beam flange width, in. (mm)
t_{bf} = beam flange thickness, in. (mm)
t_{cf} = minimum required thickness of column flange when no continuity plates are provided, in. (mm)

(3) When the beam flange is welded to the flange of the I-shape in a boxed wide-flange column having a thickness that satisfies Equations E3-10 and E3-11, continuity plates need not be provided:

$$t_{cf} \geq 0.4\sqrt{\left[1 - \frac{b_{bf}}{b_{cf}^2}\left(b_{cf} - \frac{b_{bf}}{4}\right)\right] 1.8\, b_{bf} t_{bf} \frac{F_{yb} R_{yb}}{F_{yc} R_{yc}}} \qquad (E3\text{-}10)$$

$$t_{cf} \geq \frac{b_{bf}}{12} \qquad (E3\text{-}11)$$

(4) For bolted connections, the continuity plate provisions of ANSI/AISC 358 for the specific connection type shall apply.

(2) Continuity Plate Thickness

Where continuity plates are required, the thickness of the plates shall be determined as follows:

(a) For one-sided connections, continuity plate thickness shall be at least one-half of the thickness of the beam flange.

(b) For two-sided connections, the continuity plate thickness shall be at least equal to the thicker of the two beam flanges on either side of the column.

Continuity plates shall also conform to the requirements of Section J10 of the *Specification*.

(3) Continuity Plate Welding

Continuity plates shall be welded to column flanges using CJP groove welds.

Continuity plates shall be welded to column webs using groove welds or fillet welds. The required strength of the sum of the welded joints of the continuity plates to the column web shall be the smallest of the following:

(a) The sum of the *design strengths* in tension of the contact areas of the continuity plates to the column flanges that have attached beam flanges

(b) The design strength in shear of the contact area of the plate with the column web

(c) The design strength in shear of the column panel zone

(d) The sum of the *expected yield strengths* of the beam flanges transmitting force to the continuity plates

6g. Column Splices

Column splices shall comply with the requirements of Section D2.5. Where welds are used to make the splice, they shall be complete-joint-penetration groove welds.

When bolted column splices are used, they shall have a required flexural strength that is at least equal to $R_y F_y Z_x$ (LRFD) or $R_y F_y Z_x/1.5$ (ASD), as appropriate, of the smaller column. The required shear strength of column web splices shall be at least equal to $\Sigma M_{pc}/H$ (LRFD) or $\Sigma M_{pc}/1.5H$ (ASD), as appropriate, where ΣM_{pc} is the sum of the nominal plastic flexural strengths of the columns above and below the splice.

Exception: The required strength of the column splice considering appropriate stress concentration factors or fracture mechanics stress intensity factors need not exceed that determined by a nonlinear analysis as specified in Chapter C.

E4. SPECIAL TRUSS MOMENT FRAMES (STMF)

1. Scope

Special truss moment frames (STMF) of structural steel shall satisfy the requirements in this Section.

2. Basis of Design

STMF designed in accordance with these provisions are expected to provide significant inelastic deformation capacity within a *special segment* of the truss. STMF shall be limited to span lengths between columns not to exceed 65 ft (20 m) and overall depth not to exceed 6 ft (1.8 m). The columns and truss segments outside of the special segments shall be designed to remain elastic under the forces that can be generated by the fully yielded and strain-hardened special segment.

3. Analysis

Analysis of STMF shall satisfy the following requirements.

3a. Special Segment

The *required vertical shear strength* of the special segment shall be calculated for the appropriate *load combinations* in the *applicable building code*.

3b. Nonspecial Segment

The *required strength* of nonspecial segment members and connections shall be calculated based on the load combinations in the applicable building code that include the *amplified seismic load*. In determining the amplified seismic load the effect of horizontal forces including overstrength, E_{mh}, shall be taken as the lateral forces necessary to develop the *expected vertical shear strength* of the special segment acting at mid-length and defined in Section E4.5b. Second order effects at maximum design drift shall be included.

4. System Requirements

4a. Special Segment

Each horizontal truss that is part of the SFRS shall have a special segment that is located between the quarter points of the span of the truss. The length of the special segment shall be between 0.1 and 0.5 times the truss span length. The length-to-depth ratio of any panel in the special segment shall neither exceed 1.5 nor be less than 0.67.

Panels within a special segment shall either be all Vierendeel panels or all X-braced panels; neither a combination thereof nor the use of other truss diagonal configurations is permitted. Where diagonal members are used in the special segment, they shall be arranged in an X pattern separated by vertical members. Diagonal members within the special segment shall be made of rolled flat bars of identical sections. Such diagonal members shall be interconnected at points where they cross. The interconnection shall have a required strength equal to 0.25 times the *nominal tensile strength* of the diagonal member. Bolted connections shall not be used for diagonal members within the special segment.

Splicing of chord members shall not be permitted within the special segment, nor within one-half the panel length from the ends of the special segment.

The *required axial strength* of the diagonal web members in the special segment due to dead and live loads within the special segment shall not exceed $0.03 F_y A_g$ (LRFD) or $(0.03/1.5) F_y A_g$ (ASD), as appropriate.

4b. Stability Bracing of Trusses

Each flange of the chord members shall be laterally braced at the ends of the special segment. The required strength of the lateral brace shall be

$$P_u = 0.06 R_y F_y A_f \text{ (LRFD)} \tag{E4-1a}$$

or

$$P_a = (0.06/1.5) R_y F_y A_f \text{ (ASD)} \tag{E4-1b}$$

where

A_f = gross area of the flange of the special segment chord member, in.^{2} (mm^2)

4c. Stability Bracing of Truss-to-Column Connections

The columns shall be laterally braced at the levels of top and bottom chords of the trusses connected to the columns. The lateral braces shall have a required strength of

$$P_u = 0.02 \, R_y P_{nc} \text{ (LRFD)} \tag{E4-2a}$$

or

$$P_a = (0.02/1.5) \, R_y P_{nc} \text{ (ASD)} \tag{E4-2b}$$

where

P_{nc} = nominal compressive strength of the chord member at the ends, kips (N)

4d. Stiffness of Stability Bracing

The required brace stiffness shall meet the provisions of Section 6.2 of Appendix 6 of the *Specification*, where

$$P_r = R_y P_{nc} \text{ (LRFD)} \tag{E4-3a}$$

or

$$P_r = R_y P_{nc}/1.5 \text{ (ASD)} \tag{E4-3b}$$

where

P_r = required compressive strength, kips (N)

5. Members

5a. Special Segment Members

The available shear strength of the special segment shall be calculated as the sum of the available shear strength of the chord members through flexure, and of the shear strength corresponding to the available tensile strength and 0.3 times the available compressive strength of the diagonal members, when they are used. The top and bottom chord members in the special segment shall be made of identical sections and shall provide at least 25% of the required vertical shear strength.

The *available strength*, ϕP_n (LRFD) and P_n/Ω (ASD), determined in accordance with the limit state of tensile yielding, shall be equal to or greater than 2.2 times the required strength.

$$\phi = 0.90 \text{ (LRFD)} \qquad \Omega = 1.67 \text{ (ASD)}$$

where

$$P_n = F_y A_g \tag{E4-4}$$

5b. Expected Vertical Shear Strength of Special Segment

The expected vertical shear strength of the special segment, V_{ne}, at mid-length, shall be:

$$V_{ne} = \frac{3.60 R_y M_{nc}}{L_s} + 0.036 EI \frac{L}{L_s^3} + R_y \left(P_{nt} + 0.3 P_{nc}\right) \sin \alpha \tag{E4-5}$$

where

E = modulus of elasticity of a chord member of the special segment, ksi (MPa)
I = moment of inertia of a chord member of the special segment, in.^{4} (mm^4)
L = span length of the truss, in. (mm)
L_s = length of the special segment, in. (mm)
M_{nc} = nominal flexural strength of a chord member of the special segment, kip-in. (N-mm)
P_{nt} = nominal tensile strength of a diagonal member of the special segment, kips (N)
P_{nc} = nominal compressive strength of a diagonal member of the special segment, kips (N)
R_y = ratio of the *expected yield stress* to the specified minimum yield stress
α = angle of diagonal members with the horizontal, degrees

5c. Width-to-Thickness Limitations

Chord members and diagonal web members within the special segment shall satisfy the requirements of Section D1.1b for *highly ductile members*. The width-to-thickness ratio of flat bar diagonal members shall not exceed 2.5.

5d. Built-Up Chord Members

Spacing of stitching for built-up chord members in the special segment shall not exceed $0.04Er_y/F_y$, where r_y is the radius of gyration of individual components about their weak axis.

5e. Protected Zones

The region at each end of a chord member within the special segment shall be designated as a *protected zone* meeting the requirements of Section D1.3. The protected zone shall extend over a length equal to two times the depth of the chord member from the connection with the web members. Vertical and diagonal web members from end-to-end of the special segments shall be protected zones.

6. Connections

6a. Demand Critical Welds

The following welds are *demand critical welds*, and shall satisfy the requirements of Section A3.4b and I2.3:

(1) Groove welds at column splices

(2) Welds at column-to-base plate connections

Exception: Where it can be shown that column hinging at, or near, the base plate is precluded by conditions of restraint, and in the absence of net tension under load combinations including the amplified seismic load, demand critical welds are not required.

6b. Connections of Diagonal Web Members in the Special Segment

The end connection of diagonal web members in the special segment shall have a required strength that is at least equal to the *expected yield strength* of the web member multiplied by 1.0 (LRFD) or divided by 1.5 (ASD), as appropriate. The expected yield strength of the web member shall be determined as $R_y F_y A_g$.

6c. Column Splices

Column splices shall comply with the requirements of Section D2.5. Where welds are used to make the splice, they shall be complete-joint-penetration groove welds.

When bolted column splices are used, they shall have a *required flexural strength* that is at least equal to $R_y F_y Z_x$ (LRFD) or $R_y F_y Z_x / 1.5$ (ASD), as appropriate, of the smaller column. The *required shear strength* of column web splices shall be at least equal to $\Sigma M_{pc}/H$ (LRFD) or $\Sigma M_{pc}/(1.5H)$ (ASD), as appropriate, where ΣM_{pc} is the sum of the nominal plastic flexural strengths of the columns above and below the splice.

Exception: The required strength of the column splice considering appropriate stress concentration factors or fracture mechanics stress intensity factors need not exceed that determined by a nonlinear analysis as specified in Chapter C.

E5. ORDINARY CANTILEVER COLUMN SYSTEMS (OCCS)

1. Scope

Ordinary cantilever column systems (OCCS) of structural steel shall be designed in conformance with this section.

2. Basis of Design

OCCS designed in accordance with these provisions are expected to provide minimal inelastic drift capacity through flexural yielding of the columns.

3. Analysis

There are no additional analysis requirements.

4. System Requirements

4a. Columns

Columns shall be designed using the load combinations including the *amplified seismic load*. The *required axial strength*, P_{rc}, shall not exceed 15% of the available axial strength, P_c, for these load combinations only.

4b. Stability Bracing of Columns

There are no additional stability bracing requirements for columns.

5. Members

5a. Basic Requirements

There are no additional requirements.

5b. Column Flanges

There are no additional column flange requirements.

5c. Protected Zones

There are no designated *protected zones*.

6. Connections

6a. Demand Critical Welds

No *demand critical welds* are required for this system.

6b. Column Bases

There are no additional *column base* requirements.

E6. SPECIAL CANTILEVER COLUMN SYSTEMS (SCCS)

1. Scope

Special cantilever column systems (SCCS) of structural steel shall be designed in conformance with this section.

2. Basis of Design

SCCS designed in accordance with these provisions are expected to provide limited inelastic drift capacity through flexural yielding of the columns.

3. Analysis

There are no additional analysis requirements.

4. System Requirements

4a. Columns

Columns shall be designed using the load combinations including the *amplified seismic load*. The *required strength*, P_{rc}, shall not exceed 15% of the available axial strength, P_c, for these load combinations only.

4b. Stability Bracing of Columns

Columns shall be braced to satisfy the requirements applicable to beams classified as *moderately ductile members* in Section D1.2a.

5. Members

5a. Basic Requirements

Column members shall satisfy the requirements of Section D1.1 for *highly ductile members*.

5b. Column Flanges

Abrupt changes in column flange area are prohibited in the *protected zone* as designated in Section E6.5c.

5c. Protected Zones

The region at the base of the column subject to inelastic straining shall be designated as a protected zone, and shall satisfy the requirements of Section D1.3. The length of the protected zone shall be two times the column depth, unless otherwise substantiated by testing.

6. Connections

6a. Demand Critical Welds

The following welds are *demand critical welds*, and shall satisfy the requirements of Section A3.4b and I2.3:

(1) Groove welds at column splices

(2) Welds at column-to-base plate connections

6b. Column Bases

Column bases shall be designed in accordance with Section D2.6.

CHAPTER F

BRACED-FRAME AND SHEAR-WALL SYSTEMS

This chapter provides the basis of design, the requirements for analysis, and the requirements for the system, members and connections for steel braced-frame and shear-wall systems.

The chapter is organized as follows:

- F1. Ordinary Concentrically Braced Frames (OCBF)
- F2. Special Concentrically Braced Frames (SCBF)
- F3. Eccentrically Braced Frames (EBF)
- F4. Buckling-Restrained Braced Frames (BRBF)
- F5. Special Plate Shear Walls (SPSW)

User Note: The requirements of this chapter are in addition to those required by the *Specification* and the *applicable building code*.

F1. ORDINARY CONCENTRICALLY BRACED FRAMES (OCBF)

1. Scope

Ordinary concentrically braced frames (OCBF) of structural steel shall be designed in conformance with this section. In seismically isolated structures, OCBF above the isolation system shall satisfy the requirements of Sections F1.4b, F1.5, F1.6 and F1.7 and need not satisfy the requirements of Section F1.4a.

2. Basis of Design

This section is applicable to braced frames that consist of concentrically connected members. Eccentricities less than the beam depth are permitted if they are accounted for in the member design by determination of eccentric moments using the *amplified seismic load*.

OCBF designed in accordance with these provisions are expected to provide limited inelastic deformation capacity in their members and connections.

3. Analysis

There are no additional analysis requirements.

4. System Requirements

4a. V-Braced and Inverted V-Braced Frames

Beams in V-type and inverted V-type OCBF shall be continuous at brace connections away from the beam-column connection and shall satisfy the following requirements:

(1) The *required strength* shall be determined based on the *load combinations* of the *applicable building code* assuming that the braces provide no support of dead and live loads. For load combinations that include earthquake effects, the seismic load effect, E, on the member shall be determined as follows:

   (i) The forces in braces in tension shall be assumed to be the least of the following:

      (a) The *expected yield strength* of the brace in tension, $R_y F_y A_g$

      (b) The load effect based upon the amplified seismic load

      (c) The maximum force that can be developed by the system

   (ii) The forces in braces in compression shall be assumed to be equal to $0.3 P_n$.

(2) As a minimum, one set of lateral braces is required at the point of intersection of the braces, unless the member has sufficient out-of-plane strength and stiffness to ensure stability between adjacent brace points.

4b. K-Braced Frames

K-type braced frames are not permitted for OCBF.

5. Members

5a. Basic Requirements

Braces shall satisfy the requirements of Section D1.1 for *moderately ductile members*.

5b. Slenderness

Braces in V or inverted-V configurations shall have $KL/r \leq 4\sqrt{E/F_y}$.

6. Connections

6a. Diagonal Brace Connections

The required strength of *diagonal brace* connections is the load effect based upon the amplified seismic load.

Exception: The required strength of the brace connection need not exceed the following:

(1) In tension, the expected yield strength of the brace multiplied by 1.0 (LRFD) or divided by 1.5 (ASD), as appropriate. The expected yield strength shall be determined as $R_y F_y A_g$.

(2) In compression, the expected brace strength in compression multiplied by 1.0 (LRFD) or divided by 1.5 (ASD), as appropriate. The expected brace strength in compression is permitted to be taken as the lesser of $R_y F_y A_g$ and $1.14 F_{cre} A_g$ where F_{cre} is determined from *Specification* Chapter E using the equations for F_{cr} except that the *expected yield stress* $R_y F_y$ is used in lieu of F_y. The brace length used for the determination of F_{cre} shall not exceed the distance from brace end to brace end.

(3) When oversized holes are used, the required strength for the limit state of bolt slip need not exceed a *load effect* based upon using the load combinations stipulated by the applicable building code, not including the amplified seismic load.

7. Ordinary Concentrically Braced Frames above Seismic Isolation Systems

7a. System Requirements

Beams in V-type and inverted V-type braced frames shall be continuous between columns.

7b. Members

Braces shall have a slenderness ratio, $KL/r \leq 4\sqrt{E/F_y}$.

F2. SPECIAL CONCENTRICALLY BRACED FRAMES (SCBF)

1. Scope

Special concentrically braced frames (SCBF) of structural steel shall be designed in conformance with this section.

2. Basis of Design

This section is applicable to *braced frames* that consist of concentrically connected members. Eccentricities less than the beam depth are permitted if the resulting member and connection forces are addressed in the design and do not change the expected source of inelastic deformation capacity.

SCBF designed in accordance with these provisions are expected to provide significant inelastic deformation capacity primarily through brace buckling and yielding of the brace in tension.

3. Analysis

The *required strength* of columns, beams and connections in SCBF shall be based on the *load combinations* in the *applicable building code* that include the *amplified seismic load*. In determining the amplified seismic load the effect of horizontal forces including overstrength, E_{mh}, shall be taken as the larger force determined from the following two analyses:

(i) An analysis in which all braces are assumed to resist forces corresponding to their expected strength in compression or in tension

(ii) An analysis in which all braces in tension are assumed to resist forces corresponding to their expected strength and all braces in compression are assumed to resist their expected post-buckling strength

Braces shall be determined to be in compression or tension neglecting the effects of gravity loads. Analyses shall consider both directions of frame loading.

The expected brace strength in tension is $R_y F_y A_g$.

The expected brace strength in compression is permitted to be taken as the lesser of $R_y F_y A_g$ and $1.14 F_{cre} A_g$ where F_{cre} is determined from *Specification* Chapter E using the equations for F_{cr}, except that the expected yield stress $R_y F_y$ is used in lieu of F_y. The brace length used for the determination of F_{cre} shall not exceed the distance from brace end to brace end.

The expected post-buckling brace strength shall be taken as a maximum of 0.3 times the expected brace strength in compression.

User Note: Braces with a slenderness ratio of 200 (the maximum permitted by Section F2.5b) buckle elastically for permissible materials; the value of $0.3F_{cr}$ for such braces is 2.1 ksi. This value may be used in Section F2.3(ii) for braces of any slenderness and a liberal estimate of the required strength of framing members will be obtained. Alternatively, 0 ksi may also be used to simplify the analysis.

Exceptions:

(1) It is permitted to neglect flexural forces resulting from seismic drift in this determination. Moment resulting from a load applied to the column between points of lateral support must be considered.

(2) The required strength of columns need not exceed the least of the following:
 (a) The forces determined using load combinations stipulated by the applicable building code including the amplified seismic load, applied to a building frame model in which all compression braces have been removed
 (b) The forces corresponding to the resistance of the foundation to overturning uplift
 (c) Forces determined from nonlinear analysis as defined in Section C3

4. System Requirements

4a. Lateral Force Distribution

Along any line of braces, braces shall be deployed in alternate directions such that, for either direction of force parallel to the braces, at least 30% but no more than 70% of the total horizontal force along that line is resisted by braces in tension, unless the *available strength* of each brace in compression is larger than the required strength resulting from the application of the appropriate load combinations stipulated by the applicable building code including the amplified seismic load. For the purposes of this provision, a line of braces is defined as a single line or parallel lines with a plan offset of 10% or less of the building dimension perpendicular to the line of braces.

4b. V- and Inverted V-Braced Frames

Beams that are intersected by braces away from beam-to-column connections shall satisfy the following requirements:

(1) Beams shall be continuous between columns.
(2) Beams shall be braced to satisfy the requirements for *moderately ductile members* in Section D1.2a.

As a minimum, one set of lateral braces is required at the point of intersection of the V-type (or inverted V-type) braced frames, unless the beam has sufficient out-of-plane strength and stiffness to ensure stability between adjacent brace points.

> **User Note:** One method of demonstrating sufficient out-of-plane strength and stiffness of the beam is to apply the bracing force defined in Equation A-6-7 of Appendix 6 of the *Specification* to each flange so as to form a torsional couple; this loading should be in conjunction with the flexural forces determined from the analysis required by Section F2.3. The stiffness of the beam (and its restraints) with respect to this torsional loading should be sufficient to satisfy Equation A-6-8 of the *Specification*.

4c. K-Braced Frames

K-type braced frames are not permitted for SCBF.

4d. Tension-Only Frames

Tension-only frames are not permitted in SCBF.

> **User Note:** Tension-only braced frames are those in which the brace compression resistance is neglected in the design and the braces are designed for tension forces only.

5. Members

5a. Basic Requirements

Columns and braces shall satisfy the requirements of Section D1.1 for *highly ductile members*. Beams shall satisfy the requirements of Section D1.1 for moderately ductile members.

5b. Diagonal Braces

Braces shall comply with the following requirements:

(1) Slenderness: Braces shall have a slenderness ratio, $KL/r \leq 200$.

(2) Built-up Braces: The spacing of connectors shall be such that the slenderness ratio, a/r_i, of individual elements between the connectors does not exceed 0.4 times the governing slenderness ratio of the built-up member.

The sum of the available shear strengths of the connectors shall equal or exceed the available *tensile strength* of each element. The spacing of connectors shall be uniform. Not less than two connectors shall be used in a built-up member. Connectors shall not be located within the middle one-fourth of the clear brace length.

Exception: Where the buckling of braces about their critical bucking axis does not cause shear in the connectors, the design of connectors need not comply with this provision.

(3) The brace effective net area shall not be less than the brace gross area. Where reinforcement on braces is used the following requirements shall apply:

(i) The specified minimum yield strength of the reinforcement shall be at least the specified minimum yield strength of the brace.

(ii) The connections of the reinforcement to the brace shall have sufficient strength to develop the expected reinforcement strength on each side of a reduced section.

5c. Protected Zones

The *protected zone* of SCBF shall satisfy Section D1.3 and include the following:

(1) For braces, the center one-quarter of the brace length and a zone adjacent to each connection equal to the brace depth in the plane of buckling

(2) Elements that connect braces to beams and columns

6. Connections

6a. Demand Critical Welds

The following welds are *demand critical welds*, and shall satisfy the requirements of Section A3.4b and I2.3:

(1) Groove welds at column splices

(2) Welds at column-to-base plate connections

Exception: Where it can be shown that column hinging at, or near, the base plate is precluded by conditions of restraint, and in the absence of net tension under load combinations including the amplified seismic load, demand critical welds are not required.

(3) Welds at beam-to-column connections conforming to Section F2.6b(b)

6b. Beam-to-Column Connections

Where a brace or gusset plate connects to both members at a beam-to-column connection, the connection shall conform to one of the following:

(a) The connection shall be a simple connection meeting the requirements of *Specification* Section B3.6a where the required rotation is taken to be 0.025 rad; or

(b) The connection shall be designed to resist a moment equal to the lesser of the following:

(i) A moment corresponding to the expected beam flexural strength multiplied by 1.1 (LRFD) or by 1.1/1.5 (ASD), as appropriate. The expected beam flexural strength shall be determined as $R_y M_p$.

(ii) A moment corresponding to the sum of expected column flexural strengths multiplied by 1.1 (LRFD) or by 1.1/1.5 (ASD), as appropriate. The sum of expected column flexural strengths shall be $\Sigma(R_y F_y Z)$.

This moment shall be considered in combination with the required strength of the brace connection and beam connection, including the amplified diaphragm *collector* forces.

6c. Required Strength of Brace Connections

The required strength in tension, compression and flexure of brace connections (including beam-to-column connections if part of the braced-frame system) shall be determined as required below. These required strengths are permitted to be considered independently without interaction.

(1) Required Tensile Strength

The *required tensile strength* is the lesser of the following:

(a) The *expected yield strength*, in tension, of the brace, determined as $R_y F_y A_g$ (LRFD) or $R_y F_y A_g / 1.5$ (ASD), as appropriate.

Exception: Braces need not comply with the requirements of Equation J4-1 and J4-2 of the *Specification* for this loading.

> **User Note:** This exception applies to braces where the section is reduced or where the net section is effectively reduced due to shear lag. A typical case is a slotted HSS brace at the gusset plate connection. Section F2.5b requires braces with holes or slots to be reinforced such that the effective net area exceeds the gross area.
>
> The brace strength used to check connection limit states, such as brace block shear, may be determined using expected material properties as permitted by Section A3.2.

(b) The maximum load effect, indicated by analysis, that can be transferred to the brace by the system.

When oversized holes are used, the required strength for the limit state of bolt slip need not exceed a load effect based upon using the load combinations stipulated by the applicable building code, including the amplified seismic load.

> **User Note:** For other limit states the loadings of (a) and (b) apply.

(2) Required Compressive Strength

Brace connections shall be designed for a required compressive strength based on buckling limit states that is at least equal to 1.1 times the expected brace strength in compression (LRFD) or (1.1/1.5) times the expected brace strength in compression (ASD), as appropriate, where the expected brace strength in compression is as defined in Section F2.3.

(3) Accommodation of Brace Buckling

Brace connections shall be designed to withstand the flexural forces or rotations imposed by brace buckling. Connections satisfying either of the following provisions are deemed to satisfy this requirement:

(a) Required Flexural Strength: Brace connections designed to withstand the flexural forces imposed by brace buckling shall have an available flexural strength of at least the expected brace flexural strength multiplied by 1.1

(LRFD) or by 1.1/1.5 (ASD), as appropriate. The expected brace flexural strength shall be determined as $R_y M_p$ of the brace about the critical buckling axis.

(b) Rotation Capacity: Brace connections designed to withstand the rotations imposed by brace buckling shall have sufficient rotation capacity to accommodate the required rotation at the *design story drift*. Inelastic rotation of the connection is permitted.

> **User Note:** Accommodation of inelastic rotation is typically accomplished by means of a single gusset plate with the brace terminating before the line of restraint. The detailing requirements for such a connection are described in the Commentary.

6d. Column Splices

Column splices shall comply with the requirements of Section D2.5. Where groove welds are used to make the splice, they shall be complete-joint-penetration groove welds. Column splices shall be designed to develop at least 50% of the lesser available flexural strength of the connected members.

The *required shear strength* shall be $\Sigma M_{pc}/H_c$ (LRFD) or $\Sigma M_{pc}/(1.5H_c)$ (ASD), as appropriate,

where
H_c = clear height of the column between beam connections, including a structural slab, if present, in. (mm)
ΣM_{pc} = sum of the nominal plastic flexural strengths, $F_{yc} Z_c$, of the columns above and below the splice, kip-in. (N-mm)

F3. ECCENTRICALLY BRACED FRAMES (EBF)

1. Scope

Eccentrically braced frames (EBF) of structural steel shall be designed in conformance with this section.

2. Basis of Design

This section is applicable to *braced frames* for which one end of each brace intersects a beam at an eccentricity from the intersection of the centerlines of the beam and an adjacent brace or column, forming a *link* that is subject to shear and flexure. Eccentricities less than the beam depth are permitted in the brace connection away from the link if the resulting member and connection forces are addressed in the design and do not change the expected source of inelastic deformation capacity.

EBF designed in accordance with these provisions are expected to provide significant inelastic deformation capacity primarily through shear or flexural yielding in the links.

Where links connect directly to columns, design of their connections to columns shall provide the performance required by Section F3.6e(1) and demonstrate this conformance as required by Section F3.6e(2).

3. **Analysis**

The *required strength* of *diagonal braces* and their connections, beams outside links, and columns shall be based on the load combinations in the *applicable building code* that include the *amplified seismic load*. In determining the amplified seismic load, the effect of horizontal forces including overstrength, E_{mh}, shall be taken as the forces developed in the member assuming the forces at the ends of the links correspond to the adjusted link shear strength. The adjusted link shear strength shall be taken as R_y times the link nominal shear strength, V_n, given in Section F3.5b(2) multiplied by 1.25 for I-shaped links and 1.4 for box links.

Exceptions:

(1) The effect of horizontal forces including overstrength, E_{mh}, is permitted to be taken as 0.88 times the forces determined above for the design of the following members:

   (a) The portions of beams outside links

   (b) Columns in frames of three or more stories of bracing

(2) It is permitted to neglect flexural forces resulting from seismic drift in this determination. Moment resulting from a load applied to the column between points of lateral support must be considered.

(3) The required strength of columns need not exceed the lesser of the following:

   (a) Forces corresponding to the resistance of the foundation to overturning uplift

   (b) Forces as determined from nonlinear analysis as defined in Section C3

The inelastic *link rotation angle* shall be determined from the inelastic portion of the *design story drift*. Alternatively, the inelastic link rotation angle is permitted to be determined from nonlinear analysis as defined in Section C3.

> **User Note:** The seismic load effect, E, used in the design of EBF members, such as the required axial strength used in the equations in Section F3.5, should be calculated from the analysis above.

4. **System Requirements**

4a. **Link Rotation Angle**

The link rotation angle is the inelastic angle between the link and the beam outside of the link when the total story drift is equal to the design story drift, Δ. The link rotation angle shall not exceed the following values:

(a) For links of length $1.6M_p/V_p$ or less: 0.08 rad

(b) For links of length $2.6M_p/V_p$ or greater: 0.02 rad

where

M_p = nominal plastic flexural strength, kip-in. (N-mm)
V_p = nominal shear strength of an active link, kips (N)

Linear interpolation between the above values shall be used for links of length between $1.6M_p/V_p$ and $2.6M_p/V_p$.

4b. Bracing of Link

Bracing shall be provided at both the top and bottom link flanges at the ends of the link for I-shaped sections. Bracing shall have an *available strength* and stiffness as required for expected *plastic hinge* locations by Section D1.2c.

5. Members

5a. Basic Requirements

Brace members shall satisfy width-to-thickness limitations in Section D1.1 for *moderately ductile members*.

Column members shall satisfy width-to-thickness limitations in Section D1.1b for *highly ductile members*.

Where the beam outside of the link is a different section from the link, the beam shall satisfy the width-to-thickness limitations in Section D1.1 for moderately ductile members.

> **User Note:** The diagonal brace and beam segment outside of the link are intended to remain essentially elastic under the forces generated by the fully yielded and strain hardened link. Both the diagonal brace and beam segment outside of the link are typically subject to a combination of large axial force and bending moment, and therefore should be treated as beam-columns in design, where the available strength is defined by Chapter H of the *Specification*.
>
> Where the beam outside the link is the same member as the link, its strength may be determined using expected material properties as permitted by Section A3.2.

5b. Links

Links subject to shear and flexure due to eccentricity between the intersections of brace centerlines and the beam centerline (or between the intersection of the brace and beam centerlines and the column centerline for links attached to columns) shall be provided. The link shall be considered to extend from brace connection to brace connection for center links and from brace connection to column face for link-to-column connections except as permitted by Section F3.6e.

(1) Limitations

Links shall be I-shaped cross sections (rolled wide-flange sections or built-up sections), or built-up box sections. HSS sections shall not be used as links.

Links shall satisfy the requirements of Section D1.1 for highly ductile members.

Exception: Flanges of links with I-shaped sections with link lengths, $e \leq 1.6 M_p/V_p$, are permitted to satisfy the requirements for moderately ductile members.

The web or webs of a link shall be single thickness. Doubler-plate reinforcement and web penetrations are not permitted.

For links made of built-up cross sections, complete-joint-penetration groove welds shall be used to connect the web (or webs) to the flanges.

Links of built-up box sections shall have a moment of inertia, I_y, about an axis in the plane of the EBF limited to $I_y > 0.67 I_x$, where I_x is the moment of inertia about an axis perpendicular to the plane of the EBF.

(2) Shear Strength

The *link design shear strength*, $\phi_v V_n$, and the *allowable shear strength*, V_n/Ω_v, shall be the lower value obtained in accordance with the *limit states* of *shear yielding* in the web and *flexural yielding* in the gross section. For both limit states:

$$\phi_v = 0.90 \text{ (LRFD)} \qquad \Omega_v = 1.67 \text{ (ASD)}$$

(a) For shear yielding:

$$V_n = V_p \tag{F3-1}$$

where

$$V_p = 0.6 F_y A_{lw} \text{ for } P_r/P_c \leq 0.15 \tag{F3-2}$$

$$V_p = 0.6 F_y A_{lw} \sqrt{1-(P_r/P_c)^2} \text{ for } P_r/P_c > 0.15 \tag{F3-3}$$

$$A_{lw} = (d - 2t_f) t_w \text{ for I-shaped link sections} \tag{F3-4}$$

$$= 2(d - 2t_f) t_w \text{ for box link sections} \tag{F3-5}$$

$P_r = P_u$ (LRFD) or P_a (ASD), as appropriate
P_u = required axial strength using *LRFD load combinations*, kips (N)
P_a = required axial strength using *ASD load combinations*, kips (N)
$P_c = P_y$ (LRFD) or $P_y/1.5$ (ASD), as appropriate
P_y = nominal axial yield strength = $F_y A_g$ \hfill (F3-6)

(b) For flexural yielding:

$$V_n = 2M_p/e \tag{F3-7}$$

where

$$M_p = F_y Z \text{ for } P_r/P_c \leq 0.15 \tag{F3-8}$$

$$M_p = F_y Z \left(\frac{1 - P_r/P_c}{0.85}\right) \text{ for } P_r/P_c > 0.15 \tag{F3-9}$$

e = link length, defined as the clear distance between the ends of two diagonal braces or between the diagonal brace and the column face, in. (mm)

User Note: The requirements of Section F3.5b(2) and (3) have been reformatted from the 2005 *Seismic Provisions for Structural Steel Buildings* for clarity and simplicity. However, no change to the requirements is entailed in this reformatting.

(3) Link Length

If $P_r/P_c > 0.15$, the length of the link shall be limited as follows:

When $\rho' \leq 0.5$

$$e \leq \frac{1.6M_p}{V_p} \qquad (F3\text{-}10)$$

When $\rho' > 0.5$

$$e \leq \frac{1.6M_p}{V_p}(1.15 - 0.3\rho') \qquad (F3\text{-}11)$$

where

$$\rho' = \frac{P_r/P_c}{V_r/V_c} \qquad (F3\text{-}12)$$

$V_r = V_u$ (LRFD) or V_a (ASD), as appropriate, kips (N)
V_u = required shear strength based on LRFD load combinations, kips (N)
V_a = required shear strength based on ASD load combinations, kips (N)
$V_c = V_y$ (LRFD) or $V_y/1.5$ (ASD), as appropriate, kips (N)
V_y = nominal shear yield strength, kips (N)
$= 0.6F_y A_{lw}$ (F3-13)

User Note: For links with low axial force there is no upper limit on link length. The limitations on link rotation angle in Section F3.4a result in a practical lower limit on link length.

(4) Link Stiffeners for I-Shaped Cross Sections

Full-depth web stiffeners shall be provided on both sides of the link web at the diagonal brace ends of the link. These stiffeners shall have a combined width not less than $(b_f - 2t_w)$ and a thickness not less than the larger of $0.75t_w$ or $^3/_8$ in. (10 mm), where b_f and t_w are the link flange width and link web thickness, respectively.

Links shall be provided with intermediate web stiffeners as follows:

(a) Links of lengths $1.6M_p/V_p$ or less shall be provided with intermediate web stiffeners spaced at intervals not exceeding $(30t_w - d/5)$ for a link rotation angle of 0.08 rad or $(52t_w - d/5)$ for link rotation angles of 0.02 rad or less. Linear interpolation shall be used for values between 0.08 and 0.02 rad.

(b) Links of length greater than or equal to $2.6M_p/V_p$ and less than $5M_p/V_p$ shall be provided with intermediate web stiffeners placed at a distance of 1.5 times b_f from each end of the link.

(c) Links of length between $1.6M_p/V_p$ and $2.6M_p/V_p$ shall be provided with intermediate web stiffeners meeting the requirements of (a) and (b) above.

Intermediate web stiffeners are not required in links of length greater than $5M_p/V_p$.

Intermediate web stiffeners shall be full depth. For links that are less than 25 in. (635 mm) in depth, stiffeners are required on only one side of the link web. The thickness of one-sided stiffeners shall not be less than t_w or $^3/_8$ in. (10 mm), whichever is larger, and the width shall be not less than $(b_f/2) - t_w$. For links that are 25 in. (635 mm) in depth or greater, similar intermediate stiffeners are required on both sides of the web.

The required strength of fillet welds connecting a link stiffener to the link web is $F_y A_{st}$ (LRFD) or $F_y A_{st}/1.5$ (ASD), as appropriate, where A_{st} is the horizontal cross-sectional area of the link stiffener and F_y is the yield stress of the stiffener. The required strength of fillet welds connecting the stiffener to the link flanges is $F_y A_{st}/4$ (LRFD) or $F_y A_{st}/4(1.5)$ (ASD).

(5) Link Stiffeners for Box Sections

Full-depth web stiffeners shall be provided on one side of each link web at the diagonal brace connection. These stiffeners are permitted to be welded to the outside or inside face of the link webs. These stiffeners shall each have a width not less than $b/2$, where b is the inside width of the box. These stiffeners shall each have a thickness not less than the larger of 0.75 t_w or $^1/_2$ in. (13 mm).

Box links shall be provided with intermediate web stiffeners as follows:

(a) For links of length $1.6 M_p/V_p$ or less and with web depth-to-thickness ratio, h/t_w, greater than or equal to $0.64\sqrt{E/F_y}$, full-depth web stiffeners shall be provided on one side of each link web, spaced at intervals not exceeding $20 t_w - (d - 2t_f)/8$.

(b) For links of length $1.6 M_p/V_p$ or less and with web depth-to-thickness ratio, h/t_w, less than $0.64\sqrt{E/F_y}$, no intermediate web stiffeners are required.

(c) For links of length greater than $1.6 M_p/V_p$, no intermediate web stiffeners are required.

Intermediate web stiffeners shall be full depth, and are permitted to be welded to the outside or inside face of the link webs.

The required strength of fillet welds connecting a link stiffener to the link web is $F_y A_{st}$ (LRFD) or $F_y A_{st}/1.5$ (ASD), as appropriate, where A_{st} is the horizontal cross-sectional area of the link stiffener.

User Note: Stiffeners of box links need not be welded to link flanges.

5c.   Protected Zones

Links in EBFs are a *protected zone*, and shall satisfy the requirements of Section D1.3.

6.   Connections

6a.   Demand Critical Welds

The following welds are *demand critical welds* and shall satisfy the requirements of Sections A3.4b and I2.3:

(1) Groove welds at column splices
(2) Welds at column-to-base plate connections

Exception: Where it can be shown that column hinging at, or near, the base plate is precluded by conditions of restraint, and in the absence of net tension under load combinations including the amplified seismic load, demand critical welds are not required.

(3) Welds at beam-to-column connections conforming to Section F3.6b(b)
(4) Welds attaching the link flanges and the link web to the column where links connect to columns
(5) Welds connecting the webs to the flanges in built-up beams within the link

6b. Beam-to-Column Connections

Where a brace or gusset plate connects to both members at a beam-to-column connection, the connection shall conform to one of the following:

(a) The connection shall be a simple connection meeting the requirements of *Specification* Section B3.6a where the required rotation is taken to be 0.025 radians; or

(b) The connection shall be designed to resist a moment equal to the lesser of the following:

(i) A moment corresponding to the expected beam flexural strength multiplied by 1.1 (LRFD) or by 1.1/1.5 (ASD), as appropriate. The expected beam flexural strength shall be determined as $R_y M_p$.

(ii) A moment corresponding to the sum of expected column flexural strengths multiplied by 1.1 (LRFD) or by 1.1/1.5 (ASD), as appropriate. The sum of expected column flexural strengths shall be $\Sigma(R_y F_y Z)$.

This moment shall be considered in combination with the required strength of the brace connection and beam connection, including the amplified diaphragm *collector* forces.

6c. Diagonal Brace Connections

When oversized holes are used, the required strength for the limit state of bolt slip need not exceed a load effect based upon using the load combinations stipulated by the applicable building code, including the amplified seismic load.

Connections of braces designed to resist a portion of the link end moment shall be designed as fully restrained.

6d. Column Splices

Column splices shall comply with the requirements of Section D2.5. Where groove welds are used to make the splice, they shall be complete-joint-penetration groove welds. Column splices shall be designed to develop at least 50% of the lesser available flexural strength of the connected members.

The *required shear strength* shall be $\Sigma M_{pc}/H_c$ (LRFD) or $\Sigma M_{pc}/(1.5H_c)$ (ASD), as appropriate,

where
- H_c = clear height of the column between beam connections, including a structural slab, if present, in. (mm)
- ΣM_{pc} = sum of the nominal plastic flexural strengths, $F_{yc}Z_c$, of the columns above and below the splice, kip-in. (N-mm)

6e. Link-to-Column Connections

(1) Requirements

Link-to-column connections shall be fully restrained (FR) moment connections and shall satisfy the following requirements:

(1) The connection shall be capable of sustaining the link rotation angle specified in Section F3.4a.

(2) The shear resistance of the connection, measured at the required link rotation angle, shall be at least equal to the expected shear strength of the link, $R_y V_n$, as defined in Section F3.5b(2).

(3) The flexural resistance of the connection, measured at the required link rotation angle, shall be at least equal to the moment corresponding to the nominal shear strength of the link, V_n, as defined in Section F3.5b(2).

(2) Conformance Demonstration

Link-to-column connections shall satisfy the above requirements by one of the following:

(a) Use a connection prequalified for EBF in accordance with Section K1.

> **User Note:** There are no prequalified link-to-column connections.

(b) Provide qualifying cyclic test results in accordance with Section K2. Results of at least two cyclic connection tests shall be provided and are permitted to be based on one of the following:

   (i) Tests reported in research literature or documented tests performed for other projects that are representative of project conditions, within the limits specified in Section K2.

   (ii) Tests that are conducted specifically for the project and are representative of project member sizes, material strengths, connection configurations, and matching connection material properties, within the limits specified in Section K2.

Exception: Cyclic testing of the connection is not required if the following conditions are met:

(1) Reinforcement at the beam-to-column connection at the link end precludes yielding of the beam over the reinforced length.

(2) The available strength of the reinforced section and the connection equals or exceeds the required strength calculated based upon adjusted link shear strength as described in Section F3.3.

(3) The link length (taken as the beam segment from the end of the reinforcement to the brace connection) does not exceed $1.6M_p/V_p$.

(4) Full depth stiffeners as required in Section F3.5b(4) are placed at the link-to-reinforcement interface.

F4. BUCKLING-RESTRAINED BRACED FRAMES (BRBF)

1. Scope

Buckling-restrained braced frames (BRBF) of structural steel shall be designed in conformance with this section.

2. Basis of Design

This section is applicable to frames with specially fabricated braces concentrically connected to beams and columns. Eccentricities less than the beam depth are permitted if the resulting member and connection forces are addressed in the design and do not change the expected source of inelastic deformation capacity.

BRBF designed in accordance with these provisions are expected to provide significant inelastic deformation capacity primarily through brace yielding in tension and compression. Design of braces shall provide the performance required by Section F4.5b(1) and F4.5b(2), and demonstrate this conformance as required by Section F4.5b(3). Braces shall be designed, tested and detailed to accommodate expected deformations. Expected deformations are those corresponding to a story drift of at least 2% of the story height or two times the *design story drift*, whichever is larger, in addition to brace deformations resulting from deformation of the frame due to gravity loading.

BRBF shall be designed so that inelastic deformations under the *design earthquake* will occur primarily as brace yielding in tension and compression.

2a. Brace Strength

The *adjusted brace strength* shall be established on the basis of testing as described in this section.

Where required by these Provisions, brace connections and adjoining members shall be designed to resist forces calculated based on the adjusted brace strength.

The adjusted brace strength in compression shall be $\beta\omega R_y P_{ysc}$,

where
β = compression strength adjustment factor
ω = strain hardening adjustment factor
P_{ysc} = axial yield strength of *steel core*, ksi (MPa)

The adjusted brace strength in tension shall be $\omega R_y P_{ysc}$.

Exception: The factor R_y need not be applied if P_{ysc} is established using yield stress determined from a coupon test.

The compression strength adjustment factor, β, shall be calculated as the ratio of the maximum compression force to the maximum tension force of the *test specimen* measured from the qualification tests specified in Section K3.4c for the expected deformations. The larger value of β from the two required brace qualification tests shall be used. In no case shall β be taken as less than 1.0.

The strain hardening adjustment factor, ω, shall be calculated as the ratio of the maximum tension force measured from the qualification tests specified in Section K3.4c (for the expected deformations) to the measured yield force, $R_y P_{ysc}$, of the test specimen. The larger value of ω from the two required qualification tests shall be used. Where the tested steel core material does not match that of the *prototype*, ω shall be based on coupon testing of the prototype material.

3. Analysis

Buckling-restrained braces shall not be considered as resisting gravity forces.

The *required strength* of columns, beams and connections in BRBF shall be based on the load combinations in the *applicable building code* that include the *amplified seismic load*. In determining the amplified seismic load, the effect of horizontal forces including overstrength, E_{mh}, shall be taken as the forces developed in the member assuming the forces in all braces correspond to their adjusted strength in compression or in tension.

Braces shall be determined to be in compression or tension neglecting the effects of gravity loads. Analyses shall consider both directions of frame loading.

The adjusted brace strength in tension shall be as given in Section F4.2a.

Exceptions:

(1) It is permitted to neglect flexural forces resulting from seismic drift in this determination. Moment resulting from a load applied to the column between points of lateral support must be considered.

(2) The required strength of columns need not exceed the lesser of the following:

  (a) The forces corresponding to the resistance of the foundation to overturning uplift

  (b) Forces as determined from nonlinear analysis as defined in Section C3

The brace deformation shall be determined from the inelastic portion of the design story drift and shall include the effects of beam vertical flexibility. Alternatively, the brace deformation is permitted to be determined from nonlinear analysis as defined in Section C3.

4. System Requirements

4a. V- and Inverted V-Braced Frames

V-type and inverted-V-type braced frames shall satisfy the following requirements:

(1) The required strength of beams intersected by braces, their connections and supporting members shall be determined based on the *load combinations* of the applicable building code assuming that the braces provide no support for dead and live loads. For load combinations that include earthquake effects, the vertical and horizontal earthquake effect, E, on the beam shall be determined from the adjusted brace strengths in tension and compression.

(2) Beams shall be continuous between columns. Beams shall be braced to satisfy the requirements for *moderately ductile members* in Section D1.2(a).

As a minimum, one set of lateral braces is required at the point of intersection of the V-type (or inverted V-type) braces, unless the beam has sufficient out-of-plane strength and stiffness to ensure stability between adjacent brace points.

> **User Note**: The beam has sufficient out-of-plane strength and stiffness if the beam bent in the horizontal plane meets the required brace strength and required brace stiffness for column nodal bracing as prescribed in the *Specification*. P_u may be taken as the required compressive strength of the brace.

For purposes of brace design and testing, the calculated maximum deformation of braces shall be increased by including the effect of the vertical deflection of the beam under the loading defined in Section F4.4a(1).

4b. K-Braced Frames

K-type braced frames are not permitted for BRBF.

5. Members

5a. Basic Requirements

Beam and column members shall satisfy the requirements of Section D1.1 for *highly ductile members*.

5b. Diagonal Braces

(1) Assembly

Braces shall be composed of a structural steel core and a system that restrains the steel core from buckling.

(1) Steel Core

Plates used in the steel core that are 2 in. (50 mm) thick or greater shall satisfy the minimum notch toughness requirements of Section A3.3.

Splices in the steel core are not permitted.

(2) Buckling-Restraining System

The *buckling-restraining system* shall consist of the *casing* for the steel core. In stability calculations, beams, columns and gussets connecting the core shall be considered parts of this system.

The buckling-restraining system shall limit local and overall buckling of the steel core for the expected deformations.

User Note: Conformance to this provision is demonstrated by means of testing as described in Section F4.5b(3).

(2) Available Strength

The steel core shall be designed to resist the entire axial force in the brace.

The brace design axial strength, ϕP_{ysc} (LRFD), and the brace allowable axial strength, P_{ysc}/Ω (ASD), in tension and compression, in accordance with the limit state of yielding, shall be determined as follows:

$$P_{ysc} = F_{ysc} A_{sc} \tag{F4-1}$$

$$\phi = 0.90 \text{ (LRFD)} \qquad \Omega = 1.67 \text{ (ASD)}$$

where

A_{sc} = cross-sectional area of the yielding segment of the steel core, in.^{2} (mm^2)
F_{ysc} = specified minimum yield stress of the steel core, or actual yield stress of the steel core as determined from a coupon test, ksi (MPa)

User Note: Load effects calculated based on adjusted brace strengths should not be amplified by the overstrength factor, Ω_o.

(3) Conformance Demonstration

The design of braces shall be based upon results from qualifying cyclic tests in accordance with the procedures and acceptance criteria of Section K3. Qualifying test results shall consist of at least two successful cyclic tests: one is required to be a test of a brace subassemblage that includes brace connection rotational demands complying with Section K3.2 and the other shall be either a uniaxial or a subassemblage test complying with Section K3.3. Both test types shall be based upon one of the following:

(a) Tests reported in research or documented tests performed for other projects

(b) Tests that are conducted specifically for the project

Interpolation or extrapolation of test results for different member sizes shall be justified by rational analysis that demonstrates stress distributions and magnitudes of internal strains consistent with or less severe than the tested assemblies and that considers the adverse effects of variations in material properties. Extrapolation of test results shall be based upon similar combinations of steel core and buckling-restraining system sizes. Tests are permitted to qualify a design when the provisions of Section K3 are met.

Sect. F4.] BUCKLING-RESTRAINED BRACED FRAMES (BRBF)

5c. Protected Zones

The *protected zone* shall include the steel core of braces and elements that connect the steel core to beams and columns, and shall satisfy the requirements of Section D1.3.

6. Connections

6a. Demand Critical Welds

The following welds are *demand critical welds*, and shall satisfy the requirements of Section A3.4b and I2.3:

(1) Groove welds at column splices

(2) Welds at the column-to-base plate connections

   Exception: Where it can be shown that column hinging at, or near, the base plate is precluded by conditions of restraint, and in the absence of net tension under load combinations including the amplified seismic load, demand critical welds are not required.

(3) Welds at beam-to-column connections conforming to Section F4.6b(b)

6b. Beam-to-Column Connections

Where a brace or gusset plate connects to both members at a beam-to-column connection, the connection shall conform to one of the following:

(a) The connection shall be a simple connection meeting the requirements of *Specification* Section B3.6a where the required rotation is taken to be 0.025 rad; or

(b) The connection shall be designed to resist a moment equal to the lesser of the following:

   (i) A moment corresponding to the expected beam flexural strength multiplied by 1.1 (LRFD) or by 1.1/1.5 (ASD), as appropriate. The expected beam flexural strength shall be determined as $R_y M_p$.

   (ii) A moment corresponding to the sum of expected column flexural strengths multiplied by 1.1 (LRFD) or by 1.1/1.5 (ASD), as appropriate. The sum of expected column flexural strengths shall be $\Sigma(R_y F_y Z)$.

   This moment shall be considered in combination with the required strength of the brace connection and beam connection, including the amplified diaphragm collector forces.

6c. Diagonal Brace Connections

(1) Required Strength

The required strength of brace connections in tension and compression (including beam-to-column connections if part of the braced-frame system) shall be 1.1 times the adjusted brace strength in compression (LRFD) or (1.1/1.5) times the adjusted brace strength in compression (ASD) where the adjusted brace strength is as defined in Section F4.2a.

When oversized holes are used, the required strength for the limit state of bolt slip need not exceed a load effect based upon using the load combinations stipulated by the applicable building code, including the amplified seismic load.

(2) Gusset Plate Requirements

The design of connections shall include considerations of local and overall buckling. Lateral bracing consistent with that used in the tests upon which the design is based is required.

> **User Note:** This provision may be met by designing the gusset plate for a transverse force consistent with transverse bracing forces determined from testing, by adding a stiffener to it to resist this force, or by providing a brace to the gusset plate. Where the supporting tests did not include transverse bracing, no such bracing is required. Any attachment of bracing to the steel core must be included in the qualification testing.

6d. Column Splices

Column splices shall comply with the requirements of Section D2.5. Where groove welds are used to make the splice, they shall be complete-joint-penetration groove welds. Column splices shall be designed to develop at least 50% of the lesser available flexural strength of the connected members.

The *required shear strength*, V_u or V_a, shall be determined as follows:

$$V_u = \frac{\Sigma M_{pc}}{H_c} \quad \text{(LRFD)} \qquad \text{(F4-2a)}$$

or

$$V_a = \frac{\Sigma M_{pc}}{1.5 H_c} \quad \text{(ASD)} \qquad \text{(F4-2b)}$$

as appropriate,

where
- H_c = clear height of the column between beam connections, including a structural slab, if present, in. (mm)
- ΣM_{pc} = sum of the nominal plastic flexural strengths, $F_{yc} Z_c$, of the columns above and below the splice, kip-in. (N-mm)

F5. SPECIAL PLATE SHEAR WALLS (SPSW)

1. Scope

Special plate shear walls (SPSW) of structural steel shall be designed in conformance with this section.

2. Basis of Design

This section is applicable to frames with steel web plates connected to beams and columns.

SPSW designed in accordance with these provisions are expected to provide significant inelastic deformation capacity primarily through web plate yielding and as plastic-hinge formation in the ends of *horizontal boundary elements* (HBEs).

3. Analysis

The webs of SPSW shall not be considered as resisting gravity forces.

The *required strength* of HBEs, *vertical boundary elements* (VBEs), and connections in SPSW shall be based on the load combinations in the *applicable building code* that include the *amplified seismic load*. In determining the amplified seismic load the effect of horizontal forces including overstrength, E_{mh}, shall be determined from an analysis in which all webs are assumed to resist forces corresponding to their expected strength in tension at an angle, α, as determined in Section F5.5b and HBE are resisting flexural forces at each end equal to $1.1R_yM_p$ (LRFD) or $(1.1/1.5)R_yM_p$ (ASD). Webs shall be determined to be in tension neglecting the effects of gravity loads.

The expected web yield stress shall be taken as R_yF_y. When perforated walls are used, the effective expected tension stress is as defined in Section F5.7a(4).

User Note: Shear forces per Equation E1-1 must be included in this analysis. Designers should be aware that in some cases forces from the analysis in the applicable building code will govern the design of HBE.

User Note: Shear forces in beams and columns are likely to be high and shear yielding may be a governing limit state.

4. System Requirements

4a. Stiffness of Boundary Elements

The vertical boundary elements (VBEs) shall have moments of inertia about an axis taken perpendicular to the plane of the web, I_c, not less than $0.0031t_wh^4/L$. The horizontal boundary elements (HBEs) shall have moments of inertia about an axis taken perpendicular to the plane of the web, I_b, not less than $0.0031L^4/h$ times the difference in web plate thicknesses above and below,

where
    I_b = moment of inertia of a HBE taken perpendicular to the direction of the web plate line, in.^{4} (mm^4)
    I_c = moment of inertia of a VBE taken perpendicular to the direction of the web plate line, in.^{4} (mm^4)

L = distance between VBE centerlines, in. (mm)
h = distance between HBE centerlines, in. (mm)
t_w = thickness of the web, in. (mm)

4b. HBE-to-VBE Connection Moment Ratio

The moment ratio provisions in Section E3.4a shall be met for all HBE/VBE intersections without consideration of the effects of the webs.

4c. Bracing

HBE shall be braced to satisfy the requirements for *moderately ductile members* in Section D1.2a.

4d. Openings in Webs

Openings in webs shall be bounded on all sides by *intermediate boundary elements* extending the full width and height of the panel respectively, unless otherwise justified by testing and analysis or permitted by Section F5.7.

5. Members

5a. Basic Requirements

HBE, VBE and intermediate boundary elements shall satisfy the requirements of Section D1.1 for *highly ductile members*.

5b. Webs

The panel design shear strength, ϕV_n (LRFD), and the *allowable shear strength*, V_n/Ω (ASD), in accordance with the limit state of shear yielding, shall be determined as follows:

$$V_n = 0.42 F_y t_w L_{cf} \sin 2\alpha \qquad (F5\text{-}1)$$

$$\phi = 0.90 \text{ (LRFD)} \qquad \Omega = 1.67 \text{ (ASD)}$$

where
L_{cf} = clear distance between column flanges, in. (mm)
t_w = thickness of the web, in. (mm)
α = angle of web yielding in degrees, as measured relative to the vertical. The angle of inclination, α, is permitted to be taken as 40°, or is permitted to be calculated as follows:

$$\tan^4 \alpha = \frac{1 + \dfrac{t_w L}{2 A_c}}{1 + t_w h \left(\dfrac{1}{A_b} + \dfrac{h^3}{360 I_c L} \right)} \qquad (F5\text{-}2)$$

where
A_b = cross-sectional area of an HBE, in.^{2} (mm^2)
A_c = cross-sectional area of a VBE, in.^{2} (mm^2)

5c. Protected Zone

The *protected zone* of SPSW shall satisfy Section D1.3 and include the following:

(1) The webs of SPSW

(2) Elements that connect webs to HBEs and VBEs

(3) The plastic hinging zones at each end of HBEs, over a region ranging from the face of the column to one beam depth beyond the face of the column, or as otherwise specified in Section E3.5c

6. Connections

6a. Demand Critical Welds

The following welds are *demand critical welds*, and shall satisfy the requirements of Section A3.4b and I2.3:

(1) Groove welds at column splices

(2) Welds at column-to-base plate connections

   Exception: Where it can be shown that column hinging at, or near, the base plate is precluded by conditions of restraint, and in the absence of net tension under load combinations including the amplified seismic load, demand critical welds are not required.

(3) Welds at HBE-to-VBE connections

6b. HBE-to-VBE Connections

HBE-to-VBE connections shall satisfy the requirements of Section E1.6b.

(1) Required Strength

The *required shear strength* of an HBE-to-VBE connection shall be based on the load combinations in the applicable building code that include the amplified seismic load. In determining the amplified seismic load, the effect of horizontal forces including overstrength, E_{mh}, shall be taken as the shear calculated from Equation E1-1 together with the shear resulting from the *expected yield strength* in tension of the webs yielding at an angle α.

(2) Panel Zones

The VBE panel zone next to the top and base HBE of the SPSW shall comply with the requirements in Section E3.6e.

6c. Connections of Webs to Boundary Elements

The required strength of web connections to the surrounding HBE and VBE shall equal the expected yield strength, in tension, of the web calculated at an angle α.

6d. Column Splices

Column splices shall comply with the requirements of Section D2.5. Where welds are used to make the splice, they shall be complete-joint-penetration groove welds.

Column splices shall be designed to develop at least 50% of the lesser available flexural strength of the connected members. The required shear strength, V_u or V_a, shall be determined by Equations F4-2a or F4-2b.

7. Perforated Webs

7a. Regular Layout of Circular Perforations

A perforated plate conforming to this section is permitted to be used as the web of an SPSW. Perforated webs shall have a regular pattern of holes of uniform diameter spaced evenly over the entire web-plate area in an array pattern so that holes align diagonally at a uniform angle to vertical. Edges of openings shall have a surface roughness of 500 μ-in. (13 microns) or less.

(1) Strength

The panel design shear strength, ϕV_n (LRFD), and the allowable shear strength, V_n/Ω (ASD), in accordance with the limit state of shear yielding, shall be determined as follows for perforated webs:

$$V_n = 0.42 F_y t_w L_{cf} \left(1 - \frac{0.7D}{S_{diag}}\right) \quad \text{(F5-3)}$$

$$\phi = 0.90 \text{ (LRFD)} \qquad \Omega = 1.67 \text{ (ASD)}$$

where
D = diameter of the holes, in. (mm)
S_{diag} = shortest center-to-center distance between the holes, in. (mm)

(2) Spacing

The spacing, S_{diag}, shall be at least $1.67D$.

The distance between the first holes and web connections to the HBEs and VBEs shall be at least D, but shall not exceed $(D + 0.7 S_{diag})$.

(3) Stiffness

The stiffness of such regularly perforated infill plates shall be calculated using an effective web-plate thickness, t_{eff}, given by:

$$t_{eff} = \frac{1 - \frac{\pi}{4}\left(\frac{D}{S_{diag}}\right)}{1 - \frac{\pi}{4}\left(\frac{D}{S_{diag}}\right)\left(1 - \frac{N_r D \sin\alpha}{H_c}\right)} t_w \quad \text{(F5-4)}$$

where
H_c = clear column (and web-plate) height between beam flanges, in. (mm)
N_r = number of horizontal rows of perforations
t_w = web-plate thickness, in. (mm)

α = angle of the shortest center-to-center lines in the opening array to vertical, degrees

(4) Effective Expected Tension Stress

The effective expected tension stress to be used in place of the effective tension stress for analysis per Section F5.3 is $R_y F_y (1 - 0.7\, D/S_{diag})$.

7b. Reinforced Corner Cut-Out

Quarter-circular cut-outs are permitted at the corners of the webs provided that the webs are connected to a reinforcement arching plate following the edge of the cut-outs. The plates shall be designed to allow development of the full strength of the solid web and maintain its resistance when subjected to deformations corresponding to the *design story drift*. This is deemed to be achieved if the following conditions are met.

(1) Design for Tension

The arching plate shall have the *available strength* to resist the axial tension force resulting from web-plate tension in the absence of other forces.

$$P_u = \frac{R_y F_y t_w R^2}{4e} \quad \text{(LRFD)} \tag{F5-5a}$$

or

$$P_a = \frac{R_y F_y t_w R^2 / 1.5}{4e} \quad \text{(ASD)} \tag{F5-5b}$$

as appropriate,

where
  R = radius of the cut-out, in. (mm)
  R_y = ratio of the *expected yield stress* to the specified minimum yield stress
  $e = R\left(1 - \sqrt{2}/2\right)$, in. (mm) \hfill (F5-6)

HBEs and VBEs shall be designed to resist the tension axial forces acting at the end of the arching reinforcement.

(2) Design for Beam-to-Column Connection Forces

The arching plate shall have the available strength to resist the combined effects of axial force and moment in the plane of the web resulting from connection deformation in the absence of other forces. These forces are:

$$P_u = \frac{15\, EI_y}{16 e^2}\left(\frac{\Delta}{H}\right) \quad \text{(LRFD)} \tag{F5-7a}$$

or

$$P_a = \frac{15\, EI_y}{1.5(16e^2)} \left(\frac{\Delta}{H}\right) \text{ (ASD)} \qquad \text{(F5-7b)}$$

as appropriate.

The moments are:

$$M_u = P_u e \text{ (LRFD)} \qquad \text{(F5-8a)}$$

or

$$M_a = P_a e \text{ (ASD)} \qquad \text{(F5-8b)}$$

as appropriate,

where
- E = modulus of elasticity, ksi (MPa)
- H = height of story, in. (mm)
- I_y = moment of inertia of the plate about the y-axis, in.^{4} (mm^4)
- Δ = design story drift, in. (mm)

CHAPTER G
COMPOSITE MOMENT-FRAME SYSTEMS

This chapter provides the basis of design, the requirements for analysis, and the requirements for the system, members and connections for composite moment frame systems.

The chapter is organized as follows:

    G1.   Composite Ordinary Moment Frames (C-OMF)
    G2.   Composite Intermediate Moment Frames (C-IMF)
    G3.   Composite Special Moment Frames (C-SMF)
    G4.   Composite Partially Restrained Moment Frames (C-PRMF)

User Note: The requirements of this chapter are in addition to those required by the *Specification* and the *applicable building code*.

G1. COMPOSITE ORDINARY MOMENT FRAMES (C-OMF)

1. Scope

Composite ordinary moment frames (C-OMF) shall be designed in conformance with this section. This section is applicable to moment frames with fully restrained (FR) connections that consist of either composite or reinforced concrete columns and structural steel, concrete-encased composite, or *composite beams*.

2. Basis of Design

C-OMF designed in accordance with these provisions are expected to provide minimal inelastic deformation capacity in their members and connections.

User Note: Composite ordinary moment frames, comparable to reinforced concrete ordinary moment frames, are only permitted in seismic design categories B or below in ASCE/SEI 7. This is in contrast to steel ordinary moment frames, which are permitted in higher seismic design categories. The design requirements are commensurate with providing minimal ductility in the members and connections.

3. Analysis

There are no additional analysis requirements.

4. System Requirements

There are no additional system requirements.

5. Members

There are no additional requirements for steel or composite members beyond those in the *Specification*. Reinforced concrete columns shall satisfy the requirements of ACI 318, excluding Chapter 21.

5a. Protected Zones

There are no designated *protected zones*.

6. Connections

Connections shall be fully restrained (FR). Connections shall be designed for the applicable load combinations as described in Sections B2 and B3. Beam-to-column connection *design strengths* shall be determined in accordance with the *Specification* and Section D2.7.

6a. Demand Critical Welds

There are no requirements for *demand critical welds*.

G2. COMPOSITE INTERMEDIATE MOMENT FRAMES (C-IMF)

1. Scope

Composite intermediate moment frames (C-IMF) shall be designed in conformance with this section. This section is applicable to moment frames with fully restrained (FR) connections that consist of composite or reinforced concrete columns and structural steel, concrete-encased composite or *composite beams*.

2. Basis of Design

C-IMF designed in accordance with these provisions are expected to provide limited inelastic deformation capacity through flexural yielding of the C-IMF beams and columns, and shear yielding of the column panel zones. Design of connections of beams to columns, including panel zones, *continuity plates* and diaphragms shall provide the performance required by Section G2.6b, and demonstrate this conformance as required by Section G2.6c.

> **User Note:** Composite intermediate moment frames, comparable to reinforced concrete intermediate moment frames, are only permitted in seismic design categories C or below in ASCE/SEI 7. This is in contrast to steel intermediate moment frames, which are permitted in higher seismic design categories. The design requirements are commensurate with providing limited ductility in the members and connections.

3. Analysis

There are no additional analysis requirements.

4. System Requirements

4a. Stability Bracing of Beams

Beams shall be braced to satisfy the requirements for *moderately ductile members* in Section D1.2a.

In addition, unless otherwise indicated by testing, beam braces shall be placed near concentrated forces, changes in cross section, and other locations where analysis indicates that a *plastic hinge* will form during inelastic deformations of the C-IMF.

The *required strength* of stability bracing provided adjacent to plastic hinges shall be as required by Section D1.2c.

5. Members

5a. Basic Requirements

Steel and composite members shall satisfy the requirements of Sections D1.1 for moderately ductile members.

5b. Beam Flanges

Abrupt changes in the beam flange area are prohibited in plastic hinge regions. The drilling of flange holes or trimming of beam flange width is prohibited unless testing or qualification demonstrates that the resulting configuration can develop stable plastic hinges.

5c. Protected Zones

The region at each end of the beam subject to inelastic straining shall be designated as a *protected zone*, and shall satisfy the requirements of Section D1.3.

User Note: The plastic hinge zones at the ends of C-IMF beams should be treated as protected zones. In general, the protected zone will extend from the face of the *composite column* to one-half of the beam depth beyond the plastic hinge point.

6. Connections

Connections shall be fully restrained (FR) and shall satisfy the requirements of Section D2 and this section.

User Note: All subsections of Section D2 are relevant for C-IMF.

6a. Demand Critical Welds

There are no requirements for *demand critical welds*.

6b. Beam-to-Column Connections

Beam-to-composite column connections used in the SFRS shall satisfy the following requirements:

(1) The connection shall be capable of accommodating a *story drift angle* of at least 0.02 rad.

(2) The *measured flexural resistance* of the connection, determined at the column face, shall equal at least $0.80M_p$ of the connected beam at a *story drift angle* of 0.02 rad, where M_p is defined as the nominal flexural strength of the steel, concrete-encased or *composite beams* and shall satisfy the requirements of *Specification* Chapter I.

6c. Conformance Demonstration

Beam-to-column connections used in the SFRS shall satisfy the requirements of Section G2.6b by connection testing or calculations that are substantiated by mechanistic models and component limit state design criteria consistent with these provisions.

6d. Required Shear Strength

The *required shear strength* of the connection shall be based on the load combinations in the *applicable building code* that include the *amplified seismic load*. In determining the amplified seismic load the effect of horizontal forces including overstrength, E_{mh}, shall be taken as:

$$E_{mh} = 2[1.1M_{p,exp}]/L_h \qquad (G2\text{-}1)$$

where $M_{p,exp}$ is the expected flexural strength of the steel, concrete-encased or composite beams, kip-in. (N-mm). For concrete-encased or composite beams, $M_{p,exp}$ shall be calculated using the plastic stress distribution or the strain compatibility method. Appropriate R_y factors shall be used for different elements of the cross-section while establishing section force equilibrium and calculating the flexural strength. L_h shall be equal to the distance between beam plastic hinge locations, in. (mm).

> **User Note:** For steel beams, $M_{p,exp}$ in Equation G2-1 may be taken as $R_y M_p$ of the beam.

6e. Connection Diaphragm Plates

Connection diaphragm plates are permitted for *filled composite columns* both external to the column and internal to the column.

Where diaphragm plates are used, the thickness of the plates shall be at least the thickness of the beam flange.

The diaphragm plates shall be welded around the full perimeter of the column using either complete-joint-penetration welds or two sided fillet welds. The required strength of these joints shall not be less than the *available strength* of the contact area of the plate with the column sides.

Internal diaphragms shall have circular openings sufficient for placing the concrete.

6f. Column Splices

In addition to the requirements of Section D2.5, column splices shall comply with the requirements of this section. Where groove welds are used to make the splice, they shall be complete-joint-penetration groove welds. When column splices are not made with groove welds, they shall have a *required flexural strength* that is at least equal to the nominal flexural strength, M_{pcc}, of the smaller composite column. The required shear strength of column web splices shall be at least equal to $\Sigma M_{pcc}/H$, where ΣM_{pcc} is the sum of the nominal flexural strengths of the composite columns above and below the splice. For composite columns, the nominal flexural strength shall satisfy the requirements of *Specification* Chapter I with consideration of the required axial strength, P_{rc}.

G3. COMPOSITE SPECIAL MOMENT FRAMES (C-SMF)

1. Scope

Composite special moment frames (C-SMF) shall be designed in conformance with this section. This section is applicable to moment frames with fully restrained (FR) connections that consist of either composite or reinforced concrete columns and either structural steel or concrete-encased composite or *composite beams*.

2. Basis of Design

C-SMF designed in accordance with these provisions are expected to provide significant inelastic deformation capacity through flexural yielding of the C-SMF beams and limited yielding of the column panel zones. Except where otherwise permitted in this section, columns shall be designed to be generally stronger than the fully yielded and strain-hardened beams or girders. Flexural yielding columns at the base is permitted. Design of connections of beams to columns, including panel zones, *continuity plates* and diaphragms shall provide the performance required by Section G3.6b, and demonstrate this conformance as required by Section G3.6c.

3. Analysis

There are no additional analysis requirements.

4. System Requirements

4a. Moment Ratio

The following relationship shall be satisfied at beam-to-column connections:

$$\frac{\Sigma M^*_{pcc}}{\Sigma M^*_{p,exp}} > 1.0 \tag{G3-1}$$

where

ΣM^*_{pcc} = sum of the moments in the columns above and below the joint at the intersection of the beam and column centerlines, kip-in. (N-mm). ΣM^*_{pcc} is determined by summing the projections of the nominal flexural strengths, M_{pcc}, of the columns (including haunches where used) above and below the joint to the beam centerline with a reduction for the axial force in the column. For *composite columns*, the nominal flexural strength, M_{pcc}, shall satisfy the requirements of *Specification* Chapter I with consideration of the required axial strength, P_{rc}. For reinforced concrete columns, the nominal flexural strength, M_{pcc}, shall be calculated based on the provisions of ACI 318 with consideration of the required axial strength, P_{rc}. When the centerlines of opposing beams in the same joint do not coincide, the mid-line between centerlines shall be used.

$\Sigma M^*_{p,exp}$ = sum of the moments in the steel beams or concrete-*encased composite beams* at the intersection of the beam and column centerlines, kip-in. (N-mm). $\Sigma M^*_{p,exp}$ is determined by summing the expected flexural strengths of the beams at the *plastic hinge* locations to the column centerline. It is permitted to take $\Sigma M^*_{p,exp} = \Sigma(1.1 M_{p,exp} + M_{uv})$, where $M_{p,exp}$ is calculated as specified in Section G2.6d.

M_{uv} = moment due to shear amplification from the location of the plastic hinge to the column centerline, kip-in. (N-mm).

Exception: The exceptions of Section E3.4a shall apply except that the force limit in Section E3.4a shall be $P_{rc} < 0.1 P_c$.

4b. Stability Bracing of Beams

Beams shall be braced to satisfy the requirements for *highly ductile members* in Section D1.2b.

In addition, unless otherwise indicated by testing, beam braces shall be placed near concentrated forces, changes in cross section, and other locations where analysis indicates that a plastic hinge will form during inelastic deformations of the C-SMF.

The *required strength* of stability bracing provided adjacent to plastic hinges shall be as required by Section D1.2c.

4c. Stability Bracing at Beam-to-Column Connections

Composite columns with unbraced connections shall satisfy the requirements of Section E3.4c(2).

5. Members

5a. Basic Requirements

Steel and composite members shall satisfy the requirements of Sections D1.1 for highly ductile members.

Exception: Reinforced concrete-encased beams shall satisfy the requirements for Section D1.1 for *moderately ductile members* if the reinforced concrete cover is at least 2 in. (50 mm) and confinement is provided by hoop reinforcement in regions where plastic hinges are expected to occur under seismic deformations. Hoop reinforcement shall satisfy the requirements of ACI 318 Section 21.5.3.

Concrete-encased composite beams that are part of C-SMF shall also satisfy the following requirement. The distance from the maximum concrete compression fiber to the plastic neutral axis shall not exceed:

$$Y_{PNA} = \frac{Y_{con} + d}{1 + \left(\dfrac{1,700 F_y}{E}\right)} \qquad (G3\text{-}2)$$

where
- E = modulus of elasticity of the steel beam, ksi (MPa)
- F_y = specified minimum yield stress of the steel beam, ksi (MPa)
- Y_{con} = distance from the top of the steel beam to the top of the concrete, in. (mm)
- d = overall beam depth, in. (mm)

5b. Beam Flanges

Abrupt changes in beam flange area are prohibited in plastic hinge regions. The drilling of flange holes or trimming of beam flange width is prohibited unless testing or qualification demonstrates that the resulting configuration can develop stable plastic hinges to accommodate the required *story drift angle*.

5c. Protected Zones

The region at each end of the beam subject to inelastic straining shall be designated as a *protected zone*, and shall satisfy the requirements of Section D1.3.

User Note: The plastic hinge zones at the ends of C-SMF beams should be treated as protected zones. In general, the protected zone will extend from the face of the composite column to one-half of the beam depth beyond the plastic hinge point.

6. Connections

Connections shall be fully restrained (FR) and shall satisfy the requirements of Section D2 and this section.

User Note: All subsections of Section D2 are relevant for C-SMF.

6a. Demand Critical Welds

The following welds are *demand critical welds*, and shall satisfy the requirements of Section A3.4b and I2.3:

(1) Groove welds at column splices
(2) Welds at the column-to-base plate connections

Exception: Where it can be shown that column hinging at or near the base plate is precluded by conditions of restraint, and in the absence of net tension under load combinations including the *amplified seismic load*, demand critical welds are not required.

(3) Complete-joint-penetration groove welds of beam flanges to columns, diaphragm plates that serve as a continuation of beam flanges, shear plates within the girder depth that transition from the girder to an encased steel shape, and beam webs to columns

6b. Beam-to-Column Connections

Beam-to-composite column connections used in the SFRS shall satisfy the following requirements:

(1) The connection shall be capable of accommodating a story drift angle of at least 0.04 rad.

(2) The *measured flexural resistance* of the connection, determined at the column face, shall equal at least $0.80M_p$ of the connected beam at a story drift angle of 0.04 rad, where M_p is calculated as in Section G2.6b.

6c. Conformance Demonstration

Beam-to-composite column connections used in the SFRS shall satisfy the requirements of Section G3.6b by the following:

(a) When beams are interrupted at the connection, the connections shall be qualified using test results obtained in accordance with Section K2. Results of at least two cyclic connection tests shall be provided, and shall be based on one of the following:

  (i) Tests reported in research literature or documented tests performed for other projects that represent the project conditions, within the limits specified in Section K2.

  (ii) Tests that are conducted specifically for the project and are representative of project member sizes, material strengths, connection configurations, and matching connection processes, within the limits specified by Section K2.

(b) When beams are uninterrupted or continuous through the composite or reinforced concrete column, beam flange welded joints are not used, and the connection is not otherwise susceptible to premature fracture, the performance requirements of Section G3.6b shall be demonstrated in accordance with (a) or other substantiating data.

Connections that accommodate the required story drift angle within the connection elements and provide the measured flexural resistance and shear strengths specified in Section G3.6d are permitted. In addition to satisfying the requirements noted above, the design shall demonstrate that any additional drift due to connection deformation can be accommodated by the structure. The design shall include analysis for stability effects of the overall frame, including second-order effects.

6d. Required Shear Strength

The *required shear strength* of the connection, V_u, shall be based on the load combinations in the *applicable building code* that include the amplified seismic load. In determining the amplified seismic load, the effect of horizontal forces including overstrength, E_{mh}, shall be taken as:

$$E_{mh} = 2[1.1M_{p,exp}]/L_h \qquad \text{(G3-3)}$$

where $M_{p,exp}$ is the expected flexural strength of the steel, concrete-encased, or composite beams. For concrete-encased or composite beams, $M_{p,exp}$ shall be calculated according to Section G2.6d, and L_h shall be equal to the distance between beam plastic hinge locations, in. (mm).

6e. Connection Diaphragm Plates

The continuity plates or diaphragms used for infilled column moment connections shall satisfy the requirements of Section G2.6e.

6f. Column Splices

Composite column splices shall satisfy the requirements of Section G2.6f.

G4. COMPOSITE PARTIALLY RESTRAINED MOMENT FRAMES (C-PRMF)

1. Scope

Composite partially restrained moment frames (C-PRMF) shall be designed in conformance with this section. This section is applicable to moment frames that consist of structural steel columns and *composite beams* that are connected with partially restrained (PR) moment connections that satisfy the requirements in *Specification* Section B3.6b(b).

2. Basis of Design

C-PRMF designed in accordance with these provisions are expected to provide significant inelastic deformation capacity through yielding in the ductile components of the composite PR beam-to-column moment connections. Limited yielding is permitted at other locations, such as flexural yielding of columns at the base is permitted. Design of connections of beams to columns shall be based on connection tests that provide the performance required by Section G4.6c, and demonstrate this conformance as required by Section G4.6d.

3. Analysis

Connection flexibility and composite beam action shall be accounted for in determining the dynamic characteristics, strength and drift of C-PRMF.

For purposes of analysis, the stiffness of beams shall be determined with an effective moment of inertia of the composite section.

4. System Requirements

There are no additional system requirements.

5. Members

5a. Columns

Steel columns shall satisfy the requirements of Sections D1.1 for *highly ductile members*.

5b. Beams

Composite beams shall be unencased, fully composite, and shall meet the requirements of Section D1.1 for highly ductile members. A solid slab shall be provided for a distance of 12 in. (300 mm) from the face of the column in the direction of moment transfer.

5c. Protected Zones

There are no designated *protected zones*.

6. Connections

Connections shall be partially restrained (PR) and shall satisfy the requirements of Section D2 and this section.

> **User Note:** All subsections of Section D2 are relevant for C-PRMF.

6a. Demand Critical Welds

The following welds are *demand critical welds*, and shall satisfy the requirements of Section A3.4b and I2.3:

(1) Groove welds at column splices

(2) Welds at the column-to-base plate connections

Exception: Where it can be shown that column hinging at or near the base plate is precluded by conditions of restraint, and in the absence of net tension under load combinations including the *amplified seismic load*, demand critical welds are not required.

6b. Required Strength

The *required strength* of the beam-to-column PR moment connections shall be determined considering the effects of connection flexibility and second-order moments.

6c. Beam-to-Column Connections

Beam-to-composite column connections used in the SFRS shall satisfy the following requirements:

(1) The connection shall be capable of accommodating a connection rotation of at least 0.02 rad.

(2) The measured flexural resistance of the connection determined at the column face shall increase monotonically to a value of at least $0.5M_p$ of the connected beam at a connection rotation of 0.02 rad, where M_p is defined as the nominal flexural strength of the steel beam and shall satisfy the requirements of *Specification* Chapter I.

6d. Conformance Demonstration

Beam-to-column connections used in the SFRS shall satisfy the requirements of Section G4.6c by provision of qualifying cyclic test results in accordance with

Section K2. Results of at least two cyclic connection tests shall be provided, and shall be based on one of the following:

(a) Tests reported in research literature or documented tests performed for other projects that represent the project conditions, within the limits specified in Section K2.

(b) Tests that are conducted specifically for the project and are representative of project member sizes, material strengths, connection configurations, and matching connection processes, within the limits specified by Section K2.

6e. **Column Splices**

Column splices shall satisfy the requirements of Section G2.6f.

CHAPTER H

COMPOSITE BRACED-FRAME AND SHEAR-WALL SYSTEMS

This chapter provides the basis of design, the requirements for analysis, and the requirements for the system, members and connections for composite braced frame and shear wall systems.

The chapter is organized as follows:

    H1.   Composite Ordinary Braced Frames (C-OBF)
    H2.   Composite Special Concentrically Braced Frames (C-SCBF)
    H3.   Composite Eccentrically Braced Frames (C-EBF)
    H4.   Composite Ordinary Shear Walls (C-OSW)
    H5.   Composite Special Shear Walls (C-SSW)
    H6.   Composite Plate Shear Walls (C-PSW)

User Note: The requirements of this chapter are in addition to those required by the *Specification* and the *applicable building code*.

H1. COMPOSITE ORDINARY BRACED FRAMES (C-OBF)

1. Scope

Composite ordinary braced frames (C-OBF) shall be designed in conformance with this section. Columns shall be structural steel, encased composite, filled composite or reinforced concrete members. Beams shall be either structural steel or *composite beams*. Braces shall be structural steel or filled composite members. This section is applicable to *braced frames* that consist of concentrically connected members where at least one of the elements (columns, beams or braces) is a composite or reinforced concrete member.

2. Basis of Design

This section is applicable to braced frames that consist of concentrically connected members. Eccentricities less than the beam depth are permitted if they are accounted for in the member design by determination of eccentric moments.

C-OBF designed in accordance with these provisions are expected to provide limited inelastic deformations in their members and connections. C-OBF shall satisfy the requirements of Section F1, except as modified in this section.

> **User Note:** Composite ordinary braced frames, comparable to other steel braced frames designed per the *Specification* using $R = 3$, are only permitted in seismic design categories C or below in ASCE/SEI 7. This is in contrast to steel ordinary braced frames, which are permitted in higher seismic design categories. The design requirements are commensurate with providing minimal ductility in the members and connections.

3. Analysis

There are no additional analysis requirements.

4. System Requirements

There are no additional system requirements.

5. Members

5a. Basic Requirements

There are no additional requirements.

5b. Columns

There are no additional requirements for structural steel and *composite columns*. Reinforced concrete columns shall satisfy the requirements of ACI 318, excluding Chapter 21.

5c. Braces

There are no additional requirements for structural steel and filled *composite braces*.

5d. Protected Zones

There are no designated *protected zones*.

6. Connections

Connections shall satisfy the requirements of Section D2.7.

6a. Demand Critical Welds

There are no requirements for *demand critical welds*.

H2. COMPOSITE SPECIAL CONCENTRICALLY BRACED FRAMES (C-SCBF)

1. Scope

Composite special concentrically braced frames (C-SCBF) shall be designed in conformance with this section. Columns shall be encased or filled composite. Beams shall be either structural steel or *composite beams*. Braces shall be structural steel or filled composite members. This section is applicable to *braced frames* that consist of concentrically connected members.

2. Basis of Design

This section is applicable to braced frames that consist of concentrically connected members. Eccentricities less than the beam depth are permitted if the resulting member and connection forces are addressed in the design and do not change the expected source of inelastic deformation capacity.

C-SCBF designed in accordance with these provisions are expected to provide significant inelastic deformation capacity primarily through brace buckling and yielding of the brace in tension.

3. Analysis

The analysis requirements for C-SCBF shall satisfy the analysis requirements of Section F2.3.

4. System Requirements

The system requirements for C-SCBF shall satisfy the system requirements of Section F2.4.

5. Members

5a. Basic Requirements

Composite columns and steel or *composite braces* shall satisfy the requirements of Section D1.1 for *highly ductile members*. Steel or composite beams shall satisfy the requirements of Section D1.1 for *moderately ductile members*.

> **User Note:** In order to satisfy the compactness requirement of Section F2.5a the actual width-to-thickness ratio of square and rectangular filled composite braces may be multiplied by a factor, $[(0.264 + 0.0082KL/r)]$, for KL/r between 35 and 90; KL/r being the effective slenderness ratio of the brace.

5b. Diagonal Braces

Structural steel and filled composite braces shall satisfy the requirements for SCBF of Section F2.5b. The radius of gyration in Section F2.5b shall be taken as that of the steel section alone.

5c. Protected Zones

There are no designated *protected zones*.

6. Connections

Design of connections in C-SCBF shall be based on Section D2 and the provisions of this section.

6a. Demand Critical Welds

The following welds are *demand critical welds*, and shall satisfy the requirements of Section A3.4b and I2.3:

Sect. H3.] COMPOSITE ECCENTRICALLY BRACED FRAMES (C-EBF)

(1) Groove welds at column splices
(2) Welds at the column-to-base plate connections

Exception: Where it can be shown that column hinging at, or near, the base plate is precluded by conditions of restraint, and in the absence of net tension under load combinations including the *amplified seismic load*, demand critical welds are not required.

(3) Welds at beam-to-column connections conforming to Section H2.6b(b)

6b. Beam-to-Column Connections

Where a brace or gusset plate connects to both members at a beam-to-column connection, the connection shall conform to one of the following:

(a) The connection shall be a simple connection meeting the requirements of *Specification* Section B3.6a where the required rotation is taken to be 0.025 rad; or

(b) Beam-to-column connections shall satisfy the requirements for FR moment connections as specified in Sections D2, G2.6d and G2.6e.

The required flexural strength of the connection shall be determined from analysis and be considered in combination with the *required strength* of the brace connection and beam connection, including the amplified diaphragm *collector* forces.

6c. Required Strength of Brace Connections

The required strength of brace connections shall satisfy the requirements of Section F2.6c.

6d. Column Splices

Column splices shall be designed following the requirements of Section G2.6f.

H3. COMPOSITE ECCENTRICALLY BRACED FRAMES (C-EBF)

1. Scope

Composite eccentrically braced frames (C-EBF) shall be designed in conformance with this section. Columns shall be encased composite or filled composite. Beams shall be structural steel or *composite beams*. *Links* shall be structural steel. Braces shall be structural steel or filled composite members. This section is applicable to *braced frames* for which one end of each brace intersects a beam at an eccentricity from the intersection of the centerlines of the beam and an adjacent brace or column.

2. Basis of Design

C-EBF shall satisfy the requirements of Section F3.2, except as modified in this section.

This section is applicable to braced frames for which one end of each brace intersects a beam at an eccentricity from the intersection of the centerlines of the beam and an adjacent brace or column, forming a link that is subject to shear and flexure. Eccentricities less than the beam depth are permitted in the brace connection away from the link if the resulting member and connection forces are addressed in the design and do not change the expected source of inelastic deformation capacity.

C-EBF designed in accordance with these provisions are expected to provide significant inelastic deformation capacity primarily through shear or flexural yielding in the links.

The *available strength* of members shall satisfy the requirements in the *Specification*, except as modified in this section.

3. Analysis

The analysis of C-EBF shall satisfy the analysis requirements of Section F3.3.

4. System Requirements

The system requirements for C-EBF shall satisfy the system requirements of Section F3.4.

5. Members

The member requirements of C-EBF shall satisfy the member requirements of Section F3.5.

6. Connections

The connection requirements of C-EBF shall satisfy the connection requirements of Section F3.6 except as noted below.

6a. Beam-to-Column Connections

Where a brace or gusset plate connects to both members at a beam-to-column connection, the connection shall conform to one of the following:

(a) The connection shall be a simple connection meeting the requirements of *Specification* Section B3.6a where the required rotation is taken to be 0.025 rad; or

(b) Beam-to-column connections shall satisfy the requirements for fully restrained (FR) moment connections as specified in Sections D2, G2.6d and G2.6e.

The required flexural strength of the connection shall be determined from analysis and be considered in combination with the *required strength* of the brace connection and beam connection, including the amplified diaphragm *collector* forces.

H4. COMPOSITE ORDINARY SHEAR WALLS (C-OSW)

1. Scope

Composite ordinary shear walls (C-OSW) shall be designed in conformance with this section. This section is applicable when reinforced concrete walls are composite with structural steel elements, including structural steel or composite sections acting as *boundary members* for the walls and structural steel or composite *coupling beams* that connect two or more adjacent reinforced concrete walls.

2. Basis of Design

C-OSW designed in accordance with these provisions are expected to provide limited inelastic deformation capacity through yielding in the reinforced concrete walls and the steel or composite elements. Reinforced concrete wall elements shall be designed to provide inelastic deformations at the *design story drift* consistent with ACI 318 excluding Chapter 21. Structural steel and composite coupling beams shall be designed to provide inelastic deformations at the design story drift through yielding in flexure or shear. Structural steel and composite boundary elements shall be designed to provide inelastic deformations at the design story drift through yielding due to axial force.

Reinforced concrete walls shall satisfy the requirements of ACI 318 excluding Chapter 21, except as modified in this section.

3. Analysis

Analysis shall satisfy the requirements of Chapter C as modified in this section.

(1) Uncracked effective stiffness values for elastic analysis shall be assigned in accordance with ACI 318 Chapter 10 for wall piers and composite coupling beams.

(2) When *concrete-encased shapes* function as boundary members, the analysis shall be based upon a transformed concrete section using elastic material properties.

(3) The flexibility of the connection between coupling beams and wall piers and the effect of shear distortions of the coupling beam and walls shall be taken into account.

4. System Requirements

In *coupled walls*, coupling beams are permitted to yield over the height of the structure. The coupling beam-wall connection shall develop the expected flexural and shear strengths of the coupling beam.

In coupled walls, it is permitted to redistribute coupling beam forces vertically to adjacent floors. The shear in any individual coupling beam should not be reduced by more than 20% of the elastically determined value. The sum of the coupling beam shear resistance over the height of the building shall be greater than or equal to the sum of the elastically determined values.

5. Members

5a. Boundary Members

Boundary members shall satisfy the following requirements:

(1) The *required axial strength* of the boundary member shall be determined assuming that the shear forces are carried by the reinforced concrete wall and the entire

gravity and overturning forces are carried by the boundary members in conjunction with the shear wall.

(2) When the concrete-encased structural steel boundary member qualifies as a *composite column* as defined in *Specification* Chapter I, it shall be designed as a composite column to satisfy the requirements of Chapter I of the *Specification*.

(3) Headed studs or welded reinforcement anchors shall be provided to transfer required shear strengths between the structural steel boundary members and reinforced concrete walls. Headed studs, if used, shall satisfy the requirements of *Specification* Chapter I. Welded reinforcement anchors, if used, shall satisfy the requirements of *Structural Welding Code—Reinforcing Steel* (AWS D1.4/D1.4M).

5b. Coupling Beams

(1) Structural Steel Coupling Beams

Structural steel coupling beams that are used between adjacent reinforced concrete walls shall satisfy the requirements of the *Specification* and this section. The following requirements apply to wide flange steel coupling beams.

(1) Steel coupling beams shall comply with the requirements of Section D1.1 for *moderately ductile members*.

(2) The expected shear strength, V_n, of steel coupling beams shall be computed from Equation H4-1.

$$V_n = \frac{2R_y M_p}{g} \leq R_y V_p \tag{H4-1}$$

where
A_{tw} = area of steel beam web, in.² (mm²)
$M_p = F_y Z$, kip-in. (N-mm)
V_n = expected shear strength of a steel coupling beam, kips (N)
$V_p = 0.6 F_y A_{tw}$, kips (N)
g = coupling beam clear span, in. (mm)

(3) The embedment length, L_e, shall be computed from Equations H4-2 and H4-2M. The embedment length shall be considered to begin inside the first layer of confining reinforcement in the wall boundary member.

$$V_n = 1.54\sqrt{f'_c}\left(\frac{b_w}{b_f}\right)^{0.66} \beta_1 b_f L_e \left[\frac{0.58 - 0.22\beta_1}{0.88 + \frac{g}{2L_e}}\right] \tag{H4-2}$$

$$V_n = 0.004\sqrt{f'_c}\left(\frac{b_w}{b_f}\right)^{0.66} \beta_1 b_f L_e \left[\frac{0.58 - 0.22\beta_1}{0.88 + \frac{g}{2L_e}}\right] \text{ (S.I.)} \tag{H4-2M}$$

where
L_e = embedment length of coupling beam, in. (mm)

b_w = thickness of wall pier, in. (mm)
b_f = beam flange width, in. (mm)
f'_c = concrete compressive strength, ksi (MPa)
β_1 = factor relating depth of equivalent rectangular compressive stress block to neutral axis depth, as defined in ACI 318

(4) Vertical wall reinforcement with *nominal axial strength* equal to the expected shear strength of the coupling beam shall be placed over the embedment length of the beam with two-thirds of the steel located over the first half of the embedment length. This wall reinforcement shall extend a distance of at least one tension development length above and below the flanges of the coupling beam. It is permitted to use vertical reinforcement placed for other purposes, such as for vertical boundary members, as part of the required vertical reinforcement.

(2) Composite Coupling Beams
Encased composite sections serving as coupling beams shall satisfy the requirements of Section H4.5b(1) as modified in this section:

(1) Coupling beams shall have an embedment length into the reinforced concrete wall that is sufficient to develop the expected shear strength, $V_{n,\,comp}$, computed from Equation H4-3.

$$V_{n,\,comp} = \frac{2M_{p,\,exp}}{g} \leq V_{comp} \qquad (H4\text{-}3)$$

where
$M_{p,\,exp}$ = expected flexural strength of composite coupling beam, kip-in. (N-mm). For concrete-encased or *composite beams*, $M_{p,exp}$ shall be calculated using the plastic stress distribution or the strain compatibility method. Appropriate R_y factors shall be used for different elements of the cross section while establishing section force equilibrium and calculating the flexural strength.
V_{comp} = limiting expected shear strength of an encased composite coupling beam as computed by Equations H4-4 and H4-4M, kips (N)

$$V_{comp} = R_y V_p + \left(0.0632\sqrt{f'_c}\ b_{wc} d_c + \frac{A_s F_{ysr} d_c}{s} \right) \qquad (H4\text{-}4)$$

$$V_{comp} = R_y V_p + \left(0.166\sqrt{f'_c}\ b_{wc} d_c + \frac{A_s F_{ysr} d_c}{s} \right)\ (S.I.) \qquad (H4\text{-}4M)$$

where
A_s = area of transverse reinforcement, in.² (mm²)
F_{ysr} = specified minimum yield stress of transverse reinforcement, ksi (MPa)
b_{wc} = width of concrete encasement, in. (mm)
d_c = effective depth of concrete encasement, in. (mm)
s = spacing of transverse reinforcement, in. (mm)

(2) The required embedment length shall be computed from Equations H4-2 and H4-2M by using $V_{n,comp}$ instead of V_n.

5c. Projected Zones

There are no designated *protected zones*.

6. Connections

There are no additional requirements beyond Section H4.5.

6a. Demand Critical Welds

There are no requirements for *demand critical welds*.

H5. COMPOSITE SPECIAL SHEAR WALLS (C-SSW)

1. Scope

Composite special shear walls (C-SSW) shall be designed in conformance with this section. This section is applicable when reinforced concrete walls are composite with structural steel elements, including structural steel or composite sections acting as *boundary members* for the walls and structural steel or composite *coupling beams* that connect two or more adjacent reinforced concrete walls.

2. Basis of Design

C-SSW designed in accordance with these provisions are expected to provide significant inelastic deformation capacity through yielding in the reinforced concrete walls and the steel or composite elements. Reinforced concrete wall elements shall be designed to provide inelastic deformations at the *design story drift* consistent with ACI 318 including Chapter 21. Structural steel and composite coupling beams shall be designed to provide inelastic deformations at the design story drift through yielding in flexure or shear. Coupling beam connections and the design of the walls shall be designed to account for the expected strength including strain hardening in the coupling beams. Structural steel and composite boundary elements shall be designed to provide inelastic deformations at the design story drift through yielding due to axial force.

C-SSW systems shall satisfy the requirements of Section H4 and the shear wall requirements of ACI 318 including Chapter 21, except as modified in this section.

3. Analysis

Analysis requirements of Section H4.3 shall be met with the following exceptions:

(1) Cracked effective stiffness values for elastic analysis shall be assigned in accordance with ACI 318 Chapter 10 practice for wall piers and composite coupling beams.

(2) Effects of shear distortion of the steel coupling beam shall be taken into account.

4. System Requirements

System requirements of Section H4.4 shall be satisfied with the following exceptions:

(1) In *coupled walls*, coupling beams shall yield over the height of the structure followed by yielding at the base of the wall piers.

(2) In coupled walls, the axial *design strength* of the wall at the balanced condition, P_b, shall equal or exceed the total required compressive axial strength in a wall pier, computed as the sum of the *required strengths* attributed to the walls from the gravity load components of the lateral load combination plus the sum of the expected beam shear strengths increased by a factor of 1.1 to reflect the effects of strain hardening of all the coupling beams framing into the walls.

5. Members

5a. Ductile Elements

Coupling beams are protected zones, and shall satisfy the requirements of Section D1.3. Welding on steel coupling beams is permitted for attachment of stiffeners, as required in Section F3.5b(4).

5b. Boundary Members

Unencased structural steel columns shall satisfy the requirements of Section D1.1 for *highly ductile members* and Section H4.5a(1).

In addition to the requirements of Sections H4.3(2) and H4.5a(2), the requirements in this section shall apply to walls with concrete-encased structural steel boundary members. Concrete-encased structural steel boundary members that qualify as *composite columns* in *Specification* Chapter I shall meet the highly ductile member requirements of Section D1.4b(2). Otherwise, such members shall be designed as composite compression members to satisfy the requirements of ACI 318 Section 10.13 including the special seismic requirements for boundary members in ACI 318 Section 21.9.6. Transverse reinforcement for confinement of the composite boundary member shall extend a distance of $2h$ into the wall, where h is the overall depth of the boundary member in the plane of the wall.

Headed studs or welded reinforcing anchors shall be provided as specified in Section H4.5a(3).

5c. Steel Coupling Beams

In addition to the requirements of Section H4.5b, structural steel coupling beams shall satisfy the requirements of Section F3.5b. When required in Section F3.5b(4), the coupling beam rotation shall be assumed as a 0.08 rad *link* rotation unless a smaller value is justified by rational analysis of the inelastic deformations that are expected under the design story drift. *Face bearing plates* shall be provided on both sides of the coupling beams at the face of the reinforced concrete wall. These stiffeners shall meet the detailing requirements of Section F3.5b(4).

Steel coupling beams shall comply with the requirements of Section D1.1 for highly ductile members.

The expected shear strength for which the embedment length is calculated in Equation H4-1 shall be increased by a factor of 1.1 to reflect the effects of strain hardening.

Vertical wall reinforcement as specified in Section H4.5b(1)(4) shall be confined by transverse reinforcement that meets the requirements for boundary members of ACI 318 Section 21.9.6.

Embedded steel members shall be provided with two regions of vertical transfer reinforcement attached to both the top and bottom flanges of the embedded member. The first region shall be located to coincide with the location of longitudinal wall reinforcing bars closest to the face of the wall. The second shall be placed a distance no less than $d/2$ from the termination of the embedment length. All transfer reinforcement bars shall be fully developed where they engage the coupling beam flanges. It is permitted to use straight, hooked or mechanical anchorage to provide development. It is permitted to use mechanical couplers welded to the flanges to attach the vertical transfer bars. The area of vertical transfer reinforcement required is computed by Equation H5-1:

$$A_{tb} \geq 0.03 f'_c L_e b_f / F_{ysr} \qquad (H5\text{-}1)$$

where
A_{tb} = area of transfer reinforcement required in each of the first and second regions attached to each of the top and bottom flanges, in.² (mm²)
F_{ysr} = specified minimum yield stress of transfer reinforcement, ksi (MPa)
L_e = embedment length, in. (mm)
b_f = beam flange width, in. (mm)
f'_c = concrete compressive strength, ksi (MPa)

The area of vertical transfer reinforcement shall not exceed that computed by Equation H5-2:

$$\Sigma A_{tb} < 0.08 L_e b_w - A_s \qquad (H5\text{-}2)$$

where
ΣA_{tb} = total area of transfer reinforcement provided in both the first and second regions attached to either the top or bottom flange, in.² (mm²)
A_s = area of longitudinal wall reinforcement provided over the embedment length, L_e, in.² (mm²)
b_w = wall width, in. (mm)

5d. Composite Coupling Beams

Encased composite sections serving as coupling beams shall satisfy the requirements of Section H5.5c except the requirements of Section F3.5b(4) need not be met, and Equation H5-3 shall be used instead of Equation H4-4. For all encased composite coupling beams, the limiting expected shear strength, V_{comp}, is:

$$V_{comp} = 1.1 R_y V_p + 1.56 \left(0.0632 \sqrt{f'_c} b_{wc} d_c + \frac{A_s F_{ysr} d_c}{s} \right) \qquad (H5\text{-}3)$$

$$V_{comp} = 1.1 R_y V_p + 1.56 \left(0.166 \sqrt{f'_c} b_{wc} d_c + \frac{A_s F_{ysr} d_c}{s} \right) \text{ (S.I.)} \qquad (H5\text{-}3M)$$

Sect. H6.] COMPOSITE PLATE SHEAR WALLS (C-PSW)

where
F_{ysr} = yield stress of transverse reinforcement, ksi (MPa)

5e. Protected Zones

There are no designated *protected zones*.

6. Connections

6a. Demand Critical Welds

The following welds are *demand critical welds*, and shall satisfy the requirements of Section A3.4b and I2.3:

(1) Groove welds at column splices

(2) Welds at the column-to-base plate connections

Exception: Where it can be shown that column hinging at, or near, the base plate is precluded by conditions of restraint, and in the absence of net tension under load combinations including the *amplified seismic load*, demand critical welds are not required.

6b. Column Splices

Column splices shall be designed following the requirements of Section G2.6f.

H6. COMPOSITE PLATE SHEAR WALLS (C-PSW)

1. Scope

Composite plate shear walls (C-PSW) shall be designed in conformance with this section. Composite plate shear walls consist of steel plates with reinforced concrete encasement on one or both sides of the plate, or steel plates on both sides of reinforced concrete infill, and structural steel or composite *boundary members*.

2. Basis of Design

C-PSW designed in accordance with these provisions are expected to provide significant inelastic deformation capacity through yielding in the plate webs. The *horizontal boundary elements* (HBE) and *vertical boundary elements* (VBE) adjacent to the composite webs shall be designed to remain essentially elastic under the maximum forces that can be generated by the fully yielded steel webs along with the reinforced concrete webs after the steel web has fully yielded, except that plastic hinging at the ends of HBE is permitted.

3. Analysis

3a. Webs

Steel webs shall be designed to resist the seismic load, E, determined from the analysis required by the *applicable building code*. The analysis shall account for openings in the web.

3b. Other Members and Connections

Columns, beams and connections in C-PSW shall be designed to resist seismic forces determined from an analysis that includes the expected strength of the steel webs in shear, $0.6R_yF_yA_{sp}$, and any reinforced concrete portions of the wall active at the *design story drift*. The vertical boundary elements (VBE) are permitted to yield at the base.

4. System Requirements

4a. Steel Plate Thickness

Steel plates with thickness less than $3/8$ in. (9.5 mm) are not permitted.

4b. Stiffness of Vertical Boundary Elements

The VBE shall satisfy the requirements of Section F5.4a.

4c. HBE-to-VBE Connection Moment Ratio

The beam-column moment ratio shall satisfy the requirements of Section F5.4b.

4d. Bracing

The bracing shall satisfy the requirements of Section F5.4c.

4e. Openings in Webs

Boundary members shall be provided around openings in shear wall webs as required by analysis.

5. Members

5a. Basic Requirements

Steel and composite HBE and VBE shall satisfy the requirements of Section D1.1 for *highly ductile members*.

5b. Webs

The *design shear strength*, ϕV_n, or the allowable shear strength, V_n/Ω, for the limit state of shear yielding with a composite plate conforming to Section H6.5c shall be taken as:

$$V_n = 0.6A_{sp}F_y \tag{H6-1}$$

$$\phi = 0.90 \text{ (LRFD)} \qquad \Omega = 1.67 \text{ (ASD)}$$

where
A_{sp} = horizontal area of stiffened steel plate, in.^{2} (mm^2)
F_y = specified minimum yield stress of the plate, ksi (MPa)
V_n = nominal shear strength of the steel plate, kips (N)

The *available shear strength* of C-PSW with a plate that does not meet the stiffening requirements in Section H6.5c shall be based upon the strength of the plate as given in Section F5.5 and satisfy the requirements of *Specification* Sections G2 and G3.

5c. Concrete Stiffening Elements

The steel plate shall be adequately stiffened by encasement or attachment to a reinforced concrete panel. Conformance to this requirement shall be demonstrated with an elastic plate buckling analysis showing that the composite wall can resist a nominal shear force equal to V_{ns}.

The concrete thickness shall be a minimum of 4 in. (100 mm) on each side when concrete is provided on both sides of the steel plate and 8 in. (200 mm) when concrete is provided on one side of the steel plate. Steel headed stud anchors or other mechanical connectors shall be provided to prevent local buckling and separation of the plate and reinforced concrete. Horizontal and vertical reinforcement shall be provided in the concrete encasement to meet or exceed the requirements in ACI 318 Section 14.3. The reinforcement ratio in both directions shall not be less than 0.0025. The maximum spacing between bars shall not exceed 18 in. (450 mm).

5d. Boundary Members

Structural steel and composite boundary members shall be designed to resist the *expected shear strength* of steel plate and any reinforced concrete portions of the wall active at the design story drift. Composite and reinforced concrete boundary members shall also satisfy the requirements of Section H5.5b. Steel boundary members shall also satisfy the requirements of Section F5.

5e. Protected Zones

There are no designated *protected zones*.

6. Connections

6a. Demand Critical Welds

The following welds are *demand critical welds*, and shall satisfy the requirements of Section A3.4b and I2.3:

(1) Groove welds at column splices

(2) Welds at the column-to-base plate connections

> Exception: Where it can be shown that column hinging at, or near, the base plate is precluded by conditions of restraint, and in the absence of net tension under load combinations including the *amplified seismic load*, demand critical welds are not required.

(3) Welds at HBE-to-VBE connections

6b. HBE-to-VBE Connections

HBE-to-VBE connections shall satisfy the requirements of Section F5.6b.

6c. Connections of Steel Plate to Boundary Elements

The steel plate shall be continuously welded or bolted on all edges to the structural steel framing and/or steel boundary members, or the steel component of the composite boundary members. Welds and/or slip-critical high-strength bolts required to develop the nominal shear strength of the plate shall be provided.

6d. Connections of Steel Plate to Reinforced Concrete Panel

The steel anchors between the steel plate and the reinforced concrete panel shall be designed to prevent its overall buckling. Steel anchors shall be designed to satisfy the following conditions:

(1) Tension in the Connector

The steel anchor shall be designed to resist the tension force resulting from inelastic local buckling of the steel plate.

(2) Shear in the Connector

The steel anchors collectively shall be designed to transfer the expected strength in shear of the steel plate or reinforced concrete panel, whichever is smaller.

6e. Column Splices

Column splices shall be designed following the requirements of Section G2.6f.

CHAPTER I

FABRICATION AND ERECTION

This chapter addresses requirements for fabrication and erection.

User Note: All requirements of *Specification* Chapter M also apply, unless specifically modified by these Provisions.

The chapter is organized as follows:

 I1. Shop and Erection Drawings
 I2. Fabrication and Erection

I1. SHOP AND ERECTION DRAWINGS

1. Shop Drawings for Steel Construction

Shop drawings shall indicate the work to be performed, and include items required by the *Specification*, the AISC *Code of Standard Practice for Steel Buildings and Bridges*, the *applicable building code*, the requirements of Sections A4.1 and A4.2, and the following, as applicable:

(1) Locations of pretensioned bolts
(2) Locations of Class A, or higher, faying surfaces
(3) Gusset plates drawn to scale when they are designed to accommodate inelastic rotation
(4) Weld access hole dimensions, surface profile and finish requirements
(5) Nondestructive testing (NDT) where performed by the fabricator

2. Erection Drawings for Steel Construction

Erection drawings shall indicate the work to be performed, and include items required by the *Specification*, the AISC *Code of Standard Practice for Steel Buildings and Bridges*, the applicable building code, the requirements of Sections A4.1 and A4.2, and the following, as applicable:

(1) Locations of pretensioned bolts
(2) Those joints or groups of joints in which a specific assembly order, welding sequence, welding technique or other special precautions are required

3. Shop and Erection Drawings for Composite Construction

Shop drawings and erection drawings for the steel components of composite steel-concrete construction shall satisfy the requirements of Sections I1.1 and I1.2. The shop drawings and erection drawings shall also satisfy the requirements of Section A4.3.

> **User Note:** For reinforced concrete and composite steel-concrete construction, the provisions of ACI 315 *Details and Detailing of Concrete Reinforcement* and ACI 315-R *Manual of Engineering and Placing Drawings for Reinforced Concrete Structures* apply.

I2. FABRICATION AND ERECTION

1. Protected Zone

A *protected zone* designated by these Provisions or ANSI/AISC 358 shall comply with the following requirements:

(1) Within the protected zone, holes, tack welds, erection aids, air-arc gouging, and unspecified thermal cutting from fabrication or erection operations shall be repaired as required by the *engineer of record*.

(2) Steel headed stud anchors and decking attachments that penetrate the beam flange shall not be placed on beam flanges within the protected zone. Arc spot welds as required to secure decking shall be permitted.

(3) Welded, bolted, screwed or shot-in attachments for perimeter edge angles, exterior facades, partitions, duct work, piping or other construction shall not be placed within the protected zone.

Exception: Other attachments are permitted where designated or approved by the engineer of record. See Section D1.3.

> **User Note:** AWS D1.8/D1.8M clause 6.15 contains requirements for weld removal and the repair of gouges and notches in the protected zone.

2. Bolted Joints

Bolted joints shall satisfy the requirements of Section D2.2.

3. Welded Joints

Welding and welded connections shall be in accordance with *Structural Welding Code—Steel* (AWS D1.1/D1.1M), hereafter referred to as AWS D1.1/D1.1M, and AWS D1.8/D1.8M.

Welding procedure specifications (WPSs) shall be approved by the engineer of record.

Weld tabs shall be in accordance with AWS D1.8/D1.8M clause 6.10, except at the outboard ends of continuity-plate-to-column welds, weld tabs and weld metal need not be removed closer than $1/4$ in. (6 mm) from the continuity plate edge.

AWS D1.8/D1.8M clauses relating to fabrication shall apply equally to shop fabrication welding and to field erection welding.

User Note: AWS D1.8/D1.8M was specifically written to provide additional requirements for the welding of *seismic force resisting systems*, and has been coordinated wherever possible with these Provisions. AWS D1.8/D1.8M requirements related to fabrication and erection are organized as follows, including normative (mandatory) annexes:

1. General Requirements
2. Reference Documents
3. Definitions
4. Welded Connection Details
5. Welder Qualification
6. Fabrication

Annex A. WPS Heat Input Envelope Testing of Filler Metals for *Demand Critical Welds*

Annex B. Intermix CVN Testing of Filler Metal Combinations (where one of the filler metals is FCAW-S)

Annex C. Supplemental Welder Qualification for Restricted Access Welding

Annex D. Supplemental Testing for Extended Exposure Limits for FCAW Filler Metals

AWS D1.8/D1.8M requires the complete removal of all weld tab material, leaving only base metal and weld metal at the edge of the joint. This is to remove any weld discontinuities at the weld ends, as well as facilitate magnetic particle testing (MT) of this area. At *continuity plates*, these Provisions permit a limited amount of weld tab material to remain because of the reduced strains at continuity plates, and any remaining weld discontinuities in this weld end region would likely be of little significance. Also, weld tab removal sites at continuity plates are not subjected to MT.

AWS D1.8/D1.8M clause 6 is entitled "Fabrication," but the intent of AWS is that all provisions of AWS D1.8/D1.8M apply equally to fabrication and erection activities as described in the *Specification* and in these Provisions.

4. Continuity Plates and Stiffeners

Corners of continuity plates and stiffeners placed in the webs of rolled shapes shall be detailed in accordance with AWS D1.8 clause 4.1.

CHAPTER J

QUALITY CONTROL AND QUALITY ASSURANCE

This chapter addresses requirements for quality control and quality assurance.

> **User Note:** All requirements of *Specification* Chapter N also apply, unless specifically modified by these Provisions.

The chapter is organized as follows:

- J1. Scope
- J2. Fabricator and Erector Documents
- J3. Quality Assurance Agency Documents
- J4. Inspection and Nondestructive Testing Personnel
- J5. Inspection Tasks
- J6. Welding Inspection and Nondestructive Testing
- J7. Inspection of High-Strength Bolting
- J8. Other Steel Structure Inspections
- J9. Inspection of Composite Structures
- J10. Inspection of Piling

J1. SCOPE

Quality Control (QC) as specified in this chapter shall be provided by the fabricator, erector or other responsible contractor as applicable. *Quality Assurance* (QA) as specified in this chapter shall be provided by others when required by the *authority having jurisdiction* (AHJ), *applicable building code* (ABC), purchaser, owner or *engineer of record* (EOR). Nondestructive testing (NDT) shall be performed by the agency or firm responsible for Quality Assurance, except as permitted in accordance with *Specification* Section N7.

> **User Note:** The quality assurance plan of this section is considered adequate and effective for most *seismic force resisting systems* and should be used without modification. The *quality assurance plan* is intended to ensure that the seismic force resisting system is significantly free of defects that would greatly reduce the ductility of the system. There may be cases (for example, nonredundant major transfer members, or where work is performed in a location that is difficult to access) where supplemental testing might be advisable. Additionally, where the fabricator's or erector's quality control program has demonstrated the capability to perform some tasks this plan has assigned to quality assurance, modification of the plan could be considered.

J2. FABRICATOR AND ERECTOR DOCUMENTS

1. Documents to be Submitted for Steel Construction

In addition to the requirements of *Specification* Section N3.1, the following documents shall be submitted for review by the *engineer of record* (EOR) or the EOR's designee, prior to fabrication or erection of the affected work, as applicable:

(1) Welding procedure specifications (WPS)

(2) Copies of the manufacturer's typical certificate of conformance for all electrodes, fluxes and shielding gasses to be used

(3) For *demand critical welds*, applicable manufacturer's certifications that the filler metal meets the supplemental notch toughness requirements, as applicable. Should the filler metal manufacturer not supply such supplemental certifications, the fabricator or erector, as applicable, shall have the necessary testing performed and provide the applicable test reports

(4) Manufacturer's product data sheets or catalog data for SMAW, FCAW and GMAW composite (cored) filler metals to be used

(5) Bolt installation procedures

(6) Specific assembly order, welding sequence, welding technique, or other special precautions for joints or groups of joints where such items are designated to be submitted to the engineer of record

2. Documents to be Available for Review for Steel Construction

Additional documents as required by the EOR in the contract documents shall be available by the fabricator and erector for review by the EOR or the EOR's designee prior to fabrication or erection, as applicable.

The fabricator and erector shall retain their document(s) for at least one year after substantial completion of construction.

3. Documents to be Submitted for Composite Construction

The following documents shall be submitted by the responsible contractor for review by the EOR or the EOR's designee, prior to concrete production or placement, as applicable:

(1) Concrete mix design and test reports for the mix design

(2) Reinforcing steel shop drawings

(3) Concrete placement sequences, techniques and restriction

4. Documents to be Available for Review for Composite Construction

The following documents shall be available from the responsible contractor for review by the EOR or the EOR's designee prior to fabrication or erection, as applicable, unless specified to be submitted:

(1) Material test reports for reinforcing steel

(2) Inspection procedures

(3) Nonconformance procedure

(4) Material control procedure

(5) Welder performance qualification records (WPQR) as required by AWS D1.4/D1.4M

(6) QC Inspector qualifications

The responsible contractor shall retain their document(s) for at least one year after substantial completion of construction.

J3. QUALITY ASSURANCE AGENCY DOCUMENTS

The agency responsible for quality assurance shall submit the following documents to the *authority having jurisdiction*, the *engineer of record*, and the owner or owner's designee:

(1) QA agency's written practices for the monitoring and control of the agency's operations. The written practice shall include:
  (i) The agency's procedures for the selection and administration of inspection personnel, describing the training, experience and examination requirements for qualification and certification of inspection personnel, and
  (ii) The agency's inspection procedures, including general inspection, material controls, and visual welding inspection

(2) Qualifications of management and QA personnel designated for the project

(3) Qualification records for inspectors and NDT technicians designated for the project

(4) NDT procedures and equipment calibration records for NDT to be performed and equipment to be used for the project

(5) For composite construction, concrete testing procedures and equipment

J4. INSPECTION AND NONDESTRUCTIVE TESTING PERSONNEL

In addition to the requirements of *Specification* Sections N4.1 and N4.2, visual welding inspection and nondestructive testing (NDT) shall be conducted by personnel qualified in accordance with AWS D1.8/D1.8M clause 7.2. In addition to the requirements of *Specification* Section N4.3, ultrasonic testing technicians shall be qualified in accordance with AWS D1.8/D1.8M clause 7.2.4.

> **User Note:** The recommendations of the International Code Council *Model Program for Special Inspection* should be considered a minimum requirement to establish the qualifications of a bolting inspector.

J5. INSPECTION TASKS

Inspection tasks and documentation for quality control (QC) and quality assurance (QA) for the *seismic force resisting system* (SFRS) shall be as provided in the tables in Sections J6, J7, J8, J9 and J10. The following entries are used in the tables:

1. **Observe (O)**

   The inspector shall observe these functions on a random, daily basis. Operations need not be delayed pending observations.

2. **Perform (P)**

   These inspections shall be performed prior to the final acceptance of the item.

3. **Document (D)**

   The inspector shall prepare reports indicating that the work has been performed in accordance with the contract documents. The report need not provide detailed measurements for joint fit-up, WPS settings, completed welds, or other individual items listed in the tables. For shop fabrication, the report shall indicate the piece mark of the piece inspected. For field work, the report shall indicate the reference grid lines and floor or elevation inspected. Work not in compliance with the contract documents and whether the noncompliance has been satisfactorily repaired shall be noted in the inspection report.

4. **Coordinated Inspection**

   Where a task is noted to be performed by both QC and QA, coordination of the inspection function between QC and QA is permitted in accordance with *Specification* Section N5.3.

J6. WELDING INSPECTION AND NONDESTRUCTIVE TESTING

Welding inspection and nondestructive testing shall satisfy the requirements of the *Specification*, this section and AWS D1.8/D1.8M.

> **User Note:** AWS D1.8/D1.8M was specifically written to provide additional requirements for the welding of *seismic force resisting systems*, and has been coordinated when possible with these Provisions. AWS D1.8/D1.8M requirements related to inspection and nondestructive testing are organized as follows, including normative (mandatory) annexes:
>
> 1. General Requirements
> 7. Inspection
>
> Annex F. Supplemental Ultrasonic Technician Testing
> Annex G. Supplemental Magnetic Particle Testing Procedures
> Annex H. Flaw Sizing by Ultrasonic Testing

1. **Visual Welding Inspection**

   All requirements of the *Specification* shall apply, except as specifically modified by AWS D1.8/D1.8M.

   Visual welding inspection shall be performed by both quality control and quality assurance personnel. As a minimum, tasks shall be as listed in Tables J6-1, J6-2 and J6-3.

TABLE J6-1
Visual Inspection Tasks Prior to Welding

Visual Inspection Tasks Prior to Welding	QC		QA	
	Task	Doc.	Task	Doc.
Material identification (Type/Grade)	O	–	O	–
Welder identification system	O	–	O	–
Fit-up of Groove Welds (including joint geometry) - Joint preparation - Dimensions (alignment, root opening, root face, bevel) - Cleanliness (condition of steel surfaces) - Tacking (tack weld quality and location) - Backing type and fit (if applicable)	P/O**	–	O	–
Configuration and finish of access holes	O	–	O	–
Fit-up of Fillet Welds - Dimensions (alignment, gaps at root) - Cleanliness (condition of steel surfaces) - Tacking (tack weld quality and location)	P/O**	–	O	–

** Following performance of this inspection task for ten welds to be made by a given welder, with the welder demonstrating understanding of requirements and possession of skills and tools to verify these items, the Perform designation of this task shall be reduced to Observe, and the welder shall perform this task. Should the inspector determine that the welder has discontinued performance of this task, the task shall be returned to Perform until such time as the Inspector has re-established adequate assurance that the welder will perform the inspection tasks listed.

TABLE J6-2
Visual Inspection Tasks During Welding

Visual Inspection Tasks During Welding	QC		QA	
	Task	Doc.	Task	Doc.
WPS followed - Settings on welding equipment - Travel speed - Selected welding materials - Shielding gas type/flow rate - Preheat applied - Interpass temperature maintained (min/max.) - Proper position (F, V, H, OH) - Intermix of filler metals avoided unless approved	O	–	O	–
Use of qualified welders	O	–	O	–
Control and handling of welding consumables - Packaging - Exposure control	O	–	O	–
Environmental conditions - Wind speed within limits - Precipitation and temperature	O	–	O	–
Welding techniques - Interpass and final cleaning - Each pass within profile limitations - Each pass meets quality requirements	O	–	O	–
No welding over cracked tacks	O	–	O	–

TABLE J6-3
Visual Inspection Tasks After Welding

Visual Inspection Tasks After Welding	QC		QA	
	Task	Doc.	Task	Doc.
Welds cleaned	O	–	O	–
Size, length, and location of welds	P	–	P	–
Welds meet visual acceptance criteria - Crack prohibition - Weld/base-metal fusion - Crater cross section - Weld profiles and size - Undercut - Porosity	P	D	P	D
Placement of reinforcing or contouring fillet welds (if required)	P	D	P	D
Backing removed, weld tabs removed and finished, and fillet welds added (if required)	P	D	P	D
Repair activities	P	–	P	D

2. NDT of Welded Joints

In addition to the requirements of *Specification* Section N4.5, nondestructive testing of welded joints shall be as required in this section:

2a. *k*-Area NDT

Where welding of doubler plates, continuity plates or stiffeners has been performed in the *k*-area, the web shall be tested for cracks using magnetic particle testing (MT). The MT inspection area shall include the *k-area* base metal within 3 in. (75 mm) of the weld. The MT shall be performed no sooner than 48 hours following completion of the welding.

2b. CJP Groove Weld NDT

Ultrasonic testing (UT) shall be performed on 100% of CJP groove welds in materials $^5/_{16}$ in. (8 mm) thick or greater. Ultrasonic testing in materials less than $^5/_{16}$ in. (8 mm) thick is not required. Weld discontinuities shall be accepted or rejected on the basis of criteria of AWS D1.1/D1.1M Table 6.2. Magnetic particle testing shall be performed on 25% of all beam-to-column CJP groove welds. The rate of UT and MT is permitted to be reduced in accordance with Sections J6.2g and J6.2h, respectively.

Exception: For *ordinary moment frames*, UT and MT of CJP groove welds are required only for *demand critical welds*.

2c. Base Metal NDT for Lamellar Tearing and Laminations

After joint completion, base metal thicker than $1^1/_2$ in. (38 mm) loaded in tension in the through-thickness direction in tee and corner joints, where the connected material is greater than $^3/_4$ in. (19 mm) and contains CJP groove welds, shall be ultrasonically tested for discontinuities behind and adjacent to the fusion line of such welds. Any base metal discontinuities found within $t/4$ of the steel surface shall be accepted or rejected on the basis of criteria of AWS D1.1/D1.1M Table 6.2, where t is the thickness of the part subjected to the through-thickness strain.

2d. Beam Cope and Access Hole NDT

At welded splices and connections, thermally cut surfaces of beam copes and access holes shall be tested using magnetic particle testing or penetrant testing, when the flange thickness exceeds $1^1/_2$ in. (38 mm) for rolled shapes, or when the web thickness exceeds $1^1/_2$ in. (38 mm) for built-up shapes.

2e. Reduced Beam Section Repair NDT

Magnetic particle testing shall be performed on any weld and adjacent area of the *reduced beam section* (RBS) cut surface that has been repaired by welding, or on the base metal of the RBS cut surface if a sharp notch has been removed by grinding.

2f. Weld Tab Removal Sites

At the end of welds where weld tabs have been removed, magnetic particle testing shall be performed on the same beam-to-column joints receiving UT as required under Section J6.2b. The rate of MT is permitted to be reduced in accordance with Section J6.2h. MT of continuity plate weld tabs removal sites is not required.

2g. Reduction of Percentage of Ultrasonic Testing

The reduction of percentage of UT is permitted to be reduced in accordance with *Specification* Section N5.5e, except no reduction is permitted for demand critical welds.

2h. Reduction of Percentage of Magnetic Particle Testing

The amount of MT on CJP groove welds is permitted to be reduced if approved by the *engineer of record* and the *authority having jurisdiction*. The MT rate for an individual welder or welding operator is permitted to be reduced to 10%, provided the reject rate is demonstrated to be 5% or less of the welds tested for the welder or welding operator. A sampling of at least 20 completed welds for a job shall be made for such reduction evaluation. Reject rate is the number of welds containing rejectable defects divided by the number of welds completed. This reduction is prohibited on welds in the *k*-area, at repair sites, backing removal sites, and access holes.

J7. INSPECTION OF HIGH-STRENGTH BOLTING

Bolting inspection shall satisfy the requirements of *Specification* Section N5.6 and this section. Bolting inspection shall be performed by both quality control and quality assurance personnel. As a minimum, the tasks shall be as listed in Tables J7-1, J7-2 and J7-3.

TABLE J7-1
Inspection Tasks Prior to Bolting

Inspection Tasks Prior to Bolting	QC		QA	
	Task	Doc.	Task	Doc.
Proper fasteners selected for the joint detail	O	–	O	–
Proper bolting procedure selected for joint detail	O	–	O	–
Connecting elements, including the appropriate faying surface condition and hole preparation, if specified, meet applicable requirements	O	–	O	–
Pre-installation verification testing by installation personnel observed for fastener assemblies and methods used	P	D	O	D
Proper storage provided for bolts, nuts, washers and other fastener components	O	–	O	–

TABLE J7-2
Inspection Tasks During Bolting

Inspection Tasks During Bolting	QC		QA	
	Task	Doc.	Task	Doc.
Fastener assemblies placed in all holes and washers (if required) are positioned as required	O	–	O	–
Joint brought to the snug tight condition prior to the pretensioning operation	O	–	O	–
Fastener component not turned by the wrench prevented from rotating	O	–	O	–
Bolts are pretensioned progressing systematically from the most rigid point toward the free edges	O	–	O	–

TABLE J7-3
Inspection Tasks After Bolting

Inspection Tasks After Bolting	QC		QA	
	Task	Doc.	Task	Doc.
Document accepted and rejected connections	P	D	P	D

J8. OTHER STEEL STRUCTURE INSPECTIONS

Other inspections of the steel structure shall satisfy the requirements of *Specification* Section N5.7 and this section. Such inspections shall be performed by both quality control and quality assurance personnel. Where applicable, the inspection tasks listed in Table J8-1 shall be performed.

TABLE J8-1
Other Inspection Tasks

Other Inspection Tasks	QC		QA	
	Task	Doc.	Task	Doc.
RBS requirements, if applicable – Contour and finish – Dimensional tolerances	P	D	P	D
Protected zone—no holes and unapproved attachments made by fabricator or erector, as applicable	P	D	P	D

User Note: The *protected zone* should be inspected by others following completion of the work of other trades, including those involving curtainwall, mechanical, electrical, plumbing and interior partitions.

J9. INSPECTION OF COMPOSITE STRUCTURES

Where applicable, inspections of the composite structures shall satisfy the requirements of *Specification* Section N6 and this section. These inspections shall be performed by the responsible contractor's quality control personnel and by quality assurance personnel.

Where applicable, inspection of structural steel used in composite structures shall comply with the requirements of this Chapter. Where applicable, inspection of reinforced concrete shall comply with the requirements of ACI 318, and inspection of welded reinforcing steel shall comply with the applicable requirements of Section J6.1.

Where applicable to the type of composite construction, the minimum inspection tasks shall be as listed in Tables J9-1, J9-2 and J9-3.

TABLE J9-1
Inspection of Composite Structures Prior to Concrete Placement

Inspection of Composite Structures Prior to Concrete Placement	QC Task	QC Doc.	QA Task	QA Doc.
Material identification of reinforcing steel (Type/Grade)	O	–	O	–
Determination of carbon equivalent for reinforcing steel other than ASTM A706	O	–	O	–
Proper reinforcing steel size, spacing and orientation	O	–	O	–
Reinforcing steel has not been rebent in the field	O	–	O	–
Reinforcing steel has been tied and supported as required	O	–	O	–
Required reinforcing steel clearances have been provided	O	–	O	–
Composite member has required size	O	–	O	–

TABLE J9-2
Inspection of Composite Structures During Concrete Placement

Inspection of Composite Structures During Concrete Placement	QC Task	QC Doc.	QA Task	QA Doc.
Concrete: Material identification (mix design, compressive strength, maximum large aggregate size, maximum slump)	O	D	O	D
Limits on water added at the truck or pump	O	D	O	D
Proper placement techniques to limit segregation	O	–	O	–

TABLE J9-3
Inspection of Composite Structures After Concrete Placement

Inspection of Composite Structures After Concrete Placement	QC Task	QC Doc.	QA Task	QA Doc.
Achievement of minimum specified concrete compressive strength at specified age	–	D	–	D

J10. INSPECTION OF H-PILES

Where applicable, inspection of piling shall satisfy the requirements of this section. These inspections shall be performed by both the responsible contractor's quality control personnel and by quality assurance personnel. Where applicable, the inspection tasks listed in Table J10-1 shall be performed.

TABLE J10-1
Inspection of H-Piles

Inspection of Piling	QC		QA	
	Task	Doc.	Task	Doc.
Protected zone—no holes and unapproved attachments made by the responsible contractor, as applicable	P	D	P	D

CHAPTER K

PREQUALIFICATION AND CYCLIC QUALIFICATION TESTING PROVISIONS

This chapter addresses requirements for qualification and prequalification testing.

This chapter is organized as follows:

K1. Prequalification of Beam-to-Column and Link-to-Column Connections
K2. Cyclic Tests for Qualification of Beam-to-Column and Link-to-Column Connections
K3. Cyclic Tests for Qualification of Buckling Restrained Braces

K1. PREQUALIFICATION OF BEAM-TO-COLUMN AND LINK-TO-COLUMN CONNECTIONS

1. Scope

This section contains minimum requirements for prequalification of beam-to-column moment connections in *special moment frames* (SMF), *intermediate moment frames* (IMF), and *link*-to-column connections in *eccentrically braced frames* (EBF). *Prequalified connections* are permitted to be used, within the applicable limits of prequalification, without the need for further qualifying cyclic tests. When the limits of prequalification or design requirements for prequalified connections conflict with the requirements of these Provisions, the limits of prequalification and design requirements for prequalified connections shall govern.

2. General Requirements

2a. Basis for Prequalification

Connections shall be prequalified based on test data satisfying Section K1.3, supported by analytical studies and design models. The combined body of evidence for prequalification must be sufficient to assure that the connection can supply the required *story drift angle* for SMF and IMF systems, or the required *link rotation angle* for EBF, on a consistent and reliable basis within the specified limits of prequalification. All applicable limit states for the connection that affect the stiffness, strength and deformation capacity of the connection and the *seismic force resisting system (SFRS)* must be identified. These include rupture related limit states, stability related limit states, and all other limit states pertinent for the connection under consideration. The effect of design variables listed in Section K1.4 shall be addressed for connection prequalification.

2b. Authority for Prequalification

Prequalification of a connection and the associated limits of prequalification shall be established by a connection prequalification review panel (CPRP) approved by the *authority having jurisdiction*.

3. Testing Requirements

Data used to support connection prequalification shall be based on tests conducted in accordance with Section K2. The CPRP shall determine the number of tests and the variables considered by the tests for connection prequalification. The CPRP shall also provide the same information when limits are to be changed for a previously prequalified connection. A sufficient number of tests shall be performed on a sufficient number of nonidentical specimens to demonstrate that the connection has the ability and reliability to undergo the required story drift angle for SMF and IMF and the required link rotation angle for EBF, where the link is adjacent to columns. The limits on member sizes for prequalification shall not exceed the limits specified in Section K2.3b.

4. Prequalification Variables

In order to be prequalified, the effect of the following variables on connection performance shall be considered. Limits on the permissible values for each variable shall be established by the CPRP for the prequalified connection.

4a. Beam or Link Parameters

(1) Cross-section shape: wide flange, box or other

(2) Cross-section fabrication method: rolled shape, welded shape or other

(3) Depth

(4) Weight per foot

(5) Flange thickness

(6) Material specification

(7) Span-to-depth ratio (for SMF or IMF), or link length (for EBF)

(8) Width-to-thickness ratio of cross-section elements

(9) Lateral bracing

(10) Other parameters pertinent to the specific connection under consideration

4b. Column Parameters

(1) Cross-section shape: wide flange, box, or other

(2) Cross-section fabrication method: rolled shape, welded shape or other

(3) Column orientation with respect to beam or link: beam or link is connected to column flange, beam or link is connected to column web, beams or links are connected to both the column flange and web, or other

(4) Depth

(5) Weight per foot

(6) Flange thickness

(7) Material specification

(8) Width-to-thickness ratio of cross-section elements

(9) Lateral bracing

(10) Other parameters pertinent to the specific connection under consideration

4c.   Beam-to-Column or Link-to-Column Relations

(1) Panel zone strength

(2) Doubler plate attachment details

(3) Column-to-beam (or column-to-link) moment ratio

4d.   Continuity Plates

(1) Identification of conditions under which continuity plates are required

(2) Thickness, width and depth

(3) Attachment details

4e.   Welds

(1) Location, extent (including returns), type (CJP, PJP, fillet, etc.) and any reinforcement or contouring required

(2) Filler metal classification strength and notch toughness

(3) Details and treatment of weld backing and weld tabs

(4) Weld access holes: size, geometry and finish

(5) Welding quality control and quality assurance beyond that described in Chapter J, including NDT method, inspection frequency, acceptance criteria and documentation requirements

4f.   Bolts

(1) Bolt diameter

(2) Bolt grade: ASTM A325, A325M, A490, A490M or other

(3) Installation requirements: pretensioned, snug-tight or other

(4) Hole type: standard, oversize, short-slot, long-slot or other

(5) Hole fabrication method: drilling, punching, sub-punching and reaming or other

(6) Other parameters pertinent to the specific connection under consideration

4g.   Workmanship

All workmanship parameters that exceed AISC, RCSC and AWS requirements, pertinent to specific connection under consideration, as follows:

(1) Surface roughness of thermal cut or ground edges

(2) Cutting tolerances

(3) Presence of holes, fasteners or welds for attachments

4h.   Additional Connection Details

All variables pertinent to the specific connection under consideration, as established by the CPRP.

5. Design Procedure

A comprehensive design procedure must be available for a prequalified connection. The design procedure must address all applicable limit states within the limits of prequalification.

6. Prequalification Record

A prequalified connection shall be provided with a written prequalification record with the following information:

(1) General description of the prequalified connection and drawings that clearly identify key features and components of the connection

(2) Description of the expected behavior of the connection in the elastic and inelastic ranges of behavior, intended location(s) of inelastic action, and a description of limit states controlling the strength and deformation capacity of the connection

(3) Listing of systems for which connection is prequalified: SMF, IMF or EBF

(4) Listing of limits for all prequalification variables listed in Section K1.4

(5) Listing of *demand critical welds*

(6) Definition of the region of the connection that comprises the *protected zone*

(7) Detailed description of the design procedure for the connection, as required in Section K1.5

(8) List of references of test reports, research reports and other publications that provided the basis for prequalification

(9) Summary of quality control and quality assurance procedures

K2. CYCLIC TESTS FOR QUALIFICATION OF BEAM-TO-COLUMN AND LINK-TO-COLUMN CONNECTIONS

1. Scope

This section includes requirements for qualifying cyclic tests of beam-to-column moment connections in special and *intermediate moment frames* and link-to-column connections in *eccentrically braced frames*, when required in these Provisions. The purpose of the testing described in this section is to provide evidence that a beam-to-column connection or a *link*-to-column connection satisfies the requirements for strength and *story drift angle* or *link rotation angle* in these Provisions. Alternative testing requirements are permitted when approved by the *engineer of record* and the *authority having jurisdiction*.

This section provides minimum recommendations for simplified test conditions.

2. Test Subassemblage Requirements

The *test subassemblage* shall replicate as closely as is practical the conditions that will occur in the *prototype* during earthquake loading. The test subassemblage shall include the following features:

(1) The *test specimen* shall consist of at least a single column with beams or links attached to one or both sides of the column.

(2) Points of inflection in the test assemblage shall coincide approximately with the anticipated points of inflection in the prototype under earthquake loading.

(3) Lateral bracing of the test subassemblage is permitted near load application or reaction points as needed to provide lateral stability of the test subassemblage. Additional lateral bracing of the test subassemblage is not permitted, unless it replicates lateral bracing to be used in the prototype.

3. Essential Test Variables

The test specimen shall replicate as closely as is practical the pertinent design, detailing, construction features, and material properties of the prototype. The following variables shall be replicated in the test specimen.

3a. Sources of Inelastic Rotation

The *inelastic rotation* shall be computed based on an analysis of test specimen deformations. Sources of inelastic rotation include yielding of members, yielding of connection elements and connectors, and slip between members and connection elements. For beam-to-column moment connections in special and intermediate moment frames, inelastic rotation is computed based upon the assumption that inelastic action is concentrated at a single point located at the intersection of the centerline of the beam with the centerline of the column. For link-to-column connections in eccentrically braced frames, inelastic rotation shall be computed based upon the assumption that inelastic action is concentrated at a single point located at the intersection of the centerline of the link with the face of the column.

Inelastic rotation shall be developed in the test specimen by inelastic action in the same members and connection elements as anticipated in the prototype (in other words, in the beam or link, in the column panel zone, in the column outside of the panel zone, or in connection elements) within the limits described below. The percentage of the total inelastic rotation in the test specimen that is developed in each member or connection element shall be within 25% of the anticipated percentage of the total inelastic rotation in the prototype that is developed in the corresponding member or connection element.

3b. Size of Members

The size of the beam or link used in the test specimen shall be within the following limits:

(1) The depth of the test beam or link shall be no less than 90% of the depth of the prototype beam or link.

(2) The weight per foot of the test beam or link shall be no less than 75% of the weight per foot of the prototype beam or link.

The size of the column used in the test specimen shall properly represent the inelastic action in the column, as per the requirements in Section K2.3a. In addition, the depth of the test column shall be no less than 90% of the depth of the prototype column.

Extrapolation beyond the limitations stated in this section is permitted subject to qualified peer review and approval by the authority having jurisdiction.

> **User Note:** Based upon the above criteria, beam or link depth and column depths up to and including 11% greater than that tested should be permitted for the prototype. Weight per foot of the beam or link up to and including 33% greater than that tested should be permitted for the prototype.

3c. Connection Details

The connection details used in the test specimen shall represent the prototype connection details as closely as possible. The connection elements used in the test specimen shall be a full-scale representation of the connection elements used in the prototype, for the member sizes being tested.

3d. Continuity Plates

The size and connection details of continuity plates used in the test specimen shall be proportioned to match the size and connection details of continuity plates used in the prototype connection as closely as possible.

3e. Steel Strength

The following additional requirements shall be satisfied for each member or connection element of the test specimen that supplies inelastic rotation by yielding:

(1) The yield strength shall be determined as specified in Section K2.6a. The use of yield stress values that are reported on certified material test reports in lieu of physical testing is prohibited for the purposes of this section.

(2) The yield strength of the beam flange as tested in accordance with Section K2.6a shall not be more than 15% below $R_y F_y$ for the grade of steel to be used for the corresponding elements of the prototype.

(3) The yield strength of the columns and connection elements shall not be more than 15% above or below $R_y F_y$ for the grade of steel to be used for the corresponding elements of the prototype. $R_y F_y$ shall be determined in accordance with Section A3.2.

> **User Note:** Based upon the above criteria, steel of the specified grade with a specified minimum yield stress, F_y, of up to and including 1.15 times the $R_y F_y$ for the steel tested should be permitted in the prototype. In production, this limit should be checked using the values stated on the steel manufacturer's material test reports.

3f. **Welded Joints**

Welds on the test specimen shall satisfy the following requirements:

(1) Welding shall be performed in conformance with Welding Procedure Specifications (WPS) as required in AWS D1.1/D1.1M. The WPS essential variables shall satisfy the requirements in AWS D1.1/D1.1M and shall be within the parameters established by the filler-metal manufacturer. The tensile strength and Charpy V-notch (CVN) toughness of the welds used in the *test assembly* shall be determined by tests as specified in Section K2.6c, made using the same filler metal classification, manufacturer, brand or trade name, diameter, and average heat input for the WPS used on the test specimen. The use of tensile strength and CVN toughness values that are reported on the manufacturer's typical certificate of conformance in lieu of physical testing is prohibited for purposes of this section.

(2) The specified minimum tensile strength of the filler metal used for the test specimen shall be the same as that to be used for the welds on the corresponding prototype. The tensile strength of the deposited weld as tested in accordance with Section K2.6c shall not exceed the tensile strength classification of the filler metal specified for the prototype by more than 25 ksi (172 MPa).

> **User Note:** Based upon the criteria in (2) above, should the tested tensile strength of the weld metal exceed 25 ksi (172 MPa) above the specified minimum tensile strength, the prototype weld should be made with a filler metal and WPS that will provide a tensile strength no less than 25 ksi (172 MPa) below the tensile strength measured in the *material test plate*. When this is the case, the tensile strength of welds resulting from use of the filler metal and the WPS to be used in the prototype should be determined by using an all-weld-metal tension specimen. The test plate is described in AWS D1.8/D1.8M clause A6 and shown in AWS D1.8/D1.8M Figure A.1.

(3) The specified minimum CVN toughness of the filler metal used for the test specimen shall not exceed that to be used for the welds on the corresponding prototype. The tested CVN toughness of the weld as tested in accordance with Section K2.6c shall not exceed the minimum CVN toughness specified for the prototype by more than 50%, nor 25 ft-lb (34 kJ), whichever is greater.

> **User Note:** Based upon the criteria in (3) above, should the tested CVN toughness of the weld metal in the material test specimen exceed the specified CVN toughness for the test specimen by 25 ft-lb (34 kJ) or 50%, whichever is greater, the prototype weld should be made with a filler metal and WPS that will provide a CVN toughness that is no less than 25 ft-lb (34 kJ) or 33% lower, whichever is lower, below the CVN toughness measured in the weld metal material test plate. When this is the case, the weld properties resulting from the filler metal and WPS to be used in the prototype should be determined using five CVN test specimens. The test plate is described in AWS D1.8/D1.8M clause A6 and shown in AWS D1.8/D1.8M Figure A.1.

(4) The welding positions used to make the welds on the test specimen shall be the same as those to be used for the prototype welds.

(5) Details of weld backing, weld tabs, access holes and similar items used for the test specimen welds shall be the same as those to be used for the corresponding prototype welds. Weld backing and weld tabs shall not be removed from the test specimen welds unless the corresponding weld backing and weld tabs are removed from the prototype welds.

(6) Methods of inspection and nondestructive testing and standards of acceptance used for test specimen welds shall be the same as those to be used for the prototype welds.

User Note: The filler metal used for production of the prototype is permitted to be of a different classification, manufacturer, brand or trade name, and diameter, provided that Sections K2.3f(2) and K2.3f(3) are satisfied. To qualify alternate filler metals, the tests as prescribed in Section K2.6c should be conducted.

3g. Bolted Joints

The bolted portions of the test specimen shall replicate the bolted portions of the prototype connection as closely as possible. Additionally, bolted portions of the test specimen shall satisfy the following requirements:

(1) The bolt grade (for example, ASTM A325, A325M, ASTM A490, A490M, ASTM F1852, ASTM F2280) used in the test specimen shall be the same as that to be used for the prototype, except that heavy hex bolts are permitted to be substituted for twist-off-type tension control bolts of equal minimum specified tensile strength, and vice versa.

(2) The type and orientation of bolt holes (standard, oversize, short slot, long slot or other) used in the test specimen shall be the same as those to be used for the corresponding bolt holes in the prototype.

(3) When inelastic rotation is to be developed either by yielding or by slip within a bolted portion of the connection, the method used to make the bolt holes (drilling, sub-punching and reaming, or other) in the test specimen shall be the same as that to be used in the corresponding bolt holes in the prototype.

(4) Bolts in the test specimen shall have the same installation (pretensioned or other) and faying surface preparation (no specified slip resistance, Class A or B slip resistance, or other) as that to be used for the corresponding bolts in the prototype.

4. Loading History

4a. General Requirements

The test specimen shall be subjected to cyclic loads in accordance with the requirements prescribed in Section K2.4b for beam-to-column moment connections in special and intermediate moment frames, and in accordance with the requirements prescribed in Section K2.4c for link-to-column connections in eccentrically braced frames.

Loading sequences other than those specified in Sections K2.4b and K2.4c are permitted to be used when they are demonstrated to be of equivalent or greater severity.

4b. **Loading Sequence for Beam-to-Column Moment Connections**

Qualifying cyclic tests of beam-to-column moment connections in special and intermediate moment frames shall be conducted by controlling the story drift angle, θ, imposed on the test specimen, as specified below:

(1) 6 cycles at $\theta = 0.00375$ rad
(2) 6 cycles at $\theta = 0.005$ rad
(3) 6 cycles at $\theta = 0.0075$ rad
(4) 4 cycles at $\theta = 0.01$ rad
(5) 2 cycles at $\theta = 0.015$ rad
(6) 2 cycles at $\theta = 0.02$ rad
(7) 2 cycles at $\theta = 0.03$ rad
(8) 2 cycles at $\theta = 0.04$ rad

Continue loading at increments of $\theta = 0.01$ rad, with two cycles of loading at each step.

4c. **Loading Sequence for Link-to-Column Connections**

Qualifying cyclic tests of link-to-column moment connections in eccentrically braced frames shall be conducted by controlling the total link rotation angle, γ_{total}, imposed on the test specimen, as follows:

(1) 6 cycles at $\gamma_{total} = 0.00375$ rad
(2) 6 cycles at $\gamma_{total} = 0.005$ rad
(3) 6 cycles at $\gamma_{total} = 0.0075$ rad
(4) 6 cycles at $\gamma_{total} = 0.01$ rad
(5) 4 cycles at $\gamma_{total} = 0.015$ rad
(6) 4 cycles at $\gamma_{total} = 0.02$ rad
(7) 2 cycles at $\gamma_{total} = 0.03$ rad
(8) 1 cycle at $\gamma_{total} = 0.04$ rad
(9) 1 cycle at $\gamma_{total} = 0.05$ rad
(10) 1 cycle at $\gamma_{total} = 0.07$ rad
(11) 1 cycle at $\gamma_{total} = 0.09$ rad

Continue loading at increments of $\gamma_{total} = 0.02$ rad, with one cycle of loading at each step.

5. **Instrumentation**

Sufficient instrumentation shall be provided on the test specimen to permit measurement or calculation of the quantities listed in Section K2.7.

6. Testing Requirements for Material Specimens

6a. Tension Testing Requirements for Structural Steel Material Specimens

Tension testing shall be conducted on samples taken from material test plates in accordance with Section K2.6b. The material test plates shall be taken from the steel of the same heat as used in the *test specimen*. Tension-test results from certified material test reports shall be reported, but shall not be used in lieu of physical testing for the purposes of this section. Tension testing shall be conducted and reported for the following portions of the test specimen:

(1) Flange(s) and web(s) of beams and columns at standard locations

(2) Any element of the connection that supplies inelastic rotation by yielding

6b. Methods of Tension Testing for Structural Steel Material Specimens

Tension testing shall be conducted in accordance with ASTM A6/A6M, ASTM A370, and ASTM E8, with the following exceptions:

(1) The yield strength, F_y, that is reported from the test shall be based upon the yield strength definition in ASTM A370, using the offset method at 0.002 in./in. strain.

(2) The loading rate for the tension test shall replicate, as closely as practical, the loading rate to be used for the test specimen.

6c. Testing Requirements for Weld Metal Material Specimens

Weld metal testing shall be conducted on samples extracted from the material test plate, made using the same filler metal classification, manufacturer, brand or trade name and diameter, and using the same average heat input as used in the welding of the test specimen. The tensile strength and CVN toughness of weld material specimens shall be determined in accordance with *Standard Methods for Mechanical Testing of Welds* (AWS B4.0/B4.0M). The use of tensile strength and CVN toughness values that are reported on the manufacturer's typical certificate of conformance in lieu of physical testing is prohibited for use for purposes of this section.

The same WPS shall be used to make the test specimen and the material test plate. The material test plate shall use base metal of the same grade and type as was used for the test specimen, although the same heat need not be used. If the average heat input used for making the material test plate is not within ±20% of that used for the test specimen, a new material test plate shall be made and tested.

7. Test Reporting Requirements

For each test specimen, a written test report meeting the requirements of the authority having jurisdiction and the requirements of this section shall be prepared. The report shall thoroughly document all key features and results of the test. The report shall include the following information:

(1) A drawing or clear description of the test subassemblage, including key dimensions, boundary conditions at loading and reaction points, and location of lateral braces.

(2) A drawing of the connection detail showing member sizes, grades of steel, the sizes of all connection elements, welding details including filler metal, the size and location of bolt holes, the size and grade of bolts, and all other pertinent details of the connection.

(3) A listing of all other essential variables for the test specimen, as listed in Section K2.3.

(4) A listing or plot showing the applied load or displacement history of the test specimen.

(5) A listing of all welds to be designated *demand critical*.

(6) Definition of the region of the member and connection to be designated a *protected zone*.

(7) A plot of the applied load versus the displacement of the test specimen. The displacement reported in this plot shall be measured at or near the point of load application. The locations on the test specimen where the loads and displacements were measured shall be clearly indicated.

(8) A plot of beam moment versus story drift angle for beam-to-column moment connections; or a plot of link shear force versus link rotation angle for link-to-column connections. For beam-to-column connections, the beam moment and the story drift angle shall be computed with respect to the centerline of the column.

(9) The story drift angle and the total inelastic rotation developed by the test specimen. The components of the test specimen contributing to the total inelastic rotation due to yielding or slip shall be identified. The portion of the total inelastic rotation contributed by each component of the test specimen shall be reported. The method used to compute inelastic rotations shall be clearly shown.

(10) A chronological listing of significant test observations, including observations of yielding, slip, instability, and rupture of any portion of the test specimen as applicable.

(11) The controlling failure mode for the test specimen. If the test is terminated prior to failure, the reason for terminating the test shall be clearly indicated.

(12) The results of the material specimen tests specified in Section K2.6.

(13) The welding procedure specifications (WPS) and welding inspection reports.

Additional drawings, data, and discussion of the test specimen or test results are permitted to be included in the report.

8. Acceptance Criteria

The test specimen must satisfy the strength and story drift angle or link rotation angle requirements of these Provisions for the *special moment frame*, intermediate moment frame, or eccentrically braced frame connection, as applicable. The test specimen must sustain the required story drift angle or link rotation angle for at least one *complete loading cycle*.

K3. CYCLIC TESTS FOR QUALIFICATION OF BUCKLING-RESTRAINED BRACES

1. Scope

This section includes requirements for qualifying cyclic tests of individual *buckling-restrained braces* and buckling-restrained brace subassemblages, when required in these provisions. The purpose of the testing of individual braces is to provide evidence that a buckling-restrained brace satisfies the requirements for strength and inelastic deformation by these provisions; it also permits the determination of maximum brace forces for design of adjoining elements. The purpose of testing of the brace subassemblage is to provide evidence that the brace-design can satisfactorily accommodate the deformation and rotational demands associated with the design. Further, the subassemblage test is intended to demonstrate that the hysteretic behavior of the brace in the subassemblage is consistent with that of the individual brace elements tested uniaxially.

Alternative testing requirements are permitted when approved by the *engineer of record* and the *authority having jurisdiction*. This section provides only minimum recommendations for simplified test conditions.

2. Subassemblage Test Specimen

The *subassemblage test specimen* shall satisfy the following requirements:

(1) The mechanism for accommodating inelastic rotation in the subassemblage test specimen brace shall be the same as that of the *prototype*. The rotational deformation demands on the subassemblage test specimen brace shall be equal to or greater than those of the prototype.

(2) The axial yield strength of the *steel core*, P_{ysc}, of the brace in the subassemblage test specimen shall not be less than that of the prototype where both strengths are based on the core area, A_{sc}, multiplied by the yield strength as determined from a coupon test.

(3) The cross-sectional shape and orientation of the steel core projection of the subassemblage test specimen brace shall be the same as that of the brace in the prototype.

(4) The same documented design methodology shall be used for design of the subassemblage as used for the prototype, to allow comparison of the rotational deformation demands on the subassemblage brace to the prototype. In stability calculations, beams, columns and gussets connecting the core shall be considered parts of this system.

(5) The calculated margins of safety for the prototype connection design, steel core projection stability, overall buckling and other relevant subassemblage test specimen brace construction details, excluding the gusset plate, for the prototype, shall equal or exceed those of the subassemblage test specimen construction.

(6) Lateral bracing of the subassemblage test specimen shall replicate the lateral bracing in the prototype.

(7) The *brace test specimen* and the prototype shall be manufactured in accordance with the same quality control and assurance processes and procedures.

Extrapolation beyond the limitations stated in this section is permitted subject to qualified peer review and approval by the authority having jurisdiction.

3. Brace Test Specimen

The brace test specimen shall replicate as closely as is practical the pertinent design, detailing, construction features and material properties of the prototype.

3a. Design of Brace Test Specimen

The same documented design methodology shall be used for the brace test specimen and the prototype. The design calculations shall demonstrate, at a minimum, the following requirements:

(1) The calculated margin of safety for stability against overall buckling for the prototype shall equal or exceed that of the brace test specimen.

(2) The calculated margins of safety for the brace test specimen and the prototype shall account for differences in material properties, including yield and ultimate stress, ultimate elongation, and toughness.

3b. Manufacture of Brace Test Specimen

The brace test specimen and the prototype shall be manufactured in accordance with the same quality control and assurance processes and procedures.

3c. Similarity of Brace Test Specimen and Prototype

The brace test specimen shall meet the following requirements:

(1) The cross-sectional shape and orientation of the steel core shall be the same as that of the prototype.

(2) The axial yield strength of the steel core, P_{ysc}, of the brace test specimen shall not be less than 50% nor more than 120% of the prototype where both strengths are based on the core area, A_{sc}, multiplied by the yield strength as determined from a coupon test.

(3) The material for, and method of, separation between the steel core and the buckling restraining mechanism in the brace test specimen shall be the same as that in the prototype.

Extrapolation beyond the limitations stated in this section is permitted subject to qualified peer review and approval by the authority having jurisdiction.

3d. Connection Details

The connection details used in the brace test specimen shall represent the prototype connection details as closely as practical.

3e. Materials

(1) Steel Core

The following requirements shall be satisfied for the steel core of the brace test specimen:

(a) The specified minimum yield stress of the brace test specimen steel core shall be the same as that of the prototype.

(b) The measured yield stress of the material of the steel core in the brace test specimen shall be at least 90% of that of the prototype as determined from coupon tests.

(c) The specified minimum ultimate stress and strain of the brace test specimen steel core shall not exceed those of the prototype.

(2) Buckling-Restraining Mechanism

Materials used in the buckling-restraining mechanism of the brace test specimen shall be the same as those used in the prototype.

3f. Connections

The welded, bolted and pinned joints on the *test specimen* shall replicate those on the prototype as close as practical.

4. Loading History

4a. General Requirements

The test specimen shall be subjected to cyclic loads in accordance with the requirements prescribed in Sections K3.4b and K3.4c. Additional increments of loading beyond those described in Section K3.4c are permitted. Each cycle shall include a full tension and full compression excursion to the prescribed deformation.

4b. Test Control

The test shall be conducted by controlling the level of axial or rotational deformation, Δ_b, imposed on the test specimen. As an alternate, the maximum rotational deformation is permitted to be applied and maintained as the protocol is followed for axial deformation.

4c. Loading Sequence

Loads shall be applied to the test specimen to produce the following deformations, where the deformation is the steel core axial deformation for the test specimen and the rotational deformation demand for the subassemblage test specimen brace:

(1) 2 cycles of loading at the deformation corresponding to $\Delta_b = \Delta_{by}$.

(2) 2 cycles of loading at the deformation corresponding to $\Delta_b = 0.50\Delta_{bm}$.

(3) 2 cycles of loading at the deformation corresponding to $\Delta_b = 1\Delta_{bm}$.

(4) 2 cycles of loading at the deformation corresponding to $\Delta_b = 1.5\Delta_{bm}$.

(5) 2 cycles of loading at the deformation corresponding to $\Delta_b = 2.0\Delta_{bm}$.

(6) Additional complete cycles of loading at the deformation corresponding to $\Delta_b = 1.5\Delta_{bm}$ as required for the brace test specimen to achieve a cumulative inelastic axial deformation of at least 200 times the yield deformation (not required for the subassemblage test specimen).

where
Δ_{bm} = value of deformation quantity, Δ_b, corresponding to the design story drift, in. (mm)
Δ_{by} = value of deformation quantity, Δ_b, at first significant yield of test specimen, in. (mm)

The *design story drift* shall not be taken as less than 0.01 times the story height for the purposes of calculating Δ_{bm}. Other loading sequences are permitted to be used to qualify the test specimen when they are demonstrated to be of equal or greater severity in terms of maximum and cumulative inelastic deformation.

5. Instrumentation

Sufficient instrumentation shall be provided on the test specimen to permit measurement or calculation of the quantities listed in Section K3.7.

6. Materials Testing Requirements

6a. Tension Testing Requirements

Tension testing shall be conducted on samples of steel taken from the same heat of steel as that used to manufacture the steel core. Tension test results from certified material test reports shall be reported but are prohibited in place of material specimen testing for the purposes of this Section. Tension test results shall be based upon testing that is conducted in accordance with Section K3.6b.

6b. Methods of Tension Testing

Tension testing shall be conducted in accordance with ASTM A6, ASTM A370 and ASTM E8, with the following exceptions:

(1) The yield stress that is reported from the test shall be based upon the yield strength definition in ASTM A370, using the offset method of 0.002 in./in. strain.

(2) The loading rate for the tension test shall replicate, as closely as is practical, the loading rate used for the test specimen.

(3) The coupon shall be machined so that its longitudinal axis is parallel to the longitudinal axis of the steel core.

7. Test Reporting Requirements

For each test specimen, a written test report meeting the requirements of this Section shall be prepared. The report shall thoroughly document all key features and results of the test. The report shall include the following information:

(1) A drawing or clear description of the test specimen, including key dimensions, boundary conditions at loading and reaction points, and location of lateral bracing, if any.

(2) A drawing of the connection details showing member sizes, grades of steel, the sizes of all connection elements, welding details including filler metal, the size and location of bolt or pin holes, the size and grade of connectors, and all other pertinent details of the connections.

(3) A listing of all other essential variables as listed in Sections K3.2 or K3.3, as appropriate.

(4) A listing or plot showing the applied load or displacement history.

(5) A plot of the applied load versus the deformation, Δ_b. The method used to determine the deformations shall be clearly shown. The locations on the test specimen where the loads and deformations were measured shall be clearly identified.

(6) A chronological listing of significant test observations, including observations of yielding, slip, instability, transverse displacement along the test specimen and rupture of any portion of the test specimen and connections, as applicable.

(7) The results of the material specimen tests specified in Section K3.6.

(8) The manufacturing quality control and *quality assurance plans* used for the fabrication of the test specimen. These shall be included with the welding procedure specifications and welding inspection reports.

Additional drawings, data and discussion of the test specimen or test results are permitted to be included in the report.

8. Acceptance Criteria

At least one subassemblage test that satisfies the requirements of Section K3.2 shall be performed. At least one brace test that satisfies the requirements of Section K3.3 shall be performed. Within the required protocol range all tests shall satisfy the following requirements:

(1) The plot showing the applied load vs. displacement history shall exhibit stable, repeatable behavior with positive incremental stiffness.

(2) There shall be no rupture, brace instability, or brace end connection failure.

(3) For brace tests, each cycle to a deformation greater than Δ_{by} the maximum tension and compression forces shall not be less than the *nominal strength* of the core.

(4) For brace tests, each cycle to a deformation greater than Δ_{by} the ratio of the maximum compression force to the maximum tension force shall not exceed 1.3.

Other acceptance criteria are permitted to be adopted for the brace test specimen or subassemblage test specimen subject to qualified peer review and approval by the authority having jurisdiction.

COMMENTARY
on the Seismic Provisions for Structural Steel Buildings

Seismic Provisions for Structural Steel Buildings
June 22, 2010

(The Commentary is not a part of ANSI/AISC 341-10, *Seismic Provisions for Structural Steel Buildings*, and is included for informational purposes only.)

INTRODUCTION

The Provisions is intended to be complete for normal design usage.

The Commentary furnishes background information and references for the benefit of the design professional seeking further understanding of the basis, derivations and limits of the Provisions.

The Provisions and Commentary are intended for use by design professionals with demonstrated engineering competence.

COMMENTARY PREFACE

Experience from the 1994 Northridge and 1995 Kobe earthquakes significantly expanded knowledge regarding the seismic response of structural steel building systems, particularly welded steel moment frames. Shortly after the Northridge earthquake, the SAC Joint Venture[1] initiated a comprehensive study of the seismic performance of steel moment frames. Funded by the Federal Emergency Management Agency (FEMA), SAC developed guidelines for structural engineers, building officials and other interested parties for the evaluation, repair, modification and design of welded steel moment frame structures in seismic regions. AISC actively participated in the SAC activities.

These 2010 AISC *Seismic Provisions for Structural Steel Buildings*, hereinafter referred to as the Provisions, continues the practice of incorporating recommendations from the NEHRP Provisions, most recently FEMA P-750 (FEMA, 2009), and other research. While research is ongoing, the Committee has prepared this revision of the Provisions using the best available knowledge to date. These Provisions were being developed in the same time frame as a rewrite of *Minimum Design Loads for Buildings and Other Structures*, ASCE/SEI 7 (ASCE, 2010) was being accomplished, which has subsequently been completed and published as the 2010 edition.

It is also anticipated that these Provisions will be adopted by the International Building Code, 2012 edition, and the National Fire Protection Association (NFPA) Building Construction and Safety Code, NFPA 5000, dated 2012. It is expected that both of these model building codes will reference ASCE/SEI 7 for seismic loading and neither code will contain seismic requirements.

Where there is a desire to use these Provisions with a model code that has not yet adopted these Provisions, it is essential that the AISC *Specification for Structural Steel Buildings* (AISC, 2010a), hereafter referred to as the *Specification*, be used in conjunction with these Provisions, as they are companion documents. In addition, users should also concurrently use ASCE/SEI 7 for a fully coordinated package.

---

[1] A joint venture of the Structural Engineers Association of California (SEAOC), Applied Technology Council (ATC) and California Universities for Research in Earthquake Engineering (CUREE).

CHAPTER A

GENERAL REQUIREMENTS

A1. SCOPE

In previous editions of these Provisions and the predecessor specifications to the new AISC *Specification for Structural Steel Buildings*, ANSI/AISC 360 (AISC, 2010a), the stated scope was limited to buildings. In the 2005 *Specification*, the scope was expanded to include other structures designed, fabricated and erected in a manner similar to buildings, with building-like vertical and lateral load-resisting elements. Thus the scope of the 2005 Provisions was modified for consistency with the *Specification*. For simplicity the Commentary refers to steel buildings and structures interchangeably.

However, it should be noted that these provisions were developed specifically for buildings. The Provisions, therefore, may not be applicable, in whole or in part, to some nonbuilding structures that do not have the building-like characteristics described in the paragraph above. Extrapolation of their use to such nonbuilding structures should be done with due consideration of the inherent differences between the response characteristics of buildings and these nonbuilding structures.

Structural steel systems in seismic regions are generally expected to dissipate seismic input energy through controlled inelastic deformations of the structure. These Provisions supplement the *Specification* for such applications. The seismic design loads specified in the building codes have been developed considering the energy dissipation generated during inelastic response.

The Provisions are intended to be mandatory for structures where they have been specifically referenced when defining an R factor in *Minimum Design Loads for Buildings and Other Structures*, ASCE/SEI 7 (ASCE, 2010). For steel structures, typically this occurs in seismic design category D and above, where the R factor is greater than 3. However, there are instances where an R factor of less than 3 is assigned to a system and the Provisions are still required. These limited cases occur in ASCE/SEI 7 Table 12.2-1 for cantilevered column systems and Table 15.4-1 for nonbuilding structures similar to buildings. For these systems with R factors less than 3, the use of the Provisions is required. In general, for structures in seismic design categories B and C, the designer is given a choice to either solely use the *Specification* and the R factor given for structural steel buildings not specifically detailed for seismic resistance (typically, a factor of 3) or the designer may choose to assign a higher R factor to a system detailed for seismic resistance and follow the requirements of these Provisions. Additionally, for composite steel-concrete structures, there are cases where these Provisions are required in seismic design categories B and C, as specified in Table 12.2-1 of ASCE/SEI 7. This typically occurs for composite systems designated as "ordinary" where the counterpart reinforced concrete systems have designated R factors and design requirements for seismic design categories B and C.

Previous editions of these Provisions have been limited to defining requirements for members and connections in the seismic force resisting system (SFRS). The Provisions now include requirements for columns not part of the SFRS in Section D2.5.

For buildings with composite members and/or seismic force resisting systems, an important change in these 2010 Provisions is the integration of what were formerly presented separately in Parts I (steel) and II (composite) into a combined set of provisions. The Provisions for the seismic design of composite structural steel and reinforced concrete buildings are based upon the 1994 NEHRP Provisions (FEMA, 1994) and subsequent modifications made in the 1997, 2000, 2003 and 2009 NEHRP Provisions and in ASCE/SEI 7. Because composite systems are assemblies of steel and concrete components, the portions of these Provisions pertaining to steel, the *Specification* and *Building Code Requirements for Structural Concrete and Commentary*, ACI 318-08 (ACI, 2008), form an important basis for provisions related to composite construction.

There is at present limited experience in the U.S. with composite building systems subjected to extreme seismic loads and many of the recommendations herein are necessarily of a conservative and/or qualitative nature. Extensive design and performance experience with this type of building in Japan clearly indicates that composite systems, due to their inherent rigidity and toughness, can equal or exceed the performance of reinforced concrete only or structural steel only buildings (Deierlein and Noguchi, 2004; Yamanouchi et al., 1998). Composite systems have been extensively used in tall buildings throughout the world.

Careful attention to all aspects of the design is necessary in the design of composite systems, particularly with respect to the general building layout and detailing of members and connections. Composite connection details are illustrated throughout this Commentary to convey the basic character of the force transfer in composite systems. However, these details should not necessarily be treated as design standards. The cited references provide more specific information on the design of composite connections. For a general discussion of these issues and some specific design examples, refer to Viest et al. (1997).

The design and construction of composite elements and systems continues to evolve in practice. Except where explicitly stated, these Provisions are not intended to limit the application of new systems for which testing and analysis demonstrates that the structure has adequate strength, ductility and toughness. It is generally anticipated that the overall behavior of the composite systems herein will be similar to that for counterpart structural steel systems or reinforced concrete systems and that inelastic deformations will occur in conventional ways, such as flexural yielding of beams in fully restrained (FR) moment frames or axial yielding and/or buckling of braces in braced frames. However, differential stiffness between steel and concrete elements is more significant in the calculation of internal forces and deformations of composite systems than for structural steel only or reinforced concrete only systems. For example, deformations in composite elements can vary considerably due to the effects of cracking.

When systems have both ductile and nonductile elements, the relative stiffness of each should be properly modeled; the ductile elements can deform inelastically while the nonductile elements remain nominally elastic. When using elastic analysis, member stiffness should be reduced to account for the degree of cracking at the onset of significant yielding in the structure. Additionally, it is necessary to account for material overstrength that may alter relative strength and stiffness.

A2. REFERENCED SPECIFICATIONS, CODES AND STANDARDS

The specifications, codes and standards referenced herein are listed with the appropriate revision date in this Section or in Section A2 of the *Specification*. Since the Provisions act as a supplement to the *Specification*, the references listed in Section A2 of the *Specification* are not repeated again in the Provisions.

A3. MATERIALS

1. Material Specifications

The structural steels that are explicitly permitted for use in seismic applications have been selected based upon their inelastic properties and weldability. In general, they meet the following characteristics: (1) a pronounced stress-strain plateau at the yield stress; (2) a large inelastic strain capability [for example, tensile elongation of 20% or greater in a 2 in. (50 mm) gage length]; and (3) good weldability. Other steels should not be used without evidence that the above criteria are met. For structural wide flange shapes, ASTM A992 and ASTM A913 additional supplementary requirements provide a limitation on the ratio of yield stress to tensile stress to be not greater than 0.85.

The limitation on the specified minimum yield stress for members expecting inelastic action refers to inelastic action under the effects of the design earthquake. The 50 ksi (345 MPa) limitation on the specified minimum yield stress for members was restricted to those systems in Chapters E, F, G and H expected to undergo moderate to significant inelastic action, while a 55 ksi (380 MPa) limitation was assigned to Sections E1, F1, G1, H1 and H4, since those systems are expected to undergo limited inelastic action. The listed steels conforming to ASTM A1011 with a yield of 55 ksi (380 MPa) are included as they have adequate ductility considering their limited thickness range. This steel is commonly used by the metal building industry in built-up sections.

An exception has been added to allow the yield stress limits to be exceeded where testing or rational criteria permit. An example of testing that would permit higher strength steels for elements would be cyclic tests per Sections K2 and K3 of the Provisions where the element is subject to the anticipated level of inelastic strain for the intended use.

Modern steels of higher strength, such as ASTM A913 Grade 65 (450), are generally considered to have properties acceptable for seismic column applications where limited inelastic action may occur. An exception permits structural steel with a specified minimum yield stress up to 65 ksi (450 MPa) for columns in those designated systems where the anticipated level of inelastic yielding will be minor.

Conformance with the material requirements of the *Specification* is satisfied by the testing performed in accordance with ASTM provisions by the manufacturer. Supplemental or independent material testing is only required for material that cannot be identified or traced to a material test report and materials used in qualification testing, according to Section K2 of the Provisions.

ASTM A1043/1043M Grade 36 (250) and Grade 50 (345) have been added as approved steels for the SFRS, since they meet the inelastic property and weldability requirements noted in the first paragraph above.

While ASTM A709/A709M steel is primarily used in the design and construction of bridges, it could also be used in building construction. Written as an umbrella specification, its grades are essentially the equivalent of other approved ASTM specifications. For example, ASTM A709/A709M Grade 50 (345) is essentially ASTM A572/A572M Grade 50 (345) and ASTM A709/A709M Grade 50W (345W) is essentially ASTM A588/A588M Grade 50 (345). Thus, if used, ASTM A709/A709M material should be treated as would the corresponding approved ASTM material grade.

For rotary-straightened W-shapes, an area of reduced notch toughness has been documented in a limited region of the web immediately adjacent to the flange as illustrated in Figure C-A3.1. Recommendations issued by AISC (AISC, 1997a) were followed up by a series of industry sponsored research projects (Kaufmann et al., 2001; Uang and Chi, 2001; Kaufmann and Fisher, 2001; Lee et al., 2002; Bartlett et al., 2001). This research generally corroborates AISC's initial findings and recommendations.

2. Expected Material Strength

The Provisions employ a methodology for many seismic systems (for example, special moment frames, special concentrically braced frames, and eccentrically

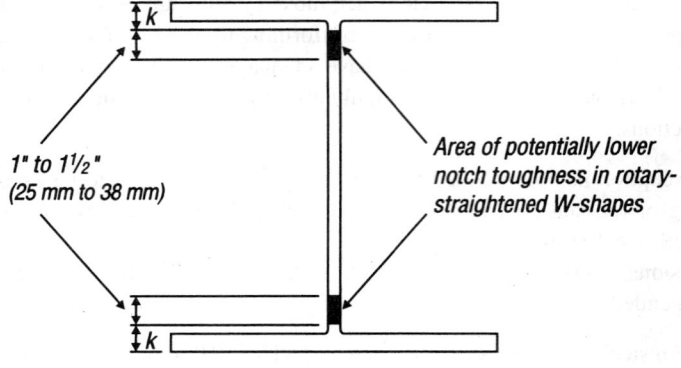

Fig. C-A3.1. "k-area."

braced frames) that can be characterized as "capacity design." That is, the required strength of most elements is defined by forces corresponding to the expected capacity (available strength) of certain designated yielding members (for example, the link in eccentrically braced frames). This methodology serves to confine ductility demands to members that have specific requirements to ensure their ductile behavior; furthermore, the methodology serves to ensure that within that member the desired, ductile mode of yielding governs and other, nonductile modes are precluded.

Such a capacity-design methodology requires a realistic estimate of the expected strength of the designated yielding members. To this end, the expected yield stresses of various steel materials have been established by a survey of mill certificates, and the ratio of expected to nominal yield stress has been included in the Provisions as R_y. The expected capacity of the designated yielding member is defined as R_y times the nominal strength of the member based on the desired yield mode; this expected strength is amplified to account for strain-hardening in some cases. For determination of the required strength of adjoining elements and their connection to the designated yielding members, neither the resistance factor (LRFD), nor the safety factor (ASD), are applied to the strength of the designated yielding members.

Where the capacity-design methodology is employed to preclude nonductile modes of failure within the designated yielding member, it is reasonable to use the expected material strength in the determination of the member capacity. For limit states based on yield, the factor R_y applies equally to the designated yielding member capacity used to compute the required strength and to the strength with respect to the limit states to be precluded. An example of this condition is yielding of the beam outside the link in an eccentrically braced frame; the required strength is based on yield of the link beam, and yield limit states, such as combined flexure and compression, can be expected to be similarly affected by increased material strength. The factor R_y is not applied to members other than the designated yielding member.

Similarly, fracture limit states within the designated yielding member are affected by increased material strength. Such limit states include block shear rupture and net section rupture of braces in special concentrically braced frames, where the required strength is calculated based on the brace expected yield strength in tension. The ratio of expected tensile strength over the specified minimum tensile strength is somewhat less than that of expected yield stress over the specified minimum yield stress, so a separate factor was created called R_t. This factor applies only to fracture limit states in designated yielding members. As is the case with R_y, R_t is applied in the determination of the capacity of designated yielding members and not the capacity of other members.

The specified values of R_y for rolled shapes are somewhat lower than those that can be calculated using the mean values reported in a survey conducted by the Structural Shape Producers Council. Those values were skewed somewhat by the inclusion of a large number of smaller members, which typically have a higher measured yield stress than the larger members common in seismic design. The given values are considered to be reasonable averages, although it is recognized that they are not maxima. The expected yield strength, $R_y F_y$, can be determined by testing conducted in accordance with the requirements for the specified grade of steel. Such an approach should only be followed in unusual cases where there is extensive evidence that the values of R_y are significantly unconservative. It is not expected that this would be the approach followed for typical building projects. Refer to ASTM A370 for testing requirements. The higher values of R_y for ASTM A36/A36M ($R_y = 1.5$) shapes are indicative of the most recently reported properties of these grades of steel. The values of R_y will be periodically monitored to ensure that current production practice is properly reflected.

A study (Liu et al., 2007) was used in determining the R_t values shown in Table A3.1. These values are based on the mean value of R_t/R_y for individual samples. Mean values are considered to be sufficiently conservative for these calculations considering that they are applied along with a ϕ factor of 0.75. An additional analysis of tensile data was carried out (Harrold, 2004) to determine appropriate R_y and R_t factors for ASTM A529 Grade 50 (345), A529 Grade 55 (380), A1011 HSLAS Grade 55 (380), and A572 Grade 55 (380) steels, that were added to Table A3.1.

While both ASTM A500 (Grades B or C) and ASTM A501 material specifications are grouped in Table A3.1, ASTM A501 material will likely have R_y values less than those specified in Table A3.1 as this material is not cold worked as is ASTM A500 material. Presently, ASTM A501 material is not as commonly used nor as readily available as ASTM A500 (Grades B or C). Due to the limited production data available for ASTM A501, these Provisions continue to conservatively use R_y and R_t values for ASTM A501 based primarily on ASTM A500 (Grades B or C) production data.

ASTM A572/A572M Grade 42 (290) shapes are no longer commonly produced and have therefore been removed from Table A3.1. However, thick plate sections of this material grade are still used for connections, built-up shapes, and column bases. Consequently, ASTM A572/A572M Grade 42 (290) has been added to Table A3.1 of these Provisions for plates. As limited production data is available for plates of this material grade, a value of R_y of 1.3 is specified corresponding to approximately the same 55 ksi (380 MPa) expected yield stress as ASTM A572/A572M Grade 50 (345) plate. The R_t value of 1.0 specified for plates of this material grade considers the expected tensile strength, R_tF_u, of the material to be the same as the specified tensile strength, F_u, which is conservative when used for determining nominal strength, R_n, limit states.

Values of R_y and R_t for ASTM 1043/1043M Grades 36 (250) and 50 (345) have been added based on a survey of production data.

3. **Heavy Sections**

The *Specification* requirements for notch toughness cover hot-rolled shapes with a flange thickness exceeding 2 in. (50 mm) and plate elements with thickness that is greater than or equal to 2 in. (50 mm) in tension applications. In the Provisions, this requirement is extended to cover: (1) shapes that are part of the SFRS with flange thickness greater than or equal to $1^1/_2$ in. (38 mm); and, (2) plate elements with thickness greater than or equal to 2 in. (50 mm) that are part of the SFRS, such as the flanges of built-up girders and connection material subject to inelastic strain under seismic loading. Because smaller shapes and thinner plates are generally subjected to sufficient cross-sectional reduction during the rolling process such that the resulting notch toughness will exceed that required above (Cattan, 1995), specific requirements have not been included herein.

Connection plates in which inelastic strain under seismic loading may be expected include, but are not limited to:

1. Gusset plates for diagonal braces that are designed to allow rotation capacity per Section F2.6c(3)(b)
2. Bolted flange plates for moment connections such as per Chapter 7 (BFP moment connection) of ANSI/AISC 358 (AISC, 2010b) and similar flange plate moment connections in OMF systems
3. Bolted end plates for moment connections such as per Chapter 6 of ANSI/AISC 358
4. Base plates of column bases designed to yield inelastically to limit forces on anchor rods or to allow column rotation

The requirements of this Section may not be necessary for members that resist only incidental loads. For example, a designer might include a member in the SFRS to develop a more robust load path, but the member will experience only an insignificant level of seismic demand. An example of such a member might include a transfer girder with thick plates where its design is dominated by its gravity load demand. It would be inconsistent with the intent of this Section if the designer were to arbitrarily exclude a member with insignificant seismic loads from the SFRS that would otherwise improve the seismic performance of the building in order to avoid the toughness requirements in this Section. The *Specification* requirements noted above would still apply in this case.

Early investigations of connection fractures in the 1994 Northridge earthquake identified a number of fractures that some speculated were the result of inadequate through-thickness strength of the column flange material. As a result, in the period immediately following the Northridge earthquake, a number of recommendations were promulgated that suggested limiting the value of through-thickness stress demand on column flanges to ensure that through-thickness yielding did not initiate in the column flanges. This limit state often controlled the overall design of these connections. However, the actual cause for the fractures that were initially thought to be through-thickness failures of the column flange are now considered to be unrelated to this material property. Detailed fracture mechanics investigations conducted as part of the FEMA/SAC project confirm that damage initially identified as through-thickness failures is likely to have occurred as a result of certain combinations of filler metal and base material strength and notch toughness, conditions of stress in the connection, and the presence of critical flaws in the welded joint. In addition to the analytical studies, extensive through-thickness testing conducted specifically to determine the susceptibility to through-thickness failures of modern column materials meeting ASTM A572 Grade 50 and ASTM A913 Grade 65 specifications did not result in significant through-thickness fractures (FEMA, 2000g).

In addition, none of the more than 100 full-scale tests on "post-Northridge" connection details have demonstrated any through-thickness column fractures. This combined analytical and laboratory research clearly shows that due to the high restraint inherent in welded beam flange to column flange joints, the through-thickness yield and tensile strengths of the column material are significantly elevated in the region of the connection. For the modern materials tested, these strengths significantly exceed those loads that can be delivered to the column by the beam flange. For this reason, no limits are suggested for the through-thickness strength of the base material by the FEMA/SAC program or in these Provisions.

The preceding discussion assumes that no significant laminations, inclusions or other discontinuities occur in regions adjacent to welded beam flange-to-column flange joints and other tee and corner joints. Section J6.2c checks the integrity of this material after welding. A more conservative approach would be to ultrasonically test the material for laminations prior to welding. A similar requirement has been included in the Los Angeles City building code since 1973; however, in practice the base material prior to welding generally passes the ultrasonic examination, and interior defects, if any, are found only after heating and cooling during the weld process. Should a concern exist, the ultrasonic inspection prior to welding should be conducted to ASTM A435 for plates and ASTM A898, level 1, for shapes.

4. Consumables for Welding

As in previous Provisions, specified levels of filler metal and weld metal Charpy V-notch (CVN) toughness are required in all member and connection welds in the load path of the SFRS. With this edition of the Provisions, the specific requirements for notch toughness are no longer directly stated, but addressed through reference to the requirements of *Structural Welding Code—Steel*, AWS D1.1/D1.1M and *Structural Welding Code—Seismic Supplement,* AWS D1.8/D1.8M (AWS, 2010) and (AWS, 2009).

The Provisions further designate certain welds as demand critical welds, and require that these welds be made with filler metals that meet minimum levels of CVN toughness using two different test temperatures and specified test protocols, unless otherwise exempted from testing. Welds designated as demand critical welds are identified in the Provisions section applicable to the specific SFRS. Demand critical welds are generally complete-joint-penetration groove (CJP) welds so designated because they are subjected to yield level or higher stress demand and located in a joint whose failure would result in significant degradation in the strength or stiffness of the SFRS.

For demand critical welds, FEMA 350 (FEMA, 2000a) and 353 (FEMA, 2000d) recommended filler metal that complied with minimum Charpy V-notch (CVN) requirements using two test temperatures and specified test protocols. Previous editions of the Provisions included the dual CVN requirement suggested in the FEMA documents but required a lower temperature than the FEMA recommendations for the filler metal classification [−20°F (−29 °C) rather than 0 °F (−18 °C)]. The use of this lower temperature was consistent with the filler metal used in the SAC/FEMA tests and matched the filler metals frequently used for such welds at the time the testing was conducted. The filler metal classification requirement was revised in this edition of the Provisions to reflect the original FEMA recommendation and AWS D1.8/D1.8M requirements because filler metals classified at either temperature ensure that some ductile tearing would occur before final fracture, and because the more critical CVN weld metal property is the minimum of 40 ft-lb (54 J) at 70 °F (21 °C), as determined in AWS D1.8/D1.8M Annex A. This change now permits the use of common welding processes and filler metals, such as GMAW and SAW filler metals that are frequently classified for 20 ft-lb (27 J) at 0 °F (−18 °C).

In a structure with exposed structural steel, an unheated building, or a building used for cold storage, the demand critical welds may be subject to service temperatures less than 50 °F (10 °C) on a regular basis. In these cases, the Provisions require that the minimum qualification temperature for AWS D1.8/D1.8M Annex A be adjusted such that the test temperature for the Charpy V-notch toughness qualification tests be no more than 20 °F (11 °C) above the lowest anticipated service temperature (LAST). For example, weld metal in a structure with a LAST of 0 °F (−18 °C) would need to be qualified at a test temperature less than or equal to 20 °F (−7 °C) and −50 °F (−46 °C) in lieu of 70 °F (21 °C) and 0 °F (−18 °C), respectively. For purposes of the Provisions, the LAST may be considered to be the lowest one-day mean temperature (LODMT) compiled from National Oceanic and Atmospheric Administration (NOAA) data.

All other welds in members and connections in the load path of the SFRS require filler metal with a minimum specified CVN toughness of 20 ft-lbs (27 J) at 0 °F (−18 °C) using the AWS A5 classification. Manufacturer certification may also be used to meet this CVN requirement. Welds carrying only gravity loads, such as filler beam connections and welds for collateral members of the SFRS such as deck welds, minor collectors, and lateral bracing, do not require filler metal meeting these notch toughness requirements.

It is not the intent of the Provisions to require project-specific CVN testing of either the welding procedure specification (WPS) or any production welds. Further, these weld notch toughness requirements are not intended to apply to electric resistance welding (ERW) and submerged arc welding (SAW) when these welding processes are used in the production of hollow structural sections and pipe, such as ASTM A500 and A53/A53M.

5. Concrete and Steel Reinforcement

The limitations on structural steel grades used in composite construction are the same as those given in Sections A3.1 and D2. The limitations in Section A3.5 on concrete and reinforcing bars are the same as those specified for the seismic design of reinforced concrete structures in the *Building Code Requirements for Structural Concrete and Commentary*, ACI 318 Chapter 21 (ACI, 2008). While these limitations are particularly appropriate for construction in seismic design categories D and higher, they apply in any seismic design category when systems are designed with the assumption that inelastic deformation will be required.

A4. STRUCTURAL DESIGN DRAWINGS AND SPECIFICATIONS

1. General

(1) and (2) To ensure proper understanding of the contract requirements and the application of the design, it is necessary to identify the specific types of seismic force resisting system (SFRS) or systems used on the project. Once this is established, those involved know the applicable requirements of the Provisions.

The special design, construction and quality requirements of the Provisions Chapter J, compared to the general requirements of the *Specification* Chapter N, are applicable to the SFRS. The additional quality control and quality assurance requirements of Chapter J are prepared to address the additional requirements for the SFRS, not the structure as a whole. Therefore, it is necessary to clearly designate which members and connections comprise the SFRS.

(3) The protected zone is immediately around the plastic hinging region. Unanticipated connections, attachments or notches may interfere with the formation of the hinge, or initiate a fracture. Because the location of the protected zone depends upon the hinge location, which may vary, the extent of the protected zone must be identified.

(4) Floor and roof decks may be designed to serve as diaphragms and transfer seismic loads, and additional connection details may be needed to provide this load transfer. Consideration should also be made for other floor and roof deck connections when the deck has not been specifically designed and detailed as a diaphragm, as the system may behave as one.

2. Steel Construction

(1) It is necessary to designate working points and connection type(s), and any other detailing requirements for the connections in the SFRS.

(2) Provide information as to the steel specification and grade of the steel elements that comprise the connection, the size and thickness of those elements, weld material size, strength classification and required CVN toughness, and bolt material diameter and grade, as well as bolted joint type.

(3) Demand critical welds are identified in the Provisions for each type of SFRS. Demand critical welds have special Charpy V-notch (CVN) toughness and testing requirements to ensure that this notch toughness will be provided.

(4) Where SCBF brace connections are designed to provide rotation capacity to accommodate buckling in accordance with Section F2.6c(3)(b), they require special detailing as illustrated in Figures C-F2.7, C-F2.8 and C-F2.9. These connections must be identified in the structural design drawings.

(6) The majority of welded connection applications in buildings are in temperature-controlled settings. Where connections are subjected to temperatures of less than 50 °F (10 °C) during service, additional requirements for welding filler metals are necessary for demand critical *welds* to ensure adequate resistance to fracture at the lower service temperatures.

(7) The presence of backing may affect the flow of stresses within the connection and contribute to stress concentrations. Therefore, backing removal may be required at some locations. Removal of backing should be evaluated on a joint specific basis, based upon connection prequalification requirements or qualification testing. AWS D1.8/D1.8M provides details for weld backing removal, additional fillet welds, weld tab removal, tapered transitions, and weld access holes.

(8) Where steel backing remains in place in tee and corner joints with the load applied perpendicular to the weld axis, a fillet weld between the backing and the flange element of the tee or corner joint reduces the stress concentration at the weld root. The requirement for this fillet weld should be evaluated on a joint specific basis, based upon connection prequalification requirements or qualification testing for moment connections, and the requirements of the Provisions for column to base plate connections. AWS D1.8/D1.8M provides details for additional fillet welds at weld backing.

(9) In tee and corner joints where loads are perpendicular to the weld axis, a reinforcing fillet weld applied to a CJP groove weld reduces the stress concentration at the corner between the weld face or root and the member. AWS D1.8/D1.8M provides details for reinforcing fillet welds. Such reinforcement is not required for most groove welds in tee or corner joints.

(10) The presence of weld tabs may affect the flow of stresses within the connection and contribute to stress concentrations. In addition, weld starts and stops made on weld tabs typically contain welds of lesser quality and are not subjected to nondestructive testing. Therefore, complete or partial weld tab removal may be required at some locations. Removal of weld tabs should be evaluated on a joint-specific basis, based upon connection prequalification requirements or qualification testing. AWS D1.8/D1.8M provides details for weld tab removal.

(11) AWS D1.8/D1.8M provides details for tapered transition when required for welded butt joints between parts of unequal thickness and width.

(12) Analysis and research regarding the use of weld access holes have shown that the shape of the weld access hole can have a significant effect on the behavior of moment connections. The selection of weld access hole configuration should be evaluated on a joint-specific basis, based upon connection prequalification requirements in ANSI/AISC 358 or qualification testing in accordance with these Provisions. The use of different weld access holes other than those prescribed by AWS D1.1/D1.1M or the *Specification* has not been found necessary for specific moment connection types, nor necessary for locations such as column splices and column-to-base plate connections. Care should be exercised to avoid specifying special weld access hole geometries when not justified. In some situations, weld access holes are undesirable, such as in end plate moment connections.

(13) In typical structural frame systems, the specification of specific assembly order, welding sequence, welding technique, or other special precautions beyond

those provided in this document should not be necessary. Such additional requirements would only be required for special cases, such as those of unusually high restraint.

3. **Composite Construction**

Structural design drawings and specifications, shop drawings and erection drawings for composite steel-concrete construction are basically similar to those given for all-steel structures. For the reinforced concrete portion of the work, in addition to the requirements in ACI 318 Section 1.2, attention is called to the ACI *Detailing Manual* (ACI, 2004b), with emphasis on Section 2.10, which contains requirements for seismic design of frames, joints, walls, diaphragms and two-way slabs.

CHAPTER B
GENERAL DESIGN REQUIREMENTS

B1. GENERAL SEISMIC DESIGN REQUIREMENTS

When designing structures to resist earthquake motions, each structure is categorized based upon its occupancy and use to establish the potential earthquake hazard that it represents. Determining the available strength differs significantly in each specification or building code. The primary purpose of these Provisions is to provide information necessary to determine the required and available strengths of steel structures. The following discussion provides a basic overview of how several seismic codes or specifications categorize structures and how they determine the required strength and stiffness. For the variables required to assign seismic design categories, limitations of height, vertical and horizontal irregularities, site characteristics, etc., the applicable building code should be consulted. In *Minimum Design Loads for Buildings and Other Structures,* ASCE/SEI 7-10 (ASCE, 2010), structures are assigned to one of four risk categories. Category IV, for example, includes essential facilities. Structures are then assigned to a seismic design category based upon the risk categories and the seismicity of the site adjusted by soil type. Seismic design categories B and C are generally applicable to structures with moderate seismic risk, and special seismic provisions like those in these Provisions are optional. However, special seismic provisions are mandatory in seismic design categories D, E and F, which cover areas of high seismic risk, unless stated otherwise in ASCE/SEI 7-10.

B2. LOADS AND LOAD COMBINATIONS

The Provisions give member and element load requirements that supplement those in the applicable building code. In the 2002 *Seismic Provisions for Structural Steel Buildings* (AISC, 2002), where element forces were defined by the strength of another element, the additional requirements of the Provisions were typically expressed as required strengths. In order to accommodate both LRFD and ASD, the 2005 edition of the Provisions (AISC, 2005b) instead gave two required "available strengths," one for LRFD and one for ASD. ["Available strength" is the term used in the *Specification* (AISC, 2010a) to cover both design strength (LRFD) and allowable strength (ASD).]

In some instances, the loads defined in the Provisions must be combined with other loads. In such cases, the Provisions simply define the seismic load, E or E_{mh}, which is combined with other loads using the appropriate load factor from the seismic load combinations in the applicable building code, and thus both LRFD and ASD are supported.

The Provisions are intended for use with load combinations given in the applicable building code. However, since they are written for consistency with the load combinations given in ASCE/SEI 7-10 and the 2009 *International Building Code* (ICC, 2009), consistency with the applicable building code should be confirmed if another building code is applicable.

The engineer is expected to use these Provisions in conjunction with the *Specification*. Typically, the Provisions do not define available strengths. In certain cases, the designer is directed to specific limit states or provisions in the *Specification*.

An amplification or overstrength factor, Ω_o, applied to the horizontal portion of the earthquake load, E, is prescribed in ASCE/SEI 7-10, the 2009 IBC, the 2009 NEHRP Provisions (FEMA, 2009) and the 2009 *Building Construction and Safety Code*, NFPA 5000 provisions (NFPA, 2009). However, these codes do not all express the load combinations that incorporate this factor in exactly the same format. In the future, if all codes adopt ASCE/SEI 7 by reference, it will be possible to directly reference the appropriate combinations within these Provisions. When used in these Provisions, the term amplified seismic load is intended to refer to the appropriate load combinations in the applicable building code that account for overstrength of members of the seismic force resisting system. The load combinations containing the overstrength factor, Ω_o, should be used where these Provisions require use of the amplified seismic load. In ASCE/SEI 7-10 these load combinations are found in Section 12.4.3.2, Load Combinations with Overstrength Factor. ASCE/SEI 7-10 provides different requirements for addressing such effects for different seismic design categories; orthogonal effects are required to be considered for all but the lowest seismic design categories.

The calculation of seismic loads for composite systems per the ASCE/SEI 7 provisions is the same as is described above for steel structures. The seismic response modification factor, R, and the deflection amplification factor, C_d, for some structural systems have been changed in ASCE/SEI 7-10 to make them more consistent with similar systems in structural steel only and reinforced concrete only systems. This is based on the fact that, when carefully designed and detailed according to these Provisions, the overall inelastic response for composite systems should be similar to comparable steel and reinforced concrete systems. Therefore, where specific loading requirements are not specified in the applicable building code for composite systems, appropriate values for the seismic response modification coefficient can be inferred from specified values for steel and/or reinforced concrete systems. These are predicated upon meeting the design and detailing requirements for the composite systems specified in these Provisions. Unlike the requirements for steel systems, for composite systems that include reinforced concrete members, the design loads and the corresponding design strengths are limited to those defined based on load and resistance factor design. This is done to ensure consistency between provisions for steel, composite and reinforced concrete members that are designed in accordance with the *Specification* and the *Building Code Requirements for Structural Concrete and Commentary,* ACI 318 (ACI, 2008).

B3. DESIGN BASIS

2. Available Strength

It is intended that nominal strengths, resistance and safety factors, and available strengths of steel and composite members in the seismic force resisting system (SFRS) be determined in accordance with the *Specification*, unless noted otherwise in the Provisions. For reinforced concrete members in the SFRS, it is intended that they be designed in accordance with ACI 318.

CHAPTER C

ANALYSIS

C1. GENERAL REQUIREMENTS

For nonseismic applications, story drift limits like deflection limits are commonly used in design to assure the serviceability of the structure. These limits vary because they depend upon the structural usage and contents. As an example, for wind loads such serviceability limit states are regarded as a matter of engineering judgment rather than absolute design limits (Fisher and West, 1990) and no specific design requirements are given in the *Specification*.

The situation is somewhat different when considering seismic effects. Research has shown that story drift limits, although primarily related to serviceability, also improve frame stability (P-Δ effects) and seismic performance because of the resulting additional strength and stiffness. Although some building codes, load standards and resource documents contain specific seismic drift limits, there are major differences among them as to how the limit is specified and applied. Nevertheless, drift control is important to both the serviceability and the stability of the structure. As a minimum, the designer should use the drift limits specified in the applicable building code.

The analytical model used to estimate building drift should accurately account for the stiffness of the frame elements and connections and other structural and nonstructural elements that materially affect the drift. Recent research on steel moment frame connections indicates that in most cases the effect of panel zone deformations on elastic drift can be adequately accounted for by modeling beams to extend between column centerlines without rigid end offsets, and that explicit panel zone modeling is not required (FEMA, 2000f). In cases where nonlinear element deformation demands are of interest, panel zone shear behavior should be represented in the analytical model whenever it significantly affects the state of deformation at a beam-to-column connection. Mathematical models for the behavior of the panel zone in terms of shear force-shear distortion relationships have been proposed by many researchers. FEMA 355C presents a good discussion of how to incorporate panel zone deformations into the analytical model (FEMA, 2000d).

Adjustment of connection stiffness is usually not required for connections traditionally considered as fixed, although FEMA 350 (FEMA, 2000a) contains recommendations for adjusting calculated drift for frames with reduced beam sections. Nonlinear models should contain nonlinear elements where plastic hinging is expected to properly capture the inelastic deformation of the frame.

For composite systems that include composite members or steel members combined with reinforced concrete, the properties of the composite and concrete members should be modeled to represent the effects of concrete cracking. For design by elastic analysis, the composite and concrete member properties should reflect the effective stiffness of the members at the onset of significant yielding in the members. The following guidance is provided for calculating effective stiffness values for design by elastic analysis:

- In concrete beam and column members, stiffness properties for elastic analysis are typically specified as a fraction of the flexural stiffness, EI_g, where E is the elastic modulus of concrete and I_g is the gross moment of inertia. For concrete frames, ACI 318 Section 8.8.2 (ACI, 2008) recommends effective stiffness values ($EI_{effective}$) in the range of 0.35 to $0.50EI_g$ for beams and 0.50 to $0.70EI_g$ for columns. More detailed recommendations that account explicitly for axial load are given in *Seismic Rehabilitation of Existing Buildings* (ASCE 41-06) including Supplement No. 1, (ASCE, 2006a and 2006b) which recommends effective stiffness values of (a) $0.70EI_g$ for columns with unfactored gravity compressive loads that are greater than $0.5A_g f'_c$) (where A_g is the gross member area and f'_c is the concrete compressive strength) and (b) $0.30EI_g$ for columns (and beams) with axial gravity loads less than $0.1A_g f'_c$. Linear interpolation of stiffness is suggested for axial loads between 0.1 to $0.5A_g f'_c$.

- For concrete walls, the cracked section properties in their assumed plastic hinge region may be taken as $0.35EI_g$ and $0.75EA_g$. The walls above the hinged region are typically expected to remain essentially elastic. For these regions and walls that are anticipated to remain in the elastic range, the cracked section properties for the walls may be taken as $0.70EI_g$ and $1.0EA_g$. ASCE 41-06 also includes recommendations, which are deemed to be conservative for new composite ordinary shear walls.

- For concrete-encased or concrete-filled beam-columns, the effective stiffness may be specified based on the use of a cracked transformed section [see, for example Ricles and Paboojian (1994); Varma et al. (2002)]. Attention should be paid to the relative values of the girder versus beam-column effective stiffnesses.

- For steel beams with composite slabs in which the shear connection is such that the contribution of the composite slab can be included in the stiffness and subject to reverse curvature due to earthquake loading, a reasonable assumption is to specify a flexural stiffness that is equal to the average of the beam stiffness in positive and negative bending. Assuming that the beams are designed to have full composite action, it is suggested to take the effective stiffness as equal to $0.5(E_s I_s + E_s I_{tr})$, where E_s is the steel modulus, I_s is the moment of inertia of the bare steel beam, and I_{tr} is the transformed moment of inertia of the beam and slab. The effective width of the slab can be determined per Chapter H of the *Specification*.

The story drift limits in *Minimum Design Loads for Buildings and Other Structures*, ASCE/SEI 7 (ASCE, 2010) and the 2009 NEHRP Provisions (FEMA, 2009) are to be compared to an amplified story drift that approximates the difference in deflection between the top and bottom of the story under consideration during a large earthquake. The amplified story drift is determined by multiplying the elastic drift caused by the horizontal component of the earthquake load, E, by a deflection amplification factor, C_d, which is dependent upon the type of building system used.

Each story of the structure should be investigated to ascertain that lateral drifts induced by earthquake response do not result in a condition of instability under gravity loads.

P-Δ effects can have a significant impact on the ability of structures to resist collapse when subjected to strong ground shaking. If earthquake induced displacements are sufficiently large to create negative instantaneous stiffness, collapse is likely to occur. For this reason, ASCE/SEI 7 Section 12.8.7 limits the ratio of secondary moment to primary moment.

Any of the elastic methods in the *Specification* Chapter C or Appendix 7 can be used to assess the stability of frames in high seismic regions. When using the equivalent lateral load procedure for seismic design and the direct analysis provisions in *Specification* Chapter C, the reduced stiffness and notional load provisions should not be included in the calculation of the fundamental period of vibration or the evaluation of seismic drift limits.

Like most of the provisions in the *Specification*, the stability requirements are intended for cases where the strength limit state is based on the nominal elastic-plastic limit in the most critical members and connections (for example, the "first hinge" limit point), not to ensure stability under seismic loads where large inelastic deformations are expected. Thus, the provisions of the *Specification* Chapter C do not alone ensure stability under seismic loads. Stability under seismic loads is synonymous with collapse prevention, which is provided for in the prescriptive design requirements given for each system, including such elements as:

- The basic determination of the seismic design force (R factors, site effects, ρ factors, etc.)
- The drift limits under the seismic lateral load (a factor of both the limiting drift and the specified C_d factor)
- The "theta" limits (sidesway stability collapse prevention)
- Other design requirements, such as strong-column weak-beam requirements, limitations on bracing configurations, etc.

C2. ADDITIONAL REQUIREMENTS

Additional analysis requirements are prescribed in the Provisions for a number of framing systems that are regarded as "highly ductile," such as SMF, EBF, BRBF, STMF, SPSW, etc. Those requirements are intended to achieve several desired performance objectives, the most important of which is to limit the inelastic activity within certain structural elements and to prevent or minimize it from occurring elsewhere during expected ground motions. The examples of intended yielding members include beams in SMF, shear links in EBF, bracing members in SCBF and BRBF, plate shear walls in SPSW, etc. The required strength of intended yielding members is determined by elastic analysis methods for the prescribed load combinations, while that of other elements which are intended to remain essentially elastic is determined by a pseudo-capacity design approach which varies from system to system.

Consequently, the intended performance goals may not always be fully achieved even after a number of design and assessment iterations.

The plastic design method is perhaps the most direct way to achieve the objective of a desired yield mechanism for structures (Goel and Chao, 2008). In the plastic design approach the desired yield mechanism is first selected by clearly identifying the intended or designated yielding members (DYM) and those which are intended to remain elastic, which may be called nondesignated yielding members (non-DYM). The required strength of the DYM is determined by using a mechanism based plastic analysis for the appropriate load combination. Caution needs to be exercised at this step to make sure that the applied lateral forces are those associated with the targeted yield mechanism. Methods have been proposed by investigators, such as the yield point spectrum method by Aschheim (2002), and an energy based method by Goel and Chao (2008). Those methods have the added advantage that drift (ductility) control is built into the determination of design lateral forces. In contrast, the design seismic forces typically specified by current building codes are intended to be used with elastic design methods, and drift and stability checks need to be performed separately which often result in rather cumbersome iterations. The second step of determining the required strength of non-DYM can be carried out by one of the following possible methods:

1. A static elastic analysis of suitably selected structural portions ("free bodies") consisting of non-DYM with lateral forces to keep them in equilibrium under the expected forces from DYM and other applicable loads.

2. A nonlinear static pushover analysis of the entire structure up to a target drift level by modeling the DYM to behave inelastically, while the non-DYM are modeled (or "forced") to behave elastically in order to be able to determine their required strength.

3. A nonlinear dynamic analysis of the structure as modeled for the pushover analysis mentioned above, using an appropriately selected ensemble of ground motions.

In the above analysis options, second-order effects should be included at expected drift levels.

The advantages of a mechanism-based design approach as outlined above include:

1. Enhanced performance and safety, especially under severe ground motions.

2. Ease and economy of repairs after an event, because the structural damage would be confined to known members (DYM) and locations. This may translate into lower overall life-cycle cost of the structures.

3. The non-DYM would not need to be detailed for as stringent ductility requirements as the DYM.

4. Innovative structural schemes can be developed by selecting from a variety of ductile energy dissipating members and devices as DYM and "ordinary" (not so ductile) members and connections for non-DYM, made of a suitable combination of materials.

C3. NONLINEAR ANALYSIS

Nonlinear analysis may be used in the Provisions in certain situations (e.g., exception in Sections E2.6g and E3.6g). Procedures such as those given in ASCE/SEI 7-10 should be followed unless a more rational method can be justified.

CHAPTER D

GENERAL MEMBER AND CONNECTION DESIGN REQUIREMENTS

D1. MEMBER REQUIREMENTS

1. Classification of Sections for Ductility

Members of the seismic force resisting system (SFRS) that are anticipated to undergo inelastic deformation have been classified as either moderately ductile members or highly ductile members. During the design earthquake, moderately ductile members are anticipated to undergo moderate plastic rotation of 0.02 rad or less, whereas highly ductile members are anticipated to undergo significant plastic rotation of 0.04 rad or more. The member rotations result from either flexure or flexural buckling. The requirements for moderately ductile and highly ductile members apply only to those members designated as such in the Provisions.

1a. Section Requirements for Ductile Members

To provide for reliable inelastic deformations in those SFRS members that require moderate to high levels of inelasticity, the member flanges must be continuously connected to the web(s). This requirement does not preclude the use of members built up from shapes. Built-up shapes shall comply with the requirements in the *Specification* and any additional requirements of these Provisions or ANSI/AISC 358 (AISC, 2010b) that are specific to the system or connection type being used.

1b. Width-to-Thickness Limitations of Steel and Composite Sections

To provide for reliable inelastic deformations in those members of the SFRS that require moderate to high levels of inelasticity, the width-to-thickness ratios of compression elements should be less than or equal to those that are resistant to local buckling when stressed into the inelastic range. Table D1.1 provides width-to-thickness ratios that correspond to the anticipated level of inelastic behavior for both moderately ductile and highly ductile members. The limiting width-to-thickness ratios for moderately ductile members generally correspond to λ_p values in Table B4.1b of the *Specification* with exceptions for round and rectangular HSS, stems of WTs, and webs in flexural compression. Although the limiting width-to-thickness ratios for compact compression elements, λ_p, given in *Specification* Table B4.1b, are sufficient to prevent local buckling before the onset of strain-hardening, the available test data suggests that these limits are not adequate for the required inelastic performance of highly ductile members in the SFRS. The limiting width-to-thickness ratios for highly ductile members, λ_{hd}, given in Table D1.1 are deemed adequate for the large ductility demands to which these members may be subjected (Sawyer, 1961; Lay, 1965; Kemp, 1986; Bansal, 1971).

For highly ductile members, the limiting width-to-thickness ratios for webs of rolled or I-shaped built-up beams and webs of built-up shapes used as beams or columns have been modified from the previous edition of these provisions for cases of combined bending and axial compression. These modifications, based on a reevaluation of available data, were made to eliminate a previous inconsistency in the specified web slenderness limits as the axial force approached zero. A review of the literature indicated little research is currently available on web buckling under cyclic axial force and bending. Consequently, the limits are based primarily on research on the effects of web slenderness on ductility under combined bending and axial compression under monotonic loading, including work by Haaijer and Thurlimann (1958), Perlynn and Kulak (1974), and Dawe and Kulak (1986). The current web slenderness limits were chosen to be consistent with those suggested by Dawe and Kulak (Dawe and Kulak, 1986) with minor modifications. The equations have been adjusted to converge to $1.49\sqrt{E/F_y}$ at $C_a = 1.0$ and to equal $2.45\sqrt{E/F_y}$ at $C_a = 0$. The latter value is consistent with the recommendations for special moment frame (SMF) beams per Uang and Fan (2001) and FEMA 350 (FEMA, 2000a) for cases where the axial force is zero. The limiting width-to-thickness ratios of stiffened webs for moderately ductile beam or column members correspond to those in Appendix 1 of the *Specification*, Design by Inelastic Analysis. For I-shaped beams in SMF and intermediate moment frames (IMF), the effects of axial compression on the limiting web slenderness ratio can be neglected when C_a is less than or equal to 0.125 (see footnote [d] of Table D1.1). This exception is provided because it is believed that small levels of axial compression, and its consequent effect on web bucking in beams, will be less detrimental to system performance than in columns.

Axial forces during the design earthquake may approach the available tensile strength of diagonal braces. In order to preclude local buckling of the webs of I-shaped members used as diagonal braces, the web width-to-thickness limitation for nonslender elements for members subject to axial compression per Table B4.1a of the *Specification* is required.

Ongoing research of special concentrically braced frame (SCBF) diagonal braces indicates that the width-to-thickness ratios of walls of round and rectangular HSS sections may not be adequate to prevent premature fracture during multiple cycles of flexural buckling of these members. (Fell et al., 2006) To reduce the possibility of fracture, the width-to-thickness values of highly ductile members comprised of these shapes have been reduced in these provisions by approximately 15%. While width-to-thickness ratios of moderately ductile members composed of unstiffened elements generally match those of Table B4.1b of the *Specification*, in order to reduce the possibility of fracture of diagonal braces of moderately ductile round and rectangular HSS members, the width-to-thickness values of these shapes correspond to the respective λ_{ps} (seismically compact) values in the 2005 Provisions (AISC, 2005b). As round and rectangular HSS members used as beams or columns designated as moderately ductile members are not anticipated to experience flexural buckling, exceptions have been added relaxing the width-to-thickness ratios to the λ_p values of Table B4.1b of the *Specification* (see footnotes [c] and [e] of Table D1.1).

A small relaxation in the width-to-thickness ratio of the stem of tees used as highly ductile members has been added for two cases (see footnote [a] of Table D1.1). The relaxed value corresponds to the λ_p value in Table B4.1b of the *Specification*. For the first case, where buckling is anticipated to occur about the plane of the stem, little inelastic deformation should occur in the stem itself. The second case takes advantage of a common practice for the connection of tees which is to bolt or weld a connection plate only to the outside of the flange of the tee with no connection to the web. Because the axial load is applied eccentrically to the neutral axis of the tee, a bending stress occurs which reduces the compressive stresses at the tip of the stem. Currently there is insufficient data or research on buckling of stems of tees to permit a more substantial relaxation for highly ductile members, nor to permit a relaxation for tees used as moderately ductile members.

During the service life of a steel H-pile, it is primarily subjected to axial compression and acts as an axially loaded column. Therefore, the b/t ratio limitations given in Table B4.1 of the *Specification* should suffice. During a major earthquake, because of lateral movements of pile cap and foundation, the steel H-pile becomes a beam-column and may have to resist large bending moments and uplift. Cyclic tests (Astaneh-Asl and Ravat, 1997) indicated that local buckling of piles satisfying the width-to-thickness limitations in Table D1.1 occurred after many cycles of loading. In addition, this local buckling did not have much effect on the cyclic performance of the pile during cyclic testing or after cyclic testing stopped and the piles were once again under only axial load.

The width-to-thickness criteria for highly ductile filled rectangular members remain unchanged from the requirements for special seismic systems in the 2005 Provisions. Provisions have been added for highly ductile filled circular members (Varma and Zhang, 2009). For moderately ductile members, the requirements are the same as for composite columns in the *Specification*, which are also unchanged from the 2005 Provisions.

In Section A3.2, the expected yield stress, $R_y F_y$, of the material used in a member is required for the purpose of determining the effect of the actual member strength on its connections to other members of the seismic force resisting system. The width-to-thickness requirements in Table D1.1, calculated using specified minimum yield stress, are expected to permit inelastic behavior without local buckling and therefore need not be computed using the expected yield strength.

2. Stability Bracing of Beams

The requirements for stability bracing of beams designated as moderately ductile members and highly ductile members are a function of the anticipated levels of inelastic yielding as discussed in Commentary Section D1.1 for members with these two designations.

2a. Moderately Ductile Members

The limiting requirement for spacing of stability bracing of $0.17 E r_y / F_y$ for moderately ductile beam members is the same limit specified in the previous provisions for

IMF beams, as the level of inelastic behavior in IMF beams is considered representative of moderately ductile beams. Since the minimum required story drift angle of a SMF system is twice that of an IMF system, the use of a less severe maximum stability spacing requirement for IMF beams that is twice that of SMF beams is appropriate. The commentary on Section D1.2b gives further discussion on stability bracing of beams.

In addition to nodal bracing, these provisions allow both torsional and relative bracing per Appendix 6 of the *Specification*. While torsional bracing is appropriate for beams with minimal or no compressive axial loads, beams with significant axial loads may require lateral bracing or lateral bracing combined with torsional bracing to preclude axial buckling.

For calculating bracing strength according to Equations A-6-5 and A-6-7 of the *Specification,* the use of $C_d = 1$ is justified because the AISC equations have an implicit assumption that the beams will be subjected to top flange loading. One can see this by comparing the *Specification* Equations A-6-5 and A-6-7 to the *Specification* Commentary Equations C-A-6-6a and C-A-6-6b, where the *Specification* equations are based on a conservative assumption of $C_t = 2$. In the case of seismic frames, where the moments are introduced via the beam-column connections, $C_t = 1$. Strictly speaking, the correct solution would be to use the commentary equation with $C_t = 1$ and $C_d = 1$ at all locations except for braces at the inflection point where $C_d = 2$. The current Provisions imply that the product of $C_t (C_d) = 2.0$ by the implied value of $C_t = 2$ and $C_d = 1$.

2b. Highly Ductile Members

Spacing of stability braces for highly ductile members is specified not to exceed $0.086 E r_y / F_y$. This limitation is identical to the requirement in the previous Provisions for beams in SMF as the degree of inelastic behavior is representative of highly ductile members. The spacing requirement for beams in SMF was originally based on an examination of lateral bracing requirements from early work on plastic design and based on limited experimental data on beams subject to cyclic loading. Lateral bracing requirements for SMF beams have since been investigated in greater detail in Nakashima et al. (2002). This study indicates that a beam lateral support spacing of $0.086 E r_y / F_y$ is appropriate, and slightly conservative, to achieve a story drift angle of 0.04 rad.

2c. Special Bracing at Plastic Hinge Locations

In addition to bracing along the beam length, the provisions of this Section call for the placement of stability bracing to be near the location of expected plastic hinges of highly ductile members. Such guidance dates to the original development of plastic design procedures in the early 1960s. In moment frame structures, many connection details attempt to move the plastic hinge a short distance away from the beam-to-column connection. Testing carried out as part of the SAC program (FEMA, 2000a) indicated that the bracing provided by typical composite floor slabs is adequate to avoid excessive strength deterioration up to the required story drift angle of 0.04 rad. Therefore, the FEMA recommendations do not require the placement of supplemental lateral bracing at plastic hinge locations adjacent to column connections for beams with composite floor construction. These provisions allow the placement of lateral or torsional braces to be consistent with the tested connections that are used to justify the design. For conditions where drifts larger than 0.04 rad are anticipated or improved performance is desired, the designer may decide to provide additional stability bracing near these plastic hinges. If lateral braces are used, they should provide an available strength of 6% of the expected capacity of the beam flange at the plastic hinge location. If a reduced beam section connection detail is used, the reduced flange width may be considered in calculating the bracing force. If torsional braces are used, they should provide an available strength of 6% of the expected bending capacity of the beam at the plastic hinge. Placement of bracing connections should consider the requirements of Section D1.3.

3. Protected Zones

The FEMA/SAC testing has demonstrated the sensitivity of regions undergoing large inelastic strains to discontinuities caused by welding, rapid change of section, penetrations, or construction caused flaws. For this reason, operations as specified in Section I2.1 that cause discontinuities are prohibited in regions subject to large inelastic strains. These provisions designate these regions as protected zones. The protected zones are designated in the Provisions in the sections applicable to the designated type of system and in ANSI/AISC 358. The protected zones include moment frame hinging zones, links of eccentrically braced frames (EBFs), the ends and the center of SCBF diagonal braces, etc.

Not all regions experiencing inelastic deformation are designated protected zones; for example, the beam-column panel zone of moment frame systems. It should be noted that yield level strains are not strictly limited to the plastic hinge zones and caution should also be exercised in creating discontinuities in these regions as well.

4. Columns

4a. Required Strength

It is imperative that columns that are part of the SFRS have adequate strength to avoid global buckling or tensile rupture. Since the late 1980s, the Seismic Provisions and other codes and standards have included requirements that are similar to those included in this section. The required forces for design of the columns are intended to represent reasonable limits on the axial forces that can be imposed. Design for these forces is expected to prevent global column failure. These axial forces are permitted to be applied without consideration of concurrent bending moments that may occur at column ends. Research has shown that columns can withstand high axial forces (up to $0.75F_y$) with significant end rotations due to story drift (Newell and Uang, 2008). The column design using these forces is typically checked using $K = 1.0$. This approach is based on the recognition that in the SFRS, column bending moments would be largest at the column ends and would normally result in reverse curvature in the column. This being the case, the bending moments would not contribute to column buckling, and the assumption of $K = 1$ would be conservative. However, bending moments resulting from a load applied between points of lateral support can contribute to column buckling and are therefore required to be considered concurrently with axial loads.

Clearly, the above described approach provides no assurance that columns will not yield and the combination of axial load and bending is often capable of causing yielding at the ends of columns. Column yielding may be caused by a combination of high bending moments and modest axial loads, as is normal in moment frames or by a combination of high axial load and bending due to the end rotations that occur in braced frame structures. While yielding of columns may result in damage that is significant and difficult to repair, it is judged that, in general, it will not result in column ruptures or global buckling, either of which would threaten life safety.

In the previous Provisions, the limits $P_u/\phi_c P_n$ (LRFD) > 0.4 or $\Omega_c P_a/P_n$ (ASD) > 0.4, as appropriate, were used as the trigger for requiring the inclusion of amplified seismic loads in the check of column strength. However, the 0.4 limit could be unconservative for columns with light gravity loads in systems with large Ω_o values. Consequently the 0.4 limit has been eliminated and the effect of amplified seismic loads shall be considered on all columns in the SFRS. A simplified check, when no load is applied to the column between points of lateral support, would be to consider Section D2.4a(2) satisfied when the ratio of *required strength* of a column without amplified seismic loads to the column's available strength is less than the value $1/\Omega_o$.

Although the provisions in Section D1.4a are believed to provide reasonable assurance of adequate performance, it should be recognized that these are minimum standards and there may be additional concerns where higher levels of performance, or greater levels of reliability are merited. For example, nonlinear analyses often indicate conditions wherein column end moments are not reversed and may contribute to buckling.

For the exception noted in Section D1.4a(2)(b), realistic soil capacities must be used when determining the limiting resistance of the foundation to overturning uplift.

4b. Encased Composite Columns

The basic requirements and limitations for determining the design strength of reinforced concrete encased composite columns are the same as those in the *Specification*. Additional requirements for reinforcing bar details of composite columns that are not covered in the *Specification* are included based on provisions in ACI 318 (ACI, 2008). Examples for determining the effective shear width, b_w, of the reinforced concrete encasement are given in Figure C-D1.1.

Composite columns can be an ideal solution for use in seismic regions because of their inherent structural redundancy (Viest et al., 1997; El-Tawil and Deierlein, 1999). For example, if a composite column is designed such that the structural steel can carry most or all of the dead load acting alone, then an extra degree of protection and safety is afforded, even in a severe earthquake where excursions into the inelastic range can be expected to deteriorate concrete cover and buckle reinforcing steel. However, as with any column of concrete and reinforcement, the designer should be aware of the constructability concerns with the placement of reinforcement and potential for congestion. This is particularly true at beam-to-column connections where potential interference between a steel spandrel beam, a perpendicular floor beam, vertical bars, joint ties, and shear stud anchors can cause difficulty in reinforcing bar placement and a potential for honeycombing of the concrete.

Seismic detailing requirements for composite columns are specified in the following two categories: moderately ductile and highly ductile. Requirements for limited ductility (ordinary) composite columns, which were included in the 2005 Provisions, have been removed from the 2010 Provisions because these requirements are now part of the basic composite column requirements in the *Specification*. The required level of detailing is specified in these Provisions for seismic systems in Chapters G and H. Moderately ductile requirements are intended for seismic systems permitted in seismic design category C, and highly ductile requirements are intended for seismic systems permitted in seismic design categories D and above. Note that the highly ductile requirements apply to members of special seismic systems permitted in seismic design category D, even if the systems are employed for use in lower seismic design categories.

(1) *Moderately Ductile Members*

The more stringent tie spacing requirements for moderately ductile encased composite columns follow those for reinforced concrete columns in regions of moderate seismicity as specified in ACI 318 Chapter 21. These requirements are applied to all composite columns for systems permitted in seismic design category C to make the composite column details at least equivalent to the minimum level of detailing for columns in intermediate moment frames of reinforced concrete (FEMA, 2000e; ICC, 2009).

(2) *Highly Ductile Members*

The additional requirements for encased composite columns used in special seismic systems are based upon comparable requirements for structural steel and reinforced concrete columns in composite systems permitted in seismic design

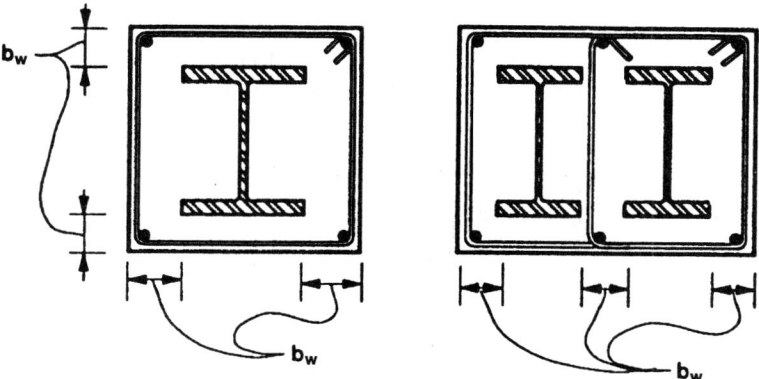

Fig. C-D1.1. Effective widths for shear strength calculation of encased composite columns.

categories D and above (FEMA, 2009; ICC, 2009). For additional explanation of these requirements, see the Commentary for Section D1.4a in these Provisions and ACI 318 Chapter 21.

The minimum tie area requirement in Equation D1-8 is based upon a similar provision in ACI 318 Chapter 21, except that the required tie area is reduced to take into account the steel core. The tie area requirement in Equation D1-8 and related tie detailing provisions are waived if the steel core of the composite member can alone resist the expected (arbitrary point in time) gravity load on the column because additional confinement of the concrete is not necessary if the steel core can inhibit collapse after an extreme seismic event. The load combination of $1.0D + 0.5L$ is based upon a similar combination proposed as loading criteria for structural safety under fire conditions (Ellingwood and Corotis, 1991).

The requirements for composite columns in composite special moment frames (C-SMF) are based upon similar requirements for steel and reinforced concrete columns in SMF (FEMA, 2009; ICC, 2009). For additional commentary, see Section E3 in these Provisions and ASCE/SEI 7 (ASCE, 2010).

The strong-column/weak-beam concept follows that used for steel and reinforced concrete columns in SMF. Where the formation of a plastic hinge at the column base is likely or unavoidable, such as with a fixed base, the detailing should provide for adequate plastic rotational ductility. For seismic design category E, special details, such as steel jacketing of the column base, should be considered to avoid spalling and crushing of the concrete.

Closed hoops are required to ensure that the concrete confinement and nominal shear strength are maintained under large inelastic deformations. The hoop detailing requirements are equivalent to those for reinforced concrete columns in SMF. The transverse reinforcement provisions are considered to be conservative since composite columns generally will perform better than comparable reinforced concrete columns with similar confinement. However, further research is required to determine to what degree the transverse reinforcement requirements can be reduced for composite columns. It should be recognized that the closed hoop and cross-tie requirements for C-SMF may require special details such as those suggested in Figure C-D1.2 to facilitate the erection of the reinforcement around the steel core.

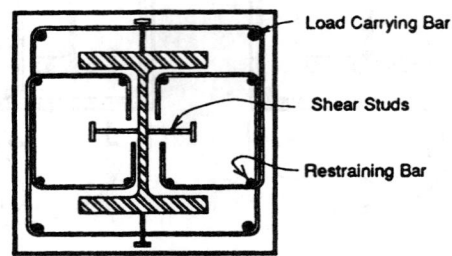

Fig. C-D1.2. Example of a closed hoop detail for an encased composite column.

Ties are required to be anchored into the confined core of the column to provide effective confinement.

4c. Filled Composite Columns

The basic requirements and limitations for detailing and determining the design strength of filled composite columns are the same as those in *Specification* Chapter I.

The shear strength of the filled member is conservatively limited to the nominal shear yield strength of the hollow structural section (HSS) because the actual shear strength contribution of the concrete fill has not yet been determined in testing. This approach is recommended until tests are conducted (Furlong, 1997; ECS, 1994). Even with this conservative approach, shear strength rarely governs the design of typical filled composite columns with cross-sectional dimensions up to 30 in. (762 mm). Alternatively, the shear strength for filled tubes can be determined in a manner that is similar to that for reinforced concrete columns with the steel tube considered as shear reinforcement and its shear yielding strength neglected. However, given the upper limit on shear strength as a function of concrete crushing in ACI 318, this approach would only be advantageous for columns with low ratios of structural steel to concrete areas (Furlong, 1997).

5. Composite Slab Diaphragms

In composite construction, floor and roof slabs typically consist of either composite or noncomposite metal deck slabs that are connected to the structural framing to provide an in-plane composite diaphragm that collects and distributes seismic loads. Generally, composite action is distinguished from noncomposite action on the basis of the out-of-plane shear and flexural behavior and design assumptions.

Composite metal deck slabs are those for which the concrete fill and metal deck work together to resist out-of-plane bending and out-of-plane shear. Flexural strength design procedures and codes of practice for such slabs are well established (ASCE, 1991a, 1991b; AISI, 2007; SDI, 2001, 2007).

Noncomposite metal deck slabs are one-way or two-way reinforced concrete slabs for which the metal deck acts as formwork during construction, but is not relied upon for composite action. Noncomposite metal deck slabs, particularly those used as roofs, can be formed with metal deck and overlaid with insulating concrete fill that is not relied upon for out-of-plane strength and stiffness. Whether or not the slab is designed for composite out-of-plane action, the concrete fill inhibits buckling of the metal deck, increasing the in-plane strength and stiffness of the diaphragm over that of the bare steel deck.

The diaphragm should be designed to collect and distribute seismic loads to the seismic force resisting system. In some cases, loads from other floors should also be included, such as at a level where a change in the structural stiffness results in redistribution. Recommended diaphragm (in-plane) shear strength and stiffness values for metal deck and composite diaphragms are available for design from industry sources that are based upon tests and recommended by the applicable building code (SDI,

2001, 2004, 2007). In addition, research on composite diaphragms has been reported in the literature (Easterling and Porter, 1994).

As the thickness of concrete over the steel deck is increased, the shear strength can approach that for a concrete slab of the same thickness. For example, in composite floor deck diaphragms having cover depths between 2 in. (51 mm) and 6 in. (152 mm), measured shear stresses on the order of $3.5\sqrt{f'_c}$ (where $\sqrt{f'_c}$ and f'_c are in units of psi) have been reported. In such cases, the diaphragm strength of concrete metal deck slabs can be conservatively based on the principles of reinforced concrete design (ACI, 2008) using the concrete and reinforcement above the metal deck ribs and ignoring the beneficial effect of the concrete in the flutes.

Shear forces are transferred through welds and/or shear anchors in the collector and boundary elements. Fasteners between the diaphragm and the steel framing should be capable of transferring forces using either welds or shear anchors. Where concrete fill is present, it is generally advisable to use mechanical devices such as steel headed stud anchors to transfer diaphragm forces between the slab and collector/boundary elements, particularly in complex shaped diaphragms with discontinuities. However, in low-rise buildings without abrupt discontinuities in the shape of the diaphragms or in the seismic force resisting system, the standard metal deck attachment procedures may be acceptable.

D2. CONNECTIONS

1. General

Adequate behavior of connections of members in various systems in the SFRS is ensured by satisfying one of the following general conditions:

(1) Connections in some systems are verified by testing to ensure adequate performance [IMF, SMF and buckling-restrained braced frames (BRBF) systems, for example].

(2) Connections of members in some systems are designed to resist the required strength of the connected member or an adjoining member and therefore the maximum connection forces are limited by yielding of a member [SCBF and BRBF diagonal braces, for example].

(3) Connections of some members must be designed to resist forces based on the load combinations including the amplified seismic load (column splices, collectors and OCBF diagonal braces for example).

A review of the requirements of these Provisions and *Mimimum Design Loads for Buildings and Other Structures* (ASCE/SEI 7-10) indicates that connections in the SFRS satisfy at least one of the above conditions. Therefore, the requirement in the 2005 Seismic Provisions that the design of a connection ensures a ductile limit state has been deleted.

2. Bolted Joints

The potential for full reversal of design load and the likelihood of inelastic deformations of members and/or connected parts necessitates that pretensioned bolts be used in bolted joints in the SFRS. However, earthquake motions are such that slip cannot and need not be prevented in all cases, even with slip-critical connections. Accordingly, the Provisions call for bolted joints to be proportioned as pretensioned bearing joints but with faying surfaces prepared as for Class A or better slip-critical connections. That is, bolted connections can be proportioned with available strengths for bearing connections as long as the faying surfaces are still prepared to provide a minimum slip coefficient, $\mu = 0.30$. The resulting nominal amount of slip resistance will minimize damage in more moderate seismic events. This requirement is intended for joints where the faying surface is primarily subjected to shear. Where the faying surface is primarily subjected to tension or compression from seismic load effects, for example, in a bolted end plate moment connection, the requirement for preparation of the faying surfaces may be relaxed.

It is an acceptable practice to designate bolted joints as slip-critical as a simplified means of specifying the requirements for pretensioned bolts with slip-critical faying surfaces. However when the fabricator is permitted to design the connections, specifying that bolted joints must be designed as slip-critical may result needlessly in additional and/or larger bolts.

To prevent excessive deformations of bolted joints due to slip between the connected plies under earthquake motions, the use of holes in bolted joints in the SFRS is limited to standard holes and short-slotted holes with the direction of the slot perpendicular to the line of force. Exceptions are provided for alternative hole types that are justified as a part of a tested assembly and for oversized holes in diagonal brace members of certain systems.

An exception allows the use of oversized holes in one ply of connections of diagonal bracing members in Sections F1, F2, F3 and F4 when the connection is designed as a slip-critical joint. The required strength for the limit state of bolt slip for the connection is specified in the applicable Section. As reported in FEMA 355D (FEMA, 2000d), bolted joints with oversized holes in tested moment connections were found to behave as full stiffness connections for most practical applications. Bolted connections of diagonal bracing with oversized holes should behave similarly. Oversized holes in diagonal bracing connections with slip-critical bolts will provide additional tolerance for field connections, yet should remain as slip-resistant for most seismic events. If the bolts did slip in the oversized holes in an extreme situation, the connections should still behave similarly to full stiffness connections. Story drifts may also increase slightly if bolts slip, and the effect of bolt slip should be considered in drift calculations. In order to minimize the amount of slip, oversized holes for bolts are limited to one ply of the connection. For large diameter bolts, the amount of slippage can also be minimized by limiting the bolt hole size to a maximum of $^{3}/_{16}$ in. greater than the bolt diameter, rather than the maximum diameter permitted by the *Specification*. The available slip resistance of bolts in oversized holes shall reflect the reduced available strength for oversized holes per Section J3.8 of the *Specification*. The reduction of pretension with bolts installed in oversized holes results in a lower static slip load, but the overall behavior of connections with oversized holes has been shown to be similar to those with standard holes (Kulak et al., 1987).

To prevent excessive deformations of bolted joints due to bearing on the connected material, the bearing strength is limited by the "deformation-considered" option in *Specification* Section J3.10 ($R_n = 2.4dtF_u$). The philosophical intent of this limitation in the *Specification* is to limit the bearing deformation to an approximate maximum of $^{1}/_{4}$ in. (6 mm). It should be recognized, however, that the actual bearing load in a seismic event may be much larger than that anticipated in design and the actual deformation of holes may exceed this theoretical limit. Nonetheless, this limit should effectively minimize damage in moderate seismic events.

Connections or joints in which bolts in combination with welds resist a common force are prohibited. Due to the potential of full load reversal and the likelihood of inelastic deformations in connecting plate elements, bolts may exceed their slip resistances under significant seismic loads. Welds that are in a common shear plane to these bolts will likely not deform sufficiently to allow the bolts to slip into bearing, particularly if subject to cyclic load reversal. Consequently, the welds will tend to resist the entire force and may fail if they are not designed as such. These provisions prohibit bolts from sharing a common force with welds in all situations. In addition to prohibiting sharing of loads on a common faying surface, sharing of a common force between different elements in other conditions is also prohibited. For example, bracing connections at beam-to-column joints are often configured such that the vertical component of the brace is resisted by a combination of both the beam web and the gusset connections to the columns (see Figures C-D2.1 and C-D2.2). Since these two elements are in a common shear plane with limited deformation capability, if one element were welded and the other bolted, the welded joint would likely resist all the force. By making the connections of these elements to the column either both bolted or both welded, both elements would likely participate in resisting the force. Similarly, wide flange bracing connections should not be designed such that bolted web connections share in resisting the axial loads with welded flanges (or vice versa).

Bolts in one element of a member may be designed to resist a force in one direction while other elements may be connected by welds to resist a force in a different direction or shear plane. For example, a beam-to-column moment connection may use welded flanges to transfer flexure and/or axial loads, while a bolted web connection transfers the beam shear. Similarly column splices may transfer axial loads and/or flexure through flange welds with horizontal shear in the column web transferred through a bolted web connection. In both of these cases there should be adequate deformation capability between the flange and web connections to allow the bolts to resist loads in bearing independent of the welds.

The Provisions do not prohibit the use of erection bolts on a field welded connection such as a shear tab in the web of a wide flange beam moment connection. In this instance the bolts would resist the temporary erection loads, but the welds would need to be designed to resist the entire anticipated force in that element.

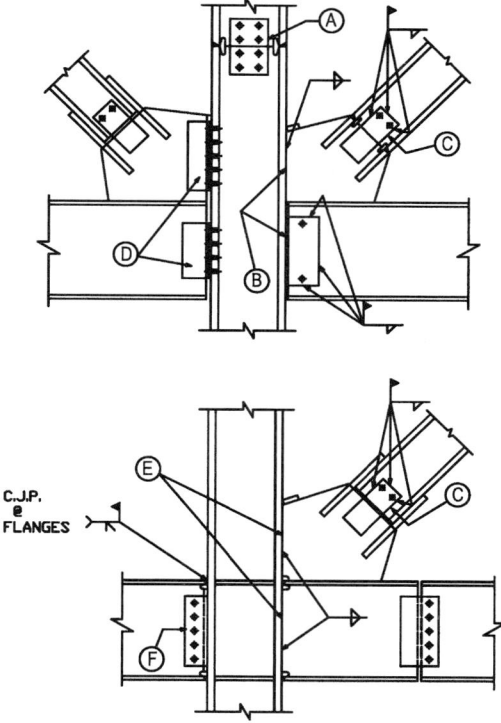

Ⓐ A bolted web connection may be designed to resist column shear while welded flanges resist axial and/or flexural forces.

Ⓑ Connection using both gusset and beam web welded to column allows both elements to participate in resisting the vertical component of the brace force. Note erection bolts may be used to support beam temporarily.

Ⓒ Flanges and web are both welded to resist axial force in combination. Bolts are for erection only.

Ⓓ Both web of beam and gussets are bolted to column allowing sharing of vertical and horizontal forces.

Ⓔ A stub detail allows both gusset and beam web to be shop welded to column. Flanges of supported beam may be welded to transfer flexural and axial forces.

Ⓕ For beam moment connections, bolted webs can resist shear while welded flanges resist flexural and axial forces. (Moment connections must meet the requirements of Chapter E of the *Provisions*, as required.)

Fig. C-D2.1. Desirable details that avoid shared forces between welds and bolts.

3. Welded Joints

The general requirements for design of welded joints are specified in Chapter J of the *Specification*. Additional design requirements for specific systems or connection types are specified elsewhere in the Provisions. Section 7.3, Welded Joints, of the 2005 Provisions also invoked certain requirements for weld filler metal toughness and welding procedures. In these provisions, the requirements are specified in Sections A3.4 and I2.3.

4. Continuity Plates and Stiffeners

The available lengths for welds of continuity plates and stiffeners to the web and flanges of rolled shapes are reduced by the detailing requirements of AWS D1.8, Clause 4.1 as specified in Section I2.4 of the Provisions. See Figures C-D2.3 (a) and (b). These large corner clips are necessary to avoid welding into the k-area of wide flange shapes. See Section A3.1 commentary and AWS D1.8, clause 4 commentary for discussion.

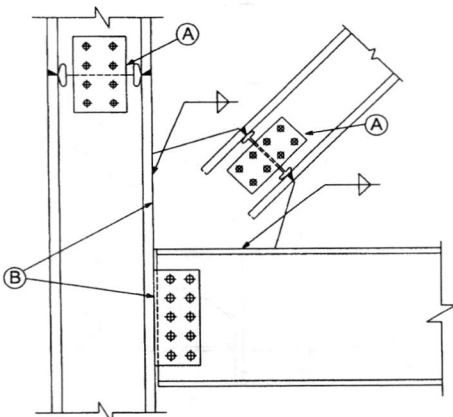

(A) Brace or column members should not be designed with a combination of bolted web and welded flanges resisting axial forces.

(B) Brace connections to columns with gussets welded to the column and the beam web bolted to the column will transfer forces differently from all-welded or all-bolted connections. The welded joint of the gusset to the column will tend to resist the entire vertical force at the column face (the vertical component of the brace force, plus the beam reaction). Also, the transfer of horizontal force through the bolted web to the column face will be precluded by the stiffer path through the welded joints of the gusset, so the gusset-to-beam joint will tend to resist the entire horizontal component of the brace force. Pass-through forces at beam-column connection will bypass the shear plate and go through the gusset. Equilibrium of the connection requires additional moments in both the beam and column, as well as higher forces in the welds of the gusset to the column and to the beam to transfer these forces.

Fig. C-D2.2. Problematic bolted/welded member connections.

5. Column Splices

5a. Location of Splices

Column splices should be located away from the beam-to-column connection to reduce the effects of flexure. For typical buildings, the 4 ft (1.2 m) minimum distance requirement will control. When splices are located 4 to 5 ft (1.2 to 1.5 m) above the floor level, field erection and construction of the column splice will generally be simplified due to improved accessibility and convenience. In general, it is recommended that the splice be within the middle third of the story height. For less typical buildings, where the floor-to-floor height is insufficient to accommodate this requirement, the splice should be placed as close as practicable to the midpoint of the clear distance between the finished floor and the bottom flange of the beam above. It is not intended that these column splice requirements be in conflict with applicable safety regulations, such as the OSHA *Safety Standards for Steel Erection* (OSHA, 2010) developed by the Steel Erection Negotiated Rulemaking Advisory Committee (SENRAC). This requirement is not intended to apply at columns that begin at a floor level, such as a transfer column, or columns that are interrupted at floor levels by cantilevered beams. However, the splice connection strength requirements of Section D2.5 still apply.

5b. Required Strength

Except for moment frames, the available strength of a column splice is required to equal or exceed both the required strength determined in Section D2.5b and the required strength for axial, flexural and shear effects at the splice location determined from load combinations stipulated by the applicable building code.

Partial-joint-penetration groove welded splices of thick column flanges exhibit virtually no ductility under tensile loading (Popov and Stephen, 1977; Bruneau et al., 1987). Consequently, column splices made with partial-joint-penetration groove welds require a 100% increase in required strength and must be made using weld metal with minimum Charpy V-notch (CVN) toughness properties.

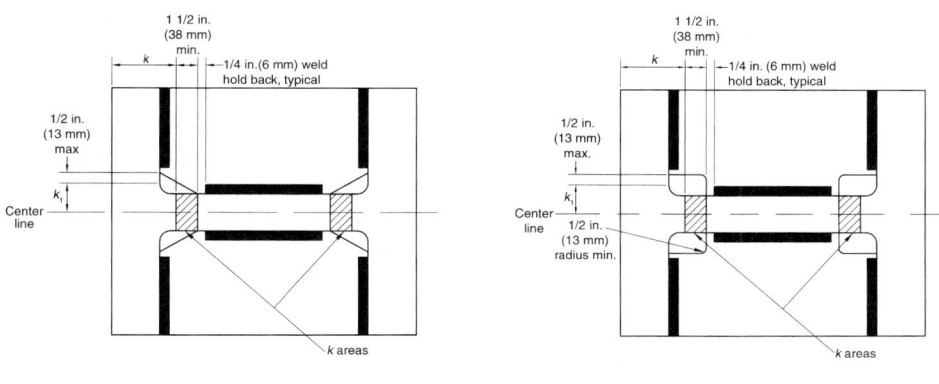

(a) Straight corner clip   *(b) Curved corner clip*

Fig. C-D2.3. Configuration of continuity plates.

The calculation of the minimum available strength in Section D2.5b(2) includes the ratio R_y. This results in a minimum available strength that is not less than 50% of the expected yield strength of the column flanges. A complete-joint-penetration (CJP) groove weld may be considered as satisfying this requirement. However, when applicable, tapered transitions are required in order to relieve stress concentrations where local yielding could occur at changes in column flange width or thickness per Section D2.5b(3). Tensile stresses are to be calculated by adding the uniform axial stress with the elastic bending stress or stresses, using the elastic section modulus, S, for both LRFD and ASD.

The possible occurrence of tensile loads in column splices utilizing partial-joint-penetration (PJP) groove welds during a maximum considered earthquake should be evaluated. When tensile loads are possible, it is suggested that some restraint be provided against relative lateral movement between the spliced column shafts. For example, this can be achieved with the use of flange splice plates. Alternatively, web splice plates that are wide enough to maintain the general alignment of the spliced columns can be used. Shake-table experiments have shown that when columns that are unattached at the base reseat themselves after lifting, the performance of a steel frame remains tolerable (Huckelbridge and Clough, 1977).

These provisions are applicable to common frame configurations. Additional considerations may be necessary when flexure dominates over axial compression in columns in moment frames, and in end columns of tall narrow frames where overturning forces can be very significant. The designer should review the conditions found in columns in buildings with tall story heights when large changes in column sizes occur at the splice, or when the possibility of column buckling in single curvature over multiple stories exists. In these and similar cases, special column splice requirements may be necessary for minimum available strength and/or detailing.

Where CJP groove welds are not used, the connection is likely to be a PJP groove weld. The unwelded portion of the PJP groove weld forms a discontinuity that acts like a notch that can induce stress concentrations. A PJP groove weld made from one side could produce an edge crack-like notch (Barsom and Rolfe, 1999). A PJP groove weld made from both sides would produce a buried crack-like notch. The strength of such crack-like notches may be computed by using fracture mechanics methodology. Depending on the specific characteristics of the particular design configuration, geometry and deformation, the analysis may require elastic-plastic or plastic finite element analysis of the joint. The accuracy of the computed strength will depend on the finite element model and mesh size used, the assumed strength and fracture toughness of the base metal, heat affected zone and weld metal, and on the residual stress magnitude and distribution in the joint.

5c. Required Shear Strength

Inelastic analyses (FEMA, 2000f) of moment frame buildings have shown the importance of the columns that are not part of the SFRS in helping to distribute the seismic shears between the floors. Even columns that have beam connections considered to be pinned connections may develop large bending moments and shears due to nonuniform drifts of adjacent levels. For this reason, it is recommended that splices of such columns be adequate to develop the shear forces corresponding to these large column moments in both orthogonal directions. Accordingly, columns that are part of the SFRS must be connected for the greater of the forces resulting from these drifts, or the requirements specific to the applicable system in Chapters E, F, G or H.

FEMA 350 (FEMA, 2000a) recommends that: "Splices of columns that are not part of the seismic force resisting system should be made in the center one-third of the column height, and should have sufficient shear capacity in both orthogonal directions to maintain the alignment of the column at the maximum shear force that the column is capable of producing." The corresponding commentary suggests that this shear should be calculated assuming plastic hinges at the ends of the columns in both orthogonal directions.

Further review (Krawinkler, 2001) of nonlinear analyses cited in FEMA 355C (FEMA, 2000d) showed that, in general, shears in such columns will be less than one-half of the shear calculated from $2M_{pc}/H$, where M_{pc} is the nominal plastic flexural strength of the column and H is the height of the story. For this reason, Section D2.5c requires that the calculated shear in the splices be M_{pc}/H (LRFD) or $M_{pc}/(1.5H)$ (ASD).

5d. **Structural Steel Splice Configurations**

Bolted web connections are preferred by many engineers and contractors because they have advantages for erection, and when plates are placed on both sides of the web, whether they are bolted or welded, they are expected to maintain alignment of the column in the event of a flange splice fracture. A one-sided web plate may be used when it is designed as a back-up plate for a CJP web weld. This plate is also commonly used as a column erection aid. Partial-joint-penetration (PJP) groove welded webs are not recommended, because fracture of a flange splice would likely lead to fracture of the web splice, considering the stress concentrations inherent in such welded joints.

Weld backing for groove welds in column splices may remain. The justification for this is that unlike beam-to-column connections, splices of column flanges and webs using weld backing result in no transversely loaded notch.

6. **Column Bases**

Column bases must have adequate strength to permit the expected ductile behavior for which the system is designed in order for the anticipated performance to be achieved.

Column bases are required to be designed for the same forces as those required for the members and connections framing into them. If the connections of the system are required to be designed for the amplified seismic loads or loads based on member strengths, the connection to the column base must also be designed for those loads.

Column bases are considered to be column splices. The required strength of column bases include the requirements prescribed in Section D2.5.

It is necessary to decompose the required tension strength of connections of diagonal brace members to determine the axial and shear forces imparted on the column base.

The requirement for removal of weld tabs and weld backing at column to base plate connections made with groove welds has been added to Section D2.6 as it is applicable to all SFRS systems in Sections E, F, G and H. The use of weld backing for a CJP weld of a column to a base plate creates a transverse notch. Consequently weld backing must be removed. For OMF, IMF and SMF systems, weld backing is allowed to remain at the CJP welds of the top flange of beam-to-column moment connections if a fillet weld is added per Chapter 3 of ANSI/AISC 358 (AISC, 2010b). Similarly, an exception has been added for column bases to permit weld backing to remain at the inside flanges and at the webs of wide flange shapes when a reinforcing fillet weld is added between the backing bar and the base plate.

6a. Required Axial Strength

The required axial (vertical) strength of the column base is computed from the column required strength in Sections D1.4a and D2.5b, in combination with the vertical component of the required connection strength of any braces present.

6b. Required Shear Strength

The required shear (horizontal) strength of the column base is computed from the required strength in Section D2.5c, in combination with the horizontal component of the required connection strength of any braces present. An exception to the shear force per Section D2.5c is allowed for single story columns with simple connections at both ends as shear from story drift will not develop in columns where flexure cannot occur at both ends.

There are several possible mechanisms for shear forces to be transferred from the column base into the supporting concrete foundation. Surface friction between the base plate and supporting grout and concrete is probably the initial load path, especially if the anchor rods have been pretensioned. Unless the shear force is accompanied by enough tension to completely overcome the dead loads on the base plate, this mechanism will probably resist some or all of the shear force. However, many building codes prescribe that friction cannot be considered when resisting code earthquake loads, and another design calculation method must be utilized. The other potential mechanisms are anchor rod bearing against the base plates, shear keys bearing on grout in the grout pocket, or bearing of the column embedded in a slab or grade beam. See Figure C-D2.4.

Anchor rod bearing is usually considered in design and is probably sufficient consideration for light shear loads. It represents the shear limit state if the base plate has overcome friction and has displaced relative to the anchor rods. The anchor rods are usually checked for combined shear and tension. Bearing on the base plate may also be considered, but usually the base plate is so thick that this is not a problem. Note that oversize holes are typically used for anchor rods, and a weld washer may be required to transmit forces from the base plate to the anchor rods. Where shear is transferred through the anchor rods, anchor rods are subject to flexure.

A shear key should be considered for heavy shear loads, although welding and construction issues must be considered. If tension and/or overturning loads are present, anchor rods must also be provided to resist tension forces.

For foundations with large free edge distances, concrete blowout strength is controlled by concrete fracture; and the concrete capacity design (CCD) method prescribed in ACI 318 Appendix D provides a relatively accurate estimate of shear key concrete strength. For foundations with smaller edge distances, shear key concrete blowout strength is controlled by concrete tensile strength; and the 45° cone method prescribed in ACI 349 (ACI, 2006) and AISC Design Guide 1, *Base Plate and Anchor Rod Design* (AISC, 2010d) provides a reasonable estimate of shear key concrete strength. In recognition of limited physical testing of shear keys, it is recommended that the shear key concrete blowout strength be estimated by the lower of these two methods (Gomez et al., 2009).

Where columns are embedded, the bearing strength of the surrounding concrete can be utilized. Note that the concrete element must then be designed to resist this force and transfer it into other parts of the foundation or into the soil.

When the column base is embedded in the foundation, it can serve as a shear key to transfer shear forces. It is sometimes convenient to transfer shear forces to concrete grade beams through reinforcing steel welded to the column. Figure C-D2.5 shows two examples of shear transfer to a concrete grade beam. The reinforcing steel must be long enough to allow a splice with the grade beam reinforcing steel, allowing transfer of forces to additional foundations.

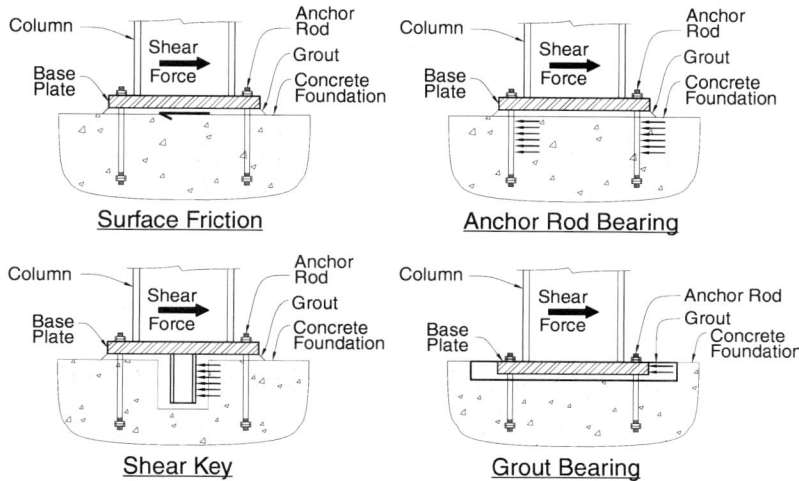

Fig. C-D2.4. Shear transfer mechanisms—column supported by foundation.

6c. **Required Flexural Strength**

Column bases for moment frames can be of several different types, as follows:

(1) A rigid base assembly may be provided which is strong enough to force yielding in the column. The designer should employ the same guidelines as given for the rigid fully restrained connections. Such connections may employ thick base plates, haunches, cover plates, or other strengthening as required to develop the column hinge. Where haunched type connections are used, hinging occurs above the haunch, and appropriate consideration should be given to the stability of the column section at the hinge. See Figure C-D2.6 for examples of rigid base

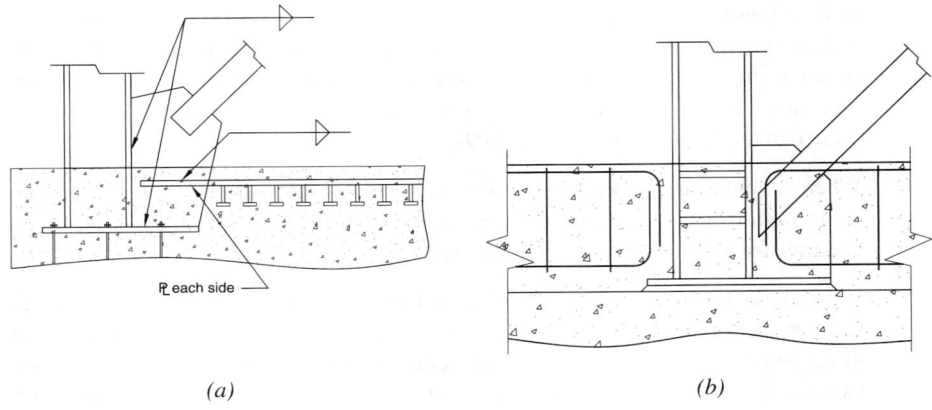

Fig. C-D2.5. Examples of shear transfer to a concrete grade beam.

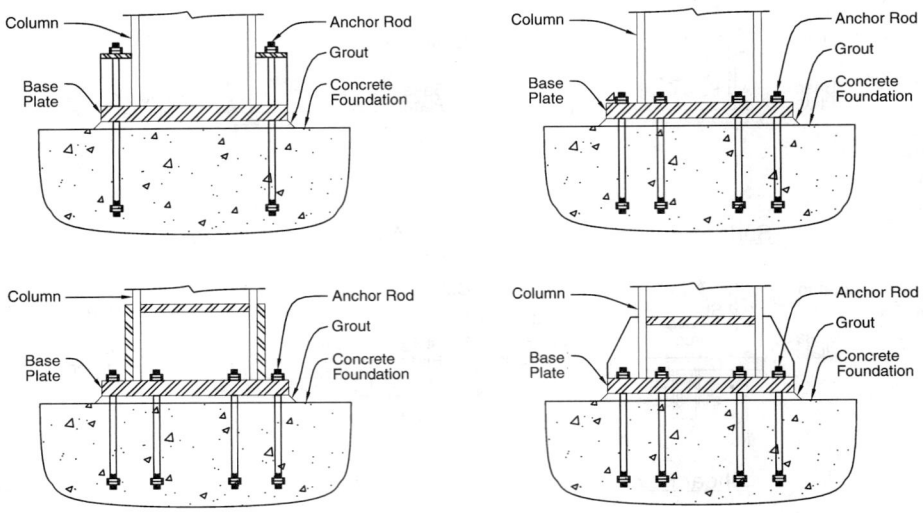

Fig. C-D2.6. Example "rigid base" plate assembly for moment frames.

assemblies that can be designed to be capable of forcing column hinging. In some cases, yielding can occur in the concrete grade beams rather than in the column. In this case the concrete grade beams should be designed in conformance with ACI 318, Chapter 21.

(2) Large columns may be provided at the bottom level to limit the drift, and a "pinned base" may be utilized. The designer should ensure that the required shear capacity of the column, base plate and anchor rods can be maintained up to the maximum rotation that may occur. It should be recognized, however, that without taking special measures, column base connections will generally provide partial rotational fixity.

(3) The column may continue below the assumed seismic base (for example, into a basement, crawl space or grade beam) in such a way that column fixity is assured without the need for a rigid base plate connection. The designer should recognize that hinging will occur in the column, just above the seismic base or in the grade beam. If hinging is considered to occur in the grade beam, then the grade beam should be designed in conformance with ACI 318, Chapter 21. The horizontal shear to be resisted at the ends of the column below the seismic base should be calculated considering the expected strength, $R_y F_y$, of the framing. See Figure C-D2.7 for examples of a column base fixed within a grade beam.

Based on experimental observations, the ultimate strength of the column base will be reached when any one of the following yielding scenarios is activated (Gomez et al., 2010):

(1) Flexural yielding of both the tension side and compression side of the base plate
(2) Axial yielding of the anchor rods on the tension side
(3) Crushing of the concrete or grout

Historically, both triangular concrete stress blocks and rectangular concrete stress blocks have been used for the analysis of column base plates, the rectangular stress blocks give the best agreement with test results (Gomez et al., 2010).

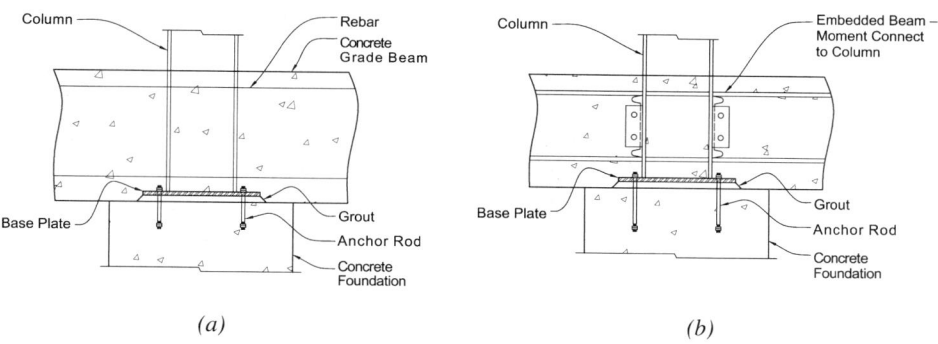

Fig. C-D2.7. Examples of column base fixity in a grade beam.

7. Composite Connections

The use of composite connections often simplifies some of the special challenges associated with traditional steel and concrete construction. For example, compared to structural steel, composite connections often avoid or minimize the use of field welding, and compared to reinforced concrete, there are fewer instances where anchorage and development of primary beam reinforcement is a problem.

Given the many alternative configurations of composite structures and connections, there are few standard details for connections in composite construction (Griffis, 1992; Goel, 1992a; Goel, 1993). However, tests are available for several connection details that are suitable for seismic design. References are given in this section of the Commentary and Commentary Chapters G and H. In most composite structures built to date, engineers have designed connections using basic mechanics, equilibrium, existing standards for steel and concrete construction, test data, and good judgment. The provisions in this section are intended to help standardize and improve design practice by establishing basic behavioral assumptions for developing design models that satisfy equilibrium of internal forces in the connection for seismic design.

General Requirements
The requirements for deformation capacity apply to both connections designed for gravity load only and connections that are part of the seismic force resisting system. The ductility requirement for gravity load only connections is intended to avoid failure in gravity connections that may have rotational restraint but limited rotation capacity. For example, Figure C-D2.8 shows a connection between a reinforced concrete wall and steel beam that is designed to resist gravity loads and is not considered to be part of the seismic force resisting system. However, this connection is required to be designed to maintain its vertical shear strength under rotations and/or moments that are imposed by inelastic seismic deformations of the structure.

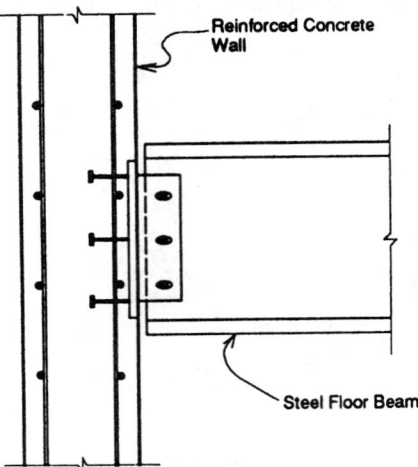

Fig. C-D2.8. Steel beam-to-reinforced concrete wall gravity load shear connection.

In calculating the required strength of connections based on the nominal strength of the connected members, allowance should be made for all components of the members that may increase the nominal strength above that usually calculated in design. For example, this may occur in beams where the negative moment strength provided by slab reinforcement is often neglected in design but will increase the moments applied through the beam-to-column connection. Another example is in filled HSS braces where the increased tensile and compressive strength of the brace due to concrete should be considered in determining the required connection strength. Because the evaluation of such conditions is case specific, these provisions do not specify any allowances to account for overstrength. However, as specified in Section A3.2, calculations for the required strength of connections should, as a minimum, be made using the expected yield strength of the connected steel member or of the reinforcing bars in the connected concrete or composite member.

Nominal Strength of Connections
In general, forces between structural steel and concrete will be transferred by a combination of bond, adhesion, friction and direct bearing. Transfers by bond and adhesion are not permitted for nominal strength calculation purposes because: (1) these mechanisms are not effective in transferring load under inelastic load reversals; and (2) the effectiveness of the transfer is highly variable depending on the surface conditions of the steel and shrinkage and consolidation of the concrete.

Transfer by friction shall be calculated using the shear friction provisions in ACI 318 where the friction is provided by the clamping action of steel ties or studs or from compressive stresses under applied loads. Since the provisions for shear friction in ACI 318 are based largely on monotonic tests, the values are reduced by 25% where large inelastic stress reversals are expected. This reduction is considered to be a conservative requirement that does not appear in ACI 318 but is applied herein due to the relative lack of experience with certain configurations of composite structures.

In many composite connections, steel components are encased by concrete that will inhibit or fully prevent local buckling. For seismic deign where inelastic load reversals are likely, concrete encasement will be effective only if it is properly confined. One method of confinement is with reinforcing bars that are fully anchored into the confined core of the member (using requirements for hoops in ACI 318, Chapter 21). Adequate confinement also may occur without special reinforcement where the concrete cover is very thick. The effectiveness of the latter type of confinement should be substantiated by tests.

For fully encased connections between steel (or composite) beams and reinforced concrete (or composite) columns such as shown in Figure C-D2.9, the panel zone nominal shear strength can be calculated as the sum of contributions from the reinforced concrete and steel shear panels (see Figure C-D2.10). This superposition of strengths for calculating the panel zone nominal shear strength is used in detailed design guidelines (Deierlein et al., 1989; ASCE, 1994; Parra-Montesinos and Wight, 2001) for composite connections that are supported by test data (Sheikh et al., 1989; Kanno and Deierlein, 1997; Nishiyama et al., 1990; Parra-Montesinos and Wight, 2001). Further information on the use and design of such connections is included in the commentary to Section G3.

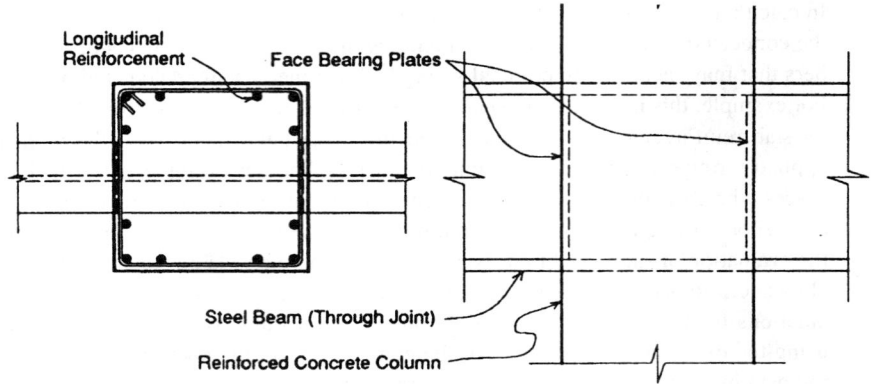

Fig. C-D2.9. Reinforced concrete column-to-steel beam moment connection.

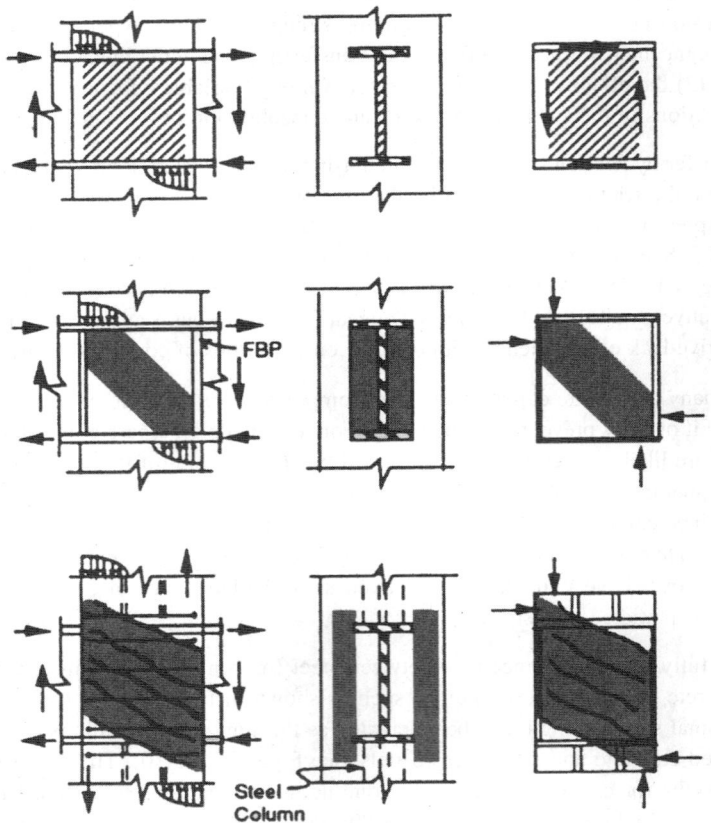

Fig. C-D2.10. Panel shear mechanisms in steel beam-to-reinforced concrete column connections (Deierlein et al., 1989)

Reinforcing bars in and around the joint region serve the dual functions of resisting calculated internal tension forces and providing confinement to the concrete. Internal tension forces can be calculated using established engineering models that satisfy equilibrium (for example, classical beam-column theory, the truss analogy, strut and tie models). Tie requirements for confinement usually are based on empirical models derived from test data and past performance of structures (ACI, 2002; Kitayama et al., 1987).

(1) In connections such as those in C-PRMF, the force transfer between the concrete slab and the steel column requires careful detailing. For C-PRMF connections (see Figure C-D2.11), the strength of the concrete bearing against the column flange should be checked (Green et al., 2004). Only the solid portion of the slab (area above the ribs) should be counted, and the nominal bearing strength should

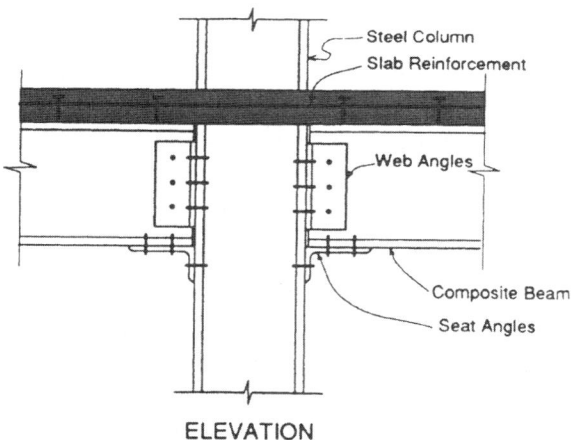

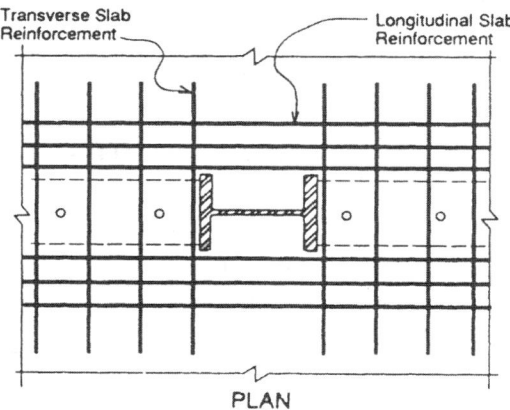

Fig. C-D2.11. Composite partially restrained connection.

be limited to $1.2 f'_c$ (Ammerman and Leon, 1990). In addition, because the force transfer implies the formation of a large compressive strut between the slab bars and the column flange, adequate transverse steel reinforcement should be provided in the slab to form the tension tie. From equilibrium calculations, this amount should be the same as that provided as longitudinal reinforcement and should extend at least 12 in. (305 mm) beyond either side of the effective slab width.

(2) Due to the limited size of joints and the congestion of reinforcement, it often is difficult to provide the reinforcing bar development lengths specified in ACI 318 for transverse column reinforcement in joints. Therefore, it is important to take into account the special requirements and recommendations for tie requirements as specified for reinforced concrete connections in ACI 318, Section 21.5 and in ACI (2002), Kitayama et al. (1987), Sheikh and Uzumeri (1980), Park et al., (1982), and Saatcioglu (1991). Test data (Sheikh et al., 1989; Kanno and Deierlein, 1997; Nishiyama et al., 1990) on composite beam-to-column connections similar to the one shown in Figure C-D2.9 indicate that the face bearing (stiffener) plates attached to the steel beam provide effective concrete confinement.

(3) As in reinforced concrete connections, large bond stress transfer of loads to column bars passing through beam-to-column connections can result in slippage of the bars under extreme loadings. Current practice for reinforced concrete connections is to control this slippage by limiting the maximum longitudinal bar sizes as described in ACI (2002).

8. Steel Anchors

At this time, there is insufficient data to generate specification requirements for the shear strength of stud anchors subjected to inelastic cyclic loads, although it is clear that some strength and stiffness reduction occurs with cycling (McMullin and Astaneh-Asl, 1994; Civjan and Singh, 2003). The degradation in behavior is particularly serious if the stud anchors are subjected to combined tension and shear (Saari et al., 2004). For composite members that are part of the SFRS in intermediate or special systems, a 25% reduction of the stud available strength given in the *Specification* is required to allow for the effect of cyclic loads if the studs are expected to yield.

D3. DEFORMATION COMPATIBILITY OF NON-SFRS MEMBERS AND CONNECTIONS

Members that are not part of the SFRS and their connections may incur forces in addition to gravity loads as a result of story deflection of the SFRS during a seismic event. Section 12.12.5 of ASCE/SEI 7 requires structural components that are not considered part of the SFRS to be able to resist the combined effects of gravity loads with any additional forces resulting from the design story drifts. The load effect due to the design story drift should be considered as an ultimate or factored load. Inelastic deformations of members and connections at these load levels are acceptable provided that instabilities do not result.

Nonuniform drifts of adjacent story levels may create significant bending moments in multistory columns. These bending moments will usually be greatest at story levels. Inelastic yielding of columns resulting from these bending moments can be accommodated when suitable lateral bracing is provided at story levels and when column shapes have adequate compactness (Newell and Uang, 2008). High shear forces at column splices resulting from these bending moments are addressed by the required shear strength requirements of Section D2.5c. The requirements for column splice location per Section D2.5a are intended to locate splices where bending moments are typically lower.

The P-Δ effect of the design story drift will also create additional axial forces due to column inclination in both single story and multistory columns. Connections of columns to beams or diaphragms should be designed to resist horizontal forces that result from the inclination of the columns. For single story columns, and multi story columns where the inclination is constant, only the effect of the beam reactions at the story level requires a horizontal thrust to create equilibrium at that story level. However for multistory columns where the column inclination changes at a level, the entire column axial force requires a horizontal thrust for equilibrium. See Figure C-D3.1 for comparison of effect of column inclination on horizontal force at story level. Likewise unequal drifts in multistory columns induce both flexure and shear in the column. Flexure will not be induced in columns with constant inclination and simple connections to beams.

Equivalent lateral force analysis methods have not been developed with an eye toward accurately estimating differences in story drift. Use of a modal response

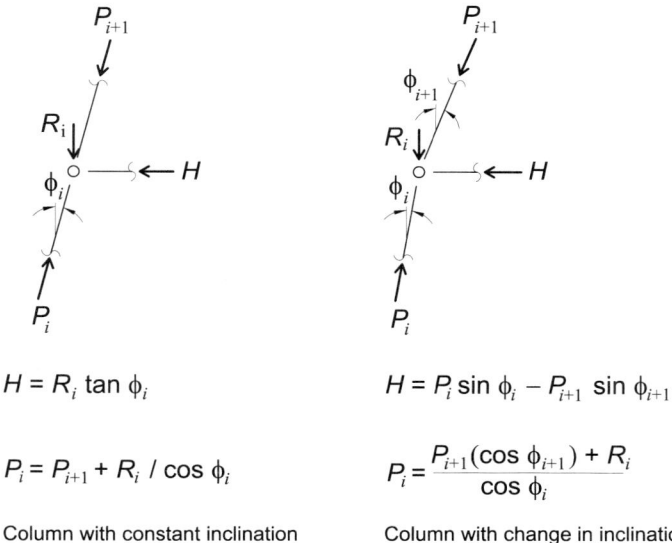

Fig. C-D3.1. *Effect of column inclination on horizontal story force.*

spectrum analysis to estimate differences in story drift is also problematic as this quantity is not tracked mode by mode in typical software. However, column shear can be tracked modally. Also, the horizontal thrust can be determined by detaching the column from the diaphragm and introducing a link element. Alternatively, thrust can be calculated from the change in column inclination, which can be estimated from the moment (and can be tracked mode by mode).

Properly designed simple connections are required at beam-to-column joints to avoid significant flexural forces. As per Section J1 of the *Specification*, inelastic deformation of the connections is an acceptable means of achieving the required rotation. Standard shear connections per Part 10 of the AISC *Steel Construction Manual* (2005c) can be considered to allow adequate rotation at the joints without significant flexural moments. Double angles supporting gravity loads have been shown to attain maximum rotations of 0.05 to 0.09 rad and are suitable for combined gravity and axial forces as are WT connections which have demonstrated rotations of 0.05 to 0.07 rad (Astaneh-Asl, 2005a). Shear tabs (single plates), while inherently more rigid than double angles, have been shown to withstand gravity rotations ranging from 0.026 to 0.103 rad, and cyclic rotations of 0.09 rad (Astaneh-Asl, 2005b). Note that reducing the number of bolts in shear plates and consequently the connection depth increases the maximum rotation. Other connections at beam-to-column joints are acceptable if they are configured to provide adequate rotational ductility. Part 9 of the AISC *Steel Construction Manual* provides guidance on rotational ductility of end plate and WT connections that can be applied to many types of connections to ensure ductile behavior.

Beams and columns connected with moment connections that develop large flexural stresses as a result of story drift should be considered as part of the SFRS and accordingly shall be subject to the requirements of the Provisions.

D4. H-PILES

The provisions on seismic design of H-piles are based on the data collected on the actual behavior of H-piles during recent earthquakes, including the 1994 Northridge earthquake (Astaneh-Asl et al., 1994) and the results of cyclic tests of full-scale pile tests (Astaneh-Asl and Ravat, 1997). In the test program, five full size H-Piles with reinforced concrete pile caps were subjected to realistic cyclic vertical and horizontal displacements expected in a major earthquake. Three specimens were vertical piles and two specimens were batter piles. The tests established that during cyclic loading for all three vertical pile specimens a very ductile and stable plastic hinge formed in the steel pile just below the reinforced concrete pile cap. When very large inelastic cycles were applied, local buckling of flanges within the plastic hinge area occurred. Eventually, low cycle fatigue fracture of flanges or overall buckling of the pile occurred. However, before the piles experienced fracture through locally buckled areas, vertical piles tolerated from 40 to 65 large inelastic cyclic vertical and horizontal displacements with rotation of the plastic hinge exceeding 0.06 rad for more than 20 cycles.

1. Design Requirements

Prior to an earthquake, piles, particularly vertical piles, are primarily subjected to gravity axial load. During an earthquake, piles are subjected to horizontal and vertical displacements as shown in Figure C-D4.1. The horizontal and vertical displacements of piles generate axial load (compression and possibly uplift tension), bending moment, and shear in the pile.

During tests of H-piles, realistic cyclic horizontal and vertical displacements were applied to the pile specimens. Figure C-D4.2 shows test results in terms of axial load and bending moment for one of the specimens. Based on the performance of test specimens, it was concluded that H-piles should be designed following the provisions of the *Specification* regarding members subjected to combined loads.

2. Battered H-Piles

The vertical pile specimens demonstrated very large cyclic ductility as well as considerable energy dissipation capacity. A case study of performance of H-piles during the 1994 Northridge earthquake (Astaneh-Asl et al., 1994) indicated excellent performance for pile groups with vertical piles only. However, the battered pile specimens did not show as much ductility as the vertical piles. The battered piles tolerated from 7 to 17 large inelastic cycles before failure. Based on relatively limited information on actual seismic behavior of battered piles, it is possible that during a major earthquake, battered piles in a pile group fail and are no longer able

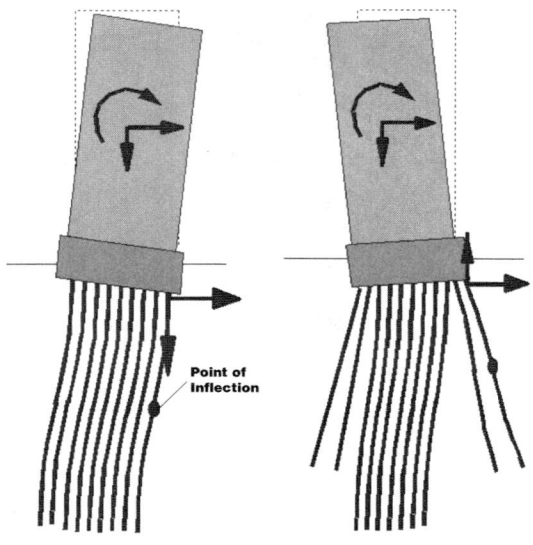

(a) Vertical Piles Only   *(b) Vertical and Battered Piles*

Fig. C-D4.1. Deformations of piles and forces acting on an individual pile.

to support the gravity load after the earthquake. Because of this possibility, the use of battered piles to carry gravity loads is discouraged. Unless, through realistic cyclic tests, it is shown that battered piles will be capable of carrying their share of the gravity loads after a major earthquake, the vertical piles in seismic design categories D, E and F should be designed to support the gravity load alone, without participation of the batter piles.

3. **Tension**

   Due to overturning moment, piles can be subjected to tension. Piles subjected to tension should have sufficient mechanical attachments within their embedded area to transfer the tension force in the pile to the pile cap or foundation.

4. **Protected Zone**

   Since it is anticipated that during a major earthquake, a plastic hinge is expected to form in the pile just under the pile cap or foundation, the use of mechanical attachment and welds over a length of pile below the pile cap equal to the depth of the pile cross section is prohibited. This region is therefore designated as a protected zone.

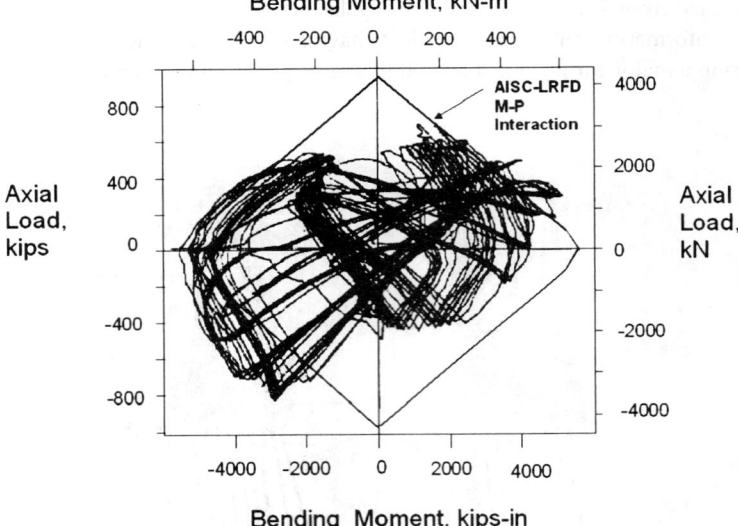

Fig. C-D4.2. Axial load-moment interaction for H-pile test.

CHAPTER E

MOMENT-FRAME SYSTEMS

E1. ORDINARY MOMENT FRAMES (OMF)

1. Scope

Ordinary moment frames (OMF) resist lateral load by rigid frame action in a system where flexure and shear dominate both the elastic and inelastic response of the beams and columns, and where moment resisting beam-to-column connections are provided. OMF must satisfy all the applicable requirements of Chapters A, B, C, D, I and J of these Provisions, as well as the requirements in Section E1. The requirements of Section E1 deal primarily with the design of beam-to-column connections.

2. Basis of Design

Compared to intermediate moment frame (IMF) and special moment frame (SMF) systems, OMF are expected to provide only minimal levels of inelastic deformation capacity. To compensate for this lower level of ductility, OMF are designed to provide larger lateral strength than IMF and SMF, and thus, are designed using a lower R factor. Systems such as OMF with high strength and low ductility have seen much less research and testing than higher ductility systems. Consequently, the design requirements for OMF are based much more on judgment than on research. Due to the limited ductility of OMF and due to the limited understanding of the seismic performance of these systems, ASCE/SEI 7 (ASCE, 2010) places significant height and other limitations on their use.

Although the design basis for OMF is to provide for minimal inelastic deformation capacity, there is no quantitative definition of "minimal inelastic deformation capacity." Despite the lack of a quantitative inelastic deformation requirement, the overall intent of OMF design is to avoid highly brittle behavior in its response to lateral load.

To provide for minimal inelastic deformation capacity, i.e., to avoid highly brittle behavior, the general intent of the OMF design provisions is that connection failure should not be the first significant inelastic event in the response of the frame to earthquake loading. This is based on the view that connection failure, in general, is one of the less ductile failure modes exhibited by structural steel frames. Thus, as lateral load is increased on an OMF, the intent is that the limit of elastic response be controlled by limit states other than connection failure, such as reaching the limiting flexural or shear strength of a beam or a column, reaching the limiting shear strength of the panel zone, etc. For higher ductility systems such as IMF and SMF, inelasticity is intended to occur in specific frame elements. For example, in SMF, inelasticity is intended to occur primarily in the form of flexural yielding of the beams. This is not the case with OMF, where the initial inelastic response is permitted to occur in any frame element.

Thus, the basic design requirement for an OMF is to provide a frame with strong connections. That is, connections should be strong enough so that, as noted above, connection failure is not the first significant inelastic event in the response of the frame to earthquake loading. This applies to all connections in the frame, including beam-to-column connections, column splices, and column base connections. Requirements for OMF column splices and column base connections are covered in Section D2. Requirements for beam-to-column connections are covered in Section E1.6.

There is an exception where initial inelastic response of an OMF is permitted to occur in beam-to-column connections. This is for OMF provided with partially restrained (PR) moment connections. Requirements for PR moment connections are covered in Section E1.6c.

Design and detailing requirements for OMF are considerably less restrictive than for IMF and SMF. The OMF provisions are intended to cover a wide range of moment frame systems that are difficult or impossible to qualify as IMF or SMF. This includes, for example, metal building systems, knee-braced frames, moment frames where the beams and/or columns are trusses (but not STMF), moment frames where the beams and/or columns are HSS, etc.

OMF Knee-Brace Systems. Knee-brace systems use an axial brace from the beam to the column to form a moment connection. Resistance to lateral loads is by flexure of the beam and column. These systems can be designed as an OMF. The knee-brace system can be considered as analogous to a moment frame with haunch-type connections. The knee brace carries axial force only, while the beam-to-column connection carries both axial force and shear. A design approach for knee-braced systems is to connect the beam end to the column and the brace ends based on the forces required to develop $1.1R_yM_p$ (LRFD) of the beam, where R_y is the ratio of the expected yield stress to the specified minimum yield stress, and M_p is the nominal plastic flexural strength, per Section E1.6b(a), at the location of the brace-to-beam work point. The beam-to-column connection, knee-brace connections, and knee-brace member design should then be designed for the resulting forces. The column and beams should be braced out of plane, either directly or indirectly at the knee brace locations, consistent with the requirements of Appendix 6 of the *Specification*.

OMF Truss Systems. In some moment frame configurations, trusses are used for the beam elements in place of rolled shapes. These systems can be designed as a special truss moment frame (STMF) following the requirements of Section E4. However, in cases where frame geometry or other restrictions preclude the use of a special truss moment frame STMF, these systems can also be designed as an OMF. As an OMF, a design approach would be to design the truss and the truss-to-column connections for the maximum force that can be transferred by the system, consistent with the requirements of Section E1.6b(b). The maximum force that can be delivered to the truss and truss-to-column connections can be based on the flexural capacity of the columns, taken as $1.1R_yM_{p\text{-}column}$ (LRFD), combined with vertical loads from the prescribed load combinations. Thus, the intent is to design a weak column system where inelasticity is expected to occur in the columns. The column should be braced out of plane, either directly or indirectly at the location of the top and bottom chord connection of the truss, consistent with the requirements of Appendix 6 of the *Specification*.

4. System Requirements

Unlike SMF, there is no beam-column moment ratio (i.e., strong column-weak beam) requirement for OMF. Consequently, OMF systems can be designed so that inelasticity will occur in the columns.

5. Members

There are no special restrictions or requirements on member width-to-thickness ratios or member stability bracing, beyond meeting the requirements of the *Specification*. Although not required, the judicious application of width-to-thickness limits and member stability bracing requirements as specified for moderately ductile members in Section D1 would be expected to improve the performance of OMF.

6. Connections

For all moment frame systems designed according to these Provisions, including SMF, IMF and OMF, the beam-to-column connections are viewed as critical elements affecting the seismic performance of the frame. For SMF and IMF systems, connection design must be based on qualification testing per Section K2 or through the use of a connection prequalified per Section K1. For OMF, connections need not be prequalified nor qualified by testing. Rather, design of beam-to-column connections can be based on strength calculations or on prescriptive requirements. Design and detailing requirements for beam-to-column connections in OMF are provided in this section.

6b. FR Moment Connections

Three options are provided in this section for design of FR moment connections. Designs satisfying any one of these three options are considered acceptable. Note that for all options, the required shear strength of the panel zone may be computed from the basic code prescribed loads, with the available shear strength computed using Equations J10-11 and J10-12 of the *Specification*. This may result in a design where initial yielding of the frame occurs in the panel zones. This is viewed as acceptable behavior due to the high ductility exhibited by panel zones.

(a) The first option permits the connection to be designed for the flexural strength of the beam, taken as $1.1R_yM_p$ (LRFD) or $(1.1/1.5)R_yM_p$ (ASD) of the beam. The 1.1 factor in the equation accounts for limited strain hardening in the beam and other possible sources of overstrength. The required shear strength of the connection is computed using the code prescribed load combinations, where the shear due to earthquake loading is computed per Equation E1-1. The available strength of the connection is computed using the *Specification*. Note that satisfying these strength requirements may require reinforcing the connection using, for example, cover plates or haunches attached to the beam. The required flexural strength of the connection specified in this section, i.e., $1.1R_yM_p$ (LRFD) or $(1.1/1.5)R_yM_p$ (ASD) of the beam, should also be used when checking if continuity plates are needed per Sections J10.1 through J10.3 in the *Specification*. However, this value of bending moment need not be used when determining the required shear strength of the column panel zone. As noted above, the required shear strength of the panel zone may be computed using the basic code prescribed loads.

(b) The second option permits design of the connection for the maximum moment and shear that can be transferred to the connection by the system. Factors that can limit the forces transferred to the connection include column yielding, panel zone yielding, foundation uplift, or the limiting earthquake force using $R = 1$. In the case of column yielding, the forces at the connection can be computed assuming the column reaches a limiting moment of $1.1R_yM_{p\text{-}column}$ for LRFD, or this value divided by 1.5 for ASD. In the case of panel zone yielding, the forces at the connection can be computed assuming the shear force in the panel zone is $1.1R_y$ times the shear given by Equations J10-11 and J10-12 in the *Specification* for LRFD, or this value divided by 1.5 for ASD. For frames with web-tapered members, as typically used in metal building systems, the flexural strength of the beam (rafter) or column will typically be first reached at some distance away from the connection. For such a case, the connection can be designed for the forces that will be generated when the flexural strength of a member is first reached anywhere along the length of the member. The nominal flexural strength of the member at the critical location should be increased by $1.1R_y$ to determine the required strength of the connection.

(c) The third option for beam-to-column connections is a prescriptive option for cases where a wide flange beam is connected to the flange of a wide flange column. The prescriptive connection specified in the section is similar to the welded unreinforced flange-bolted web (WUF-B) connection described in FEMA 350 (FEMA, 2000a). Some of the key features of this connection include the treatment of the complete-joint-penetration (CJP) beam flange to column welds as demand critical, treatment of backing bars and weld tabs using the same requirements as for SMF connections, and the use of special weld access hole geometry and quality requirements. Testing has shown that connections satisfying these requirements can develop moderate levels of ductility in the beam or panel zone prior to connection failure (Han et al., 2007).

Option (c) also permits the use of any connection in OMF that is permitted in IMF or SMF systems. Thus, any of the prequalified IMF or SMF connections in ANSI/AISC 358 can be used in OMF. However, when using ANSI/AISC 358 connections in an OMF, items specified in ANSI/AISC 358 that are not otherwise required in OMF systems are not required. For example, the WUF-W connection prequalified in ANSI/AISC 358 can be used for an OMF connection. However, items specified in ANSI/AISC 358 that would not be required when a WUF-W connection is used in an OMF include beam and column width-to-thickness limitations for IMF and SMF, beam stability bracing requirements for IMF or SMF, beam-column moment ratio requirements for SMF, column panel zone shear strength requirements for IMF or SMF, or requirements for a protected zone.

None of these items are required for OMF, and therefore are not required when the WUF-W connection is used as an OMF. Similar comments apply to all connections prequalified in ANSI/AISC 358.

6c. PR Moment Connections

Section E1.6c gives strength requirements for PR Connections, but does not provide complete prescriptive design requirements. For design information on PR connections, the reader is referred to Leon (1990); Leon (1994); Leon and Ammerman (1990); Leon and Forcier (1992); Bjorhovde et al. (1990); Hsieh and Deierlein (1991); Leon et al., (1996); and FEMA 355D (FEMA, 2000e).

E2. INTERMEDIATE MOMENT FRAMES (IMF)

1. Scope

IMF must satisfy all the applicable requirements of Chapters A, B, C, D, I and J of these Provisions, as well as the requirements in Section E2.

2. Basis of Design

IMF are intended to provide limited levels of inelastic rotation capacity and are based on tested designs. Due to the lesser rotational capacity of IMF as compared to SMF, ASCE/SEI 7 requires use of a lower seismic response modification coefficient, R, than that for SMF and places significant height and other limitations on its use.

While the design for SMF is intended to limit the majority of the inelastic deformation to the beams, the inelastic drift capability of IMF is permitted to be derived from inelastic deformations of beams, columns and panel zones.

The IMF connection is based on a tested design with a qualifying story drift angle of 0.02 rad. It is assumed that this limited connection rotation will be achieved by use of larger frame members than would be required in an SMF, because of the lower R and/or higher C_d/R values used in design.

Commentary Section E3 for SMF offers additional commentary relevant to IMF.

4. System Requirements

4a. Stability Bracing of Beams

See Commentary Section D1.2a on stability bracing of moderately ductile members and Commentary Section E3.4b for additional commentary.

5. Members

5a. Basic Requirements

This section refers to Section D1, which provides requirements for connection of webs to flanges as for built-up members and requirements for width-to-thickness ratios for the flanges and webs of the members. Because the rotational demands on IMF beams and columns are expected to be lower than for SMF, the width-to-thickness limitations for IMF are less severe than for SMF. See Commentary Section E3.5a for further discussion.

5b. **Beam Flanges**

The requirements in this Section are identical to those in Section E3.5b. See Commentary on Section E3.5b.

5c. **Protected Zones**

For commentary on protected zones see Commentary Section D1.3.

6. **Connections**

6a. **Demand Critical Welds**

The requirements in this Section are identical to those in Section E3.6a. See Commentary on Section E3.6a.

6b. **Beam-to-Column Connection Requirements**

The minimum story drift angle required for qualification of IMF connections is 0.02 rad while that for SMF connections is 0.04 rad. This level of story drift angle has been established for this type of frame based on engineering judgment applied to available tests and analytical studies, primarily those included in FEMA (2000d) and FEMA (2000f).

ANSI/AISC 358 (AISC, 2010b) describes six different connections that have been prequalified for use in both IMF and SMF systems. The prequalified connections include the reduced beam section (RBS), the bolted unstiffened extended end plate (BUEEP), the bolted stiffened extended end plate (BSEEP), the bolted flange plate (BFP), the welded unreinforced flange-welded web (WUF-W), and the Kaiser bolted bracket (KBB). In a few cases, the limitations on use of the connections are less strict for IMF than for SMF, but generally, the connections are the same.

The Commentary on the 2005 Provisions included a lengthy discussion regarding the use of a connection with welded unreinforced flanges and a bolted web, commonly referred to as the WUF-B connection. This connection is described in detail in FEMA 350, Section 3.5.1 (FEMA, 2000a). This connection is not included in ANSI/AISC 358-10 and therefore, for use in an IMF, it must be qualified in accordance with either Section K1 or K2 of the Provisions.

6c. **Conformance Demonstration**

The requirements for conformance demonstration for IMF connections are the same as for SMF connections, except that the required story drift angle is smaller. Refer to Commentary Section E3.6c.

6d. **Required Shear Strength**

The requirements for shear strength of the connection are the same for IMF as for SMF. See Commentary Section E3.6d for commentary.

6e. Panel Zone

The panel zone for IMF is required to be designed according to Section J10.6 of the *Specification*, with no further requirements in the Provisions. As noted in the commentary to Section E2.2, panel zone yielding is permitted as part of the inelastic action contributing to the drift capacity of the IMF and the requirements of the *Specification* are considered adequate for the expected performance.

6f. Continuity Plates

The requirements in this Section are identical to those in Section E3.6f. See Commentary Section E3.6f for further discussion.

6g. Column Splices

The requirements in this Section are identical to those in Section E3.6g. See Commentary Section E3.6g for further discussion.

E3. SPECIAL MOMENT FRAMES (SMF)

1. Scope

Special moment frames (SMF) must satisfy all the applicable requirements of Chapters A, B, C, D, I and J of these Provisions, as well as the requirements in Section E3.

2. Basis of Design

SMF are generally expected to experience significant inelastic deformations during large seismic events. It is expected that most of the inelastic deformation will take place as rotation in beam "hinges," with limited inelastic deformation in the panel zone of the column. The beam-to-column connections for these frames are required to be qualified based on tests that demonstrate that the connection can sustain a story drift angle of at least 0.04 rad based on a specified loading protocol. Other provisions are intended to limit or prevent excessive panel zone distortion, column hinging, and local buckling that may lead to inadequate frame performance in spite of good connection performance.

ANSI/AISC 358 (AISC, 2010b) provides requirements for six prequalified connections that are permitted to be employed in SMF systems. If connection types to be used in the structure do not meet the configurations or limitations therein, they are required to be prequalified per Section K1, or qualified per Section K2.

Since SMF and IMF connection configurations and design procedures are based on the results of qualifying tests, the configurations of connections in the prototype structure must be consistent with the tested configurations. Similarly, the design procedures used in the prototype connections must be consistent with the test specimens. Also, material properties of the test specimens must fairly represent the prototype connections. For connections included in ANSI/AISC 358, specific requirements are spelled out therein. Refer to the commentary for Sections K1 and K2 for more discussion on this topic.

4. System Requirements

4a. Moment Ratio

The strong-column weak-beam (SC/WB) concept is perhaps one of the least understood seismic provisions in steel design. It is often mistakenly assumed that it is formulated to prevent any column flange yielding in a frame, and that if such yielding occurs, the column will fail. Tests have shown that yielding of columns in moment frame sub-assemblages does not necessarily reduce the lateral strength at the expected seismic displacement levels.

The SC/WB concept is more of a global frame concern than a concern at the interconnections of individual beams and columns. Schneider et al. (1991) and Roeder (1987) showed that the real benefit of meeting SC/WB requirements is that the columns are generally strong enough to force flexural yielding in beams in multiple levels of the frame, thereby achieving a higher level of energy dissipation in the system. Weak column frames, particularly those with weak or soft stories, are likely to exhibit an undesirable response at those stories with the highest column demand-to-capacity ratios.

It should be noted that compliance with the SC/WB concept and Equation E3-1 gives no assurance that individual columns will not yield, even when all connection locations in the frame comply. Nonlinear response history analyses have shown that, as the frame deforms inelastically, points of inflection shift and the distribution of moments varies from the idealized condition. Nonetheless, yielding of the beams rather than the columns will predominate and the desired inelastic performance will, in general, be achieved in frames with members sized to meet the requirement in Equation E3-1.

Early formulations of the SC/WB relationship idealized the beam/column intersection as a point at the intersection of the member centerlines. Post-Northridge beam-to-column moment connections are generally configured to shift the plastic hinge location into the beam away from the column face and a more general formulation was needed. ANSI/AISC 358 provides procedures to calculate the location of plastic hinges for the connections included therein. For other configurations, the locations can be determined from the applicable qualifying tests. Recognition of expected beam strength (see Commentary Section A3.2) is also incorporated into Equation E3-1.

Three exceptions to Equation E3-1 are given. In the first exception, columns with low axial loads used in one-story buildings or in the top story of a multi-story building need not meet Equation E3-1 because concerns for inelastic soft or weak stories are not significant in such cases. Additionally, exception is made for a limited percentage of columns with axial loads that are considered to be low enough to limit undesirable performance while still providing reasonable design flexibility where the requirement in Equation E3-1 would be impractical, such as at large transfer girders. Finally, Section E3.4a provides an exception for columns in levels that are significantly stronger than in the level above because column yielding at the stronger level would be unlikely.

In applying Equation E3-1, recognition should be given to the location of column splices above the girder-to-column connection being checked. When the column splice is located at 4 feet or more above the top of the girder, it has been customary to base the calculation on the column size that occurs at the joint. If the column splice occurs closer to the top of the beam, or when the column above the splice is much smaller than that at the joint, consideration should be given to whether the column at the joint is capable of providing the strength assumed using the customary approach.

4b. Stability Bracing of Beams

See Commentary Section D1.2b on stability bracing of highly ductile members.

In addition to bracing along the beam length, the provisions of Section D1.2c call for the placement of lateral bracing near the location of expected plastic hinges. Such guidance dates to the original development of plastic design procedures in the early 1960s. In moment frame structures, many connection details attempt to move the plastic hinge a short distance away from the beam-to-column connection. Testing carried out as part of the SAC program (FEMA, 2000a) indicated that the bracing provided by typical composite floor slabs is adequate to avoid excessive strength deterioration up to the required story drift angle of 0.04 rad. Therefore, the FEMA recommendations do not require the placement of supplemental lateral bracing at plastic hinge locations adjacent to column connections for beams with composite floor construction. These provisions allow the placement of lateral braces to be consistent with the tested connections that are used to justify the design. For conditions where drifts larger than the anticipated 0.04 rad are anticipated or improved performance is desired, the designer may decide to provide additional lateral bracing near these plastic hinges. If lateral braces are provided, they should provide an available strength of 6% of the expected capacity of the beam flange at the plastic hinge location. If a reduced beam section connection detail is used, the reduced flange width may be considered in calculation of the bracing force. The requirements of Section E3.5c, Protected Zones, should be considered when placing bracing connections.

4c. Stability Bracing at Beam-to-Column Connections

Columns of SMF are required to be braced to prevent rotation out of the plane of the moment frame because of the anticipated inelastic behavior in, or adjacent to, the beam-to-column connection during high seismic activity.

(1) Braced Connections

Beam-to-column connections are usually braced laterally by the floor or roof framing. When this is the case and it can be shown that the column remains elastic outside of the panel zone, lateral bracing of the column flanges is required only at the level of the top flanges of the beams. If it cannot be shown that the column remains elastic, lateral bracing is required at both the top and bottom beam flanges because of the potential for flexural yielding, and consequent lateral-torsional buckling of the column.

The required strength for lateral bracing at the beam-to-column connection is 2% of the nominal strength of the beam flange. In addition, the element(s) providing lateral bracing should provide adequate stiffness to inhibit lateral movement of the column flanges (Bansal, 1971). In some cases, a bracing member will be required for such lateral bracing (direct stability bracing). Alternatively, calculations may show that adequate lateral bracing can be provided by the column web and continuity plates or by the flanges of perpendicular beams (indirect stability bracing).

The 1997 Provisions (AISC, 1997b) required column lateral bracing when the ratio in Equation E3-1 was less than 1.25. The intent of this provision was to require bracing to prevent lateral-torsional buckling for cases where it cannot be assured that the column will not hinge. Studies utilizing inelastic analyses (Gupta and Krawinkler, 1999; Bondy, 1996) have shown that, in severe earthquakes, plastic hinging can occur in the columns even when this ratio is significantly larger than 1.25. (See also discussion under Section E3.4a). The revised limit of 2.0 was selected as a reasonable cutoff because column plastic hinging for values greater than 2.0 only occurs in the case of extremely large story drifts. The intent of the revisions to this Section is to encourage appropriate bracing of column flanges rather than to force the use of much heavier columns, although other benefits may accrue by use of heavier columns, including possible elimination of continuity and doubler plates that may offset the additional material cost.

(2) Unbraced Connections

Unbraced connections occur in special cases, such as in two-story frames, at mechanical floors or in atriums and similar architectural spaces. When such connections occur, the potential for out-of-plane buckling at the connection should be minimized. Three provisions are given for the columns to limit the likelihood of column buckling.

5. Members

5a. Basic Requirements

Reliable inelastic deformation capacity for highly ductile members requires that width-to-thickness ratios of projecting elements be limited to a range that provides a cross section resistant to local buckling well into the inelastic range. Although the width-to-thickness ratios for compact elements in *Specification* Table B4.1 are sufficient to prevent local buckling before the onset of yielding, available test data suggest that these limits are not adequate for the required inelastic rotations in SMF. The limits given in Table D1.1 of the Provisions are deemed adequate for the large ductility demands to which these members may be subjected (Sawyer, 1961; Lay, 1965; Kemp, 1986; Bansal, 1971).

5b. Beam Flanges

Abrupt changes in beam flange area in locations of high strain, as occurs in plastic hinge regions of SMF, can lead to fracture due to stress concentrations. For connections such as the reduced beam section (RBS), the gradual flange area reduction, when properly configured and fabricated can be beneficial to the beam and connection performance. Such conditions are permitted when properly substantiated by testing.

5c. Protected Zones

For commentary on protected zones see Commentary Section D1.3.

6. Connections

6a. Demand Critical Welds

For general commentary on demand critical welds see Commentary Section A3.4.

The requirement to use demand critical welds for complete-joint-penetration (CJP) groove welded joints in beam-to-column connections of SMF was first included in the 2002 Provisions (AISC, 2002). The requirement for notch-tough welds with Charpy V-notch toughness of 20 ft-lb at −20 °F was introduced in the 1999 Supplement No. 1 to the 1997 Provisions. FEMA 350 and 353 (FEMA, 2000b) recommended that supplemental requirements beyond the basic toughness noted above should be applied to CJP welds in these connections. Welds for which these special requirements apply are referred to as demand critical welds.

The requirement to use demand critical welds for groove welded column splices and for welds at column base plates is new to these Provisions. Although it is likely that, in general, strain demands at near-mid-height column splice locations are less severe than those at beam-to-column joints, Shen et al. (2010) showed that bending at these locations can be large enough to cause flange yielding. This fact, coupled with the severe consequence of failure, was adequate justification for this requirement.

For the case of column-to-base plate connections at which plastic hinging is expected in the column, the condition is very similar to the condition at a beam-to-column connection. Where columns extend into a basement or are otherwise restrained in such a way that the column hinging will occur at a level significantly above the base plate, this requirement is judged to be overly conservative, and an exception is provided.

6b. Beam-to-Column Connections

Sections E3.6b and E3.6c have been rewritten to clarify the requirements and to coordinate the requirements with Sections K1 and K2. Section E3.6b gives the performance and design requirements for the connections and Section E3.6c provides the requirements for verifying that the selected connections will meet the performance requirements. These requirements have been derived from the research of the SAC Joint Venture as summarized in FEMA 350.

FEMA 350 recommends two criteria for the qualifying drift angle (QDA) for special moment frames. The "strength degradation" drift angle, as defined in FEMA 350, means the angle where "either failure of the connection occurs, or the strength of the connection degrades to less than the nominal plastic capacity, whichever is less." The "ultimate" drift angle capacity is defined as the angle "at which connection damage is so severe that continued ability to remain stable under gravity loading is uncertain." Testing to this level can be hazardous to laboratory equipment and staff, which is part of the reason that it is seldom done. The strength degradation QDA is set at 0.04 rad and the ultimate QDA is set at 0.06 rad. These values formed the basis for extensive probabilistic evaluations of the performance capability of various structural systems (FEMA, 2000f) demonstrating with high statistical confidence that frames with these types of connections can meet the intended performance goals. For the sake of simplicity, and because many connections have not been tested to the ultimate QDA, the Provisions adopt the single criterion of the strength degradation QDA. In addition, the ultimate QDA is more appropriately used for the design of high performance structures.

Although connection qualification primarily focuses on the level of plastic rotation achieved, the tendency for connections to experience strength degradation with increased deformation is also of concern. Strength degradation can increase rotation demands from P-Δ effects and the likelihood of frame instability. In the absence of additional information, it is recommended that this degradation should not reduce flexural strength, measured at a drift angle of 0.04 rad, to less than 80% of the nominal flexural strength, M_p, calculated using the specified minimum yield stress, F_y. Figure C-E3.1 illustrates this behavior. Note that 0.03 rad plastic rotation is equivalent to 0.04 rad drift angle for frames with an elastic drift of 0.01 rad.

ANSI/AISC 358 describes six different connections that have been prequalified for use in both IMF and SMF systems. The prequalified connections include the reduced beam section (RBS), the bolted unstiffened extended end plate (BUEEP), the bolted stiffened extended end plate (BSEEP), the bolted flange plate (BFP), the welded unreinforced flange-welded web (WUF-W), and the Kaiser bolted bracket (KBB). In a few cases, the limitations on use of the connections are less strict for IMF than for SMF, but generally, the connections are the same.

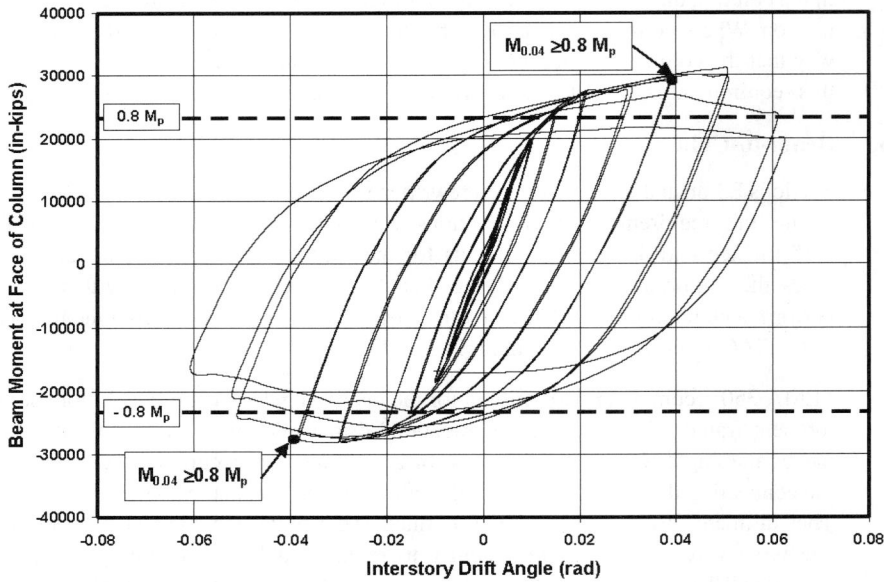

Fig. C-E3.1. Acceptable strength degradation, per Section E3.6b.

6c. Conformance Demonstration

This Section provides requirements for demonstrating conformance with the requirements of Section E3.6b. This provision specifically permits the use of prequalified connections meeting the requirements of ANSI/AISC 358 to facilitate and standardize connection design. Connections approved by other prequalification panels may be acceptable, but are subject to the approval of the authority having jurisdiction. Use of connections qualified by prior tests or project specific tests may also be used, although the engineer of record is responsible for substantiating the connection performance. Published testing, such as that conducted as part of the SAC project and reported in FEMA 350 and 355 or project-specific testing, may be used to satisfy this provision.

6d. Required Shear Strength

The required shear strength, V_u or V_a, as appropriate, of the beam-to-column joint is defined as the summation of the shear resulting from application of the factored gravity loads and the shear that results from application of the required flexural strengths on the two ends of the beam segment between the hinge points, which can be determined as $1.1 R_y F_y Z / L_h$ (LRFD) or $(1.1/1.5) R_y F_y Z / L_h$ (ASD), where Z is the plastic section modulus of the beam, and L_h is the distance between plastic hinge locations. However, in some cases, such as when large gravity loads occur or when panel zones are weak, rational analysis may indicate that lower combinations of end moments are justified.

6e. Panel Zone

(1) Required Shear Strength

Cyclic testing has demonstrated that significant ductility can be obtained through shear yielding in column panel zones through many cycles of inelastic distortion (Popov et al., 1996; Slutter, 1981; Becker, 1971; Fielding and Huang, 1971; Krawinkler, 1978). Consequently, it is not generally necessary to provide a panel zone that is capable of developing the full flexural strength of the connected beams if the available strength of the panel zone can be predicted. However, the usual assumption that the Von Mises criterion applies and the shear strength is $0.55 F_y d_c t_w$, where d_c is the depth of the column and t_w is the thickness of the column web, does not match the actual behavior observed in many tests taken into the inelastic range. Due to the presence of the column flanges, strain hardening and other phenomena, panel zone shear strengths in excess of $F_y d_c t_w$ have been observed. Accordingly, Equations J10-11 and J10-12 of the *Specification* account for the significant strength contribution of thick column flanges.

Despite the ductility demonstrated by properly proportioned panel zones in previous studies, excessive panel zone distortions can adversely affect the performance of beam-to-column connections (Englekirk, 1999; El-Tawil et al., 1999). Consequently, the provisions require that the panel zone design meet the minimum standard of the above noted equations, or match that of the successfully tested connections used to qualify the connection being used.

The application of the moments at the column face to determine the required shear strength of the panel zone recognizes that beam hinging will take place at a location away from the beam-to-column connection, which will result in amplified effects on the panel zone shear.

The 2005 Provisions required that the panel zone strength match that of tested or prequalified connections, with a minimum requirement to meet Section J10.6 of the *Specification*. ANSI/AISC 358 has adopted the simplified approach of only requiring conformance with Section J10.6. This relieves the designer of having to make determinations based on test reports. Additionally, from a practical standpoint, it is often difficult, or impossible, to match the test data, especially when the prototype column web is stronger than that used in the test. For this reason, it was judged that the requirement should be set as a minimum, rather than a match. This same approach has been adopted here, with an alternative permitting use of test data, if preferred by the designer.

The equations in Section J10.6 of the *Specification* represent the available strength in the inelastic range and, therefore, are for comparison to limiting strengths of connected members. In Section E3.6e(1) of the Provisions, ϕ_v has been set equal to unity and Ω_v set equal to 1.50, to allow a direct comparison between available strength of the beam and the column panel zone. In the *Specification*, the engineer is given the option to consider inelastic deformations of the panel zone in the analysis. Separate sets of equations are provided for use when these deformations are and are not considered. In the 2002 Seismic Provisions, only one equation was provided (Equation 9-1, which is the same as Equation J10-11 of the *Specification*) and consideration of the inelastic deformation of the panel zone in the analysis was required.

In general, analyses based on centerline dimensions of the beams and columns, and including P-Δ, can be considered as meeting the requirements to permit use of Equations J10-11 or J10-12. For further discussion on this issue, the designer is referred to Hamburger et al. (2009).

If the alternative procedure is chosen, the panel zone thickness must be determined using the same method as the one used to determine the panel zone thickness in the tested connection. The intent is that the local deformation demands on the various elements in the structure be consistent with the results of the tests that justify the use of the connection. The expected shear strength of the panel zone in relation to the maximum expected demands that can be developed by the beam(s) framing into the column should be consistent with the relative strengths that existed in the tested connection configuration. Many of the connection tests were performed with a one-sided configuration. If the structure has a two-sided connection configuration with the same beam and column sizes as a one-sided connection test, the panel zone shear demand will be about twice that of the test. Therefore, in order to obtain the same relative strength, the panel zone thickness to be provided in the structure should be approximately twice that of the test.

(2) Panel Zone Thickness

To minimize shear buckling of the panel zone during inelastic deformations, the minimum panel zone thickness is set by Equation E3-7 at one-ninetieth (1/90) of the sum of its depth and width. Thus, when the column web and web doubler plate(s) each meet the requirements of Equation E3-7 their interconnection with plug welds is not required. Otherwise, the column web and web doubler plate(s) can be interconnected with plug welds as illustrated in Figure C-E3.2 and the total panel zone thickness can be used in Equation E3-7.

When plug welds are required, Section E3.6e(2) requires a minimum of four plug welds. As a minimum, it is clear that the spacing should divide the plate into rectangular panels in such a way that all panels meet the requirements of Equation E3-7. Additionally, since a single plug weld would seem to create a boundary condition that is much different than a continuously restrained edge, it would be advisable to place the plug welds in pairs or lines, dividing the plate into approximately equally sized rectangles. Plug welds, when used, should, as a minimum, meet the requirements of Section J2.3 of the *Specification*.

An alternative detail is shown in Figure C-E3.3(c), where web doubler plates are placed symmetrically in pairs spaced away from the column web. In this configuration, both the web doubler plates and the column web are required to each independently meet Equation E3-7 in order to be considered as effective.

(3) Panel Zone Doubler Plates

There are several different conditions of use of web doubler plates depending on the need for continuity plates and on the particular design conditions. As noted in the previous section, doublers may be placed against the column web or spaced away from the web, and they may be used with or without continuity plates. When doublers are used with continuity plates, they may be located between the continuity plates, or they may be extended above and below the continuity plates. There are different requirements for welding the plates depending on the various configurations described. The most significant recent research on

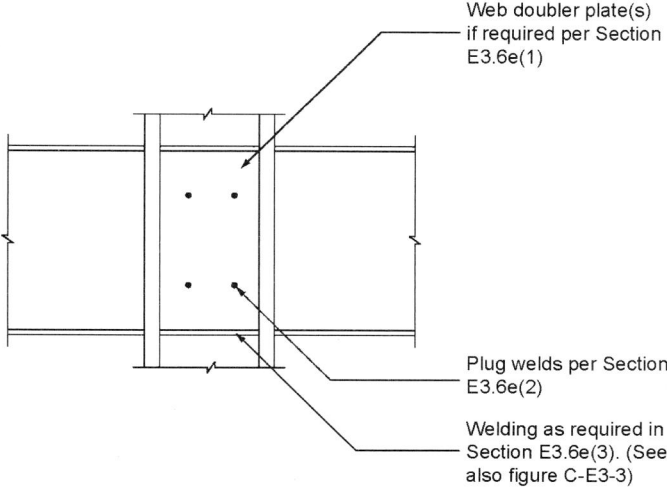

Fig. C-E3.2. Connecting web doubler plates with plug welds.

panel zone and doubler plate performance is described in the paper by Lee et al. (2005b). The research described in this paper suggests that the most critical welds for web doublers are those that connect the doubler to the column flanges. Accordingly, these welds are required to develop the available strength of the full doubler plate thickness. Either a complete-joint-penetration groove-welded joint or a fillet-welded joint can be used as illustrated in Figure C-E3.3(a) and C-E3.3(b) respectively. The plate thickness and column fillet radius should be considered in selecting the fillet-welded joint. A back bevel on the plate will be required to clear the fillet of the column, and thus a large fillet weld will be needed to develop an appropriate net section. Other weld configurations, such as a partial-joint-penetration groove weld with a reinforcing fillet, should also be considered as acceptable for this weld.

The above cited research indicates that doublers are effective even when the top and bottom edges of the doublers are not connected to the web or the continuity plates. There are two concerns that lead to the requirement to weld the tops and bottoms:

(1) If doublers do not meet the requirements of Equation E3-7, the top and bottom welds are needed to limit buckling of the doublers.

(2) Where continuity plates are used, stress concentrations at the column flange due to the discontinuity of the doubler-continuity plate interface may be undesirable.

Where continuity plates are not used and the doublers meet the requirements of Equation E3-7, the doubler top and bottom edges are not required to be welded.

When continuity plates are used, doubler plates may extend between top and bottom continuity plates and be welded directly to the column flanges and the continuity plates, or they may extend above and below the top and bottom continuity plates and be welded to the column flanges and web, and the continuity plates. In the former case, the welded joint connecting the continuity plate to the column web and web doubler plate is required to be configured to transmit the proportionate load from the continuity plate to each element of the panel zone.

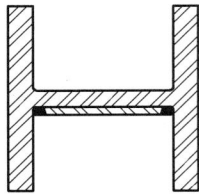
(a) Groove-welded
(see k-area discussion, Commentary Sections A3.1 and D2.4)

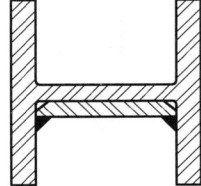
(b) Fillet-welded
(fillet weld size may be controlled by geometry, due to back-side bevel on web doubler plate)

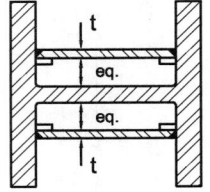
(c) Pair of equal-thickness web doubler plates, groove-welded to column

Fig. C-E3.3. Web doubler plates.

In the latter case, the welded joint connecting the continuity plate to the web doubler plate is required to be sized to transmit the load from the continuity plate to the web doubler plate and the web doubler plate thickness is required to be selected to transmit this same load.

The use of diagonal stiffeners for strengthening and stiffening of the panel zone has not been adequately tested for low-cycle reversed loading into the inelastic range. Thus, no specific recommendations are made at this time for special seismic requirements for this detail.

6f. Continuity Plates

Beam flange continuity plates serve several purposes in moment connections. They help to distribute beam flange forces to the column web, they stiffen the column web to prevent local crippling under the concentrated beam-flange forces, and they minimize stress concentrations that can occur in the joint between the beam flange and the column due to nonuniform stiffness of the column flange.

(1) Continuity Plate Requirements

When the beam flange connects to the flange of a wide-flange, built-up I-shape, or cruciform W-shaped column in which the column extends above and below the beam, and the column flange thickness satisfies Equations E3-8 and E3-9, continuity plates are not required. This is because, under these conditions, beam-flange forces can be adequately transferred to the column webs without the stiffening effects and secondary load paths provided by these plates, and the column flanges are thick enough to provide an appropriate stress distribution at the beam flange-to-column flange weld. Equation E3-8 is similar to the equation in older codes, except for the R_y factors. Justification for the use of Equations E3-8 and E3-9 is based on studies by Ricles discussed in FEMA 355D. Subsequent research by Lee et al. (2005a) confirmed the adequacy of designs based on these equations.

Equations E3-8 and E3-9 have been developed based on consideration of the behavior of columns in lower stories of buildings, where the column extends a considerable distance above the top flange of the connected beam. These equations do not apply in the top story of a building, where the column terminates at approximately the level of the top flange of the beam. In such cases, beam-flange continuity plates or column cap plates, having a thickness not less than that of the connected beam flange, should be provided. Figure C-E3.4 presents a detail for such a connection, where the beam flange is welded directly to the cap plate and the cap plate is welded to the column so as to deliver the beam-flange forces to the column web.

Alternatively, if the column projects sufficiently above the beam top flange, Equations E3-8 and E3-9 can be considered valid. Although comprehensive research to establish the necessary distance that the column must extend above the beam for this purpose has not been performed, it may be judged to be sufficient if the column is extended above the top beam flange a distance not less than $d_c/2$ or $b_f/2$, whichever is less, where b_f is the flange width of the column.

For boxed wide-flange section columns in which the beams are connected to the flange of the I-shaped section, Equations E3-10 and E3-11 have been developed to provide a similar stiffness of column flange to that provided by Equations E3-8 and E3-9 for unboxed sections. As with Equations E3-8 and E3-9, Equations E3-10 and E3-11 are not strictly valid for the case of a moment connection at the roof level of a building, in which the column does not extend significantly above the beam top flange. In these cases, a cap plate detail similar to that illustrated in Figure C-E3.4 should be used.

When beams are moment connected to the side plates of boxed wide-flange column sections, continuity plates or cap plates should always be provided opposite the beam flanges, as is required for box section columns.

(2) Continuity Plate Thickness

Requirements for thickness of continuity plates as given in Section E3.6f(2) are based on the studies by Ricles cited previously.

(3) Continuity Plate Welding

The connection of continuity plates to column webs is designed to be capable of transmitting the maximum shear forces that can be delivered to the connection. This may be limited by the beam-flange force, the shear strength of the continuity plate itself, the welded joint between continuity plate and column flange, or the strength of the column panel zone.

The Provisions require that continuity plates be attached to column flanges with CJP groove welds in order that the strength of the beam flange can be properly

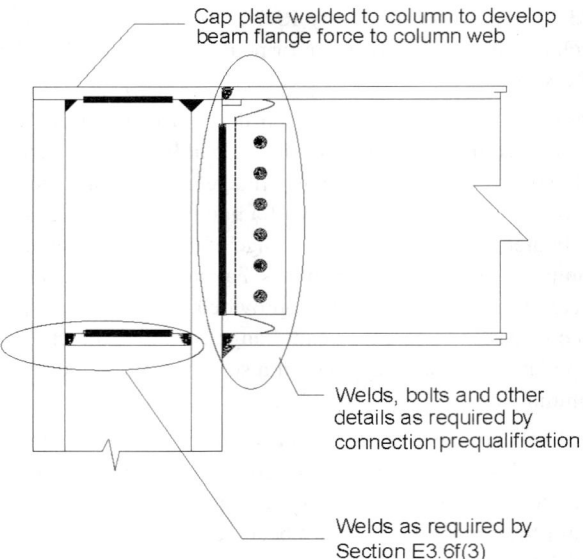

Fig. C-E3.4. Cap plate detail at column top.
(Figure C–2.2 from ANSI/AISC 358-10)

developed into the continuity plate. Research by Lee et al. (2005a, 2005b) demonstrated that properly sized fillet welded connections also performed adequately for this purpose, although this is not yet permitted by the Provisions. For single-sided connections in which a moment-connected beam attaches to only one of the column flanges, it is probably not necessary to use CJP groove welds to attach the continuity plate to the column flange that does not have a beam attached. In such cases, acceptable performance is expected if the continuity plate is attached to the column with a pair of minimum-size fillet welds.

6g. Column Splices

In the 1997 Provisions, there were no special requirements for column splices in SMF systems other than those currently given in Section D2.5. The requirement in Section D2.5a was intended to take care of column bending at the splice by requiring splices to be at least 4 ft or one-half the column clear height from the beam-to-column connection. This requirement was based on the general recognition that in elastic analyses of moment frames the columns are typically bent in double curvature with an inflection point somewhere near the middle of the column height and, therefore, little bending of the column was expected at the splice.

Nonlinear analyses performed during the FEMA/SAC project following the Northridge earthquake, and subsequently (Shen et al., 2010) clearly demonstrated that bending moments in the mid-height of columns can be substantial and that, in fact, the columns may be bent in single curvature under some conditions. Given this fact, and the recognition of the potential for severe damage or even collapse due to failure of column splices, the need for special provisions for splices of moment frame columns was apparent.

The provisions of Section E3.6g are intended to assure that the expected flexural strength of the smaller column is fully developed, either through use of CJP groove welds or another connection that provides similar strength, and that the shear strength of the splice is sufficient to resist the shear developed when M_{pc} occurs at each end of the spliced column.

The exception permits the design of splices based on appropriate inelastic analysis to determine required strength, coupled with the use of principles of fracture mechanics to determine the available strength of the connection.

E4. SPECIAL TRUSS MOMENT FRAMES (STMF)

1. Scope

Truss-girder moment frames have often been designed with little or no regard for truss ductility. Research has shown that such truss moment frames have very poor hysteretic behavior with large, sudden reductions in strength and stiffness due to buckling and fracture of web members prior to or early in the dissipation of energy through inelastic deformations (Itani and Goel, 1991; Goel and Itani, 1994a). The resulting hysteretic degradation as illustrated in Figure C-E4.1 results in excessively large story drifts in building frames subjected to earthquake ground motions with peak accelerations on the order of $0.4g$ to $0.5g$.

Research led to the development of special truss girders that limit inelastic deformations to a special segment of the truss (Itani and Goel, 1991; Goel and Itani, 1994b; Basha and Goel, 1994). As illustrated in Figure C-E4.2, the chords and web members (arranged in an X pattern) of the special segment are designed to withstand large inelastic deformations, while the rest of the structure remains elastic. Special truss moment frames (STMF) have been validated by extensive testing of full-scale subassemblages with story-high columns and full-span special truss girders. As illustrated in Figure C-E4.3, STMF are ductile with stable hysteretic behavior. The stable hysteretic behavior continues for a large number of cycles, up to 3% story drifts.

STMF must satisfy all the applicable requirements of Chapters A, B, C, D, I and J of these Provisions, as well as the requirements in Section E4.

2.   Basis of Design

Because STMF are relatively new and unique, the span length and depth of the truss girders are limited at this time to the range used in the test program.

3.   Analysis

3a. Special Segment

The design procedure of STMF is built upon the concept that the special segment of truss girders will yield in shear under the prescribed earthquake load combinations, while all other frame members and connections remain essentially elastic. Thus, for the purpose of determining the required shear strength of special segments the truss

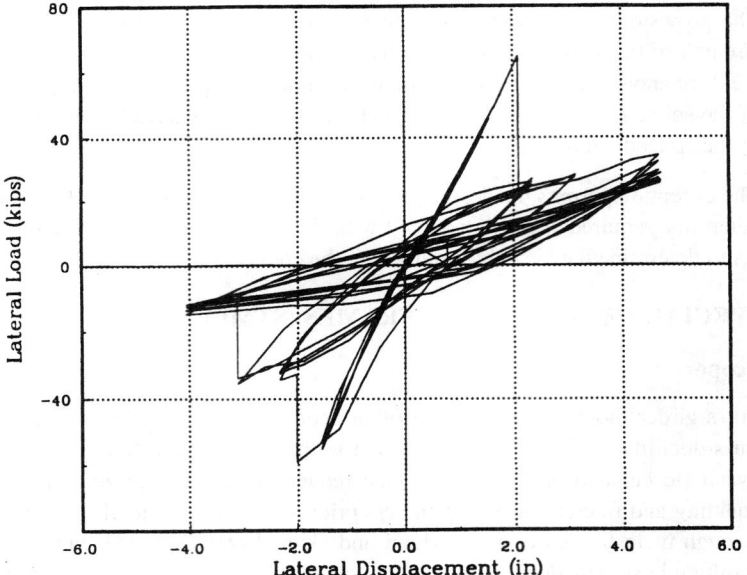

Fig. C-E4.1. Strength degradation in undetailed truss girder.

girders can be treated as analogous beams in moment frames (Rai et al., 1998). The chord and diagonal members of the special segments are then designed to provide the required shear strength as specified in Section E4.5a of the Provisions.

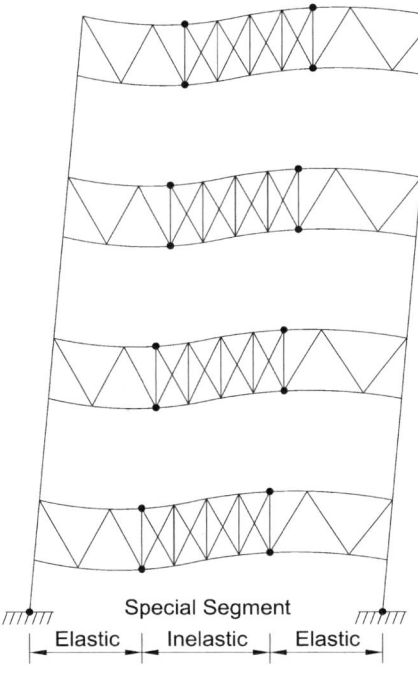

Fig. C-E4.2. Intended yield mechanism of STMF with diagonal web members in special segment.

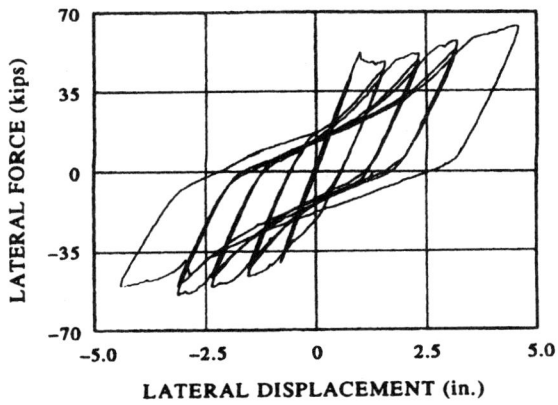

Fig. C-E4.3. Hysteretic behavior of STMF.

3b. Nonspecial Segment

All frame members and connections of STMF outside the special segments must have adequate strength to resist the combination of factored gravity loads and maximum expected shear strength of the special segments by accounting for reasonable strain-hardening and material overstrength. For this purpose, one of several analysis approaches can be used. One approach is to consider the equilibrium of properly selected elastic portions (sub-structures) of the frame and perform elastic analysis. Alternatively, a nonlinear static pushover analysis of a model of the entire frame can be carried out up to the maximum design drift. The intended yielding members of the special segments, including chord and diagonal members and column bases, are modeled to behave inelastically, while all others are modeled (or "forced") to behave elastically. Second order effects should be included in the analysis as needed.

4. System Requirements

4a. Special Segment

It is desirable to locate the STMF special segment near midspan of the truss girder because shear due to gravity loads is generally lower in that region. The lower limit on special segment length of 10% of the truss span length provides a reasonable limit on the ductility demand, while the upper limit of 50% of the truss span length represents more of a practical limit.

The required strength of interconnection for X-diagonals is intended to account for buckling over half the full diagonal length (El-Tayem and Goel, 1986; Goel and Itani, 1994b). It is recommended that half the full diagonal length be used in calculating the available compressive strength of the interconnected X-diagonal members in the special segment.

Because it is intended that the yield mechanism in the special segment form over its full length, no major structural loads should be applied within the length of the special segment. In special segments with open Vierendeel panels, in other words, when no diagonal web members are used, any structural loads should be avoided. Accordingly, a restrictive upper limit is placed on the axial load in diagonal web members due to gravity loads applied directly within the special segment.

4b. Stability Bracing of Trusses

The top and bottom chords are required to be laterally braced to provide for the stability of the special segment during cyclic yielding. The lateral bracing requirements for truss chord members have been slightly revised to make it consistent with what was used successfully in the original testing program.

4c. Stability Bracing of Truss-to-Column Connections

Columns should be laterally braced at the points of connection with the truss members in order to provide adequate stability during expected cyclic deformations of the frames. A lateral bracing requirement has been added which is partly based on what was used successfully in the original testing program.

5. Members

5a. Special Segment Members

STMF are intended to dissipate energy through flexural yielding of the chord members and axial yielding and buckling of the diagonal web members in the special segment. It is desirable to provide minimum shear strength in the special segment through flexural yielding of the chord members and to limit the axial load to a maximum value. Plastic analysis can be used to determine the required shear strength of the truss special segments under the factored earthquake load combination.

5b. Expected Vertical Shear Strength of Special Segment

STMF are required to be designed to maintain elastic behavior of the truss members, columns and all connections, except for the members of the special segment that are involved in the formation of the yield mechanism. Therefore, all members and connections outside the special segments are to be designed for calculated loads by applying the combination of gravity loads and equivalent lateral loads that are necessary to develop the maximum expected nominal shear strength of the special segment, V_{ne}, in its fully yielded and strain-hardened state. Thus, Equation E4-5, as formulated, accounts for uncertainties in the actual yield strength of steel and the effects of strain hardening of yielded web members and hinged chord members. It is based upon approximate analysis and test results of special truss girder assemblies that were subjected to story drifts up to 3% (Basha and Goel, 1994). Tests (Jain et al., 1978) on axially loaded members have shown that $0.3P_{nc}$ is representative of the average nominal post-buckling strength under cyclic loading. Based on a more recent study by Chao and Goel (2008) the first two terms of Equation E4-5 have been revised to give a more accurate estimate of contribution from the chord members.

Equation E4-5 was formulated without considering the contribution from any intermediate vertical members within the special segment other than those at the ends of the special segment. In cases where those intermediate vertical members possess significant flexural strength, their contribution should also be included in calculating the value of V_{ne}. A modified equation which includes the contribution of intermediate vertical members has been proposed by Chao and Goel (2008). However, research work to experimentally validate that equation is currently in progress.

5c. Width-to-Thickness Limitations

The ductility demand on diagonal web members in the special segment can be rather large. Flat bars are suggested at this time because of their high ductility. Tests (Itani and Goel, 1991) have shown that single angles with width-to-thickness ratios that are less than $0.18\sqrt{E/F_y}$ also possess adequate ductility for use as web members in an X-configuration. Chord members in the special segment are required to be compact cross sections to facilitate the formation of plastic hinges.

5d. Built-Up Chord Members

Built-up chord members in the special segment can be subjected to rather large rotational demands at the plastic hinges requiring close stitch spacing in order to prevent lateral-torsional buckling of the individual elements. Based on the findings from a recent experimental study (Parra-Montesinos et al., 2006) a stitch spacing requirement for chord members in the special segment has been added.

5e. Protected Zones

When special segments yield under shear, flexural plastic hinges will form at the ends of the chord members. Therefore, those regions are designated as protected zones. Also, included in the protected zones are vertical and diagonal members of the special segments, because those members are also expected to experience significant yielding.

6. Connections

6a. Demand Critical Welds

Refer to the commentary on Section E3.6a.

6b. Connections of Diagonal Web Members in the Special Segment

The diagonal members of the special segments are expected to experience large cyclic deformations in axial tension and post-buckling compression. Their end connections must possess adequate strength to resist the expected tension yield strength.

6c. Column Splices

The requirements in this Section are identical to those in Section E3.6g. See Commentary Section E3.6g for further discussion.

E5. ORDINARY CANTILEVER COLUMN SYSTEMS (OCCS)

1. Scope

Ordinary cantilever column systems (OCCS) must satisfy all the applicable requirements of Chapters A, B, C, D, I and J of these Provisions, as well as the requirements in Section E5.

2. Basis of Design

ASCE/SEI 7 (ASCE, 2010) includes two types of cantilever column systems, ordinary and special. OCCS are intended to provide a minimal level of inelastic rotation capability at the base of the column. This system is permitted in seismic design categories B and C only, and to heights not exceeding 35 ft. A low seismic response modification coefficient, R, of 1.25 is assigned due to the system's limited inelastic capacity and lack of redundancy. The OCCS has no requirements beyond those in the *Specification* except as noted in Section E5.4a.

4. System Requirements

4a. Columns

ASCE/SEI 7 (ASCE, 2010) limits the required axial load on columns in these systems under the load combinations including amplified seismic load to 15% of the available strength. This limitation is included in these provisions. Columns in OCCS would be prone to P-Delta collapse if high axial loads were permitted.

E6. SPECIAL CANTILEVER COLUMN SYSTEMS (SCCS)

1. Scope

Special cantilever column systems (SCCS) must satisfy all the applicable requirements of Chapters A, B, C, D, I and J of these Provisions, as well as the requirements in Section E6.

2. Basis of Design

ASCE/SEI 7 includes two types of cantilever column systems, ordinary and special. The SCCS is intended to provide a limited level of inelastic rotation capability at the base of the column. This system is permitted in seismic design categories B thru F, but is limited to heights not exceeding 35 ft. A relatively low seismic response modification coefficient, R, of 2.5 is assigned due to the system's limited inelastic capacity and lack of redundancy.

4. System Requirements

4a. Columns

ASCE/SEI 7 limits the required axial load on columns in these systems under the load combinations including amplified seismic load to 15% of the available strength. This limitation is included in these provisions. Columns in SCCS would be prone to P-Delta collapse if high axial loads were permitted because even modest rotations at the base of the columns can translate into significant drift at the top where the majority of the gravity load is generally applied.

4b. Stability Bracing of Columns

Stability bracing of columns at the spacing required for moderately ductile members is required. Although the columns themselves must satisfy requirements for highly ductile members, the wider spacing of braces permitted is considered to be adequate because of the relatively low inelastic demand expected and the practical difficulty in achieving bracing in many of these structures. For structures where there is no reasonable way to meet bracing requirements, need for bracing may be precluded by selecting appropriately proportioned members.

5. Members

5a. Basic Requirements

The column members are required to satisfy the width-to-thickness and other provisions for highly ductile members. The intention is to preclude local buckling at the hinging location (bottom of the column), which in this type of structure, with little redundancy, could lead rapidly to collapse.

5b. Column Flanges

Abrupt changes in beam flange area in locations of high strain, as occurs in plastic hinge regions at the base of SCCS columns, can lead to fracture due to stress concentrations.

5c. Protected Zones

For commentary on protected zones see Commentary Section D1.3.

6. Connections

6a. Demand Critical Welds

For general commentary on demand critical welds, see Commentary Section A3.4. For additional commentary appropriate to column splices and column-to-base plate connections, see Section E3.6a.

6b. Column Bases

It is apparent that a column base in the SCCS must be capable of developing the moment capacity of the column, including overstrength and strain hardening. Detailed requirements are provided in Section D2.6 and commentary is provided in the corresponding commentary section.

CHAPTER F

BRACED-FRAME AND SHEAR-WALL SYSTEMS

F1. ORDINARY CONCENTRICALLY BRACED FRAMES (OCBF)

1. Scope

Ordinary concentrically braced frames (OCBF) have minimal design requirements compared to other braced-frame systems. The Provisions assume that the applicable building code significantly restricts the permitted use of OCBF and specifies a low R factor so that ductility demands will be low. Specifically, it is assumed that the restrictions given in ASCE/SEI 7 (ASCE, 2010) govern the use of the structural system.

The scope includes OCBF above an isolation system. The provisions in Section F1.7 are intended for use in the design of OCBF for which forces have been determined using R_i equal to 1.0. R_i is defined in ASCE/SEI 7 as the "numerical coefficient related to the type of seismic force-resisting system above the isolation system." Such OCBF are expected to remain essentially elastic during design level earthquakes and, therefore, provisions that are intended to accommodate significant inelastic response, such as Section F1.4a, are not required for their design.

2. Basis of Design

OCBF are not expected to be subject to large inelastic demands due to the relatively low R factor assigned to the system in ASCE/SEI 7.

3. Analysis

Due to the expected limited inelastic demands on OCBF, an elastic analysis is considered sufficient when supplemented with use of the amplified seismic load as required by these provisions.

4. System Requirements

4a. V-Braced and Inverted V-Braced Frames

V- and inverted-V-type bracing can induce a high unbalanced force in the intersecting beam. Unlike the special concentrically braced frame (SCBF) provisions, which require that the beams at the intersections of such braces be designed for the expected yield strength of the braces to prevent a plastic hinge mechanism in the beam, the corresponding OCBF provisions permit the beam design on the basis of the maximum force that can be developed by the system. This relief for OCBF acknowledges that, unlike SCBF, the beam forces in an OCBF frame at the time of an imminent system failure mode could be less critical than those due to the expected yield strength of the connecting braces. See the commentary for Sections F2.6c(1) for techniques that may be used to determine the maximum force developed by the system.

4b. K-Braced Frames

K-bracing can have very poor post-elastic performance. After brace buckling, the action of the brace in tension induces large flexural forces on the column, possibly leading to buckling. No adequate design procedures addressing the high-consequence stability issues are available.

5. Members

5a. Basic Requirements

Only moderate ductility is expected of OCBF. Accordingly, in the 2010 Provisions, the member ductility requirement has been modified to require moderately ductile members.

5b. Slenderness

In V- and inverted V-braced frames, slender braces are not permitted. This restriction is intended to limit the unbalanced forces that develop in framing members after brace buckling; see Commentary Section F2.4c.

6. Connections

6a. Diagonal Brace Connections

Bracing connections are designed for forces corresponding to the expected brace strength, the maximum force that the system can develop (see Commentary Section F2.6 for discussion), or the amplified seismic load so as to delay the connection limit state. Net section rupture of the member is to be included with connection limit states and designed for the amplified seismic load. The Provisions permits the required strength of a brace connection in an OCBF to be limited by the load effect based on the amplified seismic load, which is considered appropriate for systems designed for limited ductility.

The Provisions permit that bolt slip be designed for a lower force level than is required for other limit states. This reflects the fact that bolt slip does not constitute connection failure and that the associated energy dissipation can serve to reduce seismic response.

7. Ordinary Concentrically Braced Frames above Seismic Isolation Systems

Above isolation, system and member ductility demands are greatly reduced compared to nonisolated OCBF. Accordingly, highly ductile members are not required, nor are beams required to resist forces corresponding to brace nonlinear behavior. However, most engineers recognize that, since the intent of the code is now to preclude collapse in the maximum credible earthquake, should an earthquake occur that is larger than those considered in the design, some ductility of the system is desirable for the survivability of the structure, and certain basic requirements remain: amplified compression strength and the elimination of the nonductile K-bracing configuration.

The requirements in this Section are similar to Section F1.5, except that the KL/r limitation is applied to all braces. Tension-only bracing is not considered to be appropriate for use above isolation systems under the conditions permitted.

The requirements of Section F1.4a are considered to be excessive for OCBFs above the isolation system because the forces on the system are limited and buckling of braces is not anticipated. The only requirement is for the beams to be continuous between columns.

F2. SPECIAL CONCENTRICALLY BRACED FRAMES (SCBF)

1. Scope

Special concentrically braced frames (SCBF) are a type of concentrically braced frame; that is, braced frames in which the centerlines of members that meet at a joint intersect at a point, thus forming a vertical truss system that resists lateral loads. A few common types of concentrically braced frames are shown in Figure C-F2.1, including diagonally braced, X-braced, and V-braced (or inverted V-braced). Use of tension-only bracing in any configuration is not permitted for SCBF. Because of their geometry, concentrically braced frames provide complete truss action with members subjected primarily to axial loads in the elastic range. However, during a moderate to severe earthquake, the bracing members and their connections are expected to undergo significant inelastic deformations into the post-buckling range.

2. Basis of Design

SCBF are distinguished from OCBF (and from braced frames designed with $R = 3$) by requirements for ductility. Accordingly, provisions were developed so that the SCBF would exhibit stable and ductile behavior in the event of a major earthquake. Earlier design provisions have been retained for OCBF in Section F1.

During a severe earthquake, bracing members in a concentrically braced frame are subjected to large deformations in cyclic tension and compression. In the compression direction flexural buckling causes the formation of flexural plastic hinges in the

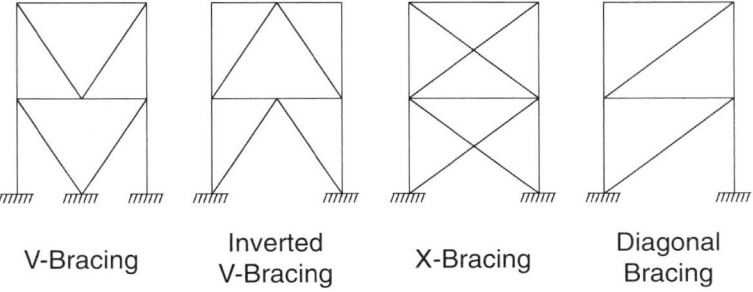

Fig. C-F2.1. Examples of concentric bracing configurations.

brace as it deforms laterally. These plastic hinges are similar to those in beams and columns in moment frames. Braces in a typical concentrically braced frame can be expected to yield and buckle at rather moderate story drifts of about 0.3% to 0.5%. In a severe earthquake, the braces could undergo post-buckling axial deformations 10 to 20 times their yield deformation. In order to survive such large cyclic deformations without premature failure, the bracing members and their connections must be properly detailed.

Damage during past earthquakes and that observed in laboratory tests of concentrically braced frames has generally resulted from the limited ductility and corresponding brittle failures, which are usually manifested in the rupture of connection elements or bracing members. The lack of compactness in braces results in severe local buckling, resulting in a high concentration of flexural strains at these locations and reduced ductility. Braces in concentrically braced frames are subject to severe local buckling, with diminished effectiveness in the nonlinear range at low story drifts. Large story drifts that result from early brace ruptures can impose excessive ductility demands on the beams and columns, or their connections.

Research has demonstrated that concentrically braced frames, with proper configuration, member design and detailing can possess ductility far in excess of that previously ascribed to such systems. Extensive analytical and experimental work by Goel has shown that improved design parameters, such as limiting width-to-thickness (to minimize local buckling), closer spacing of stitches, and special design and detailing of end connections greatly improve the post-buckling behavior of concentrically braced frames (Goel, 1992b; Goel, 1992c). The design requirements for SCBF are based on those developments.

Previous requirements for concentrically braced frames sought reliable behavior by limiting global buckling. Cyclic testing of diagonal bracing systems verifies that energy can be dissipated after the onset of global buckling if brittle failures due to local buckling, stability problems and connection fractures are prevented. When properly detailed for ductility as prescribed in the Provisions, diagonal braces can sustain large inelastic cyclic deformations without experiencing premature failures.

Analytical studies (Tang and Goel, 1987; Hassan and Goel, 1991) on bracing systems designed in strict accordance with earlier code requirements for concentrically braced frames predicted brace failures without the development of significant energy dissipation. Failures occurred most often at plastic hinges (local buckling due to lack of compactness) or in the connections. Plastic hinges normally occur at the ends of a brace and at the brace midspan. Analytical models of bracing systems that were designed to ensure stable ductile behavior when subjected to the same ground motion records as the previous concentrically braced frame designs exhibited full and stable hysteresis without fracture. Similar results were observed in full-scale tests in Wallace and Krawinkler (1985) and Tang and Goel (1989).

Since the stringent design and detailing requirements for SCBF are expected to produce more reliable performance when subjected to high energy demands imposed by severe earthquakes, model building codes have reduced the design load level below that required for OCBF.

Bracing connections should not be configured in such a way that beams or columns of the frame are interrupted to allow for a continuous brace element. This provision is necessary to improve the out-of-plane stability of the bracing system at those connections.

A zipper column system and a two-story X-braced system are illustrated in Figure C-F2.2. Two-story X- and zipper-braced frames can be designed with post-elastic behavior consistent with the expected behavior of V-braced SCBF. These configurations can also capture the increase in post-elastic axial loads on beams at other levels. It is possible to design two-story X-braced and zipper frames with post-elastic behavior that is superior to the expected behavior of V-braced SCBF by proportioning elements to discourage single-story mechanisms (Khatib et al., 1988). For more information on these configurations see Khatib et al. (1988); Yang et al. (2008); and Tremblay and Tirca (2003).

3. **Analysis**

While SCBF are typically designed on the basis of an elastic analysis, their expected behavior includes significant nonlinearity due to brace buckling and yielding, which is anticipated in the maximum credible earthquake. Braced-frame system ductility can only be achieved if beams and column buckling can be precluded. Thus there is a need to supplement the elastic analysis in order to have an adequate design.

The required strength of braces is typically determined based on the analysis required by ASCE/SEI 7. The analysis required by this section is used in determining the required strength of braced-frame beams and columns, as well as of brace

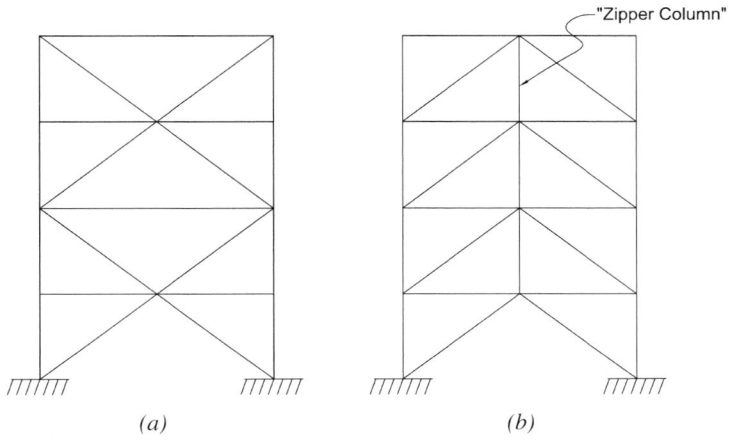

Fig. C-F2.2. (a) Two story X-braced frame, (b) "zipper column" with inverted V-bracing.

connections, as it is necessary to design these elements to resist forces corresponding to brace yielding.

In previous editions of the Provisions, the expected nonlinear behavior of SCBF was addressed through a series of design rules that defined required strengths of elements superseding those derived using elastic elements. These included:

- Forces for beams in V- and inverted V-braced frames
- Forces for the design of brace connections
- Forces for column design

These design rules were intended to approximate forces corresponding to inelastic response without requiring an inelastic analysis.

While these requirements addressed the most important shortcomings of elastic analysis, several other cases have been identified, including:

- Beams not intersected by braces in the two-story X-braced configuration (e.g., the beam at the third floor in Figure C-F2.3(a)
- Interior columns in multi-bay braced frames. See Figure C-F2.3(b)

Rather than creating new (and increasingly complicated) design rules to address these omissions in previous Provisions, it was decided to simply mandate explicit consideration of the inelastic behavior by requiring a plastic-mechanism analysis, the simplest form of inelastic analysis. It is naturally desirable that engineers performing analyses of ductile systems give some thought to the manner in which they will behave.

Because the compression behavior of braces differs substantially from the tension behavior, two separate analyses are required:

- An analysis in which all braces have reached their maximum forces
- An analysis in which tension braces are at their maximum strength level and compression braces have lost a significant percentage of their strength after buckling

The first-mode of deformation is considered when determining whether a brace is in compression or in tension. That is, the columns are considered to be inclined in one direction rather than in reverse curvature (see Figure C-F2.4). Consideration must also be given to the behavior when the columns slope the opposite direction.

Consistent with previous editions of these Provisions, when maximum axial forces are calculated for columns, the engineer is permitted to neglect the flexural forces that result from the design story drifts. This permits straightforward determination of seismic forces using spreadsheet software.

The analysis requirements utilize the expected strengths of braces in tension and compression. The full tension strength can be expected to be in the range of $R_y P_y$. The expected compressive strength of braces has been modified from the 2005 Provisions to address the following:

- Proper influence of material overstrength for all slenderness ranges
- Correct maximum value for braces of very low slenderness

Tests have shown that typical bracing members demonstrate a minimum residual post-buckling compressive strength of about 30% of the initial compressive strength (Hassan and Goel, 1991).

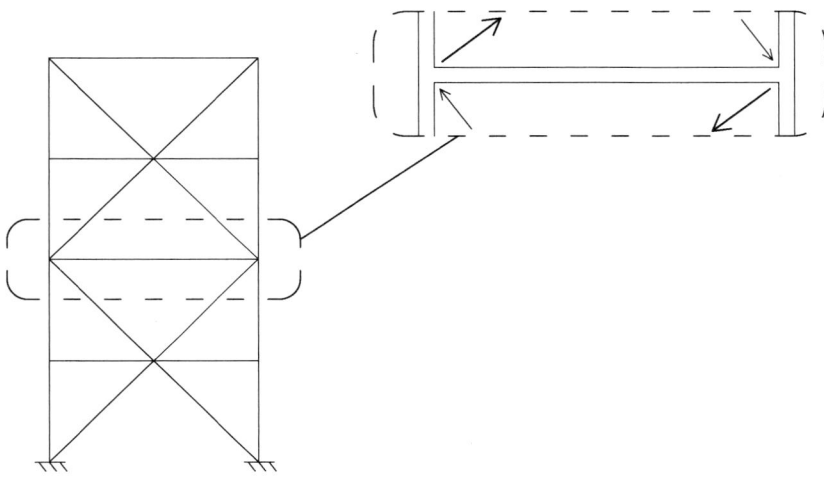

(a) Post-elastic flow of forces through braced-frame beam.

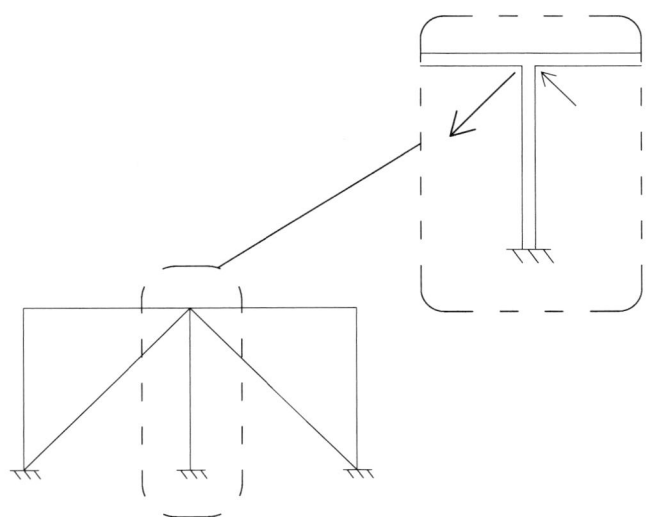

(b) Post-elastic flow of forces through interior braced-frame column.

Fig. C-F2.3. Examples of post-elastic flow of forces in braced-frame systems.

4. System Requirements

4a. Lateral Force Distribution

This provision attempts to balance the tensile and compressive resistance across the width and breadth of the building since the buckling and post-buckling strength of the bracing members in compression can be substantially less than that in tension. Good balance helps prevent the accumulation of inelastic drifts in one direction.

An exception is provided for cases where the bracing members are sufficiently oversized to provide essentially elastic response. It is envisioned that such an exception would apply to a small number of braces in the structure. It is generally preferable to have braces sized in proportion to their required strength. Where braces have vastly different overstrengths the inelastic demands may be concentrated (and amplified) in a small number of braces.

4b. V- and Inverted V-Braced Frames

V-braced and inverted V-braced (chevron) frames exhibit a special problem that sets them apart from other configurations. The expected behavior of SCBF is that upon continued lateral displacement as the brace in compression buckles, its force drops while that in the brace in tension continues to increase up to the point of yielding. In order for this to occur in these frames, an unbalanced vertical force must be resisted by the intersected beam, as well as its connections and supporting members.

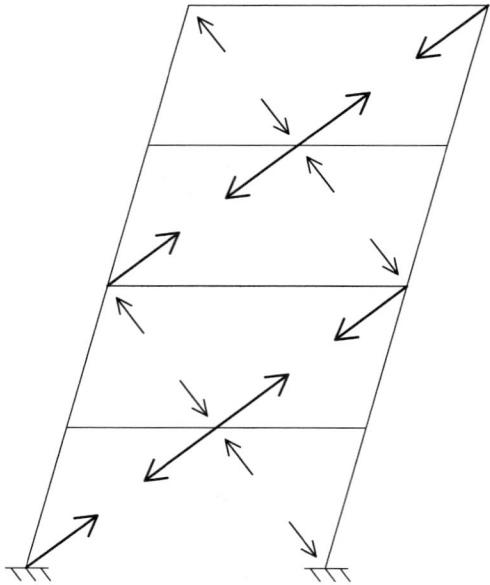

Fig. C-F2.4. Anticipated braced-frame mechanism.

The adverse effect of this unbalanced load can be mitigated by using bracing configurations, such as V- and inverted V-braces in alternate stories creating an X-configuration over two story modules (Khatib, et al., 1988). See Figure C-F2.2a.

Adequate lateral bracing at the brace-to-beam intersection is necessary in order to prevent adverse effects of possible lateral-torsional buckling of the beam. The stability of this connection is influenced by the flexural and axial forces in the beam, as well as by any torsion imposed by brace buckling or the post-buckling residual out-of-straightness of a brace. The committee did not believe that under these conditions the bracing requirements in the *Specification* are sufficient to ensure the torsional stability of this connection. Therefore a requirement based on the moment due to the flexural strength of the beam is imposed.

4c. K-Braced Frames

K-bracing is generally not considered desirable in concentrically braced frames and is prohibited entirely for SCBF because it is considered undesirable to have columns that are subjected to unbalanced lateral forces from the braces, as these forces may contribute to column failures.

4d. Tension-Only Frames

SCBF provisions have not been developed for use with braces that only act in tension. Thus tension-only braced frames are not allowed for SCBF. (Tension-only bracing is allowed for OCBF).

5. Members

5a. Basic Requirements

Traditionally, braces have shown little or no ductility after overall (member) buckling, which produces a plastic hinge at the brace midpoint. At this plastic hinge, local buckling can cause large strains, leading to fracture at low drifts. It has been found that braces with compact elements are capable of achieving significantly more ductility by forestalling local buckling (Goel, 1992b; Hassan and Goel, 1991; Tang and Goel, 1989). Width-to-thickness ratios of compression elements in bracing members have been set to be at or below the requirements for compact sections in order to minimize the detrimental effects of local buckling and subsequent fracture during repeated inelastic cycles.

Tests have shown fracture due to local buckling is especially prevalent in rectangular HSS with width-to-thickness ratios larger than the prescribed limits (Hassan and Goel, 1991; Tang and Goel, 1989). Even for square HSS braces designed to meet the seismic width-to-thickness ratios of these Provisions, local buckling leading to fracture may represent a limitation on the performance (Yang and Mahin, 2005).

The same limitations apply to columns in SCBF, as their flexural strength and rotation capacity has been shown to be a significant contributor to the stability of SCBF (Tremblay, 2001, 2003). It has also been demonstrated that SCBF can be subject to significant story drift (Sabelli et al., 2003), requiring columns to undergo inelastic rotation.

Enhanced ductility and fracture life of rectangular HSS bracing members can be achieved in a variety of ways. The HSS walls can be stiffened by using longitudinal stiffeners, such as rib plates or small angle sections in a hat configuration (Liu and Goel, 1987). Use of plain concrete infill has been found to be quite effective in reducing the severity of local buckling in the post-buckling range of the member (Liu and Goel, 1988; Lee and Goel, 1987). Based on their test results, Goel and Lee (1992) formulated an empirical equation to determine the effective width-to-thickness ratio of concrete-filled rectangular HSS bracing members. The effective width-to-thickness ratio can be calculated by multiplying the actual width-to-thickness ratio by a factor, $[(0.0082KL/r) + 0.264]$, for KL/r between 35 and 90, where KL/r is the effective slenderness ratio of the member. The purpose of concrete infill as described herein is to inhibit the detrimental effects of local buckling of the HSS walls. Use of concrete to achieve composite action of braces is covered in Section H2.5b.

As an alternative to using a single large HSS, consideration may be given to using double smaller tube sections stitched together and connected at the ends to a single gusset plate (or cross shape if needed) in much the same way as double angle or channel sections are used in a back-to-back configuration (Lee and Goel, 1990). Such double tube sections offer a number of advantages, including: reduced fit up problems, smaller width-to-thickness ratio for the same overall width of the section, in-plane buckling in most cases eliminating the problem of out-of-plane bending of gusset plates, greater energy dissipation as three plastic hinges form in the member, and greater strength because of the effective length factor, K, being close to 0.5 as opposed to $K = 1.0$ when out-of-plane buckling occurs in a single HSS and single gusset plate member.

5b. Diagonal Braces

The required strength of bracing members with respect to the limit state of net section rupture is the expected brace strength. It should be noted that some, if not all, steel materials commonly used for braces have expected yield strengths significantly higher than their specified minimum yield strengths; some have expected yield strengths almost as high as their expected tensile strength. For such cases, no significant reduction of the brace section is permissible and connections may require local reinforcement of the brace section. This is the case for knife-plate connections between gusset plates and ASTM A53 or A500 braces (for example, pipe, square, rectangular or round HSS braces), where the over-slot of the brace required for erection leaves a reduced section. If this section is left unreinforced, net section rupture will be the governing limit state and brace ductility may be significantly reduced (Korol, 1996; Cheng et al., 1998). Reinforcement may be provided in the form of steel plates welded to the tube, increasing the effective area at the reduced brace section (Yang and Mahin, 2005). Braces with two continuous welds to the gusset wrapped around its edge (instead of the more typical detail with four welds stopping short of the gusset edge) performed adequately in the tests by Cheng. However, this practice may be difficult to implement in field conditions; it also creates a potential stress riser that may lead to crack initiation.

Where there is no reduction in the section, or where the section is reinforced so that the effective net area is at least as great as the brace gross area, this requirement does not apply. The purpose of the requirement is to prevent net section rupture prior to significant ductility; having no reduction in the section is deemed sufficient to ensure this behavior. Reinforcement, if present, should be connected to the brace in a manner that is consistent with the assumed state of stress in the design. It is recommended that the connection of the reinforcement to the brace be designed for the strength of the reinforcement on either side of the reduced section.

The slenderness (KL/r) limit is 200 for braces in SCBF. Research has shown that frames with slender braces designed for compression strength behave well due to the overstrength inherent in their tension capacity. Tremblay (2000), Tang and Goel (1989) and Goel and Lee (1992) have found that the post-buckling cyclic fracture life of bracing members generally increases with an increase in slenderness ratio. An upper limit is provided to preclude dynamic effects associated with extremely slender braces.

Closer spacing of stitches and higher stitch strength requirements are specified for built-up bracing members in SCBF (Aslani and Goel, 1991; Xu and Goel, 1990) than those required for typical built-up members. This is especially critical for double-angle and double-channel braces that impose large shear forces on the stitches upon buckling. These are intended to restrict individual element bending between the stitch points and consequent premature fracture of bracing members. Typical spacing following the requirements of the *Specification* is permitted when buckling does not cause shear in the stitches. Bolted stitches are not permitted within the middle one-fourth of the clear brace length as the presence of bolt holes in that region may cause premature fractures due to the formation of a plastic hinge in the post-buckling range. Studies also showed that placement of double angles in a toe-to-toe configuration reduces bending strains and local buckling (Aslani and Goel, 1991).

5c. **Protected Zones**

Welded or shot-in attachments in areas of inelastic strain may lead to fracture. Such areas in SCBF include gusset plates and expected plastic-hinge regions in the brace.

Figures C-F2.5 and C-F2.6 show the protected zone of an inverted V- and an X-braced frame, respectively. Note that for the X-braced frame, the half-length of the brace is used and a plastic hinge is anticipated at any of the brace quarter points.

6. **Connections**

6a. **Demand Critical Welds**

Groove welds at column splices are designated as demand critical for several reasons. First, although the consequences of a brittle failure at a column splice are not clearly understood, it is believed that such a failure may endanger the safety of the frame. Second, the actual forces that will occur at a column splice during an earthquake are very difficult to predict. The locations of points of inflection in the columns during an earthquake are constantly moving, are ground motion dependent, and cannot be reliably predicted from analysis. Thus, even though analysis of the frame under code specified load combinations (with the amplified seismic load) may show that no tension will occur at a weld, such an analysis cannot be considered reliable for the prediction of these demands. Because of the critical nature of column splices and the

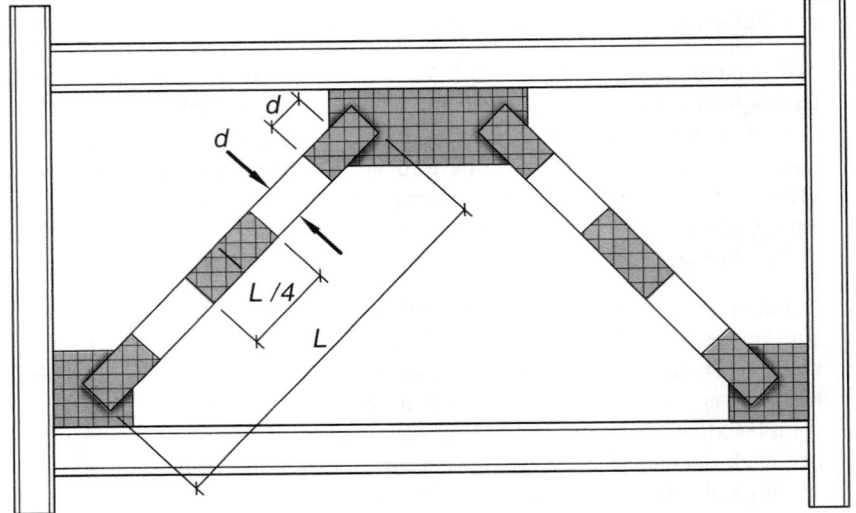

Fig. C-F2.5. Protected zone of inverted V-braced frame.

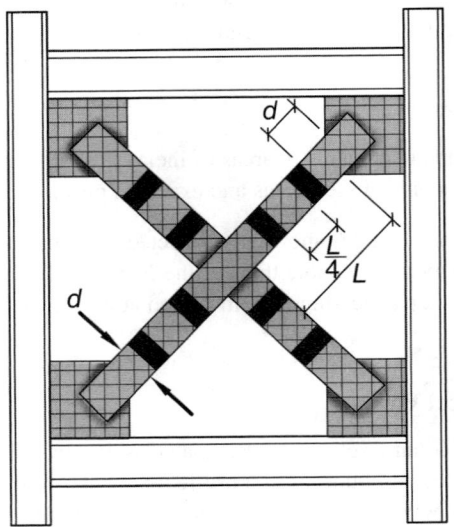

Fig. C-F2.6. Protected zone of X-braced frame.

inability to accurately predict the forces that will occur at these locations, it is the intent of the Provisions that column splices be one of the strongest elements of the frame and be designed in a conservative manner. Accordingly, in order to provide a high degree of protection against brittle failure at column splice groove welds, the use of demand critical welds is specified. PJP groove welds are included in this requirement, because the unfused portion on the weld makes PJP welds particularly prone to brittle failure.

6b. Beam-to-Column Connections

Braced frames are likely to be subject to significant inelastic drift. Thus their connections will undergo significant rotation. Connections with gusset plates can be vulnerable to rupture if they are not designed to accommodate this rotation. Recent testing at UC Berkeley (Uriz and Mahin, 2004) has indicated that designs that do not properly account for the stiffness and distribution of forces in braced frame connections may be subject to undesirable performance.

The provision allows the engineer to select from two options. The first is a simple connection (for which the required rotation is defined as 0.025 rad). An example of this would be a configuration tested at the University of Illinois (Stoakes and Fahnestock, 2010) that effectively allowed rotation between the beam and column. See Figure C-F2.7. (Note that the connection illustrated does not indicate the typical SCBF hinge zone discussed in Commentary Section F2.6c.)

Fahnestock et al. (2006) also tested a connection with rotation capacity outside the gusset plate; this connection is discussed in Commentary Section F4.6c. A similar concept was proposed by Thornton and Muir (2008). See Figure C-F2.8.

The second option is a fully restrained moment connection (for which the maximum moment can be determined from the expected strength of the connecting beam or column. Such connections must meet the same requirements for beam-to-column connections in ordinary moment frames, as specified in Section E1.6.

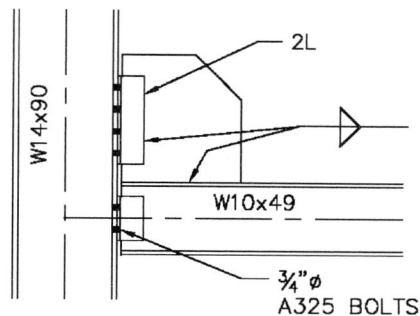

Fig. C-F2.7. Beam-to-column connection that allows rotation
(Stoakes and Fahnestock, 2010).

6c. Required Strength of Brace Connections

Many of the failures reported in concentrically braced frames due to strong ground motions have been in the connections. Similarly, cyclic testing of specimens designed and detailed in accordance with typical provisions for concentrically braced frames has produced connection failures (Astaneh-Asl et al., 1986). Although typical design practice has been to design connections only for axial loads, good post-buckling response demands that eccentricities be accounted for in the connection design, which should be based upon the maximum loads the connection may be required to resist. Good connection performance can be expected if the effects of brace member cyclic post-buckling behavior are considered.

Certain references suggest limiting the free edge length of gusset plates, including SCBF brace-to-beam connection design examples in the *Seismic Design Manual*, (AISC, 2006), and other references (Astaneh-Asl et al., 2006; ICC, 2006). However, the committee has reviewed the testing cited and has concluded that such edge stiffeners do not offer any advantages in gusset plate behavior. There is therefore no limitation on edge dimensions in these provisions.

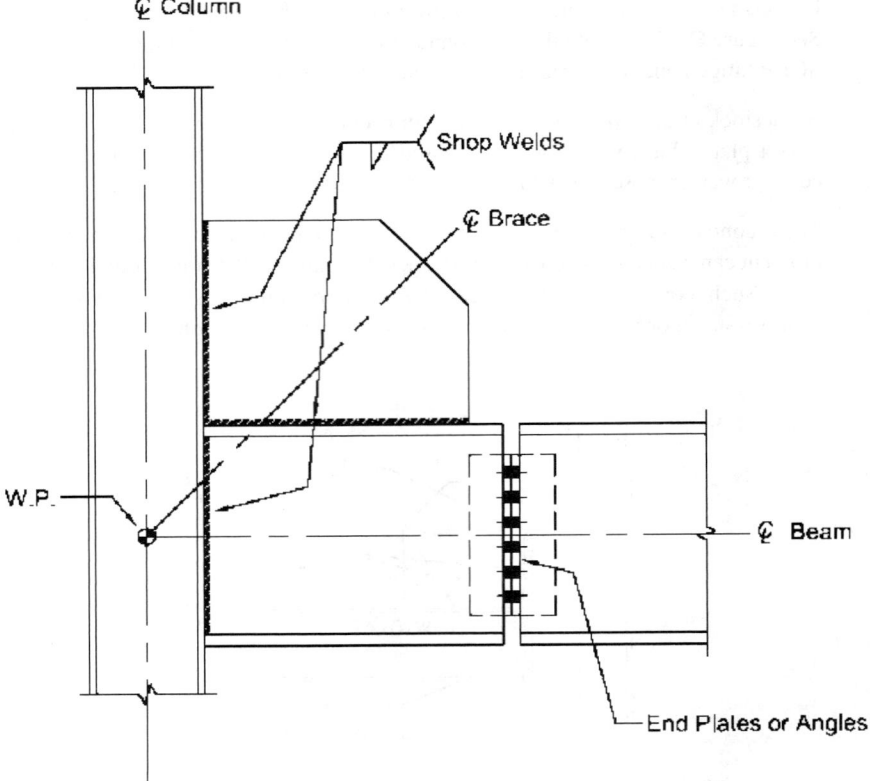

Fig. C-F2.8. Beam-to-column connection that allows rotation (Thornton and Muir, 2008).

(1) Required Tensile Strength

Braces in SCBF are required to have gross section tensile yielding as their governing limit state so that they will yield in a ductile manner. Local connection failure modes such as block shear rupture must be precluded. Therefore, the calculations for these failure modes must use the maximum load that the brace can develop.

The minimum of two criteria (the expected axial tensile strength of the bracing member and the maximum force that could be developed by the overall system) determines the required strength of both the bracing connection and the forces delivered to the beam-to-column connection. This second limit is included in the specification for structures where elements other than the tension bracing limit the system strength. Depending on the specific situation(s), there are a number of ways one can determine the maximum force transferred to the connection. They include:

(1) Perform a pushover analysis to determine the forces acting on the connections when the maximum frame capacity (leading to an imminent collapse mechanism) is reached.

(2) Determine how much force can be resisted before causing uplift of a spread footing (note that the foundation design forces are not required to resist more than the code base shear level). This type of relief is not typically applicable to a deep foundation since the determination of when uplift will occur is not easy to determine with good accuracy.

(3) Perform a suite of inelastic time history analyses and envelop the connection demands.

Calculating the maximum connection force by one of the three methods noted above is not a common practice on design projects. In some cases, such an approach could result in smaller connection demands. But, from a conceptual basis, since the character of the ground motions is not known to any great extent, it is unrealistic to expect that such forces can be accurately calculated. All three approaches rely on an assumed distribution of lateral forces which may not match reality (approach 3 probably is the best estimate, but also the most calculation intensive). In most cases, providing the connection with a capacity large enough to yield the member is needed because of the large inelastic demands placed on a structure by a major earthquake.

Requirements specific to member net section rupture are included in Section F2.5b.

Bolt slip has been removed as a limit state which must be precluded. The consequences of exceeding this limit state in the maximum credible earthquake are not considered severe if bearing failure and block-shear rupture are precluded.

(2) Required Compressive Strength

Bracing connections should be designed to withstand the maximum force that the brace can deliver in compression. A factor of 1.1 has been adopted here in part due to the use of conservative column curve equations in determining this force.

(3) Accommodation of Brace Buckling

Braces in SCBF are expected to undergo cyclic buckling under severe ground motions, forming plastic hinges at their center and at each end. To prevent fracture resulting from brace rotations, bracing connections must either have sufficient strength to confine inelastic rotation to the bracing member or sufficient ductility to accommodate brace end rotations.

For brace buckling in the plane of the gusset plates, the end connections should be designed to resist the expected compressive strength and the expected flexural strength of the brace as it transitions from pure compression to pure flexure (Astaneh-Asl et al., 1986). Note that a realistic value of K should be used to represent the connection fixity.

For brace buckling out of the plane of single plate gussets, weak-axis bending in the gusset is induced by member end rotations. This results in flexible end conditions with plastic hinges at midspan in addition to the hinges that form in the gusset plate. Satisfactory performance can be ensured by allowing the gusset plate to develop restraint-free plastic rotations. This requires that the free length between the end of the brace and the assumed line of restraint for the gusset be sufficiently long to permit plastic rotations, yet short enough to preclude the occurrence of plate buckling prior to member buckling. A length of two times the plate thickness is recommended (Astaneh-Asl et al., 1986). Note that this free distance is measured from the end of the brace to a line that is perpendicular to the brace centerline, drawn from the point on the gusset plate nearest to the brace end that is constrained from out-of-plane rotation.

This condition is illustrated in Figure C-F2.9 and provides hysteretic behavior as illustrated in Figure C-F2.10. The distance of $2t$ shown in Figure C-F2.9 should be considered the minimum offset distance. In practice, it may be advisable to specify a slightly larger distance (for example, $2t + 1$ in.) on construction documents to provide for erection tolerances. More information on seismic design of gusset plates can be obtained from Astaneh-Asl (1998).

Alternatively, connections with stiffness in two directions, such as cross gusset plates, can be detailed. Test results indicate that forcing the plastic hinge to occur in the brace rather than the connection plate results in greater energy dissipation capacity (Lee and Goel, 1987).

Where fixed end connections are used in one axis with pinned connections in the other axis, the effect of the fixity should be considered in determining the critical buckling axis.

6d. Column Splices

In the event of a major earthquake, columns in concentrically braced frames can undergo significant bending beyond the elastic range after buckling and yielding of the braces. Even though their bending strength is not utilized in the design process when elastic design methods are used, columns in SCBF are required to have adequate compactness and shear and flexural strength in order to maintain their lateral strength during large cyclic deformations of the frame. In addition, column splices are required to have sufficient strength to prevent failure under expected post-elastic forces. Analytical studies on SCBF that are not part of a dual system have shown that columns can carry as much as 40% of the story shear (Tang and Goel, 1987; Hassan and Goel, 1991). When columns are common to both SCBF and special moment frames (SMF) in a dual system, their contribution to story shear may be as high as 50%. This feature of SCBF greatly helps in making the overall frame hysteretic loops

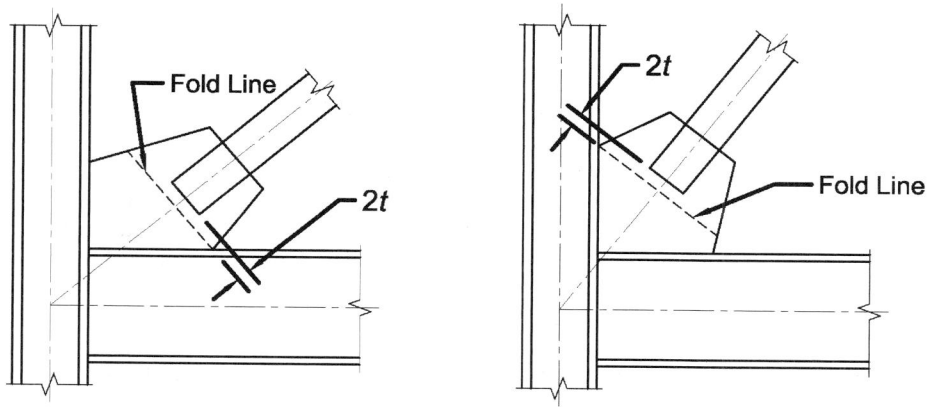

Fig. C-F2.9. Brace-to-gusset plate requirement for buckling out-of-plane bracing system.

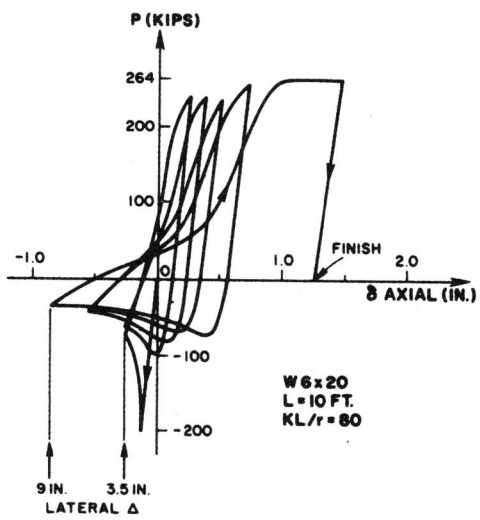

Fig. C-F2.10. P-δ diagram for a strut.

"full" when compared with those of individual bracing members which are generally "pinched" (Hassan and Goel, 1991; Black et al., 1980). See Figure C-F2.11.

F3. ECCENTRICALLY BRACED FRAMES (EBF)

1. Scope

Eccentrically braced frames (EBF) are composed of columns, beams and braces. The distinguishing characteristic of an EBF is that at least one end of every brace is connected so that the brace force is transmitted through shear and bending of a short beam segment, called the link, defined by a horizontal eccentricity between the intersection points of the two brace centerlines with the beam centerline (or between the intersection points of the brace and column centerlines with the beam centerline for links adjacent to columns). In contrast with concentrically braced frames, beams in EBF are always subject to high shear and bending forces. Figure C-F3.1 illustrates some examples of eccentrically braced frames and the key components of an EBF: the links, the beam segments outside of the links, the diagonal braces, and the columns.

These provisions are primarily intended to cover the design of EBF in which the link is a horizontal framing member located between the column and a brace or between two braces. For the inverted Y-braced EBF configuration shown in Figure C-F3.1(d), the link is attached underneath the beam. If this configuration is to be used, lateral bracing should be provided at the intersection of the diagonal braces and the vertical link, unless calculations are provided to justify the design without such bracing.

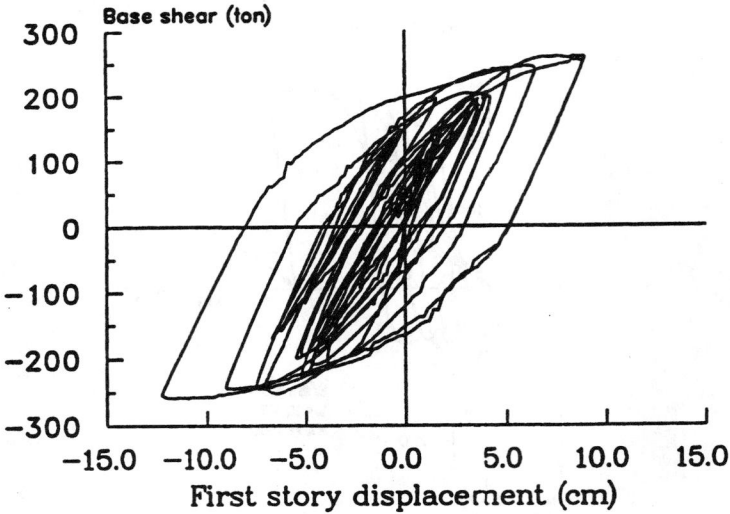

Fig. C-F2.11. Base shear vs. story drift of a SCBF.

2. Basis of Design

Research has shown that EBF can provide an elastic stiffness that is comparable to that of SCBF and OCBF, particularly when short *link* lengths are used, and excellent ductility and energy dissipation capacity in the inelastic range, comparable to that of SMF, provided that the links are not too short (Roeder and Popov; 1978; Libby, 1981; Merovich et al., 1982; Hjelmstad and Popov, 1983; Malley and Popov, 1984; Kasai and Popov, 1986a, 1986b; Ricles and Popov, 1987a, 1987b; Engelhardt and Popov, 1989a, 1989b; Popov et al., 1989). Inelastic action in EBF under seismic loading is

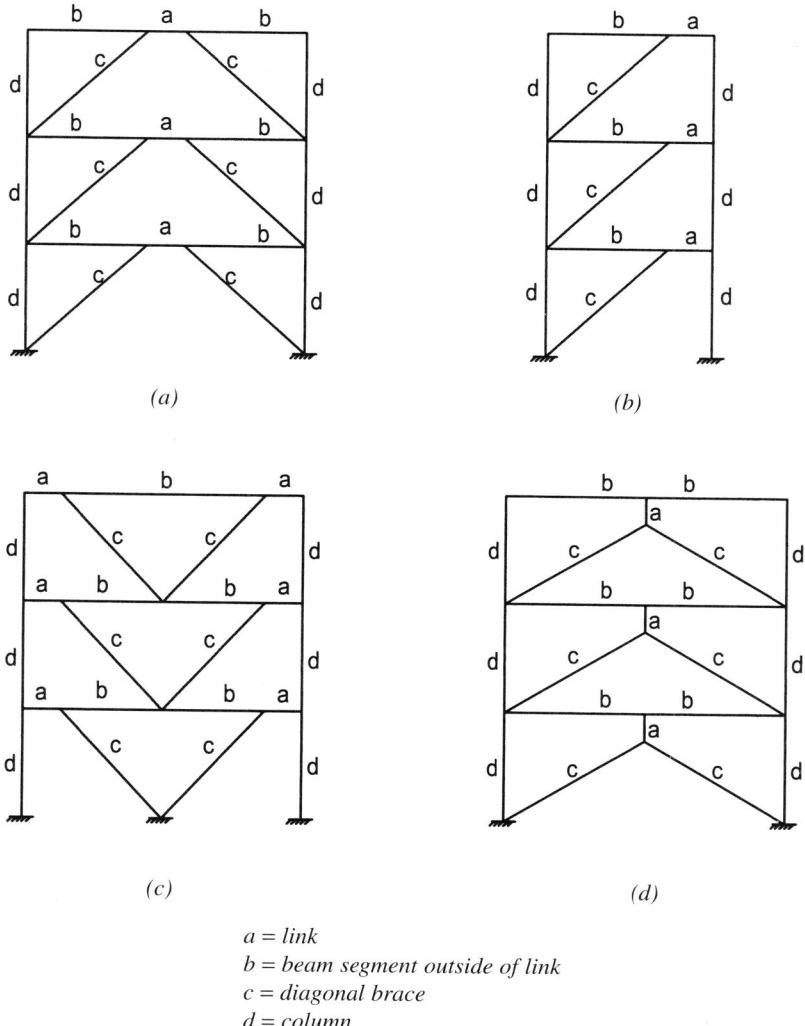

a = link
b = beam segment outside of link
c = diagonal brace
d = column

Fig. C-F3.1. Examples of eccentrically braced frames.

restricted primarily to the links. These provisions are intended to ensure that cyclic yielding in the links can occur in a stable manner while the diagonal braces, columns, and portions of the beam outside of the link remain essentially elastic under the forces that can be developed by fully yielded and strain-hardened links.

In some bracing arrangements, such as that illustrated in Figure C-F3.2 with links at each end of the brace, links may not be fully effective. If the upper link has a significantly lower design shear strength than that of the link in the story below, the upper link will deform inelastically and limit the force that can be developed in the brace and to the lower link. When this condition occurs the upper link is termed an active link and the lower link is termed an inactive link. The presence of potentially inactive links in an EBF increases the difficulty of analysis.

It can be shown with plastic frame analyses that, in some cases, an inactive link will yield under the combined effect of dead, live and earthquake loads, thereby reducing the frame strength below that expected (Kasai and Popov, 1984). Furthermore, because inactive links are required to be detailed and constructed as if they were active, and because a predictably inactive link could otherwise be designed as a pin, the cost of construction is needlessly increased. Thus, an EBF configuration that ensures that all links will be active, such as those illustrated in Figure C-F3.1, are recommended. Further recommendations for the design of EBF are available (Popov et al., 1989).

Columns in EBF are designed following capacity design principles so that the full strength and deformation capacity of the frame can be developed without failure of any individual column and without the formation of a soft story. While this does not represent a severe penalty for low-rise buildings, it is difficult to achieve for taller structures, which may have link beam sizes governed by drift-control considerations. In such cases it is anticipated that designers will adopt nonlinear analysis techniques as discussed in Chapter C.

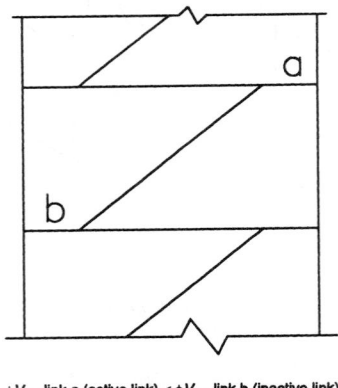

ϕV_n – link a (active link) < ϕV_n – link b (inactive link)

Fig. C-F3.2. EBF—active and inactive links.

Plastic hinge formation in columns should be avoided because, when combined with hinge formation in the links, it can result in the formation of a soft story. The requirements of Sections D1.4a and F3.3 address the required strength for column design.

Additional design requirements have been added to the Provisions to address the special case of box links (those consisting of built-up tubular cross sections). Box links are generally not susceptible to lateral-torsional buckling, and eccentrically braced frames having such links have been shown (Berman and Bruneau, 2007, 2008a, 2008b) to perform in a ductile manner without the need for lateral bracing of the link beam, provided the specified section compactness requirements are met. This can be of benefit when eccentrically braced frames are desirable in locations where such lateral bracing cannot be achieved, such as between two elevator cores, or along the facade of building atriums.

3. Analysis

The required strength of links is typically determined based on the analysis required by ASCE/SEI 7. The analysis required by this section is used in determining the required strength of braces, columns, beams outside the link and columns, as well as brace connections. The requirements presented here are essentially a reformatting of design rules for these elements into an analysis format.

The intent of the Provisions is to assure that yielding and energy dissipation in an EBF occur primarily in the links. Consequently, the columns, diagonal braces and beam segment outside of the link must be designed to resist the loads developed by the fully yielded and strain hardened link. That is, the brace and beam should be designed following capacity design principles to develop the full inelastic capacity of the links. Limited yielding outside of the links, particularly in the beams, is sometimes unavoidable in an EBF. Such yielding is likely not detrimental to the performance of the EBF, as long as the beam and brace have sufficient strength to develop the link's full inelastic strength and deformation capacity.

In most EBF configurations, the diagonal brace and the beam are subject to large axial loads combined with significant bending moments. Consequently, both the diagonal brace and the beam should be designed as beam-columns.

The diagonal brace and beam segment outside of the link must be designed for some reasonable estimate of the maximum forces that can be developed by the fully yielded and strain hardened link. For this purpose, the nominal shear strength of the link, V_n, as defined by Equation C-F3-1 is increased by two factors. First, the nominal shear strength is increased by R_y to account for the possibility that the link material may have actual yield strength in excess of the specified minimum value. Secondly, the resulting expected shear strength of the link, $R_y V_n$, is further increased to account for strain hardening in the link.

Experiments have shown that links can exhibit a high degree of strain hardening. Recent tests on rolled wide-flange links constructed of ASTM A992 steel (Arce, 2002) showed strength increases due to strain hardening ranging from 1.2 to 1.45, with an average value of about 1.30. Past tests on rolled wide-flange links constructed of

ASTM A36 steel have sometimes shown strength increases due to strain hardening in excess of 1.5 (Hjelmstad and Popov, 1983; Engelhardt and Popov, 1989a). Further, recent tests on very large welded built-up wide-flange links for use in major bridge structures have shown strain hardening factors close to 2.0 (McDaniel et al., 2002; Dusicka and Itani, 2002). These sections, however, typically have proportions significantly different from rolled shapes.

Past researchers have generally recommended a factor of 1.5 (Popov and Engelhardt, 1988) to account for expected link strength and its strain hardening in the design of the diagonal brace and beam outside of the link. However, for purposes of designing the diagonal brace, these Provisions have adopted a strength increase due to strain hardening only equal to 1.25. This factor was chosen to be less than 1.5 for a number of reasons, including the use of the R_y factor to account for expected material strength in the link but not in the brace, and the use of resistance factors or safety factors when computing the strength of the brace. Further, this value is close to but somewhat below the average measured strain hardening factor for recent tests on rolled wide-flange links of ASTM A992/A992M steel. Designers should recognize that strain hardening in links may sometimes exceed this value, and so a conservative design of the diagonal brace is appropriate. Additionally, if large built-up link sections are used with very thick flanges and very short lengths ($e < M_p/V_p$), designers should consider the possibility of strain hardening factors substantially in excess of 1.25 (Richards, 2004).

Based on the above, the required strength of the diagonal brace can be taken as the forces developed by the following values of link shear and link end moment:

For $e \leq \dfrac{2M_p}{V_p}$

    Link shear $= 1.25 R_y V_p$

    Link end moment $= \dfrac{e(1.25 R_y V_p)}{2}$

For $e > \dfrac{2M_p}{V_p}$

    Link shear $= \dfrac{2(1.25 R_y M_p)}{e}$

    Link end moment $= 1.25 R_y M_p$

The above equations assume link end moments will equalize as the link yields and deforms plastically. For link lengths less than $1.6 M_p/V_p$ attached to columns, link end moments do not fully equalize (Kasai and Popov, 1986a). For this situation, the link ultimate forces can be estimated as follows:

For links attached to columns with $e \leq \dfrac{1.6 M_p}{V_p}$

    Link shear $= 1.25 R_y V_p$

    Link end moment at column $= R_y M_p$

    Link end moment at brace $= \left[e(1.25 R_y V_p) - R_y M_p \right] \geq 0.75 R_y M_p$

The link shear force will generate axial force in the diagonal brace, and for most EBF configurations, will also generate substantial axial force in the beam segment outside of the link. The ratio of beam or brace axial force to link shear force is controlled primarily by the geometry of the EBF and is therefore not affected by inelastic activity within the EBF (Engelhardt and Popov, 1989a). Consequently, this ratio can be determined from an elastic frame analysis and can be used to amplify the beam and brace axial forces to a level that corresponds to the link shear force specified in the above equations. Further, as long as the beam and brace are designed to remain essentially elastic, the distribution of link end moment to the beam and brace can be estimated from an elastic frame analysis.

This is typically done by multiplying the beam and brace forces by the ratio of the expected, strain-hardened link shear strength to the link shear demand from the analysis. One could also use a free-body diagram to determine these forces based on the link strength, and apportion moments based on the elastic analysis. For example, if an elastic analysis of the EBF under lateral load shows that 80% of the link end moment is resisted by the beam and the remaining 20% is resisted by the brace, the ultimate link end moments given by the above equations can be distributed to the beam and brace in the same proportions. Care should be taken in this latter approach if the centerline intersections fall outside the link; see Commentary Section F3.5b.

Finally, an inelastic frame analysis can be conducted for a more accurate estimate of how link end moment is distributed to the beam and brace in the inelastic range.

As described above, these Provisions assume that as a link deforms under large plastic rotations, the link expected shear strength will increase by a factor of 1.25 due to strain hardening. However, for the design of the beam segment outside of the link, the Provisions permit reduction of the seismic force by a factor of 0.88, consistent with the 1.1 factor in the 2005 Provisions (1.25(0.88) = 1.1). This relaxation on link ultimate forces for purposes of designing the beam segment reflects the view that beam strength will be substantially enhanced by the presence of a composite floor slab, and also that limited yielding in the beam will not likely be detrimental to EBF performance, as long as stability of the beam is assured. Consequently, designers should recognize that the actual forces that will develop in the beam will be substantially greater than computed using this 1.1 factor, but this low value of required beam strength will be mitigated by contributions of the floor slab in resisting axial load and bending moment in the beam and by limited yielding in the beam. Based on this approach, a strain hardening factor of 1.25 is called for in the analysis for I-shaped links. The resulting axial force and bending moment in the beam can then be reduced by a factor of 1.1/1.25 = 0.88. In cases where no composite slab is present, designers should consider computing required beam strength based on a link strain hardening factor of 1.25.

Design of the beam segment outside of the link can sometimes be problematic in EBF. In some cases, the beam segment outside of the link is inadequate to resist the required strength based on the link ultimate forces. For such cases, increasing the size of the beam may not provide a solution because the beam and the link are typically the same member. Increasing the beam size therefore increases the link size, which in turn, increases the link ultimate forces and therefore increases the beam required strength. The relaxation in beam required strength based on the 1.1 factor on link strength was adopted by the Provisions largely as a result of such problems reported by designers, and by the view that EBF performance would not likely be degraded by such a relaxation due to beneficial effects of the floor slab and limited beam yielding, as discussed above. Design problems with the beam can also be minimized by using shear yielding links ($e \leq 1.6M_p/V_p$) as opposed to longer links. The end moments for shear yielding links will be smaller than for longer links, and consequently less moment will be transferred to the beam. Beam moments can be further reduced by locating the intersection of the brace and beam centerlines inside of the link, as described below. Providing a diagonal brace with a large flexural stiffness so that a larger portion of the link end moment is transferred to the brace and away from the beam can also substantially reduce beam moment. In such cases, the brace must be designed to resist these larger moments. Further, the connection between the brace and the link must be designed as a fully restrained moment resisting connection. Test results on several brace connection details subject to axial load and bending moment are reported in Engelhardt and Popov (1989a). Finally, built-up members can be considered for link beam design.

High axial forces in the beam outside the link can complicate beam selection if the beam outside the link and the link beam are the same member, as is typical. These axial forces can be reduced or eliminated by selection of a beneficial configuration. Frames with center links may be reconfigured to eliminate beam axial forces from levels above by adopting a two-story-X configuration as proposed by Engelhardt and Popov (1989b) and shown in Figure C-F3.3. Frames with the link at the column share the frame shear between the brace and the column at the link. Selection of beneficial bay size and link length can maximize the percentage of the frame shear resisted by the column, thus minimizing the horizontal component of the brace force and consequently minimizing the axial force in the beam outside the link of the level below. More specifically, avoiding very shallow angles (less than 40°) between the diagonal brace and the beam is recommended (Engelhardt et al., 1992).

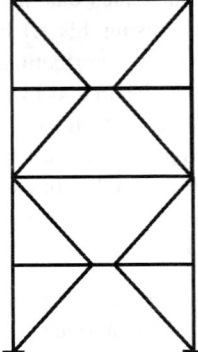

Fig. C-F3.3. Two-story X EBF configuration (Engelhardt and Popov, 1989a).

The required strength of the diagonal brace connections in EBF is the same as the required strength of the diagonal brace. Similar to the diagonal brace and beam segment outside of the link, the columns of an EBF should also be designed using capacity design principles. That is, the columns should be designed to resist the maximum forces developed by the fully yielded and strain hardened links. As discussed in Commentary Section F3.5b and in this section, the maximum shear force developed by a fully yielded and strain hardened link can be estimated as $1.25R_y$ times the link nominal shear strength, V_n, where the 1.25 factor accounts for strain hardening. For capacity design of the columns, this section permits reduction of the strain hardening factor to 1.1 (by multiplying seismic forces by 0.88; 1.25(0.88) = 1.1). This relaxation reflects the view that all links above the level of the column under consideration will not likely reach their maximum shear strength simultaneously. Consequently, applying the 1.25 strain hardening factor to all links above the level of the column under consideration is likely too conservative for a multistory EBF. For a low-rise EBF with only a few stories, designers should consider increasing the strain hardening factor on links to 1.25 for capacity design of the columns, since there is a greater likelihood that all links may simultaneously reach their maximum shear strength. For taller buildings this factor of 1.1 is likely overly conservative. No reliable methods have been developed for estimating such reduced forces on the basis of a linear analysis; designers may elect to perform a nonlinear analysis per Chapter C.

In addition to the requirements of this Section, columns in EBF must also be checked in accordance with the requirements of Section D1.4a, which are applicable to all systems.

Tests showed (Berman and Bruneau 2006, 2008a, 2008b) that strain hardening is larger for links with built-up box cross sections than for wide-flange links. Comparing the over-strength obtained for box links compared to that obtained for wide-flange links by Richards (2004), Berman and Bruneau indicated that built-up box rectangular links have a maximum strength typically 11% larger than wide-flange links. The forces to consider for the design of the braces, beams (outside the link), and columns are therefore increased accordingly.

4.   System Requirements

4a. Link Rotation Angle

The total link rotation angle is the basis for controlling tests on link-to-column connections, as described in Section K2.4c. In a test specimen, the total link rotation angle is computed by simply taking the relative displacement of one end of the link with respect to the other end, and dividing by the link length. The total link rotation angle reflects both elastic and inelastic deformations of the link, as well as the influence of link end rotations. While the total link rotation angle is used for test control, acceptance criteria for link-to-column connections are based on the link inelastic rotation angle.

To assure satisfactory behavior of an EBF, the inelastic deformation expected to occur in the links in a severe earthquake should not exceed the inelastic deformation capacity of the links. In the Provisions, the link rotation angle is the primary variable used to describe inelastic link deformation. The link rotation angle is the plastic rotation angle between the link and the portion of the beam outside of the link.

The link rotation angle can be estimated by assuming that the EBF bay will deform in a rigid-plastic mechanism as illustrated for various EBF configurations in Figure C-F3.4. In this figure, the link rotation angle is denoted by the symbol γ_p. The link rotation angle can be related to the plastic story drift angle, θ_p, using the relationships shown in the Figure C-F3.4. The plastic story drift angle, in turn, can be computed as the plastic story drift, Δ_p, divided by the story height, h. The plastic story drift is equal to the difference between the design story drift and the elastic drift. Alternatively, the link rotation angle can be determined more accurately by inelastic dynamic analyses.

The inelastic response of a link is strongly influenced by the length of the link as related to the ratio, M_p/V_p, of the link cross section. When the link length is

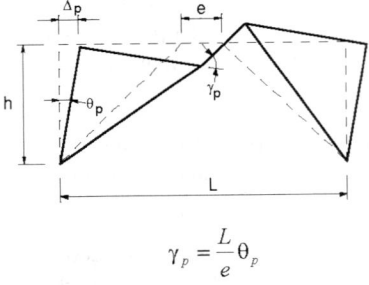

$$\gamma_p = \frac{L}{e}\theta_p$$

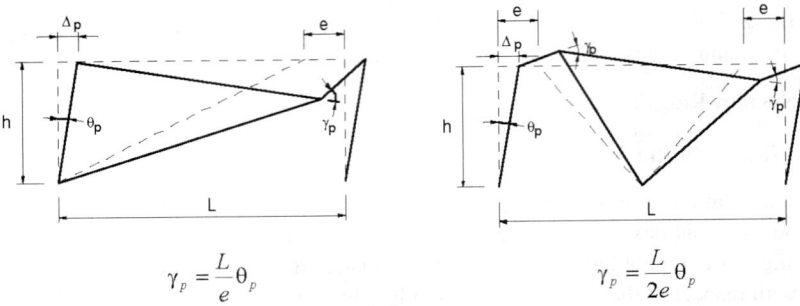

$$\gamma_p = \frac{L}{e}\theta_p \qquad \gamma_p = \frac{L}{2e}\theta_p$$

L = bay width
h = story height
Δ_p = plastic story drift
θ_p = plastic story drift angle, rad (= Δ_p/h)
γ_p = plastic link rotation angle, rad

Fig. C-F3.4. Link rotation angle.

selected not greater than $1.6M_p/V_p$, shear yielding will dominate the inelastic response. If the link length is selected greater than $2.6M_p/V_p$, flexural yielding will dominate the inelastic response. For link lengths intermediate between these values, the inelastic response will occur through some combination of shear and flexural yielding. The inelastic deformation capacity of links is generally greatest for shear yielding links, and smallest for flexural yielding links. Based on experimental evidence, the link rotation angle is limited to 0.08 rad for shear yielding links ($e \leq 1.6M_p/V_p$) and 0.02 rad for flexural yielding links ($e \geq 2.6M_p/V_p$). For links in the combined shear and flexural yielding range ($1.6M_p/V_p < e < 2.6M_p/V_p$), the limit on link rotation angle is determined according to link length by linear interpolation between 0.08 and 0.02 rad.

It has been demonstrated experimentally (Whittaker et al., 1987; Foutch, 1989) as well as analytically (Popov et al., 1989) that links in the first floor usually undergo the largest inelastic deformation. In extreme cases this may result in a tendency to develop a soft story. The plastic link rotations tend to attenuate at higher floors and decrease with the increasing frame periods. Therefore for severe seismic applications, a conservative design for the links in the first two or three floors is recommended. This can be achieved by providing links with an available shear strength at least 10% over the required shear strength.

4b.   Bracing of Link

Lateral restraint against out-of-plane displacement and twist is required at the ends of the link to ensure stable inelastic behavior. This Section specifies the required strength and stiffness of link end lateral bracing. In typical applications, a composite deck can likely be counted upon to provide adequate lateral bracing at the top flange of the link. However, a composite deck alone cannot be counted on to provide adequate lateral bracing at the bottom flange of the link and direct bracing through transverse beams or a suitable alternative is recommended.

A link with a built-up box cross section, tested without lateral bracing in a full EBF configuration, exhibited no lateral-torsional buckling (Berman and Bruneau, 2007). Slender box cross sections (significantly taller than wide) could develop lateral-torsional buckling, but the unbraced length required to do so for such sections is still considerably longer than for wide-flange links. As a result, except for unusual aspect ratios, links with built-up box cross-sections will not require lateral bracing. While no physical lateral bracing is required to ensure satisfactory seismic performance of links with built-up box sections designed as specified in the Provisions, a lateral load acting outside of the frame plane and applied at the brace-to-beam points has been conservatively specified, together with a stiffness requirement, to prevent the use of link beams that would be too weak or flexible (out-of-plane of the frame) to provide lateral restraint to the brace.

5.   Members

5a.   Basic Requirements

The ductility demands in EBF are concentrated in the links. Braces, columns and beams outside the link should have very little yielding in a properly designed EBF.

As long as the brace is designed to be stronger than the link, as is the intent of these provisions, then the link will serve as a fuse to limit the maximum load transferred to the brace, thereby precluding the possibility of brace buckling. Consequently, many of the design provisions for braces in SCBF systems intended to permit stable cyclic buckling of braces are not needed in EBF. Similarly, the link also limits the loads transferred to the beam beyond the link, thereby precluding failure of this portion of the beam if it is stronger than the link and to the columns.

For most EBF configurations, the beam and the link are a single continuous wide flange member. If this is the case, the available strength of the beam can be increased by R_y. If the link and the beam are the same member, any increase in yield strength present in the link will also be present in the beam segment outside of the link.

5b.   Links

Inelastic action in EBF is intended to occur primarily within the links. The general provisions in this Section are intended to ensure that stable inelasticity can occur in the link.

At brace connections to the link, the link length is defined by the edge of the brace connection (see Figure C-F3.5). Brace connection details employing gussets are commonly configured so that the gusset edge aligns vertically with the intersection of the brace and beam centerlines. For brace connections not employing gussets, the intersection of the brace at the link end may not align vertically with the intersection of the brace and beam centerlines; the intersection of centerlines may fall within the link (Figure C-F3.5) or outside of the link (Figure C-F3.7). Bracing using HSS members is shown in Figure C-F3.6. In either case, flexural forces in the beam outside the link and the brace may be obtained from an analysis that models the member centerline intersections, provided that the force level in the analysis corresponds to the expected strain-hardened link capacity as required by Section F3.3.

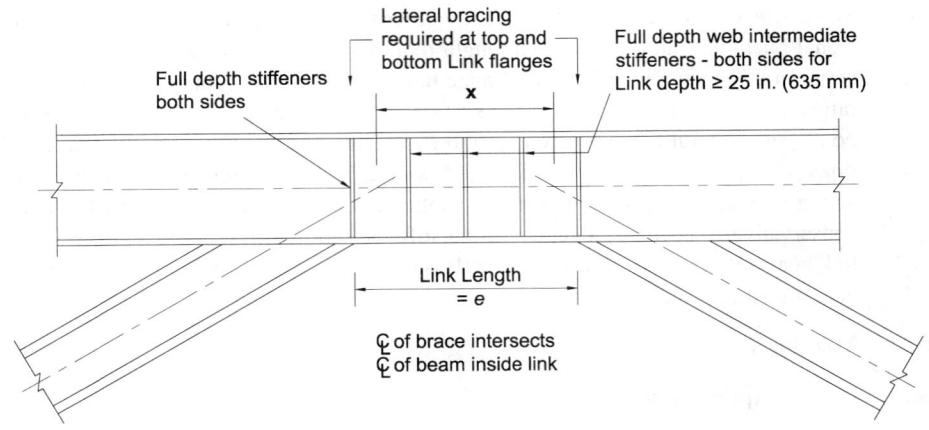

Fig.C-F3.5. EBF with W-shape bracing ($x < e$).

However, such a centerline analysis will not produce correct link end moments. See Commentary to item (1) below and Figure C-F3.5. Link end moments for either case can more accurately be obtained using the following equation:

$$M = Ve/2$$

Where V is the link beam shear in the condition under consideration (whether it be corresponding to the design base shear or to the fully yielded, strain-hardened link as required in Section F3.3).

However, link end moments are not directly used in selecting the link member in the typical design procedure. Section F3.5b(2) converts link flexural strength to an

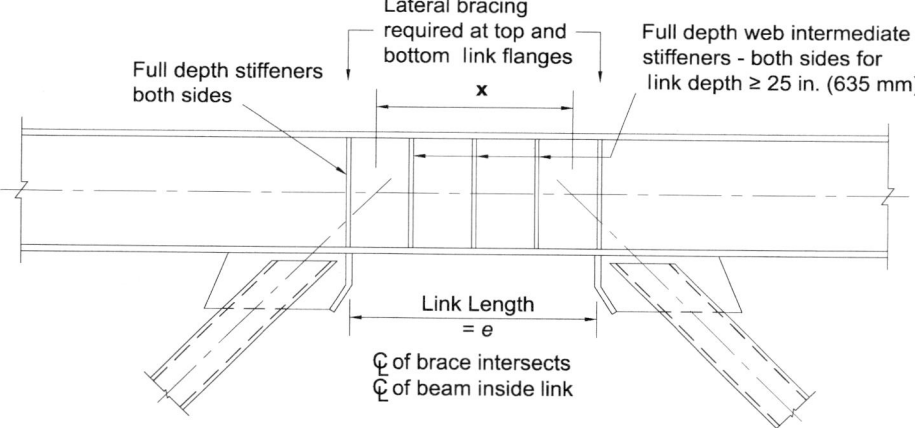

Fig. C-F3.6. EBF with HSS bracing ($x < e$).

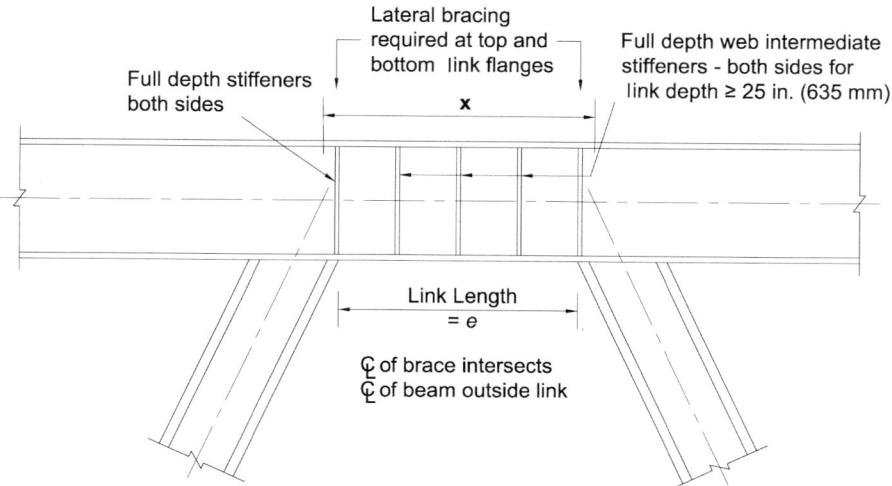

Fig. C-F3.7 EBF with W-shape bracing ($x > e$).

equivalent shear strength based on link length. Comparison of that equivalent shear strength to the required shear strength is sufficient for design and the results of a centerline model analysis can be used without modification.

(1) Limitations

Width-to-thickness limits for links are specified in Table D1.1. Previous editions of these Provisions required the link cross section to meet the same width-to-thickness criteria as is specified for beams in SMF. Based on research on local buckling in links (Okazaki et al., 2004a; Richards et al., 2004), the flange width-to-thickness limits for links are only required to meet the compactness limits for moderately ductile members. This new limit corresponds to λ_p in Table B4.1b of the *Specification*. Limits on slenderness of link built-up box cross sections are provided to prevent links that are significantly taller than wide (that could develop lateral-torsional buckling). Based on research by Berman and Bruneau (2008a, 2008b), it is recommended that, for built-up box links with link lengths $e \leq 1.6M_p/V_p$, the web width-to-thickness ratio be limited to $1.67\sqrt{E/F_y}$. For built-up box links with link lengths $e > 1.6M_p/V_p$, it is recommended that the web width-to-thickness ratio be limited to $0.64\sqrt{E/F_y}$. Specimens with links other than at mid-width of the braced bay have not been tested.

The reinforcement of links with web doubler plates is not permitted as such reinforcement may not fully participate as intended in inelastic deformations. Additionally, beam web penetrations within the link are not permitted because they may adversely affect the inelastic behavior of the link.

The 2005 Provisions required that the intersection of the beam and brace centerlines should occur at the end of the link, or inside of the link. The reason for this restriction was that when the intersection of the beam and brace centerlines occur outside of the link, additional moment is generated in the beam outside of the link. However, locating the intersection of the beam and brace centerline outside of the link is sometimes unavoidable for certain member sizes and brace connection geometries. Further, it is acceptable to locate the intersection outside of the link, as long as the additional moment in the beam is considered in the design. Consequently, the restriction has been removed to allow greater flexibility in EBF design.

When the distance between intersection of the beam and brace centerlines, x, exceeds the link length, e, as is shown in Figure C-F3.7, the total moment resisted by the beam outside the link and the brace (if moment-connected) exceeds the link end moment. Conversely if the link length, e, exceeds the distance between the intersection of the beam and brace centerlines, x, as is shown in Figures C-F3.5 and C-F3.6, the link end moment at the design level will exceed the forces indicated using a centerline model. In both conditions, care should be taken to ensure sufficient strength at the design level and proper estimation of forces in the beam outside the link and in the brace at drifts corresponding to a fully yielded, strain-hardened link.

(2) Shear Strength

The nominal shear strength of the link, V_n, is the lesser of that determined from the plastic shear strength of the link section or twice the plastic moment divided by the link length, as dictated by statics assuming equalization of end moments in the inelastic range of behavior. Accordingly, the nominal shear strength of the link can be computed as follows:

$$V_n = \begin{cases} V_p & \text{for } e \leq \dfrac{2M_p}{V_p} \\ \dfrac{2M_p}{e} & \text{for } e > \dfrac{2M_p}{V_p} \end{cases} \quad \text{(C-F3-1)}$$

The effects of axial load on the link can be ignored if the required axial strength on the link does not exceed 15% of the nominal yield strength of the link, P_y. In general, such an axial load is negligible because the horizontal component of the brace load is transmitted to the beam segment outside of the link. However, when the framing arrangement is such that larger axial forces can develop in the link, such as from drag struts or a modified EBF configuration, the available shear strength and the length of the link are reduced (according to Sections F3.5b(2) and F3.5b(3), respectively).

(3) Link Length

The rotations that can be achieved in links subject to flexural yielding with high axial forces have not been adequately studied. Consequently where high axial forces can develop in the link, its length is limited to ensure that shear yielding, rather than flexural yielding, governs and thus to ensure stable inelastic behavior.

(4) Link Stiffeners for I-Shaped Cross Sections

A properly detailed and restrained link web can provide stable, ductile and predictable behavior under severe cyclic loading. The design of the link requires close attention to the detailing of the link web thickness and stiffeners.

Full-depth stiffeners are required at the ends of all links and serve to transfer the link shear forces to the reacting elements as well as restrain the link web against buckling.

The maximum spacing of link intermediate web stiffeners in shear yielding links ($e \leq 1.6M_p/V_p$) is dependent upon the size of the link rotation angle (Kasai and Popov, 1986b) with a closer spacing required as the rotation angle increases. Intermediate web stiffeners in shear yielding links are provided to delay the onset of inelastic shear buckling of the web. Flexural yielding links having lengths greater than or equal to $2.6M_p/V_p$ but less than $5M_p/V_p$ are required to have an intermediate stiffener at a distance from the link end equal to 1.5 times the beam flange width to limit strength degradation due to flange local buckling and lateral-torsional buckling. Links of a length that are between the shear and flexural limits are required to meet the stiffener requirements for both shear and flexural yielding links. When the link length exceeds $5M_p/V_p$, link intermediate web stiffeners are not required. Link intermediate web stiffeners are required to extend full depth in order to effectively resist shear buckling of the web and to effectively limit strength degradation due to flange local buckling and lateral-torsional buckling. Link intermediate web stiffeners are required on both sides of the web for links 25 in. (635 mm) in depth or greater. For links that are less than 25 in. (635 mm) deep, the stiffener need be on one side only.

All link stiffeners are required to be fillet welded to the link web and flanges. Link stiffeners should be detailed to avoid welding in the k-area of the link. Recent research has indicated that stiffener-to-link web welds that extend into the k-area of the link can generate link web fractures that may reduce the plastic rotation capacity of the link (Okazaki et al., 2004a; Richards et al., 2004).

(5) Link Stiffeners for Box Sections

Similar to wide-flange links, the maximum spacing of stiffeners for shear yielding built-up box links ($e \leq 1.6M_p/V_p$) is dependent upon the magnitude of the link rotation angle. The equation for maximum spacing needed for the links to develop a link rotation angle of 0.08 rad [specified as $20t_w - (d - 2t_f)/8$] is derived in Berman and Bruneau (2005a). A similar equation was also derived for a 0.02 rad limit, resulting in a maximum required stiffener spacing of $37t_w - (d - 2t_f)/8$. However, experimental and analytical data is only available to support the closer stiffener spacing required for the 0.08 rad link rotation angle. Therefore, that more restrictive stiffener spacing is required for all links until other data becomes available.

The use of intermediate web stiffeners was shown (Berman and Bruneau 2006, 2008a, 2008b) to be significant on the shear yielding strength in built-up box links with h/t_w greater than $0.64\sqrt{E/F_y}$ and less than or equal to $1.67\sqrt{E/F_y}$. For shear links with h/t_w less than or equal to $0.64\sqrt{E/F_y}$, flange buckling was the controlling limit state and intermediate stiffeners had no effect. Thus, intermediate web stiffeners are not required for links with web depth-to-thickness ratios less than $0.64\sqrt{E/F_y}$. For links with lengths exceeding $1.6M_p/V_p$, compression local buckling of both webs and flanges (resulting from the compressive stresses associated with the development of the plastic moment) dominated link strength degradation. This buckling was unaffected by the presence of intermediate web stiffeners. As a result, intermediate web stiffeners are not required for links with lengths exceeding $1.6M_p/V_p$.

When intermediate stiffeners were used in the built-up box tested and simulated numerically by Berman and Bruneau (2006, 2008a, 2008b), these stiffeners were welded to both the webs and the flanges. A typical cross section is shown in Figure C-F3.8. However, presence of the stiffeners did not impact flange buckling, and these may therefore not need to be connected to the flange. This would have advantages over the detail in Figure C-F3.8. In particular, the intermediate stiffeners could be fabricated inside the built-up box link, improving resistance to corrosion and risk of accumulation of debris between the stiffeners (in cases of exterior exposures), and enhancing architectural appeal. Review of the literature (Malley and Popov, 1983; Bleich, 1952; Salmon and Johnson, 1996) showed that the derivation of minimum required areas and moment of inertia equations for sizing intermediate stiffeners did not depend on connection to the flanges. Whereas web stiffeners in I-shaped links may also serve to provide stability to the flanges (Malley and Popov, 1983), this is not the case in built-up box cross sections. Thus, welding of intermediate stiffeners to the flanges of the built-up box section links is not critical and not required.

5c. **Protected Zones**

The link, as the expected area of inelastic strain, is the protected zone.

6. **Connections**

6a. **Demand Critical Welds**

Inelastic strain in the weld material is likely at column base plates, column splices, and in moment connections in eccentrically braced frames. In addition, it is likely in welds of a built-up link member. Thus these are required to be treated as demand critical welds. See Commentary Section F2.6a.

6b. **Beam-to-Column Connections**

See Commentary on Section F2.6b.

6c. **Diagonal Brace Connections**

In the 2005 Provisions, the brace connection was required to be designed for the same forces as the brace (which is the forces generated by the fully yielded and strain hardened link). The brace connection, however, was also required to be designed for a compressive axial force corresponding to the nominal buckling strength of the brace. This second requirement has been eliminated. Braces in EBFs are designed to preclude buckling, and it is considered unnecessarily conservative to design the brace connection for the buckling strength of the brace.

Bracing connections are required to be designed to resist forces corresponding to link yielding and strain hardening. The strain hardening factors used in Section F3.3, 1.25 for I-shaped links and 1.4 for box links, are somewhat low compared to some values determined from testing; however, the reliability of connections remains sufficient due to the use of lower resistance factors for nonductile limit states.

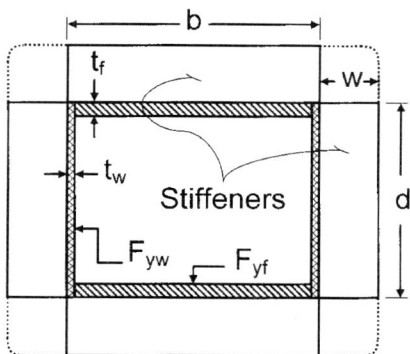

Figure C-F3.8. Built-up box link cross section with intermediate stiffener.

Bolt slip has been removed as a limit state which must be precluded. The consequences of exceeding this limit state in the maximum credible earthquake are not considered severe if bearing failure and block-shear rupture are precluded.

6d. **Column Splices**

Column splice requirements consistent with SCBF have been added. See Commentary Section F2.6d.

6e. **Link-to-Column Connections**

Prior to the 1994 Northridge earthquake, link-to-column connections were typically constructed in a manner substantially similar to beam-to-column connections in SMF. Link-to-column connections in EBF are therefore likely to share many of the same problems observed in moment frame connections. Consequently, in a manner similar to beam-to-column connections in SMF, the Provisions require that the performance of link-to-column connections be verified by testing in accordance with Section K2, or by the use of prequalified link-to-column connections in accordance with Section K1; there are no prequalified connections at the time of publication.

The load and deformation demands at a link-to-column connection in an EBF are substantially different from those at a beam-to-column connection in an SMF. Link-to-column connections must therefore be tested in a manner that properly simulates the forces and inelastic deformations expected in an EBF. Designers are cautioned that beam-to-column connections which qualify for use in an SMF may not necessarily perform adequately when used as a link-to-column connection in an EBF. Link-to-column connections must therefore be tested in a manner that properly simulates the forces and inelastic deformations expected in an EBF. For example, the reduced beam section (RBS) connection has been shown to perform well in SMF. However, the RBS is generally not suitable for link-to-column connections due to the high moment gradient in links. Similarly, recent research (Okazaki, 2004; Okazaki et al., 2004b) has demonstrated that other details that have shown good performance in moment frame beam-to-column connections (such as the WUF-W and the free flange details) can show poor performance in EBF link-to-column connections.

At the time of publication of the Provisions, development of satisfactory link-to-column connection details is the subject of ongoing research. Designers are therefore advised to consult the research literature for the latest developments. Until further research on link-to-column connections is available, it may be advantageous to avoid EBF configurations with links attached to columns.

The Provisions permit the use of link-to-column connections without the need for qualification testing for shear yielding links when the connection is reinforced with haunches or other suitable reinforcement designed to preclude inelastic action in the reinforced zone adjacent to the column. An example of such a connection is shown in Figure C-F3.9. This reinforced region should remain essentially elastic for the fully yielded and strain hardened link strength as required by Section F3.3; the exception for beams outside links does not apply. That is, the reinforced connection should be designed to resist the link shear and moment developed by the expected shear strength of the link, R_yV_n, multiplied by 1.25 to account for strain hardening. As an alternative to the reinforced link-to-column connection detail illustrated in Figure C-F3.9, preliminary testing and analysis have shown very promising performance for a reinforced connection detail wherein a pair of stiffeners is provided in the first link web panel next to the column, with the stiffeners oriented parallel to the link web. This link-to-column connection detail is described in Okazaki et al. (2009). Alternatively, the EBF can be configured to avoid link-to-column connections entirely.

The Provisions do not explicitly address the column panel zone design requirements at link-to-column connections. Based on limited research (Okazaki, 2004) it is recommended that the panel zone of link-to-column connections be designed in a manner similar to that for SMF beam-to-column connections (Section E3.6e) with the required shear strength of the panel zone determined from the analysis required by Section F3.3; the reduction in force for columns does not apply as the panel-zone shear is attributable to a single link, rather than to links at multiple levels that may not all be yielding simultaneously.

F4. BUCKLING-RESTRAINED BRACED FRAMES (BRBF)

1. Scope

Buckling-restrained braced frames (BRBF) are a special class of concentrically braced frames. Just as in SCBF, the centerlines of BRBF members that meet at a joint intersect at a point to form a complete vertical truss system that resists lateral forces. BRBF have more ductility and energy absorption than SCBF because overall brace buckling, and its associated strength degradation, is precluded at forces and deformations corresponding to the design story drift. See Section F2 for the effects of buckling in SCBF. Figure C-F2.1 shows possible concentrically braced frame configurations; note that neither X-bracing nor K-bracing is an option for BRBF. Figure C-F4.1 shows a schematic of a BRBF bracing element (adapted from Tremblay et al., 1999).

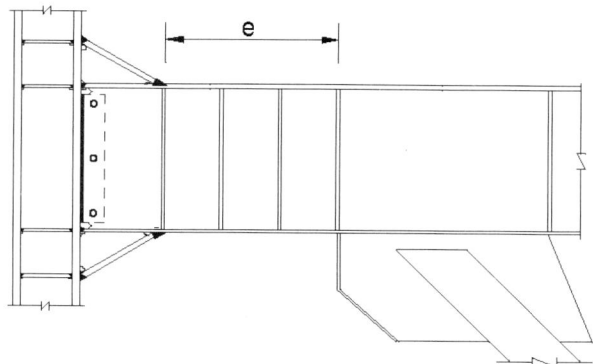

Fig. C-F3.9. Example of a reinforced link-to-column connection.

2. Basis of Design

BRBF are characterized by the ability of bracing elements to yield inelastically in compression as well as in tension. In BRBF, the bracing elements dissipate energy through stable tension-compression yield cycles (Clark et al., 1999). Figure C-F4.2 shows the characteristic hysteretic behavior for this type of brace as compared to that of a buckling brace. This behavior is achieved through limiting buckling of the steel core within the bracing elements. Axial stress is de-coupled from flexural buckling resistance; axial load is confined to the steel core while the buckling restraining mechanism, typically a casing, resists overall brace buckling and restrains high-mode steel core buckling (rippling).

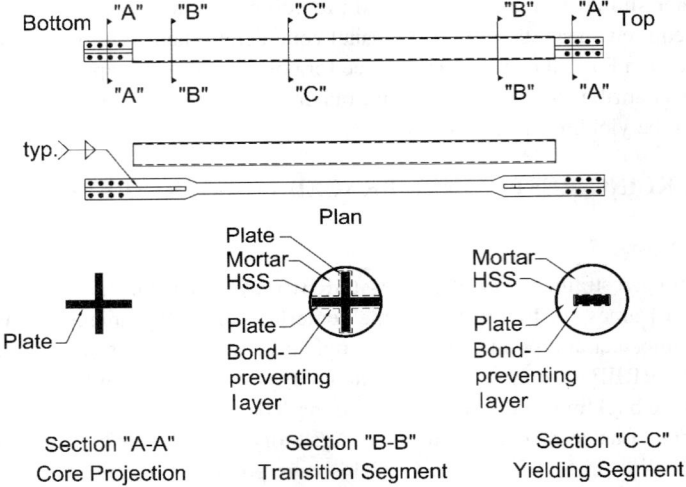

Fig. C-F4.1. Details of a type of buckling-restrained brace (courtesy of R. Tremblay).

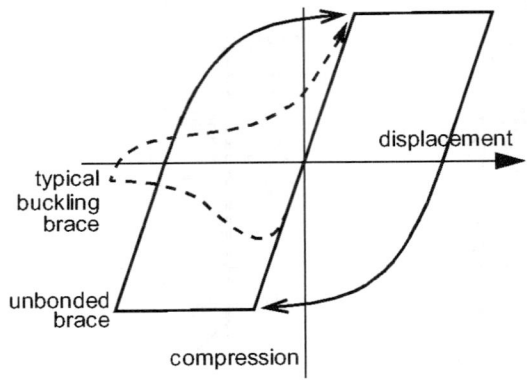

Fig. C-F4.2. Typical buckling-restrained (unbonded) brace hysteretic behavior (courtesy of Seismic Isolation Engineering).

Buckling-restrained braced frames are composed of columns, beams and bracing elements, all of which are subjected primarily to axial forces. Braces of BRBF are composed of a steel core and a buckling-restraining system encasing the steel core. In addition to the schematic shown in Figure C-F4.1, examples of BRBF bracing elements are found in Watanabe et al. (1988); Wada et al. (1994); and Clark et al. (1999). The steel core within the bracing element is intended to be the primary source of energy dissipation. During a moderate to severe earthquake the steel core is expected to undergo significant inelastic deformations.

BRBF can provide elastic stiffness that is comparable to that of EBF. Full-scale laboratory tests indicate that properly designed and detailed bracing elements of BRBF exhibit symmetrical and stable hysteretic behavior under tensile and compressive forces through significant inelastic deformations (Watanabe et al., 1988; Wada et al., 1998; Clark et al., 1999; Tremblay et al., 1999). The ductility and energy dissipation capability of BRBF is expected to be comparable to that of a SMF and greater than that of a SCBF. This high ductility is attained by limiting buckling of the steel core.

The Provisions are based on the use of brace designs qualified by testing. They are intended to ensure that braces are used only within their proven range of deformation capacity, and that yield and failure modes other than stable brace yielding are precluded at the maximum inelastic drifts corresponding to the design earthquake. For analyses performed using linear methods, the maximum inelastic drifts for this system are defined as those corresponding to 200% of the design story drift. For nonlinear time-history analyses, the maximum inelastic drifts can be taken directly from the analyses results. A minimum of 2% story drift is required for determining expected brace deformations for testing (see Section K3) and is recommended for detailing. This approach is consistent with the linear analysis equations for design story drift in ASCE/SEI 7 and the 2009 *NEHRP Recommended Provisions* FEMA P-750 (FEMA, 2009). It is also noted that the consequences of loss of connection stability due to the actual seismic displacements exceeding the calculated values may be severe; braces are therefore required to have a larger deformation capacity than directly indicated by linear static analysis.

The value of 200% of the design story drift for expected brace deformations represents the mean of the maximum story response for ground motions having a 10% chance of exceedance in 50 years (Fahnestock et al., 2003; Sabelli et al., 2003). Near-fault ground motions, as well as stronger ground motions, can impose deformation demands on braces larger than those required by these provisions. While exceeding the brace design deformation may result in poor brace behavior such as buckling, this is not equivalent to collapse. Detailing and testing braces for larger deformations will provide higher reliability and better performance.

The design engineer utilizing these provisions is strongly encouraged to consider the effects of configuration and proportioning of braces on the potential formation of building yield mechanisms. The axial yield strength of the core, P_{ysc}, can be set precisely with final core cross-sectional area determined by dividing the specified brace capacity by the actual material yield strength established by coupon testing, multiplied by the resistance factor. In some cases, cross-sectional area will be governed by brace stiffness requirements to limit drift. In either case, careful proportioning of braces can make yielding distributed over the building height much more likely than in conventional braced frames.

It is also recommended that engineers refer to the following documents to gain further understanding of this system: Uang and Nakashima (2003); Watanabe et al. (1988); Reina and Normile (1997); Clark et al. (1999); Tremblay et al. (1999); and Kalyanaraman et al. (1998).

The design provisions for BRBF are predicated on reliable brace performance. In order to assure this performance, a quality assurance plan is required. These measures are in addition to those covered in the *Code of Standard Practice* (AISC, 2010c), and *Specification* Chapters N and J. Examples of measures that may provide quality assurance are:

- Special inspection of brace fabrication. Inspection may include confirmation of fabrication and alignment tolerances, as well as nondestructive testing (NDT) methods for evaluation of the final product.
- Brace manufacturer's participation in a recognized quality certification program. Certification should include documentation that the manufacturer's quality assurance plan is in compliance with the requirements of the *Specification*, the Provisions and the *Code of Standard Practice*. The manufacturing and quality control procedures should be equivalent to, or better, than those used to manufacture brace test specimens.

2a. Brace Strength

Testing of braces is considered necessary for this system to ensure proper behavior. The applicability of tests to the designed brace is defined in Section K3. Commentary Section E3.6b, which describes in general terms the applicability of tests to designs, applies to BRBF.

Tests cited serve another function in the design of BRBF: the maximum forces that the brace can develop in the system are determined from test results. These maximum forces are used in the analysis required in Section F4.3.

The compression-strength adjustment factor, β, accounts for the compression overstrength (with respect to tension strength) noted in buckling-restrained braces in recent testing (SIE, 1999a and 1999b). The strain hardening adjustment factor, ω, accounts for strain hardening. Figure C-F4.3 shows a diagrammatic bilinear force-displacement relationship in which the compression strength adjustment factor, β, and the strain hardening adjustment factor, ω, are related to brace forces and nominal material yield strength. These quantities are defined as

$$\beta = \frac{\beta \omega F_{ysc} A_{sc}}{\omega F_{ysc} A_{sc}} = \frac{P_{max}}{T_{max}} \qquad \text{(C-F4-1)}$$

$$\omega = \frac{\omega F_{ysc} A_{sc}}{F_{ysc} A_{sc}} = \frac{T_{max}}{F_{ysc} A_{sc}} \qquad \text{(C-F4-2)}$$

where

A_{sc} = cross-sectional area of the yielding segment of steel core, in.² (mm²)
F_{ysc} = measured yield strength of the steel core, ksi (MPa)
P_{max} = maximum compression force, kips (N)
T_{max} = maximum tension force within deformations corresponding to 200% of the design story drift (these deformations are defined as $2.0\Delta_{bm}$ in Section K3.4c), kips (N)

Note that the specified minimum yield stress of the steel core, F_y, is not typically used for establishing these factors; instead, F_{ysc} is used which is determined by the coupon tests required to demonstrate compliance with Section K3. Braces with values of β and ω less than unity are not true buckling-restrained braces and their use is precluded by the provisions.

The expected brace strengths used in the design of connections and of beams and columns are adjusted upwards for various sources of overstrength, including amplification due to expected material strength (using the ratio R_y) in addition to the strain hardening, ω, and compression adjustment, β, factors discussed above. The amplification due to expected material strength can be eliminated if the brace yield stress is determined by a coupon test and is used to size the steel core area to provide the desired available strength precisely. Other sources of overstrength, such as imprecision in the provision of the steel core area, may need to be considered; fabrication tolerance for the steel core is typically negligible.

3. **Analysis**

Beams and columns are required to be designed considering the maximum force that the adjoining braces are expected to develop. In the Provisions, these requirements are presented as an analysis requirement, although they are consistent with the design requirements in the 2005 Provisions.

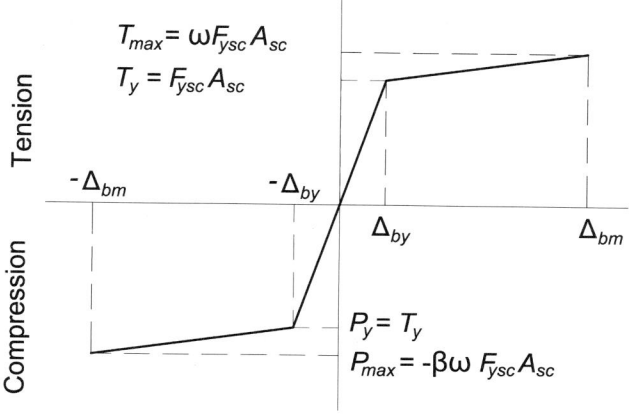

Fig. C-F4.3. Diagram of brace force-displacement.

4. System Requirements

4a. V- and Inverted V- Braced Frames

In SCBF, V-bracing has been characterized by a change in deformation mode after one of the braces buckles (see Commentary Section F2.4b). This is primarily due to the negative post-buckling stiffness, as well as the difference between tension and compression capacity, of traditional braces. Since buckling-restrained braces do not lose strength due to buckling, and have only a small difference between tension and compression capacity, the practical requirements of the design provisions for this configuration are relatively minor. Figure C-F4.4 shows the effect of beam vertical displacement under the unbalanced load caused by the brace compression overstrength. The vertical beam deflection adds to the deformation demand on the braces, causing them to elongate more than they compress (due to higher compression strength compared to tension strength). Therefore, where V-braced frames are used, it is required that a beam be provided that has sufficient strength to permit the yielding of both braces within a reasonable story drift considering the difference in tension and compression capacities determined by testing. The required brace deformation capacity must include the additional deformation due to beam deflection under this load. Since other requirements, such as the brace testing protocol (Section K3.4c) and the stability of connections (Section F4.6), depend on this deformation, engineers will find significant incentive to avoid flexible beams in this configuration. Where the special configurations shown in Figure C-F2.2 are used, the requirements of this Section are not relevant.

4b. K-Braced frames

K-braced frames are not permitted for BRBF due to the possibility of inelastic flexural demands on columns.

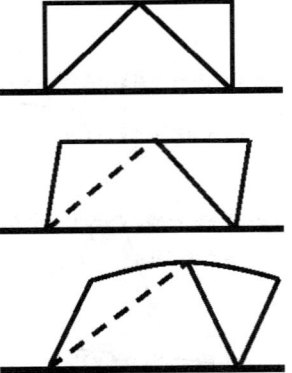

Fig. C-F4.4. Post-yield change in deformation mode for V- and inverted V-BRBF.

5. Members

5a. Basic Requirements

Highly ductile sections for beams and columns are required due to the possibility of inelastic rotations at the design story drift.

5b. Diagonal Braces

(1) Assembly

(1) Steel Core

The steel core is composed of a yielding segment and steel core projections; it may also contain transition segments between the projections and yielding segment. The cross-sectional area of the yielding segment of the steel core is expected to be sized so that its yield strength is fairly close to the demand calculated from the applicable building code. Designing braces close to the required strengths will help ensure distribution of yielding over multiple stories in the building. Conversely, over-designing some braces more than others (for example, by using the same size brace on all floors) may result in an undesirable concentration of inelastic deformations in only a few stories. The length and area of the yielding segment, in conjunction with the lengths and areas of the nonyielding segments, determine the stiffness of the brace. The yielding segment length and brace inclination also determines the strain demand corresponding to the design story drift.

In typical brace designs, a projection of the steel core beyond its casing is necessary in order to accomplish a connection to the frame. Buckling of this unrestrained zone is an undesirable failure mode and must therefore be precluded.

In typical practice, the designer specifies the core plate dimensions as well as the steel material and grade. The steel stress-strain characteristics may vary significantly within the range permitted by the steel specification, potentially resulting in significant brace overstrength. This overstrength must be addressed in the design of connections as well as of frame beams and columns. The designer may specify a limited range of acceptable yield stress in order to more strictly define the permissible range of brace capacity. Alternatively, the designer may specify a limited range of acceptable yield stress if this approach is followed in order to more strictly define the permissible range of core plate area (and the resulting brace stiffness). The brace supplier may then select the final core plate dimensions to meet the capacity requirement using the results of a coupon test. The designer should be aware that this approach may result in a deviation from the calculated brace axial stiffness. The maximum magnitude of the deviation is dependent on the range of acceptable material yield stress. Designers following this approach should consider the possible range of stiffness in the building analysis in order to adequately address both the building period and expected drift.

The strength of the steel core has been defined in terms of a new symbol, F_{ysc}, which is defined as either the specified minimum yield stress of the steel core, or actual yield stress of the steel core as determined from a coupon test. The use of coupon tests in establishing F_{ysc} eliminates the necessity of using the factor R_y in calculating the adjusted brace strength (see Commentary Section F4.2a). This is in recognition of the fact that coupon testing of the steel core material is in effect required by the similitude provisions in Section K3, and such coupon tests can provide a more reliable estimation of expected strength.

(2) Buckling-Restraining System

This term describes those elements providing brace stability against overall buckling. This includes the casing as well as elements connecting the core. The adequacy of the buckling-restraining system must be demonstrated by testing.

(2) Available Strength

The nominal strength of buckling restrained braces is simply based on the core area and the material yield strength. Buckling is precluded, as is demonstrated by testing.

(3) Conformance Demonstration

BRBF designs require reference to successful tests of a similarly sized test specimen and of a brace subassemblage that includes rotational demands. The former is a uniaxial test intended to demonstrate adequate brace hysteretic behavior. The latter is intended to verify the general brace design concept and demonstrate that the rotations associated with frame deformations do not cause failure of the steel core projection, binding of the steel core to the casing, or otherwise compromise the brace hysteretic behavior. A single test may qualify as both a subassemblage and a brace test subject to the requirements of Section K3; for certain frame-type subassemblage tests, obtaining brace axial forces may prove difficult and separate brace tests may be necessary. A sample subassemblage test is shown in Figure C-K3.1 (Tremblay et al., 1999).

5c. Protected Zones

The core, as the expected area of inelastic strain, is the protected zone.

6. Connections

6a. Demand Critical Welds

Inelastic strain in the weld material is likely at column base plates, column splices, and in moment connections in eccentrically braced frames. In addition, it is likely in welds of a built-up link member. Thus these are required to be treated as demand critical welds. See Commentary Section F2.6a.

6b. Beam-to-Column Connections

See Commentary Section F2.6b.

6c. Diagonal Brace Connections

Bracing connections must not yield at force levels corresponding to the yielding of the steel core; they are therefore designed for the maximum force that can be expected from the brace (see Commentary Section F4.5b). In addition, a factor of 1.1 is used. This factor is applied in consideration of the possibility of braces being subjected to deformations exceeding those at which the factors ω and β are required to be determined (in other words, 200% of the Δ_{bm}; see Commentary Section F4.2a). The engineer should recognize that the bolts are likely to slip at forces 30% lower than their design strength. This slippage is not considered to be detrimental to behavior of the BRBF system and is consistent with the design approach found in Section D2.2.

Recent testing in stability and fracture has demonstrated that gusset-plate connections may be a critical aspect of the design of BRBF (Tsai et al., 2003; Lopez et al., 2004). The tendency to instability may vary depending on the flexural stiffness of the connection portions of the buckling restrained brace and the degree of their flexural continuity with the casing. This aspect of BRBF design is the subject of continuing investigation and designers are encouraged to consult research publications as they become available. The stability of gussets may be demonstrated by testing, if the test specimen adequately resembles the conditions in the building. It is worth noting that during an earthquake the frame may be subjected to some out-of-plane displacement concurrent with the in-plane deformations, so a degree of conservatism in the design of gussets may be warranted.

Fahnestock et al. (2006) tested a connection, shown in Figure C-F4.5, that effectively provided a pin in the beam outside of the gusset plate via the splice with a WT

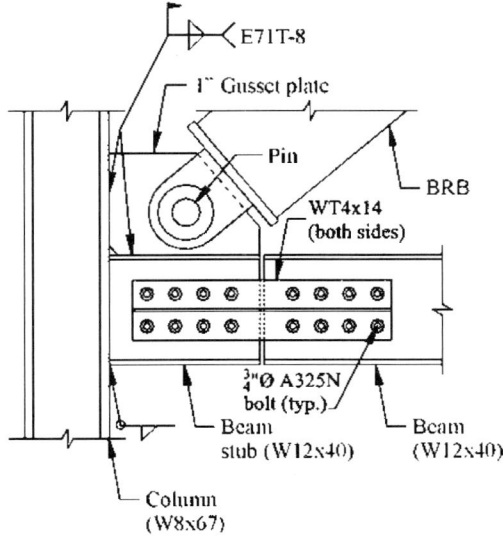

Fig. C-F4.5. Detail of connection with hinge (Fahnestock et al., 2006)

section on each side. In addition to satisfying the requirements of Section F4.6b, this connection relieves the gusset plate of in-plane moments and the related destabilization effects.

6d. **Column Splices**

See Commentary Section F2.6d.

F5. SPECIAL PLATE SHEAR WALLS (SPSW)

1. **Scope**

In special plate shear walls (SPSW), the slender unstiffened steel plates (webs) connected to surrounding horizontal and vertical boundary elements (HBE and VBE) are designed to yield and behave in a ductile hysteretic manner during earthquakes (see Figure C-F5.1). All HBE are also rigidly connected to the VBE with moment resisting connections able to develop the expected plastic moment of the HBE. Each web must be surrounded by boundary elements.

Experimental research on SPSW subjected to cyclic inelastic quasi-static and dynamic loading has demonstrated their ability to behave in a ductile manner and dissipate significant amounts of energy (Thorburn et al., 1983; Timler and Kulak, 1983; Tromposch and Kulak, 1987; Roberts and Sabouri-Ghomi, 1992; Caccese et al., 1993; Driver et al., 1997; Elgaaly, 1998; Rezai, 1999; Lubell et al., 2000;

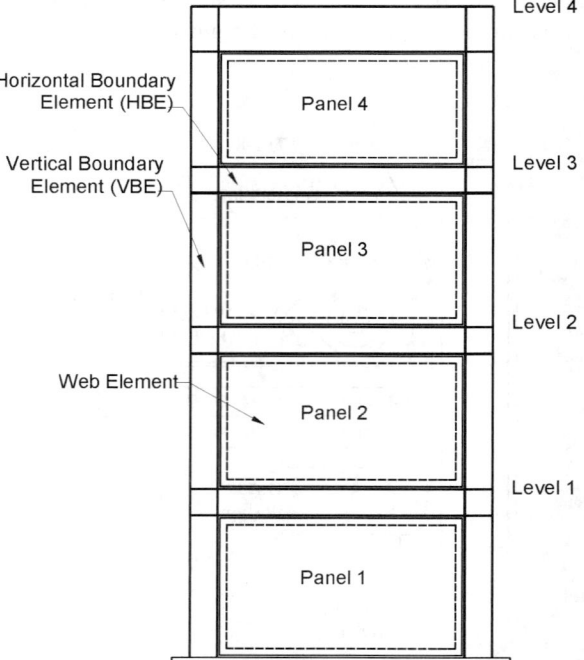

Fig. C-F5.1. Schematic of special plate shear wall.

Grondin and Behbahannidard, 2001; Berman and Bruneau, 2003a; Zhao and Astaneh-Asl, 2004; Berman and Bruneau, 2005b; Sabouri-Ghomi et al., 2005; Deng et al., 2008; Qu et al., 2008; Choi and Park, 2009; Qu and Bruneau, 2009; Vian et al., 2009a). This has been confirmed by analytical studies using finite element analysis and other analysis techniques (Sabouri-Ghomi and Roberts, 1992; Elgaaly et al., 1993; Elgaaly and Liu, 1997; Driver et al., 1997; Dastfan and Driver, 2008; Bhowmick et al., 2009; Purba and Bruneau, 2009; Shishkin et al., 2009; Vian et al., 2009b).

2. **Basis of Design**

Yielding of the webs occurs by development of tension field action at an angle close to 45° from the vertical, and buckling of the plate in the orthogonal direction. Past research shows that the sizing of VBE and HBE in an SPSW makes it possible to develop this tension field action across all of the webs. Except for cases with very stiff HBE and VBE, yielding in the webs develops in a progressive manner across each panel. Because the webs do not yield in compression, continued yielding upon repeated cycles of loading is contingent upon the SPSW being subjected to progressively larger drifts, except for the contribution of plastic hinging developing in the HBE to the total system hysteretic energy. In past research (Driver et al., 1997), the yielding of boundary elements contributed approximately 25 to 30% of the total load strength of the system.

With the exception of plastic hinging at the ends of HBEs, the surrounding HBEs and VBEs are designed to remain essentially elastic when the webs are fully yielded. Plastic hinging at the ends of HBEs is needed to develop the plastic collapse mechanism of this system. Plastic hinging in the middle of HBEs, which could partly prevent yielding of the webs, is deemed undesirable. Cases of both desirable and undesirable yielding in VBE have been observed in past testing. In the absence of a theoretical formulation to quantify the conditions leading to acceptable yielding (and supporting experimental validation of this formulation), the conservative requirement of elastic VBE response is justified.

Research literature often compares the behavior of steel plate walls to that of a vertical plate girder, indicating that the webs of an SPSW resist shears by tension field action and that the VBE of an SPSW resist overturning moments. While this analogy is useful in providing a conceptual understanding of the behavior of SPSW, many significant differences exist in the behavior and strength of the two systems. Past research shows that the use of structural shapes for the VBE and HBE in SPSW (as well as other dimensions and details germane to SPSW) favorably impacts orientation of the angle of development of the tension field action, and makes possible the use of very slender webs (having negligible diagonal compressive strength). Sizeable top and bottom HBEs are also required in the SPSW to anchor the significant tension fields that develop at the ends of the structural system. Limits imposed on the maximum web slenderness of plate girders to prevent flange buckling, or due to transportation requirements, are also not applicable to SPSW which are constructed differently. For these reasons, the use of beam design provisions in the *Specification* for the design of SPSW is not appropriate (Berman and Bruneau, 2004).

3. Analysis

Per capacity design principles, all edge boundary elements (HBE and VBE) shall be designed to resist the maximum forces developed by the tension field action of the webs fully yielding. Axial forces, shears and moments develop in the boundary elements of the SPSW as a result of the response of the system to the overall overturning and shear, and this tension field action in the webs. Actual web thickness must be considered for this calculation, because webs thicker than required may have to be used due to availability, or minimum thickness required for welding.

At the top panel of the wall, the vertical components of the tension field shall be anchored to the HBE. The HBE shall have sufficient strength to allow development of full tensile yielding across the panel width.

At the bottom panel of the wall, the vertical components of the tension field shall also be anchored to the HBE. The HBE shall have sufficient strength to allow development of full tensile yielding across the panel width. This may be accomplished by continuously anchoring the HBE to the foundation.

For intermediate HBE of the wall, the anticipated variation between the top and bottom web normal stresses acting on the HBE is usually small, or null when webs in the panel above and below the HBE have identical thickness. While top and bottom HBE are typically of substantial size, intermediate HBE are relatively smaller.

For the design of HBE, it may be important to recognize the effect of vertical stresses introduced by the tension field forces in reducing the plastic moment of the HBE. Concurrently, free-body diagrams of HBEs should account for the additional shear and moments introduced by the eccentricity of the horizontal component of the tension fields acting at the top and bottom of the HBEs (Qu and Bruneau, 2008, 2010a).

Beyond plastic-hinge formation at the ends of the HBE, in some instances the engineer may be able to justify yielding of the boundary elements by demonstrating that the yielding of a particular edge boundary element will not cause reduction on the SPSW shear capacity to support the demand and will not cause a failure in vertical gravity carrying capacity.

Forces and moments in the members (and connections), including those resulting from tension field action, may be determined from a plane frame analysis. The web is represented by a series of inclined pin-ended strips, as described in Commentary Section F5.5b. A minimum of ten equally spaced pin-ended strips per panel will be used in such an analysis.

A number of analytical approaches are possible to achieve capacity design and determine the same forces acting on the vertical boundary elements. Some example methods applicable to SPSW follow. In all cases, actual web thickness must be considered, for reasons described earlier.

Nonlinear push-over analysis. A model of the SPSW can be constructed in which bilinear elasto-plastic web elements of strength $R_y F_y A_s$ are introduced in the direction α. Bilinear plastic hinges can also be introduced at the ends of the horizontal boundary elements. Standard push-over analysis conducted with this model will provide axial forces, shears and moments in the boundary frame when the webs develop yielding. Separate checks are required to verify that plastic hinges do not develop in the horizontal boundary elements, except at their ends.

Indirect capacity design approach. The Canadian Standards Association Standard, *Limit States Design of Steel Structures* (CSA, 2001), proposes that loads in the vertical boundary members can be determined from the gravity loads combined with the seismic loads increased by the amplification factor,

$$B = \frac{V_e}{V_u} \qquad \text{(C-F5-1)}$$

where
V_e = expected shear strength, at the base of the wall, determined for the web thickness supplied, kips
    = $0.5 R_y F_y t_w L \sin 2\alpha$
V_u = factored lateral seismic force at the base of the wall

In determining the loads in VBEs, the amplification factor, B, need not be taken as greater than R.

The VBE design axial forces shall be determined from overturning moments defined as follows:

- The moment at the base is BM_u, where M_u is the factored seismic overturning moment at the base of the wall corresponding to the force V_u
- The moment BM_u extends for a height H but not less than two stories from the base
- The moment decreases linearly above a height H to B times the overturning moment at one story below the top of the wall, but need not exceed R times the factored seismic overturning moment at the story under consideration corresponding to the force V_u

The local bending moments in the VBE due to tension field action in the web shall be multiplied by the amplification factor B.

This method is capable of producing reasonable results for approximating VBE capacity design loads; however, as described above, it can be unconservative as shown in Berman and Bruneau (2008c). This procedure relies on elastic analysis of a strip model (or equivalent) for the design seismic loads, followed by amplification of the resulting VBE moments by the factor B. Therefore, it produces moment diagrams and SPSW deformations that are similar in shape to what one would obtain from a pushover analysis. Similarly, the determination of VBE axial forces from overturning calculations based on the design lateral loads amplified by B results in axial force diagrams that are of the proper shape. However, following the above procedure, the amplification factor is found only for the first story and does not include the possibly significant strength of the surrounding frame. HBEs and VBEs for SPSW are large and the portion of the base shear carried by the surrounding moment frame can be substantial. As a result, estimates of VBE demands per this method are less than those required to develop full web yielding on all stories prior to development of hinges in VBEs. In addition, in some cases, the ratio of web thickness provided to web thickness needed for the design seismic loads can be larger on the upper stories than on the lower stories. In these situations, the indirect capacity design approach would underestimate the VBE design loads for the upper stories and capacity design would not be achieved. Neglecting these effects in the determination of B will result in VBE design loads that are underestimated for true capacity design. Therefore, the full collapse mechanism should be used when determining the factor B. Such an equation is proposed in the procedure below (in Equation C-F5-15).

Combined Plastic and Linear Analysis. This procedure has been shown to give accurate VBE results compared to push-over analysis (Berman and Bruneau, 2008c). Assuming that the web plates and HBE of a SPSW have been designed according to the Provisions to resist the factored loads (or, for the case of HBE design, the maximum of the factored loads or web plate yielding), the required capacity of VBE may be found from VBE free body diagrams such as those shown in Figure C-F5.2 for a generic four-story SPSW. Those free body diagrams include distributed loads representing the web plate yielding at story i, ω_{xci} and ω_{yci}; moments from plastic hinging of HBE, M_{prli} and M_{prri}; axial forces from HBE, P_{bli} and P_{bri}; applied lateral seismic loads found from consideration of the plastic collapse mechanism, F_i; and base reactions for those lateral seismic loads, R_{yl}, R_{xl}, R_{yr} and R_{xr}. Each of these loads can then be determined as follows:

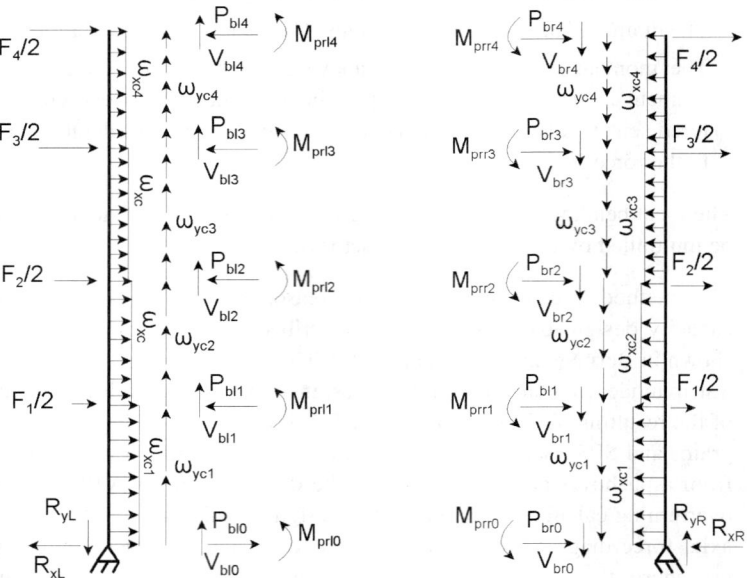

Fig. C-F5.2. VBE free body diagrams.

1. The distributed loads to be applied to the VBE (ω_{yci} and ω_{xci}) and HBE (ω_{ybi} and ω_{xbi}) from plate yielding on each story, i, may be determined as:

$$\omega_{yci} = (1/2)F_{yp}t_{wi}\sin 2\alpha \quad \text{(C-F5-2)}$$

$$\omega_{xci} = F_{yp}t_{wi}(\sin\alpha)^2 \quad \text{(C-F5-3)}$$

$$\omega_{ybi} = F_{yp}t_{wi}(\cos\alpha)^2 \quad \text{(C-F5-4)}$$

$$\omega_{xbi} = (1/2)F_{yp}t_{wi}\sin 2\alpha \quad \text{(C-F5-5)}$$

where F_{yp} and t_{wi} are the web plate yield stress and thickness at level i, respectively.

2. As part of estimating the axial load in the HBE, an elastic model of the VBE is developed as shown in Figure C-F5.3. The model consists of a continuous beam element representing the VBE which is pin-supported at the base and supported by elastic springs at the intermediate and top HBE locations. HBE spring stiffnesses at each story i, k_{bi}, can be taken as the axial stiffness of the HBE considering one half of the bay width (or HBE length for a considerably deep VBE), i.e.:

$$k_{bi} = \frac{A_{bi}E}{L/2} \quad \text{(C-F5-6)}$$

where A_{bi} is the HBE cross-sectional area, L is the bay width, and E is the modulus of elasticity. This VBE model is then loaded with the horizontal component of

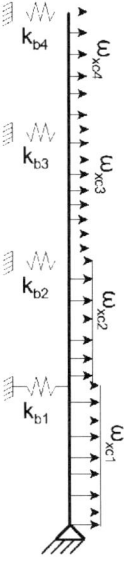

Fig. C-F5.3. Elastic VBE model with HBE springs.

the forces from the web plates yielding over each story, namely, ω_{xci} and analysis return spring forces, P_{si}.

3. The axial force component in the intermediate and top HBE resulting from the horizontal component of the plate yield forces on the HBE, ω_{xbi}, is assumed to be distributed as shown in Figure C-F5.4. Note that for the bottom HBE, this distribution is the reverse of that in the top beam. These axial force components are then combined with the spring forces from the linear VBE model, resulting in the following equations for the axial force at the left and right sides of the intermediate and top HBE (P_{bli} and P_{bri}, respectively):

$$P_{bli} = -(\omega_{xbi} - \omega_{xbi+1})\frac{L}{2} + P_{si} \qquad \text{(C-F5-7)}$$

$$P_{bri} = (\omega_{xbi} - \omega_{xbi+1})\frac{L}{2} + P_{si} \qquad \text{(C-F5-8)}$$

where the spring forces, P_{si}, should be negative indicating that they are adding to the compression in HBE. As mentioned above, the axial forces from ω_{xbi} and ω_{xbi+1} in the bottom HBE may be taken as the mirror image of those shown in Figure C-F5.4, where ω_{xbi} is zero in that particular case as there is no web below the bottom HBE. Furthermore, there are no spring forces to consider at the bottom HBE location as the horizontal component of force from web plate yielding on the lower portion of the bottom VBE is added to the base reaction determined as part of the plastic collapse mechanism analysis, as described below. Therefore, the bottom HBE axial forces on the right and left hand sides, P_{bl0} and P_{br0}, are:

$$P_{bl0} = \omega_{xb1}\frac{L}{2} \qquad \text{(C-F5-9)}$$

$$P_{br0} = -\omega_{xb1}\frac{L}{2} \qquad \text{(C-F5-10)}$$

4. The reduced plastic moment capacity at the HBE ends can be approximated by:

If $1.18\left(1 - \dfrac{|P_{bli}|}{F_{yb}A_{bi}}\right) \leq 1.0$

$$M_{prli} = 1.18\left(1 - \dfrac{|P_{bli}|}{F_{yb}A_{bi}}\right)Z_{xbi}F_{yb} \qquad \text{(C-F5-11)}$$

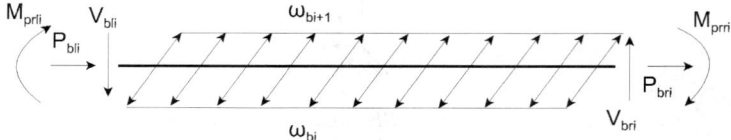

Fig. C-F5.4. HBE free body diagram.

If $1.18\left(1 - \dfrac{|P_{bli}|}{F_{yb}A_{bi}}\right) > 1.0$

$$M_{prli} = Z_{xbi} F_{yb} \qquad \text{(C-F5-12)}$$

where F_{yb} is the HBE yield strength, A_{bi} is the HBE cross-sectional area for story i, and Z_{xbi} is the HBE plastic section modulus for story i.

5. The shear forces at the left and right ends of all HBE, V_{br} and V_{bl} can be found from:

$$V_{bri} = \dfrac{M_{prri} + M_{prli}}{L} + \left(\omega_{ybi} - \omega_{ybi+1}\right)\dfrac{L}{2} \qquad \text{(C-F5-13)}$$

$$V_{bli} = V_{bri} - \left(\omega_{ybi} - \omega_{ybi+1}\right)L \qquad \text{(C-F5-14)}$$

6. The applied loads for the SPSW collapse mechanism can be found from:

$$\sum_{i=1}^{n_s} F_i H_i = \sum_{i=0}^{n_s} M_{prli} + \sum_{i=0}^{n_s} M_{prri} + \sum_{i=1}^{n_s} \dfrac{1}{2}(t_{wi} - t_{wi+1})F_{yp}LH_i \sin(2\alpha_i) \qquad \text{(C-F5-15)}$$

where F_i is the applied lateral load at each story to cause the mechanism, H_i is the height from the base to each story, and other terms are as previously defined. Note that the indices for the HBE plastic moment summations begin at zero so that the bottom HBE (denoted HBE$_0$) is included. To employ Equation C-F5-15 in calculating the applied lateral loads that cause this mechanism to form, it is necessary to assume some distribution of those loads over the height of the structure, i.e., a relationship between F_1, F_2, etc. For this purpose, a pattern equal to that of the design lateral seismic loads from the appropriate building code may be used.

7. Horizontal reactions at the column bases, R_{xL} and R_{xR}, are then determined by dividing the collapse base shear by 2 and adding the pin-support reaction from the VBE model, R_{bs}, to the reaction under the left VBE and subtracting it off the reaction under the right VBE. Vertical base reactions can be estimated from overturning calculations using the collapse loads as:

$$R_{yl} = \dfrac{\sum_{i=1}^{n_s} F_i H_i}{L} \quad \text{and} \quad R_{yr} = -R_{yl} \qquad \text{(C-F5-16)}$$

8. The moment, axial and shear force diagrams for the VBEs are established once all the components of the VBE free body diagrams are estimated. The diagrams give minimum design actions for those VBE such that they can resist full web plate yielding and HBE hinging.

VBE must be designed to remain elastic under the large shears resulting from this analysis. Existing literature shows instances of undesirable inelastic behavior when shear yielding occurred in the VBE (Qu and Bruneau, 2008; Qu and Bruneau, 2010b).

Preliminary design. For preliminary proportioning of HBE, VBE and webs, an SPSW wall may be approximated by a vertical truss with tension diagonals. Each web is represented by a single diagonal tension brace within the story. For an assumed angle of inclination of the tension field, the web thickness, t_w, may be taken as

$$t_w = \frac{2A\Omega_s \sin\theta}{L\sin 2\alpha} \qquad \text{(C-F5-17)}$$

where
- A = area of the equivalent tension brace, in.² (mm²)
- θ = angle between the vertical and the longitudinal axis of the equivalent diagonal brace
- L = distance between VBE centerlines, in. (mm)
- α = assumed angle of inclination of the tension field measured from the vertical per Section F5.5a
- Ω_s = system overstrength factor, as defined by FEMA 369 (FEMA, 2001), and taken as 1.2 for SPSW (Berman and Bruneau, 2003b)

A is initially estimated from an equivalent brace size to meet the structure's drift requirements.

4. System Requirements

Panel Aspect Ratio

The 2005 Provisions for the design of special plate shear walls (SPSW) limited their applicability to wall panels having aspect ratios of $0.8 < L/h < 2.5$. This limit was first introduced in the 2003 Edition of the *NEHRP Recommended Provisions for Seismic Regulations for New Buildings and Other Structures,* FEMA 450 (FEMA, 2003), as a most conservative measure in light of the relatively limited experience with that structural system in the U.S. at the time. Since then, SPSW designed in compliance with the Provisions and having lower aspect ratios have been observed to perform satisfactorily. For example, SPSW specimens having L/h of 0.6 (Lee and Tsai, 2008) exhibited ductile hysteretic behavior comparable to that of walls with larger aspect ratios.

No theoretical upper bound exists on L/h, but as the SPSW aspect ratio increases, progressively larger HBEs will be required, driven by the capacity design principles embodied in the design requirements. This will create a de facto practical limit beyond which SPSW design will become uneconomical and impractical, and no arbitrary limit (such as 2.5) needs to be specified provided the engineer ensures that all strips yield at the target drift response (Bruneau and Bhagwagar, 2002).

Past research has focused on walls with an L/t_w ratio ranging from 300 to 800. Although no theoretical upper bound exists on this ratio, drift limits will indirectly constrain this ratio. The requirement that webs be slender provides a lower bound on this ratio. For these reasons, no limits are specified on that ratio.

4a. Stiffness of Boundary Elements

The stiffness requirement is intended to prevent excessive in-plane flexibility and buckling of VBE. However, recent work suggests that this approach and specified

4c. Bracing

Providing stability of SPSW system boundary elements is necessary for proper performance of the system. Past experience has shown that SPSW can behave in a ductile manner with beam-to-column requirements detailed as per intermediate moment frame requirements. As such, lateral bracing requirements are specified to meet the requirements for moderately ductile members. In addition, all intersections of HBE and VBE must be braced to ensure stability of the entire panel.

4d. Openings in Webs

Large openings in webs create significant local demands and thus must have HBE and VBE in a similar fashion as the remainder of the system. When openings are required, SPSW can be subdivided in smaller SPSW segments by using HBE and VBE bordering the openings. With the exception of the structural systems described in Section F5.7, SPSW with holes in the web not surrounded by HBE/VBE have not been tested. The provisions will allow other openings that can be justified by analysis or testing.

5. Members

5a. Basic Requirements

Dastfan and Driver (2008) demonstrated that the strength of SPSW designed in compliance with current requirements is not substantially sensitive to the angle of inclination of the strips, and that using a single value of 40° throughout the design will generally lead to slightly conservative results.

Some amount of local yielding is expected in the HBE and VBE to allow the development of the plastic mechanism of SPSW systems. For that reason, HBE and VBE shall comply with the requirements in Table D1.1 for SMF.

5b. Webs

The lateral shears are carried by tension fields that develop in the webs stressing in the direction α, defined in Section F5.5a. When the HBE and VBE boundary elements of a web are not identical, the average of HBE areas may be taken in the calculation of A_b, and the average of VBE areas and inertias may be respectively used in the calculation of A_c and I_c to determine α.

Plastic shear strength of panels is given by $0.5 R_y F_y t_w L_{cf} (\sin 2\alpha)$. The nominal strength is obtained by dividing this value by a system overstrength, as defined by FEMA 369 (FEMA, 2003), and taken as 1.2 for SPSW (Berman and Bruneau, 2003b).

The above plastic shear strength is obtained from the assumption that, for purposes of analysis, each web may be modeled by a series of inclined pin-ended strips (Figure C-F5.5), oriented at angle α. Past research has shown this model provides realistic results, as shown in Figure C-F5.6 for example, provided at least 10 equally spaced strips are used to model each panel.

The specified minimum yield stress of steel used for SPSW is per Section A3.1. However, the webs of SPSW could also be of special highly ductile low yield steel having specified minimum yield in the range of 12 to 33 ksi (80 to 230 MPa).

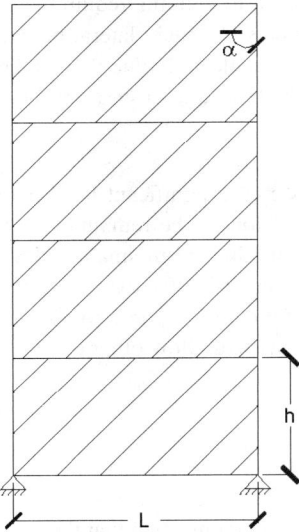

Fig. C-F5.5. Strip model of an SPSW.

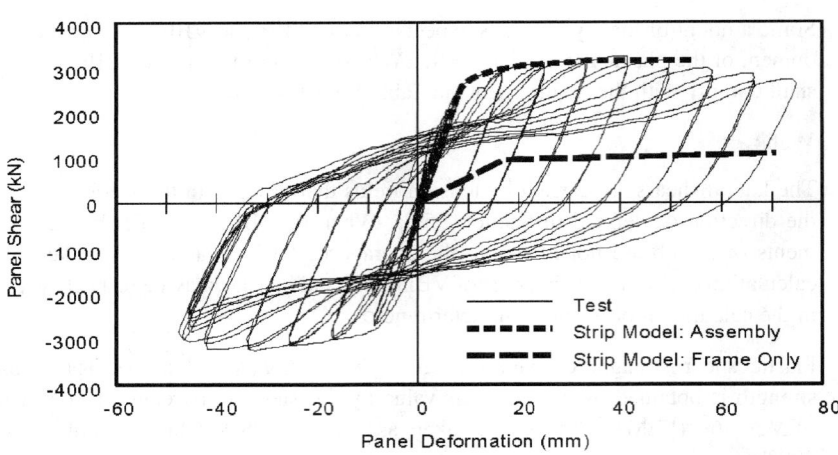

Fig. C-F5.6. Comparison of experimental results for lower panel of multi-story SPSW frame and strength predicted by strip model (after Driver et al., 1997).

5c. Protected Zone

Parts of SPSW expected to develop large inelastic deformations, and their connections, are designated as protected zones to meet the requirements of Section D1.3.

6. Connections

6a. Demand Critical Welds

Demand critical welds are required per Section A3.4b consistently with similar requirements for all SFRS.

6b. HBE-to-VBE Connections

Due to the large initial stiffness of SPSW, total system drift and plastic hinge rotation demands at the ends of HBE are anticipated to be smaller than for special moment resisting frames. The requirements of Section E6.1 for intermediate moment frames are deemed adequate for HBE-to-VBE connections.

(1) Required Strength

Connections of the HBE to VBE shall be able to develop the plastic strength of the HBE given that plastic hinging is expected at the ends of HBEs.

(2) Panel Zones

Panel zone requirements are not imposed for intermediate HBE where generally small HBE connect to sizeable VBE. The engineer should use judgment to identify special situations in which the panel zone adequacy of VBE next to intermediate HBE should be verified.

6c. Connections of Webs to Boundary Elements

Web connections to the surrounding HBE and VBE are required to develop the expected tensile strength of the webs. Net sections must also provide this strength for the case of bolted connections.

The strip model can be used to model the behavior of SPSW and the tensile yielding of the webs at angle, α. A single angle of inclination taken as the average for all the panels may be used to analyze the entire wall. The expected tensile strength of the web strips shall be defined as $R_y F_y A_s$,

where

A_s = area of a strip = $(L\cos\alpha + H\sin\alpha)/n$, in.^{2} (mm^2)
L = width of panel, in. (mm)
H = height of panel, in. (mm)
n = number of strips per panel and n shall be taken greater than or equal to 10

This analysis method has been shown, through correlation with physical test data, to adequately predict SPSW performance. It is recognized, however, that other advanced analytical techniques [such as the finite element method (FEM)] may also be used for design of SPSW. If such nonlinear (geometric and material) FEM models are used, they should be calibrated against published test results to ascertain reliability for application. Designs of connections of webs to boundary elements should also anticipate buckling of the web plate. Some minimum out-of-plane rotational restraint of the plate should be provided (Caccese et al., 1993).

6d. Column Splices

The importance of ensuring satisfactory performance of column splices is described in Commentary Section D2.5.

7. Perforated Webs

7a. Regular Layout of Circular Perforations

Special perforated steel plate walls (SPSPW) are a special case of SPSW in which a special panel perforations layout is used to allow utilities to pass through and which may be used to reduce the strength and stiffness of a solid panel wall to levels required in a design when a thinner plate is unavailable. This concept has been analytically and experimentally proven to be effective and the system remains ductile up to the drift demands corresponding to severe earthquakes (Vian and Bruneau, 2005; Vian et al., 2009a; Vian et al., 2009b; Purba and Bruneau, 2007). A typical hole layout for this system is shown in Figure C-F5.7.

Designing SPSW in low to medium rise buildings using hot-rolled steel often results in required panel thicknesses less than the minimum plate thickness available from steel producers. In such cases, using the minimum available thickness would result in large panel force over-strength, proportionally larger design demands on the surrounding VBE and HBE, and an overall less economical system. Attempts at

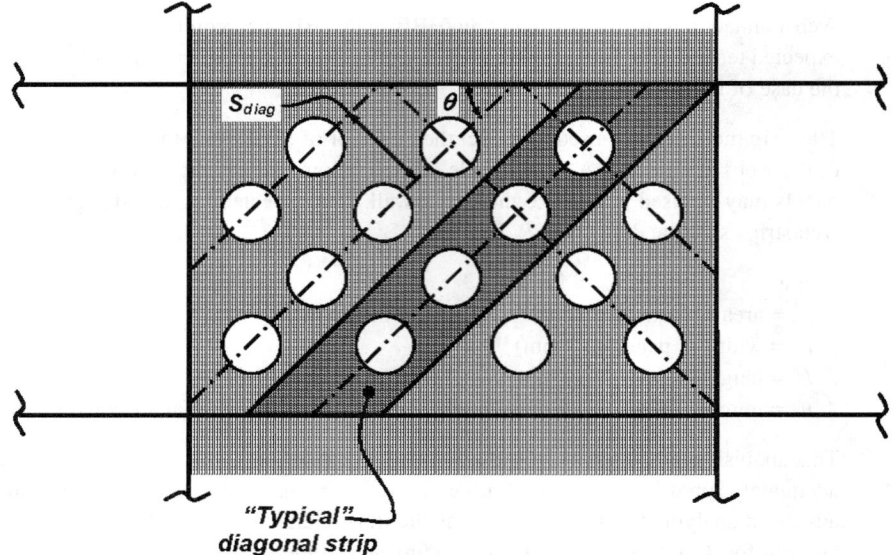

Fig. C-F5.7. Schematic detail of special perforated steel plate wall and typical diagonal strip.

alleviating this problem were addressed by the use of light-gauge, cold-formed steel panels (Berman and Bruneau, 2003a, 2005b). SPSPW instead reduce the strength of the web by adding to it a regular grid of perforations. This solution simultaneously helps address the practical concern of utility placement across SPSW. In a regular SPSW, the infill panel which occupies an entire frame bay between adjacent HBE and VBE is a protected element, and utilities that may have otherwise passed through at that location must either be diverted to another bay, or pass through an opening surrounded by HBE and VBE. This either results in additional materials (for the extra stiffening) or in labor (for the relocation of ductwork in a retrofit, for example); SPSPW provide a more economical alternative.

7b. Reinforced Corner Cut-Out

It is also possible to allow utility passage through a reinforced cutout designed to transmit the web forces to the boundary frame. While providing utility access, this proposed system provides strength and stiffness similar to a solid panel SPSW system. The openings are located immediately adjacent to the column in each of the top corners of the panel, a location where large utilities are often located. A cut-out radius as large as 19.6 in. (500 mm) for a half-scale specimen having a 6.5 ft (2000 mm) center-to-center distance between HBE has been successfully verified experimentally and analytically by Vian and Bruneau (2005) and Purba and Bruneau (2007).

Forces acting in the reinforcing arch (the curved plate at the edge of the opening) are a combination of effects due to arching action under tension forces due to web yielding, and thrusting action due to change of angle at the corner of the SPSW (Figures

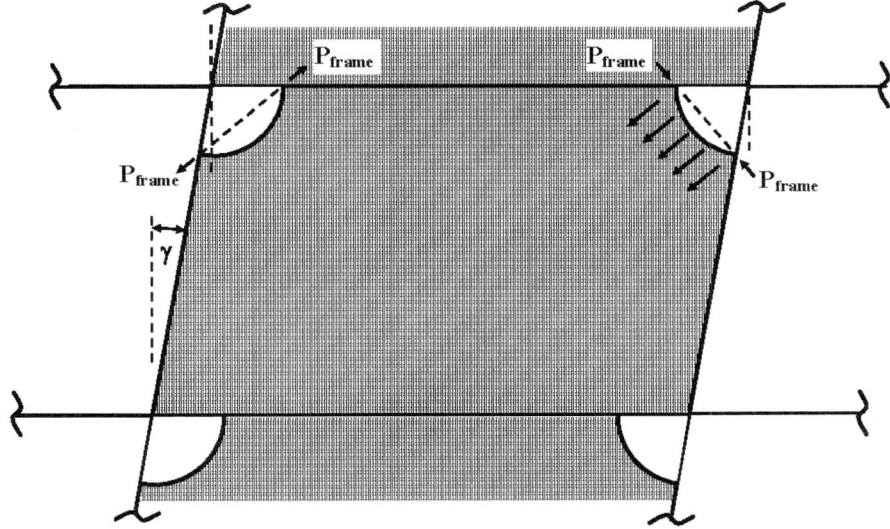

Fig. C-F5.8. Arch end reactions due to frame deformations, and infill panel forces on arches due to tension field action on reinforced cut-out corner.

C-F5.8 and C-F5.9). The latter is used to calculate the required maximum thickness of the "opening" corner arch (top left side of Figure C-F5.8, with no web stresses assumed to be acting on it). The arch plate width is not a parameter that enters the solution of the interaction equation in that calculation, and it is instead conservatively obtained by considering the strength required to resist the axial component of force in the arch due to the panel forces at the closing corner (top right side of Figure C-F5.8). Since the components of arch forces due to panel forces are opposing those due to frame corner opening (Figure C-F5.9), the actual forces acting in the arch plate will be smaller than the forces calculated by considering the components individually as is done above for design.

Note that when a plate in the plane is added to the reinforcement arch to facilitate infill panel attachment to the arch in the field, it results in a stiffer arch section that could (due to compatibility of frame corner deformation) partly yield at large drifts. However, Vian and Bruneau (2005) and Purba and Bruneau (2007) showed that the thickness of the flat plate selected per the above procedure is robust enough to withstand the loads alone, and that the presence of the stiffer and stronger T section (due to the attachment plate discussed above) is not detrimental to the system performance.

Nonlinear static pushover analysis is a tool that can be used to confirm that the selected reinforcement section will not produce an undesirable "knee-brace effect" or precipitate column yielding or beam yielding outside of the hinge region.

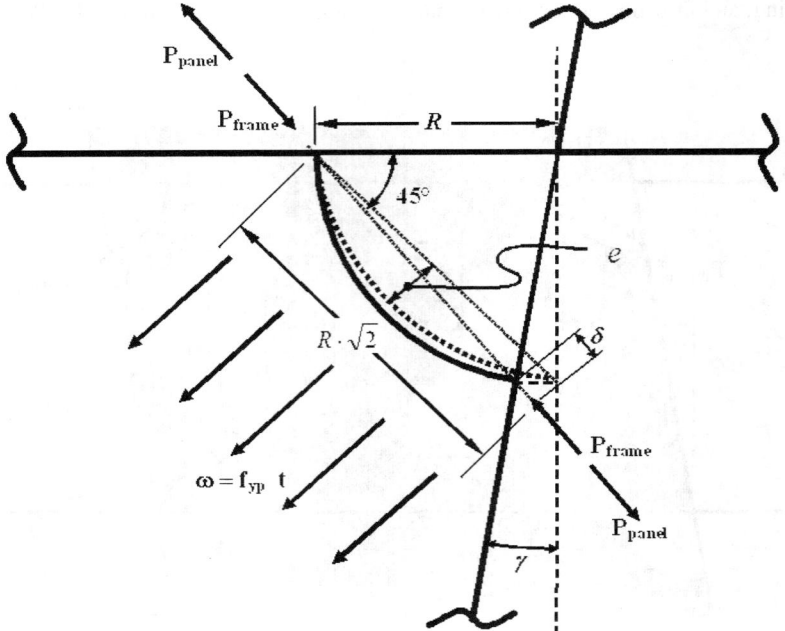

Fig. C-F5.9. Deformed configurations and forces acting on right arch.

CHAPTER G

COMPOSITE MOMENT-FRAME SYSTEMS

G1. COMPOSITE ORDINARY MOMENT FRAMES (C-OMF)

2. Basis of Design

Composite ordinary moment frames (C-OMF) represent a type of composite moment frame that is designed and detailed following the *Specification* and ACI 318 (ACI, 2008), excluding Chapter 21. ASCE/SEI 7 (ASCE, 2010) limits C-OMF to seismic design categories A and B. This is in contrast to steel ordinary moment frames, which are permitted in higher seismic design categories. The design requirements for C-OMF recognize this difference and provide minimum ductility in the members and connections. The R and C_d values for C-OMF are chosen accordingly.

G2. COMPOSITE INTERMEDIATE MOMENT FRAMES (C-IMF)

2. Basis of Design

ASCE/SEI 7 limits the use of C-IMF in seismic design category C and below. The provisions for C-IMF, as well as the associated R and C_d values in ASCE/SEI 7, are comparable to those required for reinforced concrete IMF and between those for steel IMF and OMF.

While the design of C-SMF as defined in Section G3 is intended to limit the majority of the inelastic deformation to the beams, the inelastic drift capability of C-IMF is permitted to be derived from inelastic deformations of beams, columns and panel zones.

The C-IMF connection is based on a tested design with a qualifying story drift angle of 0.02 rad.

4. System Requirements

4a. Stability Bracing of Beams

The requirement for spacing of lateral bracing in this section is less severe than that for C-SMF in Section G3.4b because of the lower required drift angle for C-IMF as compared to C-SMF. In this case, the required spacing of bracing is approximately double that of the C-SMF system.

5. Members

5a. Basic Requirements

This section refers to Section D1.1, which provides requirements for moderately ductile members. Because the rotational demands on C-IMF beams and columns are expected to be lower than C-SMF, the requirements and limitations for C-IMF members are less severe than for C-SMF.

5b. Beam Flanges

For relevant commentary on changes in cross section of beam flanges, see Commentary Section E3.5b.

5c. Protected Zones

For commentary on protected zones, see Commentary Section D1.3.

6. Connections

6a. Demand Critical Welds

There are no demand critical welds in C-IMF members because the story drift angle is 0.02 rad, which is half the value for C-SMF members, and ASCE/SEI 7 limits the use of C-IMF in seismic design category C and below.

6b. Beam-to-Column Connections

The minimum story drift angle required for qualification of C-IMF connections is 0.02 rad, which is half the value for C-SMF members, reflecting the lower level of inelastic response that is anticipated in the system.

6c. Conformance Demonstration

The requirements for conformance demonstration for C-IMF connections are the same as for C-SMF connections, except that the required story drift angle is smaller. Refer to Commentary Section G3.6c.

6d. Required Shear Strength

The requirements for shear strength of the connection for C-IMF are comparable to those of SMF, with the exception that the calculation of the expected flexural strength must account for the different constituent materials. Refer to Commentary Section E3.6d.

6e. Connection Diaphragm Plates

Connection diaphragm plates are permitted for filled composite columns both external and internal to the column. These diaphragm plates facilitate the transfer of beam flange forces into the column panel zone. These plates are required to have: (i) thickness at least equal to the beam flange, and (ii) complete-joint-penetration groove or two-sided fillet welds. They are designed with a required strength not less than the available strength of the contact area of the plate with column sides. Internal diaphragms are required to have a circular opening for placing concrete.

6f. Column Splices

The requirements for column splices for C-IMF are comparable to those of SMF, with the exception that the calculation of the expected flexural strength must account for the different constituent materials. Refer to Commentary Section E3.6g.

G3. COMPOSITE SPECIAL MOMENT FRAMES (C-SMF)

1. Scope

Composite moment frames include a variety of configurations where steel or composite beams are combined with reinforced concrete or composite columns. In particular, composite frames with steel floor framing and composite or reinforced concrete columns have been used as a cost-effective alternative to frames with reinforced concrete floors (Griffis, 1992; Furlong, 1997; Viest et al., 1997).

2. Basis of Design

Based on ASCE/SEI 7, C-SMF are primarily intended for use in seismic design categories D and above. Design and detailing provisions for C-SMF are comparable to those required for steel and reinforced concrete SMF and are intended to confine inelastic deformation to the beams and column bases. Since the inelastic behavior of C-SMF is comparable to that for steel or reinforced concrete SMF, the R and C_d values are the same as for those systems.

C-SMF are generally expected to experience significant inelastic deformations during a large seismic event. It is expected that most of the inelastic deformation will take place as rotation in beam "hinges" with limited inelastic deformation in the panel zone of the column. The beam-to-column connections for these frames are required to be qualified based on tests that demonstrate that the connection can sustain a story drift angle of at least 0.04 rad based on a specified loading protocol. Other provisions are intended to limit or prevent excessive panel zone distortion, failure of connectivity plates or diaphragms, column hinging, and local buckling that may lead to inadequate frame performance in spite of good connection performance.

C-SMF and C-IMF connection configuration and design procedures are based on the results of qualifying tests; the configuration of connections in the prototype structure must be consistent with the tested configurations. Similarly, the design procedures used in the prototype connections must be consistent with tested configurations.

4. System Requirements

4a. Moment Ratio

The strong-column weak-beam (SC/WB) mechanism implemented for composite frames is based on the similar concept for steel SMF. Refer to Commentary Section E3.4a for additional details and discussion. It is important to note that compliance with the SC/WB requirement and Equation G3-1 does not assure that individual columns will not yield, even when all connection locations in the frame comply. However, yielding of beams will predominate and the desired inelastic performance will be achieved in frames with members sized to meet the requirement of Equation G3-1.

Commentary Section E3.4a discusses the three exceptions to Equation E3-1. The same discussion applies here for Equation G3-1, with the exception that the axial force limit is $P_{rc} < 0.1 P_c$, which is done to ensure ductile behavior of composite and reinforced concrete columns.

4b. Stability Bracing of Beams

For commentary on stability bracing of beams, see Commentary Section E3.4b.

4c. Stability Bracing at Beam-to-Column Connections

The stability bracing requirements at beam-to-column connections are similar to those for unbraced connections in steel SMF. Composite columns are typically not susceptible to flexural-torsional buckling modes due to the presence of concrete. The requirements of Section E3.4c(2) are applicable because composite columns are susceptible to flexural buckling modes in the out-of-plane direction.

5. Members

5a. Basic Requirements

Reliable inelastic deformation for highly ductile members requires that width-to-thickness ratios be limited to a range that provides composite cross sections resistant to local buckling well into the inelastic range. Although the width-to-thickness ratio for compact elements in *Specification* Table I1.1 are sufficient to prevent local buckling before the onset of yielding, the available test data suggest that these limits are not adequate for the required inelastic deformations in C-SMF (Varma et al., 2002, 2004; Tort and Hajjar, 2004).

Encased composite columns classified as highly ductile members shall meet the additional detailing requirements of Sections D1.4b(1) and (2) to provide adequate ductility. For additional details, refer to Commentary Section D1.4b.

Filled composite columns shall meet the additional requirements of Section D1.4c.

When the design of a composite beam satisfies Equation G3-2, the strain in the steel at the extreme fiber will be at least five times the tensile yield strain prior to concrete crushing at strain equal to 0.003. It is expected that this ductility limit will control the beam geometry only in extreme beam/slab proportions.

5b. Beam Flanges

For relevant commentary on changes in cross section of beam flanges, see Commentary Section E3.5b.

5c. Protected Zones

For commentary on protected zones see Commentary Section D1.3.

6. Connections

While the Provisions permit the design of composite beams based solely upon the requirements in the *Specification*, the effects of reversed cyclic loading on the strength and stiffness of shear studs should be considered. This is particularly important for C-SMF where the design loads are calculated assuming large member ductility and toughness. In the absence of test data to support specific requirements in the Provisions, the following special measures should be considered in C-SMF: (1) implementation of an inspection and quality assurance plan to verify proper welding of steel headed stud anchors to the beams (see Sections A4.3 and Chapter J); and (2) use of additional steel headed stud anchors beyond those required in the *Specification* immediately adjacent to regions of the beams where plastic hinging is expected.

6a. Demand Critical Welds

For general commentary on demand critical welds see Commentary Sections A3.4 and E3.6a.

6b. Beam-to-Column Connections

Connections to Reinforced Concrete Columns: A schematic connection drawing for composite moment frames with reinforced concrete columns is shown in Figure C-D2.9, where the steel beam runs continuously through the column and is spliced away from the beam-to-column connection. Often, a small steel column that is interrupted by the beam is used for erection and is later encased in the reinforced concrete column (Griffis, 1992). Numerous large-scale tests of this type of connection have been conducted in the United States and Japan under both monotonic and cyclic loading (e.g., Sheikh et al., 1989; Kanno and Deierlein, 1997; Nishiyama et al., 1990; Parra-Montesinos and Wight, 2000; Chou and Uang, 2002; Liang and Parra-Montesinos, 2004). The results of these tests show that carefully detailed connections can perform as well as seismically designed steel or reinforced concrete connections.

In particular, details such as the one shown in Figure C-D2.9 avoid the need for field welding of the beam flange at the critical beam-to-column junction. Therefore, these joints are generally not susceptible to the fracture behavior in the immediate connection region near the column. Tests have shown that, of the many possible ways of strengthening the joint, face bearing plates (see Figure C-G3.1) and steel band plates (Figure C-G3.2) attached to the beam are very effective for both mobilizing the joint shear strength of reinforced concrete and providing confinement to the concrete. Further information on design methods and equations for these composite connections is available in published guidelines (e.g., Nishiyama et al., 1990; Parra-Montesinos and Wight, 2001). Note that while the scope of the current ASCE Guidelines (ASCE, 1994) limits their application to regions of low to moderate seismicity, recent test data indicate that the ASCE Guidelines are adequate for regions of high seismicity as well (Kanno and Deierlein, 1997; Nishiyama et al., 1990; Parra-Montesinos et al., 2003).

Connections to Encased Columns: Prior research has been conducted on the cyclic performance of encased columns and their connections (e.g., Kanno and Deierlein, 1997). Connections between steel beams and encased composite columns (see Figure C-G3.1) have been used and tested extensively in Japan. Alternatively, the connection strength can be conservatively calculated as the strength of the connection of the steel beam to the steel column. Or, depending upon the joint proportions and detail, where appropriate, the strength can be calculated using an adaptation of design models for connections between steel beams and reinforced concrete columns (ASCE, 1994). One disadvantage of this connection detail compared to the one shown in Figure C-D2.9 is that, like standard steel construction, the detail in Figure C-G3.1 requires welding of the beam flange to the steel column.

Connections to Filled Columns: Prior research has also been conducted on the cyclic performance of filled columns and their connections, and there has been substantial recent research to support design strategies (see Figure C-G3.3). (Azizinamini and Schneider, 2004; Ricles et al., 2004a; Herrera et al., 2008).

The results of these tests and the corresponding design details can be used to design the connections and prepare for the qualification according to Chapter K of the Provisions. For example, Figure C-G3.4 shows a large-scale filled composite column to steel beam connection that was tested by Ricles et al. (2004a) and demonstrated to exceed a story drift angle of 0.04 rad. In this same publication, the authors report test results for other large-scale filled composite column-to-beam connections that meet or exceed the story drift angle of 0.02 rad (for C-IMF) and 0.04 rad (for C-SMF).

For the special case where the steel beam runs continuously through the composite column, the internal load transfer mechanisms and behavior of these connections are

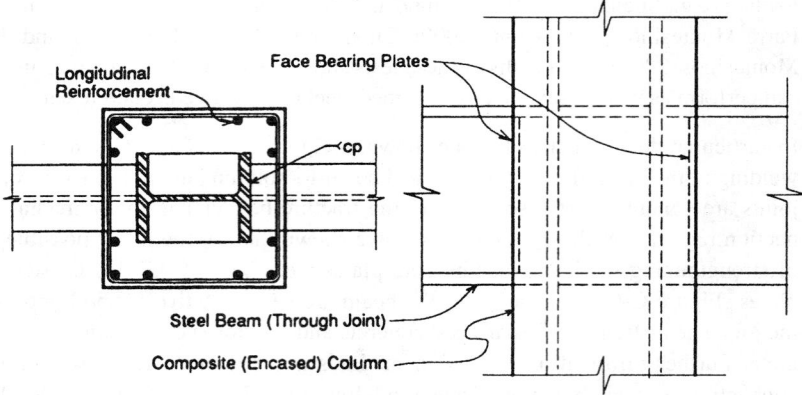

Fig. C-G3.1. Encased composite column-to-steel beam moment connection.

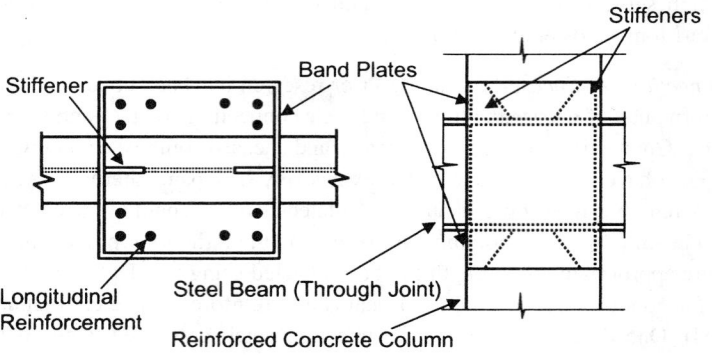

Fig. C-G3.2. Steel band plates used for strengthening the joint.

similar to those for connections to reinforced concrete columns (Figure C-G3.2). Otherwise, where the beam is interrupted at the column face, special details are needed to transfer the column flange loads through the connection (Azizinamini and Schneider, 2004).

6c. **Conformance Demonstration**

The Provisions require that connections in C-SMF meet the same story drift capacity of 0.04 rad as required for steel SMF. Section G3.6c provides conformance demonstration requirements. This provision permits the use of connections qualified by prior tests or project specific tests. The engineer is responsible for substantiating the connection.

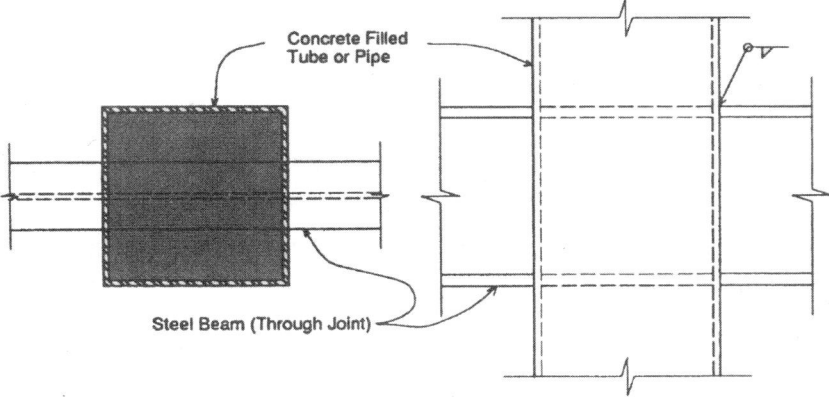

Fig. C-G3.3. Filled composite column-to-steel beam moment connection (beam flange uninterrupted).

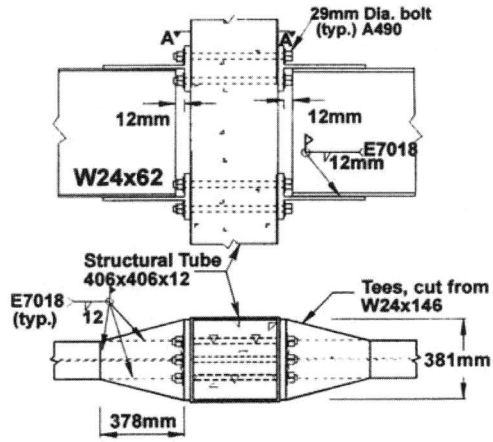

Fig. C-G3.4. Filled composite column-to-steel beam moment connection (beam flange interrupted).

For the special case when beams are uninterrupted or continuous through composite or reinforced concrete columns, and beam flange welded joints are not used, the performance requirements shall be demonstrated through large-scale testing in accordance with Section K2, or other substantiating data available in the literature (e.g., Kanno and Deierlein, 1997; Nishiyama et al., 1990; Parra-Montesinos and Wight, 2001; Parra-Montesinos et al., 2003).

6d. Required Shear Strength

The requirements for shear strength of the connection for C-SMF are comparable to those of SMF, with the exception that the calculation of the expected flexural strength must account for the different constituent materials. See Commentary Section E3.6d.

6e. Connection Diaphragm Plates

The requirements for continuity plates and diaphragms are the same for C-SMF as for C-IMF. Refer to Commentary Section G2.6e.

6f. Column Splices

The requirements for column splices are the same for C-SMF as for C-IMF. Refer to Commentary Section G2.6f.

G4. COMPOSITE PARTIALLY RESTRAINED MOMENT FRAMES (C-PRMF)

1. Scope

Composite partially restrained moment frames (C-PRMF) consist of structural steel columns and composite steel beams, connected with PR composite joints (Leon and Kim, 2004; Thermou et al., 2004; Zandonini and Leon, 1992). In PR composite joints, flexural resistance is provided by a couple incorporating a conventional steel bottom flange connection (welded or bolted plates, angles, or T-stubs) and the continuous reinforcing steel in the slab at the top of the girder (see Figure C-G4.1). The steel beam and the concrete slab are connected by steel anchors, such as studs. Shear resistance is provided through a conventional steel frame shear connection (welded or bolted plates or angles). The use of the slab reinforcing steel results in a stronger and stiffer connection, a beneficial distribution of strength and stiffness between the positive and negative moment regions of the beams, and redistribution of loads under inelastic action. In most cases, the connections in this seismic force resisting system at the roof level will not be designed as composite.

C-PRMF were originally proposed for areas of low to moderate seismicity in the eastern United States (seismic design categories C and below). However, with appropriate detailing and analysis, C-PRMF can be used in areas of higher seismicity (Leon, 1990). Tests and analyses of these systems have demonstrated that the seismically induced loads on partially restrained (PR) moment frames can be lower than those for fully restrained (FR) moment frames due to: (1) lengthening in the natural period due to yielding in the connections and (2) stable hysteretic behavior of the connections (Nader and Astaneh-Asl, 1992; DiCorso et al., 1989). Thus, in some cases, C-PRMF can be designed for lower seismic loads than ordinary moment frames (OMF).

2. Basis of Design

Design methodologies and standardized guidelines for composite partially restrained moment frames (C-PRMF) and connections have been published (Ammerman and Leon, 1990; Leon and Forcier, 1992; Leon et al., 1996; ASCE, 1998). In the design of PR composite connections, it is assumed that bending and shear forces can be considered separately.

3. Analysis

For frames up to four stories, the design of C-PRMF should be made using an analysis that, as a minimum, accounts for the partially restrained connection behavior of the connections by utilizing linear springs with reduced stiffness (Bjorhovde, 1984). The effective connection stiffness should be considered for determining member load distributions and deflections, calculating the building's period of vibration, and

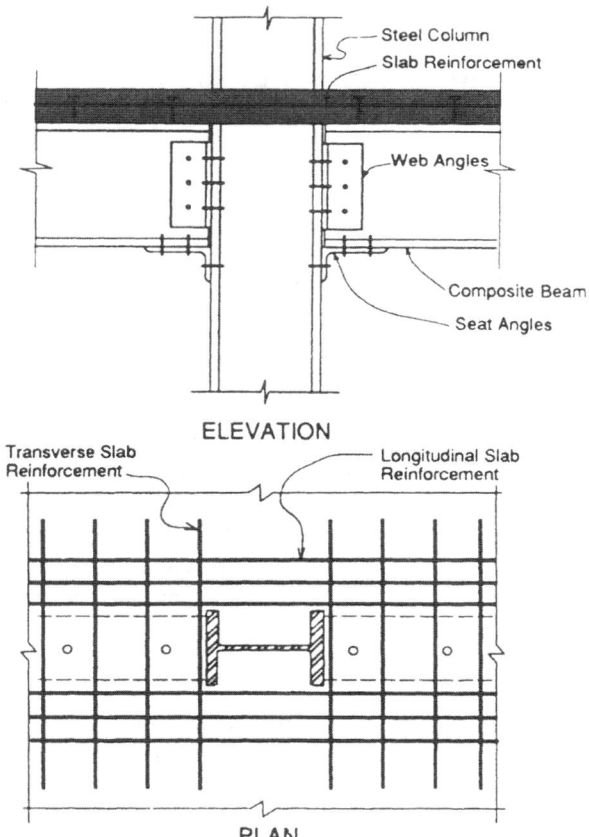

Fig. C-G4.1. Composite partially restrained connection.

checking frame stability. Different connection stiffnesses may be required for these checks (Leon et al., 1996). Frame stability can be addressed using conventional procedures. However, the connection flexibility should be considered in determining the rotational restraint at the ends of the beams. For structures taller than four stories, drift and stability need to be carefully checked using analysis techniques that incorporate both geometric and connection nonlinearities (Rassati et al., 2004; Ammerman and Leon, 1990; Chen and Lui, 1991). Because the moments of inertia for composite beams in the negative and positive regions are different, the use of either value alone for the beam members in the analysis can lead to errors. Therefore, the use of a weighted average, as discussed in the Commentary to Chapter I of the *Specification*, is recommended (Zaremba, 1988; Ammerman and Leon, 1990; Leon and Ammerman, 1990; AISC, 2010a).

4. System Requirements

The system should be designed to enforce a strong column-weak beam mechanism except for the roof level. ASCE TC (1998) suggests using the following equation to achieve this behavior, where M_{cu}^+ and M_{cu}^- refer to the connection moment strength in positive and negative bending respectively:

$$\Sigma M_{p,\,col}\left(1 - \frac{P_u}{P_0}\right) > 1.25(M_{cu}^+ + M_{cu}^-)$$

5. Members

5a. Columns

Column panel zone checks per the *Specification* should be carried out assuming the connection moment is given by concentrated forces at the bottom flange and at the center of the concrete slab.

5b. Beams

Only fully composite beams are used in this system, as the effect of partial interaction in the composite beams has not been adequately justified. Because the force transfer relies on bearing of the concrete slab against the column flange, the bearing strength of the concrete should be checked. (See Figure C-G4.2.) The full nominal slab depth should be available for a distance of at least 12 in. (152 mm) from the column flange (see Figure C-G4.3). This 12 in. requirement can be avoided if another means of load transfer is provided such as mechanically attaching the reinforcing steel directly to the column as shown in Figure C-G4.4.

6. Connections

The connecting elements are designed with a yield force that is less than that of the connected members to prevent local limit states, such as local buckling of the flange in compression, web crippling of the beam, panel zone yielding in the column, and bolt or weld failures, from controlling. When these limit states are avoided, large connection ductilities should ensure excellent frame performance under large inelastic load reversals. The 50% M_p requirement is intended to apply to both positive and negative connection strength. This requirement is intended to prevent a potential incremental collapse mechanism from developing.

6c.   **Beam-to-Column Connections**

Most PR connections do not exhibit a simple elasto-plastic behavior and thus the moment strength of the connection must be tied to a connection rotation value. A connection rotation of 0.02 rad has been used as the requirement in the *Specification*; however, for most composite PR connections, it is more appropriate to use 0.01 rad when considering the positive moment strength (tension at the bottom flange) of the connection. Most PR connections will achieve at least 80% of their ultimate strength at these rotation levels.

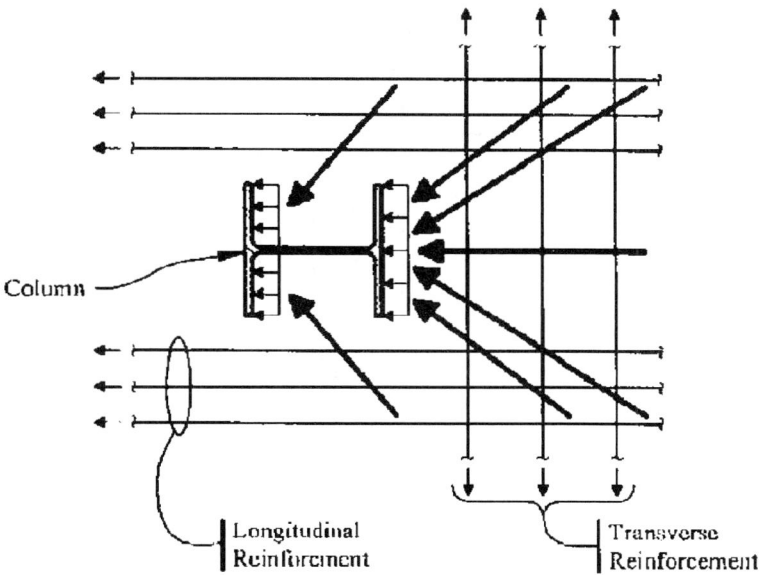

Fig. C-G4.2. Concrete slab bearing force transfer.

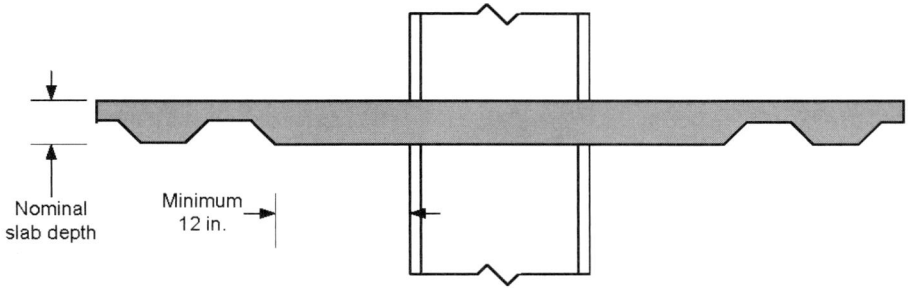

Fig. C-G4.3. Solid slab to be provided around column.

6d. Conformance Demonstration

Tests results that show general conformance with Section K2 have been reported in the literature (Leon et al., 1987; Leon, 1994). Section K2 is currently written in terms of story drift rather than in terms of connection rotation; however, the intent of Section K2 for this seismic frame system is to show that the connection is capable of sustaining cyclic strength through a connection rotation of 0.02 rad. Therefore, the loading sequence of Section K2.4b should be considered in the context of connection rotation rather than story drift and need only be taken through step (6) of the loading sequence.

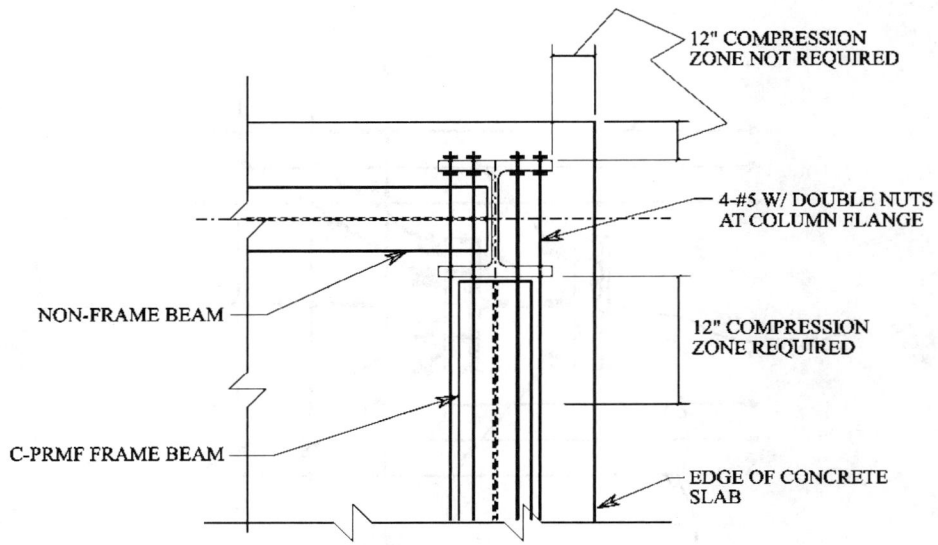

Fig. C-G4.4. Alternate means of providing slab force transfer.

CHAPTER H

COMPOSITE BRACED-FRAME AND SHEAR-WALL SYSTEMS

H1. COMPOSITE ORDINARY BRACED FRAMES (C-OBF)

Composite braced frames consisting of steel, composite and/or reinforced concrete elements have been used in low- and high-rise buildings in regions of low and moderate seismicity. The composite ordinary braced frame (C-OBF) category is provided for systems without special seismic detailing that are used in seismic design categories A, B and C. Thus, the C-OBF systems are considered comparable to structural steel systems that are designed according to the *Specification* using a seismic response factor of $R = 3$. Because significant inelastic load redistribution is not relied upon in the design, there is no distinction between frames where braces frame concentrically or eccentrically into the beams and columns.

1. Scope

The combination of steel, concrete and/or composite member types that is permitted for C-OBF is intended to accommodate any reasonable combination of member types as permitted by the *Specification* and ACI 318 (ACI, 2008).

6. Connections

Examples of connections used in C-OBF are shown in Figures C-H1.1 through C-H1.3. As with other systems designed in accordance with the *Specification* for a seismic response factor of $R = 3$, the connections in C-OBF should have design

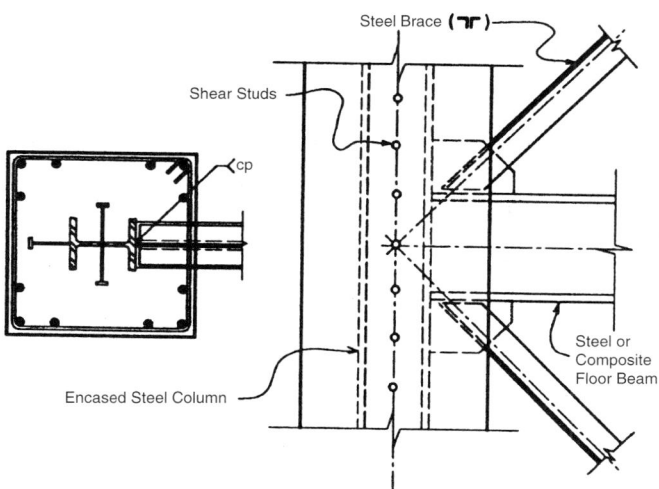

Fig. C-H1.1. Reinforced concrete (or composite) column-to-steel concentric brace.

strengths that exceed the required strengths for the earthquake loads in combination with gravity and other significant loads. The provisions of Section D2.7 should be followed insofar as they outline basic assumptions for calculating the strength of force transfer mechanisms between structural steel and concrete members and components.

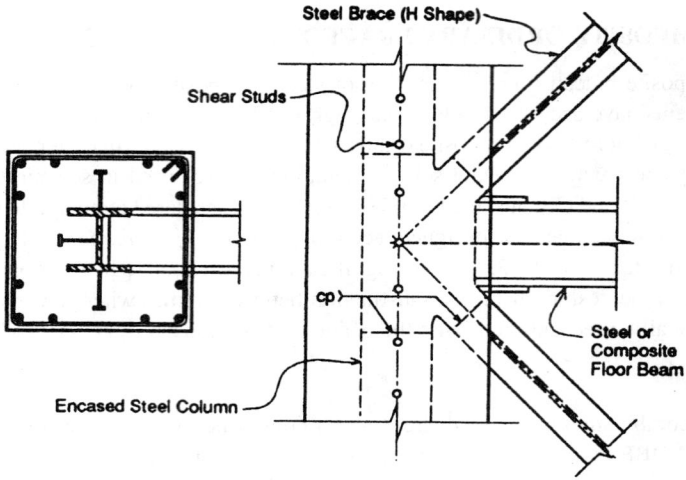

Fig. C-H1.2. Reinforced concrete (or composite) column-to-steel concentric brace.

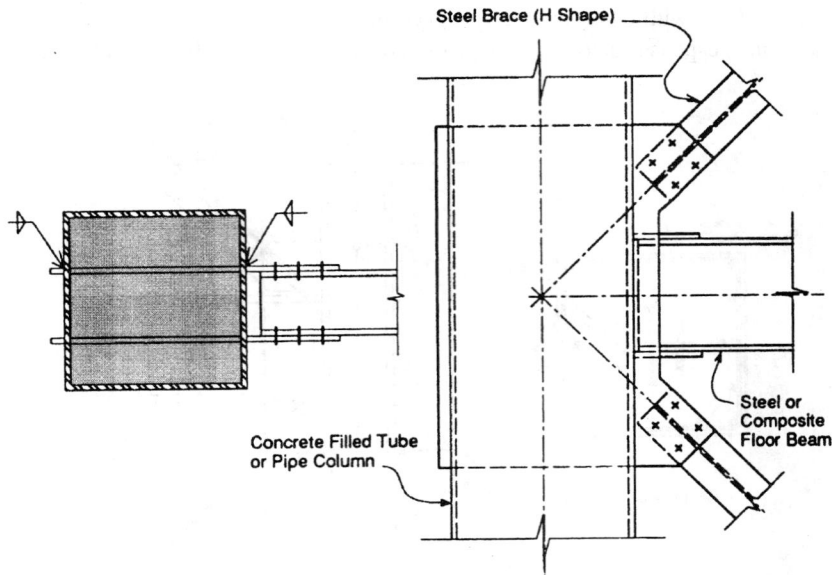

Fig. C-H1.3. Filled HSS or pipe column-to-steel concentric base.

H2. COMPOSITE SPECIAL CONCENTRICALLY BRACED FRAMES (C-SCBF)

The composite special concentrically braced frame (C-SCBF) is one of two types of composite braced frames that are specially detailed for seismic design categories D and above; the other is the composite eccentrically braced frame (C-EBF). While experience using C-SCBF is limited in high seismic regions, the design provisions for C-SCBF are intended to provide behavior that is comparable to steel SCBF, wherein the braces often are the elements most susceptible to inelastic deformations (see Commentary Section F2). The R and C_d values and usage limitations for C-SCBF are similar to those for steel SCBF.

1. Scope

Unlike C-CBF, which permit the use of concrete columns, the scope for C-SCBF is limited to systems with composite columns to help ensure reliable force transfer from the steel or composite braces and beams into the columns.

2. Basis of Design

The basis of design is comparable to steel SCBF. Thus, the provisions for analysis, system requirements, members and connections make reference to the provisions of Section F2. Refer to the associated commentary for Section F2 where reference is made to that section in the Provisions.

5. Members

Composite columns in C-SCBF are detailed with similar requirements to highly ductile composite columns in C-SMF. Special attention should be paid to the detailing of the connection elements (MacRae et al., 2004).

5b. Diagonal Braces

Braces that are all steel should be designed to meet all requirements for steel braces in Section F2.

In cases where composite braces are used (either filled or encased), the concrete has the potential to stiffen the steel section and prevent or deter brace buckling while at the same time increasing the capability to dissipate energy. The filling of hollow structural sections (HSS) with concrete has been shown to effectively stiffen the HSS walls and inhibit local buckling (Goel and Lee, 1992). For encased steel braces, the concrete should be sufficiently reinforced and confined to prevent the steel shape from buckling. To provide high ductility, the composite braces are required to be designed to meet all requirements of encased composite columns as specified in Section D1.4b. Composite braces in tension should be designed based on the steel section alone unless test data justify higher strengths.

6. Connections

Careful design and detailing of the connections in a C-SCBF is required to prevent connection failure before developing the full strength of the braces in either tension or compression. Where the brace is composite, the added brace strength afforded by the concrete should be considered in the connection design. In such cases, it would be unconservative to base the connection strength on the steel section alone. Connection design and detailing should recognize that buckling of the brace could cause excessive rotation at the brace ends and lead to local connection failure. Therefore, as in steel SCBF, the brace connection should either be designed to accommodate the inelastic rotations associated with brace buckling or to have sufficient strength and stiffness to accommodate plastic hinging of the brace adjacent to the connection.

6a. **Demand Critical Welds**

For general commentary on demand critical welds see Commentary Section A3.4.

6b. **Beam-to-Column Connections**

Ductile connections between the beam and column are required for C-SCBF. Rotation requirements for both simple and moment-resisting connections are provided. See the Commentary for the referenced sections.

6d. **Column Splices**

The requirements for column splices are comparable to those of C-IMF. Refer to Commentary Section G2.6f.

H3. COMPOSITE ECCENTRICALLY BRACED FRAMES (C-EBF)

1. **Scope**

Structural steel EBF have been extensively tested and utilized in seismic regions and are recognized as providing excellent resistance and energy absorption for seismic loads (see Commentary Section F3). While there has been little use of composite eccentrically braced frames (C-EBF), the inelastic behavior of the critical steel link region should be comparable to that of steel EBF and inelastic deformations in the encased composite or filled composite columns should be minimal as well as in the structural steel or filled composite braces. Therefore, the R and C_d values and usage limitations for C-EBF are the same as those for steel EBF. As described below, careful design and detailing of the brace-to-column and link-to-column connections is essential to the performance of the system.

2. **Basis of Design**

The basic design requirements for C-EBF are the same as those for steel EBF, with the primary energy absorption being provided by the structural steel link.

A small eccentricity of less than the beam depth is allowed for brace-to-beam or brace-to-column connections away from the link. Small eccentricities are sometimes required for constructability reasons and will not result in changing the location of predominate inelastic deformation capacity away from the link as long as the resulting secondary forces are properly accounted for.

3. Analysis

As with EBF, satisfactory behavior of C-EBF is dependent on making the braces and columns strong enough to remain essentially elastic under loads generated by inelastic deformations of the links. Since this requires an accurate calculation of the shear link nominal strength, it is important that the shear region of the link not be encased in concrete.

6. Connections

In C-EBF where the link is not adjacent to the column, the concentric brace-to-column connections are similar to those shown for C-CBF (Figures C-H2.1 through C-H2.3). An example where the link is adjacent to the column is shown in Figure C-H3.1. In this case, the link-to-column connection is similar to composite beam-to-column moment connections in C-SMF (Section G3) and to steel coupling beam-to-wall connections (Section H5).

6a. Beam-to-Column Connections

While the majority of the energy dissipation is anticipated to occur at the *link,* beam-to-column connections in C-EBF are anticipated to go through large rotations as the system undergoes large inelastic deformations. The maximum inelastic deformations are anticipated to be on the order of 0.025 rad, resulting in the requirement that when simple beam-to-column connections are used that they be capable of undergoing this rotation demand. Alternatively, fully restrained, ordinary moment connections can also be used since they have been shown to accommodate this rotation demand.

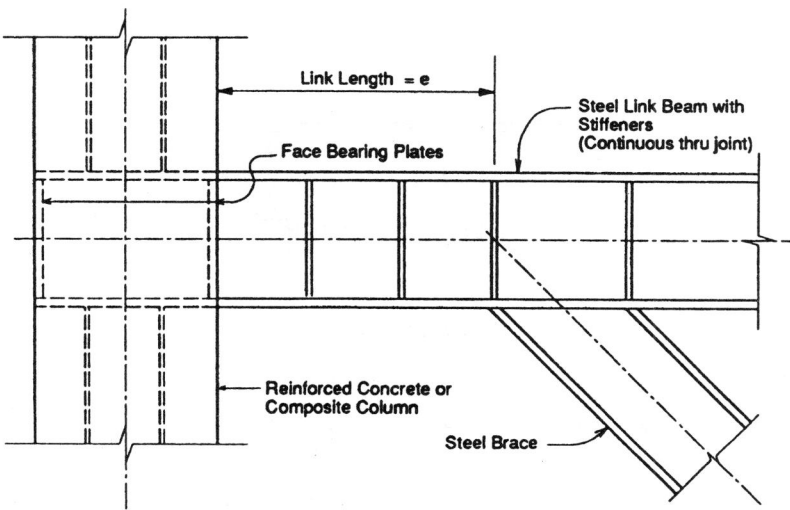

Fig. C-H3.1. *Reinforced concrete (or composite) column-to-steel eccentric brace. (Note: Stiffeners are designed according to Section F3.5a.)*

H4. COMPOSITE ORDINARY SHEAR WALLS (C-OSW)

1. Scope

The provisions in this Section apply to reinforced concrete walls with structural steel or composite sections serving as boundary elements. Examples of such systems are shown in Figure C-H4.1.

This Section also applies to coupled wall systems with steel or composite coupling beams connecting two or more adjacent walls (see Figure C-H4.2). In this case, the walls may or may not have structural steel or composite sections serving as boundary elements. Structural steel or composite boundary elements may be used as wall boundary elements or for erection purposes only. In the latter case, the structural steel members may be relatively small. The detailing of coupling beam-to-wall connections depends on whether structural shapes are embedded in the wall boundaries or the wall has conventional reinforced concrete boundary elements. If steel or composite column boundary elements are used, the coupling beams can frame into the columns and transmit the coupling forces through a moment resisting connection with the steel column (see Figure C-H4.3(a)). The use of a moment connection is, however, not preferred given the cost and difficulty of constructing ductile connections. Alternatively, the coupling beam may be connected to the embedded boundary column with a shear connection while the moment resistance is achieved by a combination of bearing along the embedment length and shear transfer provided by steel headed stud anchors along the coupling beam flanges. In such cases, special reinforcement detailing in the wall boundary region similar to that found in reinforced concrete walls is required. An example is shown in Figure C-H4.3(b).

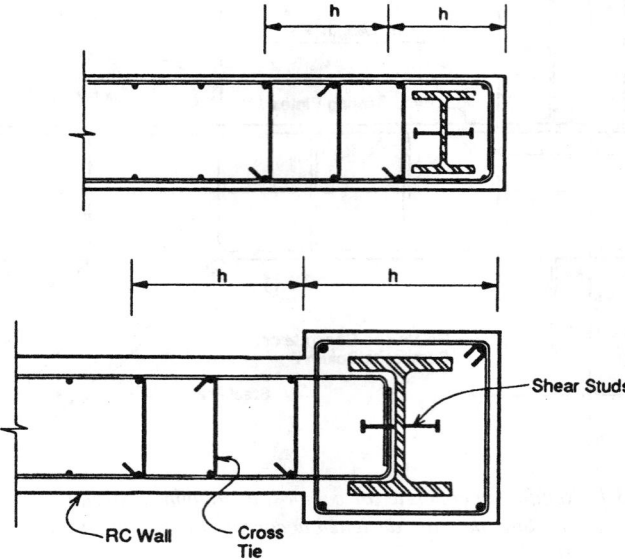

Fig. C-H4.1. Reinforced concrete walls with steel and composite boundary element.

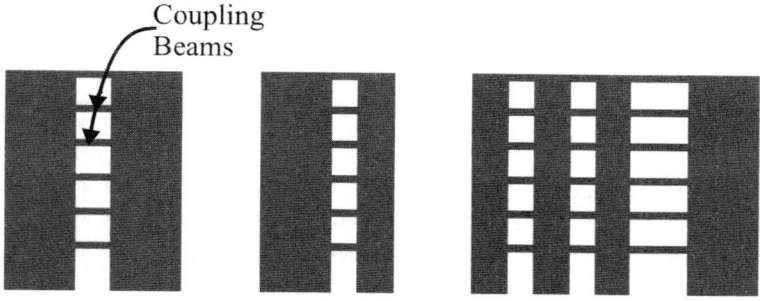

Fig. C-H4.2. Examples of coupled wall geometry.

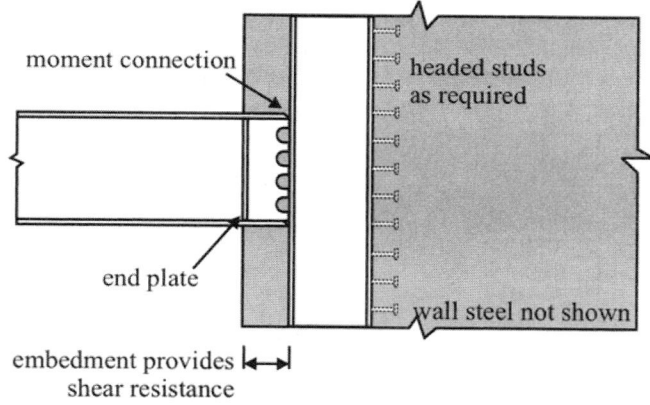

(a) Steel coupling beam attached to steel wall boundary element column

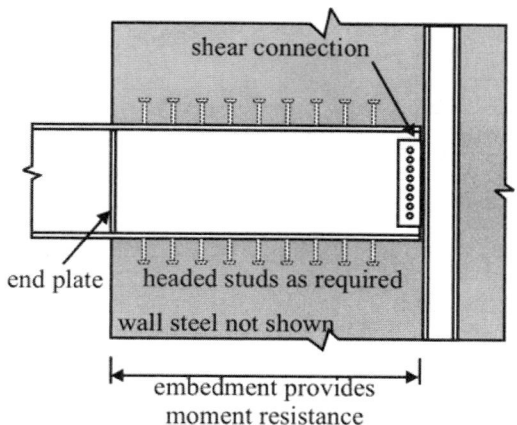

(b) Steel coupling beam attached to steel erection column

Fig. C-H4.3. Steel coupling beam details.

If structural steel or composite boundary elements are not present, the coupling beam should be embedded a sufficient distance into the wall so that the coupling forces are transmitted entirely through the interaction that occurs between the embedded coupling beam and the surrounding concrete. Examples of such embedment regions are as shown in Figure C-H4.4.

It is not necessary, nor is it typically practical, to pass wall boundary transverse reinforcing bars through the web of the embedded coupling beam. A practical alternative is to place hooked ties on either side of the web and provide short vertical bars between the flanges to anchor these ties.

2. **Basis of Design**

    The level of inelastic deformation in composite ordinary shear walls is limited. Yielding of coupling beams is not anticipated and the walls are expected to remain in the elastic range. However, the coupling beams need to be detailed to ensure that they can yield in shear or flexure. Meeting the requirements of Section D1.1 for moderately ductile members ensures yielding in flexure. Equation H4-1 for steel coupling beams and Equations H4-3, H4-4 and H4-4M allow for yielding and resulting ductility in shear implicitly in the calculations. It is thus expected that the h/t requirements of Section G2 of the *Specification* will be satisfied such that

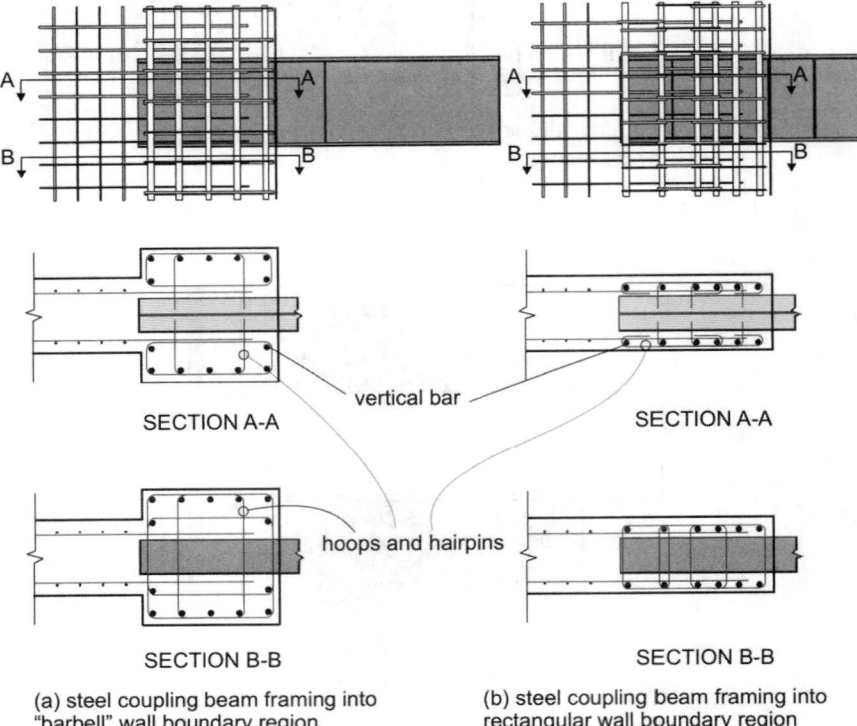

Fig. C-H4.4. *Example details of a steel coupling beam embedded in reinforced concrete wall.*

$C_v = 1.0$ in the calculation of the nominal shear strength of a steel coupling beam, or of a steel beam within a composite coupling beam. For a composite coupling beam, as discussed in Section H4.5b, the shear strengths in Equations H4-4 and H4-4M are assessed assuming the minimum shear reinforcement requirements from ACI 318 are satisfied, thus enabling the coupling beam to yield in shear. The wall piers are to be designed based on nonseismic provisions of ACI 318, i.e., the requirements of Chapter 21 do not have to be satisfied for these ordinary systems.

3. **Analysis**

In order to compute the design forces and deformations, the wall piers, coupling beam elements, and the coupling beam-wall connections need to be modeled considering cracked section properties for concrete. Guidance from ACI 318 Chapter 10 (Section 10.10.4.1) and ASCE 41 (ASCE, 2006a) is available (see Commentary Chapter C).

Modeling of the wall piers falls into three main classes (in increasing degree of complexity): 1) equivalent frame models, 2) multi-spring models, and 3) continuum finite element model (ASCE, 2009). Previous studies (Shahrooz et al., 1993; Gong and Shahrooz, 2001b; Harries et al., 1997) have demonstrated that steel or steel-concrete composite coupling beams do not behave as having a fixed boundary condition at the face of the wall. The additional flexibility needs to be taken into account in equivalent frame or multi-spring models to ensure that wall forces and lateral deflections are computed with reasonable accuracy. If the embedment length of the beam is known, the effective fixed point of steel or steel-concrete composite coupling beams may be taken at approximately one-third of the embedment length from the face of the wall (Shahrooz et al., 1993; Gong and Shahrooz, 2001b). Thus, the effective span of the equivalent fixed-end beam used for analysis, $g_{effective}$, is $g + 0.6L_e$ where g is the effective clear span and L_e is the embedment length. If the value of L_e is not available, the procedure proposed by Harries et al. (1997) may be used. In this procedure, the effective flexural stiffness (reduced to account for the presence of shear) of a steel coupling beam is reduced to 60% of its gross section value:

$$I_{eff} = 0.60I\left(1 + \frac{\lambda 12EI}{g^2 GA_w}\right)$$

where
- I = moment of inertia of steel coupling beam, in.^{4} (mm^4)
- E = modulus of elasticity of steel, ksi (MPa)
- G = shear modulus of steel, ksi (MPa)
- A_w = area of steel section assumed to resist shear, which is typically the area of the steel web, in.^{2} (mm^2)
- λ = cross section shape factor for shear (1.5 for W-shapes).

In order to account for expected spalling at the face of the wall, the effective length of the beam is increased by the wall cover dimension, c (see Figure C-H4.6 for definition of c). Therefore, the value of $g_{effective}$ becomes $g + 2c$. Both of these procedures are based on the assumption that the embedment of the coupling beam into the wall provides the necessary moment resistance at the beam end. For steel or steel-concrete composite coupling beams connected to a vertical steel member embedded in the wall boundary region (as shown in Figure C-H4.3), the effective clear span, g, should be taken as the distance between the faces of the embedded vertical "columns."

4. System Requirements

The coupling beam forces can be redistributed vertically, both up and down the structure, in order to optimize the design (Harries and McNeice, 2006). Redistribution can also help to lower the required wall overstrength and improve constructability by permitting engineers to use one beam section over larger vertical portions of the wall. Given the benefits of redistribution and the inherent ductility of steel coupling beams, a 20% redistribution of coupling beam design forces is recommended provided the sum of the resulting shear strength (e.g., the design strength, ϕV_n, for LRFD) exceeds the sum of the coupling beam design force determined from the lateral loading (e.g., the required strength, V_u, for LRFD) (CSA, 2004), i.e., $\Sigma \phi V_n / \Sigma V_u \geq 1$. This concept is schematically illustrated in Figure C-H4.5.

5. Members

5b. Coupling Beams

Coupling beam response is intended to be similar to shear link response in eccentrically braced frames. The expected coupling beam chord rotation plays an important role in how the coupling beam is detailed. This angle may be computed from $\theta_b = \dfrac{L - g_{effective}}{g_{effective}} \theta_d$ in which L is the distance between the centroids of the wall

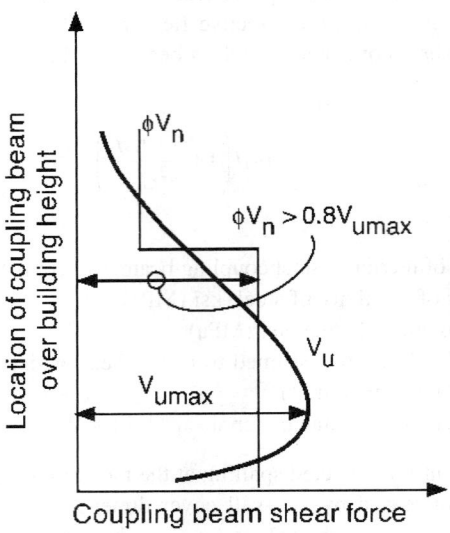

Fig. C-H4.5. *Vertical distribution of coupling beam shear.*

piers, $g_{effective}$ is the effective clear span as discussed in Commentary Section H4.3, and θ_d is the story drift angle, computed as the story drift divided by the story height (Harries et al., 2000).

For cases in which the coupling beam embedment into the wall piers is the only mechanism of moment resistance, the embedment length has to be long enough to develop the nominal shear strength of the coupling beam. Models have been developed for connections between steel brackets and reinforced concrete columns (e.g., Mattock and Gaafar, 1982). These models are used to compute an embedment length required to prevent bearing failure of concrete surrounding the flanges of the embedded steel members. A number of studies (Shahrooz et al., 1993; Gong and Shahrooz, 2001a, 2001b; Fortney, 2005) have demonstrated the adequacy of Mattock and Gaafar's model for coupling beams subjected to reversed cyclic loading. Other models (Harries et al., 1997) may also be used. Equation H4-2 is based on the model developed by Mattock and Gaafar (1982) and recommended by ASCE (2009). The strength model in this equation is intended to mobilize the moment arm between bearing forces C_f and C_b shown in Figure C-H4.6.

The Provisions stipulate that the concrete cover near the wall face spalls. As a result, the calculated value of L_e needs to be increased by the cover thickness. If the wall has a boundary member, the cover is taken as the distance from the wall face to the first layer of the confining reinforcement. For walls without boundary members, the cover is taken as the cover to the first wall longitudinal reinforcement.

A parabolic distribution of bearing stresses is assumed for C_b, and C_f is estimated by a uniform stress equal to $0.85f'_c$. The bearing stresses are distributed over the width of the beam flange, b_f.

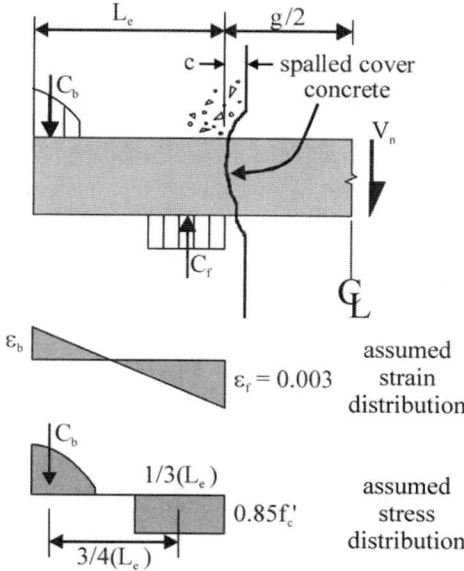

Fig. C-H4.6. Method for computing the embedment capacity.

Vertical wall reinforcement sufficient to develop the maximum shear strength of the coupling beam will provide adequate control of the gaps that open at the beam flanges under reversed cyclic loading (Harries et al., 1997). Harries et al. (1997) recommends that two-thirds of the required vertical wall reinforcement be located within a distance of one-half the embedment length from the face of the wall. The vertical bars must have adequate tension development length above and below the flanges of the coupling beam. The vertical reinforcement in wall boundary elements, if present, is typically sufficient to meet these requirements.

Steel coupling beams may be encased in reinforced concrete. Previous research (Gong and Shahrooz, 2001a, 2001b) indicates that nominal encasement significantly improves resistance to flange and web buckling, and enhances the strength of the coupling beam. The required embedment length must be computed recognizing the beneficial effects of encasement. Equations H4-4 and H4-4M for computing the shear strength of encased coupling beams have been calibrated based on meeting the ACI 318 minimum shear reinforcement requirements (ACI 318 Sections 11.4.5.1 and 11.4.6.3). Hence, minimum shear reinforcement needs to be provided regardless of the calculated value of shear force in the coupling beam.

H5. COMPOSITE SPECIAL SHEAR WALLS (C-SSW)

1. Scope

The provisions in this Section apply to coupled wall systems with steel or composite coupling beams. The reinforced concrete walls may or may not have structural steel or composite sections serving as boundary elements. Examples of such systems are discussed in Commentary Section H4.1. The focus of this Section is on composite special shear walls.

2. Basis of Design

The preferred sequence of yielding for coupled walls is for the coupling beams to yield over the entire height of the structure prior to yielding of the walls at their bases (Santhakumar, 1974). This behavior relies on coupling beam-wall connections that can develop the expected flexural and shear strengths of the coupling beams. For steel coupling beams, or steel beams embedded within composite coupling beams, satisfying the requirements of Section F3.5b ensures adequate ductility for shear yielding. For a composite coupling beam, the shear strengths in Equations H5-3 and H5-3M are assessed assuming the minimum shear reinforcement requirements are satisfied from ACI 318, thus enabling the coupling beam to yield in shear.

3. Analysis

Wall piers in special shear walls will experience significant plastic deformations. Appropriate stiffness values need to be selected to account for the differences between the cracked section properties of the walls in the plastic hinge region and regions that are expected to remain elastic. Guidance from ACI 318 Chapter 10 (Section 10.10.4.1) and ASCE 41 is available (see also Commentary Chapter C).

4. System Requirements

In order to ensure the preferred plastic mechanism in coupled walls, i.e., that the coupling beams yield prior to the wall piers, a wall overstrength factor, ω_o, is applied to the wall design forces. The required wall overstrength is taken as the ratio of the sum of the nominal shear strengths of the coupling beams, V_n, magnified by $1.1R_y$, to the sum of the coupling beam required shear strengths determined for the case of factored lateral loading, V_u, (excluding the effects of torsion) (CSA, 2004) where

$$\omega_o = \Sigma 1.1 R_y V_n / \Sigma V_u \qquad \text{(C-H5-1)}$$

This factor, therefore, includes the natural overstrength resulting from the design procedure and strength reduction factors and the overstrength resulting from designing for critical beams and using this design over a vertical cluster of beams (or all the beams) in the structure. The 20% vertical redistribution of beam forces described in Section H4.4 is permitted for special wall systems and will help to mitigate large wall overstrength factors.

The required wall overstrength can have a significant effect on wall pier design forces (Fortney, 2005; Harries and McNeice, 2006) and can adversely affect the economy of the system. Required wall overstrength will typically be greater in structures having a higher coupling ratio due to the relatively steep gradient of beam shear demand over the height of the structure (Figure C-H4.5). An advantage of a greater coupling ratio is that wall pier forces are reduced, but the larger wall overstrength factor may negate this advantage. Permitting the redistribution of beam forces as described in Section H4.4 may minimize this effect.

5. Members

5a. Ductile Elements

Coupling beams must be able to undergo substantial inelastic deformation reversals; therefore, coupling beams are designated as protected zones. Well-established guidelines for shear links in eccentrically braced frames need to be followed.

5b. Boundary Members

Concerns have been raised that walls with encased steel boundary members may have a tendency to split along planes 1 and 2 shown in Figure C-H5.1. Transverse reinforcement within a distance $2h$ (h = width of the wall) will resist splitting along plane 1 while the wall horizontal reinforcement will be adequate to prevent plane 2 failure.

5c. Steel Coupling Beams

The method described in Section H4.5b is recommended for establishing a reasonable value of coupling beam rotation. In lieu of calculating its value, the coupling beam rotation may conservatively be taken as 0.08 rad, which is the upper limit of link rotation angle in eccentrically braced frames. It should, however, be noted that

0.08 rad is conservative for coupled walls, and this rotation will result in an unnecessarily large number of stiffeners for the coupling beam.

In addition to the potential use of stiffeners along the span between the reinforced concrete walls, face-bearing plates must be provided at the face of the wall. Face bearing plates are full-width stiffeners located on both sides of the web, in effect, closing the opening in the concrete form required to install the beam. Face bearing plates provide confinement and assist in transfer of loads to the concrete through direct bearing. If it is convenient for formwork, face-bearing plates may extend beyond the flanges of the coupling beam although the plate must be installed on the inside of the form and is thereby flush with the face of the wall. The face bearing plates are detailed as a stiffener at the end of a link beam [Section F3.5b(4)]. Near the end of the embedded region, additional stiffeners similar to the face bearing plates need to be provided. These stiffeners are to be aligned with the vertical transfer bars near the end of the embedded region.

In addition to boundary element reinforcing, two regions of vertical "transfer bars" are to be provided to assist in the transfer of vertical forces and thus improve the embedment capacity (Shahrooz et al., 1993; Gong and Shahrooz, 2001a, 2001b; Fortney, 2005). Evaluation of experimental data in which transfer bars had been used (Gong and Shahrooz, 2001a, 2001b; Fortney, 2005) indicates that the minimum required area of vertical transfer reinforcement is $A_{tb} \geq 0.03 f'_c L_e b_f / F_{ysr}$ (see Figure C-H5.2). The transfer bars need to be placed close to the face of the wall and near the end of embedment length in order to develop an internal force couple that can alleviate the bearing stresses around the flanges and improve the energy dissipation characteristics of coupling beam-wall connections (Gong and Shahrooz, 2001a, 2001b). Although the required embedment length of the coupling beam may be reduced if the contribution of these bars is taken into account (Qin, 1993), to avoid excessive inelastic damage in the connection region, it is recommended by Harries et al. (1997) and Shahrooz et al. (1993) that the contribution of the transfer bars be neglected in the determination of the required embedment length. The vertical transfer bars may be attached directly to the top and bottom flanges or be passed through holes in the flanges and mechanically anchored by bolting or welding. The use of mechanical half couplers that are welded to the flanges has been successfully tested (Gong and Shahrooz, 2001a, 2001b; Fortney, 2005). U-bar hairpin reinforcement anchored by the embedded coupling beam may also be used (Figure C-H5.3). These

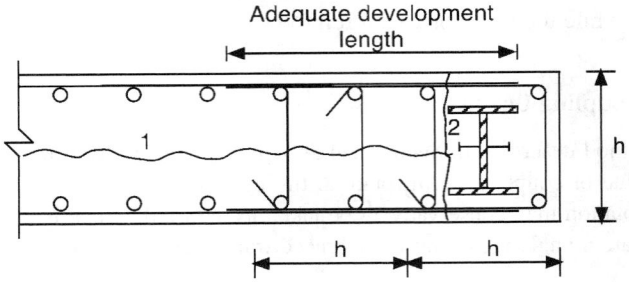

Fig. C-H5.1. Reinforcement to prevent splitting failures.

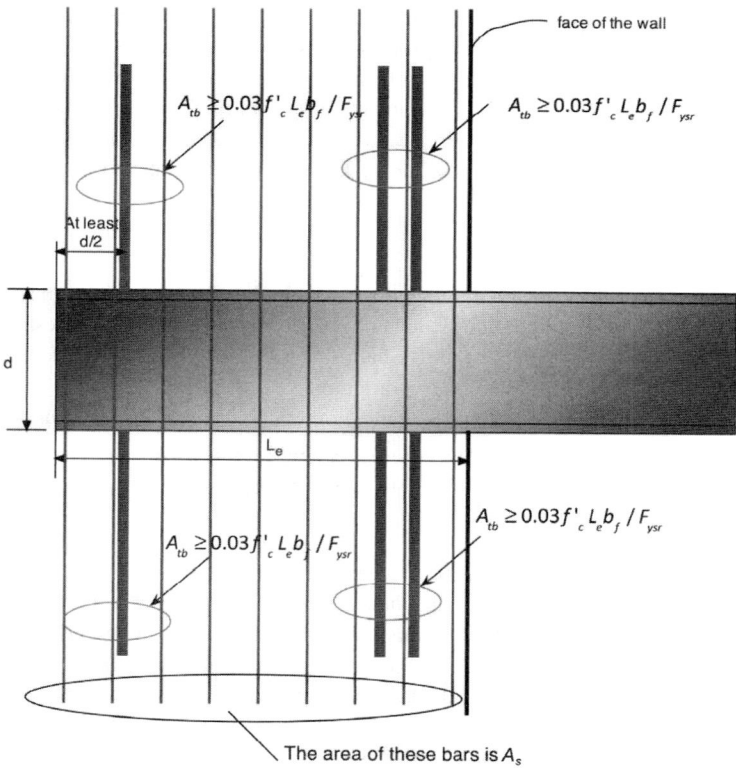

Fig. C-H5.2. *Transfer bars.*

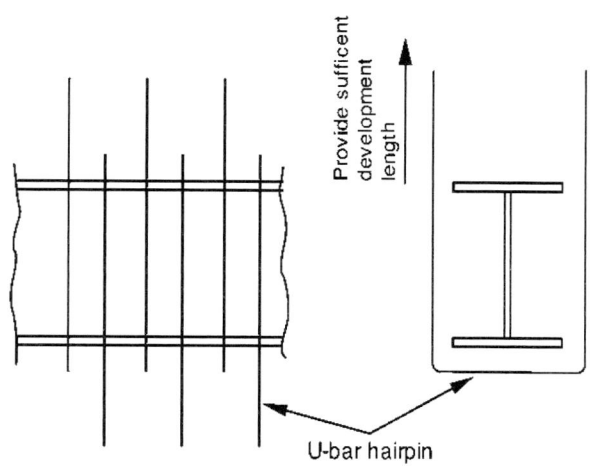

Fig. C-H5.3. *Alternating U-shaped hairpins.*

hairpins will be alternated to engage the top and bottom flanges. The transfer bars have to be fully developed in tension either by providing an adequate tension development length or through the use of headed bars. In order to prevent congestion, the sum of the areas of transfer bars and wall longitudinal bars over the embedment length (A_s shown in Figure C-H5.2 or the area of U-bar hairpins in Figure C-H5.3) is limited to 8% of the wall cross section taken as the wall width times the embedment length.

5d. Composite Coupling Beams

The required embedment length needs to be calculated to ensure that the capacity of the composite coupling beam is developed. Based on analytical studies and experimental verifications (Gong and Shahrooz, 2001a), the shear strength of the composite coupling beams may be computed based on Equation H5-3. The specified concrete compressive strength, f'_c, and nominal yield strength of transverse reinforcement, F_{ysr}, need to be used as this equation has been calibrated to account for concrete and reinforcing steel material overstrengths.

6. Connections

Structural steel sections as boundary elements in composite special shear walls are anticipated to undergo significant inelastic deformations, particularly in the plastic hinge region. The boundary columns have to be adequately anchored to the foundation system. Equally important are the splices along the boundary columns. These connections are designated as demand critical welds.

H6. COMPOSITE PLATE SHEAR WALLS (C-PSW)

1. Scope

Steel plate reinforced composite shear walls can be used most effectively where story shear loads are large and the required thickness of conventionally reinforced shear walls is excessive. Limited research on these types of systems has included configurations in which reinforced concrete is used on one side of the steel plate to mitigate the effects of local buckling (Zhao and Astaneh-Asl, 2004), and cases where two steel plates are used with reinforced concrete between them (e.g., Ozaki et al., 2004).

3. Analysis

3a. Webs

In keeping with the intended system response, the provisions of this section target having the steel webs of the C-PSW system be the primary structural elements that first attain inelastic response.

3b. Other Members and Connections

The provisions of this section target having the boundary elements of the C-PSW system remain essentially elastic under the maximum forces that can be generated by the fully yielded steel webs, along with the engaged portions of the reinforced concrete webs after the steel webs have fully yielded, except that plastic hinging at the ends of HBE and the column base are permitted.

4. System Requirements

The provisions of Section F5 are invoked for Sections H6.4b, H6.4c and H6.4d to ensure the boundary elements have adequate stiffness and strength.

4e. Openings in Webs

Careful consideration should be given to the shear and flexural strength of wall piers and of spandrels adjacent to openings. In particular, composite walls with large door openings may require structural steel boundary members attached to the steel plate around the openings.

5. Members

5b. Webs

The Provisions limit the shear strength of the wall to the yield stress of the plate because there is insufficient basis from which to develop design rules for combining the yield stress of the steel plate and the reinforced concrete panel. Moreover, since the shear strength of the steel plate usually is much greater than that of the reinforced concrete encasement, neglecting the contribution of the concrete does not have a significant practical impact. ASCE/SEI 7 assigns structures with composite walls a slightly higher R value than special reinforced concrete walls because the shear yielding mechanism of the steel plate will result in more stable hysteretic loops than for reinforced concrete walls.

5c. Concrete Stiffening Elements

Minimum reinforcement in the concrete cover or infill is required to maintain the integrity of the wall under reversed cyclic in-plane loading and out-of-plane loads. Consideration should be given to splitting of the concrete element on a plane parallel to the steel plate. Until further research data are available, the minimum required wall reinforcement is based upon the specified minimum value for reinforced concrete walls in ACI 318. Examples of such reinforcement are shown in Figures C-H6.1 through C-H6.4.

5d. Boundary Members

C-PSW systems can develop significant diagonal compressions struts, particularly if the concrete is activated directly at the design story drift. These provisions ensure that the boundary elements have adequate strength to resist this force.

6. Connections

Two examples of connections between composite walls to either steel or composite boundary elements are shown in Figures C-H6.1 and C-H6.2.

6a. Demand Critical Welds

In addition to the welds at the column splices and base plates, the welds at the connections between the boundary elements are potentially subjected to large inelastic excursions and so are designated as demand critical.

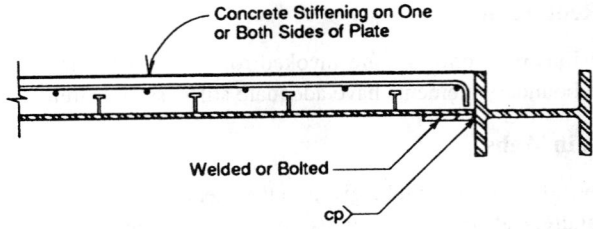

Fig. C-H6.1. Concrete stiffened steel shear wall with steel boundary member.

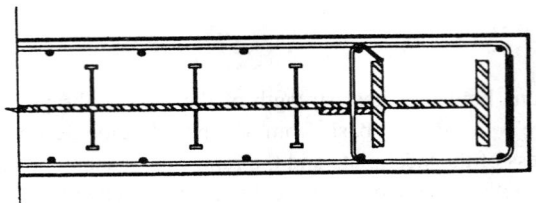

Fig. C-H6.2. Concrete stiffened steel shear wall with composite (encased) boundary member.

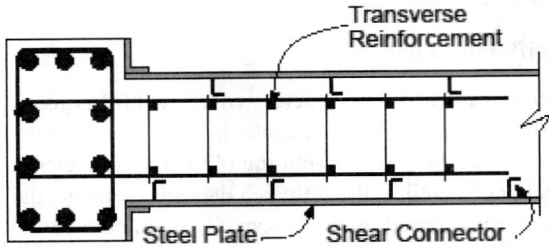

Fig. C-H6.3. Concrete filled C-PSW with a boundary element and transverse reinforcement.

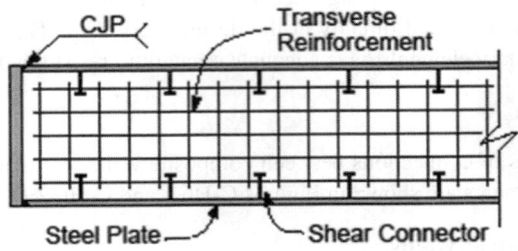

Fig. C-H6.4. Concrete filled C-PSW with transverse reinforcement to provide integrity of the concrete infill.

6b. HBE-to-VBE Connections

The provisions of Section F5 are invoked to provide adequate strength in the boundary element connections.

6c. Connections of Steel Plate to Boundary Elements

The Provisions require that the connections between the plate and the boundary members be designed to develop the full yield stress of the plate.

6d. Connections of Steel Plate to Reinforced Concrete Panel

The thickness of the concrete encasement and the spacing of shear stud connectors should be calculated to allow the steel plate to reach yield prior to overall or local buckling. It is recommended that overall buckling of the composite panel be checked using elastic buckling theory with a transformed section stiffness for the wall. It is recommended that local steel plate buckling be checked using elastic buckling theory considering steel connectors as fixed plate support points (Choi et al., 2009).

CHAPTER I

FABRICATION AND ERECTION

I1. SHOP AND ERECTION DRAWINGS

AISC 303 Section 4.2(a) (AISC, 2010c) requires the transfer of information from the contract documents (design drawings and project specifications) into accurate and complete shop and erection drawings. Therefore, relevant items in the design drawings and project specifications that must be followed in fabrication and erection should be placed on the shop and erection drawings, or in typical notes issued for the project.

3. Shop and Erection Drawings for Composite Construction

For reinforced concrete and composite steel-concrete construction, it is recommended that the following provisions be satisfied: *Details and Detailing of Concrete Reinforcement*, ACI 315 (ACI, 1999), *Manual of Structural and Placing Drawings for Reinforced Concrete Structures*, ACI 315R (ACI, 2004a), and *ACI Detailing Manual*, ACI SP-66 (ACI, 2004b), including modifications required by Chapter 21 of the *Building Code Requirements for Structural Concrete and Commentary*, ACI 318 (ACI, 2008) and *Recommendations for the Design of Beam-Column Joints in Monolithic Concrete Structures*, ACI 352 (ACI, 2002).

I2. FABRICATION AND ERECTION

1. Protected Zone

Stress concentrations could lead to fracture in regions of high plastic strain, therefore there is a prohibition on placement of welded attachments in the protected zone. Arc spot welds (puddle welds) associated with the attachment of steel deck to structural steel do not produce a high stress concentration, therefore these welds are permitted. Erection aids and attachments to meet OSHA safety requirements may be necessary in the protected zone. If erection aids or other attachments are required to be placed within the protected zone, good welding practices, including proper preheat, should be used. It may be necessary to remove the erection aid or attachment afterwards, and the surfaces of the protected zone may need to be further smoothed by grinding to remove any notch effects. In these and other such cases, the protected zone must be repaired. All such repairs must be approved by the engineer to ensure that severe stress concentrations would not cause a fracture during a seismic event.

2. Bolted Joints

The default installation requirement for high-strength bolts in the *Specification* is to the snug-tightened condition. Within the Provisions Section D2.2, the default condition for bolted connections in the SFRS is pretensioned bolts with faying surfaces of Class A slip coefficient or higher.

3. **Welded Joints**

With this edition of the Provisions, direct reference is made to AWS D1.8/D1.8M for welded connection details, replacing such details stated in Appendix W of the previous edition.

Because the selection and proper use of welding filler metals is critical to achieving the necessary levels of strength, notch toughness, and quality, the review and approval of welding procedure specifications is required. The engineer of record may use outside consultants to review these documents, if needed.

Welds are sometimes specified for the full length of a connection. Weld tabs are used to permit the starts and stops of the weld passes to be placed outside the weld region itself, allowing for removal of the start and stop conditions and their associated discontinuities. Because the end of the weld, after tab removal, is an outside surface that needs to be notch-free, proper removal methods and subsequent finishing is necessary.

At continuity plates, the end of the continuity plate to column flange weld near the column flange tip permits the use of a full weld tab, and removal is generally efficient if properly detailed. With this edition of the Provisions, it is permitted to allow $1/4$ in. (6 mm) of weld tab material to remain at the outboard end of the continuity plate-to-column weld ends because the strain demand placed on this weld is considerably less than that of a beam-to-column flange weld, and the probability of significant weld discontinuities with the distance permitted is small. Also, complete weld tab removal at beam-to-column joints is required to facilitate magnetic particle testing required by Section J6.2f, but such testing is not required for continuity plate welds. At the opposite end of the continuity plate to column flange weld, near the column radius, weld tabs are not generally desirable and may not be practicable because of clip size and k-area concerns. Weld tabs at this location, if used, should not be removed because the removal process has the potential of causing more harm than good.

CHAPTER J

QUALITY CONTROL AND QUALITY ASSURANCE

J1. SCOPE

Chapter N of the *Specification* contains requirements for Quality Control (QC) and Quality Assurance (QA) for structural steel and composite construction. Users should also refer to the Commentary of *Specification* Chapter N for additional information regarding these QC and QA requirements, which are applicable to work addressed in the *Specification*, and are also applicable to the seismic force resisting system (SFRS). These Provisions add requirements that are applicable only to the SFRS.

To assure ductile seismic response, steel framing is required to meet the quality requirements as appropriate for the various components of the structure. The applicable building code may have specific quality assurance plan (QAP) requirements, also termed a statement of special inspections. The quality assurance plan should include the requirements of Chapter J.

Section N7 of the *Specification* permits waiver of QA when the fabricator or erector is approved by the authority having jurisdiction (AHJ) to do the work without QA. Chapter 17 and Appendix Q of the prior edition of these Provisions did not contain any provisions for a waiver, but rather required the invocation of QC and QA as contained in those Provisions when required by the ABC or AHJ. Under the scope of this edition of the Provisions, QC is a requirement whether or not invoked. QA is a requirement when invoked by the AHJ, ABC, purchaser, owner or EOR.

The Provisions, *Specification*, *Code of Standard Practice* (AISC 303) (AISC, 2010c), AWS D1.1 *Structural Welding Code—Steel* (AWS, 2010) and the RCSC *Specification for Structural Joints Using High-Strength Bolts* (RCSC, 2009) provide inspection and acceptance criteria for steel building structures.

The QAP is typically prepared by the engineer of record, and is a part of the contract documents. This Chapter provides the minimum acceptable requirements for a QAP that applies to the construction of welded joints, bolted joints and other details in the SFRS. The engineer of record should evaluate what is already a part of the contractor's quality control system in determining the quality assurance needs for each project. Where the fabricator's quality control system is considered adequate for the project, including compliance with the special needs for seismic applications, the QAP may be modified to reflect this. Similarly, where additional needs are identified, such as for innovative connection details or unfamiliar construction methods, supplementary requirements should be specified, as appropriate. The QAP as contained in Chapter J is recommended for adoption without revision because consistent application of the same requirements is expected to improve reliability in the industry.

The quality assurance plan should be provided to the fabricator and erector as part of the bid documents, as any special quality control or quality assurance requirements may have substantial impact on the cost and scheduling of the work.

Structural observation at the site by the engineer of record or other design professional is an additional component of a QAP that is not addressed as part of Chapter J, and should be developed based upon the specific needs of the project.

A QAP, similar to that required for all-steel structures, should be developed for composite structures and components. For the reinforced concrete portion of the work, in addition to the requirements in ACI 318 Section 1.3, attention is called to the ACI *Detailing Manual* (ACI, 1999), with emphasis on the provisions of ACI 121R (Quality Management Systems for Concrete Construction).

J2. FABRICATOR AND ERECTOR DOCUMENTS

1. Documents to be Submitted for Steel Construction

(1) through (4): The selection and proper use of welding filler metals is critical to achieving the necessary levels of strength, notch toughness and quality, and submittal to the engineer of welding filler metal documentation and welding procedure specifications (WPS) is required. Submittal allows a thorough review on the part of the engineer, and allows the engineer to use outside consultants to review these documents, if needed.

In the *Specification*, welding filler metal documentation and WPS are to be available for review. In the Provisions, these items must be submitted because the performance of the welded joints that transfer load in the SFRS may affect overall building performance in a seismic event. Also, the engineer's approval of the WPS is a requirement of the Provisions (see Section I2.3), but is not a requirement in the *Specification*.

(5) Bolt installation procedures include instructions for pre-installation verification testing by the fabricator's or erector's personnel, and instructions for installing the bolts using the method chosen for pretensioning (commonly turn-of-nut method, twist-off type tension control bolt method, direct tension indicator method, or calibrated wrench method). In the *Specification*, these items are to be available for review. In the Provisions, these items must be submitted because the performance of the bolted joints that transfer load in the SFRS may affect overall building performance in a seismic event.

2. Documents to be Available for Review for Steel Construction

Certain items are of a nature that submittal of substantial volumes of documentation is not necessary, and it is acceptable to have these documents reviewed at the fabricator's or erector's facility by the engineer or designee, such as the QA Agency. The engineer may require submittal of these documents. The one year retention of the documents following substantial completion is to ensure their availability for further review until occupancy is permitted, and for a period following occupancy should issues arise, without placing an undue storage burden on the holder of the documents.

3. **Documents to be Submitted for Concrete Construction**

The items listed concern concrete and reinforcing steel embedded in the concrete, items that are outside the scope of the definition of structural steel as defined in AISC 303. Therefore, these documents are to be prepared and submitted by the contractor responsible for providing or installing the items.

4. **Documents to be Available for Review for Composite Construction**

The elements listed are of a nature that submittal of substantial volumes of documentation is not necessary, and it is acceptable to have these documents reviewed at the responsible contractor's facility by the engineer or designee, such as the QA Agency. The engineer may require submittal of these documents. The one year retention of these documents following substantial completion is to ensure their availability for further review until occupancy is permitted, and for a period following occupancy should issues arise, without placing an undue storage burden on the holder of the documents.

J3. QUALITY ASSURANCE AGENCY DOCUMENTS

QA Agencies should have internal procedures (written practices) that document how the Agency performs and documents inspection and testing. ASTM E329, *Standard Specification for Agencies Engaged in the Testing and/or Inspection of Materials Used in Construction*, is commonly used as a guide in preparing and reviewing written practices. ASTM E329 defines the minimum requirements for inspection agency personnel or testing agency laboratory personnel, or both, and the minimum technical requirements for equipment and procedures utilized in the testing and inspection of materials used in construction. Criteria are provided for evaluating the capability of an agency to properly perform designated tests on construction materials, and establish essential characteristics pertaining to the organization, personnel, facilities and quality systems of the agency. It can be used as a basis to evaluate an agency and is intended for use in qualifying and/or accrediting agencies, public or private, engaged in the testing and inspection of construction materials, including steel construction.

J4. INSPECTION AND NONDESTRUCTIVE TESTING PERSONNEL

Personnel performing welding inspection and nondestructive testing should be qualified to perform their designated tasks, whether functioning in a role as QC or QA. Standards are available that provide guidance for determining suitable levels of training, experience, knowledge and skill for such personnel. These standards are typically included in a written practice used by QA agencies. They may be used as a part of a fabricator's or erector's QC program.

For personnel performing bolting inspection, no standard currently exists that provides guidance as to suitable levels of training, experience, knowledge or skill in performing such tasks. Therefore, the QA agency's written practice should contain the agency's criteria for determining their personnel qualifications to perform bolting inspection. Similarly, a fabricator's or erector's QC program should contain their criteria for bolting inspector qualification.

J5. INSPECTION TASKS

Chapter J defines two inspection levels for required inspection tasks and labels them as either observe or perform. This is in contrast to common building code terminology which use or have used the terms periodic or continuous. This change in terminology reflects the multi-task nature of welding and high strength bolting operations, and the required inspections during each specific phase.

1. Observe

The *Specification* defines and uses the observe function in the same manner as used in the Provisions; however, to reflect the higher demand on and the consequence of failure of connections in the SFRS, these inspections are to be performed on a daily basis as a minimum.

2. Perform

The Specification defines and uses the perform function in the same manner as used in the Provisions. There is no requirement to make perform inspections on a daily basis, as is required for observe functions, because the perform functions are specific tasks to be completed prior to final acceptance of the designated item, and need be performed at that time.

3. Document

Inspection reports and nonconformance reports are required. The *Specification* contains limited requirements for documentation by QA of the types of inspections performed, including NDT. The Provisions require specific reporting of inspections in the same manner, but add requirements for both QC and QA reports for specific inspection tasks as described in the Document columns in the tables contained in Sections J6, J7, J8, J9 and J10.

J6. WELDING INSPECTION AND NONDESTRUCTIVE TESTING

1. Visual Welding Inspection

Visual inspection by a qualified inspector prior to, during, and after welding is emphasized as the primary method used to evaluate the conformance of welded joints to the applicable quality requirements. Joints are examined prior to the commencement of welding to check fit-up, preparation of bevels, gaps, alignment and other variables. During welding, adherence to the welding procedure specification (WPS) is maintained. After the joint is welded, it is then visually inspected to the requirements of AWS D1.1/D1.1M.

The commentary to the *Specification* Section N5.4 on welding inspection contains extensive discussion regarding the observation of welding operations, including the determination of suitable intervals for performing such inspections. Welds in the SFRS should be considered for higher levels of observation, compared to welds not in the SFRS and addressed by Chapter N in the *Specification*. Welds designated demand critical within the SFRS should be considered as warranting higher levels of observation, compared to other welds not designated demand critical within the SFRS.

2. Nondestructive Testing (NDT) of Welded Joints

The use of nondestructive testing methods as required by this Section is recommended to verify the soundness of welds that are subject to tensile loads as a part of the SFRS, or to verify that certain critical elements do not contain significant notches that could cause failure. Ultrasonic testing (UT) is capable of detecting serious embedded flaws in groove welds in all standard welded joint configurations. UT is not suitable for inspecting most fillet welds, nor should it be relied upon for the detection of surface or near-surface flaws. Magnetic particle testing (MT) is capable of detecting serious flaws on or near the surface of all types of welds, and is used for the surface examination of critical groove welds. The use of penetrant testing (PT) is not recommended for general weld inspection, but may be used for crack detection in specific locations such as weld access holes, or for the location of crack tips for cracks detected visually.

2a. *k*-Area NDT

The *k*-area of rotary straightened wide-flange sections may have reduced notch toughness. Preliminary recommendations (AISC, 1997a) discouraged the placement of welds in this area because of post-weld cracking that occurred on past projects. Where such welds are to be placed in the *k*-area, inspection of these areas is needed to verify that such cracking has not occurred.

For doubler plates, where welding in the *k*-area is performed, MT in the *k*-area should be performed on the side of the member web opposite the weld location, and at the end of the weld. If both sides of the member web receive doubler plates in the *k*-area, MT of the member web should be performed after welding of one side, prior to welding of the opposite side.

Cracking in the *k*-area is known to occur in a delayed manner, typically within 24 to 48 hours after welding. The cracks generally, but not always, penetrate the thickness of the base metal.

The *Specification* requires only visual inspection of the *k*-area after welding is performed in the *k*-area, without a designated delay period. For the SFRS, the Provisions require additional MT to be performed no sooner than 48 hours after completion of such weld.

2b. CJP Groove Weld NDT

UT is used to detect serious embedded flaws in groove welds, but is not suitable for the detection of surface or near-surface flaws. MT is used to detect serious flaws on or near the surface of these welds. Because visual inspection is also implemented for all CJP groove welds, detecting the most serious surface defects, MT is performed at a rate of 25%.

2c. Base Metal NDT for Lamellar Tearing and Laminations

Lamellar tearing is the separation (tearing) of base metal along planes parallel to a rolled surface of a member. The tearing is the result of decohesion of "weak planes," usually associated with elongated "stringer" type inclusions, from the shrinkage of large weld metal deposits under conditions of high restraint, applying stress in the through-thickness direction of the base metal.

Lamellar tears rarely occur when the weld size is less than about $^3/_4$ to 1 in. (20 to 25 mm). Typically, inclusions located deeper from the surface than $t/4$ do not contribute to lamellar tearing susceptibility.

An appropriate criterion for laminations in SFRS connections does not exist in current standards. Although AWS D1.1, Table 6.2 criteria has been written and is applicable to weld metal, not base metal, the use of Table 6.2 criteria has been deliberately selected as conservative acceptance criteria for laminations in these applications, immediately adjacent to and behind the weld.

2d. Beam Cope and Access Hole NDT

The stress flow near and around weld access holes is very complex, and the stress levels are very high. Notches serve as stress concentrations, locally amplifying this stress level which can lead to cracking. The surface of the weld access hole must be smooth, free from significant surface defects. Both penetrant testing (PT) and MT are capable of detecting unacceptable surface cracks.

2e. Reduced Beam Section Repair NDT

Because plastic straining and hinging, and potentially buckling, takes place in the thermally cut area of the reduced beam section, the area must be free of significant notches and cracks that would serve as stress concentrations and crack initiation sites. Inadvertent notches from thermal cutting, if sharp, may not be completely removed if relying solely upon visual inspection. If a welded repair is made, NDT is performed to verify that no surface or subsurface cracks have been caused by the repair.

2f. Weld Tab Removal Sites

Because weld tabs serve as locations for the starting and stopping of welds, and are therefore likely to contain a number of weld discontinuities, they are removed. To ensure that no significant discontinuities present in the tab extend into the finished weld itself, MT is performed. Any weld end discontinuities would be present at the surface of the joint, and therefore would be more detrimental to performance than an embedded discontinuity.

J7. INSPECTION OF HIGH-STRENGTH BOLTING

The commentary to *Specification* Section N5.6 on bolting inspection contains extensive discussion regarding the observation of bolting operations. Bolts in the SFRS should be considered for higher levels of observation, compared to bolts not in the SFRS and addressed by Chapter N in the *Specification*.

J8. OTHER STEEL STRUCTURE INSPECTIONS

Section N5.7 of the *Specification* provides for general inspection of the details of the steel frame, which would include those members in the SFRS, as well as anchor rods.

Section J8 of the Provisions adds inspection of specific details unique to seismic construction.

J9. INSPECTION OF COMPOSITE STRUCTURES

Section N6 of the *Specification* provides for general inspection of the steel decks and steel headed stud anchors when used in composite construction. The Provisions add inspection of the reinforcing steel and concrete materials and placement when used in a composite structural system. QC inspection of these items is performed by the contractor responsible for that portion of the work.

J10. INSPECTION OF H-PILES

The *Specification* contains no inspection requirements for piling, as piling is not considered structural steel in AISC 303. The Provisions address only steel H-pile when a part of the SFRS. The inspection is limited to verification of the protected zone, unique to seismic construction. Piling materials, pile driving, embedment, etc. are not included. Where welded joints in piling occur, inspections should be performed as for welding of other structural steel as described in Section J6.

CHAPTER K

PREQUALIFICATION AND CYCLIC QUALIFICATION TESTING PROVISIONS

K1. PREQUALIFICATION OF BEAM-TO-COLUMN AND LINK-TO-COLUMN CONNECTIONS

1. Scope

Section K1 describes requirements for prequalification of beam-to-column connections in special and intermediate moment frames (SMF and IMF) and of link-to-column connections in eccentrically braced frames (EBF). The concept of prequalified beam-to-column connections for moment frame systems, as used in the Provisions, has been adopted from FEMA 350 (FEMA, 2000a), and has been extended to include prequalified link-to-column connections for EBF.

Following observations of moment connection damage in the 1994 Northridge earthquake, these Provisions adopted the philosophy that the performance of beam-to-column and link-to-column connections should be verified by realistic-scale cyclic testing. This philosophy is based on the view that the behavior of connections under severe cyclic loading, particularly in regard to the initiation and propagation of fracture, cannot be reliably predicted by analytical means alone. Consequently, the satisfactory performance of connections must be confirmed by laboratory testing conducted in accordance with Section K2. In order to meet this requirement, designers fundamentally have two options. The first option is to provide substantiating test data, either from project specific tests or from tests reported in the literature, on connections matching project conditions within the limits specified in Section K2. The second option open to designers is to use a prequalified connection.

The option to use prequalified connections in the Provisions does not alter the fundamental view that the performance of beam-to-column and link-to-column connections should be confirmed by testing. However, it is recognized that requiring designers to provide substantiating test data for each new project is unnecessarily burdensome, particularly when the same connections are used on a repeated basis that have already received extensive testing, evaluation and review.

It is the intent of the Provisions that designers be permitted to use prequalified connections without the need to present laboratory test data, as long as the connection design, detailing and quality assurance measures conform to the limits and requirements of the prequalification. The use of prequalified connections is intended to simplify the design and design approval process by removing the burden on designers to present test data, and by removing the burden on the authority having jurisdiction to review and interpret test data. The use of prequalified connections is not intended as a guarantee against damage to, or failure of, connections in major earthquakes. The engineer of record in responsible charge of the building, based upon an understanding of and familiarity with the connection performance, behavior and limitations is responsible for selecting appropriate connection types suited to the application and implementing designs, either directly or by delegated responsibility.

The use of prequalified connections is permitted, but not required, by the Provisions.

2. General Requirements

2a. Basis for Prequalification

In general terms, a prequalified connection is one that has undergone sufficient testing, analysis, evaluation and review so that a high level of confidence exists that the connection can fulfill the performance requirements specified in Section E3.6b for special moment frames, in Section E2.6b for intermediate moment frames, or in Section F3.6e for eccentrically braced frames. Prequalification should be based primarily on laboratory test data, but supported by analytical studies of connection performance and by the development of detailed design criteria and design procedures. The behavior and expected performance of a prequalified connection should be well understood and predictable. Further, a sufficient body of test data should be available to ensure that a prequalified connection will perform as intended on a consistent and reliable basis.

Further guidance on prequalification of connections is provided by the commentary for FEMA 350, which indicates that the following four criteria should be satisfied for a prequalified connection:

There is sufficient experimental and analytical data on the connection performance to establish the likely yield mechanisms and failure modes for the connection.

Rational models for predicting the resistance associated with each mechanism and failure mode have been developed.

Given the material properties and geometry of the connection, a rational procedure can be used to estimate which mode and mechanism controls the behavior and deformation capacity (that is, story drift angle) that can be attained for the controlling conditions.

Given the models and procedures, the existing database is adequate to permit assessment of the statistical reliability of the connection.

2b. Authority for Prequalification

While the general basis for prequalification is outlined in Section K1.2a, it is not possible to provide highly detailed and specific criteria for prequalification, considering the wide variety of possible connection configurations, and considering the continually changing state-of-the-art in the understanding of connection performance. It is also recognized that decisions on whether or not a particular connection should be prequalified, and decisions on establishing limits on prequalification, will ultimately entail a considerable degree of professional engineering judgment. Consequently, a fundamental premise of these provisions is that prequalification can only be established based on an evaluation of the connection by a panel of knowledgeable individuals. Thus, the Provisions call for the establishment of a connection prequalification review panel (CPRP). Such a panel should consist of individuals with a high degree of experience, knowledge and expertise in connection behavior, design and construction. It is the responsibility of the CPRP to review all available data on a connection, and then determine if the connection warrants prequalification and determine the associated limits of prequalification, in accordance with Section K1. It is the intent of the Provisions that only a single, nationally recognized CPRP be established. To that end, AISC established the AISC connection prequalification review panel (CPRP) and developed *Prequalified Connections for Special and Intermediate Steel Moment Frames for Seismic Applications*, ANSI/AISC 358 (AISC, 2010b).

Use of connections reviewed by connection review panels other than the AISC CPRP, as permitted in Section K1.2b, and determined suitable for prequalification status in accordance with the Provisions, are subject to approval of the authority having jurisdiction.

3. **Testing Requirements**

It is the intent of the Provisions that laboratory test data form the primary basis of prequalification, and that the connection testing conforms to the requirements of Section K2. FEMA 350 specifies the minimum number of tests on nonidentical specimens needed to establish prequalification of a connection, or subsequently to change the limits of prequalification. However, in the Provisions, the number of tests needed to support prequalification or to support changes in prequalification limits is not specified. The number of tests and range of testing variables needed to support prequalification decisions will be highly dependent on the particular features of the connection and on the availability of other supporting data. Consequently, this Section requires that the CPRP determine whether the number and type of tests conducted on a connection are sufficient to warrant prequalification or to warrant a change in prequalification limits. Both FEMA 350 and the Provisions refer to "nonidentical" test specimens, indicating that a broad range of variables potentially affecting connection performance should be investigated in a prequalification test program. It may also be desirable to test replicas of nominally identical specimens in order to investigate repeatability of performance prior to and after failure and to demonstrate consistency of the failure mechanism. Individuals planning a test program to support prequalification of a connection are encouraged to consult with the CPRP, in advance, for a preliminary assessment of the planned testing program.

Tests used to support prequalification are required to comply with Section K2. That Section requires test specimens be loaded at least to a story drift angle as specified in Section E3.6b for special moment frames or in Section E2.6b for intermediate moment frames, or a link rotation angle as specified in Section F3.4a for eccentrically braced frames. These provisions do not include the additional requirement for connection rotation capacity at failure, as recommended in FEMA 350 (FEMA, 2000a). For purposes of prequalification, however, it is desirable to load specimens to larger deformation levels in order to reveal the ultimate controlling failure modes. Prequalification of a connection requires a clear understanding of the controlling failure modes for a connection; in other words, the failure modes that control the strength and deformation capacity of the connection. Consequently, test data must be available to support connection behavior models over the full range of loading, from the initial elastic response to the inelastic range of behavior, and finally through to the ultimate failure of the connection.

The story drift angle developed by a moment connection test specimen is the primary acceptance criterion for a beam-to-column moment connection in a moment frame. In an actual building, the story drift angle is computed as the story displacement divided by the story height, and includes both elastic and inelastic components of deformation. For a test specimen, story drift angle can usually be computed in a straightforward manner from displacement measurements on the test specimen. Guidelines for computing the story drift angle of a connection test specimen are provided by SAC (1997).

When a connection is being considered for prequalification by the CPRP, *all* test data for that connection must be available for review by the CPRP. This includes data on unsuccessful tests of connections that represent or are otherwise relevant to the final connection. Testing performed on a preliminary connection configuration that is not relevant to the final design need not be submitted. However, parametric studies on weak and strong panel zones of a connection that otherwise match the final connection are examples of developmental tests that should be submitted. Individuals seeking prequalification of a connection are obliged to present the entire known database of tests for the connection. Such data is essential for an assessment of the reliability of a connection. Note that unsuccessful tests do not necessarily preclude prequalification, particularly if the reasons for unsuccessful performance have been identified and addressed in the connection design procedures. For example, if ten tests are conducted on varying sized members and one test is unsuccessful, the cause for the "failure" should be determined. If possible, the connection design procedure should be adjusted in such a way to preclude the failure and not invalidate the other nine tests. Subsequent tests should then be performed to validate the final proposed design procedure.

4. Prequalification Variables

This Section provides a list of variables that can affect connection performance, and that should be considered in the prequalification of connections. The CPRP should consider the possible effects of each variable on connection performance, and establish limits of application for each variable. Laboratory tests or analytical studies investigating the full range of all variables listed in this Section are not required and would not be practical. Connection testing and/or analytical studies investigating the effects of these variables are only required where deemed necessary by the CPRP. However, regardless of which variables are explicitly considered in testing or analytical studies, the CPRP should still consider the possible effects of all variables listed in this Section, and assign appropriate limits.

5. Design Procedure

In order to prequalify a connection, a detailed and comprehensive design procedure consistent with the test results and addressing all pertinent limit states must be available for the connection. This design procedure must be included as part of the prequalification record, as required in Section K1.6. Examples of the format and typical content of such design procedures can be found in FEMA 350 (FEMA, 2000a).

6. Prequalification Record

A written prequalification record is required for a prequalified connection. As a minimum, the prequalification record must include the information listed in Section K1.6. The prequalification record should provide a comprehensive listing of all information needed by a designer to determine the applicability and limitations of the connection, and information needed to design the connection. The prequalification record need not include detailed records of laboratory tests or analytical studies. However, a list of references should be included for all test reports, research reports, and other publications used as a basis of prequalification. These references should, to the extent possible, be available in the public domain to permit independent review of the data and to maintain the integrity and credibility of the prequalification process. FEMA 350 (FEMA, 2000a) provides an example of the type and formatting of information needed for a prequalified connection.

For connections prequalified by CPRP, ANSI/AISC 358 serves as the prequalification record.

K2. CYCLIC TESTS FOR QUALIFICATION OF BEAM-TO-COLUMN AND LINK-TO-COLUMN CONNECTIONS

1. Scope

The development of testing requirements for beam-to-column moment connections was motivated by the widespread occurrence of fractures in such connections in the 1994 Northridge earthquake. To improve performance of connections in future earthquakes, laboratory testing is required to identify potential problems in the design, detailing, materials or construction methods to be used for the connection. The requirement for testing reflects the view that the behavior of connections under severe cyclic loading cannot be reliably predicted by analytical means only.

It is recognized that testing of connections can be costly and time consuming. Consequently, this Section has been written with the simplest testing requirements possible, while still providing reasonable assurance that connections tested in accordance with these Provisions will perform satisfactorily in an earthquake. Where conditions in the actual building differ significantly from the test conditions specified in this Section, additional testing beyond the requirements herein may be needed to ensure satisfactory connection performance. Many of the factors affecting connection performance under earthquake loading are not completely understood. Consequently, testing under conditions that are as close as possible to those found in the actual building will provide for the best representation of expected connection performance.

It is not the intent of these Provisions that project-specific connection tests be conducted on a routine basis for building construction projects. Rather, it is anticipated that most projects would use connection details that have been previously prequalified in accordance with Section K1. If connections are being used that have not been prequalified, then connection performance must be verified by testing in accordance with Section K2. However, even in such cases, tests reported in the literature can be used to demonstrate that a connection satisfies the strength and rotation requirements of the Provisions, so long as the reported tests satisfy the requirements of this Section. Consequently, it is expected that project-specific connection tests would be conducted for only a very small number of construction projects.

Although the provisions in this Section predominantly address the testing of beam-to-column connections in moment frames, they also apply to qualifying cyclic tests of link-to-column connections in EBF. While there are no reports of failures of link-to-column connections in the Northridge earthquake, it cannot be concluded that these similar connections are satisfactory for severe earthquake loading as it appears that few EBF with a link-to-column configuration were subjected to strong ground motion in that earthquake. Many of the conditions that contributed to poor performance of moment connections in the Northridge earthquake can also occur in link-to-column connections in EBF. Further, recent research on link-to-column connections (Okazaki et al., 2004b; Okazaki, 2004) has demonstrated that such connections, designed and constructed using pre-Northridge practices, show poor performance in laboratory testing. Consequently, in these provisions, the same testing requirements are applied to both moment connections and to link-to-column connections.

When developing a test program, the designer should be aware that the authority having jurisdiction may impose additional testing and reporting requirements not covered in this Appendix. Examples of testing guidelines or requirements developed by other organizations or agencies include those published by SAC (FEMA, 2000a; SAC, 1997), by the ICC Evaluation Service (ICC, 2008), and by the County of Los Angeles (County of Los Angeles Department of Public Works, 1996). Prior to developing a test program, the appropriate authority having jurisdiction should be consulted to ensure the test program meets all applicable requirements. Even when not required, the designer may find the information contained in the foregoing references to be useful resources in developing a test program.

2. **Test Subassemblage Requirements**

A variety of different types of subassemblages and test specimens have been used for testing moment connections. A typical subassemblage is planar and consists of a single column with a beam attached on one or both sides of the column. The specimen can be loaded by displacing either the end of the beam(s) or the end of the column. Examples of typical subassemblages for moment connections can be found in the literature, for example in SAC (1996) and Popov et al. (1996).

In the Provisions, test specimens generally need not include a composite slab or the application of axial load to the column. However, such effects may have an influence on connection performance, and their inclusion in a test program should be considered as a means to obtain more realistic test conditions. An example of test subassemblages that include composite floor slabs and/or the application of column axial loads can be found in Popov et al. (1996); Leon et al. (1997); and Tremblay et al. (1997). A variety of other types of subassemblages may be appropriate to simulate specific project conditions, such as a specimen with beams attached in orthogonal directions to a column. A planar bare steel specimen with a single column and a single beam represents the minimum acceptable subassemblage for a moment connection test. However, more extensive and realistic subassemblages that better match actual project conditions should be considered where appropriate and practical, in order to obtain more reliable test results.

Examples of subassemblages used to test *link*-to-column connections can be found in Hjelmstad and Popov (1983); Kasai and Popov (1986c); Ricles and Popov (1987b); Engelhardt and Popov (1989a); Dusicka and Itani (2002); McDaniel et al. (2002); Arce (2002); and Okazaki et al. (2004b).

3. Essential Test Variables

3a. Sources of Inelastic Rotation

This Section is intended to ensure that the inelastic rotation in the test specimen is developed in the same members and connection elements as anticipated in the prototype. For example, if the prototype moment connection is designed so that essentially all of the inelastic rotation is developed by yielding of the beam, then the test specimen should be designed and perform in the same way. A test specimen that develops nearly all of its inelastic rotation through yielding of the column panel zone would not be acceptable to qualify a prototype connection wherein flexural yielding of the beam is expected to be the predominant inelastic action.

Because of normal variations in material properties, the actual location of inelastic action may vary somewhat from that intended in either the test specimen or in the prototype. An allowance is made for such variations by permitting a 25% variation in the percentage of the total inelastic rotation supplied by a member or connecting element in a test specimen as compared with the design intent of the prototype. Thus, for the example above where 100% of the inelastic rotation in the prototype is expected to be developed by flexural yielding of the beam, at least 75% of the total inelastic rotation of the test specimen is required to be developed by flexural yielding of the beam in order to qualify this connection.

For link-to-column connections in eccentrically braced frames (EBF), the type of yielding (shear yielding, flexural yielding, or a combination of shear and flexural yielding) expected in the test specimen link should be substantially the same as for the prototype link. For example, a link-to-column connection detail which performs satisfactorily for a shear-yielding link ($e \leq 1.6M_p/V_p$) may not necessarily perform well for a flexural-yielding link ($e \geq 2.6M_p/V_p$). The load and deformation demands at the link-to-column connection will differ significantly for these cases.

Satisfying the requirements of this Section will require the designer to have a clear understanding of the manner in which inelastic rotation is developed in the prototype and in the test specimen.

One of the key parameters measured in a connection test is the inelastic rotation that can be developed in the specimen. Previously in the Provisions, inelastic rotation was the primary acceptance criterion for beam-to-column moment connections in moment frames. The acceptance criterion in the Provisions is now based on story drift angle, which includes both elastic and inelastic rotations. However, inelastic rotation provides an important indication of connection performance in earthquakes and should still be measured and reported in connection tests. Researchers have used a variety of different definitions for inelastic rotation of moment connection test specimens in the past, making comparison among tests difficult. In order to promote consistency in how test results are reported, these Provisions require that inelastic rotation for moment connection test specimens be computed based on the assumption that all inelastic deformation of a test specimen is concentrated at a single point at the intersection of the centerline of the beam with the centerline of the column. With this definition, inelastic rotation is equal to the inelastic portion of the story drift angle. Previously the Provisions defined inelastic rotation of moment connection specimens with respect to the face of the column. The definition has been changed to the centerline of the column to be consistent with recommendations of SAC (SAC, 1997; FEMA, 2000a).

For tests of link-to-column connections, the key acceptance parameter is the link inelastic rotation, also referred to in these Provisions as the link rotation angle. The link rotation angle is computed based upon an analysis of test specimen deformations, and can normally be computed as the inelastic portion of the relative end displacement between the ends of the link, divided by the link length. Examples of such calculations can be found in Kasai and Popov (1986c); Ricles and Popov (1987a); Engelhardt and Popov (1989a); and Arce (2002).

3b. **Size of Members**

The intent of this Section is that the member sizes used in a test specimen should be, as nearly as practical, a full-scale representation of the member sizes used in the prototype. The purpose of this requirement is to ensure that any potentially adverse scale effects are adequately represented in the test specimen. As beams become deeper and heavier, their ability to develop inelastic rotation may be somewhat diminished (Roeder and Foutch, 1996; Blodgett, 2001). Although such scale effects are not yet completely understood, at least two possible detrimental scale effects have been identified. First, as a beam gets deeper, larger inelastic strains are generally required in order to develop the same level of inelastic rotation. Second, the inherent restraint associated with joining thicker materials can affect joint and connection performance. Because of such potentially adverse scale effects, the beam sizes used in test specimens are required to adhere to the limits given in this Section.

This Section only specifies restrictions on the degree to which test results can be scaled up to deeper or heavier members. There are no restrictions on the degree to which test results can be scaled down to shallower or lighter members. No such restrictions have been imposed in order to avoid excessive testing requirements and because currently available evidence suggests that adverse scale effects are more likely to occur when scaling up test results rather than when scaling down. Nonetheless, caution is advised when using test results on very deep or heavy members to qualify connections for much smaller or lighter members. It is preferable to obtain test results using member sizes that are a realistic representation of the prototype member sizes.

As an example of applying the requirements of this Section, consider a moment connection test specimen constructed with a W36×150 beam. This specimen could be used to qualify any beam with a depth up to 40 in. (= 36/0.9) and a weight up to 200 lb/ft (= 150/0.75). The limits specified in this Section have been chosen somewhat arbitrarily based on judgment, as no quantitative research results are available on scale effects.

When choosing a beam size for a test specimen, several other factors should be considered in addition to the depth and weight of the section. One of these factors is the width-to-thickness ratio, b/t, of the beam flange and web. The b/t ratios of the beam may have an important influence on the performance of specimens that develop plastic rotation by flexural yielding of the beam. Beams with high b/t ratios develop local buckling at lower inelastic rotation levels than beams with low b/t ratios. This local buckling causes strength degradation in the beam, and may therefore reduce the load demands on the connection. A beam with very low b/t ratios may experience little if any local buckling, and will therefore subject the connection to higher moments. On the other hand, the beam with high b/t ratios will experience highly localized deformations at locations of flange and web buckling, which may in turn initiate a fracture. Consequently, it is desirable to test beams over a range of b/t ratios in order to evaluate these effects.

These provisions also require that the depth of the test column be at least 90% of the depth of the prototype column. Tests conducted as part of the SAC program indicated that performance of connections with deep columns may differ from the performance with W12 and W14 columns (Chi and Uang, 2002). Additional recent research on moment connections with deep columns is reported by Ricles et al. (2004b).

In addition to adhering separately to the size restrictions for beams and to the size restrictions for columns, the combination of beam and column sizes used in a test specimen should reasonably reflect the pairing of beam and column sizes used in the prototype. For example, say a building design calls for the use of a W36 beam attached to a W36 column. Say also, that for the connection type proposed for this building, successful tests have been run on specimens using a W36 beam attached to a W14 column, and on other specimens using a W24 beam attached to a W36 column. Thus, test data is available for this connection on specimens meeting the beam size limitations of Section K2.3b, and separately on specimens meeting the column size restrictions of Section K2.3b. Nonetheless, these tests would not be suitable for qualifying this connection for the case of a W36 beam attached to a W36 column, since the combination of beam and column sizes used in the test specimens does not match the combination of beam and column sizes in the prototype, within the limits of Section K2.3b.

3e. Steel Strength

The actual yield stress of structural steel can be considerably greater than its specified minimum value. Higher levels of actual yield stress in members that supply inelastic rotation by yielding can be detrimental to connection performance by developing larger forces at the connection prior to yielding. For example, consider a moment connection design in which inelastic rotation is developed by yielding of the beam, and the beam has been specified to be of ASTM A36/A36M steel. If the beam has an actual yield stress of 55 ksi (380 MPa), the connection is required to resist a moment that is 50% higher than if the beam had an actual yield stress of 36 ksi (250 MPa). Consequently, this Section requires that the materials used for the test specimen represent this possible overstrength condition, as this will provide for the most severe test of the connection.

As an example of applying these provisions, consider again a test specimen in which inelastic rotation is intended to be developed by yielding of the beam. In order to qualify this connection for ASTM A992/A992M beams, the test beam is required to have a yield stress of at least 47 ksi (324 MPa) (= $0.85 R_y F_y$ for ASTM A992/A992M). This minimum yield stress is required to be exhibited by both the web and flanges of the test beam.

The requirements of this Section are applicable only to members or connecting elements of the test specimen that are intended to contribute to the inelastic rotation of the specimen through yielding. The requirements of this Section are not applicable to members or connecting elements that are intended to remain essentially elastic.

3f. Welded Joints

The intent of the Provisions is to ensure that the welds on the test specimen replicate the welds on the prototype as closely as practicable. Accordingly, it is required that the welding variables, such as current and voltage, be within the range established by the weld metal manufacturer. Other essential variables, such as steel grade, type of joint, root opening, included angle and preheat level, are required to be in accordance with AWS D1.1/D1.1M. It is not the intent of this Section that the electrodes used to make welds in a test specimen must necessarily be the same AWS classification, diameter or brand as the electrodes to be used on the prototype.

4. Loading History

The loading sequence prescribed in Section K2.4b for beam-to-column moment connections is taken from SAC/BD-97/02, *Protocol for Fabrication, Inspection, Testing, and Documentation of Beam Column Connection Tests and Other Experimental Specimens* (SAC, 1997). This document should be consulted for further details of the loading sequence, as well as for further useful information on testing procedures. The prescribed loading sequence is not intended to represent the demands presented by a particular earthquake ground motion. This loading sequence was developed based on a series of nonlinear time history analyses of steel moment frame structures subjected to a range of seismic inputs. The maximum deformation, as well as the cumulative deformation and dissipated energy sustained by beam-to-column connections in these analyses, were considered when establishing the prescribed loading sequence and the connection acceptance criteria. If a designer conducts a nonlinear time history analysis of a moment frame structure in order to evaluate demands on the beam-to-column connections, considerable judgment will be needed when comparing the demands on the connection predicted by the analysis with the demands placed on a connection test specimen using the prescribed loading sequence. In general, however, a connection can be expected to provide satisfactory performance if the cumulative plastic deformation and the total dissipated energy sustained by the test specimen prior to failure are equal to or greater than the same quantities predicted by a nonlinear time-history analysis. When evaluating the cumulative plastic deformation, both total rotation (elastic plus inelastic) as well as inelastic rotation at the connection should be considered. SAC/BD-00/10 (SAC, 2000) can be consulted for further information on this topic.

Section K2.4c specifies the loading sequence for qualifying tests on link-to-column connections. This loading sequence has been changed from a previous edition of these Provisions. Recent research on EBF (Richards and Uang, 2003; Richards, 2004) has demonstrated that the loading protocol specified for testing of links in Section S6.3 of Appendix S in the 2002 Provisions is excessively conservative. A loading protocol for link testing was first added to Appendix S in *Supplement No. 2* to the 1997 Provisions, and remained unchanged in the 2002 Provisions. When the link loading protocol was added to Appendix S, no research was available that provided a rational basis for link testing. The loading protocol was therefore chosen on a somewhat conservative and arbitrary basis. Concerns that the loading protocol may be excessively conservative were raised when a number of shear links tested under this protocol failed somewhat prematurely due to low cycle fatigue fractures of the link web (Okazaki et al., 2004a; Arce, 2002). As a result of concerns regarding the rationality of the current link loading protocol, research was conducted to establish a rational loading protocol for link-to-column connections in EBF. This study (Richards and Uang, 2003; Richards, 2004) developed a recommended loading protocol for links, using a methodology similar to that used for moment frame connection testing, as developed under the FEMA/SAC program. The loading protocol for link-to-column connections developed in this study is the basis of the new loading sequence in Section K2.6c.

The loading sequence specified in ATC-24, *Guidelines for Cyclic Seismic Testing of Components of Steel Structures* (ATC, 1992) is considered as an acceptable alternative to those prescribed in Sections K2.4b and K2.4c. Further, any other loading sequence may be used for beam-to-column moment connections or link-to-column connections, as long as the loading sequence is equivalent to or more severe than those prescribed in Sections K2.4b and K2.4c. To be considered as equivalent or more severe, alternative loading sequences should meet the following requirements: (1) the number of inelastic loading cycles should be at least as large as the number of inelastic loading cycles resulting from the prescribed loading sequence; and (2) the cumulative plastic deformation should be at least as large as the cumulative plastic deformation resulting from the prescribed loading sequence.

Dynamically applied loads are not required by the Provisions. Slowly applied cyclic loads, as typically reported in the literature for connection tests, are acceptable for the purposes of the Provisions. It is recognized that dynamic loading can considerably increase the cost of testing, and that few laboratory facilities have the capability to dynamically load very large-scale test specimens. Furthermore, the available research on dynamic loading effects on steel connections has not demonstrated a compelling need for dynamic testing. Nonetheless, applying the required loading sequence dynamically, using loading rates typical of actual earthquake loading, will likely provide a better indication of the expected performance of the connection, and should be considered where possible.

6. **Testing Requirements for Material Specimens**

Tension testing is required for members and connection elements of the test specimen that contribute to the inelastic rotation of the specimen by yielding. These tests are required to demonstrate conformance with the requirements of Section K2.3e, and to permit proper analysis of test specimen response. Tension test results reported on certified mill test reports are not permitted to be used for this purpose. Yield stress values reported on a certified mill test report may not adequately represent the actual yield strength of the test specimen members. Variations are possible due to material sampling locations and tension test methods used for certified mill test reports.

ASTM standards for tension testing permit the reporting of the upper yield point. Yield strength may be reported using either the 0.2% offset or 0.5% elongation under load. For steel members subject to large cyclic inelastic strains, the upper yield point can provide a misleading representation of the actual material behavior. Thus, while an upper yield point is permitted by ASTM, it is not permitted for the purposes of this Section. Determination of yield stress using the 0.2% strain offset method based on independent testing using common specimen size for all members is required in this Section. This follows the protocol used during the SAC investigation.

Since this tension testing utilizes potentially different specimen geometry, testing protocol, and specimen location, differences from the material test report are to be expected. Appendix X2 of ASTM A6 discusses the variation of tensile properties within a heat of steel for a variety of reasons. Based on previous work, this appendix reports the value of one standard deviation of this variance to be 8% of the yield strength using ASTM standards.

This special testing is not required for project materials as the strength ratios in Table A3.1 were developed using standard producer material test report data. Therefore, supplemental testing of project material should only be required if the identity of the material is in question prior to fabrication.

Only tension tests are required in this Section. Additional materials testing, however, can sometimes be a valuable aid for interpreting and extrapolating test results. Examples of additional tests, which may be useful in certain cases, include Charpy

8. Acceptance Criteria

A minimum of two tests is required for each condition in the prototype in which the variables remain unchanged. The designer is cautioned, however, that two tests, in general, cannot provide a thorough assessment of the capabilities, limitations, and reliability of a connection. Thus, where possible, it is highly desirable to obtain additional test data to permit a better evaluation of the expected response of a connection to earthquake loading. Further, when evaluating the suitability of a proposed connection, it is advisable to consider a broader range of issues other than just inelastic rotation capacity.

One factor to consider is the controlling failure mode after the required inelastic rotation has been achieved. For example, a connection that slowly deteriorates in strength due to local buckling may be preferable to a connection that exhibits a more brittle failure mode such as fracture of a weld, fracture of a beam flange, etc., even though both connections achieved the required inelastic rotation.

In addition, the designer should also carefully consider the implications of unsuccessful tests. For example, consider a situation where five tests were run on a particular type of connection, two tests successfully met the acceptance criteria, but the other three failed prematurely. This connection could presumably be qualified under the Provisions, since two successful tests are required. Clearly, however, the number of failed tests indicates potential problems with the reliability of the connection. On the other hand, the failure of a tested connection in the laboratory should not, by itself, eliminate that connection from further consideration. As long as the causes of the failure are understood and corrected, and the connection is successfully retested, the connection may be quite acceptable. Thus, while the acceptance criteria in the Provisions have intentionally been kept simple, the choice of a safe, reliable and economical connection still requires considerable judgment.

K3. CYCLIC TESTS FOR QUALIFICATION OF BUCKLING-RESTRAINED BRACES

The provisions of this Section require the introduction of several new variables. The quantity Δ_{bm} represents both an axial displacement and a rotational quantity. Both quantities are determined by examining the profile of the building at the design story drift, Δ_m, and extracting joint lateral and rotational deformation demands.

Determining the maximum rotation imposed on the braces used in the building may require significant effort. The engineer may prefer to select a reasonable value (in other words, story drift), which can be simply demonstrated to be conservative for each brace type, and is expected to be within the performance envelope of the braces selected for use on the project.

Two types of testing are referred to in this Section. The first type is subassemblage testing, described in Section K3.4, an example of which is illustrated in Figure C-K3.1.

The second type of testing described in Section K3.3 as brace specimen testing is permitted to be uniaxial testing.

1. **Scope**

The development of the testing requirements in the Provisions was motivated by the relatively small amount of test data on buckling-restrained braced frame (BRBF) systems available to structural engineers. In addition, no data on the response of BRBF to severe ground motion is available. Therefore, the seismic performance of these systems is relatively unknown compared to more conventional steel-framed structures.

The behavior of a buckling-restrained braced frame differs markedly from conventional braced frames and other structural steel seismic-force-resisting systems. Various factors affecting brace performance under earthquake loading are not well understood and the requirement for testing is intended to provide assurance that the braces will perform as required, and also to enhance the overall state of knowledge of these systems.

It is recognized that testing of brace specimens and subassemblages can be costly and time-consuming. Consequently, this Section has been written with the simplest testing requirements possible, while still providing reasonable assurance that prototype BRBF based on brace specimens and subassemblages tested in accordance with these provisions will perform satisfactorily in an actual earthquake.

It is not intended that the Provisions drive project-specific tests on a routine basis for building construction projects. In most cases, tests reported in the literature or supplied by the brace manufacturer can be used to demonstrate that a brace and

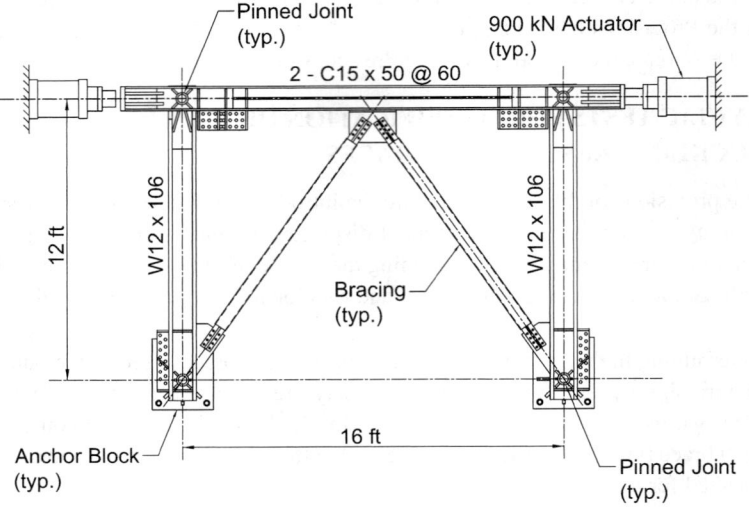

Fig. C-K3.1. Example of test subassemblage.

subassemblage configuration satisfies the strength and inelastic rotation requirements of these provisions. Such tests, however, should satisfy the requirements of this Section.

The Provisions of this Section have been written allowing submission of data on previous testing, based on similarity conditions. As the body of test data for each brace type grows, the need for additional testing is expected to diminish. The Provisions allow for manufacturer-designed braces, through the use of a documented design methodology.

Most testing programs developed for primarily axial-load-carrying components focus largely on uniaxial testing. However, these Provisions are intended to direct the primary focus of the program toward testing of a subassemblage that imposes combined axial and rotational deformations on the brace specimen. This reflects the view that the ability of the brace to accommodate the necessary rotational deformations cannot be reliably predicted by analytical means alone. Subassemblage test requirements are discussed more completely in Commentary Section K3.4.

Where conditions in the actual building differ significantly from the test conditions specified in this Section, additional testing beyond the requirements described herein may be needed to assure satisfactory brace performance. Prior to developing a test program, the appropriate regulatory agencies should be consulted to assure the test program meets all applicable requirements.

The brace deformation at first significant yield is used in developing the test sequence described in Section K3.6c. The quantity is required to determine the actual cumulative inelastic deformation demands on the brace. If the nominal yield stress of the steel core were used to determine the test sequence, and significant material overstrength were to exist, the total inelastic deformation demand imposed during the test sequence would be overestimated.

2. **Subassemblage Test Specimen**

    The objective of subassemblage testing is to verify the ability of the brace, and in particular its steel core extension and buckling restraining mechanism, to accommodate the combined axial and rotational deformation demands without failure.

    It is recognized that subassemblage testing is more difficult and expensive than uniaxial testing of brace specimens. However, the complexity of the brace behavior due to the combined rotational and axial demands, and the relative lack of test data on the performance of these systems, indicates that subassemblage testing should be performed.

    Subassemblage testing is not intended to be required for each project. Rather, it is expected that brace manufacturers will perform the tests for a reasonable range of axial loads, steel core configurations, and other parameters as required by the provisions. It is expected that this data will subsequently be available to engineers on other projects. Manufacturers are therefore encouraged to conduct tests that establish the device performance limits to minimize the need for subassemblage testing on projects.

Similarity requirements are given in terms of measured axial yield strength of both the prototype and the test specimen braces. This is better suited to manufacturer's product testing than to project-specific testing. Comparison of coupon test results is a way to establish a similarity between the subassemblage test specimen brace and the prototype braces. Once similarity is established, it is acceptable to fabricate test specimens and prototype braces from different heats of steel.

A variety of subassemblage configurations are possible for imposing combined axial and rotational deformation demands on a test specimen. Some potential subassemblages are shown in Figure C-K3-2. The subassemblage need not include connecting beams and columns provided that the test apparatus duplicates, to a reasonable degree, the combined axial and rotational deformations expected at each end of the brace.

Rotational demands may be concentrated in the steel core extension in the region just outside the buckling restraining mechanism. Depending on the magnitude of the rotational demands, limited flexural yielding of the steel core extension may occur.

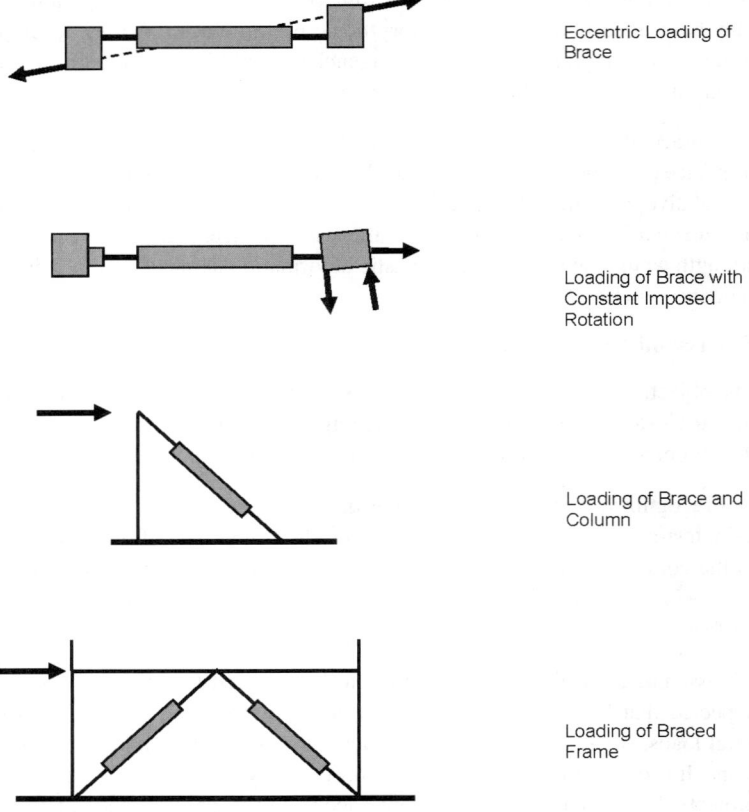

Fig. C-K3.2. Possible test subassemblages.

Rotational demands can also be accommodated by other means, such as tolerance in the buckling restraint layer or mechanism, elastic flexibility of the brace and steel core extension, or through the use of pins or spherical bearing assemblies. It is in the engineer's best interest to include in a subassemblage testing all components that contribute significantly to accommodating rotational demands.

It is intended that the subassemblage test specimen be larger in axial-force capacity than the prototype. However, the possibility exists for braces to be designed with very large axial forces. Should the brace yield force be so large as to make subassemblage testing impractical, the engineer is expected to make use of the provisions that allow for alternate testing programs, based on building official approval and qualified peer review. Such programs may include, but are not limited to, nonlinear finite element analysis, partial specimen testing, and reduced-scale testing, in combination with full-scale uniaxial testing where applicable or required.

The steel core material was not included in the list of requirements. The more critical parameter, calculated margin of safety for the steel core projection stability, is required to meet or exceed the value used in the prototype. The method of calculating the steel core projection stability should be included in the design methodology.

3. Brace Test Specimen

The objective of brace test specimen testing is to establish basic design parameters for the BRBF system.

The allowance of previous test data (similarity) to satisfy these provisions is less restrictive for uniaxial testing than for subassemblage testing. Subassemblage test specimen requirements are described in Commentary Section K3.2.

A considerable number of uniaxial tests have been performed on some brace systems and the engineer is encouraged, wherever possible, to submit previous test data to meet these provisions. Relatively few subassemblage tests have been performed. This type of testing is considered a more demanding test of the overall brace performance.

It is recognized that the fabrication tolerances used by brace manufacturers to achieve the required brace performance may be tighter than those used for other fabricated structural steel members. The engineer is cautioned against including excessively prescriptive brace specifications, as the intent of these provisions is that the fabrication and supply of the braces is achieved through a performance-based specification process. It is considered sufficient that the manufacture of the test specimen and the prototype braces be conducted using the same quality control and assurance procedures, and the braces be designed using the same design methodology.

The engineer should also recognize that manufacturer process improvements over time may result in some manufacturing and quality control and assurance procedures changing between the time of manufacture of the brace test specimen and of the prototype. In such cases reasonable judgment is required.

During the planning stages of either a subassemblage or uniaxial brace test, certain conditions may exist that cause the test specimen to deviate from the parameters established in the testing section. These conditions may include:

- Lack of availability of beam, column, and brace sizes that reasonably match those to be used in the actual building frame
- Test set-up limitations in the laboratory
- Transportation and field-erection constraints
- Actuator to subassemblage connection conditions that require reinforcement of test specimen elements not reinforced in the actual building frame

In certain cases, both the authority having jurisdiction and the peer reviewer may deem such deviations acceptable. The cases in which such deviations are acceptable are project-specific by nature and, therefore, do not lend themselves to further description in this Commentary. For these specific cases, it is recommended that the engineer of record demonstrate that the following objectives are met:

- Reasonable relationship of scale
- Similar design methodology
- Adequate system strength
- Stable buckling-restraint of the steel core in the prototype
- Adequate rotation capacity in the prototype
- Adequate cumulative strain capacity in the prototype

In many cases it will not be practical or reasonable to test the exact brace connections present in the prototype. These provisions are not intended to require such testing. In general, the demands on the steel core extension to gusset-plate connection are well defined due to the known axial capacity of the brace and the limited flexural capacity of the steel core extension. While the subsequent design of the bolted or welded gusset-plate connection is itself a complicated issue and the subject of continuing investigation, it is not intended that these connections become the focus of the testing program.

For the purposes of utilizing previous test data to meet the requirements of this Section, the requirements for similarity between the brace and subassemblage brace test specimen can be considered to exclude the steel core extension connection to frame.

The intent of the provisions is to allow test data from previous test programs to be presented where possible. See Section K3.2 for additional commentary.

The intent of this Provision is to ensure that the end connections of the brace test specimen reasonably represent those of the prototype. It is possible that due to fabrication or assembly constraints variations in fit-up, faying-surface preparation, or bolt or pin hole fabrication and size may occur. In certain cases, such variations may not be detrimental to the qualification of a successful cyclic test. The final acceptability of variations in brace-end connections rests on the opinion of the building official.

The subassemblage test specimen is required to undergo combined axial and rotational deformations similar to those in the prototype. It is recognized that identical braces, in different locations in the building, will undergo different maximum axial and rotational deformation demands. In addition, the maximum rotational and axial deformation demands may be different at each end of the brace. The engineer is expected to make simplifying assumptions to determine the most appropriate combination of rotational and axial deformation demands for the testing program.

Some subassemblage configurations will require that one deformation quantity be fixed while the other is varied as described in the test sequence above. In such a case, the rotational quantity may be applied and maintained at the maximum value, and the axial deformation applied according to the test sequence. The engineer may wish to perform subsequent tests on the same subassemblage specimen to bound the brace performance.

The loading sequence requires each tested brace to achieve ductilities corresponding to 2.0 times the design story drift and a cumulative inelastic axial ductility capacity of 200 times the yield displacement. Both of these requirements are based on a study in which a series of nonlinear dynamic analyses was conducted on model buildings in order to investigate the performance of this system. The ductility capacity requirement represents a mean of response values (Sabelli et al., 2003). The cumulative ductility requirement is significantly higher than expected for the design basis earthquake, but testing of braces has shown this value to be easily achieved. It is expected that as more test data and building analysis results become available these requirements may be revisited.

The ratio of brace yield deformation, Δ_{by}, to the brace deformation corresponding to the design story drift, Δ_{bm}, must be calculated in order to define the testing protocol. This ratio is typically the same as the ratio of the displacement amplification factor (as defined in the applicable building code) to the actual overstrength of the brace; the minimum overstrength is determined by the resistance factor (LRFD) or the safety factor (ASD) in Section F4.5b(2).

Engineers should note that there is a minimum brace deformation demand, Δ_{bm}, corresponding to 1% story drift; provision of overstrength beyond that required to so limit the design story drift may not be used as a basis to reduce the testing protocol requirements. Testing to at least twice this minimum (in other words, to 2% drift) is required.

Table C-K3.1 shows an example brace test protocol. For this example, it is assumed that the brace deformation corresponding to the design story drift is four times the yield deformation; it is also assumed that the design story drift is larger than the 1% minimum. The test protocol is then constructed from steps 1 through 4 of Section K3.4c. In order to calculate the cumulative inelastic deformation, the cycles are converted from multiples of brace deformation at the design story drift, Δ_{bm}, to multiples of brace yield deformation, Δ_{by}. Since the cumulative inelastic drift at the end of the $2.0\Delta_{bm}$ cycles is less than the minimum of $200\Delta_{by}$ required for brace tests, additional cycles to $1.5\Delta_{bm}$ are required. At the end of three such cycles, the required cumulative inelastic deformation has been reached.

TABLE C-K3.1
Example Brace Testing Protocol

Cycle	Deformation	Inelastic Deformation	Cumulative Inelastic Deformation
2 @ Δ_{by}		$= 2*4*(\Delta_{by} - \Delta_{by}) = 0\Delta_{by}$	$0\Delta_{by} = 0\Delta_{by}$
2 @ $1/2\Delta_{bm}$	$= 4 @ 2.0\Delta_{by}$	$= 2*4*(2.0\Delta_{by} - \Delta_{by}) = 8\Delta_{by}$	$0\Delta_{by} + 8\Delta_{by} = 8\Delta_{by}$
2 @ Δ_{bm}	$= 4 @ 4.0\Delta_{by}$	$= 2*4*(4.0\Delta_{by} - \Delta_{by}) = 24\Delta_{by}$	$8\Delta_{by} + 24\Delta_{by} = 32\Delta_{by}$
2 @ $1^{1}/_{2}\Delta_{bm}$	$= 2 @ 6.0\Delta_{by}$	$= 2*4*(6.0\Delta_{by} - \Delta_{by}) = 40\Delta_{by}$	$32\Delta_{by} + 40\Delta_{by} = 72\Delta_{by}$
2 @ $2\Delta_{bm}$	$= 2 @ 8.0\Delta_{by}$	$= 2*4*(8.0\Delta_{by} - \Delta_{by}) = 56\Delta_{by}$	$72\Delta_{by} + 56\Delta_{by} = 128\Delta_{by}$
4 @ $1^{1}/_{2}\Delta_{bm}$	$= 2 @ 6.0\Delta_{by}$	$= 4*4*(6.0\Delta_{by} - \Delta_{by}) = 80\Delta_{by}$	$128\Delta_{by} + 80\Delta_{by} = 208\Delta_{by}$
Cumulative inelastic deformation at end of protocol = $208\Delta_{by}$			

Dynamically applied loads are not required by the Provisions. The use of slowly applied cyclic loads, widely described in the literature for brace specimen tests, is acceptable for the purposes of these Provisions. It is recognized that dynamic loading can considerably increase the cost of testing, and that few laboratory facilities have the capability to apply dynamic loads to very large-scale test specimens. Furthermore, the available research on dynamic loading effects on steel test specimens has not demonstrated a compelling need for such testing.

If rate-of-loading effects are thought to be potentially significant for the steel core material used in the prototype, it may be possible to estimate the expected change in behavior by performing coupon tests at low (test cyclic loads) and high (dynamic earthquake) load rates. The results from brace tests would then be factored accordingly.

5. Instrumentation

Minimum instrumentation requirements are specified to permit determination of necessary data. It is expected that alternative instrumentation adequate for these purposes will be used in some cases.

6. Materials Testing Requirements

Tension testing of the steel core material used in the manufacture of the test specimens is required. In general, there has been good agreement between coupon test results and observed tensile yield strengths in full-scale uniaxial tests. Material testing required by this appendix is consistent with that required for testing of beam-to-column moment connections. For further information on this topic refer to Commentary Section K2.6.

7. Test Reporting Requirements

The results reported are necessary for conformance demonstration and for determination of strain-hardening and compression-overstrength requirements. As nonlinear modeling becomes more common, the production of test data to calibrate nonlinear elements is becoming an important secondary function. Little data exists on the behavior of braces beyond their design range; such information can be useful in verifying the reliability of the system.

8. Acceptance Criteria

The acceptance criteria are written so that the minimum testing data that must be submitted is at least one subassemblage test and at least one uniaxial test. In many cases the subassemblage test specimen also qualifies as a brace test specimen provided the requirements of Section K3.3 are met. If project specific subassemblage testing is to be performed it may be simplest to perform two subassemblage tests to meet the requirements of this Section. For the purposes of these requirements a single subassemblage test incorporating two braces in a chevron or other configuration is also considered acceptable.

Depending on the means used to connect the test specimen to the subassemblage or test apparatus, and the instrumentation system used, bolt slip may appear in the load versus displacement history for some tests. This may appear as a series of downward spikes in the load versus displacement plot and is not generally a cause for concern, provided the behavior does not adversely affect the performance of the brace or brace connection.

These acceptance criteria are intended to be minimum requirements. The 1.3 limit in Section K3.8, requirement (4), is essentially a limitation on β. These provisions were developed assuming that $\beta < 1.3$ so this provision has been included in the test requirements. Currently available braces should be able to satisfy this requirement.

REFERENCES

ACI (1999), "Details and Detailing of Concrete Reinforcement," ACI 315-99, contained within ACI Detailing Manual, SP-66(04), American Concrete Institute, Farmington Hills, MI.

ACI (2004a), *Manual of Structural and Placing Drawings for Reinforced Concrete Structures*, ACI 315R-04, contained within ACI Detailing Manual, SP-66(04), American Concrete Institute, Farmington Hills, MI.

ACI (2004b) *ACI Detailing Manual*, ACI SP-66(04), American Concrete Institute, Farmington Hills, MI.

ACI (2002), "Recommendations for the Design of Beam Column Joints in Monolithic Concrete Structures," ACI 352R-02, American Concrete Institute, Farmington Hills, MI.

ACI (2006), *Code Requirements for Nuclear Safety-Related Concrete Structures & Commentary*, ACI 349-06, American Concrete Institute, Farmington Hills, MI.

ACI (2008), *Building Code Requirements for Structural Concrete,* ACI 318-08 *& commentary*, American Concrete Institute, Farmington Hills, MI.

AISC (1997a), "K-area Advisory Statement," *Modern Steel Construction*, February, American Institute of Steel Construction, Chicago, IL.

AISC (1997b), *Seismic Provisions for Structural Steel Buildings*, April 15, American Institute of Steel Construction, Chicago, IL.

AISC (2002), *Seismic Provisions for Structural Steel Buildings*, May 21, American Institute of Steel Construction, Chicago, IL.

AISC (2005a), *Specification for Structural Steel Buildings,* ANSI/AISC 360-05, March 9, American Institute of Steel Construction, Chicago, IL.

AISC (2005b), *Seismic Provisions for Structural Steel Buildings*, ANSI/AISC 341-05, March 9, American Institute of Steel Construction, Chicago, IL.

AISC (2005c), *Steel Construction Manual,* 13th Ed., American Institute of Steel Construction, Chicago, IL.

AISC (2006), *Seismic Design Manual*, 1st Ed., American Institute of Steel Construction, Chicago, IL.

AISC (2010a), *Specification for Structural Steel Buildings,* ANSI/AISC 360-10, American Institute of Steel Construction, Chicago, IL.

AISC (2010b), *Prequalified Connections for Special and Intermediate Steel Moment Frames for Seismic Applications*, ANSI/AISC 358-10, American Institute of Steel Construction, Chicago, IL.

AISC (2010c), *Code of Standard Practice for Steel Buildings and Bridges*, AISC 303, March 18, American Institute of Steel Construction, Chicago, IL.

REFERENCES

AISC (2010d) *Design Guide 1, Base Plate and Anchor Rod Design*, 2nd Edition, American Institute of Steel Construction, Chicago, IL

AISI (2007), *North American Specification for the Design of Cold-Formed Steel Structural Members*, American Iron and Steel Institute, Washington, DC.

Ammerman, D.J. and Leon, R.T. (1990), "Unbraced Frames with Semi-Rigid Connections," *Engineering Journal,* AISC, Vol. 27, No. 1, 1st Quarter, pp. 12–21, Chicago, IL.

Arce, G. (2002), "Impact of Higher Strength Steels on Local Buckling and Overstrength in EBFs," Master's Thesis, Department of Civil Engineering, University of Texas at Austin, Austin, TX.

ASCE (1991a), *Standard for the Structural Design of Composite Slabs*, ANSI/ASCE 3-91, American Society of Civil Engineers, Reston, VA.

ASCE (1991b), *Standard Practice for Construction and Inspection of Composite Slabs*, ANSI/ASCE 9-91, American Society of Civil Engineers, Reston, VA.

ASCE (1994), "Guidelines for Design of Joints between Steel Beams and Reinforced Concrete Columns," *Journal of Structural Engineering*, American Society of Civil Engineers, Vol. 120, No. 8, August, pp. 2330–2357, Reston, VA.

ASCE Task Committee on Design Criteria for Composite Structures in Steel and Concrete (1998), "Design Guide for Partially Restrained Composite Connections," *Journal of Structural Engineering*, American Society of Civil Engineers, Vol. 124, No.10, pp. 1099–1114.

ASCE (2006a) *Seismic Rehabilitation of Existing Buildings,* ASCE 41-06, American Society of Civil Engineers, Reston, VA.

ASCE (2006b). *Seismic Rehabilitation of Existing Buildings: Supplement No.1*, ASCE 41-06, American Society of Civil Engineers, Reston, VA.

ASCE (2009), "Recommendations for Seismic Design of Hybrid Coupled Wall Systems," American Society of Civil Engineers, Reston, VA.

ASCE (2010), *Minimum Design Loads for Buildings and Other Structures*, SEI/ASCE 7-10, American Society of Civil Engineers, Reston, VA.

Aschheim, M. (2002), "Seismic Design Based on the Yield Displacement," *Earthquake Spectra*, EERI, Volume 18, No. 4, pp. 581–600.

Aslani, F. and Goel, S.C. (1991), "Stitch Spacing and Local Buckling in Seismic Resistant Double Angle Bracing Members," *Journal of Structural Engineering*, ASCE, Vol. 177, No. 8, August.

Astaneh-Asl, A., Goel, S.C., and Hanson, R.D. (1986), "Earthquake-resistant Design of Double Angle Bracing," *Engineering Journal*, AISC, Vol. 23, No. 4, 4th Quarter, Chicago, IL.

Astaneh-Asl, A., Bolt, B., McMullin, K., Donikian, R.R., Modjtahedi, D. and Cho, S-W (1994), "Seismic Performance of Steel Bridges During the 1994 Northridge Earthquake," *Report No. UCB/CEE-Steel-94/01,* Dept. of Civil and Env. Engrg., Univ. of California, Berkeley, CA.

REFERENCES

Astaneh-Asl, A. and Ravat, S. (1997), "Cyclic Tests of Steel H-Piles," *Report No. UCB/CEE-Steel-97/01*, Dept. of Civil and Env. Engrg., Univ. of California, Berkeley, CA.

Astaneh-Asl, A. (1998), "Seismic Behavior and Design of Gusset Plates for Braced Frames," *Steel Tips*, Structural Steel Education Council, Moraga, CA.

Astaneh-Asl, A. (2005a), "Notes on Design of Double-Angle and Tee Shear Connections for Gravity and Seismic Loads," *Steel Tips*, Structural Steel Education Council, Moraga, CA.

Astaneh-Asl, A. (2005b), "Design of Shear Tab Connections for Gravity and Seismic Loads," *Steel Tips*, Structural Steel Education Council, Moraga, CA.

Astaneh-Asl, A., Cochran, M., and Sabelli, R. (2006), *Seismic Detailing of Gusset Plates for Special Concentrically Braced Frames*, Steel Technical and Product Information (Steel TIPS) report, Structural Steel Educational Council, Moraga, CA.

ATC (1992), "Guidelines for Cyclic Seismic Testing of Components of Steel Structures," ATC-24, Applied Technology Council, Redwood City, CA.

AWS (2010), *Structural Welding Code—Steel*, ANSI/AWS D1.1: 2010, American Welding Society, Miami, FL.

AWS (2009), *Structural Welding Code—Seismic Supplement*, ANSI/AWS D1.8/D1.8M: 2009, American Welding Society, Miami, FL.

Azizinamini, A. and Schneider, S.P. (2004), "Moment Connections to Concrete-Filled Steel Tubes," *Journal of Structural Engineering*, ASCE, Vol. 130, No. 2, pp. 213–222.

Bansal, J.P. (1971), "The Lateral Instability of Continuous Steel Beams," CESRL Dissertation No. 71-1, University of Texas, Austin, TX.

Barsom, J.M. and Rolfe, S.T. (1999), *Fracture and Fatigue Control in Structures: Applications of Fracture Mechanics*, 3rd Ed., ASTM International MNL41, American Society for Testing and Materials, West Conshohocken, PA.

Bartlett, F.M., Jelinek, J.J., Schmidt, B.J., Dexter, R.J., Graeser, M.D., and Galambos, T.V. (2001), *Updating Standard Shape Material Properties Database for Design and Reliability*, University of Western Ontario and University of Minnesota.

Basha, H.S. and Goel, S.C. (1994), *Seismic Resistant Truss Moment Frames with Ductile Vierendeel Segment*, Research Report No. UMCEE 94-29, The University of Michigan Department of Civil and Environmental Engineering, Ann Arbor, MI.

Becker, E.R. (1971), *Panel Zone Effect on the Strength of Rigid Steel Frames*, USCOE 001, University of Southern California Structural Mechanics Laboratory, Los Angeles, CA.

Berman, J.W. and Bruneau, M. (2003a), "Experimental Investigation of Light-Gauge Steel Plate Shear Walls for the Seismic Retrofit of Buildings," *Technical Report MCEER-03-001*, Multidisciplinary Center for Earthquake Engineering Research, Buffalo, NY.

Berman, J. and Bruneau, M. (2003b), "Plastic Analysis and Design of Steel Plate Shear Walls," *Journal of Structural Engineering*, ASCE, Vol. 129, No. 11, November, pp. 1448–1456, Reston, VA.

Berman, J.W. and Bruneau, M. (2004), "Steel Plate Shear Walls are Not Plate Girders," *Engineering Journal*, AISC, 3rd Quarter, Chicago, IL.

Berman, J. W. and Bruneau, M. (2005a), "Approaches for the Seismic Retrofit of Braced Steel Bridge Piers and Proof-of-Concept Testing of a Laterally Stable Eccentrically Braced Frame," *Technical Report MCEER-05-0004*, Multidisciplinary Center for Earthquake Engineering Research, Buffalo, NY.

Berman, J., and Bruneau, M. (2005b), "Experimental Investigation of Light-Gauge Steel Plate Shear Walls," ASCE *Journal of Structural Engineering*, Vol. 131, No. 2, pp. 259–267.

Berman, J. and Bruneau, M. (2006), "Further Development of Tubular Eccentrically Braced Frame Links for the Seismic Retrofit of Braced Steel Truss Bridge Piers," *Technical Report MCEER-06-0006*, Multidisciplinary Center for Earthquake Engineering Research, Buffalo, NY.

Berman, J. and Bruneau, M. (2007), "Experimental and Analytical Investigation of Tubular Links for Eccentrically Braced Frames," *Engineering Structures*, Vol. 29, No. 8, pp. 1929–1938.

Berman, J. and Bruneau, M. (2008a), "Tubular Links for Eccentrically Braced Frames, I: Finite Element Parametric Study," *Journal of Structural Engineering*, ASCE, Vol. 134, No. 5, pp. 692–701.

Berman, J. and Bruneau, M. (2008b), "Tubular Links for Eccentrically Braced Frames, II: Experimental Verification," *Journal of Structural Engineering*, ASCE, Vol. 134, No. 5, pp. 702–712.

Berman, J. and Bruneau, M. (2008c), "Capacity Design of Vertical Boundary Elements in Steel Plate Shear Walls," *Engineering Journal*, AISC, Vol. 45, No.1, pp. 57–71.

Bhowmick, A.K., Driver, R.G., and Grondin, G.Y. (2009), Seismic analysis of steel plate shear walls considering strain rate and *P*-delta effects," *Journal of Constructional Steel Research* Vol. 65, Issue 5, pp. 1149–1159.

Bjorhovde, R. (1984), "Effects of End Restraint on Column Strength—Practical Applications," *Engineering Journal*, AISC, Vol. 21, No. 1, 1st Quarter, pp. 1–13, Chicago, IL.

Bjorhovde, R., Colson, A., and Brozzetti, J. (1990), "Classification System for Beam-to-Column Connections," *Journal of Structural Engineering*, ASCE, Vol. 116, No. 11, pp. 3059–3076.

Black, R.C., Wenger, W. A., and Popov, E.P. (1980), *Inelastic Buckling of Steel Struts Under Cyclic Load Reversals*, Report No. UCB/EERC-80-40, Earthquake Engineering Research Center, Berkeley, CA.

Bleich, F. (1952), "Buckling Strength of Metal Structures," Engineering Societies Monographs, McGraw Hill, 508p.

Blodgett, O.W. (2001), "Notes on Beam to Column Connections," *Steel Moment Frame Connection Advisory No. 3*, SAC 95-01, SAC Joint Venture, Sacramento, CA.

REFERENCES

Bondy, K.P. (1996), "A More Rational Approach to Capacity Design of Seismic Moment Frame Columns," *Earthquake Spectra*, EERI, Volume 12, No. 3, August, Oakland, CA.

Bruneau, M. and Bhagwagar, T. (2002), "Seismic Retrofit of Flexible Steel Frames using Thin Infill Panels," *Engineering Journal*, AISC, Vol. 24, No. 4, pp. 443–453, Chicago, IL.

Bruneau, M., Mahin, S.A., and Popov, E.P. (1987), *Ultimate Behavior of Butt Welded Splices in Heavy Rolled Steel Sections*, Report No. UCB/EERC-87/10, Earthquake Engineering Research Center, Berkeley, CA.

Caccese, V., Elgaaly, M., and Chen, R. (1993), "Experimental Study of Thin Steel-Plate Shear Walls Under Cyclic Load," *Journal of Structural Engineering*, ASCE, Vol. 119, No. 2, February, pp. 573–587, Reston, VA.

Cattan, J. (1995), *Statistical Analysis of Charpy V-Notch Toughness for Steel Wide-Flange Structural Shapes*, AISC, Chicago, IL.

Chao, S-H., and Goel, S. (2008), "A Modified Equation for Expected Maximum Shear Strength of the Special Segment for Design of Special Truss Moment Frames," *Engineering Journal*, AISC, 2nd Quarter, pp. 117–125.

Chen, W.F. and Lui, E.M. (1991), *Stability Design of Steel Frames*, CRC Press, Boca Raton, FL.

Cheng, J.J., Kulak, G.L., and Khoo, H.-A. (1998), "Strength of Slotted Tubular Tension Members," *Canadian Journal of Civil Engineering*, Vol. 25, pp. 982–991.

Chi, B. and Uang, C.M. (2002), "Cyclic Response and Design Recommendations of Reduced Beam Section Moment Connections with Deep Columns," *Journal of Structural Engineering*, ASCE, Vol. 128, No. 4, pp. 464–473, Reston, VA.

Choi, B. J., Kim, K. Y., Kim, C. H., and Kim, T. Y. (2009), "Experimental Compression Behavior of Stiffened Steel Plate Concrete (SSC) Structures under Compression Loading," Proceedings of 20th International Conference on Structural Mechanics in Reactor Technology (SMiRT 20), Espoo, Finland, August 9-14, Division 5, Paper 2008.

Choi, I.R. and Park, H.G. (2009), "Steel Plate Shear Walls with Various Infill Plate Designs," *Journal of Structural Engineering*, ASCE, Vol. 135, Issue 7, pp. 785–796.

Chou, C.-C. and Uang, C.-M. (2002), "Cyclic Performance of a Type of Steel Beam to Steel-Encased Reinforced Concrete Column Moment Connections," *Journal of Constructional Steel Research*, Vol. 58, pp. 637–663.

Civjan, S.A. and Singh, P. (2003), "Behavior of Shear Studs Subjected to Fully Reversed Cyclic Loading," *Journal of Structural Engineering*, ASCE, Vol. 129, No. 11, pp. 1466–1474, Reston,VA.

Clark, P., Aiken, I., Kasai, K., Ko, E. and Kimura, I. (1999), "Design Procedures for Buildings Incorporating Hysteretic Damping Devices," *Proceedings of the 68th Annual Convention*, pp. 355–371, Structural Engineers Association of California, Sacramento, CA.

County of Los Angeles Department of Public Works (1996), *County of Los Angeles Current Position on Design and Construction of Welded Moment Resisting Frame Systems (WMRF)*.

REFERENCES

CSA (2001), S16, *Limit States Design of Steel Structures,* Canadian Standards Association, Mississauga, ON.

CSA (2004), A23.3-04, *Design of Concrete Structures*, Canadian Standards Assocation, Rexdale, ON.

Dastfan, M. and Driver, R.G. (2008), "Flexural Stiffness Limits for Frame Members of Steel Plate Shear Wall Systems," *Proceedings of Annual Stability Conference*, Nashville, Structural Stability Research Council, Rolla, MO.

Dawe, J.L. and Kulak, G.L. (1986), "Local Buckling Behavior of Beam-Columns," *Journal of Structural Engineering*, ASCE, Vol. 112, No. 11, pp. 2447–2461

Deierlein, G.G. and Noguchi, H. (2004), "Overview of US-Japan Research on the Seismic Design of Composite Reinforced Concrete and Steel Moment Frame Structures," *Journal of Structural Engineering*, ASCE, Vol. 130, No. 2, pp. 361–367, Reston, VA.

Deierlein, G.G., Sheikh, T.M., and Yura, J.A. (1989), "Part 2: Beam Column Moment Connections for Composite Frames," *Journal of Structural Engineering*, ASCE, Vol. 115, No. 11, November, pp. 2877–2896, Reston, VA.

Deng, X., Dastfan, M., and Driver, R.G. (2008), "Behavior of Steel Plate Shear Walls with Composite Columns." *Proceedings*, Structures Congress, American Society of Civil Engineers, April 24-26, Vancouver, BC.

DiCorso, P.J., Reinhorn, A.M., Dickerson, J.R., Radziminski, J.B., and Harper, W.L. (1989), *Response of Frames with Bolted Semi-Rigid Connections, Part I—Experimental Study on Analytical Predictions*, Technical Report NCEER-89-0015, National Center for Earthquake Engineering Research, Buffalo, NY.

Driver, R.G., Kulak, G.L., Kennedy, D.J.L., and Elwi, A.E. (1997), "Seismic Behavior of Steel Plate Shear Walls," *Structural Engineering Report No. 215*, Department of Civil Engineering, University of Alberta, Edmonton, AB.

Dusicka, P. and Itani, A.M. (2002), "Behavior of Built-Up Shear Links Under Large Cyclic Deformations," *Proceedings of the 2002 Annual Meeting of the Structural Stability Research Council*, Structural Stability Research Council, Gainesville, FL.

Easterling, W.S. and Porter M.L. (1994), "Steel-Deck Reinforced Concrete Diaphragms I & II," *Journal of Structural Engineering*, ASCE, Vol. 120, No. 2, February, pp. 560–596, Reston, VA.

Ellingwood, B.R. and Corotis, R.B. (1991), "Load Combinations for Buildings Exposed to Fires," *Engineering Journal*, AISC, Vol. 28, No. 1, 1st Quarter, pp. 37–44, Chicago, IL.

Elgaaly M. (1998), "Thin Steel Plate Shear Walls Behavior and Analysis," *Thin Walled Structures*, Vol. 32, pp. 151–180.

Elgaaly, M., Caccese, V., and Du, C. (1993), "Postbuckling Behavior of Steel-Plate Shear Walls Under Cyclic Loads," *Journal of Structural Engineering*, ASCE, Vol. 119, No. 2, February, pp. 588–605.

Elgaaly, M. and Lui, Y. (1997), "Analysis of Thin-Steel-Plate Shear Walls," *Journal of Structural Engineering,* ASCE, Vol. 123, No. 11, November, pp. 1487–1496.

El-Tawil, S. and Deierlein, G.G. (1999), "Strength and Ductility of Encased Composite Columns," *Journal of Structural Engineering*, ASCE, Vol. 125, No. 9, pp. 1009–1019.

El-Tawil, S., Mikesell, T.D., Vidarsson, E., and Kunnath, S.K. (1999), "Inelastic Behavior and Design of Steel Panel Zones," *Journal of Structural Engineering*, ASCE, Vol. 125, No. 2, pp. 183–193, Reston, VA.

El-Tayem, A. and Goel, S.C. (1986), "Effective Length Factor for the Design of X-Bracing Systems," *Engineering Journal*, AISC, Vol. 23, No. 1, pp. 41–45, Chicago, IL.

Engelhardt, M.D. and Popov, E.P. (1989a), *Behavior of Long Links in Eccentrically Braced Frames*, Report No. UCB/EERC-89/01, Earthquake Engineering Research Center, Berkeley, CA.

Engelhardt, M.D. and Popov, E.P. (1989b), "On Design of Eccentrically Braced Frames," *Earthquake Spectra*, Vol. 5, No.3, August, Earthquake Engineering Research Institute, Oakland, CA.

Engelhardt, M.D., Tsai, K.C., and Popov, E.P. (1992), "Stability of Beams in Eccentrically Braced Frames," Chapter in *Stability and Ductility of Steel Structures Under Cyclic Loading*, Y. Fukumoto and G.C. Lee, Editors. CRC Press, pp. 99–112, 1992.

Englekirk, R.E. (1999), "Extant Panel Zone Design Procedures for Steel Frames are Questioned," *Earthquake Spectra*, EERI, Volume 15, No. 2, May, pp. 361–370, Oakland, CA.

ECS (1994), *Eurocode 4, Design of Composite Steel and Concrete Structures, Part 1-1: General Rules and Rules for Buildings*, European Committee for Standardization, Brussels, Belgium.

Fahnestock, L.A., Sause, R. and Ricles, J.M. (2003), "Analytical and Experimental Studies on Buckling-Restrained Braced Composite Frames," *Proceedings of International Workshop on Steel and Concrete Composite Construction* (IWSCCC-2003), Report No. NCREE-03-026, National Center for Research on Earthquake Engineering, Taipei, pp. 177–188.

Fahnestock, L. A. Ricles, J. M., and Sause, R. (2006), "Experimental Study of a Large-Scale Buckling Restrained Using the Pseudo-Dynamic Testing Method," *Proceedings of the 8th National Conference on Earthquake Engineering*, San Francisco, CA, April, 2006.

Fell, Benjamin V., Kanvinde, Amit M., Deierlein, Gregory G., Myers, Andrew T., and Fu, Xiangyang (August 2006), "Buckling and Fracture of Concentric Braces Under Inelastic Cyclic Loading," *Steel Tips*, Structural Steel Education Council.

FEMA (1994), *NEHRP (National Earthquake Hazards Reduction Program) Recommended Provisions for Seismic Regulations for New Buildings*, Federal Emergency Management Agency, Washington, DC.

FEMA (1997), *NEHRP Recommended Provisions for Seismic Regulations for New Buildings and Other Structures*, FEMA 302, Part 1- Provisions, Federal Emergency Management Agency, Washington, DC.

FEMA (2000a), *Recommended Seismic Design Criteria for New Steel Moment-Frame Buildings*, FEMA 350, Federal Emergency Management Agency, Washington, DC.

FEMA (2000b), *Recommended Specifications and Quality Assurance Guidelines for Steel Moment-Frame Construction for Seismic Applications*, FEMA 353, Federal Emergency Management Agency, Washington, DC.

FEMA (2000d), *State of the Art Report on Systems Performance of Steel Moment Frames Subject to Earthquake Ground Shaking*, FEMA 355C, prepared by the SAC Joint Venture for the Federal Emergency Management Agency, Washington, DC.

FEMA (2000e), *State of the Art Report on Connection Performance*, FEMA 355D, prepared by the SAC Joint Venture for the Federal Emergency Management Agency, Washington, D.C.

FEMA (2000f), *State of the Art Report on Performance Prediction and Evaluation of Steel Moment-Frame Buildings*, FEMA 355F, prepared by the SAC Joint Venture for the Federal Emergency Management Agency, Washington, DC.

FEMA (2000g), *State of the Art Report on Welding and Inspection*, Chapter 6, FEMA 355B, Federal Emergency Management Agency, Washington, DC.

FEMA (2001), *NEHRP Recommended Provisions For Seismic Regulations. Part 2: Commentary*, FEMA 369, Federal Emergency Management Agency, Washington, DC.

FEMA (2003), *NEHRP Recommended Provisions for Seismic Regulations for New Buildings and Other Structures*, FEMA 450, Federal Emergency Management Agency, Washington, DC.

FEMA (2009) *NEHRP Recommended Seismic Provisions for New Buildings and Other Structures*, FEMA P-750, Federal Emergency Management Agency, Washington, DC.

Fielding, D.J. and Huang, J.S. (1971), "Shear in Steel Beam-to-Column Connections," *Welding Journal*, AWS, Vol. 50, No. 7, Miami, FL.

Fisher, J.M. and West, M.A. (1990), *Serviceability Design Considerations for Low-Rise Buildings*, AISC, Chicago, IL.

Fortney, P.J. (2005), "The Next Generation of Coupling Beams," Department of Civil and Environmental Engineering, Ph.D. dissertation, University of Cincinnati, Cincinnati, OH.

Foutch, D.A. (1989), "Seismic Behavior of Eccentrically Braced Steel Building," *Journal of Structural Engineering*, ASCE, Vol. 115, No. 8, August, pp. 1857–1876, Reston, VA.

Furlong, R.W. (1997), "Composite Columns," *Composite Construction Design for Buildings*, Chapter 4, ASCE/McGraw Hill, New York, NY.

Goel, S.C. (1992a), *Recommendations for US/Japan Cooperative Research Program. Phase 5: Composite and Hybrid Structures*, Report UMCEE 92-29, University of Michigan, Department of Civil and Environmental Engineering, Ann Arbor, MI.

Goel, S.C. (1992b), "Cyclic Post Buckling Behavior of Steel Bracing Members," *Stability and Ductility of Steel Structures under Cyclic Loading*, pp.75–104, CRC Press, Boca Raton, FL.

Goel, S.C. (1992c), "Earthquake Resistant Design of Ductile Braced Steel Structures," *Stability and Ductility of Steel Structures under Cyclic Loading*, pp. 297–308, CRC Press, Boca Raton, FL.

Goel, S.C. (1993), *Proceedings of a US/Japan Cooperative Research Program Workshop on Composite and Hybrid Structures*, Report UMCEE 93, University of Michigan, Department of Civil and Environmental Engineering, Ann Arbor, MI.

Goel, S.C. and Chao, S-H. (2008), *Performance-Based Plastic Design: Earthquake Resistant Steel Structures*, International Code Council.

Goel, S.C. and Lee, S. (1992), "A Fracture Criterion for Concrete-Filled Tubular Bracing Members Under Cyclic Loading," *Proceedings of the 1992 ASCE Structures Congress*, ASCE, pp. 922–925, Reston, VA.

Goel, S.C. and Itani, A. (1994a), "Seismic Behavior of Open Web Truss Moment Frames," *Journal of Structural Engineering*, ASCE, Vol. 120, No. 6, June, pp.1763–1780, Reston, VA.

Goel, S.C. and Itani, A. (1994b), "Seismic Resistant Special Truss Moment Frames," *Journal of Structural Engineering*, ASCE, Vol. 120, No. 6, June, pp.1781–1797, Reston, VA.

Gomez, I., Kanvinde, A., Smith, C., and Deierlein, G. (2009), "Shear Transfer In Exposed Column Base Plates," March, 2009.

Gomez, I., Kanvinde, A., and Deierlein, G. (2010), "Exposed Column Base Connections Subjected to Axial Compression and Flexure," Final Report Presented to the American Institute of Steel Construction, April, 2010.

Gong, B. and Shahrooz, B.M. (2001a), "Concrete-Steel Composite Coupling Beams. Part I: Component Testing," *Journal of Structural Engineering*, ASCE, Vol. 127, No. 6, pp. 625–631, Reston, VA.

Gong, B. and Shahrooz, B.M. (2001b), "Concrete-Steel Composite Coupling Beams. Part II: Subassembly Testing and Design Verification," *Journal of Structural Engineering*, ASCE, Vol. 127, No. 6, pp. 632–638, Reston, VA.

Green, T.P, Leon, R.T., and Rassati, G.A. (2004), "Bidirectional Tests on Partially Restrained Composite Beam-to-Column Connections," *Journal of Structural Engineering*, ASCE, Vol. 130, No. 2, pp. 320–327, Reston, VA.

Griffis, L.G. (1992), "Composite Frame Construction," *Constructional Steel Design: An International Guide*, Elsevier Science Publishers, pp. 523–553, London, England.

Grondin and Behbahannidard (2001), "Cyclic Behaviour of Three-Storey Unstiffened Steel Plate Shear Wall," Canadian Society of Civil Engineers Conference.

Gupta, A. and Krawinkler, H. (1999), *Prediction of Seismic Demands for SMRFs with Ductile Connections and Elements*, SAC/BD-99/06, SAC Joint Venture, Sacramento, CA.

Haaijer, G. and Thurlimann, B. (1958), "On Inelastic Buckling of Steel," *Journal of the Engineering Mechanics Division*, ASCE, April.

Hamburger, R.O., Krawinkler, H., Malley, J.O., and Adan, S.M. (2009), "Seismic Design of Steel Special Moment Frames: a Guide for Practicing Engineers," NIST GCR 09-917-3, NEHRP Seismic Design Technical Brief, No. 2.

Han, S.W., Kwon, G., and Moon, K.H. (2007), "Cyclic Behavior of Post-Northridge WUF-B Connections," *Journal of Constructional Steel Research*, Vol. 63, No. 3, pp. 365–374.

Harries, K.A., Mitchell, D., Redwood, R.G., and Cook, W.D. (1997), "Seismic Design of Coupling Beams—A Case for Mixed Construction," *Canadian Journal of Civil Engineering*, Vol. 24, No. 3, pp. 448–459.

Harries, K. A., Gong, B., and Shahrooz, B.M. (2000), "Behavior and Design of Reinforced Concrete, Steel, and Steel-Concrete Coupling Beams," *Earthquake Spectra*, EERI, Vol. 16, No. 4, pp. 775–800.

Harries, K.A. and McNeice, D.S. (2006), "Performance-Based Design of High-Rise Coupled Wall Systems," *Structural Design of Tall and Special Structures*, Vol. 15, No. 3, pp. 289–306.

Hassan, O. and Goel, S.C. (1991), *Seismic Behavior and Design of Concentrically Braced Steel Structures*, Report UMCE 91-1, University of Michigan, Department of Civil and Environmental Engineering, Ann Arbor, MI.

Herrera, R.A., Ricles, J.M., and Sause, R. (2008), "Seismic Performance Evaluation of a Large-Scale Composite MRF Using Pseudo-Dynamic Testing," *Journal of Structural Engineering*, ASCE, Vol. 134 (2).

Hjelmstad, K.D. and Popov, E.P. (1983), "Cyclic Behavior and Design of Link Beams," *Journal of Structural Engineering*, ASCE, Vol. 109, No. 10, October, Reston, VA.

Hsieh, S.H. and Deierlein, G.G. (1991), "Nonlinear Analysis of Three-Dimensional Steel Frames with Semi-Rigid Connections," *Computers and Structures*, Vol. 41, No. 5, pp. 995–1009.

Huckelbridge, A.A. and Clough, R.W. (1977), *Earthquake Simulator Tests of Nine-Story Steel Frame with Columns Allowed to Uplift*, Report No. UCB/EERC-77/23, Earthquake Engineering Research Center, Berkeley, CA.

ICC (2006), *IBC Structural/Seismic Design Manual, Volume 3: Building Design Examples for Steel and Concrete*, International Code Council.

ICC (2008), *Acceptance Criteria for Steel Moment Frame Connection Systems*, AC 129, ICC Evaluation Service, Inc., Whittier, CA.

ICC (2009), *International Building Code*, International Code Council, Birmingham, AL.

Itani, A. and Goel, S.C. (1991), *Earthquake Resistant Design of Open Web Framing Systems*, Research Report No. UMCE 91-21, University of Michigan, Department of Civil and Environmental Engineering, Ann Arbor, MI.

Jain, A., Goel, S.C., and Hanson, R.D. (1978), *Hysteresis Behavior of Bracing Members and Seismic Response of Braced Frames with Different Proportions*, Research Report No. UMEE78R3, University of Michigan, Department of Civil Engineering, Ann Arbor, MI.

Kalyanaraman, V., Sridhara, B.N., and Thairani, V. (1998), "Core Loaded Earthquake Resistant Bracing Design," *Proceedings of the 2nd International Conference on Steel Construction*, San Sebastian, Spain.

Kanno, R. and Deierlein, G.G. (1997), "Seismic Behavior of Composite (RCS) Beam-Column Joint Subassemblies," *Composite Construction III*, ASCE, Reston, VA.

Kasai, K. and Popov, E.P. (1984), "On Seismic Design of Eccentrically Braced Steel Frames," *Proceedings of the 8th World Conference on Earthquake Engineering*, EERI, Vol. 5, pp.387-394, San Francisco, CA.

Kasai, K. and Popov, E.P. (1986a), "General Behavior of WF Steel Shear Link Beams," *Journal of Structural Engineering*, ASCE, Vol. 112, No. 2, February, Reston, VA.

Kasai, K. and Popov, E.P. (1986b), "Cyclic Web Buckling Control for Shear Link Beams," *Journal of Structural Engineering*, ASCE, Vol. 112, No. 3, March, Reston, VA.

Kasai, K. and Popov, E.P. (1986c), *A Study of Seismically Resistant Eccentrically Braced Frames*, Report No. UCB/EERC-86/01, Earthquake Engineering Research Center, Berkeley, CA.

Kaufmann, E.J. and Fisher, J.W. (2001), *Effect of Straightening Method on the Cyclic Behavior of K-area In Steel Rolled Shapes*, ATLSS Center, Lehigh University, December.

Kaufmann, E.J., Metrovich, B.R., and Pense, A.W. (2001), *Characterization of Cyclic Inelastic Strain Behavior On Properties of A572 Gr. 50 and A913 Gr 50 Rolled Sections*, ATLSS Center, Lehigh University, August.

Kemp, A.R. (1986), "Factors Affecting the Rotation Capacity of Plastically Designed Members," *The Structural Engineer*, Vol. 64B, No. 2, June, The Institution of Structural Engineers, London, England.

Khatib, I., Mahin, S.A., and Pister, K.S. (1988), *Seismic Behavior of Concentrically Braced Steel Frames*, Report No. UCB/EERC 88-01, Earthquake Engineering Research Center, Berkeley, CA.

Kitayama, K., Otani, S., and Aoyama, H. (1987), "Earthquake Resistant Design Criteria for Reinforced Concrete Interior Beam-Column Joints," *Proceedings of the Pacific Conference on Earthquake Engineering*, Wairakei, New Zealand.

Korol, R.M. (1996), "Shear Lag in Slotted HSS Tension Members," *Canadian Journal of Civil Engineering*, Vol. 23, pp. 1350–1354.

Krawinkler, H. (1978), "Shear in Beam-Column Joints in Seismic Design of Steel Frames," *Engineering Journal*, AISC, Vol. 15, No. 3, 3rd Quarter, Chicago, IL.

Krawinkler, H. (2001), Private Communication with Mark Saunders, November 2001.

Kulak, G.L., Fisher, J.W., and Struik, J.H.A. (1987), *Guide to Design Criteria for Bolted and Riveted Joints*, 2nd Ed., John Wiley & Sons, New York, NY.

Lay, M.G. (1965), "Flange Local Buckling in Wide-Flange Shapes," *Journal of the Structural Division*, ASCE, Vol. 91, No. 6, December, Reston, VA.

Lee, C.S. and Tsai, K.C. (2008), "Experimental Response of Four 2-Story Narrow Steel Plate Shear Walls," *Proceeding of the 2008 Structures Congress*, Vancouver, BC, ASCE, Reston, VA.

Lee, D., Cotton, S.C., Dexter, R.J., Hajjar, J.F., Ye, Y., and Ojard, S.D. (2002), *Column Stiffener Detailing and Panel Zone Behavior of Steel Moment Frame Connections*, University of Minnesota, June (Draft final).

Lee, D., Cotton, S. C., Hajjar, J. F., Dexter, R. J., and Ye, Y. (2005a), "Cyclic Behavior of Steel Moment-Resisting Connections Reinforced by Alternative Column Stiffener Details: I. Connection Performance and Continuity Plate Detailing," *Engineering Journal*, AISC, Vol. 42, No. 4, 4th Quarter, pp. 189–214.

Lee, D., Cotton, S. C., Hajjar, J. F., Dexter, R. J., and Ye, Y. (2005b), "Cyclic Behavior of Steel Moment-Resisting Connections Reinforced by Alternative Column Stiffener Details: II. Panel Zone Behavior and Doubler Plate Detailing," *Engineering Journal*, AISC, Vol. 42, No. 4, 4th Quarter, pp. 215–238.

Lee, H. and Goel, S.C. (1990), *Seismic Behavior of Steel Built-up Box-Shaped Bracing Members, and Their Use in Strengthening Reinforced Concrete Frames*, Report No. UMCE 90-7, University of Michigan, Ann Arbor, MI.

Lee, S. and Goel, S.C. (1987), *Seismic Behavior of Hollow and Concrete-Filled Square Tubular Bracing Members*, Report No. UMCE 87-11, University of Michigan, Department of Civil Engineering, Ann Arbor, MI.

Leon, R.T. (1990), "Semi-Rigid Composite Construction," *Journal of Constructional Steel Research*, Vol. 15, No. 2, pp. 99–120, Elsevier Science Publishers, London, England.

Leon, R.T. (1994), "Composite Semi-Rigid Construction," *Engineering Journal*, AISC, Vol. 31, No. 2, 2nd Quarter, pp. 57–67.

Leon, R.T. and Ammerman, D.J. (1990), "Semi-Rigid Connection for Gravity Loads," *Engineering Journal*, AISC, Vol. 27, No. 1, 1st Quarter, pp. 1–11, Chicago, IL.

Leon, R.T., Ammerman, D., Lin, J., and McCauley, R. (1987), "Semi-Rigid Composite Steel Frames," *Engineering Journal*, AISC Vol. 24, No. 4, 4th Quarter, pp. 147–156.

Leon, R.T. and Forcier, G.P. (1992), "Parametric Study of Composite Frames," *Connections in Steel Structures II*, Proceedings of the Second International Workshop on Connections in Steel Structures, AISC, pp. 152–159, Chicago, IL.

Leon, R.T., Hajjar, J.F., and Shield, C.K. (1997), "The Effect of Composite Floor Slabs on the Behavior of Steel Moment-Resisting Frames in the Northridge Earthquake," *Composite Construction in Steel and Concrete III*, ASCE, pp. 735–751, Reston, VA.

Leon, R.T., Hoffman, J.J., and Staeger, T. (1996), *Design Guide 8 Partially Restrained Composite Connections*, online content, AISC, Chicago, IL.

Leon, R.T. and Kim, D.H. (2004), "Seismic Performance of PR Frames in Zones of Infrequent Seismicity," *Proceedings of the 13th World Conference in Earthquake Engineering*, Paper 2696, IAEE, Vancouver, BC.

Liang, X. and Parra-Montesinos, G. (2004), "Seismic Behavior of RCS Beam-Column-Slab Subassemblies and Frame Systems," *Journal of Structural Engineering*, ASCE, Vol. 130, No. 2, pp. 310–319, Reston, VA.

REFERENCES

Libby, J.R. (1981), "Eccentrically Braced Frame Construction—A Case History," *Engineering Journal*, AISC, Vol. 18, No. 4, 4th Quarter, Chicago, IL.

Liu, J., Sabelli, R., Brockenbrough, R. L., and Fraser, T. P. (2007), "Expected Yield Stress and Tensile Strength Ratios for Determination of Expected Member Capacity in the 2005 AISC Seismic Provisions," *Engineering Journal*, AISC, Vol. 44, No. 1, 1st Quarter, Chicago, IL.

Liu, Z. and Goel, S.C. (1987), *Investigation of Concrete Filled Steel Tubes under Cyclic Bending and Buckling*, UMCE Report 87-3, University of Michigan, Ann Arbor, MI.

Liu, Z. and Goel, S.C. (1988), "Cyclic Load Behavior of Concrete-Filled Tubular Braces," *Journal of the Structural Division*, ASCE, Vol. 114, No. 7, Reston, VA.

Lopez, W., Gwie, D.S., Lauck, T.W., and Saunders, C.M. (2004), "UC Berkeley Stanley Hall—Structural Design and Experimental Verification of a Buckling-Restrained Braced Frame System," *Engineering Journal*, AISC, Vol. 41, No. 4, 4th Quarter, Chicago, IL.

Lubell, A.S., Prion, H.G.L., Ventura, C.E., and Rezai, M. (2000), "Unstiffened Steel Plate Shear Wall Performance under Cyclic Loading," *Journal of Structural Engineering*, ASCE, Vol. 126, No.4, April, pp. 453–460, Reston, VA.

MacRae, G., Roeder, C.W., Gunderson, C., and Kimura, Y. (2004), "Brace-Beam-Column Connections for Concentrically Braced Frames with Concrete Filled Tube Columns," *Journal of Structural Engineering*, ASCE, Vol. 130, No. 2, pp. 233–243, Reston, VA.

Malley, J.O. and Popov, E.P. (1984), "Shear Links in Eccentrically Braced Frames," *Journal of Structural Engineering*, ASCE, Vol. 110, No. 9, September, Reston, VA.

Malley, J.O. and Popov, E.P. (1983), "Design Considerations for Shear Links in Eccentrically Braced Frames," Earthquake Engineering Research Center Report EERC-83/24, University of California, Berkeley, CA., 126 p.

Mattock A. H. and Gaafar G. H. (1982), "Strength of Embedded Steel Sections as Brackets," *ACI Journal*, Vol. 79, No. 2.

McDaniel, C.C., Uang, C.M., and Seible, F. (2002), "Cyclic Testing of Built-Up Steel Shear Links for the New Bay Bridge," *Journal of Structural Engineering*, ASCE, Vol. 129, No. 6, pp. 801–809, Reston, VA.

McMullin, K.M. and Astaneh-Asl, A. (1994), "Cyclical Behavior of Welded Steel Shear Studs," Structures Congress XII, ASCE, pp. 1024–1029.

Merovich, A.T., Nicoletti, J.P., and Hartle, E. (1982), "Eccentric Bracing in Tall Buildings," *Journal of the Structural Division*, ASCE, Vol. 108, No. 9, September, Reston, VA.

Nader, M.N. and Astaneh-Asl, A. (1992), "Seismic Design Concepts for Semi-Rigid Frames," *Proceedings of the 1992 ASCE Structures Congress*, ASCE, pp. 971–975, New York, NY.

Nakashima, M., Kanao, I., and Liu, D. (2002), "Lateral Instability and Lateral Bracing of Steel Beams Subjected to Cyclic Loading," *Journal of Structural Engineering*, ASCE, Vol. 128, No. 10, October, pp. 1308–1316.

REFERENCES

Newell, James D., and Uang, Chia-Ming (2008), "Cyclic Behavior of Steel Wide-Flange Columns Subjected to Large Drift," *Journal of Structural Engineering*, ASCE, Vol. 134, No. 8, August, pp. 1334–1342.

NFPA (2009), *Building Construction and Safety Code*, NFPA 5000, Washington, D.C.

Nishiyama, I., Hasegawa, T., and Yamanouchi, H. (1990), "Strength and Deformation Capacity of Reinforced Concrete Column to Steel Beam Joint Panels," *Building Research Institute Report 71*, Ministry of Construction, Tsukuba, Japan.

Okazaki, T. (2004), "Seismic Performance of Link-to-Column Connections in Steel Eccentrically Braced Frames," Ph.D. Dissertation, Department of Civil Engineering, University of Texas at Austin, Austin, TX.

Okazaki, T., Arce, G., Ryu, H., and Engelhardt, M.D. (2004a), "Recent Research on Link Performance in Steel Eccentrically Braced Frames," *Proceedings, 13th World Conference on Earthquake Engineering*, August 1-6, 2004, Vancouver, BC.

Okazaki, T., Engelhardt, M.D., Nakashima, M., and Suita, K. (2004b), "Experimental Study on Link-to-Column Connections in Steel Eccentrically Braced Frames," *Proceedings, 13th World Conference on Earthquake Engineering*, August 1-6, 2004, Vancouver, BC.

Okazaki, T., Engelhardt, M.D., Drolias, A., Schell, E., Hong, J.K., and Uang, C.M. (2009), "Experimental Investigation of Link-to-Column Connections in Eccentrically Braced Frames," *Journal of Constructional Steel Research*, Vol. 65, pp. 1401–1412.

OSHA (2010), *Safety Standards for Steel Erection*, Part Number 1926, Subpart R, Occupational Safety and Health Administration, Washington, D.C.

Ozaki, M., Akita, S., Osuga, H., and Adachi, N. (2004), "Study on Steel Plate Reinforced Concrete Panels Subjected to Cyclic In-Plane Shear," *Nuclear Engineering and Design*, Vol. 228, pp. 225–244.

Park, R., Priestley, M.J.N., and Gill, W.D. (1982), "Ductility of Square Confined Concrete Columns," *Journal of the Structural Division*, ASCE, Vol. 108, No. ST4, April, pp. 929–950, Reston, VA.

Parra-Montesinos, G., Goel, S., and Kim, K-Y. (2006), "Behavior of Steel Double Channel Built-Up Chords of STMF under Reversed Cyclic Bending," *Journal of Structural Engineering*, ASCE, September 2006.

Parra-Montesinos, G., Liang, X., and Wight, J.K. (2003), "Towards Deformation-Based Capacity Design of RCS Beam-Column Connections," *Engineering Structures*, Vol. 25, No. 5, pp. 681–690.

Parra-Montesinos, G. and Wight, J.K. (2000), "Seismic Response of Exterior RC Column-to-Steel Beam Connections," *Journal of Structural Engineering*, ASCE, Vol. 126, No. 10, pp. 1113–1121, Reston, VA.

Parra-Montesinos, G. and Wight, J.K. (2001), "Modeling Shear Behavior of Hybrid RCS Beam-Column Connections," *Journal of Structural Engineering*, ASCE, Vol. 127, No. 1, pp. 3–11.

REFERENCES

Perlynn, M.J. and Kulak, G.L. (1974), "Web Slenderness Limits for Compact Beam-Columns," Structural Engineering Report No. 50, Department of Civil Engineering, University of Alberta.

Popov, E.P., Blondet, M., Stepanov, L., and Stojadinovic, B. (1996), "Full-Scale Beam-to-Column Connection Tests," University of California Department of Civil Engineering, Berkeley, CA.

Popov, E.P. and Engelhardt, M.D. (1988), "Seismic Eccentrically Braced Frames," *Journal of Constructional Steel Research*, Vol. 10, pp. 321–354.

Popov, E.P., Engelhardt, M.D., and Ricles, J.M. (1989), "Eccentrically Brace Frames: U. S. Practice," *Engineering Journal*, AISC, Vol. 26, No. 2, 2nd Quarter, pp. 66–80, Chicago, IL.

Popov, E.P. and Stephen, R.M. (1977), "Tensile Capacity of Partial Penetration Welds," *Journal of the Structural Division*, ASCE, Vol. 103, No. ST9, September, Reston, VA.

Purba, R. and Bruneau, M. (2007), "Design Recommendations for Perforated Steel Plate Shear Walls," Technical Report MCEER-07-0011, Multidisciplenary Center for Earthquake Engineering Research, State University of New York at Buffalo, Buffalo, NY.

Purba, R., Bruneau, M. (2009), "Finite Element Investigation and Design Recommendations for Perforated Steel Plate Shear Walls," *Journal of Structural Engineering*, ASCE, Vol.135, No. 11, pp.1367–1376.

Qin, F. (1993), "Analysis of Composite Connection between Reinforced Concrete Walls and Steel Coupling Beams," M.S. Thesis, Department of Civil and Environmental Engineering, University of Cincinnati, Cincinnati, OH.

Qu B. and Bruneau, M. (2008), "Seismic Behavior and Design of Boundary Frame Members in Steel Plate Shear Walls," Technical Report MCEER-08-0012, Multidisciplinary Center for Earthquake Engineering Research, State University of New York at Buffalo, Buffalo, NY.

Qu, B. and Bruneau, M. (2009), "Design of Steel Plate Shear Walls Considering Boundary Frame Moment Resisting Action," *Journal of Structural Engineering*, ASCE, Vol.135, No. 12, pp. 1511–1521.

Qu, B. and Bruneau, M. (2010a), "Capacity Design of Intermediate Horizontal Boundary Elements of Steel Plate Shear Walls," *Journal of Structural Engineering*, ASCE, Vol.136, No. 6, pp. 665–675.

Qu, B. and Bruneau, M. (2010b), "Behavior of Vertical Boundary Elements in Steel Plate Shear Walls," *Engineering Journal,* AISC, Vol. 47, 2nd Quarter, Chicago, IL.

Qu, B., Bruneau, M., Lin, C.H., and Tsai, K.C. (2008), "Testing of Full-Scale Two-Story Steel Plate Shear Wall with Reduced Beam Section Connections and Composite Floors," *Journal of Structural Engineering*, ASCE, Vol. 134, Issue 3, pp. 364–373.

Rai, D.C., Basha, H., and Goel, S.C. (1998), "Special Truss Moment Frames (STMF): Design Guide," Research Report No. UMCEE 98-44, Department of Civil and Environmental Engineering, University of Michigan, Ann Arbor, MI.

Rassati, G.A., Leon R.T., and Noe, S. (2004), "Component Modeling of Partially Restrained Composite Joints under Cyclic and Dynamic Loading," *Journal of Structural Engineering*, ASCE, Vol. 130, No. 2, pp. 343–351, Reston, VA.

RCSC (2009), *Specification for Structural Joints Using High-Strength Bolts*, Research Council on Structural Connections, American Institute of Steel Construction, Chicago, IL.

Reina, P. and Normile, D. (1997), "Fully Braced for Seismic Survival," *Engineering News-Record*, July 21, pp. 34–36, The McGraw Hill Companies, New York, NY.

Rezai, M. (1999), "Seismic Behavior of Steel Plate Shear Walls by Shake Table Testing," Ph.D. Dissertation, University of British Columbia, Vancouver, BC.

Richards, P. (2004), "Cyclic Stability and Capacity Design of Steel Eccentrically Braced Frames," Ph.D. Dissertation, University of California, San Diego, San Diego, CA. (Advisor: C.M. Uang).

Richards, P. and Uang, C.M. (2003), "Development of Testing Protocol for Short Links in Eccentrically Braced Frames," Report No. SSRP-2003/08, Department of Structural Engineering, University of California, San Diego, San Diego, CA.

Richards, P., Uang, C.M., Okazaki, T., and Engelhardt, M.D. (2004), "Impact of Recent Research Findings on Eccentrically Braced Frame Design," *Proceedings, 2004 SEAOC Convention*, August 25-28, 2004, Monterey, CA.

Ricles, J.M. and Paboojian, S.D. (1994), "Seismic Performance of Steel-Encased Composite," *Journal of Structural Engineering*, ASCE, 120(8), pp. 2474–2494.

Ricles, J.M., Peng, S.W., and Lu, L.W. (2004a), "Seismic Behavior of Composite Concrete Filled Steel Tube Column-Wide Flange Beam Moment Connections," *Journal of Structural Engineering*, ASCE, Vol. 130, No. 2, pp. 223–232, Reston, VA.

Ricles, J.M. and Popov, E.P. (1987a), *Dynamic Analysis of Seismically Resistant Eccentrically Braced Frames*, Report No. UCB/EERC-87/107, Earthquake Engineering Research Center, Berkeley, CA.

Ricles, J.M. and Popov, E.P. (1987b), *Experiments on EBFs with Composite Floors*, Report No. UCB/EERC-87/06, Earthquake Engineering Research Center, Berkeley, CA.

Ricles, J.M., Zhang, X., Lu, L.W., and Fisher, J. (2004b), "Development of Seismic Guidelines for Deep Column Steel Moment Connections," ATLSS Report No. 04-13, Lehigh University, Bethlehem, PA.

Roberts, T.M. and Sabouri-Ghomi, S. (1992), "Hysteretic Characteristics of Unstiffened Perforated Steel Plate Shear Walls," *Thin Walled Structures*, Elsevier Ltd., Vol. 14, pp. 139–151.

Roeder, C.W. (1987), *Inelastic Dynamic Analysis of Two Eight-Story Moment Frames*, Structural Engineers Association of Washington, Seattle, WA.

Roeder, C.W. and Foutch, D.F. (1996), "Experimental Results for Seismic Resistant Steel Moment Frame Connections," *Journal of Structural Engineering,* ASCE, Vol. 122, No. 6, June, Reston, VA.

Roeder, C.W. and Popov, E.P. (1978), "Eccentrically Braced Frames for Earthquakes," *Journal of the Structural Division*, ASCE, Vol. 104, No. 3, March, Reston, VA.

Saari, W.K., Hajjar, J.F., Schultz, A. E., and Shield, C.K. (2004), "Behavior of Shear Studs in Steel Frames with Reinforced Concrete Infill Walls," *Journal of Constructional Steel Research*, Vol. 60, pp. 1453–1480.

Saatcioglu, M. (1991), *Deformability of Steel Columns*, ACI 127.5, American Concrete Institute, Detroit, MI.

Sabelli, R., Mahin, S., and Chang, C. (2003), "Seismic Demands on Steel Braced Frame Buildings with Buckling-Restrained Braces," *Engineering Structures*, Vol. 25, pp. 655–666.

Sabouri-Ghomi, S. and Roberts, T.M. (1992), "Nonlinear Dynamic Analysis of Steel Plate Shear Walls Including Shear and Bending Deformations," *Engineering Structures*, AISC, Vol. 14, No. 3, pp. 309–317.

Sabouri-Ghomi, S., Ventura, C.E., and Kharrazi, M.H.K. (2005), Shear Analysis and Design of Ductile Steel Plate Walls," *Journal of Structural Engineering*, ASCE, Vol. 131, Issue 6, pp. 878–889.

SAC (1996), *Experimental Investigations of Beam-Column Subassemblages* SAC96-01, SAC Joint Venture, Sacramento, CA.

SAC (1997), "Protocol for Fabrication, Inspection, Testing, and Documentation of Beam-Column Connection Tests and Other Specimens," SAC/BD-97/02 Version 1.1, SAC Joint Venture, Sacramento, CA.

SAC (2000), "Loading Histories for Seismic Performance Testing of SMRF Components and Assemblies," SAC/BD-00/10, SAC Joint Venture, Sacramento, CA.

Salmon, C.G. and Johnson, J.E. (1996), *Steel Structures: Design and Behavior*, 4th Edition, HarperCollins College Publishers, New York, NY.

Santhakumar, A.R. (1974), "Ductility of Coupled Shear Walls," Ph.D. Dissertation, Department of Civil Engineering, University of Canterbury, Canterbury, New Zealand.

Sawyer, H.A. (1961), "Post-Elastic Behavior of Wide-Flange Steel Beams," *Journal of the Structural Division*, ASCE, Vol. 87, No. ST8, December, Reston, VA.

Schneider, S.P., Roeder, C.W., and Carpenter, J.E. (1991), *Seismic Performance of Weak-Column Strong-Beam Steel Moment Resisting Frames*, University of Washington Department of Civil Engineering, Seattle, WA.

SDI (2001), *Standard Practice Details*, No. SPD2, Steel Deck Institute, Fox River Grove, IL.

SDI (2004), *Diaphragm Design Manual*, 3rd Edition—No. DDMO3, Steel Deck Institute, Fox River Grove, IL.

SDI (2007), *Design Manual for Composite Decks, Form Decks and Roof Decks*, No. 31, Steel Deck Institute, Fox River Grove, IL.

Shahrooz, B. M., Remetter, M. E., and Qin, F. (1993), "Seismic Design and Performance of Composite Coupled Walls," *Journal of Structural Engineering*, ASCE, Vol. 119, No. 11, pp. 3291–3309, Reston, VA.

Sheikh, S.A. and Uzumeri, S.M. (1980), "Strength and Ductility of Tied Columns," *Journal of the Structural Division*, ASCE, Vol. 106, No. ST5, February, pp. 1079–1102, Reston, VA.

Sheikh, T.M., Deierlein, G.G., Yura, J.A., and Jirsa, J.O. (1989), "Part 1: Beam Column Moment Connections for Composite Frames," *Journal of Structural Engineering*, ASCE, Vol. 115, No. 11, November, pp. 2859–2876, Reston, VA.

Shen, J., Sabol, T., and Akbas, B. and Sutchiewcharn, N. (2010), "Seismic Demand on Column Splices in Steel Moment Frames," *Engineering Journal*, AISC, Vol. 47, No. 4, Chicago, IL.

Shishkin, J.J., Driver, R.G., and Grondin, G.Y. (2009), "Analysis of Steel Plate Shear Walls Using the Modified Strip Model," *Journal of Structural Engineering*, ASCE, Vol. 135, Issue 11, pp. 1357–1366.

SIE (1999a), "Tests of Nippon Steel Corporation Unbonded Braces," Report to Ove Arup & Partners, Seismic Isolation Engineering, CA.

SIE (1999b), "Report on Pre-Prototype Tests of Yielding Braces," Seismic Isolation Engineering, CA.

Slutter, R. (1981), "Tests of Panel Zone Behavior in Beam Column Connections," *Lehigh University Report No. 200.81.403.1*, Lehigh University, Bethlehem, PA.

Stoakes, C. D. and Fahnestock, L.A. (2010), "Flexural Behavior of Concentrically-Braced Frame Beam-Column Connections," Proceedings, 2010 Structures Congress, ASCE, Orlando, FL.

Tang, X. and Goel, S.C. (1987), *Seismic Analysis and Design Considerations of Braced Steel Structures*, Report UMCE 87-4, University of Michigan, Department of Civil and Environmental Engineering, Ann Arbor, MI.

Tang, X. and Goel, S.C. (1989), "Brace Fractures and Analysis of Phase I Structure," *Journal of Structural Engineering*, ASCE, Vol. 115, No. 8, August, pp. 1960–1976, Reston, VA.

Thermou, G.E., Elnashai, A.S., Plumier, A., and Doneaux, C. (2004), "Seismic Design and Performance of Composite Frames," *Journal of Constructional Steel Research*, Vol. 60, pp. 31–57.

Thorburn, L.J., Kulak, G.L., and Montgomery, C.J. (1983), "Analysis of Steel Plate Shear Walls," Structural Engineering Report No. 107, Department of Civil Engineering, University of Alberta, Edmonton, AB.

Thornton, W.A. and Muir, L.S. (2008), "Vertical Bracing Connections in the Seismic Regime," *Proceedings, Connections VI: Sixth International Workshop on Connections in Steel Structures*, AISC, Chicago, IL.

Timler, P.A. and Kulak, G.L. (1983), "Experimental Study of Steel Plate Shear Walls," Structural Engineering Report No. 114, Department of Civil Engineering, University of Alberta, Edmonton, AB.

Tort, C. and Hajjar, J.F. (2004), "Damage Assessment of Rectangular Concrete-Filled Steel Tubes for Performance-Based Design," *Earthquake Spectra*, EERI, Vol. 20, No. 4, pp. 1317–1348, Oakland, CA.

Tremblay, R. (2000), "Influence of Brace Slenderness on the Seismic Response of Concentrically Braced Steel Frames," *Behaviour of Steel Structures in Seismic Areas—Stessa 2000: Proceedings of the Third International Conference*, August 21-24, 2000, F.M. Mazzolani and, R. Tremblay, eds., pp. 527–534, Montreal, QU.

Tremblay, R. (2001), "Seismic Behavior and Design of Concentrically Braced Steel Frames," *Engineering Journal*, AISC, Vol. 38, No. 3, Chicago, IL.

Tremblay, R. (2003), "Achieving a Stable Inelastic Seismic Response for Multi-Story Concentrically Braced Steel Frames," *Engineering Journal*, AISC, Vol. 40, No. 2, Chicago, IL.

Tremblay, R., Degrange, G., and Blouin, J. (1999), "Seismic Rehabilitation of a Four-Storey Building with a Stiffened Bracing System," *Proceedings 8th Canadian Conference on Earthquake Engineering*, Canadian Association for Earthquake Engineering, pp. 549–554, Vancouver, B.C..

Tremblay, R., Tchebotarev, N., and Filiatrault, A. (1997), "Seismic Performance of RBS Connections for Steel Moment Resisting Frames: Influence of Loading Rate and Floor Slab," *Proceedings, STESSA '97*, Kyoto, Japan.

Tremblay R. and Tirca L. (2003), "Behavior and design of multi-storey zipper concentrically braced steel frames for the mitigation of soft-storey response," STESSA, 2003, Balkema, Naples.

Tromposch, E.W. and Kulak, G.L. (1987), "Cyclic and Static Behavior of Thin Panel Steel Plate Shear Walls," Structural Engineering Report No. 145, Department of Civil Engineering, University of Alberta, Edmonton, AB.

Tsai, K.-C., Weng, Y.-T., Lin, M.-L., Chen, C.-H., Lai, J.-W., and Hsiao, P.-C. (2003), "Pseudo Dynamic Tests of a Full-Scale CFT/BRB Composite Frame: Displacement Based Seismic Design and Response Evaluations," *Proceedings of the International Workshop on Steel and Concrete Composite Construction (IWSCCC-2003)*, Report No. NCREE-03-026, National Center for Research in Earthquake Engineering, pp. 165–176, Taipei, Taiwan.

Uang, C.M. and Chi, B. (2001), *Effect of Straightening Method on the Cyclic Behavior of k Area in Steel Rolled Shapes*, Report No. SSRP-2001/05, University of California, San Diego, CA, May.

Uang, C.M. and Fan, C.C. (2001), "Cyclic Stability Criteria for Steel Moment Connections with Reduced Beam Section," *Journal of Structural Engineering*, ASCE, Vol. 127, No. 9, September, Reston, VA.

Uang, C.M. and Nakashima, M. (2003), "Steel Buckling-Restrained Frames," *Earthquake Engineering: Recent Advances and Applications*, Chapter 16, Y. Bozorgnia and V.V. Bertero, (eds.), CRC Press, Boca Raton, FL.

Uriz, P. and Mahin, S. A. (2004), "Seismic Performance Assessment of Concentrically Braced Steel Frames," *Proceedings of 13th World Conference on Earthquake Engineering*, Vancouver, BC; August 1-6, 2004; Paper No. 1639.

Varma, A.H., Ricles, J.M., Sause, R. and Lu, L.W. (2002), "Experimental Behavior of High Strength Square Concrete Filled Tube Columns," *Journal of Structural Engineering*, ASCE, Vol. 128, No. 3, pp. 309–318, Chicago, IL.

Varma, A.H., Ricles, J.M., Sause, R., and Lu, L.W. (2004), "Seismic Behavior and Design of High Strength Square Concrete Filled Tube Beam Columns," *Journal of Structural Engineering*, ASCE, Vol. 130, No. 2, pp. 169–179, Reston, VA.

Varma, A. H., and Zhang, K. (2009), "Design of Non-Compact and Slender Concrete Filled Members," Bowen Laboratory Report No. 2009-01, School of Civil Engineering, Purdue University, West Lafayette, Indiana.

Vian, D., and Bruneau, M. (2005), "Steel Plate Walls for Seismic Design and Retrofit of Building Structures," Technical Report MCEER-05-0010, Multidisciplinary Center for Earthquake Engineering Research, State University of New York at Buffalo, Buffalo, NY.

Vian D., Bruneau, M., Tsai, K.C., and Lin, Y.C. (2009a), "Special Perforated Steel Plate Shear Walls with Reduced Beam Section Anchor Beams I: Experimental Investigation," *Journal of Structural Engineering*, ASCE, Vol. 135, No. 3, pp. 211–220.

Vian D., Bruneau, M., and Purba, R. (2009b), "Special Perforated Steel Plate Shear Walls with Reduced Beam Section Anchor Beams II: Analysis and Design Recommendations," *Journal of Structural Engineering*, ASCE, Vol. 135, No. 3, pp. 221–228.

Viest, I.M., Colaco, J.P., Furlong, R.W., Griffis, L.G., Leon, R.T., and Wyllie, L.A., Jr. (1997), *Composite Construction: Design for Buildings*, McGraw-Hill/ASCE, Reston, VA.

Wada, A., Connor, J., Kawai, H., Iwata, M., and Watanabe, A. (1994), "Damage Tolerant Structure," *ATC-15-4 Proceedings of Fifth US-Japan Workshop on the Improvement of Building Structural Design and Construction Practices*, pp. 27–39, Applied Technology Council, Redwood City, CA.

Wada, A., Saeki, E., Takeuchi, T., and Watanabe, A. (1998), "Development of Unbonded Brace," Nippon Steel's Unbonded Braces (promotional document), pp. 1–16, Nippon Steel Corporation Building Construction and Urban Development Division, Tokyo, Japan.

Wallace, B.J. and Krawinkler, H. (1985), *Small-Scale Model Experimentation on Steel Assemblies*, John A. Blume Earthquake Engineering Center Report No. 75, Stanford University Department of Civil Engineering, Palo Alto, CA.

Watanabe, A., Hitomi Y., Saeki, E., Wada, A., and Fujimoto, M. (1988), "Properties of Brace Encased in Buckling-Restraining Concrete and Steel Tube," *Proceedings of Ninth World Conference on Earthquake Engineering*, Vol. IV, pp. 719–724, Japan Association for Earthquake Disaster Prevention, Tokyo-Kyoto, Japan.

Whittaker, A.S., Uang, C.-M., and Bertero, V.V. (1987), *Earthquake Simulation Tests and Associated Studies of a 0.3 – Scale Model of a Six-Story Eccentrically Braced Steel Structure*, Report No. UBC/EERC-87/02, Earthquake Engineering Research Center, Berkeley, CA.

Xu, P. and Goel, S.C. (1990), *Behavior of Double Channel Bracing Members Under Large Cyclic Deformations*, Report No. UMCE 90-1, University of Michigan Department of Civil Engineering, Ann Arbor, MI.

Yamanouchi, H., Nishiyama, I., and Kobayashi, J. (1998), "Development and Usage of Composite and Hybrid Building Structure in Japan," *ACI SP-174*, American Concrete Institute, pp. 151–174.

Yang, C. S., Leon, R. T., and DesRoches, R. (2008), "Design and behavior of zipper-braced frames," *Engineering Structures,* Elsevier Ltd., Vol. 30, Issue 4, pp. 1092–1100.

Yang, F. and Mahin, S. (2005), "Limiting Net Section Fracture in Slotted Tube Braces," *Steel TIPS Report,* Structural Steel Education Council, Moraga, CA.

Zandonini, R. and Leon, R.T. (1992), "Composite Connections," *Constructional Steel Design: An International Guide*, pp. 501–522, Elsevier Science Publishers, London, England.

Zaremba, C.J. (1988), "Strength of Steel Frames Using Partial Composite Girders," *Journal of Structural Engineering*, ASCE, Vol. 114, No. 8, August, pp. 1741–1760, Reston, VA.

Zhao, Q. and Astaneh-Asl, A. (2004), "Cyclic Behavior of Traditional and Innovative Composite Shear Walls," *Journal of Structural Engineering*, ASCE, Vol. 130, No. 2, pp. 271–284.

ANSI/AISC 358-10
ANSI/AISC 358s1-11
An American National Standard

Prequalified Connections for Special and Intermediate Steel Moment Frames for Seismic Applications

Including Supplement No. 1

2010
(includes 2011 supplement)

Supersedes ANSI/AISC 358-05 and ANSI/AISC 358s1-09

Approved by the AISC Connection Prequalification Review Panel
and issued by the AISC Board of Directors

AMERICAN INSTITUTE OF STEEL CONSTRUCTION
One East Wacker Drive, Suite 700
Chicago, Illinois 60601-1802

AISC © 2011

by

American Institute of Steel Construction

All rights reserved. This book or any part thereof must not be reproduced in any form without the written permission of the publisher.

The AISC logo is a registered trademark of AISC and is used under license.

The information presented in this publication has been prepared in accordance with recognized engineering principles and is for general information only. While it is believed to be accurate, this information should not be used or relied upon for any specific application without competent professional examination and verification of its accuracy, suitability and applicability by a licensed professional engineer, designer or architect. The publication of the material contained herein is not intended as a representation or warranty, on the part of the American Institute of Steel Construction or of any other person named herein, that this information is suitable for any general or particular use or of freedom from infringement of any patent or patents. Anyone making use of this information assumes all liability arising from such use.

Caution must be exercised when relying upon other specifications and codes developed by other bodies and incorporated by reference herein since such material may be modified or amended from time to time subsequent to the printing of this edition. The Institute bears no responsibility for such material other than to refer to it and incorporate it by reference at the time of the initial publication of this edition.

Printed in the United States of America

PREFACE

(This Preface is not part of ANSI/AISC 358-10, *Prequalified Connections for Special and Intermediate Steel Moment Frames for Seismic Applications*, but is included for informational purposes only.)

This edition of the Standard was developed using a consensus process in concert with the *Specification for Structural Steel Buildings* (ANSI/AISC 360-10) and *Seismic Provisions for Structural Steel Buildings* (ANSI/AISC 341-10). This Standard is incorporated by reference in the *Seismic Provisions*.

The most significant modification to this edition of the Standard is the incorporation of three prequalified connections into the body of the standard that were previously part of a supplement (Chapters 7, 8, and 9). It is anticipated that future supplements will be issued as additional moment connections become prequalified.

This printing includes Supplement No. 1 to *Prequalified Connections for Special and Intermediate Steel Moment Frames for Seismic Applications* (ANSI/AISC 358s1-11) which consists of the material in Chapter 10. The Supplement was finalized shortly after the completion of this Standard.

The Symbols, Glossary, and Appendices to this Standard are an integral part of the Standard. A non-mandatory Commentary has been prepared to provide background for the provisions of the Standard and the user is encouraged to consult it. Additionally, non-mandatory User Notes are interspersed throughout the Standard to provide concise and practical guidance in the application of the provisions.

The reader is cautioned that professional judgment must be exercised when data or recommendations in this Standard are applied, as described more fully in the disclaimer notice preceding the Preface.

This Standard was approved by the AISC Connection Prequalification Review Panel (CPRP):

Ronald O. Hamburger, Chairman
Kevin Moore, Vice-Chairman
Richard Apple
Scott F. Armbrust
Michael L. Cochran
Theodore L. Droessler
Michael D. Engelhardt
Louis F. Geschwindner
Gary Glenn
W. Steven Hofmeister
Gregory H. Lynch

Brett R. Manning
Thomas M. Murray
Charles W. Roeder
Paul A. Rouis
Thomas A. Sabol
Robert E. Shaw, Jr.
James A. Stori
James A. Swanson
Chia-Ming Uang
Behnam Yousefi
Keith A. Grubb, Secretary

The CPRP gratefully acknowledges the following corresponding members and staff for their contributions to this document:

Scott M. Adan	Raymond Kitasoe
Leigh Arber	Roberto T. Leon
David Bleiman	James O. Malley
Helen Chen	Duane K. Miller
Charles J. Carter	Steven E. Pryor
Cynthia J. Duncan	Gian Andrea Rassati
Willian C. Gibb	James M. Ricles
Leonard M. Joseph	C. Mark Saunders

TABLE OF CONTENTS

SYMBOLS 9.2–xiii
GLOSSARY 9.2–xix

STANDARD

CHAPTER 1. GENERAL 9.2–1
    1.1.   Scope 9.2–1
    1.2.   References 9.2–1
    1.3.   General 9.2–2

CHAPTER 2. DESIGN REQUIREMENTS 9.2–3
    2.1.   Special and Intermediate Moment Frame Connection Types 9.2–3
    2.2.   Connection Stiffness 9.2–3
    2.3.   Members 9.2–3
        1.   Rolled Wide-Flange Members 9.2–3
        2.   Built-up Members 9.2–3
        2a.   Built-up Beams 9.2–4
        2b.   Built-up Columns 9.2–4
    2.4.   Connection Design Parameters 9.2–5
        1.   Resistance Factors 9.2–5
        2.   Plastic Hinge Location 9.2–5
        3.   Probable Maximum Moment at Plastic Hinge 9.2–6
        4.   Continuity Plates 9.2–6
    2.5.   Panel Zones 9.2–6
    2.6.   Protected Zone 9.2–6

CHAPTER 3. WELDING REQUIREMENTS 9.2–7
    3.1.   Filler Metals 9.2–7
    3.2.   Welding Procedures 9.2–7
    3.3.   Backing at Beam-to-Column and Continuity
           Plate-to-Column Joints 9.2–7
        1.   Steel Backing at Continuity Plates 9.2–7
        2.   Steel Backing at Beam Bottom Flange 9.2–7
        3.   Steel Backing at Beam Top Flange 9.2–7
        4.   Prohibited Welds at Steel Backing 9.2–8
        5.   Nonfusible Backing at Beam Flange-to-Column Joints 9.2–8
    3.4.   Details and Treatment of Weld Tabs 9.2–8
    3.5.   Tack Welds 9.2–8
    3.6.   Continuity Plates 9.2–9
    3.7.   Quality Control and Quality Assurance 9.2–9

Prequalified Connections for Special and Intermediate Steel Moment Frames
for Seismic Applications, 2010, incl. Supplement No. 1

CHAPTER 4. BOLTING REQUIREMENTS 9.2–10
    4.1.   Fastener Assemblies 9.2–10
    4.2.   Installation Requirements 9.2–10
    4.3.   Quality Control and Quality Assurance 9.2–10

CHAPTER 5. REDUCED BEAM SECTION (RBS)
MOMENT CONNECTION 9.2–11
    5.1   General 9.2–11
    5.2   Systems 9.2–11
    5.3.   Prequalification Limits 9.2–11
        1.   Beam Limitations 9.2–11
        2.   Column Limitations 9.2–12
    5.4.   Column-Beam Relationship Limitations 9.2–13
    5.5.   Beam Flange-to-Column Flange Weld Limitations 9.2–13
    5.6.   Beam Web-to-Column Flange Connection Limitations 9.2–13
    5.7.   Fabrication of Flange Cuts 9.2–14
    5.8.   Design Procedure 9.2–15

CHAPTER 6. BOLTED UNSTIFFENED AND STIFFENED EXTENDED
END-PLATE MOMENT CONNECTIONS 9.2–19
    6.1.   General 9.2–19
    6.2.   Systems 9.2–19
    6.3.   Prequalification Limits 9.2–20
    6.4.   Beam Limitations 9.2–20
    6.5.   Column Limitations 9.2–21
    6.6.   Column-Beam Relationship Limitations 9.2–22
    6.7.   Continuity Plates 9.2–22
    6.8.   Bolts 9.2–22
    6.9.   Connection Detailing 9.2–22
        1.   Gage 9.2–22
        2.   Pitch and Row Spacing 9.2–22
        3.   End-Plate Width 9.2–24
        4.   End-Plate Stiffener 9.2–24
        5.   Finger Shims 9.2–25
        6.   Composite Slab Detailing for IMF 9.2–25
        7.   Welding Details 9.2–26
    6.10. Design Procedure 9.2–27
        1.   End-Plate and Bolt Design 9.2–27
        2.   Column-Side Design 9.2–33

CHAPTER 7. BOLTED FLANGE PLATE (BFP)
MOMENT CONNECTION 9.2–39
    7.1.   General 9.2–39
    7.2.   Systems 9.2–39

7.3.	Prequalification Limits	9.2–40
	1. Beam Limitations	9.2–40
	2. Column Limitations	9.2–40
7.4.	Column-Beam Relationship Limitations	9.2–41
7.5.	Connection Detailing	9.2–41
	1. Plate Material Specifications	9.2–41
	2. Beam Flange Plate Welds	9.2–41
	3. Single-Plate Shear Connection Welds	9.2–41
	4. Bolt Requirements	9.2–41
	5. Flange Plate Shims	9.2–42
7.6.	Design Procedure	9.2–42

CHAPTER 8. WELDED UNREINFORCED FLANGE-WELDED WEB (WUF-W) MOMENT CONNECTION 9.2–46

8.1.	General	9.2–46
8.2.	Systems	9.2–46
8.3.	Prequalification Limits	9.2–46
	1. Beam Limitations	9.2–46
	2. Column Limitations	9.2–47
8.4.	Column-Beam Relationship Limitations	9.2–48
8.5.	Beam Flange-to-Column Flange Welds	9.2–48
8.6.	Beam Web-to-Column Connection Limitations	9.2–48
8.7.	Design Procedure	9.2–50

CHAPTER 9. KAISER BOLTED BRACKET (KBB) MOMENT CONNECTION 9.2–52

9.1.	General	9.2–52
9.2.	Systems	9.2–53
9.3.	Prequalification Limits	9.2–53
	1. Beam Limitations	9.2–53
	2. Column Limitations	9.2–54
	3. Bracket Limitations	9.2–55
9.4.	Column-Beam Relationship Limitations	9.2–55
9.5.	Bracket-to-Column Flange Limitations	9.2–55
9.6.	Bracket-to-Beam Flange Connection Limitations	9.2–56
9.7.	Beam Web-to-Column Connection Limitations	9.2–57
9.8.	Connection Detailing	9.2–57
9.9.	Design Procedure	9.2–58

CHAPTER 10. CONXTECH CONXL MOMENT CONNECTION 9.2–64

10.1	General	9.2–64
10.2	Systems	9.2–67
10.3	Prequalification Limits	9.2–68
	1. Beam Limitations	9.2–68

		2.	Column Limitations 9.2–68
		3.	Collar Limitations 9.2–69

10.4	Collar Connection Limitations 9.2–69
10.5	Beam Web-to-Collar Connection Limitations 9.2–71
10.6	Beam-Flange-to-Collar Flange Welding Limitations 9.2–71
10.7	Column-Beam Relationship Limitations 9.2–72
10.8	Design Procedure 9.2–72
10.9	Part Drawings 9.2–77

APPENDIX A. CASTING REQUIREMENTS 9.2–84

A1.	Cast Steel Grade 9.2–84
A2.	Quality Control (QC) 9.2–84

	1.	Inspection and Nondestructive Testing Personnel 9.2–84
	2.	First Article Inspection (FAI) of Castings 9.2–84
	3.	Visual Inspection of Castings 9.2–84
	4.	Nondestructive Testing (NDT) of Castings 9.2–84
	4a.	Procedures 9.2–84
	4b.	Required NDT 9.2–85
	5.	Weld Repair Procedures 9.2–85
	6.	Tensile Requirements 9.2–85
	7.	Charpy V-Notch (CVN) Requirements 9.2–85
	8.	Casting Identification 9.2–85

A3.	Manufacturer Documents 9.2–86

	1.	Submittal to Patent Holder 9.2–86
	2.	Submittal to Engineer of Record and Authority Having Jurisdiction 9.2–86

APPENDIX B. FORGING REQUIREMENTS 9.2–87

B1.	Forged Steel Grade 9.2–87
B2.	Bar Stock 9.2–87
B3.	Forging Temperature 9.2–87
B4.	Heat Treatment 9.2–87
B5.	Finish 9.2–87
B6.	Quality Assurance 9.2–87
B7.	Documentation 9.2–88

COMMENTARY

INTRODUCTION 9.2–89

CHAPTER 1. GENERAL 9.2–90

1.1.	Scope 9.2–90
1.2.	References 9.2–91
1.3.	General 9.2–91

CHAPTER 2. DESIGN REQUIREMENTS ... 9.2–92
2.1. Special and Intermediate Moment Frame Connection Types ... 9.2–92
2.3. Members ... 9.2–92
    2. Built-up Members ... 9.2–92
    2b. Built-up Columns ... 9.2–93
2.4. Connection Design Parameters ... 9.2–94
    1. Resistance Factors ... 9.2–94
    2. Plastic Hinge Location ... 9.2–94
    3. Probable Maximum Moment at Plastic Hinge ... 9.2–94
    4. Continuity Plates ... 9.2–94

CHAPTER 3. WELDING REQUIREMENTS ... 9.2–97
3.3. Backing at Beam-to-Column and Continuity Plate-to-Column Joints ... 9.2–97
    1. Steel Backing at Continuity Plates ... 9.2–97
    2. Steel Backing at Beam Bottom Flange ... 9.2–97
    3. Steel Backing at Beam Top Flange ... 9.2–98
    4. Prohibited Welds at Steel Backing ... 9.2–98
    5. Nonfusible Backing at Beam Flange-to-Column Joints ... 9.2–98
3.4. Details and Treatment of Weld Tabs ... 9.2–98
3.5. Tack Welds ... 9.2–99
3.6. Continuity Plates ... 9.2–99
3.7. Quality Control and Quality Assurance ... 9.2–100

CHAPTER 4. BOLTING REQUIREMENTS ... 9.2–101
4.1. Fastener Assemblies ... 9.2–101
4.2. Installation Requirements ... 9.2–101
4.3. Quality Control and Quality Assurance ... 9.2–101

CHAPTER 5. REDUCED BEAM SECTION (RBS) MOMENT CONNECTION ... 9.2–102
5.1 General ... 9.2–102
5.2 Systems ... 9.2–103
5.3. Prequalification Limits ... 9.2–103
    1. Beam Limitations ... 9.2–103
    2. Column Limitations ... 9.2–104
5.4. Column-Beam Relationship Limitations ... 9.2–106
5.5. Beam Flange-to-Column Flange Weld Limitations ... 9.2–106
5.6. Beam Web-to-Column Flange Connection Limitations ... 9.2–106
5.7. Fabrication of Flange Cuts ... 9.2–107
5.8. Design Procedure ... 9.2–107

CHAPTER 6. BOLTED UNSTIFFENED AND STIFFENED EXTENDED END-PLATE MOMENT CONNECTIONS ... 9.2–110
6.1. General ... 9.2–110

6.2.	Systems	9.2–110
6.3.	Prequalification Limits	9.2–111
6.4.	Beam Limitations	9.2–111
6.5.	Column Limitations	9.2–111
6.6.	Column-Beam Relationship Limitations	9.2–112
6.7.	Continuity Plates	9.2–112
6.8.	Bolts	9.2–112
6.9.	Connection Detailing	9.2–112
6.10.	Design Procedure	9.2–113

CHAPTER 7. BOLTED FLANGE PLATE (BFP) MOMENT CONNECTION 9.2–114

7.1.	General	9.2–114
7.2.	Systems	9.2–116
7.3.	Prequalification Limits	9.2–116
	1. Beam Limitations	9.2–116
	2. Column Limitations	9.2–117
7.4.	Column-Beam Relationship Limitations	9.2–117
7.5.	Connection Detailing	9.2–117
7.6.	Design Procedure	9.2–118

CHAPTER 8. WELDED UNREINFORCED FLANGE-WELDED WEB (WUF-W) MOMENT CONNECTION 9.2–121

8.1.	General	9.2–121
8.3.	Prequalification Limits	9.2–123
8.4.	Column-Beam Relationship Limitations	9.2–123
8.5.	Beam Flange-to-Column Flange Welds	9.2–124
8.6.	Beam Web-to-Column Connection Limitations	9.2–124
8.7.	Design Procedure	9.2–125

CHAPTER 9. KAISER BOLTED BRACKET (KBB) MOMENT CONNECTION 9.2–127

9.1.	General	9.2–127
9.2.	Systems	9.2–127
9.3.	Prequalification Limits	9.2–128
	1. Beam Limitations	9.2–128
	2. Column Limitations	9.2–128
	3. Bracket Limitations	9.2–129
9.4.	Column-Beam Relationship Limitations	9.2–129
9.5.	Bracket-to-Column Flange Limitations	9.2–129
9.6.	Bracket-to-Beam Flange Connection Limitations	9.2–131
9.7.	Beam Web-to-Column Connection Limitations	9.2–131
9.8.	Connection Detailing	9.2–131
9.9.	Design Procedure	9.2–132

CHAPTER 10. CONXTECH CONXL MOMENT CONNECTION 9.2–136
    10.1   General 9.2–136
    10.2   Systems 9.2–138
    10.3   Prequalification Limits 9.2–138
            1.   Beam Limitations 9.2–138
            2.   Column Limitations 9.2–139
            3.   Collar Limitations 9.2–139
    10.4   Collar Connection Limitations 9.2–139
    10.5   Beam Web-to-Collar Connection Limitations 9.2–140
    10.6   Beam Flange-to-Collar Flange Welding Limitations 9.2–140
    10.7   Column-Beam Relationship Limitations 9.2–140
    10.8   Design Procedure 9.2–140

APPENDIX A. CASTING REQUIREMENTS 9.2–143
    A1.   Cast Steel Grade 9.2–143
    A2.   Quality Control (QC) 9.2–143
            2.   First Article Inspection (FAI) of Castings 9.2–143
            3.   Visual Inspection of Castings 9.2–143
            4.   Nondestructive Testing (NDT) of Castings 9.2–143
            6.   Tensile Requirements 9.2–144
    A3.   Manufacturer Documents 9.2–145

APPENDIX B. FORGING REQUIREMENTS 9.2–146

REFERENCES 9.2–147

SYMBOLS

The Standard uses the following symbols in addition to the terms defined in the *Specification for Structural Steel Buildings* (ANSI/AISC 360-10) and the *Seismic Provisions for Structural Steel Buildings* (ANSI/AISC 341-10). Some definitions in the list below have been simplified in the interest of brevity. In all cases, the definitions given in the body of the Standard govern. Symbols without text definitions, used in only one location and defined at that location, are omitted in some cases. The section or table number on the right refers to where the symbol is first used.

Symbol	Definition	Section
A_c	Contact areas between the continuity plate and the column flanges that have attached beam flanges, in.^{2} (mm^2)	6.7
A_c	Area of concrete in the column, in.^{2} (mm^2)	10.8
A_s	Area of steel in the column, in.^{2} (mm^2)	10.8
C_{pr}	Factor to account for peak connection strength, including strain hardening, local restraint, additional reinforcement, and other connection conditions	2.4.3
C_t	Factor used in Equation 6.10-17	6.10
F_{EXX}	Filler metal classification strength, ksi (MPa)	9.9
F_{fu}	Factored beam flange force, kips (N)	6.10
F_{nt}	Nominal tensile strength of bolt from the AISC *Specification*, ksi (MPa)	6.10
F_{nv}	Nominal shear strength of bolt from the AISC *Specification*, ksi (MPa)	6.10
F_{su}	Required stiffener strength, kips (N)	6.10
F_u	Specified minimum tensile strength of the yielding element, ksi (MPa)	10.8
F_{ub}	Specified minimum tensile strength of beam material, ksi (MPa)	7.6
F_{uf}	Specified minimum tensile strength of flange material, ksi (MPa)	9.9
F_{up}	Specified minimum tensile strength of end-plate material, ksi (MPa)	6.10
F_{up}	Specified minimum tensile strength of plate material, ksi (MPa)	7.6
F_w	Nominal weld design strength per the AISC *Specification*, ksi (MPa)	9.9
F_y	Specified minimum yield stress of the yielding element, ksi (MPa)	10.8
F_{yb}	Specified minimum yield stress of beam material, ksi (MPa)	6.10
F_{yc}	Specified minimum yield stress of column flange material, ksi (MPa)	6.10
F_{yf}	Specified minimum yield stress of flange material, ksi (MPa)	9.9
F_{yp}	Specified minimum yield stress of end-plate material, ksi (MPa)	6.10
F_{ys}	Specified minimum yield stress of stiffener material, ksi (MPa)	6.10
H_l	Height of the story below the node, in. (mm)	10.8

Symbol	Definition	Section
H_u	Height of the story above the node, in. (mm)	10.8
L_{bb}	Length of bracket, in. (mm)	Table 9.1
L_h	Distance between *plastic hinge locations*, in. (mm)	5.8
L_{st}	Length of end plate stiffener, in. (mm)	6.9.4
M^*_{pb}	Moment at the intersection of the beam and column centerlines determined by projecting the beam maximum developed moments from the column face, kip-in. (N-mm)	10.8
M^*_{pc}	Moment at the beam and column centerline determined by projecting the sum of the nominal column plastic moment strength, reduced by the axial stress, kip-in. (N-mm)	10.8
M^*_{pcl}	Plastic moment nominal strength of the column below the node, about the axis under consideration considering simultaneous axial loading and loading about the transverse axis, kip-in. (N-mm)	10.8
M^*_{pcu}	Plastic moment nominal strength of the column above the node, about the axis under consideration considering simultaneous axial loading and loading about the transverse axis, kip-in (N-mm)	10.8
M_{bolts}	Moment at collar bolts, kip-in. (N-mm)	10.8
M_{cf}	Column flange flexural strength, kip-in. (N-mm)	6.9
M_f	Probable maximum moment at face of column, kip-in. (N-mm)	5.8
M_{np}	Moment without prying action in the bolts, kip-in. (N-mm)	Table 6.2
M_{pe}	Plastic moment of beam based on expected yield stress, kip-in. (N-mm)	5.8
M_{pr}	Probable maximum moment at plastic hinge, kip-in. (N-mm)	2.4.3
M_{uv}	Additional moment due to shear amplification from the center of the reduced beam section to the centerline of the column, kip-in. (N-mm)	5.4(2)(a)
N	Thickness of beam flange plus 2 times the *reinforcing fillet* weld size, in. (mm)	6.10
P	Axial load acting on the column at the section under consideration in accordance with the applicable load combination specified by the building code, but not considering amplified seismic load, kips (N)	10.8
P_t	Minimum specified tensile strength of bolt, kips (N)	Table 6.2
P_{uc}	Required compressive strength using LRFD load combinations, kips (N)	Symbols
R_{pt}	Minimum bolt pretension, kips (N)	10.8
R_n	Required force for continuity plate design, kips (N)	6.10
R_n	Nominal strength	7.6
R_n^{pz}	Nominal panel zone shear strength, kips (N)	10.8
R_t	Ratio of the expected tensile strength to the specified minimum tensile strength for flange material	9.9
R_u^{pz}	Required panel zone shear strength, kips (N)	10.8
R_y	Ratio of the expected yield stress to the specified minimum yield stress, F_y,	9.9

SYMBOLS

Symbol	Definition	Section
S_1	Distance from face of the column to the nearest row of bolts, in. (mm)	7.6
S_h	Distance from face of column to the plastic hinge, in. (mm)	2.3.2a
V_{bolts}	Probable maximum shear at collar bolts, kips (N)	10.8
V_{cf}	Probable maximum shear at face of collar flange, kips (N)	10.8
V_{col}	Column shear, kips (N)	10.8
V_f	Probable maximum shear at the face of column, kips (N)	10.8
$V_{gravity}$	Beam shear force resulting from $1.2D + f_1L + 0.2S$, kips (N)	5.8
V_h	Beam shear force at *plastic hinge location*, kips (N)	7.6
V_{RBS}	Larger of the two values of shear force at the center of the reduced beam section at each end of a beam, kips (N)	5.4(2)(a)
V'_{RBS}	Smaller of the two values of shear force at the center of the reduced beam section at each end of a beam, kips (N)	Commentary 5.8
V_u	Required shear strength of beam and beam web-to-column connection, kips (N)	5.8
Y_c	Column flange yield line mechanism parameter, in. (mm)	Table 6.5
Y_m	Simplified column flange yield-line mechanism parameter	9.9
Y_p	End-plate yield line mechanism parameter, in. (mm)	Table 6.2
Z_c	Plastic section modulus of the column about either axis, in.^{3} (mm^3)	10.8
Z_e	Effective plastic modulus of the section (or connection) at the location of a plastic hinge, in.^{3} (mm^3)	2.4.3
Z_{RBS}	Plastic section modulus at the center of the reduced beam section, in.^{3} (mm^3)	5.8
Z_x	Plastic section modulus about the x-axis, in.^{3} (mm^3)	5.8
a	Horizontal distance from face of column flange to the start of a reduced beam section cut, in. (mm)	5.4(2)(a)
a	Distance from the outside face of the collar to the reduced beam section cut, in. (mm)	10.8
b	Width of compression element as defined in the AISC *Specification*, in. (mm)	2.3.2b(2)
b	Length of a reduced beam section cut, in. (mm)	5.4(2)(a)
b_{bb}	Width of bracket, in. (mm)	Table 9.1
b_{bf}	Width of beam flange, in. (mm)	5.8
b_{cf}	Width of column flange, in. (mm)	9.9
b_f	Width of flange, in. (mm)	5.3.1(7)
b_p	Width of end plate, in. (mm)	Table 6.1
c	Depth of cut at center of the reduced beam section, in. (mm)	5.8
d	Depth of beam, in. (mm)	5.3.1(7)
d_b	Diameter of column flange bolts, in. (mm)	9.9
$d_{b\,req'd}$	Required bolt diameter, in. (mm)	6.10
d_c	Depth of column, in. (mm)	5.4(2)(a)
d_{col}	Depth of the column, in. (mm)	10.8
d_e	Column bolt edge distance, in. (mm)	Table 9.2
d_{eff}	Effective depth of beam, calculated as the centroidal distance between bolt groups in the upper and lower brackets, in. (mm)	9.9

Symbol	Definition	Section
d_{leg}^{CC}	Effective depth of the collar corner assembly leg, in. (mm)	10.8
f'_c	Specified compressive strength of the concrete fill, ksi (MPa)	10.8
f_1	Load factor determined by the applicable building code for live loads but not less than 0.5	5.8
g	Horizontal distance (gage) between fastener lines, in. (mm)	Table 6.1
h_1	Distance from the centerline of a compression flange to the tension-side inner bolt rows in four-bolt extended and four-bolt stiffened extended end-plate moment connections, in. (mm)	Table 6.2
h_{bb}	Height of bracket, in. (mm)	Table 9.1
h_i	Distance from centerline of compression flange to the centerline of the ith tension bolt row, in. (mm)	6.10
h_o	Distance from centerline of compression flange to the tension-side outer bolt row in four-bolt extended and four-bolt stiffened extended end-plate moment connections, in. (mm)	Table 6.2
h_p	Height of plate, in. (mm)	8.6
h_{st}	Height of stiffener, in. (mm)	6.9.4
k_1	Distance from web center line to flange toe of fillet, in. (mm)	3.6
k_c	Distance from outer face of a column flange to web toe of fillet (design value) or fillet weld, in. (mm)	6.9
k_{det}	Largest value of k_1 used in production, in. (mm)	3.6
l	Bracket overlap distance, in. (mm)	9.9
l_w	Length of available fillet weld, in. (mm)	9.9
l_w^{CC}	Total length of available fillet weld at collar corner assembly, in. (mm)	10.8
l_w^{CWX}	Total length of available fillet weld at collar web extension, in. (mm)	10.8
n	Number of bolts	7.6
n_b	Number of bolts at compression flange	6.10
n_{cb}	Number of column bolts	Table 9.1
n_{cf}	Number of collar bolts per collar flange	10.8
n_i	Number of inner bolts	6.10
n_o	Number of outer bolts	6.10
p	Perpendicular tributary length per bolt, in. (mm)	9.9
p_b	Vertical distance between the inner and outer row of bolts in an eight-bolt stiffened extended end-plate moment connection, in. (mm)	Table 6.1
p_b	Column bolt pitch, in. (mm)	Table 9.2
p_{fi}	Vertical distance from the inside of a beam tension flange to the nearest inside bolt row, in. (mm)	Table 6.1
p_{fo}	Vertical distance from the outside of a beam tension flange to the nearest outside bolt row, in. (mm)	Table 6.1
p_{si}	Distance from the inside face of continuity plate to the nearest inside bolt row, in. (mm)	6.9.2
p_{so}	Distance from the outside face of continuity plate to the nearest outside bolt row, in. (mm)	6.9.2

Symbol	Definition	Section
r_h	Radius of horizontal bracket, in. (mm)	Table 9.2
r_{ut}	Required collar bolt tension strength, kips (N)	10.8
r_v	Radius of bracket stiffener, in. (mm)	Table 9.2
s	Distance from the centerline of the most inside or most outside tension bolt row to the edge of a yield line pattern, in. (mm)	Table 6.2
s	Spacing of bolt rows in a bolted flange plate moment connection, in. (mm)	7.6
s_{bolts}	Distance from center of plastic hinge to the centroid of the collar bolts, in. (mm)	10.8
s_f	Distance from center of plastic hinge to face of column, in. (mm)	10.8
s_h	Distance from center of plastic hinge to center of column, in. (mm)	10.8
t_{bf}	Thickness of beam flange, in. (mm)	5.8
t_{bw}	Thickness of beam web, in. (mm)	6.10
t_{col}	Wall thickness of HSS or built-up box column, in. (mm)	10.8
t_{collar}	Distance from the face of the column to the outside face of the collar, in. (mm)	10.8
t_{cw}	Thickness of column web, in. (mm)	6.10
t_f^{CC}	Fillet weld size required to join collar corner assembly to column, in. (mm)	10.8
t_f^{CWX}	Fillet weld size required to join each side of the beam web to the collar web extension, in. (mm)	10.8
t_{leg}^{CC}	Effective thickness of the collar corner assembly leg, in. (mm)	10.8
t_p	Thickness of plate, in. (mm)	Table 6.1
t_s	Thickness of stiffener, in. (mm)	6.10
w	Minimum size of fillet weld, in. (mm)	Table 9.2
w	Uniform beam gravity load, kips per linear ft (N per linear mm)	Commentary 5.8
w_u	Distributed load on the beam, kips/ft (N/mm), using the load combination $1.2D + f_1 L + 0.2S$	10.8
ϕ_d	Resistance factor for ductile limit states	2.4.1
ϕ_n	Resistance factor for nonductile limit states	2.4.1

GLOSSARY

The Standard uses the following terms in addition to the terms defined in the 2010 AISC *Specification for Structural Steel Buildings* and the 2010 AISC *Seismic Provisions for Structural Steel Buildings*. Terms defined in this Glossary are *italicized* in the Glossary and where they first appear within a section or long paragraph throughout the Standard.

Air carbon arc cutting. Process of cutting steel by the heat from an electric arc applied simultaneously with an air jet.

Backing. Piece of metal or other material, placed at the weld *root* to facilitate placement of the *root* pass.

Backgouge. Process of removing by grinding or *air carbon arc cutting* all or a portion of the *root* pass of a complete-joint-penetration groove weld, from the reverse side of a joint from which a *root* was originally placed.

Cascaded weld ends. Method of terminating a weld in which subsequent weld beads are stopped short of the previous bead, producing a cascade effect.

Concrete structural slab. Reinforced concrete slab or concrete fill on steel deck with a total thickness of 3 in. (75 mm) or greater and a concrete compressive strength in excess of 2,000 psi (14 MPa).

Nonfusible backing. Backing material that will not fuse with the base metals during the welding process.

Plastic hinge location. Location in a column-beam assembly where inelastic energy dissipation is assumed to occur through the development of plastic flexural straining.

Probable maximum moment at the plastic hinge. Expected moment developed at a *plastic hinge location* along a member, considering the probable (mean) value of the material strength for the specified steel and effects of strain hardening.

Reinforcing fillet. Fillet weld applied to a groove welded T-joint to obtain a contour to reduce stress concentrations associated with joint geometry.

Root. Portion of a multi-pass weld deposited in the first pass of welding.

Thermal cutting. Group of cutting processes that severs or removes metal by localized melting, burning or vaporizing of the workpiece.

Weld tab. Piece of metal affixed to the end of a welded joint to facilitate the initiation and termination of weld passes outside the structural joint.

CHAPTER 1

GENERAL

1.1. SCOPE

This Standard specifies design, detailing, fabrication and quality criteria for connections that are prequalified in accordance with the AISC *Seismic Provisions for Structural Steel Buildings* (herein referred to as the AISC *Seismic Provisions*) for use with special moment frames (SMF) and intermediate moment frames (IMF). The connections contained in this Standard are prequalified to meet the requirements in the AISC *Seismic Provisions* only when designed and constructed in accordance with the requirements of this Standard. Nothing in this Standard shall preclude the use of connection types contained herein outside the indicated limitations, nor the use of other connection types, when satisfactory evidence of qualification in accordance with the AISC *Seismic Provisions* is presented to the authority having jurisdiction.

1.2. REFERENCES

The following standards form a part of this Standard to the extent that they are referenced and applicable:

American Institute of Steel Construction (AISC)

ANSI/AISC 341-10 *Seismic Provisions for Structural Steel Buildings* (herein referred to as the AISC *Seismic Provisions*)
ANSI/AISC 360-10 *Specification for Structural Steel Buildings* (herein referred to as the AISC *Specification*)

ASTM International (ASTM)

A354-07a *Standard Specification for Quenched and Tempered Alloy Steel Bolts, Studs, and Other Externally Threaded Fasteners*
A370-09 *Standard Test Methods and Definitions for Mechanical Testing of Steel Products*
A488/A488M-10 *Standard Practice for Steel Castings, Welding, Qualifications of Procedures and Personnel*
A574-11 *Standard Specification for Alloy Steel Socket Head Cap Screws*
A609/A609M-91(2007) *Standard Practice for Castings, Carbon, Low-Alloy, and Martensitic Stainless Steel, Ultrasonic Examination Thereof*
A781/A781M-11 *Standard Specification for Castings, Steel and Alloy, Common Requirements, for General Industrial Use*
A802/A802M-95(2010) *Standard Practice for Steel Castings, Surface Acceptance Standards, Visual Examination*
A903/A903M-99(2007) *Standard Specification for Steel Castings, Surface Acceptance Standards, Magnetic Particle and Liquid Penetrant Inspection*

A958/A958M-10 *Standard Specification for Steel Castings, Carbon and Alloy, with Tensile Requirements, Chemical Requirements Similar to Standard Wrought Grades*

B19-10 *Standard Specification for Cartridge Brass Sheet, Strip, Plate, Bar, and Disks*

B36/B36M *Standard Specification for Brass Plate, Sheet, Strip, And Rolled Bar*

E186-10 *Standard Reference Radiographs for Heavy Walled (2 to $4^1/_2$ in. (50.8 to 114 mm)) Steel Castings*

E446-10 *Standard Reference Radiographs for Steel Castings Up to 2 in. (50.8 mm) in Thickness*

E709-08 *Standard Guide for Magnetic Particle Examination*

American Welding Society (AWS)

AWS D1.1/D1.1M-2010 *Structural Welding Code—Steel*
AWS D1.8/D1.8M-2009 *Structural Welding Code—Seismic Supplement*

Manufacturers Standardization Society (MSS)

MSS SP-55-2006 *Quality Standard for Steel Castings for Valves, Flanges and Fittings and Other Piping Components—Visual Method for Evaluation of Surface Irregularities*

Research Council on Structural Connections (RCSC)

Specification for Structural Joints using High-Strength Bolts, 2009 (herein referred to as the RCSC *Specification*)

1.3. GENERAL

All design, materials and workmanship shall conform to the requirements of the AISC *Seismic Provisions* and this Standard. The connections contained in this Standard shall be designed according to the Load and Resistance Factor Design (LRFD) provisions. Connections designed according to this Standard are permitted to be used in structures designed according to the LRFD or Allowable Strength Design (ASD) provisions of the AISC *Seismic Provisions*.

CHAPTER 2
DESIGN REQUIREMENTS

2.1. SPECIAL AND INTERMEDIATE MOMENT FRAME CONNECTION TYPES

The connection types listed in Table 2.1 are prequalified for use in connecting beams to column flanges in special moment frames (SMF) and intermediate moment frames (IMF) within the limitations specified in this Standard.

2.2. CONNECTION STIFFNESS

All connections contained in this Standard shall be considered fully restrained (Type FR) for the purpose of seismic analysis.

2.3. MEMBERS

The connections contained in this Standard are prequalified in accordance with the requirements of the AISC *Seismic Provisions* when used to connect members meeting the limitations of Sections 2.3.1 or 2.3.2, as applicable.

1. Rolled Wide-Flange Members

Rolled wide-flange members shall conform to the cross section profile limitations applicable to the specific connection in this Standard.

2. Built-up Members

Built-up members having a doubly symmetric, I-shaped cross section shall meet the following requirements:

(1) Flanges and webs shall have width, depth and thickness profiles similar to rolled wide-flange sections meeting the profile limitations for wide-flange sections applicable to the specific connection in this Standard, and

(2) Webs shall be continuously connected to flanges in accordance with the requirements of Sections 2.3.2a or 2.3.2b, as applicable.

TABLE 2.1.
Prequalified Moment Connections

Connection Type	Chapter	Systems
Reduced beam section (RBS)	5	SMF, IMF
Bolted unstiffened extended end plate (BUEEP)	6	SMF, IMF
Bolted stiffened extended end plate (BSEEP)	6	SMF, IMF
Bolted flange plate (BFP)	7	SMF, IMF
Welded unreinforced flange-welded web (WUF-W)	8	SMF, IMF
Kaiser bolted bracket (KBB)	9	SMF, IMF
ConXtech ConXL moment connection (ConXL)	10	SMF, IMF

2a. Built-up Beams

The web and flanges shall be connected using complete-joint-penetration (CJP) groove welds with a pair of *reinforcing fillet* welds within a zone extending from the beam end to a distance not less than one beam depth beyond the *plastic hinge location*, S_h, unless specifically indicated in this Standard. The minimum size of these fillet welds shall be the lesser of $5/16$ in. (8 mm) and the thickness of the beam web.

Exception: This provision shall not apply where individual connection prequalifications specify other requirements.

2b. Built-up Columns

Built-up columns shall conform to the provisions of subsections (1) through (4), as applicable. Built-up columns shall satisfy the requirements of the AISC *Specification* except as modified in this Section. Transfer of all internal forces and stresses between elements of the built-up column shall be through welds.

(1) I-Shaped Columns

The elements of built-up I-shaped columns shall conform to the requirements of the AISC *Seismic Provisions*.

Within a zone extending from 12 in. (300 mm) above the upper beam flange to 12 in. (300 mm) below the lower beam flange, unless specifically indicated in this Standard, the column webs and flanges shall be connected using CJP groove welds with a pair of reinforcing fillet welds. The minimum size of the fillet welds shall be the lesser of $5/16$ in. (8 mm) and the thickness of the column web.

(2) Boxed Wide-Flange Columns

The wide-flange shape of a boxed wide-flange column shall conform to the requirements of the AISC *Seismic Provisions*.

The width-to-thickness ratio, b/t, of plates used as flanges shall not exceed, $0.6\sqrt{E/F_y}$, where b shall be taken as not less than the clear distance between plates.

The width-to-thickness ratio, h/t_w, of plates used only as webs shall conform to the requirements of the AISC *Seismic Provisions*.

Within a zone extending from 12 in. (300 mm) above the upper beam flange to 12 in. (300 mm) below the lower beam flange, flange and web plates of boxed wide-flange columns shall be joined by CJP groove welds. Outside this zone, plate elements shall be continuously connected by fillet or groove welds.

(3) Built-up Box Columns

The width-to-thickness ratio, b/t, of plates used as flanges shall not exceed, $0.6\sqrt{E/F_y}$, where b shall be taken as not less than the clear distance between web plates.

The width-to-thickness ratio, h/t_w, of plates used only as webs shall conform to the requirements of the AISC *Seismic Provisions*.

Within a zone extending from 12 in. (300 mm) above the upper beam flange to 12 in. (300 mm) below the lower beam flange, flange and web plates of box columns shall be joined by CJP groove welds. Outside this zone, box column web and flange plates shall be continuously connected by fillet welds or groove welds.

(4) Flanged Cruciform Columns

The elements of flanged cruciform columns, whether fabricated from rolled shapes or built up from plates, shall meet the requirements of the AISC *Seismic Provisions*.

Within a zone extending from 12 in. (300 mm) above the upper beam flange to 12 in. (300 mm) below the lower beam flange, the web of the tee-shaped sections shall be welded to the web of the continuous I-shaped section with CJP groove welds with a pair of reinforcing fillet welds. The minimum size of fillet welds shall be the lesser of $5/16$ in. (8 mm) or the thickness of the column web. Continuity plates shall conform to the requirements for wide-flange columns.

2.4. CONNECTION DESIGN PARAMETERS

1. Resistance Factors

Where available strengths are calculated in accordance with the AISC *Specification*, the resistance factors specified therein shall apply. When available strengths are calculated in accordance with this Standard, the resistance factors ϕ_d and ϕ_n shall be used as specified in the applicable section of this Standard. The values of ϕ_d and ϕ_n shall be taken as follows:

(a) For ductile limit states:
$\phi_d = 1.00$

(b) For nonductile limit states:
$\phi_n = 0.90$

2. Plastic Hinge Location

The distance of the plastic hinge from the face of the column, S_h, shall be taken in accordance with the requirements for the individual connection as specified herein.

3. **Probable Maximum Moment at Plastic Hinge**

The *probable maximum moment at the plastic hinge* shall be:

$$M_{pr} = C_{pr} R_y F_y Z_e \quad (2.4.3\text{-}1)$$

where

M_{pr} = probable maximum moment at plastic hinge, kip-in. (N-mm)
R_y = ratio of the expected yield stress to the specified minimum yield stress F_y as specified in the AISC *Seismic Provisions*
Z_e = effective plastic section modulus of the section (or connection) at the location of the plastic hinge, in.^{3} (mm^3)
C_{pr} = factor to account for the peak connection strength, including strain hardening, local restraint, additional reinforcement, and other connection conditions. Unless otherwise specifically indicated in this Standard, the value of C_{pr} shall be:

$$C_{pr} = \frac{F_y + F_u}{2F_y} \leq 1.2 \quad (2.4.3\text{-}2)$$

where

F_y = specified minimum yield stress of the yielding element, ksi (MPa)
F_u = specified minimum tensile strength of the yielding element, ksi (MPa)

4. **Continuity Plates**

Beam flange continuity plates shall be provided in accordance with the AISC *Seismic Provisions*.

Exceptions:
1. For bolted end-plate connections, continuity plates shall be provided in accordance with Section 6.7.
2. For the Kaiser bolted bracket connection, the provisions of Chapter 9 shall apply. When continuity plates are required by Chapter 9, thickness and detailing shall be in accordance with the AISC *Seismic Provisions*.

2.5. PANEL ZONES

Panel zones shall conform to the requirements of the AISC *Seismic Provisions*.

2.6. PROTECTED ZONE

The protected zone shall be as defined for each prequalified connection. Unless otherwise specifically indicated in this Standard, the protected zone of the beam shall be defined as the area from the face of the column flange to one-half of the beam depth beyond the plastic hinge. The protected zone shall meet the requirements of the AISC *Seismic Provisions*, except as indicated in this Standard. Bolt holes in beam webs, when detailed in accordance with the individual connection provisions of this Standard, shall be permitted.

CHAPTER 3

WELDING REQUIREMENTS

3.1. FILLER METALS

Filler metals shall conform to the requirements of the AISC *Seismic Provisions*.

3.2. WELDING PROCEDURES

Welding procedures shall be in accordance with the AISC *Seismic Provisions*.

3.3. BACKING AT BEAM-TO-COLUMN AND CONTINUITY PLATE-TO-COLUMN JOINTS

1. **Steel Backing at Continuity Plates**

   Steel *backing* used at continuity plate-to-column welds need not be removed. At column flanges, steel backing left in place shall be attached to the column flange using a continuous $^5/_{16}$-in. (8-mm) fillet weld on the edge below the CJP groove weld.

   When backing is removed, the *root* pass shall be *backgouged* to sound weld metal and backwelded with a *reinforcing fillet*. The reinforcing fillet shall be continuous with a minimum size of $^5/_{16}$ in. (8 mm).

2. **Steel Backing at Beam Bottom Flange**

   Where steel backing is used with CJP groove welds between the bottom beam flange and the column, the backing shall be removed.

   Following the removal of steel backing, the root pass shall be backgouged to sound weld metal and backwelded with a reinforcing fillet. The size of the reinforcing fillet leg adjacent to the column flange shall be a minimum of $^5/_{16}$ in. (8 mm), and the reinforcing fillet leg adjacent to the beam flange shall be such that the fillet toe is located on the beam flange base metal.

   Exception: If the base metal and weld root are ground smooth after removal of the backing, the reinforcing fillet adjacent to the beam flange need not extend to base metal.

3. **Steel Backing at Beam Top Flange**

   Where steel backing is used with CJP groove welds between the top beam flange and the column, and the steel backing is not removed, the steel backing shall be attached to the column by a continuous $^5/_{16}$-in. (8-mm) fillet weld on the edge below the CJP groove weld.

4. **Prohibited Welds at Steel Backing**

   Backing at beam flange-to-column flange joints shall not be welded to the underside of the beam flange, nor shall tack welds be permitted at this location. If fillet welds or tack welds are placed between the backing and the beam flange in error, they shall be repaired as follows:

   (1) The weld shall be removed such that the fillet weld or tack weld no longer attaches the backing to the beam flange.

   (2) The surface of the beam flange shall be ground flush and shall be free of defects.

   (3) Any gouges or notches shall be repaired. Repair welding shall be done with E7018 SMAW electrodes or other filler metals meeting the requirements of Section 3.1 for demand critical welds. A special welding procedure specification (WPS) is required for this repair. Following welding, the repair weld shall be ground smooth.

5. **Nonfusible Backing at Beam Flange-to-Column Joints**

   Where *nonfusible backing* is used with CJP groove welds between the beam flanges and the column, the backing shall be removed and the root backgouged to sound weld metal and backwelded with a reinforcing fillet. The size of the reinforcing fillet leg adjacent to the column shall be a minimum of $^5/_{16}$ in. (8 mm) and the reinforcing fillet leg adjacent to the beam flange shall be such that the fillet toe is located on the beam flange base metal.

   Exception: If the base metal and weld root are ground smooth after removal of the backing, the reinforcing fillet adjacent to the beam flange need not extend to base metal.

3.4. DETAILS AND TREATMENT OF WELD TABS

Where used, *weld tabs* shall be removed to within $^1/_8$ in. (3 mm) of the base metal surface and the end of the weld finished, except at continuity plates where removal to within $^1/_4$ in. (6 mm) of the plate edge shall be permitted. Removal shall be by *air carbon arc cutting* (CAC-A), grinding, chipping or *thermal cutting*. The process shall be controlled to minimize errant gouging. The edges where weld tabs have been removed shall be finished to a surface roughness of 500 μ-in. (13 microns) or better. The contour of the weld end shall provide a smooth transition to adjacent surfaces, free of notches, gouges and sharp corners. Weld defects greater than $^1/_{16}$ in. (1.5 mm) deep shall be excavated and repaired by welding in accordance with an applicable WPS. Other weld defects shall be removed by grinding, faired to a slope not greater than 1:5.

3.5. TACK WELDS

In the protected zone, tack welds attaching backing and weld tabs shall be placed where they will be incorporated into a final weld.

3.6. CONTINUITY PLATES

Along the web, the corner clip shall be detailed so that the clip extends a distance of at least $1^1/_2$ in. (38 mm) beyond the published k_{det} dimension for the rolled shape. Along the flange, the plate shall be clipped to avoid interference with the fillet radius of the rolled shape and shall be detailed so that the clip does not exceed a distance of $^1/_2$ in. (12 mm) beyond the published k_1 dimension. The clip shall be detailed to facilitate suitable weld terminations for both the flange weld and the web weld. When a curved corner clip is used, it shall have a minimum radius of $^1/_2$ in. (12 mm).

At the end of the weld adjacent to the column web/flange juncture, weld tabs for continuity plates shall not be used, except when permitted by the engineer of record. Unless specified to be removed by the engineer of record, weld tabs shall not be removed when used in this location.

Where continuity plate welds are made without weld tabs near the column fillet radius, weld layers shall be permitted to be transitioned at an angle of 0° to 45° measured from the vertical plane. The effective length of the weld shall be defined as that portion of the weld having full size. Nondestructive testing (NDT) shall not be required on the tapered or transition portion of the weld not having full size.

3.7. QUALITY CONTROL AND QUALITY ASSURANCE

Quality control and quality assurance shall be in accordance with the AISC *Seismic Provisions*.

CHAPTER 4

BOLTING REQUIREMENTS

4.1. FASTENER ASSEMBLIES

Bolts shall be pretensioned high-strength bolts conforming to ASTM A325/A325M, A490/A490M, F1852 or F2280, unless other fasteners are permitted by a specific connection.

4.2. INSTALLATION REQUIREMENTS

Installation requirements shall be in accordance with AISC *Seismic Provisions* and the RCSC *Specification*, except as otherwise specifically indicated in this Standard.

4.3. QUALITY CONTROL AND QUALITY ASSURANCE

Quality control and quality assurance shall be in accordance with the AISC *Seismic Provisions*.

CHAPTER 5

REDUCED BEAM SECTION (RBS) MOMENT CONNECTION

5.1. GENERAL

In a reduced beam section (RBS) moment connection (Figure 5.1), portions of the beam flanges are selectively trimmed in the region adjacent to the beam-to-column connection. Yielding and hinge formation are intended to occur primarily within the reduced section of the beam.

5.2. SYSTEMS

RBS connections are prequalified for use in special moment frame (SMF) and intermediate moment frame (IMF) systems within the limits of these provisions.

5.3. PREQUALIFICATION LIMITS

1. Beam Limitations

Beams shall satisfy the following limitations:

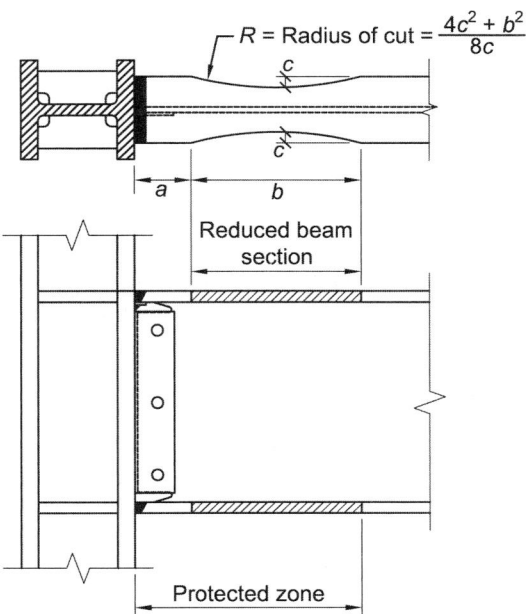

Fig. 5.1. Reduced beam section connection.

(1) Beams shall be rolled wide-flange or built-up I-shaped members conforming to the requirements of Section 2.3.

(2) Beam depth is limited to W36 (W920) for rolled shapes. Depth of built-up sections shall not exceed the depth permitted for rolled wide-flange shapes.

(3) Beam weight is limited to 300 lb/ft (447 kg/m).

(4) Beam flange thickness is limited to $1^3/_4$ in. (44 mm).

(5) The clear span-to-depth ratio of the beam shall be limited as follows:

  (a) For SMF systems, 7 or greater.

  (b) For IMF systems, 5 or greater.

(6) Width-to-thickness ratios for the flanges and web of the beam shall conform to the requirements of the AISC *Seismic Provisions*.

  When determining the width-to-thickness ratio of the flange, the value of b_f shall not be taken as less than the flange width at the ends of the center two-thirds of the reduced section provided that gravity loads do not shift the location of the plastic hinge a significant distance from the center of the reduced beam section.

(7) Lateral bracing of beams shall be provided in conformance with the AISC *Seismic Provisions*. Supplemental lateral bracing shall be provided near the reduced section in conformance with the AISC *Seismic Provisions* for lateral bracing provided adjacent to the plastic hinges.

  When supplemental lateral bracing is provided, its attachment to the beam shall be located no greater than $d/2$ beyond the end of the reduced beam section farthest from the face of the column, where d is the depth of the beam. No attachment of lateral bracing shall be made to the beam in the region extending from the face of the column to the end of the reduced section farthest from the face of the column.

  Exception: For both systems, where the beam supports a *concrete structural slab* that is connected between the protected zones with welded shear connectors spaced a maximum of 12 in. (300 mm) on center, supplemental top and bottom flange bracing at the reduced section is not required.

(8) The protected zone shall consist of the portion of beam between the face of the column and the end of the reduced beam section cut farthest from the face of the column.

2. **Column Limitations**

Columns shall satisfy the following limitations:

(1) Columns shall be any of the rolled shapes or built-up sections permitted in Section 2.3.

(2) The beam shall be connected to the flange of the column.

(3) Rolled shape column depth shall be limited to W36 (W920) maximum. The depth of built-up wide-flange columns shall not exceed that for rolled shapes. Flanged cruciform columns shall not have a width or depth greater than the depth allowed for rolled shapes. Built-up box-columns shall not have a width or depth exceeding 24 in. (610 mm). Boxed wide-flange columns shall not have a width or depth exceeding 24 in. (610 mm) if participating in orthogonal moment frames.

(4) There is no limit on the weight per foot of columns.

(5) There are no additional requirements for flange thickness.

(6) Width-to-thickness ratios for the flanges and web of columns shall conform to the requirements of the AISC *Seismic Provisions*.

(7) Lateral bracing of columns shall conform to the requirements of the AISC *Seismic Provisions*.

5.4. COLUMN-BEAM RELATIONSHIP LIMITATIONS

Beam-to-column connections shall satisfy the following limitations:

(1) Panel zones shall conform to the requirements of the AISC *Seismic Provisions*.

(2) Column-beam moment ratios shall be limited as follows:
    (a) For SMF systems, the column-beam moment ratio shall conform to the requirements of the AISC *Seismic Provisions*. The value of ΣM^*_{pb} shall be taken equal to $\Sigma(M_{pr} + M_{uv})$, where M_{pr} is computed according to Equation 5.8-5, and where M_{uv} is the additional moment due to shear amplification from the center of the reduced beam section to the centerline of the column. M_{uv} can be computed as $V_{RBS}(a + b/2 + d_c/2)$, where V_{RBS} is the shear at the center of the reduced beam section computed per Step 4 of Section 5.8, a and b are the dimensions shown in Figure 5.1, and d_c is the depth of the column.
    (b) For IMF systems, the column-beam moment ratio shall conform to the requirements of the AISC *Seismic Provisions*.

5.5. BEAM FLANGE-TO-COLUMN FLANGE WELD LIMITATIONS

Beam flange to column flange connections shall satisfy the following limitations:

(1) Beam flanges shall be connected to column flanges using complete-joint-penetration (CJP) groove welds. Beam flange welds shall conform to the requirements for demand critical welds in the AISC *Seismic Provisions*.

(2) Weld access hole geometry shall conform to the requirements of the AISC *Specification*.

5.6. BEAM WEB-TO-COLUMN FLANGE CONNECTION LIMITATIONS

Beam web to column flange connections shall satisfy the following limitations:

(1) The required shear strength of the beam web connection shall be determined according to Equation 5.8-9.

(2) Web connection details shall be limited as follows:

(a) For SMF systems, the beam web shall be connected to the column flange using a CJP groove weld extending between weld access holes. The single plate shear connection shall extend between the weld access holes as shown in Figure 5.1. The single plate shear connection shall be permitted to be used as *backing* for the CJP groove weld. The thickness of the plate shall be at least $^3/_8$ in. (10 mm). *Weld tabs* are not required at the ends of the CJP groove weld at the beam web. Bolt holes in the beam web for the purpose of erection are permitted.

(b) For IMF systems, the beam web shall be connected to the column flange as required for SMF systems.

Exception: For IMF, it is permitted to connect the beam web to the column flange using a bolted single plate shear connection. The bolted single plate shear connection shall be designed as a slip-critical connection, with the design slip resistance per bolt determined according to the AISC *Specification*. For seismic loading, the nominal bearing strength at bolt holes shall not be taken greater than the value given by Equation J3-6a of the AISC *Specification*. The design shear strength of the single plate shear connection shall be determined based on shear yielding of the gross section and on shear rupture of the net section. The plate shall be welded to the column flange with a CJP groove weld, or with fillet welds on both sides of the plate. The minimum size of the fillet weld on each side of the plate shall be 75% of the thickness of the plate. Standard holes shall be provided in the beam web and in the plate, except that short-slotted holes (with the slot parallel to the beam flanges) may be used in either the beam web or in the plate, but not in both. Bolts are permitted to be pretensioned either before or after welding.

5.7. FABRICATION OF FLANGE CUTS

The reduced beam section shall be made using *thermal cutting* to produce a smooth curve. The maximum surface roughness of the thermally cut surface shall be 500 μ-in. (13 microns) in accordance with ANSI B46.1, as measured using AWS C4.1–77 Sample 4 or a similar visual comparator. All transitions between the reduced beam section and the unmodified beam flange shall be rounded in the direction of the flange length to minimize notch effects due to abrupt transitions. Corners between the reduced section surface and the top and bottom of the flanges shall be ground to remove sharp edges, but a minimum chamfer or radius is not required.

Thermal cutting tolerances shall be plus or minus $^1/_4$ in. (6 mm) from the theoretical cut line. The beam effective flange width at any section shall have a tolerance of plus or minus $^3/_8$ in. (10 mm).

Gouges and notches that occur in the thermally cut RBS surface may be repaired by grinding if not more than $^1/_4$ in. (6 mm) deep. The gouged or notched area shall be faired in by grinding so that a smooth transition exists, and the total length of the area ground for the transition shall be no less than five times the depth of the removed gouge on each side of the gouge. If a sharp notch exists, the area shall be inspected by magnetic particle testing (MT) after grinding to ensure that the entire depth of notch has been removed. Grinding that increases the depth of the RBS cut more than $1/4$ in. (6 mm) beyond the specified depth of cut is not permitted.

Gouges and notches that exceed $1/4$ in. (6 mm) in depth, but not exceeding $1/2$ in. (12 mm) in depth, and those notches and gouges where repair by grinding would increase the effective depth of the RBS cut beyond tolerance, may be repaired by welding. The notch or gouge shall be removed and ground to provide a smooth root radius of not less than $1/4$ in. (6 mm) in preparation for welding. The repair area shall be preheated to a minimum temperature of 150 °F (65 °C) or the value specified in AWS D1.1/D1.1M, whichever is greater, measured at the location of the weld repair.

Notches and gouges exceeding $1/2$ in. (12 mm) in depth shall be repaired only with a method approved by the engineer of record.

5.8. DESIGN PROCEDURE

Step 1. Choose trial values for the beam sections, column sections and RBS dimensions a, b and c (Figure 5.1) subject to the limits:

$$0.5 b_{bf} \leq a \leq 0.75 b_{bf} \tag{5.8-1}$$

$$0.65 d \leq b \leq 0.85 d \tag{5.8-2}$$

$$0.1 b_{bf} \leq c \leq 0.25 b_{bf} \tag{5.8-3}$$

where
b_{bf} = width of beam flange, in. (mm)
a = horizontal distance from face of column flange to the start of an RBS cut, in. (mm)
b = length of an RBS cut, in. (mm)
c = depth of cut at center of the reduced beam section, in. (mm)
d = depth of beam, in. (mm)

Confirm that the beams and columns are adequate for all load combinations specified by the applicable building code, including the reduced section of the beam, and that the design story drift for the frame complies with applicable limits specified by the applicable building code. Calculation of elastic drift shall consider the effect of the reduced beam section. In lieu of more detailed calculations, effective elastic drifts may be calculated by multiplying elastic drifts based on gross beam sections by 1.1 for flange reductions up to 50% of the beam flange width. Linear interpolation may be used for lesser values of beam width reduction.

Step 2. Compute the plastic section modulus at the center of the reduced beam section:

$$Z_{RBS} = Z_x - 2 c t_{bf} (d - t_{bf}) \tag{5.8-4}$$

where

Z_{RBS} = plastic section modulus at center of the reduced beam section, in.^{3} (mm^3)
Z_x = plastic section modulus about the x-axis, for full beam cross section, in.^{3} (mm^3)
t_{bf} = thickness of beam flange, in. (mm)

Step 3. Compute the probable maximum moment, M_{pr}, at the center of the reduced beam section:

$$M_{pr} = C_{pr} R_y F_y Z_{RBS} \qquad (5.8\text{-}5)$$

Step 4. Compute the shear force at the center of the reduced beam sections at each end of the beam.

The shear force at the center of the reduced beam sections shall be determined from a free body diagram of the portion of the beam between the centers of the reduced beam sections. This calculation shall assume the moment at the center of each reduced beam section is M_{pr} and shall include gravity loads acting on the beam based on the load combination $1.2D + f_1 L + 0.2S$, where f_1 is the load factor determined by the applicable building code for live loads, but not less than 0.5.

> **User Note:** The load combination of $1.2D + f_1 L + 0.2S$ is in conformance with ASCE/SEI 7. When using the International Building Code, a factor of 0.7 must be used in lieu of the factor of 0.2 when the roof configuration is such that it does not shed snow off of the structure.

Step 5. Compute the probable maximum moment at the face of the column.

The moment at the face of the column shall be computed from a free-body diagram of the segment of the beam between the center of the reduced beam section and the face of the column, as illustrated in Figure 5.2.

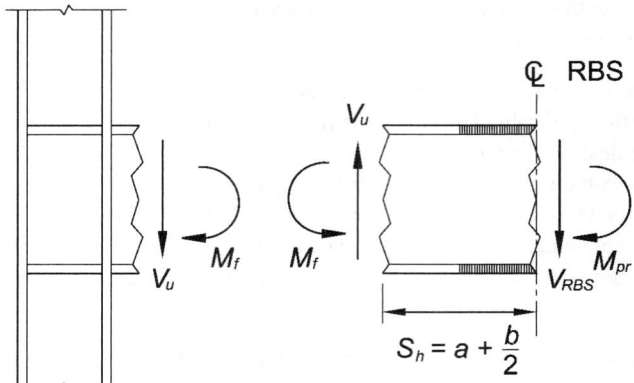

Fig. 5.2. Free-body diagram between center of RBS and face of column.

Based on this free-body diagram, the moment at the face of the column is computed as follows:

$$M_f = M_{pr} + V_{RBS}S_h \tag{5.8-6}$$

where

M_f = probable maximum moment at face of column, kip-in. (N-mm)
S_h = distance from face of the column to the plastic hinge, in. (mm)
  = $a + b/2$, in. (mm)
V_{RBS} = larger of the two values of shear force at the center of the reduced beam section at each end of the beam, kips (N)

Equation 5.8-6 neglects the gravity load on the portion of the beam between the center of the reduced beam section and the face of the column. If desired, the gravity load on this small portion of the beam is permitted to be included in the free-body diagram shown in Figure 5.2 and in Equation 5.8-6.

Step 6. Compute M_{pe}, the plastic moment of the beam based on the expected yield stress:

$$M_{pe} = R_y F_y Z_x \tag{5.8-7}$$

Step 7. Check the flexural strength of the beam at the face of the column:

$$M_f \leq \phi_d M_{pe} \tag{5.8-8}$$

If Equation 5.8-8 is not satisfied, adjust the values of c, a and b, or adjust the section size, and repeat Steps 2 through 7.

Step 8. Determine the required shear strength, V_u, of beam and beam web-to-column connection from:

$$V_u = \frac{2 M_{pr}}{L_h} + V_{gravity} \tag{5.8-9}$$

where

V_u = required shear strength of beam and beam web-to-column connection, kips (N)
L_h = distance between *plastic hinge locations*, in. (mm)
$V_{gravity}$ = beam shear force resulting from $1.2D + f_1L + 0.2S$ (where f_1 is the load factor determined by the applicable building code for live loads, but not less than 0.5), kips (N)

Check design shear strength of beam according to Chapter G of the AISC *Specification*.

User Note: The load combination of $1.2D + f_1L + 0.2S$ is in conformance with ASCE/SEI 7. When using the International Building Code, a factor of 0.7 must be used in lieu of the factor of 0.2 when the roof configuration is such that it does not shed snow off of the structure.

Step 9. Design the beam web-to-column connection according to Section 5.6.

Step 10. Check continuity plate requirements according to Chapter 2.

Step 11. Check column-beam relationship limitations according to Section 5.4.

CHAPTER 6

BOLTED UNSTIFFENED AND STIFFENED EXTENDED END-PLATE MOMENT CONNECTIONS

6.1. GENERAL

Bolted end-plate connections are made by welding the beam to an end-plate and bolting the end-plate to a column flange. The three end-plate configurations shown in Figure 6.1 are covered in this section and are prequalified under the AISC *Seismic Provisions* within the limitations of this Standard.

The behavior of this type of connection can be controlled by a number of different limit states including flexural yielding of the beam section, flexural yielding of the end-plates, yielding of the column panel zone, tension rupture of the end-plate bolts, shear rupture of the end-plate bolts, or rupture of various welded joints. The design criteria provide sufficient strength in the elements of the connections to ensure that the inelastic deformation of the connection is achieved by beam yielding.

6.2. SYSTEMS

Extended end-plate moment connections are prequalified for use in special moment frame (SMF) and intermediate moment frame (IMF) systems.

Exception: Extended end-plate moment connections in SMF systems with *concrete structural slabs* are prequalified only if:

(1) In addition to the limitations of Section 6.3, the nominal beam depth is not less than 24 in. (610 mm);

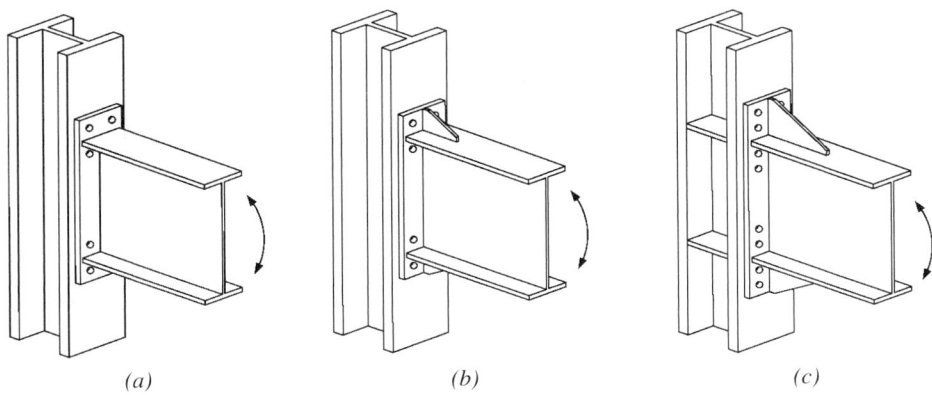

Fig. 6.1. Extended end-plate configurations: (a) four-bolt unstiffened, 4E; (b) four-bolt stiffened, 4ES; (c) eight-bolt stiffened, 8ES.

TABLE 6.1
Parametric Limitations on Prequalification

Parameter	Four-Bolt Unstiffened (4E)		Four-Bolt Stiffened (4ES)		Eight-Bolt Stiffened (8ES)	
	Maximum in. (mm)	Minimum in. (mm)	Maximum in. (mm)	Minimum in. (mm)	Maximum in. (mm)	Minimum in. (mm)
t_{bf}	3/4 (19)	3/8 (10)	3/4 (19)	3/8 (10)	1 (25)	9/16 (14)
b_{bf}	9 1/4 (235)	6 (152)	9 (229)	6 (152)	12 1/4 (311)	7 1/2 (190)
d	55 (1400)	13 3/4 (349)	24 (610)	13 3/4 (349)	36 (914)	18 (457)
t_p	2 1/4 (57)	1/2 (13)	1 1/2 (38)	1/2 (13)	2 1/2 (64)	3/4 (19)
b_p	10 3/4 (273)	7 (178)	10 3/4 (273)	7 (178)	15 (381)	9 (229)
g	6 (152)	4 (102)	6 (152)	3 1/4 (83)	6 (152)	5 (127)
p_{fi}, p_{fo}	4 1/2 (114)	1 1/2 (38)	5 1/2 (140)	1 3/4 (44)	2 (51)	1 5/8 (41)
p_b	—	—	—	—	3 3/4 (95)	3 1/2 (89)

b_{bf} = width of beam flange, in. (mm)
b_p = width of end-plate, in. (mm)
d = depth of connecting beam, in. (mm)
g = horizontal distance between bolts, in. (mm)
p_b = vertical distance between the inner and outer row of bolts in an 8ES connection, in. (mm)
p_{fi} = vertical distance from the inside of a beam tension flange to the nearest inside bolt row, in. (mm)
p_{fo} = vertical distance from the outside of a beam tension flange to the nearest outside bolt row, in. (mm)
t_{bf} = thickness of beam flange, in. (mm)
t_p = thickness of end-plate, in. (mm)

(2) There are no shear connectors within 1.5 times the beam depth from the face of the connected column flange; and

(3) The concrete structural slab is kept at least 1 in. (25 mm) from both sides of both column flanges. It is permitted to place compressible material in the gap between the column flanges and the concrete structural slab.

6.3. PREQUALIFICATION LIMITS

Table 6.1 is a summary of the range of parameters that have been satisfactorily tested. All connection elements shall be within the ranges shown.

6.4. BEAM LIMITATIONS

Beams shall satisfy the following limitations:

(1) Beams shall be rolled wide-flange or built-up I-shaped members conforming to the requirements of Section 2.3. At moment-connected ends of welded built-up sections, within at least the depth of beam or 3 times the width of flange, whichever is less, the beam web and flanges shall be connected using either a complete-joint-penetration (CJP) groove weld or a pair of fillet welds each having a size 75% of the beam web thickness but not less than $^1/_4$ in. (6 mm). For the remainder of the beam, the weld size shall not be less than that required to accomplish shear transfer from the web to the flanges.

(2) Beam depth, d, is limited to values shown in Table 6.1.

(3) There is no limit on the weight per foot of beams.

(4) Beam flange thickness is limited to the values shown in Table 6.1.

(5) The clear span-to-depth ratio of the beam shall be limited as follows:
   (a) For SMF systems, 7 or greater.
   (b) For IMF systems, 5 or greater.

(6) Width-to-thickness ratios for the flanges and web of the beam shall conform to the requirements of the AISC *Seismic Provisions*.

(7) Lateral bracing of beams shall be provided in accordance with the AISC *Seismic Provisions*.

(8) The protected zone shall be determined as follows:
   (a) For unstiffened extended end-plate connections: the portion of beam between the face of the column and a distance equal to the depth of the beam or 3 times the width of the beam flange from the face of the column, whichever is less.
   (b) For stiffened extended end-plate connections: the portion of beam between the face of the column and a distance equal to the location of the end of the stiffener plus one-half the depth of the beam or 3 times the width of the beam flange, whichever is less.

6.5. COLUMN LIMITATIONS

Columns shall satisfy the following limitations:

(1) The end-plate shall be connected to the flange of the column.

(2) Rolled shape column depth shall be limited to W36 (W920) maximum. The depth of built-up wide-flange columns shall not exceed that for rolled shapes. Flanged cruciform columns shall not have a width or depth greater than the depth allowed for rolled shapes.

(3) There is no limit on the weight per foot of columns.

(4) There are no additional requirements for flange thickness.

(5) Width-to-thickness ratios for the flanges and web of the column shall conform to the requirements of the AISC *Seismic Provisions*.

6.6. COLUMN-BEAM RELATIONSHIP LIMITATIONS

Beam-to-column connections shall satisfy the following limitations:

(1) Panel zones shall conform to the requirements of the AISC *Seismic Provisions*.

(2) Column-beam moment ratios shall conform to the requirements of the AISC *Seismic Provisions*.

6.7. CONTINUITY PLATES

Continuity plates shall satisfy the following limitations:

(1) The need for continuity plates shall be determined in accordance with Section 6.10.

(2) When provided, continuity plates shall conform to the requirements of Section 6.10.

(3) Continuity plates shall be attached to columns by welds in accordance with the AISC *Seismic Provisions*.

Exception: Continuity plates less than or equal to $3/8$ in. (10 mm) shall be permitted to be welded to column flanges using double-sided fillet welds. The required strength of the fillet welds shall not be less than $F_y A_c$, where A_c is defined as the contact areas between the continuity plate and the column flanges that have attached beam flanges and F_y is defined as the specified minimum yield stress of the continuity plate.

6.8. BOLTS

Bolts shall conform to the requirements of Chapter 4.

6.9. CONNECTION DETAILING

1. Gage

The gage, g, is as defined in Figures 6.2 through 6.4. The maximum gage dimension is limited to the width of the connected beam flange.

2. Pitch and Row Spacing

The minimum pitch distance is the bolt diameter plus $1/2$ in. (13 mm) for bolts up to 1 in. (25 mm) diameter, and the bolt diameter plus $3/4$ in. (19 mm) for larger diameter bolts. The pitch distances, p_{fi} and p_{fo}, are the distances from the face of the beam flange to the centerline of the nearer bolt row, as shown in Figures 6.2 through 6.4. The pitch distances, p_{si} and p_{so}, are the distances from the face of the continuity plate to the centerline of the nearer bolt row, as shown in Figures 6.2 through 6.4.

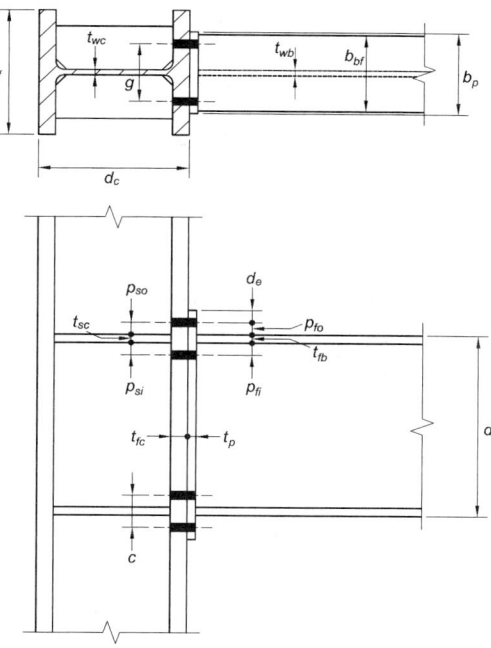

Fig. 6.2. Four-bolt unstiffened extended end-plate (4E) geometry.

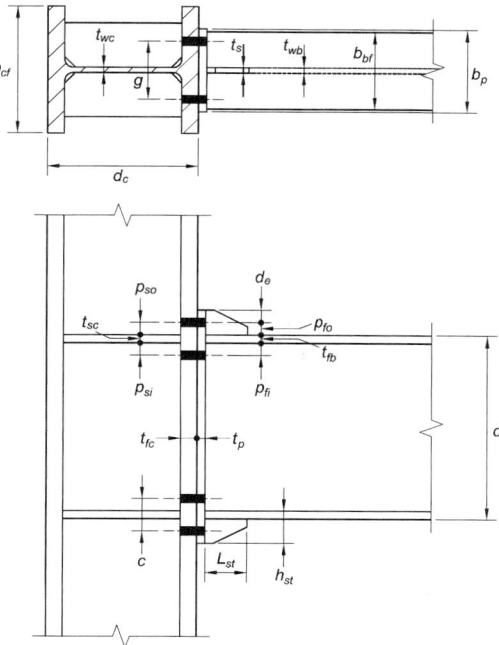

Fig. 6.3. Four-bolt stiffened extended end-plate (4ES) geometry.

The spacing, p_b, is the distance between the inner and outer row of bolts in an 8ES end-plate moment connection and is shown in Figure 6.4. The spacing of the bolt rows shall be at least $2^2/_3$ times the bolt diameter.

> **User Note:** A distance of 3 times the bolt diameter is preferred. The distance must be sufficient to provide clearance for any welds in the region.

3. **End-Plate Width**

   The width of the end-plate shall be greater than or equal to the connected beam flange width. The effective end-plate width shall not be taken greater than the connected beam flange plus 1 in. (25 mm).

4. **End-Plate Stiffener**

   The two extended stiffened end-plate connections, Figures 6.1(b) and (c), require a stiffener welded between the connected beam flange and the end-plate. The minimum stiffener length shall be:

   $$L_{st} = \frac{h_{st}}{\tan 30°} \tag{6.9-1}$$

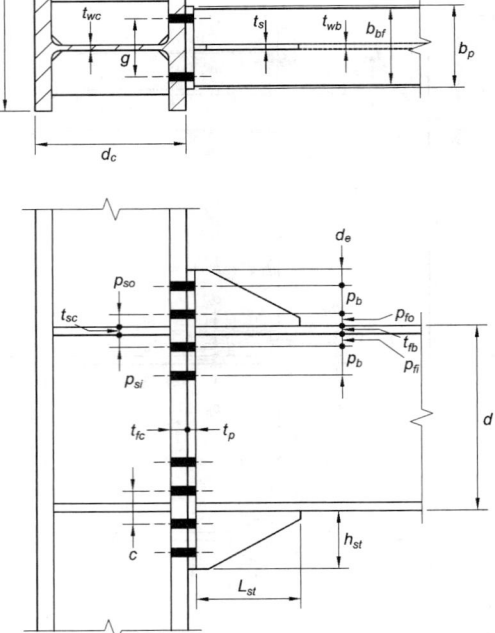

Fig. 6.4. Eight-bolt stiffened extended end-plate (8ES) geometry.

where h_{st} is the height of the stiffener, equal to the height of the end-plate from the outside face of the beam flange to the end of the end-plate as shown in Figure 6.5.

The stiffener plates shall be terminated at the beam flange and at the end of the end-plate with landings approximately 1 in. (25 mm) long. The stiffener shall be clipped where it meets the beam flange and end-plate to provide clearance between the stiffener and the beam flange weld.

When the beam and end-plate stiffeners have the same material strengths, the thickness of the stiffeners shall be greater than or equal to the beam web thickness. If the beam and end-plate stiffener have different material strengths, the thickness of the stiffener shall not be less than the ratio of the beam-to-stiffener plate material yield stresses times the beam web thickness.

5. **Finger Shims**

    The use of finger shims (illustrated in Figure 6.6) at the top and/or bottom of the connection and on either or both sides is permitted, subject to the limitations of the RCSC *Specification*.

6. **Composite Slab Detailing for IMF**

    In addition to the protected zone limitations, welded shear stud connectors shall not be placed along the top flange of the beam for a distance equal to $1^1/_2$ times the depth of the beam, measured from the face of the column.

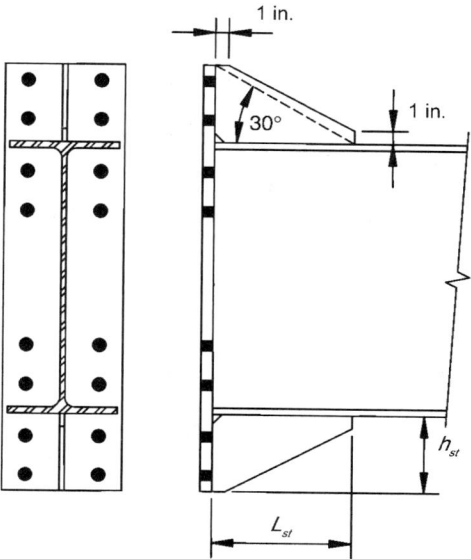

Fig. 6.5. End-plate stiffener layout and geometry for 8ES. Geometry for 4ES similar.

Compressible expansion joint material, at least 1 in. (25 mm) thick, shall be installed between the slab and the column face.

7. **Welding Details**

Welding of the beam to the end-plate shall conform to the following limitations:

(1) Weld access holes shall not be used.

(2) The beam flange to end-plate joint shall be made using a CJP groove weld without *backing*. The CJP groove weld shall be made such that the *root* of the weld is on the beam web side of the flange. The inside face of the flange shall have a 5/16-in. (8-mm) fillet weld. These welds shall be demand critical.

(3) The beam web to end-plate joint shall be made using either fillet welds or CJP groove welds. When used, the fillet welds shall be sized to develop the full strength of the beam web in tension from the inside face of the flange to 6 in. (150 mm) beyond the bolt row farthest from the beam flange.

(4) Backgouging of the root is not required in the flange directly above and below the beam web for a length equal to $1.5k_1$. A full-depth PJP groove weld shall be permitted at this location.

(5) When used, all end-plate-to-stiffener joints shall be made using CJP groove welds.

Exception: When the stiffener is 3/8 in. (10 mm) thick or less, it shall be permitted to use fillet welds that develop the strength of the stiffener.

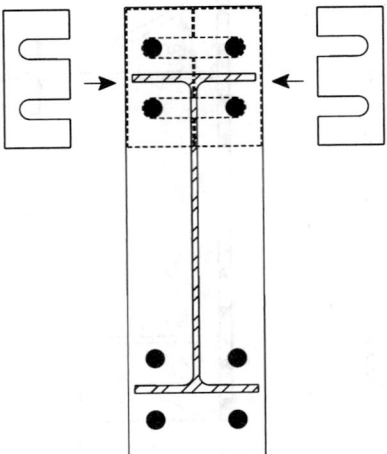

Fig. 6.6. Typical use of finger shims.

6.10. DESIGN PROCEDURE

Connection geometry is shown in Figures 6.2, 6.3 and 6.4 for the 4E, 4ES and 8ES connections, respectively.

1. End-Plate and Bolt Design

Step 1. Determine the sizes of the connected members (beams and column) and compute the moment at the face of the column, M_f.

$$M_f = M_{pr} + V_u S_h \quad (6.10\text{-}1)$$

where

M_{pr} = probable maximum moment at plastic hinge, kip-in. (N-mm), given by Equation 2.4.3-1
S_h = distance from face of column to plastic hinge, in. (mm)
  = the lesser of $d/2$ or $3b_{bf}$ for an unstiffened connection (4E)
  = $L_{st} + t_p$ for a stiffened connection (4ES, 8ES)
V_u = shear force at end of beam, kips (N)

$$= \frac{2M_{pr}}{L_h} + V_{gravity} \quad (6.10\text{-}2)$$

b_{bf} = width of beam flange, in. (mm)
d = depth of connecting beam, in. (mm)
L_h = distance between *plastic hinge locations*, in. (mm)
L_{st} = length of end-plate the stiffener, as shown in Figure 6.5, in. (mm)
t_p = thickness of end-plate, in. (mm)
$V_{gravity}$ = beam shear force resulting from $1.2D + f_1 L + 0.2S$ (where f_1 is a load factor determined by the applicable building code for live loads, but not less than 0.5), kips (N)

> **User Note:** The load combination of $1.2D + f_1 L + 0.2S$ is in conformance with ASCE/SEI 7. When using the International Building Code, a factor of 0.7 must be used in lieu of the factor of 0.2 when the roof configuration is such that it does not shed snow off of the structure.

Step 2. Select one of the three end-plate moment connection configurations and establish preliminary values for the connection geometry (g, p_{fi}, p_{fo}, p_b, g, h_i, etc.) and bolt grade.

Step 3. Determine the required bolt diameter, $d_{b\ req'd}$, using one of the following expressions.

For four-bolt connections (4E, 4ES):

$$d_{b\ req'd} = \sqrt{\frac{2 M_f}{\pi \phi_n F_{nt} (h_o + h_1)}} \quad (6.10\text{-}3)$$

For eight-bolt connections (8ES):

$$d_{b \text{ req'd}} = \sqrt{\frac{2 M_f}{\pi \phi_n F_{nt} (h_1 + h_2 + h_3 + h_4)}} \quad (6.10\text{-}4)$$

where
- F_{nt} = nominal tensile strength of bolt from the AISC *Specification*, ksi (MPa)
- h_i = distance from the centerline of the beam compression flange to the centerline of the *i*th tension bolt row.
- h_o = distance from centerline of compression flange to the tension-side outer bolt row, in. (mm)

Step 4. Select a trial bolt diameter, d_b, not less than that required in Section 6.10.1 Step 3.

Step 5. Determine the required end-plate thickness, $t_{p,\,req'd}$.

$$t_{p,\,req'd} = \sqrt{\frac{1.11 M_f}{\phi_d F_{yp} Y_p}} \quad (6.10\text{-}5)$$

where
- F_{yp} = specified minimum yield stress of the end-plate material, ksi (MPa)
- Y_p = end-plate yield line mechanism parameter from Tables 6.2, 6.3 or 6.4, in. (mm)

Step 6. Select an end-plate thickness, t_p, not less than the required value.

Step 7. Calculate F_{fu}, the factored beam flange force.

$$F_{fu} = \frac{M_f}{d - t_{bf}} \quad (6.10\text{-}6)$$

where
- d = depth of the beam, in. (mm)
- t_{bf} = thickness of beam flange, in. (mm)

Step 8. Check shear yielding of the extended portion of the four-bolt extended unstiffened end-plate (4E):

$$F_{fu}/2 \leq \phi_d R_n = \phi_d (0.6) F_{yp} b_p t_p \quad (6.10\text{-}7)$$

where b_p is the width of the end-plate, in. (mm), to be taken as not greater than the width of the beam flange plus 1 in. (25 mm).

If Equation 6.10-7 is not satisfied, increase the end-plate thickness or increase the yield stress of the end-plate material.

Step 9. Check shear rupture of the extended portion of the end-plate in the four-bolt extended unstiffened end-plate (4E):

$$F_{fu}/2 \leq \phi_n R_n = \phi_n (0.6) F_{up} A_n \quad (6.10\text{-}8)$$

TABLE 6.2
Summary of Four-Bolt Extended Unstiffened End-Plate Yield Line Mechanism Parameter

End-Plate Geometry and Yield Line Pattern	Bolt Force Model
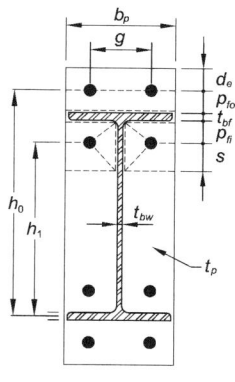	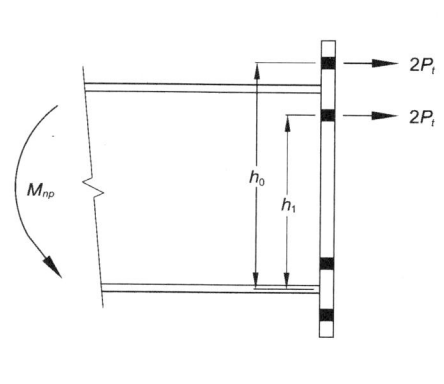

End-Plate	$Y_p = \dfrac{b_p}{2}\left[h_1\left(\dfrac{1}{p_{fi}}+\dfrac{1}{s}\right)+h_0\left(\dfrac{1}{p_{fo}}\right)-\dfrac{1}{2}\right]+\dfrac{2}{g}\left[h_1(p_{fi}+s)\right]$
	$s = \dfrac{1}{2}\sqrt{b_p g}$ Note: If $p_{fi} > s$, use $p_{fi} = s$.

TABLE 6.3
Summary of Four-Bolt Extended Stiffened End-Plate Yield Line Mechanism Parameter

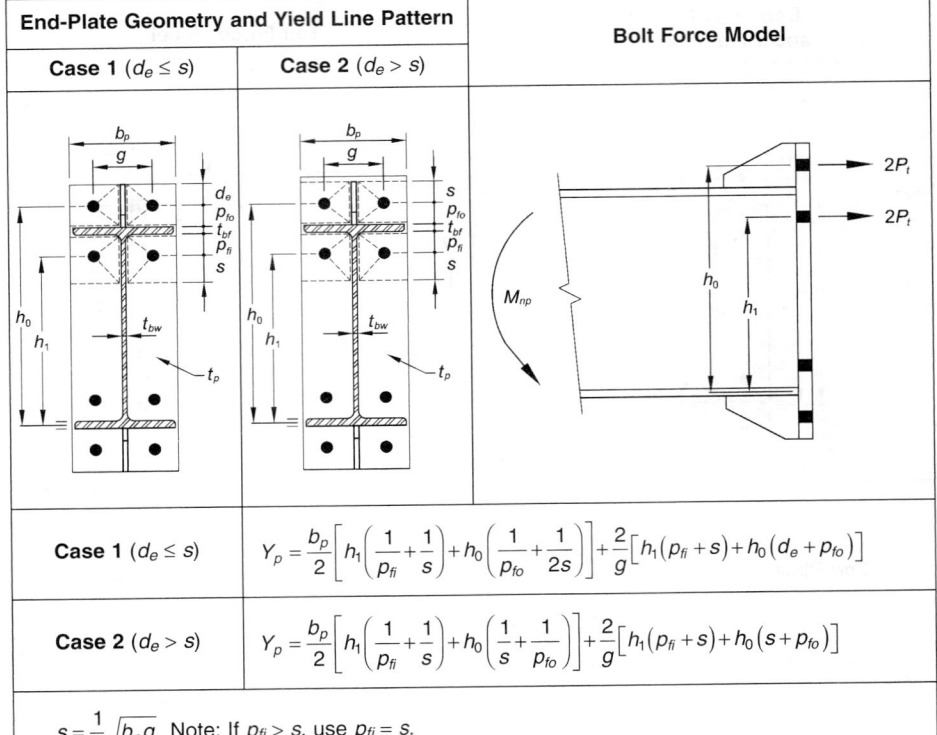

Case 1 ($d_e \leq s$)	$Y_p = \dfrac{b_p}{2}\left[h_1\left(\dfrac{1}{p_{fi}}+\dfrac{1}{s}\right)+h_0\left(\dfrac{1}{p_{fo}}+\dfrac{1}{2s}\right)\right]+\dfrac{2}{g}\left[h_1(p_{fi}+s)+h_0(d_e+p_{fo})\right]$
Case 2 ($d_e > s$)	$Y_p = \dfrac{b_p}{2}\left[h_1\left(\dfrac{1}{p_{fi}}+\dfrac{1}{s}\right)+h_0\left(\dfrac{1}{s}+\dfrac{1}{p_{fo}}\right)\right]+\dfrac{2}{g}\left[h_1(p_{fi}+s)+h_0(s+p_{fo})\right]$

$s = \dfrac{1}{2}\sqrt{b_p g}$ Note: If $p_{fi} > s$, use $p_{fi} = s$.

TABLE 6.4
Summary of Eight-Bolt Extended Stiffened End-Plate Yield Line Mechanism Parameter

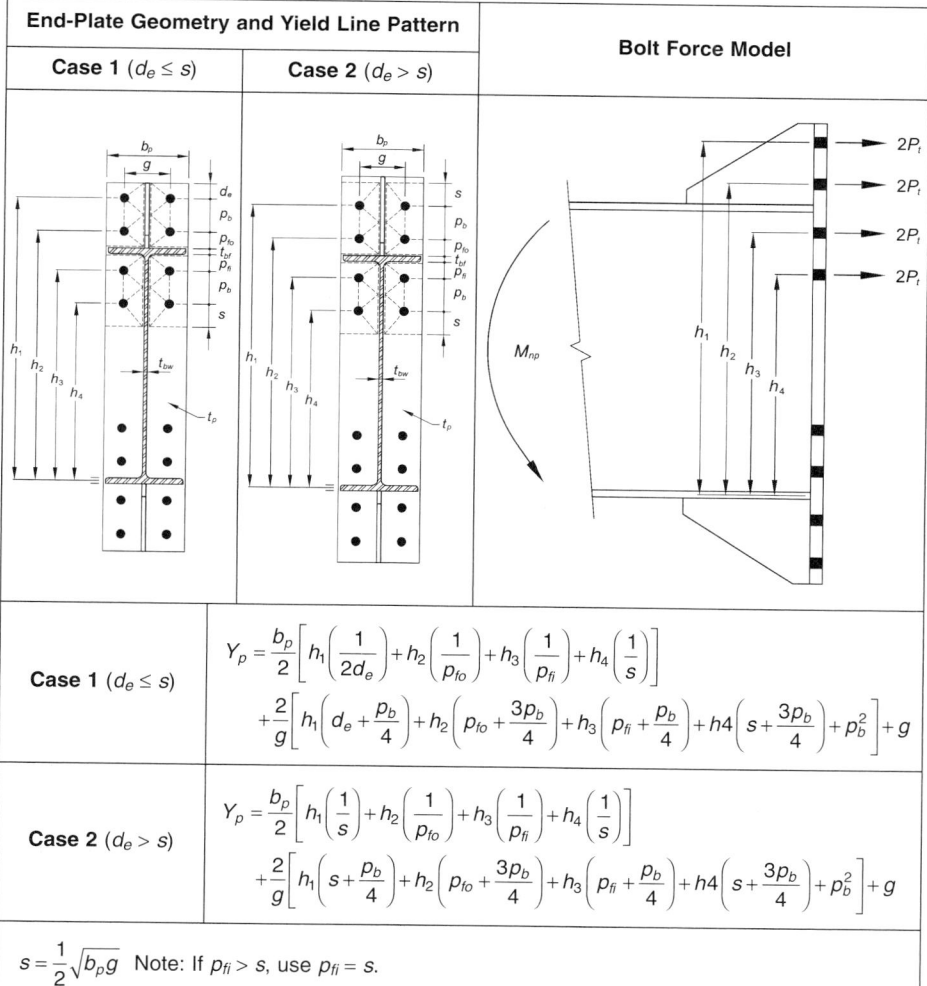

Case 1 ($d_e \leq s$):
$$Y_p = \frac{b_p}{2}\left[h_1\left(\frac{1}{2d_e}\right)+h_2\left(\frac{1}{p_{fo}}\right)+h_3\left(\frac{1}{p_{fi}}\right)+h_4\left(\frac{1}{s}\right)\right]$$
$$+\frac{2}{g}\left[h_1\left(d_e+\frac{p_b}{4}\right)+h_2\left(p_{fo}+\frac{3p_b}{4}\right)+h_3\left(p_{fi}+\frac{p_b}{4}\right)+h4\left(s+\frac{3p_b}{4}\right)+p_b^2\right]+g$$

Case 2 ($d_e > s$):
$$Y_p = \frac{b_p}{2}\left[h_1\left(\frac{1}{s}\right)+h_2\left(\frac{1}{p_{fo}}\right)+h_3\left(\frac{1}{p_{fi}}\right)+h_4\left(\frac{1}{s}\right)\right]$$
$$+\frac{2}{g}\left[h_1\left(s+\frac{p_b}{4}\right)+h_2\left(p_{fo}+\frac{3p_b}{4}\right)+h_3\left(p_{fi}+\frac{p_b}{4}\right)+h4\left(s+\frac{3p_b}{4}\right)+p_b^2\right]+g$$

$s = \frac{1}{2}\sqrt{b_p g}$ Note: If $p_{fi} > s$, use $p_{fi} = s$.

where
- F_{up} = specified minimum tensile stress of end-plate, ksi (MPa)
- A_n = net area of end-plate
  - = $t_p[b_p - 2(d_b + ^1/_8)]$ when standard holes are used, in.²
  - = $t_p[b_p - 2(d_b + 3)]$ when standard holes are used, mm²
- d_b = bolt diameter, in. (mm)

If Equation 6.10-8 is not satisfied, increase the end-plate thickness or increase the yield stress of the end-plate material.

Step 10. If using either the four-bolt extended stiffened end-plate (4ES) or the eight-bolt extended stiffened end-plate (8ES) connection, select the end-plate stiffener thickness and design the stiffener-to-beam flange and stiffener-to-end-plate welds.

$$t_s \geq t_{bw}\left(\frac{F_{yb}}{F_{ys}}\right) \qquad (6.10\text{-}9)$$

where
- t_{bw} = thickness of beam web, in. (mm)
- t_s = end plate stiffener thickness, in. (mm)
- F_{yb} = specified minimum yield stress of beam material, ksi (MPa)
- F_{ys} = specified minimum yield stress of stiffener material, ksi (MPa)

The stiffener geometry shall conform to the requirements of Section 6.9.4. In addition, to prevent local buckling of the stiffener plate, the following width-to-thickness criterion shall be satisfied.

$$\frac{h_{st}}{t_s} \leq 0.56\sqrt{\frac{E}{F_{ys}}} \qquad (6.10\text{-}10)$$

where h_{st} is the height of the stiffener, in. (mm), equal to the height of the end-plate from the outside face of the beam flange to the end of the end-plate.

The stiffener-to-beam-flange and stiffener-to-end-plate welds shall be designed to develop the stiffener plate in shear at the beam flange and in tension at the end-plate. Either fillet or complete-joint-penetration (CJP) groove welds are suitable for the weld of the stiffener plate to the beam flange. CJP groove welds shall be used for the stiffener-to-end-plate weld. If the end-plate is ³/₈ in. (10 mm) thick or less, double-sided fillet welds are permitted.

Step 11. The bolt shear rupture strength of the connection is provided by the bolts at one (compression) flange; thus

$$V_u \leq \phi_n R_n = \phi_n(n_b)F_{nv}A_b \qquad (6.10\text{-}11)$$

where

n_b = number of bolts at the compression flange
  = 4 for 4E and 4ES connections
  = 8 for 8ES connections
A_b = nominal gross area of bolt, in.² (mm²)
F_{nv} = nominal shear strength of bolt from the AISC *Specification*, ksi (MPa)
V_u = shear force at the end of the beam, kips (N), given by Equation 6.10-2

Step 12. Check bolt-bearing/tear-out failure of the end-plate and column flange:

$$V_u \leq \phi_n R_n = \phi_n(n_i)r_{ni} + \phi_n(n_o)r_{no} \tag{6.10-12}$$

where

n_i = number of inner bolts
  = 2 for 4E and 4ES connections
  = 4 for 8ES connections
n_o = number of outer bolts
  = 2 for 4E and 4ES connections
  = 4 for 8ES connections
r_{ni} = $1.2 L_c t F_u < 2.4 d_b t F_u$ for each inner bolt
r_{no} = $1.2 L_c t F_u < 2.4 d_b t F_u$ for each outer bolt
L_c = clear distance, in the direction of force, between the edge of the hole and the edge of the adjacent hole or edge of the material, in. (mm)
F_u = specified minimum tensile strength of end-plate or column flange material, ksi (MPa)
d_b = diameter of the bolt, in. (mm)
t = end-plate or column flange thickness, in. (mm)

Step 13. Design the flange to end-plate and web to end-plate welds using the requirements of Section 6.9.7.

2. **Column-Side Design**

Step 1. Check the column flange for flexural yielding:

$$t_{cf} \geq \sqrt{\frac{1.11 M_f}{\phi_d F_{yc} Y_c}} \tag{6.10-13}$$

where

F_{yc} = specified minimum yield stress of column flange material, ksi (MPa)
Y_c = unstiffened column flange yield line mechanism parameter from Table 6.5 or Table 6.6, in. (mm)
t_{cf} = column flange thickness, in. (mm)

TABLE 6.5
Summary of Four-Bolt Extended Column Flange Yield Line Mechanism Parameter

Unstiffened Column Flange Geometry and Yield Line Pattern	Stiffened Column Flange Geometry and Yield Line Pattern
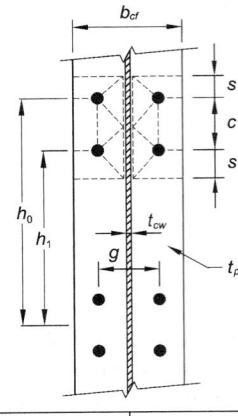	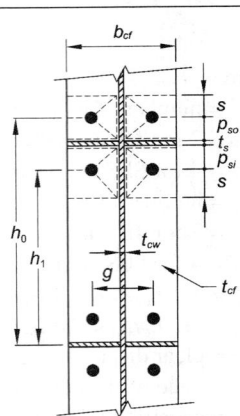

Unstiffened Column Flange	$Y_c = \dfrac{b_{cf}}{2}\left[h_1\left(\dfrac{1}{s}\right)+h_0\left(\dfrac{1}{s}\right)\right]+\dfrac{2}{g}\left[h_1\left(s+\dfrac{3c}{4}\right)+h_0\left(s+\dfrac{c}{4}\right)+\dfrac{c^2}{2}\right]+\dfrac{g}{2}$ $s = \dfrac{1}{2}\sqrt{b_{cf}\,g}$
Stiffened Column Flange	$Y_c = \dfrac{b_{cf}}{2}\left[h_1\left(\dfrac{1}{s}+\dfrac{1}{p_{si}}\right)+h_0\left(\dfrac{1}{s}+\dfrac{1}{p_{so}}\right)\right]+\dfrac{2}{g}\left[h_1(s+p_{si})+h_0(s+p_{so})\right]$ $s = \dfrac{1}{2}\sqrt{b_{cf}\,g}$ Note: If $p_{si} > s$, use $p_{si} = s$.

TABLE 6.6
Summary of Eight-Bolt Extended Column Flange Yield Line Mechanism Parameter

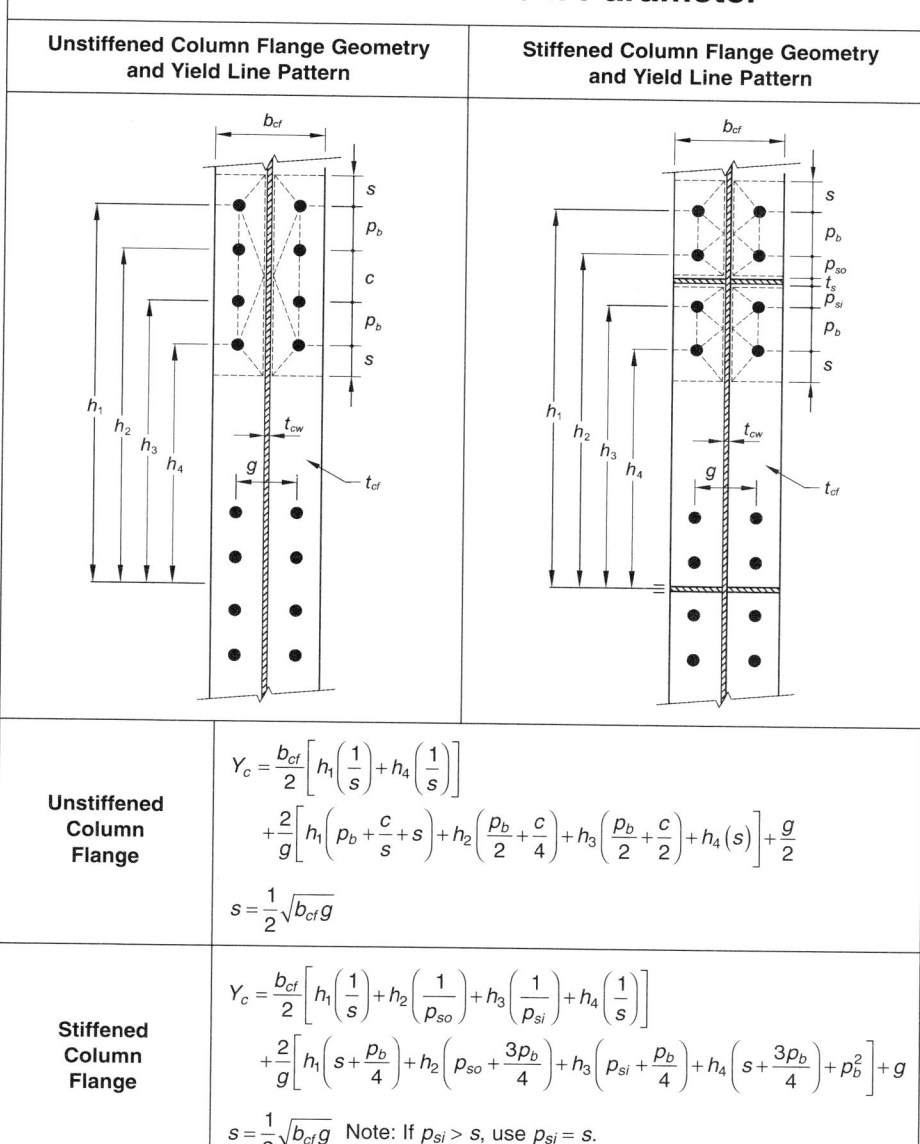

Unstiffened Column Flange

$$Y_c = \frac{b_{cf}}{2}\left[h_1\left(\frac{1}{s}\right) + h_4\left(\frac{1}{s}\right)\right]$$
$$+ \frac{2}{g}\left[h_1\left(p_b + \frac{c}{s} + s\right) + h_2\left(\frac{p_b}{2} + \frac{c}{4}\right) + h_3\left(\frac{p_b}{2} + \frac{c}{2}\right) + h_4(s)\right] + \frac{g}{2}$$

$$s = \frac{1}{2}\sqrt{b_{cf}\,g}$$

Stiffened Column Flange

$$Y_c = \frac{b_{cf}}{2}\left[h_1\left(\frac{1}{s}\right) + h_2\left(\frac{1}{p_{so}}\right) + h_3\left(\frac{1}{p_{si}}\right) + h_4\left(\frac{1}{s}\right)\right]$$
$$+ \frac{2}{g}\left[h_1\left(s + \frac{p_b}{4}\right) + h_2\left(p_{so} + \frac{3p_b}{4}\right) + h_3\left(p_{si} + \frac{p_b}{4}\right) + h_4\left(s + \frac{3p_b}{4}\right) + p_b^2\right] + g$$

$$s = \frac{1}{2}\sqrt{b_{cf}\,g}\quad \text{Note: If } p_{si} > s, \text{ use } p_{si} = s.$$

If Equation 6.10-13 is not satisfied, increase the column size or add continuity plates.

If continuity plates are added, check Equation 6.10-13 using Y_c for the stiffened column flange from Tables 6.5 and 6.6.

Step 2. If continuity plates are required for column flange flexural yielding, determine the required stiffener force.

The column flange flexural design strength is

$$\phi_d M_{cf} = \phi_d F_{yc} Y_c t_{cf}^2 \qquad (6.10\text{-}14)$$

where Y_c is the unstiffened column yield line mechanism parameter from Table 6.5 or Table 6.6, in. (mm). Therefore, the equivalent column flange design force is

$$\phi_d R_n = \frac{\phi_d M_{cf}}{(d - t_{bf})} \qquad (6.10\text{-}15)$$

Using $\phi_d R_n$, the required force for continuity plate design is determined in Section 6.10.2 Step 6.

Step 3. Check the local column web yielding strength of the unstiffened column web at the beam flanges.

Strength requirement:

$$F_{fu} \leq \phi_d R_n \qquad (6.10\text{-}16)$$

$$R_n = C_t(6k_c + t_{bf} + 2t_p) F_{yc} t_{cw} \qquad (6.10\text{-}17)$$

where
C_t = 0.5 if the distance from the column top to the top face of the beam flange is less than the depth of the column
= 1.0 otherwise
F_{yc} = specified yield stress of column web material, ksi (MPa)
k_c = distance from outer face of the column flange to web toe of fillet (design value) or fillet weld, in. (mm)
t_{cw} = column web thickness, in. (mm)

If the strength requirement of Equation 6.10-16 is not satisfied, column web continuity plates are required.

Step 4. Check the unstiffened column web buckling strength at the beam compression flange.

Strength requirement:

$$F_{fu} \leq \phi R_n \qquad (6.10\text{-}18)$$

where $\phi = 0.75$

(a) When F_{fu} is applied at a distance greater than or equal to $d_c/2$ from the end of the column

$$R_n = \frac{24t_{cw}^3\sqrt{EF_{yc}}}{h} \qquad (6.10\text{-}19)$$

(b) When F_{fu} is applied at a distance less than $d_c/2$ from the end of the column

$$R_n = \frac{12t_{cw}^3\sqrt{EF_{yc}}}{h} \qquad (6.10\text{-}20)$$

where h is the clear distance between flanges less the fillet or corner radius for rolled shapes; clear distance between flanges when welds are used for built-up shapes, in. (mm)

If the strength requirement of Equation 6.10-18 is not satisfied, then column web continuity plates are required.

Step 5. Check the unstiffened column web crippling strength at the beam compression flange.

Strength requirement:

$$F_{fu} \leq \phi R_n \qquad (6.10\text{-}21)$$

where $\phi = 0.75$

(a) When F_{fu} is applied at a distance greater than or equal to $d_c/2$ from the end of the column

$$R_n = 0.80\, t_{cw}^2 \left[1 + 3\left(\frac{N}{d_c}\right)\left(\frac{t_{cw}}{t_{cf}}\right)^{1.5}\right]\sqrt{\frac{EF_{yc}t_{cf}}{t_{cw}}} \qquad (6.10\text{-}22)$$

(b) When F_{fu} is applied at a distance less than $d_c/2$ from the end of the column

(i) for $N/d_c \leq 0.2$,

$$R_n = 0.40\, t_{cw}^2 \left[1 + 3\left(\frac{N}{d_c}\right)\left(\frac{t_{cw}}{t_{cf}}\right)^{1.5}\right]\sqrt{\frac{EF_{yc}t_{cf}}{t_{cw}}} \qquad (6.10\text{-}23)$$

(ii) for $N/d_c > 0.2$,

$$R_n = 0.40\, t_{cw}^2 \left[1 + \left(\frac{4N}{d_c} - 0.2\right)\left(\frac{t_{cw}}{t_{cf}}\right)^{1.5}\right]\sqrt{\frac{EF_{yc}t_{cf}}{t_{cw}}} \qquad (6.10\text{-}24)$$

where
- N = thickness of beam flange plus 2 times the groove weld reinforcement leg size, in. (mm)
- d_c = overall depth of the column, in. (mm)

If the strength requirement of Equation 6.10-21 is not satisfied, then column web continuity plates are required.

Step 6. If stiffener plates are required for any of the column side limit states, the required strength is

$$F_{su} = F_{fu} - \min(\phi R_n) \qquad (6.10\text{-}25)$$

where $\min(\phi R_n)$ is the minimum design strength value from Section 6.10.2 Step 2 (column flange bending), Step 3 (column web yielding), Step 4 (column web buckling), and Step 5 (column web crippling).

The design of the continuity plates shall also conform to Chapter E of the AISC *Seismic Provisions*, and the welds shall be designed in accordance with Section 6.7(3).

Step 7. Check the panel zone in accordance with Section 6.6(1).

CHAPTER 7

BOLTED FLANGE PLATE (BFP) MOMENT CONNECTION

7.1. GENERAL

Bolted flange plate (BFP) moment connections utilize plates welded to column flanges and bolted to beam flanges. The top and bottom plates must be identical. Flange plates are welded to the column flange using complete-joint-penetration (CJP) groove welds and beam flange connections are made with high-strength bolts. The beam web is connected to the column flange using a bolted shear tab with bolts in short-slotted holes. Details for this connection type are shown in Figure 7.1. Initial yielding and plastic hinge formation are intended to occur in the beam in the region near the end of the flange plates.

7.2. SYSTEMS

Bolted flange plate connections are prequalified for use in special moment frame (SMF) and intermediate moment frame (IMF) systems within the limitations of these provisions.

Exception: Bolted flange plate connections in SMF systems with *concrete structural slabs* are only prequalified if the concrete structural slab is kept at least 1 in. (25 mm) from both sides of both column flanges. It is permissible to place compressible material in the gap between the column flanges and the concrete structural slab.

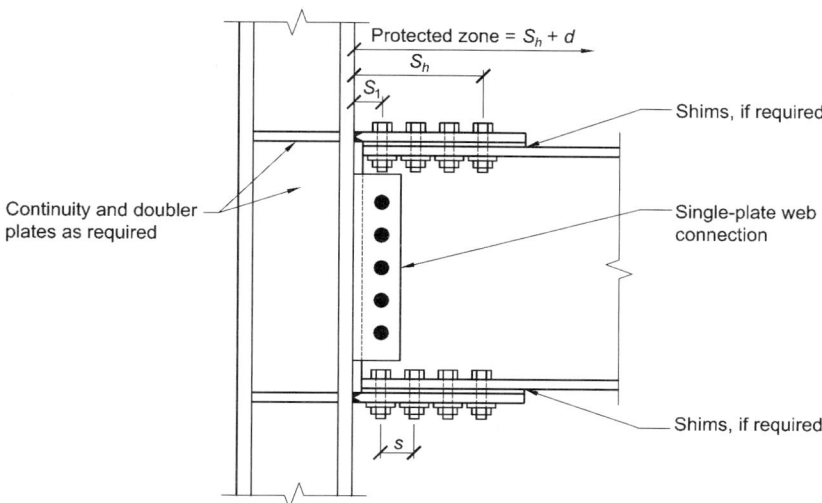

Fig. 7.1. Bolted flange plate moment connection.

7.3. PREQUALIFICATION LIMITS

1. Beam Limitations

Beams shall satisfy the following limitations:

(1) Beams shall be rolled wide-flange or welded built-up I-shaped members conforming to the requirements in Section 2.3.

(2) Beam depth is limited to a maximum of W36 (W920) for rolled shapes. Depth of built-up sections shall not exceed the depth permitted for rolled wide-flange shapes.

(3) Beam weight is limited to a maximum of 150 lb/ft (224 kg/m).

(4) Beam flange thickness is limited to a maximum of 1 in. (25 mm).

(5) The clear span-to-depth ratio of the beam is limited as follows:
   (a) For SMF systems, 9 or greater.
   (b) For IMF systems, 7 or greater.

(6) Width-to-thickness ratios for the flanges and web of the beam shall conform to the requirements of the AISC *Seismic Provisions*.

(7) Lateral bracing of beams shall be provided as follows:

   Lateral bracing of beams shall conform to the requirements of the AISC *Seismic Provisions*. To satisfy the requirements of Chapter E of the AISC *Seismic Provisions* for lateral bracing at plastic hinges, supplemental lateral bracing shall be provided at both the top and bottom beam flanges, and shall be located a distance of d to $1.5d$ from the bolt farthest from the face of the column. No attachment of lateral bracing shall be made within the protected zone.

   Exception: For both SMF and IMF systems, where the beam supports a concrete structural slab that is connected along the beam span between protected zones with welded shear connectors spaced at a maximum of 12 in. (300 mm) on center, supplemental top and bottom flange bracing at plastic hinges is not required.

(8) The protected zone consists of the flange plates and the portion of the beam between the face of the column and a distance equal to the beam depth beyond the bolt farthest from the face of the column.

2. Column Limitations

(1) Columns shall be any of the rolled shapes or welded built-up sections permitted in Section 2.3.

(2) The beam shall be connected to the flange of the column.

(3) Rolled shape column depth shall be limited to W36 (W920) maximum when a concrete structural slab is provided. In the absence of a concrete structural slab, the rolled shape column depth is limited to W14 (W360) maximum. Flanged

cruciform columns shall not have a width or depth greater than the depth allowed for rolled shapes. Built-up box columns shall not have a width or depth exceeding 24 in. (610 mm). Boxed wide-flange columns shall not have a width or depth exceeding 24 in. (610 mm) if participating in orthogonal moment frames.

(4) There is no limit on weight per foot of columns.

(5) There are no additional requirements for flange thickness.

(6) Width-to-thickness ratios for the flanges and web of columns shall conform to the requirements of the AISC *Seismic Provisions*.

(7) Lateral bracing of columns shall conform to the requirements of the AISC *Seismic Provisions*.

7.4. COLUMN-BEAM RELATIONSHIP LIMITATIONS

Beam-to-column connections shall satisfy the following limitations:

(1) Panel zones shall conform to the requirements of the AISC *Seismic Provisions*.

(2) Column-beam moment ratios shall conform to the requirements of the AISC *Seismic Provisions*.

7.5. CONNECTION DETAILING

1. Plate Material Specifications

All connection plates shall conform to one of the following specifications: ASTM A36/A36M or A572/A572M Grade 50 (345).

2. Beam Flange Plate Welds

Flange plates shall be connected to the column flange using CJP groove welds and shall be considered demand critical. *Backing*, if used, shall be removed. The *root* pass shall be *backgouged* to sound weld metal and back welded.

3. Single-Plate Shear Connection Welds

The single-plate shear connection shall be welded to the column flange. The single-plate to column-flange connection shall consist of CJP groove welds, two-sided partial-joint-penetration (PJP) groove welds, or two-sided fillet welds.

4. Bolt Requirements

Bolts shall be arranged symmetrically about the axes of the beam and shall be limited to two bolts per row in the flange plate connections. The length of the bolt group shall not exceed the depth of the beam. Standard holes shall be used in beam flanges. Holes in flange plates shall be standard or oversized holes. Bolt holes in beam flanges and in flange plates shall be made by drilling or by sub-punching and reaming. Punched holes are not permitted.

User Note: Although standard holes are permitted in the flange plate, their use will likely result in field modifications to accommodate erection tolerances.

Bolts in the flange plates shall be ASTM A490 or A490M or ASTM F2280 assemblies. Threads shall be excluded from the shear plane. Bolt diameter is limited to $1^{1}/_{8}$ in. (28 mm) maximum.

5. **Flange Plate Shims**

Shims with a maximum overall thickness of $^{1}/_{4}$ in. (6 mm) may be used between the flange plate and beam flange as shown in Figure 7.1. Shims, if required, may be finger shims or may be made with drilled or punched holes.

7.6. **DESIGN PROCEDURE**

Step 1. Compute the *probable maximum moment at the plastic hinge*, M_{pr}, in accordance with Section 2.4.3.

Step 2. Compute the maximum bolt diameter to prevent beam flange tensile rupture.

For standard holes with two bolts per row:

$$d_b \leq \frac{b_f}{2}\left(1 - \frac{R_y F_y}{R_t F_u}\right) - {}^{1}/_{8} \text{ in.} \qquad (7.6\text{-}2)$$

$$d_b \leq \frac{b_f}{2}\left(1 - \frac{R_y F_y}{R_t F_u}\right) - 3 \text{ mm} \qquad \text{(S.I.)} \qquad (7.6\text{-}2\text{M})$$

Select a bolt diameter. Check that the edge distance for the beam flange holes satisfies the AISC *Specification* requirements.

Step 3. Assume a flange plate thickness, t_p. Estimate the width of the flange plate, b_{fp}, considering bolt gage, bolt edge distance requirements, and the beam flange width. Determine the controlling nominal shear strength per bolt considering bolt shear and bolt bearing:

$$r_n = \min\begin{cases} 1.0 F_{nv} A_b \\ 2.4 F_{ub} d_b t_f \\ 2.4 F_{up} d_b t_p \end{cases} \qquad (7.6\text{-}3)$$

where
 A_b = nominal unthreaded body area of bolt, in.² (mm²)
 F_{nv} = nominal shear strength of bolt from the AISC *Specification*, ksi (MPa)
 F_{ub} = specified minimum tensile strength of beam material, ksi (MPa)
 F_{up} = specified minimum tensile strength of plate material, ksi (MPa)
 d_b = nominal bolt diameter, in. (mm)
 t_f = beam flange thickness, in. (mm)
 t_p = flange plate thickness, in. (mm)

Step 4. Select a trial number of bolts.

User Note: The following equation may be used to estimate the trial number of bolts.

$$n \geq \frac{1.25 M_{pr}}{\phi_n r_n (d + t_p)} \quad (7.6\text{-}4)$$

where
n = number of bolts rounded to the next higher even number increment
d = beam depth, in. (mm)

Step 5. Determine the beam *plastic hinge location*, S_h, as dimensioned from the face of the column.

$$S_h = S_1 + s\left(\frac{n}{2} - 1\right) \quad (7.6\text{-}5)$$

where
S_1 = distance from face of column to nearest row of bolts, in. (mm)
s = spacing of bolt rows, in. (mm)

The bolt spacing between rows, s, and the edge distance shall be sufficiently large to ensure that L_c, as defined in the AISC *Specification*, is greater than or equal to $2d_b$.

Step 6. Compute the shear force at the beam plastic hinge location at each end of the beam.

The shear force at the hinge location, V_h, shall be determined from a free body diagram of the portion of the beam between the plastic hinge locations. This calculation shall assume the moment at the plastic hinge location is M_{pr} and shall include gravity loads acting on the beam based on the load combination $1.2D + f_1 L + 0.2S$, where f_1 is the load factor determined by the applicable building code for live loads, but not less than 0.5.

User Note: The load combination of $1.2D + f_1 L + 0.2S$ is in conformance with ASCE/SEI 7. When using the International Building Code, a factor of 0.7 must be used in lieu of the factor of 0.2 when the roof configuration is such that it does not shed snow off of the structure.

Step 7. Calculate the moment expected at the face of the column flange.

$$M_f = M_{pr} + V_h S_h \quad (7.6\text{-}6)$$

where V_h is the larger of the two values of shear force at the beam hinge location at each end of the beam, kips (N).

Equation 7.6-6 neglects the gravity load on the portion of the beam between the plastic hinge and the face of the column. If desired, the gravity load on this small portion of the beam is permitted to be included.

Step 8. Compute F_{pr}, the force in the flange plate due to M_f.

$$F_{pr} = \frac{M_f}{(d + t_p)} \quad (7.6\text{-}7)$$

where
d = depth of beam, in. (mm)
t_p = thickness of flange plate, in. (mm)

Step 9. Confirm that the number of bolts selected in Step 4 is adequate.

$$n \geq \frac{F_{pr}}{\phi_n r_n} \quad (7.6\text{-}8)$$

Step 10. Check that the thickness of the flange plate assumed in Step 3 is adequate:

$$t_p \geq \frac{F_{pr}}{\phi_d F_y b_{fp}} \quad (7.6\text{-}9)$$

where
F_y = specified minimum yield stress of flange plate, ksi (MPa)
b_{fp} = width of flange plate, in. (mm)

Step 11. Check the flange plate for tensile rupture.

$$F_{pr} \leq \phi_n R_n \quad (7.6\text{-}10)$$

where R_n is defined in the tensile rupture provisions of Chapter J of the AISC *Specification*.

Step 12. Check the beam flange for block shear.

$$F_{pr} \leq \phi_n R_n \quad (7.6\text{-}11)$$

where R_n is as defined in the block shear provisions of Chapter J of the AISC *Specification*.

Step 13. Check the flange plate for compression buckling.

$$F_{pr} \leq \phi_n R_n \quad (7.6\text{-}12)$$

where R_n is defined in the compression buckling provisions of Chapter J of the AISC *Specification*.

> **User Note:** When checking compression buckling of the flange plate, the effective length, KL, may be taken as $0.65 S_1$.

Some iteration from Steps 3 through 13 may be required to determine an acceptable flange plate size.

Step 14. Determine the required shear strength, V_u, of the beam and the beam-web-to-column connection from:

$$V_u = \frac{2M_{pr}}{L_h} + V_{gravity} \qquad (7.6\text{-}13)$$

where
- L_h = distance between plastic hinge locations, in. (mm)
- $V_{gravity}$ = beam shear force resulting from $1.2D + f_1L + 0.2S$ (where f_1 is a load factor determined by the applicable building code for live loads, but not less than 0.5), kips (N)

User Note: The load combination of $1.2D + f_1L + 0.2S$ is in conformance with ASCE/SEI 7. When using the International Building Code, a factor of 0.7 must be used in lieu of the factor of 0.2 when the roof configuration is such that it does not shed snow off of the structure.

Check design shear strength of beam according to the AISC *Specification*.

Step 15. Design a single-plate shear connection for the required shear strength, V_u, calculated in Step 14 and located at the face of the column, meeting the requirements of the AISC *Specification*.

Step 16. Check the continuity plate requirements according to Chapter 2.

Step 17. Check the column panel zone according to Section 7.4.

The required shear strength of the panel zone shall be determined from the summation of the moments at the column faces as determined by projecting moments equal to $R_y F_y Z_e$ at the plastic hinge points to the column faces. For d, add twice the thickness of the flange plate to the beam depth.

CHAPTER 8

WELDED UNREINFORCED FLANGE–WELDED WEB (WUF-W) MOMENT CONNECTION

8.1. GENERAL

In the welded unreinforced flange-welded web (WUF-W) moment connection, inelastic rotation is developed primarily by yielding of the beam in the region adjacent to the face of the column. Connection rupture is controlled through special detailing requirements associated with the welds joining the beam flanges to the column flange, the welds joining the beam web to the column flange, and the shape and finish of the weld access holes. An overall view of the connection is shown in Figure 8.1.

8.2. SYSTEMS

WUF-W moment connections are prequalified for use in special moment frame (SMF) and intermediate moment frame (IMF) systems within the limits of these provisions.

8.3. PREQUALIFICATION LIMITS

1. Beam Limitations

Beams shall satisfy the following limitations:

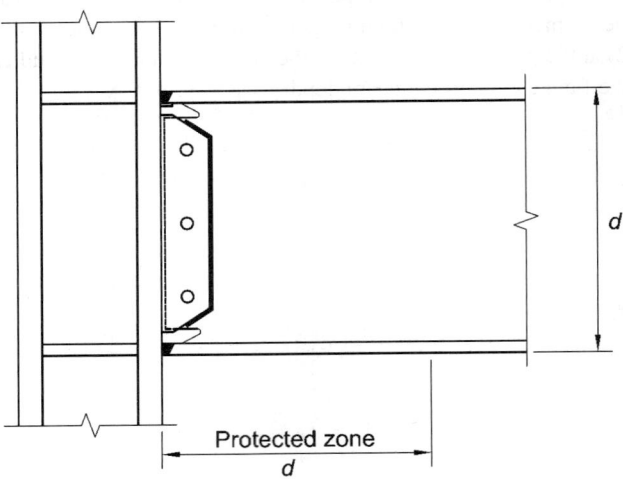

Fig. 8.1. WUF-W moment connection.

(1) Beams shall be rolled wide-flange or built-up I-shaped members conforming to the requirements of Section 2.3.

(2) Beam depth is limited to a maximum of W36 (W920) for rolled shapes. Depth of built-up sections shall not exceed the depth permitted for rolled wide-flange shapes.

(3) Beam weight is limited to a maximum of 150 lb/ft (224 kg/m).

(4) Beam flange thickness is limited to a maximum of 1 in. (25 mm).

(5) The clear span-to-depth ratio of the beam is limited as follows:
  (a) For SMF systems, 7 or greater.
  (b) For IMF systems, 5 or greater.

(6) Width-to-thickness ratios for the flanges and web of the beam shall conform to the requirements of the AISC *Seismic Provisions*.

(7) Lateral bracing of beams shall be provided as follows:

  Lateral bracing of beams shall conform to the requirements of the AISC *Seismic Provisions*. To satisfy the requirements of the AISC *Seismic Provisions* for lateral bracing at plastic hinges, supplemental lateral bracing shall be provided at both the top and bottom beam flanges, and shall be located at a distance of d to $1.5d$ from the face of the column. No attachment of lateral bracing shall be made to the beam in the region extending from the face of the column to a distance d from the face of the column.

  Exception: For both SMF and IMF systems, where the beam supports a *concrete structural slab* that is connected along the beam span between protected zones with welded shear connectors spaced at a maximum of 12 in. (300 mm) on center, supplemental top and bottom flange bracing at plastic hinges is not required.

(8) The protected zone consists of the portion of beam between the face of the column and a distance one beam depth, d, from the face of the column.

2. Column Limitations

Columns shall satisfy the following limitations:

(1) Columns shall be any of the rolled shapes or built-up sections permitted in Section 2.3.

(2) The beam shall be connected to the flange of the column.

(3) Rolled shape column depth shall be limited to a maximum of W36 (W920). The depth of built-up wide-flange columns shall not exceed that for rolled shapes. Flanged cruciform columns shall not have a width or depth greater than the depth allowed for rolled shapes. Built-up box columns shall not have a width or depth exceeding 24 in. (610 mm). Boxed wide-flange columns shall not have a width or depth exceeding 24 in. (610 mm) if participating in orthogonal moment frames.

(4) There is no limit on the weight per foot of columns.

(5) There are no additional requirements for flange thickness.

(6) Width-to-thickness ratios for the flanges and web of columns shall conform to the requirements of the AISC *Seismic Provisions*.

(7) Lateral bracing of columns shall conform to the requirements of the AISC *Seismic Provisions*.

8.4. COLUMN-BEAM RELATIONSHIP LIMITATIONS

Beam-to-column connections shall satisfy the following limitations:

(1) Panel zones shall conform to the requirements of the AISC *Seismic Provisions*.

(2) Column-beam moment ratios shall be limited as follows:

   (a) For SMF systems, the column-beam moment ratio shall conform to the requirements of the AISC *Seismic Provisions*. The value of ΣM^*_{pb} shall be taken equal to $\Sigma(M_{pr} + M_{uv})$, where M_{pr} is computed according to Step 1 in Section 8.7 and M_{uv} is the additional moment due to shear amplification from the plastic hinge to the centerline of the column. M_{uv} is permitted to be computed as $V_h(d_c/2)$, where V_h is the shear at the plastic hinge computed per Step 3 of Section 8.7, and d_c is the depth of the column.

   (b) For IMF systems, the column-beam moment ratio shall conform to the requirements of the AISC *Seismic Provisions*.

8.5. BEAM FLANGE-TO-COLUMN FLANGE WELDS

Beam flange-to-column flange connections shall satisfy the following limitations:

(1) Beam flanges shall be connected to column flanges using complete-joint-penetration (CJP) groove welds. Beam flange welds shall conform to the requirements for demand critical welds in the AISC *Seismic Provisions*.

(2) Weld access hole geometry shall conform to the requirements of AWS D1.8/D1.8M Section 6.10.1.2. Weld access hole quality requirements shall conform to the requirements of AWS D1.8.

8.6. BEAM WEB-TO-COLUMN CONNECTION LIMITATIONS

The overall details of the beam web-to-column flange connection are shown in Figure 8.2. Single-plate shear connection shall conform to the requirements shown in Figure 8.2. Beam web-to-column flange connections shall satisfy the following limitations:

(1) A single-plate shear connection shall be provided with a thickness equal at least to that of the beam web. The height of the single plate shall allow a $^1/_4$-in. (6-mm) minimum and $^1/_2$-in. (12-mm) maximum overlap with the weld access hole at the top and bottom as shown in Figure 8.3. The width shall extend 2 in. (50 mm) minimum beyond the end of the weld access hole.

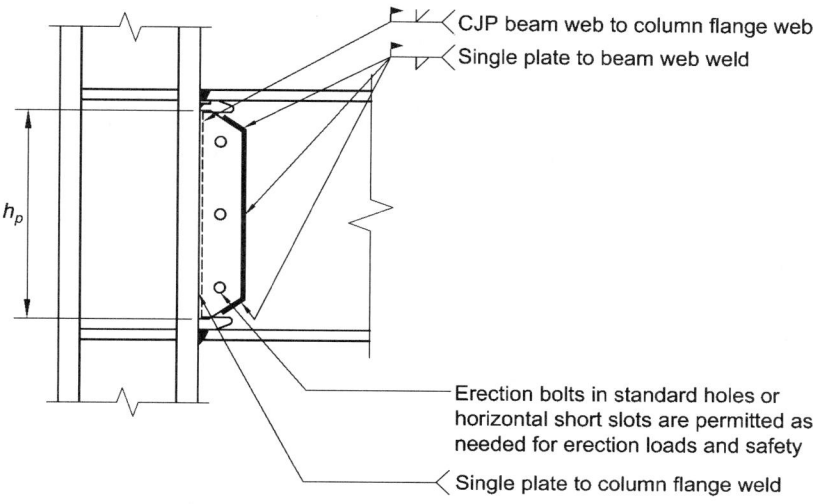

Fig. 8.2. General details of beam web-to-column flange connection.

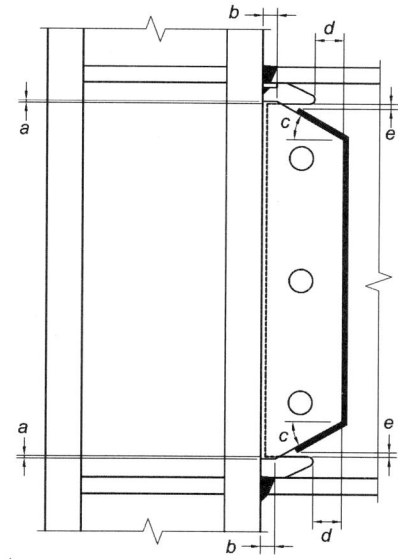

Notes

a = ¼ in. (6 mm) minimum, ½ in. (12 mm) maximum
b = 1 in. (25 mm) minimum
c = 30° (±10°)
d = 2 in. (50 mm) minimum
e = ½ in. (12 mm) minimum distance, 1 in. (25 mm) maximum distance from end of fillet weld to edge of access hole

Fig. 8.3. Details at top and bottom of single-plate shear connection.

(2) The single-plate shear connection shall be welded to the column flange. The design shear strength of the welds shall be at least $h_p t_p (0.6 R_y F_{yp})$, where h_p is defined as the length of the plate, as shown in Figure 8.2, and t_p is the thickness of the plate.

(3) The single-plate shear connection shall be connected to the beam web with fillet welds, as shown in Figures 8.2 and 8.3. The size of the fillet weld shall equal the thickness of the single plate minus $1/16$ in. (2 mm). The fillet welds shall extend along the sloped top and bottom portions of the single plate, and along the vertical single plate length, as shown in Figures 8.2 and 8.3. The fillet welds on the sloped top and bottom portions of the single plate shall be terminated at least $1/2$ in. (12 mm) but not more than 1 in. (25 mm) from the edge of the weld access hole, as shown in Figure 8.3.

(4) Erection bolts in standard holes or horizontal short slots are permitted as needed.

(5) A CJP groove weld shall be provided between the beam web and the column flange. This weld shall be provided over the full length of the web between weld access holes, and shall conform to the requirements for demand critical welds in the AISC *Seismic Provisions* and AWS D1.8/D1.8M. *Weld tabs* are not required. Weld tabs, if used, must be removed after welding in accordance with the requirements of Section 3.4. When weld tabs are not used, the use of *cascaded weld ends* within the weld groove shall be permitted at a maximum angle of 45°. Nondestructive testing (NDT) of cascaded weld ends need not be performed.

8.7. DESIGN PROCEDURE

Step 1. Compute the *probable maximum moment at the plastic hinge*, M_{pr}, in accordance with Section 2.4.3. The value of Z_e shall be taken as equal to Z_x of the beam section and the value of C_{pr} shall be taken as equal to 1.4.

> **User Note:** The C_{pr} value of 1.4 for WUF-W moment connections is based on experimental data that shows a high degree of strain hardening.

Step 2. The *plastic hinge location* shall be taken to be at the face of the column; that is, $S_h = 0$.

Step 3. Compute the shear force, V_h, at the plastic hinge location at each end of the beam.

The shear force at the plastic hinge locations shall be determined from a free body diagram of the portion of the beam between the plastic hinges. This calculation shall assume the moment at each plastic hinge is M_{pr} and shall include gravity loads acting on the beam between the hinges based on the load combination $1.2D + f_1 L + 0.2S$.

User Note: The load combination of $1.2D + f_1L + 0.2S$ is in conformance with ASCE/SEI 7. When using the International Building Code, a factor of 0.7 must be used in lieu of the factor of 0.2 when the roof configuration is such that it does not shed snow off of the structure.

Step 4. Check column-beam relationship limitations per Section 8.4. For SMF, the required shear strength of the panel zone, per the AISC *Seismic Provisions*, shall be determined from the summation of the probable maximum moments at the face of the column. The probable maximum moment at the face of the column shall be taken as M_{pr}, computed per Step 1. Provide doubler plates as necessary.

Step 5. Check beam design shear strength:

The required shear strength, V_u, of the beam shall be taken equal to the larger of the two values of V_h computed at each end of the beam in Step 3.

Step 6. Check column continuity plate requirements per Section 2.4.4. Provide continuity plates as necessary.

CHAPTER 9

KAISER BOLTED BRACKET (KBB) MOMENT CONNECTION

The user's attention is called to the fact that compliance with this chapter of the standard requires use of an invention covered by patent rights. By publication of this standard, no position is taken with respect to the validity of any claim(s) or of any patent rights in connection therewith. The patent holder has filed a statement of willingness to grant a license under these rights on reasonable and nondiscriminatory terms and conditions to applicants desiring to obtain such a license. The statement may be obtained from the standards developer.

9.1. GENERAL

In a Kaiser bolted bracket (KBB) moment connection, a cast high-strength steel bracket is fastened to each beam flange and bolted to the column flange as shown in Figure 9.1. The bracket attachment to the beam flange is permitted to be either welded (Figure 9.1a) or bolted (Figure 9.1b). When welded to the beam flange, the

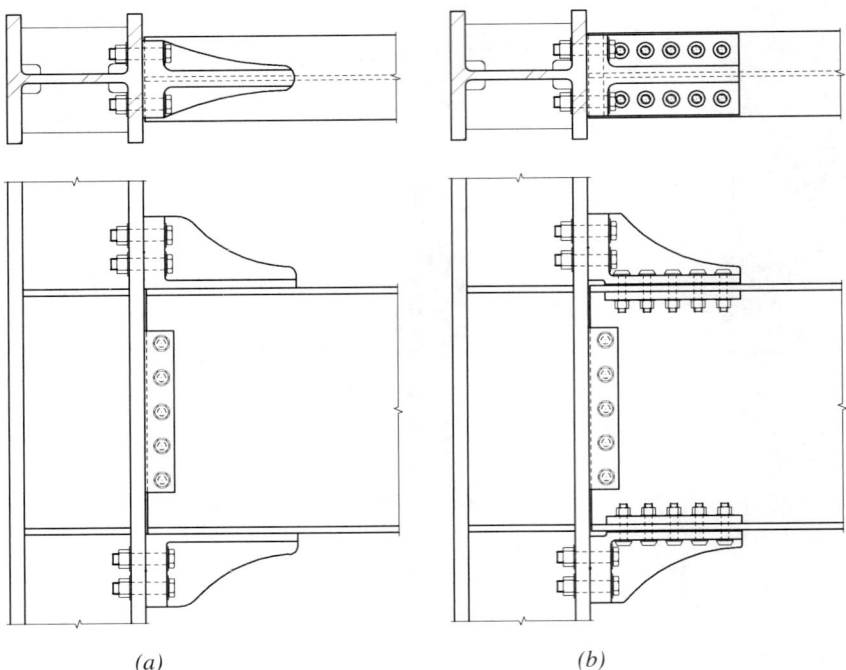

Fig. 9.1. Kaiser bolted bracket connection. (a) W-series connection; (b) B-series connection.

five W-series bracket configurations available are shown in Figure 9.2. When bolted to the beam flange, the two B-series bracket configurations available are shown in Figure 9.3. The bracket configuration is proportioned to develop the probable maximum moment strength of the connected beam. Yielding and plastic hinge formation are intended to occur primarily in the beam at the end of the bracket away from the column face.

9.2. SYSTEMS

KBB connections are prequalified for use in special moment frame (SMF) and intermediate moment frame (IMF) systems within the limits of these provisions.

Exception: KBB SMF systems with *concrete structural slabs* are prequalified only if the concrete structural slab is kept at least 1 in. (25 mm) from both sides of both column flanges and the vertical flange of the bracket. It is permitted to place compressible material in the gap in this location.

9.3. PREQUALIFICATION LIMITS

1. Beam Limitations

Beams shall satisfy the following limitations:

(1) Beams shall be rolled wide-flange or built-up I-shaped members conforming to the requirements of Section 2.3.

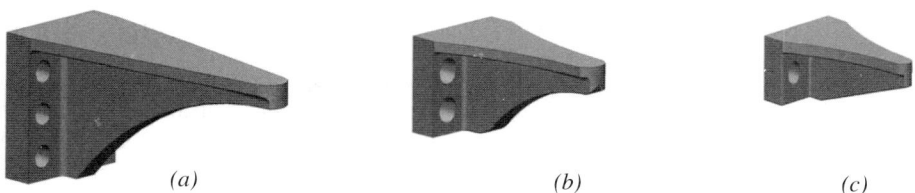

Fig. 9.2. Kaiser bolted bracket W-series configurations:
(a) six column bolts, W1.0; (b) four column bolts, W2.0 and W2.1; and
(c) two column bolts, W3.0 and W3.1.

Fig. 9.3. Kaiser bolted bracket B-series configurations:
(a) six column bolts, B1.0, and (b) four column bolts, B2.1.

(2) Beam depth is limited to a maximum of W33 (W840) for rolled shapes. Depth of built-up sections shall not exceed the depth permitted for rolled wide-flange shapes.

(3) Beam weight is limited to a maximum of 130 lb/ft (195 kg/m).

(4) Beam flange thickness is limited to a maximum of 1 in. (25 mm).

(5) Beam flange width shall be at least 6 in. (152 mm) for W-series brackets and at least 10 in. (250 mm) for B-series brackets.

(6) The clear span-to-depth ratio of the beam shall be limited to 9 or greater for both SMF and IMF systems.

(7) Width-to-thickness ratios for the flanges and web of the beam shall conform to the requirements of the AISC *Seismic Provisions*.

(8) Lateral bracing of beams shall be provided as follows:
   (a) For SMF systems, in conformance with the AISC *Seismic Provisions*. Supplemental lateral bracing shall be provided at the expected plastic hinge in conformance with the AISC *Seismic Provisions*.

   When supplemental lateral bracing is provided, attachment of supplemental lateral bracing to the beam shall be located at a distance d to $1.5d$ from the end of the bracket farthest from the face of the column, where d is the depth of the beam. No attachment of lateral bracing shall be made to the beam in the region extending from the face of the column to a distance d beyond the end of the bracket.

   (b) For IMF systems, in conformance with the AISC *Seismic Provisions*.

   Exception: For both systems, where the beam supports a concrete structural slab that is connected between the protected zones with welded shear connectors spaced at maximum of 12 in. (300 mm) on center, supplemental top and bottom flange bracing at the expected hinge is not required.

(9) The protected zone consists of the portion of beam between the face of the column and one beam depth, d, beyond the end of the bracket farthest from the face of the column.

2. **Column Limitations**

The columns shall satisfy the following limitations:

(1) Columns shall be any of the rolled shapes or built-up sections permitted in Section 2.3.

(2) The beam shall be connected to the flange of the column.

(3) The column flange width shall be at least 12 in. (305 mm).

(4) Rolled shape column depth shall be limited to W36 (W920) maximum when a concrete structural slab is provided. In the absence of a concrete structural slab,

rolled shape column depth is limited to W14 (W360) maximum. The depth of built-up wide-flange columns shall not exceed that for rolled shapes. Flanged cruciform columns shall not have a width or depth greater than the depth allowed for rolled shapes. Built-up box columns shall not have a width or depth exceeding 16 in. (406 mm). Boxed wide-flange columns shall not have a width or depth exceeding 16 in. (406 mm) if participating in orthogonal moment frames.

(5) There is no limit on the weight per foot of columns.

(6) There are no additional requirements for flange thickness.

(7) Width-to-thickness ratios for the flanges and web of columns shall conform to the requirements of the AISC *Seismic Provisions*.

(8) Lateral bracing of the columns shall conform to the requirements of the AISC *Seismic Provisions*.

3. Bracket Limitations

The high strength cast-steel brackets shall satisfy the following limitations:

(1) Bracket castings shall conform to the requirements of Appendix A.

(2) Bracket configuration and proportions shall conform to Section 9.8.

(3) Holes in the bracket for the column bolts shall be vertical short-slotted holes. Holes for the beam bolts shall be standard holes.

(4) Material thickness, edge distance and end distance shall have a tolerance of $\pm^1/_{16}$ in. (2 mm). Hole location shall have a tolerance of $\pm^1/_{16}$ in. (2 mm). The overall dimensions of the bracket shall have a tolerance of $\pm^1/_8$ in. (3 mm).

9.4. COLUMN-BEAM RELATIONSHIP LIMITATIONS

Beam-to-column connections shall satisfy the following limitations:

(1) Panel zones shall conform to the requirements in the AISC *Seismic Provisions*.

(2) Column-beam moment ratios shall conform to the requirements of the AISC *Seismic Provisions*.

9.5. BRACKET-TO-COLUMN FLANGE LIMITATIONS

Bracket-to-column flange connections shall satisfy the following limitations:

(1) Column flange fasteners shall be pretensioned ASTM A490, A490M, A354 Grade BD bolts, or A354 Grade BD threaded rods, and shall conform to the requirements of Chapter 4.

(2) Column flange bolt holes shall be $^1/_8$ in. (3 mm) larger than the nominal bolt diameter. Bolt holes shall be drilled or subpunched and reamed. Punched holes are not permitted.

(3) The use of finger shims on either or both sides at the top and/or bottom of the bracket connection is permitted, subject to the limitations of the RCSC *Specification*.

(4) When bolted to a box column, a steel washer plate shall be inserted between the box column and the bracket on both faces of the column. The washer plate shall be ASTM A572/A572M Grade 50 (345) or better and shall be designed to transfer the bolt forces to the outside edges of the column. Where required, the vertical plate depth may extend beyond the contact surface area by up to 4 in. (102 mm). The plate thickness shall not exceed 3 in. (76 mm). The fasteners shall pass through the interior of the box column and be anchored on the opposite face. The opposite face shall also have a steel washer plate.

(5) When connecting to the orthogonal face of a box column concurrent with a connection on the primary column face, a $1^3/_4$-in. (44-mm) steel spacer plate shall be inserted between the beam flanges and the brackets of the orthogonal connection. The spacer plate shall be made of any of the structural steel materials included in the AISC *Specification* and shall be the approximate width and length matching that of the bracket contact surface area.

9.6. BRACKET-TO-BEAM FLANGE CONNECTION LIMITATIONS

Bracket-to-beam-flange connections shall satisfy the following limitations:

(1) When welded to the beam flange, the bracket shall be connected using fillet welds. Bracket welds shall conform to the requirements for demand critical welds in the AISC *Seismic Provisions* and AWS D1.8/D1.8M, and to the requirements of AWS D1.1/D1.1M. The weld procedure specification (WPS) for the fillet weld joining the bracket to the beam flange shall be qualified with the casting material. Welds shall not be started or stopped within 2 in. (51 mm) of the bracket tip and shall be continuous around the tip.

(2) When bolted to the beam flange, fasteners shall be pretensioned ASTM A490 or A490M bolts with threads excluded from the shear plane and shall conform to the requirements of Chapter 4.

(3) Beam flange bolt holes shall be $1^5/_{32}$ in. (29 mm) and shall be drilled using the bracket as a template. Punched holes are not permitted.

(4) When bolted to the beam flange, a $^1/_8$-in. (3-mm)-thick brass washer plate with an approximate width and length matching that of the bracket contact surface area shall be placed between the beam flange and the bracket. The brass shall be a half-hard tempered ASTM B19 or B36/B36M sheet.

(5) When bolted to the beam flange, a 1-in. (25-mm)-thick by 4-in. (102-mm)-wide ASTM A572/A572M Grade 50 (345) plate washer shall be used on the opposite side of the connected beam flange.

9.7. BEAM WEB-TO-COLUMN CONNECTION LIMITATIONS

Beam web-to-column flange connections shall satisfy the following limitations:

(1) The required shear strength of the beam web connection shall be determined according to Section 9.9.

(2) The single-plate shear connection shall be connected to the column flange using a two-sided fillet weld, two-sided PJP groove weld or CJP groove weld.

9.8. CONNECTION DETAILING

If welded to the beam flange, Figure 9.4 shows the connection detailing for the W-series bracket configurations. If bolted to the beam flange, Figure 9.5 shows the connection detailing for the B-series bracket configurations. Table 9.1 summarizes

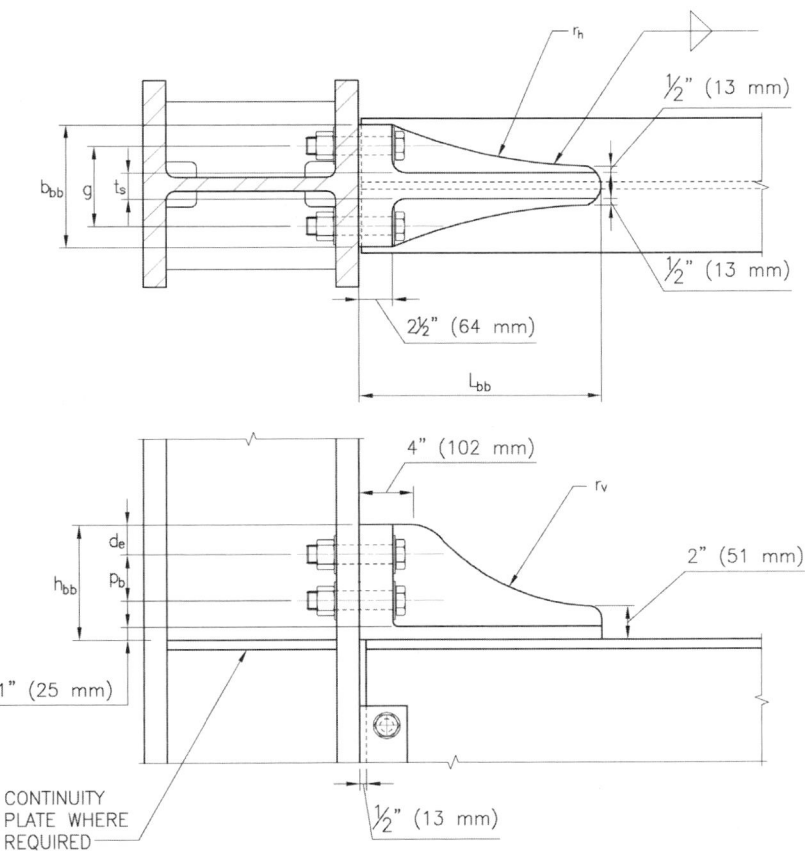

Fig. 9.4. W-series connection detailing.

the KBB proportions and column bolt parameters. Table 9.2 summarizes the design proportions for the W-series bracket configuration. Table 9.3 summarizes the design proportions for the B-series bracket configurations.

9.9. DESIGN PROCEDURE

Step 1. Select beam and column elements which satisfy the limits of Section 9.3.

Step 2. Compute the probable maximum moment, M_{pr}, at the location of the plastic hinge according to Section 2.4.3.

Step 3. Select a trial bracket from Table 9.1.

Step 4. Compute the shear force at the beam hinge location at each end of the beam. The shear force at the hinge location, V_h, shall be determined from a free-body diagram of the portion of the beam between the hinge locations. This calculation shall assume the moment at the hinge location is M_{pr} and shall include gravity loads acting on the beam based on the load combination $1.2D + f_1 L + 0.2S$, kips (N) where f_1 is the load factor determined by the applicable building code for live loads, but not less than 0.5.

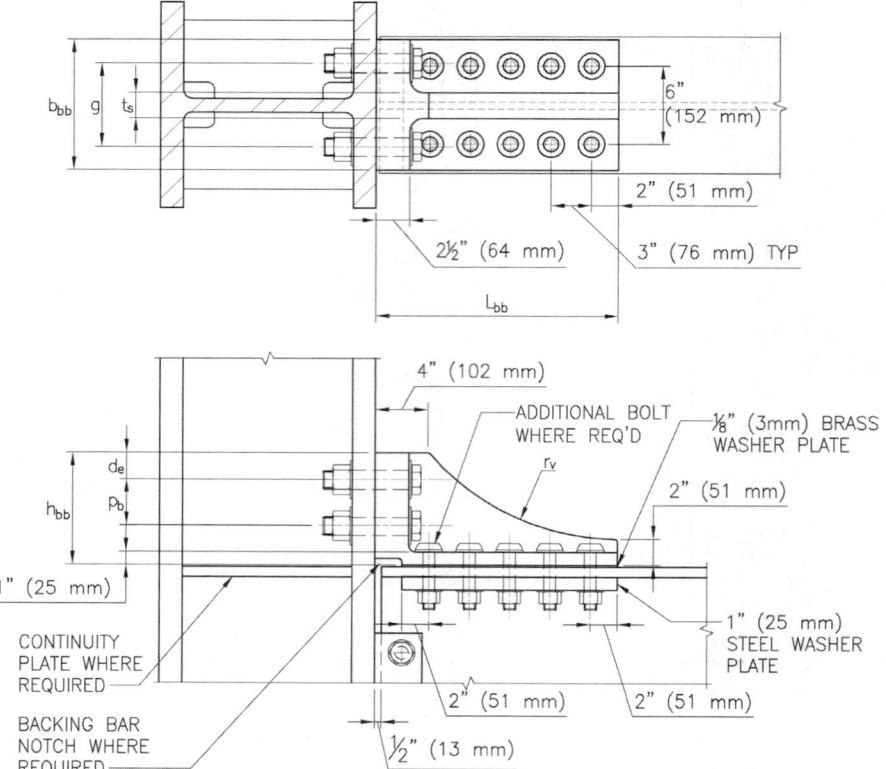

Fig. 9.5. B-series connection detailing.

Prequalified Connections for Special and Intermediate Steel Moment Frames for Seismic Applications, 2010, incl. Supplement No. 1
AMERICAN INSTITUTE OF STEEL CONSTRUCTION

TABLE 9.1
Kaiser Bolted Bracket Proportions

Bracket Designation	Bracket Length, L_{bb} in. (mm)	Bracket Height, h_{bb} in. (mm)	Bracket Width, b_{bb} in. (mm)	Number of Column Bolts, n_{cb}	Column Bolt Gage, g in. (mm)	Column Bolt Diameter in. (mm)
W3.0	16 (406)	5 1/2 (140)	9 (229)	2	5 1/2 (140)	1 3/8 (35)
W3.1	16 (406)	5 1/2 (140)	9 (229)	2	5 1/2 (140)	1 1/2 (38)
W2.0	16 (406)	8 3/4 (222)	9 1/2 (241)	4	6 (152)	1 3/8 (35)
W2.1	18 (457)	8 3/4 (222)	9 1/2 (241)	4	6 1/2 (165)	1 1/2 (38)
W1.0	25 1/2 (648)	12 (305)	9 1/2 (241)	6	6 1/2 (165)	1 1/2 (38)
B2.1	18 (457)	8 3/4 (222)	10 (254)	4	6 1/2 (165)	1 1/2 (38)
B1.0	25 1/2 (648)	12 (305)	10 (254)	6	6 1/2 (165)	1 1/2 (38)

TABLE 9.2
W-Series Bracket Design Proportions

Bracket Designation	Column Bolt Edge Distance, d_e in. (mm)	Column Bolt Pitch, p_b in. (mm)	Bracket Stiffener Thickness, t_s in. (mm)	Bracket Stiffener Radius, r_v in. (mm)	Bracket Horizontal Radius, r_h in. (mm)	Minimum Fillet Weld Size, w in. (mm)
W3.0	2 1/2 (64)	n.a.	1 (25)	n.a.	28 (711)	1/2 (13)
W3.1	2 1/2 (64)	n.a.	1 (25)	n.a.	28 (711)	5/8 (16)
W2.0	2 1/4 (57)	3 1/2 (89)	2 (51)	12 (305)	28 (711)	3/4 (19)
W2.1	2 1/4 (57)	3 1/2 (89)	2 (51)	16 (406)	38 (965)	7/8 (22)
W1.0	2 (51)	3 1/2 (89)	2 (51)	28 (711)	n.a.	7/8 (22)

TABLE 9.3
B-Series Bracket Design Proportions

Bracket Designation	Column Bolt Edge Distance, d_e in. (mm)	Column Bolt Pitch, p_b in. (mm)	Bracket Stiffener Thickness, t_s in. (mm)	Bracket Stiffener Radius, r_v in. (mm)	Number of Beam Bolts, n_{bb}	Beam Bolt Diameter in. (mm)
B2.1	2 (51)	3 1/2 (89)	2 (51)	16 (406)	8 or 10	1 1/8 (29)
B1.0	2 (51)	3 1/2 (89)	2 (51)	28 (711)	12	1 1/8 (29)

Prequalified Connections for Special and Intermediate Steel Moment Frames for Seismic Applications, 2010, incl. Supplement No. 1
AMERICAN INSTITUTE OF STEEL CONSTRUCTION

User Note: The load combination of $1.2D + f_1L + 0.2S$ is in conformance with ASCE/SEI 7. When using the International Building Code, a factor of 0.7 must be used in lieu of the factor of 0.2 when the roof configuration is such that it does not shed snow off of the structure.

Step 5. Compute the probable maximum moment at the face of the column:

$$M_f = M_{pr} + V_h S_h \tag{9.9-1}$$

where

M_f = probable maximum moment at the face of the column, kip-in. (N-mm)
S_h = distance from the face of the column to the plastic hinge, in. (mm)
  = L_{bb} per Table 9.1, in. (mm)
V_h = larger of the two values of shear force at the beam hinge location at each end of the beam, kips (N)

Equation 9.9-1 neglects the gravity load on the portion of the beam between the plastic hinge and the face of the column. If desired, the gravity load on this small portion of the beam is permitted to be included.

Step 6. The following relationship shall be satisfied for the bracket column bolt tensile strength:

$$r_{ut} \leq \phi_n F_{nt} A_b \tag{9.9-2}$$

where

$$r_{ut} = \frac{M_f}{d_{eff} n_{cb}} \tag{9.9-3}$$

A_b = bolt nominal cross-sectional area, in.² (mm²)
F_{nt} = nominal tensile strength of bolt from the AISC *Specification*, ksi (MPa)
d_{eff} = effective beam depth, calculated as the centroidal distance between bolt groups in the upper and lower brackets, in. (mm)
n_{cb} = number of column bolts per Table 9.1

Step 7. Determine the minimum column flange width to prevent flange tensile rupture:

$$b_{cf} \geq \frac{2\left[d_b + \frac{1}{8}\text{ in.}\right]}{\left(1 - \frac{R_y F_{yf}}{R_t F_{uf}}\right)} \tag{9.9-4}$$

$$b_{cf} \geq \frac{2\left[d_b + 3\text{ mm}\right]}{\left(1 - \frac{R_y F_{yf}}{R_t F_{uf}}\right)} \quad \text{(S.I.)} \tag{9.9-4M}$$

where
- b_{cf} = width of column flange, in. (mm)
- d_b = diameter of column flange bolts, in. (mm)
- F_{yf} = specified minimum yield stress of the flange material, ksi (MPa)
- F_{uf} = specified minimum tensile strength of the flange material, ksi (MPa)
- R_y = ratio of expected yield stress to the specified minimum yield stress for the flange material
- R_t = ratio of expected tensile strength to the specified minimum tensile strength for the flange material

Step 8. Check the minimum column flange thickness to eliminate prying action:

$$t_{cf} \geq \sqrt{\frac{4.44 r_{ut} b'}{\phi_d p F_y}} \tag{9.9-5}$$

where
$$b' = 0.5\,(g - k_1 - 0.5 t_{cw} - d_b) \tag{9.9-6}$$
- g = column bolt gage, in. (mm)
- k_1 = column web centerline distance to the flange toe of the fillet, in. (mm)
- p = perpendicular tributary length per bolt, in. (mm)
  - = 3.5 in. (89 mm) for W1.0/B1.0
  - = 5.0 in. (127 mm) for all other brackets
- t_{cf} = minimum column flange thickness required to eliminate prying action, in. (mm)
- t_{cw} = column web thickness, in. (mm)

If the selected column flange thickness is less than that required to eliminate prying action, select a column with a satisfactory flange thickness or include the bolt prying force in Equation 9.9-2 per Part 9 of the AISC *Steel Construction Manual*.

Step 9. The column flange thickness shall satisfy the following requirement to eliminate continuity plates:

$$t_{cf} \geq \sqrt{\frac{M_f}{\phi_d F_{yf} d_{\mathit{eff}} Y_m}} \tag{9.9-7}$$

where
- Y_m = simplified column flange yield line mechanism parameter
  - = 5.9 for W3.0/W3.1
  - = 6.5 for W2.0/W2.1/B2.1
  - = 7.5 for W1.0/B1.0
- t_{cf} = minimum column flange thickness required to eliminate continuity plates, in. (mm)

Step 10. Continuity Plate Requirements

For W14 and shallower columns, continuity plates are not required if Equation 9.9-7 is satisfied. For column sections deeper than W14, continuity plates shall be provided.

Step 11. If the bracket is welded to the beam flange proceed to Step 14; otherwise, determine the minimum beam flange width to prevent beam flange tensile rupture:

$$b_{bf} \geq \frac{2\left[d_b + {}^1/_{32} \text{ in.}\right]}{\left(1 - \dfrac{R_y F_{yf}}{R_t F_{uf}}\right)} \quad (9.9\text{-}8)$$

$$b_{bf} \geq \frac{2\left[d_b + 1 \text{ mm}\right]}{\left(1 - \dfrac{R_y F_{yf}}{R_t F_{uf}}\right)} \quad \text{(S.I.)} \quad (9.9\text{-}8M)$$

where

b_{bf} = width of beam flange, in. (mm)
d_b = diameter of beam flange bolts, in. (mm)

Step 12. The following relationship shall be satisfied for the beam bolt shear strength:

$$\frac{M_f}{\phi_n F_{nv} A_b d_{eff} n_{bb}} < 1.0 \quad (9.9\text{-}9)$$

where

F_{nv} = nominal shear strength of bolt from the AISC *Specification*, ksi (MPa)
n_{bb} = the number of beam bolts per Table 9.3

Step 13. Check the beam flange for block shear per the following:

$$\frac{M_f}{d_{eff}} \leq \phi_n R_n \quad (9.9\text{-}10)$$

where R_n is as defined in the block shear provisions of Chapter J of the AISC *Specification*.

Step 14. If the bracket is bolted to the beam flange proceed to Step 15. Otherwise, the following relationship shall be satisfied for the fillet weld attachment of the bracket to the beam flange:

$$\frac{M_f}{\phi_n F_w d_{eff} l_w (0.707 w)} < 1.0 \quad (9.9\text{-}11)$$

where

F_w = nominal weld design strength per the AISC *Specification*
     = $0.60 F_{EXX}$
F_{EXX} = filler metal classification strength, ksi (MPa)
L_{bb} = bracket length per Table 9.3, in. (mm)

l = bracket overlap distance, in. (mm)
  = 0 in. (0 mm) if $b_{bf} \geq b_{bb}$
  = 5 in. (125 mm) if $b_{bf} < b_{bb}$
l_w = length of available fillet weld, in. (mm)
  = $2(L_{bb} - 2.5 \text{ in.} - l)$ (9.9-12)
  = $2(L_{bb} - 64 \text{ mm} - l)$ (S.I.) (9.9-12M)
w = minimum fillet weld size per Table 9.2, in. (mm)

Step 15. Determine the required shear strength, V_u, of the beam and beam web-to-column connection from:

$$V_u = \frac{2M_{pr}}{L_h} + V_{gravity} \qquad (9.9\text{-}13)$$

where
  L_h = distance between *plastic hinge locations*, in. (mm)
  $V_{gravity}$ = beam shear force resulting from $1.2D + f_1 L + 0.2S$ (where f_1 is a load factor determined by the applicable building code for live loads, but not less than 0.5), kips (N)

User Note: The load combination of $1.2D + f_1 L + 0.2S$ is in conformance with ASCE/SEI 7. When using the International Building Code, a factor of 0.7 must be used in lieu of the factor of 0.2 when the roof configuration is such that it does not shed snow off of the structure.

Check design shear strength of beam according to Chapter G of the AISC *Specification*.

Step 16. Design the beam web-to-column connection according to Section 9.7.

Step 17. Check column panel zone according to Section 9.4. Substitute the effective depth, d_{eff}, of the beam and brackets for the beam depth, d.

Step 18. (Supplemental) If the column is a box configuration, determine the size of the steel washer plate between the column flange and the bracket such that:

$$Z_x \geq \frac{M_f(b_{cf} - t_{cw} - g)}{4\phi_d F_y d_{eff}} \qquad (9.9\text{-}14)$$

where
  F_y = specified minimum yield stress of the washer material, ksi (MPa)
  Z_x = plastic section modulus of the washer plate, in.^{3} (mm^3)
  g = column bolt gage, in. (mm)

CHAPTER 10

CONXTECH CONXL MOMENT CONNECTION

The user's attention is called to the fact that compliance with this chapter of the standard requires use of an invention covered by patent rights.* By publication of this standard, no position is taken with respect to the validity of any claim(s) or of any patent rights in connection therewith. The patent holder has filed a statement of willingness to grant a license under these rights on reasonable and nondiscriminatory terms and conditions to applicants desiring to obtain such a license, and the statement may be obtained from the standards developer.

10.1 GENERAL

The ConXtech® ConXL™ moment connection permits full-strength, fully restrained connection of wide flange beams to concrete-filled 16 in. (406 mm) square HSS or built-up box columns using a high-strength, field-bolted collar assembly. Beams are shop-welded to forged flange and web fittings (collar flange assembly) and are field-bolted together through forged column fittings (collar corner assembly) that are shop welded to the columns. Beams may be provided with reduced beam section (RBS) cutouts if necessary to meet strong-column/weak-beam criteria. ConXL connections may be used to provide moment connections to columns in orthogonal frames. All moment beams connecting to a ConXL node (intersection of moment beams and column) must be of the same nominal depth.

Figure 10.1 shows the connection geometry and major connection components. Each ConXL collar assembly is made up of forged collar corners and collar flanges conforming to the material requirements of ASTM A572/A572M Grade 50 (Grade 345). At each ConXL node there are four collar corner assemblies (Figure 10.2), one at each corner of the square built-up or HSS column. Each ConXL node also contains four collar flange assemblies (Figure 10.3), one for each face of the square column. Each collar flange assembly can contain the end of a moment beam that is shop-welded to the collar flange assembly. The combination of collar corner assemblies, collar flange assemblies, and square concrete-filled column create the ConXL node.

Figure 10.2 shows the collar corner assemblies. The collar corner assembly is made up of a collar corner top (CCT) piece, a collar corner bottom (CCB) piece; and for beam depths greater than 18 in. (460 mm), a collar corner middle (CCM) piece. The

---

* The connectors and structures illustrated are covered by one or more of the following U.S. and foreign patents: U.S. Pat. Nos.: 7,941,985; 6,837,016; 7,051,917; 7,021,020; Australia Pat. Nos. 2001288615; 2004319371; Canada Pat. Nos. 2,458,706; 2,564,195; China Pat. Nos. ZL 01 8 23730.4; ZL 2004 8 0042862.5; Japan Pat. Nos. 4165648; 4427080; Mexico Pat. Nos. 262,499; 275284; Hong Kong Pat. No. 1102268. Other U.S. and foreign patent protection pending.

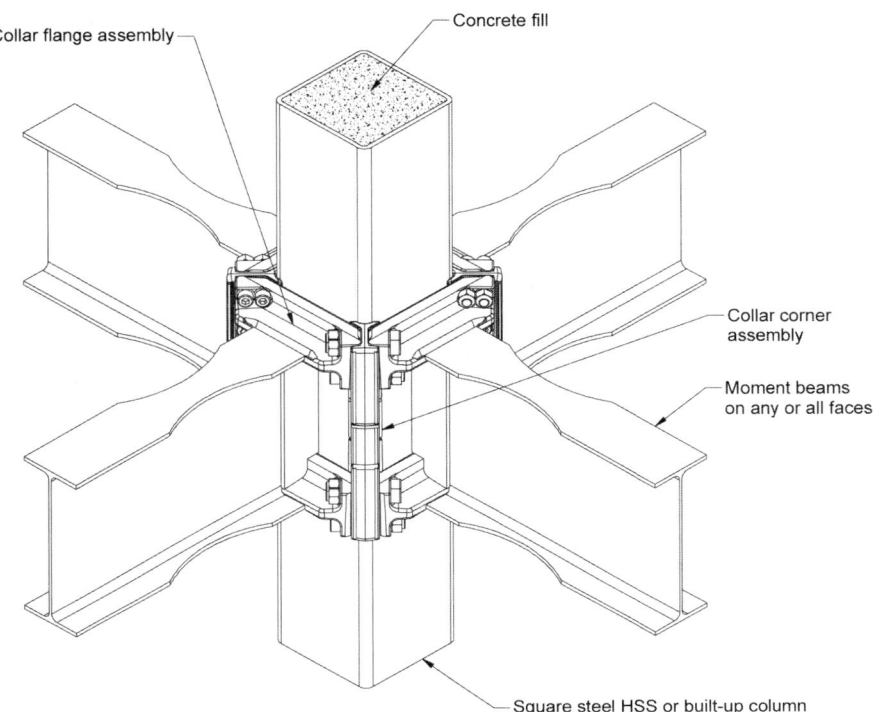

Fig. 10.1. Assembled ConXL moment connection.

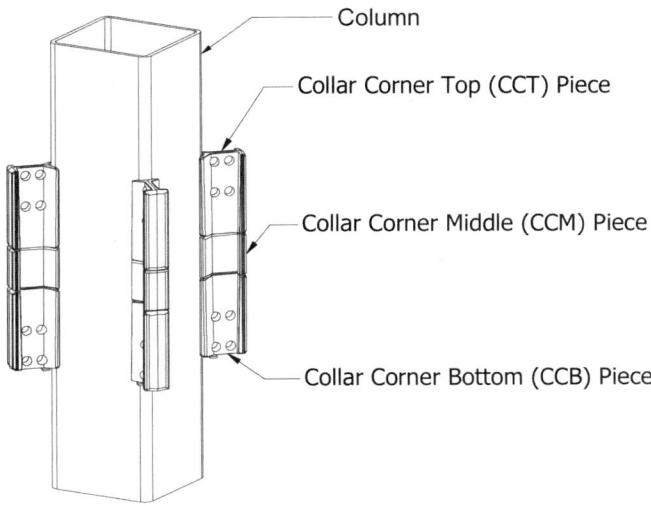

Fig. 10.2. Column with attached collar corner assemblies.

CCT, CCB and CCM are partial joint penetration- (PJP-) welded together to create the collar corner assembly; they are then shop fillet welded to the corners of the square column.

Figure 10.3 shows the collar flange assembly. Each collar flange assembly is made up of a collar flange top (CFT), collar flange bottom (CFB) and a collar web extension (CWX).

If a beam at the node requires a moment connection, the CFT (or CFB) is aligned with and shop-welded to the top (or bottom) flange of the beam.

Moment-connected beam webs are also shop-welded to the CWX. If a beam at the node does not require a moment connection, the size of the CWX remains unchanged and a shear plate connection is shop-welded to the CWX to accommodate a non-moment beam that does not need to match the nominal depth of the moment-connected beam(s).

If no beams exist on a node at a particular column face, the CFT and CFB are aligned at the nominal depth of the moment beam, and the CWX shall be permitted to be optionally omitted.

Section 10.9 contains drawings indicating the dimensions of individual pieces.

Columns are delivered to the job site with the collar corner assemblies shop-welded to the column at the proper floor framing locations. Beams are delivered to the job site with the collar flange assemblies shop-welded to the ends of the beams. During

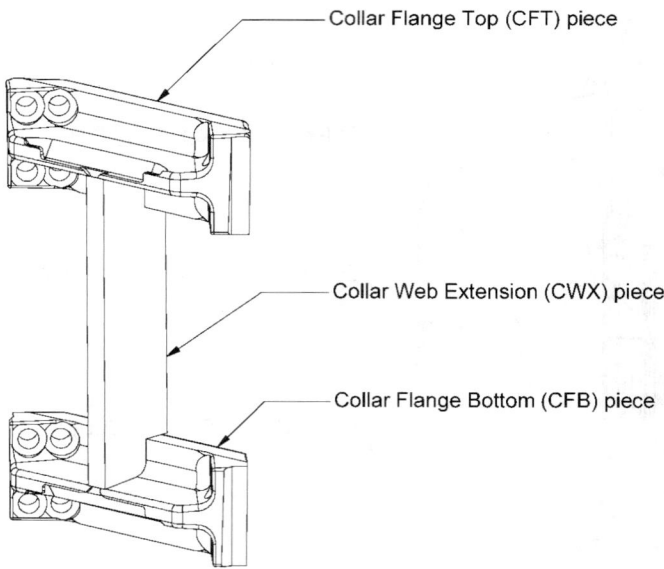

Fig. 10.3. Collar flange assembly.

frame erection the collar flange assemblies with or without beams are lowered into the column collar corner assemblies. When all four faces of the column are filled with collar flanges the collar bolts are inserted and pre-tensioned, effectively clamping and compressing the collar flange assemblies around the collar corner assemblies and square column.

Beam flange flexural forces in moment beams are transferred to the collar flange assemblies via complete joint penetration (CJP) groove welds. Collar flanges transfer compressive beam flange forces to the collar corners through flexure of the collar flange and direct bearing onto the collar corners. The collar flange transfers beam flange tensile forces in flexure to the pre-tensioned collar bolts. The collar bolts transfer these forces in tension through the orthogonal collar flanges, which then transfer the forces through the rear collar bolts attached to the collar flange on the opposite face of the column. These combined forces are then transferred to the column walls through a combination of bearing and the fillet welds attaching the collar corners to the column. Finally, a portion of these forces are transferred to the concrete fill which is in direct contact with the column walls.

The behavior of this connection is controlled by flexural hinging of the beams adjacent to the collar assembly. When RBS cutouts are provided yielding and plastic hinge formation primarily occur within the reduced beam section.

10.2 SYSTEMS

The ConXL moment connection is prequalified for use in special moment frame (SMF) and intermediate moment frame (IMF) systems within the limits of these provisions. The ConXL moment connection is prequalified for use in planar moment-resisting frames or in orthogonal intersecting moment-resisting frames.

ConXL SMF systems with concrete structural slabs are prequalified only if a vertical flexible joint at least 1 in. (25 mm) thick is placed in the concrete slab around the collar assembly and column similar to that shown in Figure 10.4.

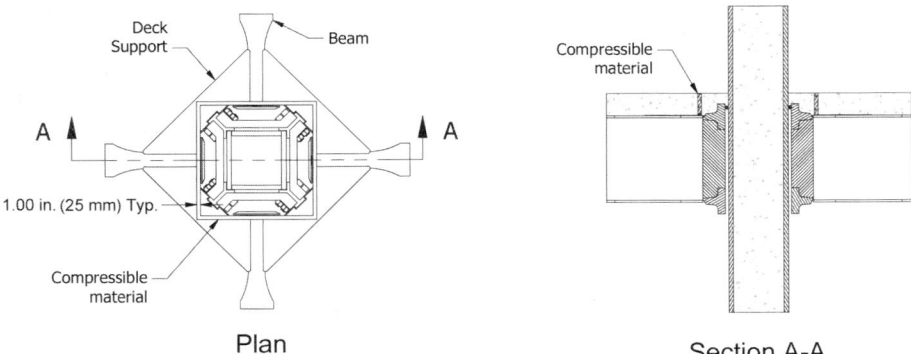

Fig. 10.4. Use of compressible material to isolate structural slab from connection.

10.3 PREQUALIFICATION LIMITS

1. Beam Limitations

Beams shall satisfy the following limitations:

(1) Beams shall be rolled wide-flange or built-up I-shaped members conforming to the requirements of Section 2.3.

(2) Beam depths shall be limited to the following beam shapes or their built-up equivalents: W30, W27, W24, W21 and W18.

(3) Beam flange thickness shall be limited to a maximum of 1 in. (25 mm).

(4) Beam flange width shall be limited to a maximum of 12 in. (300 mm).

(5) The clear span-to-depth ratio of the beam shall be limited as follows:

(a) For SMF systems, 7 or greater.

(b) For IMF systems, 5 or greater.

(6) Width-thickness ratios for beam flanges and webs shall conform to the limits of the AISC *Seismic Provisions*. The value of b_f used to determine the width-thickness ratio of beams with RBS cutouts shall not be less than the flange width at the center two-thirds of the reduced section provided that gravity loads do not shift the location of the plastic hinge a significant distance from the center of the reduced beam section.

(7) Lateral bracing of beams shall conform to the applicable limits of the AISC *Seismic Provisions*.

Exception: For SMF and IMF systems, where the beam supports a concrete structural slab that is connected between the protected zones with welded shear connectors spaced at a maximum of 12 in. (300 mm) on center, supplemental top and bottom flange bracing at the expected hinge is not required.

(8) For RBS connections the protected zone consists of the portion of the connection assembly and beam between the column face and the farthest end of the reduced beam section. For beams without reduced beam sections the protected zone consists of the portion of the connection assembly and beam extending from the column face to a distance of d from the outside face of the collar flange.

2. Column Limitations

Columns shall satisfy the following limitations:

(1) Columns shall be square 16 in. (406 mm) HSS sections or square 16 in. (406 mm) built-up box sections permitted in Section 2.3.

(2) There is no limit on column weight per foot.

(3) Column wall thickness shall not be less than $3/8$ in. (10 mm). Column wall thickness for HSS columns shall not be less than $3/8$ in. (10 mm) nominal.

(4) Width-thickness ratios for columns shall conform to the applicable limits for filled composite columns in the AISC *Seismic Provisions*.

(5) Lateral bracing of columns shall conform to the applicable limits in the AISC *Seismic Provisions*.

(6) Columns shall be completely filled with structural concrete having unit weight not less than 110 pounds per cubic foot (17 kN/m^3). Concrete shall have 28-day compressive strength not less than 3,000 psi (21 MPa).

3. **Collar Limitations**

   Collar forgings shall satisfy the following limitations:

   (1) Collar forgings shall conform to the requirements of Appendix B, Forging Requirements.

   (2) Collar configuration and proportions shall conform to Section 10.9, ConXL Part Drawings.

   (3) Collar flange bolt holes shall be $^1/_8$ in. (3 mm) larger than the nominal bolt diameter. Bolt holes shall be drilled.

   (4) Collar corner bolt holes shall be $^1/_8$ in. (3 mm) larger than the nominal bolt diameter. Bolt holes shall be drilled.

   (5) Material thickness, edge distance, end distance and overall dimension shall have a tolerance of $\pm\ ^1/_{16}$ in. (2 mm).

   (6) Faying surfaces shall be machined and meet the requirements for Class A slip-critical surfaces.

10.4. COLLAR CONNECTION LIMITATIONS

Collar connections shall satisfy the following limitations:

(1) Collar bolts shall be pretensioned $1^1/_4$-in.- (31.8-mm-) diameter high-strength bolts conforming to ASTM A574 with threads excluded from the shear plane and shall conform to the requirements of Sections 4.2 and 4.3.

(2) The collar bolts shall be pretensioned to the requirements for ASTM A490 bolts in the RCSC *Specification*.

(3) Welding of CCT, CCM and CCB pieces to form collar corner assemblies shall consist of partial joint penetration groove welds per Figure 10.5

(4) Welding of collar corner assemblies to columns shall consist of flare bevel groove welds with $^3/_8$ in. (10 mm) fillet reinforcing per Figure 10.6

(5) Collar flanges shall be welded to CWX pieces with $^5/_{16}$ in. (8 mm) fillet welds, each side per Figure 10.7

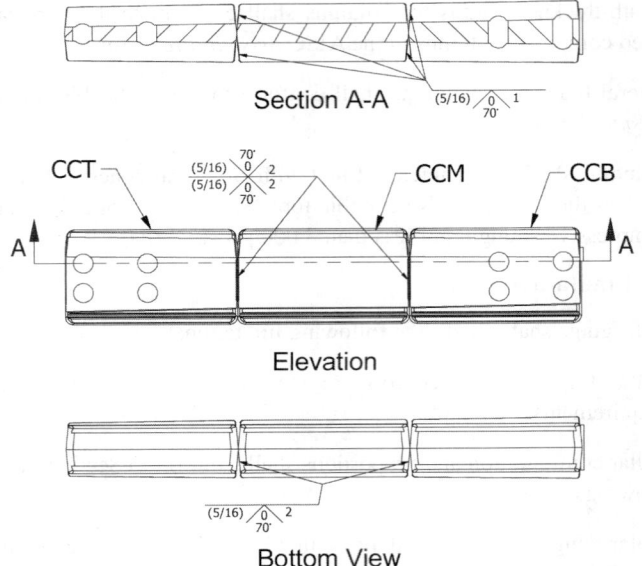

Fig. 10.5. Collar corner assembly welding.

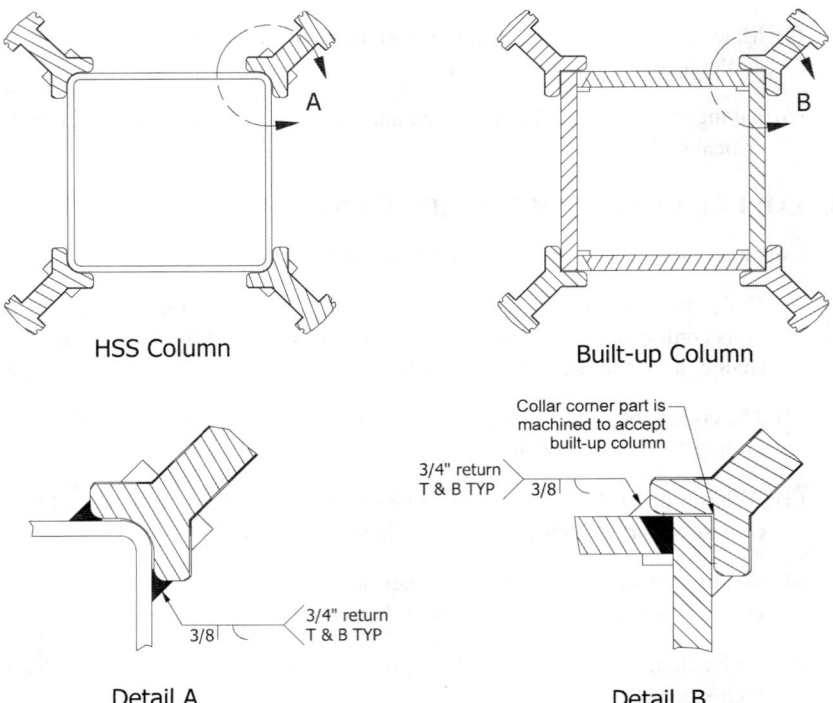

Fig. 10.6. Collar-corner-assembly-to-column weld, plan view.

(6) Beams shall be welded to collar flange assemblies with complete joint penetration groove welds per Figure 10.8

10.5 BEAM WEB-TO-COLLAR CONNECTION LIMITATIONS

Beam-web-to-collar connections shall satisfy the following limitations:

(1) The required shear strength of the beam web connection shall be determined according to Section 10.8.

(2) The beam web is welded to the collar web extension (CWX) with a two-sided fillet weld. The fillet welds shall be sized to develop the required shear strength of the connection.

10.6 BEAM FLANGE-TO-COLLAR FLANGE WELDING LIMITATIONS

Welding of the beam to the collar flange shall conform to the following limitations:

(1) Weld access holes are not allowed. Welding access to top and bottom flanges shall be made available by rotating the beam to allow a CJP weld in the flat position (Position 1G per AWS D1.1).

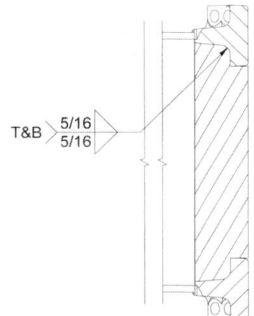

Fig. 10.7. Collar-web-extension-to-collar-flange welds, elevation.

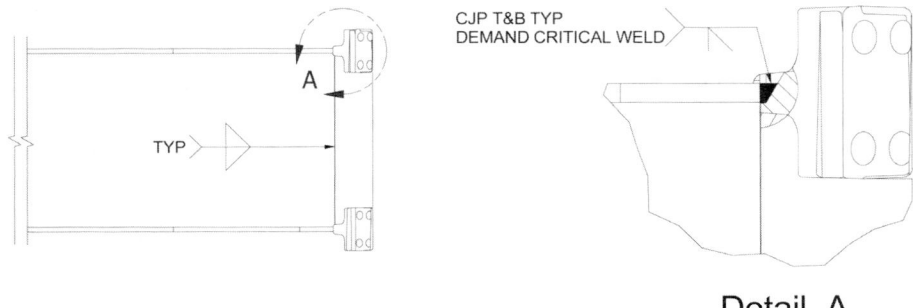

Fig. 10.8. Collar-flange-assembly-to-beam welds, elevation.

(2) The beam-flange-to-collar-flange weld shall be made with a CJP groove weld within the weld prep area of the collar flange. Reinforcing 5/16-in. (8 mm) fillet welds shall be placed on the back side of the CJP groove welds. The CJP flange weld shall conform to the requirements for demand critical welds in the AISC *Seismic Provisions* and AWS D1.8 and to the requirements of AWS D1.1.

10.7 COLUMN-BEAM RELATIONSHIP LIMITATIONS

Beam-to-column connections shall satisfy the following limitations:

(1) Panel zones shall conform to the applicable requirements of the AISC *Seismic Provisions*.

(2) Column-beam moment ratios shall be limited as follows:

(a) For SMF systems, the column-beam moment ratio about each principal axis shall conform to the requirements of the AISC *Seismic Provisions* considering simultaneous development of the expected plastic moments in the moment-connected beams framing into all sides of the ConXL node.

(b) For IMF systems, the column-beam moment ratio shall conform to the requirements of the AISC *Seismic Provisions*.

10.8 DESIGN PROCEDURE

Step 1. Compute the probable maximum moment at the plastic hinge, M_{pr}, in accordance with Section 2.4.3.

$$M_{pr} = C_{pr} R_y F_y Z_e \tag{2.4.3-1}$$

where

$$C_{pr} = \frac{F_y + F_u}{2F_y} \leq 1.2 \text{ (for RBS beams)} \tag{2.4.3-2}$$

$C_{pr} = 1.1$ (for non-RBS beams)
R_y = ratio of the expected yield stress to the specified minimum yield stress, F_y, as specified in the AISC *Seismic Provisions*
F_y = specified minimum yield stress of the yielding element, ksi (MPa)
F_u = specified minimum tensile strength of the yielding element, ksi (MPa)
Z_e = effective plastic section modulus of the section at the location of the plastic hinge, in.^{3} (mm^3)

For beams with an RBS cutout, the plastic hinge shall be assumed to occur at the center of the reduced section of beam flange. For beams without an RBS cutout, the plastic hinge shall be assumed to occur at a distance $d/2$ from the outside face of the collar (see Figure 10.9) where d is the beam depth.

Step 2. Compute the shear force, V_h, at the location of the plastic hinge at each end of the beam.

The shear force at each plastic hinge location shall be determined from a free body diagram of the portion of the beam between the plastic hinge locations. This calculation shall assume that the moment at the center of the plastic hinge is M_{pr} and shall consider gravity loads acting on the beams between plastic hinges in accordance with the equation:

$$V_h = \frac{2M_{pr}}{L_h} + V_{gravity} \qquad (10.8\text{-}1)$$

where
L_h = distance between plastic hinge locations, in. (mm)
$V_{gravity}$ = beam shear force resulting from $1.2D + f_1L + 0.2S$ (where f_1 is the load factor determined by the applicable building code for live loads, but not less than 0.5), kips (N)

User Note: The load combination of $1.2D + f_1L + 0.2S$ is in conformance with ASCE 7. When using the International Building Code, a factor of 0.7 must be used in lieu of the factor of 0.2 when the roof configuration is such that it does not shed snow off of the structure.

When concentrated loads are present on the beam between the points of plastic hinging they must be considered using standard considerations of statics when calculating the beam shear and using the same load combination.

Step 3. Confirm that columns are adequate to satisfy biaxial strong column-weak beam conditions. For the purpose of satisfying this requirement, it shall be permitted to take the yield strength of the column material as the specified F_y and to consider the full composite behavior of the column for axial load and flexural action.

User Note: The specified value of F_y need not be the minimum value associated with the grade of steel if project specifications require a higher minimum yield strength.

The value of ΣM^*_{pb} about each axis shall be taken equal to $\Sigma(M_{pr} + M_v)$, where M_{pr} is computed according to Equation 2.4.3-1, and where M_v is the additional moment due to the beam shear acting on a lever arm extending from the assumed point of plastic hinging to the centerline of the column. M_v on each side of the column can be computed as the quantity $V_h s_h$, where V_h is the shear at the point of theoretical plastic hinging, computed in accordance with Equation 10.8-1 and s_h is the distance of the assumed point of plastic hinging to the column centerline.

For beams with reduced beam section (RBS) cutout, the distance s_h shall be taken as the distance from the center of the column to the center of the reduced section of beam flange. For beams without an RBS cutout, the distance s_h shall be taken as the distance from the center of the column to a point one-half the beam depth ($d/2$) from the outside face of collar (see Figure 10.9).

The value of ΣM^*_{pc} about each axis shall be taken as:

$$\Sigma M^*_{pc} = M^*_{pcu} + M^*_{pcl} + \frac{\Sigma M^*_{pb}}{(H_u + H_l)} d \qquad (10.8\text{-}2)$$

where
- M^*_{pcu} = plastic moment nominal strength of the column above the node, about the axis under consideration considering simultaneous axial loading and loading about the transverse axis, kip-in. (N-mm)
- M^*_{pcl} = plastic moment nominal strength of the column below the node, about the axis under consideration considering simultaneous axial loading and loading about the transverse axis, kip-in. (N-mm)
- H_u = height of the story above the node, in. (mm)
- H_l = height of the story below the node, in. (mm)

For sections with equal properties about both axes, It shall be permitted to take M^*_{pcu} and M^*_{pcl} as:

$$M^*_{pcu} = M^*_{pcl} = 0.67 Z_c F_y \left(1 - \frac{P_u}{A_s F_y + 0.85 A_c f'_c}\right) \qquad (10.8\text{-}3)$$

where
- A_c = area of concrete in the column, in.² (mm²)
- A_s = area of steel in the column, in.² (mm²)
- f'_c = specified compressive strength of the concrete fill, ksi (MPa)
- P_u = axial load acting on the column at the section under consideration in accordance with the applicable load combination specified by the building code, but not considering amplified seismic load, kips (N)
- Z_c = plastic section modulus of the column about either axis, in.³ (mm³)

Step 4. Compute the moment at the collar bolts for each beam:

$$M_{bolts} = M_{pr} + V_h s_{bolts} \qquad (10.8\text{-}4)$$

where
- M_{bolts} = moment at collar bolts, kip-in. (N-mm)
- s_{bolts} = distance from center of plastic hinge to the centroid of the collar bolts, in. (mm) as given by the equations:

$$s_{bolts} = \frac{t_{collar}}{2} + a + \frac{b}{2} \quad \text{(for RBS beams)} \qquad (10.8\text{-}5)$$

$$s_{bolts} = \frac{t_{collar}}{2} + \frac{d}{2} \quad \text{(for non-RBS beams)} \qquad (10.8\text{-}6)$$

where
- a = distance from the outside face of the collar to the RBS cut, in. (mm)
- b = length of the RBS cut, in. (mm)
- t_{collar} = distance from the face of the column to the outside face of the collar, taken as 7 1/8 in. (181 mm) as illustrated in Figure 10.9

Step 5. Verify that the beam flange force does not exceed the available tensile strength of the bolts at the flange connection. The following relationship shall be satisfied for the collar bolts tensile strength:

$$\frac{r_{ut}}{\phi_d R_{pt}} = \frac{r_{ut}}{102} \leq 1.0 \qquad (10.8\text{-}7)$$

$$\frac{r_{ut}}{\phi_d R_{pt}} = \frac{r_{ut}}{454,000} \leq 1.0 \qquad \text{(S.I.)} \qquad (10.8\text{-}7\text{M})$$

where
r_{ut} = required collar bolt tension strength, kips (N)
$$= \frac{M_{bolts}}{n_{cf} d \sin 45°} = 0.177 \frac{M_{bolts}}{d} \qquad (10.8\text{-}8)$$
n_{cf} = number of collar bolts per collar flange
   = 8
R_{pt} = minimum bolt pretension, kips (N)

Step 6: Compute V_{bolts}, the probable maximum shear at the collar bolts, equal to the shear at the plastic hinge, V_h, plus any additional gravity loads between the plastic hinge and center of the collar flange, using the load combination of Step 2. Confirm that V_{bolts} is less than the slip critical, Class A bolt available slip resistance in accordance with the AISC *Specification* and using a resistance factor, ϕ, of unity.

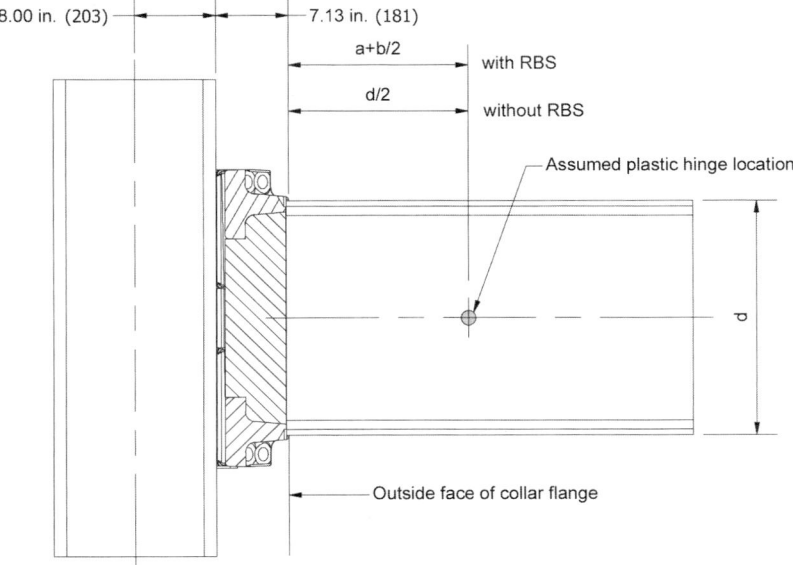

Fig. 10.9. Assumed plastic hinge location.

User Note: Note that for 1¹/₄ in. (31.8 mm) ASTM A574 bolts the value of T_b is the same as for 1¹/₄ in. (31.8 mm) ASTM A490 bolts and has a value of 102 kips (454 kN).

Step 7: Compute V_{cf}, the probable maximum shear at the face of collar flange, equal to the shear at the plastic hinge, V_h, plus any additional gravity loads between the plastic hinge and the outside face of the collar flange using the load combination of Step 2.

Check the design shear strength of the beam according to the requirements of the AISC *Specification* against V_{cf}.

Step 8: Determine required size of the fillet weld connecting the beam web to the collar web extension (CWX) using the following relationship:

$$t_f^{CWX} \geq \frac{\sqrt{2}V_{cf}}{\phi_n F_w l_w^{CWX}} \qquad (10.8\text{-}9)$$

where
t_f^{CWX} = fillet weld size required to join each side of the beam web to the CWX, in. (mm)
F_w = nominal weld design strength per the AISC *Specification*
    = $0.60 F_{EXX}$, ksi (MPa)
l_w^{CWX} = total length of available fillet weld at CWX, in. (mm), taken as 54 in. (1370 mm) for W30 (W760) sections; 48 in. (1220 mm) for W27 (W690) sections; 42 in. (1070 mm) for W24 (W610) sections; 36 in. (914 mm) for W21 (W530) sections; and 30 in. (762 mm) for W18 (W460) sections

Step 9: Compute V_f, the probable maximum shear at the face of column, equal to the shear at the plastic hinge, V_h, plus any additional gravity loads between the plastic hinge and the face of the column using the load combination of Step 2.

Determine size of fillet weld connecting collar corner assemblies to column using the following relationship:

$$t_f^{CC} \geq \frac{\sqrt{2}V_f}{\phi_n F_w l_w^{CC}} \qquad (10.8\text{-}10)$$

where
t_f^{CC} = fillet weld size required to join collar corner assembly to column, in. (mm)
l_w^{CC} = total length of available fillet weld at collar corner assembly, in. (mm), taken as 72 in. (1830 mm) for W30 (W760) sections; 66 in. (1680 mm) for W27 (W690) sections, 60 in. (1520 mm) for W24 (W610) sections; 54 in. (1370 mm) for W21 (W530) sections, and 48 in. (1220 mm) for W18 (W460) sections

Step 10: Determine the required shear strength of the column panel zone, R_u^{pz}, using the following relationship:

$$R_u^{pz} = \frac{\sum(M_{pr} + V_h s_f)}{d} - V_{col} \tag{10.8-11}$$

where
V_{col} = column shear, kips (N)
$$= \frac{\sum(M_{pr} + V_h s_h)}{H} \tag{10.8-12}$$
R_u^{pz} = required panel zone shear strength, kips (N)

s_f = distance from center of plastic hinge to face of column, in. (mm)

$$= t_{collar} + a + \frac{b}{2} \quad \text{(RBS beam)} \tag{10.8-13}$$

$$= t_{collar} + \frac{d}{2} \quad \text{(non-RBS beam)} \tag{10.8-14}$$

s_h = distance from the center of plastic hinge to center of column, in. (mm), given by the equations:

$$= \frac{d_{col}}{2} + t_{collar} + a + \frac{b}{2} \quad \text{(RBS beam)} \tag{10.8-15}$$

$$= \frac{d_{col}}{2} + t_{collar} + \frac{d}{2} \quad \text{(non-RBS beam)} \tag{10.8-16}$$

$$H = \frac{H_u + H_l}{2} \tag{10.8-17}$$

d_{col} = depth of the column, in. (mm)

Step 11: Determine the nominal design panel zone shear strength, ϕR_n^{pz}, using the following relationship:

$$\phi R_n^{pz} = \phi_d 0.6 F_y A_{pz} \tag{10.8-18}$$

where

$$A_{pz} = 2d_{col}t_{col} + 4\left(d_{leg}^{CC} t_{leg}^{CC}\right) \tag{10.8-19}$$

d_{leg}^{CC} = effective depth of the collar corner assembly leg, taken as $3^1/_2$ in. (89 mm)
t_{col} = wall thickness of HSS or built-up box column, in. (mm)
t_{leg}^{CC} = effective thickness of the collar corner assembly leg, taken as $^1/_2$ in. (12 mm)

User Note: If required strength exceeds available strength the designer may increase the column section and/or decrease the beam section strength assuring that all other design criteria are met.

10.9 PART DRAWINGS

Figures 10.10 through 10.15 provide indicate the dimensions of the various components of the ConXtech ConXL moment connection.

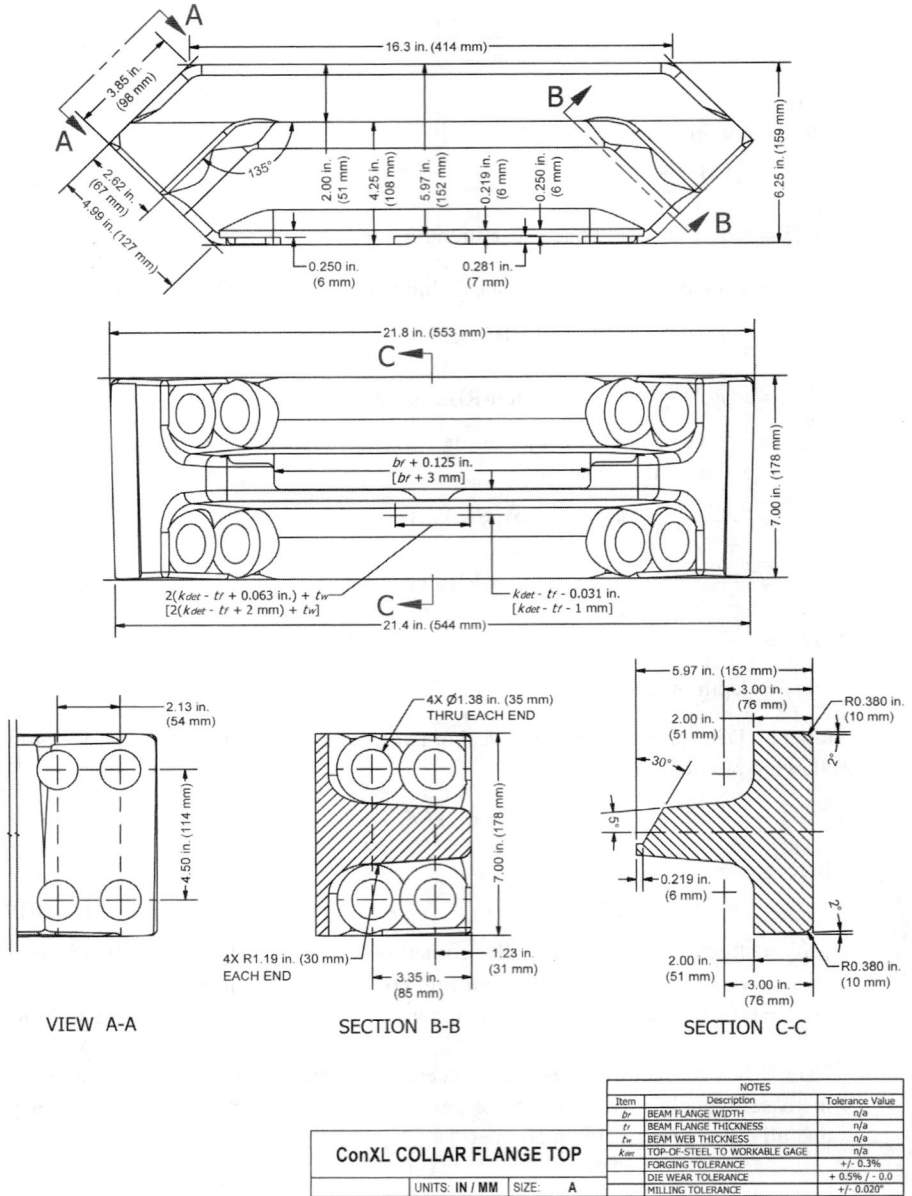

Fig. 10.10. Collar flange top (CFT).

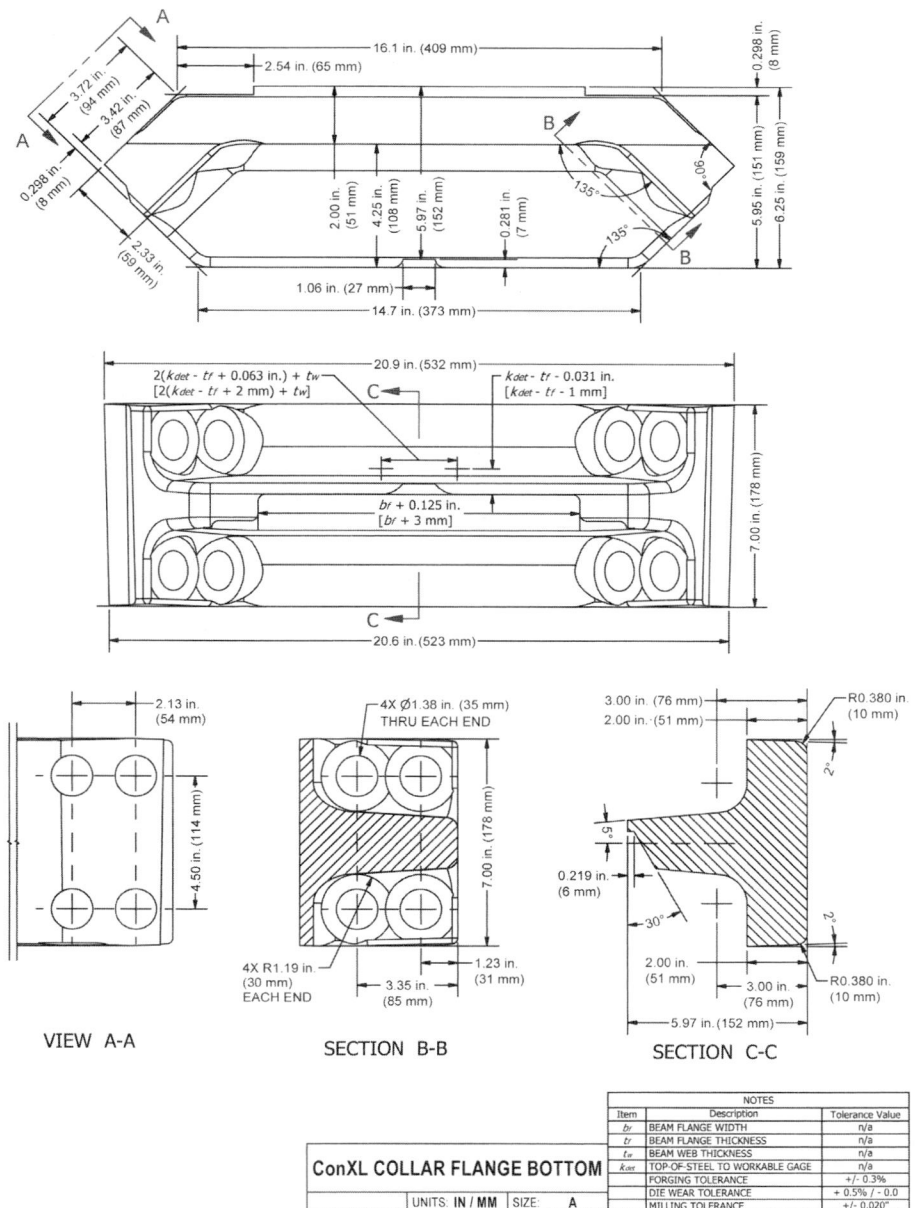

Fig. 10.11. Collar flange bottom (CFB).

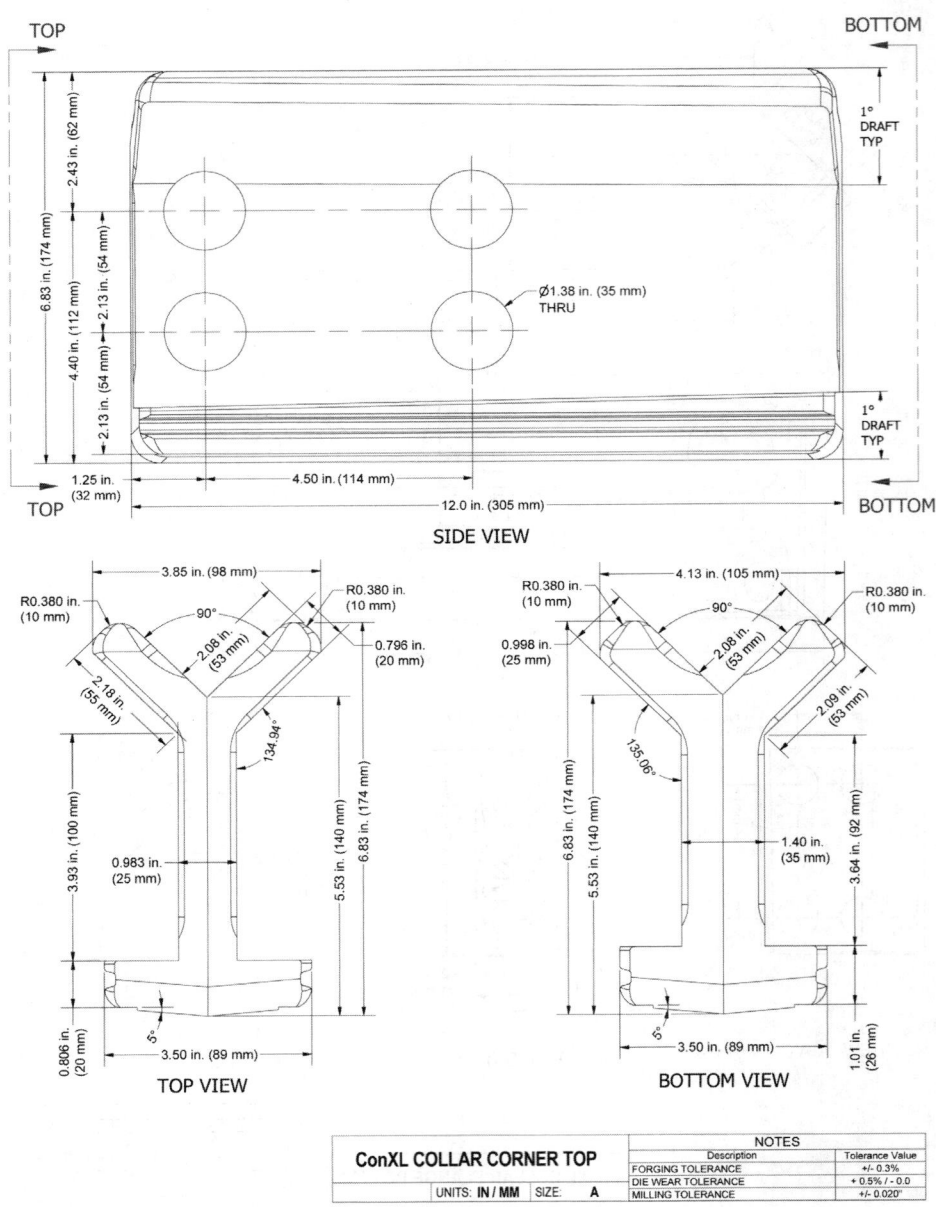

Fig. 10.12. Collar corner top (CCT).

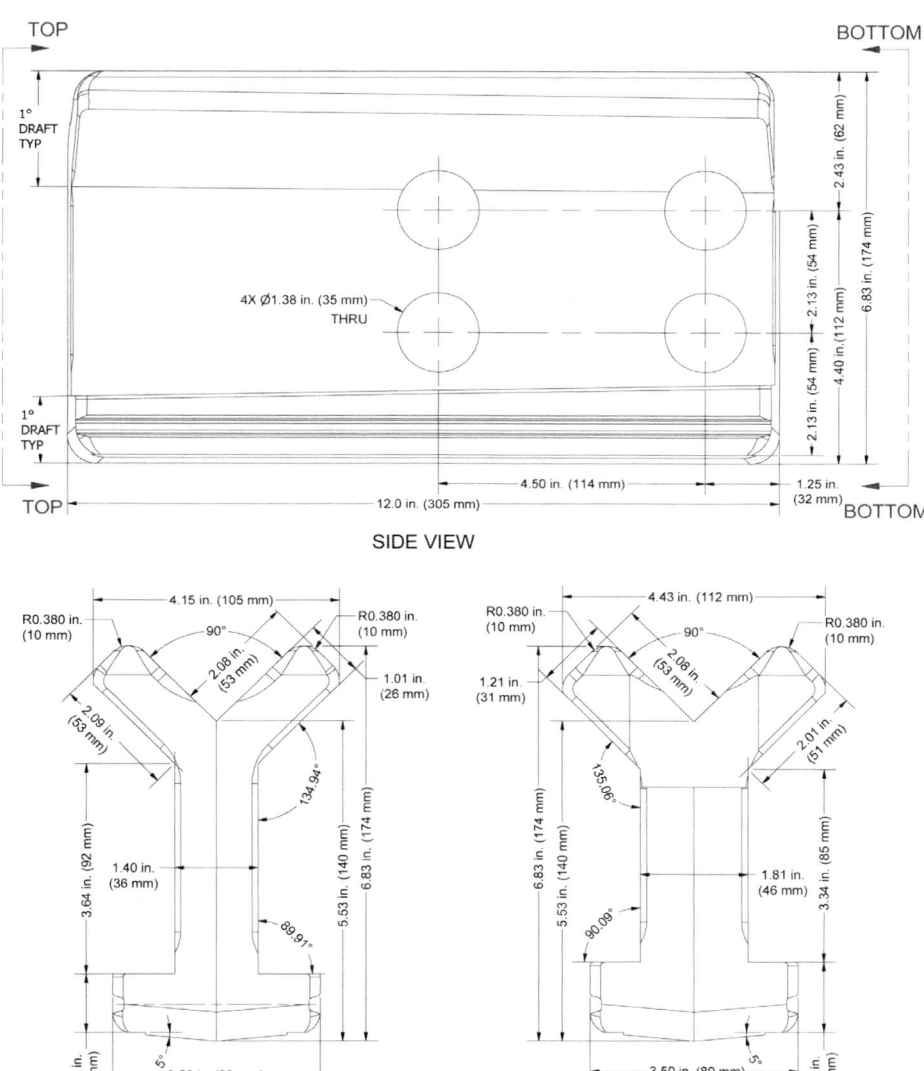

Fig. 10.13. Collar corner bottom (CCB).

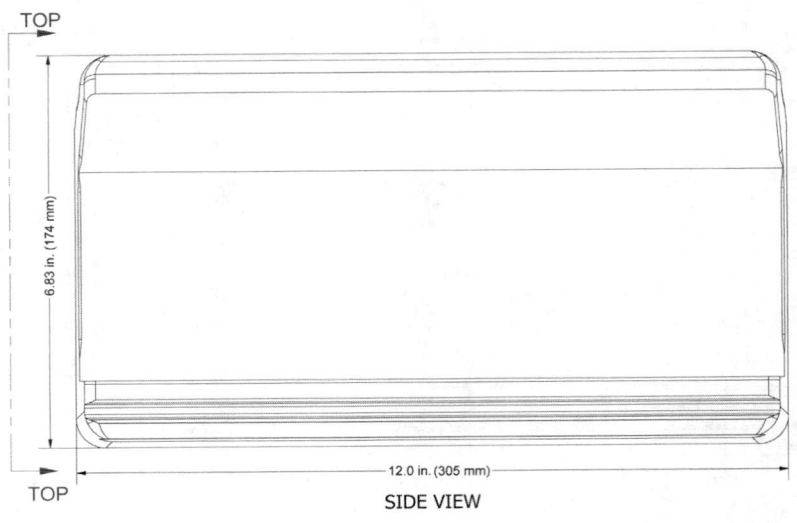
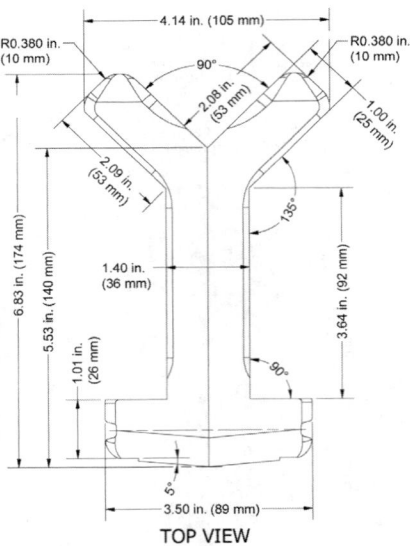

Fig. 10.14. Collar corner middle (CCM).

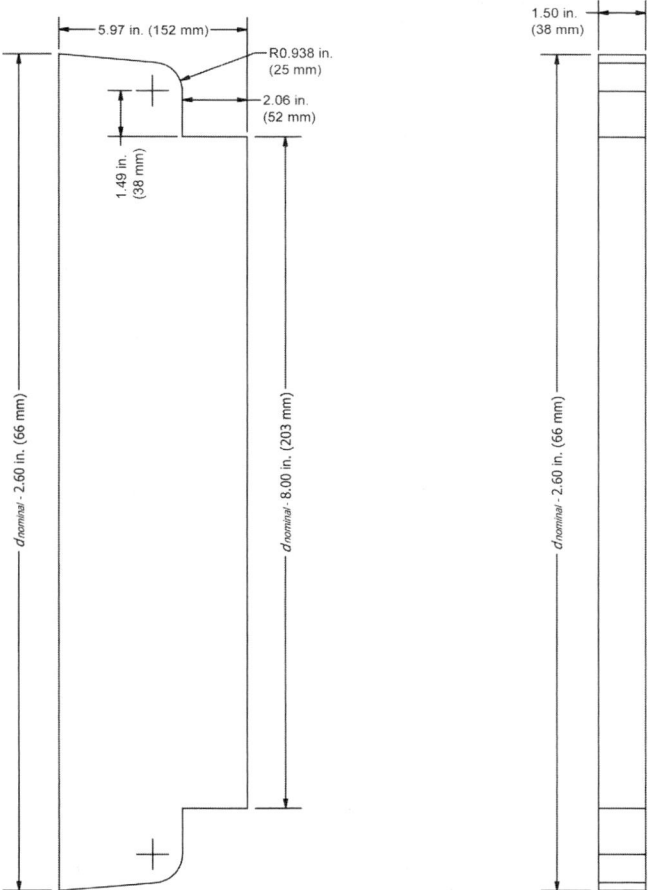

Fig. 10.15. Collar web extension (CWX).

APPENDIX A
CASTING REQUIREMENTS

A1. CAST STEEL GRADE

Cast steel grade shall be in accordance with ASTM A958/A958M Grade SC8620 class 80/50.

A2. QUALITY CONTROL (QC)

1. Inspection and Nondestructive Testing Personnel

Visual inspection and nondestructive testing shall be conducted by the manufacturer in accordance with a written practice by qualified inspectors. The procedure and qualification of inspectors is the responsibility of the manufacturer. Qualification of inspectors shall be in accordance with ASNT-TC-1a or an equivalent standard. The written practice shall include provisions specifically intended to evaluate defects found in cast steel products. Qualification shall demonstrate familiarity with inspection and acceptance criteria used in evaluation of cast steel products.

2. First Article Inspection (FAI) of Castings

The first article is defined as the first production casting made from a permanently mounted and rigged pattern. FAI shall be performed on the first casting produced from each pattern. The first article casting dimensions shall be measured and recorded. FAI includes visual inspection in accordance with Section A2.3, nondestructive testing in accordance with Section A2.4, tensile testing in accordance with Section A2.6, and Charpy V-notch testing in accordance with Section A2.7.

3. Visual Inspection of Castings

Visual inspection of all casting surfaces shall be performed to confirm compliance with ASTM A802/A802M and MSS SP-55 with a surface acceptance Level I.

4. Nondestructive Testing (NDT) of Castings

4a. Procedures

Radiographic testing (RT) shall be performed by quality assurance (QA) according to the procedures prescribed in ASTM E446 and ASTM E186 with an acceptance Level III or better.

Ultrasonic testing (UT) shall be performed by QA according to the procedures prescribed by ASTM A609/A609M Procedure A with an acceptance Level 3, or better.

Magnetic particle testing (MT) shall be performed by QA according to the procedures prescribed by ASTM E709 with an acceptance Level V, or better, in accordance with ASTM A903/A903M.

4b. **Required NDT**

(1) First Article
RT and MT shall be performed on the first article casting.

(2) Production Castings
UT shall be performed on 100% of the castings.
MT shall be performed on 50% of the castings.

(3) Reduction of Percentage of UT
The UT rate is permitted to be reduced if approved by the engineer of record and the authority having jurisdiction. The UT rate may be reduced to 25%, provided the number of castings not conforming to Section A2.4a is demonstrated to be 5% or less. A sampling of at least 40 castings shall be made for such reduction evaluation. This reduction is not permitted for castings with weld repairs.

(4) Reduction of Percentage of MT
The MT rate is permitted to be reduced if approved by the engineer of record and the authority having jurisdiction. The MT rate may be reduced to 10%, provided the number of castings not conforming to Section A2.4a is demonstrated to be 5% or less. A sampling of at least 20 castings shall be made for such reduction evaluation. This reduction is not permitted for castings with weld repairs.

5. **Weld Repair Procedures**

Castings with discontinuities that exceed the requirements of Section A2.4a shall be weld repaired. Weld repair of castings shall be performed in accordance with ASTM A488/A488M. The same testing method that discovered the discontinuities shall be repeated on repaired castings to confirm the removal of all discontinuities that exceed the requirements of Section A2.4a.

6. **Tensile Requirements**

Tensile tests shall be performed for each heat in accordance with ASTM A370 and ASTM 781/A781M.

7. **Charpy V-Notch (CVN) Requirements**

CVN testing shall be performed in accordance with ASTM A370 and ASTM 781/A781M. Three notched specimens shall be tested with the first heat, and with each subsequent 20th ton (18,100 kg) of finished material. The specimens shall have a minimum CVN toughness of 20 ft-lb (27 J) at 70 °F (21 °C).

8. **Casting Identification**

The castings shall be clearly marked with the pattern number and a unique serial number for each individual casting providing traceability to heat and production records.

A3. MANUFACTURER DOCUMENTS

1. Submittal to Patent Holder

The following documents shall be submitted to the patent holder, prior to the initiation of production as applicable:

(1) Material chemical composition report

(2) First article inspection report

2. Submittal to Engineer of Record and Authority Having Jurisdiction

The following documents shall be submitted to the engineer of record and the authority having jurisdiction, prior to, or with shipment as applicable:

(1) Production inspection and NDT reports

(2) Tensile and CVN test reports

(3) Weld repair reports

(4) Letter of approval by the patent holder of the manufacturer's FAI report

APPENDIX B

FORGING REQUIREMENTS

B1. FORGED STEEL GRADE

Raw material shall conform to the requirements of ASTM A572/A572M, Gr. 50 (345). Forging process shall conform to the requirements of ASTM A788 and ASTM A668. Mechanical properties shall conform to the requirements of Table B1.1.

B2. BAR STOCK

Bar stock shall be cut to billets appropriate to the part being forged. All billets shall be marked with heat number.

B3. FORGING TEMPERATURE

Billets shall be forged at a minimum temperature of 2150 °F (1180 °C) and a maximum temperature of 2250 °F (1230 °C).

B4. HEAT TREATMENT

Immediately following impression forging, part the part being forged shall be normalized for one hour at 1650 °F (900 °C) then air cooled.

B5. FINISH

Finished forgings shall have shot blast finish, clean of mill scale.

B6. QUALITY ASSURANCE

One sample of bar stock from each heat shall be cut to a length of 6 in. (152 mm) and forged to a 5 in. by 2 in.-thick bar (127 mm by 50 mm). Samples shall be marked with longitudinal and transverse directions. Chemistry and physical properties per Table B1.1 shall be verified to ASTM A572/A572M Gr. 50 (345) for both longitudinal and transverse direction on each sample.

TABLE B1.1
Required Mechanical Properties

Yield strength	50 ksi (345 MPa) minimum
Tensile strength	65 ksi (450 MPa) minimum
Elongation in 2 in. (50 mm)	22% minimum
Reduction of area	38% minimum
Charpy V-notch toughness	20 ft-lb at 70 °F (27 J at 21 °C)

Magnetic particle testing shall be conducted on the initial 12 pieces from each run to verify tooling and forging procedures. Cracks shall not be permitted. If cracks are found, the tooling or forging procedure shall be modified and an additional 12 initial pieces shall be tested. This process shall be repeated until 12 crack-free samples are obtained prior to production.

B7. DOCUMENTATION

Laboratory test data documenting chemistry, strength, elongation, reduction of area, and Charpy requirements for the samples tested in accordance with Section B6 shall be submitted.

Inspection reports documenting satisfactory performance of magnetic particle tests per Section B6 shall be submitted.

Certification of conformance with the requirements of this Appendix shall be submitted to the purchaser.

COMMENTARY
on Prequalified Connections for Special and Intermediate Steel Moment Frames for Seismic Applications

Including Supplement No. 1

(The Commentary is not part of ANSI/AISC 358-10, *Prequalified Connections for Special and Intermediate Steel Moment Frames for Seismic Applications*, or ANSI/AISC 358s1-11, *Supplement No. 1* to ANSI/AISC 358-10, but is included for informational purposes only.)

INTRODUCTION

The Standard is intended to be complete for normal design usage.

The Commentary furnishes background information and references for the benefit of the design professional seeking further understanding of the basis, derivations and limits of the Standard.

The Standard and Commentary are intended for use by design professionals with demonstrated engineering competence.

CHAPTER 1
GENERAL

1.1. SCOPE

Design of special moment frames (SMF) and intermediate moment frames (IMF) in accordance with the AISC *Seismic Provisions* and applicable building codes includes an implicit expectation that they will experience substantial inelastic deformations when subjected to design-level earthquake ground shaking, generally concentrated at the moment-resisting beam-to-column connections. In the 1994 Northridge earthquake, a number of steel moment frame buildings were found to have experienced brittle fractures that initiated at the welded beam flange-to-column flange joints of moment connections. These brittle fractures were unexpected and were quite different from the anticipated behavior of ductile beam flexural yielding in plastic hinge zones. Where they occurred, these brittle fractures prevented the formation of ductile plastic hinge zones and resulted in frame behavior substantially different from that upon which the design requirements for these systems were based.

Following this discovery, the Federal Emergency Management Agency (FEMA) provided funding to a coalition of universities and professional associations, known as the SAC Joint Venture. Over a period of six years, the SAC Joint Venture, with participation from AISC, AISI, AWS and other industry groups, conducted extensive research into the causes of the damage that had occurred in the Northridge earthquake and effective means of reducing the possibility of such damage in future earthquakes.

Numerous issues were identified in the SAC studies as contributing causes of these brittle fractures. This Standard specifically addresses the following four causes that were identified in the SAC study:

(1) Connection geometries that resulted in large stress concentrations in regions of high triaxiality and limited ability to yield;

(2) Use of weld filler metals with low inherent notch toughness and limited ductility;

(3) High variability in the yield strengths of beams and columns resulting in unanticipated zones of weakness in connection assemblies; and

(4) Welding practice and workmanship that fell outside the acceptable parameters of the AWS D1.1/D1.1M *Structural Welding Code* at that time.

A more complete listing of the causes of damage sustained in the Northridge earthquake may be found in a series of publications (FEMA 350, FEMA 351, FEMA 352, FEMA 353, FEMA 355C, and FEMA 355D) published in 2000 by the SAC Joint Venture that presented recommendations for design and construction of moment resisting frames designed to experience substantial inelastic deformation during design ground shaking. These recommendations included changes to material specifications for base metals and welding filler metals, improved quality assurance procedures during construction and the use of connection geometries that had been demonstrated by testing and analysis to be capable of resisting appropriate levels of inelastic deformation without fracture. Most of these recommendations have been incorporated into the AISC *Seismic Provisions for Structural Steel Buildings* (AISC, 2010) as well as into AWS D1.8/D1.8M *Structural Welding Code—Seismic Supplement* (AWS, 2009).

Following the SAC Joint Venture recommendations, the AISC *Seismic Provisions* require that moment connections used in special or intermediate steel moment frames be demonstrated by testing to be capable of providing the necessary ductility. Two means of demonstration are acceptable. One means consists of project-specific testing in which a limited number of full-scale specimens, representing the connections to be used in a structure, are constructed and tested in accordance with a protocol prescribed in Chapter K of the AISC *Seismic Provisions*. Recognizing that it is costly and time consuming to perform such tests, the AISC *Seismic Provisions* also provide for prequalification of connections consisting of a rigorous program of testing, analytical evaluation and review by an independent body, the Connection Prequalification Review Panel (CPRP). Connections contained in this Standard have met the criteria for prequalification when applied to framing that complies with the limitations contained herein and when designed and detailed in accordance with this Standard.

1.2. REFERENCES

References for this Standard are listed at the end of the Commentary.

1.3. GENERAL

Connections prequalified under this Standard are intended to withstand inelastic deformation primarily through controlled yielding in specific behavioral modes. To obtain connections that will behave in the indicated manner, proper determination of the strength of the connection in various limit states is necessary. The strength formulations contained in the LRFD method are consistent with this approach.

CHAPTER 2
DESIGN REQUIREMENTS

2.1. SPECIAL AND INTERMEDIATE MOMENT FRAME CONNECTION TYPES

Limitations included in this Standard for various prequalified connections include specification of permissible materials for base metals, mechanical properties for weld filler metals, member shape and profile, and connection geometry, detailing and workmanship. These limitations are based on conditions, demonstrated by testing and analytical evaluation, for which reliable connection behavior can be attained. It is possible that these connections can provide reliable behavior outside these limitations; however, this has not been demonstrated. When any condition of base metal, mechanical properties, weld filler metals, member shape and profile, connection geometry, detailing or workmanship falls outside the limitations specified herein, project-specific qualification testing should be performed to demonstrate the acceptability of connection behavior under these conditions.

Limited testing of connections of wide-flange beams to the webs of I-shaped columns had been conducted prior to the Northridge earthquake by Tsai and Popov (1986, 1988). This testing demonstrated that these "minor-axis" connections were incapable of developing reliable inelastic behavior even at a time when major axis connections were thought capable of developing acceptable behavior. No significant testing of such minor axis connections following the Northridge earthquake has been conducted. Consequently, such connections are not currently prequalified under this Standard.

Similarly, although there has been only limited testing of connections in assemblies subjected to biaxial bending of the column, the judgment of the CPRP was that as long as columns are designed to remain essentially elastic and inelastic behavior is concentrated within the beams, it would be possible to obtain acceptable behavior of beam-column connection assemblies subjected to biaxial loading. Flanged cruciform section columns, built-up box columns, and boxed wide-flange columns are permitted to be used in assemblies subjected to bi-axial loading for those connections types where inelastic behavior is concentrated in the beam, rather than in the column. It should be noted that the strong column—weak beam criteria contained in AISC 341 are valid only for planar frames. When both axes of a column participate in a moment frame, columns should be evaluated for the ability to remain essentially elastic while beams framing to both column axes undergo flexural hinging.

2.3. MEMBERS

2. Built-up Members

The behavior of built-up I-shaped members has been extensively tested in bolted end-plate connections and has been demonstrated to be capable of developing the necessary inelastic deformations. These members have not generally been tested in other prequalified connections; however, the conditions of inelastic deformation imposed on the built-up shapes in these other connection types are similar to those tested for the bolted end-plate connections.

2b. Built-up Columns

Four built-up column cross section shapes are covered by this Standard. These are illustrated in Figure C-2.1 and include:

(1) I-shaped welded columns that resemble standard rolled wide-flange shapes in cross section shape and profile.

(2) Cruciform W-shape columns, fabricated by splitting a wide-flange section in half and welding the webs on either side of the web of an unsplit wide-flange section at its mid-depth to form a cruciform shape, each outstanding leg of which terminates in a rectangular flange.

(3) Box columns, fabricated by welding four plates together to form a closed box-shaped cross section.

(4) Boxed W-shape columns constructed by adding side plates to the sides of an I-shaped cross section.

The preponderance of connection tests reviewed as the basis for prequalifications contained in this Standard consisted of rolled wide-flange beams connected to the flanges of rolled wide-flange columns. A limited number of tests of connections of wide-flange beams to built-up box section columns were also reviewed.

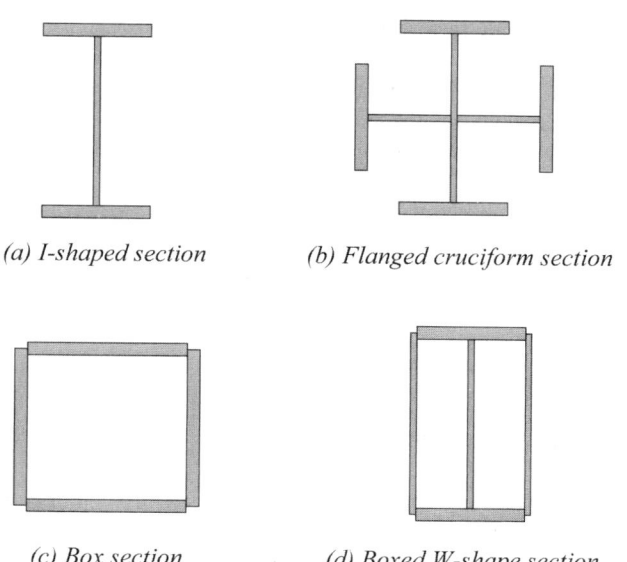

(a) I-shaped section   (b) Flanged cruciform section

(c) Box section   (d) Boxed W-shape section

Figure C-2.1. Column shapes. Plate preparation and welds are not shown.

The flanged cruciform column and boxed wide-flange columns have not specifically been tested. However, it was the judgment of the CPRP that as long as such column sections met the limitations for I-shaped sections and box-shaped sections, respectively, and connection assemblies are designed to ensure that most inelastic behavior occurred within the beam as opposed to the column, the behavior of assemblies employing these sections would be acceptable. Therefore, prequalification has been extended to these cross sections for connection types where the predominant inelastic behavior is in the beam rather than the column.

2.4. CONNECTION DESIGN PARAMETERS

1. Resistance Factors

A significant factor considered in the formulation of resistance factors is the occurrence of various limit states. Limit states that are considered brittle (non-ductile) and subject to sudden catastrophic failure are typically assigned lower resistance factors than those that exhibit yielding (ductile) failure. Because, for the prequalified connections, design demand is determined based on conservative estimates of the material strength of weak elements of the connection assembly, and materials, workmanship and quality assurance are more rigorously controlled than for other structural elements, resistance factors have been set somewhat higher than those traditionally used. It is believed that these resistance factors, when used in combination with the design, fabrication, erection and quality-assurance requirements contained in the Standard, will provide reliable service in the prequalified connections.

2. Plastic Hinge Location

This Standard specifies the presumed location of the plastic hinge for each prequalified connection type. In reality, inelastic deformation of connection assemblies is generally distributed to some extent throughout the connection assembly. The plastic hinge locations specified herein are based on observed behavior during connection assembly tests and indicate the locations of most anticipated inelastic deformation in connection assemblies conforming to the particular prequalified type.

3. Probable Maximum Moment at Plastic Hinge

The probable plastic moment at the plastic hinge is intended to be a conservative estimate of the maximum moment likely to be developed by the connection under cyclic inelastic response. It includes consideration of likely material overstrength and strain hardening.

4. Continuity Plates

Beam flange continuity plates serve several purposes in moment connections. They help to distribute beam flange forces to the column web, they stiffen the column web to prevent local crippling under the concentrated beam-flange forces and they minimize stress concentrations that can occur in the joint between the beam flange and column due to nonuniform stiffness of the attached elements.

Almost all connection assembly testing has been conducted on specimens that include a significant length (typically one half story height) of column above and below the beam or beams framing into the column. Thus, the condition that typically exists in a structure's top story, where the column terminates at the level of the beam top flange has not specifically been tested to demonstrate acceptable detailing. A cap plate detail similar to that illustrated in Figure C-2.2 is believed to be capable of providing reliable performance when connection elements do not extend above the beam top flange. In some connections, e.g. extended end plate and Kaiser bolted bracket connections, portions of the connection assembly extend above the column top flange. In such cases, the column should be extended to a sufficient height above the beam flange to accommodate attachment and landing of those connection elements. In such cases, stiffener plates should be placed in the column web, opposite the beam top flange, as is done at intermediate framing levels.

The attachment of continuity plates to column webs is designed to be capable of transmitting the maximum shear forces that can be delivered to the continuity plate. This may be limited by the beam-flange force, the shear strength of the continuity plate itself, or the welded joint between continuity plate and column flange.

The AISC *Seismic Provisions* require that continuity plates be attached to column flanges with CJP groove welds so the strength of the beam flange can be properly developed into the continuity plate. For single-sided connections in which a moment-connected beam attaches to only one of the column flanges, it is generally not necessary to use CJP groove welds to attach the continuity plate to the column flange

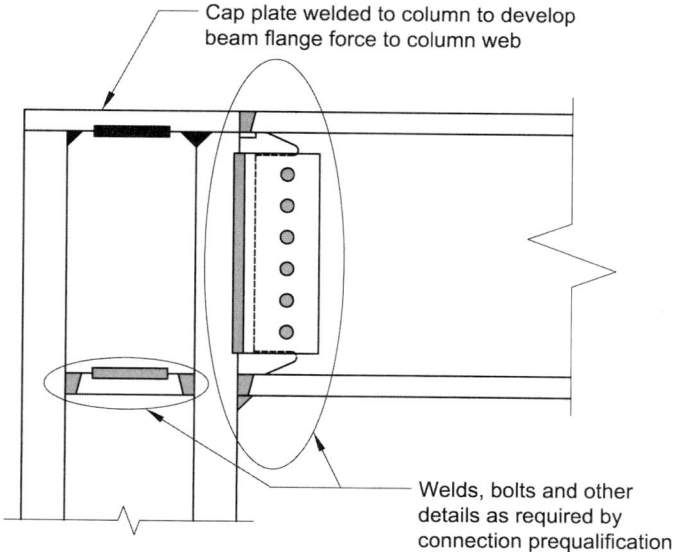

Figure C-2.2. Example cap plate detail at column top for RBS connection.

that does not have a beam attached. In such cases, acceptable performance can often be obtained by attaching the continuity plate to the column with a pair of minimum-size fillet welds.

When beams are moment connected to the side plates of boxed wide-flange column sections, continuity plates or cap plates should always be provided opposite the beam flanges, as is required for box section columns.

CHAPTER 3

WELDING REQUIREMENTS

3.3. BACKING AT BEAM-TO-COLUMN AND CONTINUITY PLATE-TO-COLUMN JOINTS

At the root of groove welds between beam flanges or continuity plates and column flanges, the inherent lack of a fusion plane between the left-in-place steel backing and the column flange creates a stress concentration and notch effect, even when the weld has uniform and sound fusion at the root. Further, when ultrasonic testing is performed, this left-in-place backing may mask significant flaws that may exist at the weld root. These flaws may create a more severe notch condition than that caused by the backing itself (Chi et al., 1997).

1. Steel Backing at Continuity Plates

The stress and strain level at the groove weld between a continuity plate and column flange is considerably different than that at the beam flange-to-column flange connection; therefore it is not necessary to remove the backing. The addition of the fillet weld beneath the backing makes the inherent notch at the interface an internal notch, rather than an external notch, reducing the notch effect. When backing is removed, the required reinforcing fillet weld reduces the stress concentration at the right-angle intersection of the continuity plate and the column flange.

2. Steel Backing at Beam Bottom Flange

The removal of backing, whether fusible or nonfusible, followed by backgouging to sound weld metal, is required so that potential root defects within the welded joint are detected and eliminated, and the stress concentration at the weld root is eliminated.

The influence of left-in-place steel backing is more severe on the bottom flange, as compared to the top flange, because at the bottom flange, the stress concentration from the backing occurs at the point of maximum applied and secondary tensile stresses in the groove weld, at the weld root, and at the outer fiber of the beam flange.

A reinforcing fillet weld with a $^5/_{16}$-in. (8-mm) leg on the column flange helps to reduce the stress concentration at the right-angle intersection of the beam flange and column flange, and is placed at the location of maximum stress. The fillet weld's horizontal leg may need to be larger than $^5/_{16}$ in. (8 mm) to completely cover the weld root area, eliminating the potential for multiple weld toes at the root that serve as small stress concentrations and potential fracture initiation points. When grinding the weld root and base metal area, previously deposited weld toe regions and their associated fracture initiation sites are removed, and the horizontal leg of the fillet weld need not be extended to base metal.

3. **Steel Backing at Beam Top Flange**

   Because of differences in the stress and strain conditions at the top and bottom flange connections, the stress/strain concentration and notch effect created by the backing/column interface at the top flange is at a lower level, compared to that at the bottom flange. Therefore, backing removal is not required. The addition of the reinforcing fillet weld makes the inherent notch at the interface an internal notch, rather than an external notch, further reducing the effect. Because backing removal, backgouging and backwelding would be performed through an access hole beneath the top flange, these operations should be avoided whenever possible.

4. **Prohibited Welds at Steel Backing**

   Tack welds for beam flange-to-column connections should be made within the weld groove. Tack welds or fillet welds to the underside of beam at the backing would direct stress into the backing itself, increasing the notch effect at the backing/column flange interface. In addition, the weld toe of the tack weld or fillet weld on the beam flange would act as a stress concentration and a potential fracture initiation site.

   Proper removal of these welds is necessary to remove the stress concentration and potential fracture initiation site. Any repair of gouges and notches by filling with weld metal must be made using filler metals with the required notch toughness.

5. **Nonfusible Backing at Beam Flange-to-Column Joints**

   After nonfusible backing is removed, backgouging to sound metal removes potential root flaws within the welded joint. A reinforcing fillet weld with a $5/16$-in. (8-mm) leg on the column flange helps reduce the stress concentration at the right-angle intersection of the beam flange and column flange.

   The fillet weld's horizontal leg may need to be larger than $5/16$ in. (8 mm) to completely cover the weld root area, eliminating the potential for small stress concentrations and potential fracture initiation points. When grinding the weld root and base metal area, previously deposited weld toe regions and their associated fracture initiation sites are removed, therefore the horizontal leg of the fillet weld need not be extended to base metal.

3.4. DETAILS AND TREATMENT OF WELD TABS

Weld tabs are used to provide a location for initiation and termination of welds outside the final weld location, improving the quality of the final weld. The removal of weld tabs is performed to remove the weld discontinuities and defects that may be present at these start and stop locations. Because weld tabs are located at the ends of welds, any remaining weld defects at the weld-end removal areas may act as external notches and fracture initiation sites and are therefore removed. A smooth transition is needed between base metal and weld to minimize stress concentrations.

3.5. TACK WELDS

Tack welds outside weld joints may create unintended load paths and may create stress concentrations that become crack initiation sites when highly strained. By placing tack welds within the joint, the potential for surface notches and hard heat affected zones (HAZs) is minimized. When placed within the joint, the HAZ of a tack weld is tempered by the subsequent passes for the final weld.

Tack welds for beam flange-to-column connections are preferably made in the weld groove. Tack welds of backing to the underside of beam flanges would be unacceptable, and any tack welds between weld backing and beam flanges are to be removed in accordance with Section 3.3.4. Steel backing may be welded to the column under the beam flange, where a reinforcing fillet is typically placed.

When tack welds for the attachment of weld tabs are placed within the weld joint, they become part of the final weld.

3.6. CONTINUITY PLATES

The rotary straightening process used by steel rolling mills to straighten rolled sections cold works the webs of these shapes in and near the k-area. This cold working can result in an increase in hardness, yield strength, ultimate tensile strength, and yield-to-tensile ratio; and a decrease in notch toughness. In some instances, Charpy V-notch toughness has been recorded to be less than 2 ft-lb at 70 °F [3 J at 20 °C] (Barsom and Korvink, 1998). These changes do not negatively influence the in-service behavior of uncracked shapes. However, the potential for post-fabrication k-area base metal cracking exists in highly restrained joints at the weld terminations for column continuity plates, web doublers, and thermal cut coped beams.

When the minimum clip dimensions are used along the member web, the available continuity plate length must be considered in the design and detailing of the welds to the web. For fillet welds, the fillet weld should be held back one to two weld sizes from each clip. For groove welds, weld tabs should not be used in the k-area since they could cause base metal fracture from the combination of weld shrinkage, the stress concentration/notch effect at the weld end, and the low notch-toughness web material.

When the maximum clip dimensions are used along the member flange, the width, hence the capacity, of the continuity plate is not reduced substantially. Care must be used in making quality weld terminations near the member radius, as the use of common weld tabs is difficult. If used, their removal in this region may damage the base metal, necessitating difficult repairs. The use of cascaded ends within the weld groove may be used within the dimensional limits stated. Because of the incomplete filling of the groove, the unusual configuration of the weld, and the relatively low level of demand placed upon the weld at this location, nondestructive testing of *cascaded weld ends* in groove welds at this location are not required.

3.7. QUALITY CONTROL AND QUALITY ASSURANCE

Chapter J of the AISC *Seismic Provisions* specifies the minimum requirements for a quality assurance plan for the seismic load resisting system. It may be appropriate to supplement the Chapter J provisions with additional requirements for a particular project based on the qualifications of the contractor(s) involved and their demonstrated ability to produce quality work. Contract documents are to define the quality control (QC) and quality assurance (QA) requirements for the project.

QC includes those tasks to be performed by the contractor to ensure that their materials and workmanship meet the project's quality requirements. Routine welding QC items include personnel control, material control, preheat measurement, monitoring of welding procedures, and visual inspection.

QA includes those tasks to be performed by an agency or firm other than the contractor. QA includes monitoring of the performance of the contractor in implementing the contractor's QC program, ensuring that designated QC functions are performed properly by the contractor on a routine basis. QA may also include specific inspection tasks that are included in the contractor's QC plan, and may include nondestructive testing of completed joints.

CHAPTER 4

BOLTING REQUIREMENTS

4.1. FASTENER ASSEMBLIES

ASTM F1852 twist-off type tension-control fastener assemblies are appropriate equivalents for ASTM A325 bolts. ASTM F2280 twist-off type tension control fastener assemblies are appropriate substitutes for ASTM A490. Such assemblies are commonly produced and used, and are addressed by the RCSC *Specification* (RCSC, 2009).

4.2. INSTALLATION REQUIREMENTS

Section D2 of the AISC *Seismic Provisions* designates all bolted joints to be pretensioned joints, with the additional requirement that the joint's faying surfaces meet Class A conditions for slip-critical joints. Some connection types designate the bolted joint to be designed as slip-critical, and others waive the faying surface requirements of the AISC *Seismic Provisions*.

4.3. QUALITY CONTROL AND QUALITY ASSURANCE

See Commentary Section 3.7.

CHAPTER 5

REDUCED BEAM SECTION (RBS) MOMENT CONNECTION

5.1. GENERAL

In a reduced beam section (RBS) moment connection, portions of the beam flanges are selectively trimmed in the region adjacent to the beam-to-column connection. In an RBS connection, yielding and hinge formation are intended to occur primarily within the reduced section of the beam, and thereby limit the moment and inelastic deformation demands developed at the face of the column.

A large number of RBS connections have been tested under a variety of conditions by different investigators at institutions throughout the world. A listing of relevant research is presented in the references at the end of this document. Review of available test data indicates that RBS specimens, when designed and constructed according to the limits and procedures presented herein, have developed interstory drift angles of at least 0.04 rad under cyclic loading on a consistent basis. Tests on RBS connections show that yielding is generally concentrated within the reduced section of the beam and may extend, to a limited extent, to the face of the column. Peak strength of specimens is usually achieved at an interstory drift angle of approximately 0.02 to 0.03 rad. Specimen strength then gradually reduces due to local and lateral-torsional buckling of the beam. Ultimate failure typically occurs at interstory drift angles of approximately 0.05 to 0.07 rad, by low cycle fatigue fracture at local flange buckles within the RBS.

RBS connections have been tested using single-cantilever type specimens (one beam attached to column), and double-sided specimens (specimen consisting of a single column, with beams attached to both flanges). Tests have been conducted primarily on bare-steel specimens, although some testing is also reported on specimens with composite slabs. Tests with composite slabs have shown that the presence of the slab provides a beneficial effect by helping to maintain the stability of the beam at larger interstory drift angles.

Most RBS test specimens were tested pseudo-statically, using a loading protocol in which applied displacements are progressively increased, such as the loading protocol specified in ATC-24 (ATC, 1992) and the loading protocol developed in the FEMA/SAC program and adopted in Chapter K of the AISC *Seismic Provisions*. Two specimens were tested using a loading protocol intended to represent near-source ground motions that contain a large pulse. Several specimens were also tested dynamically. Radius-cut RBS specimens have performed well under all of these loading conditions. See Commentary Section 5.7 for a discussion of other shapes of RBS cuts.

5.2. SYSTEMS

Review of the research literature presented in the reference section at the end of this document and summarized in Commentary Section 5.1 indicates that the radius-cut RBS connection meets the prequalification requirements in Chapter K, Section K1 of the AISC *Seismic Provisions* for special and intermediate moment frames.

5.3. PREQUALIFICATION LIMITS

1. Beam Limitations

A wide range of beam sizes has been tested with the radius-cut RBS. The smallest beam size reported in the literature was a Canadian W530×82, roughly equivalent to a W21×50. The heaviest beam reported was a W36×300 (W920×446) (FEMA, 2000e), no longer produced. Although the AISC *Seismic Provisions* permit limited increases in beam depth and weight compared to the maximum sections tested, the prequalification limits for maximum beam depth and weight were established based on the test data for W36×300 (W920×446). It was the judgment of the CPRP that for the purposes of establishing initial prequalification limits, adherence to the maximum tested specimen would be appropriately conservative. There is no evidence that modest deviations from the maximum tested specimen would result in significantly different performance, and the limit on maximum flange thickness is approximately 4% thicker than the 1.68 in. (43 mm) flange in a W36×300 (W920×446).

Beam depth and beam span-to-depth ratio are significant in the inelastic behavior of beam-to-column connections. For the same induced curvature, deep beams will experience greater strains than shallower beams. Similarly, beams with shorter span-to-depth ratio will have a sharper moment gradient across the beam span, resulting in reduced length of the beam participating in plastic hinging and increased strains under inelastic rotation demands. Most of the beam-to-column assemblies that have been tested had configurations approximating beam spans of about 25 ft (7.6 m) and beam depths varying from W30 (W760) to W36 (W920) so that beam span-to-depth ratios were typically in the range of eight to ten (FEMA, 2000e). Given the degree to which most specimens significantly exceeded the minimum interstory drift demands, it was judged reasonable to set the minimum span-to-depth ratio at seven for SMF and five for IMF.

Local buckling requirements for members subjected to significant inelastic rotation are covered in the AISC *Seismic Provisions*. For the purposes of calculating the width-to-thickness ratio, it is permitted to take the flange width at the two-thirds point of the RBS cut. This provision recognizes that the plastic hinge of the beam forms within the length of the RBS cut where the width of the flange is less than at the uncut section. This provision will result in a lower width-to-thickness ratio when taken at the RBS cut compared to that at the uncut section. Many of the RBS tests conducted as a part of the FEMA/SAC program used a W30×99 (W760×147) beam that does not quite satisfy the flange width-to-thickness ratio at the uncut section.

Nevertheless, the tests were successful. For these reasons, it was judged reasonable to permit the calculation of the width-to-thickness ratio a reasonable distance into the RBS cut.

In developing this prequalification, the CPRP also reviewed lateral bracing requirements for beams with RBS connections. Some concerns were raised in the past that the presence of the RBS flange cuts might make the beam more prone to lateral-torsional buckling and that supplemental lateral bracing should be provided at the RBS. The issue of lateral bracing requirements for beams with RBS connections was subsequently investigated in both experimental and analytical studies (FEMA, 2000f; Yu et al., 2000). These studies indicated that for bare steel specimens (no composite slab), interstory drift angles of 0.04 rad can be achieved without a supplemental lateral brace at the RBS, as long as the normal lateral bracing required for beams in SMF systems is provided, per Section D1.2b of the AISC *Seismic Provisions*.

Studies also indicated that although supplemental bracing is not required at the RBS to achieve 0.04 rad interstory drift angles, the addition of a supplemental brace can result in improved performance. Tests on RBS specimens with composite slabs indicated that the presence of the slab provided a sufficient stabilizing effect that a supplemental brace at the RBS is not likely to provide significantly improved performance (FEMA, 2000f; Engelhardt, 1999; Tremblay et al., 1997). Based on the available data, beams with RBS connections that support a concrete structural slab are not required to have a supplemental brace at the RBS.

In cases where a supplemental brace is provided, the brace should not be connected within the reduced section (protected zone). Welded or bolted brace attachments in this highly strained region of the beam may serve as fracture initiation sites. Consequently, if a supplemental brace is provided, it should be located at or just beyond the end of the RBS that is farthest from the face of the column.

The protected zone is defined as shown in Figure 5.1 and extends from the face of the column to the end of the RBS farthest from the column. This definition is based on test observations that indicate yielding typically does not extend past the far end of the RBS cut.

2. **Column Limitations**

   Nearly all tests of RBS connections have been performed with the beam flange welded to the column flange (i.e., strong-axis connections). The limited amount of weak-axis testing has shown acceptable performance. In the absence of more tests, the CPRP recommended limiting prequalification to strong-axis connections only.

   The majority of RBS specimens were constructed with W14 (W360) columns. However, a number of tests have also been conducted using deeper columns, including W18, W27 and W36 (W460, W690 and W920) columns. Testing of deep-column RBS specimens under the FEMA/SAC program indicated that stability problems may occur when RBS connections are used with deep columns (FEMA, 2000f). In FEMA 350 (FEMA, 2000b), RBS connections were only prequalified for W12 (W310) and W14 (W360) columns.

The specimens in the FEMA/SAC tests conducted showed a considerable amount of column twisting (Gilton et al., 2000). However, two of the three specimens tested achieved 0.04-rad rotation, albeit with considerable strength degradation. The third specimen just fell short of 0.04-rad rotation and failed by fracture of the column web near the k-area. Subsequent study attributed this fracture to column twisting.

Subsequent to the FEMA/SAC tests, an analytical study (Shen et al., 2002) concluded that boundary conditions used in these tests may not be representative of what would be found in an actual building. Consequently, the large-column twisting (and presumably resultant k-area column fracture) seen in the FEMA/SAC tests would not be expected in real buildings. The study also concluded that deep columns should not behave substantially different from W14 (W360) columns and that no special bracing is needed when a slab is present. This was followed by a more extensive analytical and large-scale experimental investigation on RBS connections with columns up to W36 (W920) in depth (Ricles et al., 2004). This investigation showed that good performance can be achieved with deep columns when a composite slab is present or when adequate lateral bracing is provided for the beam and/or column in the absence of a slab. Based on a review of this recent research, the prequalification of RBS connections is extended herein to include W36 (W920) columns.

The behavior of RBS connections with cruciform columns is expected to be similar to that of a rolled wide-flange column because the beam flange frames into the column flange, the principal panel zone is oriented parallel to that of the beam and the web of the cut wide-flange column is to be welded with a CJP groove weld to the continuous web one foot above and below the depth of the frame girder. Given these similarities and the lack of evidence suggesting behavioral limit states different from those associated with rolled wide-flange shape, cruciform column depths are limited to those imposed on wide-flange shapes.

Successful tests have also been conducted on RBS connections with built-up box columns. The largest box column for which test data was available was 24 in. by 24 in. (610 mm by 610 mm). Consequently, RBS connections have been prequalified for use with built-up box columns up to 24 in. (610 mm). Limits on the width-to-thickness ratios for the walls of built-up box columns are specified in Section 2.3.2b(3) and were chosen to reasonably match the box columns that have been tested.

The use of box columns participating in orthogonal moment frames, that is, with RBS connections provided on orthogonal beams, is also prequalified. Although no data were available for test specimens with orthogonal beams, this condition should provide ostensibly the same performance as single-plane connections, since the RBS does not rely on panel zone yielding for good performance, and the column is expected to remain essentially elastic for the case of orthogonal connections.

Based on successful tests on wide-flange columns and on built-up box columns, boxed wide-flange columns would also be expected to provide acceptable performance. Consequently, RBS connections are prequalified for use with boxed wide-flange columns. When moment connections are made only to the flanges of the wide-flange portion of the boxed wide-flange, the column may be up to W36 in depth. When the boxed wide-flange column participates in orthogonal moment frames, then neither the depth nor the width of the column is allowed to exceed 24 in. (610 mm), applying the same limits as for built-up boxes.

5.4. COLUMN-BEAM RELATIONSHIP LIMITATIONS

Column panel zone strength provided on RBS test specimens has varied over a wide range. This includes specimens with very strong panel zones (no yielding in the panel zone), specimens with very weak panel zones (essentially all yielding in the panel zone and no yielding in the beam), and specimens where yielding has been shared between the panel zone and the beam. Good performance has been achieved for all levels of panel zone strength (FEMA, 2000f), including panel zones that are substantially weaker than permitted in Section E3.6e of the AISC *Seismic Provisions*. However, there are concerns that very weak panel zones may promote fracture in the vicinity of the beam-flange groove welds due to "kinking" of the column flanges at the boundaries of the panel zone. Consequently, the minimum panel zone strength specified in Section E3.6e of the AISC *Seismic Provisions* is required for prequalified RBS connections.

5.5. BEAM FLANGE-TO-COLUMN FLANGE WELD LIMITATIONS

Complete-joint-penetration groove welds joining the beam flanges to the column flanges provided on the majority of RBS test specimens have been made by the self-shielded flux cored arc welding process (FCAW-S) using electrodes with a minimum specified Charpy V-notch toughness. Three different electrode designations have commonly been used in these tests: E71T-8, E70TG-K2 and E70T-6. Further, for most specimens, the bottom flange backing was removed and a reinforcing fillet added, top flange backing was fillet welded to the column, and weld tabs were removed at both the top and bottom flanges.

Test specimens have employed a range of weld access-hole geometries, and results suggest that connection performance is not highly sensitive to the weld access-hole geometry. Consequently, prequalified RBS connections do not require specific access-hole geometry. Weld access holes should satisfy the requirements of Section 6.10 of AWS D1.8/D1.8M (AWS, 2009). The alternative geometry for weld access holes specified in Section 6.10.1.2 of AWS D1.8/D1.8M is not required for RBS connections.

5.6. BEAM WEB-TO-COLUMN CONNECTION LIMITATIONS

Two types of web connection details have been used for radius-cut RBS test specimens: a welded and a bolted detail. In the welded detail, the beam web is welded directly to the column flange using a complete-joint-penetration groove weld. For the bolted detail, pretensioned high-strength bolts are used. Specimens with both types of web connections have achieved at least 0.04-rad interstory drift angles, and consequently both types of web connection details were permitted for RBS connections in FEMA 350 (2000b).

Previous test data (Engelhardt et al., 2000) indicate that beyond an interstory drift angle of 0.04 rad, specimens with bolted web connections show a higher incidence of fracture occurring near the beam-flange groove welds, as compared to specimens with welded web connections. Thus, while satisfactory performance is possible with a bolted web connection, previous test data indicate that a welded web is beneficial in reducing the vulnerability of RBS connections to fracture at the beam-flange groove welds.

Subsequent to the SAC/FEMA testing on RBS connections, a test program (Lee et al., 2004) was conducted that directly compared nominally identical RBS connections except for the web connection detail. The RBS specimens with welded web connections achieved 0.04-rad interstory drift angle, whereas RBS specimens with bolted web connections failed to achieve 0.04 rad.

Thus, while past successful tests have been conducted on RBS connections with bolted web connections, recent data has provided contradictory evidence, suggesting bolted web connections may not be suitable for RBS connections when used for SMF applications. Until further data is available, a welded web connection is required for RBS connections prequalified for use in SMF. For IMF applications, bolted web connections are acceptable.

The beam web-to-plate CJP groove weld is intended to extend the full distance between the weld access holes to minimize the potential for crack-initiation at the ends of the welds, hence the requirement for the plate to extend from one weld access hole to the other. All specimens were tested with the full-depth weld configuration.

5.7. FABRICATION OF FLANGE CUTS

Various shapes of flange cutouts are possible for RBS connections, including a constant cut, a tapered cut, and a radius cut. Experimental work has included successful tests on all of these types of RBS cuts. The radius cut avoids abrupt changes of cross section, reducing the chances of a premature fracture occurring within the reduced section. Further, the majority of tests reported in the literature used radius-cut RBS sections. Consequently, only the radius-cut RBS shape is prequalified.

An issue in the fabrication of RBS connections is the required surface finish and smoothness of the RBS flange cuts. No research data was found that specifically addressed this issue. Consequently, finish requirements for RBS cuts were chosen by the CPRP based on judgment and are consistent with those specified in FEMA 350 (2000b).

5.8. DESIGN PROCEDURE

Dimensions of the RBS cuts for the test specimens reported in the literature vary over a fairly small range. The distance from the face of the column to the start of the RBS radius cut (designated as a in Figure 5.1) ranged from 50 to 75% of the beam-flange width. The length of the cuts (designated as b in Figure 5.1) varied from approximately 75 to 85% of the beam depth. The amount of flange width removed at the minimum section of the RBS varied from about 38 to 55%. Flange removal for prequalified RBS connections is limited to a maximum of 50%, to avoid excessive loss of strength or stiffness.

The design procedure presented herein for prequalified RBS connections is similar to that presented in FEMA 350 (2000b). The overall basis for sizing the RBS radius cut in this design procedure is to limit the maximum beam moment that can develop at the face of the column to the actual plastic moment (based on expected yield stress) of the beam when the minimum section of the RBS is fully yielded and strain hardened. Test data indicate that connecting the beam at the face of the column in accordance with the requirements herein allows the connection to resist this level of moment while minimizing the chance of fracture at the beam-flange groove welds.

Step 4 of the design procedure requires computation of the shear force at the center of the RBS radius cut. This shear force is a function of the gravity load on the beam

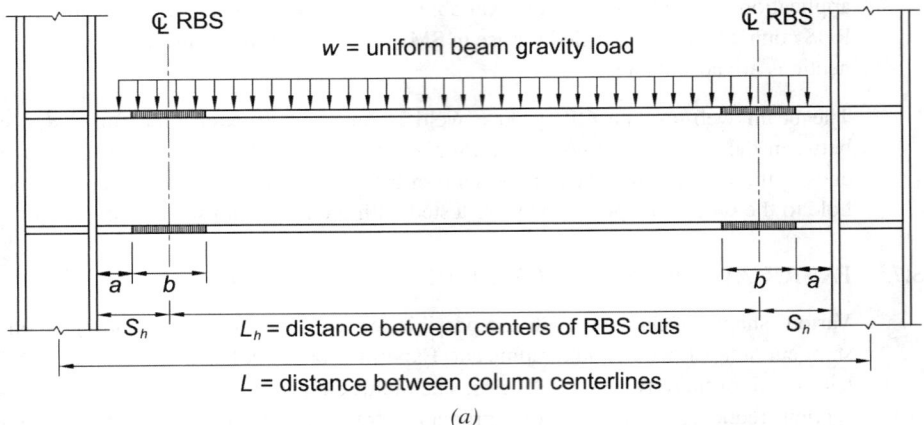

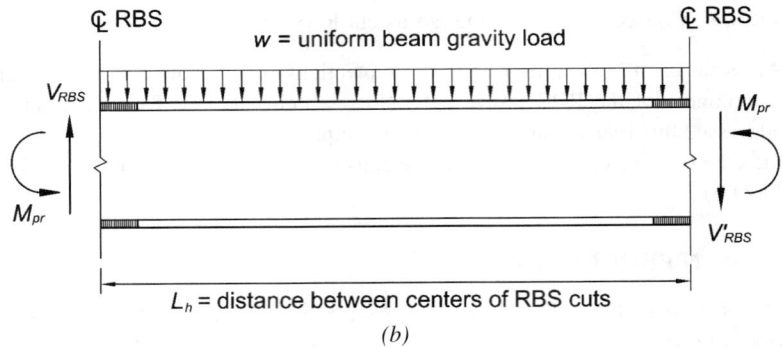

Figure C-5.1. Example calculation of shear at center of RBS cuts.
(a) Beam with RBS cuts and uniform gravity load;
(b) Free-body diagram of beam between RBS cuts and calculation of shear at RBS.

and the plastic moment capacity of the RBS. An example calculation is shown in Figure C-5.1 for the case of a beam with a uniformly distributed gravity load.

For gravity load conditions other than a uniform load, the appropriate adjustment should be made to the free-body diagram in Figure C-5.1 and to Equations C-5.8-1 and C-5.8-2.

$$V_{RBS} = \frac{2M_{pr}}{L_h} + \frac{wL_h}{2} \tag{C-5.8-1}$$

$$V'_{RBS} = \frac{2M_{pr}}{L_h} - \frac{wL_h}{2} \tag{C-5.8-2}$$

Equations C-5.8-1 and C-5.8-2 assume that plastic hinges will form at the RBS at each end of the beam. If the gravity load on the beam is very large, the plastic hinge at one end of the beam may move toward the interior portion of the beam span. If this is the case, the free-body diagram in Figure C-5.1 should be modified to extend between the actual plastic hinge locations. To determine whether Equations C-5.8-1 and C-5.8-2 are valid, the moment diagram for the segment of the beam shown in Figure C-5.1(b)—that is, for the segment of the beam between the centers of the RBS cuts—is drawn. If the maximum moment occurs at the ends of the span, then Equations C-5.8-1 and C-5.8-2 are valid. If the maximum moment occurs within the span and exceeds M_{pe} of the beam (see Equation 5.8-7), then the modification described above will be needed.

CHAPTER 6

BOLTED UNSTIFFENED AND STIFFENED EXTENDED END-PLATE MOMENT CONNECTIONS

6.1. GENERAL

The three extended end-plate moment configurations currently addressed in this chapter are the most commonly used end-plate connection configurations in steel moment frames. AISC Design Guide 4, *Extended End-Plate Moment Connections, Seismic and Wind Applications* (Murray and Sumner, 2003) provides background, design procedures, and complete design examples for the three configurations. The guide was developed before this Standard was written, and there are small differences between the design procedures in the guide and in Commentary Section 6.10. The primary differences are in the resistances factors. The Standard supersedes the design guide in all instances.

Prequalification test results for the three extended end-plate moment connections are found in FEMA (1997); Meng (1996); Meng and Murray (1997); Ryan and Murray (1999); Sumner et al. (2000a); Sumner et al. (2000b); Sumner and Murray (2001); and Sumner and Murray (2002). Results of similar testing but not used for prequalification are found in Adey et al. (1997); Adey et al. (1998); Adey et al. (2000); Castellani et al. (1998); Coons (1999); Ghobarah et al. (1990); Ghobarah et al. (1992); Johnstone and Walpole (1981); Korol et al. (1990); Popov and Tsai (1989); and Tsai and Popov (1990).

The intent of the design procedure in Section 6.10 is to provide an end-plate moment connection with sufficient strength to develop the strength of the connected flexural member. The connection does not provide any contribution to inelastic rotation. All inelastic deformation for an end-plate connection is achieved by beam yielding and/or column panel zone deformation.

The design procedure in Section 6.10 is based on Borgsmiller and Murray (1995) and is similar to the "thick plate" procedure in AISC Design Guide 16 (Murray and Shoemaker, 2002). The procedure is basically the same as that in FEMA 350 (2000b), but with much clarification. Applicable provisions in FEMA 353 (2000d) are incorporated into the procedure as well.

6.2. SYSTEMS

The three extended end-plate moment connections in Figure 6.1 are prequalified for use in IMF and SMF systems, except in SMF systems where the beam is in direct contact with concrete structural slabs. The exception applies only when shear studs are used to attach the concrete slab to the connected beam and is because of the lack of test data to date. Prequalification testing has generally been performed with bare steel specimens. Sumner and Murray (2002) performed one test in which a slab was present. In this test, headed studs were installed from near the end-plate moment connection to the end of the beam, and the concrete was in contact with the column flanges and web. The lower bolts failed prematurely by tension rupture because of the increase in the distance from the neutral axis due to the presence of the composite slab. In later testing, Murray repeated this test but placed a flexible material between the vertical face of the end plate and the slab to inhibit slab participation in transfer of load to the column. This specimen performed acceptably and resulted in provisions for using concrete structural slabs when such flexible material is placed between the slab and the plate.

6.3. PREQUALIFICATION LIMITS

The parametric limitations in Table 6.1 were determined from reported test data in the prequalification references. Only connections that are within these limits are prequalified.

Beams may be either hot-rolled or built-up. If built-up sections are used, the web-to-flange weld may be a one-sided fillet weld, except within the beam, depth, or three times the flange width of the face of the end-plate. Within this length, fillet welds on both sides are required of a size at least $1/4$ in. (6 mm) for constructability or 0.75 times the beam-flange web thickness to develop the web material, whichever is greater. Complete-joint-penetration (CJP) groove welds may be used in lieu of fillet welds.

For tapered members, the depth of the beam at the connection is used to determine the limiting span-to-depth ratio.

6.4. BEAM LIMITATIONS

The beam size limitations in Table 6.1 are directly related to connection testing. Since many of the tested beam sections were built-up members, the limitations are in cross section dimensions instead of rolled-beam designations. There is no evidence that modest deviations from these dimensions will result in significantly different performance.

Similar to RBS testing, most of the tested beam-column assemblies had configurations approximating beam span-to-depth ratios in the range of eight to ten. However, it was judged reasonable to set the minimum span-to-depth ratio at 7 for SMF and 5 for IMF.

The protected zone is based on test observations.

6.5. COLUMN LIMITATIONS

Extended end-plate moment connections may be used only with rolled or built-up I-shaped sections and must be flange connected. There are no other specific column requirements for extended end-plate moment connections.

6.6. COLUMN-BEAM RELATIONSHIP LIMITATIONS

There are no specific column-to-beam relationship limitations for extended end-plate moment connections.

6.7. CONTINUITY PLATES

Continuity plate design must conform to the requirements of Section 2.4.4. The design procedure in Section 6.10 contains provisions specific to extended end-plate moment connections, and the procedure is discussed generally in AISC Design Guide 13, *Wide-Flange Column Stiffening at Moment Connections* (Carter, 1999).

6.8. BOLTS

Prequalification tests have been conducted with both pretensioned ASTM A325 and A490 bolts. Bolt length should be such that at least two complete threads are between the unthreaded portion of the shank and the face of the nut after the bolt is pretensioned. Slip-critical connection provisions are not required for end-plate moment connections.

6.9. CONNECTION DETAILING

Maximum gage, that is, the horizontal distance between outer bolt columns, is limited to the width of the beam flange to ensure a stiff load path. Monotonic tests have shown that the stiffness and strength of an end-plate moment connection are decreased when the bolt gage is wider than the beam flange.

Inner bolt pitch, the distance between the face of the beam flange and the first row of inside or outside bolts, must be sufficient to allow bolt tightening. The minimum pitch values specified have been found to be satisfactory. An increase in pitch distance can significantly increase the required end-plate thickness.

The end-plate can be wider than the beam flange, but the width used in design calculations is limited to the beam flange width plus 1 in. (25 mm). This limitation is based on the CPRP's assessment of unpublished results of monotonic tests of end-plate connections.

The requirements for the length of beam-flange-to-end-plate stiffeners are established to ensure a smooth load path. The 30° angle is the same as used for determining the Whitmore section width in other types of connections. The required 1-in. (25-mm) land is needed to ensure the quality of the vertical and horizontal weld terminations.

Tests have shown that the use of finger shims between the end-plate and the column flange do not affect the performance of the connection (Sumner et al., 2000a).

Design procedures are not available for connections of beams with composite action at an extended end-plate moment connection. Therefore, careful composite slab detailing is necessary to prevent composite action that may increase tension forces in the lower bolts. Welded steel stud anchors are not permitted within $1^1/_2$ times the beam depth, and compressible material is required between the concrete slab and the column face (Sumner and Murray, 2002; Yang et al., 2003).

Cyclic testing has shown that use of weld access holes can cause premature fracture of the beam flange at extended end-plate moment connections (Meng and Murray, 1997). Short to long weld access holes were investigated with similar results. Therefore, weld access holes are not permitted for extended end-plate moment connections.

Strain gage measurements have shown that the web plate material in the vicinity of the inside tension bolts generally reaches the yield strain (Murray and Kukreti, 1988). Consequently, it is required that the web-to-end-plate weld(s) in the vicinity of the inside bolts be sufficient to develop the strength of the beam web.

The beam-flange-to-end-plate and stiffener weld requirements equal or exceed the welding that was used to prequalify the three extended end-plate moment connections. Because weld access holes are not permitted, the beam-flange-to-end plate weld at the beam web is necessarily a partial-joint-penetration (PJP) groove weld. The prequalification testing has shown that these conditions are not detrimental to the performance of the connection.

6.10. DESIGN PROCEDURE

The design procedure in this section, with some modification, was used to design the prequalification test specimens. The procedure is very similar to that in the AISC Design Guide 4 (Murray and Sumner, 2003) except that different resistance factors are used. Example calculations are found in the design guide. Column stiffening example calculations are found in AISC Design Guide 13 (Carter, 1999).

CHAPTER 7

BOLTED FLANGE PLATE (BFP) MOMENT CONNECTION

7.1. GENERAL

The bolted flange plate (BFP) connection is a field-bolted connection. The fundamental seismic behaviors expected with the BFP moment connection include:

(1) initial yielding of the beam at the last bolt away from the face of the column

(2) slip of the flange plate bolts, which occurs at similar resistance levels to the initial yielding in the beam flange, but the slip does not contribute greatly to the total deformation capacity of the connection

(3) secondary yielding in the column panel zone, which occurs as the expected moment capacity and strain hardening occur

(4) limited yielding of the flange plate, which may occur at the maximum deformations

This sequence of yielding has resulted in very large inelastic deformation capacity for the BFP moment connection, but the design procedure is somewhat more complex than some other prequalified connections.

The flange plates and web shear plate are shop-welded to the column flange and field-bolted to the beam flanges and web, respectively. ASTM A490 or A490M bolts with threads excluded from the shear plane are used for the beam flange connections because the higher shear strength of the A490 or A490M bolts reduces the number of bolts required and reduces the length of the flange plate. The shorter flange plates that are therefore possible reduce the seismic inelastic deformation demands on the connection and simplify the balance of the resistances required for different failure modes in the design procedure. Flange plate connections with A325 or A325M bolts may be possible, but will be more difficult to accomplish because of the reduced bolt strength, greater number of bolts, and longer flange plates required. As a result, the connection is not prequalified for use with A325 or A325M bolts.

Prequalification of the BFP moment connection is based upon 20 BFP moment connection tests under cyclic inelastic deformation (FEMA, 2000e; Schneider and Teeraparbwong, 1999; Sato, Newall and Uang, 2008). Additional evidence supporting prequalification is derived from bolted T-stub connection tests (FEMA, 2000e; Swanson et al., 2000), since the BFP moment connection shares many yield mechanisms, failure modes, and connection behaviors with the bolted T-stub connection. The tests were performed under several deformation-controlled test protocols, but most use variations of the ATC-24 (ATC, 1992) or the SAC steel protocol (Krawinkler et al., 2000), which are both very similar to the prequalification test protocol of Chapter K of the AISC *Seismic Provisions* (AISC 2010a). The 20

BFP tests were performed on connections with beams ranging in depth from W8 (W200) to W36 (W920) sections, and the average total demonstrated ductility capacity exceeded 0.057 rad. Hence, the inelastic deformation capacity achieved with BFP moment connections is among the best achieved from seismic testing of moment frame connections. However, the design of the connection is relatively complex because numerous yield mechanisms and failure modes must be considered in the design process. Initial and primary yielding in the BFP moment connection is flexural yielding of the beam near the last row of bolts at the end of the flange plate. However, specimens with the greatest ductility achieve secondary yielding through shear yielding of the column panel zone and limited tensile yielding of the flange plate. Hence, a balanced design that achieves yielding from multiple yield mechanisms is encouraged.

Most past tests have been conducted on specimens with single-sided connections, and the force-deflection behavior is somewhat pinched as shown in Figure C-7.1. Because plastic hinging at the end of the flange plate is the controlling yield mechanism, the expected plastic moment at this location dominates the connection design. The pinching is caused by a combination of bolt slip and the sequence of yielding and strain hardening encountered in the connection. Experiments have shown that the expected peak moment capacity at the plastic hinge is typically on the order of 1.15 times the expected M_p of the beam, as defined in the AISC *Seismic Provisions*, and the expected moment at the face of the column is on the order of 1.3 to 1.5 times the expected M_p of the beam depending upon the span length, number of bolts, and length of the flange plate. The stiffness of this connection is usually slightly greater than 90% of that anticipated with a truly rigid, fully restrained (FR) connection. This reduced stiffness is expected to result in elastic deflection no more than 10% larger than computed with an FR connection, and so elastic calculations with rigid connections are considered to be adequate for most practical design purposes.

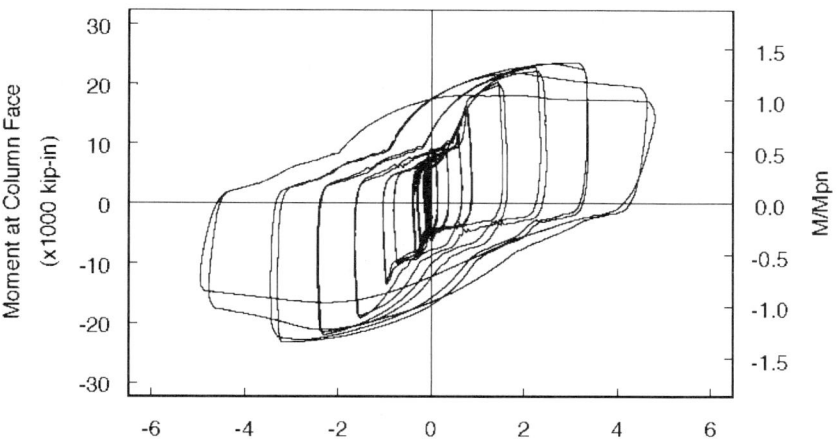

Fig. C-7.1. Moment at face of column vs. total connection rotation for a BFP moment connection with a W30×108 (W760×161) beam and a W14×233 (W360×347) column.

7.2. SYSTEMS

Review of the research literature shows that BFP moment connections meet the qualifications and requirements of both SMF and IMF frames. However, no test data are available for BFP moment connections with composite slabs, and so the BFP moment connection is not prequalified with reinforced concrete structural slabs that contact the face of the columns. Reinforced concrete structural slabs that make contact with the column may:

- significantly increase the moment at the face of the column;
- cause significant increases of the force and strain demands in the bottom flange plate; and
- result in reduced inelastic deformation capacity of the connection.

Therefore, prequalification of the BFP moment connection is restricted to the case where the concrete structural slab has a minimum separation or isolation from the column. In general, isolation is achieved if steel stud anchors are not included in the protected zone and if the slab is separated from all surfaces of the column by an open gap or by use of compressible foam-like material.

7.3. PREQUALIFICATION LIMITS

1. Beam Limitations

The SMF prequalification limits largely reflect the range of past testing of the BFP moment connection. Limits for IMF connections somewhat exceed these limits because 18 of the past 20 tests used to prequalify the connection developed plastic rotations larger than those required to qualify as a SMF connection, and all 20 tests greatly exceed the rotation required to qualify as an IMF connection.

BFP moment connections have been tested with beams as large as the W36×150 (W920×223) while achieving the ductility required for qualification as an SMF. Consequently, the W36 (W920) beam depth, 150 lb/ft weight limit (223 kg/m mass limit), and 1 in. (25 mm) flange thickness limits are adopted in this provision. Past tests have shown adequate inelastic rotation capacity to qualify as an SMF in tests with span-to-depth ratios less than 5 and greater than 16, and so lower bound span-to-depth ratio limits of 7 and 9 are conservatively adopted for the IMF and SMF applications, respectively. Inelastic deformation is expected for approximately one beam depth beyond the end of the flange plate, and limited yielding is expected in the flange plate. As a result, the protected zone extends from the column face to a distance equal to the depth of the beam beyond the bolt farthest from the face of the column.

Primary plastic hinging of the BFP moment connection occurs well away from the face of the column, and lateral-torsional deformation will occur as extensive yielding develops in the connection. As a result, lateral bracing of the beam is required at the end of the protected zone. The bracing is required within the interval between 1 and 1.5 beam depths beyond the flange bolts farthest from the face of the column. This permits some variation in the placement of the lateral support to allow economical use of transverse framing for lateral support where possible. As with other moment frame connections, supplemental lateral bracing at the column flange connection can typically be accommodated by the stiffness of the diaphragm and transverse framing.

As for other prequalified connections, the BFP moment connection requires compact flanges and webs as defined by the AISC *Seismic Provisions*, and built-up I-shaped beams conforming to Section 2.3 are permitted. It should be noted, however, that the BFP and most other prequalified connections do not have specific seismic test data to document the prequalification of built-up beam sections. This prequalification is provided, because long experience shows that built-up steel sections provide similar flexural behavior as hot-rolled shapes with comparable materials and proportions.

2.   Column Limitations

BFP moment connections have been tested with wide flange columns up to W14×233 (W360×347) sections. The SMF prequalification limits largely reflect the range of past testing of the BFP moment connection. All 20 tests were completed with strong axis bending of the column, and the prequalification of the BFP moment connections is limited to connections made to the column flange.

As with most other prequalified connections, the BFP moment connection has not been tested with columns deeper than W14 (W360) sections or with built-up column sections. It was the judgment of the CPRP that the BFP moment connection places similar or perhaps smaller demands on the column than other prequalified connections. The demands may be smaller because of the somewhat smaller strain-hardening moment increase achieved with the BFP moment connection as compared to the welded web-welded flange and other FR connections. The location of yielding of the BFP moment connection is somewhat analogous to the RBS connection, and therefore, prequalification limits for the column are comparable to those used for the RBS connection.

7.4.   COLUMN-BEAM RELATIONSHIP LIMITATIONS

The BFP moment connection is expected to sustain primary yielding in the beam starting at the last flange plate bolt line away from the face of the column. Secondary yielding is expected in the column panel zone and very limited subsequent yielding is expected in the flange plate. Yielding in the column outside the connection panel zone is strongly discouraged. Therefore, the BFP moment connection employs a similar weak beam-strong column check and panel zone resistance check as used for other prequalified connections.

7.5.   CONNECTION DETAILING

The BFP moment connection requires plate steel for the flange plate, shear plate, and possibly panel zone doubler plates. Past tests have been performed with plates fabricated both from ASTM A36/A36M and A572/A572M Grade 50 (Grade 345) steels. Therefore, the prequalification extends to both plate types. The designer should be aware of potential pitfalls with the material selection for the flange plate design. The flange plate must develop tensile yield strength over the gross section and ultimate tensile fracture resistance over the effective net section. A36/A36M steel has greater separation of the nominal yield stress and the minimum tensile strength, and this may simplify the satisfaction of these dual requirements. However, variation in expected yield stress is larger for A36/ A36M steel, and design calculations may more accurately approximate actual flange plate performance with A572/A572M steel.

The flange plate welds are shop welds, and these welds are subject to potential secondary yielding caused by strain hardening at the primary yield location in the beam. As a result, the welds are required to be demand-critical complete-joint-penetration (CJP) groove welds. If backing is used, it must be removed and the weld must be backgouged to sound material and back welded to assure that the weld can sustain yielding of the flange plate. Since the welds are shop welds, considerable latitude is possible in the selection of the weld process as long as the finished weld meets the demand critical weld requirements stipulated in the AISC *Seismic Provisions*. In the test specimens used to prequalify this connection, electroslag, gas shielded metal arc, and flux cored arc welding have been used.

The BFP moment connection places somewhat less severe demands on the web connection than most FR connections, because of the somewhat greater flexibility of the bolted flange connection. As a result, the shear plate may be welded with CJP groove welds or partial-joint-penetration (PJP) groove welds or fillet welds.

Bolts in the flange plate are limited to two rows of bolts, and the bolt holes must be made by drilling or sub-punching and reaming. These requirements reflect testing used to prequalify the BFP moment connection, but they also reflect practical limitations in the connection design. Net section rupture is a clear possibility in the beam flange and flange plates, and it is very difficult to meet the net section rupture criteria if more than two rows of bolts are employed.

A single row of bolts causes severe eccentricity in the connection and would lead to an excessively long connection. Punched bolt holes without reaming are not permitted, because punching may induce surface roughness in the hole that may initiate cracking of the net section under high tensile stress. As noted earlier, the connection is prequalified only for A490 or A490M bolts with threads excluded from the shear plane. Bolt diameter is limited to a maximum of $1^1/_8$ in. (28 mm), because larger bolts are seldom used and the $1^1/_8$ in. (28 mm) diameter is the maximum used in past BFP tests. The bolt diameter must be selected to ensure that flange yielding over the gross area exceeds the net section capacity of the beam flange.

Oversized bolt holes were included in some past tests, because the oversized holes permit easier alignment of the bolts and erection of the connection and resulted in good performance of the connection. Further, the beam must fit between two welded flange plates with full consideration of rolling and fabrication tolerances. As a result, shims may be used to simplify erection while ensuring a tight connection fit.

7.6. DESIGN PROCEDURE

The BFP moment connection is somewhat more complex than some other connections, because a larger number of yield locations and failure modes are encountered with this connection. **Step 1** of this procedure defines the maximum expected moment, M_{pr}, at the last bolt away from the face of the column in the flange plate. The beam flange must have greater net section fracture resistance than its yield resistance, because tensile yield of the flange is a ductile mechanism and net section rupture is a brittle failure. **Step 2** establishes the maximum bolt diameter that can meet this balanced criterion. While this requirement is rational, it should be noted that net section rupture of the beam flange has not occurred in any past BFP tests, since the beam web clearly reduces any potential for flange rupture.

The shear strength of the flange bolts is the smallest strength permitted by bolt shear with threads excluded from the shear plane, the bolt bearing on the flange plate, bolt bearing on the beam flange and block shear considerations. **Step 3** provides this evaluation. **Step 4** is an approximate evaluation of the number of bolts needed to develop the BFP moment connection. The moment for the bolts is larger than M_{pr} because the centroid of the bolt group is at a different location than the primary hinge location. However, this moment cannot be accurately determined until the geometry of the flange plate and bolt spacing are established. The 1.25 factor is used as an empirical increase in this moment to provide this initial estimate for the number of bolts required. The bolts are tightened to meet slip-critical criteria, but the connection is not slip-critical: the bolts are designed as bearing bolts.

Once the required number of bolts is established, bolt spacing and an initial estimate of the flange plate length can be established. This geometry is illustrated and summarized in Figure 7.1, and **Step 5** defines critical dimensions of this geometry for later design checks.

Step 6 is similar to other connection types in that the shear force at the plastic hinge is based upon the maximum shear achieved with maximum expected moments at the plastic hinges at both ends of the beam plus the shear associated with appropriate gravity loads on the beam.

Step 7 uses the geometry established in Step 5 and the maximum shear force established in Step 6 to determine the maximum expected moment at the face of the column flange, M_f. The maximum expected force in the flange plate, F_{pr}, is determined from M_f in **Step 8**.

The flange plate bolts cannot experience a tensile force larger than F_{pr}, and so **Step 9** checks the actual number of bolts required in the connection. If this number is larger or smaller than that estimated in Step 4, it may be necessary to change the number of bolts and repeat Steps 5 through 9 until convergence is achieved.

Steps 10 and 11 check the flange plate width and thickness to ensure that tensile yield strength and tensile rupture strength, respectively, exceed the maximum expected tensile force in the flange. The net section rupture check of Step 11 employs the nonductile resistance factor, while the flange yielding check of Step 10 employs the ductile resistance factor; this check also allows limited yielding in the flange plate and ensures ductility of the connection. **Step 12** checks block shear of the bolt group in the flange plate, and **Step 13** checks the flange plate for buckling, when F_{pr} is in compression. Both block shear and buckling of the flange plate are treated as non-ductile behaviors.

Step 14 is somewhat parallel to Step 6 except that the beam shear force at the face of the column is established, and this shear force is then used to size and design the single shear plate connection is **Step 15**.

Continuity plates and panel zone shear strength are checked in **Steps 16 and 17**, respectively. These checks are comparable to those used for other prequalified connections.

As previously noted, the BFP moment connection has provided quite large inelastic rotational capacity in past research. It has done this by attaining primary yielding in the beam at the end of the flange plate away from the column and through secondary yielding as shear yielding in the column panel zone and tensile yielding in the flange plate. Bolt slip occurs but does not contribute greatly to connection ductility. This rather complex design procedure attempts to achieve these goals by balancing the resistances for different yield mechanisms and failure modes in the connection and by employing somewhat greater conservatism for brittle behaviors than for ductile behaviors.

CHAPTER 8

WELDED UNREINFORCED FLANGE-WELDED WEB (WUF-W) MOMENT CONNECTION

8.1. GENERAL

The welded unreinforced flange–welded web (WUF-W) moment connection is an all-welded moment connection, wherein the beam flanges and the beam web are welded directly to the column flange. A number of welded moment connections that came into use after the 1994 Northridge earthquake, such as the reduced beam section and connections provided with beam flange reinforcement, were designed to move the plastic hinge away from the face of the column. In the case of the WUF-W moment connection, the plastic hinge is not moved away from the face of the column. Rather, the WUF-W moment connection employs design and detailing features that are intended to permit the connection to achieve SMF performance criteria without fracture. Key features of the WUF-W moment connection that are intended to control fracture are as follows:

- The beam flanges are welded to the column flange using CJP groove welds that meet the requirements of demand critical welds in the AISC *Seismic Provisions*, along with the requirements for treatment of backing and weld tabs and welding quality control and quality assurance requirements, as specified in Chapter 3 of this Standard.

- The beam web is welded directly to the column flange using a CJP groove weld that extends the full-depth of the web—that is, from weld access hole to weld access hole. This is supplemented by a single-plate connection, wherein a single plate is welded to the column flange and is then fillet welded to the beam web. Thus, the beam web is attached to the column flange with both a CJP groove weld and a welded single-plate connection. The single-plate connection adds stiffness to the beam web connection, drawing stress toward the web connection and away from the beam flange to column flange connections. The single plate also serves as backing for the CJP groove weld connecting the beam web to the column flange.

- Instead of using a conventional weld access hole detail as specified in Section J1.6 of the AISC *Specification* (AISC, 2010b), the WUF-W moment connection employs a special seismic weld access hole with requirements on size, shape and finish that reduce stress concentrations in the region around the access hole detailed in AWS D1.8/D1.8M (AWS, 2009).

Prequalification of the WUF-W moment connection is based on the results of two major research and testing programs. Both programs combined large-scale tests with extensive finite element studies. Both are briefly described herein.

The first research program on the WUF-W moment connection was conducted at Lehigh University as part of the SAC-FEMA program. Results are reported in several publications (Ricles et al., 2000, 2002). This test program formed the basis of prequalification of the WUF-W moment connection in FEMA 350 (FEMA, 2000e). As part of the Lehigh program, tests were conducted on both interior and exterior type specimens. The exterior specimens consisted of one beam attached to a column. The interior specimens consisted of a column with beams attached to both flanges. One of the interior specimens included a composite floor slab. All specimens used W36×150 (W920×223) beams. Three different column sizes were used: W14×311, W14×398 and W27×258 (W360×463, W360×592 and W690×384). All WUF-W moment connection specimens tested in the Lehigh program satisfied the rotation criteria for SMF connections (±0.04 rad total rotation). Most specimens significantly exceeded the qualification criteria. Considering that the interior type specimens included two WUF-W moment connections each, 12 successful WUF-W moment connections were tested in the Lehigh program. This research program included extensive finite element studies that supported the development of the special seismic weld access hole and the details of the web connection.

The second major research program on the WUF-W moment connection was conducted at the University of Minnesota. The purpose of this research program was to examine alternative doubler plate details, continuity plate requirements, and effects of a weak panel zone. All test specimens used the WUF-W moment connection. Results are reported in several publications (Lee et al., 2002, 2005a, 2005b). Six interior type specimens were tested in the Minnesota program. All specimens used W24×94 beams. Three column sizes were used: W14×283, W14×176 and W14×145. All specimens were designed with panel zones weaker than permitted by the AISC *Seismic Provisions*. Two of the test specimens, CR1 and CR4, were inadvertently welded with low-toughness weld metal. This resulted in premature weld failure in specimen CR4 (failure occurred at about 0.015 rad rotation). With the exception of CR4, all specimens achieved a total rotation of ±0.04 rad, and sustained multiple cycles of loading at ±0.04 rad prior to failure. All successful specimens exhibited substantial panel zone yielding, due to the weak panel zone design. This test program was also supported by extensive finite element studies.

Considering the WUF-W moment connection research programs at both Lehigh and the University of Minnesota, WUF-W moment connection specimens have shown excellent performance in tests. There is only one reported failed test, due to the inadvertent use of low-toughness weld metal for beam flange CJP groove welds (Minnesota Specimen CR4). Of all of the WUF-W moment connection specimens that showed good performance (achieved rotations of at least ±0.04 rad), approximately one-half had panel zones weaker than permitted by the AISC *Seismic Provisions*. The other half satisfied the panel zone strength criteria of the AISC *Seismic Provisions*. This suggests that the WUF-W moment connection performs well for both strong and weak panel zones; therefore the connection is not highly sensitive to panel zone strength.

The protected zone for the WUF-W moment connection is defined as the portion of the beam extending from the face of the column to a distance d from the face of the column, where d is the depth of the beam. Tests on WUF-W moment connection specimens show that yielding in the beam is concentrated near the face of the column, but extends to some degree over a length of the beam approximately equal to its depth.

8.3. PREQUALIFICATION LIMITS

The WUF-W moment connection is prequalified for beams up to W36 (W920) in depth, up to 150 lb/ft in weight (223 kg/m mass limit), and up to a beam flange thickness of 1 in. (25 mm). This is based on the fact that a W36×150 (W920×223) is the deepest and heaviest beam tested with the WUF-W moment connection. The 1-in. (25-mm) flange thickness limitation represents a small extrapolation of the 0.94-in. (23.9-mm) flange thickness for the W36×150 (W920×223). Limits are also placed on span-to-depth ratio based on the span-to-depth ratios of the tested connections and based on judgment of the CPRP.

Beam lateral bracing requirements for the WUF-W moment connection are identical to those for the RBS moment connection. The effects of beam lateral bracing on cyclic loading performance have been investigated more extensively for the RBS moment connection than for the WUF-W moment connection. However, the available data for the WUF-W moment connection suggests that beams are less prone to lateral-torsional buckling than with the RBS moment connections. Consequently, it is believed that lateral bracing requirements established for the RBS moment connection are satisfactory, and perhaps somewhat conservative, for the WUF-W moment connection.

Column sections used in WUF-W moment connection test specimens were W14 (W360) and W27 (W690) sections. However, column limitations for the WUF-W moment connection are nearly the same as for the RBS moment connection, which includes wide-flange shapes up to W36 (W920) and box columns up to 24 in. by 24 in. (610 mm by 610 mm). A primary concern with deep columns in moment frames has been the potential for twisting and instability of the column driven by lateral-torsional buckling of the beam. Because beams with WUF-W moment connections are viewed as somewhat less prone to lateral-torsional buckling than beams with RBS moment connections, the column limitations established for the RBS moment connection were judged as appropriate for the WUF-W moment connection.

8.4. COLUMN-BEAM RELATIONSHIP LIMITATIONS

WUF-W moment connection test specimens have shown good performance with a range of panel zone shear strengths, ranging from very weak to very strong panel zones. Tests conducted at the University of Minnesota (Lee et al., 2005b) showed excellent performance on specimens with panel zones substantially weaker than required in the AISC *Seismic Provisions*. However, there are concerns that very weak panel zones may contribute to premature connection fracture under some circumstances, and it is believed further research is needed before weak panel zone designs can be prequalified. Consequently, the minimum panel zone strength required in Section E3.6e of the AISC *Seismic Provisions* is required for prequalified WUF-W moment connections for SMF. For IMF systems, the AISC *Seismic Provisions* have no special panel zone strength requirements, beyond the AISC *Specification*. This may lead to designs in which inelastic action is concentrated within the panel zone. As described earlier, based on successful tests on WUF-W moment connection specimens with weak panel zones, this condition is not viewed as detrimental for IMF systems.

8.5. BEAM FLANGE-TO-COLUMN FLANGE WELDS

Beam flanges are required to be connected to column flanges with CJP groove welds. The welds must meet the requirements of demand critical welds in the AISC *Seismic Provisions*, as well as the detailing and quality control and quality assurance requirements specified in Chapter 3 of this Standard. These beam flange-to-column flange weld requirements reflect the practices used in the test specimens that form the basis for prequalification of the WUF-W moment connection and reflect what are believed to be best practices for beam flange groove welds for SMF and IMF applications.

A key feature of the WUF-W moment connection is the use of a special weld access hole. The special seismic weld access hole has specific requirements on the size, shape and finish of the access hole. This special access hole was developed in research on the WUF-W moment connection (Ricles et al., 2000, 2002) and is intended to reduce stress concentrations introduced by the presence of the weld access hole. The size, shape and finish requirements for the special access hole are specified in AWS D1.8/D1.8M, Section 6.10.1.2 (AWS, 2009).

8.6. BEAM WEB-TO-COLUMN CONNECTION LIMITATIONS

The beam web is connected to the column flange with a single plate that is welded to the column flange and fillet welded to the beam web and with a full-depth (weld access hole to weld access hole) CJP groove weld. The single plate serves as backing for the beam web CJP groove weld. The use of the CJP groove weld combined with the fillet-welded single plate is believed to increase the stiffness of the beam web connection. The stiffer beam web connection serves to draw stress away from the beam flanges and therefore reduces the demands on the beam flange groove welds.

Most of the details of the beam web-to-column connection are fully prescribed in Section 8.6; thus few design calculations are needed for this connection. An exception to this is the connection of the single plate to the column. This connection must develop the shear strength of the single plate, as specified in Section 8.6(2). This can be accomplished by the use of CJP groove welds, PJP groove welds, fillets welds or combinations of these welds. The choice of these welds is left to the discretion of the designer. In developing the connection between the single plate and the column flange, designers should consider the following issues:

- The use of a single-sided fillet weld between the single plate and the column flange should be avoided. If the single plate is inadvertently loaded or struck in the out-of-plane direction during erection, the fillet weld may break and may lead to erection safety concerns.

- The end of the beam web must be set back from the face of the column flange a specified amount to accommodate the web CJP root opening dimensional requirements. Consequently, the single plate-to-column weld that is placed in the web CJP root opening must be small enough to fit in that specified root opening. For example, if the CJP groove weld is detailed with a $^1/_4$-in. (6-mm) root opening, a fillet weld between the single plate and the column flange larger than $^1/_4$ in. (6 mm) will cause the root of the CJP groove weld to exceed $^1/_4$ in. (6 mm).

- Placement of the CJP groove weld connecting the beam web to the column flange will likely result in intermixing of weld metal, with the weld attaching the single plate to the column flange. Requirements for intermix of filler metals specified in AWS D1.8/D1.8M (AWS, 2009) should be followed in this case.

The CJP groove weld connecting the beam web to the column flange must meet the requirements of demand critical welds. Note that weld tabs are permitted, but not required, at the top and bottom ends of this weld. If weld tabs are used, they should be removed after welding according to the requirements of Section 3.4. If weld tabs are not used, the CJP groove weld should be terminated in a manner that minimizes notches and stress concentrations, such as with the use of cascaded ends.

The fillet weld connecting the beam web to the single plate should be terminated a small distance from the weld access hole, as shown in Figure 8.3. This is to avoid introducing notches at the edge of the weld access hole.

8.7. DESIGN PROCEDURE

For the WUF-W moment connection, many of the details of the connection of the beam to the column flange are fully prescribed in Sections 8.5 and 8.6. Consequently, the design procedure for the WUF-W moment connection largely involves typical checks for continuity plates, panel zone shear strength, column-beam moment ratio and beam shear strength.

With the WUF-W moment connection, yielding of the beam (i.e., plastic hinge formation) occurs over the portion of the beam extending from the face of the column to a distance of approximately one beam depth beyond the face of the column. For purposes of the design procedure, the location of the plastic hinge is taken to be at the face of the column. That is, $S_h = 0$ for the WUF-W moment connection. It should be noted that the location of the plastic hinge for design calculation purposes is somewhat arbitrary, since the plastic hinge does not occur at a single point but instead occurs over some length of the beam. The use of $S_h = 0$ is selected to simplify the design calculations. The value of C_{pr} was calibrated so that when used with $S_h = 0$, the calculated moment at the column face reflects values measured in experiments.

Note that the moment in the beam at the column face is the key parameter in checking panel zone strength, column-beam moment ratio, and beam shear strength.

The value of C_{pr} for the WUF-W moment connection is specified as 1.4, based on an evaluation of experimental data. Tests on WUF-W moment connections with strong panel zones (Ricles et al., 2000) showed maximum beam moments, measured at the face of the column, as high as $1.49M_p$, where M_p was based on measured values of F_y. The average maximum beam moment at the face of the column was $1.33M_p$. Consequently, strain hardening in the beam with a WUF-W moment connection is quite large. The value of C_{pr} of 1.4 was chosen to reflect this high degree of strain hardening. Combining the value of $C_{pr} = 1.4$ with $S_h = 0$ results in a moment at the face of the column, $M_f = M_{pr} = 1.4R_yF_yZ$, that reasonably reflects maximum column face moments measured in experiments.

CHAPTER 9

KAISER BOLTED BRACKET (KBB) MOMENT CONNECTION

9.1. GENERAL

The Kaiser bolted bracket (KBB) moment connection is designed to eliminate field welding and facilitate frame erection. Depending on fabrication preference, the brackets can be either fillet welded (W-series) or bolted (B-series) to the beam. The B-series can also be utilized to improve the strength of weak or damaged connections, although it is not prequalified for that purpose. Information on the cast steel and the process used to manufacture the brackets is provided in Appendix A.

The proprietary design of the brackets is protected under U.S. patent number 6,073,405 held by Steel Cast Connections LLC. Information on licensing rights can be found at http://www.steelcastconnections.com. The connection is not prequalified when brackets of an unlicensed design and/or manufacture are used.

Connection prequalification is based on 21 full-scale bolted bracket tests representing both new and repaired applications (Kasai and Bleiman, 1996; Gross et al., 1999; Newell and Uang, 2006; and Adan and Gibb, 2009). These tests were performed using beams ranging in depth from W16 to W36 (W410 to W920) and columns using W12, W14 and W27 (W310, W360 and W690) sections. Built-up box columns have also been tested. The test subassemblies have included both single cantilever and double-sided column configurations. Concrete slabs were not present in any tests. During testing, inelastic deformation was achieved primarily through the formation of a plastic hinge in the beam. Some secondary yielding was also achieved in the column panel zone. Peak strength typically occurred at an interstory drift angle between 0.025 and 0.045 rad. Specimen strength then gradually decreased with additional yielding and deformation. In the KBB testing reported by Adan and Gibb (2009), the average specimen maximum interstory drift angle exceeded 0.055 rad.

9.2. SYSTEMS

Review of the research literature and testing referenced in this document indicates that the KBB moment connection meets the prequalification requirement for special and intermediate moment frames.

The exception associated with concrete structural slab placement at the column and bracket flanges is based on testing conducted on the stiffened extended end-plate moment connection (Seek and Murray, 2008). While bolted bracket testing has been conducted primarily on bare-steel specimens, some limited testing has also been performed on specimens with a concrete structural slab. In these tests, the presence of the slab provided a beneficial effect by maintaining the stability of the beam at larger interstory drift angles (Gross et al., 1999; Newell and Uang, 2006). However, in the absence of more comprehensive testing with a slab, the placement of the concrete is subject to the exception.

9.3. PREQUALIFICATION LIMITS

1. **Beam Limitations**

    A wide range of beam sizes was tested with bolted brackets. The lightest beam size reported in the literature was a W16×40 (W410×60). The heaviest beam reported was a W36×210 (W920×313). In the W36×210 test, the specimen met the requirements, but subsequently experienced an unexpected nonductile failure of the bolts connecting the bracket to the column. The next heaviest beams reported to have met the requirements were W33×130 and W36×150 (W840×193 and W410×60). Based on the judgment of the CPRP, the maximum beam depth and weight was limited to match that of the W33×130 (W840×193). The maximum flange thickness was established to match a modest increase above that of the W36×150 (W410×60).

    The limitation associated with minimum beam flange width is required to accommodate fillet weld attachment of the W-series bracket and to prevent beam flange tensile rupture when using the B-series bracket.

    Bolted bracket connection test assemblies used configurations approximating beam spans between 24 and 30 ft (7310 and 9140 mm). The beam span-to-depth ratios were in the range of 8 to 20. Given the degree to which most specimens significantly exceeded the requirement, it was judged reasonable to set the minimum span-to-depth ratio at 9 for both SMF and IMF systems.

    As with other prequalified connections, beams supporting a concrete structural slab are not required to have a supplemental brace near the expected plastic hinge. If no floor slab is present, then a supplemental brace is required. The brace may not be located within the protected zone.

2. **Column Limitations**

    Bolted bracket connection tests were performed with the brackets bolted to the column flange (i.e., strong-axis connections). In the absence of additional testing with brackets bolted to the column web (weak-axis connections), the prequalification is limited to column flange connections.

    Test specimen wide flange column sizes ranged from a W12×65 to a W27×281 (W310×97 to W690×418). Testing performed by Ricles et al. (2004) of deep-column RBS connections demonstrated that deep columns do not behave substantially different from W14 (W360) columns when a slab is present or when adequate lateral bracing is provided for the beam and/or column in the absence of a slab. Based on the similarity in performance to that of the RBS connection, the KBB is prequalified to include column sizes up to W36 (W920).

The behavior of a flanged cruciform column in KBB connections is expected to be similar to that of a rolled wide-flange. Therefore, flanged cruciform columns are prequalified, subject to the limitations imposed on rolled wide-flange shapes.

Two of the tests were successfully conducted using a built-up box column. In the first box column test, connections were made on two opposing column faces. Then, in the second test, a connection was made to the orthogonal face of the same column. These two tests were intended to prequalify a box column participating in orthogonal moment frames. The tested box column was $15^5/_8$ in. (397 mm) square (Adan and Gibb, 2009). Consequently, bolted bracket connections are prequalified for use with built-up box columns up to 16 in. (406 mm) square.

Based on both successful wide-flange and built-up box column testing, acceptable performance would also be expected for boxed wide-flange columns. Therefore, the use of boxed wide-flange columns is also prequalified. When moment connections are made only to the flanges of the wide-flange portion of the boxed wide-flange, subject to the bracing limitations mentioned previously, the column may be as deep as a W36 (W920). When the boxed wide-flange column participates in orthogonal moment frames, neither the depth nor the width of the column is allowed to exceed 16 in. (406 mm), applying the same limit as a built-up box.

3. Bracket Limitations

The ASTM cast steel material specification used to manufacture the brackets is based on recommendations from the Steel Founders' Society of America (SFSA).

The cast brackets are configured and proportioned to resist applied loads in accordance with the limit states outlined by Gross et al. (1999). These limit states include column flange local buckling, bolt prying action, combined bending and axial loading on the bracket, shear, and for the B-series, bolt bearing deformation and block shear rupture.

In tests representing new applications, the bracket column bolt holes were cast vertically short-slotted. The vertically slotted holes provide field installation tolerance. In tests representing a repair application, the holes were cast standard diameter. There has been no difference in connection performance using either type of cast hole (Adan and Gibb, 2009).

9.4. COLUMN-BEAM RELATIONSHIP LIMITATIONS

The reduction of column axial and moment strength due to the column bolt holes need not be considered when checking column-beam moment ratios. Research performed by Masuda et al. (1998) indicated that a 30 to 40% loss of flange area due to bolt holes showed only a corresponding 10% reduction in the yield moment strength.

9.5. BRACKET-TO-COLUMN FLANGE CONNECTION LIMITATIONS

In the prequalification tests, fasteners joining the bracket to the column flange were pretensioned ASTM A490 or A490M bolts. The column bolt head can be positioned on either the column or bracket side of the connection. Where possible, the column bolts are tightened prior to the bolts in the web shear tab.

When needed, finger shims between the bracket and column face allow for fit between the bracket and column contact surfaces. Tests indicated that the use of finger shims does not affect the performance of the connection.

Because the flanges of a box column are stiffened only at the corners, tightening of the column bolts can cause excessive local flange bending. Therefore, as shown in Figure C-9.1, a washer plate is required between the box column flange and the bracket.

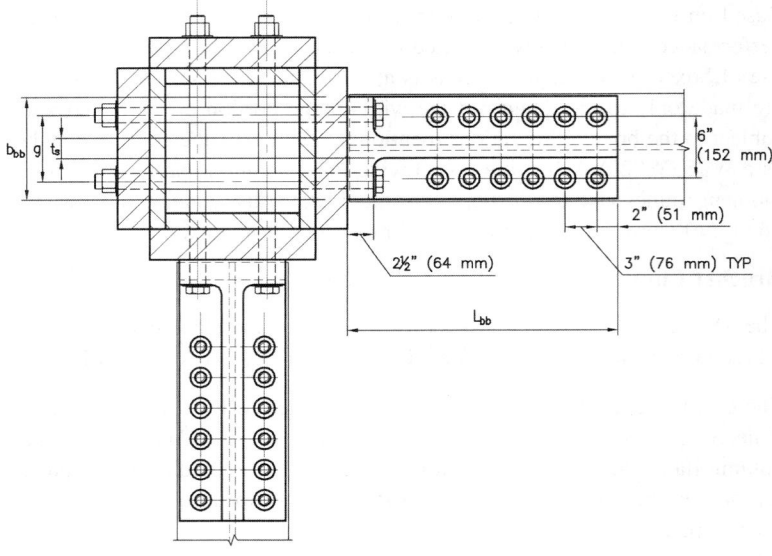

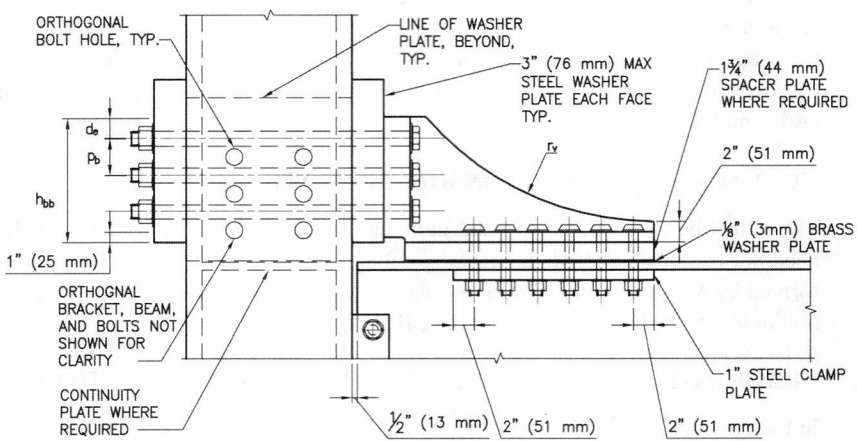

Fig. C-9.1. Box column connection detailing for KBB.

As shown in Figure C-9.1, orthogonally connected beams framing into a box column are raised one-half of the column bolt spacing distance to avoid overlapping the column bolts.

9.6. BRACKET-TO-BEAM FLANGE CONNECTION LIMITATIONS

The cast steel brackets are not currently listed as a prequalified material in AWS D1.1/D1.1M (AWS, 2010). Therefore, the weld procedure specification (WPS) for the fillet weld joining the bracket to the beam flange is required to be qualified by test with the specific cast material.

Bolts joining the bracket to the beam flange in prequalification tests have been conducted with pretensioned ASTM A490 or A490M bolts with the threads excluded from the shear plane. The beam bolt head can be positioned on either the beam or bracket side of the connection. Given the beam bolt pattern and hole size, it is necessary to use the bracket as a template when drilling the beam bolt holes. The holes must be aligned to permit insertion of the bolts without undue damage to the threads.

The brass washer plate prevents abrading of the beam and bracket contact surfaces. In the initial developmental stages of the connection, several specimens configured without the brass plate experienced flange net section rupture through the outermost bolt holes. Observation of the failed specimens indicated that fracture likely initiated at a notch created by the abrading contact surfaces near the hole. Furthermore, energy released through the beam-bracket slip-stick mechanism caused loud, intermittent bursts of noise, particularly at high levels of inelastic drift (Kasai and Bleiman, 1996). To overcome these problems, the brass plate was inserted between the bracket and the beam flange. The idea is based on the use of a brass plate as a special friction-based seismic energy dissipator (Grigorian et al., 1992). Although not intended to dissipate energy in the bolted bracket connection, the brass plate provides a smooth slip mechanism at the bracket-to-beam interface.

When bolting the bracket to a beam flange, a steel washer or clamp plate is positioned on the opposite side of the connected flange. The restraining force of the clamp plate prevents local flange buckling from occurring near the outermost bolt holes. In tests performed without the clamp plates, flange distortion increased the strains near the holes. The increased strain caused necking and fracture through the flange net area. In similar tests performed with the clamp plates, yielding and fracture occurred outside the connected region through the flange gross area (Kasai and Bleiman, 1996).

9.7. BEAM WEB-TO-COLUMN CONNECTION LIMITATIONS

All of the bolted bracket connection tests were performed with a bolted web connection where pretensioned high-strength bolts were used. Therefore, the KBB is prequalified for a bolted beam web-to-column connection.

9.8. CONNECTION DETAILING

Both Figures 9.4 and 9.5 show the connection configured with continuity plates where required. The use of continuity plates is dictated by the need to satisfy prescribed limit states for the flange and web of the column. In a bolted connection, the configuration of the fasteners can impede the ability of the continuity plates to effectively address these limit states. The design intent for the KBB is to satisfy the prescribed limit states without continuity plates. In tests of wide flange columns without continuity plates, the absence of the continuity plates did not appear to promote local flange bending or lead to other detrimental effects (Adan and Gibb, 2009). However, in the absence of additional tests on deeper column sections, prequalification without continuity plates is limited to W12 and W14 sections.

9.9. DESIGN PROCEDURE

The design procedure for prequalified KBB connections is intended to develop the probable maximum moment capacity of the connecting beam. Test data indicate that connecting the brackets to the column and beam in accordance with the requirements herein allows the connection to resist this level of moment.

Tables C-9.1, C-9.1M, C-9.2, and C-9.2M can be used as a guide in selecting trial bracket-beam combinations in conjunction with **Steps 1 and 3**. The tables are based on beams that satisfy the limitations of Section 9.3.1 for ASTM A992/A992M or A572/A572M Grade 50 (Grade 345) wide flange shapes.

Step 4 of the procedure requires computation of the shear force at the expected plastic hinge. This shear force is a function of the gravity load on the beam and the plastic moment strength. A calculation similar to that for the RBS moment connection is required for the case of a beam with a uniformly distributed gravity load as shown in Figure C-5.1. For the KBB, L_h is the distance between the expected plastic hinge locations and S_h is the distance from the face of the column to the hinge. The explanation associated with Equations C-5.8-1 and C-5.8-2 also applies to the KBB.

Step 6 is based on the limit state of bolt tensile rupture as defined in Section J3.6 of the AISC *Specification* (AISC, 2010b), where the required bolt tensile strength is determined in Equation 9.9-3.

Steps 7 and 11 of the procedure apply to rolled or built-up shapes with flange holes, proportioned based on flexural strength of the gross section. The flexural strength is limited in accordance with the limit state of flange tensile rupture as defined in Section F13.1(a) of the AISC *Specification*. When the flange width is adequate, the tensile rupture limit state does not apply.

Step 8 of the procedure requires a column flange prying action check as outlined in Chapter 9 of the AISC *Manual*. The computations include provisions from the research performed by Kulak, Fisher and Struik (1987).

Step 9 of the procedure is based on the limit state of column flange local bending as defined in Section J10.1 of the AISC *Specification*. The limit state determines the strength of the flange using a simplified yield line analysis. Yield line analysis is a method that determines the flexural load at which a collapse mechanism will form in a flat plate structure and employs the principle of virtual work to develop an upper bound solution for plate strength. Given the bolted bracket configuration, the

TABLE C-9.1
Recommended W-Series Bracket-Beam Combinations

Bracket Designation	Beam Designations
W1.0	W33×130, W30×124, W30×116, W24×131, W21×122, W21×111
W2.1	W30×108, W27×114, W27×102, W24×103, W21×93, W18×106, W18×97
W2.0	W27×94, W24×94, W24×84, W24×76, W21×83, W21×73, W21×68, W21×62, W18×86, W18×71, W18×65
W3.1	W24×62, W24×55, W21×57, W18×60, W18×55, W16×57
W3.0	W21×50, W21×44, W18×50, W18×46, W18×35, W16×50, W16×45, W16×40, W16×31

TABLE C-9.1M
Recommended W-Series Bracket-Beam Combinations

Bracket Designation	Beam Designations
W1.0	W840×193, W760×185, W760×173, W610×195, W530×182, W530×165
W2.1	W760×161, W690×170, W690×152, W610×153, W530×138, W460×158, W460×144
W2.0	W690×140, W610×140, W610×125, W610×113, W530×123, W530×109, W530×101, W530×92, W460×128, W460×106, W460×97
W3.1	W610×92, W610×82, W530×85, W460×89, W460×82, W410×85
W3.0	W530×74, W530×66, W460×74, W460×68, W460×52, W410×75, W410×67, W410×60, W410×46.1

Prequalified Connections for Special and Intermediate Steel Moment Frames for Seismic Applications, 2010, incl. Supplement No. 1
AMERICAN INSTITUTE OF STEEL CONSTRUCTION

TABLE C-9.2
Recommended B-Series Bracket-Beam Combinations

Bracket Designation	Beam Designations
B1.0	W33×130, W30×124, W30×116, W24×131, W21×122, W21×111
B2.1	W30×108, W27×114, W27×102, W27×94, W18×106, W18×97

TABLE C-9.2M
Recommended B-Series Bracket-Beam Combinations

Bracket Designation	Beam Designations
B1.0	W840×193, W760×185, W760×173, W610×195, W530×182, W530×165
B2.1	W760×161, W690×170, W690×152, W690×140, W460×158, W460×144

solution can be simplified to determine the controlling yield line pattern that produces the lowest failure load. Because a continuity plate would interfere with the installation of the connecting bolts, the procedure requires that the column flange thickness adequately satisfies the limit state without the requirement to provide continuity plates.

Although **Step 9** requires a flange thickness that will adequately satisfy the column flange local bending limit state, the limit states of web local yielding, web crippling, and web compression buckling as defined in Sections J10.2, J10.3 and J10.5 of the AISC *Specification*, respectively, may also be applicable. In shallow seismically compact W12 (W310) and W14 (W360) sections these additional limit states will not control. However, in some deeper sections, the additional limit states may govern. Therefore, **Step 10** requires continuity plates in the deeper sections to adequately address the limit states and to stabilize deep column sections. The plates are positioned at the same level as the beam flange as shown in Figures 9.4 and 9.5.

Step 12 of the procedure is based on the limit state of bolt shear rupture as defined in Section J3.6 of the AISC *Specification*. When this connection first appeared in the 2009 Supplement No. 1 to AISC 358-05, a bolt shear overstrength factor of 1.1 was included in the denominator of Equation 9.9-9 based on research subsequently reported by Tide (2010). The 2010 AISC *Specification* has since incorporated that factor into the tabulated shear strengths of bolts, necessitating its removal here.

The procedure outlined in Step 12 omits a bolt-bearing limit state check per Section J3.10 of the AISC *Specification* because the provisions of Sections 9.3.1(5) and 9.3.1(7) preclude the use of beams where the bolt bearing would limit the strength of the connection.

Step 14 of the procedure is based on the limit state of weld shear rupture as defined in Section J2.4 of the AISC *Specification*. The procedure assumes a linear weld group loaded through the center of gravity.

Step 18 of the procedure is supplemental if the column is a built-up box configuration. The procedure is based on the limit state of yielding (plastic moment) as defined in Section F11.1 of the AISC *Specification*. The design assumes a simply supported condition with symmetrical point loads applied at the bolt locations.

CHAPTER 10

CONXTECH CONXL MOMENT CONNECTION

10.1. GENERAL

The ConXtech® ConXL™ moment connection is designed to provide robust cost effective moment framing, while eliminating field welding and facilitating fast frame erection. The patented ConXL fabrication and manufacturing process utilizes forged parts, welding fixtures and robotic welders to produce a standardized connection.

The collars and collar assemblies illustrated, and methodologies used in their fabrication and erection, are covered by one or more of the U.S. and foreign patents shown at the bottom of the first page of Chapter 10. Additional information on the ConXL connection can also be found at http://www.conxtech.com.

Prequalification of the ConXL moment connection is based on the 17 qualifying cyclic tests listed in Table C10.1, as well as nonlinear finite element modeling of the connection. The test database includes five bi-axial moment connection tests. These unprecedented bi-axial moment connection tests subjected the framing in the orthogonal plane to a constant shear creating a moment across the column-beam joint equivalent to that created by the probable maximum moment at the plastic hinge of the primary beams, while the framing in the primary plane was simultaneously subjected to the qualifying cyclic loading specified by AISC 341-05 Appendix S until failure occurred. Tests were conducted using a variety of column to beam strength ratios. Many tests were conducted with an intentionally reinforced column, consisting of a concrete-filled HSS with an embedded W12 (W310) inside the HSS, forcing all inelastic behavior out of the column. In one of the biaxial tests, simultaneous flexural yielding of the column was initiated during cycling. Typically, failures consist of low-cycle fatigue of a beam flange in the zone of plastic hinging, following extensive rotation and local buckling deformation.

The ConXL connection is a true bi-axial moment connection capable of moment-connecting up to four beams to a column. All moment-connected columns require a full set of four collar flange top (CFT) pieces and four collar flange bottom (CFB) pieces at every beam-column moment connected joint, even if a column face has no beam present. Each column face with either a moment-connected beam or simply supported beam will have the full collar flange assembly [CFT, CFB and collar web extension (CWX)] with the simply connected beam connected to the CWX with a standardized bolted connection.

Unlike more conventional moment frame design, which focuses on keeping the number of moment-resisting frames to a minimum for reasons of economy, the efficient ConXL system distributes the bi-axial moment connection to nearly every beam-column-beam joint throughout the structure creating a distributed moment-resisting space frame. Thus, instead of a less redundant structure with more concentrated

TABLE C10.1
Summary of ConXL Tests

Test No.	Test Condition	Column Size	Primary Axis Beam	Secondary Axis Beam	Rotation (radians)
1101	Planar	HSS 16×16×$^5/_8$*	W18×76 RBS	N/A	0.05
1102	Planar	HSS 16×16×$^5/_8$*	W18×119	N/A	0.05
1103	Planar	HSS 16×16×$^5/_8$*	W24×84 RBS	N/A	0.06
1104	Planar	HSS 16×16×$^5/_8$*	W24×104	N/A	0.05
1105	Planar	HSS 16×16×$^5/_8$*	W24×117×6[†]	N/A	0.04
1106	Planar	HSS 16×16×$^5/_8$*	W24×117×9[†]	N/A	0.04
1107	Planar	HSS 16×16×$^5/_8$	W21×62 RBS	N/A	0.04
1108	Planar	HSS 16×16×$^5/_8$	W21×62 RBS	N/A	0.06
1201	Planar	HSS 16×16×$^5/_8$*	W30×108 RBS	N/A	0.05
1202	Planar	HSS 16×16×$^5/_8$*	W30×108 RBS	N/A	0.05
1203	Planar	HSS 16×16×$^5/_8$*	W30×90	N/A	0.04
1204	Planar	HSS 16×16×$^5/_8$*	W30×90	N/A	0.04
2102	Bi-Axial	BU 16×16×1.25	W30×108 RBS	W30×148	0.05
2103	Bi-Axial	BU 16×16×1.25	W30×108 RBS	W30×148	0.06
2105	Bi-Axial	HSS 16×16×$^1/_2$	W21×55 RBS	W21×83	0.06
2106	Bi-Axial	BU 16×16×1.25	W30×108 RBS	W30×148	0.05
2107	Bi-Axial	BU 16×16×1.25	W30×108 RBS	W30×148	0.05

* Column consisted of HSS 16 with supplementary W12×136 housed within concrete fill.
[†] Beam flanges were trimmed to the indicated width in order to test the ability of the collar to withstand (a) narrow-flange beams [6 in. (150 mm) flange] and (b) maximum forces [9 in. (230 mm) flange].
BU indicates built-up box section columns.

lateral force resistance, all or almost all beam-column connections are moment-resisting creating extensive redundancy. The distribution of moment connections throughout the structure also allows for reduced framing sizes and provides excellent floor vibration performance due to fixed-fixed beam end conditions. The highly distributed lateral force resistance also provides for reduced foundation loads and an inherently robust resistance to progressive collapse.

Finite element models of tested beam-column assemblies confirm that the contribution of concrete column fill can be accounted for using the gross transformed properties of the column. Beams and columns should be modeled without rigid end offsets. Prescriptive reductions in beam stiffness to account for reduced beam section (RBS) property reductions are conservative for ConXL framing, as the RBS is located farther away from the column centerline than is typical of standard RBS connections. Therefore, modeling of ConXL assemblies employing RBS beams should model the reduced beam sections explicitly, rather than using prescriptive reductions in stiffness to account for the beam flange reduction.

Because ConXL systems have their lateral force resistance distributed throughout the structure, torsional resistance can be less than structures with required lateral force resistance concentrated on exterior lines. It is possible to minimize this effect by selecting stiffer members towards the building perimeter, to increase the torsional inertia.

10.2. SYSTEMS

The ConXL moment connection is unique in that it meets the prequalification requirements for special and intermediate moment frames in orthogonal intersecting moment-resisting frames. It can also be used in more traditional plane frame applications. These requirements are met with a single standardized connection.

The exception associated with concrete structural slab placement at the column and collar assembly is based on testing conducted on the stiffened extended end-plate moment connection (Seek and Murray, 2005). Early testing by Murray of a bolted-end-plate specimen with a concrete slab in place failed by tensile rupture of the bolts. This was postulated to be the result of composite action between the beam and slab, resulting in increased beam flexural strength and increased demands on the bolt relative to calculated demands neglecting composite effects. Later testing referenced above demonstrated that placement of a flexible material in the slab adjacent to the column sufficiently reduced this composite action and protected the bolts. Although ConXL connections have not been tested with slabs present, it is believed that the same protective benefits of the flexible material apply to this connection.

ConXL's highly distributed lateral force resistance reduces the need for metal deck/concrete fill to act as a diaphragm and drag forces to a limited number of moment resisting frames. Each moment-resisting column and connected beams resist a tributary lateral load and typically minimal concrete reinforcement or deck attachment is required.

10.3. PREQUALIFICATION LIMITS

1. Beam Limitations

Minimum beam depth is controlled by the collar dimensions and is 18 in. (460 mm). Maximum beam depth is controlled by strong-column weak-beam considerations and is limited to 30 in. (760 mm) for practical purposes. The flange width and thickness requirements are limited by the ability of the collar flange to accommodate the beam flange weld and also by the strength of the bolts. A key ConXL requirement for allowable beam sections is limiting the force delivered by the beam to the bolts connecting the collar flange/beam to the collar corner assemblies/column so as to not overcome the pretension load applied to the bolts. This requirement is covered in detail in Section 10.8.

ConXL connections have been successfully tested without reduced beam section reductions in flange width and are qualified for use without such reductions. However, RBS cuts in beam flanges can be a convenient way to achieve strong column weak beam limitations without increasing column weight.

Lateral bracing of beams is per the AISC *Seismic Provisions*. During the bi-axial moment connection tests, the test beams (W30×108 with 50% RBS, W21×55 with 50% RBS) were not braced at the RBS and were braced at the beam ends, 10 ft. (3050 mm) from the column center.

All moment-connected beams are required to meet seismic compaction requirements of the AISC *Seismic Provisions*, if RBS beams are used, the width-thickness ratio is taken within its reduced flange width as permitted for RBS connections (Section 5.3.1(6)).

2. Column Limitations

The key requirement for ConXL moment columns is a square sectional dimension of 16 in. (406 mm). Section type (built-up box or HSS) can vary, as can steel strength and wall thickness. All columns used in ConXL moment connections are concrete-filled with either normal or lightweight concrete, having minimum compressive strength of 3,000 psi (21 MPa). Columns are typically filled with concrete at the job site, after erection and bolting is complete. The concrete is pumped to the top of column and allowed to free-fall the full height of column, using the column as a tremie. There are no obstructions, stiffener plates, etc. within the column, thus the column is similar to a tremie-pipe allowing the concrete an unobstructed path to its placement with excellent consolidation (Suprenant, 2001).

3. Collar Limitations

Appendix B describes the forged steel material specification used to manufacture the collars. The forging process produces an initial collar (blank collar) slightly larger than the final overall dimensions. The collar is then machined to their manufacturing dimensions within the required tolerances.

10.4. COLLAR CONNECTION LIMITATIONS

The collars are the key elements of the ConXL connection. They are standardized components and no further design or sizing of these components are required. The same components are used for all beams and columns. The same is true for the collar bolts, where the specification, size and number of bolts always remain the same. The design procedure ensures that column-beam combinations used in the ConXL connection fall within the code requirements of these standard connection components.

The bolts used in the ConXL connection are $1^1/_4$ in. diameter ASTM A574 bolts. These bolts are similar in chemistry and mechanical properties to ASTM A490 bolts, but have socket heads to accommodate their use in this connection. Metric bolts conforming to ASTM A574M have not been tested and are not prequalified for use in this connection. Pretensioning is performed to the requirements for $1^1/_4$ in. diameter ASTM A490 bolts [102 kips (454 kN) per Table J3.1 of the 2010 AISC *Specification*].

10.5. BEAM WEB-TO-COLLAR CONNECTION LIMITATIONS

The collar web extension (CWX) is $1^1/_2$ in. (38 mm) thick, thus the minimum sized fillet weld between the CWX and beam web is a $^5/_{16}$ in. (8 mm) fillet weld. This weld size for a two-sided fillet weld (each side of the web) should be sufficient for all allowable beams; this should be confirmed during the design procedure calculations.

10.6. BEAM FLANGE-TO-COLLAR FLANGE WELDING LIMITATIONS

Weld access holes are not permitted in the ConXL connection. Welding of the beam flange to the collar flange is performed in a proprietary ConXtech beam weld fixture, which rotates the beam to allow access to the bottom flange for welding in the flat position. The beam weld fixture enables the manufacturing of the moment beam within ConXL tolerances.

10.7. COLUMN-BEAM RELATIONSHIP LIMITATIONS

The ConXL moment connection is a bi-axial connection. Strong-column weak beam requirements specified by the AISC *Seismic Provisions* were formulated considering the typical planar framing prevalent in moment-frame construction following the 1994 Northridge earthquake. Because the ConXL connection is primarily used in intersecting moment frames, with biaxial behavior an inherent part of the design, the committee felt that it was imperative to require that columns have sufficient strength to develop expected simultaneous flexural hinging in beams framing into all column faces. The bi-axial calculation considers all moment beams attached to the column. This calculation is covered in detail in Section 10.8.

10.8. DESIGN PROCEDURE

Step 1. As with other connections, the first step in the design procedure is to compute the probable maximum moment at the plastic hinge. Note that differing C_{pr} factors are applied for RBS and non-RBS beams. The factor for non-RBS beams is compatible with the standard requirements in the AISC *Seismic Provisions* while that for RBS beams is compatible with the requirements of this standard for RBS connections.

Step 2. As with other connections, the equation given for computation of shear forces has to include consideration of gravity loads that are present. The equations presented in the design procedure assume uniform gravity loading. Modifications to these equations are necessary for cases with concentrated loads present. These modifications must satisfy static equilibrium requirements.

Step 3. The ConXL moment connection is a true bi-axial moment connection, thus the committee determined that columns must be sufficiently strong to permit simultaneous development of flexural hinging in all beams framing to a column, not just beams along a single plane. This bi-axial column-beam moment evaluation is more conservative than current AISC *Seismic Provisions* requirements which considers plastic hinging of beams in a single plane only, even though columns supporting moment frames in orthogonal directions are possible with other connections using built-up box sections or other built-up column sections. In calculating the ConXL bi-axial column-beam moment ratio it is permitted to take the actual yield strength of the column material in lieu of the specified minimum yield stress, F_y, and to consider the full composite behavior of the column for axial load and flexural action (inter-story drift analysis). The default formula for column strength provided in the design procedure assumes that equal strength beams are present on all faces of the connection. When some beams framing to a column are stronger than others, it is permitted to use basic principals of structural mechanics to compute the actual required flexural strength.

The design procedure also considers the critical beam strength as it relates to the column strength at locations just above the beam's top flange and just below the beam's bottom flange, where flexural demand on the columns are greatest. Flexural demand on the column within the panel zone is less than at these locations.

Step 5. The available tensile strength for the bolts used in the ConXL connection is specified as the minimum bolt pretension load. The purpose of assigning the minimum pretension load as the available bolt tensile strength is to prevent overcoming of bolt pretension, at least up to the bolt loading subjected by the probable maximum moment. The minimum bolt pretension load is 102 kips (454 kN). Bolts are checked for tension only because the frictional force developed by the bolt pretension will resist beam shear (see Steps 6 and 7).

Steps 6 and 7. Beam shear is resisted by the friction developed between the collar flanges and the collar corners. The collar flanges are clamped against the collar corner assemblies and column when the collar bolts are pretensioned. This pretension clamping force creates friction between the machined surfaces of the collar flanges and collar corners. The machined surfaces are classified as a Class B Surface (unpainted blast-cleaned steel surfaces). The design frictional resistance per bolt is:

$R_n = \mu D_u h_{sc} T_b N_s$
$\phi = 0.85$
$\mu = 0.5$
$D_u = 1.13$
$h_{sc} = 1.0$
$T_b = 102$ kips (454 kN)
$N_s = 1$
$\phi R_n = 0.85 \times 0.5 \times 1.13 \times 1.0 \times 102 \times 1 = 49.0$ kips/bolt (218 kN/bolt)

Prequalified Connections for Special and Intermediate Steel Moment Frames for Seismic Applications, 2010, incl. Supplement No. 1
AMERICAN INSTITUTE OF STEEL CONSTRUCTION

There are 16 bolts per beam end providing a total of 784 kips (3487 kN) of frictional resistance against shear. This frictional force is significantly greater than any beam shear developed by an allowable beam.

Steps 8 and 9. The available length of weld for the collar web extension and collar corner assemblies allow for minimum sized fillet welds to resist beam shear.

Steps 10 and 11. The collar corner assemblies provide additional strength to the column walls to resist panel zone shear. Without taking into consideration the contribution of the concrete fill, the column section along with the collar corner assemblies should provide sufficient strength for anticipated panel zone shear; this should be confirmed during the design procedure calculations.

APPENDIX A

CASTING REQUIREMENTS

A1. CAST STEEL GRADE

The cast steel grade is selected for its ability to provide ductility similar to that of rolled steel. The material has a specified yield and tensile strength of 50 ksi (354 MPa) and 80 ksi (566 MPa), respectively. The ASTM specification requires the castings be produced in conjunction with a heat treatment process that includes normalizing and stress relieving. It also requires each heat of steel meet strict mechanical properties. These properties include the specified tensile and yield strengths, as well as elongation and area reduction limitations.

A2. QUALITY CONTROL (QC)

See Commentary Section 3.7.

2. First Article Inspection (FAI) of Castings

The intent of this section is that at least one casting of each pattern undergo FAI. When a casting pattern is replaced or when the rigging is modified, FAI is to be repeated.

3. Visual Inspection of Castings

All casting surfaces shall be free of adhering sand, scales, cracks, hot tears, porosity, cold laps, and chaplets. All cored holes in castings shall be free of flash and raised surfaces. The ASTM specification includes acceptance criteria for the four levels of surface inspection. Level I is the most stringent criteria. The Manufacturers Standardization Society (MSS) specification includes a set of reference comparators for the visual determination of surface texture, surface roughness and surface discontinuities.

4. Nondestructive Testing (NDT) of Castings

These provisions require the use of nondestructive testing to verify the castings do not contain indications that exceed the specified requirements.

Radiographic testing (RT) is capable of detecting internal discontinuities and is specified only for the FAI. The ASTM specifications contain referenced radiographs and five levels of RT acceptance. The lower acceptance levels are more stringent and are typically required on high-performance aerospace parts such as jet engine turbine blades or on parts that may leak such as valves or pumps. Level III is considered the industry standard for structurally critical components.

Ultrasonic testing (UT) is also capable of detecting internal discontinuities and is specified for production castings. The ASTM specification includes seven levels of

A3. MANUFACTURER DOCUMENTS

Submittal documents allow a thorough review on the part of the patent holder, engineer of record, the authority having jurisdiction and outside consultants, if required.

APPENDIX B
FORGING REQUIREMENTS

There is no Commentary for this Appendix.

REFERENCES

The following references have been reviewed as a basis for the prequalification of the connections described in this Standard. Although some references are not specifically cited in this Standard, they have been reviewed by the AISC Connection Prequalification Review Panel and are listed here to provide an archival record of the basis for this Standard in accordance with the requirements of Chapter K of the AISC *Seismic Provisions*.

ALL CONNECTIONS

AISC (2005a), *Seismic Provisions for Structural Steel Buildings*, AISC/ANSI 341-05, American Institute of Steel Construction, Inc., Chicago, IL.

AISC (2005b), *Specification for Structural Steel Buildings*, AISC/ANSI 360-05, American Institute of Steel Construction, Inc., Chicago, IL.

AISC (2005c), *Steel Construction Manual*, 13th ed., American Institute of Steel Construction, Chicago, IL.

AISC (2010a), *Seismic Provisions for Structural Steel Buildings*, AISC/ANSI 341-10, American Institute of Steel Construction, Chicago, IL.

AISC (2010b), *Specification for Structural Steel Buildings*, AISC/ANSI 360-10, American Institute of Steel Construction, Chicago, IL.

ATC (1992), *Guidelines for Cyclic Seismic Testing of Components of Steel Structures*, ATC-24, Applied Technology Council, Redwood City, CA.

AWS (2010), *Structural Welding Code—Steel*, AWS D1.1/D1.1M:2010, American Welding Society, Miami, FL.

AWS (2009), *Structural Welding Code—Seismic Supplement*, AWS D1.8/D1.8M:2009, American Welding Society, Miami, FL.

Barsom, J. and Korvink, S. (1998), "Effects of Strain Hardenening and Strain Aging on the K-Region of Structural Shapes," Report No. SAC/BD-98/02, SAC Joint Venture, Sacramento, CA.

Chi, W.M., Deierlein, G. and Ingraffea, A. (1997), "Finite Element Fracture Mechanics Investigation of Welded Beam-Column Connections," Report No. SAC/BD-97/05, SAC Joint Venture, Sacramento, CA.

FEMA (1997), *NEHRP Recommended Provisions for Seismic Regulations for New Buildings and Other Structures*, FEMA 302, Part 1—Provisions, Federal Emergency Management Agency, Washington, DC.

FEMA (2000a), *Recommended Post-Earthquake Evaluation and Repair Criteria for Welded Steel Moment-Frame Buildings*, FEMA 352, Federal Emergency Management Agency, Washington, DC.

REFERENCES

FEMA (2000b), *Recommended Seismic Design Criteria for New Steel Moment-Frame Buildings*, FEMA 350, Federal Emergency Management Agency, Washington, DC.

FEMA (2000c), *Recommended Seismic Evaluation and Upgrade Criteria for Existing Welded Steel Moment-Frame Buildings*, FEMA 351, Federal Emergency Management Agency, Washington, DC.

FEMA (2000d), *Recommended Specifications and Quality Assurance Guidelines for Steel Moment-Frame Construction for Seismic Applications*, FEMA 353, Federal Emergency Management Agency, Washington, DC.

FEMA (2000e), *State of the Art Report on Connection Performance*, FEMA 355D, prepared by the SAC Joint Venture for the Federal Emergency Management Agency, Washington, DC.

FEMA (2000f), *State of the Art Report on Systems Performance of Steel Moment Frames Subject to Earthquake Ground Shaking*, FEMA 355C, prepared by the SAC Joint Venture for the Federal Emergency Management Agency, Washington, DC.

RCSC (2009), *Specification for Structural Joints Using High-Strength Bolts*, Research Council on Structural Connections, Chicago, IL.

Tsai, K.C. and Popov, E.P. (1986), "Two Beam-to-Column Web Connections," Report No. UCB/EERC-86/05, Earthquake Engineering Research Center, University of California at Berkeley, April.

Tsai, K.C. and Popov, E.P. (1988), "Steel Beam-Column Joints in Seismic Moment Resisting Frames," Report No. UCB/EERC-88/19, Earthquake Engineering Research Center, University of California at Berkeley, November.

CHAPTER 5
REDUCED BEAM SECTION (RBS) MOMENT CONNECTION

Chambers, J.J., Almudhafer, S. and Stenger, F. (2003), "Effect of Reduced Beam Section Frames Elements on Stiffness of Moment Frames," *Journal of Structural Engineering*, American Society of Civil Engineers, Vol. 129, No. 3.

Chen, S.J. and Chao, Y.C. (2001), "Effect of Composite Action on Seismic Performance of Steel Moment Connections with Reduced Beam Sections," *Journal of Constructional Steel Research*, Elsevier Science Publishers, Vol. 57.

Chen, S.J., Yeh, C.H. and Chu, J.M. (1996), "Ductile Steel Beam-to-Column Connections for Seismic Resistance," *Journal of Structural Engineering*, American Society of Civil Engineers, Vol. 122, No. 11, pp. 1292–1299.

Engelhardt, M.D. (1999), "The 1999 T.R. Higgins Lecture: Design of Reduced Beam Section Moment Connections," *Proceedings 1999 North American Steel Construction Conference, Toronto, Canada*, pp. 1-1 to 1-29, American Institute of Steel Construction, Inc., Chicago, IL.

Engelhardt, M.D., Fry, G., Jones, S., Venti, M. and Holliday, S. (2000), "Behavior and Design of Radius-Cut Reduced Beam Section Connections," Report No. SAC/BD-00/17, SAC Joint Venture, Sacramento, CA.

REFERENCES

Engelhardt, M.D., Winneberger, T., Zekany, A.J. and Potyraj, T.J. (1998), "Experimental Investigation of Dogbone Moment Connections," *Engineering Journal*, American Institute of Steel Construction, Vol. 35, No. 4, pp. 128–139.

FEMA (2000g), *State of the Art Report on Past Performance of Steel Moment-Frame Buildings in Earthquakes*, FEMA 355e, prepared by the SAC Joint Venture for the Federal Emergency Management Agency, Washington, DC.

Gilton, C., Chi, B. and Uang, C.M. (2000), "Cyclic Response of RBS Moment Connections: Weak Axis Configuration and Deep Column Effects," Report No. SAC/BD-00/23, SAC Joint Venture, Sacramento, CA.

Grubbs, K.V., "The Effect of the Dogbone Connection on the Elastic Stiffness of Steel Moment Frames" (1997). M.S. Thesis, Department of Civil Engineering, The University of Texas at Austin.

Iwankiw, N.R. and Carter, C. (1996), "The Dogbone: A New Idea to Chew On," *Modern Steel Construction*, American Institute of Steel Construction, April 1996.

Lee, C.H., Kim, J.H., Jeon, S.W., and Kim, J.H. (2004), "Influence of Panel Zone Strength and Beam Web Connection Method on Seismic Performance of Reduced Beam Section Steel Moment Connections," *Proceedings of the CTBUH 2004 Seoul Conference—Tall Buildings for Historical Cities*, Council on Tall Buildings and Urban Habitat, Bethlehem, PA.

Moore, K.S., Malley, J.O. and Engelhardt, M.D. (1996), "Design of Reduced Beam Section (RBS) Moment Connections," *Steel Tips*, Structural Steel Education Council, Moraga, CA.

Okahashi, Y. (2003), "Reduced Beam Section Connection without Continuity Plates," M.S. Thesis, Department of Civil and Environmental Engineeering, University of Utah.

Plumier, A. (1990), "New Idea for Safe Structures in Seismic Zones," *IABSE Symposium—Mixed Structures Including New Materials*, Brussels, Belgium.

Plumier, A. (1997), "The Dogbone: Back to the Future," *Engineering Journal*, American Institute of Steel Construction, Vol. 34, No. 2, pp. 61-67.

Popov, E.P., Yang, T.S. and Chang, S.P. (1998), "Design of Steel MRF Connections Before and After 1994 Northridge Earthquake," *International Conference on Advances in Steel Structures*, Hong Kong, December 11–14, 1996. Also in: *Engineering Structures*, Elsevier Science Publishers, Vol. 20, No. 12, pp.1030–1038.

Ricles, J.M., Zhang, X., Lu, L.W., and Fisher, J. (2004), "Development of Seismic Guidelines for Deep Column Steel Moment Connections," ATLSS Report No. 04-13, Lehigh University, Bethlehem, PA.

Shen, J., Kitjasateanphun, T. and Srivanich, W. (2000), "Seismic Performance of Steel Moment Frames with Reduced Beam Sections," *Journal of Constructional Steel Research*, Elsevier Science Publishers, Vol. 22

Shen, J., Astaneh-Asl, A. and McCallen, D.B. (2002), "Use of Deep Columns in Special Steel Moment Frames," *Steel Tips*, Structural Steel Education Council, Moraga, CA.

Suita, K., Tamura, T., Morita, S., Nakashima, M. and Engelhardt, M.D. (1999), "Plastic Rotation Capacity of Steel Beam-to-Column Connections Using a Reduced Beam Section and No Weld Access Hole Design—Full Scale Tests for Improved Steel Beam-to-Column Subassemblies—Part 1," *Structural Journal*, Architectural Institute of Japan, No. 526, pp. 177–184, December 1999 (in Japanese).

Tremblay, R., Tchebotarev, N. and Filiatrault, A. (1997), "Seismic Performance of RBS Connections for Steel Moment Resisting Frames: Influence of Loading Rate and Floor Slab," *Proceedings, Stessa '97*, Kyoto, Japan.

Tsai, K.C., Chen, W.Z. and Lin, K.C. (1999), "Steel Reduced Beam Section to Weak Panel Zone Moment Connections," *Proceedings: Workshop on Design Technologies of Earthquake-Resistant Moment-Resisting Connections in Steel Buildings*, May 17–18, 1999, Taipei, Taiwan (in Chinese).

Uang, C.M. and Fan, C.C. (1999) "Cyclic Instability of Steel Moment Connections with Reduced Beam Section," Report No. SAC/BD-99/19, SAC Joint Venture, Sacramento, CA.

Uang, C.M. and Richards, P. (2002), "Cyclic Testing of Steel Moment Connections for East Tower of Hoag Memorial Hospital Presbyterian," Third Progress Report, University of California, San Diego, CA.

Yu, Q.S., Gilton, C. and Uang, C.M. (2000), "Cyclic Response of RBS Moment Connections: Loading Sequence and Lateral Bracing Effects," Report No. SAC/BD-00/22, SAC Joint Venture, Sacramento, CA, 2000.

Zekioglu, A., Mozaffarian, H., Chang, K.L., Uang, C.M. and Noel, S. (1997), "Designing after Northridge," *Modern Steel Construction*, American Institute of Steel Construction, Inc., Chicago, IL.

Zekioglu, A., Mozaffarian, H. and Uang, C.M. (1997), "Moment Frame Connection Development and Testing for the City of Hope National Medical Center," *Building to Last—Proceedings of Structures Congress XV*, ASCE, Portland, American Society of Civil Engineers, Reston, VA.

CHAPTER 6
BOLTED UNSTIFFENED AND STIFFENED EXTENDED END-PLATE MOMENT CONNECTIONS

Abel, M.S. and Murray, T.M. (1992a), "Multiple Row, Extended Unstiffened End-Plate Connection Tests," Research Report CE/VPI-ST-92/04, Department of Civil Engineering, Virginia Polytechnic Institute and State University, Blacksburg, VA.

Abel, M.S. and Murray, T.M. (1992b), "Analytical and Experimental Investigation of the Extended Unstiffened Moment End-Plate Connection with Four Bolts at the Beam Tension Flange," Research Report CE/VPI-ST-93/08, Department of Civil Engineering, Virginia Polytechnic Institute and State University, Blacksburg, VA.

Adey, B.T., Grondin, G.Y. and Cheng, J.J.R. (1997), "Extended End Plate Moment Connections under Cyclic Loading," Structural Engineering Report No. 216, Department of Civil and Environmental Engineering, University of Alberta, Edmonton, Alberta, Canada.

REFERENCES

Adey, B.T., Grondin, G.Y. and Cheng, J.J.R. (1998), "Extended End Plate Moment Connections under Cyclic Loading," *Journal of Constructional Steel Research*, Elsevier Science Publishers, Vol. 46, pp. 1–3.

Adey, B.T., Grondin, G.Y. and Cheng, J.J.R. (2000), "Cyclic Loading of End Plate Moment Connections," *Canadian Journal of Civil Engineering*, National Research Council of Canada, Vol. 27, No. 4, pp. 683–701.

Agerskov, H. (1976), "High Strength Bolted Connections Subject to Prying," *Journal of the Structural Division*, American Society of Civil Engineers, Vol. 102, No. ST1, pp. 161–175.

Agerskov, H. (1977), "Analysis of Bolted Connections Subject to Prying." *Journal of the Structural Division*, American Society of Civil Engineers, Vol. 103, No. ST11, pp. 2145–2163.

Ahuja, V. (1982), "Analysis of Stiffened End-Plate Connections Using Finite Element Method," M.S. Thesis, School of Civil Engineering and Environmental Science, University of Oklahoma, Norman, OK.

Bahaari, M.R. and Sherbourne, A.N. (1993), "Modeling of Extended End-plate Bolted Connections." *Proceedings of the National Steel Structures Congress*, pp. 731–736, American Institute of Steel Construction, Inc., Chicago, IL.

Bjorhovde, R., Brozzetti, J. and Colson, A. (1987) "Classification of Connections," *Connections in Steel Structures—Behaviour, Strength and Design*, Elsevier Science Publishers, London, U.K., pp. 388–391.

Bjorhovde, R., Colson, A. and Brozzetti, J. (1990), "Classification System for Beam-to-Column Connections," *Journal of Structural Engineering*, American Society of Civil Engineers, Vol. 116, No. 11, pp. 3059–3076.

Borgsmiller, J.T. and Murray, T.M. (1995), "Simplified Method for the Design of Moment End-Plate Connections," Research Report CE/VPI-ST-95/19, Department of Civil Engineering, Virginia Polytechnic Institute and State University, Blacksburg, VA.

Bursi, O.S. and Leonelli, L. (1994), "A Finite Element Model for the Rotational Behavior of End Plate Steel Connections," SSRC Proceedings 1994 Annual Task Group Technical Session, pp. 162–175, Lehigh University, Bethlehem, PA.

Carter, C.J. (1999), *Stiffening of Wide-Flange Columns at Moment Connections: Wind and Seismic Applications, Design Guide No. 13*, American Institute of Steel Construction, Chicago, IL.

Castellani, A., Castiglioni, C.A., Chesi, C., and Plumier, A. (1998), "A European Research Program on the Cyclic Behaviour of Welded Beam to Column Connections," *Proceedings of the NEHRP Conference and Workshop on Research on the Northridge, California Earthquake of January 17, 1994*, Vol. III-B, pp. 510–517, National Earthquake Hazards Reduction Program, Washington, DC.

Coons, R.G. (1999), "Seismic Design and Database of End Plate and T-stub Connections," M.S. Thesis, University of Washington, Seattle, WA.

REFERENCES

Disque, R.O. (1962), "End-Plate Connections," *National Engineering Conference Proceedings,* American Institute of Steel Construction, 1962, pp. 30–37.

Douty, R.T. and McGuire, S. (1965), "High Strength Bolted Moment Connections," *Journal of the Structural Division,* American Society of Civil Engineers, Vol. 91, No. ST2, pp. 101–126.

Fleischman, R.B., Chasten, C.P., Lu, L-W and Driscoll, G.C. (1991), "Top-and-Seat Angle Connections and End-Plate Connections: Snug vs. Fully Pretensioned Bolts," *Engineering Journal,* American Institute of Steel Construction, Vol. 28, pp. 18–28.

Ghassemieh, M. (1983), "Inelastic Finite Element Analysis of Stiffened End-Plate Moment Connections," M.S. Thesis, School of Civil Engineering and Environmental Science, University of Oklahoma, Norman, OK.

Ghobarah, A., Korol, R.M. and Osman, A. (1992), "Cyclic Behavior of Extended End-Plate Joints," *Journal of Structural Engineering,* American Society of Civil Engineers, Vol. 118, No. 5, pp. 1333–1353.

Ghobarah, A., Osman, A. and Korol, R.M. (1990), "Behaviour of Extended End-Plate Connections under Cyclic Loading," *Engineering Structures,* Elsevier Science Publishers, Vol. 12, pp. 15–26.

Granstrom, A. (1980), "Bolted End-Plate Connections," Stalbyggnads Institute SBI Report 86.3, pp. 5–12.

Griffiths, J.D. (1984), "End-Plate Moment Connections—Their Use and Misuse," *Engineering Journal,* American Institute of Steel Construction, Vol. 21, No. 1, pp. 32–34.

Hasan, R., Kishi, N. and Chen, W.F. (1997), "Evaluation of Rigidity of Extended End-Plate Connections," *Journal of Structural Engineering,* American Society of Civil Engineers, Vol. 123, No. 12, pp. 1595–1602.

Hendrick, D., Kukreti, A. and Murray, T. (1984), "Analytical and Experimental Investigation of Stiffened Flush End-Plate Connections with Four Bolts at the Tension Flange," Research Report FSEL/MBMA 84-02, Fears Structural Engineering Laboratory, University of Oklahoma, Norman, OK.

Hendrick, D., Kukreti, A. and Murray, T. (1985), "Unification of Flush End-Plate Design Procedures," Research Report FSEL/MBMA 85-01, Fears Structural Engineering Laboratory, University of Oklahoma, Norman, OK.

Johnstone, N.D. and Walpole, W.R. (1981), "Bolted End-Plate Beam-to-Column Connections Under Earthquake Type Loading," Report 81-7, Department of Civil Engineering, University of Canterbury, Christchurch, New Zealand.

Kato, B. and McGuire, W.F. (1973), "Analysis of T-Stub Flange-to-Column Connections," *Journal of the Structural Division,* American Society of Civil Engineers, Vol. 99 No. ST5, pp. 865–888.

Kennedy, N.A., Vinnakota, S. and Sherbourne, A.N. (1981), "The Split-Tee Analogy in Bolted Splices and Beam-Column Connections," *Proceedings of the International Conference on Joints in Structural Steelwork,* pp. 2.138–2.157.

Kline, D., Rojiani, K. and Murray, T. (1989), "Performance of Snug Tight Bolts in Moment End-Plate Connections," MBMA Research Report, Department of Civil Engineering, Virginia Polytechnic Institute and State University, Blacksburg, VA. Revised July 1995.

Korol, R.M., Ghobarah, A. and Osman, A. (1990), "Extended End-Plate Connections Under Cyclic Loading: Behaviour and Design," *Journal of Constructional Steel Research*, Elsevier Science Publishers, Vol. 16, No. 4, pp. 253–279.

Krishnamurthy, N. (1978), "A Fresh Look at Bolted End-Plate Behavior and Design," *Engineering Journal*, American Institute of Steel Construction, Vol. 15, No. 2, pp. 39–49.

Krishnamurthy, N. and Graddy, D. (1976), "Correlation between 2- and 3-Dimensional Finite Element Analysis of Steel Bolted End Plate Connections," *Computers and Structures*, Vol. 6, No. 4/5, pp. 381–389.

Kukreti, A.R., Ghassemieh, M. and Murray, T.M. (1990), "Behavior and Design of Large-Capacity Moment End-Plates," *Journal of Structural Engineering*, American Society of Civil Engineers, Vol. 116, No. 3, pp. 809–828.

Kukreti, A.R., Murray, T.M. and Abolmaali, A. (1987), "End-Plate Connection Moment-Rotation Relationship," *Journal of Constructional Steel Research*, Elsevier Science Publishers, Vol. 8, pp. 137–157.

Mann, A.P. and Morris, L.J. (1979), "Limit Design of Extended End-Plate Connections," *Journal of the Structural Division*, American Society of Civil Engineers, Vol. 105, No. ST3, pp. 511–526.

Meng, R.L. (1996), "Design of Moment End-Plate Connections for Seismic Loading," Doctoral Dissertation, Virginia Polytechnic Institute and State University, Blacksburg, VA.

Meng, R.L. and Murray, T.M. (1997), "Seismic Performance of Bolted End-Plate Moment Connections," *Proceedings of the 1997 National Steel Construction Conference*, American Institute of Steel Construction, pp. 30–1 to 30–14.

Morrison, S.J., Astaneh-Asl, A. and Murray, T. (1985), "Analytical and Experimental Investigation of the Extended Stiffened Moment End-Plate Connection with Four Bolts at the Beam Tension Flange," Research Report FSEL/MBMA 85-05, Fears Structural Engineering Laboratory, University of Oklahoma, Norman, OK.

Morrison, S.J., Astaneh-Asl, A. and Murray, T. (1986), "Analytical and Experimental Investigation of the Multiple Row Extended 1/3 Moment End-Plate Connection with Eight Bolts at the Beam Tension Flange," Research Report FSEL/MBMA 86-01, Fears Structural Engineering Laboratory, University of Oklahoma, Norman, OK.

Murray, T.M. (1986), "Stability of Gable Frame Panel Zone Plates," *Proceedings of the Structural Stability Research Council Annual Technical Session*, pp. 317–325, Structural Stability Research Council, Bethlehem, PA.

Murray, T.M. (1988), "Recent Developments for the Design of Moment End-Plate Connections," *Journal of Constructional Steel Research*, Vol. 10, pp. 133–162.

Murray, T.M. (1990), *Extended End-Plate Moment Connections, Design Guide No. 4*, American Institute of Steel Construction, Chicago, IL.

REFERENCES

Murray, T.M., Kline, D.P. and Rojiani, K.B. (1992), "Use of Snug-Tightened Bolts in End-Plate Connections," *Connections in Steel Structures II: Behavior, Strength and Design*, Edited by R. Bjorhovde et al., pp. 27–34, American Institute of Steel Construction, Chicago, IL.

Murray, T.M. and Kukreti, A.R. (1988), "Design of 8-Bolt Stiffened Moment End Plates," *Engineering Journal*, American Institute of Steel Construction, Vol. 25, No. 2, pp. 45–52, American Institute of Steel Construction, Chicago, IL.

Murray, T.M. and Shoemaker, W.L. (2002), *Flush and Extended Multiple Row Moment End Plate Connections, Design Guide No. 16,* American Institute of Steel Construction, Inc., Chicago, IL.

Murray, T.M. and Sumner, E.A. (2003), *Extended End-Plate Moment Connections: Seismic and Wind Applications, Design Guide No. 4,* 2nd Ed., American Institute of Steel Construction, Chicago, IL.

Nair, R., Birkemoe, P. and Munse, W. (1974), "High Strength Bolts Subject to Tension and Prying." *Journal of the Structural Division*, American Society of Civil Engineers, Vol. 100, No. ST2, pp. 351–372.

Packer, J. and Morris, L. (1977), "A Limit State Design Method for the Tension Region of Bolted Beam-Column Connections," *The Structural Engineer*, Vol. 55, No. 10, pp. 446–458.

Popov, E. and Tsai, K.C. (1989), "Performance of Large Seismic Steel Moment Connections under Cyclic Loads," *Engineering Journal*, American Institute of Steel Construction, Vol. 12, pp. 51–60.

Ryan, J.C. and Murray, T.M. (1999), Evaluation of the Inelastic Rotation Capability of Extended End-Plate Moment Connections, Research Report No. CE/VPI-ST-99/13, submitted to Metal Building Manufacturers Association and American Institute of Steel Construction, Virginia Polytechnic Institute and State University, Blacksburg, VA.

Salmon, C. and Johnson, J. (1980), *Steel Structures, Design and Behavior*, 2nd Ed., Harper & Row, New York, NY.

Seek, M. W. and T. M. Murray (2008), "Seismic Strength of Moment End-Plate Connections with Attached Concrete Slab," Proceedings, Connections VI, American Institute of Steel Construction, Chicago, IL, June 23-25.

Srouji, R., Kukreti, A.R. and Murray, T.M. (1983a), "Strength of Two Tension Bolt Flush End-Plate Connections," Research Report FSEL/MBMA 83-03, Fears Structural Engineering Laboratory, University of Oklahoma, Norman, OK.

Srouji, R., Kukreti, A.R. and Murray, T.M. (1983b), "Yield-Line Analysis of End-Plate Connections With Bolt Force Predictions," Research Report FSEL/MBMA 83-05, Fears Structural Engineering Laboratory, University of Oklahoma, Norman, OK.

Sumner, E.A., Mays, T.W. and Murray, T.M. (2000a), Cyclic Testing of Bolted Moment End-Plate Connections, Research Report No. CE/VPI-ST-00/03, SAC Report No.

SAC/BD00/21, submitted to the SAC Joint Venture, Virginia Polytechnic Institute and State University, Blacksburg, VA.

Sumner, E.A., Mays, T.W. and Murray, T.M. (2000b), "End-Plate Moment Connections: Test Results and Finite Element Method Validation," *Connections in Steel Structures IV, Proceedings of the Fourth International Workshop*, pp. 82–93, American Institute of Steel Construction, Chicago, IL.

Sumner, E.A. and Murray, T.M. (2001), "Experimental Investigation of the MRE 1/2 End-Plate Moment Connection," Research Report No. CE/VPI-ST-01/14, Department of Civil Engineering, Virginia Polytechnic Institute and State University, Blacksburg, VA.

Sumner, E.A. and Murray, T.M. (2002), "Behavior of Extended End-Plate Moment Connections Subject to Cyclic Loading," *Journal of Structural Engineering*, American Society of Civil Engineers, Vol. 128, No. 4, pp. 501–508.

Tsai, K.C. and Popov, E.P. (1990), "Cyclic Behavior of End-Plate Moment Connections," *Journal of Structural Engineering*, American Society of Civil Engineers, Vol. 116, No. 11, pp. 2917–2930.

Yang, H., Tagawa, Y. and Nishiyama, I. (2003). "Elasto-Plastic Behavior of 'New Composite Beam System'," *Steel Structures*, Vol. 3, pp. 45–52.

Young, J. and Murray, T.M. (1996), "Experimental Investigation of Positive Bending Moment Strength of Rigid Knee Connections," Research Report No. CE/VPI-ST 9617, Department of Civil Engineering, Virginia Polytechnic Institute and State University, Blacksburg, VA.

CHAPTER 7
BOLTED FLANGE PLATE (BFP) MOMENT CONNECTION

Krawinkler, H., (1992), "Guidelines for Cyclic Seismic Testing of Components of Steel Structures" Report ATC-24, Applied Technology Council, Redwood City, CA.

Krawinkler, H., Gupta, A., Medina, R. and Luco, N. (2000), "Loading Histories for Seismic Performance Testing of SMRF Components and Assemblies," Report SAC/BD-00/10, SAC Joint Venture, Sacramento, CA.

Sato, A., Newell, J., and Uang, C.M. (2008), "Cyclic Behavior and Seismic Design of Bolted Flange Plate Steel Moment Connections," *Engineering Journal*, American Institute of Steel Construction, Vol. 45, No. 4.

Schneider, S.P. and Teeraparbwong, I. (1999), "SAC Task 7.09: Bolted Flange Plate Connections," report submitted to the SAC Joint Venture by the University of Illinois, Urbana, IL.

Seek, M. W. and T. M. Murray (2008), "Seismic Strength of Moment End-Plate Connections with Attached Concrete Slab," Proceedings, Connections VI, American Institute of Steel Construction, Chicago, IL, June 23-25.

Swanson, J., Leon, R.D. and Smallridge, J. (2000), "Tests on Bolted Connections," Report SAC/BD-00/04, SAC Joint Venture, Sacramento, CA.

CHAPTER 8
WELDED UNREINFORCED FLANGE–WELDED WEB (WUF-W) MOMENT CONNECTION

Lee, D., Cotton, S.C., Dexter, R.J., Hajjar, J.F., Ye, Y. and Ojard, S.D. (2002), "Column Stiffener Detailing and Panel Zone Behavior of Steel Moment Frame Connections." Structural Engineering Report No. ST-01-3.2, Department of Civil Engineering, University of Minnesota.

Lee, D., Cotton, S.C., Hajjar, J., Dexter, R.J. and Ye, Y. (2005a), "Cyclic Behavior of Steel Moment-Resisting Connections Reinforced by Alternative Column Stiffener Details I. Connection Performance and Continuity Plate Detailing," *Engineering Journal*, American Institute of Steel Construction, Vol. 42, No. 4.

Lee, D., Cotton, S.C., Hajjar, J., Dexter, R.J. and Ye, Y. (2005b), "Cyclic Behavior of Steel Moment-Resisting Connections Reinforced by Alternative Column Stiffener Details II. Panel Zone Behavior and Doubler Plate Detailing," *Engineering Journal*, American Institute of Steel Construction, Vol. 42, No. 4.

Ricles, J.M., Mao, C., Lu, L.W. and Fisher, J.W. (2000), "Development and Evaluation of Improved Details for Ductile Welded Unreinforced Flange Connections." Report No. SAC/BD-00-24, SAC Joint Venture, Sacramento, CA.

Ricles, J.M., Mao, C., Lu, L.W. and Fisher, J.W. (2002), "Inelastic Cyclic Testing of Welded Unreinforced Moment Connections," *Journal of Structural Engineering*, American Society of Civil Engineers, Vol. 128, No. 4.

CHAPTER 9
KAISER BOLTED BRACKET (KBB) MOMENT CONNECTION

Adan, S.M. and Gibb, W. (2009), "Experimental Evaluation of Kaiser Bolted Bracket Steel Moment Resisting Connections," *Engineering Journal*, American Institute of Steel Construction, Vol. 46, No. 3, pp. 181-195.

Grigorian, C.E., Yang, T.S. and Popov, E.P. (1992), "Slotted Bolted Connection Energy Dissipators," EERC Report No. 92/10, Earthquake Engineering Research Center, University of California, Berkeley, CA.

Gross, J.L., Engelhardt, M.D., Uang, C.M., Kasai, K. and Iwankiw, N.R. (1999), *Modification of Existing Welded Steel Moment Frame Connections for Seismic Resistance, Design Guide No. 12,* American Institute of Steel Construction, Chicago, IL.

Kasai, K. and Bleiman, D. (1996), "Bolted Brackets for Repair of Damaged Steel Moment Frame Connections," *7th US–Japan Workshop on the Improvement of Structural Design and Construction Practices: Lessons Learned from Northridge and Kobe,* Kobe, Japan.

Kulak, G.L., Fisher, J.W. and Struik, J.H.A. (1987), *Guide to Design Criteria for Bolted and Riveted Joints,* 2nd Ed., Wiley, New York, NY.

Masuda, H., Tamaka, A., Hirabayashi, K. and Genda, I. (1998), "Experimental Study on the Effect of Partial Loss of Sectional Area on the Static Characteristics of H-Beams," *Journal of Structural and Construction Engineering* (Transaction of AIJ), Architectural Institute of Japan, No. 512, pp. 157–164, October (in Japanese).

Newell, J. and Uang, C.M. (2006), "Cyclic Testing of Steel Moment Connections for the CALTRANS District 4 Office Building Seismic Rehabilitation," UCSD Report No. SSRP-05/03, University of California, San Diego, CA.

Ricles, J.M., Zhang, X., Lu, L.W. and Fisher, J. (2004), "Development of Seismic Guidelines for Deep Column Steel Moment Connections," ATLSS Report No. 04-13, Lehigh University, Bethlehem, PA.

Seek, M. W. and T. M. Murray (2008), "Seismic Strength of Moment End-Plate Connections with Attached Concrete Slab," Proceedings, Connections VI, American Institute of Steel Construction, Chicago, IL, June 23-25.

Tide, R.H.R. (2010), "Bolt Shear Design Considerations," *Engineering Journal*, American Institute of Steel Construction, Vol. 47, No. 1, pp. 47-63.

CHAPTER 10
CONXTECH CONXL MOMENT CONNECTION

AISC (2005a), *Seismic Provisions* for Structural Steel Buildings, AISC/ANSI 341-05, American Institute of Steel Construction, Inc., Chicago, IL.

AISC (2005b), *Specification for Structural Steel Buildings*, AISC/ANSI 360-05, American Institute of Steel Construction, Inc., Chicago, IL.

AISC (2010a), *Seismic Provisions for Structural Steel Buildings*, AISC/ANSI 341-10, American Institute of Steel Construction, Chicago, IL.

AISC (2010b), *Specification for Structural Steel Buildings*, AISC/ANSI 360-10, American Institute of Steel Construction, Chicago, IL.

Seek, M.W. and Murray, T.M. (2005), "Cyclic Test of 8-Bolt Extended Stiffened Steel Moment End Plate Connection with Concrete Structural Slab," report submitted to the American Institute of Steel Construction, AISC, Virginia Polytechnic Institute and State University, Blacksburg, VA.

Suprenant, Bruce (2001), "Free Fall of Concrete, ASCC Position Statement No. 17," *Concrete International*, ACI, Vol. 23, No. 6.

APPENDIX A
CASTING REQUIREMENTS

Briggs, C.W. (1967), "The Evaluation of Discontinuities in Commercial Steel Castings by Dynamic Loading to Failure in Fatigue," Steel Founders' Society of America (SFSA), Rocky River, OH.

REFERENCES

Carlson, K.D., Lin, Z., Hardin, R.A., Beckermann, C., Mazurkevich, G. and Schneider, M.C. (2003), "Modeling of Porosity Formation and Feeding Flow in Steel Casting," *Proceedings of Modeling of Casting, Welding and Advanced Solidification Processes X*, May 25–30, Destin, FL.

Hardin R.A., Ou, S., Carlson, K. and Beckermann, C. (1999), "Relationship between Casting Simulation and Radiographic Testing: Results from the SFSA Plate Casting Trials," *Proceedings of the 1999 SFSA Technical and Operating Conference*, November 4–6, Chicago, IL.

Niyama E., Nchida T., Marikawa M. and Shigeki, S. (1982), "A Method of Shrinkage Prediction and its Application to Steel Castings Practice," paper presented at *49th International Foundry Congress*, Chicago, IL.

PART 10
ENGINEERED DAMPING SYSTEMS

10.1 SCOPE 10–2
10.2 INTRODUCTION TO ENGINEERED DAMPING SYSTEMS 10–2
10.3 VISCOUS DAMPERS 10–2
10.4 VISCOELASTIC DAMPERS 10–4
10.5 FRICTION DAMPERS 10–5
10.6 YIELDING METALLIC DAMPERS 10–5
10.7 APPLICATION TO ANALYSIS AND DESIGN 10–6
PART 10 REFERENCES 10–7

10.1 SCOPE

Although not covered in the AISC *Seismic Provisions*, damping can be used to reduce seismic response and is thus an important methodology that may be utilized in the design of steel structures. This discussion provides an overview of types of engineered damping systems, but does not provide design guidance. A more thorough treatment may be found in Constantinou et al. (1998), Hanson and Soong (2001), Liang et al. (2011), Ramirez et al. (2002), and Whittaker et al. (2003).

10.2 INTRODUCTION TO ENGINEERED DAMPING SYSTEMS

Damping provides a means of dissipating a segment of input seismic energy. The inherent damping from structural and nonstructural sources is approximated as an equivalent viscous effect. For steel structures, the bare framing typically provides 1 to 2% equivalent viscous damping. The origin of this structural damping is from sources such as slip of bolts, minor yielding of members, and rocking of base plates. When nonstructural components such as partitions and other tenant improvements are added, the equivalent viscous damping can be as high as 5%. This additional damping is from the interaction of various nonstructural components and slip of components. The magnitude of damping is highly dependent on type and extent of nonstructural elements.

Such inherent damping is difficult to estimate and is too small to provide an effective means for energy dissipation. By contrast, supplementary damping can be incorporated in both new construction and seismic retrofit applications to produce much larger equivalent damping, typically in the range of 10 to 40%. This supplementary damping is more reliable and readily predictable. Supplementary damping can be counted on to dissipate significant seismic energy and thus reduce demand on the structural system. This reduction of drift, acceleration and inelastic behavior will provide significantly higher performance than that of conventional construction which relies heavily on inelastic energy dissipation by the frame elements.

For design, the expected level of damage is approximately proportional to the seismic input intensity. This implies acceptance of a certain level of damage based on the seismic input intensity. Research has shown (Miyamoto et al., 2011) that steel moment frame buildings designed per the applicable building code and using supplementary damping will have a significantly lower collapse rate than a building designed without supplementary damping.

Supplementary damping devices used for seismic design fall into four broad categories: viscous dampers, viscoelastic dampers, friction dampers and metallic dampers. Each of these four broad groups of damper devices also includes subsets. For example, both solid viscoelastic dampers and fluid viscoelastic dampers are available. Damping devices are manufactured by a number of vendors; see Figure 10-1 for examples of the dampers. The various damper types have distinct force-deformation (F-u) constitutive relations as shown in Figure 10-2.

10.3 VISCOUS DAMPERS

Viscous dampers are velocity-proportional devices. Figure 10-2(a) illustrates the force-displacement relationship of a viscous damper. The constitutive force-velocity relationship for viscous dampers can be written as:

10.4 VISCOELASTIC DAMPERS

(a) When $\dot{u} \geq 0$

$$F = C\dot{u}^\alpha \qquad (10\text{–}1)$$

(b) When $\dot{u} < 0$

$$F = -C \, | \dot{u} |^\alpha \qquad (10\text{–}2)$$

where
C = damping constant
$\dot{u}$ = velocity
α = velocity exponent

The choice of the parameters C and α depends on the structural system. In general, for buildings with higher story stiffness, larger values of C would be required. For viscous

(a) Viscous damper

(b) Viscoelastic damper

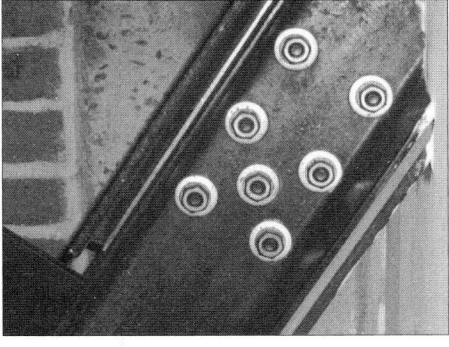

(c) Friction damper

(d) Yielding metallic damper

Fig. 10-1. Images of viscous, viscoelastic, friction and yielding metallic dampers.
(a) New medical office complex (Miyamoto International)
(b) Retrofit of an existing commercial building (Miyamoto International)
(c) Retrofit of an existing commercial building (Miyamoto International)
(d) Retrofit of an existing building on campus of University of California Berkeley
(NISEE e-library, Pacific Earthquake Engineering Research Center)

dampers, the brace used to connect the damper to the structural members must be stiff enough to ensure that the nearly pure viscous behavior is reproduced in the damper. In commercially available structural analysis programs, viscous dampers are modeled as Maxwell's model of a dashpot in series with a spring. A value of α equal to 1 denotes a purely viscous response. This type of damper is referred to as a linear damper. Dampers with α of smaller than unity are commonly referred to as nonlinear dampers. In general, viscous dampers do not increase stiffness or strength of the structural system and serve as purely damping elements. In other words, if the static pushover curves for a system with and without dampers are plotted, the curves will be identical so long as the demand on the viscous damper is within its operational capacity. Viscous dampers are ideal for structures with longer periods, such as moment frame buildings. They have been used extensively in new construction and for retrofitting steel moment frame buildings, resulting in reduced inelastic demand on the members and beam-to-column connections. The most common type of viscous damper is the fluid viscous damper which is similar to the shock absorbers used in automobiles. Figure 10-1(a) shows a typical viscous damper.

10.4 VISCOELASTIC DAMPERS

Viscoelastic dampers are displacement- and velocity-proportional devices. Figure 10-2(b) illustrates the force-displacement relationship of a viscoelastic damper. The constitutive force-displacement relation for viscoelastic dampers can be written as:

$$F = Ku + C\dot{u} \qquad (10\text{-}3)$$

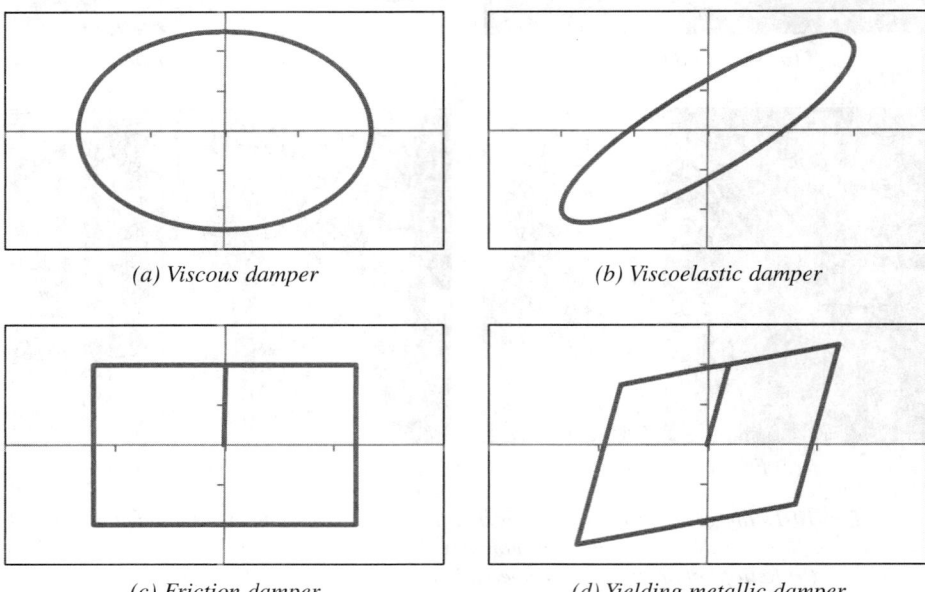

Fig. 10-2. Idealized force-displacement relation for various types of dampers.

where
  K = effective stiffness of the device
  u = displacement

The damper's effective stiffness, K, and damping constant, C, depend on both the amplitude of excitation and its frequency.

Viscoelastic dampers increase both the stiffness and strength of the structural system. In other words, if the static pushover curve for a system with and without dampers is plotted, the curve with a viscoelastic damper will have a higher capacity and a larger stiffness. Viscoelastic dampers are ideal for structures for which an increase in both damping and stiffness is desired. For example, they can be used to mitigate torsional response and increase damping for vulnerable structures or to reduce soft story response in buildings. An example of the viscoelastic damper is illustrated in Figure 10-1(b).

In commercially available structural analysis programs, solid viscous dampers can be modeled as Kelvin's model of a dashpot in parallel with a spring.

10.5 FRICTION DAMPERS

Friction dampers exhibit elastic-perfectly-plastic response. Figure 10-2(c) illustrates the force-displacement relationship of a friction damper. They are displacement-proportional devices and increase the strength of the structural system. The constitutive force-displacement relation for friction dampers can be written as:

(a) When $u < u_{slip}$

$$F = Ku \tag{10-4}$$

(b) When $u \geq u_{slip}$

$$F = 0 \tag{10-5}$$

where
  K = initial stiffness of the device
  u_{slip} = displacement at which the slip initiates

Friction dampers are ideal for structures for which damping and an increase in initial stiffness is desired. They can also be used to limit the force that is delivered to a structural member placed in series with them. Figure 10-1(c) shows a typical friction damper.

In commercially available structural analysis programs, friction dampers are modeled as Coulomb's model producing bilinear response.

10.6 YIELDING METALLIC DAMPERS

Yielding metallic dampers exhibit multi-linear force displacement response. Figure 10-2(d) illustrates the force-displacement relationship of a yielding metallic damper. They are displacement-dependent devices. The constitutive force-displacement relation for the yielding metallic dampers can be written as:

$$F = f(u) \tag{10-6}$$

where $f(u)$ denotes the nonlinear dependence of the device force on its displacement.

Yielding metallic dampers increase both stiffness and strength of the structural system. In other words, if the static pushover curves for a system with and without dampers are plotted, the curve with dampers will have a higher capacity and larger stiffness.

Yielding dampers are ideal for structures for which both damping and stiffness increases are desired. In U.S. practice, yielding metallic elements such as buckling-restrained braces are not typically treated as dampers, but as primary lateral load resisting elements. See Figure 10-1(d).

In commercially available structural analysis programs, yielding metallic dampers are modeled as multi-linear one-dimensional elements.

10.7 APPLICATION TO ANALYSIS AND DESIGN

Each type of damper has specific characteristics, advantages and disadvantages for structural applications, and each type is made by several different manufacturers. To design with dampers, it is critical to capture the dynamic characteristics and the nonlinear behavior of the structure. This permits the designer to determine the amount of additional damping and stiffness required to achieve the desired performance. The equivalent viscous damping ratio can be estimated by computing the area of the force-displacement curve for one full cycle of loading.

Engineered damping systems inherently require a performance-based design approach. Engineered damping systems are effective for almost all but the most rigid buildings. A thorough discussion of engineered, supplemental damping system design is beyond the scope of this Manual, but ASCE/SEI 41-06 (ASCE, 2006) is considered state-of-the-art for existing structures, and FEMA P-750 (FEMA, 2009) and ASCE/SEI 7-10 provide information for new structures.

PART 10 REFERENCES

ASCE (2006), *Seismic Rehabilitation of Existing Buildings,* ASCE 41-06, American Society of Civil Engineers, Reston, VA.

Constantinou, M.C., Soong, T.T. and Dargush, G.F. (1998), *Passive Energy Dissipation Systems for Structural Design and Retrofit,* Monograph No. 1, Multidisciplinary Center for Earthquake Engineering Research, University of Buffalo, State University of New York, Buffalo, NY.

FEMA (2009), *NEHRP Recommended Seismic Provisions For New Buildings And Other Structures*, FEMA P-750, Federal Emergency Management Agency, Washington, DC.

Hanson, R.D. and Soong, T.T. (2001), *Seismic Design with Supplemental Energy Dissipation Devices*, MNO-8, Earthquake Engineering Research Institute, Oakland, CA.

Liang, Z, Lee, G.C., Dargush, G.F. and Song, J. (2011), *Structural Damping: Applications in Seismic Response Modification*, CRC Press.

Miyamoto, H.K., Gilani, A.S.J., Wada, A. and Ariyaratana, C. (2011), "Identifying the Collapse Hazard of Steel Special Moment-Frame Buildings with Viscous Dampers Using the FEMA P695 Methodology," *Earthquake Spectra*, Earthquake Engineering Research Institute, Vol. 27, Issue 4, pp. 1,147-1,168.

Ramirez, O.M., Constantinou, M.C., Whittaker, A.S., Kircher, C.A. and Chrysosotomou, C.Z. (2002), "Elastic and Inelastic Seismic Response of Buildings with Damping Systems," *Earthquake Spectra*, Earthquake Engineering Research Institute, Vol. 18, Issue 3, pp. 531-547.

Whittaker, A.S., Constantinou, M.C., Ramirez, O.M., Johnson, M.W. and Chrysostomou, C.Z. (2003), "Equivalent Lateral Force and Modal Analysis Procedures of the 2000 NEHRP Provisions for Buildings with Damping Systems," *Earthquake Spectra*, Earthquake Engineering Research Institute, Vol. 19, Issue 4, pp. 959-980.

INDEX

Analysis 2–2, 9.1–10
Applicable building code 1–6
ASCE/SEI 7
    Analysis methods 2–6
    Design coefficients and factors for SFRS 1–30
    Systems 1–10
Beam design
    Buckling-restrained braced frame 5–430
    Collectors and chords 8–1
    Composite ordinary shear wall steel coupling beam 7–21
    Composite ordinary shear wall composite coupling beam 7–30
    Composite special shear wall steel coupling beam 7–39, 7–41
    Eccentrically braced frame 5–353
    Embedment length 7–14
    Moment frame ($R = 3$) 3–10
    Ordinary concentrically braced frames 5–18
    Ordinary moment frames 4–12
    Special concentrically braced frames 5–104, 5–119
    Special moment frames 4–46
Beam embedment length 7–14
Bolted flange plate moment connection 9.2–39
Bolted joints 9.1–20
Bracing connection
    Braced frames ($R = 3$) 3–25
    Eccentrically braced frames 5–372, 5–379
    Ordinary concentrically braced frames 5–25
    Special concentrically braced frames 5–140, 5–178, 5–202, 5–269, 5–299
Brace design
    Braced frames ($R = 3$) 3–22
    Buckling-restrained braced frames 5–419
    Eccentrically braced frames 5–362
    Ordinary concentrically braced frames 5–6, 5–75
    Special concentrically braced frames 5–87
Braced frame systems
    $R = 3$ systems 3–21
    $R > 3$ systems 5–1

    Buckling-restrained braced frames. 5–413, 9.1–65
    Composite braced frames. 7–3, 9.1–88
    Eccentrically braced frames. 5–334, 9.1–57
    Ordinary concentrically braced frames 5–3, 9.1–50
    Special concentrically braced frames. 5–82, 9.1–52
Buckling-restrained braced frames 5–413, 9.1–65
    Adjusted brace strength 9.1–65
    Analysis. 9.1–66
    Beam design 5–430, 9.1–67
    Bracing connections 9.1–69
    Brace design 5–419, 9.1–67
    Column design 5–425, 9.1–67
    Column splices 9.1–70
    Connections. 9.1–69
        Beam-to-column connections. 9.1–69
        Demand critical welds 9.1–69
        Diagonal brace connections 9.1–69
        Column splices 9.1–70
    Protected zone. 9.1–69
    Qualifying cyclic testing. 9.1–129
    System requirements. 9.1–67
        V-and inverted V-braced frames 9.1–67
        K-braced frames 9.1–67
    Width-to-thickness limitations. 1–36
Building drift 1–18
Building separations. 1–17
Capacity design 2–2
Chord design 8–2, 8–8
Classification of sections for local buckling. 1–25, 9.1–11
Collector design 8–4, 8–8
Column design 9.1–16
    Braced frames ($R = 3$) 3–24
    Buckling-restrained braced frames 5–425
    Eccentrically braced frames. 5–367
    Ordinary concentrically braced frames 5–15
    Ordinary moment frames. 4–7
    Moment frames ($R = 3$) 3–6
    Special concentrically braced frames 5–98
    Special moment frames 4–42
Column splice design 9.1–21

Gravity column splice	4–76
Special concentrically braced frames	5–129
Special moment frames	4–90
Composite eccentrically braced frames	7–7, 9.1–91
Analysis	9.1–92
Members	9.1–92
Connections	9.1–92
Composite intermediate moment frames	6–4, 9.1–78
Analysis	9.1–78
Beam flanges	9.1–79
Members	9.1–79
Connections	9.1–79
Beam-to-column connections	9.1–79
Conformance demonstration	9.1–79
Connection diaphragm plates	9.1–80
Column splices	9.1–81
Demand critical welds	9.1–79
Required shear strength	9.1–80
Protected zones	9.1–79
Composite ordinary braced frames	7–3, 9.1–88
Analysis	9.1–89
Beams	9.1–89
Braces	9.1–89
Columns	9.1–89
Connections	9.1–89
Demand Critical Welds	9.1–89
Protected zones	9.1–89
Composite ordinary moment frames	6–2, 9.1–77
Analysis	9.1–77
Connections	9.1–78
Demand critical welds	9.1–78
Members	9.1–78
Protected zones	9.1–78
Composite ordinary shear walls	7–16
Analysis	9.1–93
Beam embedment length	7–19, 7–21, 9.1–94
Boundary members	9.1–93
Composite coupling beam design	7–19, 7–30, 9.1–94
Connections	9.1–96
Demand critical welds	9.1–96

    Protected zones 9.1–96
    Steel coupling beam design 7–19, 7–21, 9.1–94
Composite partially restrained moment frames 6–11, 9.1–85
    Analysis. 9.1–85
    Columns 9.1–85
    Beams 9.1–86
    Protected Zones. 9.1–86
    Connections. 9.1–86
        Demand Critical Welds. 9.1–86
        Required Strength. 9.1–86
        Beam-to-Column Connections. 9.1–86
        Conformance Demonstration 9.1–86
        Column Splices. 9.1–87
Composite plate shear walls 9.1–99
    Analysis. 9.1–99
    Boundary members. 9.1–101
    Column splices 9.1–102
    Concrete stiffening elements. 9.1–101
    Connections. 9.1–101
    Demand critical welds 9.1–101
    Protected zones 9.1–101
    Webs 9.1–100
Composite shear walls 7–10
Composite slab diaphragms 9.1–19
Composite special concentrically braced frames 7–5, 9.1–89
    Analysis. 9.1–90
    Beams 9.1–90
    Braces 9.1–90
    Columns 9.1–90
    Connections. 9.1–90
        Beam-to-column connections. 9.1–91
        Column splices 9.1–91
        Demand critical welds 9.1–90
        Required strength of brace connections. 9.1–91
    Protected zones 9.1–90
Composite special moment frames. 6–7, 9.1–81
    Analysis. 9.1–81
    Beam flanges. 9.1–83
    Members 9.1–82
    Connections. 9.1–83

    Beam-to-column connections 9.1–84
    Conformance demonstration. 9.1–84
    Connection diaphragm plates 9.1–85
    Column splices 9.1–85
    Demand critical welds 9.1–83
    Required shear strength 9.1–84
  Protected zones 9.1–83
  System requirements. 9.1–81
    Moment ratio 9.1–81
    Stability bracing of beams 9.1–82
    Stability bracing at beam-to-column connections 9.1–82
Composite special shear walls 7–36, 9.1–96
  Analysis. 9.1–96
  Beam embedment length 7–40, 9.1–97
  Boundary members 9.1–97
  Column splices 9.1–99
  Composite coupling beams. 9.1–98
  Connections. 9.1–99
  Demand critical welds 9.1–99
  Ductile elements 9.1–97
  Face bearing plates 7–41, 9.1–97
  Steel coupling beams 7–39, 7–41, 9.1–97
  Vertical transfer bars 7–41, 9.1–98
  Wall overstrength 7–40
  Web stiffeners. 7–40
Composite structural steel and reinforced concrete buildings 1–23, 9.1–77, 9.1–88
(also see individual composite systems)
  Composite connections 9.1–24
  Composite members
    Composite slab diaphragms 9.1–19
    Encased composite columns 9.1–17
    Filled composite columns 9.1–19
  Composite moment frames 6–1
  Composite shear walls 7–10
Connection design
  Bolted joints; see Bolted joints
  Braced frame ($R = 3$) 3–25
  Column bases 9.1–22
  Column splice design 4–67, 9.1–21
  Collector connection 8–22

Composite connections ... 9.1–24
Continuity plates and stiffeners ... 9.1–21
Eccentrically braced frames ... 5–372, 5–379
Moment frames ($R = 3$) ... 3–13
Ordinary concentrically braced frames ... 5–25
Ordinary moment frames ... 4–16
Prequalified ... 9.2–i
Rectangular filled composite column-to-steel beam ... 6–21
Reinforced concrete column-to-steel beam ... 6–15
Round filled composite column-to-steel beam ... 6–19
Special concentrically braced frames ... 5–140, 5–178, 5–202, 5–269, 5–299
Special moment frames ... 4–57
Steel anchors ... 9.1–26
Welded joints; see Welded joints
Continuity plates and stiffeners ... 9.1–21
ConXtech ConXL moment connection ... 9.2–64
Coupling beams ... 7–11, 7–19, 7–21, 7–30, 7–39, 7–41
Damping systems; see Engineered damping systems
Deflection amplification factor ... 1–14
Deflection compatibility ... 1–18
Deformation compatibility ... 9.1–26
Demand critical welds ... 1–21
Design basis earthquake ... 1–10
Design drawings ... 1–21, 1–22, 9.1–6
Diaphragm design ... 8–2, 8–8
Direct analysis method ... 2–2
Drift limits ... 1–18, 2–19; see also Story drift
Ductile design mechanism ... 2–2
Ductility requirements ... 1–25, 1–34, 9.1–11
Earthquake ground motion ... 1–7
Eccentrically braced frames ... 5–334, 9.1–57
    Analysis ... 9.1–58
    Beam outside of link design ... 5–353, 9.1–59
    Brace design ... 5–362, 9.1–59
    Brace-to-link connections ... 5–372
    Column design ... 5–367, 9.1–59
    Connections ... 9.1–62
        Beam-to-column connections ... 9.1–63
        Column splice design ... 9.1–63
        Demand critical welds ... 9.1–62

Diagonal brace connections	5–379, 9.1–63
Link-to-column connections	9.1–64
Links	5–340, 9.1–59
Protected zone	9.1–62
Story drift	5–339
System requirements	9.1–58
Link rotation angle	9.1–58
Bracing of link	9.1–59
Width-to-thickness limitations	1–36, 5–338, 9.1–59
Effective length method	2–5
Encased W-shapes, plastic capacities	7–52
End-plate moment connection	9.2–19
Engineered damping systems	10–1
Analysis and design	10–6
Friction dampers	10–5
Viscous dampers	10–2
Viscoelastic dampers	10–4
Yielding metallic dampers	10–5
Equivalent lateral force analysis	2–6
Erection drawings	9.1–103
Expansion joints	1–16
Expected material strength	9.1–3
Filled HSS, plastic capacities	7–54
First-order analysis method	2–6
Friction dampers	10–5
H-pile	9.1–26
Battered H–piles	9.1–26
Protected zone	9.1–27
Tension in H-piles	9.1–26
Heavy section CVN requirements	9.1–5
Inspection	
Inspection of bolting	9.1–113
Inspection of composite structures	9.1–115
Inspection of H-piles	9.1–117
Inspection of welding	9.1–109
Inspection tasks	9.1–108
Other steel structure inspections	9.1–115
Personnel	9.1–108
Intermediate moment frames	4–33, 9.1–30
Analysis	9.1–31

Beam flanges. 9.1–31
Comparison to SMF and OMF 4–122
Connections. 9.1–32
    Beam-to-column connections. 9.1–32
    Column splices 9.1–34
    Conformance demonstration. 9.1–33
    Continuity plates. 9.1–34
    Demand critical welds 9.1–32
    Required shear strength 9.1–33
    Panel zone. 9.1–33
Stability bracing of beams 9.1–31
Protected zone. 9.1–32
Width-to-thickness limitations 1–36, 9.1–31
Kaiser bolted bracket moment connection 9.2–52
K-braced frames 5–443, 9.1–54
Link design. 5–340
Load combinations 9.1–8
Local buckling. 1–25, 9.1–11
    Pipe 1–29, 1–57
    Rectangular HSS 1–28, 1–53
    Round HSS 1–29, 1–55
    Single angles 1–27, 1–52
    Square HSS 1–28, 1–54
    W-Shapes. 1–26, 1–36
Lowest anticipated service temperature 1–18, 9.1–7
Low-seismic design ($R = 3$). 1–13, 3–2
Material specifications 1–21, 9.1–2
Maximum considered earthquake 1–10
Maximum force delivered by system 1–16, 5–136
Modal response spectrum analysis. 2–7
Moment connections
    Composite moment frames 6–14
    Moment frame ($R = 3$). 3–13
    Ordinary moment frames. 4–16
    Prequalified. 9.1–118, 9.2–i
    Qualification of. 9.1–121
    Special moment frames 4–57
Moment frames
    $R = 3$ systems. 3–4
    $R > 3$ systems. 4–1

Intermediate moment frames	4–33, 9.1–30
Ordinary moment frames	4–2, 9.1–28
Special moment frames	4–33, 9.1–34
Nonbuilding structures	5–443
Ordinary cantilever column systems	9.1–47
Ordinary concentrically braced frames	5–3, 9.1–50
Above Seismic Isolation Systems	9.1–52
Analysis	9.1–50
Brace members	5–6, 5–75, 9.1–51
Brace connections	5–25, 9.1–51
Beam design	5–18
Column design	5–15
System requirements	9.1–50
V-braced and inverted V-braced frames	9.1–50
K-braced frames	5–443, 9.1–51
Nonbuilding structures	5–443
Width-to-thickness limitations	1–36, 9.1–51
Ordinary moment frames	4–2, 9.1–28
Analysis	9.1–28
Beam design	4–12, 9.1–28
Beam-to-column connections	4–16, 9.1–29
Column design	4–7, 9.1–28
Comparison to IMF and SMF	4–122
Demand critical welds	9.1–29
FR moment connections	9.1–29
PR moment connections	9.1–30
Story drift and stability check	4–5
Overstrength factor	1–14, 9.1–8
Performance goals	1–5
Period	1–7
Prequalified connection standard	9.2–i
Prequalification	9.1–118
Beam-to-column connections	9.1–118, 9.2–i
Link-to-column connections	9.1–118
Protected zone	1–22, 9.1–16
Qualification	
Buckling-restrained braced frames	9.1–129
Beam-to-column connections	9.1–121
Link-to-column connections	9.1–121
Quality assurance	1–19, 9.1–106

Quality control 1–19, 9.1–106
Reduced beam section moment connection 4–57, 9.2–11
Redundancy factor 1–15
Response modification coefficient 1–12, 9.1–1
    $R = 3$ applications 1–13, 3–2
Response spectra 1–7
R-factor (see Response modification coefficient)
Risk category 1–7
Seismic design category 1–7
Seismic joints 1–17
Seismic Provisions for Structural Steel Buildings 9.1–i
Shear wall coupling 7–10
Shop drawings 9.1–103
Special cantilever column systems 9.1–48
    Width-to-thickness limitations 1–36
Special concentrically braced frames 5–82, 9.1–52
    Analysis 5–93, 9.1–52
    Beam design 5–104, 5–119, 9.1–54
    Brace design 5–83, 5–87, 9.1–54
    Column design 5–98, 9.1–54
    Column splice design 5–129, 9.1–57
    Connection design 5–140, 5–178, 5–202, 5–269, 5–299, 9.1–55
    Demand critical welds 9.1–55
    System requirements 9.1–53
        Lateral force distribution 9.1–53
        V-braced and inverted V-braced frames 5–83, 9.1–53
        K-braced frames 5–443, 9.1–54
        Tension-only frames 9.1–54
    Width-to-thickness limitations 1–36, 5–82, 9.1–54
Special moment frame 4–33, 4–76, 9.1–34
    Analysis 9.1–34
    Beam design 4–46, 9.1–38
    Beam flanges 9.1–38
    Column base 4–94, 4–113
    Column design 4–42, 9.1–38
    Column-beam moment ratio 9.1–34
    Comparison to OMF and IMF 4–122
    Connections 9.1–38
        Demand critical welds 9.1–38
        Beam-to-column connections 4–57, 9.1–39

      Conformance demonstration. 9.1–39
      Required shear strength 9.1–40
      Panel zone 9.1–40
      Continuity plates. 9.1–41
      Column splices 4–90, 9.1–43
      Stability bracing at beam-to-column connections 9.1–37
   Stability bracing of beams 9.1–36
   Protected zones 9.1–38
   Story drift and stability check 4–36
   Width-to-thickness limitations. 1–36, 9.1–38
Special plate shear walls 9.1–70
   Analysis. 9.1–71
   Connections. 9.1–73
      Column splices. 9.1–73
      Connections of webs to boundary elements. 9.1–73
      Demand critical welds 9.1–73
      HBE-to-VBE connections 9.1–73
   Perforated webs. 9.1–74
   Protected zone. 9.1–73
   System requirements. 9.1–71
      Stiffness of boundary elements 9.1–71
      HBE-to-VBE connection moment ratio 9.1–72
      Bracing. 9.1–72
      Openings in webs 9.1–72
   Webs 9.1–72
   Width-to-thickness limitations 1–26, 1–36, 9.1–72
Special truss moment frames 9.1–43
   Analysis. 9.1–43
      Special segment 9.1–43
      Nonspecial segment 9.1–44
   Built-up chord members 9.1–46
   Connections. 9.1–46
      Demand critical welds 9.1–46
      Connections of diagonal web members in the special segment. 9.1–47
      Column splices 9.1–47
   Protected zones 9.1–46
   Strength of special segment members 9.1–45
   System requirements. 9.1–44
      Special segment 9.1–44
      Stability bracing of trusses 9.1–44

Stability bracing of truss-to-column connections	9.1–45
Stiffness of stability bracing	9.1–45
Width-to-thickness limitations	1–36, 9.1–46
Specifications, codes and standards	1–4, 9.1–i
Stability bracing	9.1–14
Stability design methods	2–5
Steel headed stud anchors	1–30, 1–58
Steel plate shear walls; see Special plate shear walls	
Stiffeners; see Continuity plates and stiffeners	
Story drift	1–18, 3–5, 4–36
Structural modeling	2–7
Gravity loads	2–13
Stiffness of structural elements	2–8
Strength of structural elements	2–7
Viscoelastic dampers	10–4
Viscous dampers	10–2
Wall overstrength	7–15
Weld access hole	1–25, 1–33
Welded joints	9.1–21
Welded unreinforced flange-welded web moment connection	9.2–46
Welding provisions	9.1–5, 9.2–7
AWS D1.8	1–23
Demand critical welds	1–21, 9.1–5
Inspection	9.1–109
Notch toughness verification test	9.1–5
Width-to-thickness limitations; see Local buckling and specific systems	
Yielding metallic dampers	10–5